W0018436

MEMBRANE FUSION

MEMBRANE FUSION

edited by

Jan Wilschut • Dick Hoekstra

University of Groningen
Groningen, The Netherlands

CRC Press
Taylor & Francis Group
Boca Raton London New York

CRC Press is an imprint of the
Taylor & Francis Group, an **informa** business

First published 1991 by Marcel Dekker, Inc.

Published 2019 by CRC Press
Taylor & Francis Group
6000 Broken Sound Parkway NW, Suite 300
Boca Raton, FL 33487-2742

© 1991 by Taylor & Francis Group, LLC
CRC Press is an imprint of Taylor & Francis Group, an Informa business

No claim to original U.S. Government works

ISBN 13: 978-0-8247-8301-3 (hbk)

This book contains information obtained from authentic and highly regarded sources. Reasonable efforts have been made to publish reliable data and information, but the author and publisher cannot assume responsibility for the validity of all materials or the consequences of their use. The authors and publishers have attempted to trace the copyright holders of all material reproduced in this publication and apologize to copyright holders if permission to publish in this form has not been obtained. If any copyright material has not been acknowledged please write and let us know so we may rectify in any future reprint.

Except as permitted under U.S. Copyright Law, no part of this book may be reprinted, reproduced, transmitted, or utilized in any form by any electronic, mechanical, or other means, now known or hereafter invented, including photocopying, microfilming, and recording, or in any information storage or retrieval system, without written permission from the publishers.

For permission to photocopy or use material electronically from this work, please access www.copyright.com (http://www.copyright.com/) or contact the Copyright Clearance Center, Inc. (CCC), 222 Rosewood Drive, Danvers, MA 01923, 978-750-8400. CCC is a not-for-profit organization that provides licenses and registration for a variety of users. For organizations that have been granted a photocopy license by the CCC, a separate system of payment has been arranged.

Trademark Notice: Product or corporate names may be trademarks or registered trademarks, and are used only for identification and explanation without intent to infringe.

Visit the Taylor & Francis Web site at
http://www.taylorandfrancis.com

and the CRC Press Web site at
http://www.crcpress.com

Library of Congress Cataloging-in-Publication Data

Membrane fusion/edited by Jan Wilschut, Dick Hoekstra.
        p.   cm.
     Includes bibliographical references.
     Includes index.
     ISBN 0-8247-8301-8 (alk. paper)
     1. Membrane fusion. I. Wilschut, Jan. II. Hoekstra, Dick.
     [DNLM: 1. Cell Membrane--physiology. 2. Cell Transformation,
  Viral--physiology. 3. Membrane Fusion--physiology.   QH 601 M53251]
  QH601.M467 1990
  574.87'5--dc 20
  DNLM/DLC
  for Library of Congress   90-13828
                                                          CIP

Preface

The relevance and importance of cellular membrane fusion are perhaps best illustrated by the notion that current interest in this phenomenon is shared by scientists in quite varied disciplines. Biophysicists and biochemists are intrigued by the complex molecular rearrangements that must occur during the merging of two lipid bilayer membranes. Cell biologists, on the other hand, appreciate the importance of the frequent and highly specific intracellular membrane fusion events that occur in the processing of substances taken into the cell by endocytosis and in the trafficking of cellular components from the site of their synthesis to their ultimate destination in the cell. Furthermore, after early observations of the formation of multinucleated cells in various viral infections, virologists now recognize membrane fusion as the mechanism by which enveloped viruses infect their host cells. Finally, many scientists have come to appreciate the potential of the application of membrane fusion techniques, both in the study of fundamental cell-biological processes as well as in the area of biotechnology.

The purpose of this book is to provide the reader with an overview of recent progress in research on membrane fusion as it relates to these diverse areas of scientific interest. We hope that the book will show interrelationships, that it will bind together, rather than dig into a limited number of specific topics presenting these in every possible detail. The reader will note that there is occasional overlap between chapters. This has been our deliberate policy. Not only was it our intention to show interrelationships, we have also encouraged the authors to present their own personal views. As a result, some topics are discussed from quite different perspectives. It is our hope that the book, by presenting membrane fusion viewed from these different perspectives, will provide a coherent picture of the ubiquitous significance of the fusion process in cell biology.

The book contains six parts, subdivided in various chapters, each discussing a particular topic. To introduce the reader and to provide a basis for an optimal understanding of the rest of the book, Part I presents three rather general chapters on membrane structure, lipid polymorphism, and intermembrane forces. Part II covers fusion in model membrane systems. Most of the work presented in this part is aimed at elucidation of the molecular mechanisms involved in membrane fusion. Several chapters address the question of the relevance of the study of membrane fusion in simple model systems as a basis for a better understanding of biological membrane fusion, where molecular mechanisms involved in the actual fusion event as well as in its modulation and control still remain largely obscure. Viral membrane fusion represents the only example of a biological membrane fusion event of which the characteristics are relatively well understood, in some cases even at the molecular level. Therefore, a separate part, Part III, is devoted to the fusogenic properties of enveloped viruses, studied in both cellular and model systems. This part, in particular, illustrates the importance and relevance of the convergence of various scientific disciplines.

As to membrane fusion occurring intracellularly in the trafficking of cellular components, fascinating attempts are being made to dissect the pathways of endocytosis and exocytosis in reconstituted cell-free systems, to identify the molecular components involved in the fusion reactions, and to unravel control mechanisms. Part IV of this book is devoted to the various aspects of these intracellular, or endoplasmic, fusion processes. Exoplasmic fusion, occurring in cell-cell fusion processes, is discussed in Part V. Cell-cell fusion is not limited to pathological conditions such as those induced by certain viral infections; on the contrary, the controlled cell-cell fusion during fertilization constitutes the basis of life in higher organisms, and cell-cell fusion is also involved in the formation of multi-nucleated muscle cells.

Part VI focuses on *applications* of membrane fusion techniques in cell-biological research, biotechnology, and medicine. Topics include the generation of hybrid cells in, for example, the production of monoclonal antibodies, the introduction of foreign molecules into cells or specific cell organelles, and targeted drug delivery.

Finally, we wish to thank all the authors for their excellent contributions to this book. We also wish to thank many of them for their patience: finalizing this large project has taken longer than we initially anticipated, and perhaps this book would never have been published were it not for the continuous support and help of the Marcel Dekker personnel, the production editor Elaine Grohman in particular. Also, the secretarial assistance of Rinske Kuperus in our own department has been invaluable.

Jan Wilschut
Dick Hoekstra

Contributors

William E. Balch Department of Molecular Biology, Scripps Clinic and Research Foundation, La Jolla, California

Patricia A. Baldwin[a] Cancer Research Institute, School of Medicine, University of California—San Francisco, San Francisco, California

Nutrit Ballas Department of Biological Chemistry, Institute of Life Sciences, The Hebrew University of Jerusalem, Jerusalem, Israel

J. J. M. Bergeron Department of Anatomy, Faculty of Medicine, University of Montreal and McGill University, Montreal, Quebec, Canada

Robert Blumenthal Section on Membrane Structure and Function, Laboratory of Mathematical Biology, National Cancer Institute, National Institutes of Health, Bethesda, Maryland

Jan Bondeson Department of Medical Physiological Chemistry, University of Lund, Lund, Sweden

Lawrence T. Boni[b] Department of Physiology and Biophysics, Harvard Medical School, Boston, Massachusetts

Francois Boulay[c] Department of Cell Biology, Yale University School of Medicine, New Haven, Connecticut

---

Present affiliation:
[a]California Biotechnology, Inc., Mountain View, California.
[b]The Liposome Company, Inc., Princeton, New Jersey.
[c]Centre National de la Recherche Scientifique, Grenoble, France.

Brad Chazotte Department of Cell Biology and Anatomy, Laboratories for Cell Biology, School of Medicine, University of North Carolina—Chapel Hill, Chapel Hill, North Carolina

Vitaly Citovsky Department of Biological Chemistry, Institute of Life Sciences, The Hebrew University of Jerusalem, Jerusalem, Israel

Pieter R. Cullis Department of Biochemistry, University of British Columbia, Vancouver, British Columbia, Canada

John Davey[a] Department of Biochemistry, University of Dundee, Dundee, Scotland

Johannes de Gier Centre for Biomembranes and Lipid Enzymology, University of Utrecht, Utrecht, The Netherlands

Toon de Kroon Centre for Biomembranes and Lipid Enzymology, University of Utrecht, Utrecht, The Netherlands

Ben de Kruijff Centre for Biomembranes and Lipid Enzymology, University of Utrecht, Utrecht, The Netherlands

Lou de Leij Department of Clinical Immunology, University Hospital Groningen, Groningen, The Netherlands

Robert W. Doms[b] Department of Cell Biology, Yale University School of Medicine, New Haven, Connecticut

Arnold J. M. Driessen Department of Microbiology, University of Groningen, Haren, The Netherlands

Nejat Düzgüneş Cancer Research Institute, School of Medicine, and Department of Pharmaceutical Chemistry, School of Pharmacy, University of California—San Francisco, San Francisco, California

Ofer Eidelman[c] Section on Membrane Structure and Function, Laboratory of Mathematical Biology, National Cancer Institute, National Institutes of Health, Bethesda, Maryland

Charles G. Glabe Department of Molecular Biology and Biochemistry, University of California—Irvine, Irvine, California

Manfred Gratzl Department of Anatomy and Cell Biology, University of Ulm, Ulm, Federal Republic of Germany

---

Present affiliation:
[a]School of Biochemistry, University of Birmingham, Birmingham, England.
[b]Laboratory of Viral Diseases, National Institute of Allergy and Infectious Diseases, National Institutes of Health, Bethesda, Maryland.
[c]Department of Biological Chemistry, Institute of Life Sciences, The Hebrew University of Jerusalem, Jerusalem, Israel.

Charles R. Hackenbrock Department of Cell Biology and Anatomy, School of Medicine, University of North Carolina—Chapel Hill, Chapel Hill, North Carolina

Anne M. Haywood Departments of Pediatrics, Microbiology, and Medicine, University of Rochester Medical Center, Rochester, New York

Ari Helenius Department of Cell Biology, Yale University School of Medicine, New Haven, Connecticut

Dick Hoekstra Laboratory of Physiological Chemistry, University of Groningen, Groningen, The Netherlands

Keelung Hong Cancer Research Institute, School of Medicine, University of California—San Francisco, San Francisco, California

Michael J. Hope The Canadian Liposome Co. Ltd., Vancouver, British Columbia, Canada

Sek-Wen Hui Department of Biophysics, Roswell Park Memorial Institute, and Department of Biophysics, State University of New York—Buffalo, Buffalo, New York

Alan M. Kleinfeld[a] Harvard Medical School, Boston, Massachusetts

Karen A. Knudsen Department of Cell Biology, Lankenau Medical Research Center, Philadelphia, Pennsylvania

Wilhelmus N. Konings Department of Microbiology, University of Groningen, Haren, The Netherlands

Peter I. Lelkes Department of Medicine, University of Wisconsin Medical School, Milwaukee Clinical Campus, Milwaukee, Wisconsin

John Lenard Department of Physiology and Biophysics, University of Medicine and Dentistry of New Jersey—Robert Wood Johnson Medical School at Rutgers, Piscataway, New Jersey

Dov Lichtenberg Department of Physiology and Pharmacology, Sackler School of Medicine, Tel Aviv University, Ramat Aviv, Israel

Abraham Loyter Department of Biochemistry, Institute of Life Sciences, The Hebrew University of Jerusalem, Jerusalem, Israel

Mark Marsh Chester Beatty Laboratories, Institute of Cancer Research, London, England

Paul R. Meers[b] Cancer Research Institute, School of Medicine, University of California—San Francisco, San Francisco, California

Present affiliation:
[a]Medical Biology Institute, La Jolla, California.
[b]Boston University School of Medicine, Boston, Massachusetts.

Shlomo Nir Seagram Center for Soil and Water Sciences, Faculty of Agriculture, The Hebrew University of Jerusalem, Rehovot, Israel

Michel Ollivon[a] Section on Membrane Structure and Function, Laboratory of Mathematical Biology, National Cancer Institute, National Institutes of Health, Bethesda, Maryland

Jacques M. Paiement Department of Anatomy, University of Montreal and McGill University, Montreal, Quebec, Canada

Demetrios Papahadjopoulos Cancer Research Institute, School of Medicine, University of California—San Francisco, San Francisco, California

V. Adrian Parsegian Physical Sciences Laboratory, National Institute of Diabetes, Digestive, and Kidney Diseases and Division of Computer Research and Technology, National Institutes of Health, Bethesda, Maryland

Helmut Plattner Faculty of Biology, University of Konstanz, Konstanz, Federal Republic of Germany

Harvey B. Pollard Laboratory of Cell Biology and Genetics, National Institute of Diabetes, Digestive, and Kidney Diseases, National Institutes of Health, Bethesda, Maryland

Paul Quinn Department of Experimental Pathology, University College, London, England

R. Peter Rand Department of Biological Sciences, Brock University, St. Catharines, Ontario, Canada

David S. Roos Department of Biological Sciences, University of Pennsylvania, Philadelphia, Pennsylvania

Robert M. Straubinger[b] Department of Pharmacology, University of California—San Francisco, San Francisco, California

Roger Sundler Department of Medical and Physiological Chemistry, University of Lund, Lund, Sweden

Francis C. Szoka, Jr. Department of Pharmaceutical Chemistry, School of Pharmacy, University of California—San Francisco, San Francisco, California

Ted F. Taraschi[c] Centre for Biomembranes and Lipid Enzymology, University of Utrecht, Utrecht, The Netherlands

T. Hauw The Department of Clinical Immunology, University Hospital Groningen, Groningen, The Netherlands

Present affiliation:
[a]Centre Nationale de la Recherce Scientifique, Thiais, France.
[b]Department of Pharmaceutics, State University of New York—Buffalo, Buffalo, New York.
[c]Department of Biochemistry, Jefferson College, Philadelphia, Pennsylvania.

Thomas E. Thompson   Department of Biochemistry, University of Virginia
Health Sciences Center, Charlottesville, Virginia

Colin P. Tilcock   Department of Radiology, University of British Columbia,
Vancouver, British Columbia, Canada

Victor D. Vacquier   Marine Biology Research Division, Scripps Institution
of Oceanography, University of California—San Diego, La Jolla, California

Nol van der Steen[a]   Centre for Biomembranes and Lipid Enzymology,
University of Utrecht, Utrecht, The Netherlands

Peter van Hoogevest[b]   Centre for Biomembranes and Lipid Enzymology,
University of Utrecht, Utrecht, The Netherlands

Gerrit van Meer   Department of Cell Biology, Medical School, University
of Utrecht, Utrecht, The Netherlands

A. J. Verkleij   Department of Molecular Cell Biology and Institute of
Molecular Biology and Medical Biotechnology, University of Utrecht,
Utrecht, The Netherlands

Anne Walter[c]   Section on Membrane Structure and Function, Laboratory
of Mathematical Biology, National Cancer Institute, National Institutes of
Health, Bethesda, Maryland

Graham Barry Warren[d]   Department of Biochemistry, University of Dundee,
Dundee, Scotland

Judy White[e]   Yale University School of Medicine, New Haven, Connecticut

Jan Wilschut   Department of Physiological Chemistry, University of
Groningen, Groningen, The Netherlands

Joshua Zimmerberg   Molecular Forces and Assembly Group, Laboratory
of Biochemistry and Metabolism, National Institute of Diabetes, Digestive,
and Kidney Diseases, National Institutes of Health, Bethesda, Maryland

U. Zimmermann   Department of Biotechnology, University of Würzburg,
Würzburg, Federal Republic of Germany

Present affiliation:
[a]Organon Teknika, Turnhout, Belgium.
[b]CIBA-GEIGY, Ltd., Basel, Switzerland.
[c]Department of Physiology and Biophysics, Wright State University, Dayton
Ohio.
[d]Department of Cell Biology, Imperial Cancer Research Fund, London,
England.
[e]Department of Pharmacology, School of Medicine, University of California—
San Francisco, San Francisco, California.

Contents

Introduction

CELLULAR MEMBRANE FUSION: A LOCAL CATASTROPHE
WITH TEMPORAL AND SPATIAL CONTROL

Cellular membrane fusion represents an incongruity as a concept: cellular membranes are by definition the structures that completely separate two aqueous compartments. They therefore define and maintain the individuality of each cell and each organelle within the cell. Their fusion abrogates that very individuality by allowing the components of two separate membranes to coalesce, resulting in mixing of the two aqueous compartments enclosed by these two membranes. Obviously, membrane fusion would be incompatible with cellular evolution unless it was highly controlled with respect to both time and space. Each cell maintains its individuality as a replicating unit, by not fusing with adjoining cells, and yet the egg cell will fuse with one sperm cell that has managed to reach it, while subsequently developing resistance to thousands of other newly arriving sperm cells. This is an example of exquisite time control for membrane fusion. Myoblast fusion into giant myotubes can be timed by maintaining the individual cells at low Ca^{2+} and then triggering their fusion by additions of Ca^{2+}! Membrane fusion is occurring continuously within the interior of a cell, as part of endocytosis, secretion, membrane recycling, and so forth, and yet an exquisite order is maintained by each type of fusion responding to a unique "stimulus" at specific "recognition" sites.

The diversity of cellular membrane fusion phenomena could provoke one to think that there must be a multitude of mechanisms that can account for such diversity. Yet, it is quite possible that the diversity is due to the superimposed controls necessary for the proper timing and positioning of a particular fusion event, and that beyond these controls, the mechanism for the fusion reaction itself could be very similar in many, or even all, cases. One is struck by the analogy of the key and lock, where the diversity in shapes is reduced to the same action once the "fit" is accomplished. I suggest that this "fit" in cellular membrane fusion could simply be the "hydrophobic" contact between the two candidate contacting membranes. This is not normally possible, because of the highly hydrated nature of the molecules aligning the surface of cell membranes. These are not only the parts of membrane proteins protruding from the bilayer, but also the headgroups of the majority of membrane lipids, such as most of the phospholipids and glycolipids. Dehydration at the point of close contact between two membranes and the coincidence of packing defects among lipid molecules at this point could be sufficient requirements for a hydrophobic contact and initiation of fusion.

Such dehydration and generation of defects at the point of contact has been suggested for the Ca^{2+}-induced fusion of liposomes (1-3; Chapter 4), and, by extrapolation, for other Ca^{2+}-requiring processes, such as exocytosis during secretion. Such a simple mechanism could also account for PEG-induced fusion (Chapters 11,28,30), and perhaps for "electrofusion" (Chapter 29). It is even more tenuous but still possible to make this extrapolation for cell-cell fusion phenomena, such as fertilization (Chapter 27) and myogenesis (Chapter 26), but much more difficult for other fusion processes that do not require Ca^{2+}, such as virus-cell fusion, or endoplasmic reticulum-Golgi and other miscellaneous intracellular trafficking.

Having made an attempt to generalize, I will now point out some of the unique features that distinguish some fusion phenomena from others. The requirement for Ca^{2+}, although prevalent for many fusion reactions, does not appear to be so for virus-cell fusion (Chapter 13) or other intracellular fusion phenomena not related to exocytosis (Chapter 21). Low pH has been established as the triggering factor for some of the virus-cell fusion events (Chapter 15) and may also be involved in intracellular "trafficking" in situations where fusion is initiated in the lumen side of vesicular membranes (Chapter 22). Very little is known about other natural intracellular triggering signals for fusion apart from Ca^{2+} and low pH. Polyamines may certainly be a possibility (4). Other striking differences between various cellular events are topological, such as ectoplasmic (cell to cell) as opposed to endoplasmic (vesicle to plasma membrane) and planar (cell to cell, vesicle to plasma membrane) as opposed to tubular (endocytosis) (5).

Model membranes such as liposomes and planar lipid bilayers have their own peculiar characteristics with respect to fusion, and their value comes from the observed differences as well as similarities to the fusion characteristics of cellular membranes. It would be naive to dismiss them as too simple compared to cell membranes, or as too dissimilar in reference to specific fusion characteristics (Chapter 3). Their importance as a tool toward understanding the mechanism of membrane fusion comes exactly from their simplicity and their potential to imitate individual facets of membrane fusion (Chapter 4). Beyond that, different levels of sophistication can be imposed by increasing the complexity of the components and their control by other elements (Chapters 9,10,18). This approach, the synthetic one represented by model membranes, is complementary, and not a substitute for the more direct analytical approach applied to biological membranes (Chapter 1). Because of the complexity of cellular membranes, both approaches are necessary for reaching an understanding of the mechanism of membrane fusion at the molecular level.

The value of this book derives precisely from the diversity of the views and approaches of the various contributors toward the phenomenon of membrane fusion. These vary from the most basic kinetic modeling of artificial membranes (Chapter 5) to the clinically relevant question of viral infectivity (Chapter 13) or, from a theoretical understanding of the physical forces between two membranes in close approach (Chapter 3), to the engineering of efficient drug delivery vehicles (Chapters 31,32,36).

Demetrios Papahadjopoulos

REFERENCES

1. Portis, A., Newton, C., Pangborn, W., and Papahadjopoulos, D. (1979). *Biochemistry* 18:780-790.
2. Düzgüneş, N., Paiement, J., Freeman, K. B., Lopez, N., Wilschut, J., and Papahadjopoulos, D. (1984). *Biochemistry* 23:3486-3494.
3. Papahadjopoulos, D., Nir, S., and Düzgüneş, N. (1990). *J. Bioenergetics Biomembranes* 22:157-179.
4. Hong, K., Schuber, F., and Papahadjopoulos, D. (1983). *Biochim. Biophys. Acta* 732:469-472.
5. Membrane Dynamics Group Report. In *Transport of Macromolecules, Life Sciences Research Report 11* (S. Silverstein, ed.), Dahlem Conferences, Berlin, pp. 503-516 (1978).

MEMBRANE FUSION

GENERAL ASPECTS OF MEMBRANE STRUCTURE AND FUNCTION

1

Lipid and Protein Structure of Biological Membranes

ALAN M. KLEINFELD*

Harvard Medical School, Boston, Massachusetts

I. INTRODUCTION

Biological membranes are composed of a wide variety of lipids, proteins, and carbohydrates. The complex topological and compositional arrangement of biological membranes no doubt reflects the functional diversity of these organelles. A complete explanation of the many functions performed by biological membranes requires that the structure and dynamics of the constituent elements, as well as the interactions among the constituents, be understood at the molecular level.

At the present time membrane structure is reasonably well described by the "fluid mosaic" model developed by Singer and Nicolson (1). In the context of this model, the organization of transmembrane and peripheral proteins with respect to the lipid bilayer gives the membrane its mosaic character. Free diffusion of lipids and proteins, both laterally and rotationally, gives the membrane its fluid character. Guided by this general outline, research during the past 15 years has focused on membrane structure and dynamics. This chapter presents an overview of recent developments in the structure of membrane components and certain of their interactions.

Phospholipids represent the major components of what in most membranes is a complex mixture of many different varieties of lipid molecules. High-resolution structures of phospholipid molecules have been determined by X-ray diffraction and nuclear magnetic resonance (NMR). There is, therefore, a reasonably complete understanding of the molecular structure of the constituents of the lipid bilayer. The situation with regard to protein is much less complete, if for no other reason than the much greater diversity of membrane proteins as compared to lipids. Application of DNA techniques has yielded the sequences of a number of major membrane proteins. As far as high-resolution tertiary structure is concerned, however, there is still relatively little known. With the exception of the recently solved structure of the photosynthetic reaction center of *Rhodopseudomonas viridis*

Present affiliation: Medical Biology Institute, La Jolla, California.

3

at 3 Å (2), no protein has been solved at a resolution high enough to
identify individual amino acids (~2 Å is needed). The best resolution obtained for a membrane protein with most of its structure within the bilayer
is 7 Å, and this has been achieved only for a single protein, bacteriorhodopsin (3).

Studies involving the interactions between lipids in model membranes
have clarified the role of acyl chain length, degree of acyl chain saturation,
and the nature of the head group. These lipid characteristics have also
been correlated with phospholipid packing and mobility, bilayer width,
vesicle size, and the physical state of the membrane. Interactions between
different lipids, especially binary mixtures of phospholipids, have been
shown to lead to phase separation. Biological membranes, although vastly
more complex, exhibit many of the properties of model membranes.

Protein-protein interactions of the kind that lead to oligomeric associations of proteins within the plane of the membrane have been more difficult
to study than lipid-lipid interactions. Two-dimensional crystals of proteins
as well as transmembrane complexes of several subunits are observed in
a number of membranes. A review of this subject has appeared (9), and
it will not be discussed further in this chapter.

The interaction between lipids and proteins is central to the understanding of membrane function. Membrane lipid must significantly affect protein
function, since membrane proteins generally remain tightly bound to the
lipid bilayer and it is likely that the protein structure is at least partially
determined by its dissolution in the bilayer. The question, then, is whether
this dissolution is passive, so that any medium of appropriate dielectric
constant would achieve the same structure, or whether the interaction
is specific to the nature of the lipid. The past several years have witnessed
a considerable advance in our understanding of this issue and a significant
section of this chapter is devoted to its discussion.

This chapter has as its focus recent developments in membrane structure,
and it represents an expanded and more current review than in Kleinfeld
(5). Membrane dynamics, such as lipid and protein mobility, have been
reviewed elsewhere and will be touched upon only briefly in this review.

II. LIPID STRUCTURE

A. Properties of Individual Lipid Molecules

Lipid constituents of most biological membranes can be classed as either
glycero, sphinganine, or sterol. The molecules in each of these classes
are amphipathic, with a specific polar head group aligned near the lipid-
water interface, followed by an intermediate region which may also have
some polar character; finally, there is a portion of the molecule that extends
into the hydrocarbon region of the bilayer. To varying degrees, the lipid
structure of these molecules is reasonably well understood at the angstrom
level of resolution. X-ray diffraction studies on single crystals and X-ray
diffraction and NMR of aqueous lipid dispersions provide a reasonably
complete picture of the molecular structure of the major lipids (see Ref. 6
for review and Ref. 7 for a brief review of lipid structure in biological
membranes). Although it has not yet been possible to use this structural
information to derive a complete picture of the bilayer, these results,
as discussed below, place severe restrictions on the possible forms of
lipid associations.

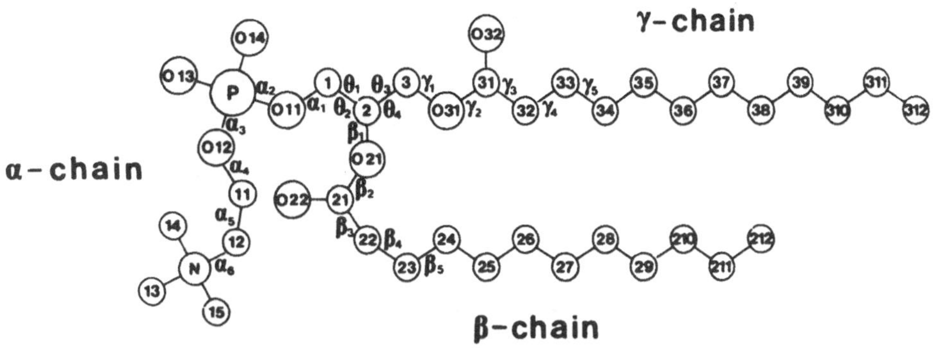

FIGURE 1 X-ray crystallographic structure of phosphatidylcholine. A similar structure was obtained for phosphatidylethanolamine. [Reprinted with permission from Hauser et al. (6).]

1. Glycerophospholipids

Molecular structures of phosphatidylcholine (PC) and phosphatidylethanolamine (PE) obtained from single crystals are shown in Figure 1, which illustrates some of the general features of lipid structure. The axis formed by the phosphorylcholine or phosphorylethanolamine groups is nearly perpendicular to the gamma chain so that the head group lies parallel to the membrane surface. These results obtained by X-ray diffraction and NMR on single crystals (6,8,9) have been confirmed by X-ray, neutron diffraction, and NMR on fully hdyrated dispersions of phospholipids (10-13). The surface parallel conformation of the head group is not affected by hydration or by the presence of cholesterol (11). Thus, the stability of the head group orientation is determined by intramolecular interactions, and therefore, this configuration probably prevails in biological as well as model membranes. Although the orientation of the phosphorylcholine or phosphorylethanolamine axis relative to the plane of the membrane is rigid, the axis itself undergoes rapid rotation ($>10^4$/s) about the c(1)-c(2) axis and, of course, the molecule as a whole rotates about an axis normal to the surface.

According to the crystal structure, a polar region can be defined which extends from the head group to the carbonyls of the glycerol backbone and has a thickness, depending on head group and acyl chain length, of between 8 and 10 Å. These results are in excellent agreement with the electron density profile obtained by X-ray diffraction on aqueous dispersions (14). Thus, a significant fraction of the bilayer thickness may form a region of relatively high dielectric constant (also suggested by EPR and fluorescence studies, see below). The acyl chain orientation is parallel or tilted to the normal in the gel state. In either case, however, the phospholipid crystal structure (6), as well as NMR studies of model and biological membranes (12), indicate that the first two carbons of the beta chain are parallel to the surface and therefore the beta chain does not extend as far into the bilayer as the gamma chain.

An important consequence of these studies is that structures determined from the solid crystal can be applied to the fully hydrated state. Even the rotational mobility of the head group about the c(1)-c(2) axis can be inferred from the single-crystal structures. On the other hand, the strong hydrogen bonding between head groups that is observed in the crystal study is probably not as significant in the fully hydrated state, at temperatures above the phase transition, where the average separation of head groups exceeds hydrogen bonding distances. Thus, while the crystal structure provides an essential starting point for using constituent structures to predict the form of the physiologically interesting aggregate, information from other sources will be necessary to completely understand membrane structure.

2. Sterols

Cholesterol is one of the major constituents of the plasma membrane of most eukaryotic cells. Single-crystal, high-resolution studies have been carried out on a large number of different sterol molecules (15). These studies clearly show the planar configuration of the steroid ring and suggest that the hydrocarbon tail exhibits some folding, although the extended conformation is energetically most favorable. The configuration of cholesterol in membranes composed of cholesterol-phospholipid mixtures has been studied by X-ray diffraction (16) and neutron diffraction (11). These studies indicate that cholesterol is oriented in membranes with its hydroxyl group (at the 3 position of the steroid ring) located in the polar region of the bilayer, its steroid ring plane parallel to the lipid acyl chains, and its hydrocarbon tail extended into the bilayer. The steroid ring is found by ^{2}H-NMR studies (see Ref. 13 for review) to be highly ordered ($S_{mol} = 0.87$) relative to the lipid acyl chains of egg PC.

3. Sphingolipids

Sphingolipids, especially the sphingomyelins (sphingosine with a phosphorylcholine head group), are major components of animal cell membranes (see reviews in Refs. 17,18). The structure of the sphingolipids follows the same general pattern as that of glycerophospholipids and sterols; sphingolipids have a polar head group, a region of intermediate polarity, and a nonpolar hydrocarbon tail region. The ceramide moiety is that portion of the molecule which does not include the head group and is composed of a sphingosine (a C_{18} amino alcohol) and a fatty acid linked to the 2-amino position of the sphingoside. These fatty acids are generally saturated and range from C_{16} to C_{28}. Single-crystal studies of a neutral glycosphingolipid (19), as well as studies of several structures (20 of the component moieties (the sphinosine base and fatty acid), have yielded a fairly complete picture of the atomic structure. This image is quite similar in many respects to that of the glycerophospholipids. Pascher, however, has noted that sphingolipids have a considerably different hydrogen bonding configuration. It is suggested that these interactions are necessary to maintain bilayer stability in the face of the potentially disruptive interactions of the highly polar hydroxyl and amide groups of the sphingosine moiety.

An important characteristic of sphingolipids is that the ceramide moiety forms the base of the major glycolipids, the glycosphingolipids. Although glycosphingolipids constitute a fairly minor membrane component, they are believed to play major roles in a number of important cellular functions. It is also not surprising, given their probable role in cell-cell interactions, that the glycolipids tend to be concentrated in the plasma membrane. Interestingly, the sphingomyelins seem also to be concentrated in the plasma membrane. The carbohydrate moieties of glycolipids exhibit a wide variety of configurations, from simple single-sugar neutral and charged head groups (cerebrosides and sulfatides) to multisugar complexes which are charged and highly branched (gangliosides).

III. LIPID-LIPID INTERACTIONS

A. Single Lipid Species

1. Bilayer Structure

Aqueous dispersions of phospholipid give rise to a number of distinct structures [21], including the bilayer, hexagonal tubes, and micelles, as shown in Figure 2. The nature of the phase depends on lipid species, lipid concentration, temperature, pH, and ionic strength. Nonbilayer or hexagonal tube phases have been found to coexist with the bilayer phase for special lipid mixtures in model systems (22,23). Since in biological membranes the majority of the lipids are organized in a lamellar arrangement, the following section deals primarily with the bilayer phase. The possible existence and role of nonbilayer structures in membrane function are discussed in Chapters 2, 3, 4, 6, and 10.

Specific conditions are required for bilayer formation. PCs with acyl chain lengths greater than 12 carbons, for example, are capable of forming bilayer phases over a wide range of conditions, including those relevant to the physiological state. Bilayers of pure PE, however, form only within a restricted range of temperatures and at pH greater than 11. Structural studies on single-lipid crystals provide insight into conditions necessary for bilayer formation (6). These results indicate that the PE head group occupies an area of about 38 Å^2 both in the anhydrous state and in the

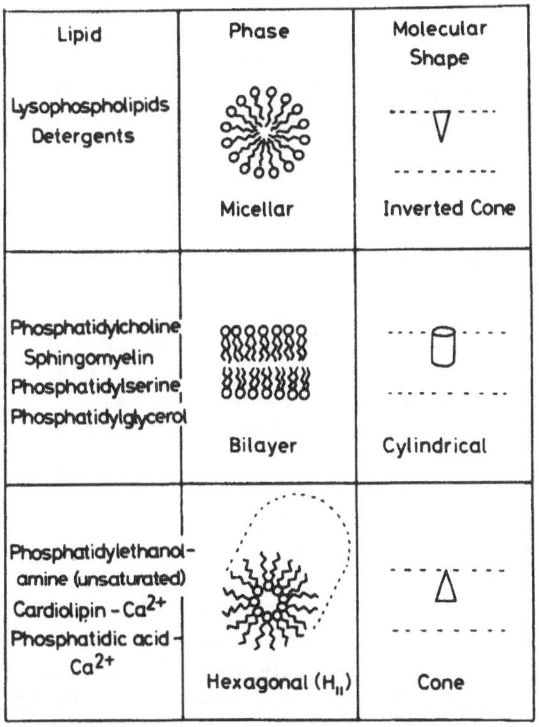

Lipid	Phase	Molecular Shape
Lysophospholipids Detergents	Micellar	Inverted Cone
Phosphatidylcholine Sphingomyelin Phosphatidylserine Phosphatidylglycerol	Bilayer	Cylindrical
Phosphatidylethanolamine (unsaturated) Cardiolipin - Ca^{2+} Phosphatidic acid - Ca^{2+}	Hexagonal (H$_\text{II}$)	Cone

FIGURE 2 Phospholipid phases in aqueous dispersions. The structure of each phase is shown in the center panel, examples of phospholipids that form such phases are given in the left panel, and the overall shape of the phospholipid molecule is shown in the right panel. [Reprinted with permission from Cullis and de Kruijff (22).]

presence of excess water. The PC head group, on the other hand, occupies an area of about 50 $Å^2$ in the anhydrous state and about 60-70 $Å^2$ in the presence of excess water. Below the transition temperature the sum of the acyl chain cross-sectional areas (for saturated chains) is about 20 $Å^2$, small enough for these chains to pack either perpendicular or tilted to the surface of both PE and PC bilayers.

As the temperature is raised above the transition temperature, the acyl chain area expands to 50 $Å^2$. The acyl chains will therefore curve outward from the head group, forming the inverted micelle shown in Figure 2. For PE where head groups maintain tight packing even in excess water, acyl chain-head group packing constraints are not consistent with a bilayer phase. The inverted micelles that are formed under these conditions are not stable in isolation and combine to form the hexagonal tube phase (Fig. 2). At elevated pH, PE acquires a net charge and the resulting electrostatic repulsion creates a sufficiently large head group area to accommodate a bilayer phase. In contrast, the head groups of hydrated PC occupy a suifficiently large area for acyl chain expansion to occur within a bilayer phase. In the liquid crystalline state, in fact, the PC head groups occupy a considerably greater area than is necessary to accommodate the acyl chains. Over a wide range of conditions, therefore, PC is able to form bilayers.

X-ray diffraction studies on dispersions of PCs demonstrate that the thickness of the glycerohydrocarbon region increases from about 26 Å for 12 carbon acyl chains to about 30 Å for the 18-carbon lipid (24). Since this increase is less than expected for fully extended acyl chains and since, for the same variation in acyl chain length, the area/phospholipid is found to increase from 60 $Å^2$ to 75 $Å^2$, appreciable folding is suggested for the central portion of the acyl chains. The head-to-head width of biological membranes is between 40 and 45 Å, according to X-ray diffraction studies on multilayer preparations (14). This value is consistent with the 30-Å width of the glycerohydrocarbon region of an 18-carbon lipid and a head group to glycerol backbone thickness of between 5 and 7 Å from the lipid single-crystal results (6).

The studies of the PC bilayer thickness indicate that the terminal ends of the acyl chains fold rather than intercalate between the acyl chains of the opposite bilayer leaflet. This suggests that lipid-mediated coupling of the two bilayer leaflets (a possible transbilayer signaling mechanism) is not a significant factor in PC bilayers. X-ray diffraction and electron microscopy of PCs with mixed 10- and 18-carbon acyl chains suggest, however, that interdigitation, and therefore possible bilayer coupling, occurs when the bilayer is in the gel state (25,26).

2. *Properties of the Bilayer Normal to the Surface*

The interior of the membrane is not well approximated by a uniform isotropic hydrocarbon fluid, as witnessed by several important properties of the bilayer that exhibit significant variation in the direction normal to the surface. The degree of order as a function of bilayer depth has been studied by ^{2}H or ^{19}F NMR using lipids selectively labeled along the acyl chain (12,27). The molecular order parameter S_{mol} reflects the average orientation of each segment along the acyl chain, having a value of zero for the completely disordered isotropic chain and a value of unity for a chain in the all-*trans* state. Seelig and his co-workers have found that except for a slight dip at the 2 position, S_{mol} is relatively constant between carbons 1 and 9 and then falls rapidly for positions deeper in the

bilayer, reaching a minimum at the center. The value at the minimum is
about threefold smaller than the maximum value at the surface of the bilayer.
Similar profiles were obtained in the ^{19}F studies of Macdonald et al. (27).
In the liquid crystalline state the profiles are relatively unaffected by
fatty acyl chain substitutions. For *A. laidlawii B* membranes in the gel state,
however, enrichment in palmitate, but not unsaturated fatty acids, abolishes
the order profile gradient (27). The behavior observed in the ^{2}H and [19]F
studies is characteristic of model membranes, biological membranes, and
whole cells.

The order parameter profile has also been investigated by measuring
the fluorescence anisotropy decay of the n-(9-anthroyloxy) fatty acids
(28-30). These fatty acids labeled at one of eight different positions along
the chain are added exogenously to the membranes and can therefore be
used to study a wide variety of membranes. In Figure 3 the order parameter
profiles determined by fluorescence anisotropy are shown for red cell ghosts,
liposomes, and paraffin oil. These profiles are in good agreement with
those obtained by NMR. The larger S values for ghosts presumably reflect
the high cholesterol content of these membranes while, as expected, the
isotropic solvent paraffin oil exhibits an almost flat profile.

The order parameter reflects the ensemble-averaged orientation and
may not be simply related to the rotational rate or segmental mobility.
Information on the acyl chain mobility has been obtained from ^{13}C NMR
and fluorescence decay anisotropy measurements (28,29,31; Kleinfeld,

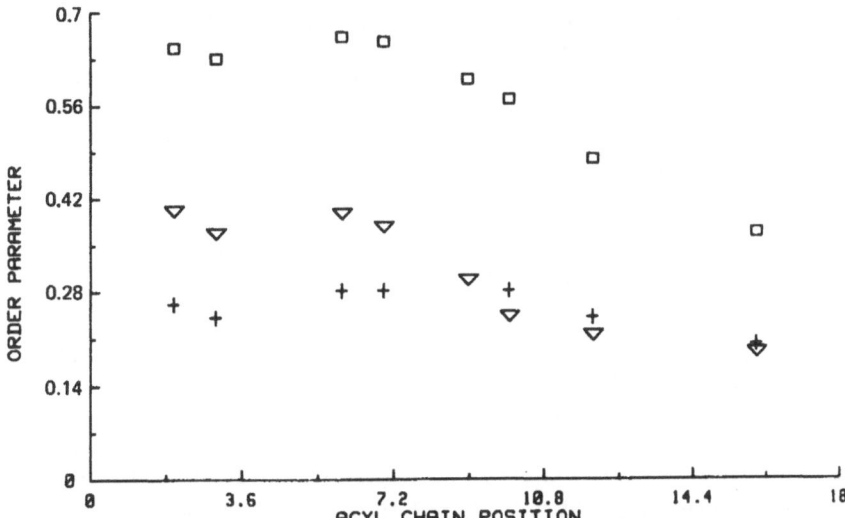

FIGURE 3   Order parameter variation along fatty acyl chains as determined
by the decay anisotropy of the anthroyloxy fatty acids. Fluorescence decay
anisotropies of the n-(9-anthroyloxy) fatty acids (n between 2 and 16)
were measured in red cell ghosts (□), small unilamellar vesicles of egg
phosphatidylcholine (▽) and paraffin oil (+). Differential polarized phase
lifetimes were measured at an excitation wavelength of 383 nm and the
results were analyzed according to Kutchai et al. (1983). The order parameter was calculated as $S = (r/r_0)^{1/2}$, where r is the measured value at 383
nm and $r_0 = 0.3$. (A. M. Kleinfeld, unpublished observations.)

unpublished observations). The results of these studies indicate that the mobility profile does, in fact, parallel the order parameter variation.

The polarity of the bilayer displays appreciable variation with position along the acyl chain and probably reflects the degree of water penetration into the bilayer. The first measurements of the polarity profile were performed using the n-doxylstearic spin-labeled fatty acids (32). These electron spin resonance (ESR) measurements, in fully hydrated membranes, demonstrated that the polarity at the C_5 position was greater than ethanol, while the polarity between the C_{12} and C_{16} positions was intermediate between mineral oil and ethanol. In dehydrated membranes, however, all probes sensed a similar polarity, suggesting that the observed gradient was due to water penetration. Confirmation of the hydrated membrane profile has been obtained by fluorescence lifetime studies of the anthroyloxy fatty acid probes (33). The degree of water penetration appears to be modulated by lipid composition (34). In membranes composed of a 1:1 mixture of PE and cholesterol, the depth to which a significant amount of water can penetrate appears to decrease by about 2.5 Å, compared to pure PE.

3. *Phase Transitions*

Phase transitions in lipid bilayers have provided an important tool for the investigation of membrane structure. The phenomenon is a collective one and is thought to hold the key to many important cellular events. It is readily observed in model bilayers and its occurrence (in the form of lateral phase separation, discussed below) has been observed in some cellular membranes (35,36).

Lipid bilayer phase transitions involve changes primarily in the acyl chains from a less to a more disordered state. There are a number of bilayer phases through which the transitions can proceed, and the occurrence of these phases depends primarily on the lipid head group. Using the nomenclature of Ranck et al. (37), the phases below the main phase transition are designated with a beta and those above with an alpha. The main transition is accompanied by a significant enthalpy change (~9 kcal/mole) and is therefore considered to be first order although, since it occurs with a finite width, it is not rigorously a phase transition (36). In phosphatidylcholine with homogenous fully saturated acyl chains there is a state in which the chains are all *trans*, tilted to the bilayer, and in a monoclinic form; this state is designated $L_{\beta'}$. In addition, there is another state (the $P_{\beta'}$ state) in which the chains are also all *trans* and tilted but, in addition, a long-range ripple is superimposed on the monoclinic gel state. This state is more disordered than the $L_{\beta'}$ and therefore occurs at higher temperature (and is sometimes referred to as pretransition). Recently it has been observed (38-40) that in dipalmitoylphosphatidylcholine (DPPC), the gel state is not an equilibrium state and that after prolonged incubation at 0°C the gel state slowly (>3.5 days) transforms into a more ordered state. Upon heating, the transition from this state (crystalline $L_{\beta'}$) to $L_{\beta'}$ is accompanied by a net enthalpy change. Thus, a full description of the transitions in DPPC is $L_{\beta'}$ (crystalline) → $L_{\beta'}$ (gel) → $P_{\beta'}$ → L_α. Only the main transition is observed in phosphatidylethanolamine, and this occurs at approximately 20°C above the corresponding PC transition (36).

In addition to changes in the acyl chains, the bilayer width decreases and the area/lipid increases for a $L_{\beta'}$ to L_α transition. The increase in lipid area is also associated with an increase in hydration from 11 to 15 molecules H_2O/DPPC molecule below the main transition to about 27 above.

IV. LIPID MIXTURES AND DOMAINS

It was implicit in the original formulation of the fluid mosaic model (1) that the membrane constituents formed a well-mixed fluid. Numerous studies now indicate that membrane components exhibit varying degrees of immiscibility. Rates of protein diffusion are generally inconsistent, with movement restricted solely by a homogeneous lipid viscosity, and in some instances may be consistent with zero mobility. Studies on the lipid composition of biological membranes generally demonstrate compositional asymmetry between the inner and outer bilayer leaflets and therefore an immiscibility of the lipids across the bilayer. Our focus here is on lateral immiscibility of lipids.

Lateral immiscibility in the form of lateral phase separation has been amply demonstrated in model systems (41-43). Phase separation has also been demonstrated in specialized biological membranes or those perturbed by enrichment or depletion of a particular component (fatty acids supplementation or sterol depletion). A more difficult issue, which has received considerable attention recently, is whether lipid immiscibility is a general phenomenon in biological membranes, resulting in small and consequently difficult to detect regions of lipid phase separation. As these issues have recently been the subject of reviews elsewhere, only a brief discussion of this subject is presented here (44-48).

A. Model Membranes

Phase behavior has been studied in model membranes composed primarily of binary mixtures. Investigations have focused on mixtures of PC using virtually every biophysical as well as many biochemical methods. To a lesser degree, PE, sterols, and sphingolipids have also been studied. It is generally accepted that there are four different phases or phase mixtures in binary systems. (1) For lipids with nearly identical structure and/or phase behavior, bilayers exhibit nearly ideal mixing, so that above some critical temperature the composite system is entirely fluid, while below this temperature the acyl chains are all *trans*. This is virtually indistinguishable from a single species phase transition discussed above. True ideal mixing, therefore, implies near homogeneity in composition. (2) Binary mixtures of PC, such as DMPC and DPPC, and PC-sphingomyelin (49-51), in which the components are similar, exhibit nearly ideal mixing but differ somewhat in their phase transition temperature. Thus, above a critical temperature the lipids form an ideally mixed liquid phase. As the temperature is reduced, a value is reached where the presence of the higher-melting-point species causes solid domains to form in equilibrium with the fluid lipid. The solid domains thus formed are not homogeneous in composition, but have a greater proportion of the higher melting component at higher temperatures. When the temperature is lowered sufficiently, the entire phase becomes a well-mixed solid phase in which both components cocrystallize. (3) For lipids sufficiently different from one another the liquid phase will approach ideality. Lowering the temperature yields a coexisting fluid-solid phase. Finally, as the temperature is reduced further, two solid phases form in which the lower melting point crystallizes separately from a phase composed primarily of the higher-melting component. (4) In a few special instances liquid-liquid immiscibility is exhibited. This behavior appears to depend on the nature of the lipids (52,53) or the radius of curvature of the bilayer (49,50). A recent study employing rapid-freezing differential scanning calorimetry (DSC) indicates that liquid-liquid immisci-bility may be an important characteristic even of binary mixtures that appear to be well mixed by conventional DSC (54).

The nature of the liquid-solid phase separation has been investigated by combining freeze fracture electron microscopy, electron diffraction, and ESR (41,42). In all cases examined, phase separation appears to occur laterally within the plane of a single hemileaflet. Thus the separation does not correspond to separation into distinct membranes (fission), nor does it appear that the phases of the two leaflets of the bilayer are correlated. Lateral phase separations can be induced isothermally, as, for example, in the case of addition of Ca^{2+} to phosphatidylserine (PS)-PC mixtures. The detailed topology of the phase separation is more difficult to assess, although some evidence suggests patches with dimensions of the order of 0.2 to several μm (55,169,170).

Cholesterol's prevalence in the plasma membrane of most eukaryotic cells as well as its importance in disease has led to considerable interest in the physical properties of cholesterol-phospholipid mixtures. This interest has been rewarded by a rich variety of observed phenomena. Cholesterol almost certainly plays an important role in lateral phase separation. How-ever, the complexity of the interaction of cholesterol and other lipids has limited our ability to predict the lipid topology of membranes containing cholesterol (see, for example, Ref. 48). Indeed, even the relatively straightforward issue of spontaneous cholesterol exchange between lipid vesicles is the subject of continuing controversy. Studies of Bar et al. (56,171) suggest that as much as 90% of cholesterol in small unilamellar vesicles composed of various phospholipids is nonexchangeable. It is pro-posed that these mixtures are extremely nonideal and that stable domains exist in certain phospholipid mixtures. In particular, there appears to be a highly preferential association of cholesterol and sphingomyelin (171-173).

Several studies indicate that glycosphingolipids can exhibit lateral phase separation as admixtures in phospholipid membranes (18,57). Because of their relatively long and often saturated fatty acid chains these lipids are expected to be good candidates for formation of gel domains. It has been suggested that addition of Ca^{2+} to membranes containing GM_1 gangli-osides would cross-link the negative head groups and thereby induce clustering. A recent study by Masserini and Freire (57) suggests that clustering of GM_1 is driven by acyl chain interactions and is only in-directly a function of Ca^{2+}. Thus, in a mixture of phospholipids and GM_1, the major effect of Ca^{2+} is to bind to the PC head groups (with greater affinity than to sialic acid). This in some way alters the phospholipid-GM_1 interaction to cause exclusion of GM_1 from the mixture. The thermotropic behavior of the excluded GM_1 is characteristic of its saturated long acyl chain.

B. Biological Membranes

Biological membranes are obviously more complicated than the binary model membranes discussed above. The increased heterogeneity manifests itself both in head group and in acyl chain diversity; a considerable fraction of the acyl chains are *cis*-unsaturated and would therefore be expected to undergo phase transitions at very low temperatures compared to 37°C. In spite of these factors, membranes obtained from cells specifically en-riched in a particular fatty acid, or depleted in sterols, exhibit thermotropic behavior similar to that of binary model membranes (48,58). In addition, a number of cells exhibit macroscopic lateral phase separation (59-61,174). These large-scale segregations of certain lipid classes are probably a result of lipid-lipid interactions as well as interactions of lipids with other cellular constituents.

In addition to these relatively large effects, there appears to be a class of phenomena, suggesting domains or clusters of lipids on a microscopic scale, for which the evidence is more indirect. The evidence for such domains has been obtained from calorimetry, photobleaching, and spectroscopy. Many of these effects have been discovered at temperatures appreciably below the physiologically important ones. Clearly, a central issue is whether membrane lipid domains are significant structural factors at physiological temperatures. Several studies do, in fact, indicate domain formation under physiological conditions (62,63,175-178).

A number of cellular phenomena have been interpreted in terms of lipid domains. These phenomena generally involve the observation that altering fatty acid composition, either by changing the lipid acyl chain composition (64,65) or by adding exogenous fatty acids (46-48,66), alters a specific cellular function although the fatty acids remain unesterified. Modulation of cellular function by sterol alteration also suggests domains in lymphocytes (67).

These observations suggest a number of physiological roles for domains in cellular function. For example, depending on their function, proteins may be located in regions where the lipids are in a solid phase and partitioning of lipophilic molecules (fatty acids, drugs, hormones) into this region may serve as a trigger for a particular function. There are also a number of studies which suggest that domain formation itself may occur in response to physiological stimulus or growth (68-70,174).

V. MEMBRANE PROTEIN STRUCTURE

Complete understanding of cellular function and the effects of lipids on this function must ultimately be based on the molecular structure of membrane proteins. Unfortunately, our knowledge of the molecular structure of membrane proteins is quite primitive. There is only one integral membrane protein (the photosynthetic reaction center) whose structure is known to high resolution. As a consequence, and to a much greater extent than for water-soluble proteins, there has been a great tendency to rely on predictive methods to understand the structure of membrane proteins. This section, therefore, summarizes both the experimental and predictive results. The emphasis is on proteins with major portions associated with lipid. Other recent reviews of membrane protein structure include those of Senior (71), Benga and Holmes (72), and Eisenberg (73).

A. Primary Structure

A benchmark of the difficulty in determining membrane protein structure is that before about 1980 very few membrane proteins had even been sequenced. Only a single protein having a majority of its structure associated with the lipid bilayer was sequenced directly (bacteriorhodopsin) (74,75). This paucity of primary sequence information is a direct reflection of the difficulty of purifying and solubilizing proteins that are tightly associated with the lipid bilayer. With the advent of nucleic acid sequencing techniques this situation has changed significantly. Nucleic acid sequences have now been determined for the following major membrane proteins: lactose permease (76), bacteriorhodopsin (77), Na^+ channel (78), band 3 (79), glucose transporter of the human erthrocyte (80), and the NaK-ATPase (81).

B. Secondary Structure

Secondary structure can be studied by a variety of techniques, including circular dichroism (CD), infrared spectroscopy (IR), Raman spectroscopy, and ultimately X-ray diffraction or electron microscopy reconstruction. As a practical matter, the bulk of secondary structure information has been obtained by the spectroscopic techniques. Each of the spectroscopic techniques has its own advantages and disadvantages. Although CD is sensitive to membrane scattering effects and relatively insensitive to β-sheet structure, membrane scattering effects can be accounted for, and β structure has been determined by CD (82,83). Until recently, IR absorption bands in water limited membrane studies to dried samples (84,85). Recent studies have presented methods to overcome this problem and measurements have been carried out in hydrated samples (86). Raman spectroscopy appears not to suffer from either water or membrane interference; it is, however, appreciably less sensitive than CD or IR (87).

Secondary structure of a fairly wide variety of membrane proteins has been studied by the spectroscopic methods. A good many of these studies indicate that the major secondary structure is α-helix (71,88,89). These results, together with the image of bacteriorhodopsin obtained by electron microscopy (3) and the X-ray image of the photosynthetic reaction center, suggested that, in general, the transmembrane region is formed by α-helices whose axes are roughly parallel to the lipid acyl chains. In addition, a number of theoretical arguments have been advanced to support this view (71,90).

Even in the case of bacteriorhodopsin, however, several dissenting reports have suggested that at least some of the transmembrane region may be composed of β strands parallel to the lipid chains (86,91). Much less controversial evidence for β structure has been obtained in a number of studies which demonstrate that the major portion of the secondary structure is composed of β sheet: the nonpolar peptide of cytochrome b_5 (92), the gap junction (93), three outer-membrane proteins of *Escherichia coli* (94), the α toxin of *Staphylococcus aureus* (83), and other examples cited in Wallace et al. (95). Even turns or bends within the membrane which were thought to be energetically unfavorable are indicated in at least two cases (96,97). In summary, it does not appear that α-helices parallel to the lipid acyl chains constitute the universal secondary structure of the transmembrane region of membrane proteins.

C. Tertiary Structure

High-resolution images of membrane proteins have been obtained by X-ray diffraction of three-dimensional crystals (photosynthetic reaction center) and image reconstruction of two-dimensional electron micrographs. In addition to these methods, which may ultimately yield a complete molecular solution, a variety of experimental methods have been used to determine the location of particular portions of membrane-bound proteins. These approaches can be used to complement diffraction methods that are as yet unable to distinguish individual residues. They also offer the possibility of monitoring at least part of the protein structure in situ and possibly in different physiological states.

Chemical labeling and proteolysis with selected proteases have been used to determine the number of polypeptide loops across the membrane as well as the location and general topology of the extramembranous segments (Band 3) (98), bacteriorhodopsin (99), and histocompatibility antigen (97). Regions within the bilayer have been mapped out using ferritin labeling (100), photoactivatible reagents attached to fatty acids (101,102), and fluorescence quenching of tryptophan with spin-labeled fatty acids (103,104). Resonance energy transfer between tryptophan and lipid-associated acceptors has been used successfully to determine the distribution of tryptophan residues (105-109).

Although portions of membrane proteins that can be removed from the membrane and can be crystallized have been solved by X-ray diffraction (110,111,168), only a single complete membrane protein has been solved at high resolution. Crystallization of the photosynthetic reaction center of *Rhodopseudomonas viridis* has allowed this complex of four membrane proteins to be solved to 3-Å resolution (2). Although at this resolution it is not possible to uniquely identify all amino acid side chains, a complete structure should be available once the primary sequence is fitted to the 3 Å image. Figure 4 shows the image of the polypeptide backbone at 3-Å resolution. Not seen in this figure are the 14 chromophores which serve as prosthetic groups responsible for the transformation of light energy into electron movement across the bacterial membrane. This image is obtained from three-dimensional protein-detergent crystals and, therefore, its transmembrane region cannot be identified with certainty. It is, however, likely that the 10 α-helices connecting the two large globular regions are the transmembrane segments.

It is likely that the crystallization of the photosynthetic reaction center was facilitated by the large fraction of the protein complex that extends out of the lipid bilayer. In contrast, the structure of proteins that are primarily lipid-associated is at a much less refined level. Purification and crystallization of these proteins present much greater difficulties than either water-soluble proteins or membrane proteins with large extramembranous regions. The partial structure of a protein having most of its mass associated with the lipid bilayer was obtained using the naturally occurring two-dimensional crystals of bacteriorhodopsin, formed in the purple membrane of *Halobacterium halobium* (3). Using electron diffraction and microscopy, an image of the protein was obtained at a resolution of 7 Å parallel and 14 Å perpendicular to the plane of the membrane. More recently, this resolution has been improved to 6.5 and 12 Å, respectively (112).

Although at this resolution the location and orientation of individual amino acid residues cannot be discerned, a general outline of the protein shape and secondary structure can be inferred. The model of Figure 5 indicates that the protein is composed of seven rods, roughly arranged as a cylinder of radius 15 Å. Approximately 70% of the cylinder, whose length is about 40 Å, is contained within the membrane. Based on the analysis of the electron density, measurements of the circular dichroism, the primary sequence, and theoretical arguments, it has generally been concluded that the seven rods are α-helices and that the segments connecting

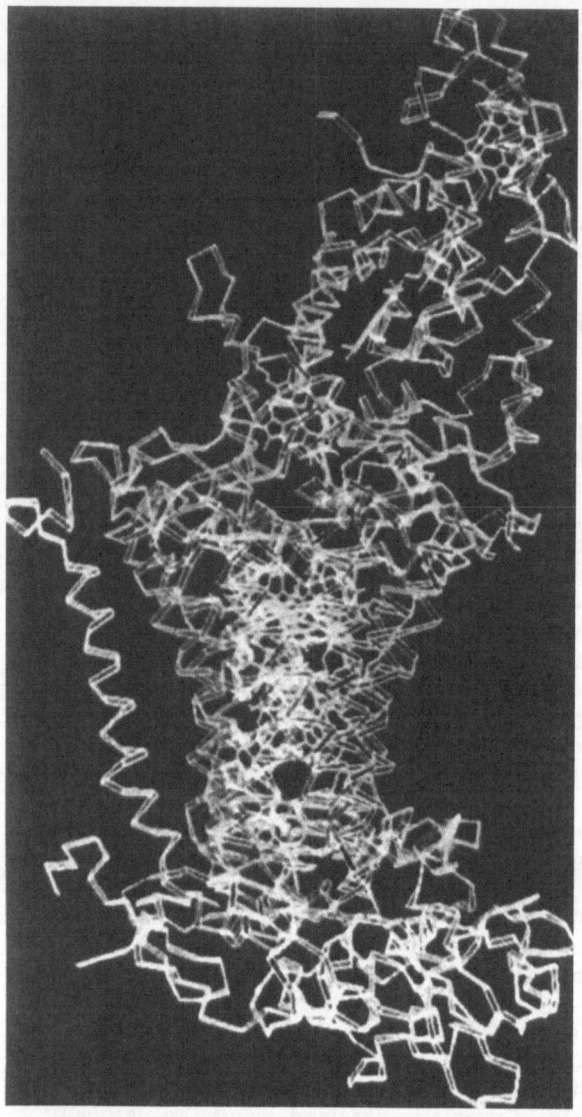

FIGURE 4   Structure of the photosynthetic reaction center of *R. viridis*,
obtained by X-ray crystallography at 3-Å resolution. This view shows
the protein complex in the orientation it is likely to have in the membrane.
The central helical region is the putative transmembrane portion. [Reprinted
with permission from Deisenhofer et al. (2).]

the loops are extramembranous and nonhelical (3,71,74,75,89,113). Plausible
models based on the the α-helical configuration have been proposed for
the spatial arrangement of the primary sequence through the membrane
(75,99,114). These models are consistent with the electron diffraction results and with labeling and proteolytic digestion studies which define the
topology of the extramembranous regions (99,107,115,116).

Outlines of the membrane-associated portions of several other proteins
have been obtained at lower resolution. These include: gap junctions

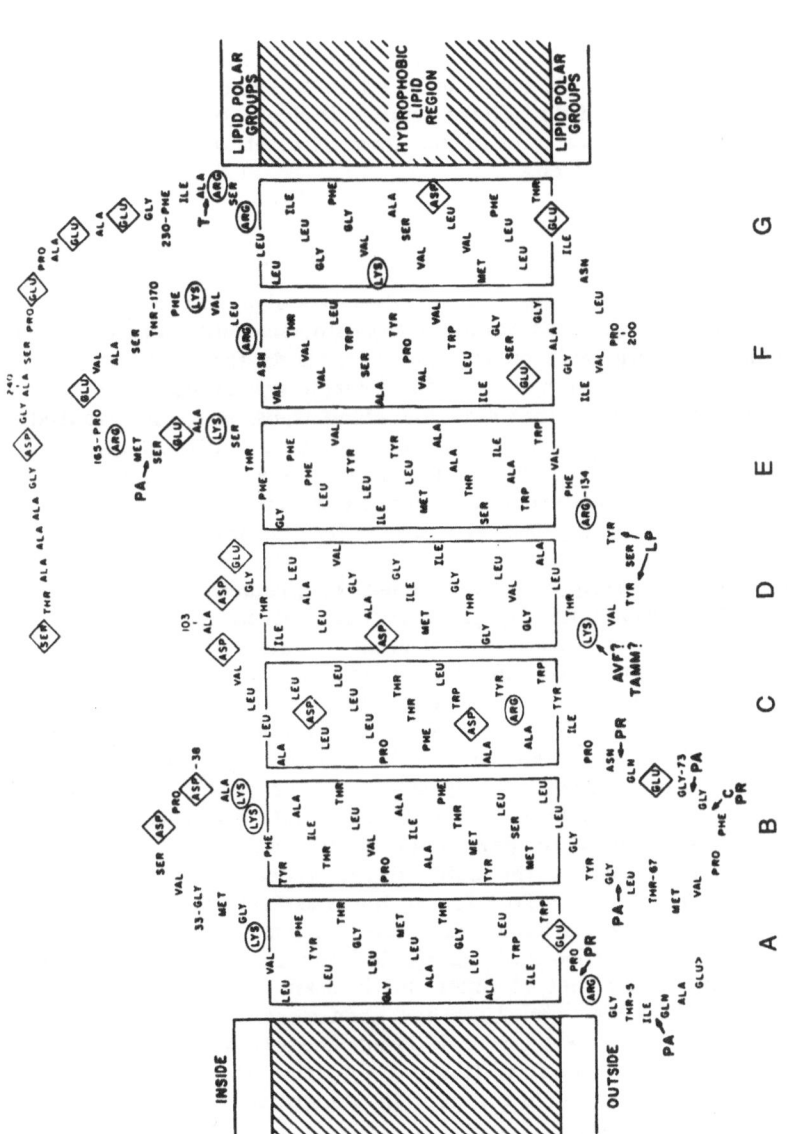

FIGURE 5 A model for the structure of bacteriorhodopsin. This proposed arrangement of the primary sequence through seven alpha helices is based on electron diffraction, energy minimization, and biochemical evidence. [Reprinted with permission from Engelman et al. (99).]

(117,118), ubiquinol:cytochrome c reductase (119), cytochrome oxidase
(120), and the acetylcholine receptor (121). These studies are generally
consistent with a cylindrical shape for the membrane-spanning portion
of the protein. In addition, high-resolution studies have been conducted
on small membrane-associated peptides such as alamethicin (122) and melittin
(123). These studies have solved the structures at resolutions between
1.5 and 2.5 Å, sufficient to determine the location of the amino acid side
chains. The results are important for indicating how side chains interact
with lipids and should be useful for analyzing larger proteins. In the
case of melittin, at least, the recent study of Vogel and Jähnig (124)
suggests that the crystallographic image may not be sufficient to infer
the membrane form of this peptide. These studies indicate that mellitin
may associate with the membrane only in the form of a tetramer whose
axis is parallel to the lipid acyl chains.

D. Predictive Methods

Limited high-resolution membrane protein structures and an increasingly
large number of primary sequences obtained from nucleic acid techniques
have placed considerable emphasis on the use of methods for predicting
the secondary structure of membrane proteins. Predictive methods have
been developed for water-soluble proteins for which there are a substantial
number of high-resolution structures determined by X-ray diffraction.
The most successful of these methods are primarily statistical; the probability of a given amino acid occurring in a particular type of secondary
structure is determined from the database formed by the known high-
resolution structures.

A similar approach has frequently been used for predicting membrane
protein structure: The probability for placement of a given amino acid
in the secondary structure of a membrane protein is obtained from the
database of *water-soluble proteins*. Intrinsic deficiencies in this approach
must be expected since the different environmental constraints imposed
on membrane proteins must certainly lead to different kinds of structures.
Noncovalent interactions that are responsible for the particular secondary
and tertiary structures of a given primary sequence are essentially electrical
in nature. These are certain to be different in the relatively isotropic and
highly polar environment formed by permanent water dipoles and the very
anisotropic environment formed by the lipid bilayer whose polarity decreases
markedly from the lipid/water interface to the hydrocarbon interior.

The validity of predictive methods has recently been reviewed by Wallace
and her collaborators (95). These authors applied several of the statistical
predictive methods, including that of Chou and Fasman (125,126), to membrane proteins whose secondary structure had been measured and to the
photosynthetic reaction center and crambin, membrane proteins whose
structure is known to high resolution. It was found that these methods
have little value in predicting membrane protein structure, at least when
no additional information is used.

Another, increasingly popular approach to predicting membrane protein
structure is to make use of the hydrophobicity of individual amino acids
(127). Thus, using an appropriate scale to gauge the hydrophobicity of
each amino acid side chain (the choice of scale is complex and controversial;
see for example, Refs. 128 and 129), this procedure forms a running average of the hydrophobicity along the primary sequence. The object of this
method is to identify those regions along the sequence that are exceptionally

hydrophobic or hydrophilic. It is assumed that these regions are, respectively, inside and outside the bilayer. This procedure seems to correctly predict the membrane-spanning sequences of the well-understood membrane proteins such as bacteriorhodopsin and the photosynthetic reaction center. Not surprisingly, given the simplicity of this technique, it is often unclear what level to choose as hydrophobic and how short a segment can be considered to be integral to the lipid bilayer. The latter issue is especially relevant if non-α-helical regions and/or turns are allowed within the bilayer.

To overcome some of the deficiencies of a simple hydrophobicity analysis as a predictor of membrane protein structure, increasing use is being made of the amphipathic helix concept (78,113,130,131). Very often a substantial fraction of the amino acid side chains in the membrane-spanning segments are polar or charged. Forcing such residues to abut the hydrocarbon lipid interface is clearly energetically unfavorable. Thus, it is expected that to accommodate both polar and nonpolar residues within the membrane the polarity of the side chains should exhibit a regularity (period of 3-4 for an α-helix and 2 for a β-strand) which would allow an α-helix or β-strand to be oriented with polar or charged residues facing away and nonpolar residues abutting the lipid protein interface. The polar or charged residues can fulfill hydrogen bonding or salt bridge requirements by lining an aqueous channel or interacting with another protein surface. Structures of varying degrees of detail have been predicted using these methods, and in some cases, experimental evidence for secondary structure has been obtained, for example, in case of Lac permease from *E. coli* (87,132), melittin (124), Omp A and porin from the outer membrane of *E. coli* (94), and the sodium channel (133).

E. Environment and the Structure of Membrane Proteins

As membrane proteins are distinguished by lipid solubility, it is not surprising that their form and function are sensitive to environment. What may be surprising is the degree to which environment can alter the secondary and tertiary structure. A large number of studies have established the lipid specificity or sensitivity of membrane protein function (see, for example, Ref. 134). Such studies do not generally establish a correlation between the observed alterations in function and specific structural alteration. It is quite possible that significant functional alterations may be associated with rather subtle structural changes. Demonstrating differences in membrane protein structure, although easier than determining the structure itself, is generally difficult for the same reasons associated with determining membrane protein structure. Several recent studies that have examined this issue in some detail serve as an important caveat in interpreting structural determinations of membrane proteins (see Section VI for additional studies). These results indicate that structural determinations in crystals must be, at least partially, verified by noncrystallographic methods.

Gramicidin is a 15-amino-acid, alternating L and D peptide which forms monovalent cation channels in membranes. The structure has been solved at 1.7 Å in crystals formed in methanol (135). In these crystals the peptide forms a dimer composed of two intertwined antiparallel β helices of 7.2 residues per turn. CD measurements of gramicidin in organic solutions are also consistent with an intertwined helix configuration (135,136). In contrast, the structure in lipid bilayers as determined by CD (135), NMR (137), and fluorescence (108) is composed of two $\beta^{6.3}$ helices in an end-to-end configuration.

The M13 coat protein also exhibits major secondary structural changes in response to environmental alterations. The protein transforms from virtually all α-helix to virtually all β-sheet, simply by changing the lipid composition or the lipid-to-protein ratio upon reconstituting the protein into lipid vesicles by detergent dialysis (138). Recently it has been demonstrated that significant alterations in tertiary structure are possible by quite subtle alterations in the environment of the M13 coat protein (139). In these studies ^{19}F NMR studies of fluorotyrosine indicate that two different M13 conformations, separated by 10 kcal/mole, exist in deoxycholate micelles, whereas only a single conformation was observed in octylglucoside.

VI. INTERACTIONS BETWEEN LIPIDS AND PROTEINS

Interactions between lipids and proteins have been studied by observing the effect of lipid on protein and, separately, the effect of protein on lipid structure. Proteins have been found to alter the order and mobility of the lipid acyl chain, the translational mobility of the whole lipid, and the lateral organization of the bulk lipid. The effect of lipid on protein structure and dynamics is less well characterized. A number of studies do, however, suggest alterations in specific protein segments in response to changes in lipid structure. In this section these two effects are discussed separately since they are studied by different methods and they represent potentially different physiological functions.

A. Effects of Lipids on Proteins

Lipid-protein interactions involve the alteration in secondary, tertiary, or quaternary protein structure induced by changes in the composition or physical state of the lipid. Several studies in reconstituted systems using magnetic resonance have demonstrated that the mobility of individual amino acid side chains can be altered by changes in the lipid phase (140-142). Secondary structural changes in response to alterations in lipid composition and lipid-to-protein ratio have been observed in reconstituted glycosyltransferase (143) and fd coat protein (138). Evidence also exists that fatty acid perturbation in plasma membranes may affect the conformation of individual amino acid residues (144,145). Interactions of small peptides with membranes indicate that binding occurs with a specific orientation relative to the membrane surface (73), but that the actual conformational changes are quite subtle (146). A large body of indirect evidence for lipid-protein interactions has been obtained from functional changes observed in response to lipid and fatty acid alterations (72,147-149).

B. Effects of Proteins on Lipid

Protein-lipid interactions have attracted a great deal of attention in the past decade. This interest has been stimulated by the availability of a rich variety of lipid probes and by increased understanding of reconstitution of proteins into lipid bilayers. The central issue in these investigations has been whether and how the presence of protein in membranes affects lipid structure and dynamics. Proteins may affect the local properties of individual lipid molecules, such as the acyl chain order or mobility, or more global properties, such as the lateral and transmembrane distribution and mobility. In this section we do not deal with interactions affecting the transmembrane lipid properties and touch only briefly on lateral dynamics.

The issue that has generated the most interest (and controversy) is whether a tightly bound lipid boundary is formed around transmembrane proteins. Tightness of binding is directly proportional to the residence time of a particular lipid in the protein-lipid interface. There are two important time scales with which to evaluate this residence time: the time characteristic of the protein's function and the bulk lipid residence time. As times that are important for protein function are probably greater than 10^{-6} s, the lipid residence time should be greater than this in order for the binding to be functionally important. The bulk lipid residence time, the average time one lipid spends near another, is about 10^{-7} s (150). This, therefore, is the shortest residence time in which any specific lipid can play a structural role. In this section we pay particular attention to the issue of the lipid boundary, using these time limits to gauge the significance of protein-lipid interactions.

The first direct physical evidence that lipid at the protein-lipid interface might exhibit special properties was obtained using cytochrome c oxidase reconstituted into liposomes composed of mitochondrial lipid and a spin-labeled fatty acid (151). The ESR spectra from such preparations suggested two environments, one in which the spin label exhibits nearly the same isotropic tumbling that it does in pure lipid liposomes (only a single environment is detected in pure lipid) and one in which the probe is motionally restricted. Variation of the lipid-to-protein ratio demonstrated that over a broad range of values the immobilized component was proportional to the protein concentration. At sufficiently high concentrations virtually all the probe was immobilized. From the observed saturation ratio it was estimated that the boundary lipid constituted one lipid layer or annulus around the protein. It was suggested that in normal membranes an annulus formed around integral membrane proteins and that the lipids in this annulus were tightly bound. Subsequent studies using ESR, fluorescence, and Raman spectroscopy on several other reconstituted and plasma membrane preparations have confirmed the existence of immobilized lipid related to the presence of protein (152-159).

The observation of an immobilized lipid fraction by these techniques does not, however, constitute evidence for laterally immobilized lipid on the time scale of protein function or the bulk lipid exchange rate. The techniques have time windows ranging from 10^{-8} to 10^{-12} s. In the case of ESR, for example, it is only necessary for the spin label to exchange between the annulus and the bulk lipid at rates slower than about 10^8/s for an immobilized component to appear in the ESR spectrum. NMR, which is sensitive to much slower events (exchange rates longer than 10^5/s), can be used in the same way as ESR to detect regions of immobilized lipid. In most NMR studies no evidence of an immobile lipid component is observed, and therefore these studies establish a lower limit to the exchange rate of 10^5/s (160-162). The constraints set by the ESR and NMR results therefore require that if bound lipid exists, its mean residence time in the annulus is between 10^{-8} and 10^{-5} s. Indeed, there is some evidence obtained from the analysis of the position and line shape of the ESR spectra that the fatty acid spin label exchange rate is about 10^7/s (150). This value is about the same as the bulk lipid exchange rate, suggesting that the exchange of lipid in the annulus with bulk lipid is unhindered and, therefore, the annular lipid is *not* tightly bound to the protein.

Spectroscopic techniques generally reveal a component that is rotationally immobile for times longer than the temporal resolution of the method
and, therefore, only indirectly reflects lipid-to-protein binding affinities.
Measurements of the effect of protein on lipid lateral mobility and the lipid-
protein association constants represent more direct approaches to the question
of how tightly lipid is bound to the protein surface. Several investigations
using fluorescence photobleaching recovery to measure the lipid lateral
mobility in the presence and absence of protein have been carried out
in reconstituted and native membranes (163). Results in red cells indicate
that the rate of lipid diffusion is slower by a factor of 4 in whole membranes
as compared to liposomes formed from extracted lipid (163). This decrease,
however, is entirely explicable in terms of lipid collisions, with proteins
acting as simple-hard surfaces, suggesting, therefore, that the protein-
lipid interaction does not exhibit a significant attractive interaction.

The association constant for the binding of lipid at the protein-lipid
interface may be determined from the partition of lipid between the annulus
and bulk regions, using spectroscopic markers to indicate the fraction
in each region. In addition to the ESR method (164), spin-labeled fatty
acids and brominated phospholipids have been used as short-ranged
quenchers of intrinsic protein fluorescence, to determine the fraction of
these probes in close apposition to the protein surface (103,104,165). The
results of these studies indicate that the partition of neutral lipid between
the boundary and bulk phases is not significantly different from unity.
Fluorescence energy transfer between tryptophan and fluorescent fatty
acids suggests that the fatty acid is excluded from a region in immediate
apposition to the protein. This is consistent with a preferential association
of the phospholipid with the protein as compared to the fatty acid and
with a tightly bound lipid annulus (105,106,166). It is possible, however,
that this represents not tight binding of the lipids to the protein surface,
but rather decreased partition of the lipid probes into a region in which
the acyl chain free energy is reduced by restrictions to rotation or by
electrostatic repulsion between the protein and the charged probes.

Although there is little evidence for specific binding of neutral lipid
to the lipid-protein boundary, a number of studies suggest that tight
binding may be exhibited by charged lipids and will, therefore, probably
be protein specific. Proteins from myelin, studied by a number of techniques, appear to induce formation of domains, enriched in charged lipids,
from mixtures of neutral and charged lipids (155). The lipids in these
domains as well as in the bulk phase retain their unmixed phase behavior,
suggesting that the segregation occurs without significantly affecting the
acyl chain properties. Studies on cytochrome c oxidase and the matrix
protein of vesicular stomatitis virus also indicate the segregation of charged
lipids and demonstrate an approximately twofold increase in partition preference of the charged lipids for the boundary, in comparison to neutral lipid
(153,164,167). The Ca^{2+}-ATPase from sarcoplasmic reticulum, on the other
hand, displays no binding specificity for charged lipids (165).

Although proteins may not retard lateral diffusion of lipids other than
by providing simple barriers, a number of studies have demonstrated an
effect on acyl chain order and mobility. Deuterium NMR studies in reconstituted systems of several proteins demonstrate that the degree of acyl chain
order decreases in response to protein incorporation (160,162). This may
reflect the influence of the uneven, relatively convoluted surface of the
protein on the packing of the neighboring lipid acyl chains (162). Measurements of the fluorescence decay anisotropy of the fatty acid analog parinaric

acid confirm the decrease in order, but also indicate that the rotational mobility of the lipid acyl chains is decreased by protein (157). The decrease in rotational mobility is consistent with the protein side chains acting to retard the *gauche-trans* rotation of the lipid methylene groups. Thus, it appears that virtually all measured protein interactions are consistent with the protein acting to retard movement by virtue of the slow diffusion of the protein mass as compared to the lipid.

These results can, in fact, be understood in terms of an "infinite" barrier presented by the protein. A lipid molecule adjacent to a protein differs from one in the bulk lipid phase since its possible jump positions are limited by the protein. As the relative protein concentration increases, some of the lipids will also be trapped within aggregates of protein and display further immobility.

Although these results suggest that, on average, there is no specific protein-neutral lipid interaction, it should be emphasized that only a small fraction of membrane proteins have been studied thus far.

ACKNOWLEDGMENTS

This work was supported by a grant-in-aid from the American Heart Association and with funds contributed in part by the Massachusetts affiliate (86-1253) and a grant from the National Science Foundation (PCM-830268). This work was done during the tenure of an Established Investigatorship of the American Heart Association and with funds contributed in part by the Massachusetts affiliate (82-174).

REFERENCES

1. Singer, S. J., and Nicolson, G. L. (1972). The fluid mosaic model of the structure of cell membranes. *Science* 175:720-731.
2. Deisenhofer, J., Epp, O., Miki, K., Huber, R., and Michel, H. (1985). Structure of the protein subunits in the photosynthetic reaction centre of *Rhodopseudomonas viridis* at 3 Å resolution. *Nature* 318:618-624.
3. Henderson, R., and Unwin, P. N. T. (1975). Three-dimensional model of purple membrane obtained by electron microscopy. *Nature* 257:23-32.
4. Klingenberg, M. (1981). Membrane protein oligomeric structure and transport function. *Nature* 290:449-454.
5. Kleinfeld, A. M. (1987). Current views of membrane structure. *Curr. Topics Membranes Transport* 29, pp. 1-27.
6. Hauser, H., Pascher, I., Pearson, R. H., and Sundell, S. (1981). Preferred conformation and molecular packing of phosphatidylethanolamine and phosphatidylcholine. *Biochim. Biophys. Acta* 650:21-51.
7. Storch, J., and Kleinfeld, A. M. (1985). The lipid structure of biological membranes. *Trends Biochem. Sci.* 10:418-420.
8. Griffin, R. G., Powers, L., and Pershan, P. S. (1978). Head-group conformation in phospholipids: A phosphorus-31 nuclear magnetic resonance study of oriented monodomain dipalmitoylphosphatidylcholine bilayers. *Biochemistry* 17:2718-2722.
9. Herzfeld, J., Griffin, R. G., and Haberkorn, R. A. (1978). Phosphorus-31 chemical shift tensors in barium diethyl phosphate and urea-phosphoric acid: Model compounds for phospholipid head-group studies. *Biochemistry* 17:2711-2718.

10. Franks, N. P. (1976). Structural analysis of hydrated egg lecithin and cholesterol bilayers. I. X-ray diffraction. *J. Mol. Biol.* 100:345-358.

11. Worcester, D. L., and Franks, N. P. (1976). Structural analysis of hydrated egg lecithin and cholesterol bilayers. II. Neutron diffraction. *J. Mol. Biol.* 100:359-378.

12. Seelig, J., and Seelig, A. (1980). Lipid conformation in model membranes and biological membranes. *Q. Rev. Biophys.* 13:19-61.

13. Davis, J. H. (1983). The description of membrane lipid conformation, order, and dynamics by ^{2}H-NMR. *Biochim. Biophys. Acta* 737:117-171.

14. Blaurock, A. E. (1982). Evidence of bilayer structure and of membrane interactions from X-ray diffraction analysis. *Biochim. Biophys. Acta* 650:167-207.

15. Duax, W. L., Griffin, J. F., Rohrer, D. C., and Weeks, C. M. (1980). Conformational analysis of sterols: Comparison of X-ray crystallographic observations with data from other sources. *Lipids* 15:783-792.

16. Caspar, D. L. D., and Kirschner, D. A. (1971). *Nature New Biol.* 231:46.

17. Barenholz, Y., and Thompson, T. E. (1980). Sphingomyelins in bilayers and biological membranes. *Biochim. Biophys. Acta* 604:129-158.

18. Thompson, T. E., and Tillack, T. W. (1985). Organization of glyco-sphingolipids in bilayers and plasma membranes of mammalian cells. *Annu. Rev. Biophys. Chem.* 14:361-386.

19. Pascher, I., and Sundell, S. (1985). Interaction and space requirements of the phosphate headgroups in membrane lipids. The crystal structure of disodium lysophosphatidate dihydrate. *Chem. Phys. Lipids* 37:241-250.

20. Pascher, I. (1976). Molecular arrangements in sphingolipids: Conformation and hydrogen bonding of ceramide and their implication on membrane stability and permeability. *Biochim. Biophys. Acta* 455:433-451.

21. Luzzati, V. (1968). X-ray diffraction studies of lipid-water systems. In: *Biological Membranes* (D. Chapman, ed.). Academic Press, London, pp. 71-123.

22. Cullis, P. R., and de Kruijff, B. (1979). Lipid polymorphism and the functional roles of lipids in biological membranes. *Biochim. Biophys. Acta* 559:399-420.

23. Hui, S. W., Stewart, T. P., Yeagle, P. L., and Albert, A. D. (1981). Bilayer to non-bilayer transition in mixtures of phosphatidylethanolamine and phosphatidylcholine: Implications for membrane properties. *Arch. Biochem. Biophys.* 207:227-240.

24. Cornell, A., and Separovic, F. (1983). Membrane thickness and acyl chain length. *Biochim. Biophys. Acta* 733:189-193.

25. McIntosh, T. J., Simon, S. A., Ellington, J. C., Jr., and Porter, N. A. (1984). New structural model for mixed-chain phosphatidylcholine bilayers. *Biochemistry* 23:4038-4044.

26. Hui, S. W., Mason, J. T., and Huang, C. (1984). Acyl chain inter-digitation in saturated mixed-chain phosphatidylcholine bilayer dispersions. *Biochemistry* 23:5570-5577.

27. Macdonald, P. M., Sykes, B. D., and McElhaney, R. N. (1984). Fatty acyl chain structure, orientational order, and the lipid phase transition in *Acholeplasma laidlawii* B membranes. A review of recent ^{19}F nuclear magnetic resonance studies. *Can. J. Biochem.* 62:1134-1150.

28. Vincent, M., Foresta, B., Gallay, J., and Alfsen, A. (1982). Nano-second fluorescence decays of n-(9-anthroyloxy) fatty acids in dipalmitoylphosphatidylcholine vesicles with regard to isotropic solvents. *Biochemistry* 21:708-716.

29. Kutchai, H., Chandler, L. H., and Zavoico, G. B., (1983). Effects

of cholesterol on acyl chain dynamics in multilamellar vesicles of various phosphatidylcholines. *Biochim. Biophys. Acta* 736:137-149.

30. Storch, J., and Schacter, D. (1985). Calcium alters the acyl chain composition and lipid fluidity of rat hepatocyte plasma membranes in vitro. *Biochim. Biophys. Acta* 812:473-484.

31. Lee, A. G., Birdsall, N. J. M., Metcalf, J. C., Toon, P. A., and Warren, G. B. (1974). Clusters in lipid bilayers and the interpretation of thermal effects in biological membranes. *Biochemistry* 13:3699-3765.

32. Griffith, O. H., Dehlinger, P. J., and Van, S. P. (1974). Shape of the hydrophobic barrier of phospholipid bilayers (evidence for water penetration in biological membranes). *J. Membrane Biol.* 15:159-192.

33. Chalpin, B., and Kleinfeld, A. M. (1983). Interaction of fluorescent quenchers with the n-(9-anthroyloxy) fatty acid membrane probes. *Biochim. Biophys. Acta* 731:465-474.

34. Simon, S. A., McIntosh, T. J., and Latorre, R. (1982). Influence of cholesterol on water penetration into bilayers. *Science* 216:65-67.

35. Chapman, D. (1975). Phase transitions and fluidity characteristics of lipids and cell membranes. *Q. Rev. Biophys.* 8:185-235.

36. Nagle, J. F. (1980). Theory of the main lipid bilayer phase transition. *Annu. Rev. Phys. Chem.* 31:157-195.

37. Ranck, J. L., Mateu, L., Sadler, D. M., Tardieu, A., Gulik-Krzywicki, T., and Luzzati, V. J. (1974). Order-disorder conformational transitions of the hydrocarbon chains of lipids. *Mol. Biol.* 85:249-277.

38. Chen, S. C., Sturtevant, J. M., and Gaffney, B. J. (1980). Scanning calorimetric evidence for a third phase transition in phosphatidylcholine bilayers. *Proc. Natl. Acad. Sci. USA* 77:5060-5063.

39. Ruocco, M. J., and Shipley, G. G. (1982). Characterization of the sub-transition of hdyrated dipalmitoylphosphatidylcholine bilayers. *Biochim. Biophys. Acta* 684:59-66.

40. Fuldner, H. H. (1981). Characterization of a third lipid transition in multilamellar dipalmitoyllecithin liposomes. *Biochemistry* 20:5707-5710.

41. Shimshick, E. J., and McConnell, H. M. (1973). Lateral phase separation in phospholipid membranes. *Biochemistry* 12:2351-2360.

42. Grant, C. W. M., Wu, S. H-W., and McConnell, H. M. (1974). Lateral phase separations in binary lipid mixtures: Correlation between spin label and freeze-fracture electron microscope studies. *Biochim. Biophys. Acta* 363:151-158.

43. Luna, E. J., and McConnell, H. M. (1977). Lateral phase separations in binary mixtures of phospholipids having different charges and different crystalline structures. *Biochim. Biophys. Acta* 470:303-316.

44. Bach, D., Bursuker, I., and Goldman, R. (1977). Differential scanning calorimetry and enzyme activity of rat liver microsomes in the presence and absence of 1-tetrahydrocannabinol. *Biochim. Biophys. Acta* 469:171-179.

45. Jain, M. K., and White, H. B. (1978). Long-range order in biomembranes. *Adv. Lipid Res.* 15:1-60.

46. Karnovsky, M. J., Kleinfeld, A. M., Hoover, R. L., Dawidowicz, E. A., MacIntyre, D. E., Salzman, E. A., and Klausner, R. D. (1982). Lipid domains in membranes. *Ann. NY Acad. Sci.* 401:61-75.

47. Karnovsky, M. J., Kleinfeld, A. M., Hoover, R. L., and Klausner, R. D. (1982). The concept of lipid domains in membranes. *J. Cell Biol.* 49:1-6.

48. Klausner, R. D., and Kleinfeld, A. M. (1984). Lipid domains in membranes. In: *Cell Surface Dynamics* (A. S. Perelson, C. H. DeLisi, and F. W. Wiegel, eds.). Marcel Dekker, New York, pp. 23-58.

49. Lentz, B. R., Barenholz, Y., and Thompson, T. E. (1976). Fluorescence depolarization studies of phase transitions and fluidity in phospholipid bilayers. 1. Single component phosphatidylcholine liposomes. *Biochemistry* 15:4521–4528.

50. Lentz, B. R. Barenholz, Y., and Thompson, T. E. (1976). Fluorescence depolarization studies of phase transitions and fluidity in phospholipid bilayers. 2. Two component phosphatidylcholine liposomes. *Biochemistry* 15:4529–4537.

51. Lentz, B. R., Hoechli, M., and Barenholz, Y. (1981). Acyl chain order and lateral domain formation in mixed phosphatidylcholine-sphingomyelin multilamellar and unilamellar vesicles. *Biochemistry* 20:6803–6809.

52. Wu, S. H., and McConnell, H. M. (1975). Phase separations in phospholipid membranes. *Biochemistry* 14:847–854.

53. Galla, H. J., and Sackmann, E. (1975). Chemically induced lipid phase separation in model membranes containing charged lipids: A spin label study. *Biochim. Biophys. Acta* 401:509–529.

54. Melchior, D. L. (1986). Lipid domains in fluid membranes—A quick-freeze DSC study. *Science* 234:1577–1580.

55. Hui, S. W. (1981). Geometry of phase-separated domains in phospholipid bilayers by diffraction-contrast electron microscopy. *Biophys. J.* 34:383–395.

56. Bar, L. K., Barenholz, Y., and Thompson, T. E. (1986). Fraction of cholesterol undergoing spontaneous exchange between small unilamellar phosphatidylcholine vesicles. *Biochemistry* 25:6701–6705.

57. Masserini, M., and Freire, E. (1986). Thermotropic characterization of phosphatidylcholine vesicles containing ganglioside G_{M1} with homogeneous ceramide chain length. *Biochemistry* 25:1043–1049.

58. Melchior, D. L. (1982). Lipid phase transitions and regulation of membrane fluidity in prokaryotes. *Curr. Topics Membrane Transport* 17:263–316.

59. Bearer, E. L., and Freind, D. S. (1980). Anionic lipid domains: Correlation with functional topography in a mammalian cell membrane. *Proc. Natl. Acad. Sci. USA* 77:6601–6605.

60. Wolf, D. E. (1983). The plasma membrane in early embryogenesis. In: *Development of Mammals*, Vol. 5 (M. H. Johnson, ed.). Elsevier, New York, pp. 187–208.

61. Wolf, D. E., and Voglmayr, J. K., (1984). Diffusion and regionalization in membranes of maturing rat spermatozoa. *J. Cell Biol.* 98:1678–1684.

62. Brasitus, T. A., Tall, A. R., and Schachter, D. (1980). Thermotropic transitions in rat intestinal plasma membranes studied by differential scanning calorimetry and fluorescence polarization. *Biochemistry* 19:1256–1261.

63. Mutsch, B., Gains, N., and Hauser, H. (1983). Order-disorder phase transition and lipid dynamics in rabbit small intestinal brush border membranes. Effect of proteins. *Biochemistry* 22:6326–6333.

64. Horwitz, A. F., Hatten, M. E., and Berger, M. M. (1974). Membrane fatty acid replacements and their effect on growth and lectin-induced agglutinability. *Proc. Natl. Acad. Sci. USA* 71:3115–3119.

65. Hatten, M. E., Scandella, C. J., Horwitz, A. F., and Burger, M. M. (1978). Similarities in the membrane fluidity of 3T3 and SV101-3T3 cells and its relation to concanavalin A and wheat germ agglutinin-induced agglutination. *J. Biol. Chem.* 253:1972–1977.

66. Hill, D. J., Dawidowicz, E. A., Andrews, M. L., and Karnovsky, M. J. (1983). Modulation of microsomal glucose-6-phosphate translocase activity by free fatty acids: Implications for lipid domain structure in microsomal membranes. *J. Cell Physiol.* 115:1-8.

67. Hoover, R. L., Dawidowicz, E. A., Robinson, J. M., and Karnovsky, M. J. (1983). Role of cholesterol in the capping of surface immunoglobulin receptors on murine lymphocytes. *J. Cell Biol.* 97:73-80.

68. Curtain, C. C., Looney, F. D., and Smelstorius, J. A. (1980). Lipid domain formation and ligand-induced lymphocyte membrane changes. *Biochim. Biophys. Acta* 596:43-56.

69. De Laat, S. W., Van Der Saag, P. T., Elson, E. L., and Schlessinger, J. (1979). Lateral diffusion of membrane lipids and proteins is increased specifically in neurites of differentiating neuroblastoma cells. *Biochim. Biophys. Acta* 558:247-250.

70. Packard, B. S., Saxton, M. J., Bissell, M. J., and Klein, M. P. (1984). Plasma membrane reorganization induced by tumor promoters in an epithelial cell line. *Proc. Natl. Acad. Sci. USA* 81:449-452.

71. Senior, A. E. (1983). Secondary and tertiary structure of membrane proteins involved in proton translocation. *Biochim. Biophys. Acta* 726:81-95.

72. Benga, G., and Holmes, R. P. (1984). Interactions between components in biological membranes and their implications for membrane function. *Prog. Biophysics* 43:195-257.

73. Eisenberg, D. (1984). Three-dimensional structure of membrane and surface proteins. *Annu. Rev. Biochem.* 53:595-623.

74. Khorana, H. G., Gerber, G. E., Herlihy, W. C., Gray, C. P., Anderegg, R. J., Nihei, K., and Biemann, K. (1979). Amino acid sequence of bacteriorhodopsin. *Proc. Natl. Acad. Sci. USA* 76:5046-5050.

75. Ovchinnikov, Yu. A., Abdulaev, N. G., Feigina, M. Yu., Kiselev, A. V., and Lobanov, N. A. (1979). The structural basis of the functioning of bacteriorhodopsin: An overview. *FEBS Lett.* 100:219-224.

76. Buchel, D. E., Gronenborg, B., and Muller-Hill, B. (1980). Sequence of the lactose permease gene. *Nature* 283:541.

77. Dunn, R., McCoy, J., Simsek, M., Majumdar, A., Chang, S. H., RajBhandary, U. L., and Khorana, G. (1981). The bacteriorhodopsin gene. *Proc. Natl. Acad. Sci USA* 78:6744-6748.

78. Noda, M., Shimizu, S., Tanabe, T., Takai, T., Kayano, T., Ikeda, T., Takahashi, H,m Nakayama, H., Kanoka, Y., Minamino, N., Kangawa, K., Miyata, T., and Numa, S. (1984). Primary structure of *Electochorus electricus* sodium channel deduced from cDNA sequence. *Nature* 312:121.

79. Kopito, R. R. and Lodish, H. F. (1985). Primary structure and transmembrane orientation of the murine anion exchange protein. *Nature* 316:234.

80. Mueckler, M., Caruso, C., Baldwin, S. A., Panico, M., Blench, I., Morris, H. R., Allard, W. J., Lienhard, G. E., and Lodish, H. F. (1985). Sequence and structure of a human glucose transporter. *Science* 229:941.

81. Shull, G. E., Schwartz, A., and Lingrel, J. B. (1985). Amino-acid sequence of the catalytic subunit of the $(Na^+ + K^+)ATPase$ deduced from a complementary DNA. *Nature* 316:691.

82. Wallace, B. A., and Mao, D. (1984). Circular dichroism analyses of membrane proteins: An examination of differential light scattering and absorption flattening effects in large membrane vesicles and membrane sheets. *Analy. Biochem.* 142:317-328.

83. Tobkes, N., Wallace, B. A., and Bayley, H. (1985). Secondary structure and assembly mechanism of an oligomeric channel protein. *Biochemistry* 24:1915-1920.

84. Rothschild, K. J., and Clark, N. A. (1979). Polarized infrared spectroscopy of oriented purple membrane. *Biophys. J.* 25:473-488.

85. Nabedryk, E., Bardin, A. M., and Breton, J. (1985). Further characterization of protein secondary structures in purple membrane by circular dichroism and polarized infrared spectroscopies. *Biophys. J.* 48:873-876.

86. Lee, D. C., Hayward, J. A., Restall, C. J., and Chapman, D. (1985). Second derivative infrared spectroscopic studies of the secondary structures of bacteriorhodopsin and Ca^{2+}-ATPase. *Biochemistry* 24:4364-4373.

87. Vogel, H., Wright, K. J., and Jähnig, F. (1985). The structure of the lactose permease derived from Raman spectroscopy and prediction methods. *EMBO J.* 4:3625-3631.

88. Guidotti, G. (1977). The structure of intrinsic membrane proteins. *J. Supramol. Struct.* 7:489-497.

89. Long, M. M., Urry, D. W., and Stoeckenius, W. (1977). Circular dichroism of biological membranes: Purple membrane of *Halobacterium halobium*. *Biochem. Biophys. Res. Commun.* 75:725-731.

90. Engelman, D. M., and Steitz, T. A. (1981). The spontaneous insertion of proteins into and across membranes: The helical hairpin hypothesis. *Cell* 23:411-422.

91. Jap, B. K., Maestre, M. F., Hayward, S. B., and Glaeser, R. M. (1983). Peptide-chain secondary structure of bacteriorhodopsin. *Biophys. J.* 43:81-89.

92. Dailey, H. A., and Stittmatter, P. (1978). Structural and functional properties of the membrane binding segment of cytochrome b_5. *J. Biol. Chem.* 253:8203-8209.

93. Makowski, L., Caspar, D. L. D., Goodenough, D. A., and Phillips, W. C. (1982). Gap junction structures. III. The effect of variations in the isolation procedure. *Biophys. J.* 37:189-191.

94. Vogel, H., and Jähnig, F. (1986). Models for the structure of outer-membrane proteins of *Escherichia coli* derived from Raman spectroscopy and prediction methods. *J. Mol. Biol.* 190:191-199.

95. Wallace, B. A., Cascio, M., and Mielke, D. E. (1986). Evaluation of prediction methods for membrane protein secondary structures. *Proc. Natl. Acad. Sci. USA* 83:9423-9427.

96. Dailey, H. A., and Strittmatter, P. (1981). Orientation of the carboxyl and NH2 termini of the membrane-binding segment of cytochrome b_5 on the same side of phospholipid bilayers. *J. Biol. Chem.* 256:3951-3955.

97. Cardoza, J. D., Kleinfeld, A. M., Stallcup, K. C., and Mescher, M. F. (1984). Hairpin configuration of H-2K^k in liposomes formed by detergent dialysis. *Biochemistry* 23:4401-4409.

98. Brock, C. J., Tanner, M. J. A., and Kemph, C. (1983). The human erythrocyte anion-transport protein. *Biophys. J.* 213:577-586.

99. Engelman, D. M., Goldman, A., and Steitz, T. A. (1982). The identification of helical segments in the polypeptide chain of bacteriorhodopsin. *Methods Enzymol.* 88:81-88.

100. Henderson, R., Jubb, J. S., and Whytock, S. Specific Labelling of the protein and lipid on the extracellular surface of purple membrane. *J. Mol. Biol.* 123:259-274.

101. Robson, R. J., Radhakrishnan, R., Ross, A. H., Takagaki, Y., and Khorana, H. G. Photochemical cross-linking in studies of lipid-

protein interactions. In: *Lipid-Protein Interactions*, Vol. 2 (P. C. Jost and O. H. Griffith, eds.), Wiley, New York, 1982, pp. 149-192.

102. Takagaki, Y., Radhakrishnan, R., Gupta, C. M., and Khorana, G. (1983. The membrane-embedded segment of cytochrome b_5 as studied by cross-linking with photoactivatable phospholipids. *J. Biol. Chem.* 258:9128-9135.

103. London, E., and Feigenson, G. W. (1981a). Fluorescence quenching in model membranes. 1. Characterization of quenching caused by a spin-labeled phospholipid. *Biochemistry* 20:1932-1938.

104. London, E., and Feigenson, G. W. (1981b). Fluorescence quenching in model membranes. 2. Determination of the local lipid environment of the calcium adenosinetriphosphatase from sarcoplasmic reticulum. *Biochemistry* 20:1939-1948.

105. Fleming, P. J., Koppel, D. E., Lan, A. L. Y., and Stritmatter, P. (1979). Intramembrane position of the fluorescent tryptophanyl residue in membrane-bound cytochrome b5. *Biochemistry* 18:5458-5464.

106. Kleinfeld, A. M., and Lucakovic, M. L. (1985). Energy transfer study of cytochrome b_5 using the anthroyloxy fatty acid membrane probes. *Biochemistry* 24:1883-1890.

107. Kleinfeld, A. M., LeGrange, J. D., and Caplan, S. R. (1986). Tryptophan imaging of bacterio-opsin reconstituted into lipid vesicles. *Biophys. J.* 49:477a.

108. Boni, L. T., Connolly, A. J., and Kleinfeld, A. M. (1986). Transmembrane distribution of gramicidin by tryptophan energy transfer. *Biophys. J.* 49:122-123.

109. Kleinfeld, A. M. Tertiary structure of membrane proteins by resonance energy transfer. In: *Spectroscopic Membrane Probes* (L. M. Loew, ed.), CRC Press, Boca Raton, FL, 1987.

110. Mathews, F. S., Czerwiski, E. W., and Argos, P. The X-ray crystallographic structure of calf liver cytochrome b5. In: *The Porphyrins*, Vol. 3 (D. Dolphin, ed.), Academic Press, New York, 1979, pp. 107-145.

111. Wilson, I. A., Skehel, J. J., and Wiley, D. C. (1981). Structure of the haemagglutin membrane glycoprotein of influenza virus at 3A resolution. *Nature* 289:366-373.

112. Leifer, D., and Henderson, R. J. (1983). Three-dimensional structure of orthorhombic purple membrane at 6.5 A resolution. *J. Mol. Biol.* 163:451-466.

113. Engelman, D. M., and Zacci, G. (1980). Bacteriorhodopsin is an inside-out protein. *Proc. Natl. Acad. Sci. USA* 77:5894-5898.

114. Stoeckenius, W., and Bogomolni, R. A., (1982). Bacteriorhodopsin and related pigments of halobacteria. *Annu. Rev. Biochem.* 52:587-616.

115. Agard, D. A., and Stroud, R. M. (1982). Linking regions between helices in bacteriorhodopsin revealed. *Biophys. J.* 37:589-602.

116. Huang, K-S., Liao, M-J., Gupta, C. M., Royal, N., Beimann, K., and Khorana, H. G. (1982). The site of attachment of retinal in bacteriorhodopsin. *J. Biol. Chem.* 257:8596-8599.

117. Makowski, L., Caspar, D. L. D., Phillips, W. C., Baker, T. S., and Goodenough, D. A. (1984). Gap junction structures. VI. Variation and conservation in connexon conformation and packing. *Biophys. J.* 45:208-218.

118. Unwin, P. T., and Zampighi, G. (1980). Structure of the junction between communicating cells. *Nature* 283:545-549.

119. Leonard, K., Wingfield, P., Arad, T., and Weiss, H. (1981). Three-dimensional structure of ubiquinol:cytochrome c reductase from Neurospora mitochondria determined by electron microscopy of membrane crystals. *J. Mol. Biol.* 149:259–274.

120. Deatherage, J. F., Henderson, R., and Capaldi, R. A. (1982). Relationship between membrane and cytoplasmic domains in cytochrome c oxidase by electron microscopy in media of different density. *J. Mol. Biol.* 158:501–514.

121. Changeux, J., Devillers-Thiery, A., and Chemouilli, P. (1984). Acetylcholine receptor: An allosteric protein. *Science* 225:1335–1345.

122. Fox, R. O., Jr., and Richards, F. M. (1982). A voltage-gated ion channel model inferred fromthe crystal structure of alamethicin at 1.5-A resolution. *Nature* 300:325–330.

123. Terwilliger, T. C., Weissman, L., and Eisenberg, D. (1982). The structure of melittin in the form I crystals and its implication for melittin's lytic and surface activities. *Biophys. J.* 37:353–361.

124. Vogel, H., and Jähnig, F. (1986b). The structure of melittin in membranes. *Biophys. J.* 50:573–582.

125. Chou, P. Y., and Fasman, G. D. (1974a). Conformational parameters for amino acids in helical, β-sheet, and random coil regions calculated from proteins. *Biochemistry* 13:211–222.

126. Chou, P. Y., and Fasman, G. D. (1974b). Prediction of protein conformation. *Biochemistry* 13:222–245.

127. Kyte, J., and Doolittle, R. F. (1982). A simple method for displaying the hydropathic character of a protein. *J. Mol. Biol.* 157:105–132.

128. Wolfenden, R., and Radzicka, A. (1986). How hydrophobic is tryptophan, and rejoinder. *Trends Biochem. Sci.* 11:401–402.

129. Fauchere, J. L., and Pliska, V. (1986). Reply to Wolfenden and Radzicka. *Trends Biochem. Sci.* 11:402.

130. Finer-Moore, J., and Stroud, R. M. (1984). Amphipathic analysis and possible formation of the ion channel in an acetylcholine receptor. *Proc. Natl. Acad. Sci. USA* 81:155–159.

131. Anantharamaiah, G. M., Jones, J. L., Brouillette, C. G., Schmidt, C. F., Chung, B. H., Hughes, T. A., Bhown, A. S., and Segrest, J. P. (1984). Studies of synthetic peptide analogs of the amphipathic helix; structure of complexes with dimyristoyl phosphatidylcholine. *J. Biol. Chem.* 160:10248–10255.

132. Kaback, R. H. (1986). Active transport in *Escherichia coli*: Passage to permease. *Annu. Rev. Biophys. Biophys. Chem.* 15:279–319.

133. Guy, H. R., and Seetharamulu, P. (1986). Molecular model of the action potential of the sodium channel. *Proc. Natl. Acad. Sci. USA* 83:508–512.

134. Seto-Young, D., Chen, C. C., and Wilson, T. H. (1986). Effect of different phospholipids on the reconstitution of two functions of the lactose carrier of *Escherichia coli*. *J. Memb. Biol.* 84:259–267.

135. Wallace, B. A. (1986). Structure of gramicidin A. *Biophys. J.* 49:295–306.

136. Naik, V. M., and Krimm, S. (1984). The structure of crystalline and membrane bound gramicidin A by vibrational analysis. *Biochem. Biophys. Res. Commun.* 125:919–925.

137. Weinstein, S., Durkin, J. T., Veatch, W. R., and Blout, E. R. (1985). Conformation of the gramicidin A channel in phospholipid vesibles: A fluorine-19 nuclear magnetic resonance study. *Biochemistry* 24:4374–4382.

138. Dunker, A. K., Fodor, S. P. A., and Williams, R. W. (1982). Lipid-dependent structural changes of an amphomorphic membrane protein. *Biophys. J.* 37:201-203.

139. Wilson, L. M., and Dahlquist, F. W. (1985). Membrane protein conformational change dependent on the hydrophobic environment. *Biochemistry* 24:1920-1928.

140. Hagen, D. S., Weiner, J. H., and Sykes, B. D. (1978). Fluorotyrosine M13 coat protein: Fluorine-19 nuclear magnetic resonance study of the motional properties of an integral membrane protein in phospholipid vesicles. *Biochemistry* 17:3860-3866.

141. Stollery, J. G., Boggs, J. M., Moscarello, M. A., and Deber, C. M. (1980). Direct observation by carbon-13 nuclear magnetic resonance of membrane-bound human myelin basic protein. *Biochemistry* 19:2391-2396.

142. Boggs, J. M., Stollery, J. G., and Moscarello, M. A. (1980). Effect of lipid environment on the motion of a spin-label covalently bound to myelin basic protein. *Biochemistry* 19:1226-1234.

143. Beadling, L., and Rothfield, L. I. (1978). Modulation of the conformation of a membrane glycosyltransferase by specific lipids. *Natl. Acad. Sci. USA* 75:3669-3672.

144. Esfahani, M., and Devlin, T. M. (1982). Effects of lipid fluidity on quenching characteristics of tryptophan fluorescence in yeast plasma membrane. *J. Biol. Chem.* 257:9919-9921.

145. Pjura, W. J., Kleinfeld, A. M., Klausner, R. D., and Karnovsky, M. J. (1982). Fatty acid perturbation of a membrane protein-lipid interaction: A Tb fluorescence study. *Biophys. J.* 37:69-71.

146. Deber, C. M., and Behnam, B. A. (1984). Role of membrane lipids in peptide hormone function: Binding of enkephalins to micelles. *Proc. Natl. Acad. Sci. USA* 81:61-65.

147. Melchior, D. L., and Steim, J. M. (1976). Thermotropic transitions in biomembranes. *Annu. Rev. Biophys. Bioeng.* 5:205-238.

148. Criado, M., Eibl, H., and Barrantes, F. J. (1984). Functional properties of the acetylcholine receptor incorporated in model lipid membranes. *J. Biol. Chem.* 259:9188-9198.

149. Stubbs, C. D., and Smith, A. D. (1984). The modification of mammalian membrane polyunsaturated fatty acid composition in relation to membrane fluidity and function. *Biochim. Biophys. Acta* 779:89-137.

150. Marsh, D., Watts, A., Pates, D., Uhl, R., Knowles, P. F., and Esmann, M. (1982). ESR spin-label studies of lipid-protein interactions in membranes. *Biophys. J.* 37:265-274.

151. Jost, P. C., Griffith, O. H., Capaldi, R. A., and Vanderkooi, G. (1973). Evidence for boundary lipid in membranes. *Proc. Natl. Acad. Sci. USA* 70:480-484.

152. Hesketh, T. R., Smith, G. A., Houslay, M. D., McGill, K. A., Birdsall, N. J. M., Metcalfe, J. C., and Warren, G. B. (1976). Annular lipids determine the ATPase activity of a calcium transport protein complexed with dipalmitoyllecithin. *Biochemistry* 15:4145-4151.

153. Cable, M. B., and Powell, G. L. (1980). Spin-labeled cardiolipin: Preferential segregation in the boundary layer of cytochrome c oxidase. *Biochemistry* 19:5679-5686.

154. Marsh, D., and Watts, A. Spin labeling and lipid-protein interactions in membranes. In: *Lipid-Protein Interactions*, Vol. 2 (P. C. Jost and O. H. Griffith, eds.), Wiley, New York, 1982, pp. 53-126.

155. Boggs, J. M., Moscarello, M. A., and Papahadjopoulos, D. Structural organization of myelin—role of lipid-protein interactions determined in model systems. In: *Lipid-Protein Interactions*, Vol. 2 (P. C. Jost and O. H. Griffith, eds.), Wiley, New York, 1982, pp. 1-51.

156. Thomas, D. D., Bigelow, D. J., Squier, T. C., and HIldago, C. (1982). Rotational dynamics of protein and boundary lipid in sarcoplasmic reticulum membrane. *Biophys. J.* 37:217-225.

157. Wolber, P. K., and Hudson, B. S. (1982). Bilayer acyl chain dynamics and lipid-protein interaction. The effect of the M13 bacteriophage coat protein on the decay of the fluorescence anisotropy of parinaric acid. *Biophys. J.* 37:253-262.

158. Taraschi, T., and Mendelsohn, R. (1980). Lipid-protein interaction in the glycophorin-dipalmitoylphosphatidylcholine system: Raman spectroscopic investigation. *Proc. Natl. Acad. Sci. USA* 77:2362-2366.

159. Levin, I. W., Lavialle, F., and Mollay, C. (1982). Comparative effects of melittin and its hydrophobic and hydrophilic fragments on bilayer organization by Raman spectroscopy. *Biophys. J.* 37:339-349.

160. Oldfield, E., Gilmore, R., Glaser, M., Gutowsky, H. S., Hshung, J. C., Kang, S. Y., King, T. E., and Meadows, M. (1978). Deuterium nuclear magnetic resonance investigation of the effects of proteins and polypeptides on hydrocarbon chain order in model membrane systems. *Proc. Natl. Acad. Sci. USA* 75:4657-4660.

161. Seelig, J., Tamm, L., Hymel, L., and Fleischer, S. (1981). Deuterium and phosphorus nuclear magnetic resonance and fluorescence depolarization studies of functional reconstituted sarcoplasmic reticulum membrane vesicles. *Biochemistry* 20:3922-3932.

162. Seelig, J., Seelig, A., and Tamm, L. Nuclear magnetic resonance and lipid-protein interactions. In: *Lipid-Protein Interactions*, Vol. 2 (P. C. Jost and O. H. Griffith, eds.), Wiley, New York, 1982, pp. 127-148.

163. Golan, D. E., Alecio, M. R., Veatch, W. R., and Rando, R. R. (1984). Lateral mobility of phospholipid and cholesterol in the human erythrocyte membrane: Effects of protein-lipid interactions. *Biochemistry* 23:332-339.

164. Griffith, O. H., Brotherus, J. R., and Jost, P. C. Equilibrium constants and number of binding sites for lipid-protein interactions in membranes. In: *Lipid-Protein Interactions*, Vol. 2 (P. C. Jost and O. H. Griffith, eds.), Wiley, New York, 1982, pp. 225-281.

165. East, J. M., and Lee, A. G. (1982). Lipid selectivity of the calcium and magnesium ion dependent adenosinetriphosphatase, studied with fluorescence quenching by a brominated phospholipid. *Biochemistry* 21:4144-4151.

166. Lee, A. G., East, J. M., Jones, O. T., McWhirter, J., and Simmonds, A. C. (1982). Interaction of fatty acids with the calcium-magnesium ion dependent adenosinetriphosphatase from sarcoplasmic reticulum. *Biochemistry* 21:6441-6446.

167. Wiener, J. R., Pal, R., Barenholz, Y., and Wagner, R. R. (1983). Influence of the peripheral matrix protein of vesicular stomatitis virus on the membrane dynamics of mixed phospholipid vesicles: Fluorescence studies. *Biochemistry* 22:2162-2170.

168. Bjorkman, P. J., Saper, M. A., Samraoui, B., Bennett, W. S., Strominger, J. L., and Wiley, D. C. (1987). Structure of the human class I histocompatibility antigen, HLA-A2. *Nature* 329:506-512.

169. Haverstick, D. M., and Glaser, M. (1988). Visualization of domain formation in the inner and outer leaflets of a phospholipid bilayer. *J. Cell Biol.* 106:1885-1892.

170. Eklund, K. K., Vuorinen, J., Mikkola, J., Virtanene, J. A., and Kinnunen, P. K. J. (1988). Ca^{2+}-induced lateral phase separation in phosphatidic acid/phosphatidylcholine monolayers as revealed by fluorescence microscopy. *Biochemistry* 27:3433-3437.

171. Bar, L. K. Barenholz, Y., and Thompson, T. E. (1987). Dependence on phospholipid composition of the fraction of cholesterol undergoing spontaneous exchange between small unilamellar vesicles. *Biochemistry* 26:5460-5465.

172. Lund-Kat, S., Laboda, H. M., McLean, L. R., and Phillips, M. C. (1988). Influence of molecular packing and phospholipid type on rates of cholesterol exchange. *Biochemistry* 27:3416-3423.

173. Lange, Y., Swaisgood, M. H., Ramos, B. V., and Steck, T. L. (1989). Plasma membranes contain half the phospholipid and 90% of the cholesterol and sphingolmyelin in cultured human fibroblasts. *J. Biol. Chem.* 264:3786-3793.

174. Wolf, D. E., Scott, B. K., and Millette, C. F. (1986). The development of regionalized lipid diffusibility in the germ cell plasma membrane during spermatogenesis in the mouse. *J. Cell Biol.* 103:1745-1750.

175. Goppelt, M., Eichorn, R., and Resch Krebs, G. (1986). Lipid composition of functional domains of the lymphocyte plasma membrane. *Biochim. Biophys. Acta* 854:184-190.

176. Yechiel, E., and Edidin, M. (1987). Micrometer-scale domains in fibroblast plasma membranes. *J. Cell Biol.* 105:755-760.

177. Yechiel, E., Barenholz, Y., and Henis, Y. I. (1985). Lateral mobility and organization of phospholipids and proteins in rat myocyte membranes. *J. Biol. Chem.* 260:9132-9136.

178. Yechiel, E., and Barenholz, Y. (1985). Relationships between membrane lipid composition and biological properties of rat myocytes. *J. Biol. Chem.* 260:9123-9131.

2

Lipid Polymorphism

PIETER R. CULLIS and COLIN P. TILCOCK

University of British Columbia, Vancouver, British Columbia, Canada

MICHAEL J. HOPE

The Canadian Liposome Co. Ltd., Vancouver, British Columbia, Canada

I. INTRODUCTION

The ability of aqueous dispersions of liquid-crystalline lipids to adopt a variety of structures in addition to the bilayer organization is well established (1,2). It is also becoming generally recognized that these polymorphic capabilities may be directly related to many functional capacities of membranes, including membrane fusion (3,4). In this chapter we present a synopsis of the polymorphic capabilities of lipids, discuss the possible theoretical basis for polymorphism, and indicate the implications for the structure and function of certain biological membranes.

II. LIPID POLYMORPHISM AND LIPID DIVERSITY: AN OVERVIEW

Biological membranes contain a large variety of different molecular species of lipids. This diversity has naturally led to questions regarding the functional roles of individual lipid components. As yet, this has not resulted in a general framework within which the functions of lipids with differing headgroups and/or fatty acid components can be understood. However, a major theme of this chapter is the proposal that the properties inherent in the polymorphic abilities of lipids offer important insights, which may lead to such basic understanding. A particular point is that lipid polymorphism appears to offer more basic insight into lipid function in membranes than can be achieved through rationales of lipid function relying on membrane fluidity arguments.

It is useful to outline the background leading to this statement. Briefly, the recognition that membrane lipids provide a fluid, liquid-crystalline bilayer structure in membranes (5) was combined with the realization that acyl-chain composition and headgroup type can strongly influence the gel (frozen) or liquid-crystalline nature of the resulting membrane. It was therefore proposed that the presence of different lipids is required to provide appropriate fluidity characteristic in a given membrane. In addition, local domains of appropriate lipid composition could possibly modulate local fluidity characteristics, possibly influencing protein function. However, for reasons summarized here and elsewhere (1,2,6), this rationale for lipid diversity has proved unsatisfactory. The primary reasons for this include the fact that lipids do not appear to be present in a gel state in the vast majority of biological membranes at physiological temperatures. In addition, there is little evidence to support the contention that local domains of differing fluidity can be readily achieved in liquid-crystalline bilayer membranes in response to physiological stimuli such as ionic strength, pH, divalent cations, or proteins. Finally, the membrane fluidity parameter itself is loosely defined and can lead to confusion. For example, it is commonly assumed that more saturated lipids or the presence of cholesterol makes membranes less "fluid." This is not necessarily the case. Membrane fluidity is rigorously defined as the reciprocal of the membrane viscosity, which in turn is inversely proportional to the rotational and lateral diffusion rates (D_R and D_T, respectively) of membrane components (7). Thus, a linear relation between membrane fluidity and D_R and D_T would be expected, which is not observed. Incorporation of cholesterol into phosphatidylcholine (PC) model membranes (at temperatures above the gel to liquid-crystalline transition temperature) has little or no influence on the lateral diffusion rates observed (8,9) and can actually increase the rotational diffusion rates (10). The major influence of cholesterol or decreased unsaturation is to increase the order in the hydrocarbon matrix (11).

Lipid polymorphism appears to offer a more acceptable framework within which to characterize the physical properties of lipids and their functional roles in biological membranes. Reasons for this are detailed at length in subsequent sections of this chapter. Briefly, there are four major points. First, an appreciable fraction (30 mol% or more) of lipids in biological membranes either adopt or induce nonbilayer structures in various model systems, and there is strong evidence to suggest that under appropriate conditions, the large majority of membrane lipids can adopt nonbilayer phases. Second, the structural preferences of lipids in pure and mixed lipid systems can be modulated by factors such as ionic strength, pH, divalent cations, and membrane proteins, indicating the possibility of regulated roles in membrane-mediated phenomena. Third, certain membrane-mediated processes, such as fusion, clearly require a local departures from lamellar organization in order to proceed. Finally, detailed consideration of the factors leading to different phase preferences of lipids is leading to an appreciation of parameters such as lipid "shape" or intrinsic membrane "curvature." These parameters can strongly influence bilayer membrane properties such as the order in the hydrocarbon region, which may reflect basic conserved properties in membranes.

III. STRUCTURAL PREFERENCES OF LIPIDS

A. Introduction

The subject of lipid polymorphism has been the topic of many reviews (1,2,6) and it is difficult to avoid repetition. However, for the sake of completeness, certain points must be reiterated. First, the two major polymorphic phases adopted by pure aqueous dispersions of membrane lipids are the bilayer or lamellar phase and the hexagonnal (H_{II}) phase. The hexagonal (H_{II}) phase consists of hexagonally packed arrays of lipid cylinders where the polar headgroups are oriented toward an aqueous pore of ~20 Å diameter. Nonbilayer structures such as the hexagonal H_{II} phase are liquid-crystalline structures; gel-state lipids invariably adopt the bilayer organization. Further, the hexagonal phase per se is not likely to be a major structure available to membrane lipids in vivo, as a permeability barrier could not be maintained. Indeed, the observation of H_{II} structure in vivo appears to be correlated with pathological consequences (12). As indicated below, it is more likely that intermediates between bilayer and H_{II} organization, such as inverted micelles, could provide local, discrete, and transitory departures from bilayer structure compatible with overall membrane integrity.

Techniques commonly employed to monitor lipid phase structure include X-ray procedures, as well as freeze-fracture and nuclear magnetic resonance (NMR) techniques. The advantages and limitations of these techniques have been extensively discussed elsewhere (1,2,13). X-ray analyses provide, in principle, unambiguous information on lipid phase structure. The ^{31}P-NMR technique provides a rapid diagnostic procedure, giving rise to phase assignments entirely consistent with X-ray studies (14). Briefly, bilayer phospholipids give rise to an asymmetrical lineshape with a low-field shoulder, where the dominant motional averaging arises due to rapid axial rotation of the phospholipid about its long axis. In the H_{II} phase, additional motional averaging occurs owing to the ability of the lipids to diffuse laterally around the cylinders characteristic of this phase. This gives rise to lineshapes with reversed asymmetry which are a factor of two narrower (1). Finally, phospholipids in small bilayer systems or in inverted micellar or other nonbilayer alternatives give rise to narrow symmetrical ^{31}P NMR spectra due to the isotropic motional averaging processes available in these structures. Examples of the three types of ^{31}P-NMR response are given in Figure 1.

The application of freeze-fracture techniques to the study of lipid polymorphism has been reviewed elsewhere (15) and provides an ability to visualize local structure in a lipid dispersion. Bilayer systems give rise to flat, relatively featureless fracture planes, whereas hexagonal H_{II} phase systems display corrugated fracture faces (see Fig. 1) arising as the fracture plane cleaves between hexagonally packed lipid cylinders. The greatest utility of the freeze-fracture technique arises from an ability to detect nonbilayer lipid structure giving rise to "lipidic particle" morphology (15) (Fig. 1). NMR and X-ray techniques cannot be employed to detect these systems. This is due to the absence of a regular lattice, which precludes observation by X-ray procedures, and the fact that a variety of structures (e.g., small bilayer systems, micelles) can give rise to the narrow NMR resonances that would be expected to arise from lipidic particles. Limitations of the freeze-fracture procedure include the fact that nonbilayer structures formed only at relatively high temperature ($\geq 30°C$) are often difficult to capture via freeze-fracture owing to their tendency to revert to a lamellar structure during the freezing process. Also, to avoid formation of ice crystals, cryoprotectants such as glycerol are commonly employed. In some cases, this can influence lipid morphology (16). Finally, nonbilayer structures that have very short lifetimes, including those expected to occur as intermediates in fusion, may well be difficult to observe by freeze-fracture techniques (17).

Thus far we have indicated that isolated species of hydrated, liquid-crystalline membrane lipids adopt the bilayer or H_{II} organization; that

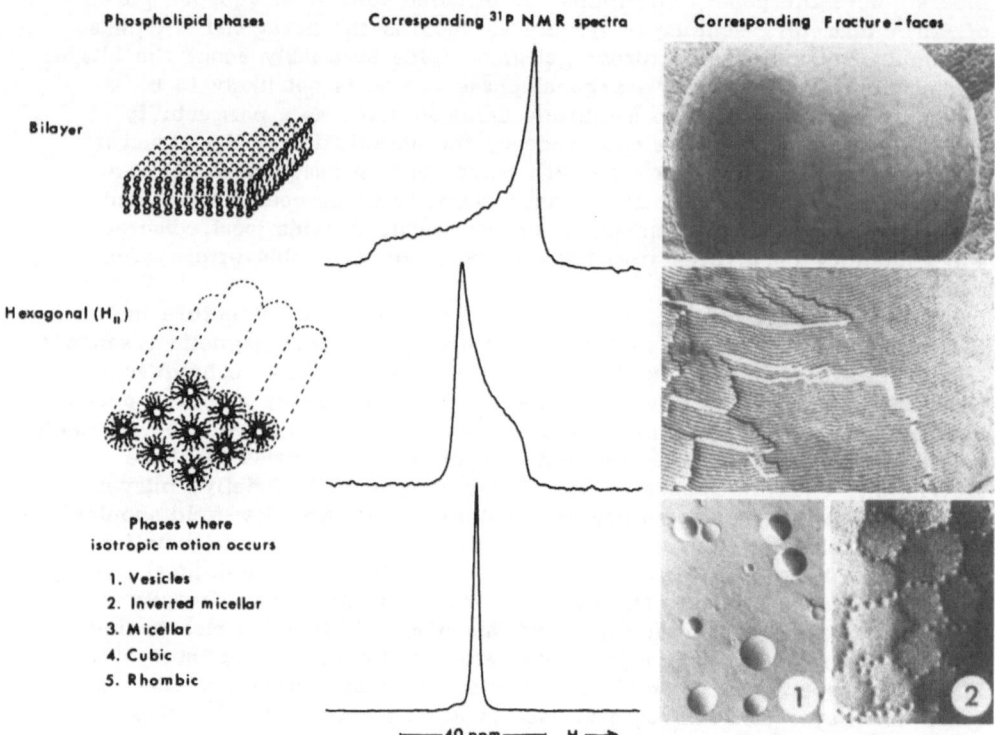

FIGURE 1 ^{31}P NMR and freeze-fracture characteristic of phospholipids
in various phases. The bilayer ^{31}P NMR spectrum was obtained from
aqueous dispersions of egg yolk phosphatidylcholine, and the hexagonal
(H_{II}) phase spectrum from phosphatidylethanolamine (prepared from soybean
phosphatidylcholine). The ^{31}P NMR spectrum representing isotropic motion
was obtained from a mixture of 70 mol% soya phosphatidylethanolamine and
30% egg yolk phosphatidylcholine after heating to 90°C for 15 min. All
preparations were hydrated in 10 mM Tris-acetic acid (pH 7.0) containing
100 mM NaCl, and the ^{31}P NMR spectra were recorded at 30°C in the
presence of proton decoupling. The freeze-fracture micrographs represent
typical fracture faces obtained from bilayer and H_{II} phase systems as well
as structures giving rise to isotropic motional averaging. The bilayer configuration (total erythrocyte lipids) gives rise to a smooth fracture face,
whereas the hexagonal (H_{II}) configuration is characterized by ridges displaying a periodicity of 6–15 mm. Common conformations that give rise
to isotropic motion are represented in the bottom micrograph: (1) bilayer
vesicles (~100 nm diameter) of egg phosphatidylcholine prepared by extrusion techniques and (2) large lipid structures containing lipidic particles.
The latter system was generated by fusing SUVs composed of egg phosphatidylethanolamine and 20 mol% egg phosphatidylserine which were prepared
at pH 7 and then incubated at pH 4 for 15 min to induce fusion.

X-ray, ^{31}P-NMR, and freeze-fracture are useful techniques to visualize
these structures; and that intermediate "lipidic particle" arrangements
can be detected by freeze-fracture protocols. From the point of view of
fusion and other membrane contact phenomena [e.g., tight junctions (18)]
the latter structures are of particular interest. Lipidic particles are commonly
observed in mixed systems composed of lipids which adopt bilayer structure
in isolation and lipids preferring the H_{II} arrangement. There is now considerable evidence to suggest that these structures are intermediates between
bilayer and H_{II} arrangements of lipids (15) and correspond to inverted
micelles formed at the nexus of intersecting bilayers (see Fig. 2). It may
be noted that the morphology of lipidic particles can vary considerably
(15). This can be attributed to the evolution of inverted-micellar contact
sites to form inverted tubes [e.g., "line defects" (19)] or to form interlamellar attachment sites (20), or other possibilities as discussed in
Section IV.

B. Structural Preferences of Lipids

The polymorphic phase preferences of lipids have been the subject of extensive investigations. A sensible summary of the large amount of data
obtained is difficult to achieve in text form. In Table 1 we present a synopsis of these investigations and confine our written remarks to providing
an appropriate overview. We address, in turn, the properties of pure
lipid systems, mixed lipid systems, and, subsequently, factors modulating
these structural preferences.

The phase behavior of the major classes of phospholipids, including
phosphatidylcholine (PC), phosphatidylglycerol (PG), phosphatidylethanolamine (PE), phosphatidylserine (PS), phosphatidic acid (PA), phosphatidylinositol (PI), cardiolipin (CL), and sphingomyelin (SM), have all received
detailed attention. The results summarized in Table 1 support the following
general points. First, effectively all lipids with fully saturated fatty acid
constituents only adopt the bilayer organization (either liquid-crystalline
or gel state) over the temperature interval 0-100°C, irrespective of hydration, pH, ionic strength, or divalent cation concentration. Second, among
the unsaturated lipids, only PE adopts the hexagonal H_{II} phase at "physiological" temperatures, pH values, and salt concentrations. More unsaturated
species of PE adopt the hexagonal phase more readily, as indicated by
a progressive lowering of the bilayer to H_{II} transition temperature (T_H)
as the unsaturation is increased. For example, the molecular species 1-
palmitoyl-2-oleoyl-PE exhibits a $T_H \simeq 75°C$, whereas for dilinoleoyl-PE
T_H is less than -10°C. PE's of eukaryotic origin exhibit T_H values in the
range of 10°C. Among the other phospholipids, unsaturated varieties of
PS, PA, and CL can prefer the H_{II} organization at low pH values and/or
at Ca^{2+} concentrations of ~2 mM or higher.

The phase properties of mixed lipid systems are of more direct interest
to the properties of biological membranes. Again, the results of Table 1
support a number of general observations. First, all species of phospholipid
that adopt the bilayer phase in isolation can stabilize hexagonal-preferring
lipids (e.g., unsaturated PE) into an overall bilayer organization in mixed
systems. The proportions of bilayer lipid required to achieve this can
vary substantially (20-50 mol%). Second, cholesterol has the general ability
to induce H_{II} phase structure in mixtures of unsaturated PE with bilayer-
stabilizing phospholipids, such as PC and PS. Fatty acids and other fusogenic compounds exhibit a similar ability to induce H_{II} organization. A third

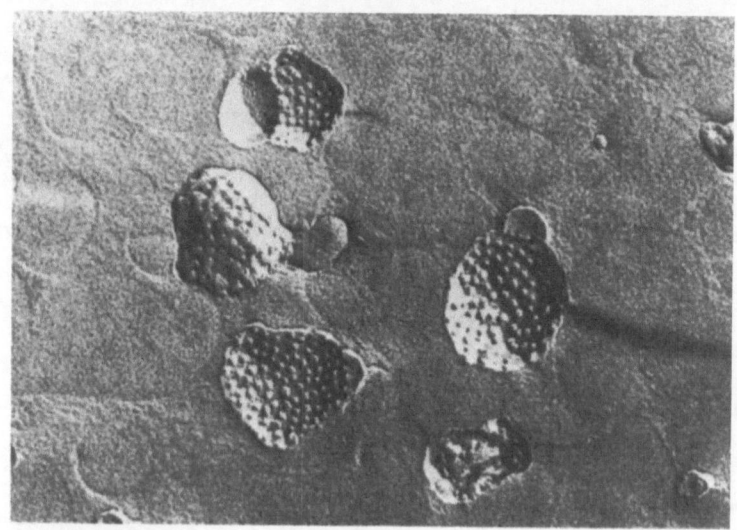

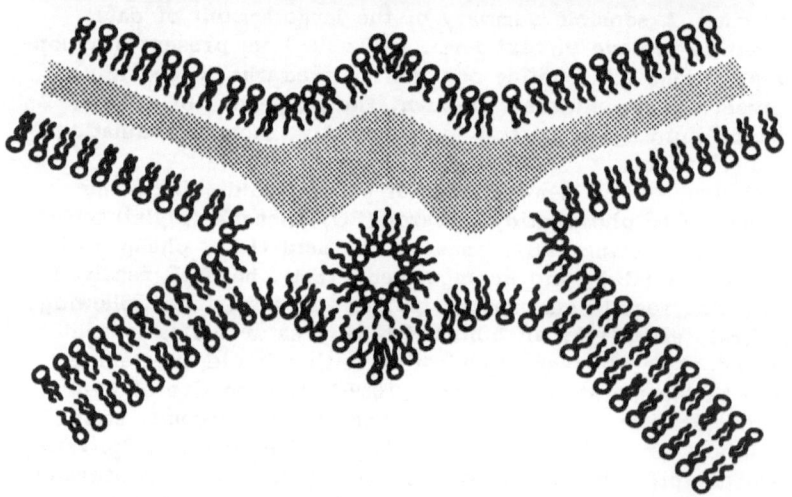

FIGURE 2   Freeze-fracture micrograph of lipidic particles induced by Ca^{2+} in a lipid system consisting of cardiolipin and soya phosphatidylethanolamine in the molar ratio of 1:4 (magnification ×80,000). A model of the lipidic particle as an inverted micelle is depicted below the micrograph. The shaded area represents the fracture region.

point concerns the hysteresis effects often observed in the temperature-dependent phase behavior of certain mixed lipid systems. For example, unsaturated PE-PC systems which adopt the bilayer phase at ambient temperatures can often be converted to systems exhibiting H_{II} phase and "isotropic" motional averaging components (as detected by ^{31}P NMR) on heating (1,2). Subsequent cooling does not necessarily result in conversion back to bilayer phase structure. Indeed, the bilayer phase can often only be reset by freezing the sample. Freeze-fracture studies commonly reveal lipidic particles in such systems, and these and other data (1,21) can be interpreted to suggest formation of "honeycomb" structures resulting from

TABLE 1 Phase Behavior of Various Lipids

Species	Phase	Conditions	References
Phosphatidylcholine[a]			
Egg	C, H_{II}	5% water, 50°C	46,48
	L	10% water	46,48,49
16:0/16:0	L_α	20% water, 41°C	50,51
	P_β	36–41°C	51,52,53
	L_β	35°C	50,51
18:1$_c$/18:1$_c$	L_α	0°C	54
20:4/20:4	L_α	0–90°C	b
Sphingomyelin			
16:0/16:0	L	10–50°C	58
Egg	L	20°C	59
Bovine brain	L	40% water, 25–50°C	49,59–61
Phosphonolipids			
16:0/16:0	L	−20–20°C	62
Tetrahymena	L	30°C	63–65
	H_{II}	45°C	63–65
Phosphatidylethanol-			
amines[c]			
Diacyl species			
20:0/20:0	H_{II}	96°C	66
18:0/18:0	H_{II}	110°C	66,67
16:0/16:0	H_{II}	109–123°C	66,67
14:0/14:0	H_{II}	85°C	66,68,69
16:0/18:1$_c$	H_{II}	75°C	70
18:1$_c$/16:0	H_{II}	70°C	b
18:1$_c$/18:1$_c$	H_{II}	60°C	68,70,71
18:1$_c$/18:1$_c$	H_{II}	10°C	68,72
18:2/18:2	H_{II}	−15°C	68,72
18:3/18:3	H_{II}	−15°C	70
20:4/20:4	H_{II}	−30°C	70
22:6/22:6	H_{II}	−30°C	70
Egg	L	pH 79	71
Egg	H_{II}	25–35°C	71
	L	pH 8.5, pressure	73,74
From egg PC	H_{II}	40–45°C	75
E. coli	H_{II}	55–60°C	71
Human erythrocyte	H_{II}	8°C	71
Porcine erythrocyte	L + H_{II}	20–40°C, 10–90% water	76
Rat liver e.r.	H_{II}	7°C	77
Rabbit s.r.	H_{II}	0°C	78
Soya bean	H_{II}	−10°C	79
Rat mitochondrial	H_{II}	10°C	80
Dialkyl species			
18:1/18:0	H_{II}	80°C	66
16:0/16:0	H_{II}	86°C	66,67

(continued)

(Table 1, continued)

Species	Phase	Conditions	References
14:0/14:0	H_{II}	93°C, excess water	66,67
		78°C, salt NaCl	66
		100°C, excess water	66
		70°C, low water	66
Effect of acyl chain linkage			
Vinyl ether	H_{II}	30°C	80,81
Alkyl ether	H_{II}	53°C	80
Acyl ester	H_{II}	68°C	80
Phosphatidylserines			
16:0/16:0	L	60°C, EDTA	82–85
	H_{II}	Anhydrous	57
	L	70°C + Ca	83
14:0/14:0	L	50°C, EDTA	82,86,87
$18:1_t/18:1_t$	L	—	88
$18:1_c/18:1_c$	L	-7°C, EDTA	82,89
Bovine brain	L	EDTA	90
	L	Na, Mg or Ca salt	25,90–93
Egg	L	± Ca	94
Egg	H_{II}	pH 3 at 40°C	94
Human erythrocyte	L	± Ca	94
Phosphatidylglycerol			
18:0/18:0	L	pH 9.5, Na salt	95
16:0/16:0	L	pH 9.5, Na salt	95–97
14:0/14:0	L	100°C	97–100
14:0/14:0	H_{II}	90°C, 1 M $CaCl_2$	101
12:0/12:0	L	Na,K,NH_4 or Ca salt	95
$18:1_c/18:1_c$	L	Ca salt	95
Egg	L	30°C	102
E. coli	L	30°C	102
Cardiolipin			
Bovine heart	L	50% water	103–107
	H_{II}	50% water	103
	H_{II}	50% water + Ca	103,104
	H_{II}	50°C + Ca, pH 3, high salt	107,108
Dilyso	M	20°C, 0.5 M NaCl	108
	L	20°C, 3 M NaCl	108
Monolyso	L	20°C, 3 M NaCl	108
Acyl	H_{II}	20°C, 3 M NaCl	108
B. subtilis	L	25°C, Na salt	105
	H_{II}	0°C Ca, Mg salt	105
	H_{II}	25°C Ba salt	105
Phosphatidic acid			
16:0/16:0	L	pH 3.5 12	109
	H_{II}	104°C, pH 4.6, 1 M NaCl	110
14:0/14:0	$P_{\beta'}$	5°C, pH 13	110
	L	20-55°C	109,112

Species	Phase	Conditions	References
$18:1_c/18:1_c$	L	pH 4-8	113
	H_{II}	pH 6 + Ca, Mg, Mn	113
	L	pH 8, 30°C	114
	H_{II}	pH 5.5 + Ca, pH 4	114
Egg	L	pH 8-12	115
	H_{II}	pH 6 + Ca	97
Phosphatidylinositol			
Soya	L	25°C ± Ca	116,117
	L	+ Ca	103
Glycosyldiglycerides			
A. laidlawii			
MGluDG	H_{II}	30°C	118-120
DGluDG	L	30°C	118-120
	C	MGDG/DGDG 1.2/1	120
18:0/18:0 MGalDG	L	20°C	121,122
16:0/16:0 MGalDG	H_{II}	70°C	123
16:1/16:1 MGalDG	H_{II}	38°C	123
18:3/18:3 MGalDG	H_{II}	20°C	121
Maize galactolipid	H_{II}	-20-100°C, 10% water	124
	C	60-100°C, 10-20% water	124
	L	0-40°C, 20% water	124
Perlargonium leaves			
MGalDG	H_{II}	0-80°C	125
DGalDG	L	0-80°C	125
Sulfoquinovosyl DG	L	20-80°C	125
Spinach MGalDG	H_{II}	-15°C	126
Wheat cloropast			
MGalDG	H_{II}	-10-80°C	127
DGalDG	L	-10-80°C	127
Cerebrosides and gangliosides			
Psychosine	Coagel	20% water, 20-70°C	128
	L	20-45% water, 20-70°C	128
	H	50-605 water, 20-70°C	128
	M	70% water, 20-70°C	128
Bovine brain			
Cerebroside	L	20-40% water, 70°C	49,128,129
Palmitoyl-Car	L	80°C	130
Sulfatide	L	20-40% water, 40°C	128,129
	C	50-60% water, 40°C	128
	M	70% water, 20-80°C	128
Bovine brain			
Ganglioside	H_{II}	18-50% water, 50°C	132
	M	50% water	132
GalCer, GlcCer LacCer, GM_3, GM_1 GD_1, GT_1	L	Mixtures with PC or monoolein	133,134

(continued)

(Table 1, continued)

Species	Phase	Conditions	References
Lysolipids			
18:0 PC	L_β	25°C	135,136
	M	27°C	136
16:0 PC	L	-10°C	137,138
	M	25°C	137
12:0 PC	M	25°C	139
18:3 PE	L	-10°C	140
	H_{II}	0°C	140
	M	10°C (inverted)	140
18:2 PE	L	-10°C	140
	M	20°C (inverted)	140
18:1 PE	L	0-90°C	140
Lyso-PC	L	+ equimolar cholesterol	141
Lyso-PC	L	+ equimolar fatty acid	142
Effect of polypeptides and proteins			
CL	L	+ Ca and polylysine	143,144
CL/PE	H_{II}	+ polylysine	143,144
CL	H_{II}	+ cytochrome c	145
PA	H_{II}	+ myelin basic protein	146
Axon lipids	H_{II}	+ cardiotoxin V4	147
PE	H_{II}	+ gramicidin 1:200 PE	148-150
PC	H_{II}	+ gramicidin 1:25 PC	148-150
PE	L	+ glycophorin	151
PE	L	+ cytochrome oxidase	152
PE	L	+ chlorophyllase	153
Effect of anesthetics and other lipophilic compounds			
PA	H_{II}	pH 6 + chloropromazine	113
CL	H_{II}	Dibucaine or chlorpromazine	104
Egg-PE	L	2 mM chlorpromazine	154
	L	% MM dibucaine	154
	L	10 mM tetracaine	154
	L	100 mM procaine	154
Egg-PE	L	Triton X-100, 20 mol%	155
	L	Deoxycholate, 5 mol%	155
	L	Octylglucoside, 10 mol%	155
	L	Lyso-PC 5 mol%	155
PE/PS	H_{II}	+Ca^{2+}	
	L	+Ca^{2+} + dibucaine	156
Erythrocyte lipids	H_{II}	+ palmitoleic acid	36
	H_{II}	+ retinol	36
	H_{II}	+ oleic acid	36
	H_{II}	+ glycerylmonooleate	36
	L	+ glycerylmonostearate	36
PC,PE	H_{II}	+ glycerylmonooleate	157,158
Egg-PC	H_{II}	30 mol% diacylglycerol	159

Species	Phase	Conditions	References
Egg-PE	H_{II}	5 mol% diacylglycerol	159
CL/PC (1:1)	L	+ Ca and adriamycin	160
PE/CL (2:1)	H_{II}	+ adriamycin	160
PE/PS (1:1)	H_{II}	+ adriamycin	160
Mixed lipid systems			
PC/PS	L	± Ca, phase separation	25,46,86
PC/PA	L	± Ca, phase separation	161,162
PC/PG	L	± Ca, phase separation	86
PE/DOPS	L	30 mol% PS, + Mg	163,164
	L + H_{II}	+ Ca, phase separation, pH 5	164
PE/DLPS	L	± Ca	165
PE/PG	L	30 mol% PG	102
	H_{II}	+ Ca, no phase separation	102
PE/PI	L	15 mol% PI	117
	H_{II}	+ Ca	117
PE/CL	L	30 mol% CL	145
PE/PC	L	20 mol% PC	
PE/PC/Chol	H_{II}	+ equimolar cholesterol	79,166
PE/DOPS/Chol	L	30°C	165
	H_{II}	+ Ca, no phase separation	165
PE/DLPS/Chol	L	30°C	165
	H_{II}	+ Ca, no phase separation	165
PE/PS/Chol	H_{II}	+ NaCl	167
PC/SM/Chol (erythrocyte outer monolayer)	L		168
PC/PE/PS/Chol (erythrocyte inner monolayer)	H_{II}	+ Ca^{2+}	168

Unless otherwise stated, the lipids are assumed to be fully hydrated and at neutral pH. Phases are indicated as follows: L = lamellar; H_I = hexagonal H_I; H_{II} = yexagonal H_{II}; C = cubic; M = micellar.

[a]All PC's adopt only lamellar gel or liquid-crystal phases except at low hydration. For reviews of gel-liquid crystal behavior see Refs. 55,56. The mesomorphism of anhydrous and monohydrated PC's has been discussed elsewhere [57].

[b]C. P. S. Tilcock (unpublished).

[c]General trends are that T_H decreases with increasing acyl chain unsaturation, high salt or low hydration. Alkaline pH or high pressures raise T_H. For saturated chains, decreasing chain length increases T_H. Plasmalogens (vinyl ethers) exhibit lower T_H values than the corresponding alkyl or acyl lipids.

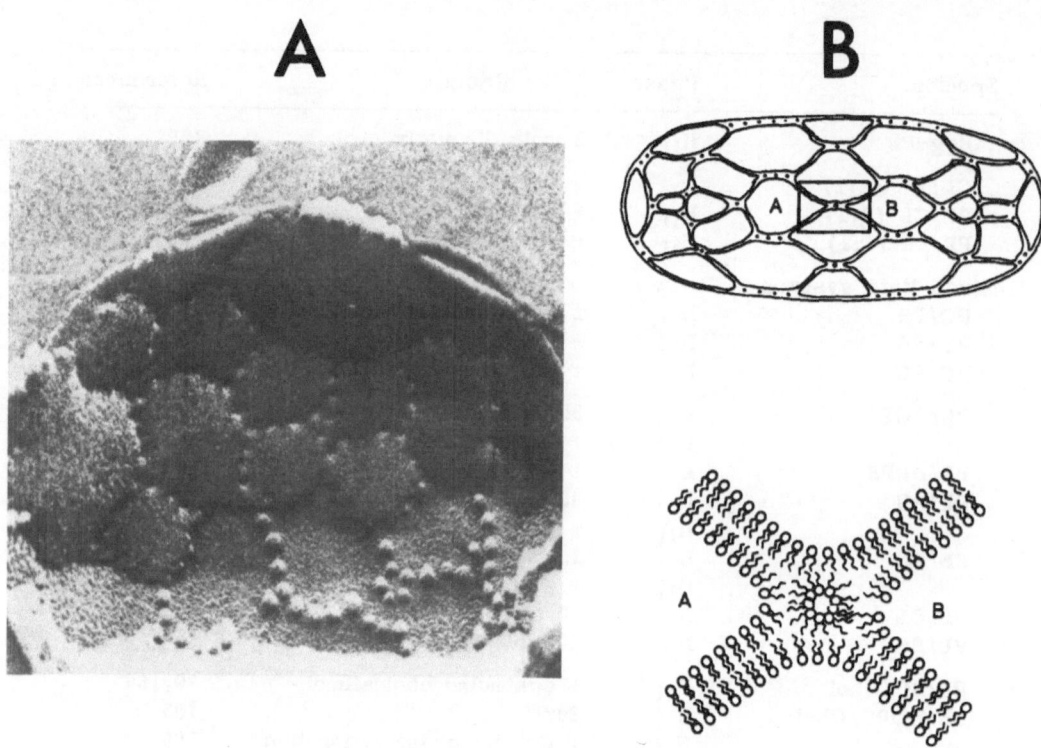

FIGURE 3   (A) Freeze-fracture micrograph of aggregation produced on dialysis of sonicated vesicles (soya PE/egg PS, 5:1) against a pH = 3.0 buffer. (B) Honeycomb structure interpretation.

fusion of apposed bilayers in multilamellar systems, as shown in Figure 3 (see also Chapter 6). The importance of these observations concerns their possible relation to relatively stable interbilayer connections such as occur in tight junctions (18).

As pointed out previously, in order for nonbilayer lipid structures to play regulated roles in membrane-mediated phenomena, isothermal mechanisms for the generation of such structures in vivo must exist. It is, therefore, particularly gratifying that a large variety of biologically relevant variables, such as ionic strength, pH, divalent cation, and protein, strongly influence the lipid polymorphism. This is apparent in Table 1 and has already been indicated in part by the ability of low pH values (pH $\leq$ 4.0) to induce H_{II} phase organization in unsaturated PS and PA dispersions as well as the ability of Ca^{2+} to trigger the H_{II} phase in CL systems. However, as shown in Table 1, the ability of exogenous factors to modulate the polymorphism of mixed lipid systems is even more pronounced. In (bilayer) mixtures of unsaturated PE stabilized by acidic (negatively charged) lipids, for example, the addition of Ca^{2+} can trigger H_{II} formation. Alternatively, in similar systems stabilized by unsaturated PS and PA, lower pH values again lead to H_{II} phase formation. Increased ionic strength can give rise to H_{II} structures in previously bilayer PE/PS/cholesterol liposomes. Proteins can have similar bilayer-destabilizing (or bilayer-stabilizing) effects, as indicated by the ability of cytochrome c to induce nonbilayer lipid organization in CL/PE dispersions. Similar observations have been made for the A_1 basic protein from myelin (30), as well as cardiotoxin and mellitin. Other proteins, such as cytochrome oxidase and glycophorin, can stabilize the bilayer (see Table 1).

C. Mixing Properties of Lipids

As indicated in the previous section, most membrane lipids can adopt H_{II} phase structure under appropriate circumstances. However, in mixed lipid systems, questions concerning the mixing properties of component lipids can be raised, particularly in a multicomponent system where different phase structures (e.g., bilayer, H_{II}, and lipidic particle) are observed to coexist. In such systems it may be expected that lipids preferring the bilayer organization would be predominantly in the bilayer component, whereas H_{II}-preferring lipid would be in the H_{II} component. However, results from this laboratory (22) indicate that ideal lipid mixing is maintained in PC/PE/cholesterol systems for which bilayer, H_{II}, and "isotropic" components are observed by ^{31}P NMR. Similar conclusions can be drawn from results obtained for CL/PC systems (23). However, there is some contention in this area, as effects apparently consistent with enrichment of PE in H_{II} phase components existing in PE/PC systems have been observed (24).

In mixed lipid systems where H_{II} phase structure is induced by factors such as pH, ionic strength, or Ca^{2+}, two major types of phenomena can occur. The first of these, which has been observed in unsaturated PE/PS systems, concerns the ability of Ca^{2+} to sequester PS into crystalline ("cochleate") domains (25). For example, in DOPS/DOPE systems, the addition of Ca^{2+} can result in lateral phase separation of PS to form cochleate domains, which allows the PE to adopt the H_{II} phase it prefers in isolation (26). However, this behavior is not observed in systems stabilized by more unsaturated varieties of PS or in systems containing extra components such as cholesterol (27). In bilayer DOPS/DOPE/cholesterol (molar ratio 1:1:1) systems, for example, the addition of Ca^{2+} results in complete H_{II}

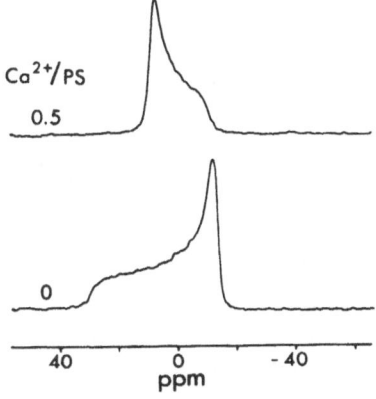

FIGURE 4 81.0 MHz ^{31}P NMR spectra at 30°C of a DOPE-DOPC-DOPS-cholesterol (1:1:1:3) mixture in the absence and presence of Ca^{2+}. The Ca^{2+} was added to achieve a Ca^{2+}/PS molar ratio of 0.5. For further details see Ref. 165.

phase structure where both the PE and the PS (and presumably the choles-terol as well) are contained in the H_{II} organization. The major point of these remarks is to emphasize the conclusion that the large majority of membrane lipids can adopt nonbilayer structure and that stimulation of H_{II} and other nonbilayer phases does not usually result in lateral segregation of component lipids. A dramatic example of this is given in Figure 4, where the ^{31}P NMR characteristics of a DOPC/DOPE/DOPS/cholesterol (molar ratio 1:1:1:3) in the presence and absence of Ca^{2+} are noted. In the absence of Ca^{2+}, a ^{31}P NMR spectrum consistent with lamellar organization is observed, whereas in the presence of Ca^{2+}, complete H_{II} phase formation is observed, as indicated by ^{31}P NMR. This and other (27) evidence show unequivocally that all the DOPC, DOPE, and DOPS adopt the H_{II} organization when Ca^{2+} is present.

IV. THE MOLECULAR BASIS OF LIPID POLYMORPHISM

The ability of lipids to adopt the H_{II} phase and other nonbilayer structures on hydration has naturally stimulated considerable interest in the factors that drive these remarkable structural transitions and the mechanisms involved. The literature in these areas is starting to provide real insight, which, as emphasized in Section II, has general implications for the proper-ties and roles of lipids in membranes.

A. Factors Determining Lipid Phase Structure: The Shape and Curvature Concepts

Progress in this area basically stems from the work of Israelachvili and co-workers (28), who have examined the molecular properties of amphiphiles which form spherical and nonspherical micelles on dispersion in water. A basic packing property that has proved useful to explain these properties is a dimensionless shape parameter defined as $S = V/A_OL_C$. Here A_O is an "optimum" area per molecule at the lipid-water interface, V is the volume per molecule, and L_C is the length of the fully extended acyl chain. That the S parameter relates to a molecular shape property is easily realized from Figure 5, which defines an additional parameter A_H as the cross-sectional area subtended at the hydrophobic end of the molecule. It is straightforward to show that for $A_H/A_O < 1$, $S > 1$; for A_H/A_O, $S = 1$; and for $A_H/A_O > 1$, $S < 1$. Using the language introduced previously (1), lipids that have a preferred shape corresponding to $S < 1$ are referred to as having a "cone" shape, whereas lipids where $S \simeq 1$ are cylindrical and lipids where $S > 1$ have an "inverted" cone shape. The relationship between these shape properties and the geometry of the macroscopic lipid aggregate is clear, as lipids in a micellar phase must exhibit an inverted cone shape in order to satisfy geometrical packing constraints, bilayer lipids must be roughly cylindrical, whereas lipids in H_{II} or inverted micellar structures must have a net cone shape.

    The shape concept of lipids might appear unduly simplistic. However, it has proven remarkably successful in providing a qualitative but predictive framework for understanding the polymorphic phase preferences of lipid dispersions. Before detailing this success, it is important to realize that lipid shape is an inclusive phenomenological concept which lumps together a large variety of complex molecular forces. For example, A_O, the optimum cross-sectional area at the lipid-water interface, would be expected to be

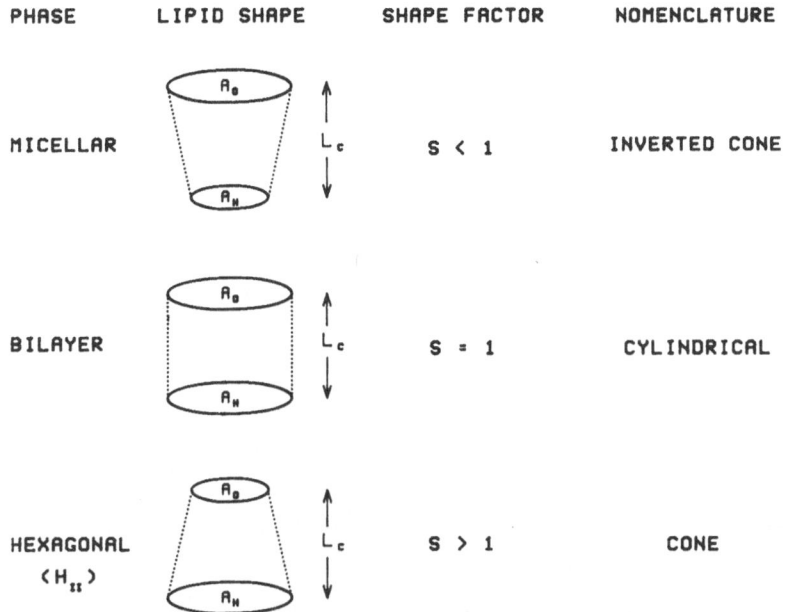

FIGURE 5 Shape features exhibited by membrane lipids. A_O refers to the area subtended by the polar region at the lipid-water interface, whereas A_H refers to the area subtended at the intermonolayer hydrophobic interface. The shape factor $S = V_O/A_O L_C = 1/3[1 + (A_H/A_O)^{1/2} + A_H/A_O]$, where V_O is the volume of the lipid.

sensitive to the size of the lipid headgroup (large headgroup leading to large A_O values), the charge on the headgroup (charged headgroups giving larger effective A_O's due to inter-headgroup electrostatic repulsion effects), the hydration of the headgroup (lower hydration, smaller A_O), and so on. Alternatively, A_H will be sensitive to factors that modulate the splay at the end of the hydrocarbon chains. Thus, increased acyl chain unsaturation, increased temperature, and increased acyl chain length (for liquid-crystalline lipids) would all be expected to increase the preferred value of A_H, leading to increased cone shape and possible H_{II} phase formation.

The features influencing lipid shape and their predicted influence on the bilayer (L_α) to H_{II} phase transition are summarized in Figure 6. All of these factors modulate lipid polymorphism in the predicted manner. As previously indicated (see Table 1), particularly for PE systems, increased acyl-chain unsaturation leads to increased proclivity for H_{II} structure, and increased temperature induces bilayer to H_{II} transitions. The smaller headgroup of PE (as compared to PC) is consistent with H_{II} organization. The proclivity of unsaturated PE's for the bilayer phase at pH $\geqslant 9$ is consistent with deprotonation of the primary amine, resulting in a charged headgroup and thus a larger effective A_O. Similarly, protonation of the PS carboxyl and PA phosphate at lower pH values (≤ 4, see Table 1) leads to reduced interheadgroup electrostatic repulsion, smaller A_O values, and H_{II} structure. The ability of Ca^{2+} to trigger bilayer H_{II} transitions in CL systems as well as unsaturated PS/PE, PG/PE, PA/PE, and PI/PE systems can be rationalized on a similar basis, as can the ability of high salt concentrations to induce H_{II} organization in PE/PS/cholesterol systems. With

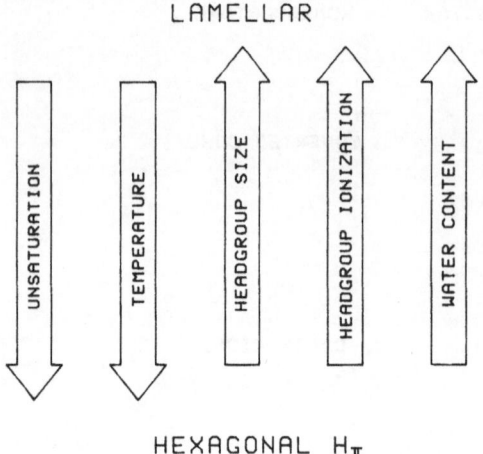

LAMELLAR

HEXAGONAL H_{II}

FIGURE 6 Factors influencing the bilayer to hexagonal (H_{II}) phase transition for membrane lipids.

regard to hydration, lower water content has been shown to reduce the T_H in a variety of systems, most notably egg PC systems where H_{II} phase structure can be observed under conditions of low water content and high temperature (Table 1):

In summary, the shape concept offers gratifying correlation between predictions and experiment. With regard to fusion, it is interesting to note that all factors promoting cone-shape character in component lipids (i.e., promoting H_{II} organization) also promote fusion (4).

However, the shape parameter is not a well-defined, measurable quantity. For example, as discussed in Section IV.B, the lipid shape can vary according to the environment. Theoretical work has therefore been directed toward obtaining a more quantitative measure of forces driving H_{II} phase formation, leading to the "curvature" concept introduced by Gruner and co-workers (29,30). In this approach each monolayer is viewed independently, and the presence of cone-shaped lipid is expressed as a tendency of each monolayer to curl up into cylinders with an equilibrium radius of curvature R_o. The binding energy for monolayers expressing a radius of curvature R is then given (to least significant order in 1/R) by $E = k_c (1/R - 1/R)^2/2$ where k_c is the elastic compressibility modulus. The tendency of the monolayers to form cylinders is countered by the requirement to fill the inter-cylinder spaces which requires acyl-chain stretching. Thus, the actual structure assumed depends on a competition between curvature and hydrocarbon stretching forces.

B. Lipid Shape and Bilayer Packing Properties

The fact that lipid molecules exhibit certain preferred shapes which strongly influence the polymorphic phase adopted (bilayer or H_{II}) leads to interesting possibilities concerning the properties of membranes containing a proportion of lipids that prefer H_{II} organization, but where bilayer structure is maintained. Clearly, this directly relates to the situation in biological membranes. Assuming a reasonably flat bilayer, the distributed presence of cone-shaped molecules would result in a smaller "optimum" surface area as compared to the optimum hydrophobic area assumed at the hydrophobic (intermonolayer) interface. Least energy considerations would then suggest a lateral compression of the bilayer to counter the large hydration energies associated with increased water penetration into the hydrophobic region. In turn, this compression would be expected to result in larger-order parameters in the hydrocarbon and correspondingly reduced membrane permeability. Such increased hydrocarbon order has been observed (employing ^{2}H NMR) in bilayer PE/PC systems as compared to PC systems (31), and decreases in K^+ permeability have been observed as PE is titrated into PC bilayers (32). Similar effects may be related to the ability of cholesterol, a cone-shaped molecule, to increase acyl-chain order and reduce membrane permeability (33). The presence of cone-shaped lipids in bilayer membranes therefore appears to lead to increased order in the hydrocarbon. In turn, factors modulating the shape preferences of lipdis would be expected to lead to corresponding modulations of hydrocarbon order, assuming bilayer structure is maintained. This could be expressed as changes in membrane permeability or membrane protein function. The appeal of these observations lies in the resulting ability to relate the quantitative shape concepts to the well-defined, measurable hydrocarbon order parameters, which may well represent a conserved, regulated quantity in membranes.

C. Mechanisms and Dynamics of Bilayer Nonbilayer Transitions

The results indicated in the previous sections identify likely forces and stresses that result in an overall preference for bilayer, H_{II}, or other nonbilayer phases. However, they do not address the actual mechanism whereby such reorganizations proceed or the dynamics involved.

An initial observation relating to the mechanism of the bilayer-to-H_{II} transition indicated that the inverted cylinders characteristic of the H_{II} organization form parallel to the planes of closely apposed bilayers (21). Concurrently, the observation was also made that lipidic particles appeared to be intermediate structures between bilayer and H_{II} phases (34), leading to the possibility that lipidic particles (interpreted as inverted micelles) are intermediates in these bilayer-to-H_{II} transitions as well as intermediates in fusion processes. In bilayer-to-H_{II} transitions, a general intermediate role of inverted micelles did not appear warranted, however, as no lipidic particles or narrow isotropic ^{31}P NMR resonances could be observed during such transitions occurring, for example, in unsaturated PE systems. However, Siegel (17,19,20), in an interesting series of papers, provides a thermodynamic analysis of the dynamics of inverted micellar intermediates (IMI) and concludes that the lifetime of IMI is likely to be very short (e.g., $< 10^{-4}$ s) and the steady-state number of IMI will contain only a small fraction of the total lipid. Thus, detection of IMI in bilayer-to-H_{II} transitions would be very difficult by NMR or freeze-fracture techniques. A general model of bilayer-to-H_{II} transitions results, where IMI form between bilayers and provide nucleation points for formation of the cylinders characteristic of the H_{II} phase as the lipid dispersion progressively converts to the H_{II} organization.

Siegel (20) has also provided an elegant analysis of events that may be expected when formation of inverted cylinders from IMI is reduced or inhibited. This may arise in systems incubated just below the bilayer-to-H_{II} transition temperature, in small unilamellar systems where a limited interface area prevents formation of long interbilayer inverted tubes, as well as in lipid systems that evolve into cubic rather than H_{II} phases. These structures exhibit isotropic NMR resonances and are usually translucent. In such situations, an evolution of the IMI into interlamellar attachment sites (ILA) is favored. These interlamellar attachment sites basically represent a completed fusion event.

In summary, the theoretical basis for lipid polymorphism is becoming increasingly well established. The success of the phenomenological shape concept has led to identification of membrane curvature as a quantitative and measurable parameter influencing lipid polymorphism, as well as to increasing insight into the relation between lipid shape properties and hydrocarbon order. Finally, detailed thermodynamic analyses are leading to recognition of the interbilayer inverted micelle as a fundamental intermediate in membrane-membrane interactions, as a precursor to either H_{II} organization, cubic structure, or fusion processes.

V. LIPID POLYMORPHISM AND BIOLOGICAL MEMBRANES

The polymorphic properties of lipids summarized in the previous sections arise from studies performed on model membrane systems. The aim of such studies is, of course, to provide a basic understanding of the physical properties and functional roles of lipids in biological membranes. This has not yet been achieved. However, certain aspects of biological membrane lipid composition and behavior are instructive.

More than 95% of the phospholipid in the erythrocyte is organized in a bilayer organization (35), and similar conclusions may be drawn for most other biological membranes. Chemical fusogens can induce H_{II} structure for erythrocyte "ghost" membranes (3,36). However, no other form of disruption [phospholipase treatment, proteolytic digestion (37)] causes reorganization of membrane bilayer structure, although isolated dispersions of "inner monolayer" lipid exhibit H_{II} organization in the presence of Ca^{2+} (38).

In the case of organelle membranes, the lipid composition and unsaturation result in a much more delicate balance between bilayer structure and other structural alternatives. Limited dehydration of endoplasmic reticulum (microsomal) membranes results in H_{II} organization (41), for example, whereas total lipid extracts of the mitochondrial membrane exhibit a proportion of H_{II} phase structure (40). Similarly, lipid extracts from retinal disk membranes adopt H_{II} organization (39) (indicating a bilayer-stabilizing role for rhodopsin) and lipid extracts from chloroplast membranes exhibit similar behavior (42). It may therefore be suggested that the lipid composition exhibited by these membranes is compatible with the presence of localized nonbilayer lipid structure. This may find expression in fusion processes as well as nonbilayer adhesion points between membranes, such as between the inner and outer mitochondrial membranes (15) and structures related to tight junctions (18).

The observed balance between bilayer and H_{II} phase lipids in membranes has led to the suggestion that maintenance of appropriate proportions of cylindrical and cone-shaped lipids represents conserved quantities in membranes (6). Evidence in support of this hypothesis has been generated employing certain prokaryotes that have rather limited biosynthetic abilities for lipid biosynthesis, allowing appreciable manipulation of lipid composition and corresponding elucidation of factors regulating lipid composition. *Acholeplasma laidlawii* is perhaps the best example, as inhibition of endogenous fatty acid synthesis has led to the development of strains with an essentially homogeneous fatty acid composition (43). Studies on these systems have led to interesting observations regarding a balance between different molecular species of lipid, which can be interpreted in terms of a requirement for a balance between cone-shaped (H_{II} phase) lipids and cylindrical (bilayer) species. In particular, it is observed that as the acyl-chain length or unsaturation is increased, the ratios of endogenous monoglucosyldiglyceride (MGluDG) to diglucosyldiglyceride (DGluDG) decrease dramatically (43). As MGluDG is a cone-shaped (H_{II} phase) lipid and DGluDG a cylindrical (bilayer phase) lipid, and as increases in chain length and unsaturation give rise to increased cone-shaped character, the changes in the MGluDG/DGluDG ratio are consistent with a need to conserve lipid shape distributions. Similarly, other workers have shown that higher levels of cholesterol (a cone-shaped molecule) result in lower MGluDG/DGluDG ratios (44), whereas the inclusion of low levels of inverted cone (anesthetic) molecules in the growth medium has reverse effects (45). These observations have been used to support the contention (6) that conservation of lipid shape properties may be more basic than maintenance of membrane lipid "fluidity" per se. However, it is also possible that the basic conserved quantity is the order in the hydrocarbon, which is modulated (increased) by the presence of cone-shaped lipids (see Section IV.B).

VI. CONCLUDING REMARKS

Our phenomenological understanding of the structural properties of lipids found in membranes and factors regulating these polymorphic capabilities is rapidly becoming a mature body of knowledge. This is having two important consequences. First, the ability of membrane lipids to adopt transitory or long-lived nonbilayer alternatives has direct application to phenomena, such as fusion, that require local departures from bilayer structure. Second, the observation that biological membranes maintain a balance between bilayer and nonbilayer lipids of different shapes may lead to an understanding of conserved quantities, such as curvature or hydrocarbon packing properties, which dictate membrane lipid composition.

NOTE ADDED IN PROOF

Since submission of this chapter, a number of additional findings have been made in this research area. Unfortunately, we have not been able to include all of them. Some recent references are cited at the end of the reference list.

ACKNOWLEDGMENTS

The research programs of Drs. Cullis, Hope, and Tilcock are supported by the Canadian Medical Research Council. PRC is an MRC Scientist.

REFERENCES

1. Cullis, P. R., and De Kruijff, B. (1979). Lipid polymorphism and the

functional roles of lipids in biological membranes. *Biochim. Biophys. Acta* 559:399–420.

2. Cullis, P. R., Hope, M. J., De Kruijff, B., Verkleij, A. J., and Tilcock, C. P. S. (1985). Structural properties and functional roles of phospholipids in biological membranes. In *Phospholipids and Cellular Regulation* (J. F. Kuo, ed.). CRC Press, Boca Raton, FL, Chapter 1.

3. Cullis, P. R., and Hope, M. J. (1978). Effects of fusogenic agents on the membrane structure of erythrocyte ghosts: The mechanism of membrane fusion. *Nature* 271:672–674.

4. Cullis, P. R., De Kruijff, B., Verkleij, A. J., and Hope, M. J. (1986). Lipid polymorphism and membrane fusion. *Biochem. Soc. Trans.* 14:242–245.

5. Singer, S. J., and Nicolson, G. L. (1972). The fluid mosaic model of the structure of cell membranes. *Science* 175:720–736.

6. Rilfors, L., Lindblom, A., Wieslander, A., and Christiansson, A. (1984). Lipid bilayer stability in biological membranes. In: *Biomembranes, Vol. 12, Membrane Fluidity* (M. A. Manson and M. Kates, eds.). Plenum Press, New York, Chapter 6.

7. Safman, P. G., and Delbruck, M. (1975). Brownian motion in biological membranes. *Proc. Natl. Acad. Sci. USA* 72:3111–3114.

8. Cullis, P. R. (1976). Lateral diffusion rates of phosphatidylcholine in vesicle membranes. Effects of cholesterol and hydrocarbon phase transitions. *FEBS Lett.* 70:233–239.

9. Lindblom, G., Johansson, L. B. A., and Arvidson, G. (1981). Effects of cholesterol on membranes. *Biochemistry* 20:2204–2210.

10. Ghosh, R., and Seelig, J. (1982). The interaction of cholesterol with bilayers of phosphatidylethanolamine. *Biochim. Biophys. Acta* 691:151–160.

11. Davis, J. H. (1983). The description of membrane lipid conformation, order and dynamics by ^{2}H NMR. *Biochim. Biophys. Acta* 737:117–172.

12. Bucheim, W., Drenckhahn, D., and Lullmann-Rauch, R. (1979). Freeze-fracture studies of cytoplasmic inclusions occurring in experimental lipidosis as induced by amphiphilic cationic drugs. *Biochim. Biophys. Acta* 575:71–80.

13. Gruner, S. M., Cullis, P. R., Hope, M. J., and Tilcock, C. P. S. (1985). Lipid polymorphism: The molecular basis of nonbilayer phases, *Annu. Rev. Biophys. Biophys. Chem.* 14:211–238.

14. Tilcock, C. P. S., Cullis, P. R., and Gruner, S. M. (1986). On the validity of ^{31}P NMR determinations of phospholipid polymorphic phase behavior. *Chem. Phys. Lipids* 40:47–56.

15. Verkleij, A. J. (1984). Lipidic intramembranous particles. *Biochim. Biophys. Acta* 779:43–92.

16. Seu, A., Brain, A. P. R., Quinn, P. J., and Williams, W. P. (1982). Formation of inverted lipid micelles in aqueous dispersions of mixed sn-3-galactosyldiacylglycerols induced by heat and ethylene glycol. *Biochim. Biophys. Acta* 686:215–224.

17. Siegel, D. P. (1984). Inverted micellar structure in bilayer membranes. Formation rates and half-lives. *Biophys. J.* 45:399–421.

18. Kachar, B., and Reese, T. S. (1982). Evidence for the lipidic nature of tight junction strands. *Nature* 296:464–466.

19. Siegel, D. P. (1986). Inverted micellar intermediates and the transitions between lamellar, cubic and inverted hexagonal lipid phases: I. Mechanism of the L_α-H_{II} phase transitions. *Biophys. J.* 49, in press.

20. Siegel, D. P. (1986). Inverted micellar intermediates and the transitions between lamellar, cubic and inverted hexagonal lipid Phases: II. Impli-

cations for membrane-membrane interactions and membrane fusion. *Biophys. J.* 49, in press.

21. Cullis, P. R., De Kruijff, B., Hope, M. J., Nayer, R., and Schmid, S. L. (1980). Phospholipids and membrane transport. *Can. J. Biochem.* 58:1091-1100.

22. Tilcock, C. P. S., Bally, M. B., Farren, S. B., and Cullis, P. R. (1982). Influence of cholesterol on the structural preferences of dioleoylphosphatidylethanolamine-dioleoylphosphatidylcholine systems: A ^{31}P and ^{2}H NMR study. *Biochemistry* 21:4596-4601.

23. De Kruijff, B., Verkleij, A. J., van Echteld, C. J. A., Gerritsen, W. J., Mombers, C., Noordam, P. C., and de Gier, J. (1979). The observation of lipidic particles in lipid bilayers as seen by P NMR and freeze-fracture electron microscopy. *Biochim. Biophys. Acta* 555: 200-209.

24. Eriksson, P. O., Rilfors, L., Lindblom, G., and Arvidson, G. (1985). Multicomponent spectra from ^{31}P NMR studies of the phase equilibria in the system dileoylphosphatidylcholine-dioleoylphosphatidylethanolamine-water. *Chem. Phys. Lipids* 37:357-372.

25. Jacobson, K., and Papahadjopoulos, D. (1975). Phase transitions and phase separations in phospholipid membranes induced by changes in temperatures, pH and concentration of divalent cations. *Biochemistry* 14:153-161.

26. Tilcock, C. P. S., and Cullis, P. R. (1981). The polymorphic phase behaviour of mixed phosphatidylserine-phosphatidylethanolamine model systems as detected by ^{31}P NMR. Effects of divalent cations and pH. *Biochim. Biophys. Acta* 641:189-201.

27. Tilcock, C. P. S., Bally, M. B., Farren, S. B., Cullis, P. R., and Gruner, S. M. (1984). Cation-dependent segregation phenomena and phase behaviour in model membrane systems containing phosphatidyl-serine: Influence of cholesterol and acyl chain unsaturation. *Biochemistry* 23:2696-2703.

28. Israelachvili, J. N., Marcelja, S., and Horn, R. G. (1980). Physical principles of membrane organization. *Q. Rev. Biophys.* 13:121-160.

29. Kirk, G. L., Gruner, S. M., and Stein, D. L. (1984). A thermodynamic model of the lamellar to inverse hexagonal phase transition of lipid membrane-water systems. *Biochemistry* 23:1093-1201.

30. Gruner, S. M. (1985). Intrinsic curvature hypothesis for biomembrane lipid composition: A role for nonbilayer lipids. *Proc. Natl. Acad. Sci. USA* 82:3665-3669.

31. Cullis, P. R., Hope, M. J., and Tilcock, C. P. S. (1986). Lipid polymorphism and the roles of lipids in membranes. *Chem. Phys. Lipids* 40:127-144.

32. Papahadjopoulos, D., and Watkins, J. C. (1967). Phospholipid model membranes. II. Permeability properties of hydrated liquid crystals. *Biochim. Biophys. Acta* 135:269-281.

33. De Gier, J., Mandersloot, J. S., and Van Deenen, L. L. M. (1968). Lipid composition and permeability of liposomes. *Biochim. Biophys. Acta* 150:166-175.

34. Verkleij, A. J., Van Echteld, C. J. A., Gerritsen, W. J., Cullis, P. R., and De Kruijff, B. (1980). The lipidic particle as an intermediate structure in membrane fusion processes and bilayer to hexagonal (H$_{II}$) transitions. *Biochim. Biophys. Acta* 600:620-625.

35. Cullis, P. R., and Grathwohl, Ch. (1977). Hydrocarbon phase transitions and lipid protein interactions in the erythrocyte membrane. *Biochim. Biophys. Acta* 471:213-224.

36. Hope, M. J., and Cullis, P. R. (1981). The role of non-bilayer lipid structures in the fusion of human erythrocytes induced by lipid fusogens. *Biochim. Biophys. Acta* 640:82–90.

37. Van Meer, G., De Kruijff, B., Op den Kamp, J. A. F., and Van Deenen, L. L. M. (1980). Preservation of bilayer structures in human erythrocytes and erythrocyte ghosts after phospholipase treatment. *Biochim. Biophys. Acta* 600:1–12.

38. Hope, M. J., and Cullis, P. R. (1979). The bilayer stability of inner monolayer lipids from the human erythrocyte. *FEBS Lett.* 107:323–327.

39. De Grip, W. J., Drenthe, E. H. S., Van Echteld, C. J. A., De Kruijff, B., and Verkleij, A. J. (1979). A possible role of rhodopsin in maintaining bilayer structure in the photoreceptor membrane. *Biochim. Biophys. Acta* 558:330–338.

40. Cullis, P. R., De Kruijff, B., Hope, M. J., Nayer, R., and Rietveld, A. (1980). Structural properties of phospholipids in rat liver mitochondrial membranes. *Biochim. Biophys. Acta* 600:625–635.

41. Crowe, L. M., and Crowe, J. H. (1982). Hydration dependent hexagonal phase lipid in a biological membrane. *Arch. Biochem. Biophys.* 271:582–591.

42. Sprague, S. G., and Staehelin, L. A. (1984). A rapid reverse phase evaporation method for the reconstitution of uncharged thylakoid membrane lipids that resist hydration. *Plant Physiol.* 75:502–570.

43. Silvius, J. R., Mak, N., and McElhaney, R. N. (1980). Lipid and protein composition and thermotropic phase transitions in fatty acid homogeneous membranes of *Acholeplama laidlawii* B. *Biochim. Biophys. Acta* 597:199–212.

44. Wieslander, A., Christiansson, A., Rilfors, L., and Lindblom, G. (1980). Lipid bilayer stability in membranes. *Biochemistry* 19:3650–3660.

45. Christiansson, A., Gutman, H., Wieslander, A., and Lindblom, G. (1981). Effects of anesthetics on water permeability and lipid metabolism in *Acholeplasma laidlawii* membranes. *Biochim. Biophys. Acta* 645:24–31.

46. Luzzatti, V. (1968). X-ray diffraction studies of lipid-water systems. In: *Biological Membranes*, Vol. 1 (D. Chapman, ed.). Academic Press, New York, pp. 71–123.

47. Fontell, K. (1978). Liquid-crystalline behaviour in lipid-water systems. *Prog. Chem. Fats Other Lipids* 16:145–162.

48. Small, D. M. (1967). Phase equilibria and structure of dry and hdyrated egg lecithin. *J. Lipid Res.* 8:551–557.

49. Reiss-Husson, F. (1967). Structure des phases liquide-cristallines des differents phospholipides, monoglycerides, sphingolipides, anhydres ou en presence d'eau. *J. Mol. Biol.* 25:363–382.

50. Chapman, D., Williams, R. M., and Ladbrooke, B. D. (1967). Physical studies of phospholipids VI. Thermotropic and lyotropic mesomorphism of some 1,2-diacylphosphatidylcholines. *Chem. Phys. Lipids* 1:445–473.

51. Janiak, M. J., Small, D. M., and Shipley, G. G. (1976). Nature of the thermal pretransition of synthetic phospholipid: Dimyristoyl and dipalmitoyllecithin. *Biochemistry* 15:4575–4082.

52. Boroske, E., and Trahms, L. (1983). A ^{1}H and ^{13}C NMR study of the motional changes of dipalmitoyllecithin associated with the pretransition. *Biophys. J.* 42:275–281.

53. Heutschel, M., Hosemann, R., and Helfrich, W. (1980). Direct X-ray study of the molecular tilt in dipalmitoyllecithin bilayers. *Z. Naturforsch.* 35:643–646.

54. Phillips, M. C., Hauser, H., and Paltauf, F. (1972). The inter- and

intra-molecular mixing of hydrocarbon chains in lipid/water systems. *Chem. Phys. Lipids* 8:127-133.

55. Silvius, J. R. (1982). Thermotropic phase transitions of pure lipids in model membranes and their modification by membrane proteins. In: *Lipid-Protein Interactions*, Vol. 2, Chapter 7 (P. C. Jost and O. H. Griffith, eds.). Wiley, New York.

56. Lytz, R. K., Reinert, J. C., Church, S. E., and Wickman, H. H. (1984). Structural properties of a monobrominated analog of 1,2-dipalmitoyl-sn-glycero-3-phosphorylcholine. *Chem. Phys. Lipids* 35:63-76.

57. Williams, R. M., and Chapman, D. (1970). Phospholipids, liquid-crystals and cell membranes. *Prog. Chem. Fats Other Lipids* 11:1-79.

58. Calhoun, W. I., and Shipley, G. G. (1979). Sphingomyelin-lecithin bilayers and their interaction with cholesterol. *Biochemistry* 18:1717-1722.

59. Hui, S. W., Stewart, T. P., and Yeagle, P. L. (1980). Temperature-dependent morphological and phase behaviour of sphingomyelin. *Biochim. Biophys. Acta* 601:271-281.

60. Mackay, A. L., Wassall, S. R., Valic, M. I., Gorrissen, H., and Cushley, R. J. (1980). ^2H and ^{31}P NMR studies of cholesteryl palmitate in sphingomyelin dispersions. *Biochim. Biophys. Acta* 601:22-33.

61. Cullis, P. R., and Hope, M. J. (1980). The bilayer stabilizing role of sphingomyelin in the presence of cholesterol. A ^{31}P NMR study. *Biochim. Biophys. Acta* 597:533-542.

62. Jarrell, H. C., Byrd, R. A., Deslauriers, R., Ekiel, L., and Smith, I. C. P. (1981). Characterization of the phase behaviour of phospholipids in model and biological membranes by ^{31}P NMR. *Biochim. Biophys. Acta* 648:80-86.

63. Deslauriers, R., Ekiel, L., Byrd, R. A., Jarrell, H., and Smith, I. C. P. (1982). A ^{31}P NMR study of the structural and functional aspects of phosphate and phosphonate distribution in Tetrahymena. *Biochim. Biophys. Acta* 720:329-338.

64. Ferguson, K. A., Hui, S. W., Stewart, T. P., and Yeagle, P. L. (1982). Phase behaviour of the major lipids of Tetrahymena ciliary membranes. *Biochim. Biophys. Acta* 684:179-186.

65. Hill, R. J., Deslauriers, R., Butler, K. W., Colvin, R., and Smith, I. C. P. (1984). The bilayer properties of the ciliary membranes of Tetrahymena Thermophilia as revealed by ^{31}P NMR. *Biochim. Biophys. Acta* 773:74-82.

66. Seddon, J. M., Cevc, G., and Marsh, D. (1982). Calorimetric studies of the gel-fluid (Lb-La) and lamellar-inverted hexagonal (La-H$_{II}$) phase transition of dialkyl and diacyl phosphatidylethanolamines. *Biochemistry* 22:1280-1289.

67. Harlos, K., and Eibl, H. (1980). Hexagonal phases in phospholipids with saturated chains. *Biochemistry* 20:2888-2892.

68. Tilcock, C. P. S., and Cullis, P. R. (1982). The polymorphic phase behaviour and miscibility properties of synthetic phosphatidylethanolamines. *Biochim. Biophys. Acta* 684:212-222.

69. Wen, S. Y. K., Hess, D., Kaufman, J. W., Collins, J. M., and Lis, L. J. (1983). Raman spectroscopic and X-ray diffraction studies of the effect of temperature and Ca^{2+} on phosphatidylethanolamine. *Chem. Phys. Lipids* 32:165-173.

70. Dekker, C. J., Geurts van Kessel, W. S. M., Klomp, J. P. G., Pieters, J., and De Kruijff, B. (1983). Synthesis and polymorphic phase behaviour of polyunsaturated phosphatidylcholines and phosphatidylethanolamines. *Chem. Phys. Lipids* 33:93-106.

71. Cullis, P. R., and De Kruijff, B. (1978). The polymorphic phase behaviour of phosphatidylethanolamines of natural and synthetic origin. A ^{31}P NMR study. *Biochim. Biophys. Acta* 513:31-41.

72. Cullis, P. R., and De Kruijff, B. (1976). ^{31}P NMR studies of unsonicated aqueous dispersions of neutral and acidic phospholipids: Effects of phase transition, pH and divalent cations on the motion in the phosphate region of the polar headgroup. *Biochim. Biophys. Acta* 436:523-533.

73. Hardman, P. D. (1982). Spin-label characterization of lamellar-to-hexagonal (H_{II}) phase transition in egg phosphatidylethanolamine aqueous dispersions. *Eur. J. Biochem.* 124:95-103.

74. Yager, P., and Chang, E. L. (1983). Destabilization of a lipid non-bilayer phase by high pressure. *Biochim. Biophys. Acta* 731:491-494.

75. Mantsch, H. H., Martin, A., and Cameron, D. G. (1981). Characterization by infrared spectroscopy of the bilayer to non-bilayer phase transition of phosphatidylethanolamines. *Biochemistry* 20:3138-3145.

76. Rand, R. P., Tinker, D. O., and Fast, P. G. (1971). Polymorphism of phosphatidylethanolamines from two natural sources. *Chem. Phys. Lipids* 6:333-342.

77. De Kruijff, B., Rietveld, A., and Cullis, P. R. (1980). ^{31}P NMR studies on membrane phospholipids in microsomes, rat liver slices and intact perfused rat liver. *Biochim. Biophys. Acta* 600:343-356.

78. Cullis, P. R., De Kruijff, B., Hope, M. J., Verkleij, A. J., Nayar, R., Farren, S. B., Tilcock, C. P. S., Madden, T. D., and Bally, M. B. (1983). Structural properties and functional roles of phospholipids in biological membranes. In: *Membrane Fluidity in Biology*, Vol. 1 (R. C. Aloia, ed.). Academic Press, New York.

79. Cullis, P. R., Van Dijck, P. W. M., De Kruijff, B., and De Gier, J. (1978). Effect of cholesterol on the properties of equimolar mixtures of synthetic phosphatidylethanolamine and phosphatidylcholine. A ^{31}P NMR and differential scanning calorimetry study. *Biochim. Biophys. Acta* 513:21-32.

80. Lohner, K., Hermetter, A., and Paltauf, F. (1984). Phase behavior of ethanolamine plasmalogen. *Chem. Phys. Lipids* 34:163-170.

81. Paltauf, F. (1983). Ether lipids in biological and model membranes. In: *Ether Lipids. Biochemical and Biomedical Aspects* (H. K. Mangold and F. Paltauf, eds.). Academic Press, New York.

82. Browning, J. L., and Seelig, J. (1980). Bilayers of phosphatidylserine: A deuterium and phosphorus nuclear magnetic resonance study. *Biochemistry* 19:1262-1270.

83. Luna, E., and McConnell, H. (1977). Lateral phase separations in binary mixtures of phospholipids having different charges and different crystalline structures. *Biochim. Biophys. Acta* 470:303-316.

84. MacDonald, R., Simon, S., and Baer, E. (1976). Ionic influences on the phase transition of dipalmitoylphosphatidylserine. *Biochemistry* 15:885-891.

85. Cevc, G., Watts, A., and Marsh, D. (1981). Titration of the phase transition of phosphatidylserine bilayer membranes. *Biochemistry* 20:4955-4965.

86. Van Dijck, P. W. M., De Kruijff, B., Verkleij, A. J., Van Deenen, L. L. M., and De Gier, J. (1978). Comparative studies of the effects of pH and Ca^{2+} on bilayers on various negatively charged phospholipids and their mixtures with phosphatidylcholine. *Biochim. Biophys. Acta* 512:84-96.

87. Mombers, C. Verkleij, A. J., De Gier, J., and Van Deenen, L. L. M. (1979). The interaction of spectrin-actin and synthetic phospholipids. II. The interaction with phosphatidylserine. *Biochim. Biophys. Acta* 551:271-281.

88. Comfurius, P., and Zwaal, R. (1977). The enzymatic synthesis of phosphatidylserine and purification by CM-cellulose column chromatography. *Biochim. Biophys. Acta* 448:36-42.

89. Van Dijck, P. W. M. (1979). Negatively charged phospholipids and their position in the cholesterol affinity sequence. *Biochim. Biophys. Acta* 555:89-101.

90. Portis, A., Newton, C., Pangborn, W., and Papahadjopoulos, D. (1979). Studies on the mechanism of membrane fusion: Evidence for an intermembrane Ca^{2+}-phospholipid complex, synergism with Mg^{2+} and inhibition by spectrin. *Biochemistry* 18:780-790.

91. Papahadjopoulos, D., Vail, W. J., Jacobson, K., and Poste, G. (1975). Cochleate lipid cylinders: Formation by fusion of unilamellar lipid vesicles. *Biochim. Biophys. Acta* 394:483-491.

92. Hauser, H., Finer, E., and Darke, A. (1977). Crystalline anhydrous Ca-phosphatidylserine bilayers. *Biochem. Biophys. Res. Commun.* 76:267-274.

93. Hark, S. K., and Ho, J. T. (1980). Raman study of calcium-induced fusion and molecular segregation of phosphatidylserine-dimyrisotylphosphatidylcholine membranes. *Biochim. Biophys. Acta* 601:54-62.

94. Hope, M. J., and Cullis, P. R. (1980). Effects of divalent cations and pH on phosphatidylserine model membranes. A ^{31}P NMR study. *Biochem. Biophys. Res. Commun.* 92:846-852.

95. Findlay, E. J., and Barton, P. G. (1978). Phase behaviour of synthetic phosphatidylglycerols and binary mixtures with phosphatidylcholines in the presence and absence of calcium ions. *Biochemistry* 17:2400-2405.

96. Ranck, J. L., Keira, T., and Luzatti, V. (1977). A novel packing of the hydrocarbon chains in lipids. The low temperature phases of dipalmitoylphosphatidylglycerol. *Biochim. Biophys. Acta* 488:432-441.

97. Papahadjopoulos, D., Vail, W. J., Pangborn, W. A., and Poste, G. (1976). Studies on membrane fusion. II. Induction of fusion in pure phospholipid membranes by calcium ions and other divalent metals. *Biochim. Biophys. Acta* 448:265-283.

98. Van Dijck, P. W. M., Ververgaert, P. H. J. Th., Verkleij, A. J., Van Deenen, L. L. M., and De Gier, J. (1975). Influence of Ca^{2+} and Mg^{2+} on the thermotropic behaviour and permeability properties of liposomes prepared from dimyristoylphosphatidylglycerol and mixtures of dimyristoylphosphatidylglycerol and dimyristoylphosphatidylcholine. *Biochim. Biophys. Acta* 406:465-478.

99. Papahadjopoulos, D., Jacobson, K., Nir, S., and Isac, T. (1973). Phase transitions in phospholipid vesicles. Fluorescence polarization and permeability measurements concerning the effect of temperature and cholesterol. *Biochim. Biophys. Acta* 311:330-348.

100. Sacre, M. M., Hoffman, W., Turner, M., Tocanne, J. F., and Chapman, D. (1979). Differential scanning calorimetry of some phosphatidylglycerol lipid-water systems. *Chem. Phys. Lipids* 69:69-83.

101. Harlos, K., and Eibl, H. (1980). Influence of calcium on phosphatidylglycerol. Two separate lamellar structures. *Biochemistry* 19:895-899.

102. Farren, S. B., and Cullis, P. R. (1980). Polymorphism of phosphatidylglycerol-phosphatidylethanolamine model membrane systems. *Biochem. Biophys. Res. Commun.* 97:182-191.

103. Rand, R. P., and Sengupta, S. (1972). Cardiolipin forms hexagonal structures with cations. *Biochim. Biophys. Acta* 225:484-492.

104. Cullis, P. R., Verkleij, A. J., and Ververgaert, P. H. J. Th. (1978). Polymorphic phase behaviour of cardiolipin as detected by ^{31}P NMR and freeze-fracture techniques. Effects of calcium, dibucaine and chlorpromazine. *Biochim. Biophys. Acta* 513:11-20.

105. Vasilenko, I., De Kruijff, B., and Verkleij, A. J. (1982). Polymorphic phase behaviour of cardiolipin from bovine heart and from *Bacillus subtilis* as detected by ^{31}P NMR and freeze-fractured techniques. *Biochim. Biophys. Acta* 684:282-286.

106. De Kruijff, B., Verkleij, A. J., Leunissen-Bijvelt, J., Van Echteld, C. J. A., Hille, J., and Rijnbout, H. (1982). Further aspects of the Ca^{2+}-dependent polymorphism of bovine heart cardiolipin. *Biochim. Biophys. Acta* 693:1-12.

107. Seddon, J. M., Kaye, R. D., and Marsh, D. (1983). Induction of the lamellar-inverted hexagonal phase transition in cardiolipin by protons and monovalent cations. *Biochim. Biophys. Acta* 734:347-352.

108. Powell, G. L., and Marsh, D. (1985). Polymorphic phase behaviour of cardiolipin derivatives studied by ^{31}P NMR and X-ray diffraction. *Biochemistry* 24:2902-2908.

109. Blume, A., and Eibl, H. (1979). The influence of charge on bilayer membranes. Calorimetric investigations of phosphatidic acid bilayers. *Biochim. Biophys. Acta* 558:13-21.

110. Harlos, K., and Eibl, H. (1981). Hexagonal phases in lipids with saturated chains. *Biochemistry* 20:2880-2890.

111. Harlos, K., Stumpel, J., and Eibl, H. (1979). Influence of pH on phosphatidic acid multilayers. A rippled structure at high pH values. *Biochim. Biophys. Acta* 555:409-416.

112. Elamrani, K., and Blume, A. (1984). Phase transition kinetics of phosphatidic acid bilayers. A stopped-flow study of the electrostatically induced transition. *Biochim. Biophys. Acta* 769:578-584.

113. Verkleij, A. J., De Maagd, R., Leunissen-Bijvelt, J., and De Kruijff, B. (1982). Divalent cations and chlorpromazine can induce non-bilayer structures in phosphatidic acid-containing model membranes. *Biochim. Biophys. Acta* 684:255-262.

114. Farren, S. B., Hope, M. J., and Cullis, P. R. (1983). Polymorphic phase preferences of phosphatidic acid. A ^{31}P and ^{2}H NMR study. *Biochem. Biophys. Res. Commun.* 111:675-682.

115. Gains, N., and Hauser, H. (1983). Characterization of small unilamellar vesicles produced in unsonicated phosphatidic acid and phosphatidyl-choline phosphatidic acid dispersions by pH adjustment. *Biochim. Biophys. Acta* 731:31-39.

116. Sundler, R., and Papahadjopoulos, D. (1981). Control of membrane fusion by phospholipid headgroups. I. Phosphatidate/phosphatidylinositol specificity. *Biochim. Biophys. Acta* 649:743-750.

117. Nayer, R., Schmid, S. L., Hope, M. J., and Cullis, P. R. (1982). Structural preferences of phosphatidylinositol and phosphatidylinositol-phosphatidylethanolamine model membranes. Influence of Ca^{2+} and pH. *Biochim. Biophys. Acta* 688:169-176.

118. Wieslander, A., Ulmius, J., Lindblom, G., and Fontell, K. (1978). Water binding and phase structure for different *Acholeplasma laidlawii* membrane lipids studied by deuteron magnetic resonance and X-ray diffraction. *Biochim. Biophys. Acta* 512:241-253.

119. Wieslander, A., Christiansson, A., Rilfors, L., and Lindblom, G. (1980). Lipid bilayer stability in membranes. Regulation of lipid composition in *Acholeplasma laidlawii* as governed by molecular shape. *Biochemistry* 19:3650-3655.

120. Wieslander, A., Rilfors, L., Lennart, B. J., and Lindblom, G. (1981). Reversed cubic phase with membrane glucolipids from *Acholeplasma laidlawii*. ¹H, ²H and diffusion nuclear magnetic resonance measurements. *Biochemistry* 20:730-735.

121. Sen, A., Williams, W. P., and Quinn, P. J. (1981). The structure and thermotropic properties of pure 1,2-diacylgalactosylglycerols in aqueous systems. *Biochim. Biophys. Acta* 663:380-389.

122. Quinn, P. J., and Williams, W. P. (1983). The structural role of lipids in photosynthetic membranes. *Biochim. Biophys. Acta* 737:223-266.

123. Mannock, D. A., Brain, A. P. R., and Williams, W. P. (1985). The phase behaviour of 1,2-diacyl-3-monogalactosyl-sn-glycerol derivatives. *Biochim. Biophys. Acta* 817:289-298.

124. Rivas, E., and Luzzatti, V. (1969). Polymorphisme des lipides polaires et des galactolipides de chloroplastes de maïs, en presence d'eau. *J. Mol. Biol.* 41:261-275.

125. Shipley, G. G., Green, J. P., and Nichols, B. W. (1973). The phase behaviour of monogalactosyl, digalactosyl and sulphoquinovosyl diglycerides. *Biochim. Biophys. Acta* 311:531-544.

126. Mansourian, A., and Quinn, P. J. (1986). Phase properties of binary mixtures of monogalactosyldiacylglycerols differing in hydrocarbon chain substituents dispersed in aqueous systems. *Biochim. Biophys. Acta* 855:169-178.

127. Brentel, I., Selstam, E., and Lindblom, G. (1985). Phase equilibria of mixtures of plant galactolipids. The formation of bicontinuous cubic phase. *Biochim. Biophys. Acta* 812:816-826.

128. Abrahamsson, S., Pascher, I., Larsson, K., and Karlsson, K. (1972). Molecular arrangements in glycolipids. *Chem. Phys. Lipids* 8:152-179.

129. Bunow, M. R., and Levin, I. W. (1980). Molecular conformation of cerebrosides in bilayers determined by Raman spectroscopy. *Biophys. J.* 32:1007-1021.

130. Ruocco, M. J., Atkinson, D., Small, D. M., Skarjune, R. P., Oldfield, E., and Shipley, G. G. (1981). X-ray diffraction and calorimetric study of anhydrous and hydrated -palmitoylgalactosylsphingosine (cerebroside). *Biochemistry* 20:5957-5966.

131. Boggs, J. M., Koshy, K. M., and Rangaraj, G. (1984). Effect of fatty acid chain length, fatty acid hydroxylation and various cations on the phase behaviour of synthetic cerebroside sulfate. *Chem. Phys. Lipids* 36:65-89.

132. Curatolo, W., Small, D. M., and Shipley, G. G. (1977). Phase behaviour and structural characteristics of hydrated bovine brain gangliosides. *Biochim. Biophys. Acta* 468:11-20.

133. McDaniel, R., and McLaughlin, S. (1985). The interaction of calcium with gangliosides in bilayer membranes. *Biochim. Biophys. Acta* 819:153-160.

134. Maggio, B., Ariga, T., Sturtevant, J. M., and Ku, R. K. (1985). Thermotropic behaviour of binary mixtures of dipalmitoyl phosphatidylcholine and glycosphingolipids in aqueous dispersions. *Biochim. Biophys. Acta* 818:1-12.

135. Jain, M. K., Crecely, R. W., Hille, J. D. R., De Haas, G. H., and Gruner, S. M. (1985). Phase properties of aqueous dispersions of n-octadecylphosphatidylcholine. *Biochim. Biophys. Acta* 813:68-76.

136. Wu, W., Huang, C., Conley, T. G., Martin, R. B., and Levin, I. W. (1982). Lamellar-micellar transition of 1-stearyllysophosphatidylcholine assemblies in excess water. *Biochemistry* 21:5957-5961.

137. Van Echteld, C. J. A., De Kruijff, B., Mandersloot, J. G., and De Gier, J. (1981). Effects of lysophosphatidylcholines on phosphatidylcholine and phosphatidylcholine/cholesterol liposome systems as revealed by ^{31}P NMR, electron microscopy and permeability studies. *Biochim. Biophys. Acta* 649:211-220.

138. Allegrini, P. R., Van Scharrenburg, G., De Haas, G. H., and Seelig, J. (1983). ^{2}H and ^{31}P NMR studies of bilayers composed of 1-acyllysophosphatidylcholine and fatty acids. *Biochim. Biophys. Acta* 731:448-455.

139. Hauser, H., Guyer, W., Spiess, M., Pascher, I., and Sundell, S. (1980). The polar headgroup conformation of a lysophosphatidylcholine analogue in solution. A high-resolution nuclear magnetic resonance study. *J. Mol. Biol.* 137:265-282.

140. Tilcock, C. P. S., Cullis, P. R., Hope, M. J., and Gruner, S. M. (1986). The polymorphic phase behaviour of unsaturated 1-acyl phosphatidylethanolamines. A ^{31}P NMR and X-ray diffraction study. *Biochemistry* 25:816-822.

141. Ramsammy, L. S., Volwerk, H., Lipton, L. C., and Brockerhoff, H. (1983). Association of cholesterol with lysophosphatidylcholine. *Chem. Phys. Lipids* 32:83-89.

142. Jain, M. K., Van Echteld, C. J. A., Ramirez, F., De Gier, J., De Haas, G. H., and van Deenen, L. L. M. (1980). Association of lysophosphatidylcholine with fatty acids in aqueous phase to form bilayers. *Nature* 284:486-487.

143. De Kruijff, B., and Cullis, P. R. (1980). The influence of poly-L-lysine on phospholipid polymorphism. Evidence that electrostatic polypeptide-phospholipid interactions can modulate bilayer-non-bilayer transitions. *Biochim. Biophys. Acta* 601:235-240.

144. De Kruijff, B., Rietveld, A., Telders, N., and Vaandrager, G. (1985). Molecular aspects of the bilayer stabilization induced by poly(L-lysines) of varying sizes in cardiolipin liposomes. *Biochim. Biophys. Acta* 820:295-304.

145. De Kruijff, B., and Cullis, P. R. (1980). Cytochrome c specifically induces nonbilayer structures in cardiolipin-containing model membranes. *Biochim. Biophys. Acta* 602:477-490.

146. Smith, R., and Cornell, B. A. (1985). Myelin basic protein induces hexagonal phase formation in dispersions of diacylphosphatidic acid. *Biochim. Biophys. Acta* 818:275-279.

147. Gulik-Krzywicki, T., Balerna, M., Vincent, J. P., and Lazdunski, M. (1981). Freeze-fracture study of cardiotoxin action on axonal membrane and axonal membrane lipid vesicles. *Biochim. Biophys. Acta* 643:101-114.

148. Van Echteld, C. J. A., van Stight, R., De Kruijff, B., Leunissen-Bijvelt, J., Verkleij, A. J., and De Gier, J. (1981). Gramicidin promotes formation of the hexagonal H_{II} phase in aqueous dispersions of phosphatidylethanolamine and phosphatidylcholine. *Biochim. Biophys. Acta* 648:287-291.

149. Van Echteld, C. J. A., De Kruijff, B., Verkleij, A. J., Leunissen-Bijvelt, J., and De Gier, J. (1982). Gramicidin induces the formation of non-bilayer structures in phosphatidylcholine dispersions in a fatty-acid-chain-length-dependent way. *Biochim. Biophys. Acta* 692:126-138.
150. Killian, J. A., and De Kruijff, B. (1985). *Biochemistry* 24:7890-7898.
151. Taraschi, T. F., De Kruijff, B., Verkleij, A. J., and Van Echteld, C. J. A. (1982). Effect of glycophorin on lipid polymorphism. *Biochim. Biophys. Acta* 685:153-161.
152. Rietveld, A., and De Kruijff, B. (1986). *Eur. J.Biochem.*, in press.
153. Lambero, J. W. J., Verkleij, A. J., and Terpstra, W. (1984). Reconstitution of chlorophyllase with mixed plant lipids in the presence and absence of Mg^{2+} influence of single and mixed plant on enzyme stability. *Biochim. Biophys. Acta* 786:1-8.
154. Hornby, A. P., and Cullis, P. R. (1981). Influence of local and neutral anaesthetics on the polymorphic phase preferences of egg phosphatidylethanolamine. *Biochim. Biophys. Acta* 647:285-292.
155. Madden, T. D., and Cullis, P. R. (1982). Stabilization of bilayer structure for unsaturated phosphatidylethanolamines by detergents. *Biochim. Biophys. Acta* 684:149-153.
156. Cullis, P. R., and Verkleij, A. J. (1979). Modulation of membrane structure by Ca^{2+} and dibucaine. *Biochim. Biophys. Acta* 552:546-552.
157. Tilcock, C. P. S., and Fisher, D. (1982). Interactions of glycerol-monooleate and dimethylsulphoxide with phospholipids. A differential scanning calorimetry and P NMR study. *Biochim. Biophys. Acta* 685:340-346.
158. Howell, J. I., Fisher, D., Goodall, A. H., Verrinder, M., and Lucy, J. A. (1973). Interactions of membrane phospholipids with fusogenic lipids. *Biochim. Biophys. Acta* 332:1-10.
159. Das, S., and Rand, R. P. (1985). Diacylglycerol causes major structural transitions in phospholipid bilayer membranes. *Biochem. Biophys. Res. Commun.* 124:491-496.
160. Nicolay, K., Van der Neut, R., Fok, J. J., and De Kruijff, B. (1985). *Biochim. Biophys. Acta* 819:55-65.
161. Ohnishi, S., and Ito, T. (1974). Calcium-induced phase separations in phosphatidylserine-phosphatidylcholine membranes. *Biochemistry* 13:881-887.
162. Galla, H. J., and Sackmann, E. (1975). Chemically induced lipid phase separation in model membranes containing charged lipids: A spin label study. *Biochim. Biophys. Acta* 401:509-529.
163. Tokutomi, S., Lew, R., and Ohnishi, S. (1981). Ca^{2+}-induced phase separation in phosphatidylserine, phosphatidylethanolamine and phosphatidylcholine mixed membranes. *Biochim. Biophys. Acta* 643:276-282.
164. Tilcock, C. P. S., and Cullis, P. R. (1981). The polymorphic phase behaviour of mixed phosphatidylserine-phosphatidylethanolamine model systems as detected by ^{31}P NMR. Effects of divalent cations and pH. *Biochim. Biophys. Acta* 641:189-201.
165. Tilcock, C. P. S., Bally, M. B., Farren, S. B., Cullis, P. R., and Gruner, S. M. (1984). Cation-dependent segregation phenomena and phase behaviour in model membrane systems containing phosphatidylserine: Influence of cholesterol and acyl-chain composition. *Biochemistry* 23:2696-2703.
166. Tilcock, C. P. S., Bally, M. B., Farren, S. B., and Cullis, P. R. (1982). Influence of cholesterol on the structural preferences of dioleoylphosphatidylethanolamine-dioleoylphosphatidylcholine systems: A ^{31}P NMR and ^{2}H NMR study. *Biochemistry* 21:4596-4601.

167. Bally, M. B., Tilcock, C. P. S., Hope, M. J., and Cullis, P. R.
     (1983). Polymorphism of phosphatidylethanolamine-phosphatidylserine
     systems: Influence of cholesterol and Mg^{2+} on Ca^{2+}-triggered bilayer
     to hexagonal (H_{II}) transitions. *Can. J. Biochem.* 61:346-352.
168. Hope, M. J., and Cullis, P. R. (1979). The bilayer stability of the
     inner monolayer lipids of the human erythrocytes. *FEBS Lett.* 107:
     323-326.

RECENT REFERENCES ADDED IN PROOF

Ellens, H., Siegel, D. P., Alford, D., Yeagle, P. L., Boni, L., Lis,
L. J., Quinn, P. J., and Bentz, J. (1989). Membrane fusion and inverted
phases. *Biochemistry* 28:3692-3703.

Kumar, V. V., Malewicz, B., and Baumann, W. J. (1989). Lysophosphatidylcholine stabilizes small unilamellar phosphatidylcholine vesicles.
Phosphorus-31 NMR evidence for the "wedge" effect. *Biophys. J.* 55:789-
792.

Lewis, R. N. A. H., Mannock, D. A., McElhaney, R. N., Turner, D. C.,
and Gruner, S. M. (1989). Effect of fatty acyl chain length and structure
on the lamellar gel to liquid-crystalline and lamellar to reversed hexagonal
phase transitions of aqueous phosphatidylethanolamine dispersions. *Biochemistry* 28:541-548.

Mingeot-Leclercq, M. P., Schanck, A., Ronveaux-Dupal, M. F., Deleers, M.,
Brasseur, R., Ruysschaert, J. M., Laurent, G., and Tulkens, P. M.
(1989). Ultrastructural, physiochemical and conformational study of the
interaction of gentamicin and bis(beta-diethylaminoethylether) hexestrol
with negatively charged phospholipid layers. *Biochem. Pharmacol.* 38:729-
741.

3

Forces Governing Lipid Interaction and Rearrangement

V. ADRIAN PARSEGIAN

National Institute of Diabetes, Digestive, and Kidney Diseases and Division of Computer Research and Technology, National Institutes of Health, Bethesda, Maryland

R. PETER RAND

Brock University, St. Catharines, Ontario, Canada

I. INTRODUCTION*

It is tempting to begin a review of this sort piously raising the hope that understanding of physical forces between bilayer membranes will explain cell membrane fusion. We don't believe it (1). Real cellular or viral fusion involves too many changing and unidentified components to allow one to build a purely physical model with any confidence. Artificial fusion is still too specialized, ill defined, and poorly controlled to warrant exhaustive physical thinking. Nevertheless, the physical phenomena of cellular fusion must eventually be related to biochemical triggers that control them.

Having scaled down our expectations, we can concentrate on the many things to be learned from the energetics of lipid assembly and of forces between approaching membranes. For example, fusing membranes must come near contact. The work required for bilayers to do so is primarily a work of dehydration, a quantity that can change from prohibitive to negligible magnitude with well-defined, one-step changes in polar group identity.

Fusion must involve transient rearrangements that deviate widely from bilayer conformation. The energies of such rearrangements, seen as transitions in the structure of liquid-crystalline phases, are sensitive not only to differences in polar groups with different tendencies for hydration, but also to the presence of small amounts of aliphatic components or of long-chain lipids.

The lipid turnover that accompanies cell membrane fusion is a biochemical process whose triggers at the enzymatic level show up at the organelle level by virtue of changing physical properties of component

---

*The interval between the 1986 submission of this chapter and its appearance has been a time of rapid progress. We regret that the minor modifications allowed near the date of publication prohibit proper updating of this text. A review article dedicated specifically to hydration forces between phospholipid bilayers has recently appeared in *BBA Biomembranes* (68).

lipids. Conversely, physical causes such as osmotic activity may create stresses on membranes or vesicular contents to set off metabolic events.

The physical forces we refer to must be seen, then, in the context of dynamic cellular activity. These forces are an essential feature of the design of fusing systems but insufficient on their own to provide the full explanation of fusion that is sought.

After briefly reviewing the principal features of interlamellar forces, we consider their action in vesicular systems of the type often used as models for fusion. We find it useful, then, to consider the energetics of nonlamellar structures that might be related to fusion intermediates and to see how the occurrence of such arrangements reflects biochemical changes along known reaction paths.

II. LIPID ASSEMBLIES

Lipid polymorphism (see also Chapter 2) is a consequence of an internal tension built into every lipid molecule, conflicting drives to hydrate and dissolve polar groups and to isolate nonpolar parts (Fig. 1). The first accommodation to this conflict is an aggregation to merge hydrocarbons in such a way as to expose polar groups to the surrounding medium. At

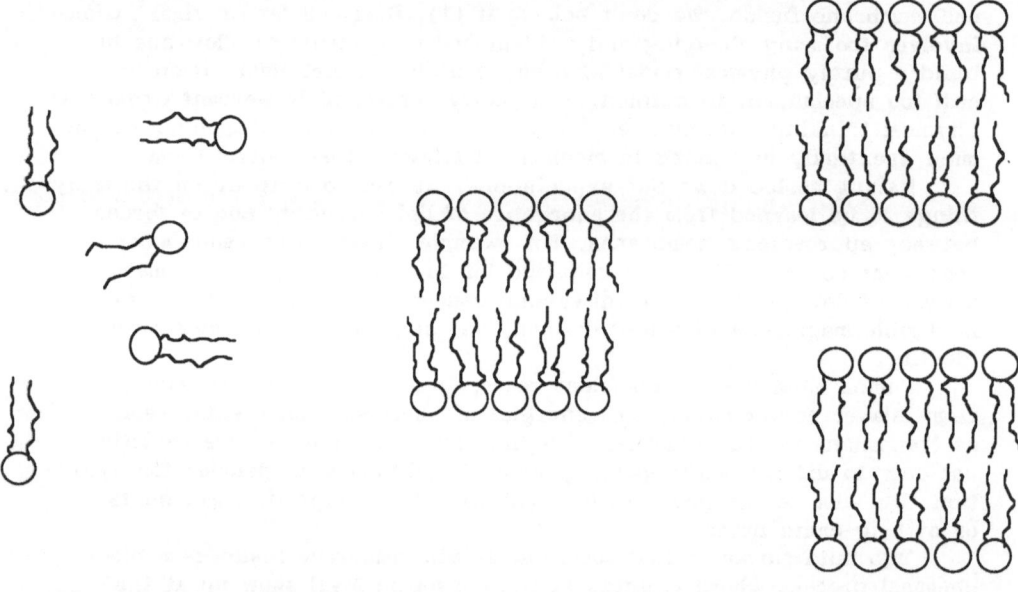

aggregation   assembly

FIGURE 1 Individual amphiphilic molecules aggregate in water, driven by conflicting tendencies to dissolve the polar groups in water and to isolate the nonpolar parts from that solvent. At a higher level of organization the aggregates assemble, driven by interactions that optimize conflicting attraction and repulsion energies. The two levels are coupled in that changes in interaction energies of either aggregation of assembly affect structure at the other level.

the next level of organization, aggregates assemble to optimize the energies of interaction between them. What one sees at this higher level are long-range versions of the forces that initially stabilize the lipid aggregate. But the two levels are not disjoint. The energies encountered when lipid aggregates come together are automatically coupled to forces of aggregate deformation or even of drastic rearrangements in packing seen as the structural phase transitions of the assemblies: hence, the expectation that the structural rearrangements we call membrane fusion must be related to membrane interaction.

III. BILAYER FORCE MEASUREMENTS

More than 10 years ago we succeeded in combining osmotic stress with X-ray diffraction from multilayers to measure directly the repulsive forces between egg phosphatidylcholine (egg PC) bilayers (2,3) (Fig. 2). A description of the techniques, recently given in detail elsewhere (4), will not be repeated here. Bilayer repulsion measured from the spontaneous separation in excess water down to virtually anhydrous contact surprised us with its strength and exponential growth. Because of the secondary importance of salts in the bathing medium and because of the insensitivity of forces to bilayer electric charge as bilayers approach contact, we concluded that the primary work of bringing bilayers together was a work

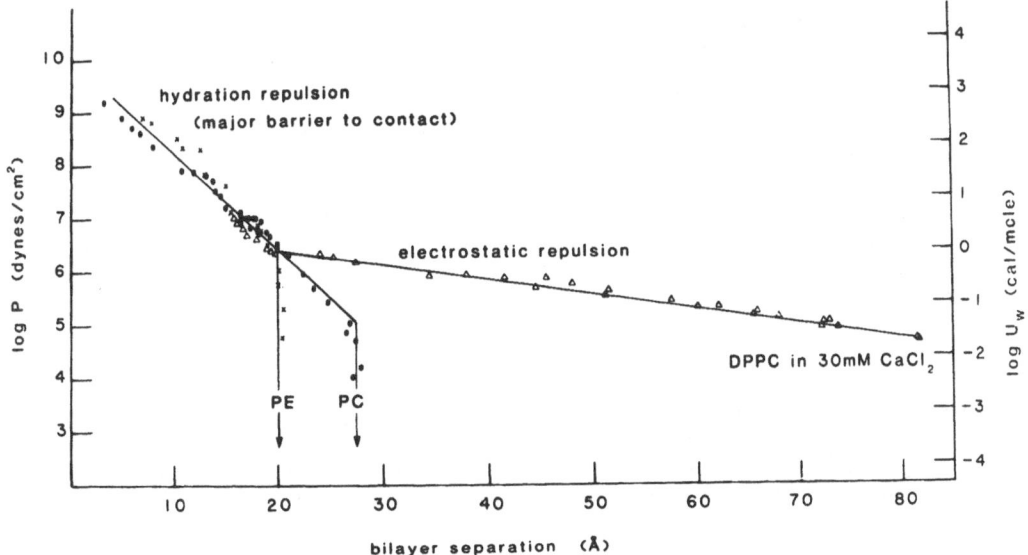

FIGURE 2 Typical examples, measured by the osmotic stress technique (4), of net repulsive pressures between phospholipid bilayers, either neutral or electrostatically charged. All lipids at close range show strong exponentially decaying hydration repulsion. A major difference among lipids is the limit of swelling. For neutral lipids this is apparently determined by the coefficient of the hydration repulsion (PE vs. PC). In the presence of net surface charge, in this case produced by the binding of Ca^{2+} to DPPC, the bilayers separate indefinitely.

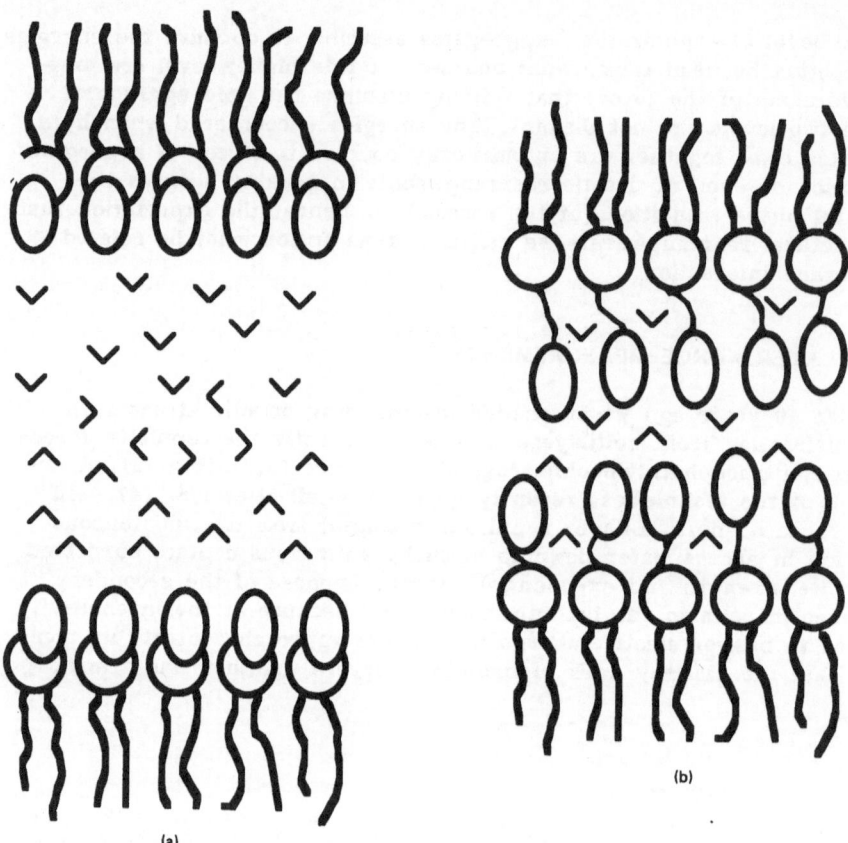

FIGURE 3   The polar group layer and its water. Without details of the structure of this region, the dimensions and composition of multilayer phases unambigously give the thicknesses of the water and bilayer regions, assuming their complete separation (a). More realistically, the water and polar groups mix (b), in ways yet to be determined. Bilayer "contact" is defined by the nature of that determination.

of dehydrating the polar groups (Fig. 2). Since that time, with vast amounts of new data collected on macromolecular, bilayer, and nonbilayer systems the idea of a hydration (or dehydration!) force has become well established (5-9).

The range of hydration forces is some 25 Å between electrically zwitterionic PC bilayers in the liquid-crystalline state. The actual definition of bilayer separation depends on convention. Most of the time we follow the popular volume-average picture according to Luzzati, which imagines a lipid-water interface that puts all the lipid on one side and all the water on the other (10). Then the repeat distance d in a multilayer is divided into lipid bilayer thickness $d_l = \phi d$ and bilayer separation $d_w = (1 - \phi)\, d$, where ϕ is the independently measured volume fraction of lipid. Of course, the lipid polar region is not a mathematically planar sheet (Fig. 3). Polar groups and water probably mix over a diffuse region. Other conventions of separation (2,3,11) assign a thickness to a polar group layer and thereby derive a smaller separation distance. Different conventions of separation can suggest "contact" at different degrees of dehydration of the interbilayer space. Differences either in the estimated change in bilayer thickness and molecular area as dehydration proceeds (11) or in what one decides to include in the polar group region (5) will affect the derived decay rates of the hydration force. However, the major portion of the work in removing interbilayer water is in the last half of the water, not the first. In addition, osmotic pressures of interbilayer spaces do not change discontinuously anywhere over the full range of dehydration. This emphasizes the necessity of treating the interbilayer contents as a mixture of polar groups and water and makes it difficult to conceive that bilayer "contact" is made at any discrete degree of dehydration. The primary advantages, to us, of the volume-average separation are twofold. First, there is no need for ad hoc assumptions about polar group size (11), only a need for component weight density and repeat spacing vs. water content, or vs. osmotic stress. Second, a physically unambiguous procedure relates unambiguous procedure relates osmotic stress P to mean molecular cross-section A in order to estimate the total work done per molecule, $P\Delta v_w$, when a value of water, Δv_w, is removed from the lamellar lattice. This is done to extract the force, as a perpendicular interlamellar stress, PA or as a lateral intra-lamellar stress $Pd_w/2$ on bilayers (3). In addition, we have recently used independently measured bilayer compressibility moduli (67) to estimate the changes in molecular areas, bilayer thickness, and interbilayer separations as the multilayers are osmotically stressed. We now believe this procedure provides the best measures of structural and hydration repulsion data. A summary of results is provided in Table 1; a more comprehensive compilation is found in Ref. 68.

Empirically the hydration force appears to vary exponentially as $P(d_w) = P_o \exp(-d_w/\lambda)$, where λ is approximately 2.0 Å for most PC's and somewhat smaller for phosphatidylethanolamine (PE) (5,6) (Table 1). We use the intuitive convention here that repulsive forces are positive. Strictly speaking, a repulsion implies a negative outward normal stress on a surface.

Despite its commanding strength in membrane contact, the hydration force is still poorly understood theoretically. The fact that repulsive forces of surprisingly similar properties are seen between parallel double helices of DNA (7) and between parallel molecules of the polysaccharide xanthan (unpublished results) and between myelin membranes (12) suggests a common mechanism to do with the properties of water near a polar surface. A model for progressively weaker perturbation of water molecules with increasing distance from a surface (schematically given in Fig. 3A) automatically produces the required exponential force (13,14) but does not seem to be easily reproduced in computer simulation (15). Alternatively, since there is good evidence that some of the distance between bilayers is a mixture of water and zwitterionic polar groups (11), one might imagine that at least part of the repulsive hydration force comes from crowding together extended polar groups (Fig. 3B). Some further consideration of the basis of the observed hydration repulsion may be found in Ref. 16.

Because they have a finite bending modulus, bilayers at a finite temperature in a multilayer array will bend and sweep out an occupation volume greater than they would if they were perfectly stiff (17-19). The mutual interference of such motion by neighboring bilayers is seen as a repulsive force due to the progressive loss of possible configurations as bilayers are forced together. Since bilayers feel each other through long-range forces before they actually collide, the mechanical motions of

TABLE 3   Interaction Parameters for a Variety of Neutral Phospholipids

Lipid	d_W	λ	P_O	E_t
Phosphatidylcholines				
Dilauryl-	27.4	2.0	10.6	
Dimyristoyl-	26.5	2.2	10.5	
Dipalmitoyl- 25°C	16.7	1.2	12.3	
50°C	31.1	2.1	11.0	
Distearyl-	19.6	1.3	12.9	
Dioleoyl-	28.1	2.1	10.6	
Stearyl,oleoyl-	24	2.0	10.5	.02
Egg-	24.9	2.1	10.6	.03
Phosphatidylethanolamines				
Egg				
Extracted	19.1	1.3	12.5	
Transesterified	14.6	1.1	12.3	
Monomethylated	21	1.8	10.3	
Dimethylated	22.7	1.8	10.4	
Palmitoyl,oleoyl-	11.4	.8	12.5	0.14
Didodecyl- (66)	11			
Diarachinoyl- (66)	13.3			

d_W (Å) = bilayer separation for lipid in excess aqueous solution; λ, P_O = decay distance and coefficient of exponential hydration force; E_t = adhesion energies (ergs/cm^2). The parameters given here are empirical and have not been modified to take account of the contribution of steric undulatory repulsion (19).

which we speak actually occur within the force field established by the hydration, electrostatic, and Van der Waals interactions described below. The weaker these underlying forces, the greater the allowed mechanical excursions and the stronger the contribution of steric entropic interactions.

In multilayer systems, over most of the range of bilayer separations, hydration repulsion is too strong to allow any significant mechanical undulation and consequent steric repulsion. Only near the point of balance between hydration and Van der Waals attraction in electrically neutral systems, where the net force is weak, are these undulation forces evident (19). They might expand the lattice by about 5 Å.

Between zwitterionic lipids hydration and steric repulsion are opposed by a Van der Waals attraction of the approximate form (3)

$$\frac{H}{6\pi d_W^3} \text{ or } \frac{H}{6\pi d_W^3} \left(1 - \frac{2}{(1 + d_l/d_W)^3} + \frac{1}{(1 + 2d_l/d_W)^3} \right)$$

and these forces balance at bilayer separations from 32 Å to 10 Å, depending on the kind and state of the lipid (5,6) (Table 1).

By manipulating giant vesicles held on the tips of aspirated pipettes, Evans and co-workers have directly measured bilayer contact energies for several kinds of lipids at the point of force balance (20-22). Because Van der Waals attraction is so much more slowly varying with interbilayer distance than hydration repulsion, the depth of this energy minimum is almost completely determined by the energy of the attractive component.

Table 1 lists several of these contact energies which show about an order of magnitude difference between PC's (0.01 erg/cm^2) and PE's (0.1 erg/cm^2), in good correlation with their interlamellar separation distances. The depths of these energies and the corresponding "hamaker" coefficient H in the Van der Waals interaction agree well with values indirectly inferred from osmotic force measurement (23).

Recently, Evans and coworkers measured the energies of interaction between vesicles composed of various ratios of palmitoyl,oleoyl-PE (POPE) and stearoyl,oleoyl-PC (SOPC) and derived the remarkable result that the coefficient of the hydration force, P_0, is the geometrical mean of the coefficients of POPE-POPE and SOPC-SOPC interactions (22). To this extent, then, one may think of the hydration interaction in terms of effective surface potentials formally analogous to what is used in electrostatic interactions (16). This result also opens up the possibility of attractive hydration forces between lipid bilayers (69), as have been observed between DNA double helices (9).

With the addition of sufficient electrostatic repulsion between charged lipid layers, energy minima and stable adhesion can disappear and lipid repulsion can extend many tens of angstroms with a range and rate of decay sensitive to the ionic activity of the bathing medium (24-29). Ion binding to the bilayer surface can be as important a determinant of apparent surface charge density as is the polar group charge itself (Fig. 2). Forces in the electrostatic regime (bilayer separations greater than about 20 Å) decay roughly in parallel but seem to reflect different degrees of surface charge in the coefficient of the electrostatic interaction (25,28). Figure 4

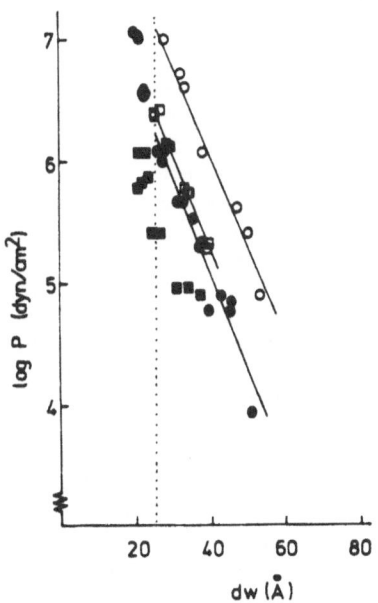

FIGURE 4 Interbilayer pressure as it varies with bilayer separation for phosphatidylserine (PS) in chloride solutions of 0.4 M Li$^+$ (□), Cs$^+$ (■), Na$^+$ (●), and TMA$^+$ (○). Solid lines are least-square fits. The dotted line represents the boundary where hydration repulsion at greater separations becomes negligible (27).

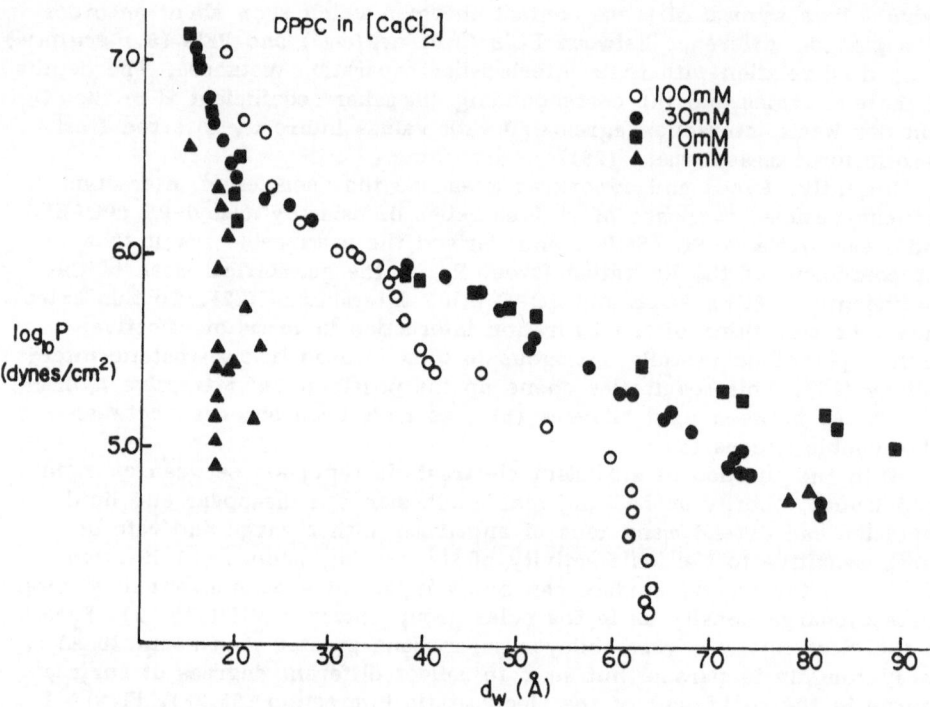

FIGURE 5 Comparison of interbilayer force as it varies with bilayer separation for DPPC in solutions of different $CaCl_2$ concentrations; 100 mM shows binding sufficient to give electrostatic repulsion but limited by sufficient screening at this concentration (26).

shows data from a set of measurements on phosphatidylserine (PS) in a number of different ionic solutions at the same ionic strength. Figure 5 shows a family of force curves for electrically zwitterionic dipalmitoyl-PC (DPPC), charged by the adsorption of Ca^{2+} ions in $CaCl_2$ solutions of varying ionic strength. The apparent surface charge also varies with the activity of the charging ion (26).

As expected, the rate of force decay increases with increasing salt concentration in both these kinds of charged bilayers. But, looking further, we find an unexpected feature of these ion-sensitive forces. The rate of exponential decay is consistently slower than that predicted by electrostatic double-layer theory (26,27). A plausible explanation for this slower decay comes from the steric interactions due to mechanical bilayer undulations (17-19). In contrast to repulsive forces between neutral bilayers, electrostatic double-layer forces between charged bilayers usually are weak enough to allow significant mechanical fluctuations. Then, as the electrostatic force decreases with bilayer separation, a progressively growing steric repulsion is added to the electrostatic forces. This coupling is nicely seen in Fig. 4, showing forces between PS bilayers in 0.4-M solutions. The theoretical Debye length is 4.8 Å. The observed decay length in tetramethylammonium chloride is 5.4 Å, while in NaCl, where the electrostatic force is weakened by apparent Na^+ binding, the decay constant is 8.4 Å. Weaker electrostatic forces lead to slower decay. Importantly, therefore, mechanical undulatory forces and electrostatic interactions appear to enhance each other. Consequently, even in this long-distance electrostatic regime bilayer interactions differ significantly from what had once been expected.

Hydration repulsion, even unenhanced with electrostatic or undulation repulsion, precludes contact between bilayers. Even for those less hydrated and more strongly adherent bilayers, such as bilayers consisting of PE, considerable work is required to remove intervening water. The only strategy for overcoming hydration repulsion is to combine anionic phospholipid surfaces with divalent cation solutions. The dimensions of many such combinations indicate the potency of these ions to dehydrate (28,30). In all, most of the intervening water is displaced as the cations bind to the bilayer surfaces, which then collapse together. The collapse is always completed without intermediate stages of bilayer separation (28). In some PS/PC mixtures, the Ca-Ps interaction is strong enough to cause dehydration also of neighboring PC polar groups (70). Often, segregation of lipid species in mixed lipid systems or a complete repacking of the lipid into a nonlamellar phase results (31,32). The combination of Ca^{2+} with PS is particularly potent; no detectable water is left and the hydrocarbon chains of the PS molecules freeze. Ca^{2+}-triggered interaction of PS-containing vesicles is commonly used in systems that attempt to model membrane fusion (see also Chapter 4). We describe below the reaction of vesicles to such modification of their interaction.

IV. FORCES MEASURED BETWEEN LIPID-COATED, CURVED MICA SHEETS

Over part of the force range, the intermembrane forces first seen by osmotic stress have recently been confirmed by use of a surface-force apparatus (SFA) (33-35). In this system, phospholipid monolayers or bilayers are deposited on mica surfaces either from vesicular suspensions (34) or from monolayers (35). Forces between crossed mica cylinders are determined through a spring system. Distances between lipid layers are inferred from interference fringes between silvered surfaces on the backs of the mica sheets. A "zero" distance, or contact, is defined from the point of contact between bare sheets, or between sheets coated with lipid molecules anchored at their polar ends, from which an estimated thickness of the additional lipid coating is subtracted.

For neutral layers, experiencing attractive Van der Waals forces of the form $H/6\pi d_w^3$ and repulsive hydration forces of the form $P_0 \exp(-d_w/\lambda)$, the zero force position between curved surfaces is shifted in to the zero energy position between planes (33). Because the measurements are made between curved rather than flat surfaces, one must transform them to compare with or to predict forces between parallel surfaces (5,33). The method usually chosen is the Derjaguin Approximation or Derjaguin Transform (36), which assumes that the interaction of two curved surfaces is the sum of interactions between facing pieces treated as planar and parallel. This approximation says that the force F_c between crossed cylinders both of radius R at minimal separation d is related to E_p the energy of interaction per unit area between two parallel planes of the same material, by

$$\frac{F_c(d)}{2\pi R} = E_p(d)$$

In effect, the force between curved surfaces is an integral of forces between plane surfaces since

$$E_p(d) = -\int_{\infty}^{d} P_{planar}(x)dx$$

Thus, for example, the interaction of charged bilayers (Figs. 2, 4, and 5) of the form

$$P = P_{es} \, e^{-d_w/\lambda_{Debye}} + P_{hyd} \, e^{-d_w/\lambda_{hyd}}$$

will be seen as

$$\frac{F_c}{2\pi R} = \lambda_{Debye} \, P_{es}e^{-d_w/\lambda_{Debye}} + \lambda_{hyd} \, P_{hyd}e^{-d_w/\lambda_{hyd}}$$

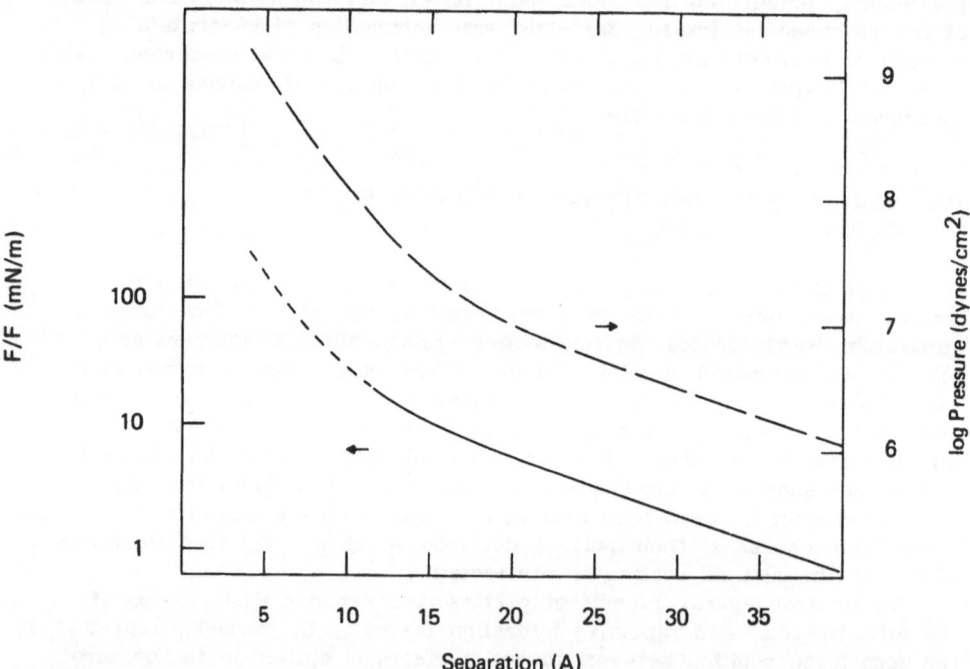

FIGURE 6 Comparison of forces measured by osmotic stress on multilayers and their expected values when immobilized onto curved mica surfaces. Dashed line: PS electrostatic forces (27) plus egg-PC hydration forces (3) after subtracting steric fluctuation contribution. Solid line: expected measurements of force/radius for the same forces between crossed cylinders of radius R over the range where forces are currently measured, i.e., with no deformation of the mica surface. Dotted line: continuation into the region where mica bending currently prevents quantitative measurement. This comparison shows that with the curved mica surfaces it is difficult to get into the region where hydration forces dominate for surfaces that are charged.

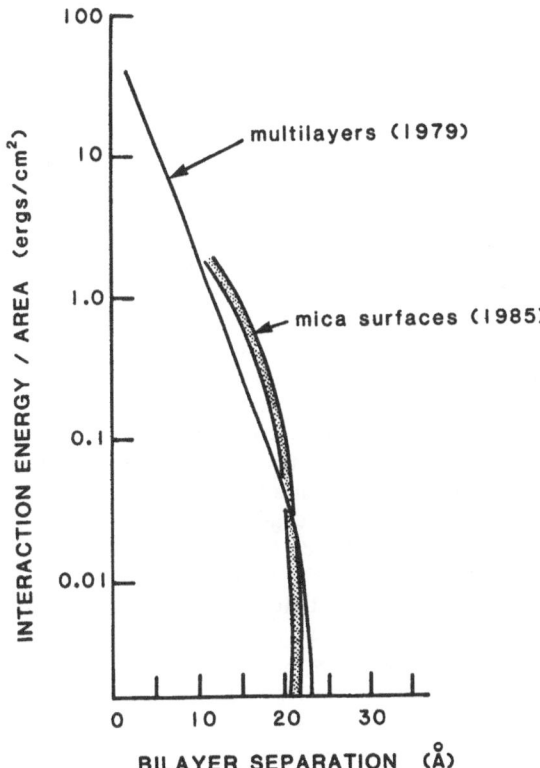

FIGURE 7 Direct comparison of the interbilayer forces measured osmotically and measured between mica surfaces, for fluid lipids over the range experimentally feasible. The agreement is remarkably good in the low-force region. See discussion in the text for more detailed quantitative comparison.

Importantly, where $\lambda_{hyd} \ll \lambda_{Debye}$ the interaction seen between curved surfaces will be dominated by the more slowly decaying electrostatic double-layer force.

Figure 6 shows this graphically for the case of combined hydration and electrostatic repulsion. The dashed line gives the measured interlamellar pressure P between parallel planes. The solid line gives the expected force between crossed cylinders as $F/2\pi R$. For curved surfaces the "elbow" that denotes a change from electrostatically dominated to hydration-dominated repulsion is shifted inward. Since mica deformation limits these measurements to $F/2\pi R = 20$ erg/cm^2, there will be charged lipid systems where underlying hydration forces are not seen.

A direct comparison of the osmotic stress and mica surface force measurements is shown in Figure 7. Plots of data for liquid-crystalline PC's on mica surfaces (shaded band, Ref. 35) agree remarkably well with data from parallel multilamellae to which the Derjaguin Transform has been applied. Data are shown over the range for which results have been reported.

At higher pressures mica surfaces deform, perhaps explaining the slight bending to the left at the upper end of the shaded band before measurements were terminated. Under some high-pressure conditions the

lipid layer is said to thin by squeezing out the half-layer facing the intervening aqueous phase (cf. Fig. 2 of Ref. 34). It has been suggested (34)
that this deformation is related to membrane fusion (see also Chapter 7).
The nature of the bilayer stress exerted through the mica sheet and the
observation that the energy apparently must include bending of the mica
make problematic any immediate connection with fusion, as it is understood
to occur either between artificial lipid bilayers or between natural cell
membranes.

One puzzling discrepancy has to do with the depth of the energy
minimum in E_p. Both the osmotic stress estimates (23) and the direct
measurement of bilayer-bilayer contact energy (37) give a value of 10^{-2}
erg/cm^2 between fluid PC bilayers. The reported estimate from coated
mica sheets is nearly 10 times higher (35), so high that it is necessary
to say that the plane at which Van der Waals attractions originate lies
at the very tip of the hydrated polar groups, so that the dielectric properties of this polar region would have to be identical to those of the hydrocarbon core. Part of the deeper energy minimum inferred in the mica
system may be removal of mechanical undulatory repulsions as bilayers
are immobilized onto the fixed surface. Part of the large attraction may
be in the method of analysis: Marra and Israelachvilli found a factor of
7 difference in Van der Waals force coefficient for the same mica-system
data treated two different ways (35). Part of the difficulty in locating
the plane of action of Van der Waals forces may stem from ambiguities
of a few angstroms in definition of the zero of bilayer separation.

Electrostatic interactions between mica surfaces correlate well with
expectations. Decay constants agree quantitatively with predicted Debye
lengths, as they should for rigid bodies (33). The magnitudes of interaction
and inferred surface potentials are respectively slightly higher or lower
than expected from zeta potential measurements for positive and negative
surfaces. One is thereby reassured on the accuracy of the electrostatic
double-layer theory. Consideration of electrostatic double-layer forces
in any fusion process, though, must recognize their enhancement by
mechanical undulations.

V. VESICLE INTERACTIONS

How does one translate the measured forces and energies of interacting
bilayers into interactions of bilayer vesicles, especially those used to probe
bilayer fusion?

Energetically, vesicular fusion in vitro is a "downhill process." The
mix is made, the reaction goes. The nature of the vesicle preparation,
the lipid composition, the salt concentration, the temperature—all contribute
to how and why two membranes become one and their enclosed volumes
merge. However, one may expect that the strains built into synthetic
vesicles, which drive their reorganization, may not be easily related to
the meticulously crafted natural vesicles and plasma membranes. For example, very small unilamellar vesicles (SUV), made by sonication, are at
the very limit of mechanical strain. Any smaller curvature and these shells
would be less strained if they exposed a raw edge of bilayer (38). Lipids
on the interior and exterior halves have different properties. The artificial
strain energy built in during their preparation is probably what drives
them into lower energy forms, often spontaneously (39).

In the study of larger, presumably less strained, vesicles the mechanical properties of the bilayer should also be taken into account. Evidence points to the importance of these properties. Bilayer vesicles made to adhere, by manipulating in various ways the forces just described, necessarily deform and become stressed as they flatten against each other (37). What appears to be critically important is the consequences of making such mechanically fragile vesicles interact strongly. Vesicle deformation is opposed by their mechanical rigidity, and if equilibrium could be achieved, the adhesion energy G and membrane tension T are related as (23)

$$\cos \theta = 1 + \frac{G}{2T}$$

where θ is the included angle between the membranes. This crucially important relation together with the measured mechanical properties of bilayers (20,37) allows prediction of the kind of response to expect from vesicles when they are stressed consequent to adhesion. Since the maximum tension and area increase that bilayers can withstand without breaking are about 3 dynes/cm and 3 ergs/cm^2, respectively (20), and θ max is 180, adhesion energies then are restricted to the 0-6 ergs/cm^2 range. Stronger adhesion at constant vesicle volume would be expected to rupture the bilayer. In addition, such stresses may cause volume loss, which can relieve bilayer tension but will depend on the osmotic content of the vesicle and the permeability of its stressed bilayer. Furthermore, because of different surface-to-volume ratios, responses of large (0.05-0.5 µm) and giant (>1 µm) vesicles will differ; large will respond preferentially by volume loss and then by area dilation, giant vesicles by the reverse (37).

These physical considerations suggest that a number of alternative responses could result after triggering vesicle adhesion. In fact, in the three systems we have investigated, all three responses have been observed directly (40,41) (Fig. 8) and correlate well with estimated adhesion energies. Pure PS vesicles interact so strongly on addition of Ca^{2+} (we estimated 27 dynes/cm^2 in 1985, but we now believe that number is even higher) that no adhering equilibrium configuration can be obtained (23). As adhesion takes place, we observe fusion and rupture with about equal probability (but see also Chapter 4) and destruction of the vesicles rapidly ensues (40-42). If the PS is diluted with the neutral lipid PE and interaction triggered by Ca^{2+}, the response is less violent and either vesicle fusion or volume loss, but not vesicle rupture, results. Eventually, the vesicles are destroyed and the lipids segregate. If the PS is diluted with PC, a neutral lipid with even stronger hydration repulsion than PE and an order of magnitude less adhesion energy, only vesicle adhesion is observed with no subsequent reaction (42). So as one would predict from the mechanics of the interaction, as the net attraction and adhesion energy is scaled down by the addition of neutral lipids with increasing hydration repulsion, the vesicle response changes from their rapid destruction to their stable aggregation. We believe that the response of the vesicles to induced adhesion is a response to bilayer mechanical stress: Where fusion occurs, it results from bilayer rupture incidental to the collapse process.

The response of bilayers to mechanical stress accounts for the qualitative differences observed in the behavior of interacting spheres (42) compared to spheres interacting with planar bilayers (43) (see also Chapter 8). Upon adhesion, bilayers of mutually adhering spheres are necessarily stressed on contact, but during adhesion of spheres to planar bilayers,

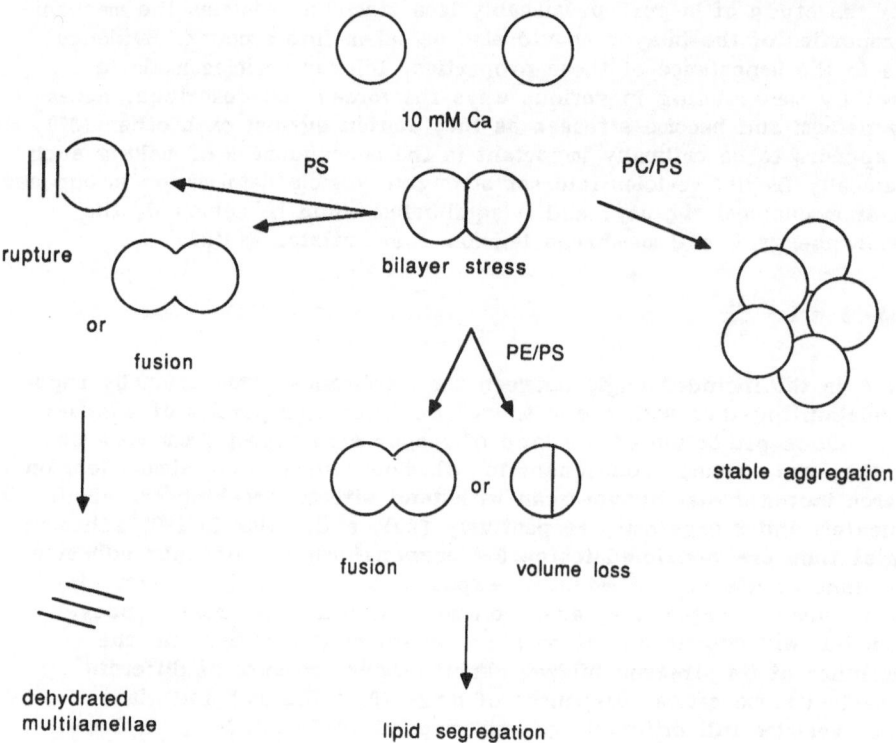

FIGURE 8 Schematic summary of directly observed responses of bilayer vesicles to their mutual interaction induced by added calcium (41,42). All three vesicle species respond in different ways. PS vesicles either fuse or break, with about equal probability, and then collapse to complete dehydration. PS/PE vesicles either fuse or lose volume, do not break, but eventually collapse and the lipids segregate. PS/PC vesicles simply adhere. The different responses are attributed to differences in the Ca^{2+}-induced adhesion energies which are reduced successively by the increasing hydration repulsion of PE and then PC.

stressed on contact, but during adhesion of spheres to planar bilayers, the latter are not stressed, since the contact can be made without significant increase in bilayer tension. Hence, unlike mutually adhering spheres, there is requirement for osmotic swelling of spheres adhered to planar bilayers in order to elicit a response (43). This interpretation is consistent as well with these indirect measurements of fusion if one accepts that volume loss without loss of fluorescent label (40), or that vesicles rupture without fusion (43), is not detected in these assays.

What, in our view, a number of phospholipid vesicle systems fail to mimic in duplicating membrane fusion, probably because of the stochastic nature of the mechanical strain, is the critical characteristic of focusing membrane destabilization to the contact area so that only fusion results and no subsequent reactions occur. What could provide this fine control? Mechanisms have been sought in those processes and protein-free membrane systems that produce nonbilayer structures (44-46) (see also Chapters 2, 4, and 6). It is well to remember that these structures are not two-dimensional and significant formation of them in cells would lead to self-destruction. On the other hand, conditions tending toward their formation within membranes might produce the transient destabilization required for fusion. One should expect such rearrangements to be restricted to the area of close membrane apposition.

VI. ENERGETICS OF NONLAMELLAR PHASES

Many fusion scenarios describe transient nonlamellar arrangements of membrane components. Only recently have direct energetic measurements been made on the inverted hexagonal (H_{II}) phase (47), an arrangement often thought related to fusion transients (44–46).

In H_{II}, polar groups face inward to parallel tubes of water whose axes form a hexagonal grid (Fig. 9). Rather than osmotic stress vs. separation, one measures the stress required to hold the lattice at a particular cylinder radius r or water volume V_W per lipid molecule.

Measurements made on three systems—pure dioleoyl-PE (DOPE), a mixture of DOPE and 5 mol% tetradecane (TD), and a mixture of DOPE and dioleyl-PC (DOPC) at a molar ratio of 3:1, containing 20 mol% TD— are shown in Fig. 10. These data have been interpreted (47) to show several features of nonlamellar packing. Different lipids or mixtures of lipids seem to assume a most-favored radius of curvature R_0 under zero osmotic stress in excess water (48,49).

Realization of this radius of curvature can be frustrated by the inability of the amphiphile acyl chain to fill the hydrocarbon part of the lattice

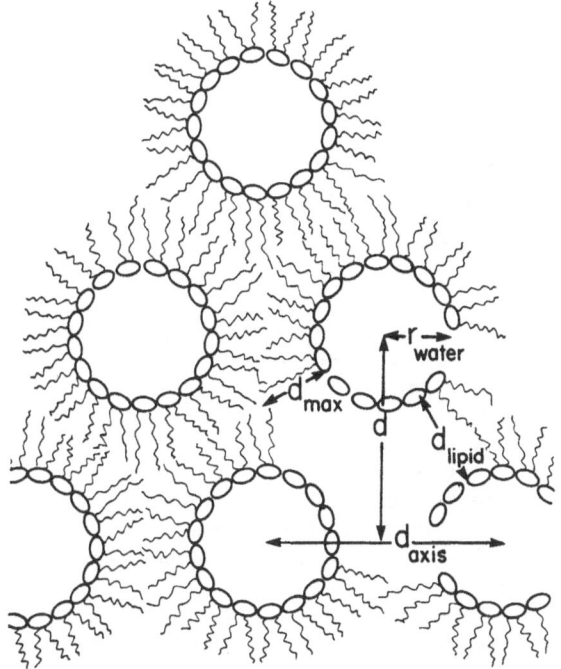

FIGURE 9 Parameters used to describe the inverted hexagonal structure, measured from X-ray diffraction and composition of the phase.

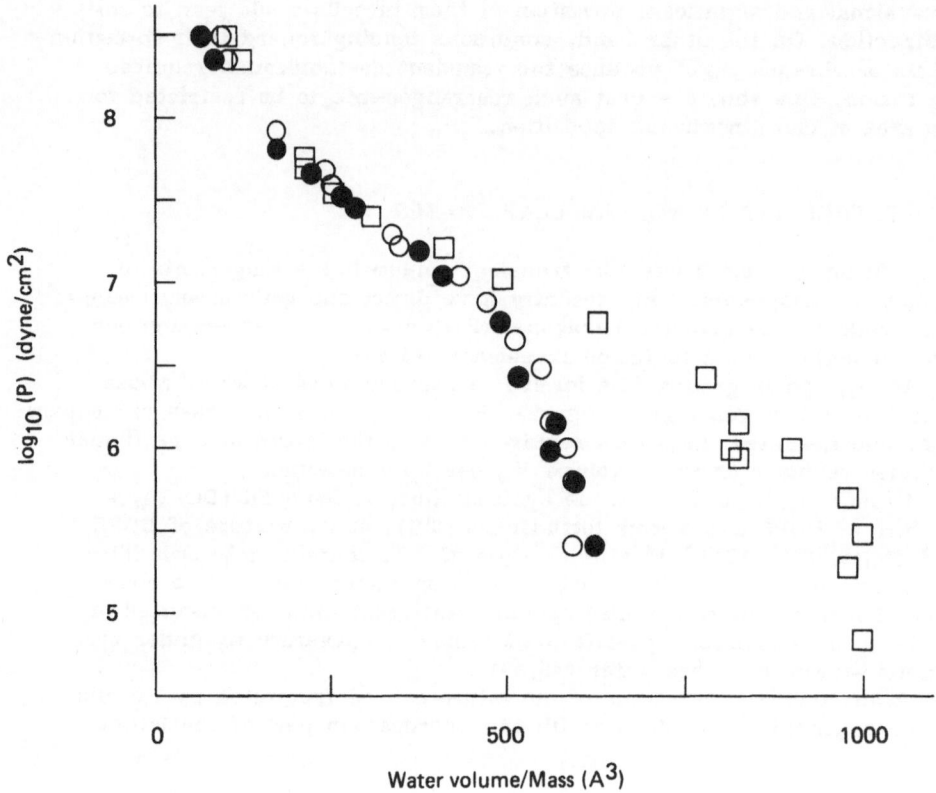

FIGURE 10 Osmotic stress vs. water volume per polar group mass equivalent to that of PE. ○ , DOPE; ● , DOPE plus 5% tetradecane; □ , DOPE/DOPC (3:1) plus 20% tetradecane.

(direction d_{max} in Fig. 9), a frustration that is removed with addition of alkane. For example, DOPE/DOPC (3:1) forms only lamellar structures at 25°C but transforms to H_{II} in the presence of dodecane or tetradecane.

The work of dehydrating H_{II} lattices near their spontaneous radius R_0 seems to be described by a bending or curvature modulus K such that the energy is given by $1/2K (1/r - 1/R_0)^2$. The value of this modulus seems to depend very little on R_0 (47). In fact, recognizing the fact that two layers are simultaneously bent in bilayers, the curvature moduli of the three H_{II} studied are effectively the same as for bilayers (50-54).

It appears that hydrocarbon chains are under little stress in the H_{II} phase. Addition of dodecane or tetradecane to DOPE, which forms H_{II} on its own, makes little or no detectable difference in the P vs. r or P vs. V_w curves. The same noneffect is seen in studies of H_{II} dimensions vs. temperature in excess water (49). If there are strains in the hydrocarbon such as clearly exist to prohibit DOPE/DOPE to form H_{II}, they do not seem to occur in H_{II}.

Siegel (55-57) has produced a series of studies examining the correlation between the tendency for the L_α - H_{II} transition and the likelihood of membrane fusion, aggregation-induced vesicle leakage, and intervesicle lipid exchange. The central idea is that if lipids are allowed to aggregate under conditions near those for transition to an H_{II} structure, there is an appreciable probability for inverted spherical or cylindrical structures ("inverted micellular intermediates") to occur between facing lamellae (see also Chapters 2, 4, and 6. It is possible to estimate the probability of such inversions and to speculate on their ability to seed metamorphosis into other lamellar structures. With this approach it should be possible to incorporate information on the energetic factors underlying L_α - H_{II} transitions to create more precise models of fusion intermediates.

VII. COUPLING THE PHYSICAL AND THE BIOCHEMICAL

The influence of alkanes or other nonpolar components aiding the occurrence of inverted structures might be an unrecognized factor in cell membrane deformation or fusion (48,58). Dolichol known to anchor oligosaccharide synthesis in endoplasmic reticulum also enables H_{II} formation (58). This material, known for its biochemical activity, might well act too to lubricate the many vesiculation and fusion activities seen in the membranes with which it is associated. Indeed, models of stalk formation in membrane transients seem to require some nonpolar filler substance inside the bending lipid membrane (54).

Structural rearrangements during fusion are likely under biochemical control in cells. One very suggestive link between the structural and the biochemical lies in the turnover of derivatives of phosphatidylinositol (PI) with the transient production of diacylglycerol (DAG). That biochemical cycle has been tightly correlated with the fusion of membranes in several different systems (59-61). While most attention in this scheme has been placed on the role of intracellular messengers that mobilize calcium (59), DAG remains in the membrane. With phospholipids and calcium it activates protein kinase C (62), whose roles are yet to be established, and phospholipases in lipid systems (63). But DAG may play another structural role. It modifies phospholipid bilayers and their interactions (64,65). It lacks significant headgroup and alone does not interact with water. When incorporated to maximum possible levels in PC, PE, or PS bilayers, it spreads apart the phospholipid polar groups but it does not enhance bilayer apposition. Above these levels it induces the lamellar to H_{II} transition in these bilayers, i.e., at concentrations that may exist at sites of DAG production in membranes.

If DAG plays a direct structural role in the fusion process, it would appear to do so in a way common to all the other models of fusion invoking nonlamellar structures. Importantly, however, it would do so under the transient biochemical control of its production. Layered on that control may be the control of dolichol activity to aid in the transformation process. We believe it is this kind of strategy the cell must use in regulating the fusion process. We end this chapter with a plea to recognize the biochemical factors, particularly involving proteins, that control physical forces governing cell membrane organization.

REFERENCES

1. Rand, R. P., and Parsegian, V. A. (1986). Mimicry and mechanism in phospholipid models of membrane fusion, *Annu. Rev. Physiol.* 48: 201-212

2. LeNeveu, D., Rand, R. P., Gingell, D., and Parsegian, V. A. (1977). Measurement and modification of foeces between lecithin bilayers, *Biophys. J.* 18:209-230.

3. Parsegian, V. A., Fuller, N. L., and Rand, R. P. (1979). Measured work of deformation and repulsion of lecithin bilayers, *Proc. Natl. Acad. Sci. USA* 76:2750–2754.

4. Parsegian, V. A., Rand, R. P., and Rau, D. C. (1986). Osmotic stress for the direct measurement of intermolecular forces. In: *Methods in Enzymology*. Vol. 127. *Biomembranes; Protons and Water: Structure and Translocation* (L. Packer, ed.). Academic Press, New York.

5. Rand, R. P. (1981). Interacting phospholipid bilayers: Measured forces and induced structural changes. *Annu. Rev. Biophys. Bioeng.* 10:277–314.

6. Lis, L. J., McAlister, M., Fuller, N., Rand, R. P., and Parsegian, V. A. (1982). Interaction between neutral phospholipid bilayer membranes. *Biophys. J.* 37:657–666.

7. Rau, D. C., Lee, B., and Parsegian, V. A. (1984). Measurement of the repulsive force between polyelectrolyte molecules in ionic solution: Hydration forces between parallel DNA double helices. *Proc. Natl. Acad. Sci. USA* 81:2621–2625.

8. Rand, R. P., Das, S., and Parsegian, V. A. (1985). The hydration force: Its character, universality and applications: Some current issues. *Chem. Scripta* 25:15–21.

9. Parsegian, V. A., Rand, R. P., and Rau, D. C. (1985). Hydration forces: What next? *Chem. Scripta* 25:28–31.

10. Luzzati, V., and Husson, F. (1962). The structure of the liquid-crystalline phases of lipid water systems. *J. Cell. Biol.* 12:207–219.

11. McIntosh, T. J., and Simon, S. A. (1986). Hydration force and bilayer deformation: A reevaluation. *Biochemistry* 25:4058–4066.

12. Rand, R. P., Fuller, N. L., and Lis, L. J. (1900). Myelin swelling and measurement of forces between myelin membranes. *Nature (London)* 279:258–260.

13. Marcelja, S., and Radic, N. (1976). Repulsion of surfaces due to boundary water. *Chem. Phys. Lett.* 42:129–130.

14. Gruen, D. W. R., Marcelja, S., and Parsegian, V. A. (1984). Water structure near the membrane surface. In: *Cell Surface Dynamics; Concepts and Models* (A. S. Perelson, C. DeLisi, and F. W. Wiegel, eds.). Marcel Dekker, New York and Basel, pp. 59–91.

15. Kjellander, R., and Marcelja, S. (1985). Perturbation of hydrogen bonding in water near polar surfaces. *Chem. Phys. Lett.* 120:393–396.

16. Cevc, G. (1985). The mean-field solvation model. Its implications and examples from lipid/water mixtures. *Chem. Scripta* 25:25–35.

17. Helfrich, W. (1978). Steric interaction of fluid membranes in multilayer systems. *Z. Naturforsch.* 33a:305–315.

18. Ostrowsky, N., and Sornette, D. (1985). Role of membrane fluidity on bilayers short range interactions. *Chem. Scripta* 25:38–40.

19. Evans, E. A., and Parsegian, V. A. (1986). The action of thermal fluctuations on forces within lamellar arrays. *Proc. Natl. Acad. Sci. USA* 83:7132–7136.

20. Evans, E., and Kwok, R. (1982). Mechanical calorimetry of large dimyristoylphosphatidylcholine vesicles in the phase transition region. *Biochemistry* 21:4874–4879.

21. Evans, E. A., and Metcalfe, M. (1984). Free energy potential for aggregation of giant, neutral lipid bilayer vesicles by van der Waals attraction. *Biophys. J.* 46:423–426.

22. Evans, E., and Needham, D. (1986). Giant vesicle bilayers composed of mixtures of lipids, cholesterol and polypeptides. *Faraday Discuss. Chem. Soc.* 81:267–280.

23. Parsegian, V. A., and Rand, R. P. (1983). Membrane interaction and deformation. *Ann. NY Acad. Sci.* 416:1-12.

24. Inoko, Y., Yamaguchi, T., Furuya, K., and Mitsui, T. (1975). Effects of cations on dipalmitoylphosphatidylcholine/cholesterol/water systems. *Biochim. Biophys. Acta* 413:24-32.

25. Cowley, A. C., Fuller, N., Rand, R. P., and Parsegian, V. A. (1978). Measurement of repulsive forces between charged phospholipid bilayers. *Biochemistry* 17:3163-3168.

26. Lis, L. J., Parsegian, V. A., and Rand, R. P. (1981). Detection of the binding of divalent cations to dipalmitoylphosphatidylcholine bilayers by its effect on bilayer interaction. *Biochemistry* 20:1761-1770.

27. Loosley-Millman, M., Rand, R. P., and Parsegian, V. A. (1982). Effects of monovalent ion binding and screening on measured electrostatic forces between charged phospholipid bilayers. *Biophys. J.* 40: 221-232.

28. Loosley-Millman, M. E. (1980). Ph.D. thesis, Guelph University, Guelph, Canada.

29. Lis, L. J., Lis, W. T., Parsegian, V. A., and Rand, R. P. (1981). Adsorption of divalent cations to a variety of phosphatidylcholine bilayers. *Biochemistry* 20:1771-1777.

30. Portis, A., Newton, C., Pangborn, W., and Papahadjopolous, D. (1979). Studies on the mechanism of membrane fusion: evidence for an intermembrane ca-phospholipid complex, synergism with mg, and inhibition by spectrin. *Biochemistry* 18:780-790.

31. Rand, R. P., and Parsegian, V. A. (1984). Physical force considerations in model and biological membranes. *Can. J.Biochem. Cell. Biol.* 62:752-759.

32. Gruner, S. M., Cullis, P.R., Hope, M. J., and Tilcock, C. P. S. (1985). Lipid polymorphism: the molecular basis of nonbilayer phases. *Annu. Rev. Biophys. Biophys. Chem.* 14:211-238.

33. Israelachvili, J. N., and Adams, G. E. (1978). Measurement of forces between two mica surfaces in aqueous electrolyte solutions in the range 0-100 nm. *J. Chem. Soc.* 74:975-1001.

34. Horn, R. (1984). Direct measurement of the force between two lipid bilayers and observation of their fusion. *Biochim. Biophys. Acta* 778: 224-228.

35. Marra, J., and Israelachvili, J. N. (1985). Direct measurements of attractive, adhesive, and repulsive forces between phosphatidylcholine and phosphatidylethanolamine bilayers in aqueous electrolyte solutions. *Biochemistry* 24:4608-4618.

36. Verwey, E. J., and Overbeek, J. Th. (1948). *The Theory of Stability of Lyophobic Colloids*. Elsevier, Amsterdam.

37. Evans, E. A., and Parsegian, V. A. (1983). Energetics of membrane deformation and adhesion in cell and vesicle aggregation. *Ann. NY Acad. Sci.* 416:13-33.

38. Cornell, B. A., Middlehurst, J., and Sepanovic, F. (1986). Small unilamellar phospholipid vesicles and the theories of membrane formation. *Faraday Discuss. Chem. Soc.* 81.

39. Evans, E. A., Gershfeld, N. L., Ginsberg, L., and Parsegian, V. A. (1982). Caveats against the use of thermodynamically unstable vesicles as models for biological systems. *Biophys. J.* 37:164a.

40. Duzgunes, N., Wilshut, J., Fraley, R., and Papahadjopoulos, D. (1981). Studies on the mechanism of membrane fusion: role of head group composition in calcium- and magnesium-induced fusion of mixed phospholipid vesicles. *Biochim. Biophys. Acta* 642:182-195.

41. Rand, R. P., Kachar, B., and Reese, T. S. (1985). Dynamic morphology of interacting phosphatidylserine vesicles. *Biophys. J.* 47:483-489.

42. Kachar, B., Fuller, N., and Rand, R. P. (1986). Morphological responses to calcium-induced interaction of phosphatidylserine-containing vesicles. *Biophys. J.* 50:778-779.

43. Cohen, F. S., Akabas, M. H., Zimmerberg, J., and Finkelstein, A. (1984). Parameters affecting the fusion of unilamellar phospholipid vesicles with planar bilayer membranes. *J. Cell. Biol.* 98:1054-1062.

44. Verkleij, A. J., Mombers, C., Gerritsen, W. J., Leunissen-Bijvelt, L., and Cullis, P. R. (1979). Fusion of phospholipid vesicles in association with the appearance of lipidic particles as visualized by freeze-fracturing. *Biochem. Biophys. Acta* 555:358-361.

45. Hui, S. W., Stewart, T. P., Yeagle, P. L., and Albert, A. D. (1981). Bilayer to non-bilayer transition in mixtures of phosphatidylethanolamine and phosphatidylcholine: Implications for membrane properties. *Arch. Biochem. Biophys.* 207:227-240.

46. Rand, R. P., Reese, T. S., and Miller, R. G. (1981). Phospholipid bilayer deformations associated with interbilayer contact and fusion. *Nature* 293:237-238.

47. Gruner, S., Parsegian, V. A., and Rand, R. P. (1986). Lipid vesicles and membranes. *Faraday Discuss. Chem. Soc.* 81:29-37.

48. Gruner, S. M. (1985). Intrinsic curvature hypothesis for biomembrane lipid composition: a role for non-bilayer lipids. *Proc. Natl. Acad. Sci. USA* 82:3665-3669.

49. Kirk, G. L., and Gruner, S. M. (1984). Lyotropic effects of alkanes and headgroup composition on the L_α-H_{II} lipid liquid crystal phase transition: hydrocarbon packing versus intrinsic curvature. *J. Phys.* 46:761-769.

50. Servuss, R. M., Harbich, W., and Helfrich, W. (1976). Measurement of the curvature-elastic modulus of egg lecithin bilayers. *Biochim. Biophys. Acta* 436:900-903.

51. Evans, E. A., and Skalak, R. (1980). *Mechanics and Thermodynamics of Biomembranes.* CRC Press, Boca Raton, FL.

52. Schneider, M. B., Jenkins, J. T., and Webb, W. W. (1984). Thermal fluctuations of large cylindrical phospholipid vesicles. *Biophys. J.* 45:891-899.

53. Engelhardt, H., Duwe, H. P., and Sackmann, E. (1985). Bilayer bending elasticity measured by Fourier analysis of thermally excited surface undulations of flaccid vesicles. *J. Phys. Lett.* 46:395-400.

54. Chernomordik, L. U., Kozlov, M. M., Melikyan, G. B., Abidor, I. G., Markin, V. S., and Chizmadzhev, Yu. A. (1985). The shape of lipid molecules and monolayer membrane fusion. *Biochim. Biophys. Acta* 812:643-655.

55. Siegel, D. P. (1984). Inverted micellar structures in bilayer membranes. *Biophys. J.* 45:399-420.

56. Siegel, D. P. (1986). Inverted micellar intermediates and the transitions between lamellar, cubic and inverted hexagonal lipid phases I. Mechanism of the L_α to H_{II} phase transition II. Implications for membrane-membrane interactions and membrane fusion. *Biophys. J.*, submitted.

57. Siegel, D. P. (1986). Membrane interactions via intermediates in lamellar-to-inverted hexagonal phase transitions. In: *Membrane Fusion* (A. E. Sowers, eds.). Plenum Press, New York.

58. Valtersson, C., van Duyn, G., Verkleij, A. J., Chojnacki, T., de Kruijff, B., and Dallner, G. (1985). The influence of dolichol, dolichol

esters, and dolichyl phosphate on phospholipid polymorphism and fluidity in model membranes. *J. Biol. Chem.* 260:2742-2751.

59. Berridge, M. J. (1984). Inositol and diacylglycerol as second messengers. *Biochem. J.* 220:345-360.
60. Wakelam, M. J. O. (1983). Inositol phospholipid metabolism and myoblast fusion. *Biochem. J.* 214:77-83.
61. Whitaker, M., and Aitchison, M. (1985). Calcium-dependent polyphosphoinstide Hydrolysis is associated with exocytosis in vitro. *FEBS Lett.* 182:119-124.
62. Nishizuka, Y. (1984). The role of protein kinase c in cell surface signal transduction and tumor promotion. *Nature (London)* 308:693-698.
63. Dawson, R. M. C., Irvine, R. F., Bray, J., and Quinn, P. J. (1984). Long-chain unsaturated diacylglycerols cause a perturbation in the structure of phospholipid bilayers rendering them susceptible to phospholipase attack. *Biochem. Biophys. Res. Commun.* 125:836-842.
64. Das, S., and Rand, R. P. (1984). Diacylglycerol causes major structural transitions in phospholipid bilayer membranes. *Biochem. Biophys. Res. Commun.* 124:491-496.
65. Das, S., and Rand, R. P. (1986). Modification by diacylglycerol of the structure and interaction of various phospholipid bilayers. *Biochemistry* 25:2882-2889.
66. Deleted in proof.
67. Evans, E. A., and Needham, D. (1987). Physical properties of surfactant bilayer membranes composed of lipids, cholesterol, and polypeptides: Thermal transitions, elasticity, cohesion and colloidal interactions. *J. Phys. Chem.* 91:4219-4228.
68. Rand, R. P., and Parsegian, V. A. (1989). Hydration forces between phospholipid bilayers, *BBA Biomembranes* 988:351-376.
69. Rand, R. P., Fuller, N. L., Parsegian, V. A., and Rau, D. C. (1988). Variation in hydration forces between neutral phospholipid bilayers: Evidence for hydration attraction. *Biochemistry* 27:7711-7722.
70. Coorssen, J. and Rand, R. P. (1988). Competitive forces between lipid membranes. *Stud. Biophys.* 127:53-60.

II
MEMBRANE FUSION IN MODEL SYSTEMS

4

Membrane Fusion in Lipid Vesicle Systems
An Overview

JAN WILSCHUT

University of Groningen, Groningen, The Netherlands

I. INTRODUCTION

The basic structural element of biological membranes is the lipid bilayer.
This bimolecular leaflet configuration is quite stable, and, obviously,
membranes resist a gross departure from this organization at any cost,
as it would defy the nature of their very being. Membranes are designed
to serve the function of a barrier, separating the cell from its surroundings
and separating intracellular compartments from the rest of the cell interior.
Any structural change that would severely impede this basic barrier function would be untolerable. On the other hand, in cell biology numerous
membrane fusion and fission events occur in order to allow the necessary
flow of mass and information between cells and from one intracellular
compartment to another. The development of higher organisms begins with
a cell-cell fusion event, while cell-cell fusion is also essential in the formation of multinucleated muscle cells. Within the cell, many transport processes
critically depend on specific membrane fission and fusion reactions involving
shuttle vesicles trafficking between different intracellular compartments.
This vesicular trafficking appears to be a general mechanism in the sorting
of cellular components during the assembly of cell organelles. Fission and
fusion of shuttle vesicles also plays a central role in the internalization
of compounds from the extracellular environment during endo- or phagocytosis, and in their subsequent intracellular processing, or, conversely,
during the exocytotic secretion of specific products synthesized by the cell.
Finally, membrane fusion is a critical step in the infectious entry of enveloped viruses into cells.

Obviously, during a membrane fusion event the two membranes involved
have to transiently leave the stable bimolecular leaflet configuration without
compromising their prime function as a relatively impermeable barrier.
This is exactly where the first and most important challenge in the investigation of biological membrane fusion lies: How does a pair of closely apposed
membranes achieve this complex merging operation within the framework
of the restrictions posed by their barrier function? This question not only

relates to the molecular mechanisms involved in fusion processes, it is
also a question regarding the *control* of these fusion events. Clearly,
the stimulus inducing a certain fusion reaction will be restricted to a very
short period of time and to a specific localization. For example, studies
on stimulus-response coupling in secretory cells have indicated that exocytosis is usually preceded by a highly transient rise in the intracellular free
Ca^{2+} concentration (1,2), which may even be limited to restricted parts
of the cytoplasm. Although the precise function of Ca^{2+} in exocytosis is
not known, it likely constitutes an essential element in the control of the
exocytotic process. Also, in other intracellular fusion processes, highly
transient fusion signals seem to operate. Many intracellular fusion reactions,
on both the endocytic and exocytic pathways, appear to involve an *N*-
ethylmaleimide-sensitive fusion protein (NSF) (3). In addition, these
intracellular fusion reactions may also require Ca^{2+} (4). With respect to
the extreme selectivity of biological membrane fusion, which must involve
highly specific and intricate recognition, molecular mechanisms remain largely
obscure as yet, although the involvement of GTP-binding and -hydrolyzing
proteins (5) has been proposed.

In this chapter we focus on membrane fusion in model systems consisting
of lipid vesicles (liposomes). Much of our current knowledge regarding
the possible molecular mechanisms of membrane fusion has been derived
from investigation of these model systems (for reviews, see Refs. 6-16).
The use of liposomal model systems in studies on membrane fusion has
several obvious advantages. Liposomes are relatively simple and relatively
well-defined structures. They can be easily made, either as multilamellar
vesicles (MLV) of different sizes or as small (<50 nm diameter) or large
(>50 nm diameter) unilamellar vesicles (SUV and LUV, respectively). The
membrane composition of liposomes as well as their aqueous contents can
be easily manipulated.

This versatility of liposomal systems, in turn, has greatly facilitated
the development of a set of fusion assays based on the incorporation of
fluorescent probes in the lipid bilayer of liposomes or on the encapsulation
of fluorescent reporter molecules in their aqueous interior, thus enabling
the registration of the mixing of membrane lipids, the coalescence of internal
compartments, or the release of aqueous contents during vesicle fusion
(17-19). This is perhaps the most important advantage of the use of liposomal
model systems, these fluorescent assays allowing the investigation of the
kinetics of the fusion reaction under well-defined conditions. The kinetics
of a particular fusion process can then be correlated with data on the
structure of lipid systems of the same composition at equilibrium so as
to provide, perhaps, insight into the possible structural nature of fusion
intermediates. Model membranes of a well-defined composition are the only
systems available in which such a comparative analysis of lipid structure,
on the one hand, and fusion kinetics in vesicle systems, on the other,
can be made.

However, this is also exactly where pitfalls may arise. First, dynamic
fusion intermediates are, almost by definition, short-lived structures,
which are not necessarily comparable to the structure of the particular
lipid at equilibrium. In addition, fusion intermediates are likely to involve
just a small fraction of the components of the apposed membranes, whereas,
again almost by definition, the equilibrium structure of a lipid dispersion
involves large cooperative units, which may well be different from fusion
intermediates, not only in terms of unit size, but also in terms of structure. Finally, membranes interacting during fusion processes are inherent
parts of closed vesicular structures that are osmotically active, whereas

often the equilibrium structure of a lipid dispersion under fusion-promoting conditions is nonvesicular.

There are other limitations to the use of liposomal model systems, which ought to be recognized as well. First, in their simplest form, lipid vesicles lack specialized recognition molecules; therefore, their interaction by necessity is nonspecific, implying that, using a liposomal model system, one cannot expect to mimic the initial specific adhesion step in the interaction between biological membranes. For example, the initial adhesion between phosphatidylserine (PS) vesicles in the presence of Ca^{2+} is the result of neutralization of the negative surface charge on the vesicles due to binding of Ca^{2+}, allowing the vesicles to aggregate under the influence of attractive Van der Waals forces. Second, fusion in most liposomal model systems lacks the control that is inherent to biological membrane fusion reactions. This implies that, unless the stimulus to fusion (e.g. Ca^{2+}) is removed, the process will simply continue until all the vesicles have been transformed into large, nonvesicular, structures. Therefore one cannot normally expect to see *postfusion stability* of the fusion products. This indicates that in the study of lipid vesicle fusion it is imperative, as indicated above, to employ methodologies that allow a kinetic and quantitative assessment of the initial stages of the process.

In the next section of this chapter we briefly present methodologies for determining fusion in model systems, particularly emphasizing kinetic fluorescence fusion assays. In Section III examples of applications of these assays in a number of widely studied systems are given and evidence is presented indicating that the vesicle-vesicle interactions occurring in these systems meet a rigorous criterion for membrane fusion. Sections IV and V discuss the factors involved in the close apposition of membranes and in their merging, and possible molecular mechanisms are presented. Finally, in Sections VI and VII an attempt is made to relate fusion in model systems to biological membrane fusion reactions.

II. FUSION ASSAYS

Various methodologies have been utilized to study membrane fusion reactions. Among the techniques used in lipid vesicle systems (for a review, see Ref. 6) are nuclear magnetic resonance (NMR) and electron spin resonance (ESR) spectroscopy, fluorescence techniques, differential scanning calorimetry, electron microscopy, gel filtration, and turbidity measurements. The criteria underlying the use of some of these techniques do not always meet a rigorous definition of membrane fusion. For example, mixing of membrane constituents or increase in vesicle size may be the result of fusion, but can also be due to exchange or transfer of individual molecules. The most rigorous criterion for lipid vesicle fusion is coalescence of the internal compartments of the vesicles concomitant with or following the mixing of membrane lipids. The fluorescence fusion assays described below are based on this stringent criterion. In addition, they permit one to assess the fusion reaction in a kinetic and quantitative fashion, which, as discussed above, is critical for a proper evaluation of the process (for a recent review on fluorescence fusion assays, see Ref. 17).

A. Mixing of Aqueous Vesicle Contents

The two most commonly used assays for mixing of aqueous vesicle contents are the Tb/DPA assay and the ANTS/DPX assay (Fig. 1a). Details of the

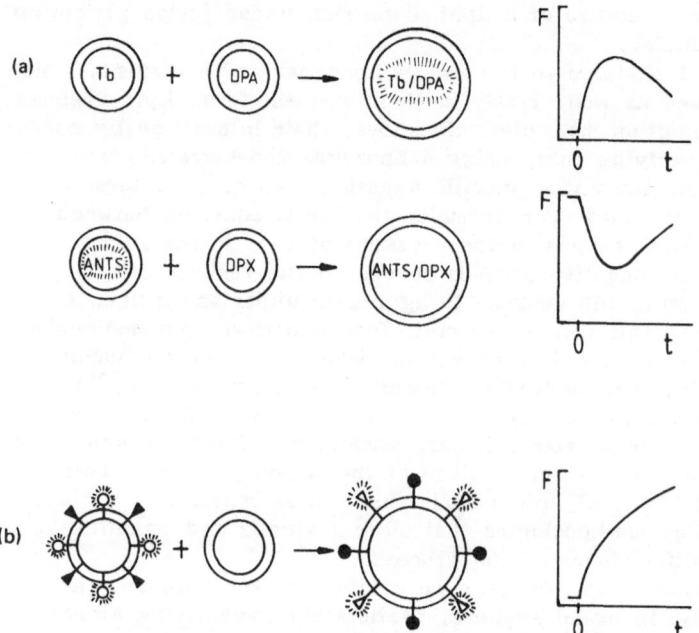

FIGURE 1   Schematic representation of the Tb/DPA (Refs. 20-23) and
ANTS/DPX (Ref. 25) assays (a) for mixing of aqueous contents and the
RET assay (b) for mixing of bilayer lipids during fusion of phospholipid
vesicles (26). In the Tb/DPA assay the mixing of aqueous contents during
vesicle fusion results in an increase of fluorescence intensity, whereas
in the ANTS/DPX assay the ANTS fluorescence is quenched. In b, the
triangles represent the donor probe molecules (*N*-NBD-PE) and the circles
the acceptor probe molecules (*N*-Rh-PE); fusion is monitored as an increase
of the donor fluorescence.

Tb/DPA assay have been described before (20-23). Briefly, it involves
encapsulation of Tb^{3+} ions in one population of vesicles and the anion
of dipicolinic acid (DPA^{2-}) in another. To prevent the Tb^{3+} from binding
to negatively charged phospholipids, it is entrapped in the presence of a
relatively weak chelator, usually citrate. Fusion of Tb- with DPA-containing
vesicles results in the fast formation of the fluorescent $Tb(DPA)_3{}^{3-}$ complex
(Fig. 1a). EDTA, included in the external medium, effectively prevents
the formation of the fluorescent complex outside the vesicles, implying
that any fluorescence observed is due to Tb/DPA complex formation in
fusing vesicles sequestered from the external medium. With a modification
of the Tb/DPA assay the kinetics of release of vesicle contents can be
determined (24).

The ANTS/DPX assay (11,25) relies on the quenching of aminonaphtha-
lenetrisulfonic acid (ANTS) by *p*-xylelene bis(pyridinium) bromide (DPX),
encapsulated in separate populations of vesicles (Fig. 1a). Fusion results
in a decrease of the fluorescence intensity. Release of vesicle contents
can be monitored by coencapsulation of ANTS and DPX in one population
of vesicles. The results obtained with this assay appear to be, in general,
comparable to those obtained with the Tb/DPA assay (11,18,25).

Obviously, the most important question to be asked with respect to the above fusion assays is whether they indeed reliably register the coalescence of internal aqueous vesicle contents rather than an artifact, such as the intermixing of aqueous contents between aggregated vesicles as a result of leakage within a large conglomerate of vesicles. In Section III of this chapter we provide evidence showing that, even though leakage does eventually occur in most vesicle fusion systems, the assays do report the initial coalescence of the internal aqueous vesicle contents.

B. Mixing of Membrane Lipids

Several fluorescence assays for monitoring the mixing of bilayer lipids during vesicle fusion have been developed. Ordinarily, these assays are based on fluorescence resonance energy transfer (RET) between a fluorescence donor and acceptor pair (Fig. 1b). A widely used RET couple consists of two derivatives of phosphatidylethanolamine (PE), labeled in their polar groups: N-NBD-PE and N-Rh-PE, the fluorescence donor and acceptor, respectively (26). These fluorophores have been demonstrated to be non-exchangeable between phospholipid vesicles, even when the vesicles are aggregated. In the most commonly used variant of the RET assay, shown in Figure 1b, the two probes are incorporated together in one vesicle bilayer. Fusion of such labeled liposomes with an unlabeled target membrane results in a decrease of the resonance energy transfer efficiency due to dilution of the probes into the newly formed membrane and, hence, in an increase of the donor fluorescence, which is usually monitored (Fig. 1b); concomitantly, the acceptor fluorescence decreases. Starting from fluorophore concentrations of 0.6–0.7 mol% each, the relative increase of the N-NBD-PE fluorescence intensity is proportional to the dilution of the probes, permitting an accurate quantitation of the fusion process (27). The RET assay can be used not only to determine fusion among phospholipid vesicles, but also for monitoring fusion between (fluorescently labeled) phospholipid vesicles and (unlabeled) biological membranes.

Besides the possibility of the occurrence of probe exchange between aggregated membrane vesicles, obviously, a major concern with lipid mixing assays is the possibility that the physical state of the membrane lipids during the overall fusion process affects the fluorescence characteristics or quantum yield of the probes that are used (17–19). For example, it has been demonstrated that the fluorescence quantum yield of N-NBD-PE increases when its surrounding lipid undergoes a transition from the lamellar to the H_{II} configuration (28,29). Also in systems that are not undergoing such a transition, such as the Ca^{2+}/PS system, effects of the physical state of the lipid on the N-NBD-PE fluorophore have been noted (19,30). Therefore, with the application of a particular lipid mixing assay, it is essential that these possibilities be considered before the assay can be reliably employed. A readily accessible control involves the use of just labeled vesicles, rather than a mixture of labeled and unlabeled vesicles as used in the fusion assays, eliminating the effects of probe dilution. Recently, Silvius et al. (30) have developed a lipid mixing assay that appears less sensitive than the N-NBD-PE/N-Rh-PE assay to lamellar to H_{II} transitions, and possibly also to changes, occurring in the Ca^{2+}/PS system. This assay utilizes two derivatives of PC, labeled in the acyl-chain moiety with a coumarin fluorophore (CPS), and a nonfluorescent dimethyl-aminophenylazophenol quencher (DABS).

III. EXAMPLES OF FUSION IN LIPID VESICLE SYSTEMS

In this section we present a number of representative examples of widely
studied systems, in which the fusion assays discussed above have been
applied extensively. Rather than discussing possible mechanisms of the
vesicle interactions occurring in these systems (for this discussion, see
Section V) we emphasize here the question as to whether the phenomenon
as it is observed meets a rigorous criterion for "fusion," as opposed to
other modes of vesicle-vesicle interaction. As indicated above, in practically
all lipid vesicle systems in which fusion can be induced, leakage of vesicle
contents occurs as well. This is not surprising, since the fusion process
in these uncontrolled systems usually represents but one stage of an overall
transformation from a vesicular condition to a condition in which the lipid
is present in the form of large nonvesicular conglomerates. Critical, therefore, is the question of whether the *initial* interaction between pairs of
vesicles results in a relatively nonleaky coalescence of the internal compartments of the vesicles, concomitant with or following mixing of bilayer lipids.

Probably the most widely studied model system for membrane fusion
is the Ca^{2+}/PS system. More than a decade ago, Papahadjopoulos and co-
workers showed that addition of Ca^{2+} to PS vesicles causes the vesicles
to aggregate, to release their internal aqueous contents, and to form large
structures, called "cochleate" cylinders (31,32). These cochleates would
transform into large unilamellar vesicles upon addition of excess EDTA,
indicating that at some stage of the Ca^{2+}-induced PS vesicle interaction
membrane fusion occurs. Yet, the Ca^{2+}/PS system has been criticized extensively as an appropriate model for membrane fusion, largely on the
basis of the occurrence of release of vesicle contents, indicating the loss
of vesicular integrity (33-37). Parsegian and Rand have argued that the
adhesion force between PS bilayers in the presence of Ca^{2+} is so strong
that, as a result of the flattening of the vesicles against one another,
the membranes break, which may, largely as a matter of coincidence, occur
at the diaphragm between the interacting vesicles, resulting in mixing of
aqueous vesicle contents; however, according to their view, breakage
of the membranes could just as well occur outside the area of vesicle-
vesicle interaction, resulting in lysis and release of vesicle contents (see
Chapter 3).

Figure 2 shows a typical result obtained by application of the fluorescence assays, discussed in the previous section, to the Ca^{2+}/PS system.
The figure presents the case of large unilamellar PS vesicles at 3 mM Ca^{2+}
(21,23,38). Clearly, the mixing of aqueous vesicle contents (drawn line)
precedes that of the release of vesicle contents to the external medium
(dotted line). Importantly, during the initial stages of the process there
is no detectable release at all (23). In later stages of the process leakage
does occur, as is evident from the secondary decrease of the Tb/DPA
signal observed during the contents mixing experiment (drawn line) as
well as from the direct measurement of release (dotted line). This secondary
leakage marks the first stages of the collapse of fused vesicles into large
nonvesicular structures, an event that will inevitably occur as long as
the process is allowed to proceed, the final equilibrium condition in the
Ca^{2+}/PS system being one of a tightly packed multilamellar structure.
If, however, the fusion reaction is interrupted in an early stage by taking
away the fusion stimulus (i.e., Ca^{2+}), the contents of the vesicles remain
sequestered from the external medium, apparently in fused vesicles (arrow).

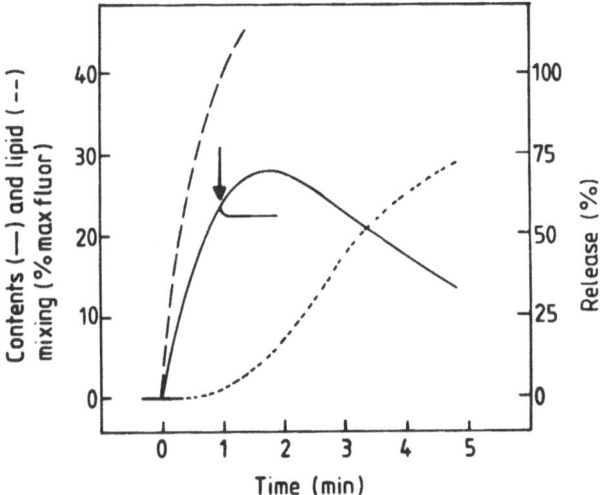

FIGURE 2   Relative kinetics of mixing and release of vesicle contents during
Ca^{2+}-induced fusion of bovine brain PS large unilamellar vesicles (100 nm
diameter). Ca^{2+} concentration was 3 mM. Fusion was measured in a 1:1
mixture of Tb- and DPA-containing vesicles (solid line); in parallel experiments the release of vesicle contents was monitored with a modification
of the Tb/DPA assay (dotted line). The arrow indicates, in a separate
experiment, the interruption of the fusion process by addition of excess
EDTA. The dashed line represents mixing of bilayer lipids monitored with
the RET assay. Calibration of the fluorescence scales is such that the
curves can be compared directly. (Data from Refs. 23 and 38.)

Application of the N-NBD-PE/N-Rh-PE RET assay for monitoring the mixing
of bilayer lipids to the same system demonstrates (dashed line) that this
process proceeds slightly faster than the mixing of aqueous vesicle contents
(38). In conclusion, in the presence of Ca^{2+}, PS LUV communicate their
contents in a nonleaky fashion, very shortly after the mixing of bilayer
lipids.

Figure 3 shows the results of a typical fusion experiment with vesicles
made of cardiolipin (CL) (39). As in the Ca^{2+}/PS system, the coalescence
of the internal vesicle compartments (drawn line) is a much faster process
than the release of aqueous contents to the external medium (dotted line).
And, again similar to the Ca^{2+}/PS results, the rate of mixing of bilayer
lipids measured with the RET assay is slightly faster than that of the
mixing of aqueous vesicle contents. With respect to the results in Figure 3,
it is of interest to refer briefly to the results in Figure 8 (which will be
discussed in more detail later), showing that in the presence of Sr^{2+} the
fluorescence signal recorded with the Tb/DPA fusion assay in the CL LUV
system is dramatically different from that in the presence of Ca^{2+}, particularly at elevated temperatures. With Sr^{2+} there is a fast and extensive
release of vesicle contents, translating, in the assay for mixing of aqueous
vesicle contents, into an only limited and highly transient fluorescence
signal; with Ca^{2+}, on the other hand, due to a low and initially even
negligible rate of leakage, extensive mixing of aqueous contents can be
observed. This result clearly demonstrates the reliability of the Tb/DPA

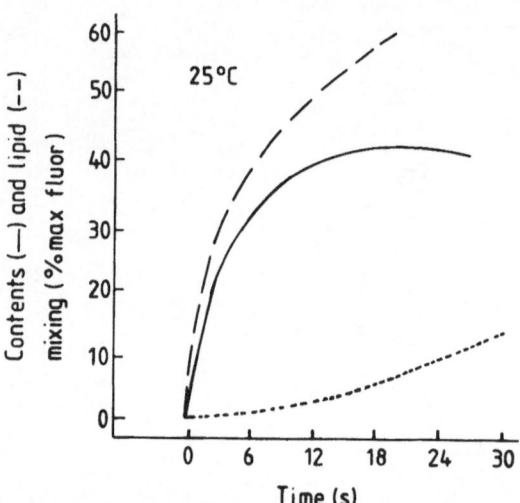

FIGURE 3   Relative kinetics of mixing and release of vesicle contents and
of mixing of bilayer lipids during Ca^{2+}-induced fusion of large unilamellar
vesicles composed bovine heart CL. Ca^{2+} concentration was 10 mM. Fusion
was measured in a 1:1 mixture of Tb- and DPA-containing vesicles (solid
line); in parallel experiments the release of vesicle contents was monitored
with a modification of the Tb/DPA assay (dotted line). The dashed line
represents mixing of bilayer lipids monitored with the RET assay; it was
assured that on the time scale of the experiment the fluorescence quantum
yield of the N-NBD-PE was not affected by a lamellar to H_{II} phase transition
in the system (see Ref. 29). Calibration of the fluorescence scale is such
that the curves can be compared directly. (Data from Ref. 39.)

assay: It does not artifactually register release of vesicle contents as
"fusion."

Besides the negatively charged phospholipid vesicle systems, the only
pure phospholipid systems in which fusion, monitored as mixing of aqueous
vesicle contents, has been convincingly demonstrated are vesicle systems
consisting of phosphatidylethanolamine (PE) or PE analogs. Having already
presented two examples of the application of fluorescent assays to fusion
in model systems, for fusion of PE-containing vesicles, we refer to a recent
excellent review, in which Bentz and Ellens discuss their extensive work
in this area (11), and to a number of specific research papers (28,40-44).
In the next section of this chapter, dealing with the possible molecular
mechanisms of fusion of lipid vesicles, we will come back to the PE systems.

Apparently, the initial interactions between PS or CL vesicles, induced
by Ca^{2+}, as well as, under certain conditions, those of PE-containing
vesicles, meet a rigorous criterion for membrane fusion, i.e., mixing of
bilayer lipids followed by a nonleaky coalescence of internal vesicle compartments. This is precisely the basis for the value of liposomal systems as
models for biological membrane fusion, and therefore, from studying these
systems we may expect to gain a fundamental insight into the basic molecular
requirements for lipid bilayer fusion, which is the key event in all membrane
fusion reactions including those occurring in cell biology.

IV. MEMBRANE APPOSITION: VESICLE AGGREGATION AND THE ESTABLISHMENT OF MOLECULAR CONTACT

The fusion process in vesicle systems proceeds in a number of distinct steps. First, the vesicles will have to adhere or aggregate; usually, in simple model systems, this aggregation is achieved by nonspecific interactions. Second, direct molecular contact, beyond the "contact" that is achieved during vesicle aggregation, has to be established. Third, a local perturbation of the packing of the bilayer lipids at the site of contact seems required; this point defect serves as a focal point of hydrophobic interactions, initiating the merging of the outer monolayers of the two bilayer membranes. Finally, the aqueous vesicle interiors have to coalesce with the concomitant mixing of the inner-monolayer lipids. In this section we briefly discuss the factors involved in the aggregation of vesicles and in the establishment of direct molecular contact; these two steps in the overall fusion process are governed primarily by *inter*bilayer forces. In Section V factors affecting the actual bilayer merging are discussed; here *intra*bilayer forces appear to play a major role.

A. Intermembrane Forces

While little is known about specific recognition and adhesion mechanisms operating between biological membranes, knowledge on nonspecific interactions between lipid bilayer membranes, and the forces involved, is quite extensive. This knowledge is based not only on studies regarding the aggregation behavior of phospholipid vesicles (for reviews, see Refs. 6 and 11), but also on direct measurement of forces between lipid bilayers adsorbed onto a mica surface (45,46) or in stacked multibilayer systems (47,48; see also Chapter 3).

Upon mutual approach, lipid bilayer membranes are subject to a number of different interaction forces (6,47-49). These forces comprise: (1) electrostatic or Coulombic interactions, (2) attractive Van der Waals interactions, and (3) hydration forces. As the forces operating between lipid bilayer membranes are discussed extensively in Chapter 3 of this volume, we give only a short account of these interactions here, relating them specifically to the phenomenon of vesicle aggregation.

Electrostatic and Van der Waals interactions both represent long-range forces: They act at surface separations much greater than, say, 1 nm. For particles of like charge electrostatic forces are repulsive and they decay exponentially with increasing surface separation. Ions in the surrounding medium screen the electrostatic surface potential and, thus, affect the decay distance of the repulsion. The Van der Waals (or electrodynamic) force is an attractive force. Van der Waals forces between large bodies in condensed media fall off much more slowly than with the classical inverse sixth power of the distance of separation and may, in fact, decay even more slowly than electrostatic repulsion. It is for this reason that Van der Waals forces, acting at long range, can stabilize aggregates of lipid vesicles in the so-called secondary minimum (50), i.e., under conditions where no direct surface contact occurs (Fig. 4).

When two phospholipid bilayers are forced to close approach, strong repulsive hydration forces arise, which are due to the work involved in the removal of the tightly bound water from the phospholipid headgroups (47,51). Hydration forces become significant at surface separations of

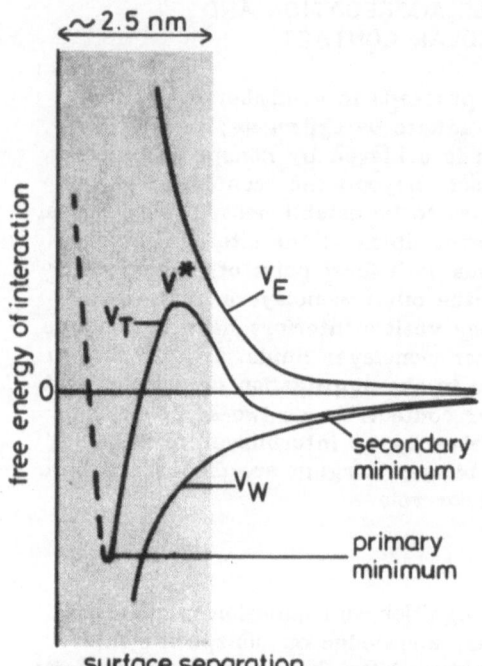

FIGURE 4 Free energy of interaction between two vesicles of like charge
as a function of their surface separation according to the DLVO theory
ignoring hydration repulsion. The total potential energy, V_T, is the sum
of the electrostatic repulsion, V_E, and the Van der Waals attraction, V_W.
The indicated separation of 2.5 nm represents the approximate range of
the repulsive hydration forces.

about 2-3.5 nm (Fig. 4), depending on the nature of the lipid involved.
With further decreasing distance, the force rises steeply and largely over-
whelms both Van der Waals attraction and electrostatic forces. Hydration
repulsion occurs between phospholipid bilayers of any composition, but
its magnitude may vary with the nature of the particular lipid involved.
For example, for neutral phosphatidylcholine (PC) bilayers the equilibrium
separation, where hydration repulsion is balanced by Van der Waals attraction, is about 2.1-2.5 nm, whereas for PE the equilibrium separation is
about 1 nm smaller and the adhesive energy almost an order of magnitude
larger (45,47,48,51,52). This difference is due to the different conformation
(53) and the lower degree of hydration of the headgroup of PE as compared
to that of PC. Recently, an attractive hydration force in PE systems has
even been postulated (54). This attraction would originate from the formation of interbilayer hydrogen-bonded water bridges (52,54). Finally, the
conformation of the choline headgroup (53) may also contribute a steric
component to the observed repulsion between PC bilayers (55).

B. Aggregation of Lipid Vesicles

A theory of the interaction between lyophobic colloids, the DLVO (Derjaguin-
Landau-Verwey-Overbeek) theory (56), in most cases provides an adequate

description of the aggregation of lipid vesicles under various ionic conditions. A sufficient condition for the formation of stable vesicle aggregates is the occurrence of a local minimum in the free energy of interaction between the vesicles, provided that the energy well is significantly deeper than kT (where k is Boltzmann's constant and T the absolute temperature). For vesicles of like charge, the DLVO theory, which takes into account electrostatic repulsion and Van der Waals attraction, predicts that there are two types of energy minimum, depending on the distance of surface separation of the interacting particles. The primary minimum would occur at distances of separation of 1 nm or less and the secondary minimum, mentioned above, at separations of 3-10 nm (Fig. 4).

Obviously, aggregation in the primary minimum is of particular importance to membrane fusion. The rate of aggregation in the primary minimum is dependent on the height of the energy barrier, V^*, which in turn depends on the interplay between electrostatic repulsion and Van der Waals attraction (Fig. 4). This energy barrier, in fact, represents the activation energy for aggregation. In a medium containing physiological salt concentrations, negatively charged phospholipid vesicles, composed of, for example, PS, will not aggregate owing to electrostatic repulsion: The activation energy V^* is too high. Divalent cations induce aggregation of the vesicles primarily by direct binding to the phospholipid headgroups, resulting in charge neutralization, but also by screening of the surface charge. Thus, the vesicles start to aggregate at certain specific divalent cation concentrations, which are dependent on the binding affinities of the cations for the anionic lipid. For example, PS vesicles in a 100-mM NaCl medium start to aggregate on a time scale of seconds in the presence of 1-2 mM Ca^{2+} or 4-5 mM Mg^{2+} (6,57). These observations are in perfect agreement with predictions from the DLVO theory, provided that the physical binding of the cations to PS is taken into account.

The DLVO theory does not consider hydration repulsion. Yet, in most cases the theory provides an adequate description of the aggregation behavior of negatively charged phospholipid vesicles under various ionic conditions, suggesting that hydration forces do not normally affect the aggregation of such vesicles to a significant extent. It is likely that, in the overall process of vesicle fusion, hydration repulsion has its prominent effect in the stage *after* vesicle aggregation, i.e., during the establishment of molecular contact.

C. Establishment of Molecular Contact

It appears that repulsive hydration forces, as discussed above and in Chapter 3, pose the major barrier to the establishment of direct intermembrane contact between aggregated vesicles. The evidence to indicate this has been discussed extensively before in several review articles (6-16); therefore, we limit the discussion here to a short summary.

First, fusion of vesicles consisting of mixtures of negatively charged and zwitterionic phospholipids has been observed to be sustained or sometimes even enhanced when PE is the zwitterionic lipid, whereas PC or sphingomyelin is almost always strongly inhibitory (58-62). This is likely to be due, at least in part, to the difference in the degree of hydration of the ethanolamine vs. choline headgroups of these lipids, as discussed above (45,47,48,51-55).

Second, the cation specificity of PS vesicle fusion correlates with the degree to which the ions are dehydrated and with their ability to form

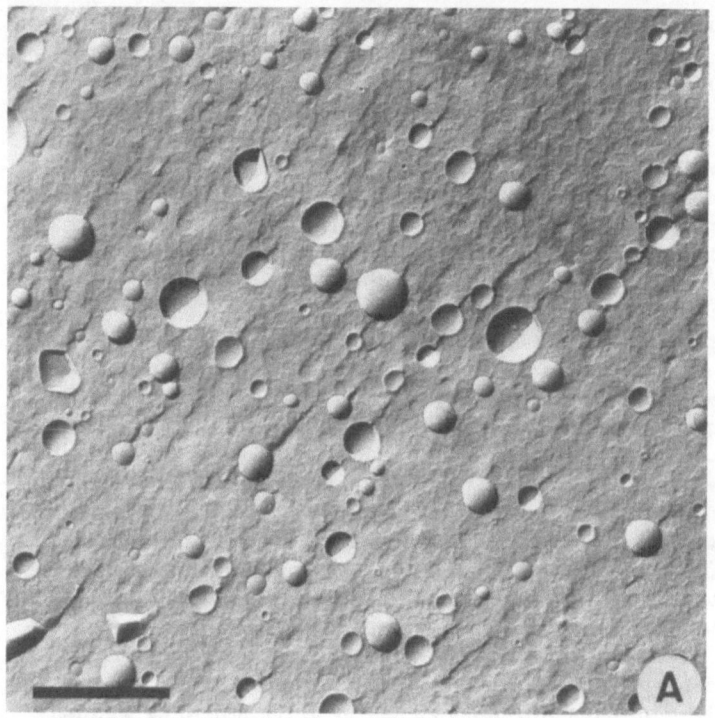

FIGURE 5 Freeze-fracture electron micrographs of PS LUV before and after interaction with Ca^{2+} or Mg^{2+}. Vesicles were incubated in the presence of $CaCl_2$ or $MgCl_2$ for 10 min at 25°C. The vesicle aggregates were spun down and, to disaggregate the vesicles, the pellets were treated with an excess EDTA. (A) Control; (B) 10 mM Mg^{2+}; (C) 5 mM Ca^{2+}. Clearly, the Mg^{2+}-treated vesicles have not increased in size, whereas the Ca^{2+}-treated vesicles have fused to considerably larger structures. Bars represent 0.5 µm. (Reproduced, with permission, from Ref. 22.)

dehydrated complexes with PS (63). The most extreme case is given by Mg^{2+}, which does not induce any fusion of PS LUV, even though the vesicles massively aggregate (Fig. 5, Ref. 22). Under comparable conditions Ca^{2+} induces efficient fusion. It has been demonstrated that Ca^{2+} has the capacity to completely dehydrate the space between two PS bilayers by forming a *trans* complex involving PS molecules on the apposed surfaces (57,64-66). On the other hand, with Mg^{2+} a *cis* complex is formed and a layer of water remains between the bilayers, thus preventing direct interbilayer contact and fusion (57).

Finally, it is well established that dehydrating agents may have the capacity to induce membrane fusion (67). For example, the cell-cell fusion activity of poly(ethylene glycol) (PEG) is exploited extensively in the production of hybrid cells (see Chapters 28 and 30). PEG also induces fusion of phospholipid vesicles, as discussed in detail in Chapter 11. The fusogenic capacity of PEG points to a key role of bilayer dehydration in membrane fusion (16). Apart from inducing a direct intermembrane contact, PEG-induced membrane dehydration appears to induce local fluctuations in lipid packing (16), a requirement for membrane fusion that is discussed more extensively in the next section of this chapter.

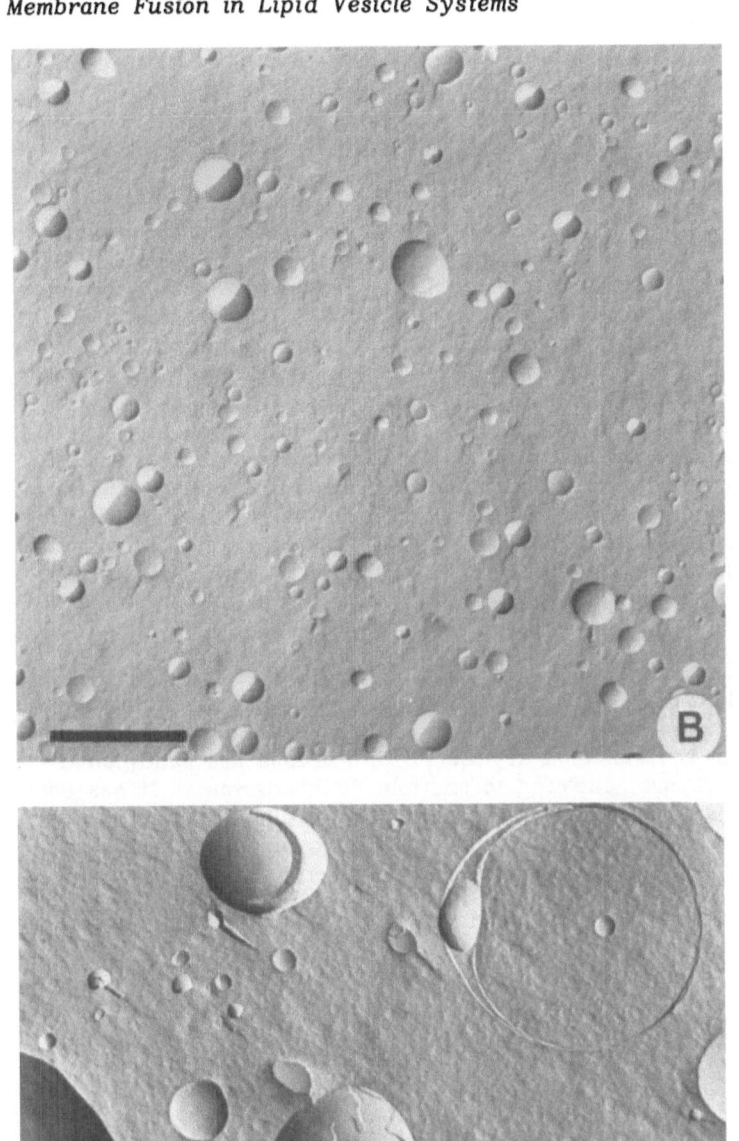

V. FACTORS AFFECTING THE ACTUAL FUSION PROCESS IN LIPID VESICLE SYSTEMS: CLUES WITH RESPECT TO MOLECULAR MECHANISMS

A. A Local Disturbance of Lipid Packing: The Focal Point for Fusion

The establishment of direct interbilayer contact is a necessary, but in itself insufficient, condition for fusion. A local disturbance of lipid packing seems required as well. The concept of bilayer fusion proceeding through point defects has been proposed by Hui et al. (68), on the basis of a study of the fusion of lipid vesicles induced by freezing and thawing. In experiments involving phospholipid bilayers adsorbed onto mica surfaces, Marra and Israelachvili (45) and Horn (46) have observed occasional "fusion" events under conditions of very large interbilayer pressures. In these cases fusion was always initiated at one point via a local thinning of the closely apposed, largely dehydrated bilayers. Horn (46) noted that fusion was facilitated by the presence of impurities in the lipid, causing imperfections in the bilayer structure. How does Ca^{2+} as the fusogenic agent induce a disturbance of lipid packing in negatively charged phospholipid vesicle systems?

1. The Ca^{2+}/PS System

It is well established that Ca^{2+} has the capacity to induce the formation of a rigid dehydrated complex with PS (57,64-66,69). Within this high-affinity complex, of the stoichiometry $Ca(PS)_2$, the ions are sandwiched between apposed bilayers, chelated to multiple PS headgroups. It has been proposed that the ability of Ca^{2+} to induce the formation of this characteristic phase forms the basis for its ability to induce fusion of PS vesicles (32,57). Although in later studies, involving Ba^{2+}- or Sr^{2+}-induced inter-

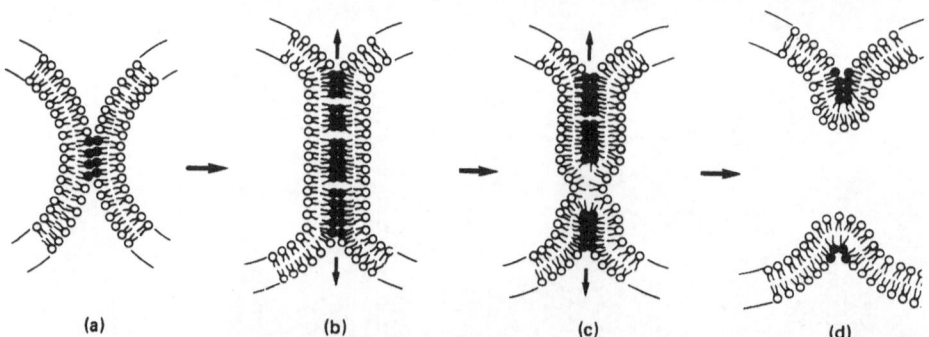

(a)   (b)   (c)   (d)

FIGURE 6 Hypothetical mechanism of divalent-cation-induced fusion of negatively charged phospholipid vesicles. (a) Vesicle aggregation and the formation of a dehydrated area of contact. (b) Expansion of the area of contact results in an increased surface to volume ratio of the vesicles and generates a tension on the vesicle bilayers; lipid packing defects develop at the area of contact due to specific interbilayer complex formation. (c) A point defect in the apposed bilayers produces a local thinning and formation of a single-bilayer diaphragm. (d) The diaphragm breaks and the bilayers merge. (Reproduced, with permission, from Ref. 15.)

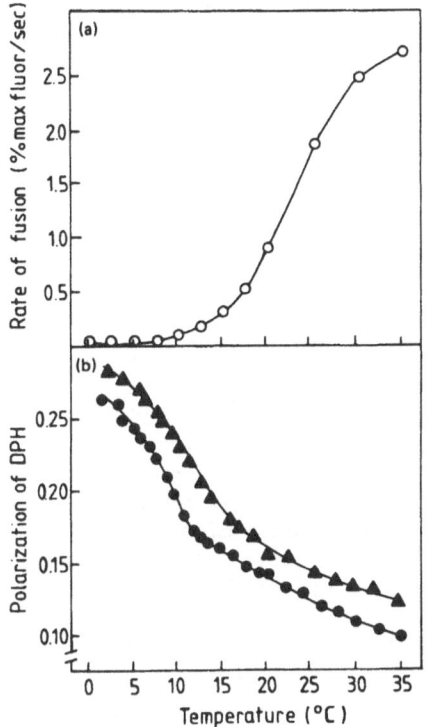

FIGURE 7 Effect of bilayer fluidity on the fusion of PS LUV. (a) Fusion of PS LUV, measured with the Tb/DPA assay as described in Figure 2, at 5 mM Ca^{2+} and different temperatures. (b) Fluidity of PS LUV, measured by fluorescence polarization of diphenyl hexatriene (DPH), in the absence of Ca^{2+} (circles) or in the presence of 0.5 mM Ca^{2+} (triangles), which does not induce vesicle aggregation on the time scale of the experiment. (Data taken from Ref. 70.)

actions between PS vesicles, it was demonstrated that an ion-induced phase transition from a fluid to a gellike state may not be essential for induction of vesicle fusion (59,63), the induction of a tighter lipid packing at the site of intermembrane contact as an important feature of the mechanism of fusion remained attractive. Indeed, if a rigid *trans* $Ca(PS)_2$ complex is formed between interacting PS vesicles (see scheme in Fig. 6), it is likely to not only, as discussed above, induce a virtually complete dehydration of the interbilayer space (Fig. 6a), it would also induce defects in the packing of the lipid. Importantly, these packing constraints would, in fact, result primarily from the *asymmetrical* interaction of the ion with just the outer monolayer of the vesicle bilayer, leaving the fluid inner monolayer unaffected (Fig. 6b), and would be relieved by merging of the outer vesicle monolayers and formation of a more or less extended single-bilayer diaphragm (Fig. 6c). In this respect, it is noteworthy that Ca^{2+}-induced fusion of PS vesicles will occur only when the vesicle bilayer, before vesicle-vesicle interaction, is in the fluid state (Fig. 7; 70). The formation of a *trans* ion/lipid complex as a focal point for fusion is particularly attractive as it necessarily limits the initial destabilization of the vesicle bilayers to the site of vesicle-vesicle interaction. In addition, it appears on the basis of recent studies of Feigenson (65,66) that the complex can be formed in the biologically relevant (sub)micromolar Ca^{2+} concentration range.

2. Inverted Structures as Fusion Intermediates: The Ca^{2+}/CL System

Another example of contact-induced destabilization of lipid bilayers is given by those systems that are prone to undergoing a lamellar to inverted-hexagonal (H_{II}) phase transition. It is well established that many, if not all, biological membranes contain significant amounts of lipids that, in isolation, do not adopt a lamellar organization but, rather, prefer an inverted configuration, such as the HII phase (71,72). Prominent examples are unsaturated PE (73), and CL in the presence of Ca^{2+} or other divalent cations (74). Early experimental evidence for the possible role of inverted structures in membrane fusion has been obtained from freeze-fracture electron microscopic observations on model systems. In these studies, so-called "lipidic particles" were observed at the interface of interacting vesicles, suggesting that inverted micelles might function as fusion intermediates (75,76; see also Chapter 6). However, subsequently, it was appreciated that the lipidic particles observed in these early studies presumably represented equilibrium structures rather than dynamic fusion intermediates (77). Yet, the inverted micellar structure as an essential feature of dynamic fusion intermediates remained an attractive model.

The discussion on the role of inverted structures in membrane fusion processes has gained considerable impetus through contributions of Siegel (78-82). His theoretical work has shown that vesicle systems that can undergo a lamellar to inverted hexagonal (L_α-H_{II}) phase transition, rapidly develop so-called inverted micellar intermediates (IMI) between the apposed bilayers at sites of vesicle contact (Fig. 8b). These structures are *very* short-lived and in dynamic equilibrium with the bilayers from which they are formed. IMI can either assemble into H_{II} phase precursors (not shown in Fig. 8) or they may evolve into so-called interlamellar attachment sites (ILA), which, in fact, represent fused structures (Fig. 8c). H_{II} phase precursor formation and ILA formation compete for the steady-state pool of IMI formed at the site of interbilayer contact. The formation of H_{II} phase precursors is a process of second-order with respect to the concentration of IMI per unit area, whereas the transformation of an IMI into an ILA is a first-order process. Therefore, under conditions of abundant IMI formation, i.e., when the temperature is above the L_α-H_{II} phase transition temperature of the lipid at equilibrium (T_H), a fast "zippering down" of the apposed bilayers into H_{II} phase precursors may be induced, with the concomitant loss of vesicular integrity. Obviously, this transformation will result in mixing of bilayer lipids, but there will be little coalescence of internal vesicle compartments; rather, the aqueous vesicle contents will be rapidly and completely released to the external medium.

Only under specific conditions IMI may break into ILA (79-82). Usually this breakage is an energetically unfavorable event; the probability of its occurrence appears to be governed, not only by the density of IMI, but also by the parameter $Z = a_L/a_H$, where a_L is the average area per molecule in the lamellar phase and a_H the area per molecule in the H_{II} phase. In fact, the probability of ILA formation is even more sensitive to the parameter Z than to the IMI density. Large Z values (say, 1.4-1.8) correspond to systems with small intrinsic radii of curvature (83); these

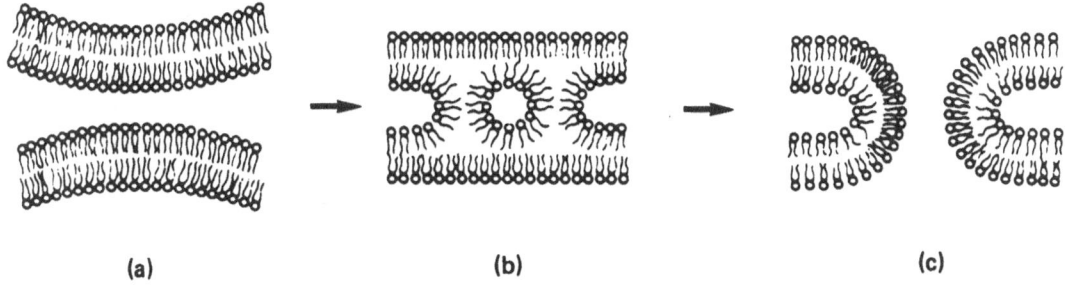

(a)　　　　　　　　　　　　(b)　　　　　　　　　　　　(c)

FIGURE 8 Fusion of lipid vesicles via the formation of (b) inverted micellar intermediates (IMI) transforming into (c) interlamellar attachment sites (ILA). (Adapted from Ref. 11.)

systems exhibit rapid L_α to H_{II} transitions, and hardly any ILA will form. In systems with smaller Z values (≤ 1.2), corresponding to larger intrinsic radii of curvature and larger-diameter H_{II} tubes at equilibrium, the transition is more hysteretic owing to abundant ILA formation.

Recently, compelling evidence has been presented indicating that fusion of vesicles containing PE or PE derivatives may well proceed via the rapid formation of inverted micellar intermediates (IMI) at the site of vesicle-vesicle contact, transforming into ILA (28,41,42,84,85). Fusion, monitored as mixing of aqueous vesicle contents, occurred only at temperatures just below the T_H of the lipid or the lipid mixture, under conditions where the lipid at equilibrium exhibits an isotropic ^{31}P-NMR signal, presumably arising from motional averaging due to lipid diffusion through accumulated interconnected ILA. These accumulated ILA are likely to be responsible for the observed cubic X-ray diffraction pattern of the lipid at equilibrium under these specific conditions (86-88).

With respect to the possible involvement of inverted structures in the fusion of CL vesicles induced by Ca^{2+}, the situation seems somewhat more complex. Clearly, in the absence of divalent cations, CL is organized in a bilayer arrangement, while in the presence of Ca^{2+} or Mg^{2+} it adopts the H_{II} configuration, the lamellar to H_{II} transition temperature of the final cation/CL complex being below 0°C (74,89). The Ca^{2+}-induced fusion of liposomes composed of mixtures of CL and PC has been studied extensively by application of kinetic fluorescence assays, and it appears that these vesicles fuse in a largely nonleaky manner (90,91). Recently, we have characterized the interaction between CL LUV in the presence of different divalent cations in the temperature range from 10 to 50°C (39). The L_α to H_{II} transition temperatures of the Sr^{2+}/CL and Ba^{2+}/CL complexes (25 and 23°C, respectively; Refs. 39,89) lie within this range. The behavior of the Sr^{2+}/CL system, in terms of fusion and destabilization, seems consistent with predictions made by Siegel (79-82); i.e., at temperatures above the T_H of the final ion-phospholipid complex, abundant formation of IMI at the site of contact between aggregated vesicles transforming very rapidly into H_{II} phase precursors results in loss of vesicular integrity and release of vesicle contents (Fig. 9b); the limited apparent mixing of vesicle contents seen above the T_H (Fig. 9a) is of such a transient nature, that it may well be incidental to the collapse process. In sharp contrast to the behavior of the Sr^{2+}/CL system, a remarkable result is obtained in the Ca^{2+}/CL system. As already shown above in Figure 3, Ca^{2+} induces a relatively nonleaky fusion of CL LUV at 25°C. Furthermore, when the

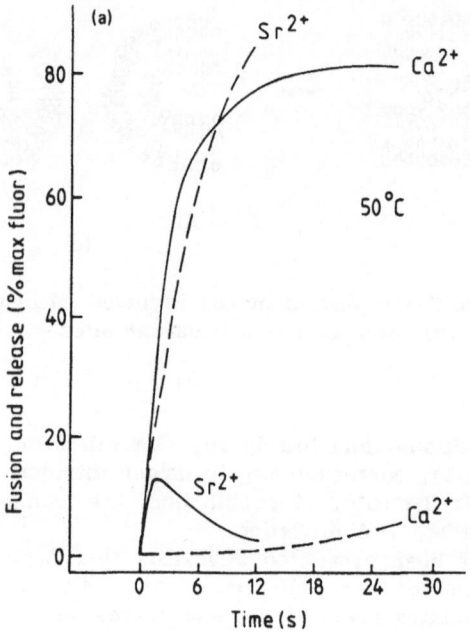

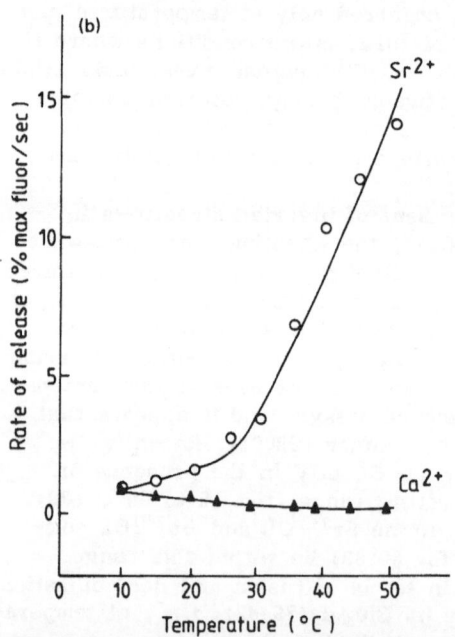

FIGURE 9 Different effects of Ca^{2+} and Sr^{2+} on CL LUV with increasing temperature. (a) Fusion of CL LUV at 50°C, measured with the Tb/DPA assay as described in Figure 3, in the presence of either 10 mM Ca^{2+} or 10 mM Sr^{2+} (solid lines); the dashed lines represent the corresponding release curves. (b) Initial rate of release of vesicle contents in the presence of either 10 mM Ca^{2+} or 10 mM Sr^{2+} as a function of temperature. The lamellar to H_{II} phase transition temperature of the Sr^{2+}/CL complex is approx. 25 °C, while the T_H for the Ca^{2+}/CL complex is below 0°C. (Data taken from Ref. 39.)

temperature is raised up to 50°C, the rate of leakage *decreases* (Fig. 9). Therefore, if the fusion of CL LUV in the presence of Ca $^+$ does proceed via the formation of IMI and ILA, apparently, in the vesicle system the L_α to H_{II} transition is kinetically retarded, in favor of extensive nonleaky fusion of the vesicles, even at temperatures much *higher* than the T_H of the final complex. It is not unlikely that the metastable phase formed in this case, in fact, represents one of accumulated ILA, a structure reminiscent of an inverted cubic phase (84-88).

B. The Energetics of Membrane Fusion: Considerations on the Driving Force for the Fusion Process

In this section we discuss the important question of the driving force for membrane fusion reactions in vesicle systems. In very general terms, the answer to this question is as simple as it is obvious: There should be energy to be gained from the transformation process. This obvious statement, however, immediately implies that large (say, 100 nm or larger) vesicles (LUV) have virtually no intrinsic tendency to fuse, as the surface energy of such vesicles is not appreciably different from that of flat bilayer membranes (92). This is different for small vesicles (approx. 25 nm), such as the SUV obtained by sonication.

1. Fusion of SUV and the Role of Bilayer Curvature

SUV, in contrast to LUV, do have a tendency to fuse spontaneously, as there is appreciable energy to be gained from their transformation to larger vesicles. SUV are strained structures forced to the very minimum of vesicle size: any further reduction would no longer be compatible with an arrangement of the constituting lipid molecules in a lamellar organization. This strain in the packing of the lipid molecules in SUV translates into an increased average area per lipid molecule in the outer and a reduced area per molecule in the inner monolayer of the vesicles relative to the area per molecule in a flat bilayer, resulting in a relatively high surface energy of the vesicles due to the partial exposure of the hydrophobic membrane interior to the aqueous environment. This high surface energy may contribute to a considerable extent the driving force for the fusion of SUV (93,94), provided that the vesicles can overcome the activation energy for aggregation. In many systems, SUV have been shown to fuse much more avidly than LUV (6,21-23,63,70,95). Often, the vesicles will go through a limited number of rounds of fusion until the strain in the packing of the lipid molecules is relieved as a result of the increase in vesicle size. An example of such a fusion process, driven by a high degree of bilayer curvature, is given by the fusion of PS SUV in the presence of Mg^{2+}. That this reaction stops spontaneously after the vesicles have reached a certain size is entirely consistent with the total lack of fusion of PS LUV in the presence of Mg^{2+} (Fig. 5, Ref. 22). Another example of SUV fusion driven by the curvature of the vesicle bilayer is discussed in Chapter 7. Fusion of SUV is instructive for a better understanding of the basic requirements of lipid bilayer fusion. On the other hand, it may not be an appropriate model for biological membrane fusion, as biological membranes do not usually exhibit a degree of curvature typical of SUV.

2. The Driving Force for Fusion of Large Vesicles

As indicated above, large vesicles (LUV) have virtually no tendency to fuse spontaneously. They have no reason to go from the situation of two stable bilayer vesicles to one stable bilayer vesicle, unless, importantly, an external osmotic stress is applied to the vesicles. For dispersed vesicles, an osmotic gradient, hypotonic outside, will eventually lead to rupture and lysis. On the other hand, for vesicles that are aggregated, it may drive the fusion of the vesicles, as fusion of two spherical vesicles into one larger spherical vesicle, with the contomitant influx of water, changes the surface to volume ratio of the vesicles in a way to diminish the osmotic stress.

If, however, a homogeneous population of LUV fuses into larger structures in the *absence* of applied osmotic stress, which does occur under certain conditions, as demonstrated above, the driving force for such a process must originate entirely in the energy of interaction of the fusion-inducing agent, such as Ca^{2+}, with the vesicles. For example, in the Ca^{2+}/PS system there is considerable energy to be gained from the transformation of the fluid bilayer structure to the final dehydrated Ca^{2+}/PS complex arranged in a tightly packed multilamellar configuration (57,65,66,69). Or, in the CL system, there is energy to be gained from the Ca^{2+}-induced lamellar to H_{II} phase transition (71,72,74,89). Only under certain conditions during the course of these ion-induced phase transitions (again, in the absence of an applied osmotic gradient) may the vesicles be transiently "trapped" in a larger vesicular configuration, be it *not* a spherical configuration. The transient nature of this kinetic trap implies that, unless the stimulus to fusion (Ca^{2+} in these examples) is removed, the *final* condition in these systems will never be one of stable fused vesicles. The best one can hope for is that the transition to the final configuration is sufficiently hysteretic so that fused vesicles can be observed for a reasonable period of time during the course of the transformation process. A nonleaky fusion event between two large vesicles in these simple systems in the absence of osmotic pressure, therefore, cannot possibly represent more than a relatively sluggish step of the system on the way to a final nonvesicular condition.

The question for pure lipid systems then becomes: Which factors are favorable to drive the system into a kinetic trap in which the vesicles may transiently mix their internal contents sequestered from the external medium? We believe that keys in this regard lie in (1) the initial interaction of the ions with the vesicle bilayer being confined to the outer lipid monolayer, (2) the initial stages of the ion-induced phase transition being confined to the site of contact between the interacting membranes, and (3) the interacting membranes being part of a closed vesicular, and therefore osmotically active, structure.

In the CL system, inverted micellar intermediates can only be formed between interacting bilayers, and their formation is likely to be facilitated by the asymmetrical interaction of Ca^{2+} with the outer surface of the vesicles only. Formation of IMI is a relatively facile process with a very low enthalpy and the IMI is still in rapid equilibrium with the two bilayers from which it is formed. However, if an IMI would happen to "break" into an ILA, the system is effectively trapped in a condition from which it can neither revert to the original bilayers nor readily proceed to H_{II} phase formation. When an ILA is formed between two interacting vesicles, the system is trapped in a fused vesicular configuration. Abundant formation of ILA between multiple vesicles may lock the system temporarily in a metastable inverted cubic configuration, assembled by accumulated ILA (81-88).

3. A Possible Role for a Hydrostatic, or Osmotic, Driving Force

Like the formation of IMI in the CL system, the formation of a high-affinity *trans* complex in the Ca^{2+}/PS system confines the instability focus to the site of contact between the interacting vesicles. How might this system be temporarily trapped in a configuration of larger fused vesicles? Unlike the IMI in the CL system, which, as an inherently unstable structure, has a tendency to "spontaneously" evolve into an ILA (79-85), the *trans* Ca^{2+}/PS complex, being perfectly compatible with a lamellar organization, will, in itself, have virtually no tendency to "break," such that a fused membrane would result; on the contrary, the structure, in itself, is extremely stable. Perhaps the asymmetrical interaction of the Ca^{2+}, resulting in a condensation of the area per lipid molecule in the outer monolayer of the vesicles and, thus, a packing mismatch with the unaffected fluid inner monolayer (Fig. 6b), is sufficient to drive the merging of the outer monolayers of the vesicles and the formation of a more or less extended single-bilayer diaphragm (Fig. 6c). However, the packing mismatch is very unlikely to induce the breakage of the single bilayer in the diaphragm. Therefore, it seems, in the Ca^{2+}/PS or related systems, an additional force is required to drive the breakage of the apposed membranes at the site of vesicle contact (92). We believe this additional force specifically orginates in a lateral stretching of the vesicle bilayers, as would occur if an osmotic gradient were applied. Undoubtedly, in the Ca^{2+}/PS system this stretching is induced by the flattening deformation of the vesicles owing to their tendency to enlarge the area of mutual contact. This flattening deformation is a direct result of the very strong adhesion between the vesicles in the presence of Ca^{2+} (35-37,47,92; see also Chapter 3). Since with the flattening deformation the volume to surface ratio of the vesicles will decrease, the encapsulated solute concentration will increase, which as a consequence will create an osmotic hydrostatic pressure, opposing the deformation. This pressure will translate into a lateral tension on the vesicle bilayers. Therefore, the lateral stretching as a result of the flattening deformation is, in fact, comparable to the stretching that would be induced by an osmotic gradient applied externally. This lateral tension is likely to drive the breakage of the diaphragm, resulting in coalescence of the internal vesicle compartments and in a transient formation of a larger vesicular structure. However, obviously, the system is far from being stable at that point. The bound Ca^{2+} will remain bound (Fig. 6d), and as the tension is relieved after breakage of the diaphragm, the remaining area of interbilayer contact will tend to increase again, possibly involving other vesicles and resulting in a renewed development of hydrostatic pressure, etc.

Summarizing, it appears as though the transformation of IMI into fused ILA does not require a lateral stretching of the vesicle bilayer, although, significantly, it may well be facilitated by it (84). On the other hand, a contact region involving a *trans* Ca^{2+}/PS complex is very unlikely to break spontaneously; here an additional lateral stretching of the bilayers appears to be required. This lateral tension is induced by the osmotic stress caused by the flattening deformation of the vesicles. Fundamentally, the critical difference between the "spontaneous" evolution of IMI into ILA, on the one hand, and the Ca^{2+}-induced fusion of PS LUV requiring an additional driving force, on the other, may well be related to a difference in the intrinsic radius of curvature of the lipid at equilibrium (83). The difference can also be understood if one considers the hypothetical interaction of two pieces of bilayer under conditions where these do *not* constitute part of a vesicle, i.e., in the absence of any hydrostatic pressure due to vesicle deformation. There is no reason to assume that IMI and ILA formation would not occur under such conditions. However, under the same conditions two PS bilayers would obviously not fuse, but rather convert rapidly to a rigid dehydrated bilamellar structure, In this respect it is interesting that fusion of lipid vesicles with planar bilayers (see Chapter 8), a system in which one of the fusion partners is not part of a closed vesicular structure, does require an externally applied osmotic gradient (96-98). Also of interest is the observation made by Bondeson and Sundler (99) on hemifusion of LUV, composed of PE and phosphatidic acid (PA), induced by pentalysine. In this system merging of the outer bilayer halves of the vesicles was observed, but neither mixing nor release of aqueous vesicle contents occurred. It may be speculated that, in this system, due to insufficient flattening deformation of the vesicles and in the absence of an external osmotic pressure gradient, breakage of the diaphragm did not occur. Finally, it is noteworthy that Ohki (93) has observed a stimulatory effect of an applied osmotic gradient on phospholipid vesicle fusion.

4.   *Energetics of Biological Membrane Fusion*

It is perhaps instructive at this point to make a comparison with the energetics of fusion occurring in biological systems. Obviously, contrary to fusion in simple lipid vesicle systems, in which fusion can at best be observed as a hysteretic step in an overall transformation of the system to a nonvesicular configuration, biological membrane fusion cannot possibly be driven by such a gross transformation of the fusion partners. In all likelihood, the susceptibility to fusion of biological membranes will be limited to a very short period of time and to highly specific areas on the interacting membranes. Possibly, in these fusion-prone areas a relative concentration of fusion-susceptible lipids, for example, the ones studied in simple vesicle systems, is induced transiently. In principle, such a relative concentration of fusion-susceptible lipids to the site of intermembrane contact could evoke a "spontaneous" (i.e., *not* driven by hydrostatic pressure) evolution of an IMI-like structure into a fused ILA. This would require the fusogen, whether it be an ion or a protein, to induce an intrinsic curvature of the membrane lipids at the site of intermembrane contact compatible with such a spontaneous transformation. With respect to protein-induced fusion, one could think in this regard of, for example, a wedgelike penetration of a hydrophobic part of the protein molecule into the lipid bilayer structure. Several hydrophobic peptides have been shown to possess the capacity to induce the formation of H_{II} phase structures in an otherwise lamellar lipid system (100). On the other hand, it is quite likely that as in the Ca^{2+}/PS system, in many cases of biological fusion a local perturbation of lipid packing at the site of intermembrane contact will not spontaneously result in fusion. In these cases, an osmotic driving force may be required for taking the fusion event to completion (101,102). As a flattening deformation of fusing vesicles, which in the Ca^{2+}/PS system generates the hydrostatic pressure, is unlikely to occur in biological systems, such an osmotic gradient will have to be generated independently, perhaps through the activation of ion pumps in the membrane of the fusing vesicles. In fact, this may constitute part of the control of the fusion process. In this respect, it is interesting that in several reconstituted systems, intracellular membrane fusion reactions have been shown to require an external input of energy, usually in the form of ATP (3,4).

Finally, apart from the above considerations and irrespective of whether the actual fusion event is dependent on an osmotic force or not, clearly an osmotic force is required for the formation of one larger *spherical* vesicle as a product of fusion of two smaller ones.

VI. TOWARD BIOLOGICAL MEMBRANES

Numerous attempts have been made to modify liposomal model systems such that they would more closely mimic the characteristics of certain particular biological membrane fusion processes. These attempts have included (1) providing lipid vesicles with recognition markers to achieve specificity in the initial interaction between the vesicles, (2) lowering the required Ca^+ concentration to the micromolar levels that are biologically relevant, (3) changing the lipid composition of the vesicles to one that more closely resembles the lipid composition of biological membranes, and (4) investigation of the fusogenic potency of peptides and proteins. In addition, lipid vesicles have been used extensively as target membranes for fusion-competent biological membranes, such as, for example, enveloped viruses or secretory vesicles. These latter studies, however, will be reviewed in various chapters in later parts of this volume and, therefore, will not be discussed here.

A. Conferring Specificity to Vesicle–Vesicle Adhesion

We have already emphasized that the nonspecific interactions involved in the aggregation of simple phospholipid vesicles, such as Van der Waals attraction and electrostatic repulsion, differ fundamentally from the highly specific recognition mechanisms that must be operating in biological systems. In a number of studies, therefore, lectin/glycolipid or sometimes lectin/glycoprotein pairs have been employed in order to confer specificity to the initial adhesion between lipid vesicles. As this area of research will be reviewed in Chapter 10, we will not discuss it here in detail. Pertinent to our present discussion, however, is the observation that lectin-induced agglutination of glycolipid-containing vesicles, consisting otherwise of negatively charged phospholipids mainly, will often cause a significant reduction of the threshold Ca^{2+} concentration required for initiating fusion of the vesicles (103,104). This important result indicates that lectin-induced adhesion of glycolipid-containing vesicles may trigger the formation of interbilayer Ca^{2+}/phospholipid complexes under conditions of Ca^{2+} concentrations in the medium which, because of electrostatic repulsion between the vesicles, would not normally allow such complex formation.

Lectin-induced agglutination of glycolipid-containing vesicles does not in all cases lower the threshold Ca^{2+} concentration required for fusion. Inhibition of fusion due to steric constraints, related to the relative bulkiness of the carbohydrate moiety of the particular glycolipid employed, has also been observed (104). Likewise, the inhibition of lipid vesicle fusion by the integral membrane protein glycophorin, as discussed in detail in Chapter 10, has been attributed to steric constraints (105).

B. Threshold Ca^{2+} Concentration

The fact that normally Ca^{2+} concentrations in the millimolar range are required to induce fusion in simple phospholipid vesicle systems has been the basis for extensive criticism, questioning the validity of these systems as a proper model for biological membrane fusion phenomena. In this respect, it is important to note that, in these simple systems, Ca^{2+} plays a dual role in the induction of the overall fusion process. First, by straightforward charge neutralization it allows Van der Waals attractions to induce vesicle aggregation; charge neutralization normally requires concentrations of the ion in the millimolar range, as dictated by the binding affinity of the ion for the lipid in dispersed vesicles. Second, Ca^{2+} triggers the actual fusion reaction by inducing the formation of interbilayer phospholipid structures. From the studies on lectin-induced vesicle agglutination, discussed above, it appears that, if the first, nonphysiological, function of Ca^{2+} is eliminated, the second, perhaps more physiological, role of the ion may not necessarily require concentrations in the millimolar range. It has been clearly demonstrated that the binding affinity of Ca^{2+} for PS in the *trans* Ca^{2+}/PS complex is higher by several orders of magnitude than the binding of the ion in a single PS bilayer (64-66). Recently, Feigenson, in a very careful study, has provided evidence to indicate that the *trans* complex can be formed at Ca^{2+} concentrations in the low-micromolar or sometimes even submicromolar range (65). Such concentrations are highly relevant physiologically, as similar levels of Ca^{2+} are required for intracellular fusion reactions or a process like exocytosis. Interestingly, the author also showed that the formation of the *trans* complex exhibits supersaturation with respect to the Ca^{2+} concentration prior to the onset of high-affinity binding, suggesting that, also in the multibilayer model system he investigated, relatively high concentrations of Ca^{2+} may be required to bring the bilayers into a close enough apposition to allow nucleation of the *trans* complex. Similarly, in Figure 7b it is shown that PS LUV in the presence of 0.5 mM Ca^{2+}, a condition that does not result in rapid vesicle aggregation, exhibit an only slightly higher gel to liquid-crystalline phase transition temperature than PS LUV in the absence of Ca^{2+}, indicating the absence of *trans* Ca^{2+}/PS complex formation due to the inability of the vesicles to interact with one another. This and other observations then raise the intriguing possibility that, if two membranes can be brought into close apposition by some independent mechanism, micromolar Ca^{2+} concentrations may suffice to induce a local destabilization at the site of contact through direct interaction with membrane lipids.

A possibly physiologically relevant mechanism for induction of close membrane apposition is given by the capacity of the protein synexin to induce aggregation of membrane vesicles (see Chapter 9). Synexin is a soluble Ca^{2+}-binding protein which has been isolated originally from adrenal medulla (see Chapter 23). It promotes aggregation of chromaffin granules or other membrane vesicles, presumably as a result of its bipolar nature, which allows it to act as a cross-linker. Synexin has been shown to reduce the Ca^{2+} threshold for phospholipid vesicle fusion to values as low as 10 µM (106,107). Rather than facilitating the fusion reaction per se, synexin exerts its stimulatory effect on the fusion process exclusively at the level of the initial aggregation of the vesicles (108), apparently inducing a membrane apposition close enough to allow interbilayer Ca^{2+}/phospholipid complexes to be formed. Interestingly, the stimulatory effect of synexin on lipid vesicle fusion is highly dependent on the lipid composition of the vesicles (106,107). This suggests that synexin, in addition to promoting membrane-membrane adhesion as such, may also contribute to the specificity of this interaction.

Like synexin, polyamines such as spermine have also been found to promote fusion of acidic phospholipid vesicles through stimulation of vesicle aggregation (109,110). Again, in these studies the actual fusion reaction remained strictly Ca^{2+}-dependent, but much lower Ca^{2+} concentrations sufficed than the threshold levels needed in the absence of the polyamines. Similar behavior has been observed for the polycation polylysine, which promotes aggregation of CL-containing vesicles, thereby substantially reducing the Ca^{2+} concentration required for fusion of the vesicles (111).

C. Effects of the Lipid Composition of the Vesicles

Animal cell membranes do contain acidic phospholipids, but these usually represent a minority among zwitterionic phospholipids, such as PC, PE, and sphingomyelin. In addition, often cholesterol is present in considerable amounts.

1. Mixtures of Negatively Charged and Zwitterionic Phospholipids

In model systems, in general Ca^{2+}-induced fusion of vesicles composed of mixtures of negatively charged and zwitterionic phospholipids turns out to be strongly inhibited by the choline phospholipids, PC or sphingomyelin, whereas in mixtures with PE fusion is usually sustained or even stimulated (44,58,59,62). This relative "fusogenic" character of PE is likely due to the low degree of hydration of its polar headgroup, allowing close membrane apposition (45,47,48,52,54,55), as well as to the tendency of PE to adopt nonlamellar configurations (11,40-43,71,72,84,85). Conversely, PC and shingomyelin appear to inhibit fusion because of the high degree of hydration of the choline headgroup (45-48,51-55) and the strong tendency of these lipids to remain in a lamellar organization.

2. Lateral Lipid Phase Separations

An important question regarding both fusion in mixed phospholipid vesicle systems and biological membrane fusion is the occurrence of lateral phase separations between the various lipid constituents. There is no question that Ca^{2+} has the capacity to induce lateral phase separations between negatively charged and zwitterionic phospholipids in mixed systems (112, 113). Recently, these phase separations have been visualized in single giant unilamellar vesicles by fluorescence microscopy (114,115). Important with respect to fusion in lipid vesicle systems, however, is the question of whether lateral phase separations are induced rapidly before vesicle aggregation, i.e., at the level of dispersed vesicles after interaction of Ca^{2+} with the outer surface of the vesicles only, and whether such phase separations are required for fusion of the vesicles to occur. Clearly, for the case of PS/PC vesicles a Ca^{2+}-induced phase separation on dispersed vesiclés does *not* occur on the time scale of vesicle aggregation and fusion, since, for example, PS/PC (molar ratio, 1:1) LUV do not fuse at all in the presence of Ca^{2+}, even though the vesicles aggregate massively (58, 59,61). In addition, Hoekstra (116) has shown for PS/PC SUV, which do fuse in the presence of Ca^{2+} because of the strong bilayer curvature, that the kinetics of fusion are much faster than the kinetics of lipid phase separation, indicating that the phase separation in this case follows rather than precedes the fusion reaction. For PS/PE vesicles, it has been demonstrated that fusion occurs even under conditions where a lateral phase separation between the two lipid constituents in the mixture is not induced at equilibrium, indicating that a phase separation does not occur in dispersed vesicles either and, therefore, is not required for fusion of the vesicles (58-60). On the other hand, Leventis et al. (117) have observed a dependence of Ca^{2+}-induced PA/PC vesicle fusion on the induction of a lateral phase separation at the stage prior to vesicle aggregation. In summary, it appears that in many cases of divalent-cation-induced aggregation and fusion of negatively charged phospholipid vesicles, a lateral lipid phase separation preceding the aggregation and fusion of the vesicles does not occur, and, therefore, it is not an absolute prerequisite for fusion. If a prior phase separation does occur, however, it may well stimulate the fusion process.

In fact, from a biological perspective the latter possibility, i.e., the induction of lateral lipid phase separations *prior to* induction of fusion, seems highly relevant. A lateral segregation of fusion-susceptible lipids, induced independently of the actual fusion stimulus, would add a powerful dimension to the control of the process. There are several ways in which a heterogeneous lipid distribution may be maintained or transiently induced in biological membranes. As an example, the asymmetrical distribution of different phospholipid species across the two halves of biological membranes (118), or a possible transient perturbation of this asymmetry, is likely to play an essential role in the control and selectivity of many cellular fusion reactions. In addition, lateral, perhaps transient, redistributions of lipids, induced by membrane proteins, could be involved in the modulation of fusion processes.

In an attempt to investigate the effect of *preexisting* lateral lipid phase separations on fusion in a lipid vesicle system, we have studied the Ca^{2+}-induced fusion of LUV, composed of an equimolar mixture of CL and distearoylphosphatidylcholine (DSPC). Within this lipid mixture a thermotropic phase separation between fluid CL-enriched domains and gel-phase DSPC-enriched domains occurs at temperatures below approximately 45°C. Figure 10 shows that fusion of these vesicles in the presence of Ca^{2+} is strongly stimulated under these conditions. The stimulatory effect of the phase separation even results, in this particular system, in a fusion capacity of the phase-separated vesicles that is considerably higher than the fusion seen with pure CL vesicles. Therefore, this observation indicates that the fusion capacity of lipid vesicles in the presence of Ca^{2+} may be drastically stimulated by a preexisting lipid phase separation. Similar mechanisms might be operating in biological membrane fusion processes. The recent observation of Feigenson (66) that, under conditions of lateral phase separation, micromolar Ca^{2+} concentrations induce the formation of the high-affinity intermembrane $Ca(PS)_2$ complex in PS/PC systems containing only 20% PS is also suggestive of a possible role of lipid phase separations in biological membrane fusion.

3. Effects of Cholesterol and Other Nonphospholipid Membrane Constituents

In a number of phospholipid vesicle systems, the inclusion of cholesterol in the lipid bilayer has been shown to stimulate fusion (62,119,120). This effect can be attributed to a decrease of the interbilayer hydration repulsion in the presence of cholesterol (121). Under certain conditions cholesterol may also facilitate the formation of nonbilayer structures and, thus, promote vesicle fusion (120). The effects of cholesterol on membrane fusion in these vesicle systems are likely to be fundamentally different from the role of

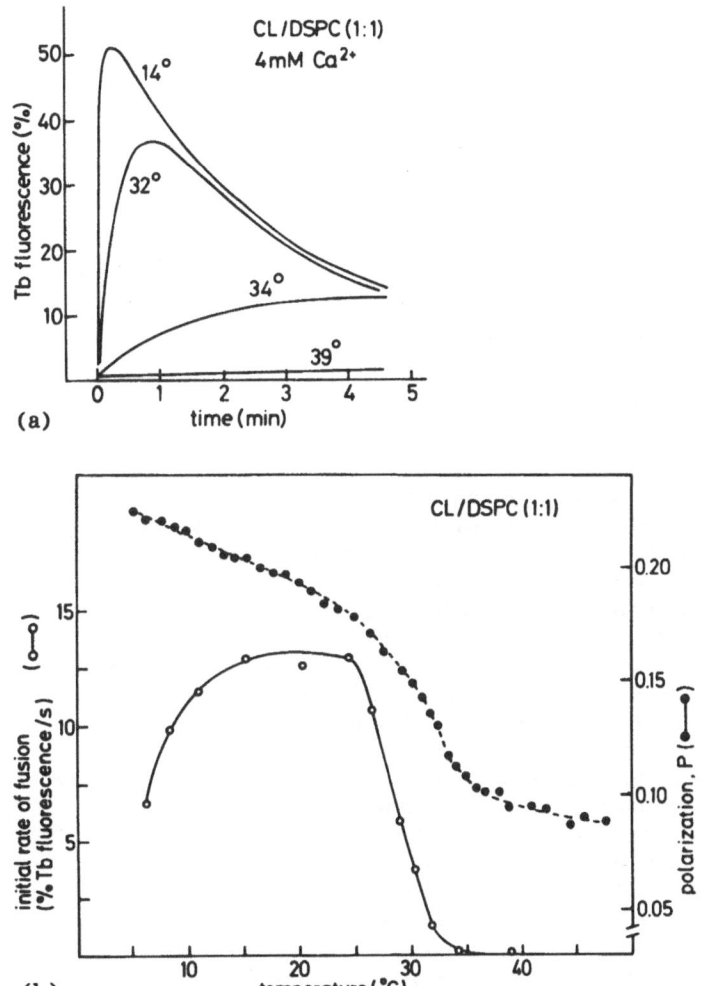

FIGURE 10 Effect of a thermotropic phase separation in CL/DSPC vesicles on the fusion of the vesicles in the presence of Ca^{2+}. (a) Fusion of CL/DSPC (molar ratio, 1:1) LUV, measured with the Tb/DPA assay as described in Figure 3, in the presence of 4 mM Ca^{2+} as a function of temperature. (b) Initial rate of fusion, measured as in panel A, as a function of temperature (open circles) and the fluidity of the vesicle bilayer in the absence of Ca^{2+} measured by fluorescence polarization (P, closed circles) of diphenyl hexatriene (DPH), as in Figure 7. Clearly, with decreasing temperature the fusion rate increases in the temperature range in which a lateral phase separation in the vesicles occurs (J. Wilschut, unpublished observations).

cholesterol in certain viral fusion reactions. For example, the absolute requirement of Semliki Forest virus fusion for the presence of cholesterol in its target membrane is due to a specific interaction between the viral fusion protein and the sterol (122).

As a minor nonphospholipid molecule in biological membranes, diacylglycerol has recently been shown to have dramatic effects on phospholipid vesicle fusion (123). Diacylglycerol is produced in the cellular plasma membrane as a second messenger during the phosphatidylinositol cycle after external stimulation of the cell (124). One of the known functions of diacylglycerol is its activation of protein kinase C. Besides this effect, it may be speculated that diacylglycerol is directly involved in the stimulus-response coupling during exocytosis by enhancing the local susceptibility of the cellular plasma membrane or secretory granule membranes to fusion. It has been demonstrated that diacylglycerol lowers the L_α to H_{II} phase transition temperature in phospholipid systems (123). It, thus, has the capacity to mediate the fusion of PE-containing vesicles (123) through the IMI/ILA mechanism, discussed above. Importantly, diacylglycerol exerts this stimulatory effect at a level of about 2 mol% relative to the membrane phospholipid, which is within the physiologically relevant range.

Finally, we mention briefly that several nonphospholipid constituents have been utilized in PE-containing vesicles in order to render the vesicles susceptible to destabilization under certain conditions. The aim in these cases was, rather than to mimic a biological fusion event, to design a vesicle system that could be used as a vehicle for the introduction of foreign molecules into cells or as a diagnostic tool. In all these cases the basic concept is to stabilize PE in a lamellar configuration by addition of a bilayer-stabilizing membrane constituent, a stabilization that would be abolished in response to a specific change in the medium. For example, free fatty acids in their unprotonated form stabilize PE in a lamellar configuration such that vesicles can be made. These vesicles exhibit pH sensitivity, since protonation of the fatty acid at a mildly acidic pH results in destabilization of the vesicles (125), and can be used to deliver encapsulated macromolecules to the cytoplasm of cells via the endocytic pathway (126-128). Likewise, membrane-associated acylated immunoglobulins can stabilize PE-containing vesicles. In this case, cross-linking of the immunoglobulin molecules by specific antigens may cause vesicle destabilization and release of contents (129). For a more extensive discussion of these and similar vesicle systems, and their applications in cell biology and biotechnology, we refer to Chapter 31 of this volume. Free fatty acids are likely to be involved at very low concentrations in certain biological membrane fusion events. For example, arachidonic acid has been shown to induce the fusion of chromaffin granules in the presence of synexin (130). Recently, it has been demonstrated that low levels of free fatty acid also stimulate the fusion of phospholipid vesicles (131).

D. Fusion Induced by Peptides and Proteins

It is well established that many biological membrane fusion processes are mediated by proteins (for a recent review, see Ref. 132). Among the best-characterized protein-mediated fusion reactions are those that are involved in the infectious entry of enveloped viruses into cells, as discussed extensively later in this volume. But also for intracellular fusion events, the involvement of specific protein components becomes increasingly apparent (3-5). For this reason many proteins and peptides, including putative fusion sequences of viral fusion proteins (133,134), have been investigated for their potential capacity to induce fusion in phospholipid vesicle systems. As this topic is discussed extensively in Chapter 9 of this volume, we will not present specific examples here. However, we do wish to discuss one aspect of peptide-induced fusion in lipid vesicle systems, which relates to our above considerations on the driving force for membrane fusion.

In most studies of peptide-induced fusion, SUV have been employed and often positive results have been obtained. This is not surprising in view of the intrinsic tendency of these vesicles to fuse. However, positive fusion results in LUV systems are rare, which, in fact, is not surprising either, as a basic requirement for fusion is not easily met under the experimental conditions. Mere destabilization of vesicle bilayers by penetration of hydrophobic peptides, not restricted to the site of vesicle-vesicle contact, is not sufficient to trigger fusion of LUV, as a driving force for the fusion reaction will be lacking in most cases. In the Ca^{2+}/PS system this driving force is provided by the flattening deformation of the vesicles, due to strong vesicle-vesicle adhesion. Such flattening deformation is not likely to occur in peptide/vesicle systems. Furthermore, if flattening were to occur, the *global* perturbation of the lipid bilayer by penetration of a hydrophobic peptide might well abolish any hydrostatic pressure response to the deformation and, thus, the necessary lateral bilayer tension. Therefore, in studies of peptide-induced vesicle fusion, which are no doubt highly relevant for the mechanism of biological membrane fusion, intricate designs must be developed, in which the destabilization of the vesicle bilayer is confined to the site of vesicle contact. In addition, with the possible exception of fusion proceeding via an IMI/ILA-like mechanism, an externally applied osmotic gradient might turn out to be essential.

VII. CONCLUDING REMARKS

Investigation of lipid vesicle systems has contributed a great deal to a better understanding of the molecular events involved in membrane fusion. It is obviously because of their relative simplicity that lipid vesicles can be studied in considerable molecular detail. On the other hand, it ought to be recognized that fusion of lipid vesicles, because of this very same simplicity of the system, does not necessarily constitute a close and proper reflection of biological membrane fusion. Therefore, conclusions derived from the investigation of model systems cannot always be directly extrapolated to biological membrane fusion. First, the nonspecific interactions that govern aggregation of phospholipid vesicles differ fundamentally from the specific interaction and recognition mechanisms operating between biological membranes. Also, with respect to the actual fusion reaction, the characteristics of fusion in a model system may deviate from those a seemingly comparable counterpart in biology. For example, fusion of influenza virus has been studied extensively in systems involving lipid vesicles as fusion targets of the virus (see Chapters 14 and 15). From these studies it has become apparent that the virus fuses efficiently with CL vesicles. However, the characteristics of this fusion process, particularly its rate and pH dependence, differ from those of the biological fusion activity of the virus (135,136). Thus, while fusion of enveloped viruses with CL vesicles may reveal important *general* requirements for membrane fusion, it is essential to choose a composition for target liposomes that more closely resembles the composition of the biological target membrane in order to be able to draw specific conclusions with respect to the viral fusion activity involved in host cell entry.

    Despite the above limitations of liposomal model systems, the critical step in the fusion of lipid vesicles remains the merging of two closely apposed lipid bilayers, as occurs during fusion in cell biology. Therefore, model systems do reveal the molecular requirements for membrane fusion, be it in general terms. However, it can only be through an increased refinement of lipid vesicle systems, necessarily involving the inclusion of components that are known to play a key role in biological membrane fusion, that one may expect to more closely mimic specific cellular fusion processes. It will be a major challenge for the coming years to reconstitute in a lipid vesicle system the fusion machinery involved in intracellular fusion of shuttle vesicles, as the constituents of this fusion machinery are being identified and characterized (3).

ACKNOWLEDGMENTS

I thank all those who have contributed to our work on lipid vesicle fusion, discussed in this chapter, particularly Michal Bental, Joe Bentz, Nejat Düzgüneş, Dick Hoekstra, Shlomo Nir, Antonio Ortiz, Demetrios Papahad-jopoulos, Janny Scholma, and Toon Stegmann. I also thank Harma Ellens and Dave Siegel for many helpful discussions and Gerrit Scherphof for his support and his interest in our fusion work. I acknowledge the financial support from the Netherlands Organization for Scientific Research (NWO), NATO (Research Grant 151.81), EMBO (short-term fellowships to J. W. and S. Nir; long-term fellowship to A. Ortiz), and the National Institutes of Health (Research Grant no. AI 25534 to N. Duzgunes and J. W.).

REFERENCES

1. Burgoyne, R. D. (1984). Mechanisms of secretion from adrenal chromaffin cells. *Biochim. Biophys. Acta* 779:201-216.
2. Baker, P. F., and Knight, D. E. (1984). Calcium control of exocytosis in bovine adrenal medullary cells. *Trends Neurosci.* 7:120-126.
3. Wilson, D. W., Wilcox, C. A., Flynn, G. C., Chen, E., Kuang, W-J., Henzel, W. J., Block, M. R., Ullrich, A., and Rothman, J. E. (1989). A fusion protein required for vesicle-mediated transport in both mammalian cells and yeast. *Nature* 339:355-359.
4. Beckers, C. J. M., and Balch, W. E. (1989). Calcium and GTP: Essential components in vesicular trafficking between the endoplasmic reticulum and Golgi apparatus. *J. Cell Biol.* 108:1245-1256.
5. Bourne, H. R. (1988). Do GTPases direct membrane traffic in secretion? *Cell* 53:669-671.
6. Nir, S., Bentz, J., Wilschut, J., and Düzgüneş, N. (1983). Aggregation and fusion of phospholipid vesicles. *Prog. Surf. Sci.* 13:1-124.
7. Düzgüneş, N. (1985). Membrane fusion. In: *Subcellular Biochemistry*, Vol. 11 (D. B. Roodyn, ed.). Plenum Press, New York, pp. 195-286.
8. Wilschut, J., and Hoekstra, D. (1984). Membrane fusion: From liposomes to biological membranes. *Trends Biochem. Sci.* 9:479-483.
9. Wilschut, J., and Hoekstra, D. (1986). Membrane fusion: Lipid vesicles as a model system. *Chem. Phys. Lipids* 40:145-166.

10. Düzgüneş, N., Hong, K., Baldwin, P. A., Bentz, J., Nir, S., and Papahadjopoulos, D. (1987). Fusion of phospholipid vesicles induced by divalent cations and protons: Modulation by phase transitions, free fatty acids, monovalent cations and polyamines. In: *Cell Fusion* (A. E. Sowers, ed.). Plenum Press, New York, pp. 241-267.

11. Bentz, J., and Ellens, H. (1988). Membrane fusion: Kinetics and mechanisms. *Colloids Surfaces* 30:65-112.

12. Papahadjopoulos, D., Meers, P. R., Hong, K., Ernst, J. D., Goldstein, I. M., and Düzgüneş, N. (1988). Calcium-induced membrane fusion: From liposomes to cellular membranes. In: *Molecular Mechanisms of Membrane Fusion* (S. Ohki et al., eds.). Plenum Press, New York, pp. 1-16.

13. Szoka, F. C. (1987). Lipid vesicles: Model systems to study membrane-membrane destabilization and fusion. In: *Cell Fusion* (A. E. Sowers, ed.). Plenum Press, New York, pp. 209-240.

14. Wilschut, J., Scholma, J., and Stegmann, T. (1988). Molecular mechanisms of membrane fusion and applications of membrane fusion techniques. In: *Technological Applications of Lipid Microstructures*, Advances in Experimental Medicine, Vol. 238 (B. P. Gaber et al., eds.). Plenum Press, New York, pp. 105-126.

15. Wilschut, J. (1988). Membrane interactions and fusion. In: *Energetics of the Secretion Response* (J. W. N. Akkerman, ed.). CRC Press, Boca Raton, FL, pp. 63-80.

16. Hoekstra, D., and Wilschut, J. (1989). Membrane fusion of artificial and biological membranes. In: *Water Transport in Biological Membranes*, Vol. I (G. Benga, ed.). CRC Press, Boca Raton, FL, pp. 143-176.

17. Düzgüneş, N., and Bentz, J. (1988). Fluorescence assays for membrane fusion. In: *Spectroscopic Membrane Probes* (L. M. Loew, ed.). CRC Press, Boca Raton, FL, pp. 117-159.

18. Düzgüneş, N., Allen, T. M., Fedor, J., and Papahadjopoulos, D. (1988). Lipid mixing during membrane aggregation and fusion: Why fusion assays disagree. *Biochemistry* 26:8435-8442.

19. Silvius, J. R., Leventis, R., and Brown, P. M. (1988). Slow artifacts in assays of lipid mixing between membranes. In: *Molecular mechanisms of Membrane Fusion* (S. Ohki et al., eds.). Plenum Press, New York, pp. 531-542.

20. Wilschut, J., and Papahadjopoulos, D. (1979). Ca^{2+}-induced fusion of phospholipid vesicles monitored by mixing of aqueous contents. *Nature* 281:690-692.

21. Wilschut, J., Düzgüneş, N., Fraley, R., and Papahadjopoulos, D. (1980). Studies on the mechanism of membrane fusion: Kinetics of calcium ion-induced fusion of phosphatidylserine vesicles followed by a new assay for mixing of aqueous vesicle contents. *Biochemistry* 19: 6011-6021.

22. Wilschut, J., Düzgüneş, D., and Papahadjopoulos, D. (1981). Calcium/magnesium specificity in membrane fusion: Kinetics of aggregation and fusion of phosphatidylserine vesicles and the role of bilayer curvature. *Biochemistry* 20:3126-3133.

23. Wilschut, J., Düzgüneş, N., Hong, K., Hoekstra, D., and Papahadjopoulos, D. (1983). Retention of aqueous contents during divalent cation-induced fusion of phospholipid vesicles. *Biochim. Biophys. Acta* 734:309-318.

24. Bentz, J., Düzgüneş, D., and Nir, S. (1983). Kinetics of divalent cation-induced fusion of phosphatidylserine vesicles: Correlation between fusogenic capacities and binding affinities. *Biochemistry* 22:3320-3330.

25. Ellens, H., Bentz, J., and Szoka, F. C. (1985). H^+- and Ca^{2+}-induced fusion and destabilization of phosphatidylethanolamine liposomes. *Biochemistry* 24:3099–3106.

26. Struck, D. K., Hoekstra, D., and Pagano, R. E. (1981). Use of resonance energy transfer to monitor membrane fusion. *Biochemistry* 20:4093–4099.

27. Driessen, A. J. M., Hoekstra, D., Scherphof, G., Kalicharan, R. D., and Wilschut, J. (1985a). Low pH-induced fusion of liposomes with membrane vesicles derived from *Bacillus subtilis*. *J. Biol. Chem.* 260: 10880–10887.

28. Bentz, J., Ellens, H., and Szoka, F. C. (1987). Destabilization of phosphatidylethanolamine-containing liposomes: Hexagonal phase and asymmetric membranes. *Biochemistry* 26:2105–2116.

29. Hong, K., Baldwin, P., Allen, T. M., and Papahadjopoulos, D. (1988). Fluorometric detection of the bilayer-to-hexagonal phase transition in liposomes. *Biochemistry* 27:3947–3955.

30. Silvius, J. R., Leventis, R., Brown, P. M., and Zuckermann, M. (1987). Novel fluorescent phospholipids for assays of lipid mixing between membranes. *Biochemistry* 26:4279–4287.

31. Papahadjopoulos, D., Vail, W. J., Jacobson, K., and Poste, G. (1975). Cochleate lipid cylinders: Formation by fusion of unilamellar vesicles. *Biochim. Biophys. Acta* 394:483–491.

32. Papahadjopoulos, D., Vail, W. J., Newton, C., Nir, S., Jacobson, J., Poste, G., and Lazo, R. (1977). Studies on membrane fusion. III: The role of Ca^{2+}-induced phase changes. *Biochim. Biophys. Acta* 465: 579–598.

33. Ginsberg, L. (1978). Does calcium cause fusion or lysis of unilamellar vesicles? *Nature* 275:758–760.

34. Kendall, D. A., and McDonald, R. C. (1982). A fluorescence assay to monitor vesicle fusion and lysis. *J. Biol. Chem.* 257:13892–13895.

35. Rand, R. P., and Parsegian, V. A. (1984). Physical force considerations in model and biological membranes. *Can. J. Biochem. Cell Biol.* 62:752–759.

36. Rand, R. P., Kachar, B., and Reese, T. S. (1985). Dynamic morphology of interacting phosphatidylserine vesicles. *Biophys. J.* 47:483–489.

37. Kachar, B., Fuller, N., and Rand, R. P. (1986). Morphological responses to calcium-induced interaction of phosphatidylserine vesicles. *Biophys. J.* 50:779–788.

38. Wilschut, J. Scholma, J. Bental, M., Hoekstra, D., and Nir, S. (1985). Ca^{2+}-induced fusion of phosphatidylserine vesicles: Mass action kinetic analysis of membrane lipid mixing and aqueous contents mixing. *Biochim. Biophys. Acta* 821:45–55.

39. Ortiz, A., Killian, J. A., Verkleij, A. J., and Wilschut, J. (1990). Fusion of cardiolipin vesicles induced by divalent cations: Relationship between lipid polymorphic behavior and fusion kinetics. *Biochemistry*, in press.

40. Ellens, H., Bentz, J., and Szoka, F. C. (1986). Destabilization of phosphatidylethanolamine liposomes at the hexagonal phase transition temperature. *Biochemistry* 25:285–294.

41. Ellens, H., Bentz, J., and Szoka, F. C. (1986). Fusion of phosphatidylethanolamine liposomes and the mechanism of the L_α-H_{II} phase transition. *Biochemistry* 25:4141–4147.

42. Ellens, H., Siegel, D. P., Alford, D., Yeagle, P. L., Boni, L., Lis, L. J., Quinn, P. J., and Bentz, J. (1989). Membrane fusion and inverted phases. *Biochemistry* 28:3692–3703.

43. Gagné, J., Stamatatos, L., Diacovo, T., Hui, S. W., Yeagle, P. L., and Silvius, J. R. (1985). Physical properties and surface interactions of bilayer membranes containing N-methylated phosphatidylethanolamines. *Biochemistry* 24:4400-4408.

44. Brown, P. M., and Silvius, J. R. (1989). Stability and fusion of lipid vesicles containing headgroup-modified analogues of phosphatidylethanolamine. *Biochim. Biophys. Acta* 980:181-190.

45. Marra, J., and Israelachvili, J. (1985). Direct measurements of forces between phosphatidylcholine and phosphatidylethanolamine in aqueous electrolyte solution. *Biochemistry* 24:4608-4618.

46. Horn, R. G. (1984). Direct measurement of the force between two bilayers and observation of their fusion. *Biochim. Biophys. Acta* 778:224-228.

47. Rand, R. P. (1981). Interacting phospholipid bilayers: Measured forces and induced structural changes. *Annu. Rev. Biophys. Bioeng.* 10:277-314.

48. McIntosh, T. J., and Simon, S. A. (1986). Hydration force and bilayer deformation: A reevaluation. *Biochemistry* 25:4058-4066.

49. Nir, S., and Bentz, J. (1978). On the forces between phospholipid bilayers. *J. Colloid Interface Sci.* 65:399-414.

50. Nir, S., Bentz, J., and Düzgünes, N. (1981). Two modes of reversible aggregation: Particle size and the DLVO theory. *J. Colloid Interface Sci.* 84:266-269.

51. LeNeveu, D. M., Rand, R. P., and Parsegian, V. A. (1976). Measurements of forces between lecithin bilayers. *Nature* 259:601-603.

52. McIntosh, T. J., and Simon, S. A. (1986). Area per molecule and distribution of water in fully hydrated dilauroylphosphatidylethanolamine bilayers. *Biochemistry* 25:4948-4952.

53. Hauser, H., Pascher, I., Pearson, R. H., and Sundell, S. (1981). Preferred conformation and molecular packing of phosphatidylethanolamine and phosphatidylcholine. *Biochim. Biophys. Acta* 650:21-51.

54. Rand, R. P., Fuller, N., Parsegian, V. A., and Rau, D. C. (1988). Variation in hydration forces between neutral phospholipid bilayers: Evidence for hydration attraction. *Biochemistry* 27:7711-7722.

55. McIntosh, T. J., Magid, A. D., and Simon, S. A. (1987). Steric repulsion between phosphatidylcholine bilayers. *Biochemistry* 26:7325-7332.

56. Verwey, E. J., and Overbeek, J. Th. G. (1948). *Theory of the Stability of Lyophobic Colloids*. Elsevier, Amsterdam.

57. Portis, A., Newton, C., Pangborn, W., and Papahadjopoulos, D. (1979). Studies on the mechanism of membrane fusion: Evidence for an intermembrane Ca^{2+}-phospholipid complex, synergism with Mg^{2+} and inhibition by spectrin. *Biochemistry* 18:780-790.

58. Düzgünes, N., Wilschut, J., Fraley, R., and Papahadjopoulos, D. (1981). Studies on the mechanism of membrane fusion: Role of headgroup composition in calcium- and magnesium-induced fusion of mixed phospholipid vesicles. *Biochim. Biophys. Acta* 642:182-195.

59. Düzgünes, N., Paiement, J., Freeman, K. B., Lopez, N. G., Wilschut, J., and Papahadjopoulos, D. (1984). Modulation of membrane fusion by ionotropic and thermotropic phase transitions. *Biochemistry* 23:3486-3494.

60. Silvius, J. R., and Gagné, J. (1984). Lipid phase behavior and calcium-induced fusion of phosphatidylethanolamine-phosphatidylserine vesicles. Calorimetric and fusion studies. *Biochemistry* 23:3232-3240.

61. Silvius, J. R., and Gagné, J. (1984). Calcium-induced fusion and lateral phase separations in phosphatidylcholine-phosphatidylserine vesicles. Correlation by calorimetric and fusion measurements. *Biochemistry* 23:3241-3247.

62. Stamatatos, L., and Silvius, J. R. (1987). Effects of cholesterol on the divalent cation-mediated interactions of vesicles containing amino and choline phospholipids. *Biochim. Biophys. Acta* 905:81-90.

63. Bentz, J., and Düzgüneş, N. (1985). Fusogenic capacities of divalent cations and the effect of liposome size. *Biochemistry* 24:5436-5443.

64. Ekerdt, R., and Papahadjopoulos, D. (1982). Intermembrane contact affects calcium binding to phospholipid vesicles. *Proc. Natl. Acad. Sci. USA* 79:2273-2277.

65. Feigenson, G. W. (1986). On the nature of calcium ion binding between phosphatidylserine lamellae. *Biochemistry* 25:5819-5825.

66. Feigenson, G. W. (1989). Calcium ion binding between lipid bilayers: The four-component system of phosphatidylserine, phosphatidylcholine, calcium chloride and water. *Biochemistry* 28:1270-1278.

67. McDonald, R. I. (1985). Membrane fusion due to dehydration by polyethylene glycol, dextran or sucrose. *Biochemistry* 24:4058-4066.

68. Hui, S. W., Stewart, T. P., Boni, L. T., and Yeagle, P. L. (1981). Membrane fusion through point defects in bilayers. *Science* 212:921-923.

69. Mattai, J., Hauser, H., Demel, R. A., and Shipley, G. G. (1989). Interactions of metal ions with phosphatidylserine bilayer membranes: Effect of hydrocarbon chain unsaturation. *Biochemistry* 28:2322-2330.

70. Wilschut, J., Düzgüneş, N., Hoekstra, D., and Papahadjopoulos, D. (1985). Modulation of membrane fusion by membrane fluidity: Temperature dependence of divalent cation-induced fusion of phosphatidylserine vesicles. *Biochemistry* 24:8-14.

71. Gruner, S., Cullis, P. R., Hope, M. J., and Tilcock, C. P. S. (1985). Lipid polymorphism: The molecular basis of nonbilayer phases. *Annu. Rev. Biophys. Biophys. Chem.* 14:211-238.

72. Cullis, P. R., Hope, M. J., and Tilcock, C. P. S. (1986). Lipid polymorphism and the roles of lipids in membranes. *Chem. Phys. Lipids* 40:127-144.

73. Cullis, P. R., and De Kruijff, B. (1978). The polymorphic phase behavior of phosphatidylethanolamines of natural and synthetic origin. A ^{31}P-NMR study. *Biochim. Biophys. Acta* 513:31-42.

74. De Kruijff, B., Verkleij, A. J., Leunissen-Bijvelt, J., Van Echteld, C. J. A., Hille, J., and Rijnbout, H. (1982). Further aspects of the Ca^{2+}-dependent polymorphism of bovine heart cardiolipin. *Biochim. Biophys. Acta* 693:1-12.

75. Verkleij, A. J., Mombers, C., Gerritsen, W. J., Leunissen-Bijvelt, J., and Cullis, P. R. (1979). Fusion of phospholipid vesicles in association with the appearance of lipidic particles as visualized by freeze-fracturing. *Biochim. Biophys. Acta* 555:358-361.

76. Verkleij, A. J. (1984). Lipidic intramembranous particles. *Biochim. Biophys. Acta* 779:43-64.

77. Bearer, E. L., Düzgüneş, N., Friend, D. S., and Papahadjopoulos, D. (1982). Fusion of phospholipid vesicles arrested by quick-freezing. The question of lipidic particles as intermediates in membrane fusion. *Biochim. Biophys. Acta* 693:93-98.

78. Siegel, D. P. (1984). Inverted micellar structures in bilayer membranes: Formation rates and half-lifes. *Biophys. J.* 45:399-420.

79. Siegel, D. P. (1986). Inverted micellar intermediates and the transitions between lamellar, cubic and inverted hexagonal phases. I. Mechanism of the L_α-H_{II} phase transition. *Biophys. J.* 49:1155-1170.

80. Siegel, D. P. (1986). Inverted micellar intermediates and the transitions between lamellar, cubic and inverted hexagonal phases. II. Implications for membrane-membrane interactions and membrane fusion. *Biophys. J.* 49:1171-1183.

81. Siegel, D. P. (1987). Inverted micellar intermediates and the transitions between lamellar, cubic and inverted hexagonal phases. III. Isotropic and inverted cubic phase formation via intermediates in the transitions between L_α and H_{II} phases. *Chem. Phys. Lipids* 42: 279-301.

82. Siegel, D. P. (1987). Membrane-membrane interactions via intermediates in lamellar-to-inverted hexagonal phase transitions. In: *Cell Fusion* (A. E. Sowers, ed.). Plenum Press, New York, pp. 181-207.

83. Gruner, S. M. (1985). Intrinsic curvature hypothesis for biomembrane lipid composition: A role for non-bilayer lipids. *Proc. Natl. Acad. Sci. USA* 82:3665-3669.

84. Siegel, D. P., Ellens, H., and Bentz, J. (1988). Membrane fusion via intermediates in L_α/H_{II} phase transitions. In: *Molecular Mechanisms of Membrane Fusion* (S. Ohki, et al., eds.). Plenum Press, New York, pp. 53-71.

85. Siegel, D., Burns, J. L., Chestnut, M. H., and Talmon, Y. (1989). Intermediates in membrane fusion and bilayer/nonbilayer phase transitions imaged by time-resolved cryo-transmission electron microscopy. *Biophys. J.* 56:161-169.

86. Gruner, S. M., Tate, M. W., Kirk, G. L., So, P. T. C., Turner, D. C., Teane, D. T., Tilcock, C. P. S., and Cullis, P. R. (1988). X-ray diffraction study of the polymorphic behavior of N-monomethylated dioleoylphosphatidylethanolamine. *Biochemistry* 27:2853-2866.

87. Shyamsunder, E., Gruner, S. M., Tate, M. W., Turner, D. C., So, P. T. C., and Tilcock, C. P. S. (1988). Observation of inverted cubic phase in hydrated dioleoylphosphatidylethanolamine membranes. *Biochemistry* 27:2332-2336.

88. Lindblom, G., and Rilfors, L. (1989). Cubic phases and isotropic structures formed by membrane phospholipids. Possible biological relevance. *Biochim. Biophys. Acta* 988:221-256.

89. Vasilenko, I., De Kruijff, B., and Verkleij, A. J. (1982). Polymorphic phase behaviour of cardiolipin from bovine heart and from *Bacillus subtilis* as detected by P-NMR and freeze-fracture techniques. Effects of Ca^{2+}, Mg^{2+}, Ba^{2+} and temperature. *Biochim. Biophys. Acta* 684:282-286.

90. Wilschut, J., Holsappel, M., and Jansen, R. (1982). Ca^{2+}-induced fusion of cardiolipin/phosphatidylcholine vesicles monitored by mixing of aqueous vesicle contents. *Biochim. Biophys. Acta* 690:297-301.

91. Wilschut, J., Nir, S., Scholma, J., and Hoekstra, D. (1985). Kinetics of Ca^{2+}-induced fusion of cardiolipin-phosphatidylcholine vesicles: Correlation between vesicle aggregation, bilayer destabilization and fusion. *Biochemistry* 24:4630-4636.

92. McDonald, R. C. (1988). Mechanisms of membrane fusion in acidic lipid-cation systems. In: *Molecular Mechanisms of Membrane Fusion* (S. Ohki et al., eds.). Plenum Press, New York, pp. 101-112.

93. Ohki, S. (1984). Effects of divalent cations, temperature, osmotic pressure gradient and vesicle curvature on phosphatidylserine vesicle fusion. *J. Membrane Biol.* 77:265-275.

94. Ohki, S. (1988). Surface tension, hydration energy and membrane fusion. In: *Molecular Mechanisms of Membrane Fusion* (S. Ohki et al., eds.). Plenum Press, New York, pp. 123-138.

95. Nir, S., Wilschut, J., and Bentz, J. (1982). The rate of fusion of phospholipid vesicles and the role of bilayer curvature. *Biochim. Biophys. Acta* 688:275-278.

96. Cohen, F. S., Akabas, M. H., and Finkelstein, A. (1982). Osmotic swelling of phospholipid vesicles causes them to fuse with a planar phospholipid bilayer membrane. *Science* 217:458-460.

97. Cohen, F. S., Niles, W. D., and Akabas, M. H. (1989). Fusion of phospholipid vesicles with a planar membrane depends on the membrane permeability of the solute used to create the osmotic pressure. *J. Gen. Physiol.* 93:201-210.

98. Niles, W. D., Cohen, F. S., and Finkelstein, A. (1989). Hydrostatic pressures developed by osmotically swelling vesicles bound to planar membranes. *J. Gen. Physiol.* 93:211-244.

99. Bondeson, J., and Sundler, R. (1985). Lysine peptides induce lipid intermixing but not fusion between phosphatidic acid-containing vesicles. *FEBS Lett.* 190:283-287.

100. Killian, J. A., and De Kruijff, B. (1986). The influence of proteins and peptides on the phase properties of lipids. *Chem. Phys. Lipids* 40:259-284.

101. Lucy, J. A., and Ahkong, Q. F. (1986). An osmotic model for the fusion of biological membranes. *FEBS Lett.* 199:1-11.

102. Lucy, J. A., and Ahkong, Q. F. (1988). Osmotic forces and the fusion of biomembranes. In: *Molecular Mechanisms of Membrane Fusion* (S. Ohki et al., eds.). Plenum Press, New York, pp. 163-179.

103. Sundler, R., and Wijkander, J. (1983). Protein-mediated intermembrane contact specifically enhances Ca^{2+}-induced fusion of phosphatidate-containing membranes. *Biochim. Biophys. Acta* 730:391-394.

104. Hoekstra, D., and Düzgüneş, N. (1986). Ricinus communis agglutinin-mediated agglutination and fusion of glycolipid-containing phospholipid vesicles: Effect of carbohydrate head group size, calcium ions and spermine. *Biochemistry* 25:1321-1330.

105. De Kroon, A. I. P. M., Van Hoogevest, P., Geurts van Kessel, W. S. M., and De Kruijff, B. (1985). Influence of glycophorin incorporation on Ca^{2+}-induced fusion of phosphatidylserine vesicles. *Biochemistry* 24:6382-6389.

106. Hong, K., Düzgünes, N., and Papahadjopoulos, D. (1982). Modulation of membrane fusion by calcium-binding proteins. *Biophys. J.* 37:297-305.

107. Hong, K., Düzgüneş, N., Ekerdt, R., and Papahadjopoulos, D. (1982). Synexin facilitates fusion of specific phospholipid vesicles at divalent cation concentrations found intracellularly. *Proc. Natl. Acad. Sci. USA* 79:4942-4944.

108. Meers, P., Bentz, J., Alford, D., Nir, S., Papahadjopoulos, D., and Hong, K. (1988). Synexin enhances the aggregation rate but not the fusion rate of liposomes. *Biochemistry* 26:4430-4439.

109. Schuber, F., Hong, K., Düzgüneş, N., and Papahadjopoulos, D. (1983). Polyamines as modulators of membrane fusion: Aggregation and fusion of liposomes. *Biochemistry* 22:6134-6140.

110. Meers, P., Hong, K., Bentz, J., and Papahadjopoulos, D. (1986). Spermine as a modulator of membrane fusion: Interaction with acidic phospholipids. *Biochemistry* 25:3109-3118.

111. Gad, A. E., Bental, M., Elyashiv, G., Weinberg, H., and Nir, S. (1985). Promotion and inhibition of vesicle fusion by polylysine. *Biochemistry* 24:6277-6282.

112. Ohnishi, S-I., and Ito, T. (1974). Calcium-induced phase separations in phosphatidylserine-phosphatidylcholine membranes. *Biochemistry* 13:881-887.

113. Tokutomi, S., Lew, R., and Ohnishi, S-I. (1981). Ca^{2+}-induced phase separation in phosphatidylserine, phosphatidylethanolamine and phosphatidylcholine mixed membranes. *Biochim. Biophys. Acta* 643:276-282.

114. Haverstick, D. M., and Glaser, M. (1987). Visualization of Ca^{2+}-induced phospholipid domains. *Proc. Natl. Acad. Sci. USA* 84:4475-4479.

115. Haverstick, D. M., and Glaser, M. (1989). Visualization of domain formation in the inner and outer leaflets of a phospholipid bilayer. *J. Cell Biol.* 106:1885-1892.

116. Hoekstra, D. (1982). Role of lipid phase separation and membrane hydration in phospholipid vesicle fusion. *Biochemistry* 21:2833-2840.

117. Leventis, R., Gagne, J., Fuller, N., Rand, R. P., and Silvius, J. R. (1986). Divalent cation induced fusion and lipid lateral segregation in phosphatidylcholine-phosphatidic acid vesicles. *Biochemistry* 25:6978-6987.

118. Op den Kamp, J. A. F. (1979). Lipid asymmetry in membranes. *Annu. Rev. Biochem.* 48:47-71.

119. Bental, M., Wilschut, J., Scholma, J., and Nir, S. (1987). Ca^{2+}-induced fusion of large unilamellar phosphatidylserine/cholesterol vesicles. *Biochim. Biophys. Acta* 898:239-247.

120. Verkleij, A. J., Van Echteld, C. J. A., Gerritsen, W. J., Cullis, P. R., and De Kruijff, B. (1980). The lipidic particle as an intermediate structure in membrane fusion processes and bilayer to hexagonal H_{II} transitions. *Biochim. Biophys. Acta* 600:620-624.

121. McIntosh, T. J., Magid, A. D., and Simon, S. A. (1989). Cholesterol modifies the short-range repulsive interactions between phosphatidylcholine membranes. *Biochemistry* 28:17-25.

122. Kielian, M., and Helenius, A. (1984). Role of cholesterol in fusion of Semlike Forest virus with membranes. *J. Virol.* 52:281-283.

123. Siegel, D. P., Banschbach, J., Alford, D., Ellens, H., Lis, L. J., Quinn, P. J., Yeagle, P. L., and Bentz, J. (1989). Physiological levels of diacylglycerol in phospholipid membranes induce membrane fusion and stabilize inverted phases. *Biochemistry* 28:3703-3709.

124. Berridge, M. J. (1987). Inositol triphosphate and diacylglycerol: Two interacting second messengers. *Annu. Rev. Biochem.* 56:159-193.

125. Düzgünes, N., Straubinger, R. M., Baldwin, P. A., Friend, D. S., and Papahadjopoulos, D. (1985). Proton-induced fusion of oleic acid-phosphatidylethanolamine liposomes. *Biochemistry* 24:3091-3098.

126. Straubinger, R. M., Düzgünes, N., and Papahadjopoulos, D. (1985). pH-sensitive liposomes mediate cytoplasmic delivery of encapsulated macromolecules. *FEBS Lett.* 179:148-154.

127. Collins, D., and Huang, L. (1987). Delivery of Diphteria toxin A fragment to the cytoplasm of toxin-resistant cells by pH-sensitive immunoliposomes. *Cancer Res.* 47:735-739.

128. Wang, C-Y., and Huang, L. (1987). pH-sensitive immunoliposomes mediate target-cell-specific delivery and controlled expression of a foreign gene in mouse. *Proc. Natl. Acad. Sci. USA* 84:7851-7855.

129. Ho, R. J. Y., Rouse, B. T., and Huang, L. (1986). Target-sensitive immunoliposomes: Preparation and characterization. *Biochemistry* 25: 5500-5506.

130. Creutz, C. E. (1981). *Cis*-unsaturated fatty acids induce the fusion of chromaffin granules aggregated by synexin. *J. Cell Biol.* 91:247-256.

131. Meers, P., Hong, K., and Papahadjopoulos, D. (1988). Free fatty acid enhancement of cation-induced fusion of liposomes: Synergism with synexin and other promoters of vesicle aggregation. *Biochemistry* 27:6784-6794.

132. Stegmann, T., Doms, R. W., and Helenius, A. (1989). Protein-mediated membrane fusion. *Annu. Rev. Biophys. Biophys. Chem.* 18:187-211.

133. Lear, J. D., and DeGrado, W. F. (1987). Membrane binding and conformational properties of peptides representing the NH_2 terminus of influenza HA-2. *J. Biol. Chem.* 262:6500-6505.

134. Wharton, S. A., Martin, S. R., Ruigrok, R. W. H., Skehel, J. J., and Wiley, D. C. (1988). Membrane fusion by peptide analogues of influenza virus hemagglutinin. *J. Gen. Virol.* 69:1847-1857.

135. Stegmann, T., Hoekstra, D., Scherphof, G., and Wilschut, J. (1986). Fusion activity of influenza virus: A comparison between artificial and biological target membranes. *J. Biol. Chem.* 261:10966-10969.

136. Stegmann, T., Nir, S., and Wilschut, J. (1989). Membrane fusion activity of influenza virus: Effects of gangliosides and negatively charged phospholipids in target liposomes. *Biochemistry* 28:1698-1704.

5

Modeling of Aggregation and Fusion of Phospholipid Vesicles

SHLOMO NIR

The Hebrew University of Jerusalem, Rehovot, Israel

I. INTRODUCTION

A. Outline

This chapter describes a model used in analysis of the kinetics of phospholipid vesicle fusion. The approach that is followed has already been extended to studies on vesicle-virus or virus-cell fusion. Analysis and examples elucidate the effects of vesicle aggregation, fusion, and deaggregation processes on the overall kinetics of vesicle fusion. The final part of the chapter is a review of the effects of several fusogens, such as cations, in particular Ca^{2+}, and the effect of vesicle size on vesicle fusion.

B. General Features and Stages in the Development of the Model

The process of membrane fusion has been considered (1,2) to consist of the following three steps: (1) a close approach or point contact between two membranes; in a system of vesicles this is the aggregation step; (2) membrane destabilization; the two apposed membranes, which are ordered structures, must undergo a transient disturbance so that a new structure can be formed later; (3) membrane merging. The mathematical procedure based on a mass action model has made it possible to distinguish between two steps in the overall fusion reaction, the close approach or aggregation, which involves a collision between particles and is of second order with respect to the particle concentrations, and the subsequent first-order fusion reaction, which involves membrane destabilization and merging (3-7).

Studies on vesicle aggregation (8-10) have shown that it can be a reversible process and emphasized the effect of particle concentration,

This work was done during a sabbatical leave at the Laboratory of Cell Biology and Genetics, National Institutes of Health, and Uniformed Services University of the Health Sciences, Bethesda, Maryland.

in addition to the effect of forces between particles. These studies have culminated in formulation of a model of dynamical aggregation and produced numerical procedures based on mass action for the simulation of aggregation processes, resulting also in closed-form expressions for equilibrium distributions (11,12). In certain cases the model (11) gave results similar to those of the analytical solution of Smoluchowski (13), but the approach of dynamical aggregation was essential to explain phenomena such as the reversible aggregation of PS vesicles in the presence of NaCl and the decrease in the overall rate of their aggregation with temperature.

While this chapter deals with the modeling of vesicle-vesicle fusion, the description given stems from a new method of representation, which was originally designed for studies on liposome-virus fusion (14). A pictorial scheme of the formation of aggregation-fusion products is shown in Figure 1. The molar concentration of an aggregate of I particles of one type and J particles of another type is denoted by A(I,J). Examples of systems investigated that consist of two types of particles are liposome-virus fusion systems (14) and systems consisting of two types of liposomes, as employed in fluorescence fusion assays. Details of these fusion assays are given in Chapter 4. One of the most commonly used assays for mixing of aqueous vesicle contents is the Tb/DPA assay, involving the formation of a fluorescent complex upon fusion of Tb- and DPA-encapsulating vesicles (15,16).

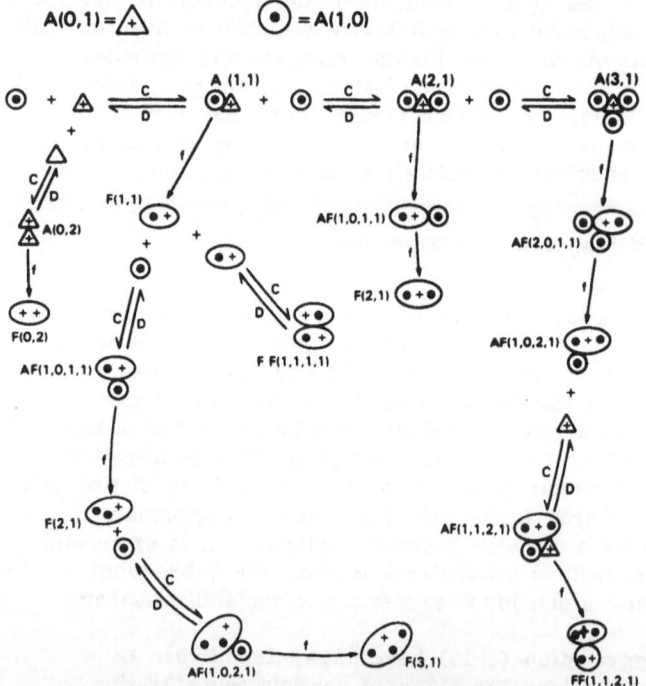

FIGURE 1   Aggregation and fusion products in a binary system that can consist of, e.g., Tb- and DPA-encapsulating vesicles, membrane-labeled and unlabeled vesicles, or liposomes and virus particles. In the latter case, structures consisting solely of one of the two types of particle, e.g., A(2,0) or F(2,0), do not usually exist; furthermore, the fusion products normally consist of a single virus particle but may contain more than one liposome.

Mixing of bilayer lipids can be conveniently monitored as a decrease in resonance energy transfer (RET) between fluorescent derivatives of phospholipids (17,18) or a decrease in self-quenching of a fluorescent probe (19). The latter assays are based on dilution of the fluorophores from a labeled vesicle into an unlabeled one during fusion. Usually, the initial state in these systems is described by a certain concentration of L = A(1,0) and V = A(0,1), where L and V represent liposomes and virus particles, Tb- and DPA-vesicles, or membrane-labeled and unlabeled vesicles, respectively. A fusion product consisting of I particles of type L and J particles of type V is denoted by F(I,J). The terms $AF(I_1, J_1, I_2, J_2)$ denote the concentration of products consisting of I_1 and J_1 unfused particles of types L and V and I_2 and J_2 fused particles of these same types, respectively. The term $FF(I_1, J_1, I_2, J_2)$ denotes an aggregate of $F(I_1, J_1)$ and $F(I_2, J_2)$, respectively. The structures A(2,0), A(0,2), F(2,0), F(0,2), etc., in Figure 1, which represent aggregation-fusion products of particles of the same type, do exist in vesicle-vesicle interactions, but were not found in the particular liposome-virus systems studied.

Figure 1 outlines a few possible pathways and products of the aggregation-fusion reaction. If we limit the number of particles in an aggregation-fusion product to $N = 8$, or up to four particles of each type, the number of nonlinear differential equations describing the reaction is 390, if a Taylor expansion up to third derivatives is employed. In actual calculations the current programs are set for $N = 8$.

II. APPROXIMATE EXPRESSIONS IN LIMITING CASES: AN ILLUSTRATION FOR THE CASE $N = 2$

The essential features of the model are elucidated by focusing on the case $N = 2$. In this case the constructs AF and FF (see Fig. 1) are avoided. The aggregation products are dimers, which are formed with a forward rate constant, $C(M^{-1} \cdot s^{-1})$. The dimers can fuse or can dissociate with the respective rate constants f and $D(s^{-1})$. The analysis of such a system has been presented in Refs. 3 and 7. The equations presented here are simpler than in these articles, because we consider here vesicles of type L interacting with vesicles of type V, where the aggregation-fusion products consist of just one L and one V particle. The schematic chain of reactions is given by

$$L + V \underset{D}{\overset{C}{\rightleftharpoons}} A(1,1) \xrightarrow{f} F(1,1) \tag{1}$$

The kinetic equations are

$$\frac{dL}{dt} = -C \cdot L \cdot V + D \cdot A \tag{2}$$

$$\frac{dV}{dt} = -C \cdot L \cdot V + D \cdot A \tag{3}$$

$$\frac{dA}{dt} = C \cdot L \cdot V - f \cdot A - D \cdot A \tag{4}$$

$$\frac{dF}{dt} = F \cdot A \tag{5}$$

In Eq. (4) the term $C \cdot L \cdot V$ is of second order in the particle concentration, and gives the rate of generation of aggregates A, whereas the first-order terms, $-fA$ and $-DA$, give the rate of annihilation of A due to fusion and dissociation. For the case where the rate of dissociation can be ignored ($D = 0$, or $D \ll f$), a simple solution has been found (14) whose form depends on the initial molar concentrations of particles L_0 and V_0. When $L_0 = V_0$, the solution for L is

$$L(t) = \frac{L_0}{1 + CL_0 t} \tag{6}$$

When $L_0 \neq V_0$ the solution is

$$L(t) = \frac{QL_0/V_0}{L_0/V_0 - \exp(-CQt)} \tag{7}$$

where $Q = L_0 - V_0$.

The solution for the fused doublet F can be given in an implicit form:

$$F(t) = f \exp(-ft) \int [L_0 - L(s)] \exp(fs) ds \tag{8}$$

where L is given by Eq. (6) or by Eq. (7). Now we will examine the results in certain limiting cases.

A. Fusion Is the Rate-Limiting Step

The analysis of this case will relate the fusion rate constant f to the duration of the actual fusion reaction. Let us consider the situation where particle aggregation has been achieved, but the onset of fusion requires a certain signal, such as a sudden injection of Ca^{2+}. Such cases have been reported in Refs. 20 and 21, employing polylysine or spermine, respectively, to preaggregate the vesicles, followed by addition of Ca^{2+}. Another example is the aggregation of small phosphatidylserine (PS) vesicles (4), or PS/cholesterol vesicles (22), the rate of which is enhanced by large concentrations (>500 mM) of NaCl, while this condition at the same time results in a reduction of the fusion rate per se. Such cases can be simulated by the following relations. Initially, say in a 1:1 mixture of L and V particles, $F = L = V = O$ and $A = A_0$. Furthermore, for simplicity we will ignore dissociation of aggregates. At the onset of fusion, Eq. (4) and Eq. (5) become

$$\frac{dA}{dt} = -fA \tag{9}$$

$$\frac{dF}{dt} = fA \tag{10}$$

and their solution is

$$F(t) = A_0 [1 - \exp(-ft)] \tag{11}$$

The fraction fused is $F(t)/A_0 = 1 - \exp(-ft)$ (12)

At short times, $ft \ll 1$,

$$\frac{F(t)}{A_0} = ft \tag{13}$$

Here the fraction fused is *independent* of particle concentration and initially it increases linearly with time. The time t required for the occurrence of a certain degree of fusion is given by

$$T = \frac{F(T)/A_0}{f} \tag{14}$$

Thus, if $f = 0.2 \text{ s}^{-1}$, it takes 500 ms until 10% fusion of preformed aggregates has been attained.

B. Aggregation Is Rate-Limiting

Here fusion is assumed to occur instantaneously following formation of aggregates. From an inspection of Eq. (4), it is clear that a sufficient condition for this limiting case is $f \gg CL_0$, if $L_0 \geqslant V_0$, or $f \gg CV_0$, otherwise. In this case the solution for F(t) is readily given by

$$F(t) = L_0 - L(t) \tag{15}$$

where L(t) is given by Eq. (6) or by Eq. (7). At short times, such that $CL_0 t \ll 1$ (or $CQt \ll 1$), the employment of Eq. (15) and Eq. (6) or Eq. (7) gives

$$F(t) = CL_0 V_0 t \tag{16}$$

The corresponding fractions of fused L or V particles are

$$\frac{L_0 - L(t)}{L_0} = CV_0 t \tag{17}$$

and

$$\frac{V_0 - V(t)}{V_0} = CL_0 t \tag{18}$$

Hence, initially, there is a linear increase in the fraction of particles fused with particle concentration. In the case of liposomes fusing with virus particles, the fraction of fused liposomes is independent of liposome concentration and it increases with virus concentration. The linearity with time can be observed if fusion follows aggregation with no delay, which means that although t is short, satisfying the condition $CL_0 t \ll 1$, yet $ft > 1$.

Table 1 illustrates this limiting case. The table presents calculated values for I, i.e., the fluorescence intensity observed in fusion assays, for two total lipid concentrations of 1 and 10 μM. The relationship between the quantity I and the percent fusion in a particular fusion assay will be given in Section III. The vesicles are assumed to be unilamellar and of radii of 50 nm. Since a lipid concentration of 1 μM corresponds to a vesicle concentration of 1.2×10^{-11} M, the lipid concentrations considered imply that $L_0 = V_0 = 6 \times 10^{-12}$ or 6×10^{-11} M, respectively. A large,

TABLE 1   Kinetics of Vesicle Aggregation and Fusion in a System Where Aggregation Is Rate-Limiting to the Overall Fusion Reaction: Effects of the Vesicle Concentration and the Fusion Rate Constant[a]

| | Lipid conc. (1 µM[b]) | | | Lipid conc. (10 µM[b]) | | |
| | I (%)[c] | | Monomer depletion for f = 5 s (%) | I (%)[c] | | Monomer depletion for f = 5 s (%) |
Time (s)	$f = 5 \ s^{-1}$	$f = 1 \ s^{-1}$		$f = 5 \ s^{-1}$	$f = 1 \ s^{-1}$	
3	0.33	0.25	0.7	3.3	2.4	6.7
6	0.69	0.6	1.4	6.6	5.7	13
9	1.05	0.95	2.1	9.7	8.9	19
15	1.7	1.7	3.5	15.5	14.8	28
30	3.5	3.4	6.8	27.4	26.8	46
60	6.8	6.7	13	43.1	42.8	66

[a]This example applies to the increase in fluorescence intensity in a fusing vesicle system consisting initially of a 1:1 mixture of Tb- and DPA-vesicles of 100-nm diameter. For details of the Tb/DPA assay, see Chapter 4.
[b]At a total lipid concentration of 1 µM the vesicle concentration is 1.2×10^{-11} M; therefore, the concentration of L_0 vesicles is 6×10^{-12} M.
[c]The I values were calculated employing the values $C = 2 \times 10^8 \ M^{-1} \cdot s^{-1}$, D = 0, and the f values indicated, while N was taken equal to 8. The relationship between the value of I and the percent fusion in the Tb/DPA assay is discussed in Section III.

but realistic value of f was used, $f = 5 \ s^{-1}$, and the value of C employed was $2 \times 10^8 \ M^{-1} \cdot s^{-1}$. Hence, for a lipid concentration of 10 µM, $C \cdot L_0 = 2 \times 10^8 \times 6 \times 10^{-11} = 1.2 \times 10^{-2} \ll f$. Now consider a 1-µM suspension and set t = 30 s. Eq. (17) yields for the percent of fused L particles a value of $1.2 \times 10^{-3} \times 30 \times 100 = 3.6$. Table 1 gives I = 3.5. For the 10-µM suspension I = 3.3 at t = 3 s. As indicated above, the linear increase of the percent fusion with time occurs if $CL_0 t \ll 1$, $CL_0 \ll f$, and $ft \gg 1$. This condition is satisfied in the case of the more dilute suspension, whereas it holds approximately for the 10-µM suspension. As long as the conditions $ft \gg 1$ and $f \gg CL_0$ hold (say $f/CL_0 \geqslant 100$), a change in the f value does not affect the I values. This point is illustrated by comparing I values generated by employing the values $f = 1$ and $5 \ s^{-1}$ (Table 1).

Table 1 illustrates the linear increase of I values with vesicle concentration, when I values are smaller than 10. Table 2 illustrates that this linearity is not preserved when the lipid concentration is further increased and a smaller value of f is employed. At later stages the simple relations presented here do not hold; eventually the overall rate of fusion decreases, because the number of particles in suspension is continuously decreasing.

Table 2 also illustrates that for the same total lipid concentration the kinetics of fluorescence increase are faster when going from a 1:1 to 1:9 ratio of L to V vesicles. Note that according to Eq. (17) the percent fusion of L particles is independent of their concentration, but it is proportional to the concentrations of V particles. Indeed, an increase of the concentration of V by 9/5 when going from a 1:1 to a 1:9 mixture yields a corresponding increase by a factor of 1.8 in the initial I values.

TABLE 2 Kinetics of Fluorescence Development During Fusion of Phospholipid Vesicles: Effects of the Total Vesicle Concentration and of the Ratio of L- to V-Type Vesicles, and the Effect of Reversible Vesicle Aggregation

Time (s)	Fluorescence intensity I (%)[a]								
	Lipid concentration (μM)								
	1(0.5,0.5)	10(5,5)		10(1,9)		100(50,50)		100(10,90)	
		D = 0	D = f	D = 0	D = f	D = 0	D = f	D = 0	D = f
3	0.04	0.4	0.4	0.8	0.8	3.9	3.4	6.8	5.6
6	0.15	1.5	1.4	2.6	2.5	11.8	9.5	19.8	16.1
9	0.3	2.8	2.7	4.9	4.8	20.7	15.9	33.2	26.3
15	0.6	5.8	5.5	10.1	9.6	36.9	27.8	54.7	43.4
30	1.5	13.3	12.6	22.2	21	59.9	48.3	79.1	68.8

[a]The table presents calculated I values (%), simulating the fluorescence intensity as observed in, e.g., the Tb/DPA assay. A concentration of 1 μM total lipid corresponds to a total vesicle concentration of 1.2×10^{-10} M. The numbers in parentheses are the micromolar concentrations of L- and V-type vesicles, respectively, corresponding to Tb- and DPA-vesicles in the Tb/DPA assay. A value of $N = 8$ was employed in the calculations. Two sets of rate constants were employed: $C = 10^8$ $M^{-1} \cdot s^{-1}$, $f = 0.2$ s^{-1}, $D = 0$ and $C = 2 \times 10^8$ $M^{-1} \cdot s^{-1}$, $f = 0.1$ s^{-1}, $D = 0.1$ s^{-1}. In the most dilute lipid concentration both sets gave identical results.

III. RELATION BETWEEN FLUORESCENCE INTENSITY AND FUSION

The model calculations provide the conversion between the degree of increase in fluorescence intensity and the extent of fusion. The examples below refer to vesicle fusion monitored by the Tb/DPA assay (15,16) or by membrane lipid mixing assays (17,18). Let us assume that leakage of material from fusing vesicles and entry of material into the vesicles in the application of the Tb/DPA assay has been accounted for (3-7,23). Similarly, it will be assumed that in the application of a membrane mixing assay, such as the RET assay (18), the dilution of probes occurs only due to fusion.

Let us first treat the case of a 1:1 population, where $L_O = V_O$; i.e., initially there are equal numbers of Tb- and DPA-vesicles, or labeled and unlabeled vesicles. When fused doublets are formed, the possible constructs are F(2,0), F(1,1), and F(0,2) with weights of 1/4, 1/2, and 1/4, respectively. The Tb/DPA complex is produced only within the products F(1,1), which comprise one-half of the total fused doublets. Hence, in the application of the Tb/DPA assay the fluorescence intensity is maximal, or I = 100%, when all the Tb has reacted with DPA; I = 50% in the hypothetical case where all vesicles have fused to doublets. In the application of the RET assay the fluorescence of the donor (N-NBD-PE) is maximal under conditions of infinite probe dilution. If in this case all L and V vesicles fuse to form a homogeneous mixture, then in a 1:1 population the ultimate value of I will be 50%, and in the above hypothetical situation where all vesicles fused to form doublets, the value of I would be 25%. For the sake of comparison between the results of the two assays, it is advantageous to multiply both of the above I values in the RET assay by a factor of two. The fused triplets are F(3,0), F(2,1), F(1,2), and F(0,3), with weights of 1/8, 3/8, 3/8, and 1/8, respectively. Hence, in the application of the Tb/DPA assay (where the DPA concentration in the vesicles is much larger than Tb concentration), a complete fusion to triplets will result in I = 75%, and, similarly, in quadruplets the I value will be 87.5%, etc. (5). When the RET assay is employed in the linear regime (see below), a complete fusion to triplets results in I = 66.7% (2/3), i.e., less than the corresponding value with the Tb/DPA assay (24). The value of 2/3 is obtained by taking into account the fraction of N-NBD-PE in each of the triplet fusion products and multiplying it with the factor of intensity increase due to dilution. The increase in intensity I due to dilution is given in the linear regime by

$$I = 100(1 - X) \qquad\qquad (19)$$

in which X is the surface density of the N-NBD-PE molecules relative to their initial concentration. Thus, initially I = 0%, whereas at an infinite dilution X = 0 and I = 100%.

In the case of fusion products consisting of n particles, the value of I for a population consisting of labeled-to-blank vesicles in ratios of L/V is

$$I = 2 \times 100[V/L + V)][(n - 1)/n] \qquad\qquad (20)$$

The factor of 2 is introduced in order to have I = 50% for a complete fusion to doublets in a 1:1 population. The model calculations (14) avoid combinatorial computations and provide directly the I values corresponding to the evolved distribution of aggregation-fusion products.

Comparison between the Tb/DPA and RET assays indicates that in the ideal case the kinetics of fluorescence increase are initially similar

TABLE 3   Kinetics of Fluorescence Development During Fusion of Phospholipid Vesicles: A Comparison Between Calculated I Values in the Application of the Tb/DPA and RET Assays

Time (s)	Fluorescence intensity I (%)[a]			
	Lipid conc. (10 μM)		Lipid conc. (100 μM)	
	Tb/DPA	RET	Tb/DPA	RET
3	0.4	0.4	3.4	3.3
6	1.4	1.4	9.5	9.2
15	5.5	5.4	27.8	26.2
30	12.6	12.2	48.3	44.1

[a]The table presents calculated I values (%); the 100% value corresponds to complete mixing of all of the vesicle contents in the Tb/DPA assay and to complete mixing of all of the bilayer lipids in the RET assay. At a total lipid concentration of 10 μM, the vesicle concentration is 1.2×10^{-10} M, and the concentration of L-type vesicles (Tb-vesicles in the Tb/DPA assay or membrane-labeled vesicles in the RET assay) is 6×10^{-11} M. The following values were employed in the calculations: $C = 2 \times 10^8$ $M^{-1} \cdot s^{-1}$, $f = 0.1$ s^{-1} and $D = 0.1$ s^{-1}.

in both cases, and later as higher-order fusion products become more abundant, the kinetics are slightly faster with the Tb/DPA assay. This comparison is illustrated in Table 3. Actual comparisons between the results of the application of both assays have been reported (24-26). In certain cases the overall rates were identical, whereas in other cases the overall rate was faster with the RET assay. A discussion of these studies is given in Section VII.A.

IV. EFFECT OF AGGREGATE DISSOCIATION

Dissociation of aggregates, or reversible aggregation, results in slower kinetics of the overall fusion reaction. It has been shown (3) that calculations employing the set (C, f, D, with D = 0) yield approximately the same I values generated with other sets (C_1, f_1, D_1) when

$$C_1 = C(1 + D_1/f_1) \qquad (21)$$

$$f_1 = f - D \qquad (22)$$

This point is illustrated in Table 2, where the calculations employed the parameter values $C = 10^8 \ M^{-1} \cdot {}^{-1}$, $f = 0.2s^{-1}$ and D = 0, in one case, and the values $C = 2 \times 10^8 \ M^{-1} \cdot {}^{-1}$, $D_1 = f_1 = 0.1s^{-1}$ in the other case. Initially, the differences between the calculated values are indeed small, and for the 1- and 10-μM suspensions, the calculations with the two sets yield practically identical I values. However, with the 100-μM suspensions the calculated I values differ by ΔI > 2%, which is more than the experimental errors. The resolution between the calculated values given by the two sets is larger with the 1:9 population. Therefore, it is desirable to employ several ratios of L and V vesicles in the analyses.

V. DISTRIBUTION OF AGGREGATION-FUSION PRODUCTS

An illustration is given in Table 4, which gives just a sample from the whole particle distribution at a certain instance, t = 6 s, in a 1:9 population of Tb- and DPA-vesicles. In the case of the more dilute (10 μM) vesicle suspension, most of the contribution to I = 2.6% arises from the term $100F(1,1)/L_0 = 2.5\%$. In comparison, I = 19.8%, and the corresponding term [F(1,1)] yields 11.7 in the case of the 100-μM suspension. In both cases F(1,1) is about 18 times larger than F(2,0), in accord with the expected relative weights of 81:18:1 of the terms F(0,2), F(1,1), and F(2,0) in a 1:9 population. The relative weight of the higher-order aggregation products is significantly smaller in the case of the more dilute suspension.

VI. DETERMINATION OF RATE CONSTANTS

A. Simulation and Prediction of Experimental Results

The analysis starts by finding the rate constants C, f, and D that yield the best simulation of the data; then additional predictions are tested. The calculations give the distribution of aggregation-fusion products, providing predictions that can be tested by measurements that reflect the increase in size of vesicles and/or the extent of aggregation. Such

TABLE 4 An Example of the Distribution of Aggregation-Fusion Products in an Initial 1:9 Population of Tb- and DPA- Vesicles at 5 = 6 s after Initiation of Aggregation and Fusion[a]

	Lipid concentration (μM)	
	100	10
I	19.8	2.6
A(1,0)	54.1	93.2
A(1,1)	14.2	3.4
A(2,0)	0.8	0.2
A(2,1)	0.3	0.01
A(3,1)	0.008	4×10^{-5}
A(4,4)	3.5×10^{-5}	0
F(1,1)	11.7	2.5
F(2,0)	0.65	0.14
F(3,0)	0.007	2×10^{-4}
F(4,4)	10^{-8}	0
AF(1,0,1,1)	0.3	0.009
AF(2,2,1,1)	0.002	10^{-7}
FF(1,1,1,1)	0.04	1.3×10^{-4}
FF(2,2,2,2)	10^{-7}	0

[a]The numbers opposite A(1,0), F(1,1), and so on represent the % of the number of those aggregation-fusion products relative to the initial number of Tb-vesicles. A(1,0) represents Tb-vesicles. The calculation employed the same rate constants as in several cases in Table 2, $C = 10^8$ $M^{-1} \cdot s^{-1}$, $f = 0.2$ s^{-1}, $D = 0$. The value 0 indicates that the value is less than 10^{-12}.

measurements may include light scattering (5,15), dynamic light scattering, or electron microscopy.

Ordinarily the rate constant C is first determined from the results with the most dilute suspensions, where aggregation is rate-limiting. Under those conditions it is possible to set $D = 0$ and variations in the value of f have little effect, provided that $f \gg CL_0$ (see Sections II, IV, and Table 2). Next, the value of f is determined from the results obtained from measurements on more concentrated vesicle suspensions, where aggregation is faster and the delay due to the fusion reaction itself becomes significant. In the initial stages of the simulation, Eqs. (6), (7), and (16-18) can be utilized, as well as the approximate formulas given in Ref. 3.

The programs can use many rate constants, describing all the different aggregation-fusion reactions as pictured in Figure 1. However, our point of view has been to reduce the number of parameters to the smallest number that can yield fair simulations and predictions and provide an adequate description of the processes. Our primary aim is to assess rate constants of the initial fusion reaction, where aggregated dimers and fused doublets are the dominant products.

As indicated above, variation of the ratio f/CL_0 with total vesicle or lipid concentration is an essential part of the procedure. However, occasionally the condition $f \gg CL_0$ may still hold even at a relatively high vesicle concentration, which means that only a lower bound can be given for the value of f. This problem may be circumvented by performing pre-aggregation experiments. Such an approach has been pursued in fusion studies employing small PS vesicles, where high concentrations of Na^+, or other monovalent cations that by themselves do not induce vesicle fusion, were employed to enhance the rate of vesicle aggregation in the presence of Ca^{2+} and reduce the rate of fusion (4,22,23). Similarly, in systems consisting of large PS vesicles, Mg^{2+}, which induces aggregation but no fusion of these vesicles (53), has been employed to preaggregate the vesicles prior to induction of fusion by Ca^{2+} (27).

Variation in the ratio of Tb- to DPA-vesicles, while keeping total vesicle concentration constant, is another means to extend the range of values of f/CL_0.

B. Correction for Leakage in the Tb/DPA Assay

In the application of the Tb/DPA assay the analysis also includes a preliminary step of correcting the observed Tb/DPA fluorescence for the dissociation of the complex, and the concomitant quenching of its fluorescence (15,16) owing to release of vesicle contents and entry of the medium into the interior of the vesicles. A detailed expression for the correction due to leakage is given in Ref. 5, but for simplicity a sufficient correction is given, for a 1:1 population of Tb- and DPA-vesicles, by Refs. 4 and 23:

$$I(t) = F(t) + 0.5 \ Q(t) \tag{23}$$

in which $F(t)$ is the measured fluorescence intensity expressed as percent of the maximal value, and $Q(t)$ is the percent quenching observed in a separate experiment in which the Tb/DPA complex is preencapsulated. The factor 0.5 accounts for the fact that the Tb/DPA complex is produced in 1/2 of the fused doublets, those of the type $F(1,1)$. At later stages, beyond $I = 20\text{-}25\%$, when triplets and higher-order products become more abundant, the correction term should be increased.

VII. DISCUSSION

Table 5 provides a summary of rate constants found in several studies on vesicle fusion. In all cases, the model gave adequate simulations and predictions of experimental results.

We illustrate just one example in Figure 2, which shows a simulation of the fusion of vesicles composed of cardiolipin (CL) and phosphatidylcholine (PC), induced by Ca^{2+} and monitored with the Tb/DPA assay. The drawn lines represent the calculated values and the data points the experimental results. The upper curve was obtained after preincubation of the vesicles for 15 s with polylysine, which by itself does not induce fusion, followed by addition of 8 mM Ca^{2+}; the lower curve represents the corresponding result without preincubation.

TABLE 5 Rate Constants of Vesicle–Vesicle Aggregation and Fusion

Composition, size, and references[a]	Fusion-inducing agents and conditions[a]	Rate[a] constant of aggregation C_{11} (= C/2)(M$^{-1} \cdot$ s^{-1})	Rate[a] constant of fusion f(s^{-1})	Rate constant of deaggregation D (s^{-1})	Comments
PS SUV (5)	1.25 mM Ca^{2+}	1.3×10^6	—[b]	—[b]	Aggregation is rate limiting; studies included measurement and calculations of intensity of light scattered at 90°
	1.5 mM Ca^{2+}	5.8×10^6	—	—	
PS SUV (3,6,7)	2.0 mM Ca^{2+}	4.5×10^7	5	—	Fusion per se is rate limiting
PS SUV (23)	2.5 mM Ca^{2+} (+ 300 mM Na$^+$)	10^7	0.01–0.1	—	Aggregation is rate limiting
	1.25 mM Ca^{2+}	$(1–1.5) \times 10^6$	>1	—	
	1.0 mM Ca^{2+} (+ 20 mM Na$^+$)	6×10^5	>5	—	
	2.5 mM Ca^{2+} + 300 mM K$^+$	$>10^6$	>1	—	
	1.25 mM Ca^{2+} +100 mM K$^+$	$(1–2) \times 10^6$	>1	—	
	1.0 mM Ca^{2+} + 20 mM K$^+$	6×10^5	>5	—	
	2.5 mM Ca^{2+} + 300 mM Li$^+$	$>10^6$	0.001–0.002	—	
	1.5 mM Ca^{2+} + 100 mM Li$^+$	$10^5 – 10^7$	0.001–0.5	—	
	1.0 mM Ca^{2+} + 20 mM Li$^+$	6×10^5	>5	—	

PS SUV (4)	3.5 mM Ba^{2+}	—	0.013	—	The medium contained 500 mM Na$^+$ and it was assumed that the actual fusion determined the overall rate [see Eq. (11) or Eq. (4) in Ref. 4]
	3.0 mM Ca^{2+}	—	0.006		
	6.0 mM Sr^{2+}	—	0.005		
	15 mM Mg^{2+}	—	0.006		
PS LUV (3,6,7)	5 mM Ca^{2+}	6.5 × 10^7	0.08	S	
PS LUV (37)	3 mM Ca^{2+}, 15°C	10^7	0.05	0	These vesicles have diameters smaller by 20% than the other LUV in the table (Z-average diameter as determined by light scattering); note the increased values of f relative to other PS LUV with Ca^{2+}
	25°C	2 × 10^7	0.3	0	
	35°C	3.5 × 10^7	> 1	—	
	3 mM Ba^{2+}, 15°C	3.5 × 10^7	0.03	0	
	25°C	5 × 10^7	0.1	0.1	
	35°C	1.5 × 10^8	0.2	0.4	
	5 mM Sr^{2+}, 15°C	> 5 × 10^6	0.001	—	
	25°C	> 2 × 10^7	0.01	—	
	35°C	7 × 10^7	0.05	0.05	
PS LUV (26)	3 mM Ca^{2+}, 15°C	1.3 × 10^7 (1.0 × 10^7)	0.006 (0.03)	0	See Ref. 26 for additional cases
	5 mM Ca^{2+}, 15°C	3.9 × 10^7 (3.5 × 10^7)	0.012 (0.12)	0	
	3 mM Ca^{2+}, 25°C	1.7 × 10^7 (3.3 × 10^7)	0.03 (0.17)	0	
	5 mM Ca^{2+}, 25°C	5.5 × 10^7 (1.3 × 10^8)	0.1 (0.6)	S	At 35°C D is slightly less than f
	3 mM Ca^{2+}, 35°C	2.2 × 10^7 (8 × 10^7)	0.12 (>1)	S	
	5 mM Ca^{2+}, 35°C	8 × 10^7 (3.4 × 10^8)	0.3 (>5)	S	

(continued)

(Table 5, continued)

Composition, size, and references	Fusion-inducing agents and conditions[a]	Rate[a] constant of aggregation C_{11} (= C/2)($M^{-1} \cdot s^{-1}$)	Rate[a] constant of fusion $f(s^{-1})$	Rate constant of deaggregation D (s^{-1})	Comments
PS SUV (22)	1.1 mM Ca^{2+}	2×10^6	>1	—	See Ref. 22 for additional cases of PS and PS/cholesterol mixtures
PS/cholesterol (9:1)	1.1 mM Ca^{2+}	1.5×10^6	>1	—	The effect of cholesterol is summarized by stating that it reduces the aggregation rate constant and enhances the fusion rate constant; a similar conclusion has been reached in PS/cholesterol mixtures up to 2:1 molar ratios with LUV (99)
PS	2.2 mM Ca^{2+}	5×10^7	>5	—	
PS/cholesterol (9:1)	2.2 mM Ca^{2+}	1.6×10^7	>5	—	
PS/cholesterol (3:1)	2.2 mM Ca^{2+}	1.3×10^7	>5	—	
PS	2.5 mM Ca^{2+} 600 mM NaCl	—	0.003	—	
PS/cholesterol (9:1)	2.5 mM Ca^{2+} 600 mM NaCl	—	0.0034		

		C	f	D	
CL/DOPC LUV (1:1) (24)	9.5 mM Ca^{2+} (25°C)	3.5 × 10^6	0.1	S	At 25°C the parameters
	10 mM Ca^{2+}	9 × 10^6	0.3	S	obtained from analysis
	11 mM Ca^{2+}	8 × 10^7	4	S	of RET assay results are
	9.5 mM Ca^{2+} (37°C)	1.3 × 10^7	0.4	S	identical with those shown;
	10 mM Ca^{2+}	3.5 × 10^7	1	S	at 37°C the RET C values
	11 mM Ca^{2+}	1.6 × 10^8	10	S	are 3-4 times larger
CL/PC 1:1 (egg PC) LUV (20)	10 mM Ca^{2+}	1.4 × 10^8	0.3	—	See Ref. 20 for more cases
	10 mM Ca^{2+} + 1:2800 PL	5 × 10^8	0.35	0.7	PL - polylysine (positively charged) MW 35000
	10 mM Ca^{2+} + 1:1400 PL	8 × 10^8	0.29	0.35	1:1400 denotes the molar ratio of PL to phosphate
	8 mM Ca^{2+} + 1:2800 PL	5 × 10^8	0.25	0.8	The charge ratio in this case is 0.2 (PL/CL)
	10 mM Ca^{2+} + 1:420 PL	—	0.3	—	Preincubation with PL for 15 s; the charge ratio in this case is 1.5

aThe Analysis was based on experiments employing the Tb/DPA fusion assay (16). Rate constants determined from experiments employing the RET fusion assay (18) are enclosed in parentheses. The uncertainties in the estimates of the rate constants are about 20, 25, and 50% for C, f, and D, respectively, but more details can be found in the cited studies. Whenever not stated explicitly, the measurements were performed at room temperature and the medium included 100 mM NaCl.

b— indicates that the corresponding rate constant was not determined. The value 0 for D indicates that D << f. S indicates that D < f/2.

A. Effect of Divalent and Monovalent Cations on the
    Aggregation and Fusion of Negatively Charged
    Phospholipid Vesicles

Table 5 shows that both C and f values increase dramatically with an
increase in the concentration of divalent cations above a certain threshold
value. The monovalent cations Li^+, Na^+, and K^+ induce reversible and
nonleaky aggregation of acidic phospholipid vesicles (4,7,8,10,23,28-30).
H^+ is the only known monovalent cation that promotes membrane destabilization and fusion (2,31-33). When vesicle fusion is induced by divalent cations,
an increase in monovalent cation concentration results in a reduced value
of the rate constant of fusion f due to competition between monovalent
and divalent cations for binding to the acidic headgroups. Conversely,
a reduction in monovalent cation concentration can enhance the rate of
fusion induced by divalent cations, up to the state that aggregation is
rate-limiting to the overall fusion process. In this case, a further reduction
in monovalent cation concentration results in slowing down the overall
fusion rate, because of the increased electrostatic repulsion between vesicles
which prevents their close approach.

The value for C_{11} (which equals $C/2$), the rate constant of aggregation,
is approximated by (34,35)

$$C_{11} = 4kTN_A \ 10^{-3} \ \exp[-V^*/(kT)]/(3\eta) \tag{24}$$

where V^* is the height of the potential barrier for close approach of particles, N_A is Avogadro's number, k is Boltzmann's constant, T is the absolute
temperature, and η is the viscosity of the medium.

In Table 5 C_{11} values vary from 10^5 to 10^9 $M^{-1} \cdot s^{-1}$, whereas the
value of C_{11} for a diffusion-controlled process ($V^* = 0$) is 3×10^9 $M^{-1} \cdot s^{-1}$
at room temperature, or 5×10^9 $M^{-1} \cdot s^{-1}$ at 37°C. Hence, the potential
barriers, V^*, are in the range of 1-10 kT. These values of V^* are significantly smaller than the values expected in a medium without added fusogens.

Calculations in the framework of the Derjaguin-Landau-Verwey-Overbeek
(DLVO) theory (35) considered V^* to consist of a sum of a negative or
attractive Van der Waals term and a positive term accounting for the electrostatic repulsion between negatively charged surfaces. The calculations of
C values for PS vesicles aggregating in the presence of combinations of
Na^+, Ca^{2+}, and Mg^{2+} extended the DLVO expressions by explicitly considering the binding of cations to the negatively charged surfaces (9,10,36).
The dramatic increase in C values with increasing Ca^{2+} concentrations,
which was deduced from fusion experiments on PS vesicles in the presence
of 100 mM Na^+ and 1-5 mM Ca^{2+}, could be predicted by the calculations
(5,10). These calculations also showed that larger Ca^{2+} concentrations
would be required for aggregation of large unilamellar vesicles (LUV) than
for aggregation of small unilamellar vesicles (SUV), because the electrostatic
repulsion energy is approximately proportional to vesicle radii.

Theoretical calculations of C values predicted (10) their increase with
temperature. This increase is due to a decrease in viscosity of the medium
with increasing temperature and due to the inverse exponential dependence
of C on V^*/kT in Eq. (24). Studies on the aggregation of PS SUV by
≥500 mM NaCl show that the kinetics of the process slow down with an
increase in temperature (8,10). The explanation is that the deaggregation
rate constant D increases more steeply with temperature than the forward
aggregation rate constant C (8,10-12). On the other hand, the overall
fusion rate increases with temperature (24,26,37,38), and Table 5 illustrates

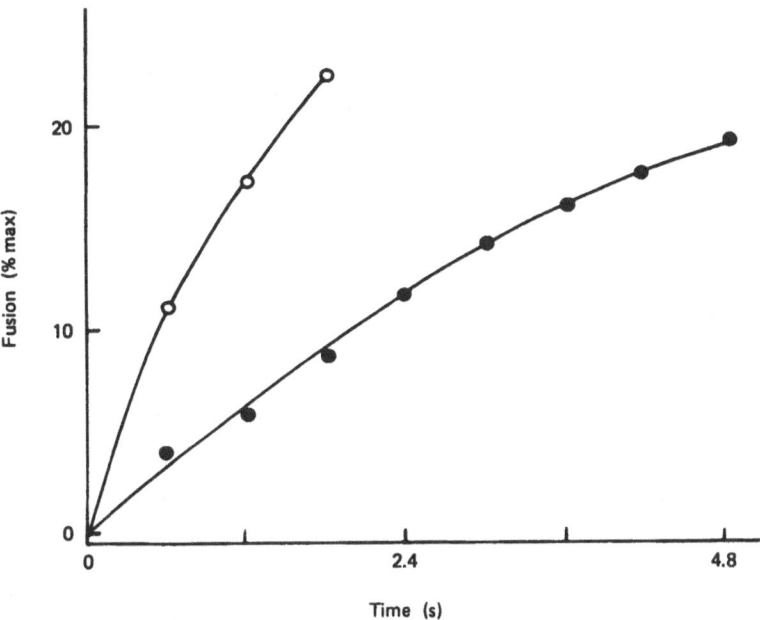

FIGURE 2 Simulation of Ca^{2+}-induced fusion of CL/PC large unilamellar vesicles, and the effect of preincubation of the vesicles with polylysine (PL). The drawn lines denote the theoretical simulations, calculated with the rate constants given in Table 5; the data points represent experimental results obtained with the Tb/DPA assay. The lower curve (solid circles) represents fusion of the vesicles in a medium containing 8 mM $CaCl_2$ and 90 nM PL; the upper curve (open circles) shows the effect of preincubation of the vesicles with the PL, 15 s prior to the addition of $CaCl_2$. (Reproduced, with permission, from Ref. 20.)

that there is indeed an increase in C values with temperature. A detailed examination reveals that there is a certain decrease in the value of V^* with temperature which is not fully accounted for by the temperature dependence of Van der Waals and electrostatic interactions (10-12,37). Because of this small decrease of V^* with temperature, it is not recommended to deduce conclusions from activation energies found by means of Arrhenius plots, because the activation energies obtained are expected to have apparent values, significantly larger than the true values (10).

The f values listed in Table 5, varying from 0.001 to 5 s^{-1}, reflect membrane destabilization and merging rates. Our view is that membrane merging is a much faster process than membrane destabilization (23). The argument is based on the fact that an increase of the Ca^{2+} concentration is expected to produce a decrease in membrane fluidity (2,39-43), whereas f values increase steeply. The steep increase in values of f with an increase in solution concentrations of divalent cations correlates well with the amount of divalent cations bound to the anionic phospholipid headgroups (4,23,27,37). Thus, an increase in the concentration of monovalent cations results in a decrease in f values, and vice versa (4,23). Monovalent cations cause a reduction in binding of divalent cations to membranes due to competition for binding sites and due to a reduction in the magnitude of the negative surface potential, which in turn results in a smaller enhancement in the concentration of the divalent cations at the interface (10,36,44). Experimental binding studies (44-48) have verified the predictions of calculations of amounts of cations bound. Consequently, calculated values of binding ratios could be used in correlations with f values for various combinations of divalent and monovalent cations.

PS vesicles undergoing aggregation and fusion exhibit an enhanced binding affinity of Ca^{2+} to PS (43,49-51). When the actual fusion is the rate-limiting step, the effectiveness of divalent cations for the fusion of PS vesicles as expressed by f values is $Ca^{2+} > Ba^{2+} > Sr^{2+} > Mg^{2+}$ (4,27). In the presence of 100 mM Na^+ it was found (4) that Ba^{2+} concentrations were smaller than those of Ca^{2+} for production of certain initial fusion rates of PS SUV. This interchange between Ca^{2+} and Ba^{2+} was in accord with aggregation studies (28) and binding studies on isolated vesicles (52). In view of the fact that fusing PS vesicles exhibit a higher affinity for Ca^{2+} than isolated vesicles, it is still possible that the same sequences of divalent cations reflect both fusogenic activity and binding strength. The author's view is that binding affinity and fusogenic capacity are reflections of interactions between divalent cations and phospholipids. When the amount of divalent cation bound to the membranes exceeds a certain critical ratio, optimization of the phospholipid-cation interaction (in a 2:1 binding mode or other modes) is achieved at a new structure of the membrane. During the transition to a new structure, the hydrocarbon core may be exposed to water (2) and the membrane becomes unstable. In the Ca^{2+}/PS system this instability is attained (2,41-43,51). In the Mg^{2+}/PS system (at room temperature) the process of fusion terminates when the vesicles reach the size of LUV, although they aggregate (53).

In other studies it has been shown that initial fusion rates correlate best with the increase in surface tension of monolayers (4-10 dynes/cm) in the presence of divalent cations. The method was applied to small PS (48) and PA (54) vesicles. It was also pointed out (55) that H^+ is the only monovalent cation that causes a significant increase in the surface tension of acidic monolayers and that this cation also induces fusion (2,31-33). In the latter study (55) the authors succeeded in predicting the increase in surface tension of a PS monolayer in the presence of Ca^{2+}, which was attributed to a 2:1 rather than a 1:1 binding mode. These findings are indeed interesting, but the authors admit (55) that their model might be limited to certain fusion mechanisms. Clearly, it cannot explain the fact that PS LUV do not fuse in the presence of Mg^{2+}. Furthermore, initial fusion rates reflect, in general, both the aggregation and the fusion processes, and their relative importance can be modulated by varying vesicle concentrations. Hence, it might be of interest to correlate the increase in surface tension with f values.

B. Membrane Instability and Membrane Contact

Membrane instabilities can be produced by various defects in the structure, as in the case of SUV consisting of dipalmitoylphosphatidylcholine (DPPC) or distearoylphosphatidylcholine (DSPC) below the phase transition temperature (56-59; see also Chapter 7) or in the La/PS system at the phase transition of the complex (60). The membrane is expected to be less ordered, and thus more susceptible to fusion, in the boundary regions between domains (2) in the membrane, although phase separations or isothermal transitions (2,61-66) need not be a necessary condition for membrane instabilities (37,38,67,68). Another example of a means of producing membrane instabilities is that of electric pulses which can induce the fusion of cells or vesicles previously brought in close proximity (69,70; see also Chapter 29).

Release of vesicle contents is a convenient criterion for membrane instability. When vesicle fusion is induced by divalent or trivalent cations, by protons or by proteins, leakage increases with vesicle concentration, indicating that membrane instability is promoted by vesicle contact, or by vesicle collisions (3,5,7,10,71,72). Although it is possible that enhanced permeability can occur at the level of isolated vesicles in the presence of divalent cations, vesicles in contact exhibit a significantly larger rate of leakage. Other studies (32,71,73) demonstrated contact-mediated leakage without fusion (in the sense of mixing of contents) in vesicles containing phosphatidylethanolamine (PE). Recently it has been reported (92) that the rate of flipflop in vesicles containing phosphatidylglycerol (PG) is also promoted by vesicle contact. It has also been reported (74-78) that if membranes can be brought into contact by various means, such as lectins, then low concentrations of Ca^{2+} are sufficient to phase-separate the lipids in the zone of contact (67). In that zone point defects are more likely to be produced (51,68,79-82). Hence, while membrane contact or vesicle aggregation is not a sufficient condition for membrane destabilization, it promotes this process, and in most cases is appears to be a necessary condition, if we exclude external sources of energy, such as electric or acoustic pulses.

On the other hand, studies on the fusion of vesicles have revealed that the aggregation rate is increased in cases where the surfaces are susceptible to structural changes. A first observation of this effect was noted in the case of PS SUV in the presence of low NaCl concentrations and 1-5 mM Ca^{2+} or Mg^{2+} (10). The experimentally found C values are several orders of magnitude larger than the calculated values [Eq. (24)] based on the sum of Van der Waals and electrostatic interactions (10,23). The inclusion of repulsive hydration forces (83) in the calculations would even further reduce the C values. The suggestion has been that due to membrane instability, the surfaces of the vesicles are distorted (10). Now, the potential barrier for the close approach of a spike to another surface is much smaller than that for the approach of two spheres, or two flattened areas. Indeed, this proposal is speculative and should be tested, but at this stage it enables us to explain qualitatively the observations mentioned above. Our proposal for the observed enhanced aggregation rates resembles a mutual feedback mechanism. Schematically, this process is as follows: collision → instability and deformation → closer contact → enhanced instability → fusion. This sequence can occur with a significant probability only if the vesicles can be destabilized, and in addition can approach each other to within 2-3 nm (further away the potential barrier is too small), and the probability of its occurrence rises when the intact vesicles can approach each other to a closer distance by a reduction of the electrostatic repulsion. This can be achieved by means of divalent cations, or proteins, and also by means of monovalent cations (8,10,29,30), although monovalent cations, by reducing the binding of divalent cations, may enhance membrane stability (4,23,45).

Contact-induced destabilization has also been observed (24) with CL/PC LUV, induced to fuse by Ca, where a steep increase of the rate constants of aggregation and fusion was found when the Ca^{2+} concentration was increased from 9.5 to 11 mM. The increase in C values could not be explained on the basis of electrostatic considerations which would give a

twofold increase rather than the 20-fold increase found. In this context
we will also mention results from studies on liposomes rich in PE, for which
only a few rate constants have been determined. It has been reported
(84-86) that the rate of fusion in these systems is enhanced when the
lipids can undergo a lamellar-to-inverted-hexagonal (L_α-H_{II}) phase transition. It has been suggested (87,88) that the above transition requires
close apposition of the bilayers in order to occur. The above and later
studies (89,90) suggest that if the lipids in the aggregating vesicles prefer
the H_{II} phase or are close to the L_α-H_{II} phase boundary, then intermediates
in this phase transition may form, which could result in fusion. It has
been found that under these conditions the liposomes undergo contact-
mediated lysis with enhanced lipid mixing but no mixing of aqueous contents
(71-73). While it was not established that inverted micelles (84-86), as
opposed to point defects (82,91), were indeed involved, predictions based
on theoretical considerations (87-89) have suggested that inverted micellar
intermediates are short-lived structures (lifetimes of 1 msec or less), which
therefore could remain undetected. By the same reasoning, it was suggested
(90) that the sharp threshold for the aggregation and fusion of CL/PC,
and the fact that the rate of lipid mixing was faster than the rate of aqueous contents mixing at 37°C (24), could imply the formation of inverted
micellar intermediates in this system. At this stage it is difficult to confirm
this suggestion by direct experiments, but the general idea is in line with
our proposal here and with the study on the CL/PC system (24) regarding
the promotion of membrane instability by aggregation and vice versa.

C. Effect of Vesicle Size

Table 5 provides a comparison of C and f values found in studies on SUV
with those found in LUV. In accord with Eq. (24), an increase in vesicle
size results in an enhanced potential barrier for close approach, which
can be overcome by an increase in the concentrations of fusogens, e.g.,
cations in the case of anionic phospholipids (5,16). In the range of vesicle
diameters of 25-200 nm, where the effect of vesicle size has been studied,
an increase in vesicle size results in a decrease in fusion activity, as
measured by the rate constant f.

The spontaneous and slow fusion of DPPC SUV (57-59; see also Chapter 7) results, in a period of days, in fused vesicles of diameters in the
range of about 70 nm, which further fuse to vesicles of about 95 nm diameter in 3-4 weeks. Corresponding diameters found in DSPC vesicle systems
are 60 and 100 nm, respectively (56). These fusion events are promoted
by lowering the temperature below the phase transition, in order to promote
defects in the structure of SUV. In the case of PS, Mg^{2+} cannot destabilize
LUV, although the vesicles aggregate, whereas SUV fuse until they reach
the size of LUV (53). Similarly, PS/PC (molar ratio 1:1) LUV aggregate
but do not fuse with Ca^{2+}, although SUV do fuse (79). The tendency
of fusion products to reach a certain limiting size has also been reported
for other systems (93,94). In the case of neutral phospholipids, the spontaneous and slow fusion of SUV terminates when they transform to LUV
as final stable products, similar to the case of PS SUV with Mg^{2+}. In the
case of PS as well as other acidic phospholipids induced to fuse by Ca^{2+}
(2,40-43,51) or La^{3+} (60), the fusion products are not stable, because
the intermolecular interactions in these systems are optimized by nonvesicular structures (e.g., cochleates), which arise as products of many fusion
events.

A detailed comparison of the fusion rate constants of LUV with SUV was carried out for PS vesicles induced to fuse with Ca^{2+} (3,6,7). The rate constants of aggregation were about the same when the medium contained 2 mM and 5 mM Ca^{2+} with SUV or LUV, respectively, but the respective f values were almost two orders of magnitude larger with SUV (see Table 5). Table 5 also illustrates that f values of smaller LUV (20% smaller Z-average diameter as determined by dynamic light scattering) are larger than those of corresponding larger LUV (37). The same significant increase of f values with lower vesicle radii has been observed with other divalent cations (27,37). In one of the latter studies (27) it was claimed that the size dependence of f values manifests itself only above a certain threshold of bound divalent cations. This conclusion was reached by determining f values for PS LUV in the presence of 5 or 10 mM Mg^{2+}, and using calculated values for bound divalent cations by employing binding constants deduced from isolated vesicles (52). Thus, at a "threshold" binding ratio (~0.16 Ca^{2+}/PS), the same value of f emerged for both SUV (4) and LUV. However, in fusion experiments the vesicles are not isolated and the experimental values of bound Ca^{2+} are significantly larger than the above calculated values. For instance, in the presence of a total Ca^{2+} concentration of 0.8 mM, the experimental value of Ca^{2+} bound per PS is 0.21 (49), in agreement with calculations for isolated vesicles (50). Immediately upon addition of 6 mM Mg^{2+} the binding ratio drops to 0.05, as expected for isolated vesicles. However, 1 min later the binding ratio becomes 0.23, i.e., above the initial value, and 2 min later it is 0.35, corresponding to a significant increase in the binding affinity of Ca^{2+} to PS (49-51). Although the comparison between calculated and experimental results of binding ratios of Ca^{2+}/PS indicates some enhancement of the binding affinity of Ca^{2+} to PS SUV in the presence of large Na^{+} concentrations (44,54), this effect is not as dramatic as with Mg^{2+} in the medium, in which case we encounter conditions where the ratio Ca^{2+}/PS is larger in the presence of Mg^{2+} than in its absence. Thus, when the increase in binding affinity of Ca^{2+} to PS LUV during aggregation and fusion is taken into account, the general result may hold that f values for SUV are larger than the corresponding values (at the same binding ratios) for LUV. This result can be summarized by saying that SUV are more unstable than LUV under all conditions, irrespective of the threshold values.

The larger instability of SUV arises from the smaller radius of curvature which poses packing constraints (and asymmetry between monolayers) on the molecules composing the membranes (95,96). Consequently, the intermolecular interactions between the phospholipids in SUV are less optimal than in the larger and symmetrical LUV. This means that SUV are structures of higher free-energy content, which are in a metastable state. In this connection we note that SUV are more permeable than LUV (16,97). The same reasoning leads us to expect larger stability and resistance to fusion in vesicles composed of phospholipids of longer acyl chains, in which case stronger intermolecular interactions are expected. This proposition (98) has to be tested directly by the application of fusion assays to kinetic studies.

ACKNOWLEDGMENTS

This work was supported in part by NIH Grant GM 31506 (J. Bentz, University of California—San Francisco, and S. Nir). Ms. Andrea Mazel (UCSF) is acknowledged for the expert typing of the manuscript. Dr. Joe Bentz is acknowledged for critically reading the manuscript and useful comments.

REFERENCES

1. Nir, S. (1977). Van der Waals interactions between surfaces of biological interest. *Prog. Surface Sci.* 8:1-58.
2. Papahadjopoulos, D., Vail, W. J., Newton, C., Nir, S., Jacobson, K., Poste, G., and Lazo, R. (1977). Studies on membrane fusion. III. The role of calcium-induced phase changes. *Biochim. Biophys. Acta* 465:579-598.
3. Bentz, J., Nir, S., and Wilschut, J. (1983). Mass action kinetics of vesicle aggregation and fusion. *Colloids Surfaces* 6:333-363.
4. Bentz, J., Düzgünes, N., and Nir, S. (1983). Kinetics of divalent cation induced fusion of phosphatidylserine vesicles: Correlation between fusogenic capacities and binding affinities. *Biochemistry* 22:3320-3330.
5. Nir, S., Bentz, J., and Wilschut, J. (1980). Mass action kinetics of phosphatidylserine vesicle fusion as monitored by coalescence of internal vesicle volumes. *Biochemistry* 19:6030-6036.
6. Nir, S., Wilschut, J., and Bentz, J. (1982). The rate of fusion of phospholipid vesicles and the role of bilayer curvature. *Biochim. Biophys. Acta* 688:275-278.
7. Nir, S., Bentz, J., Wilschut, J., and Düzgünes, N. (1983). Aggregation and fusion of phospholipid vesicles. *Prog. Surface Sci.* 13:1-124.
8. Day, E. P., Kwok, A. Y. W., Hark, S. K., Ho, J. T., Vail, W. J., Bentz, J., and Nir, S. (1980). Reversibility of sodium-induced aggregation of sonicated phosphatidylserine vesicles. *Proc. Natl. Acad. Sci. USA* 77:4026-4029.
9. Nir, S., and Bentz, J. (1978). On the forces between phospholipid bilayers. *J. Coll. Interface Sci.* 65:399-414.
10. Nir, S., Bentz, J., and Portis, A. R., Jr. (1980). Effect of cation concentrations and temperature on the rates of aggregation of acidic phospholipid vesicles: Application to fusion. *Adv. Chem. Ser.* 188:75-106.
11. Bentz, J., and Nir, S. (1981). Kinetic and equilibrium aspects of reversible aggregation. *J. Chem. Soc. Faraday I* 77:1249-1275.
12. Bentz, J., and Nir, S. (1981). Aggregation of colloidal particles modeled as a dynamical process. *Proc. Natl. Acad. Sci. USA* 78:1634-1637.
13. Smoluchowski, M. (1917). Investigation into a mathematical theory of the kinetics of coagulation of colloidal solutions. *Z. Phys. Chem. Abt.* A92:129-168.
14. Nir, S., Stegmann, T., and Wilschut, J. (1986). Fusion of influenza virus and cardiolipin liposomes at low pH: Mass action analysis of kinetics and extent. *Biochemistry* 25:257-266.
15. Wilschut, J., and Papahadjopoulos, D. (1979). Ca^{2+}-induced fusion of phospholipid vesicles monitored by mixing of aqueous contents. *Nature* 281:690-692.
16. Wilschut, J., Düzgünes, N., Fraley, R., and Papahadjopoulos, D. (1980). Studies on the mechanism of membrane fusion: Kinetics of calcium ion induced fusion of phosphatidylserine vesicles followed by a new assay for mixing of aqueous vesicle contents. *Biochemistry* 19:6011-6021.
17. Hoekstra, D. (1982). Fluorescence method of measuring the kinetics of Ca^{2+}-induced phase separations in phosphatidylserine-containing lipid vesicles. *Biochemistry* 21:1055-1061.

18. Struck, D. K., Hoekstra, D., and Pagano, R. E. (1981). Use of resonance energy transfer to monitor membrane fusion. *Biochemistry* 20:4093-4099.

19. Hoekstra, D., De Boer, T., Klappe, K., and Wilschut, J. (1984). Fluorescence method for measuring the kinetics of fusion between biological membranes. *Biochemistry* 23:5675-5681.

20. Gad, A. E., Bental, M., Elyashiv, G., Weinberg, H., and Nir, S. (1985). Promotion and inhibition of vesicle fusion by polylysine. *Biochemistry* 24:6277-6282.

21. Schuber, F., Hong, K., Düzgüneş, N., and Papahadjopoulos, D. (1983). Polyamines as modulators of membrane fusion: Aggregation and fusion of liposomes. *Biochemistry* 22:6134-6140.

22. Braun, G., Lelkes, P. I., and Nir, S. (1985). Effect of cholesterol on Ca^{2+}-induced aggregation and fusion of sonicated phosphatidylserine/cholesterol vesicles. *Biochim. Biophys. Acta* 812:688-694.

23. Nir, S., Düzgüneş, N., and Bentz, J. (1983). Binding of monovalent cations to phosphatidylserine and modulation of Ca^{2+}- and Mg^{2+}-induced vesicle fusion. *Biochim. Biophys. Acta* 735:160-172.

24. Wilschut, J., Nir, S., Scholma, J., and Hoekstra, D. (1985). Kinetics of Ca-induced fusion of cardiolipin/phosphatidylcholine vesicles: Correlation between vesicle aggregation, bilayer destabilization and fusion. *Biochemistry* 24:4630-4636.

25. Rosenberg, J., Düzgüneş, N., and Kayalar, C. (1983). Comparison of two liposome fusion assays monitoring the intermixing of aqueous contents and of membrane components. *Biochim. Biophys. Acta* 735:173-180.

26. Wilschut, J., Scholma, J., Bental, M., Hoekstra, D., and Nir, S. (1985). Ca-induced fusion of phosphatidylserine vesicles: Mass action kinetic analysis of lipid mixing and aqueous contents mixing. *Biochim. Biophys. Acta* 821:45-55.

27. Bentz, J., and Düzgüneş, N. (1985). Fusogenic capacities of divalent cations and effect of liposome size. *Biochemistry* 24:5436-5443.

28. Ohki, S., Düzgünes, N., and Leonards, K. (1984). Monovalent cation-induced phospholipid vesicle aggregation: Effect of ion binding. *Biochemistry* 21:2127-2133.

29. Ohki, S., Roy, S., Ohshima, H., and Leonards, K. (1984). Monovalent cation-induced phospholipid vesicle aggregation: Effect of ion binding. *Biochemistry* 23:6126-6132.

30. Yoshimura, T., and Aki, K. (1985). Sodium-induced aggregation of phosphatidic acid and mixed phospholipid vesicles. *Biochim. Biophys. Acta* 815:167-173.

31. Düzgünes, N., Straubinger, R. M., Baldwin, P. A., Friend, D. S., and Papahadjopoulos, D. (1985). Proton-induced fusion of oleic acid-phosphatidylethanolamine liposomes. *Biochemistry* 24:3091-3098.

32. Ellens, H., Bentz, J., and Szoka, F. C. (1985). H^{+}- and Ca^{2+}-induced fusion and destabilization of liposomes. *Biochemistry* 24:3099-3106.

33. Ohki, S., and Düzgüneş, N. (1979). Divalent cation induced interaction of phospholipid vesicle and monolayer membranes. *Biochim. Biophys. Acta* 552:438-449.

34. Fuchs, V. (1934). Über die Stabilität und Aufladung der Aerosole. *Z. Physik.* 89:736-743.

35. Verwey, E. J. W., and Overbeek, J. Th. G. (1948). *Theory of the Stability of Lyophobic Colloids.* Elsevier, Amsterdam.

36. Nir, S., Bentz, J., and Düzgüneş, N. (1981). Two modes of reversible vesicle aggregation: Particle size and the DLVO theory. *J. Coll. Interface Sci.* 84:266-269.

37. Bentz, J., Düzgüneş, N., and Nir, S. (1985). Temperature dependence of divalent cation induced fusion of phosphatidylserine liposomes: Evaluation of the kinetic rate constants. *Biochemistry* 24:1064-1072.

38. Wilschut, J., Düzgüneş, N., Hoekstra, D., and Papahadjopoulos, D. (1985). Modulation of membrane fusion by membrane fluidity: Temperature dependence of divalent cation-induced fusion of phosphatidylserine vesicles. *Biochemistry* 24:8-14.

39. Düzgüneş, N., and Papahadjopoulos, D. (1983). Ionotropic effects on phospholipid membranes: Calcium-magnesium specificity in binding, fluidity, and fusion. In: *Membrane Fluidity in Biology*, Vol. 2, *General Principles* (R. C. Aloia, ed.). Academic Press, New York, pp. 187-216.

40. Papahadjopoulos, D. (1978). Calcium-induced phase changes and fusion in natural and model membranes. In: *Membrane Fusion* (G. Poste and G. L. Nicolson, eds.). Elsevier/North Holland Biomedical Press, Amsterdam, pp. 765-790.

41. Papahadjopoulos, D., Portis, A., and Pangborn, W. (1978). Calcium-induced lipid phase transitions and membrane fusion. *Ann. NY Acad. Sci.* 308:50-66.

42. Papahadjopoulos, D., Portis, A., Pangborn, W., and Newton, C. (1978). Fusion of artificial membranes with special emphasis on the role of calcium-induced lipid phase transitions. In: *Transport of Macromolecules in Cellular Systems* (S. C. Silverstein, ed.). Dahlem Konferenzen, Berlin, pp. 413-430.

43. Papahadjopoulos, D., Poste, G., and Vail, W. J. (1979). Studies on membrane fusion with natural and model membranes. *Methods Memb. Biol.* 10:1-121.

44. Nir, S., Newton, C., and Papahadjopoulos, D. (1978). Binding of cations to phosphatidylserine vesicles. *Bioelectrochem. Bioenerg.* 5: 116-133.

45. Düzgüneş, N., Nir, S., Wilschut, J., Bentz, J., Newton, C., Portis, A., and Papahadjopoulos, D. (1981). Calcium- and magnesium-induced fusion of mixed phosphatidylserine/phosphatidylcholine vesicles: Effect of ion binding. *J. Memb. Biol.* 59:115-125.

46. Kurland, R., Newton, C., Nir, S., and Papahadjopoulos, D. (1979). Specificity of Na^+ binding to phosphatidylserine vesicles from a $^{23}Na^+$ NMR relaxation rate study. *Biochim. Biophys. Acta* 551:137-147.

47. Newton, C., Pangborn, W., Nir, S., and Papahadjopoulos, D. (1978). Specificity of Ca^{2+} and Mg^{2+} binding to phosphatidylserine vesicles and resultant phase changes of bilayer membrane structure. *Biochim. Biophys. Acta* 506:281-287.

48. Ohki, S. (1982). A mechanism of divalent ion-induced phosphatidylserine membrane fusion. *Biochim. Biophys. Acta* 689:1-11.

49. Ekerdt, R., and Papahadjopoulos, D. (1982). Intermembrane contact affects calcium binding to phospholipid vesicles. *Proc. Natl. Acad. Sci. USA* 79:2273-2277.

50. Nir, S. (1984). A model for cation adsorption in closed systems: Application to calcium binding to phospholipid vesicles. *J. Coll. Interface Sci.* 102:313-321.

51. Portis, A., Newton, C., Pangborn, W., and Papahadjopoulos, D. (1979). Studies on the mechanism of membrane fusion: Evidence for an intermembrane Ca^{2+}-phospholipid complex, synergism with Mg^{2+}, and inhibition by spectrin. *Biochemistry* 18:780-790.

52. McLaughlin, S., Mulrine, N., Gresalfi, T., Vaio, G., and McLaughlin, A. (1981). The adsorption of divalent cations to bilayer membranes containing phosphatidylserine. *J. Gen. Physiol.* 77:445-473.

53. Wilschut, J., Duzgunes, N., and Papahadjopoulos, D. (1981). Calcium/magnesium specificity in membrane fusion: Kinetics of aggregation and fusion of phosphatidylserine vesicles and the role of bilayer curvature. *Biochemistry* 20:3126-3133.

54. Ohki, S., and Ohshima, H. (1985). Divalent cation-induced phosphatidic acid membrane fusion: Effect of ion binding and membrane surface tension. *Biochim. Biophys. Acta* 812:147-154.

55. Ohki, S., and Ohshima, H. (1984). Divalent cation-induced surface tension increase in acidic phospholipid membranes: Ion binding and membrane fusion. *Biochim. Biophys. Acta* 776:177-182.

56. Larrabee, A. L. (1979). Time-dependent changes in the size distribution of distearoylphosphatidylcholine vesicles. *Biochemistry* 18:3321-3326.

57. Lichtenberg, D., and Schmidt, C. F. (1981). Molecular packing and stability in the gel phase of curved phosphatidylcholine vesicles. *Lipids* 16:555-557.

58. Lichtenberg, D., Freire, E., Schmidt, C. F., Barenholz, Y., Felgner, P. L., and Thompson, T. E. (1981). Effect of surface curvature on stability, thermodynamic behavior, and osmotic activity of dipalmitoylphosphatidylcholine single lamellar vesicles. *Biochemistry* 20:3462-3467.

59. Wong, M., Anthony, F. H., Tillack, T. W., and Thompson, T. E. (1982). Fusion of dipalmitoylphosphatidylcholine vesicles at 4°C. *Biochemistry* 21:4126-4132.

60. Hammoudah, M. M., Nir, S., Bentz, J., Mayhew, E., Stewart, T. P., Hui, S. W., and Kurland, R. J. (1981). Interactions of La^{3+} with phosphatidylserine vesicles. Binding, phase transition, leakage, ^{31}P-NMR and fusion. *Biochim. Biophys. Acta* 645:102-114.

61. Hui, S. W., Boni, L. T., Stewart, T. P., and Isaac, T. (1983). Identification of phosphatidylserine and phosphatidylcholine in calcium-induced phase separated domains. *Biochemistry* 22:3511-3516.

62. Ito, T., and Ohnishi, S-I. (1974). Ca^{2+}-induced lateral phase separations in phosphatidic acid-phosphatidylcholine membranes. *Biochim. Biophys. Acta* 352:29-37.

63. Jacobson, K., and Papahadjopoulos, D. (1975). Phase transitions and phase separations in phospholipid membranes induced by changes in temperature, pH and concentration of bivalent cations. *Biochemistry* 14:152-161.

64. Ohnishi, S-I., and Ito, T. (1974). Calcium-induced phase separations in phosphatidylserine-phosphatidylcholine membranes. *Biochemistry* 13:881-887.

65. Ohnishi, S-I., and Tokutomi, S. (1981). ESR studies of calcium- and proton-induced phase separations in phosphatidylserine-phosphatidylcholine mixed membranes. In: *Biological Magnetic Resonance*, Vol. 3 (L. J. Berliner and J. Reuben, eds.). Plenum Press, New York, pp. 121-153.

66. Shimshick, E. J., and McConnell, H. M. (1973). Lateral phase separation in phospholipid membranes. *Biochemistry* 12:2351-2359.

67. Düzgüneş, N. (1985). Membrane fusion. Subcell. *Biochemistry* 11:195-286.

68. Düzgüneş, N., Paiement, J., Freeman, K. B., Lopez, N. G., Wilschut, J., and Papahadjopoulos, D. (1984). Modulation of membrane fusion by ionotropic and thermotropic phase transition. *Biochemistry* 23:3486-3494.

69. Zimmerman, U. (1982). Electric field-mediated fusion and related electrical phenomena. *Biochim. Biophys. Acta* 694:227-277.

70. Stenger, D. A., and Hui, S. W. (1986). Kinetics of ultrastructural changes during electrically-induced fusion of human erythrocytes. *J. Memb. Biol.* 93:43-53.

71. Bentz, J., Ellens, H., Lai, M-Z., and Szoka, F. C., Jr. (1985). On the correlation between H_{II} phase and the contact-induced destabilization of phosphatidylethanolamine-containing membranes. *Proc. Natl. Acad. Sci. USA* 82:5742-5745.

72. Ellens, H., Bentz, J., and Szoka, F. C. (1984). pH-induced destabilization of phosphatidylethanolamine-containing liposomes. Role of bilayer contact. *Biochemistry* 23:1532-1538.

73. Ellens, H., Bentz, J., and Szoka, F. C. (1986). Destabilization of phosphatidylethanolamine liposomes at the hexagonal phase transition temperature. *Biochemistry* 25:285-294.

74. Düzgünes, N., and Hoekstra, D. (1986). Agglutination and fusion of glycolipid-phospholipid vesicles mediated by lectins and calcium ions. *Stud. Biophys.* 111:5-10.

75. Düzgünes, N., Hoekstra, D., Hong, K., and Papahadjopoulos, D. (1984). Lectins facilitate calcium-induced fusion of phospholipid vesicles containing glycosphingolipids. *FEBS Lett.* 173:80-84.

76. Hoekstra, D., Düzgüneş, N., and Wilschut, J. (1985). Agglutination and fusion of globoside GL-4 containing phospholipid vesicles mediated by lectins and Ca^{2+}. *Biochemistry* 24:565-572.

77. Hoekstra, D., and Düzgünes, N. (1986). Ricinus communis agglutinin-mediated agglutination and fusion of glycolipid-containing phospholipid vesicles. Effect of carbohydrate-headgroup size, calcium ions and spermine. *Biochemistry* 25:1321-1330.

78. Tokutomi, S., Lew, R., and Ohnishi, S-I. (1981). Ca^{2+}-induced phase separation in phosphatidylserine, phosphatidylethanolamine and phosphatidylcholine mixed membranes. *Biochim. Biophys. Acta* 643:276-282.

79. Düzgünes, N., Wilschut, J., Fraley, R., and Papahadjopoulos, D. (1981). Studies on the mechanism of membrane fusion: Role of headgroup composition in calcium- and magnesium-induced fusion of mixed phospholipid vesicles. *Biochim. Biophys. Acta* 642:182-195.

80. Düzgünes, N., Wilschut, J., and Papahadjopoulos, D. (1985). Control of membrane fusion by divalent cations, phospholipid head-groups and proteins. In: *Physical Methods on Biological Membranes and Their Model Systems* (F. Conti, W. E. Blumberg, J. DeGier, and F. Pocchiari, eds.). Plenum Press, New York, pp. 193-218.

81. Hoekstra, D. (1982). Role of lipid phase separations and membrane hydration in phospholipid vesicle fusion. *Biochemistry* 21:2833-2840.

82. Hui, S. W., Stewart, T. P., Boni, L. T., and Yeagle, P. L. (1981). Membrane fusion through point defects in bilayers. *Science* 212:921-923.

83. Lis, L. J., McAlister, M., Fuller, N., Rand, R. P., and Parsegian, V. A. (1982). Interactions between neutral phospholipid bilayer membranes. *Biophys. J.* 37:657-666.

84. Hope, M. J., Walker, D. C., and Cullis, P. R. (1983). Calcium and pH-induced fusion of small unilamellar vesicles consisting of phosphatidylethanolamine and negatively charged phospholipids: A freeze-fracture study. *Biochem. Biophys. Res. Commun.* 110:15-22.

85. Verkleij, A. J. (1984). Lipidic intramembranous particles. *Biochim. Biophys. Acta* 779:43-63.

86. Verkleij, A. J., van Echteld, C. J. A., Gerritsen, W. J., Cullis, P. R., and de Kruijff, B. (1980). The lipidic particle as an intermediate structure in membrane fusion processes and bilayer to hexagonal H$_{II}$ transitions. *Biochim. Biophys. Acta* 600:620–624.
87. Siegel, D. P. (1984). Inverted micellar structures in bilayer membranes: Formation rates and half-lives. *Biophys. J.* 45:399–420.
88. Siegel, D. P. (1986). Inverted micellar intermediates and the transitions between lamellar, cubic and inverted hexagonal lipid phase. I. Mechanism of the L$_\alpha$ - H$_{II}$ phase transition. *Biophys. J.* 49:1155–1170.
89. Siegel, D. P. (1986). Inverted micellar intermediates and the transitions between lamellar, cubic and inverted hexagonal lipid phases. II. Implications for membrane-membrane interactions and membrane fusion. *Biophys. J.* 49:1171–1183.
90. Siegel, D. P. (1987). Membrane-membrane interactions via intermediates in lamellar-to-inverted hexagonal phase transitions. In: *Membrane Fusion* (A. E. Sowers, ed.). Plenum Press, New York, pp. 181–207.
91. Ornberg, R. L., and Reese, T. S. (1981). Beginning of exocytosis captured by rapid-freezing of *Limulus* amebocytes. *J. Cell Biol.* 90:40–54.
92. Lentz, B. R., Alford, D. R., and Whitt, N. A. (1986). The kinetic mechanism of cation catalyzed phosphatidylglycerol transbilayer migration implies close contact between vesicles as an intermediate state. *Biophys. J.* 49:509a.
93. Liao, M. J., and Prestegard, J. H. (1979). Fusion of phosphatidic acid-phosphatidylcholine mixed lipid vesicles. *Biochim. Biophys. Acta* 550:157–173.
94. Miller, C., Arvan, P., Telford, J. N., and Racker, E. (1976). Ca^{2+}-induced fusion of proteoliposomes: Dependence on transmembrane osmotic gradient. *J. Memb. Biol.* 30:271–282.
95. Cornell, B. A., Middlehurst, J., and Separovic, F. (1980). The molecular packing and stability within highly curved phospholipid bilayers. *Biochim. Biophys. Acta* 598:405–410.
96. Huang, C., and Mason, J. T. (1978). Geometric packing constraints in egg phosphatidylcholine vesicles. *Proc. Natl. Acad. Sci. USA* 73:308–310.
97. Szoka, F., Jr., and Papahadjopoulos, D. (1980). Comparative properties and methods of preparation of lipid vesicles (liposomes). *Annu. Rev. Biophys. Bioeng.* 9:467–508.
98. Gaber, B. P., and Sheridan, J. P. (1982). Kinetic and thermodynamic studies of the fusion of small unilamellar vesicles. *Biochim. Biophys. Acta* 685:87–93.
99. Bental, M., Wilschut, J., Scholma, J., and Niz, S. (1987). Ca^{2+}-induced fusion of large unilamellar phosphatidylserine/cholesterol vesicles. *Biochim. Biophys. Acta* 898:239–247.

6

Role of Nonbilayer Lipids in Membrane Fusion

A. J. VERKLEIJ

University of Utrecht, Utrecht, The Netherlands

I. INTRODUCTION

Membrane fusion is a ubiquitous event in cell biology. In fact, every biological membrane has the potential to fuse, but this potentiality may be revealed more in one membrane than the other. At present, many effectors are known to be involved in fusion of biological membranes, such as Ca^{2+} in exocytosis, hormones, growth factors, and antibodies in receptor-mediated endocytosis, and pH during the infectious entry of many enveloped viruses into their host cells. Moreover, the involvement of other substances, such as ATP, cAMP, GTP, drugs, and Ca^{2+}-binding proteins, has been reported. Also, membrane proteins appear to be involved in fusion. The state of the cytoskeleton, the extent of glycosylation, and the distribution of membrane-spanning proteins all can affect the process (see, e.g., Ref. 1). With respect to the role of membrane proteins in fusion, it was generally assumed that both the extrinsic or membrane skeleton proteins and the intrinsic membrane-spanning proteins had to be cleared from the fusion site before actual fusion could occur. However, clearance of membrane proteins or their lateral reorganization has not been visualized using rapid freezing. Thus, since no alterations in the distribution of intramembranous particles could be seen using reliable fast freezing fixation (2-4; see also Chapter 25), only a small area of lipid bilayer appears to be sufficient for fusion (local point fusion). Also, it is important to note that membrane fusion is extremely fast, in the order of milliseconds, and under strict control (5,6).

Whereas it may be clear that proteins and many factors are crucial in the modulation of membrane fusion, it is the lipid part of the membrane that actually fuses. This implies that the lipids of the fusing membranes have to come into close apposition, which requires a reduction in electrostatic repulsion and in the hydration forces between the two approaching surfaces. Subsequently, the lipids of the two membranes have to join, which requires a local bilayer destabilization as the lipids must temporarily leave the bilayer configuration at the point of fusion. Finally, the membranes fuse and a stable bilayer configuration is reestablished.

It has been postulated that special fusogenic lipids such as lysophosphatidylcholine (7), monoglycerides (8), or phosphatidylserine (PS) play a pivotal role in membrane fusion. However, even if there were a universal mechanism for membrane fusion, in view of the high variability of lipid compositions in biological membranes, it is highly unlikely that one particular lipid would be required for this process. PS, for instance, is not present in all membranes (compare bacterial, chloroplast, and eukaryotic membranes). Moreover, the role of PS in membrane fusion can only be relevant in exocytosis or other fusion events from within the cell, as this phospholipid appears to be exclusively located in the cytoplasmic monolayer of the plasma membrane (9) and coated vesicles (10).

In an alternative hypothesis it has been proposed that the property of phospholipids to adopt the hexagonal (H_{II}) phase and/or an inverted micellar conformation is crucial for membrane fusion (11,12; see also Chapter 2). This hypothesis is based on the facts that (1) in any membrane lipids are present which upon isolation can adopt the H_{II} phase at physiological conditions, (2) fusion occurs between artificial membranes in which part of the lipid in isolation prefers the H_{II} phase, (3) all effectors known as fusion factors, including Ca^{2+}, pH, temperature, and apolar peptides, can trigger H_{II} or inverted micellar lipid configurations, and (4) the fusion intermediate to be expected in this concept, the inverted micelle, has actually been visualized by electron microscopy (the so-called lipidic particle). In this chapter all these points are summarized, discussed, and related to membrane fusion in biological and artificial systems.

II. HEXAGONAL (H_{II}) PHASE AND INVERTED MICELLES

Lipids that prefer to adopt the hexagonal II (H_{II}) phase at physiological conditions can be found in almost any biological membrane. In this configuration, the lipids are organized in hexagonally arranged cylinders in which the polar headgroups of the lipid molecules surround a narrow aqueous channel (see Chapter 2). The organization of lipids in the H_{II} phase can be detected with X-ray diffraction (13), freeze fracture electron microscopy (14,15), and ^{31}P nuclear magnetic resonance (^{31}P NMR) (16).

Examples of lipids that adopt the H_{II} phase at physiological conditions are unsaturated phosphatidyethanolamine (PE), monogalactosyldiglyceride from chloroplasts, and monoglucosyldiglyceride from *Acholeplasma laidlawii*. The tendency of other naturally occurring lipids, such as cardiolipin (CL) and phosphatidic acid (PA), to form the H_{II} phase is dependent on the pH and the presence of divalent cations. These negatively charged phospholipids can also adopt the H_{II} phase in the presence of local anesthetics, such as dibucaine and chlorpromazine. CL adopts the H_{II} phase in the presence of cytochrome c. Moreover, the H_{II} phase can be induced by gramicidin, can be modulated by cholesterol, and appears in complex mixtures of synthetic lipids and in lipid extracts of *Escherichia coli*, mitochondria, and rod outer segment membranes (for reviews, see refs. 17 and 18 and Chapter 2). Recently, it has been reported that diacylglycerol (DAG) can induce the H_{II} phase in PE, PC, or PS bilayers (19; see also Chapter 3). This is interesting in view of the role of DAG in the phosphoinositide cascade involved in certain cellular signal transduction processes. Moreover, dolichols are capable of promoting H_{II} phase formation (20).

The transition from bilayer to H_{II} phase in systems of pure PE is remarkably abrupt and occurs within a temperature range of only a few degrees. The latter point can be deduced from the fact that one has to use fast freezing rates (>1000°/s) to prevent the H_{II} to bilayer transition such that the H_{II} phase is preserved for freeze fracturing (21). It has also been shown by differential scanning calorimetry (DSC) that the enthalpy change involved in the structural rearrangement is very small compared to the heat uptake needed to melt the solid bilayer. Because the polymorphic transition occurs above the gel to liquid-crystalline transition of the bilayers, it can be concluded that the H_{II} phase is a liquid-crystalline or fluid phase (16).

Next to the pure H_{II} phase an alternative analogous nonbilayer configuration can be found, i.e., the inverted micelle. This inverted micelle is visible as a well-defined lipidic particle (15,22). Lipidic particles have been observed in pure PE systems or in CL systems in the presence of divalent cations, such as Ca^{2+}, as well as in lipid mixtures in which one of the components prefers the H_{II} phase, such as PE-containing lipid systems, CL/PC mixtures in the presence of Ca^{2+}, or CL in the presence of local anesthetics or cardiotoxins (23,24; for a review, see Ref. 14). Inverted micelles or lipidic particles can be seen as intermediary structures between the bilayer and the H_{II} phase, and they may in fact serve as intermediates in bilayer fusion.

III. FUSION IN ARTIFICIAL MEMBRANE SYSTEMS

The hypothesis that lipids preferring the H_{II} phase and/or an inverted micellar configuration play a crucial role in membrane fusion has been strongly supported by experiments with model membrane systems. Inverted micellar intermediates were first demonstrated with vesicles composed of an equimolar mixture of CL and egg-PC (Fig. 1). These vesicles, in fact, fuse in a nonleaky fashion in the presence of Ca^{2+} (25). More relevant for biological fusion events are mixtures of PE and negatively charged phospholipids. Ca^{2+}, which is known to be involved in certain biological membrane fusion events, triggers bilayer-to-nonbilayer (H_{II} and/or inverted micelle) transitions isothermally in these mixtures (17). An increase in temperature also promotes such a transition. It is therefore logical to suppose that the presence of Ca^{2+} or an increase in temperature allows the nonbilayer tendency of endogenous lipids to be expressed, thus promoting the fusion event. These predictions hold for a variety of mixtures when Ca^{2+} is added, such as mixtures of, on the one hand, PE and, on the other, PS, or PA, or CL, or phosphatidylinositol (PI), PE/PC/cholesterol mixtures (for reviews, see Refs. 15 and 17). In all these mixtures the formation of larger structures is induced, consistent with the occurrence of bilayer fusion.

IV. LIPIDIC PARTICLES

In all the above mixtures undergoing fusion, the fusion event is associated with the appearance of lipidic particles, sometimes localized in regions corresponding in the fusion interfaces. In a recent review (15), it has been argued that the different particle types, which may have confused

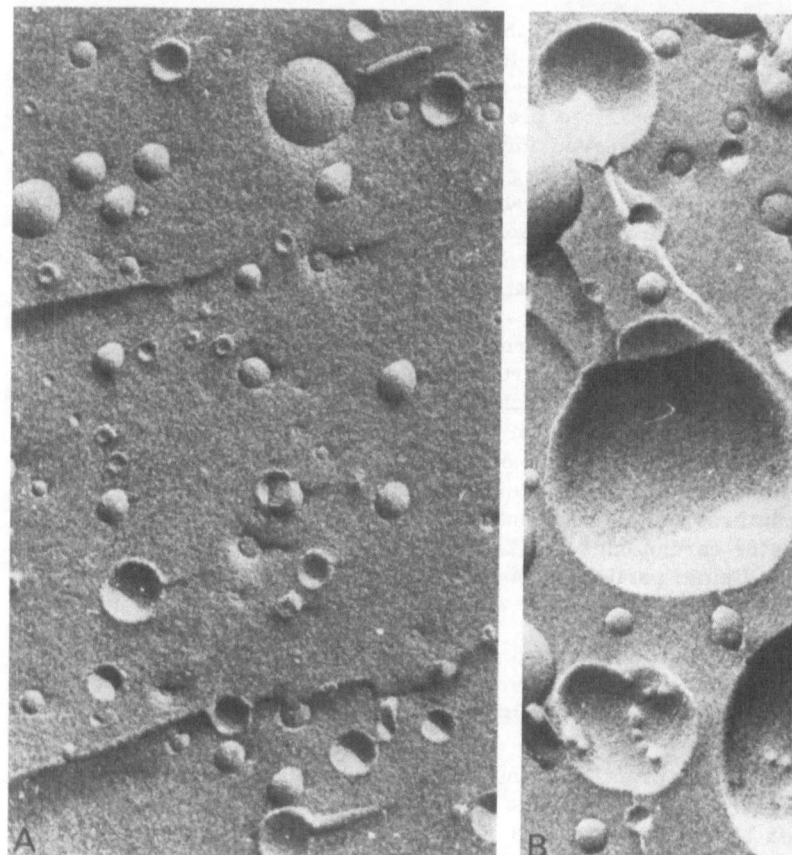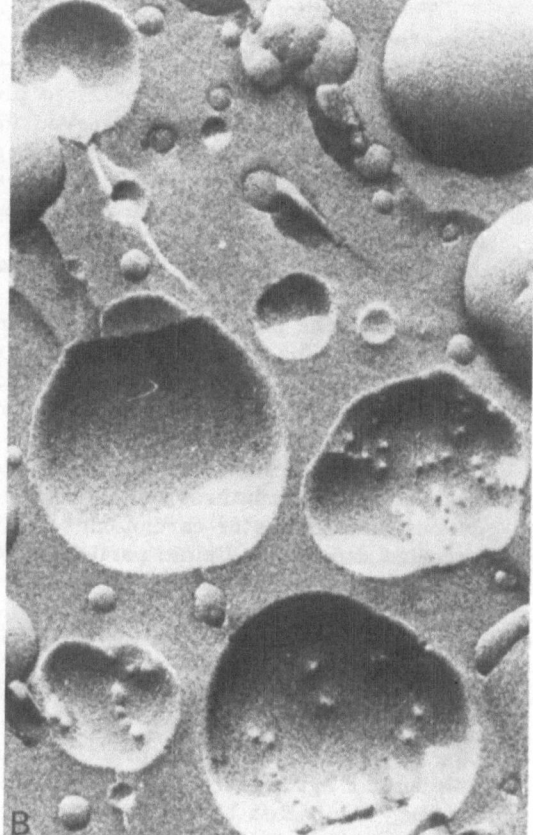

FIGURE 1   Freeze-fracture preparations of lipid vesicles composed of an equimolar mixture of CL and egg-PC before (A) and after (B) addition of Ca²⁺. Magnification 70,000×.

the issue in the early days of the discovery of lipidic particles, can be interpreted as a reflection of different stages during the fusion event.

Figure 2 shows a tentative drawing and the corresponding particles at the different fusion stages. In the "adhesion" stage, two neighboring bilayers form polar contact points requiring local dehydration and charge neutralization which is characteristic for H_{II}-preferring lipids. The bilayers are still intact but the contact points rise to deflections in the freeze-fractured membranes because of the locally high curvature. Such deflections are most likely not homogeneous in size (diameter) and are not well defined. Cusplike particles may reflect this stage and indeed represent intermembrane attachment sites (IMAS model and cusp model (26,27). In the second stage the bilayers "join," which enables intermixing of their lipids. This may proceed with either an inverted micelle or an extended micelle as an intermediate. Fracturing of this joining stage reveals well-defined particles and complementary pits. This joining stage could also be visualized by freeze-substitution after rapid freezing. In a system with multiple joining sites one can see a honeycomb network with local fusion points (Figure 3) (28). Finally, there will be a fission step which can give rise either to

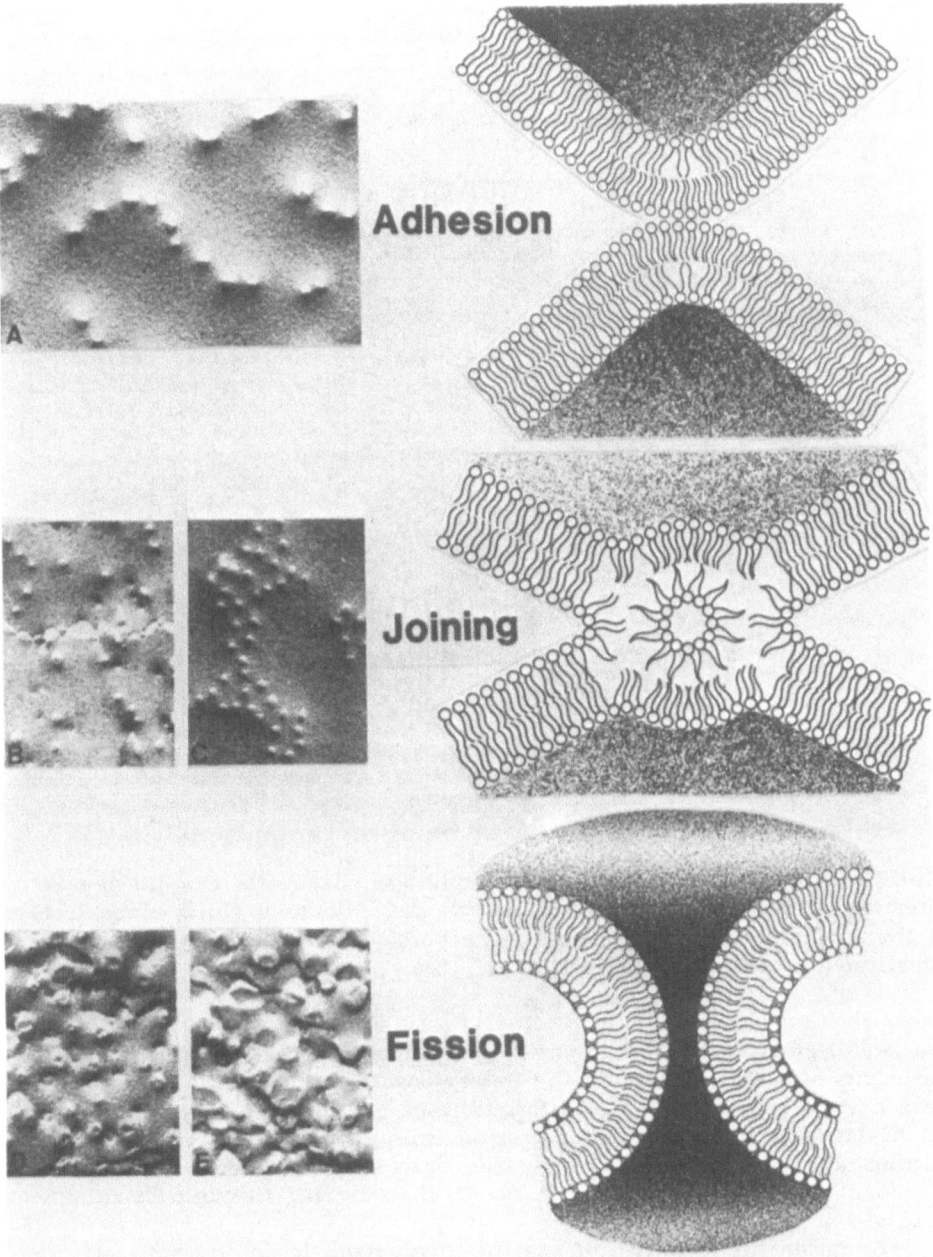

FIGURE 2 Tentative drawings of membrane fusion intermediates and their possible corresponding features as visualized by freeze fracturing. (A-E) Micrographs taken from a sample consisting of dioleoyl-PE (DOPE), dioleoyl-PC (DOPC), and cholesterol (molar ratio 3:1:2) heated up to 60°C for 10 min, cooled down to 4°C, and subsequently frozen at 4°C. Magnification 75,000×.

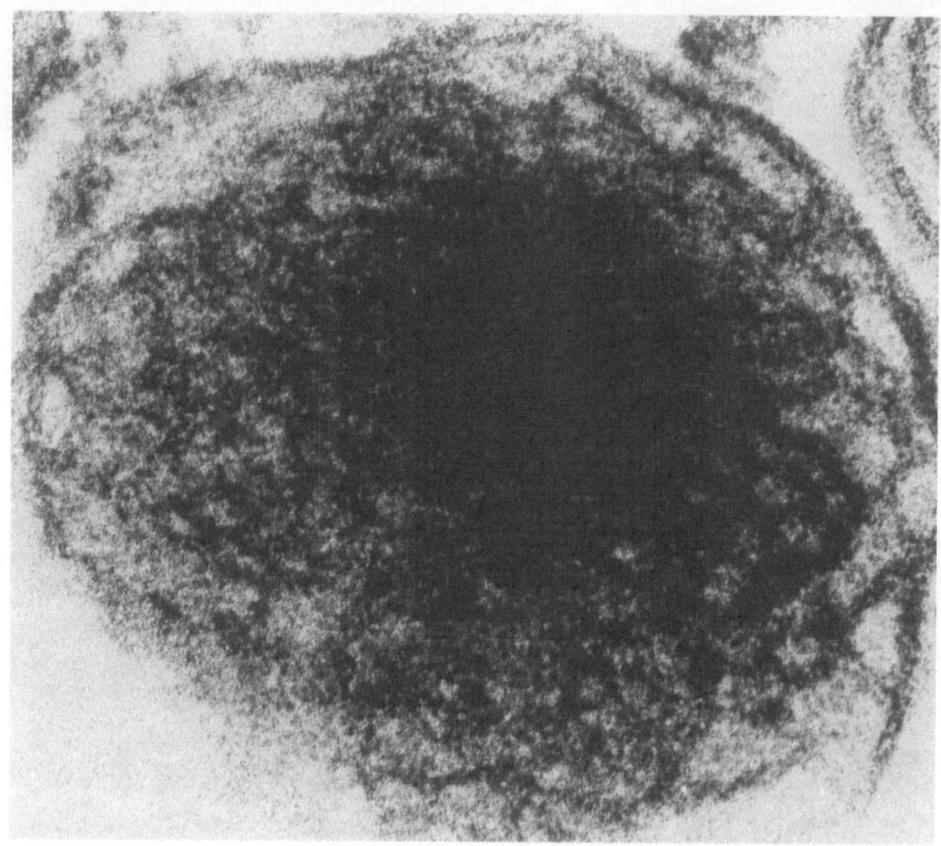

FIGURE 3 Electron micrograph of an equimolar mixture of CL and egg-PC after the addition of an aliquot of 100 mM $CaCl_2$ solution (final concentration in the sample 10 mM). Thin section of cryofixed (fast freezing) and freeze substituted according to Verkleij et al. (28).

the two original vesicles or to fusion and intermixing of the aqueous compartments of both vesicles. In the former case the bilayer will temporarily pass a stage similar to that of intermembrane attachment sites, whereas in the latter case the fracturing passing through the newly formed aqueous channel will show a marked change from cusp shapes to volcanoes with flat tops. The flat tops arise as a result of fracturing through ice in aqueous pores.

For fusion the formation of the H_{II} phase itself is not relevant. First, the H_{II} phase is a stable structure and the production of it by necessity implies the disruption of the aqueous compartment in vesicle systems. The lifetime of the inverted micelle may not be long enough for fusion to occur with other inverted micelles into tubules and H_{II} phase, which is seen under conditions where there is formation of H_{II} phase (15). The inverted micelle in the fusion process is thought to be metastable with a very short lifetime. In that respect, lipidic particles could not be found during the initial rounds of fusion and are seen at later stages (18,29). The lipidic particles seen at later stages likely have a longer lifetime. The reason for not seeing the lipidic particles during the initial fusion rounds is probably due to the fact that fusion, which is not arrested at the joining stage but proceeds as expected, is so fast that the transient intermediates (the inverted micelles) escape detection by freeze fracture electron microscopy.

Recent theoretical work of Siegel (30,31) has indicated that vesicle systems that can undergo a lamellar to H_{II} transition, rapidly develop short-lived inverted micellar intermediates (IMI) between the apposed bilayers at the sites of vesicle contact. When the temperature is above the L_α-H_{II} phase temperature of the lipid at equilibrium (T_H), abundant formation of IMI may induce a fast "zippering down" of the apposed bilayers into inverted hexagonal tubes, with the concomitant loss of vesicular integrity. This transformation seems irrelevant for membrane fusion as it will result in little, if any, coalescence of internal vesicle compartments; rather, the aqueous vesicle contents will be rapidly and completely released to the external medium. Only under specific conditions, when the IMI remain isolated, the IMI may break into so-called interlamellar attachment sites (ILA), which in effect represent fused structures. Using PE-containing vesicles, Ellens, Bentz, and co-workers (32,33) have presented evidence for fusion through transformation of IMI to ILA. Fusion occurred only at temperatures just below the T_H of the lipid or the lipid mixture, under conditions where the lipid at equilibrium exhibits an isotropic ^{31}P-NMR signal, presumably arising from motional averaging due to lipid diffusion through accumulated interconnected ILA.

The presence of nonbilayer lipids is not a prerequisite for fusion of artificial membranes. For instance, PS vesicles fuse upon addition of Ca^{2+} (34-36). It has been noted that the gel to liquid-crystalline phase transition temperature of PS is increased in the presence of divalent cations. This has led to the hypothesis that the divalent-cation-induced fusion of PS liposomes is driven by an isothermal phase transition from liquid-crystalline to the gel phase (34). However, it has been shown that PS vesicles fuse in the presence of Ba^{2+} or Sr^{2+} at temperatures where the lipid acyl chains remain in the liquid-crystalline state throughout the process, from the initial large unilamellar vesicles to the final equilibrium structures (37,38). Therefore, an ionotropic shift in the transition temperature of the lipids and the formation of solid domains are not required for fusion. Rather, in view of the fact that H_{II}-preferring lipids are present in any biological membrane (which is not the case for PS), the involvement of nonbilayer lipids seems more attractive as a general mechanism for biological membrane fusion.

V. FUSION OF BIOLOGICAL MEMBRANES

Taken together, the notions discussed above, showing that (1) biological membranes contain lipids that can adopt the H_{II} phase or inverted micellar organizations, (2) both fusion and bilayer/H_{II} phase transitions can be modulated by similar factors, such as Ca^{2+}, pH, and also apolar peptides, (3) bilayer/H_{II} transitions occur at the same time scale as membrane fusion (on the order of milliseconds), and (4) biological membrane fusion is a local point fusion at which site the lipids have to leave temporarily the bilayer configuration, all corroborate the universal hypothesis that H_{II}-phase-preferring lipids by virtue of their ability to adopt nonbilayer structures (inverted micelles) are pivotal for fusion of biological membranes.

As discussed above, in artificial systems lipidic particles as dynamic fusion intermediates might escape detection because of their short lifetime.

In biological systems it will be even more difficult to trap such intermediary structures, even with fast-freezing, because of (1) again the short lifetime, (2) the problem of synchronizing the fusion events, and (3) the limited number of fusion events per unit surface area. Even with exocytosis, at most only a few hundred vesicles per cell are involved in fusion. However, recently Schmidt et al. (3) have been able to catch exocytosis in chromaffin cells using fast-freezing devices. After stimulation of the exocytotic activity in a controlled way by addition of carbachol, the plasma membrane displayed structural features that strongly resembled the "lipidic particles" found in model systems.

Figure 4 (see also Ref. 3 and Chapter 25) shows some of the structural intermediates of this biological fusion. One can observe undefined semispherical protrusions or cusplike particles, a well-defined lipidic particle, and volcano-like particles with a flat top, which likely represent the adhesion, the joining, and the fission stage, respectively. From this study it is also clear that no particle aggregation (clearance) is visible before or during fusion and that the fusion is a local event. On the other hand, one has to realize that it is very difficult, perhaps even impossible, to make a direct correlation between fusion and these morphological features in other systems where fusion cannot be controlled.

How is fusion then modulated in biological systems? There are in principle different possibilities. One is that intrinsic proteins prevent fusion if they are laterally homogeneously distributed over the surface. Support for this idea is provided by the study of Tarashi et al. (39), who showed that glycophorin keeps PE in the lamellar phase rather than in the H_{II} phase which it prefers in isolation. Addition of a lectin which induces aggregation of the glycophorin (although not visible) allows fusion and the formation of H_{II} phases.

Alternatively, extrinsic proteins may function to prevent membrane fusion. Examples are the cytoskeleton of the erythrocyte membrane (40), clathrin coats in the fusion of caoted vesicles and the endosome (10), and probably cytoskeleton elements involved in exocytosis. These membrane skeletons have to leave the fusion site. As a result of Ca^{2+} binding to PS (which is exclusively in the cytoplasmic leaflet of the membrane (9), the bilayer stabilization is lost (17) and fusion will occur.

Fusion may also be induced by the local accumulation of a nonbilayer lipid. Of special interest is DAG, which has been shown to induce the H_{II} phase in PE, PC, or PS systems (19). Exocytosis in many cell types (41; see also Chapter 23) and myoblast fusion (42; see also Chapter 26) are closely linked to the phosphoinositide cascade of which DAG is an essential element.

Finally, it has been proposed that fusion proteins of a number of enveloped viruses expose apolar peptides in the mildly acidic environment of the endosome, thus activating the membrane fusion capacity of these viruses as an essential step in the infectious entry into their host cells (see Chapter 15). This has led Lucy to suggest that hydrophobic peptides possibly induce membrane fusion (43). In this hypothesis, hydrophobic peptides should be cleaved off from either intrinsic or extrinsic proteins by activation of endogenous cellular proteinases. Other examples mentioned by Lucy (43) are melittin and alamethicin. In this respect, it is interesting to note that several peptides, including cytochrome c, gramicidin (15), melittin (Batenburg and co-workers, unpublished observations), cardiotoxin (23,24), and possibly alamethicin [as suggested by Lau and Chan (44)] can induce nonbilayer structures like inverted micelles or H_{II} cylinders.

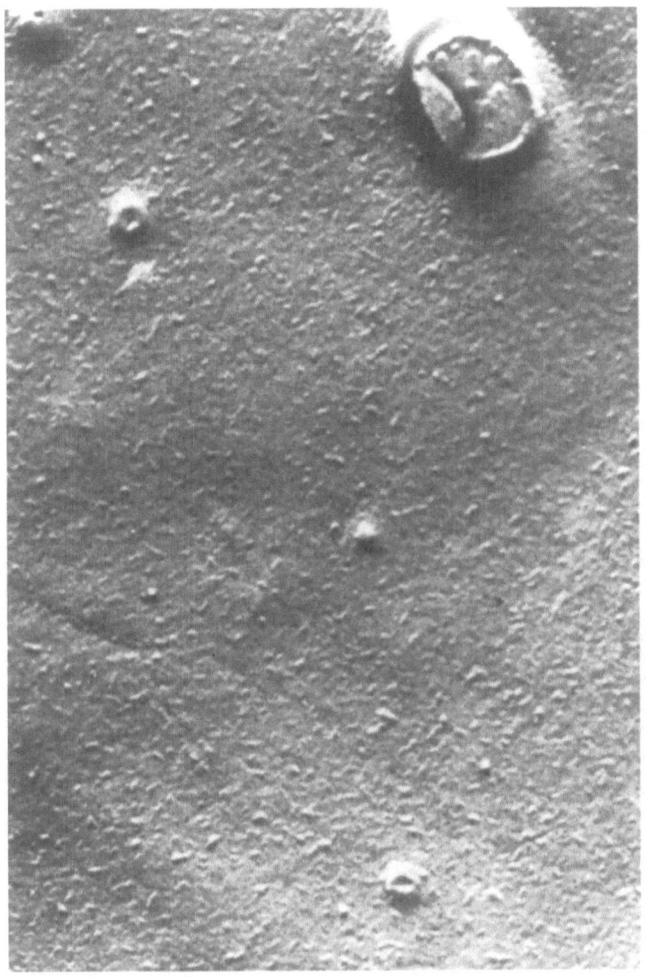

FIGURE 4 Freeze-fractured plasma membrane of chromaffin cells during stimulation with carbachol. This exoplasmic fracture face exhibits features associated with exocytosis. Arrowhead indicates a "lipidic particle"-like structure. Exocytotic openings are visible similar to the volcano-like protrusion found in model systems which represents a fission stage. Magnification 104,000×. This micrograph was kindly provided by Prof. Dr. H. Plattner. (See also Chapter 25.)

In summary, fusion may proceed after removal of limiting factors, which prevent fusion from taking place, or by local changes of the membranes, such as local accumulation of lipids or hydrophobic peptides. These then may allow adhesion and induce local nonbilayer configurations of lipids, resulting in membrane fusion.

REFERENCES

1. Schramm, M., Oates, J., Papahadjopoulos, D., and Loyter, A. (1982). Fusion and implantation in biological membranes. *Trends Pharmacol. Sci.* 3:221-229.

2. Chandler, D. E., and Hauser, J. E. (1980). Arrest of membrane fusion events in mast cells by quick freezing. *J. Cell Biol.* 86:666-674.

3. Schmidt, A., Patzak, A., Lingg, A., Winkler, H., and Plattner, H. (1983). Membrane events in adrenal chromaffin cells during exocytosis: a freeze-etching analysis after rapid cryofixation. *Eur. J. Cell Biol.* 32:31-37.

4. Chandler, D. E. (1984). Comparison of quick frozen and chemically fixed sea urchin eggs: Structural evidence that cortical granule exocytosis is preceded by a local increase in membrane mobility. *J. Cell Sci.* 72:23-36.

5. Douglas, W. W. (1974). Involvement of calcium in exocytosis and the exocytosis-vesiculation sequence. *Biochem. Soc. Symp.* 39:1-28.

6. Hauser, J. E., Riese, T. S., Dennis, M. J., Jan, Y. N., and Evans, L. (1979). Synaptic vesicle exocytosis captured by quick freezing and correlated with wuantal transmitter release. *J. Cell Biol.* 81:275-300.

7. Lucy, J. A. (1970). The fusion of biological membranes. *Nature* 227: 814-817.

8. Ahkong, Q. F., Fisher, D., Tampion, W., and Lucy, J. A. (1973). The fusion of erythrocytes by fatty acids, esters, retinol and α-tocopherol. *Biochem. J.* 136:147-155.

9. Op den Kamp, J. A. F. (1979). Lipid asymmetry in membranes. *Annu. Rev. Biochem.* 48:47-71.

10. Altstiel, L., and Branton, D. (1983). Fusion of coated vesicles with lysosomes: Measurement with a fluorescence assay. *Cell* 32:921-929.

11. Cullis, P. R., and Hope, M. J. (1978). Effects of fusogenic agents on membrane structure of erythrocyte ghosts and the mechanism of membrane fusion. *Nature (London)* 271:672-674.

12. Verkleij, A. J., Mombers, C., Gerritsen, W. J., Leunissen-Bijvelt, J., and Cullis, P. R. (1979a). Fusion of phospholipid vesicles in association with the appearance of lipidic particles as visualized by freeze fracturing. *Biochim. Biophys. Acta* 555:358-361.

13. Luzzati, V., Gulik-Krzywicki, T., and Tardieu, A. (1968). Polymorphism of lecithins. *Nature* 218:1031-1034.

14. Deamer, D. W., Leonard, R., Tardieu, A., and Branton, D. (1970). Lamellar and hexagonal lipid phases visualized by freeze etching. *Biochim. Biophys. Acta* 219:47-60.

15. Verkleij, A. J. (1984). Lipidic intramembraneous particles. *Biochim. Biophys. Acta* 779:43-63.

16. Cullis, P. R., and De Kruijff, B. (1979). Lipid polymorphism and the functional roles of lipids in biological membranes. *Biochim. Biophys. Acta* 559:399-420.

17. Cullis, P. R., De Kruijff, B., Hope, M. J., Verkleij, A. J., Nayar, R., Farren, S. B.,Tilcock, C., Madden, T. D., and Bally, M. B. (1982). Structural properties of lipids and their functional roles on biological membranes. In: *Membrane Fluidity*, Vol. 2 (R. C. Aloia, ed.). Academic Press, New York, pp. 40-79.

18. Verkleij, A. J., Leunissen-Bijvelt, J., De Kruijff, B., Hope, M., Cullis, P. R. (1984). Non-bilayer structures in membrane fusion. In: *Cell Fusion*, Ciba Foundation Symposium 103. Pitman, London, pp. 45-59.

19. Das, S., and Rand, R. P. (1984). Diacylglycerol causes major structural transitions in phospholipid bilayer membranes. *Biochem. Biophys. Res. Commun.* 124:491-496.

20. Valtersson, C., Van Duin, G.,Verkleij, A. J., Chojnacki, T., De Kruijff, B., and Dallner, G. (1985). The influence of dolichol, dolichol esters and dolichyl-phosphate on phospholipid polymorphism and fluidity in model membranes. *J. Biol. Chem.* 260:2742-2751.

21. Van Venetië, R., Hage, W. J., Bleumink, J. G., and Verkleij, A. J. (1981). Propane jet-freezing: a valid ultra-rapid freezing method for preservation of temperature-dependent lipid phases. *J. Microsc.* 123: 287-292.

22. Verkleij, A. J., Mombers, C., Leunissen-Bijvelt, J., and Ververgaert, P. H. J. (1979). Lipidic intramembraneous particles. *Nature* 279:162-163.

23. Gulik-Krzywicki, T., Balerna, M., Vincent, J. P., and Luzdanski, M. (1981). Freeze-fracture study of cardiotoxin action on axonal membrane and axonal membrane lipid vesicles. *Biochim. Biophys. Acta* 643:101-114.

24. Batenburg, A. M., Rochat, H., Verkleij, A. J., and De Kruijff, B. (1985). The penetration of a cardiotoxin into cardiolipin model membranes and its applications on lipid organization. *Biochemistry* 24:7102-7110.

25. Wilschut, J., Holsappel, M., and Jansen, R. (1982). Ca^{2+}-induced fusion of cardiolipin/phosphatidylcholine vesicles monitored by mixing of aqueous contents. *Biochim. Biophys. Acta* 690:297-301.

26. Miller, R. G. (1980). Do "lipidic particles" represent intermembrane attachment sites? *Nature* 287:166-167.

27. Hui, S. W., and Stewart, T. P. (1981). "Lipidic particles" are intermembrane attachment sites. *Nature* 290:427-428.

28. Verkleij, A. J., Humbel, B., Studer, D., and Muller, M. (1985). Lipidic particles systems as visualized by thin section electron microscopy. *Biochim. Biophys. Acta* 812:591-594.

29. Bearer, E. L.,Düzgüneş, N., Friend, D. S., and Papahadjopoulos, D. (1982). Fusion of phospholipid vesicles arrested by quick freezing. The question of lipidic particles as intermediates in membrane fusion. *Biochim. Biophys. Acta* 693:93-102.

30. Siegel, D. (1986). Inverted micellar intermediates and the transition between lamellar, cubic and inverted hexagonal lipid phases. II. Implications for membrane-membrane interactions and membrane fusion. *Biophys. J.* 49:1171-1183.

31. Siegel, D. (1987). Membrane-membrane interactions via intermediates in lamellar to inverted-hexagonal phase transitions. In: *Cell Fusion* (A. E. Sowers, ed.). Plenum Press, New York, pp. 181-207.

32. Ellens, H., Bentz, J., and Szoka, E. C. (1986). Fusion of phosphatidylethanol micelle containing liposomes and the mechanism of the L_α-H_{II} phase transition. *Biochemistry* 25:4141-4147.

33. Bentz, J., and Ellens, H. (1988). Membrane fusion: Kinetics and mechanisms. *Colloids Surfaces* 30:65-112.

34. Papahadjopoulos, D., Portis, A., and Pangborn, W. (1978). Calcium-induced lipid phase transitions and membrane fusion. *Ann. NY Acad. Sci.* 308:50-66.

35. Wilschut, J., Düzgüneş, N., Fraley, R., and Papahadjopoulos, D. (1980). Studies on the mechanism of membrane fusion: Kinetics of calcium ion-induced fusion of phosphatidylserine vesicles followed by a new assay for mixing of aqueous vesicle contents. *Biochemistry* 19: 6011-6021.

36. Wilschut, J., Düzgüneş, D., and Papahadjopoulos, D. (1981). Calcium/magnesium specificity in membrane fusion: Kinetics of aggregation and fusion of phosphatidylserine vesicles and the role of bilayer curvature. *Biochemistry* 20:3126-3133.

37. Düzgüneş, N., Paiement, J., Freeman, K., Lopez, N. G., Wilschut, J., and Papahadjopoulos, D. (1984). Modulation of membrane fusion by ionotropic and thermotropic phase transitions. *Biochemistry* 23:3486-3494.

38. Bentz, J., Düzgüneş, N., and Nir, S. (1985). Temperature dependence of divalent cation induced fusion of phosphatidylserine liposome. Evaluation of the kinetic rate constants. *Biochemistry* 24:1064-1072.

39. Taraschi, T. F., Van der Steen, A. T. M., De Kruijff, B., Tellier, C., and Verkleij, A. J. (1982). Lectin-receptor interactions in liposomes: evidence that binding of wheat-germ agglutinin to glycoprotein phosphatidylethanolamine vesicles induces nonbilayer structures. *Biochemistry* 21:5756-5764.

40. Portis, A., Newton, C., Pangborn, W., and Papahadjopoulos, D. (1979). Studies on the mechanism of membrane fusion: Evidence for an inter-membrane Ca^{2+}-phospholipid complex, synergism with Mg^{2+} and inhibition by spectrin. *Biochemistry* 18:780-790.

41. Baker, P. F., and Knight, D. E. (1984). Calcium control of exocytosis in bovine adrenal medullary cells. *Trends Neurosci.* 7:120-127.

42. Wakelam, M. J. O. (1983). Inositol phospholipid metabolism and myoblast fusion. *Biochem. J.* 214:77-82.

43. Lucy, J. A. (1984). Do hydrophobic sequences cleaved from cellular polypeptides induce membrane fusion reactions in vivo? *FEBS Lett.* 166:223-231.

44. Lau, A. L. Y., and Chan, S. J. (1975). Alamethecin-mediated fusion of lecithin vesicles. *Proc. Natl. Acad. Sci. USA* 72:2170-2174.

7

Fusion of Zwitterionic Phospholipid Vesicles

DOV LICHTENBERG

Sackler School of Medicine, Tel Aviv University, Ramat Aviv, Israel

THOMAS E. THOMPSON

University of Virginia Health Sciences Center, Charlottesville, Virginia

I. INTRODUCTION

Fusion of biological membranes is fundamental to a number of important physiological and pharmacological processes. In an attempt to gain an understanding of these complex processes, much work has been devoted to studies of the interactions between relatively simple model membranes made of phospholipids. In spite of the possible role of proteins in the fusion of biological membranes (1), studies of fusion in model membrane systems can give relevant information, since in both protein-mediated biological membrane fusion and model membrane fusion eventually the core structures of the apposing membranes, i.e., the lipid bilayers, have to merge to form one membrane.

Several types of fusion processes have been studied. This chapter reviews the work on the spontaneous fusion of the uncharged bilayer membranes made of zwitterionic phospholipids. The effects of charged lipids, various ions, and other compounds that may enhance or reduce the tendency of phospholipid vesicles to fuse are discussed in other chapters in this book (e.g., Chapters 4-6, 9). This review is not meant to be comprehensive. In fact, the selection of literature described in this chapter was made on the basis of our subjective evaluation of the concepts that currently prevail in the field and the facts that support these concepts.

II. EARLY STUDIES

The first studies on fusion of liposomes employed small unilamellar vesicles (SUV) made by ultrasonic irradiation of aqueous dispersions of phosphatidylcholines with saturated acyl chains (2,3). While such vesicles are stable for long periods of time when incubated in the liquid-crystalline state (at temperatures above the transition temperature of the lipid, T_m), size growth of the vesicles occurs when the vesicles are incubated in the gel state (below T_m) (4-10).

This rather simple description of the stability of SUV made of saturated phosphatidylcholines has become common knowledge only recently. In the early stages of research in which these vesicles were used as models for biological membranes, it was not clear how the size growth occurred and what the optimal conditions for this size growth were. Kantor and Prestegard (11) attributed the size growth of dimyristoylphosphatidylcholine (DMPC) small vesicles in the phase transition range to the presence of myristic acid (12), while Papahadjopoulos and co-workers (13) provided evidence that indicated that the vesicle size growth observed in the presence of myristic acid occurred by transfer of lipid molecules rather than by vesicle fusion. More confusion came from the finding that although vesicles made by sonication of saturated phosphatidylcholines below T_m were very unstable and in the gel state spontaneously underwent fusion into larger vesicles, these vesicles could be annealed above T_m to form stable small vesicles (14).

Sonication above the T_m of the lipid in fact results in the formation of SUV (15) that are very stable above T_m, but undergo a slow size increase in the gel phase. This was first shown in the calorimetric study of Suurkuusk et al. (6), who attributed the growth in size of dipalmitoyl-phosphatidylcholine (DPPC) SUV to incubation in the gel state, but also suggested that cycling through the phase transition might play a role. Similar claims were made by Van Dijck et al. (16).

Only later was it recognized that cycling through the phase transition actually reduces the overall rate of vesicle size growth as aggregated gel-phase vesicles undergo deaggregation when the lipid is brought into the liquid-crystalline state (7). In 1977 Kantor et al. showed that SUV made of *pure* DMPC transform into larger vesicles at an appreciable rate only below T_m (17). In 1979 it was demonstrated by Larrabee that SUV made of distearoylphosphatidylcholine (DSPC) are more stable above or in the phase transition range than in the gel phase, where the vesicles grow into larger unilamellar structures (4). One year later, it became clear that the same is true for DPPC SUV (5). More detailed work indicated that the vesicle size growth occurs through a fusion mechanism (7) and provided data on the temperature dependence of the resultant vesicle size (8-10) as well as on the size dependence of gel-phase vesicle stability (18,19).

III. PROBING SLOW VESICLE FUSION PROCESSES

Generally speaking, membrane fusion is that process in which two membranes merge into one. If, in addition, the two membranes define two internal aqueous compartments, then fusion results in the joining of these two compartments. An alternative mechanism by which the mean diameter of vesicles can grow involves net transfer of lipid molecules from some vesicles to other vesicles. While the end result of such a process will not necessarily be different from that of fusion, the two mechanisms are quite different. Proving the involvement of a fusion mechanism in vesicle size growth requires demonstration of mixing of membrane lipids and mixing of the entrapped material from the fused membranes. Thus, for rapid fusion processes, such as the calcium-induced fusion of negatively charged vesicles, a large number of techniques have been developed to monitor the mixing of entrapped volumes (see Chapter 4). However, to date no data are available that unambiguously prove the mixing of contents of small vesicles made of zwitterionic synthetic phospholipids during the course of the very slow size growth of these vesicles.

The conclusion that gel-phase SUV fuse is therefore based on direct measurement of the size and structure of the resultant vesicles and on indirect lines of evidence based on kinetic studies of the size-growth process. Studies of vesicle sizes have employed various techniques such as electron microscopy, static and dynamic light scattering, molecular sieve chromatography, differential scanning calorimetry (DSC), nuclear magnetic resonance (NMR) spectroscopy, and determination of the ratio between the number of phospholipid molecules on the outer and inner vesicle mono-layers. The last parameter, which can be evaluated from the interaction of the lipid molecules on the outer surface of the lipid bilayer with various reagents, including fluorescently labeled molecules, spin labels, and NMR shift reagents, is particularly sensitive to the number of concentric lamellae in each vesicle. It has been demonstrated very clearly with such techniques, in conjunction with other methods, that the large vesicles formed from gel-phase SUV are unilamellar (7).

Another parameter that has been followed during the size growth of DPPC SUV is the total aqueous volume entrapped within vesicles. The surface area of a vesicle formed by fusion of two identical small vesicles is about equal to twice the surface area of each of the original vesicles. The radius of the resultant vesicle is therefore larger than that of the parent vesicles by a factor of $2^{1/2}$; the volume entrapped within the resultant vesicle is larger than the volume of a parent vesicle by a factor of $2^{3/2}$ and the volume entrapped within the resultant vesicle is larger than the sum of volumes entrapped within the two parent vesicles by a factor of $2^{1/2}$. Thus, fusion results in an increase of 41% in the total volume entrapped within vesicles. If the fusion of two spherical vesicles is to result in a larger vesicle that is also spherical, fusion must be accompanied or followed by a flow of water into the resultant vesicle. For this inflow of water to occur without resulting in an osmotic gradient across the vesicle membrane, the membrane permeability to solutes must increase at least transiently.

A study of the aqueous volume trapped within growing vesicles made of DPPC (5) has shown that the increase in trapped volume is accompanied by equilibration of entrapped solutes with the ambient aqueous phase. In addition, this increase of volume was as expected on the basis of the size increase, as derived from more direct measurements. The major conclusion of this study was that vesicles of an average radius of about 110 Å grow during several weeks of incubation at room temperature to vesicles with a mean radius of about 350 Å. If the mechanism responsible for this size growth is fusion, the formation of each large vesicle requires 13-18 small vesicles (5,9).

Cooling a vesicle dispersion to a temperature below T_m results in a rapid aggregation of the vesicles (7,20-22). This process expresses itself by an increase in the turbidity of the dispersion, but it has only a slight effect on the characteristics of ^{1}H and ^{31}P NMR spectra (7). An important feature of the turbidity increase is that heating the dispersion to a temperature above T_m results in a decrease of turbidity to its original value, suggesting that vesicle aggregation is reversible. In contrast, the characteristics of the NMR resonances vary over much longer time scales and cannot be reversed by reheating the dispersion (7). Reheating, however, slows the rate of size transformation. We have interpreted this to mean that size transformation occurs by interactions between aggregated gel-phase vesicles (7). We found that the mechanism responsible for the size growth is apparently of second order (7). This finding, by itself, is not sufficient to derive a conclusion about the mechanism, but the result appears to be inconsistent with a lipid transfer mechanism (23).

IV. STABILITY OF SELF-AGGREGATES OF ZWITTERIONIC PHOSPHOLIPIDS IN AQUEOUS MEDIA AND ITS DEPENDENCE ON THE LIPID PHASE

Phosphatidylcholines (PC's) with very short acyl chains are quite soluble in water. As the length of the acyl chains is extended to eight carbon atoms, the solubility of the monomeric phospholipid becomes very limited, and above a critical concentration (cmc) micelles are formed. Lipids with still longer acyl chains cannot be packed along curved surfaces such as those of spherical micelles. Thus, the most stable state of aggregation of these PC's is the bilayer. Hydration of synthetic PC's with two saturated acyl chains longer than C_{13} results in spontaneous formation of concentric multibilayers. This, by itself, does not imply that these aggregates are the most stable form of the phospholipid in aqueous solution. In fact, the multilamellarity is a function of the ionic strength of the aqueous media. In very dilute dispersions of DPPC in distilled water the resultant large liposomes are made of only one to three bilayers. Thus, it is possible that the multilamellar structure of "swollen phospholipid" reflects a long-lived metastable state of aggregation. The formation of this state is a result of the preparation method, and the lipid is kinetically trapped in these multilamellar liposomes by the absence of an efficient mechanism by which it can segregate into unilamellar vesicles of smaller sizes. Of course, it is possible that interlamellar interactions stabilize this structure and make it the equilibrium state of aggregation, but this has not been proven as yet.

Sonication results in the formation of relatively stable SUV only if it is done above T_m. If sonication is then followed by cooling the vesicle dispersion to the gel phase, the spherical vesicles probably become polygonal or faceted (24); i.e., the vesicles are composed of areas of planar packing. We have suggested that the formation of this structure is the result of the gel-phase structural parameters and the packing asymmetry in the two monolayers of the vesicle bilayer (19). More specifically, since the transition-induced simultaneous decrease of the total surface area and increase of the bilayer thickness do not alter the general spherical morphology of the vesicles, these changes must result in packing strains. The alignment of the phospholipid fatty acid chains within a facet in a nonradial fashion provides a way of partially relieving these packing strains, but not all the curvature-induced gel-phase packing strains are relieved so that the vesicles transform slowly to larger, more planar structures.

V. TEMPERATURE DEPENDENCE OF THE FUSION OF SMALL UNILAMELLAR VESICLES

In a relatively recent report, Gaber and Sheridan (25) claim that SUV made of pure saturated PC with an acyl chain length of 14–18 carbon atoms fuse spontaneously to larger vesicles when incubated at, as well as below, the phase transition temperature. This result is in contrast to Larrabee's result with DSPC (4) as well as ours with DPPC vesicles (5,7,9). SUV made of pure saturated PC are clearly quite stable above T_m in the liquid-crystalline state. In fact, spontaneous size growth occurs only when the vesicles are cooled to a temperature below the *pretransition temperature* of the lipid (T_p). This is concluded from a number of considerations based on work from various laboratories. First, a pretransition of small enthalpic change has been observed for SUV made of DPPC in the temperature range 29-32°C (26). Second, this as well as pretransitions in other phospholipid systems may reflect changes in the ordering of the polar head groups of the lipids in the bilayer, as previously suggested by Petersen et al. (22,27). Third and most important, recent work of McConnell and Schullery (10) has shown that SUV made of DPPC fuse only below 30°C, the pre-transition temperature, T_p. Similarly, the work of Ostrowsky and Sornette (28) on fusion of small DMPC vesicles also shows that the rate of size growth changes faster in the vicinity of T_p (13-17°C) than near the main transition (21-23°). The data on the temperature dependence of vesicles aggregation suggest that the principal effect of cooling the vesicle dispersion to a temperature below T_p is to promote aggregation of the vesicles (22).

It is also important to note that the apparent equilibrium size of vesicles made by incubation of the SUV below T_p depends on the temperature of incubation in a stepwise fashion. At 13°C or above, the mean diameter of the resultant vesicles is 700 Å, independent of temperature, whereas at 8°C and below, the mean diameter of the resultant vesicles is 950 Å. In addition to this temperature effect on the vesicle size, the rate of fusion also depends on the temperature, but this effect is continuous rather than stepwise (10).

In order to understand the critical nature of the temperature dependence of the size of fused vesicles, it is necessary to note the following. First, in addition to the main transition at 41.4°C and the pretransition at 35.5°C, fully hydrated DPPC multilamellar dispersions undergo a broad endothermic phase transition at 14-18°C, denoted "the subtransition" (29,30). It can only be observed when the lipid is incubated at a low temperature for an extended period (29) and is attributed to the formation of a lamellar crystalline phase, denoted L_c, which is characterized by an ordered two-dimensional orthorhombic subcell structure (31). Second, the polar head-groups in lamellar crystalline-phase phospholipids are less hydrated than in the gel phase (30-33). Third, vesicular systems formed upon incubation of small unilamellar vesicles of DPPC at 4°C also give rise to a "subtransi-tion" at about 11°C (26).

Since in the crystalline phase (L_c) the effect of curvature is likely to persist to larger vesicle radii than in the gel (L_β) phase, the relatively stable vesicles of 700 Å diameter formed by a process similar to that occur-ring in the gel phase may be quite unstable in the L_c phase. The decreased hydration of the polar headgroups in the L_c phase makes it easier for these crystalline vesicles to fuse into larger unilamellar vesicles, 950 Å in diameter.

For any given lipid phase, the temperature-dependence of the size transformation can be explained in terms of the gradual decrease of the disorder within the lipid bilayer that accompanies the decrease of the temperature below T_p (34-37). If the lipid molecules are considered to be rigid rods, the decrease of disorder means that decreasing the temperature below T_p results in an increase in the rigidity of the molecules. Conse-quently, the distortions necessary to create planar packing areas will become larger as the temperature decreases. This will thus destabilize the vesicles further and increase the rate of fusion.

As the vesicles grow in size, the packing asymmetry becomes lesser and the vesicles become more stable in the gel phase. Thus, in the rather

heterogeneous preparation of DPPC vesicles made by the French press
method, only those vesicles with a diameter smaller than 400 Å are unstable
below T_p and fuse spontaneously to form larger single lamellar vesicles
(18). This result is consistent with the marked sensitivity of the phase
transition parameters to the vesicle size within the range of 200-400 Å,
which can be attributed to the decrease in the effective bilayer curvature
due to packing rearrangement of the lipid molecules. Although the instability
is very small for gel-phase vesicles larger than 400 Å diameter, SUV of
DPPC fuse into 700 Å diameter vesicles (5), which is larger than is required
to relieve the packing strains. This may be the result of mechanistic details
of the vesicle-vesicle fusion process as described in the following sections.

VI. THEORETICAL APPROACHES TO MEMBRANE AGGREGATION AND FUSION AND THEIR APPLICABILITY TO THE FUSION OF GEL-PHASE SMALL UNILAMELLAR VESICLES

The major forces that play a role in the aggregation of uncharged vesicles
are the London dispersion (van der Waals) attractive forces between the
two surfaces and the hydration force (38), which come from the requirement that approaching vesicles push away water molecules in their collision
path (see also Chapter 3). In addition, for elastic vesicles such as those
in the liquid crystalline state, the repulsive undulatory force (39) may
also play a role.

Based on the results of Parsegian and his colleagues (40), Ostrowsky
and Sornette (28) concluded that the close approach of two vesicles is
hindered by "the tail of the short-range repulsive interaction" at a distance
of about 50 Å between the bilayers. The repulsive forces that appear at
about this distance constitute a potential barrier of a height V_c which the
approaching vesicle must overcome in order to aggregate. For elastic liquid-
crystalline bilayers, the repulsive undulatory force makes a major contribution to V_c, thus lowering the probability of aggregation. In contrast, for
vesicles of low elasticity, such as the gel-phase bilayers, the only significant
contribution to V_c comes from bilayer hydration which is lower than that
of the liquid-crystalline bilayers. Consequently, the probability that gel-
phase vesicles will aggregate is higher.

The force between two apposing egg PC liquid-crystalline bilayers
absorbed onto mica surfaces has been measured by Horn (41) as a function
of the distance between the two mica surfaces. At large separations *between
the mica surface* (about 107 Å), the interaction energy was found to be
negative, reflecting the major role of van der Waals attraction between
the bilayers. Forcing the membranes into closer apposition required energy.
Ultimately, the increasing pressure resulted in the formation of a single
bilayer of 44 Å thickness. This process is diagrammed in Figure 1. At
the point where the separation is 107 Å, the distance between the headgroups of the lipids in the apposing bilayers is 19 Å [107 - (2 × 44)].
This value for the distance at which the attractive Van der Waals forces
and the repulsive hydration force balance each other is in a reasonable
agreement with that given by Lis et al. (42) based on their studies with
multilamellar vesicles. Thus, if the kinetic energy of the vesicles is sufficient to overcome the long-range repulsive forces so that the vesicles
approach each other at a distance smaller than about 30 Å, the van der
Waals forces can cause the membranes to come together to a distance of
about 20 Å, where the hydration force prevents closer contact. If liquid-

crystalline vesicles aggregate, the distance between the lipid headgroups on opposing bilayers will therefore be about 20 Å. In aggregates of the less hydrated gel-phase (L_β) bilayers the distance between apposing headgroups is probably smaller, and in aggregated crystalline phase (L_c) bilayers, still smaller distances may be expected.

The major forces that play a role in the fusion of two aggregated vesicles into a single larger vesicle are the repulsive hydration force, short- and long-range interactions due to steric repulsion forces, and the very short-range intermolecular interbilayer force (28,43-45) that may strongly depend on the order parameters of the interacting membranes (46). The crucial step in the fusion of vesicles is the approach of two vesicles to a distance at which the attractive very short-range interactions overcome the hydration forces. Owing to the strong hydration repulsion at short range (41,42,47), considerable energy (>10 mJ/m^2) is required to bring the membranes closer together. However, application of a sufficiently large force will cause the membranes to thin, then rupture, and fuse together into a single bilayer, as shown by Horn (41). This process is illustrated in Figure 1. However, a much smaller force is needed if the surface is dehydrated or the membrane is irregular or nonlamellar. Thus, in the experiments of Horn (41), impurities such as small amounts of n-hexane help trigger fusion.

The effect of larger amounts of n-hexane is comparable to that of alamethicin (48) and other compounds that have been claimed to induce fusion by promoting a concave curvature in the region indicated by the circle in Figure 1C. This structure resembles that of a micelle (48). More important, in molecular terms, the encircled region in Figure 1C, shown also in Figure 1D, can easily lead to the formation of a structure such as the one in Figure 1F. The latter structure is, of course, equivalent to that of Figure 1G, which is that of a fused membrane shown in Figure 1H. Large structural defects occur in unannealed vesicles made below T_p and cause them to fuse very rapidly. Annealed gel-phase vesicles, however, also have irregular molecular packing which probably plays an important role in their fusion.

In the context of their study of the kinetics of the size growth of DMPC vesicles, Ostrowsky and Sornette (28) have shown that the actual kinetics can be predicted with some accuracy on the basis of a phenomenological model, which assumes that the vesicles undergo a partially reversible, diffusion-controlled, aggregation that may lead to fusion between bilayers. The experimental results of these authors, however, are difficult to reconcile with other studies on the fusion of gel-phase SUV. Ostrowsky and Sornette studied dilute dispersions with phospholipid concentrations two to three orders of magnitude lower than those used by Larrabee (4), Schullery et al. (5), and Wong et al. (9), yet the rate of size growth appeared to be at least as fast as in these other studies, even though second-order kinetics for the fusion of DPPC SUV at 21°C have been clearly demonstrated (7,9). This discrepancy may have arisen because the aggregation that precedes fusion may occur in two distinct steps, as suggested by Petersen and Chan (22). These two steps, denoted flocculation and coagulation, differ in terms of the distances between the aggregated vesicles and are therefore likely to show different rates of deaggregation. The light-scattering results of Ostrowsky and Sornette may reflect a transformation of vesicle flocculates into coagulates in which the vesicles are more closely packed. This speculation appears to be consistent with the disagreement, observed after long periods of vesicle incubation, between the

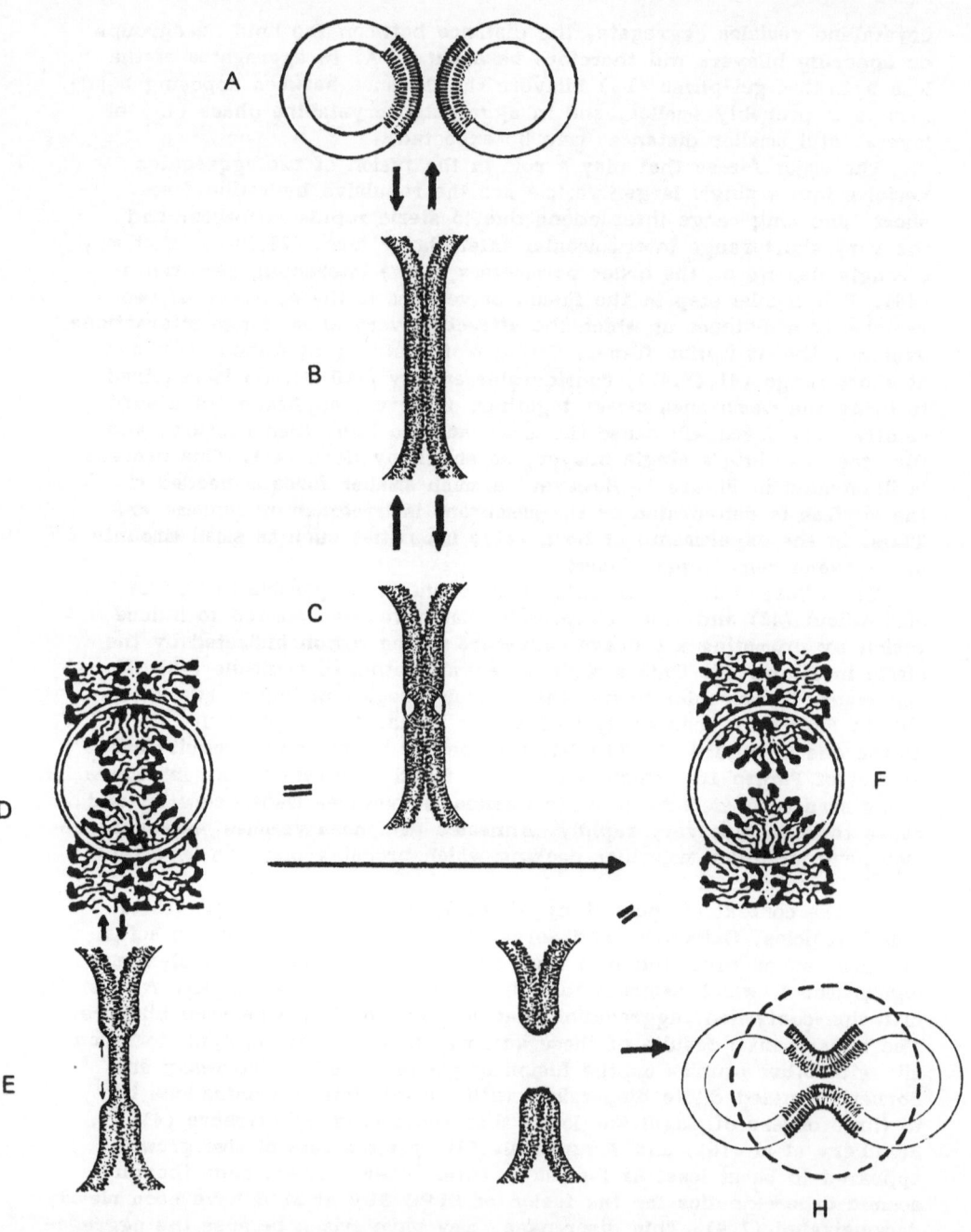

experimental scattering intensities and those predicted on the basis of the assumption that "the second step" (following aggregation) involves fusion of the aggregated vesicles (28). More important, it is consistent with the results of other laboratories on the concentration dependence of the rate of vesicle fusion. However, more experimental work is needed in this area.

A completely different approach to understanding the adhesion, instability, and fusion of membranes has been developed by Dimitrov and his collaborators (49). These authors adopted an analysis of the dynamics of thin films and carried out a series of theoretical and experimental investigations to support the validity of their general, but oversimplified model. In molecular terms, this model postulates a deformation of two approaching membranes to result in membrane destabilization and rupture of the intervening aqueous film, ultimately yielding "holes" in the membranes. These are equivalent to fusion of the two membranes and mixing of the two aqueous compartments. Intuitively, this model can be modified to describe the fusion of gel-phase SUV, but such analysis has not yet been carried out.

In conclusion, the existing theoretical attempts to understand the kinetics and structural details of membrane aggregation and fusion need to be refined in order to explain quantitatively the experimental data. In the next section we offer a qualitative formulation of the mechanisms of aggregation and fusion of SUV in the gel phase.

FIGURE 1 Model for fusion of lipid vesicles. If two SUV approaching each other (A) are forced to within less than 20 Å of each other, such as in the experiment of Horn (41) using bilayers associated with mica surfaces, the resultant associated bilayers (B) are likely to fuse only if additional energy causes them to approach very closely. Then the very short-range interactions overrule the hydration force and a structural irregularity appears (C) due to either an external force, an internal irregularity in the membrane, or an impurity residing in the membrane. Under the influence of an external force, the intermediate C (also shown in D) will result in the single bilayer shown in E, which then grows rapidly until it occupies the entire planar region under pressure. This is not likely to happen if an external force is not applied. Instead, the irregularity in C will disappear by reformation of B, or the intermediate D will undergo a slight structural change to form the intermediate F, or its equivalent (G). Flattening of this intermediate will result in the fused vesicle (H), which upon further flattening will occupy the area represented by the dashed circle in H.

Note that in gel-phase vesicles the surfaces are flat and dehydrated; thus a facet-to-facet approach described in B is likely to occur. Furthermore, the faceted surfaces are likely to form intermediates such as C, D and F, G and fuse. Also note that if the pair of fusing vesicles are a part of a larger aggregate of vesicles, flattening of the fused vesicle will cause it to collide with neighboring vesicles, thus propagating the fusion process.

VII. MECHANISTIC ASPECTS OF AGGREGATION AND FUSION OF SMALL UNILAMELLAR VESICLES MADE OF SATURATED PHOSPHATIDYLCHOLINE IN THE GEL PHASE

The process by which large unilamellar vesicles form through the fusion of SUV below the phase transition may be as follows. In the gel phase, SUV of saturated diacyl phosphatidylcholines take the form of irregular polyhedrons (24). Each facet of a polyhedron is a flat bilayer of gel-phase phosphatidylcholine. The facets are joined to each other by disorganized PC molecules in a glasslike gel phase. The minimum energy for any one polyhedron is obtained as a balance between the lower free energy of the gel-phase facets and higher free energy of the joints between facets. An additional important feature of these polyhedral vesicles is the fact that the energy required to move the joint between facets by increasing the size of one facet and decreasing the size of an adjacent facet is small. Thus, as far as the energetics of the bilayer itself is concerned, the shape of any one polyhedron can be changed by the application of small distortion forces.

An aqueous dispersion of these polyhedrons will form aggregates of limited size. The aggregates are stabilized by the energy minimum that exists between two bilayers at some critical distance from each other, which is likely to be less than 20 Å for gel-phase PC below T_p. Thus, the aggregation process can be imagined to begin when two vesicles find themselves with one facet of one polyhedron at a distance of about 20 Å or less from one facet of a second polyhedron. Additional vesicles then adhere to this initial "dimer" by the same process except that each additional vesicle can have two or more facets in apposition rather than just one facet per vesicle, as in the initial dimer. Therefore, the stabilization energy for the "trimer," "tetramer," and so forth will be larger per vesicle than it was initially in the starting vesicle "dimer." This will be the case until the first shell of 12-15 vesicles surrounding the initial "dimer" is filled. When the first shell is filled, the stabilization per vesicle for additional vesicles will be lower since only one facet of each added vesicle can be apposed by a facet of a vesicle already in the aggregate. This, in conjunction with geometrical considerations, suggests that the resultant aggregates will tend to consist of about 15 small vesicles.

Thus, the aggregate of SUV from which a fused vesicle forms is comprised of about 15 rather regular polyhedral vesicles, the arrangement of which is such that the free energy of the aggregate is minimal even if the distortions of some individual vesicles in the aggregate put them at a higher free energy than they would have if they were not part of the aggregate. The compliance characteristic of each individual vesicle would cause the aggregate as a whole to be easily deformed. The ease of deformation is, however, reduced by the fact that changes in shape of individual vesicles require concomitant changes in the aqueous volumes of the vesicle interiors which are resisted by the low water permeability of gel-phase phospholipid.

During the process of forming an aggregate, the compliance of the polyhedrons acts to maximize the facet-to-facet apposition, so that the final configuration of the aggregate may be quite regular. However, this shape can be changed by thermal fluctuations and interaggregate collisions.

If collisions between aggregates occur with sufficient energy to force a large shape change in a vesicle pair within the aggregate, the shape change will be accompanied by a transient bilayer rupture that permits bulk water flow to accommodate the internal volume change coupled to the shape change. If the transient rupture occurs in the two adjacent vesicles in the same place at the same time, fusion of these two vesicles is likely, especially when a fault in one of the vesicles occurs near a fault in a neighboring vesicle in the aggregate. The structure at the point where the two faulted bilayers meet resembles that of an inverted micelle, which previously has been suggested to be a transient step in the mechanism of vesicle-vesicle fusion (see Chapter 2,6). The fusion product will now tend to round up while being leaky. That is, the whole bilayer of this new fusion vesicle will be transiently faulted in many places. If in adjacent vesicles transient faults occur near the fusion product, a new fusion event is likely to occur, especially if two faults occur next to each other. In this way the fusion process will be propagated throughout the whole aggregate. Since the process involves stepwise fusion of unilamellar vesicles only, the final product will be a large unilamellar vesicle.

The fusion process will depend on the concentration of lipid, i.e., the concentration of aggregates, because the initial nucleating event is a collision between two aggregates. The size of the product vesicle will depend on the number of vesicles in the aggregate. This number may be larger in the crystalline phase than in the gel phase, thus resulting in a larger fused vesicle below the subtransition temperature, as compared to the size of the fused vesicle formed in the gel phase.

VIII.   EFFECTS OF VARIOUS ADDITIVES ON VESICLE SIZE GROWTH PROCESSES

As noted above, fusion is accompanied by an increase of the aqueous volume entrapped within the product vesicle. If this volume increase is not accompanied by permeation of solutes contained in the medium, an osmotic gradient will result. This gradient may inhibit further fusion, as suggested by McConnell and Schullery (10) in the context of rationalizing the dramatic inhibition of vesicle fusion observed in the presence of high concentrations of membrane impermeant solutes. The rate of fusion of gel-phase SUV is also reduced by the presence of polyvalent ions such as calcium or lanthanides in the medium. Presumably, binding of such ions to the bilayers confers to the membranes a net charge that causes them to repel each other and thus prevents aggregation followed by fusion (7).

Other types of solutes likely to affect the size growth of vesicles include surfactants, charged lipids, and alcohols. In spite of rather limited data on the effects of such additives on the rate of fusion, we suggest the following speculative generalizations. First, small vesicles are unstable in their gel phase; thus a driving force for size growth exists. The pathways that these small vesicles may follow for their size to grow involve either collisional mechanisms (fusion or lipid transfer between aggregated vesicles) or a mechanism of exchange of lipid molecules through the ambient aqueous phase. In a pure lipid system size growth occurs spontaneously, but slowly. Any additive that can distribute into the membrane and stabilize it by annealing the structural irregularities of the vesicle bilayer reduces the driving force for size growth and thus reduces the rate of fusion of gel-phase small vesicles. This probably is the mechanism by which very small mole fractions of cholesterol confer stability to DPPC SUV (10). Similar considerations may explain the instability of liquid-crystalline small vesicles made of PC and cholesterol at mole ratios of cholesterol to PC greater than 1 (50).

In general terms, other mechanisms by which additives can change the stability of vesicles are by altering a process that operates in their absence and/or by introducing a new route for size growth. The mechanism by which SUV fuse requires vesicle aggregation. Any charged molecule that distributes into the membrane and thereby charges it can be expected to interfere with vesicle aggregation by introducing electrostatic repulsion between vesicles. It is therefore not surprising that gangliosides stabilize SUV (51). On the other hand, a surface-active molecule, whether charged or not, is likely to make lipid transfer easier, by increasing the desorption rate of phospholipid molecules from the small vesicle bilayer and/or by increasing the steady-state concentration of phospholipid in the ambient aqueous phase. This may be the mechanism by which bile salts (52-55), lysolecithin (56), and other detergents such as octylglucoside (57) induce vesicle size growth. Since these size growth processes occur above, as well as below, T_m and yield sizes that are dependent on the final ratio of bile salt to lipid in the bilayers (54,55), they are not very likely to share much in common mechanistically with the fusion of gel-phase SUV. The size growth observed for SUV in the presence of various alcohols [such as ethanol, ethylene glycol, propylene glycol, and glycerol (10)] also does not appear to be mechanistically similar to the fusion of gel-phase SUV, since this process does not exhibit the bimodal size distribution kinetics observed with pure DPPC. It is possible that alcohol-induced size growth is indeed similar to the size growth caused by various surfactants, in that it makes it easier for lipid to transfer from one vesicle to another. This then may result in the growth of some of the larger vesicles at the expense of the disintegration of other SUV.

REFERENCES

1. Lucy, J. A. (1984). Do hydrophobic sequences cleaved from cellular polypeptides induce membrane fusion? *FEBS Lett.* 166:223-231.
2. Taupin, C., and McConnell, H. M. (1972). Membrane fusion. In: *Mitochondria; Biomembranes* (S. G. van den Burgh, P. Borst, L. L. M. van Deenen, J. C. Riemersma, E. C. Slater, and J. M. Tager, eds.). Elsevier/North-Holland Biomedical Press, Amsterdam, pp. 219-229.
3. Metcalfe, J. C., Birdsall, N. J. M., and Lee, A. G. (1972). NMR studies of dynamic features of membrane structure. In: *Mitochondria; Biomembranes* (S. G. van den Burgh, P. Borst, L. L. M. van Deenen, J. C. Riemersma, E. C. Slater, and J. M. Tager, eds.). Elsevier/North-Holland Biomedical Press, Amsterdam, pp. 197-217.
4. Larrabee, A. L. (1979). Time-dependent changes in the size distribution of distearoylphosphatidylcholine vesicles. *Biochemistry* 18:3321-3326.
5. Schullery, S. E., Schmidt, C. F., Felgner, P., Tillack, T. W., and Thompson, T. E. (1980). Fusion of dipalmitorylphosphatidylcholine vesicles. *Biochemistry* 19:3919-3923.
6. Suurkuusk, J., Lentz, B. R., Barenholz, Y., Biltonen, R. L., and Thompson, T. E. (1980). Fusion of dipalmitoylphosphatidylcholine vesicles. *Biochemistry* 15:1393-1401.
7. Schmidt, C. F., Lichtenberg, D., and Thompson, T. E. (1981). Vesicle-vesicle interactions in sonicated dispersions of dipalmitoylphosphatidylcholine. *Biochemistry* 20:4792-4797.
8. Wong, M., and Thompson, T. E. (1982). Aggregation of dipalmitoylphosphatidylcholine vesicles. *Biochemistry* 21:4133-4139.

9. Wong, M., Anthony, F. H., Tillack, T. W., and Thompson, T. E. (1982). Fusion of dipalmitoylphosphatidylcholine vesicles at 4°C. *Biochemistry* 21:4126-4132.

10. McConnell, D. S., and Schullery, S. E. (1985). Phospholipid vesicle Fusion and drug loading: Temperature, solute and cholesterol effects, and, a rapid preparation for solute-loaded vesicles, *Biochim. Biophys. Acta* 818:13-22.

11. Kantor, H. L., and Prestegard, J. H. (1978). Fusion of phosphatidylcholine bilayer vesicles: Role of free fatty acid. *Biochemistry* 17:3592-3597.

12. Kantor, H. L., and Prestegard, J. H. (1975). Fusion of fatty acid containing lecithin vesicles. *Biochemistry* 14:1790-1795.

13. Papahadjopoulos, D., Hui, S., Vail, W. J., and Poste, G. (1976). Studies on membrane fusion. I. Interactions of pure phospholipid membranes and the effect of myristic acid, lysolecithin, proteins and DMSO. *Biochim. Biophys. Acta* 448:245-264.

14. Lawaczeck, R., Kainosho, M., Girardet, J-L., and Chan, S. I. (1975). Effects of structural defects in sonicated phospholipid vesicles on fusion and ion permeability. *Nature* 256:584-586.

15. Huang, C. (1969). Studies on phosphatidylcholine vesicles. Formation and physical characteristics. *Biochemistry* 8:344-352.

16. Van Dijck, P. W. M., De Kruijff, B., Aarts, P. A. M. M., Verkleij, A. J., and DeGier, J. (1978). Phase transition in phospholipid model membranes of different curvature. *Biochim. Biophys. Acta* 506:183-191.

17. Kantor, H. L., Mabrey, S., Prestegard, J. H., and Sturtevant, J. M. (1977). A calorimetric examination of stable and fusing lipid bilayer vesicles. *Biochim. Biophys. Acta* 466:402-410.

18. Lichtenberg, D., Freire, E., Schmidt, C. F., Barenholz, Y., Felgner, P. L., and Thompson, T. E. (1981). Effect of surface curvature on stability, thermodynamic behavior, and osmotic activity of dipalmitoylphosphatidylcholine single lamellar vesicles. *Biochemistry* 20:3462-3467.

19. Lichtenberg, D., and Schmidt, C. F. (1981). Molecular packing and stability in the gel phase of curved phosphatidylcholine vesicles. *Lipids* 16:555-557.

20. Chong, C. S., and Colbow, K. (1976). Light Scattering and turbidity measurements on lipid vesicles. *Biochim. Biophys. Acta* 436:260-282.

21. Yi, P. N., and MacDonald, R. C. (1973). Temperature dependence of optical properties of aqueous dispersions of phosphatidylcholine. *Chem. Phys. Lipids* 11:114-134.

22. Petersen, N. O. and Chan, S. I. (1978). The effect of the thermal prephase transition and salts on the coagulation and flocculation of phosphatidylcholine bilayer vesicles. *Biochim. Biophys. Acta* 509:111-128.

23. Lawaczeck, R. (1978). Intervesicular lipid transfer and direct fusion of phospholipid vesicles: A comparison, on a kinetic basis. *J. Coll. Inter. Sci.* 66:247-256.

24. Blaurock, A. E., and Gamble, R. C. (1979). Small phosphatidylcholine vesicles appear to be faceted below the thermal phase transition. *J. Memb. Biol.* 50:187-204.

25. Gaber, B. P., and Sheridan, J. P. (1982). Kinetic and thermodynamic studies of the fusion of small unilamellar phospholipid vesicles. *Biochim. Biophys. Acta* 685:87-93.

26. Lichtenberg, D., Menashe, M., Donaldson, S., and Biltonen, R. L. (1984). Thermodynamic characterization of the pretransition of unilamellar dipalmitoylphosphatidylcholine vesicles. *Lipids* 19:395-400.

27. Petersen, N. O., Kroon, P. A., Kainosho, M., and Chan, S. I. (1975). Thermal phase transitions in deuterated lecithin bilayers. *Chem. Phys. Lipids* 14:343-349.

28. Ostrowsky, N., and Sornette, D. (1985). Interaction between nonionic vesicles. Importance of the aliphatic-chain phase transition. In: *Physics of Amphiphiles: Micelles, Vesicles and Microemulsions* (Course XC of the international school of physics "Enrico Fermi," edited by V. Degiorgio and M. Corti), North-Holland, Amsterdam, Oxford, New York, Toronto, pp. 563-586.

29. Chen, S. C., Sturtevant, J. M., and Gaffney, B. J. (1980). Scanning calorimetric evidence for a third phase transition in phosphatidylcholine bilayers. *Proc. Natl. Acad. Sci. USA* 77:5060-5063.

30. Cameron, D. G., and Mantsch, H. H. (1982). Metastability and polymorphism in the gel phase of 1,2-dipalmitoyl-3-sn-phosphatidylcholine: A Fourier transform infrared study of the subtransition. *Biophys. J.* 38:175-184.

31. Ruocco, M. J., and Shipley, G. G. (1982). Characterization of the subtransition of hydrated dipalmitoyl phosphocholine bilayers. Kinetic, hydration and structural study. *Biochim. Biophys. Acta* 691:309-320.

32. Füldner, H. H. (1981). Characterization of a third phase transition in multilamellar dipalmitoyllecithin liposomes. *Biochemistry* 20:5707-5710.

33. Wu, W.-G., Chong, P. L.-G., and Huang, C. (1985). Pressure effect on the rate of crystalline phase formation of L-α-dipalmitoylphosphatidylcholine in multilamellar dispersions. *Biophys. J.* 47:237-242.

34. Janiak, J. M., Small, D. M., and Shipley, G. G. (1979). Temperature and compositional dependence of the structure of hydrated dimyristoyl lecithin. *J. Biol. Chem.* 254:6068-6078.

35. Gaber, B. P., and Peticolas, W. L. (1977). On the quantitative interpretation of biomembrane structure by Raman spectroscopy. *Biochim. Biophys. Acta* 465:260-274.

36. Cameron, D. G., Casal, H. L., and Mantsch, H. H. (1980). Characterization of the pretransition in 1,2-dipalmitoyl-sn-glycero-3-phosphocholine by Fourier transform infrared spectroscopy. *Biochemistry* 19:3665-3672.

37. Huang, T. H., Skarjune, R. P., Wittebort, R. J., Griffin, R. G., and Oldfield, E. (1980). Restricted rotational isomerization in polymethylene chains. *J. Am. Chem. Soc.* 102:7377-7379.

38. LeNeveu, D. M., Rand, R. P., and Parsegian, V. A. (1976). Measurement of forces between lecithin bilayers. *Nature (London)* 259:601-603.

39. Marcelja, S., and Radic, N. (1976). Repulsion of interfaces due to boundary water. *Chem. Phys. Lett.* 42:129-130.

40. Parsegian, V. A., Fuller, N., and Rand, R. P. (1979). Measured work of deformation and repulsion of lecithin bilayers. *Proc. Natl. Acad. Sci. USA* 76:2750-2754.

41. Horn, R. G. (1984). Direct measurement of the force between two lipid bilayers and observation of their fusion. *Biochim. Biophys. Acta* 778:224-228.

42. Lis, L. J., McAlister, M., Fuller, N., Rand, R. P., and Parsegian, V. A. (1982). Interactions between neutral phospholipid bilayer membranes. *Biophys. J.* 37:657-666.

43. Ninham, B. W. (1980). Long range vs. short range forces. The present state of play. *J. Phys. Chem.* 84:1423-1430.

44. Nir, S. (1977). Van der Waals interactions between surfaces of biological interest. *Prog. Surf. Sci.* 8:1-58.

45. Nir, S., Bentz, J., Wilschut, J., and Düzgüneş, N. (1983). Aggregation and fusion of phospholipid vesicles. *Prog. Surf. Sci.* 13:1-24.

46. Jönssön, B., and Wennerström, H. (1983). Image-charge forces in phospholipid bilayer systems. *J. Chem. Soc. Faraday Trans.* 2 79:19-35.

47. LeNeveu, D., Rand, R. P., Gingell, D., and Parsegian, V. A. (1977). Measurement and modification of forces between lecithin bilayers. *Biophys. J.* 18:209-230.

48. Lau, A. L. Y., and Chan, S. I. (1974). Alamethicin-mediated fusion of lecithin vesicles. *Biochemistry* 13:4942-4948.

49. Dimitrov, D. S., Zhelev, D. V., and Jain, R. K. (1985). Stability of membrane systems modeled as multilayered viscoelastic films. *J. Theor. Biol.* 113:353-377.

50. Collins, J. J., and Phillips, M. C. (1982). The stability and structure of cholesterol-rich co-dispersions of cholesterol and phosphatidylcholine. *J. Lipid Res.* 23:291-298.

51. Felgner, P. L., Friere, E., Barenholz, Y., and Thompson, T. E. (1981). Asymmetric incorporation of trisialoganglioside into dipalmitoyl-phosphatidylcholine vesicles. *Biochemistry* 20:2168-2172.

52. Enoch, H. G., and Stritmatter, P. (1979). Formation and properties of 1000 Å-diameter single-bilayer phospholipid vesicles. *Proc. Nat. Acad. Sci. USA* 76:145-149.

53. Lichtenberg, D., Tamir, I., Cohen, R., and Peled, Y. (1984). Cholesterol solubilization and supersaturation in bile: Dependence on total lipid concentration and formation of metastable dispersions. In: *Surfactants in Solution*, Vol. 2 (K. L. Mittal, and B. Lindman, eds.). Plenum Press, New York, London, pp. 981-997.

54. Almog, S., Kushnir, T., Nir, S., and Lichtenberg, D. (1986). Kinetic and structural aspects of reconstitution of phosphatidylcholine vesicles by dilution of phosphatidylcholine-sodium cholate mixed micelles. *Biochemistry* 25:2597-2605.

55. Schurtenberger, P., Mazer, N. A., and Kanzig, W. (1985). Micelle to vesicle transition in aqueous solutions of bile salt and lecithin. *J. Phys. Chem.* 89:1042-1049.

56. Poole, A. R., Howell, J. I., and Lucy, J. A. (1970). Lysolecithin and cell fusion. *Nature* 227:810-813.

57. Jackson, M., Schmidt, C. F., Lichtenberg, D., Litman, B. J., and Albert, A. D. (1982). Solubilization of phosphatidylcholine bilayers by octyl glucoside. *Biochemistry* 21:4576-4582.

8

Fusion of Phospholipid Vesicles with Planar Phospholipid Membranes

JOSHUA ZIMMERBERG

*National Institute of Diabetes, Digestive, and Kidney Diseases,
National Institutes of Health, Bethesda, Maryland*

I. INTRODUCTION

Biological membrane fusion is implicated in a large number of cellular
processes, but there are few examples of surface fusion in our everyday
experience.* Implicit in studies involving fusion of membranes composed
of defined components is the hope of developing some understanding about
membrane fusion on the physicochemical level and relating that understanding
to biological membrane fusion. In this chapter we limit our discussion to
investigations studying fusion of synthetic phospholipid vesicles with planar
phospholipid bilayers and consider the difference between vesicle-vesicle
fusion and fusion of vesicles with a planar bilayer. We will examine the
relevance of these studies to biological exocytosis and the applications
of techniques involving the fusion of vesicles with a planar bilayer to
other questions of interest.

II. EVIDENCE FOR FUSION OF PHOSPHOLIPID VESICLES WITH PLANAR PHOSPHOLIPID BILAYERS

The evidence for fusion is based on either measuring the sequelae of
incorporation of the vesicular membrane into the planar film or measuring
the transfer of vesicular contents across the planar film (Fig. 1). Fusion
must be distinguished from nonfusional incorporation of material from vesicle
into planar membrane. In a number of studies it was demonstrated that
lipid, protein, or antibiotics can transfer from one membrane to another
without fusion (3-5; Marty and Finkelstein, unpublished observations).

---

*Lord Rayleigh found that water droplets in a fountain will rebound after
collision but coalesce if the water was soapy, dirty, or an electric field
was present (1). C. V. Boys reported that soap bubbles will fuse when
a charged rod of sealing wax is brought near (2).

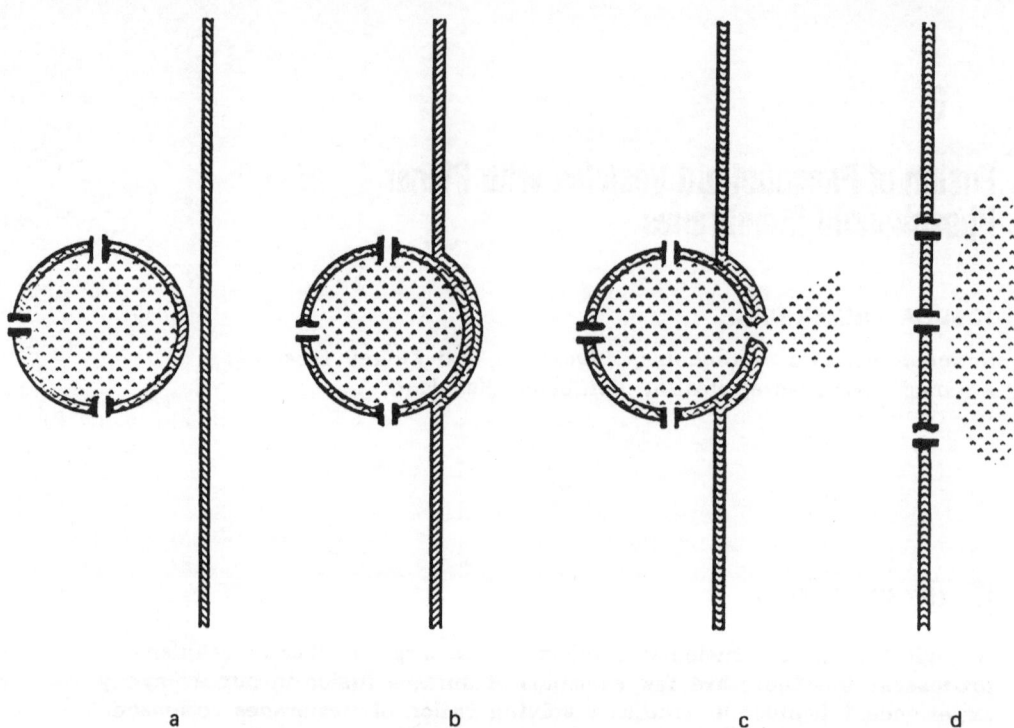

a   b   c   d

FIGURE 1 Some of the events thought to occur during fusion of vesicles
with planar bilayers. The vesicle and the planar membrane are first (a)
and then adhere after collision with the planar membrane wrapping around
the taut vesicle (b). After osmotic swelling of the vesicle, either fusion
or rupture occurs. Fusion (c) results in the transfer of many channels
to the membrane simultaneously and dye transfer across the membrane (d).
Rupture results in dye release to the same side to which the vesicles were
added.

In other studies of the interaction of vesicles with planar films such exchange was not ruled out (6-9).

   In nonfusional incorporation, presumably membrane components would
incorporate into the planar film randomly as single insertions, but in fusion
the entire vesicle membrane is incorporated into the planar film at once.
Thus if a channel-forming protein is incorporated into vesicle membrane,
fusion can be distinguished from nonfusional incorporation by observation
of the simultaneous insertion of many channels into the planar membrane
(10-12). As expected for fusion, the number of channels simultaneously
inserted increases as one increases the ratio of protein molecules to lipid
molecules in the vesicle preparation, and the size distribution of vesicles
correltes with a histogram of the number of channels inserted per event
(10). It is possible in these studies that an aggregate of channels inserts
via nonfusional incorporation instead of fusion. However, when care was
taken to disperse the channels into monomers, simultaneous incorporation
of many channels was observed (10). It is still possible, however, that
channels reaggregated once incorporated into the vesicle.

   The same conditions that promote simultaneous channel insertion lead
to the transfer of fluorescent material packaged within phospholipid vesicles

across the planar bilayer with the same time course (13). These findings have recently been confirmed with a new assay system for studying fusion of phospholipid vesicles with planar phospholipid membranes (14). In this system, large vesicles filled with a fluorescent dye at self-quenching concentrations are added to the solution bathing the planar phospholipid membrane and visualized by fluorescence microscopy. Flashes of fluorescence presumably reflecting release of dye from vesicles are recorded at the same time as simultaneous channel insertion. Similar flashes are seen during fusion of dye-filled synaptic vesicles with a planar bilayer (15). In summary, there is substantial evidence for the fusion of phospholipid vesicles with planar phospholipid membranes.

III. CONDITIONS LEADING TO FUSION

Two conditions must be met in order for vesicles to fuse with planar membranes. First, there must be an intimate contact between the two membranes. This stable adhesion has been termed a "prefusion" state and is empirically demonstrated by adding vesicles to the solution bathing a planar bilayer, washing out the free vesicles by perfusing with fresh solution, and observing fusion in the absence of free vesicles (12,16). Second, there must be an osmotic swelling of the adherent vesicle, which leads to membrane fusion. The role of osmotic swelling is the subject of a recent review (17).

The lipid and ionic requirements leading to the prefusion state (18) are in reasonable agreement with independent measurements of the equilibrium spacing between phospholipid bilayers (19); i.e., those conditions leading to vesicles adhering and then fusing with planar phospholipid membranes also lead to an intimate (<5 Å) contact between bilayers as determined by X-ray diffraction. In particular, the combination of negatively charged lipids and divalent cations, or phosphatidylethanolamine (PE) without divalent cations, leads to adhesion (18). PE-containing vesicles fuse at a higher rate than vesicles containing phosphatidylcholine (PC). The presence of hydrocarbon in the planar bilayer increases the rate of fusion but is not needed for fusion to occur.

IV. VESICLE-PLANAR BILAYER FUSION VS. VESICLE-VESICLE FUSION

The requirement for osmotic swelling in vesicle-planar membrane fusion, but not in vesicle-vesicle fusion, is the major difference between these systems. Experimentally, conditions that lead to vesicle-vesicle fusion (20,21), using identical lipids and solutions, do not lead to fusion of vesicles with planar membranes (18). What is the role of osmotic stress in fusion? It is thought that stretching adherent membranes in the region of contact exposes acyl chains to water, forcing a molecular rearrangement (fusion or rupture) to prevent further hydration (12). In vesicle-vesicle fusion, adhesion leads to stretching because the energy of adhesion deforms the vesicle sufficiently to increase pressure (22-24; see Chapter 3). Since the planar membrane, not having an internal volume, can wrap around an adherent vesicle without causing a pressure change, this stretching must be provided by osmotic swelling of the vesicle.

A perspective is required that includes not only the individual molecular interactions of membrane components and the ionic constituents of the medium

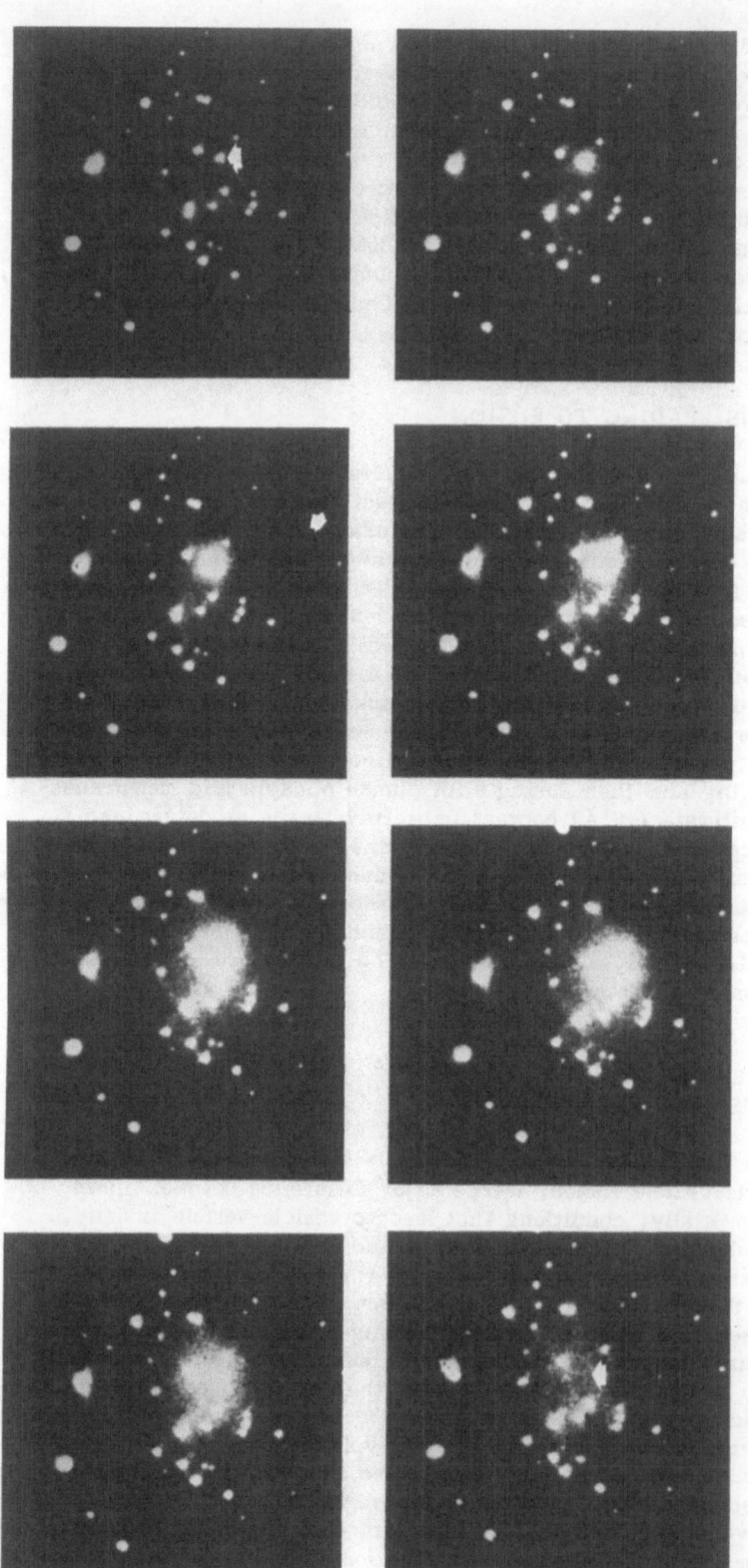

but also the contribution of the macroscopic topological sequelae of such interactions. Recent work, both theoretical and experimental, suggests the following scenario. The first step, adhesion and molecular contact, leads to a geometrical disruption—a flattening of a surface. This, in turn, not only directly alters surface curvature, but also changes enclosed volumes. This volume change, depending on the preexisting surface-to-volume ratio, may lead to a change in intravesicular pressure (12,23,24), leading to an additional mechanical stress oncomponent molecular species. This added stress will alter the free energy of the species, which may change nearest-neighbor interactions to either increase or decrease net adhesion, and thus surface deformation. These events may either progress to a stable adhesion of two membranes or lead toward rupture or fusion (12,23).

By studying the interaction of closed surfaces (vesicles) with open surfaces (planar bilayers) of defined chemical composition, surface tension, and flaccidity, we can control or eliminate the energetic contribution due to the deformation of the vesicle that follows that vesicle's adhesion to a planar membrane. With spherical vesicles we expect the planar membrane to wrap around the taut vesicle (12,24) without flattening the vesicle, but we have no evidence for this. The surprising experimental result of removing vesicular deformation is that conditions known to lead either to fusion or to rupture of phospholipid vesicles with each other lead instead to a steady-state adhesion of vesicles to planar phospholipid bilayers. Fusion occurs only when osmotic swelling of the vesicle is induced (12,16,17). Assuming a reasonable area of membrane contact, one computer model for water flow into the vesicle estimates a fusion tension of 2 dyne/cm (Niles and Cohen, unpublished results). However, the structure of the contact zone between a vesicle and the planar membrane and its area must be known before such calculations can be useful.

The requirement for osmotic swelling of the vesicle in fusion of vesicles with a planar bilayer has recently been confirmed in bilayer visualization experiments (Niles and Cohen, manuscript in preparation). As shown in Figure 2, the stable prefusion state is observed as the vesicle hits the

---

FIGURE 2 Release of vesicular contents as a result of fusion. Large (1-10 μm) unilamellar vesicles were loaded with 200 mM of the membrane-impermeant fluorescent dye calcein. Vesicles were added to one side (*cis*) of a planar phospholipid membrane, which was bathed by a buffer adjusted to be isosmotic with the vesicular contents. Cobalt, a quencher of calcein fluorescence, was also present in the *cis* compartment. Vesicles adsorbed to and fusing with the planar membrane were viewed by fluorescence video microscopy. The temporal sequence is from left to right, running top to bottom. The first panel (upper left corner) shows the planar membrane just prior to fusion. Note that all the vesicles are within the depth of focus, establishing that they are adsorbed to the planar membrane. Fusion was induced by causing the vesicles to swell. Upon fusion the dye was released to the *trans* side of the planar membrane and a flash of light appeared, as shown in the subsequent panels. The vesicle that fused in the sequence shown is indicated by an arrow. The second (upper right corner) through the eighth panels occurred 83, 125, 167, 208, 292, 333, and 417 msec, respectively, after the first panel. Note that the fused vesicle has disappeared (its previous position is marked by the arrow) in the last panel. (Figure provided by Drs. W. D. Niles and F. S. Cohen.)

membrane and adheres. Dye release is seen with osmotic swelling. By differentially applying to the two aqueous compartments compounds that quench the dye, vesicle rupture and fusion can be distinguished. Both rupture and fusion are seen, with roughly equal amounts. Lysis tensions for phospholipid bilayer membranes are known to be on the order of 3 dyne/cm (25). Since fusion and lysis are both seen, it follows that fusion tension is also on that same order. The visualization experiments point out yet another difference between biological exocytosis and model membrane fusion: specifically, the strict control that prevents lysis of secretory granules during exocytosis.

V. RELEVANCE TO BIOLOGICAL FUSION

Fusion of vesicles with a planar membrane is geometrically similar to, for example, exocytotic secretion in that the radius of curvature of a secretory vesicle is smaller than that of a cell. However, the composition of biological membranes is much different than that of phospholipid bilayers, containing proteins, phospholipases, glycoproteins, gangliosides, and sterols. In a number of systems there is evidence for morphological specializations at the relatively small area of adhesion between the secretory granule to the cellular plasma membrane (26,27; Chapter 0). The fusion event is strictly controlled, and there is no evidence for any intracellular rupture. While hyperosmotic solutions inhibit secretion in many systems (17,28), flaccid biological vesicles seem to fuse in chromaffin cells (29,30). Enveloped viruses also seem to fuse without osmotic swelling (31,32).

In a recent study on the beige mouse mast cell (33), membrane fusion as measured by capacitance changes clearly preceded any vesicle swelling, even when the vesicles were clearly shrunken (Fig. 3). In this system the swelling associated with biological exocytosis is a postfusional event, perhaps related to secretory product dispersal. Thus, in one biological system, a mechanism exists for driving fusion without obvious lateral mechanical stresses on membrane components. This mechanism driving biological fusion may be mediated by membrane components which, upon proper biological stimulus, suddenly hydrate. The important feature is macromolecular surfaces whose hydration potential is changed by a biological triggering event such that contact and adhesion with subunits within the same bilayer are altered to promote adhesion with subunits in the opposing bilayer while the original adhered surfaces hydrate.

VI. APPLICATIONS OF FUSION IN MEMBRANE RESEARCH

A. Reconstitution

Fusion of vesicles with a planar bilayer fusion is useful for channel reconstitution because one need not subject the channel to detergent solubilization in order to study it in a defined milieu. Adding either native vesicles of biological origin or phospholipid vesicles to which channels have been inserted to one side of a planar phospholipid membrane often leads to that channel's insertion into the planar membrane. Here voltage dependence, ionic selectivity, agonist binding, and lipid dependence can be easily investigated (recently reviewed in Refs. 34,35). In many of these studies it is probable that channels insert into the planar membrane via nonfusional incorporation, as discussed above. Transfer of channels to the planar

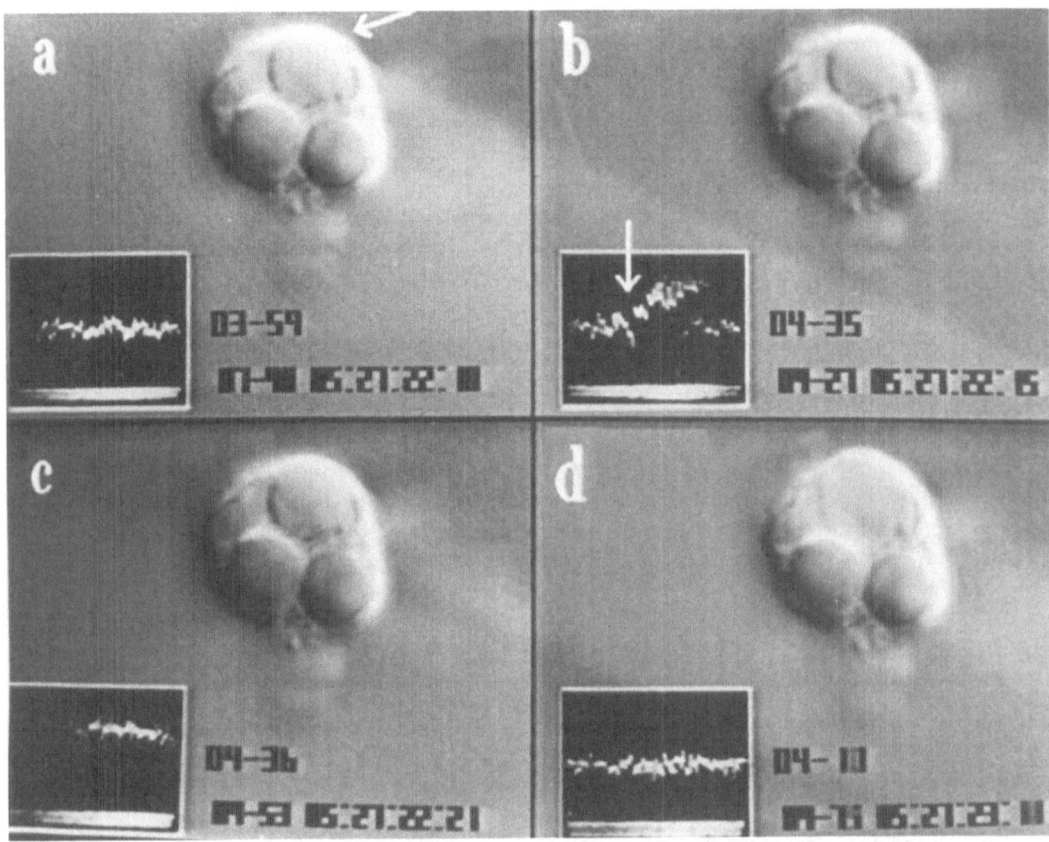

FIGURE 3 Membrane fusion precedes vesicle swelling in beige mouse mast
cells. Sequential fields of one experiment are shown. Cells are patch-clamped
in the whole-cell configuration and capacitance is measured with a phase-
sensitive detector (41). The cell is simultaneously observed with differential
interference contrast and video-enhanced microscopy. The top of the border
to the oscilloscope tracing is 10 μ; magnification to screen was 13,400.
The top oscilloscope tracing (A.D.-coupled, 50 msec/sweep) is capacitance,
the bottom is current. The top digits in the middle are conductance and
the bottom capacitance, 100 units = 200 femptofarads. Time is at the bottom
right; the changing times are seconds and hundredths of seconds.

In panel a, the resting capacitance of the cell is seen with an arrow
at the granule which will soon fuse. In panel b, the capacitance change
due to fusion of the secretory granule to the plasma membrane is evident
(arrow). In panel c, the capacitance change is almost finished but there
is no vesicular swelling. Swelling is evident in panel d.

bilayer is sometimes more desirable than fusion of native vesicles since it leads to single channels in the planar bilayer. Thus, one does not always wish to impose an osmotic swelling of the vesicle in reconstitution since doing so increases the rate of incorporation of channels.

B. Lipid Jump Experiments

By fusing a vesicle with a planar membrane, one abruptly alters the lipid environment of the proteins in this vesicle. Thus, any lipid-dependent kinetic changes or alteration of equilibrium of conformational states of a channel can be detected electrically as the lipid of the planar membrane diffuses into the new patch of recently vesicular lipid (36).

C. Biochemical Assay for Channel Protein Purification

During protein purification, it is useful to determine the purity of the preparation. For enzymes, this is accomplished with an enzymatic assay or by gel electrophoresis. For channels, the amount of material is often too small to be detected on gels. The planar bilayer offers a sensitive assay, but incorporation from detergent or monolayers is highly variable. By incorporating a known amount of protein into a known amount of lipid, one can quantify the purity of the protein by measuring the number of channels simultaneously inserted during a fusion jump (10). This is a reflection of the number of channels per vesicle of defined dimensions. In one study, the correlation between the number of channels per vesicle as measured by fusion events and the number of intramembraneous particles seen in freeze-fracture replicas was used as proof of identity of the channel (37).

D. Assay System for Fusion Molecules

One promising direction for research in membrane fusion is investigation of macromolecules responsible for the specificity and sensitivity of biological membrane fusion. By incorporating such molecules into the system, one could then test their effect on fusion of vesicles with planar bilayers. A calcium-binding protein has been shown to decrease the threshold concentration of calcium needed to promote fusion, and a calcium vs. magnesium specificity similar to that of physiological systems is observed (36). In another system fibronectin was incorporated, which again increased the rate of fusion by increasing adhesion (38). In contrast, the soluble factor synexin only aggregated vesicles in the bathing medium, thereby decreasing fusion to the planar bilayer (F. S. Cohen and S. Morris, personal communication). Synexin will promote binding of vesicles leading to fusion in the presence of an osmotic gradient if no greater than micromolar concentrations of calcium are used in the bathing solution (W. D. Niles, personal communication). In none of these systems was the requirement for osmotic swelling of the vesicles lifted. The planar bilayer system is clearly preferable to vesicle-vesicle fusion to distinguish adhesion molecules (which may cause vesicles to fuse with each other) from fusion molecules (which should lead to fusion of vesicles with planar bilayers without osmotic swelling).

Another approach that may be useful in this regard would be experiments whereby biological membranes would fuse with chemically defined films. These hybrid fusions could involve secretory vesicles fusing with planar bilayers or phospholipid vesicles fusing with planar plasma membranes, e.g., isolated planar cortices of sea urchin eggs (26). Once fusion has been verified, membrane proteins could be selectively extracted until fusion halts. This fraction could then be added back and ultimately important fractions may be identified and tested on pure phospholipid systems. Reconstitution of biological fusion may permit fusion of vesicles with planar membranes without experimentally imposed osmotic gradients.

VII. FUTURE WORK

It is likely that vesicles can fuse with planar membranes. The structure of the prefusion state is not known. Further speculation on the role of proteins in fusion would be aided by direct measurements of this structure. When two hemispherical bilayer membranes are pushed into contact, they can form a region of adhesion which is a single phospholipid bilayer membrane (39,40). When this central zone bursts, the membranes are fused. It is hoped that experiments involving energy transfer between fluorescent lipids and creative use of antibiotics will lead to measurements of both the area of contact between vesicles and planar bilayers and the number of bilayers in the prefusion state. Detailed understanding of the fusion process depends on elucidation of this structure.

ACKNOWLEDGMENTS

I thank Anne Walter and Andrew Harris for a critical reading of this manuscript and Fredric S. Cohen for many useful conversations.

REFERENCES

1. Rayleigh, J. W. S. (1964). *Scientific Papers*, Chapter 4, pp. 421–425.
2. Boys, C. V. (1959). *Soap Bubbles*, pp. 101–103.
3. Pohl, G. W., Stark, G., and Trissl, H. W. (1973). Interaction of liposomes with black lipid membranes. *Biochim. Biophys. Acta* 318:478–481.
4. Cohen, J. A., and Moronne, M. M. (1976). Interaction of charged lipid vesicles with planar bilayer lipid membranes: Detection by antibiotic membrane probes. *J. Supramol. Struc.* 5:409–416.
5. Roseman, M. A., Holloway, P. W., Calabro, M. A., and Thompson, T. E. (1977). Exchange of cytochrome b5 between phospholipid vesicles. *J. Biol. Chem.* 252:4842–4849.
6. Moore, M. R. (1976). Fusion of liposomes containing conductance probes with black lipid films. *Biochim. Biophys. Acta* 426:765–771.
7. Sokolov, Y. V., and Lishko, V. K. (1979). Study of fusion of planar bilayer phospholipid membranes with liposomes. *Biokhimia* 44:317–323.
8. Yegorova, Y. M., Chernomordik, L. V., Abidos, I. G., and Chizmadzhev, Y. A. (1981). Fusion of liposomes with a flat lipid membrane. *Biofizika* 26:145–147.
9. Yegorova, Y. M., Chernomordik, L. V., Abidor, I. G., and Chizmadzhev, Y. A. (1981). Investigation of the interaction of liposomes with BLM by a potentiodynamic method. *Bifizika* 26:363–365.
10. Cohen, F. S., Zimmerberg, J., and Finkelstein, A. (1980). Fusion of phospholipid vesicles with planar phospholipid bilayer membranes.

II. Incorporation of a vesicular membrane marker into the planar membrane. *J. Gen. Physiol.* 75:251-270.

11. Miller, C. and Racker, E. (1976). Ca++-induced fusion of fragmented sarcoplasmic reticulum with artificial planar bilayers. *J. Membrane Biol.* 30:283-300.

12. Cohen, F. S., Akabas, M. H., and Finkelstein, A. (1982). Osmotic swelling of phospholipid vesicles causes them to fuse with a planar phospholipid bilayer membranes. *Science* 217:458-460.

13. Zimmerberg, J., Cohen, F. S., and Finkelstein, A. (1980). Fusion of phospholipid vesicles with planar phospholipid bilayer membranes. I. Discharge of vesicular contents across the planar membrane. *J. Gen. Physiol.* 75:251-270.

14. Woodbury, D. J. (1986). Correlation of channel incorporation and content release during fusion of unilamellar vesicles with a planar bilayer. *Biophys. J.* 49:132a.

15. Perrin, M. S., and MacDonald, R. C. (1986). Osmotic influences on fusion of synaptic vesicles to planar bilayers. *Biophys. J.* 49:133a.

16. Akabas, M. H., Cohen, F. S., and Finkelstein, A. (1984). Separation of the osmotically driven fusion event from vesicle-planar membrane attachment in a model system for exocytosis. *J. Cell Biol.* 98:1063-1071.

17. Finkelstein, A., Zimmerberg, J., and Cohen, F. S. (1986). Osmotic swelling of vesicles: Its role in the fusion of vesicles with planar phospholipid bilayer membranes and its possible role in exocytosis. *Annu. Rev. Phys.* 48:163-174.

18. Cohen, F. S., Akabas, M. H., Zimmerberg, J., and Finkelstein, A. (1984). Parameters affecting the fusion of unilamellar phospholipid vesicles with planar bilayer membrane. *J. Cell Biol.* 98:1054-1062.

19. Rand, R. P., and Parsegian, V. A. (1986). Mimicry and mechanism in phospholipid models of membrane fusion. In: *Annual Review of Physiology*, 48th ed. (R. M. Berne and J. F. Hoffman, eds.), pp. 201-212.

20. Düzgünes, N. (1985). Membrane fusion. In: *Subcellular Biochemistry*, 11th ed. (D. B. Roodyn, ed.). Chapter 5, pp. 195-286.

21. Düzgünes, N., Wilschut, J., Fraley, R., and Papahadjopoulos, D. (1981). Studies on the mechanism of membrane fusion: Role of head group composition in calcium- and magnesium-induced fusion of mixed phospholipid vesicles. *Biochim. Biophys. Acta* 642:182-195.

22. Rand, R. P., Kachar, B., and Reese, T. S. (1985). Dynamic morphology of interacting phosphatidylserine vesicles. *Biophys. J.* 47:483-489.

23. Kachar, B., Fuller, N., and Rand, R. P. (1986). Morphological responses to calcium-induced interaction of phosphatidylserine-containing vesicles. *Biophys. J.* 50:779-788.

24. Evans, E. A., and Parsegian, V. A. (1983). Energetics of membrane deformation and adhesion in cell and vesicle aggregation. *Ann. NY Acad. Sci.* 416:13-33.

25. Evans, E., and Kwok, R. (1982). Mechanical calorimetry of large dimyristoylphosphatidylcholine vesicles in the phase transition region. *Biochemistry* 21:4874-4879.

26. Zimmerberg, J., Sardet, C., and Epel, D. (1985). Exocytosis of sea urchin egg cortical vesicles in vitro is retarded by hyperosmotic sucrose: Kinetics of fusion monitored by quantitative light-scattering microscopy. *J. Cell Biol.* 101:2398-2410.

27. Satir, B. (1972). Membrane reorganization during secretion in Tetrahymena. *Nature* 235:53-54.

28. Hampton, R. Y., and Holz, R. W. (1983). Effects of changes in osmality on the stability and function of cultured chromaffin cells and the possible role of osmotic forces in exocytosis. *J. Cell Biol.* 96:1082-1088.

29. Holz, R. W., and Senter, R. A. (1986). The effects of osmalality and ionic strength on secretion from adrenal chromaffin cells permeabilized with digitonin. *J. Neurochem.* 46:1835-1842.

30. Holz, R. W. (1986). The role of osmotic forces in exocytosis from adrenal chromaffin cells. In: *Annual Review of Physiology*, 48th ed. (R. M. Berne and J. F. Hoffman, eds.), pp. 175-190.

31. White, J., Kielian, M., and Helenius, A. (1983). Membrane fusion proteins of enveloped animal viruses. *Q. Rev. Biophys.* 16:151-195.

32. Blumenthal, R., Bali-Puri, A., Walter, A., Corell, D., and Eidelman, O. (1986). pH-dependent fusion of vesicular stomatitis virus with Vero cells: Measurement by dequenching of octadecylrhodamine fluorescence. *J. Biol. Chem.*, submitted.

33. Zimmerberg, J., Curran, M., Cohen, F. S., and Brodwick, M. (1986). Simultaneous electrical and optical measurements show that membrane fusion precedes secretory granule swelling during exocytosis of beige mouse mast cells. *Proc. Natl. Acad. Sci. USA* 84:1585-1589.

34. Cohen, F. S. (1986). Fusion of liposomes to planar bilayers. In: *Ion Channel Reconstitution* (C. Miller, ed.). Chapter 6, pp. 131-139.

35. Hanke, W. (1986). Incorporation of ion channels by fusion. In: *Ion Channel Reconstitution* (C. Miller, ed.). Chapter 7, pp. 141-153.

36. Zimmerberg, J., Cohen, F. S., and Finkelstein, A. (1980). Micromolar concentrations of calcium stimulate fusion of phospholipid vesicles with planar phospholipid bilayer membranes when a calcium-binding protein is present in the planar membrane. *Science* 210:906-908.

37. Zampighi, G. A., Hall, J. E., and Kreman, M. (1985). Purified lens junctional protein forms channels in planar lipid films. *Proc. Natl. Acad. Sci. USA* 82:8468-8472.

38. Young, T. M., and Young, D.-E. (1984). Protein-mediated intermembrane contact facilitates fusion of lipid vesicles with planar bilayers. *Biochem. Biophys. Acta* 775:441-445.

39. Neher, E. (1974). Asymmetric membranes resulting from the fusion of two black lipid bilayers. *Biochim. Biophys. Acta* 373:327-336.

40. Fisher, L. R., and Parker, N. S. (1984). Osmotic control of bilayer fusion. *Biophys. J.* 46:253-258.

41. Neher, E., and Marty, A. (1982). Discrete changes of cell membrane capacitance observed under conditions of enhanced secretion in bovine adrenal chromaffin cells. *Proc. Natl. Acad. Sci. USA* 79:6712-6716.

9

Fusion of Liposomes Induced and Modulated by Proteins and Polypeptides

KEELUNG HONG, PAUL R. MEERS,* NEJAT DÜZGÜNEŞ, and DEMETRIOS PAPAHADJOPOULOS

University of California—San Francisco, San Francisco, California

I. INTRODUCTION

Liposomes represent the lipid bilayer backbone of biological membranes and have been used extensively as a model system to study the fusion of biological membranes (1). Early studies of membrane fusion in exocytosis suggested that the bare lipid patches in the region of adhesion of the membranes might be directly involved in fusion. However, subsequent observation with rapidly frozen tissues revealed that fusion can take place in an area not entirely free of intramembranous particles (2,3). Even when a fusion event is shown to occur in the bare lipid domains of a biological membrane (4), the possibility cannot be ruled out that certain peripheral membrane proteins or small polypeptides are involved in fusion. This conclusion follows the generally accepted view that freeze fracture can only reveal the integral (transmembrane) membrane proteins (5-7). Therefore, in considering the possible components of a fusion model system, it is important to recognize that the interactions of peripheral membrane proteins or polypeptides with liposomes could be as critical as those of the intrinsic membrane proteins. In this chapter we review studies on membrane fusion induced or modulated by peripheral proteins and polypeptides. In addition, we identify a few general structural requirements for liposome fusion. Apart from building the conceptual framework for understanding fusion mechanisms, the identified features of protein-lipid interactions could be helpful for the future design of polypeptides which may facilitate cytoplasmic delivery of agents encapsulated in liposomes through fusion between liposomes and cellular membranes.

II. FUSION REACTION STEPS

Experimental observations of liposomes suggest that at least two requirements must be met in order for the lipid bilayers of two different vesicles

---

Present affiliation: Boston University School of Medicine, Boston, Massachusetts.

to fuse: (1) aggregation of liposomes with close intermembrane contact, and (2) transitional destabilization of the closely apposed membranes leading to the mixing of the membrane components. The main potential barrier to liposome aggregation is the charge repulsion of the bilayer surface. Surface hydration is a subsidiary barrier relating more specifically to close intermembrane contact. Therefore, proteins or polypeptides must be able to promote charge neutralization and dehydration of the lipid surface (at least locally) in order to facilitate liposome aggregation and close contact of the two adjacent lipid bilayers. The charge distribution on the bilayer surface and the extent of hydration are intrinsic properties of the membrane components which must be considered in order to predict liposome fusion.

A fusion reaction is not complete until the second step, which involves a transient destabilization, mixing of the membrane components, and the eventual coalescence of the liposome contents (1,8-10). This fusion step is a dynamic process requiring that the lipid molecules at the fusion area must be sufficiently mobile to rearrange themselves to reach the final stable state. The overall topology of the bilayers will dictate how this transient stage is resolved. Since bilayer continuity is a low-energy state of the interacting membranes, the process of lipid rearrangement should bring the fusion system back to another stable state with no bilayer discontinuity. With insufficient lipid mobility, a delay in lipid rearrangement at the fusion site will prolong bilayer discontinuity. Consequently, the leakage of liposome contents associated with the fusion event may increase (11,12). In general, this dynamic bilayer reorganization at the fusion site can be relatively fast in liposome fusion, where it has been shown that intermembrane close contact can be the rate-limiting step. The initiation of bilayer destabilization and the consequent lipid reorganization and leakage of contents is still the least understood step in the overall fusion reaction. Lipid packing defects and liquid-crystalline phase transitions (13), bilayers to hexagonal (H_{II}) phase transitions and/or the formation of inverted micelles (14,15), protein penetration in the hydrophobic membrane region (16), and lipid dehydration (17,18) have all been proposed as important parameters in this step (see also Chapters 2, 3, 4, and 8). Moreover, leakage of liposome contents during fusion (9), which has been a controversial point (12), has been found to vary in different experimental systems from negligible (12) to total (19). Careful assessment of the characteristics of each system will help in defining the importance of the various parameters underlying these phenomena.

Ca^{2+} is the best-known divalent cation that can perform both of the above steps during fusion of negatively charged liposomes. However some cases clearly demonstrate that liposome fusion can occur independently of Ca^{2+} in the presence of some polypeptides or proteins. The involvement of polypeptides and proteins in inducing or modulating liposome fusion is discussed below, categorized according to the requirement for Ca^{2+}.

III. Ca^{2+}-DEPENDENT FUSION

There are several known cases in which Ca^{2+} is mandatory for activating the protein to interact with liposomes (71-75). The common feature of these proteins is their Ca^{2+}-binding capability. These Ca^{2+}-binding proteins often bind to liposomes in the presence of Ca^{2+}. Even though calmodulin, parvalbumin, synhibin, prothrombin, and the proteolytic fragment of pro-thrombin are able to bind to liposomes in the presence of Ca^{2+}, liposome aggregation is not observed under these conditions (20-23). Clearly, these proteins do not have the capability to form stable bonds for bridging adjacent liposomes. We suggest that these proteins may only possess a single membrane binding site and lack a second site for membrane or protein binding which would accomplish bridging (20).

In contrast, certain proteins appear capable of Ca^{2+}-dependent cross-linking of liposomes. Thus, tubulin, when incorporated into dipalmitoyl phosphatidylcholine (DPPC) liposomes, is able to facilitate the aggregation and fusion of the resulting tubulin-liposomes in the presence of Ca^{2+} (24). Since phosphatidylcholine (PC) is not susceptible to fusion in the presence of Ca^{2+}, the tubulin-facilitated fusion of DPPC liposomes is attributable to the Ca^{2+}-binding feature of tubulin. It is conceivable that protein-protein binding is the cause of the aggregation of tubulin-liposomes.

Calelectrin of different molecular weights, isolated from different tissues, is able to aggregate synaptic vesicles, chromaffin granules, and liposomes in the presence of Ca^{2+} (25,26). Its liposome fusion activity is not yet known. In the presence of Ca^{2+}, calelectrin binds to hydrophobic chromatography columns, indicating that exposure of a hydrophobic domain increases through Ca^{2+} binding. Since calelectrin is not a basic protein (P_I is < 6.0), it is unlikely that surface charge interaction between calelectrin and negatively charged liposomes is the main mode of protein-lipid interaction. Ca^{2+}-induced hydrophobic bridging between liposomes by calelectrin appears more likely. At present, there is no clear indication that Ca^{2+}-induced polymerization of calelectrin is required for the aggregation of liposomes (25).

Synexin, originally isolated from adrenal medulla (27), facilitates membrane fusion in a number of liposome systems (20,21,28). In general, synexin increases the initial rate of fusion of liposomes containing phosphatidylserine (PS) and/or phosphatidic acid (PA) and lowers the Ca^{2+} threshold. The facilitation of fusion by synexin is drastically reduced when PC is included in liposomes. Ca^{2+} is an absolute requirement for the action of synexin in liposome fusion. Two distinct features of synexin must be considered in explaining its effect on fusion (see also Chapter 23). In the absence of liposomes, Ca^{2+} induces the self-association of synexin and the formation of large extended rods (29). This bipolar nature of synexin has been proposed as an explanation for why synexin can bridge adjacent liposomes and facilitate close intermembrane contact (20). Considering polymer as mechanistically equivalent to one synexin entity, kinetic analysis reveals that prepolymerization by Ca^{2+} always decreases the activity of synexin such that it is less than the activity of an equal amount of untreated monomers. However, the activity of synexin monomers polymerized to an average hexameric size is greater than that of one-sixth as many untreated monomers, with respect to the liposome aggregation rate constant (30). Furthermore, the analysis of the kinetics of synexin-facilitated fusion based on specific selected conditions that favor either aggregation or fusion as the overall rate-limiting step reveals that the enhancement in the overall fusion process by synexin is due to the ability of synexin to promote liposome aggregation at low Ca^{2+} concentrations and not due to an increase of the rate of the actual fusion step (30,31).

Even though the presence of lectins in animal tissues is not common, the possibility that animal lectins may be involved in intra- and extracellular adhesion and fusion is worth consideration. Recent studies on liposome fusion show that preagglutination of glycolipid-containing liposomes substantially reduces the Ca^{2+} threshold required for fusion (32-34; see also

Chapter 12). Lectins in this case are playing the role of a strong bridging device for promoting liposome aggregation via specific receptors while Ca^{2+} is still required to complete the second fusion step. In this sense, the lectins act by a mechanism similar to that of synexin.

IV. Ca^{2+}-INDEPENDENT FUSION

Small hydrophobic peptides, e.g., gramicidin and alamethicin, have been found to induce fusion of small DPPC liposomes (35,36). Fusion takes place very slowly and is temperature-dependent. The creation of small mixed-phase domains by either gramicidin or alamethicin in gel-phase vesicles is considered a source of unstable membrane regions and eventually leads to mixing of lipids and an increase in liposome size, most likely following liposome collision.

Polylysine has been used as a model polypeptide in liposome fusion (37,38). Polylysine-induced fusion of PS-containing liposomes is optimal when the number of sites for vesicle-peptide-vesicle interaction is maximized and when the net charge on the aggregated system is near zero (39). Therefore, it is clear that a major function of polylysine in liposome fusion is the promotion of liposome aggregation through charge neutralization due to the polycationic feature of polylysine. Elucidation of the additional factors that promote the crucial next step of membrane destabilization requires further study.

Small natural polypeptides such as melittin or polymyxin B can also induce fusion of acidic liposomes (40,41). The amphipathic nature of these polypeptides enables them to be incorporated into liposomes while the polycationic hydrophilic end of the molecule could attract another liposome. Similar to polylysine, the net charge of melittin liposomes has to be minimized for optimal effect (41). However, polylysine is less effective than melittin in terms of the concentration required for inducing fusion (40). Higher concentrations of polymyxin B and melittin than those required for fusion produce disruption of the bilayers. It appears that the membrane-destabilizing effect of these polypeptides may be an additional factor that promotes fusion more effectively.

Fusion-promoting capability has also been ascribed to myelin basic protein (42). This protein is capable of organizing phosphatidylglycerol (PG) bilayers into the multilamellar structure characteristic of myelin (43) and of aggregating acidic liposomes. To demonstrate the importance of the charge distribution on myelin basic protein for liposome aggregation, the overall number of positive charges on the protein molecule can be altered. Decreasing the positive charge by either phosphorylation or removal of C-terminal arginyl residues of myelin basic protein decreases its ability to induce vesicle aggregation (44). Myelin basic protein promotes aggregation of PS/PC liposomes, but fusion is observed only when an amphipathic molecule, lysolecithin, is added (24). A limited and slow fusion induced by myelin basic protein alone has been observed in PC/phosphatidylethanolamine (PC/PE) liposomes (45). Addition of palmitoyl aldehyde to PC/dioleoylPE (PC/DOPE) liposomes dramatically increases their fusion susceptibility in the presence of myelin basic protein (45).

On the same subject of Ca^{2+}-independent fusion of liposomes, glyceraldehyde-3-phosphate dehydrogenase has been observed to induce fusion of PA/PC liposomes (46). How this water-soluble enzyme interacts with lipid bilayers is not known. These experiments were carried out with sonicated vesicles. The strained bilayer of these vesicles is likely to contribute to the interaction of the protein with the membrane. It will be of interest to investigate the behavior of large unilamellar vesicles under the same conditions.

Two sperm proteins have been implicated in membrane fusion, based on evidence from liposome fusion studies (41,47,48). Bindin, a protein isolated from sea urchin sperm, associates with liposomes in the gel phase and induces aggregation of liposomes that contain both gel-phase and fluid-phase domains. When bindin is added to these PS-containing, mixed-phase liposomes, the liposomes aggregate and fuse. Ca^{2+} does not have any effect on fusion. Lysin, an acrosomal protein from abalone sperm, also interacts with phospholipids and induces fusion of negatively charged liposomes (41). In contrast to bindin, lysin appears to bind to liposomes of various composition when the membrane is in the fluid phase and induces aggregation of negatively charged liposomes. The bridging between adjacent liposomes by lysin is clearly due to the polycationic nature of lysin.

Fibronectin, a phospholipid-binding protein, mediates membrane-membrane contact and is thought to bridge adjacent liposomes through hydrophobic insertion into the bilayers. This fibronectin-mediated intermembrane contact facilitates fusion of lipid vesicles with planar bilayers. Ca^{2+} is not required in this fusion reaction, but its presence enhances the fusion rate (49).

V. CELLULAR PERSPECTIVE

The cell maintains its individuality and intracellular compartmentation by using membranes as barriers. Most biologically relevant molecules are distributed asymmetrically along the membranes. When membrane fusion takes place inside the cell, it does not necessarily have the same requirements as fusion originating outside the cell. These two types of fusion events are obviously initiated in different biological environments and the fusion stimuli could be entirely different. Fusion is a dynamic process which abrogates cellular individuality and intracellular compartmentation. Therefore, the fusion process has to be highly regulated to complete a specific cellular activity in a nondisruptive, nondestructive fashion. In view of these facts, it is unlikely that a simple membrane model or one general fusion mechanism can provide a satisfactory basis for studying all types of membrane fusion. It is conceivable, however, that different stimuli, promoters, and controlling factors may be superimposed on a basically simple mechanism, thus providing for the required spatial and temporal specificity.

A. Ectoplasmic Fusion

The target for the initial intermembrane contact in ectoplasmic fusion is the outer layers of the adjacent plasma membranes. Fusion of egg with sperm during fertilization and fusion of myoblasts during the development of skeletal muscle are typical examples of ectoplasmic fusion. Ca^{2+} is regarded as an essential element in both cases (50,51). Although red blood cells are clearly designed to be fusion resistant, they have been used for constructing a cell-fusion model. Studies on chemically induced fusion of erythrocytes suggest that proteolytic degradation of membrane-associated proteins facilitates such fusion (52). This fusion model calls for high concentrations of membrane-active fusogens, such as chlorpromazine, oleoylglycerol, and benzyl alcohol (16). High concentrations of poly(ethylene glycol) can also induce fusion of erythrocytes, in the absence of proteolysis of membrane-associated proteins (53-55). That the chemically induced erythrocyte fusion is slow can be attributed to the strong fusion resistance of the plasma membrane. The entry of Ca^{2+} into the cytoplasm in the presence of fusogens is thought to initiate or facilitate events that lead to erythrocyte fusion in the presence of fusogens (16). The combined action of calcium and phosphate also induces the fusion of chicken erythrocytes or human erythrocyte ghosts (56,57).

Erythrocytes and various cultured cells can be induced to fuse by the application of an electric field pulse after the cells are aligned by dielectrophoresis (58; see also Chapter 29). The fusogenic state of the erythrocyte membrane induced by the electric pulse appears to be long-lived; thus, establishing intermembrane contact minutes after the application of the pulse to initially nonaligned cells is sufficient to induce membrane fusion (59). Ectoplasmic fusion between cells can also be induced by viruses (see Part III on "Fusogenic Properties of Viruses") either at neutral pH, as in the case of Sendai virus (60), or at mildly acidic pH, as with influenza, Semliki Forest, and vesicular stomatitis virus (61).

B. Cytoplasmic Fusion

Initial cytoplasmic intermembrane contact is a common feature in exocytosis, endocytosis, and the transport processes between intracellular organelles. A rise of the intracellular Ca^{2+} concentration is well established for exocytic secretion in different cells. Since intracellular vesicles can be isolated, their in vitro fusion is a popular model for fusion studies (62). A great number of studies on Ca^{2+}-dependent fusion of liposomes has shown that Ca^{2+} not only effectively neutralizes the lipid surface charge, but also promotes close contact by removing the surface-bound water. The local dehydration and fluctuations in lipid packing have been emphasized as a fusion mechanism in a recent review (63). However, fusion requirements for these simple systems are often far from those known to be required for biological fusion. For example, the difference in the Ca^{2+} threshold required for fusion is generally cited as an important discrepancy (62,64,65). Other cytoplasmic components have been studied for their ability to increase the Ca^{2+} sensitivity of liposome fusion (28). Ca^{2+}-binding proteins are considered to be likely candidates (21).

VI. PROPOSED STRUCTURAL REQUIREMENTS

As pointed out earlier, we have classified protein- or polypeptide-mediated membrane fusion into Ca^{2+}-dependent and Ca^{2+}-independent fusion events. We have also considered fusion reactions on the basis of their locality with respect to the plasma membrane as either ectoplasmic or endoplasmic. Such classification takes into account the importance of the asymmetrical distribution of the membrane components and the different ionic environments surrounding the fusion site. Because of the limited information about protein structure and fusion mechanisms, we restrict our discussion to some identifiable requirements for the various steps of fusion in each of these categories.

A. Liposome Aggregation

Proteins can induce liposome aggregation by several mechanisms. Generally speaking, proteins have to overcome both the electrostatic repulsion and surface hydration which are the main forces keeping the liposomes from aggregating. In any case, the eventual aggregation will be affected either by lipid-protein interactions (where a single protein molecule will be sandwiched between two lipid membranes) or by protein-protein interactions (where two protein molecules both bound to individual membranes have increased affinity for each other). It is also conceivable that in certain cases the interacting protein or polypeptide may affect the surface characteristics of the lipid in such a way that lipid-lipid interactions are the dominant forces promoting aggregation. The following discussion analyzes these possibilities; however, a detailed assessment will have to await the accumulation of sufficient data.

1. Calcium-Dependent Protein-Lipid Interactions

Two possible modes of protein-lipid interaction can cause liposome aggregation. (1) Ca^{2+}-induced hydrophobic interaction of protein with the lipid bilayer: This type of interaction is plausible when the Ca^{2+} concentration required for promoting the protein-dependent aggregation of liposomes is generally not high enough to neutralize the surface charge of liposomes. (2) Ca^{2+} bridging between protein and lipid: This type of interaction can be either ion-ion or ion-dipole interaction where protein and lipid are ligands to Ca^{2+} while Ca^{2+} functions as a bridge between protein and lipid molecules.

If the protein molecule has only one binding site available, once Ca^{2+} bridges the protein to the liposome surface, the exposed surface of the protein may not be able to interact with another liposome. The Ca^{2+}-induced interaction of such a protein with liposomes may not result in liposome aggregation. Some Ca^{2+}-binding proteins, e.g., prothrombin and its fragment 1, synhibin, calmodulin, and parvalbumin, can be classified in this group of proteins. Studies on liposome fusion show that synexin or calelectrin self-aggregates in the presence of low concentration of Ca^{2+} (25,29). In the presence of acidic liposomes, these two proteins promote Ca^{2+}-dependent aggregation of liposomes. This indicates that synexin or calelectrin may have two or more Ca^{2+}-binding sites so that Ca^{2+} can bridge liposomes through a lipid-Ca^{2+}-$(protein)_n$-Ca^{2+}-lipid interaction, where $n \geq 1$. The exact number of n in protein-induced liposome aggregation remains to be determined. Our recent studies on the kinetics of synexin modulation of the aggregation and fusion of liposomes suggest that small synexin aggregates are still active but slightly less active than monomeric synexin (30).

2. Charge Neutralization

A protein or polypeptide with some distinctive positively charged domain can decrease the surface charge density and surface potential of negatively charged lipid bilayers. Melittin, lysin, polylysine, and polymyxin B are examples of such polypeptides which have been found to induce aggregation of negatively charged liposomes. Aggregation of liposomes can also occur simply by polypeptide bridging of adjacent liposomes through multiple binding sites or by protein-protein bonding of liposome-associated proteins. Liposome aggregation mediated by myelin basic protein is an example of the latter type of interaction. The aggregation of liposomes only occurs after binding of myelin basic protein to a liposome is saturated (66). In this case protein-protein binding may be the cause of liposome aggregation. Some polypeptides that integrate readily into lipid bilayers, as melittin does, can reverse the surface charge of liposomes when more than a charge-equivalent number of melittin molecules incorporate into bilayers. Under such conditions, liposome aggregation will not occur. In all cases it appears that charge neutralization plays a major role.

3. Specific Lipid Ligands

Proteins like lectins or fibronectin can be very effective in cross-linking adjacent liposomes containing specific ligands when charge repulsion can be overcome (32,34). It appears that the presence of Ca^{2+} does not affect this type of protein-lipid interaction, which depends on the presence of specific ligands on the liposome surface.

4. Hydrophobic Bridging

Hydrophobic bridging is not as clearly defined as charge neutralization. However, it appears to be very important in membrane fusion. The incorporation of tubulin (24) or bindin (47) to mixed-phase lipid bilayers may be attributed to the hydrophobic attachment to the lipid surface. Whether monodisperse proteins or protein aggregates initiate liposome aggregation is not clear at present. Although it is possible that some hydrophobic polypeptides produced by proteolytic cleavage may promote membrane close contact by hydrophobic insertion (16), their possible role in the various fusion steps (including both close contact and destabilization) remains to be elucidated after the polypeptides can be isolated and characterized.

B. Bilayer Destabilization and Fusion of Liposomes

A possible major role of proteins or polypeptides in the actual membrane fusion step is the penetration of the protein into the lipid bilayer to increase the potential of forming a mobile intermembrane intermediate and consequently lead to fusion. The extent of penetration of the protein into the bilayer is dependent on bilayer packing (67) and bilayer packing is dependent on the headgroups of the lipids. Thus, headgroup structure can control the extent of protein penetration into bilayers. However, the mechanism for such interactions at the molecular level is still not understood. The following discussion refers only to the few systems that have been studied extensively.

In some cases, aggregation of liposomes does not necessarily lead to fusion unless some membrane-active agents are added. The incorporation of these membrane-active agents into bilayers is expected to change the bilayer packing and to increase the mobility of bilayer components leading to bilayer reorganization. A clear example is that of lysophosphatidylcholine or aliphatic aldehydes which are required for fusion to occur when liposome aggregation is mediated by myelin basic protein (42,45).

Apart from mediating liposome aggregation, proteins like melittin, lysin, and bindin can also produce bilayer destabilization and fusion of liposomes without added membrane-active agents. Hydrophobic insertion by these proteins into bilayers, although not necessarily to the same extent, may be their common feature in inducing membrane fusion. When the perturbation of bilayer continuity by peptide insertion is above a threshold at the contact site, it is conceivable that lipid reorganization leads to fusion at points where a close membrane contact has been achieved, possibly by charge neutralization. This provides some rationalization for the action of melittin, lysin, or bindin which can mediate Ca$^+$-independent fusion of liposomes containing a large amount of hydrated phosphatidylcholine (40,41,48). In contrast, the chemical structure and the extent of hydration of the lipid group are critical in Ca^{2+}-dependent, protein-mediated fusion of liposomes (20,21,28-34). In such cases a bulky and/or highly hdyrated lipid headgroup probably inhibits liposomes from close intermembrane contact, even when aggregation is enhanced by proteins such as synexin and lectins. Occasionally, fusion is enhanced by membrane-active agents such as *cis* fatty acids, in liposome aggregates induced by proteins (68,69). As described above, the incorporation of fatty acids in lipid bilayers may enhance the lipid mobility for reorganization of apposed bilayers.

VII. CONCLUDING REMARKS

Our goal in this chapter is to call attention to the identifiable structural characteristics of various nonmembrane proteins and polypeptides known to be involved in membrane fusion. It is hoped that this will help in constructing a molecular model of the fusion reaction. The structural requirements of a protein involved in membrane fusion are closely related to the membrane components which are asymmetrically distributed on each half of the membrane. The structural requirements suggested in this review are not necessarily equally important in fusion reactions. Meeting one requirement may be enough to initiate slow fusion of liposomes, for instance, creation of micro-mixed-phase domains by gramicidin or alamethicin incorporated into the bilayer below the gel to liquid-crystalline transition, leading to an increase of liposome size, possibly through liposome collision. Similarly, studies on fusion induced by membrane-active compounds, e.g., benzyl alcohol, aliphatic halides, or chlorpromazine, as well as fusion induced by poly(ethylene glycol) can provide useful, but limited information on bilayer packing and on water binding at the membrane surface (16,52,70). When a protein can fulfill multiple requirements simultaneously, its effect will be manifested by the speed of fusion and/or its independence from additional components. On the other hand, when a protein is lacking certain structural requirements, its effect on fusion is often facilitated by an additional factor. Divalent cations or small-molecular-weight membrane-active agents are commonly introduced. Every fusion model has its own emphasis on known requirements and such emphases are not mutually exclusive. It is clear that because of the variety of biological reactions, all the examples and factors involving liposome fusion should be taken as possible elements helping to elucidate a very complex phenomenon. In this regard we view the function of a model membrane fusion system as not to imitate all the requirements of a biological fusion event, but to identify the importance of specific requirements.

ACKNOWLEDGMENTS

Our research has been supported by NIH grants GM 28117, GM 26379, and AI25534 and a Grant-in-Aid from the American Heart Association. Paul Meers is supported by postdoctoral fellowships from the American Cancer

Society (PF-2398) and the Arthritis Foundation. We thank Stephen Murray for editorial comments.

REFERENCES

1. Papahadjopoulos, D., Poste, G., and Vail, W. J. (1979). Studies on membrane fusion with natural and model membranes. *Methods Memb. Biol.* 10:1-121.
2. Plattner, H. (1981). Membrane behaviour during exocytosis. *Cell Biol. Int. Rep.* 5:435-459.
3. Chandler, D. E., and Heuser, J. E. (1980). Arrest of membrane fusion events in the mast cells by quick-freezing. *J. Cell Biol.* 86:666-674.
4. Ahkong, Q. F., Fisher, D., Tampion, W., and Lucy, J. A. (1975). Mechanism of cell fusion. *Nature* 253:194-195.
5. da Silva, P. P., Douglas, S. D., and Branton, D. (1971). Localization of antigen sites on human erythrocyte ghosts. *Nature* 232:194-196.
6. da Silva, P. P., and Branton, D. (1970). Membrane splitting in freeze-etching: Covalently bound ferritin as a membrane marker. *J. Cell Biol.* 45:598-605.
7. Chen, Y. S., and Hubbell, W. L. (1973). Temperature- and light-dependent structural changes in rhodopsin-lipid membranes. *Exp. Eye Res.* 17:517-532.
8. Wilschut, J., and Papahadjopoulos, D. (1979). Ca^{2+}-induced fusion of phospholipid vesicles monitored by mixing of aqueous contents. *Nature* 281:690-692.
9. Wilschut, J., Düzgüneş, N., Fraley, R., and Papahadjopoulos, D. (1980). Studies on the mechanism of membrane fusion: Kinetics of calcium ion induced fusion of phosphatidylserine vesicles followed by a new assay for mixing of aqueous contents. *Biochemistry* 19:6011-6021.
10. Nir, S., Bentz, J., Wilschut, J., and Düzgüneş, N. (1983). Aggregation and fusion of phospholipid vesicles. *Prog. Surface Sci.* 13:1-124
11. Wilschut, J., Düzgüneş, N., and Papahadjopoulos, D. (1981). Calcium/magnesium specificity in membrane fusion: Kinetics of aggregation and fusion of phosphatidylserine vesicles and the role of bilayer curvature. *Biochemistry* 20:3126-3133.
12. Wilschut, J., Düzgüneş, N., Hong, K., Hoekstra, D., and Papahadjopoulos, D. (1983). Retention of aqueous contents during divalent cation-induced fusion of phospholipid vesicles. *Biochim. Biophys. Acta* 734:309-318.
13. Papahadjopoulos, D., Portis, A., and Pangborn, W. (1978). Calcium-induced lipid phase transitions and membrane fusion. *Ann. NY Acad. Sci.* 308:50-66.
14. Verkleij, A. J., Mombers, C., Gerritsen, W. J., Leunissen-Bijvelt, J., and Cullis, P. R. (1979). Fusion of phospholipid vesicles in association with the appearance of lipidic particles as visualized by freeze-fracturing. *Biochim. Biophys. Acta* 555:358-361.
15. Cullis, P. R., and De Kruijff, B. (1979). Lipid polymorphism and functional roles of lipids in biological membranes. *Biochim. Biophys. Acta* 559:399-420.
16. Lucy, J. A. (1984). Do hydrophobic sequences cleaved from cellular polypeptides induce membrane fusion reactions in vivo? *FEBS Lett.* 166:223-231.

17. Düzgüneş, N., Wilschut, J., Fraley, R., and Papahadjopoulos, D. (1981). Studies on the mechanism of membrane fusion. Role of head-group composition in calcium- and magnesium-induced fusion of mixed phospholipid vesicles. *Biochim. Biophys. Acta* 642:2-195.

18. Hoekstra, D. (1982). Role of lipid phase separations and membrane hydration in phospholipid vesicle fusion. *Biochemistry* 21:2833-2840.

19. Sundler, R., and Papahadjopoulos, D. (1981). Control of membrane fusion by phospholipid head groups. *Biochim. Biophys. Acta* 649:743-750.

20. Hong, K., Düzgüneş, N., and Papahadjopoulos, D. (1982). Modulation of membrane fusion by calcium-binding proteins. *Biophys. J.* 37:297-305.

21. Hong, K., Düzgüneş, N., Ekerdt, R., and Papahadjopoulos, D. (1982). Synexin facilitates fusion of specific phospholipid vesicles at divalent cation concentrations found intracellularly. *Proc. Natl. Acad. Sci. USA* 79:4942-4944.

22. Morris, S. J., Hughes, J. M. X., and Whittaker, V. P. (1982). Purification and mode of action of synexin: A protein enhancing calcium-induced membrane aggregation. *J. Neurochem.* 39:529-536.

23. Pollard, H. B., and Scott, J. H. (1982). Synhibin: A new calcium-dependent membrane-binding protein that inhibits synexin-induced chromaffin granule aggregation and fusion. *FEBS. Lett.* 150:201-206.

24. Kumar, N., Blumenthal, R., Henkart, M., Weinstein, J. N., and Klausner, R. D. (1982). Aggregation and calcium-induced fusion of phosphatidylcholine-vesicle-tubulin complexes. *J. Biol. Chem.* 257: 15137-15144.

25. Südhof, T. C., Walker, J. H., and Obrocki, J. (1982). Calelectrin self-aggregates and promotes membrane aggregation in the presence of calcium. *EMBO J.* 1:1167-1170.

26. Südhof, T. C., Ebbecke, M., Walker, J. H., Fritsche, U., and Boustead, C. (1984). Isolation of mammalian calelectrins: A new class of ubiquitous Ca^{2+}-regulated proteins. *Biochemistry* 23:1103-1109.

27. Creutz, C. E., Pazoles, C. J., and Pollard, H. B. (1978). Identification and purification of an adrenal medullary protein (synexin) that causes calcium-dependent aggregation of isolated chromaffin granules. *J. Biol. Chem.* 253:2858-2866.

28. Hong, K., Düzgüneş, N., and Papahadjopoulos, D. (1981). Role of synexin in membrane fusion. *J. Biol. Chem.* 256:3641-3644.

29. Creutz, C. E., Pazoles, C. J., and Pollard, H. B. (1979). Self-association of synexin in the presence of calcium: Correlation with synexin-induced membrane fusion and examination of the structure of synexin aggregates. *J. Biol. Chem.* 254:553-558.

30. Meers, P., Bentz, J., Alford, D., Nir, S., Papahadjopoulos, D., and Hong, K. (1988). Synexin enhances the aggregation rate but not the fusion rate of liposomes. *Biochemistry* 27:4430-4439.

31. Hong, K., Ekerdt, R., Bentz, J., Nir, S., and Papahadjopoulos, D. (1983). Kinetics of synexin-facilitated membrane fusion. *Biophys. J.* 41:31a.

32. Düzgüneş, N., Hoekstra, D., Hong, K., and Papahadjopoulos, D. (1984). Lectins facilitate calcium-induced fusion of phospholipid vesicles containing glycosphingolipids. *FEBS. Lett.* 173:80-84.

33. Sundler, R., and Wijkander, J. (1983). Protein-mediated intermembrane contact specifically enhances Ca^{2+}-induced fusion of phosphatidate-containing membranes. *Biochim. Biophys. Acta* 730:391-394.

34. Hoekstra, D., and Düzgüneş, N. (1986). *Ricinus communis* agglutinin-mediated agglutination and fusion of glycolipid-containing phospholipid

vesicles: Effect of carbohydrate head group size, calcium ions, and spermine. *Biochemistry* 25:1321-1330.

35. Massari, S., and Colonna, R. (1986). Gramicidin induced aggregation and size increase of phosphatidylcholine vesicles. *Chem. Phys. Lipids* 39:203-220.

36. Lau, A. L. Y., and Chan, S. I. (1975). Alamethicin-mediated fusion of lecithin vesicles. *Proc. Natl. Acad. Sci. USA* 72:2170-2174.

37. Stollery, J. G. and Vail, W. J. (1977). Interactions of divalent or basic proteins with phosphatidylethanolamine vesicles. *Biochim. Biophys. Acta* 471:372-390.

38. Gad, A. E., Silver, B. L., and Eytan, G. D. (1982). Polycation-induced fusion of negatively-charged vesicles. *Biochim. Biophys. Acta* 690:124-132.

39. Walter, A., Steer, C. J., and Blumenthal, R. Polylysine induces pH-dependent fusion of acidic phospholipid vesicles: A model for polycation-induced fusion. Unpublished.

40. Eytan, G. D., and Almary, T. (1983). Melittin-induced fusion of acidic liposomes. *FEBS Lett.* 156:29-32.

41. Hong, K., and Vacquier, V. D. (1986). Fusion of liposomes induced by a cationic protein from the acrosome granule of abalone spermatozoa. *Biochemistry* 25:543-549.

42. Lampe, P. D., and Nelsestuen, G. L. (1982). Myelin basic protein-enhanced fusion of membranes. *Biochim. Biophys. Acta* 693:320-325.

43. Brady, G. W., Murphy, N. S., Fein, D. B., Wood, D. D., and Moscarello, M. A. (1981). The effect of basic myelin protein on multilayer membrane formation. *Biophys. J.* 34:345-350.

44. Cheifetz, S., and Moscarello, M. A. (1985). Effect of bovine basic protein charge microheterogeneity on protein-induced aggregation of unilamellar vesicles containing a mixture of acidic and neutral phospholipids. *Biochemistry* 24:1909-1914.

45. Surewicz, W. K., Epand, R. M., Vail, W. J., and Moscarello, M. A. (1985). Aliphatic aldehydes promote myelin basic protein-induced fusion of phospholipid vesicles. *Biochim. Biophys. Acta* 820:319-323.

46. Morero, R. D., Vinals, A. L., Bloj, B., and Farias, R. V. (1985). Fusion of phospholipid vesicles induced by muscle glyceraldehyde-3-phosphate dehydrogenase in the absence of calcium. *Biochemistry* 24:1904-1909.

47. Glabe, C. B. (1985). Interaction of the sperm adhesive protein, bindin, with phospholipid vesicles. I. Specific association of bindin with gel-phase phospholipid vesicles. *J. Cell Biol.* 100:794-799.

48. Glabe, C. G. (1985). Interaction of the sperm adhesive protein, bindin, with phospholipid vesicles. II. Bindin induces the fusion of mixed-phase vesicles that contain phosphatidylcholine and phosphatidylserine in vitro. *J. Cell Biol.* 100:800-806.

49. Young, T. M., and Young, J. D-E. (1984). Protein-mediated inter-membrane contact facilitates fusion of lipid vesicles with planar bilayers. *Biochim. Biophys. Acta* 775:441-445.

50. Yanagimachi, R. (1988). Sperm-egg fusion. In: *Membrane Fusion in Fertilization, Cellular Transport, and Viral Infection* (N. Düzgünes and F. Bronner, eds.), Academic Press, San Diego, pp. 3-43.

51. Wakelam, M. J. O. (1985). The fusion of myoblasts. *Biochem. J.* 228, 1-12.

52. Lang, R. D. A., Wickenden, C., Wynne, J., and Lucy, J. A. (1984). Proteolysis of ankyrin and of band 3 protein in chemically induced cell fusion. *Biochem. J.* 218:295-305.

53. Knutton, S. (1979). Studies on membrane fusion. III. Fusion of erythrocytes with polyethylene glycol. *J. Cell Sci.* 36:61-72.

54. Smith, C. L., Ahkong, Q. F., Fisher, D., and Lucy, F. A. (1982). Is purified poly(ethylene glycol) able to induce cell fusion? *Biochim. Biophys. Acta* 692:109-114.

55. Hui, S. W., Isac, T., Boni, L. T., and Sen, A. (1985). Action of polyethylene glycol on the fusion of human erythrocyte membranes. *J. Membrane Biol.* 84:137-146.

56. S. Majumdar, and Baker, R. F. (1980). Phosphate-calcium induced fusion of chicken erythrocytes. *Exp. Cell Res.* 126:175-182.

57. Zakai, N., Kulka, R. G., and Loyter, A. (1976). Fusion of human erythrocyte ghosts promoted by the combined action of calcium and phosphate ions. *Nature* 263:696-699.

58. Zimmerman, U. (1982). Electric field-mediated fusion and related electrical phenomena. *Biochim. Biophys. Acta* 694:227-277.

59. Sowers, A. E. (1986). A long-lived fusogenic state is induced in erythrocyte ghosts by electric pulses. *J. Cell Biol.* 102:1359-1362.

60. Okada, Y. (1988). Sendai virus-mediated cell fusion. In: *Membrane Fusion in Fertilization, Cellular Transport, and Viral Infection* (N. Düzgüneş and F. Bronner, eds.). Academic Press, San Diego, pp. 297-336.

61. White, J., Matlin, K., and Helenius, A. (1981). Cell fusion by Semliki Forest, influenza and vesicular stomatitis viruses. *J. Cell. Biol.* 89: 674-679.

62. Düzgüneş, N. (1985). Membrane fusion. In: *Subcellular Biochemistry* Vol. 11 (D. B. Roodyn, ed.), Plenum Press, New York, pp. 195-286.

63. Wilschut, J., and Hoekstra, D. (1984). Membrane fusion: From liposomes to biological membranes. *Trend. Biochem. Sci.* 9:479-483.

64. Strittmatter, W. J., Couch, C. B., and Mundy, D. I. (1985). Role of proteins in the fusion of biological membranes. In: *Membrane Fluidity in Bioloty*, Vol. 4, *Cellular Aspects* (R. C. Aloia and J. M. Boggs, eds.). Academic Press, New York, pp. 259-291.

65. Hong, K., Düzgüneş, N., Meers, P. R., and Papahadjopoulos, D. (1987). Protein modulation of liposome fusion. In: *Cell Fusion* (A. E. Sowers, ed.). Plenum Press, New York, pp. 269-284.

66. Lampe, P. D., Wei, G. J., and Nelsestuen, G. L. (1983). Stopped-flow studies of myelin basic protein association with phospholipid vesicles and subsequent vesicle aggregation. *Biochemistry* 22:1594-1599.

67. Kimelberg, H. K., and Papahadjopoulos, D. (1971). Phospholipid-protein interactions: Membrane permeability coordinated with monolayer "penetration." *Biochim. Biophys. Acta* 233:805-809.

68. Creutz, C. E. (1981). *Cis*-unsaturated fatty acids induce the fusion of chromaffin granules aggregated by synexin. *J. Cell. Biol.* 91:247-256.

69. Meers, P., Hong, K., and Papajadjopoulos, D. (1988). Free fatty acid enhancement of cation-induced fusion of liposomes: Synergism with synexin and other promoters of vesicle aggregation. *Biochemistry* 27:6784-6794.

70. Roos, D. S., Robinson, J. M., and Davidson, R. L. (1983). Cell fusion and intramembrane particle distribution in poly(ethylene glycol)-resistant cells. *J. Cell Biol.* 97:909-917.

71. Kretsinger, R. and Creutz, C. E. (1986). Consensus in exocytosis. *Nature* 320:573.

72. Geisow, M. J. (1986). Common domain structure of Ca^{2+} and lipid-binding proteins. *FEBS Lett.* 203:99-103.

73. Geisow, M. J. and Walker, J. H. (1986). New proteins involved in cell regulation by Ca^{2+} and phospholipids. *Trends Biochem. Sci.* 11:420-423.
74. Pepinsky, R. B., Tizard, R., Mattaliano, R. J., Sinclair, L. K., Miller, G. T., Browning, J. L., Chow, E. P., Burne, C., Huang, K.-S., Pratt, D., Wachter, L., Hession, C., Frey, A. Z., and Wallner, B. P. (1988). Five distinct Ca^{2+} and phospholipid binding proteins share homology with lipocortin I. *J. Biol. Chem.* 263:10799-10811.
75. Klee, C. B. (1988). Ca^{2+}-dependent phospholipid- (and membrane-) binding proteins. *Biochemistry* 27:6645-6653.

10

Effects of an Integral Membrane Glycoprotein on Phospholipid Vesicle Fusion

BEN de KRUIJFF, JOHANNES de GIER, PETER van HOOGEVEST,*
NOL van der STEEN,† TED F. TARASCHI,‡ and TOON de KROON

University of Utrecht, Utrecht, The Netherlands

I. INTRODUCTION

Membrane fusion is a process in which two membranes fuse together to form one new membrane. During the actual fusion event, which in general is short-lived, dramatic reorganizations in lipid structure must take place in order to allow for the maintenance of favorable energetics of lipid organization and retainment of barrier functions. Despite the important role of lipids in membrane fusion, additional factors must be involved in order to regulate and localize the process. Divalent cations (in particular Ca^{2+}) and (membrane) proteins are now well-known examples of such factors.

The evidence for the involvement of proteins in membrane fusion phenomena is most clear for virus-membrane fusion in which viral membrane proteins can expose a hydrophobic fusion site under the appropriate conditions (1; see also Chapters 13 and 14). Also, in fusion between exocytotic storage vesicles and the plasma membrane, proteins such as synexin appear to fulfill an important function (2). In numerous model membrane studies, using lipid vesicles, the fusogenic capacity of a large variety of different proteins and peptides has been reported. These include both water-soluble, very basic peptides, such as polylysines (3), and amphipathic peptides and proteins, such as cardiotoxin (4) and mellitin (5). In some of these cases a molecular picture of the involvement of the peptide or protein in the fusion process has emerged (4). However, such pictures are only of low resolution because either insufficient knowledge of the precise structure of the protein or peptide is available or, more often, the interactions between the protein and its surrounding lipids are not well enough understood. The aim of this chapter is to provide a comprehensive review at the molecular level of lipid-protein interactions in relation to the effect

Present affiliations:
*CIBA-GEIGY, Ltd., Basel, Switzerland.
†Organon Teknika, Turnhout, Belgium.
‡Jefferson College, Philadelphia, Pennsylvania.

of an integral membrane protein (glycophorin) on lipid vesicle fusion. This protein is not only one of the best characterized membrane proteins, but because of its receptor functions it also provides an attractive model for receptor-mediated endocytosis, a process that can be considered to be a reversal of membrane fusion.

II. PROPERTIES OF GLYCOPHORIN AND RECONSTITUTION PROCEDURES

Glycophorin is the major sialoglycoprotein of the human erythrocyte membrane. Its primary structure is known (Fig. 1) and is characterized by: (1) a large N-terminal part, which carries 16 threonine- or serine-linked carbohydrate chains which are highly negatively charged owing to the presence of a total of some 32 sialic acid residues, and (2) a stretch of 23 hydrophobic amino acids (positions 73-96) flanked at its C-terminal side by some basic residues (positions 96-100). The protein most likely forms oligomers in the membrane (even in the presence of SDS it forms dimers) and is (in part) in interaction with the other major integral membrane protein band 3 (6). No clear function has yet been described for this protein but it carries receptor properties for lectins (7), influenza virus (8), and the malaria parasite (9) and it might be involved in cytoskeleton-membrane interactions via some strongly bound phosphatidylinositol diphosphate (10). The known topology of the protein in the red cell membrane (11) is shown in Figure 2. The large glycosylated N-terminus is localized at the outside of the red cell. The protein traverses the membrane once with the hydrophobic segment, probably in an α-helical conformation. The C-terminus is localized in the cytoplasm.

Because of its high charge and carbohydrate content this protein is rather hydrophilic and can be isolated from the membrane without the use of detergents (12). This is a remarkable property for an intrinsic membrane protein and has contributed to its popularity in studies on lipid-protein interactions.

In such studies the protein is reconstituted into a model membrane system of a defined lipid composition. This can be achieved in a number of ways, and the resulting recombinant has properties that depend greatly on the method of reconstitution. Figure 3 illustrates the major reconstitution procedures used for glycophorin. Most often used is the so-called "MacDonald" organic-solvent procedure (13,14) in which a dry mixed glycophorin-lipid film (prepared from a mixed organic solution of both molecules) is hydrated with buffer. This results in the spontaneous formation of large unilamellar vesicles (LUV) containing the protein which can be isolated from protein-poor multilayered structures. The protein is incorporated in a transmembrane orientation and predominantly has a topology similar to that of the erythrocyte in that the majority (±90%) of the N-terminus is located at the outside of the vesicle (14). Using this technique the lipid-protein ratio can be varied only within a limited range (15). However, using the detergent octylglucoside LUV can be prepared which have a lower protein content and in which the protein is more symmetrically incorporated (16,17). Cosonication of glycophorin and lipid results in the formation of small unilamellar vesicles (SUV) in which now 100% of the sialic acid is accessible for external neuraminidase. In general, the lipid-protein ratio in these SUV can be made to be >300:1 (mol/mol). It is possible to reach higher protein contents using cosonication procedures (15),

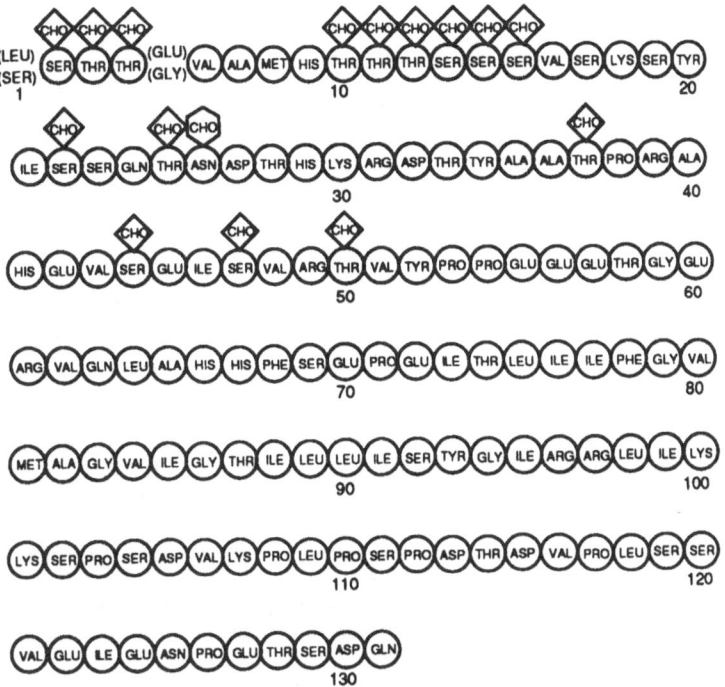

FIGURE 1 Primary structure of human erythrocyte glycophorin A (11).

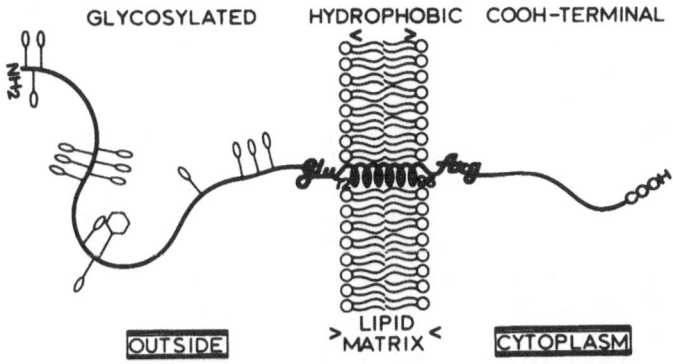

FIGURE 2 Topology of glycophorin in the erythrocyte membrane.

but the vesicular nature of these recombinants has not always been established. Before discussing the effect of glycophorin on lipid polymorphism and vesicle fusion it is useful to first summarize the consequences of the glycophorin-lipid interaction for acyl chain packing and barrier function of lipid vesicles.

Methods

	McDonald (organic solvents)	Detergent (OG: 0.5 mol %)	Cosonication
Diameter	LUV (1000-5000 A)	LUV (2500-4000 A)	SUV (250-300 A)
PC/Gly (m/m)	100-600	300-∞	300-∞

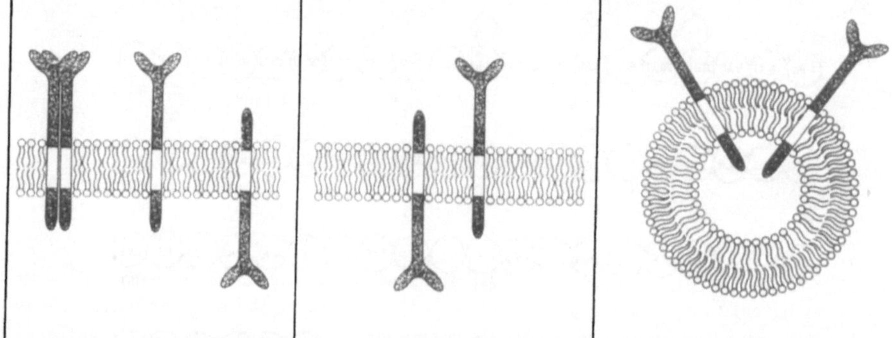

FIGURE 3 Different methods to reconstitute glycophorin (gly) into phosphatidylcholine (PC) vesicles. The Y-shaped end of glycophorin represents the carbohydrate-containing N-terminus of the protein.

III. PROPERTIES OF GLYCOPHORIN-CONTAINING LIPID VESICLES

Incorporation of glycophorin into LUV prepared with phosphatidylcholines (PC) via the organic solvent procedure results in a pertubation of acyl chain packing (disordering effect), as can be inferred from spectroscopic studies (18,19), and causes a strong reduction in the heat content of the gel to liquid-crystalline phase transition (15). The pertubation extends down to at least three layers of surrounding lipid molecules. PC in extended liquid-crystalline bilayers such as those found in LUV cannot be degraded by pancreatic phospholipase A_2 owing to insufficient penetration of the enzyme (20), whereas the presence of glycophorin leads to rapid degradation of the phospholipids by the phospholipase A_2. Apparently, the defects in lipid packing induced by glycophorin facilitate the penetration of the lipase into the bilayer.

Lipid packing defects can greatly affect the general barrier function of a bilayer. This is well established for the defects at the interface of gel and liquid-crystalline lipid domains, which not only lead to rapid permeation of polar molecules (21,22) [the "pore" radius of these defects in dimyristoylphosphatidylcholine (DMPC) vesicles determined with solutes of different size was found to be ~10 Å (23)], but also can cause a rapid transbilayer movement of lipids (24). The glycophorin-induced packing defects in unsaturated PC bilayers (prepared via the organic solvent reconstitution method) show a similar behavior. Transbilayer movement of both PC (25) and lyso-PC (26,27) is greatly increased by incorporation of the protein. Figure 4 illustrates this for the lyso component, which can be almost completely degraded by exogenous lysophospholipase, despite the fact that it is symmetrically distributed over the two halves of the

glycophorin-containing bilayers of dioleoylphosphatidylcholine (DOPC) LUV
(27). In the absence of the protein, only 50% of the lyso-PC can be degraded, showing the lack of transbilayer movement in these lipid vesicles.
Most interesting, the lipid flip-flop in this system is highly dependent
on the lipid composition of the bilayer. Incorporation of a similar amount
of glycophorin into large unilamellar vesicles made of a total lipid extract
of the human red cell membrane does not result in an increase of the lyso-PC
flip-flop (27).

More detailed insight into the way lipids modulate the presence of
packing defects in the glycophorin-containing bilayer has been obtained
from permeability experiments (16,27-30). In the absence of glycophorin
LUV of DOPC have a good barrier for glucose and dextran (Fig. 5), whereas
the presence of the protein results in a rapid loss of entrapped glucose
but not of the much larger dextran molecule. A variety of other solutes
with molecular weights of up to 800 kDa, can pass the glycophorin-
containing DOPC bilayer. The size of the glycophorin-induced packing
defect corresponds to a "pore" with a radius of 10 Å (29) similar to that
seen in the case of bilayers with coexisting gel and liquid-crystalline DMPC
domains. Although variations in the acyl chain composition of the PC molecule
have no dramatic influence on the protein-induced permeation pathway
(30), the presence of other lipid classes has a pronounced "annealing"
effect. Increasing the dioleoylphosphatidylethanolamine (DOPE) or dioleoylphosphatidylserine (DOPS) content in glycophorin-containing DOPC vesicles
results in a gradual decrease in glucose permeability (30). Figure 6 shows
this for DOPE. Incorporation of DOPE up to a mol fraction of 0.5 completely
restores the barrier function, which is even more striking since the glyco-

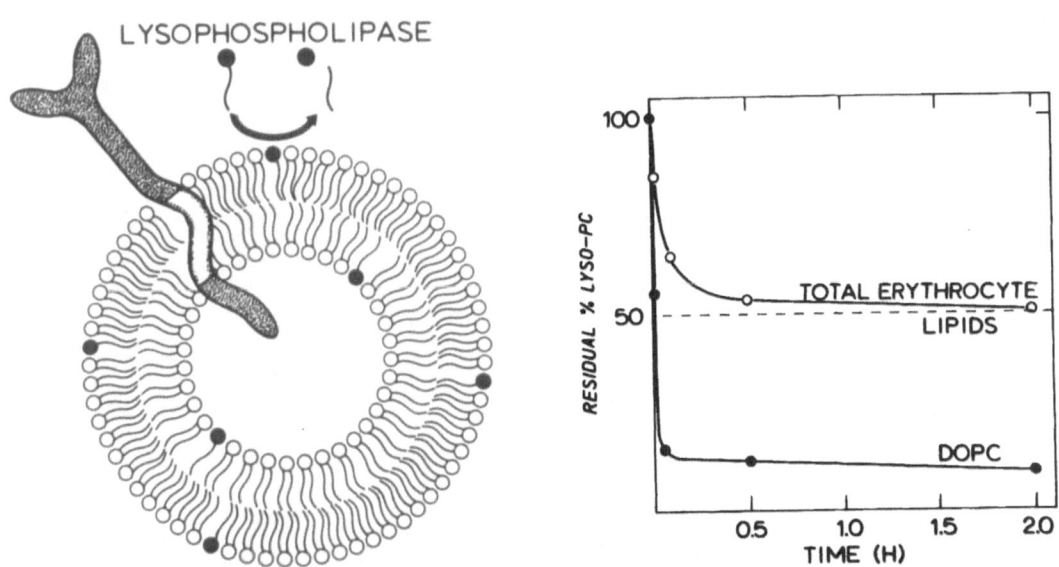

FIGURE 4 Transbilayer movement of lysophosphatidylcholine (LPC) in
glycophorin-containing LUV of DOPC (glycophorin/DOPC, molar ratio 1:400)
or total erythrocyte lipids as assayed by the degradation of preincorporated
lyso-PC (5 mol%) by externally added lysophospholipase. (For details see
Ref. 27.) The molar ratio of lipid to glycophorin was 400.

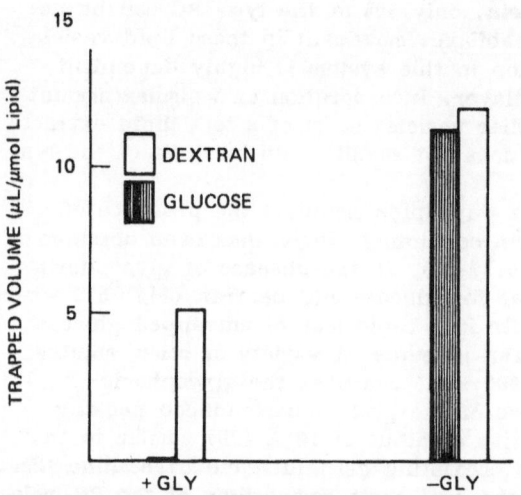

FIGURE 5 Effect of glycophorin incorporation on the barrier properties
of DOPC LUV. Presented are the trapped volumes measured via determination of enclosed dextran and glucose in the vesicles after separation of
the vesicles from the nonenclosed solutes. Glycophorin/lipid molar ratio
1:500. (For details see Ref. 28.)

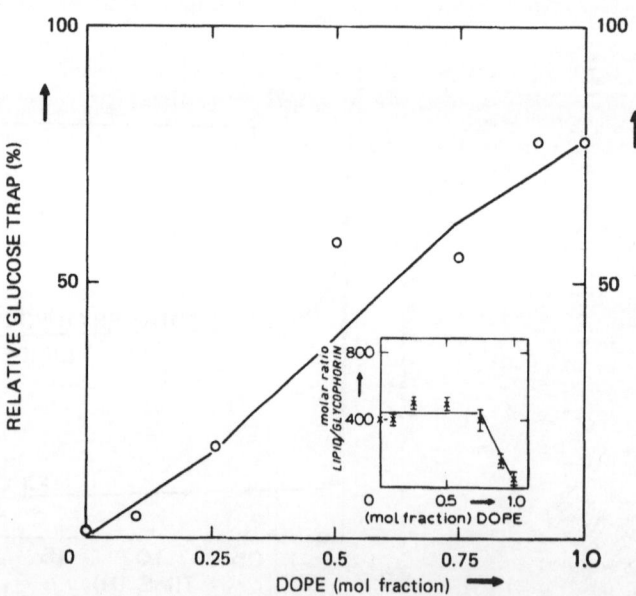

FIGURE 6 Glucose permeability of glycophorin-containing DOPC/DOPE
LUV of varying composition. The inserts show the relationships between
the DOPE and the glycophorin content of these vesicles which were prepared via the "organic solvent" procedure. (For details see Ref. 30.)

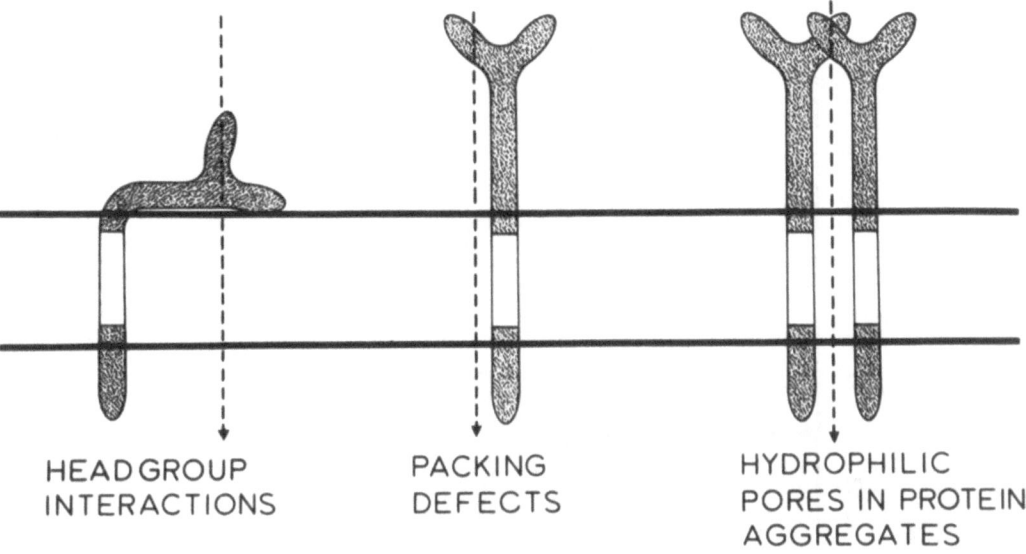

HEADGROUP
INTERACTIONS

PACKING
DEFECTS

HYDROPHILIC
PORES IN PROTEIN
AGGREGATES

FIGURE 7   Possible permeation pathways in glycophorin-containing PC bilayer.

phorin content of these vesicles slightly increases with increasing PE content. There are at least three possible explanations for the observed increases in lipid flip-flop and bilayer permeability: headgroup interactions, packing defects in the acyl chain, and hydrophilic pores present in protein aggregates (Fig. 7). Interactions between the carbohydrate-rich headgroup of glycophorin and the lipid headgroups have been inferred from spectroscopic studies (31). However, they are most likely not determining the loss of barrier function since proteolytic removal of the protein headgroup does not affect bilayer permeability (29). Furthermore, incorporation of only the hydrophobic domain of glycophorin into lipid vesicles results in a large increase in bilayer permeability (32,33). Defects in acyl chain packing are a more attractive mechanism, as already suggested by the similarity between the barrier properties of PC bilayer containing glycophorin and protein-free bilayers containing coexisting gel and liquid-crystalline lipid domains. The phospholipid class dependency of the increase in permeability and flip-flop can then be understood in terms of either specific electrostatic and/or hydrogen bonding interactions between lipid headgroups and particular amino acid residues (30) or in terms of the irregular outer contour of the membrane-spanning part of the protein which has to be matched with lipids of a particular geometrical shape in order to minimize unfavorable packing defects (28,30,34). The results on the effect of DOPE are especially relevant in this respect as this lipid prefers to organize in an H_{II} phase and thus will be cone shaped, as will be discussed in more detail in Section IV.

Most likely hydrophilic channels present in glycophorin aggregates also contribute to the increased permeability (but probably not to the lipid flip-flop). This possibility is attractive because: (1) glycophorin reconstituted in PC vesicles by the organic-solvent procedure does aggregate in the bilayer, as evidenced by the particles seen by freeze-fracture

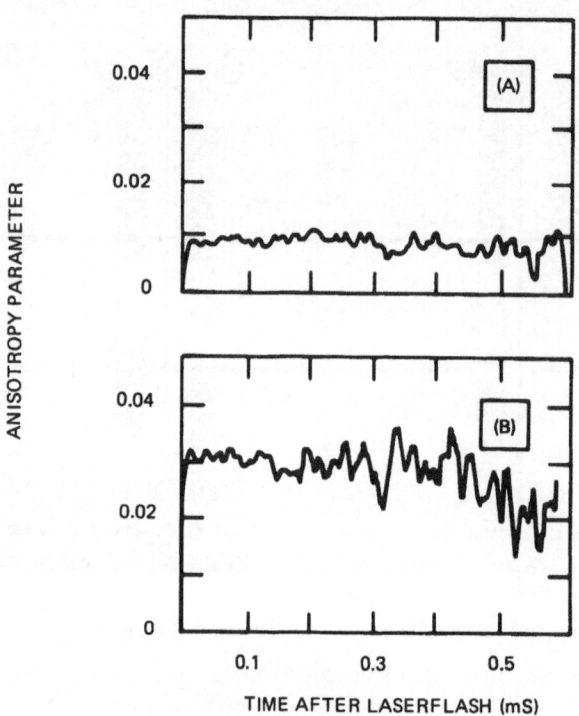

FIGURE 8   Phosphorescence anisotropy decay traces at 0°C of erythrosin-
labeled glycophorin reconstituted in DOPC (A) or DOPE (B) vesicles. (For
details see Ref. 37.)

on the fracture face of the bilayer (14); (2) in the octylglucoside reconstitution method the protein is not visible on the fracture face and the bilayer
permeability is less increased (16); (3) there is a marked correlation between
the aggregate size in reconstitutes of the hydrophobic part of glycophorin
and the bilayer permeability (32,33,35); and (4) black-lipid membrane
studies have given good evidence that other α-helix-forming hydrophobic
polypeptides such as alamethicin can form aqueous channels present within
peptide aggregates (36).

Analysis of the phosphorescence anisotropy decay of erythrosin-labeled
glycophorin reconstituted into lipid vesicles showed that the rotational
diffusion of the probe, and also the protein, is dependent on the aggregate
state of the protein, the phase state of the lipid, and the composition
of the bilayer (37). For example, the addition of the multivalent lectin
wheat germ agglutinin (WGA), which has a high affinity for the carbohydrate moiety of glycophorin (7), decreases the rate of rotational diffusion
due to protein aggregation. Rigidifying the bilayer either by lowering
the temperature, causing the acyl chains to "freeze," or by adding Ca^{2+}
in case of beef brain phosphatidylserine (PS) results in a decrease in
rotational mobility of the protein. The marked differences in the anisotropy
parameter observed for glycophorin in DOPC and DOPE vesicles (Fig. 8)
strongly suggest a higher state of aggregation of the protein in the PE
bilayer. This is in full agreement with the observation that the protein
particles visualized by freeze-fracturing on the fracture face of the bilayer
are much larger in PE than in PC systems (12,14). Most likely, the strong

intermolecular headgroup interactions in PE are the driving force behind the clustering of glycophorin. Similar differences in peptide aggregation behavior between PC and PE model membranes have been reported (38). As the bilayer permeability is very low in the glycophorin-containing PE vesicles (see previous sections), it must be concluded that the proposed channels in glycophorin oligomers are closed because of strong intermolecular PE interactions.

With the background knowledge on the properties of glycophorin-containing lipid vesicles described in this section it is possible to understand the effect of glycophorin on lipid vesicle fusion.

IV. INFLUENCE OF GLYCOPHORIN ON LIPID POLYMORPHISM IN RELATION TO VESICLE FUSION

As discussed in other chapters of this book (Chapters 2 and 8), lipids that have a tendency to adopt "inverted" nonbilayer structures like the H_{II} phase and/or lipidic particles greatly promote fusion either by virtue of their low headgroup hydration, which allows for close bilayer apposition, or because of the ability of such lipids to temporarily form nonlamellar lipid structures as fusion intermediates, or by a combination of both.

Therefore, in order to understand the effect of a protein on vesicle fusion it is necessary to know how this protein affects the macroscopic organization of the lipids. In the ideal case one would like to follow the structural organization of the lipids during the fusion process. In view of the time resolution of the process and the expected low number of participating lipids at the site of bilayer fusion, this has not been possible up to now owing to technical limitations. Therefore, the common approach normally employed in this area has been to characterize the effect of the protein on the macroscopic organization of the lipids in aqueous dispersions under equilibrium conditions and then to correlate the results from such studies with fusion measurements on unilamellar vesicles made of the same components. This approach has also been followed for glycophorin. Aqueous dispersions of unsaturated PE's undergo a temperature-dependent bilayer to H_{II} phase transition (for review see Ref. 39 and Chapter 3). This transition temperature depends on the acyl chain composition, increasing chain unsaturation resulting in lower transition temperature. ^{31}P NMR is a convenient technique to discriminate between the different lipid organizations. Because the motional possibilities of a lipid molecule depend on the macroscopic structure in which the lipid is present, the chemical shift anisotropy of the lipid phosphates is differently averaged leading to characteristic ^{31}P NMR line shapes. Thus, under conditions of high-power proton decoupling which reduces or eliminates the dipolar (^{1}H-^{31}P) interaction, the ^{31}P NMR spectrum of lipids in the liquid-crystalline phase is characterized by a high-field peak and a low-field shoulder separated by ~40 ppm. In the H_{II} phase the lateral diffusion of the lipid molecules around the tubes of which this phase is composed gives rise to a reversal of sign in the asymmetry of the spectrum (now low-field peak and high-field shoulder) and a reduction in peak-to-shoulder distance by a factor of two.

Figure 9 shows that DOPE in an aqueous dispersion is organized in an aqueous dispersion is organized in an H_{II} phase at 0°C. When glycophorin is incorporated in this PE via the organic solvent reconstitution procedure, part of the lipids form unilamellar bilayer vesicles (300-1500 Å

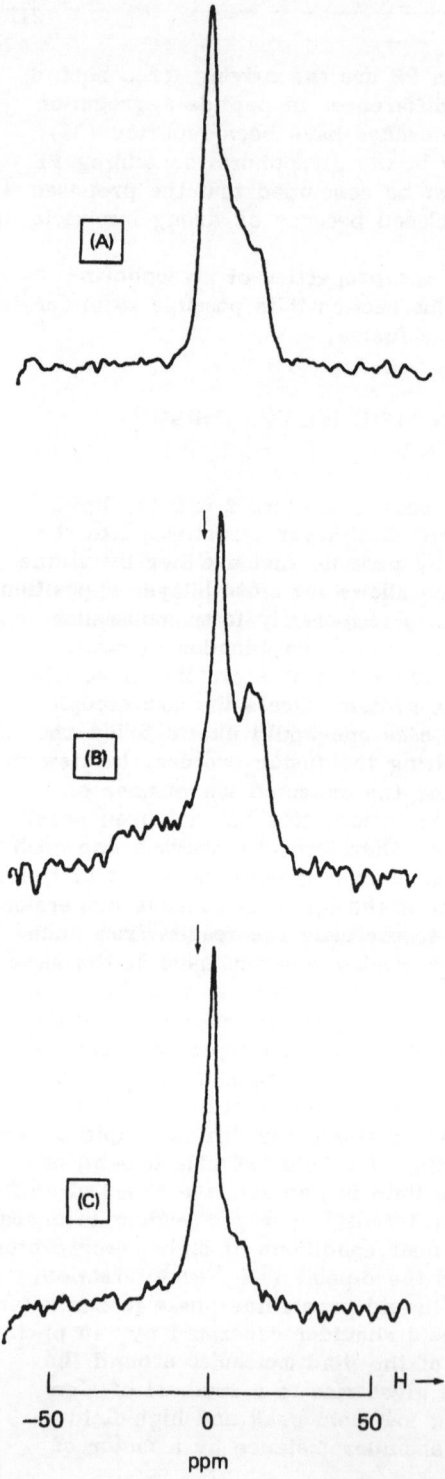

$$H \rightarrow$$

$$-50 \quad\quad 0 \quad\quad 50$$
ppm

FIGURE 9 36.4 MHz ^{31}P NMR spectra of DOPE in the absence (0°C, A) and presence (0°C, B) and (25°C, C) of glycophorin. Arrow indicates the position of the main spectral component of DOPE in the H_{II} phase. (Reproduced with permission from Ref. 12. For details see Ref. 12.)

diameter) with a fixed stoichiometry of 25 molecules of PE per glycophorin molecule, independent of the initial glycophorin/PE molar ratio in the mixed lipid-protein film (12). In these vesicles the protein appears to be more aggregated since the protein particles observed by freeze-fracture electron microscopy are much larger than those observed in the vesicles made of DOPC (see also Section III). The profound stabilization by glycophorin of DOPE is also evident by ^{31}P NMR. Figure 9B shows that most of the spectral intensity is present in a component with the characteristic "bilayer" line shape. Part of the signal is present in a narrow line shape with a chemical shift position characteristic of phospholipids undergoing rapid isotropic motion. This signal arises from the smallest vesicles in the preparation in which the chemical shift anisotropy is reduced due to vesicle tumbling and lateral diffusion of the lipid molecules. Increasing the temperature to 20°C results in an increase in this isotropic component due to the faster tumbling of the vesicles and an increased rate of lateral diffusion. No signal intensity of lipids organized in the H_{II} phase is observed despite the stronger expression of the H_{II} character due to the increase in chain motion.

Since in small unilamellar vesicles prepared by ultrasonication from DOPE and glycophorin the lipid to protein ratio can be varied more extensively, such vesicles are more suitable to get insight in the molecular details of the bilayer stabilization (40). Freeze-fracture electron microscopy analysis of such cosonicated vesicles demonstrated that in the presence of one or more glycophorin molecules per 200 DOPE molecules stable small bilayer vesicles are formed, which have an isotropic ^{31}P NMR line shape (Fig. 10A,D). Thus, one glycophorin molecule can stabilize 200 DOPE molecules into a bilayer configuration. Addition of neuraminidase, an enzyme that removes the sialic acid residues which are all externally exposed in these vesicles, has virtually no effect on the vesicle morphology or on the ^{31}P NMR spectrum at low (Fig. 10E) or high (Fig. 10B) glycophorin/PE molar ratios. Thus, the negative charge on the protein does not appear to contribute to the bilayer stabilization. However, when the entire carbohydrate-rich N-terminal part of the protein is cleaved off by trypsin, the 1:200 (glycophorin/PE, mol/mol) vesicles fuse and form the hexagonal H_{II} phase as evidenced by freeze-fracturing and the characteristic ^{31}P NMR line shape shown in Figure 10F. When the glycophorin content of the vesicles is increased, trypsin treatment also results in immediate fusion of the vesicles as judged by an increase in vesicle size (40), which is paralleled by an increase in ^{31}P NMR line width. However, no H_{II} phase is formed in this case, demonstrating that the high content of membrane-spanning hydrophobic parts of glycophorin still exert significant bilayer stabilization. Figure 11 models these data and emphasizes the two main bilayer stabilizing forces of the protein: (1) the sterical hindrance of the large glycosylated protein headgroup which inhibits close contact between the vesicles and H_{II} phase formation and (2) the direct bilayer stabilization caused by the hydrophobic segment of the protein. The profound bilayer stabilization of glycophorin is not limited to PC systems. The presence of the protein also very effectively blocks the bilayer to hexagonal H_{II} phase transition induced by Ca^{2+} in beef heart cardiolipin dispersions (41).

The multivalent plant lectin wheat-germ agglutinin (WGA) has a very high affinity for the carbohydrate part of glycophorin and can cause both clustering of the protein in a membrane and aggregation of glycophorin-containing vesicles (42). The consequences of this behavior for membrane lipid structure can be dramatic, as shown in the experiment depicted in

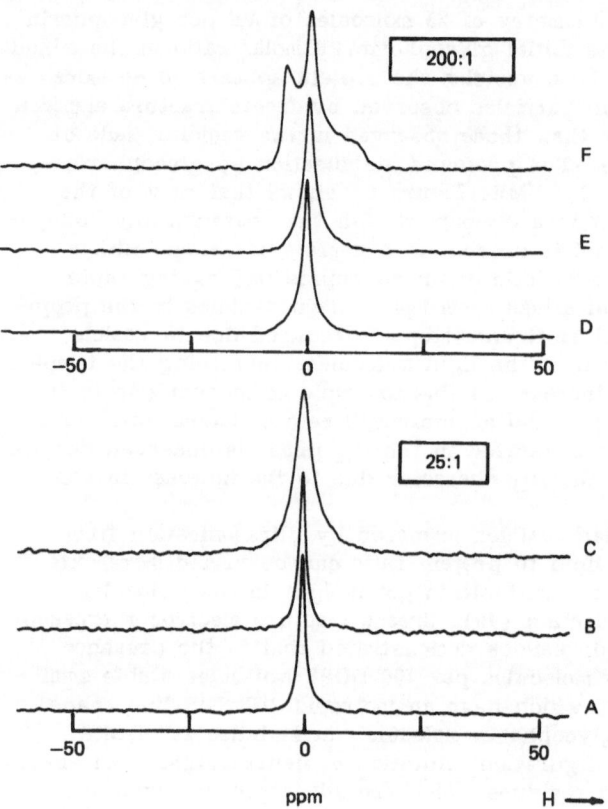

FIGURE 10   Effect of neuraminidase and trypsin treatment on the 81.0
MHz ^{31}P NMR spectra of cosonicated DOPE/glycophorin vesicles. The
molar ratios of DOPE/glycophorin are indicated. Vesicles were treated with
(B,E) neuraminidase (50 units/mg of protein, 2 h, 37°C), followed by
(C,F) trypsin treatment (5%, w/w, with respect to glycophorin, 2 h, 37°C).
(Reproduced with permission from Ref. 40.)

Figure 12. Addition of excess WGA to DOPE/glycophorin (molar ratio 200:1)
vesicles immediately results in fusion and the formation of lipidic particles
and the H$_{II}$ phase. This process, which can be completely inhibited by
an excess of N-acetylglucosamine, a sugar that competes with glycophorin
for the binding to WGA, can be interpreted as being due to the local
increase in concentration of glycophorin as a result of the WGA-glycophorin
interaction, thereby removing the bilayer stabilization of the excess DOPE
molecules (40). In fact, consistent with the trypsin data, no such bilayer
to H$_{II}$ transition is observed when WGA clusters the glycophorin molecules
in DOPE/glycophorin (molar ratio 25:1) vesicles (40). Freeze-fracture
electron microscopy showed that these vesicles did undergo aggregation
and some vesicle fusion leading to an increase in ^{31}P NMR line width.
A schematic representation of the effect of WGA on the structure of
DOPE/glycophorin recombinants is given in Figure 13. As in Figure 11,
only the final lipid structures are modeled. No information is available
on lipid structures during fusion. Although these data do not give direct
insight in the nature of the relation between vesicle fusion and lipid

organization in the glycophorin-containing vesicles, they do unambiguously demonstrate the profound parallel influence of the protein on fusion and the formation of nonbilayer structures.

The development of fluorescence techniques that monitor both the mixing of aqueous contents and lipids of vesicles, which are described in detail elsewhere in this book (Chapter 4), has greatly contributed to our present knowledge of lipid vesicle fusion. With such techniques the effect of glycophorin on vesicle fusion has also been investigated. As a test system large unilamellar vesicles prepared from beef brain PS have been used. Such vesicles undergo isothermal fusion in the presence of 3 mM or more Ca^{2+} (43,44). It is believed that lipid headgroup dehydration and the formation of a "trans" Ca^{2+}/PS complex involving PS molecules on the apposed bilayers are important parameters in this Ca^{2+}-mediated vesicle fusion process (45). In Figure 14 the kinetics of the fusion of such PS vesicles is represented as the rate of mixing of the aqueous compartments by the addition of 25 mM Ca^{2+}. This is being monitored by measuring the formation of the fluorescent $Tb(DPA)_3$ complex upon mixing of the aqueous contents of vesicles containing either Tb or DPA (43). After the initial rapid rise in fluorescence intensity a decrease is observed which is due to the collapse of the vesicles, the release of vesicle contents, and the formation of so-called cochleated structures (43). Incorporation of one glycophorin molecule per 400 PS molecules completely inhibits the

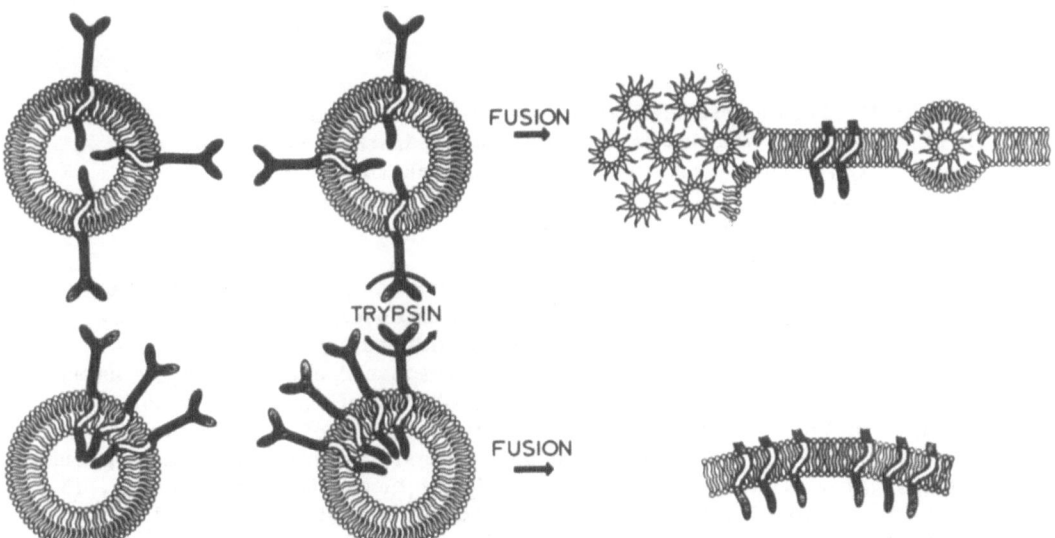

FIGURE 11 Schematic representation of the structural consequences of the removal of the headgroup of glycophorin by trypsin in DOPE vesicles. Upper panel: DOPE/glycophorin 200:1 (mol/mol); lower panel: DOPE/glycophorin 200:1 (mol/mol); lower panel: DOPE/glycophorin 25:1 molar ratio. No exact structural information is available on the transitional area between the H_{II} phase (shown here in cross-section) and the lamellar phase containing the glycophorin remnants. The visualization of an intrabilayer inverted micelle is solely meant to indicate that other nonbilayer lipid structures such as lipidic particles are present in the system after trypsin treatment.

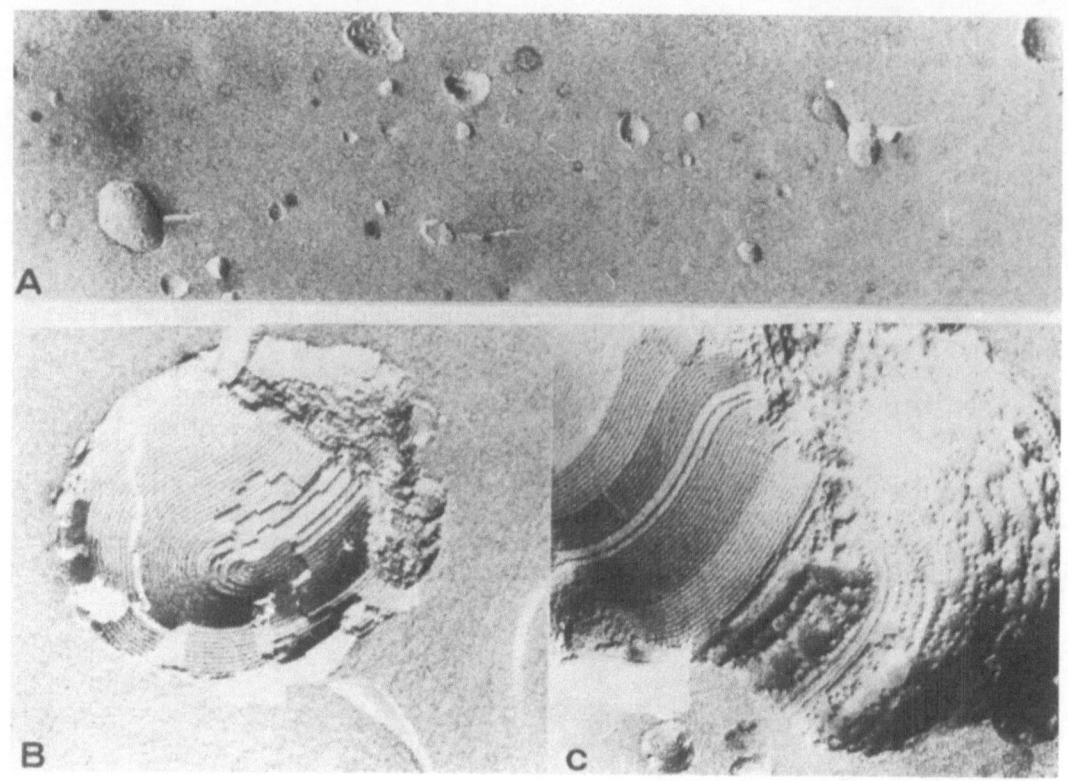

FIGURE 12   Freeze-fracture electron micrographs of cosonicated
DOPE/glycophorin vesicles (molar ratio 200:1) in the absence (A) and
presence (B,C) of 200 nmol WGA (WGA/glycophorin molar ratio 1:1).
Final magnification 100,000×. (Reproduced with permission from Ref. 40.)

Ca^{2+}-induced formation of the Tb-DPA complex in the vesicles (44), but
a time-dependent release of vesicle contents is observed which is mainly
due to the increase in bilayer permeability induced by the presence of
the glycophorin (see Section III). Using a resonance energy transfer assay
for lipid mixing (46), the strong fusion-inhibiting effect of glycophorin
could be further substantiated. Whereas the addition of 5 mM $CaCl_2$ causes
rapid lipid mixing of two PS vesicle populations, the presence of glycophorin
(PS/glycophorin; molar ratio 400:1) completely blocks this process (Fig. 15).
   Interestingly, the addition of WGA prior to the addition of Ca^{2+}, thus
clustering the glycophorin molecules, makes the glycophorin-containing
PS vesicles again fusion sensitive (Fig. 15), in full agreement with the
results described for the glycophorin-containing PE vesicles. The large
carbohydrate-bearing headgroup of glycophorin is the most crucial factor
in the fusion inhibitory effect of glycophorin, as demonstrated in Figure 16,
where it is shown that treatment of the vesicles with trypsin restores,
to a large extent, the Ca^{2+}-induced fusion of the vesicles. That the fusion
still is not as effective as in the control situation without the protein
suggests that either the extending remnants of the protein prohibit close
proximity or, alternatively and more likely, that the membrane-spanning

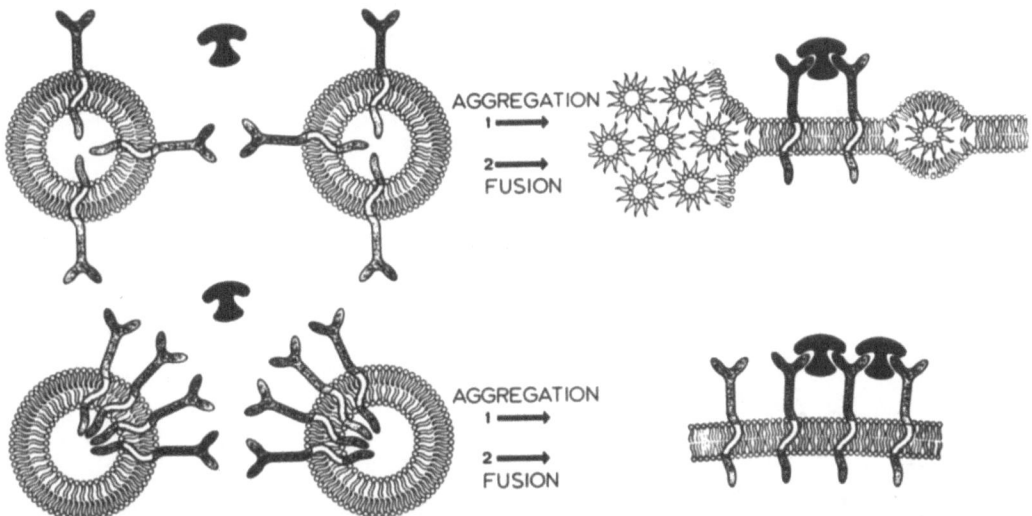

FIGURE 13 Schematic representation of the effect of WGA–glycophorin interaction on the structure of DOPE/glycophorin vesicles. Upper panel: DOPE/glycophorin (molar ratio, 200:1); lower panel: DOPE/glycophorin (molar ratio, 25:1). In this model WGA is proposed to laterally aggregate glycophorin molecules. For visualization of the nonbilayer structures see the legend of Figure 11.

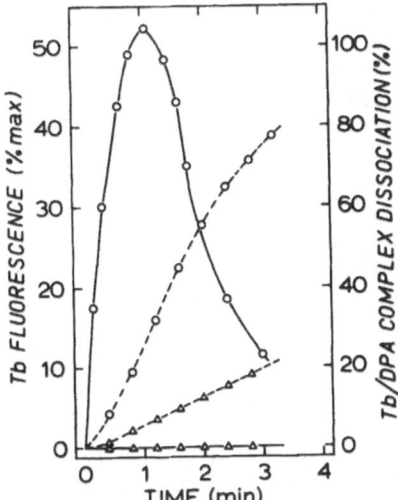

FIGURE 14 Calcium-induced fusion (——) and release of contents (---) of PS vesicles (o) and PS glycophorin (molar ratio 400:1) vesicles (△), as measured by the Tb/DPA assay. $CaCl_2$ was injected as an aliquot of a 1.0 M solution at t = 0 to a final concentration (not corrected for the 0.1 mM EDTA, present in the medium) of 25 mM. Fusion was measured as the mixing of aqueous contents of separate vesicle populations (each with a lipid concentration of 25 µM) containing either Tb or DPA and is presented as the percentage of encapsulated Tb complexed to DPA. The release of vesicle contents was measured with vesicles (with a lipid concentration of 50 µM) containing the Tb/DPA complex and shows the percentage of the fluorescent complex that has dissociated due to leakage from, and the entry of medium into, the vesicles. (Reproduced with permission from Ref. 44. For experimental details see Ref. 44.)

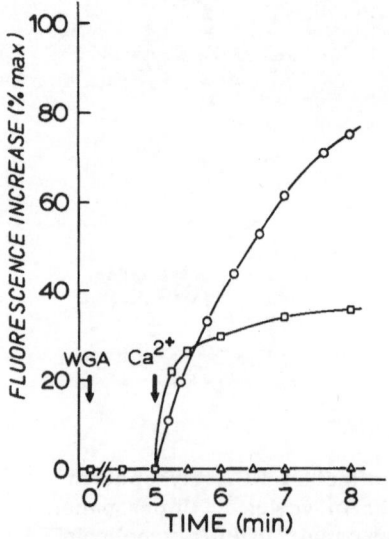

FIGURE 15   Calcium-induced fusion of PS vesicles (o), PS glycophorin
vesicles (△), and PS glycophorin vesicles after incubation with WGA (□),
as measured by the RET assay. CaCl$_2$ was introduced as an aliquot of
a 1.0 M solution to a final concentration of 5 mM. The total lipid concentration was 50 μM. The incubation of PS glycophorin vesicles with WGA
(45 μg/mL, corresponds to a WGA/glycophorin molar ratio of 10) was
carried out for 5 min prior to the injection of calcium. In the RET assay
equal amounts of two populations of vesicles are used, one containing 0.5%
N-NBD-PE and 1% N-Rh-PE and the other devoid of fluorescent lipids.
Fusion results in lipid mixing of the two vesicle populations, thus decreasing
the efficiency of the resonance energy transfer between donor (N-NBD-PE)
and acceptor (N-Rh-PE). The extent of fusion was measured by following
the emission signal of N-NBD-PE (530 nm) upon excitation 450 nm. (Reproduced with permission from Ref. 44. For further details see Ref. 44.)

part of the protein acts to stabilize the vesicle bilayer thereby inhibiting
fusion. Removal of the sialic acid residues by neuraminidase treatment
also restores some of the fusion capability of the glycophorin-containing
PS vesicles (Fig. 16), although the effect of this treatment is less than
that observed with trypsin. These and other data (44) demonstrate that
there is a striking similarity between the inhibitory effect of glycophorin
on the Ca$^+$-induced fusion of PS vesicles and the influence of glycophorin
on lipid polymorphism in PE and cardiolipin model membranes and reinforces
the conclusion that the fusion inhibitory effect of glycophorin is due to
a combination of steric hindrance of close apposition of vesicle membranes
and a direct bilayer stabilization by the protein molecule.

V.   CONCLUSIONS AND BIOLOGICAL SIGNIFICANCE

The results reviewed in this chapter demonstrate that interactions between
an integral membrane glycoprotein and its surrounding lipids can have

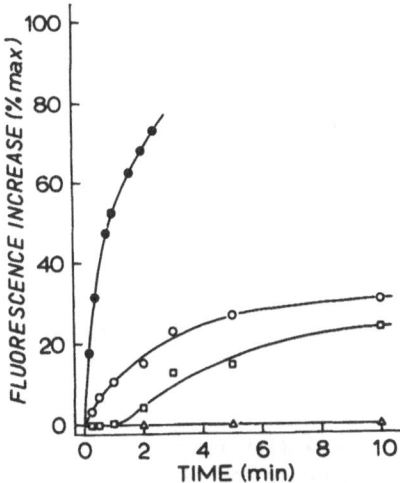

FIGURE 16 Calcium-induced fusion of PS vesicles (●) and PS/glycophorin vesicles after incubation without enzyme (△), with trypsin (○), or with neuraminidase (□), as measured by the RET assay. $CaCl_2$ was introduced as aliquots of a 1.0 M solution to a final concentration of 25 mM. The total lipid concentration was 50 µM. The PS/glycophorin vesicles were incubated with trypsin (10% w/v, glycophorin) or neuraminidase (0.04 unit/mg glyco-phorin) during 1 h at 37°C. (Reproduced with permission from Ref. 44. For experimental details see Ref. 44.)

pronounced structural and functional consequences. In the liquid-crystalline bilayer of an unsaturated typical membrane phospholipid such as DOPC, the protein causes both lipid-packing defects and aqueous pores present within protein aggregates, which can result in an enhanced transbilayer movement of both lipids and polar solutes. The formation of these transport possibilities is strongly lipid dependent, as is the aggregation state of the proteins in the bilayer. Unsaturated PE molecules play a particular role in this respect as they not only can restore the normal barrier function of the bilayer, but they also cause the protein to laterally aggregate in the plane of the bilayer. The glycophorin molecule has a pronounced bilayer stabilizing effect in model membrane systems prepared of lipids which prefer on their own an organization in the H_{II} phase. Parallel to this bilayer stabilization the protein has a strong vesicle fusion inhibitory action. For both effects the large outwardly oriented glycosylated headgroup of the protein plays a major role as it interferes by steric hindrance with the close apposition of the bilayers, which is a prerequisite for H_{II} phase formation and vesicle fusion. In addition, the hydrophobic membrane-spanning segment of the protein most likely contributes to the inhibition of H_{II} phase formation and vesicle fusion by a direct bilayer stabilizing effect on the surrounding lipids.

Clustering of the glycophorin molecules by a multivalent lectin can induce both H_{II} phase formation and vesicle fusion. Extrapolation of the data summarized in this chapter to biomembrane structure, in particular biomembrane fusion, leads to the following speculations.

1. Clustering of integral membrane proteins can be the result of (changes in) the lipid composition of the membrane.
2. Local transbilayer transport of lipids and polar solutes is facilitated by an integral protein in a lipid-dependent manner.
3. Lateral redistribution of glycoproteins by means of effector-receptor interactions could result in bilayer destabilization around the clustered glycoprotein, which could result in membrane fusion of the type encountered in receptor-mediated endocytosis.

REFERENCES

1. Lucy, J. A. (1984). Do hydrophobic sequences cleaved from cellular polypeptides induce membrane fusion reactions in vivo? *FEBS Lett.* 166:223-231.
2. Hong, K., Düzgüneş, N., and Papahadjopoulos, D. (1981). Role of synexin in membrane fusion. *J. Biol. Chem.* 256:3641-364.
3. Gad, A. E. (1983). Cationic polypeptide-induced fusion of acidic liposomes. *Biochim. Biophys. Acta* 728:377-382.
4. Batenburg, A. M., Bougis, P. E., Rochat, H., Verkleij, A. J., and De Kruijff, B. (1985). Penetration of a cardiotoxin into model membranes and its implications on lipid organization. *Biochemistry* 24:7101-7110.
5. Batenburg, A. M., Hibbeln, J. C. L., Verkleij, A. J., and de Kruijff, B. (1987). Melittin induces H_{II} phase formation in cardiolipin model membranes. *Biochim. Biophys. Acta* 903:162-156.
6. Nigg, E. A., Bron, C., Giradet, M., and Cherry, R. J. (1980). Band 3-glycophorin A association in erythrocyte membranes demonstrated by combining protein diffusion measurements with antibody-induced cross-linking. *Biochemistry* 19:1887-1893.
7. Verpoorte, J. A. (1975). Purification and characterization of glycoprotein from human erythrocyte membranes. *Int. J. Biochem.* 6:855-862.
8. Marchesi, V. T., Tillack, T. W., Jackson, R. L., Segrest, J. P. and Scott, R. E. (1972). Chemical characterization and surface orientation of the major glycoprotein of the erythrocyte membrane. *Proc. Natl. Acad. Sci. USA* 69:1445-1449.
9. Pasvol, G., Wainscoat, J. S., and Weatherall, D. J. (1982). Erythrocytes deficient in glycophorin resist invasion by the malarial parasite *Plasmodium falciparum*. *Nature* 297:64-66.
10. Anderson, R. A., and Marchesi, V. T. (1985). Regulation of the association of membrane skeletal protein 4.1 with glycophorin by a polyphosphoinositide. *Nature* 314:472-474.
11. Tomita, M., and Marchesi, V. T. (1975). Amino acid sequence and oligosaccharide attachment sites of human erythrocyte glycophorin. *Proc. Natl. Acad. Sci. USA* 72:2964-2968.
12. Taraschi, T. F., De Kruijff, B., Verkleij, A. J., and Van Echteld, C. J. A. (1982). Effect of glycophorin on lipid polymorphism. A ^{31}P NMR study. *Biochim. Biophys. Acta* 685:153-161.
13. MacDonald, R. J., and MacDonald, R. C. (1975). Assembly of phospholipid vesicles bearing sialoglycoprotein from erythrocyte membrane. *J. Biol. Chem.* 250:9206-9214.
14. Van Zoelen, E. J. J., Verkleij, A. J., Zwaal, R. F. A., and Van Deenen, L. L. M. (1978). Incorporation and asymmetric orientation of glycophorin in reconstituted protein-containing vesicles. *Eur. J. Biochem.* 86:539-546.

15. Van Zoelen, E. J. J., Van Dijck, P. W. M., De Kruijff, B., Verkleij, A. J., and Van Deenen, L. L. M. (1978). Effects of glycophorin incorporation on the physicochemical properties of phospholipid bilayers. *Biochim. Biophys. Acta* 514:9-24.

16. Van der Steen, A. T. M., Taraschi, T. F., Voorhout, W. F., and De Kruijff, B. (1983). Barrier properties of glycophorin-phospholipid systems prepared by different methods. *Biochim. Biophys. Acta* 733: 51-64.

17. Mimms, L. T., Zampighi, G., Nozaki, Y., Tanford, C., and Reynolds, J. A. (1981). Phospholipid vesicle formation and transmembrane protein incorporation using octylglycoside. *Biochemistry* 20:833-840.

18. Taraschi, T. F., and Mendelsohn, R. (1980). Lipid-protein interaction in the glycophorin-dipalmitoylphosphatidylcholine system: Raman spectroscopic investigation. *Proc. Natl. Acad. Sci. USA* 77:2362-2366.

19. Mendelsohn, R., Dluhy, R., Taraschi, T., Cameron, D. G., and Mantsch, H. H. (1981). Raman and Fourier transform infrared spectroscopic studies of the interaction between glycophorin and dimyristoyl-phosphatidylcholine. *Biochemistry* 20:6699-6706.

20. Op den Kamp, J. A. F., De Gier, J., and Van Deenen, L. L. M. (1974). Hydrolysis of phosphatidylcholine liposomes by pancreatic phospholipase A2 at the transition temperature. *Biochim. Biophys. Acta* 345:253-256.

21. Haest, C. W. M., De Gier, J., Van Es, G. A., Verkleij, A. J., and Van Deenen, L. L. M. (19). Fragility of the permeability barrier of *E. coli. Biochim. Biophys. Acta* 288:43-53.

22. Marsch, D., Walls, A., and Knowles, P. F. (1977). Cooperativity of the phase transition in single and multilayered vesicles. *Biochim. Biophys. Acta* 465:500-515.

23. Van Hoogevest, P., De Gier, J., and De Kruijff, B. (1984). Determination of the size of the packing defects in dimyristoylphosphatidylcholine bilayers, present at the phase transition temperature. *FEBS Lett.* 171:160-164.

24. De Kruijff, B., and Van Zoelen, E. J. J. (1978). Effect of the phase transition on the transbilayer movement of dimyristoylphosphatidylcholine in unilamellar vesicles. *Biochim. Biophys. Acta* 511:105-115.

25. De Kruijff, B., Van Zoelen, E. J. J., and Van Deenen, L. L. M. (1978). Glycophorin facilitates the transbilayer movement of phosphatidyl-choline in vesicles. *Biochim. Biophys. Acta* 509:537-542.

26. Van Zoelen, E. J. J., De Kruijff, B., and Van Deenen, L. L. M. (1978). Protein-mediated transbilayer movement of lysophosphatidyl-choline in glycophorin-containing vesicles. *Biochim. Biophys. Acta* 508:97-108.

27. Van der Steen, A. T. M., De Jong, W. A. C., De Kruijff, B., and Van Deenen, L. L. M. (1981). Lipid dependency of glycophorin-induced transbilayer movement of lysophosphatidylcholine in large unilamellar vesicles. *Biochim. Biophys. Acta* 647:63-72.

28. Van der Steen, A. T. M., De Kruijff, B., and De Gier, J. (1982). Glycophorin incorporation increases the bilayer permeability of large unilamellar vesicles in a lipid-dependent manner. *Biochim. Biophys. Acta* 691:13-23.

29. Van Hoogevest, P., Du Maine, A. P. M., and De Kruijff, B. (1983). Characterization of the permeability increase induced by the incorporation of glycophorin in phosphatidylcholine vesicles. *FEBS Lett.* 157:41-45.

30. Van Hoogevest, P., Du Maine, A. P. M., De Kruijff, B., and

De Gier, J. (1984). The influence of lipid composition on glycophorin-induced bilayer permeability. *Biochim. Biophys. Acta* 771:119-126.

31. Rüppel, D., Kapitza, H. G., Galla, H. J., Sixl, F., and Sackmann, E. (1982). On the micro-structure and phase diagram of dimyristoylphosphatidylcholine-glycophorin bilayers. The role of defects and the hydrophilic lipid-protein interaction. *Biochim. Biophys. Acta* 692:1-17.

32. Segrest, J. P., Gulik-Krzywicki, T., and Sardet, C. (1974). Association of the membrane-penetrating polypeptide segment of the human erythrocyte MN-glycophorin with phospholipid bilayers. I. Formation of freeze-etch intramembraneous particles. *Proc. Natl. Acad. Sci. USA* 71:3294-3298.

33. Romans, A. Y., Allen, T. M., Meckes, W., Cionelli, R., Sheng, L., Kercet, H., and Segrest, J. P. (1981). Incorporation of the trans-membrane hydrophobic domain of glycophorin into small unilamellar phospholipid vesicles. Ion flux studies. *Biochim. Biophys. Acta* 642: 135-148.

34. Israelachvili, J. N. (1977). Refinement of the fluid-mosaic model of membrane structure. *Biochim. Biophys. Acta* 469:221-225.

35. Van der Steen, A. T. M., unpublished observations.

36. Boheim, G., Hanke, W., and Jung, G. (1983). Alamethicin pore formation: Voltage-dependent flip-flop of α-helix dipoles. *Biophys. Struct. Mech.* 9:181-191.

37. Van Hoogevest, P., De Kruijff, B., and Garland, P. B. (1985). The influence of lipid composition and lectin-glycophorin interaction on the rotational diffusion of glycophorin in vesicles, as measured by time-resolved phosphorescence depolarization. *Biochim. Biophys. Acta* 813:1-9.

38. Killian, J. A., and De Kruijff, B. (1985). Thermodynamic, motional and structural aspects of gramicidin-induced hexagonal H_{II} phase formation in phosphatidylethanolamine. *Biochemistry* 24:7881-7890.

39. De Kruijff, B., Cullis, P. R., Verkleij, A. J., Hope, M. J., Van Echteld, C. J. A., and Taraschi, T. F. (1985). Lipid polymorphism and membrane function. In: *The Enzymes of Biological Membranes*, 2nd ed., Vol. 1, *Membrane Structure and Dynamics* (A. N. Martonosi, ed.). Plenum Press, New York, pp. 131-204.

40. Taraschi, T. F., Van der Steen, A. T. M., De Kruijff, B., Tellier, C., and Verkleij, A. J. (1982). Lectin-receptor interactions in liposomes: Evidence that binding of wheat-germ agglutinin to glycoprotein-phosphatidylethanolamine vesicles induces non-bilayer structures. *Biochemistry* 21:5756-5764.

41. Taraschi, T. F., De Kruijff, B., and Verkleij, A. J. (1983). The effect of an integral membrane protein on lipid polymorphism in the cardiolipin-calcium system. *Eur. J. Biochem.* 129:621-625.

42. Grant, C. W. M., and Peters, M. W. (1984). Lectin-membrane interactions. Information from model systems. *Biochim. Biophys. Acta* 779: 403-423.

43. Wilschut, J., Düzgüneş, N., Fraley, R., and Papahadjopoulos, D. (1980). Studies on the mechanism of membrane fusion: kinetics of calcium ion-induced fusion of phosphatidylserine vesicles followed by a new assay for mixing of aqueous contents. *Biochemistry* 19:6011-6021.

44. De Kroon, A. I. P. M., Van Hoogevest, P., Geurts van Kessel, W. S. M., and De Kruijff, B. (1985). Influence of glycophorin incorporation on calcium-induced fusion of phosphatidylserine vesicles. *Biochemistry* 24:6382-6389.

45. Nir, S., Bentz, J., Wilschut, J., and Düzgünes, N. (1983). Aggregation and fusion of phospholipid vesicles. *Prog. Surf. Sci.* 13:1-124.

46. Struck, D. K., Hoekstra, D., and Pagano, R. E. (1981). Use of resonance energy transfer to monitor membrane fusion. *Biochemistry* 20:4093-4099.

11

Membrane Fusion Induced by Polyethylene Glycol

SEK-WEN HUI

Roswell Park Memorial Institute and State University of New York—Buffalo, Buffalo, New York

LAWRENCE T. BONI*

Harvard Medical School, Boston, Massachusetts

I. INTRODUCTION

Polyethylene glycol (PEG) is a commonly used cell fusogen (1-6). It has been employed to fuse a wide variety of cells, including interspecific and interkingdom cell types. PEG exhibits a very low degree of cytotoxicity; it is a water-soluble fusogen, which is relatively easy to remove by washing. Its fusion efficiency is high and reproducible. In addition, it is inexpensive and commonly available.

A number of similar protocols are used in PEG-induced fusion. The protocols typically employ a short (1-3 min) incubation of cells in a concentrated PEG solution (usually 50% w/w of molecular weight 6 k), followed by washing away the PEG by dilution. Fusion between cells is only observed following the removal of PEG. These fusion protocols have been employed with success for over a decade without a proper understanding of the physicochemical mechanism of PEG-induced fusion.

A knowledge of the factors that govern membrane stability is essential to understanding the mechanism of PEG-induced fusion. The existence of an intact biological membrane depends on the balance between the constituent lipids and proteins and their aqueous environment (7). Disturbing the organization of these membrane components and/or changing the nature of the aqueous phase can create a structurally unstable membrane. The manner by which PEG induces reorganization of membrane components, facilitates cell or vesicle contact, and disrupts bilayer structure and continuity resulting in fusion is the subject of this chapter.

It is known that during the fusion process induced by PEG, patches of the plasma membrane are "bared" of intramembrane particles (IMP), the membrane proteins seen by freeze-fracture electron microscopy (3-6,8,9). The site of cell contact and subsequent fusion is associated with these IMP-denuded regions (Fig. 1). It seems that the site of action is the lipid bilayer. Therefore, in searching for the possible

---

*Present affiliation: The Liposome Company, Inc., Princeton, New Jersey.

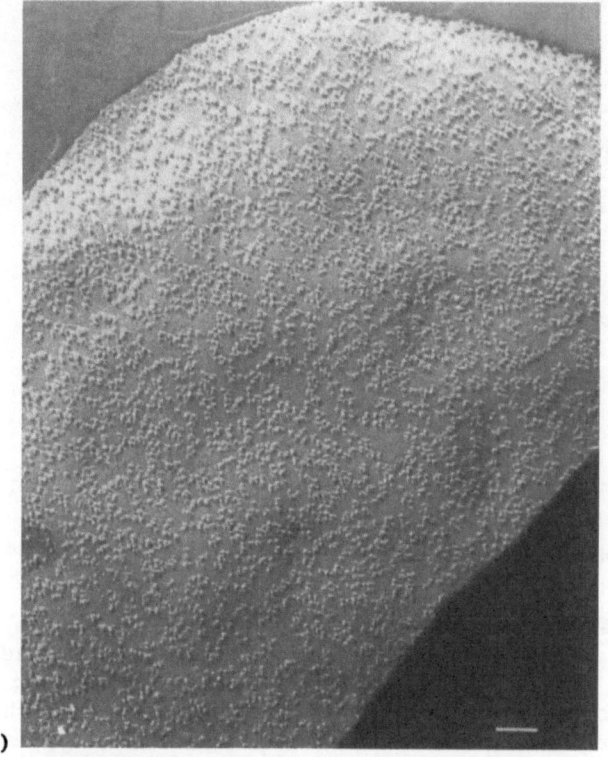

(a)

(b)

mechanisms of PEG-induced membrane fusion, we first examine its interaction with pure lipid vesicles. In subsequent sections, we address the
additional factors in biological membranes that affect the fusion efficiency
of PEG.

II. THE PHYSICAL CHEMISTRY OF POLYETHYLENE GLYCOL

PEG of molecular weight 1000 and higher has the ability to bind and structure water. High viscosities are observed for aqueous PEG solutions of
molecular weight 6 k or greater due to the unusual hydration of the polymer
along with the possible existence of a tightly held gellike network (10).
Baran et al. (11) used high-resolution proton nuclear magnetic resonance
(NMR) to determine the mobility of water molecules at various PEG concentrations. They concluded that each PEG monomer unit bound three to four
water molecules are bound at 38-45% PEG and all of the water is structured
by 13% PEG.

Through the use of differential scanning calorimetry (DSC), Blow
et al. (12) observed two endothermic peaks for solutions less than 45%
(w/w) PEG 6 k. The first peak was attributed to the melting of free water
while the second peak was believed to be due to a polymer-hydrate melt.
There was no free water peak at 45% PEG 6 k. DSC studies by Tilcock
and Fisher (13) revealed an increase in the number of water molecules
bound per polymer unit from 1.8 for PEG 200 to 2.7 for PEG 6 k. No
free water was detectable in 48.1% w/w solution of PEG 6 k. The greater
amount of bound water with increasing molecular weight PEG is attributed
to the random coiling of the polymer in aqueous solution, providing additional sites for water entrapment while decreasing the total volume of solution available for other macromolecules.

Orienting or structuring water leads to a decrease in polarity. Ingham
(14) observed an increase in the fluorescence of 8-anilinonaphthalene
1-sulphonate (ANS) with increasing molecular weight PEG (from ethylene
glycol or PEG 20 k) and increasing concentration. This is caused by a
decrease in the ability of water to reorient its dipoles, making the water
less capable of quenching the fluorescence. These results are consistent
with those of Arnold et al. (15). A gradual decrease in the dielectric
constant of water from 80 to 50 for 10-50% w/w for PEG (400, 6 k, 20 k,
and 40 k) has been reported by Arnold et al. (16), again due to the
decrease of the polarity of aqueous solutions. However, the PEG-induced
decrease in dielectric constant is still greater than the threshold allowed
for the formation of lipid bilayers in a polar solvent (16).

Commercial PEG differs according to manufacturer and from batch to
batch with respect to heterogeneity of polymerization and purity. Impurities,
such as catalysts or terminators of polymerization, antioxidants, and
oxidative decomposition products (17), may have an impact on cell fusion
and viability (18). Fusion between erythrocytes (17) and of erythrocytes
to culture cells was only observed using unpurified PEG (6). Purified
PEG was still found to be as effective as unpurified PEG in fusing liposomes
(19) and erythrocyte ghosts (9,18).

FIGURE 1 (a) Freeze-fracture electron micrograph of a human erythrocyte
showing the irregularly distributed intramembraneous particles (IMP).
(b) Clusters of IMP and bare patches revealed in 12% PEG. Bar = 0.1 μm.

Thus, two important actions of pure PEG impart perturbing effects on lipids—competition for the water necessary to maintain the bilayer structure and a decrease in the polar nature of the aqueous phase. In both senses, PEG is more effective than other common nontoxic fusogenic polymers.

III.   PEG EFFECT ON LIPID BILAYERS

A.   Modulation of the Bilayer Surface

The water-binding and water-structuring properties of PEG alters the forces required for the stability of lipid bilayers. These forces include the cohesive forces between phospholipids and water, the electrostatic repulsive forces between adjacent bilayers, and the Van der Waals attraction between the hydrocarbon chains. A decrease in the surface potentials of dipalmitoylphosphatidyl choline (DPPC) and dipalmitoylphosphatidylethanolamine (DPPE) monolayers has been observed with increasing concentrations of PEG 6 k from 0.5 to 5% (20,21). This is the result of a diminished electrostatic field perpendicular to the surface of the lipid layer. Since the decrease of surface potential was observed without changes in the area per lipid molecule, a change in the properties of the hydrated polar headgroup region must be involved.

Altering the hydration and electrostatic states of the surface leads to lipisome aggregation. Egg PC small unilamellar vesicles (SUV) have been shown to aggregate in as low as 2.5% PEG 6 k (22). The dependence of lipid concentration on the aggregation is shown in Figure 2, where the higher concentration of DMPC vesicles aggregate at lower PEG concentrations within the range of 2-5% PEG (13). Reversibility of aggregation was exhibited for PEG below 20%, as shown by the nonlinear decrease in turbidity upon dilution (22). Additional evidence for the reversibility of aggregation without fusion was shown by dynamic light scattering, electron microscopy, and ^{1}H and ^{31}P NMR (19). The results on vesicle aggregation are consistent with the ability of PEG to bind water and remove the hydration shell surrounding lipid molecules.

PEG may also bind directly to lipid molecules. A binding constant of 0.088 mol PEG/mol PC has been determined using radiolabeled PEG 6 k (19). This is higher than found for dextran binding to lipids (18.4 g dextran/mol DPPC for dextran 150 k (23). Since PEG binds to and dehydrates lipid to a greater extent than dextran, a greater effect on the packing of the bilayer could be anticipated. In a nuclear Overhauser enhancement effect experiment, a correlation peak between the protons of low-molecular-weight PEG and those of the glycerol moiety of the phospholipid was noted, which indicates a close proximity of the polymer to the lipid (Hui and Alderfer, unpublished results). Infrared spectroscopy has also revealed a strong influence of PEG on the carbonyl region of the phospholipid molecules (Hui and Mantsch, unpublished results). These findings are consistent with the particular perturbing nature of PEG at the lipid-water interface of the bilayer. NMR studies of PEG interactions with sodium dodecyl sulfate (SDS) revealed a change in chemical shift with PEG concentration for the C1, or carboxyl, carbon resonance of the SDS molecule (24). This implies that the hydrated part of the SDS molecule is bound to the polymer and that the polymer does not enter the micelle core. Measurements of ^{23}Na$^+$ NMR relaxation rates (24) suggest the involvement of sodium ions in PEG binding. The partitioning of multilamellar vesicles

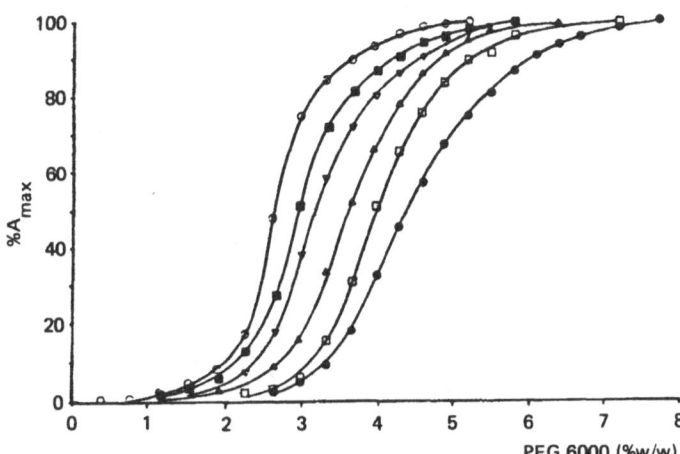

FIGURE 2   The aggregation of dimyristoylphosphatidylcholine (DMPC) vesicles by PEG 6 k. Values for the normalized absorbance (% A_{max}) of sonicated dispersions of DMPC in the presence of PEG 6 k are shown for six concentrations of lipid: 4.75 mg/ml (○), 1.58 mg/ml (▪), 0.810 mg/ml (▾), 0.372 mg/ml (▲), 0.352 mg/ml (□), and 0.176 mg/ml (●). [Reproduced with permission of Tilcock and Fisher (13).]

(MLV) in aqueous two-phase systems of PEG/dextran has shown that vesicles favor the PEG phase and act as if they possessed a net positive surface charge and PEG a negative charge (25). These results agree with the model proposed in Ref. 1, in which a cation forms a salt bridge between the polymer unit and lipid. An enhancement in the fusion efficiency of PEG due to the presence of calcium (1) or sodium (5) has been noted.

On the other hand, a recent study by Arnold et al. (75) indicates that high-molecular-weight PEG is excluded from the membrane surfaces. This conclusion was derived from the fact that the addition of PEG to the medium did not impede the electrophoretic movements of charged liposomes and erythrocytes as expected from the increase in viscosity. This argument is supported by X-ray diffraction studies of the effect of PEG and PEG-containing surfactants on interbilayer spacing (32,76).

In proton NMR studies, Ohno et al. (26) observed a decrease in motion of the choline methyl moiety over the methylene resonances of the hydrophobic region for DPPC SUV upon incubation with as low as 0.2% PEG. This is consistent with a strong interaction between PEG and the liposome surface, particularly at the headgroup region. PEG-induced broadening of egg PC SUV ^{31}P NMR resonances was shown to be much more than that induced by dextran and glycerol of the same viscosity (19). The broadening, a consequence of restricted lipid motion, is most likely due to a decrease in the lateral diffusion of the lipid. In most cases, the PEG-lipid headgroup interaction also led to a decrease in lipid fluidity as measured by fluorescence (27), by ESR (28,29), and by ^{1}H NMR spin-lattice relaxation times (19).

DSC results of Tilcock and Fisher (30) indicate both a broadening and an upward shift of the main gel to liquid-crystalline transition (T_c) for DPPC dispersions in PEG from molecular weight 400 to 6000. An upward shift and broadening of the T_c are characteristic of dehydrated lipid. Changes in both enthalpy and entropy of the transition above 30% PEG are consequences of a decrease in the cooperativity of the transition. This is consistent with the existence of packing distortions in the bilayer. Similar DSC effects have been observed for egg PC, DMPC, soy PE, and bovine phosphatidylserine (PS) dispersions (31,32). The single peak for the DSC endotherms implies that the dehydration affects all layers of the MLV equally.

X-ray diffraction shows the effect of dehydration on the lamellar repeat spacings. DMPC dispersions exhibited a reduction in spacings from 6.4 nm to 5.1 nm when incubated in 50% PEG 6 k at 40°C (32). Similar PEG-induced decreases in lamellar repeat spacings have been observed for dispersions of egg PC, bovine PS, soy PE, and egg PC/cholesterol (31,32). Parsegian et al. (77) and Arnold et al. (75) measured the osmotic pressure created by PEG and found it may reach 10^8 dyne/cm^2, higher than most other polymers. With increasing osmotic pressure, the X-ray diffraction spacing between lipid bilayers reduced to less than 0.5 nm. These findings exclude the possibility that PEG molecules intercalate between lipid bilayers during fusion-inducing treatments.

Low PEG concentrations (less than 10%) interact with lipid bilayers both by changing their surface potential and competing with the available water. The interaction with the polar headgroups changes bilayer packing, weakening the association between the phospholipids in the bilayer (23) and creating phase separated domains with packing defects along domain boundaries (33). A shearing force may then rupture the bilayer and facilitate the exposure of the hydrophobic region of the bilayer (30). It has been proposed that fusion might occur at the boundaries between these domains (13,34-36).

B. Bilayer Destabilization and the Formation of Defects

PEG facilitates two essential steps in fusion—promoting bilayer contact and subsequent bilayer destabilization. Ordered water decreases the thermodynamic barrier that prevents transfer of apolar groups to the aqueous phase (37). Destabilization of the bilayer structure must occur by allowing the lipid hydrocarbon moieties to merge without removing water from the polar headgroup (30). This contact-mediated transfer of lipid is indicative of the exposure of the hydrophobic core of the bilayer, a necessary step in the fusion process (39).

Lipid release and lipid exchange were indeed observed when PEG was used to induce cell or vesicle fusion. Studies have shown that cells treated with 40% PEG 1 k for 5 min released twice the amount of ^{3}H-glycerol (a lipid precursor) into the culture medium as that released from control cells (40). The release of the apolar fluorescent probe pyrene from erythrocyte ghosts in varying concentrations of PEG was recorded by Arnold et al. (15).

PEG also causes a decrease in the partitioning of the ESR spin label 1-oxyl-2,2,6,6-tetramethyl peperidine (TEMPO) between DPPC vesicles and the external aqueous phase (28), suggesting a reduction in the polarity of the aqueous PEG phase. A complete mixing of lipids between DMPC and DPPC MLV was shown by DSC to occur within 25 min in 45% PEG 400

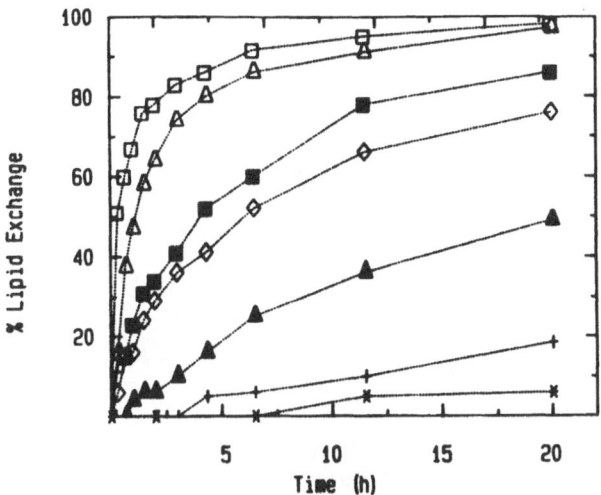

FIGURE 3 The time dependence of transfer of fluorescent phospholipid probes from 1 mM egg PC/NBD-PE/Rhodamine-PE (98.8/0.4/0.8) SUV to 9 mM egg PC SUV at 20°C in the following reagents: 2% PEG 6 k (*), 3% PEG 6 k (+), 4% PEG 6 k (▲), 5% PEG 6 k (■), 6% PEG 6 k (△), 10% PEG 6 k (□), and 10% dextran 70 k (◇). The buffer used was 7 mM Tris-HCl, pH 7.4.

(13). This could be a result of exchange through the PEG media via direct contact of the aggregated vesicles, by breakage and resealment of vesicles, or by fusion. Morgan et al. (41) observed the transfer of the fluorescent triacylglycerol and possibly phospholipid probes between DPPC vesicles upon incubation in 5% PEG 6 k.

A direct observation of transfer of phospholipid between SUV is shown in Figure 3 (Boni, unpublished results). The lipid mixing assay of Struck et al. (42) was utilized. Phospholipid vesicles containing the fluorescent phospholipids N-7-nitrobenz-2-oxa-1,3 diazol-4-yl)-PE (N-Rh-PE) were at concentrations such that the N-Rh-PE partially quenched the fluorescence of the N-NBD-PE. A release of quenching occurs upon incubation with SUV devoid of probe in various concentrations of both PEG and dextran due to lipid transfer. This is seen to occur only at aggregating concentrations of PEG 6 k. When the PEG incubation buffers contained an additional 25 mM sodium chloride, up to a twofold enhancement of lipid transfer efficiency was observed. Transfer also occurred when the vesicles were incubated in 10% and 14% dextran 70 k. Fusion of SUV to larger entities is not observed under these conditions by light scattering (19,22) and gel filtration (Boni, unpublished results). These results are consistent with the observations by light microscopy for the spread of fluorescent lipid probes from membranes of erythrocytes to culture cells in the presence of fusogenic and purified nonfusogenic 44% PEG 8 k (6), where transfer occurred in less than 30 s. The proteins of the erythrocyte membranes did not diffuse into the culture cell membranes.

Various structural defects have been observed in lipid bilayers dispersed in 50% w/w PEG 6 k by freeze fracture electron microscopy (31,32). The wormlike texture has been noted for egg PC dispersions that were

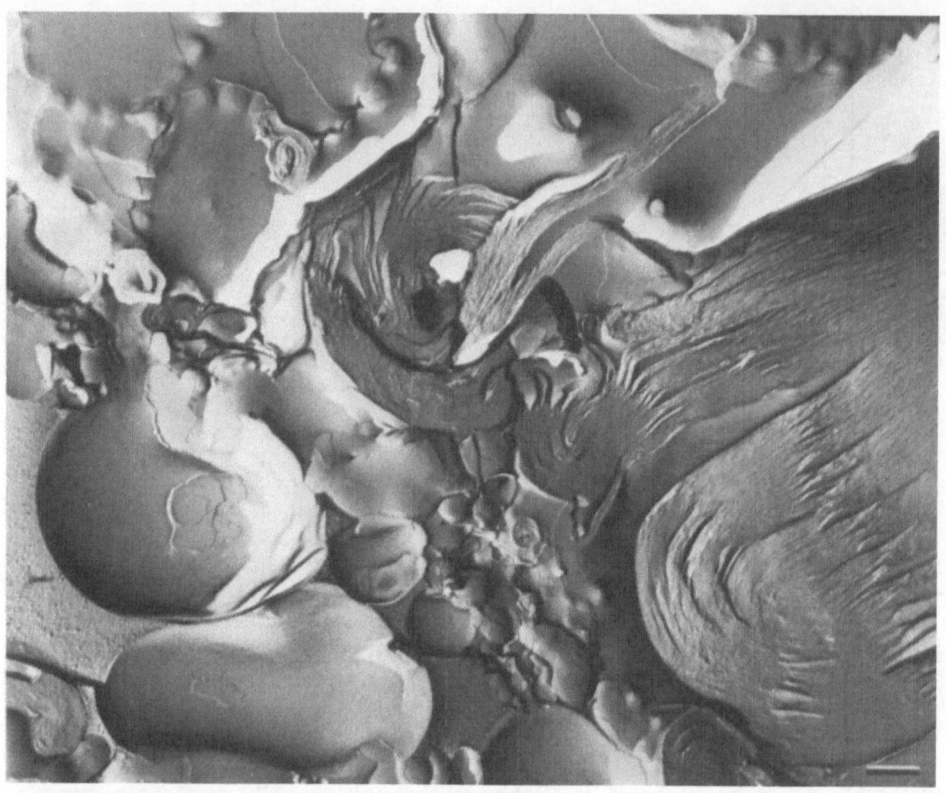

FIGURE 4 Freeze-fracture electron micrograph of egg PC MLV in 50%
PEG 6 k at 20°C exhibiting scalloped deformations. Bar = 0.2 μm.

10-20% hydrated (43-45). Scalloped deformations with protrusions along
the edges, possible sites of interlamellar contact and exposure of hydrophilic portions of the bilayer, are shown in Figure 4. Evenly spaced rows
and pits visible in Figure 4 for egg PC MLV have also been observed for
DMPC and bovine PS dispersions above their liquid-crystalline phase
transition (31,32). Similar defects have been reported by Kleman et al.
(44) for egg PC MLV partially hydrated (16% water). These structures
are caused by relaxation strain between large confocal domains, deformations
in liquid crystals (smectic mesophases) that are adjustments to uneven
surface conditions, such as varying degrees of local hydration (46). The
core or dimple of these domains connects distant lamellae. Pits and protrusions are not seen in DMPC and bovine PS MLV below their phase transitions
(32) where the lamellae may be too rigid to bend to accommodate this
external strain. Screw dislocations or links between adjacent bilayers were
also observed (32). These connections between hydrophobic regions of
neighboring lamellae allow for the rapid diffusion of lipids. The existence
of this defect structure could explain the rapid PEG-induced exchange
between MLV observed by Tilcock and Fisher (13).
      The ultrastructural bilayer defects are indications of exposed hydro-
phobic portions of the bilayer. These defects may be formed between
neighboring membranes and thus provide the next stage of the fusion

process, the bilayer contact stage, with the actual breakdown of a discrete bilayer structure.

C. Fusion

Fusion of lipisomes by PEG was first reported by Boni et al. (22). Egg PC or bovine PS SUV were incubated in 50% PEG 6 k for 1 min and washed by dilution. Fusion to MLV took place during the incubation in PEG, as shown in Figure 5. A linear decrease in turbidity after dilution of SUV in greater than 20% PEG indicated an inability of vesicles to deaggregate, implying fusion to larger entities. A more detailed size measurement was obtained by dynamic light scattering. Vesicles were incubated in varying concentrations of different molecular weight PEG, dextran, glycerol, and sucrose, which are all dehydrating agents. Light-scattering measurements were taken following dilution to allow for vesicle deaggregation. As shown in Figure 6, significant fusion occurred between 20 to 30% PEG 6 k. A decrease in fusion with a decrease in molecular weight is also observed. Impurities in PEG do not affect SUV fusion. This finding is contrary to the results of some PEG cell fusion studies that depended on PEG purity (6,17). Minimal fusion was observed when other dehydrating reagents were used (Fig. 6). Fusion of SUV in 20-40% PEG 6 k has also been observed by Saez et al. (47), Morgan et al. (41), Yoshihara and Nakae (48), and MacDonald (49). MacDonald performed parallel experiments in which SUV were mixed directly with PEG or dextran or dialyzed against these reagents. Lipid mixing was observed in all cases, indicating that direct contact of PEG with the vesicles was not required for fusion. The high degree of dehydration in the dialyzed samples could have promoted fusion or lipid exchange without fusion, thus making interpretations of the results difficult.

The kinetics of PEG-induced fusion of liposomes is more rapid than that for cells. Egg PC SUV were seen by light scattering to fuse to completion within 5 min following dilution from 25 or 45% PEG 6 k (19). This is consistent with the PEG-induced rapid lipid mixing observed by Morgan et al. (41) and MacDonald (49). A detailed kinetic study was reported by Parente and Lentz (50). They found that the initial fusion rate increased with PEG concentration, with a remarkable enhancement at the PEG concentration range of 20-30% (Figure 7). This value corresponds to that of a fusion threshold (19,49) in steady-state experiments. These authors also found that osmotic pressure plays little role in the fusion process, consistent with the fact that the vesicles fuse before the dilution step.

The lipid phase also has a significant effect on the fusion efficiency. A high degree of fusion is seen below the phase transition of DMPC SUV at as low as 6% PEG 6 k (19). This is interpreted to be due to the instability of bilayers below the phase transition where structural defects are thought to exist (51). The result also agrees with the enhanced degree of fusion or lipid exchange for vesicles below the phase transition observed by Tilcock and Fisher (13). This could also be an explanation for the lipid mixing observed by Morgan et al. (41) for DPPC vesicles in 5% PEG 6 k. Fusion is also enhanced by the bilayer/hexagonal II phase transition. When vesicles of mixtures of unsaturated PE/PC were subject to PEG-induced fusion, the extent of fusion was significantly higher if the lipid was in the neighborhood of the bilayer/hexagonal II phase transition.

The effect of vesicle size on fusion efficiency has been studied by Aldwinckle et al. (52). Fusion of MLV of PC/phosphatidic acid/cholesterol

FIGURE 5   Freeze-fracture electron micrographs of egg PC at various
stages of PEG-induced fusion: (top) SUV before the addition of PEG and
(bottom) after mixing in 50% PEG 6 k for 1 min. Bar = 0.1 μm.

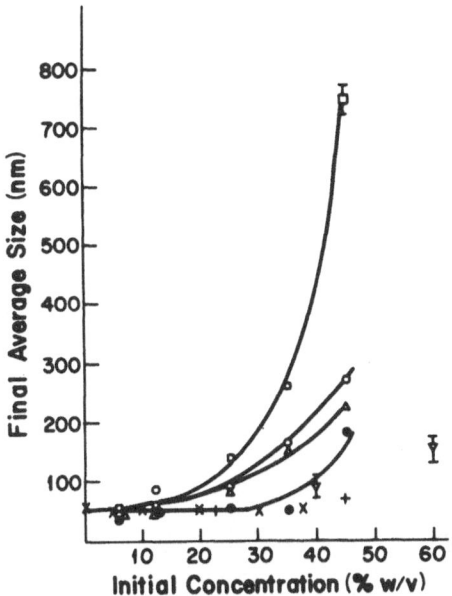

FIGURE 6 Fusion induced by poly(ethylene glycol) and analogs, as shown by dynamic light scattering following dilution of 10 mM egg PC SUV from the given percentage (w/v) of PEG 6 k (□), PEG 1 k (○), PEG 600 (△), PEG 200 (●), ethylene glycol (+), dextran 200k-300k (×) and glycerol (▽). No fusion was observed for up to 60% sucrose. All data were recorded at 20°C. [Reproduced with permission of Boni et al. (19).]

(7:2:1) to larger vesicles took place in 30% PEG. This, however, occurred over a much slower time course than that seen for SUV. A slow lipid exchange in MLV was also noted by MacDonald (49). This could be due to fusion or a result of structural defects. Parente and Lentz (50) also noted different fusion kinetics for SUV and REV (made by the reverse-phase evaporation method). In a freeze-fracture study employing LUV prepared by the ether injection technique (19), incubation in 45% PEG 6 k resulted in a shift in the size distribution of the LUV from a mean diameter of 120 to 200 nm and median diameter from 140 to 220 nm. The diameter increased by a factor of the square root of 2 corresponding to a predominant fusion between dimers. Thus, fusion between larger vesicles of low curvature is different from that between SUV; the former case is less susceptible to the curvature-strain factor and is more similar to cell membrane fusion.

Fusion of vesicles inevitably involves, besides the fusion of their membranes and the intermixing of their internal contents, a minimal leakage of their contents into the external medium.

Ca^{2+} leakage into erythrocyte ghosts upon a 1-min incubation with PEG 1.5 k and 6 k was shown to increase rapidly above 30% PEG (12). This increase in permeability follows the increase in fusion efficiency. PEG-induced leakage was observed for egg PC SUV that contained the water-soluble probe 6-carboxyfluorescein at self-quenching concentrations (47). An increase of fluorescence, indicating leakage of probe, occurred at concentrations as low as 10% PEG 1 k. Leakage from SUV was much

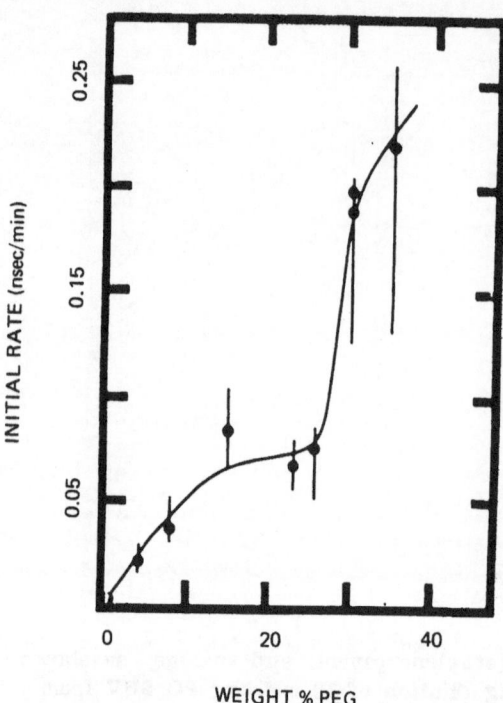

FIGURE 7 Initial rate of lipid mixing as a function of weight percent of PEG. The rate of mixing is determined from fluorescence lifetime measurements. [Reproduced with permission of Parente and Lentz (50).]

greater than from MLV. It was suggested that the fusion process in SUV is by a mechanism of vesicle lysis and reassembly. Liposomal permeability to calcium, shown in Figure 8, increased with time and concentration of PEG 6 k. A large increase occurred between 20 and 25% PEG. Aldwinkle et al. (52) proposed that the leakage was a result of local discontinuities in bilayers rather than partial or total lysis of the membrane, because the activation energy for loss was too high to be attributed to a lytic effect. The rate and the relative amount of content mixing and leakage were reported by Parente and Lentz (50). Leakage was detected much earlier than the fusion event after the addition of PEG.

Thus, a transient bilayer instability is created during the fusion process. The increase in permeability may allow molecules from the external medium to enter cells and the contents to leak out. Such loss of internal contents can be minimized by a shorter incubation time in PEG (53).

IV. PEG EFFECTS ON RECONSTITUTED AND CELL MEMBRANES

The process of PEG-induced fusion of lipid vesicles differs from that of cell membranes. The concentrations of PEG used for the fusion of biological

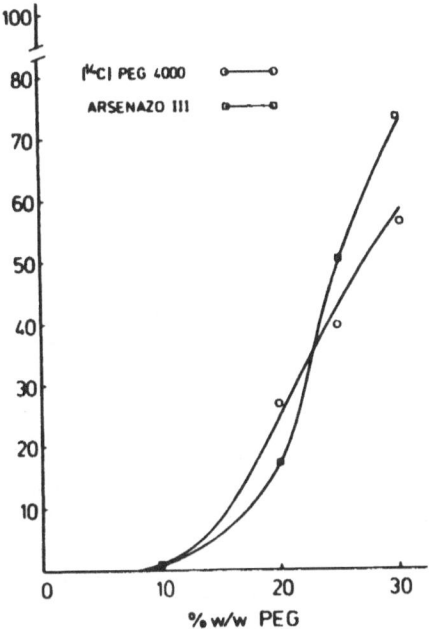

FIGURE 8 A comparison of the simultaneous extraliposomal release of isotopically labelled poly(ethylene glycol) 4 k and of Arsenazo III from multilamellar vesicles that were incubated with poly(ethylene glycol) 6 k.
[Reproduced with permission of Aldwinckle et al. (52).]

membranes are usually in the 35-50% range (2). This is higher than that
required to fuse pure lipid bilayers. Lipid vesicles fuse in the presence
of PEG. The complete fusion of biological membranes or cells requires the
removal of PEG during the posttreatment incubation period. Additional
factors must exist in biological membranes that impede the PEG-induced
fusion of lipid bilayers. These factors have been identified in several
studies. The approach has been to study the fusion process in reconstituted
or simplified membranes. The erythrocyte ghost membrane and its variations
offer models for studying PEG-induced fusion process in biological membranes.

A. The Influence of Membrane Proteins

There have been several freeze-fracture electron microscopy studies showing
that during and even after the PEG treatment, large areas free of intramembranous particles (IMP) in the plasma membrane are common (4,5,9).
The creation of IMP-free areas seems to be an essential step in the PEG-
induced membrane fusion processes. Figure 1 shows that the initially irregular distribution of IMP in human erythrocyte membranes becomes highly
patchy; the contact areas between membranes of adjacent cells are without
exception devoid of IMP. Roos et al. (54) found that cells resistant to
PEG-induced fusion also lack the ability to form IMP-free patches on their
plasma membranes at reduced temperatures. If the IMP-free areas are indeed
protein-free lipid bilayers, one function of PEG must be the clearing away

of proteins to facilitate bilayer-bilayer contact. Fusion between these lipid-rich regions may then proceed as discussed in the previous section.

To verify that the creation of IMP-free areas by PEG is a physicochemical process and not a cellular, cytoskeleton-dependent process, Hui et al. (9) treated dilute suspensions of proteoliposomes reconstituted with egg PC and glycophorin, an erythrocyte glycoprotein that spans the bilayer. At low PEG 6 k concentrations (less than 12.5%), the IMP representing glycophorin aggregated and gave rise to IMP-free areas. No visible evidence of membrane-membrane contact between vesicles existed. At higher vesicle or PEG concentrations, the vesicles were seen to be in contact at the IMP-free regions. Similar vesicle behavior was observed when cytoskeleton-free membrane vesicles budded from erythrocytes were used (Fig. 9). Patching disappears when PEG (35%) is removed by dilution. Since IMP-free areas are created prior to membrane contact, this is contrary to the theory of electrostatic displacement of IMP upon close apposition of cells proposed by Knutton and Pasternak (55). When dextran (molecular weight 500 k) or glycerol of equivalent or greater osmolarity was used isntead of 35% PEG, no extensive IMP-free region was observed (9). Thus, the creation of IMP-free areas is not based solely on the dehydration of the membrane. Other properties, such as the change in polarity of the medium, may be important.

The IMP-free regions may be created by a number of factors, including PEG-lipid binding, direct PEG-protein interactions, or by PEG displacing or structuring the water at the membrane surface. An irreversible immobilization of erythrocyte membrane proteins has been shown by ESR experiments due to an increase in the external osmotic pressure (56) or as a result of PEG treatment (28). Protein conformational changes and structural dissociation from the membrane can be caused by osmotic pressure (56) or by different solubilities of amino acids in PEG (14). In addition, a more hydrophobic solvent may increase the sensitivity of proteins to electrolytes and can thus induce their coagulation (57).

Human erythrocytes fuse in 35% or higher concentrations of PEG. When IMP-free patches were induced prior to or during the PEG treatment, the threshold PEG concentration were considerably lowered. Sublytic solutions of spermidine, trichloroacetic acid, and ethanol induce the formation of IMP-free regions either prior to or during the PEG cotreatment and lead to cell fusion in 25% PEG 8 k. DMSO, lysolecithin, and polylysine do not affect the distribution of IMP and fail to induce cell fusion in 25% PEG. This stresses the fact that the availability of IMP-free areas is an indispensable step in PEG-induced fusion (58). When turkey erythrocytes were treated with concanavalin A or wheat germ agglutinin, they fused even at PEG concentrations as low as 5%. It has been shown that both lectins cause the formation of IMP-free patches. Figure 10 shows the interdependence of Con-A and PEG concentrations in causing the fusion between turkey erythrocytes. The shifting of the fusion threshold (steep slope) from high to low concentrations of PEG as the Con-A concentration increases corresponds to the formation of IMP-free patches by Con-A (58).

It appears that PEG-induced fusion of biomembranes is a two-step process. First, IMP-free areas of "bare" lipid bilayers are exposed and then the lipid bilayers are attached and fused in the PEG environment. For most biological membranes, the first step requires a higher percentage of PEG (35%), while the latter needs 25% or less PEG, which is the same concentration required for the fusion of model lipid bilayers (19,22).

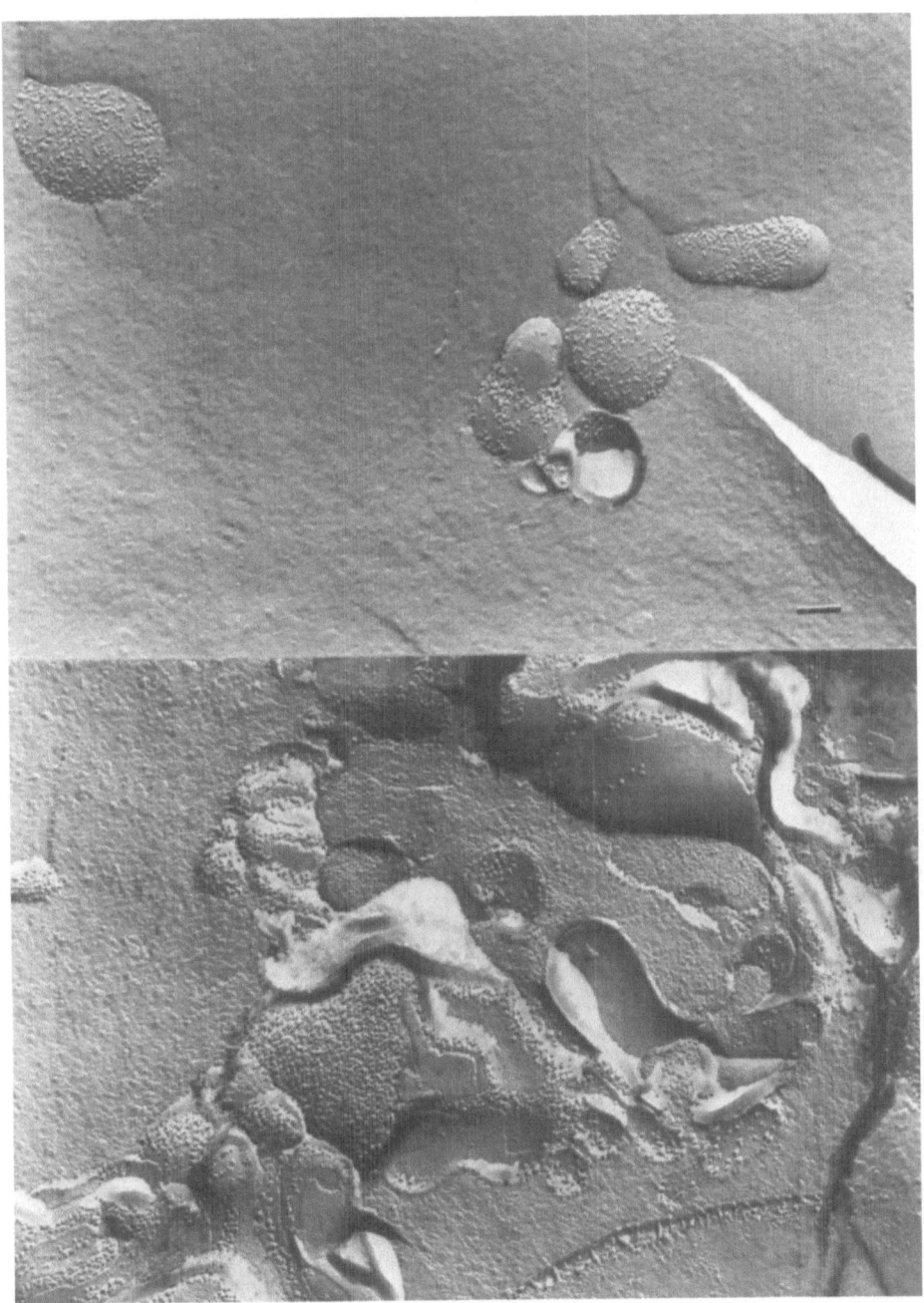

FIGURE 9 Freeze-fracture electron micrographs of (top) cytoskeleton-free erythrocyte vesicles in 6% PEG 6 k and (bottom) cytoskeleton-free vesicles in 35% PEG 6 k. All samples were frozen from 20°C. Bar = 0.1 μm.

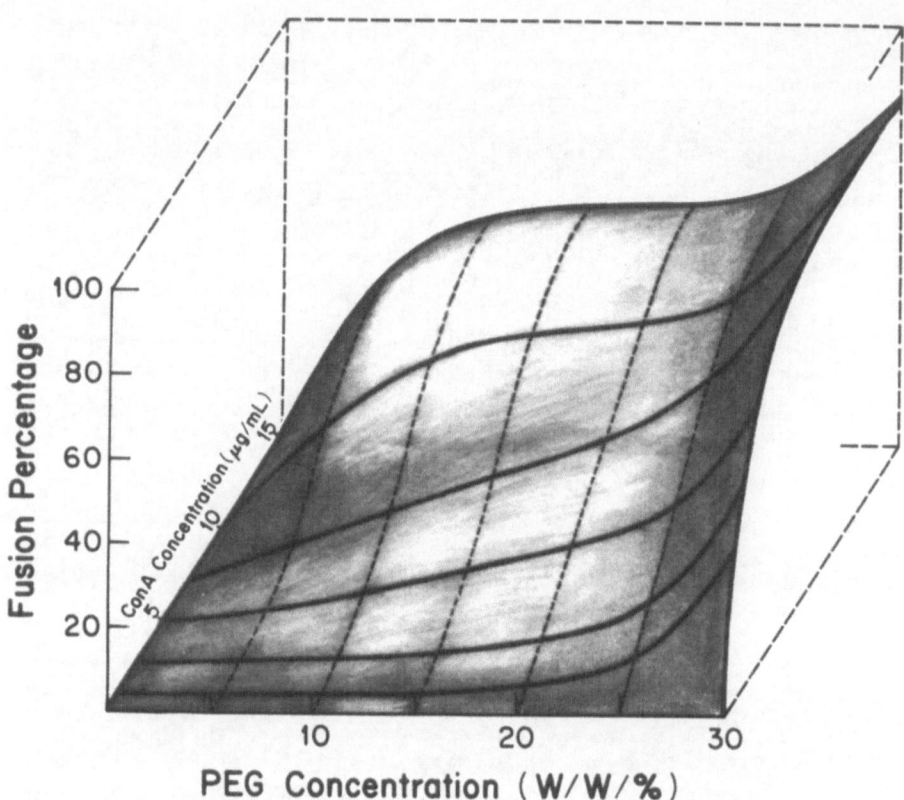

FIGURE 10   The three-variable plot showing the combined effect of Con-A and PEG on the fusion efficiency of turkey erythrocytes. [Reproduced with permission of Huang and Hui (58).]

B.   Influence of Membrane Lipids

The evidence so far, although by no means conclusive, points to the fact that most membrane proteins play only a passive role in the PEG-induced fusion process. Once the spatial hindrance of membrane proteins is removed, fusion of the bared lipid bilayer may proceed in a manner similar to that of fusion between model membranes. We then expect the composition of the membrane lipids to become a major determining factor in the fusion efficiency.

Correlating the PEG-induced fusion efficiency of various cell types with their lipid compositions is not simple, since many other factors may be involved. Roos and Choppin (59) studied the resistance to PEG-induced fusion using a series of mutant mouse fibroblasts which are morphologically similar to their parental cells. They found excellent correlation between their susceptibility to PEG-induced fusion and their membrane lipid composition. Fusion-resistant mutants contain elevated levels of neutral lipids, particularly triglycerides, as well as an ether-link diacylglycerol. Their cells also contain a higher percentage of saturated fatty acyl chains. By increasing the saturated fatty acid contents of fusion-susceptible parental cells with growth medium supplement, these cells were rendered highly resistant to PEG-induced fusion (59). The susceptibility of a variety of mutants to PEG-induced fusion was also controllable by fatty acid alterations (78). Since both *cis*- and *trans*-unsaturated fatty acids are equally effective in promoting fusion, the change in the overall phase transition temperature is probably not the cause.

The spatial distribution of lipids is highly heterogeneous in biological membranes. Cell polarization, transmembrane asymmetry, and lateral micro-domains are all important considerations with regard to lipid effects. The influence of these factors has not been critically examined in PEG-induced fusion. Preliminary results (Huang and Hui, unpublished) indicated that in erythrocyte membranes, the introduction of unsaturated PE, which resides mainly in the outer lipid layer due to the function of phospholipid exchange priteins, led to an increase in the fusion efficiency. Induced lipid microdomain may also play a role.

Fluorescently labeled lipids were used to detect any exogenous lipid domains remaining after the fusion event. With few exceptions, most com-ponents diffuse rapidly after the PEG-induced fusion (6,60,61). However, Schlegel et al. (61) reported that those lipids stained by merocyanine-540 remained in discrete domains 24 h after fusion. These more fluid lipids may be responsible for confining exogenous membrane proteins in discrete areas, or vice versa (6).

C. Osmotic Swelling and Cytoskeletal Proteins

Unsealed (leaky) human erythrocyte ghosts cannot be fused simply by PEG treatment. No fusion was observed when resealed human erythrocyte ghosts were incubated in a physiological isotonic medium after treated in PEG (9). Incubation of ghosts in a hypotonic solution following the PEG treatment enhanced the fusion efficiency. Increasing the osmolarity of the incubation medium reduced the PEG fusion efficiency. Similar results were obtained for fusion induced by Sendai virus (62). This supports the results of Zimmerberg et al. (63) where fusion between lipid vesicles and a planar bilayer was facilitated by an osmotic gradient. An additional "push" by the osmotic pressure can thus reduce the fusion barrier between biological membranes. Osmotic pressure was also shown to be critical in the fusion of protoplasts (64,65). Although lipid mixing between fusion partners occurs in PEG without the osmotic force from dilation (6,79), the mixing of contents occurs only after the dilution step (79).

Woicieszyn et al. (6) showed that an intact cytoskeleton may be the structural force that hinders the completion of cell fusion in PEG. However, when erythrocyte vesicles were incubated in isotonic or hypotonic media following PEG treatment, the response to osmotic-assisted fusion was the same regardless of whether the vesicles contained cytoskeletal elements (9). This could, however, be due to a perturbed cytoskeletal network. A possi-ble role of the cytoskeletal network is preservation of the "random" dis-tribution of the intramembrane proteins. This would impede the close contact between lipid bilayers. Cytoskeletal elements seem to play a passive role in the PEG-induced fusion process and are shown to remain intact throughout the PEG treatment (3). Fuseler et al. (66) have shown, how-ever, an increase in microtubule and stress fibers formation following PEG treatment.

V. CONCLUSION

The membrane destabilizing effect of PEG is associated with its properties to bind and structure water. PEG dehydrates membranes by competing for free water while it alters the dielectric properties of water. By making the aqueous phase "less polar," PEG facilitates the "solvation" of lipid molecules and leads to the creation of bilayer defects. Dehydration alone is not sufficient to induce fusion of membranes, as indicated by the inability of dextrans and other dehydrating agents to fuse vesicles or cells effectively. The uniqueness of PEG as compared to other dehydrating agents lie in its abilities to structure water, disrupt bilayers, form salt complexes, and aggregate membrane proteins.

The mode of action of PEG may be depicted from studies of model lipid bilayers and cellular membranes and the various factors that either enhance or inhibit their fusion efficiencies. Cell or vesicle contact occurs due to the exclusion of the cells or vesicles from the PEG-water network. Local dehydration and the change in polar and ionic environment may lead to a clustering of membrane proteins. This removes the steric hindrance to close approach (67) while also leaving bare membrane patches of high lipid content. Local dehydration and the changing of the dielectric constant of water create defects in bilayer packing and permit exposure of the hydrocarbon chains, an effect that may be facilitated by the formation of domains of different phase and composition. With the reduction of the electrostatic and hydration repulsion between bilayers due to the decrease in polarity of the medium, contact and fusion of the bilayers will proceed.

IMP-free regions are a necessity for PEG-induced fusion. This is consistent with the enhanced PEG fusion efficiency affected by a pretreatment of cells with a protease preparation (68). Protein-denuded areas have also been observed in cell fusion induced by calcium phosphate (69), Sendai virus (70), the divalent cation ionophore A23187 (71), uranyl acetate (72), and electrical breakdown (73).

The final step of removing PEG by dilution has two functions, the "sealing" of bilayer defects at fusion sites and the osmotic swelling of the cell or vesicle which expands the contact area and the fusion lumen (9,62,74). Cells with extensive contact points would then fuse with complete cytoplasmic mixing. Thus, we may say that PEG promotes membrane fusion by providing membranes with a physical environment necessary for natural fusion, in a simple treatment protocol.

ACKNOWLEDGMENT

The authors acknowledge the support of the National Institutes of Health, through grant GM-30969 to S.W.H.

REFERENCES

1. Kao, K. N., and Michayluk, M. R. (1974). A method for high-frequency intergeneric fusion of plant protoplasts. *Planta* 115:355-367.
2. Davidson, R. L., and Gerald, P. S. (1977). Mammalian somatic cell hybridization by polyethylene glycol. In: *Methods in Cell Biology* (D. Prescott, ed.). Academic Press, London and New York, p. 325.
3. Robinson, J. M., Roos, D. S., Davidson, R. L., and Karnovsky, M. J. (1979). Membrane alterations and other morphological features associated with polyethylene glycol-induced cell fusion. *J. Cell Sci.* 40:63-75.

4. Knutton, S. (1979). Studies of membrane fusion. III. Fusion of erythrocytes with polyethylene glycol. *J. Cell Sci.* 36:61–72.

5. Krahling, H. (1981). Discrimination between two fusogenic properties of aqueous polyethylene glycol solutions. *Z. Naturforsch.* 36:593–596.

6. Wojcieszyn, J. W., Schlegel, R. A., Lumley-Sapanski, K., and Jacobson, K. A. (1983). Studies on the mechanism of polyethylene glycol-mediated cell fusion using fluorescent membrane and cytoplasmic probes. *J. Cell Biol.* 96:151–159.

7. Poste, G., and Allison, A. C. (1973). Membrane fusion. *Biochim. Biophys. Acta* 300:421–465.

8. Maul, G. G., Steplewski, Z., Weibel, J., and Koprowski, H. (1976). Time sequence and morphological evaluations of cells fused by polyethylene glycol 6000. *In Vitro* 12:787–796.

9. Hui, S. W., Isac, T., Boni, L. T., and Sen, A. (1985). Action of polyethylene glycol on the fusion of human erythrocyte membranes. *J. Membrane Biol.* 84:137–146.

10. Bailey, F. E., and Koleske, J. V. (1967). Configuration and hydrodynamic properties of the polyethylene chain in solution. In: *Nonionic Surfactants* (M. J. Schick, ed.). Marcel Dekker, New York.

11. Baran, A. A., Solomentseva, I. M., Mank, V. V., and Kurilenko, O. D. (1972). Role of the solvation factor in stabilizing disperse systems containing water-soluble polymers. *Dolk. Akad. Nauk SSSR* 207:363–366.

12. Blow, A. M. J., Botham, G. M., Fisher, D., Goodall, A. H., Tilcock, C. P. S., and Lucy, J. A. (1978). Water and calcium ions in cell fusion induced by poly(ethylene glycol). *FEBS Lett.* 94:305–310.

13. Tilcock, C. P. S., and Fisher, D. (1982). The interaction of phospholipid membranes with poly(ethylene glycol) vesicle aggregation and lipid exchange. *Biochim. Biophys. Acta* 688:645–652.

14. Ingham, K. C. (1977). Polyethylene glycol in aqueous solution: solvent perturbation and gel filtration studies. *Arch. Biochem. Biophys.* 184: 59–68.

15. Arnold, K., Pratsch, L., and Grawrisch, K. (1983). Effect of poly(ethylene glycol) on phospholipid hydration and polarity of the external phase. *Biochim. Biophys. Acta* 728:120–128.

16. Arnold, K., Herrmann, A., Pratsch, L., and Gawrisch, K. (1985). The dielectric properties of aqueous solutions of poly(ethylene glycol) and their influence on membrane structure. *Biochim. Biophys. Acta* 815:515–518.

17. Honda, K., Maeda, Y., Sasakawa, S., Ohno, H., and Tsuchida, E. (1981). The components contained in polyethylene glycol of commercial grade (PEG-6,000) as cell fusogen. *Biochem. Biophys. Res. Commun.* 101:165–171.

18. Smith, C. L., Ahkong, Q. F., Fisher, D., and Lucy, J. A. (1982). Is purified poly(ethylene glycol) able to induce cell fusion? *Biochim. Biophys. Acta* 692:109–114.

19. Boni, L. T., Hah, J. S., Hui, S. W., Mukherjee, P., Ho, J. T., and Jung, C. Y. (1984). Aggregation and fusion of unilamellar vesicles by poly(ethylene glycol). *Biochim. Biophys. Acta* 775:409–418.

20. Maggio, B., Ahkong, Q. F., and Lucy, J. A. (1976). Poly(ethylene glycol), surface potential and cell fusion. *Biochem. J.* 158:647–650.

21. Maggio, B., and Lucy, J. A. (1978). Interactions of water-soluble fusogens with phospholipids in monolayers. *FEBS Letts.* 94:301–304.

22. Boni, L. T., Stewart, T. P., Alderfer, J. L., and Hui, S. W. (1981). Lipid-polyethylene glycol interactions: I. Induction of fusion between liposomes. *J. Membrane Biol.* 62:65–70.

23. Minetti, M., Aducci, P., and Viti, V. (1979). Interaction of neutral polysaccharides with phosphatidylcholine multilamellar liposomes. Phase transitions studied by binding of fluorescein-conjugated dextrans. *Biochemistry* 18:2541-2548.

24. Cabane, B. (1977). Structure of some polymer-detergent aggregates in water. *J. Phys. Chem.* 81:1639-1645.

25. Eriksson, E. and Albertsson, P-A. (1978). The effect of the lipid composition on the partition of liposomes in aqueous two-phase systems. *Biochim. Biophys. Acta* 507:425-432.

26. Ohno, H., Maeda, Y., and Tsuchida, E. (1981). 1-H NMR study of the effect of synthetic polymers on the fluidity, transition temperature and fusion of dipalmitoyl phosphatidylcholine small vesicles. *Biochim. Biophys. Acta* 642:27-36.

27. Ohno, H., Sakai, T., Tsuchida, E., Honda, K., and Sasakawa, S. (1981). The interaction of human erythrocyte ghost or liposomes with polyethylene glycol detected by fluorescence polarization. *Biochem. Biophys. Res. Commun.* 102:426-431.

28. Herrmann, A., Pratsch, L., Arnold, K., and Lassmann, G. (1983). Effect of poly(ethylene glycol) on the polarity of aqueous solutions and on the structure of vesicle membranes. *Biochim. Biophys. Acta* 733:87-94.

29. Boss, W. F., and Mott, R. L. (1980). Effects of divalent cations and polyethylene glycol on the membrane fluidity of protoplast. *Plant Physiol.* 66:835-837.

30. Tilcock, C. P. S., and Fisher, D. (1979). Interaction of phospholipid membranes with poly(ethylene glycol)s. *Biochim. Biophys. Acta* 577:53-61.

31. Boni, L. T., Stewart, T. P., Alderfer, J. L., and Hui, S. W. (1981). Lipid-polyethylene glycol interactions: II. Formation of defects in bilayers. *J. Membrane Biol.* 62:71-77.

32. Boni, L. T., Stewart, T. P., and Hui, S. W. (1984). Alterations in phospholipid polymorphism by polyethylene glycol. *J. Membrane Biol.* 80:91-104.

33. Hui, S. W. (1981). Geometry of phase-separated domains in phospholipid bilayers. *Biophys. J.* 34:383-395.

34. Papahadjopoulos, D., Portis, A., and Pangborn, W. (1978). Calcium induced lipid phase transitoons and membrane fusion. *Ann. NY Acad. Sci.* 308:50-66.

35. Jain, M. K., and White, H. B. (1978). Long-range order in biomembranes. *Adv. Lipid Res.* 15:1-60.

36. McIver, D. J. L. (1979). Control of membrane fusion by interfacial water: A model for the actions of divalent cation. *Physiol. Chem. Physics* 11:289-302.

37. Hafeti, Y., and Hanstein, W. G. (1974). Destabilization of membranes with chaotropic ions. In: *Methods in Enzymology*, Vol. XXXI, *Biomembranes*, Part A (S. Fleischer and L. Packer, eds.). Academic Press, New York, pp. 770-790.

38. Cowley, A. C., Fuller, N. L., Rand, R. P., and Parsegian, V. A. (1978). Measurement of repulsive forces between charged phospholipid bilayers. *Biochemistry* 17:3163-3168.

39. Hui, S. W., Stewart, T. P., Boni, L. T., and Yeagle, P. L. (1981). Membrane fusion through point defects in bilayers. *Science* 212:921-923.

40. McCammon, J. R., and Fan, V. S. C. (1979). Release of membrane constituents following polyethylene glycol treatment of HEp-2 cells. *Biochim. Biophys. Acta* 551:67-73.

41. Morgan, C. G., Thomas, E. W., and Yianni, Y. P. (1983). The use of fluorescence energy transfer to distinguish between poly(ethylene glycol)-induced aggregation and fusion of phospholipid vesicles. *Biochim. Biophys. Acta* 728:356-362.

42. Struck, D. K., Hoekstra, D., and Pagano, R. E. (1981). Use of resonance energy transfer to monitor membrane fusion. *Am. Chem. Soc.* 20:4093-4099.

43. Ranck, J. L., Mateu, L., Sadler, D. M., Tardieu, A., Gulik-Krzywicki, T., and Luzzati, V. (1974). Order-disorder conformational transitions of the hydrocarbon chains of lipids. *J. Mol. Biol.* 85:249-277.

44. Kleman, M., Williams, C. E., Costello, M. J., and Gulik-Krzywicki, T. (1977). Defect structures in lyotropic smectic phases revealed by freeze-fracture electron microscopy. *Philosophical Mag.* 35:33-56.

45. Costello, M. J., and Gulik-Krzywicki, T. (1976). Correlated x-ray diffraction and freeze-fracture studies on membrane model systems. Perturbations induced by freeze-fracture preparative procedures. *Biochim. Biophys. Acta* 455:412-432.

46. De Gennes, P. G. (1974). *The Physics of Liquid Crystals*. Carendon Press, Oxford.

47. Saez, R., Alonso, A., Villena, A., and Goni, F. M. (1982). Detergent-like properties of polyethyleneglycols in relation to model membranes. *FEBS Lett.* 137:323-326.

48. Yoshihara, E., and Nakae, T. (1984). Quantitative measurement of membrane fusion induced by calcium and PEG using porin function. *FEBS Lett.* 166:49-52.

49. MacDonald, R. I. (1985). Membrane fusion due to dehydration by polyethylene glycol, dextran, or sucrose. *Biochemistry* 24:4058-4066.

50. Parente, R. A., and Lentz, B. R. (1986). Rate and extent of PEG-induced large vesicle fusion monitored by bilayer and internal contents mixing. *Biochemistry* 25:6678-6682.

51. Larrabee, A. L. (1979). Time-dependent changes in the size distribution of distearoylphosphatidylcholine vesicles. *Biochemistry* 18:3321-3326.

52. Aldwinckle, T. J., Ahkong, Q. F., Bangham, A. D., Fisher, D., and Lucy, J. A. (1982). Effects of poly(ethylene glycol) on liposomes and erythrocytes. Permeability changes and membrane fusion. *Biochim. Biophys. Acta* 689:548-560.

53. Lane, R. D. (1985). A short duration PEG fusion technique for increasing production of monoclonal antibody-secreting hybridoma. *J. Immun. Methods* 81:223-228.

54. Roos, D. S., Robinson, J. M., and Davidson, R. L. (1983). Cell fusion and IMP distribution in PEG-resistant cells. *J. Cell Biol.* 97:909-917.

55. Knutton, S., and Pasternak, C. A. (1979). The mechanism of cell-cell fusion. *Trends Biochem. Sci.* 4:220-223.

56. d'Avila Nunes, M. (1981). A spin label study of erythrocyte membranes during simulation of freezing. *J. Membrane Biol.* 60:155-162.

57. Glasstone, S., and Lewis, D. (1960). Surface chemistry and colloids. In: *Elements of Physical Chemistry*. D. Van Nostrand, New York, pp. 559-599.

58. Huang, C., and Hui, S. W. (1986). Chemical co-treatments and IMP patching in PEG-induced fusion of turkey and human erythrocytes. *Biochim. Biophys. Acta* 860:539-548.

59. Roos, D. S., and Choppin, P. W. (1985). Biochemical studies on cell fusion. *J. Cell Biol.* 101:1578-1598.

60. Szoka, F., Magnusson, K-E., Wojcieszyn, J., Hou, Y., Derzko, Z., and Jacobson, K. (1981). Use of lectins and polyethylene glycol for fusion of glycolipid containing liposomes with eukaryotic cells. *Proc. Natl. Acad. Sci. USA* 78:1685–1689.

61. Schlegel, R. A., Lumley-Sapanski, K., and Williamson, P. (1985). Insertion of lipid domains into plasma membranes by fusion with erythrocytes. *Biochim. Biophys. Acta* 846:234–241.

62. Knutton, S., and Bachi, T. (1980). The role of cell swelling and haemolysis in Sendai virus-induced cell fusion and the diffusion of incorporated viral antigens. *J. Cell Sci.* 42:153–167.

63. Zimmerberg, J., Cohen, F. S., and Finkelstein, A. (1980). Micromolar Ca^{2+} stimulates fusion of lipid vesicles with planar bilayers containing a calcium-binding protein. *Science* 210:906–908.

64. Kanchanapoon, K., and Boss, W. F. (1986). Osmoregulation of fusogenic protoplast fusion. *Biochim. Biophys. Acta* 861:429–439.

65. Ahkong, Q. F. and Lucy, J. A. (1986). Membrane fusion induced by osmotic pressure. *Biochim. Biophys. Acta* 858, 206–216.

66. Fuseler, J. W., Miller, C. L., and Fuller, G. M. (1978). Alteration of cytoskeletal morphologies and growth patterns in human fibroblasts treated with PEG. *Cytobiologie* 18:345–359.

67. Maroudas, N. G. (1975). Polymer exclusion, cell adhesion and membrane fusion. *Nature* 254:695.

68. Hartmann, J. X., Galla, J. D., Emma, D. A., Kao, K. N., and Ganborg, O. L. (1976). The fusion of erythrocytes by treatment with proteolytic enzymes and polyethylene glycol. *Can. J. Genet. Cytol.* 18:503–512.

69. Zakai, N., Kulka, R. G., and Loyter, A. (1977). Membrane ultrastructural changes during calcium phosphate-induced fusion of human erythrocyte ghosts. *Proc. Natl. Acad. Sci. USA* 74:2417–2421.

70. Bachi, T., Deas, J. E., and Howe, C. (1977). Virus-erythrocyte membrane interaction. In: *Cell Surface Reviews*, Vol. 2 (G. Poste and G. L. Nicolson, eds.). North-Holland, Amsterdam, pp. 83–127.

71. Vos, J., Ahkong, Q. F., Botham, G. M., Quirk, S. J., and Lucy, J. A. (1976). Changes in the distribution of intramembranous particles in hen erythrocytes during cell fusion induced by the bivalent-cation ionophore A23187. *Biochem. J.* 158:651–653.

72. Majumdar, S., Baker, R. F., and Kalfra, V. K. (1980). Fusion of human erythrocytes induced by uranyl acetate and rare earth metals. *Biochim. Biophys. Acta* 598:411–416.

73. Stenger, D. A., and Hui, S. W. (1986). Kinetics of ultrastructural changes during electrically-induced fusion of erythrocytes. *J. Membrane Biol.* 93:43–53.

74. Pinto da Silva, P., Kazufumi, S., and Parkison, C. (1980). Fusion of human erythrocytes by Sendi virus: freeze-fracture aspects. *J. Cell Sci.* 43:419–432.

75. Arnold, K., Herrmann, A., Fawrich, K., and Pratsch, L. (1988). Water-mediated effects of PEG on membrane properties and fusion. In: *Molecular Mechanism of Membrane Fusion* (S. Ohki, D. Doyle, T. Flanagan, S. W. Hui, and E. Mayhew, eds.). Academic Press, New York, pp. 255–272.

76. Gawrisch, K., Arnold, K., Dietze, K., and Schulze, U. (1988). Hydration forces between phospholipid membranes and the polyethylene glycol induced membrane approach. In: *Electromagnetic Fields and Biomembranes* (M. Markov and M. Blank, eds.), Plenum Press, New York, pp. 9–18.

77. Parsegian, V. A., Rand, R. P., and Rau, D. C. (1986). Osmotic stress for the direct measurement of intermolecular forces. In: *Methods of Enzymology*, 127. Academic Press, New York).

78. Ross, D. (1988). Control of cell membrane fusion by lipid composition. In: *Molecular Mechanism of Membrane Fusion* (S. Ohki, D. Doyle, T. Flanagan, S. W. Hui, and E. Mayhew, eds.). Academic Press, New York, pp. 273-288.

79. Lucy, J. A. and Ahkong, Q. F. (1988). Osmotic forces and the fusion of biomembranes. In: *Molecular Mechanism of Membrane Fusion* (S. Ohki, D. Doyle, T. Flanagan, S. W. Hui, and E. Mayhew, eds.). Academic Press, New York, pp. 163-179.

12

Modulation of Membrane Fusion by Glycolipids and Lectin-Mediated Intermembrane Contact

ROGER SUNDLER and JAN BONDESON

University of Lund, Lund, Sweden

I. INTRODUCTION

Establishment of intermembrane contact is an early and critical step in the process of membrane fusion. In biomembrane fusion this part of the process is likely to involve specific recognition, whereby the appropriate target membrane and the site for the fusion event are selected. In studies on model systems, such as anionic phospholipid vesicles, intermembrane contact has often been established by aggregation/precipitation of the vesicles through increases in the concentration of monovalent or divalent counterions (1). More recently, such model systems have been improved by incorporation of glycolipids in the vesicles. Lectins can then be used for specific recognition between the vesicles, thereby allowing independent control over the establishment of intermembrane contact and the triggering of the actual fusion event (see Sections IV and V).

In mammalian cells glycolipids, as well as the carbohydrate groups of membrane glycoproteins, appear to be specifically localized to the cell surface and the corresponding luminal side of those intracellular membranes that communicate most frequently with the surface membrane. Apart from the possible informational content of these oligosaccharides, their presence in the outer half of the plasma membrane bilayer might contribute to the maintenance of cellular integrity and prevent unwanted cell-cell fusion. Because of their localization it is obvious that membrane glycolipids and lectinlike "receptor" proteins are unlikely to play a role in the intracellular fusion processes that occur between the cytoplasmic side of intracellular membranes and/or the plasma membrane. On the other hand, every instance of endocytosis requires fusion between sites on the extracellular aspect of the plasma membrane, and there are also certain physiological situations where intercellular fusion occurs, such as in oocyte fertilization and myoblast-myotube transformation (see Chapters 26 and 27). Most likely, special mechanisms have developed by which cell-cell fusion can be achieved in such instances and they might involve cell surface carbohydrates and lectinlike structures.

Regardless of the immediate relevance of glycolipid-protein recognition
in fusion of biological membranes, glycolipid-containing vesicles and oligovalent plant lectins provide a useful model system, in which certain basic
aspects of membrane-membrane interaction and the potential role of protein-
mediated intermembrane contact in the control of membrane fusion can
be investigated. By selection of appropriate glycolipids, this system can
also provide information about steric effects within membrane surfaces
(see Section III), relevant not only for intermembrane contact, but also
for the accessibility to soluble enzymes and other proteins.

II. EFFECT OF GLYCOLIPIDS ON LIPID VESICLE FUSION

Although the detailed molecular transformations involved in the process
of membrane fusion remain a matter of debate and may never be possible
to capture in a single unifying description, an inherent feature of the
process is the requirement for close apposition of the fusing membranes

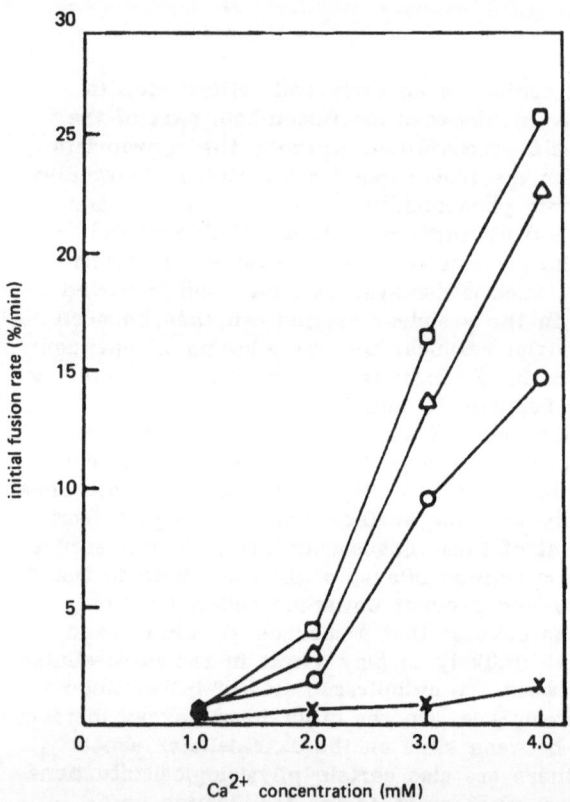

FIGURE 1 Effect of glycolipids on initial rates of Ca^{2+}-induced vesicle
fusion, assessed by the terbium/dipicolinic-acid assay for intermixing of
vesicle contents (64). Vesicles were composed of PA (20 mol%), PE (70 mol%),
and one of the following glycolipids (at 10 mol%): galactosylceramide (□),
lactosylceramide (△), trihexosylceramide (○), or globoside Gb_4 (×).
(Reproduced from Ref. 15.)

(1,2). Thus, phospholipids containing the bulky and well-hydrated head-groups phosphocholine or phosphoinositol that counteract close apposition also counteract Ca^{2+}-induced fusion of phospholipid vesicles (3,4), while phosphatidylethanolamine (PE), containing a smaller and less hydrated headgroup, instead promotes such fusion (4,5). Also, the dehydrating counterion Ca^{2+} is a more effective inducer of membrane fusion than the less dehydrating Mg^{2+} ion when they interact with lipid vesicles containing phosphatidylserine (PS) (2,6-8). These and a series of other experimental findings in model membrane systems demonstrate a crucial role for close intermembrane contact, at least in ionotropic fusion, and have been dealt with extensively elsewhere (1,9,10).

Glycolipids containing a single monosaccharide as headgroup, e.g., galactosylceramide, remain hidden among the phospholipid headgroups, while those containing a di- or oligosaccharide protrude out of the phospholipid headgroup layer in a mixed glycolipid-phospholipid membrane. This statement reflects the conditions in a phospholipid-water interface and is substantiated by molecular modeling as well as by the experimental finding that lectins do not bind to galactosylceramide, whereas they do bind to lactosylceramide (11).

In accordance with the reasoning presented above, the incorporation of several naturally occurring glycolipids or a synthetic glycophospholipid, phosphatidylethanol-N-lactobionamide, in PS or PE/phosphatidic acid (PA) vesicles leads to a certain inhibition of Ca^{2+}-induced aggregation, and also of the fusion, of such vesicles (2,12-16). The degree of interference with Ca^{2+}-induced fusion has been found to increase progressively to complete inhibition, as the size of the glycolipid headgroup increases from a mono- to a tetrasaccharide (Fig. 1). The aggregation of vesicles induced by the polyamine spermine is also counteracted by incorporation of a glycolipid (15). Other means of interference with the establishment of close intermembrane contact also lead to inhibition of Ca^{2+}-induced fusion. Thus, both spectrin binding to the surface of PS vesicles (2) and reconstitution of glycophorin into such vesicles (17) abolish Ca^{2+}-induced fusion. As will be seen below, the inhibitory effect on fusion exerted by the glycolipids themselves can be drastically changed by agglutinating lectins.

III. LECTIN INTERACTION WITH GLYCOLIPID-PHOSPHOLIPID VESICLES

The interaction of an oligovalent lectin with specific carbohydrate groups in glycolipid-containing vesicles can be conveniently monitored as vesicle agglutination (11,18). This experimental system has been utilized in numerous studies as a model for ligand binding to membrane-bound receptors (see Ref. 19). One basic feature of lectin-induced agglutination, shared also by other agglutination processes, is the need for a certain "threshold" glycolipid density in the vesicles (usually 2-5 mol%, but varying with the size of the vesicles). The "threshold" amount of glycolipid is larger than that which actually binds to the lectin during agglutination (12) and apparently does not reflect a discontinuity in lectin binding since this increases in a linear fashion with the amount of glycolipid in this range (19). Thus, the use of agglutination as a measure of lectin binding requires that the glycolipid be present in amounts exceeding the "threshold."

Another important feature of lectin-mediated agglutination, which is dictated by the polar headgroup layer of the vesicles, is the necessity for a sufficient protrusion of the lectin-binding saccharide from the vesicle surface to allow access to it by the lectin. This was clearly demonstrated by Curatolo and co-workers (11), who showed that lactosylceramide/phosphatidylcholine (PC) vesicles, but not those containing galactosylceramide, could be agglutinated by *Ricinus communis* agglutinin I. Rando and co-workers (20,21) have extended this observation, using synthetic glycolipids in which the lectin-binding group was attached to cholesterol via spacer arms differing in length. A four-membered spacer arm was insufficient to support agglutination of glycolipid/PC vesicles, but agglutination occurred when the spacer arm contained seven or more atoms. By the use of another series of synthetic glycolipids (*N*-alkylaldobionamides containing four- or six-membered spacer arms) it was found that the agglutinability with either concanavalin A or *R. communis* agglutinin I was dramatically affected also by the phospholipid composition of the vesicles (22).

The latter kind of synthetic glycolipids, complemented with analogous glycophospholipids having spacer arms of 10 or 12 atoms, were employed to investigate whether the varying effects of phospholipids on the agglutination might be exerted by steric hindrence, differing according to headgroup size (23,24). The results strongly support such an interpretation. Thus, the modulating effect of phospholipid headgroups was weakened when the spacer arm was extended from four to six atoms and was eliminated when the spacer arm was 10- or 12-membered. PE promoted agglutination, in contrast to PC, and agglutination of anionic vesicles was enhanced by Ca^{2+} and Mg^{2+} in a pattern that agreed with their expected effects on headgroup hydration. Also, the findings were quite similar whether concanavalin A or *R. communis* agglutinin I was utilized in combination with appropriate glycolipids. An enhancing effect of Ca^{2+} on the lectin agglutinability of vesicles containing lactosylceramide in mixture with PA and PE has also been reported (15). These findings point to the possibility of using lectin-glycolipid interaction as a tool to determine not only differences in hydrated size of lipid headgroups pertaining to differences in chemical structure, but also changes in anionic headgroups induced by changes in the species, valency, and concentration of counterions (9,24). The hydrated size of phospholipid headgroups is likely to affect a wide range of other interactions between soluble proteins and membrane surfaces, such as between antibodies and lipid antigens and between enzymes and membrane lipid substrates. Thus, the activity of soluble phospholipases C cleaving phosphatidylinositol (PI) and its phosphorylated derivatives is enhanced when the substrate is in mixture with PE, but not with PC (25,26). Also, the kinase phosphorylating PI-4-phosphate, to generate PI-4,5-bisphosphate, is affected in a similar way by admixture of its substrate with phospholipids differing in headgroup size (27).

IV. EFFECT OF LECTIN-MEDIATED INTERMEMBRANE CONTACT ON CALCIUM-INDUCED FUSION

In contrast to the inhibitory effect of incorporated glycolipid on Ca^{2+}-induced aggregation and fusion of vesicles, Ca^{2+}-induced fusion is more or less enhanced by prior, lectin-mediated, aggregation of the vesicles (12-16). The degree of enhancement, manifested as an increased sensitivity to Ca^{2+}, varies depending on both phospholipid composition and glycolipid structure. Thus, the Ca^{2+}-dependence for fusion of vesicles containing PA as the anionic lipid (in mixture with PE) is shifted by more than an

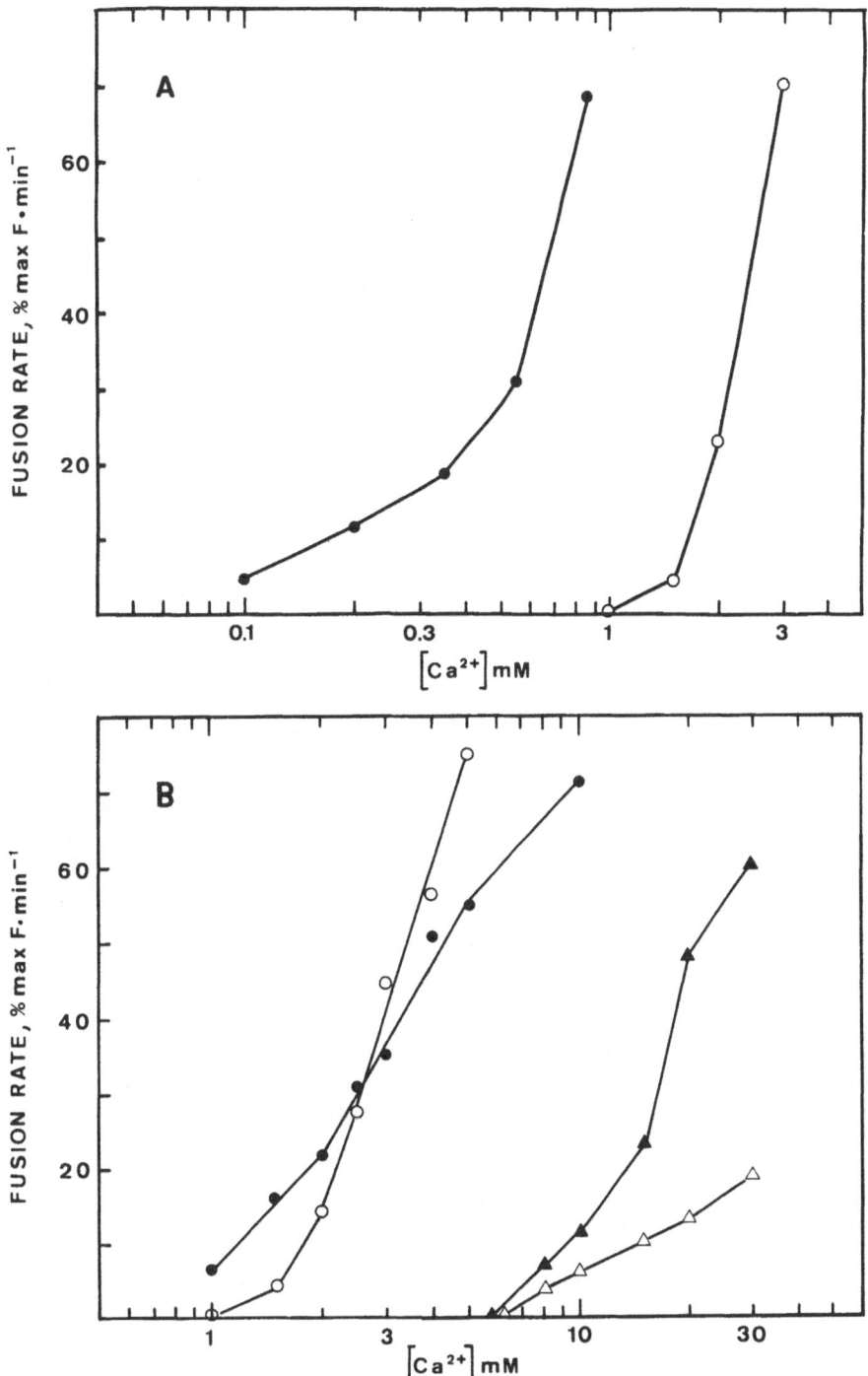

FIGURE 2 Effect of lectin-mediated intervesicle contact on initial rates of Ca^{2+}-induced vesicle fusion, assessed essentially as in Figure 1. Vesicles were composed of phosphatidylethanolamine (70 mol%), phosphatidylethanol-N-lactobionamide (10 mol%) and either of the following (at 20 mol%): (A) PA, (B) PI (△, ▲), or PI-4,5-bisphosphate (○, ●). Open symbols: no lectin; filled symbols: Ca^{2+} added after preincubation (for 1 min) with *Ricinus communis* agglutinin I (40 µg/ml). (Reproduced from Ref. 12.)

order of magnitude (Fig. 2a), indicating that intervesicle contact is strongly
rate-limiting for fusion in this system (12,15). On the other hand, only
minimal changes in Ca^{2+} sensitivity occur when PI, PI-4-phosphate, or
PI-4,5-bisphosphate is the anionic lipid in such mixed lipid vesicles (Fig. 2b),
although the rate of fusion increases for vesicles containing either PI or
PS (12). Interestingly, in the case of PA-containing vesicles, the sensitivity
to Mg^{2+}-induced fusion does not increase to any comparable extent (12).
Increase in the size of the glycolipid headgroup has been found to reduce
the enhancing effect of lectin-mediated intervesicle contact, probably by
increasing the distance between lectin-bonded vesicles (16).

There are, admittedly, potential complications in the use of lectin-
glycolipid interaction to establish intervesicle contact, in particular when
several different glycolipids or lectins are compared: An electrostatic interaction between *R. communis* agglutinin I and vesicles largely composed
of PA occurs at low ionic strength, but is overcome at higher ionic strength
(24). Evidence for apparent hydrophobic interactions between concanavalin
A and phospholipid vesicles (28), and between soybean agglutinin and
glycolipid/phospholipid vesicles subsequent to lectin-glycolipid interaction
(14), has also been presented. Furthermore, lectins differ in the degree
of monosaccharide and anomer specificity as well as the size of the carbohydrate structure recognized, and naturally occurring glycolipids may
differ widely in conformation and stereochemical configuration. Such complications may have a bearing on some of the differences reported between
various lectin-glycolipid combinations (13,14).

It has been suggested that lectins may, in addition to effecting intermembrane contact, also facilitate membrane fusion by causing lateral segregation of the fusion-inhibiting glycolipids (13,16) and that this would explain
why lectin addition before the fusogen (Ca^{2+}) is more effective than the
reverse order of addition. Naturally occurring glycolipids are often rich
in very long and saturated hydrocarbon chains, a structural feature that
could promote lateral segregation and formation of glycolipid clusters in
membranes. This mechanism could explain the enhancement of fusion between
glycolipid-containing and glycolipid-free phospholipid vesicles by wheat
germ agglutinin (13), since a molar excess of lectin (in terms of carbohydrate-
binding sites) to glycolipid was used in these experiments. However, in
other systems the amount of lectin required for vesicle aggregation and
facilitation of fusion is about an order of magnitude lower than the amount
of glycolipid in the vesicles (12,15). There are other plausible ways to
explain why the susceptibility to fusion also here decreases in a time-
dependent manner after Ca^{2+} addition, whether it is made before or after
the addition of lectin (15). One possibility is that Ca^{2+}-phospholipid interaction causes a more perpendicular orientation of the glycolipid headgroups,
thus increasing intervesicle distance (14-16); another is that it causes
a lateral reorganization that makes the areas of lectin-induced intermembrane
contact enriched in glycolipid and thereby less susceptible to fusion.

V. LECTINS AND PROTON-INDUCED FUSION

In the cytosol of mammalian cells the pH is maintained close to 7, while
the luminal compartment of lysosomes, endosomes, and probably also secretory
vesicles and part of the Golgi is considerably more acidic (pH 4.5-6) (29-31).
There is a well-known functional role for the acidic environment in releasing
receptor-bound ligands during endocytosis (32), but the pathway of endo-

cytosis may also involve membrane fusion processes that depend on a low luminal pH (33). It is also well known that the fusion of certain enveloped viruses with cellular membranes can be triggered by acidification (34).

Recent studies have shown that a low pH also can induce fusion between phospholipid vesicles under certain conditions. Highly concentrated suspensions of vesicles composed of PE (80 mol%) and either PA or PS undergo fusion upon acidification to pH 3 (35) and more dilute suspensions of PS vesicles also do so, after very drastic acidification (to pH 2) (36). Phospholipid vesicles containing PE as a major component can be made more sensitive to changes in pH by inclusion of 20-30 mol% of a protonizable component. Thus, acidification to pH 6 or lower is reported to induce extensive fusion between vesicles containing palmitoylhomocysteine (37) in mixture with PE (38). The extent of fusion (at pH 4.8) is strongly reduced when PC or PS is used instead of PE. Vesicles consisting of oleic acid (30 mol%) and PE also fuse at or below pH 6.5 (39), and substitution of PC for PE, or PS for oleic acid also prevents fusion. In some pH-sensitive vesicles described, notably those containing cholesterol-hemisuccinate (36,40,41) or N-succinyl-PE (42), destabilization and release of vesicle content is claimed to occur without prior fusion of the vesicles.

To investigate how lectin-mediated intermembrane contact would affect the pH dependence for proton-induced fusion of phospholipid vesicles,

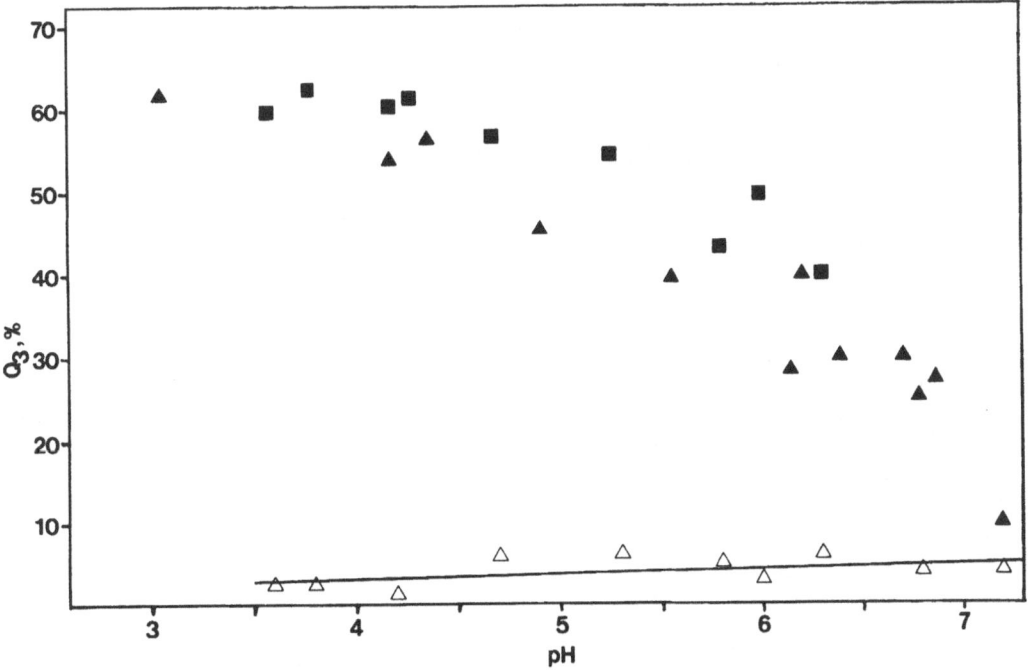

FIGURE 3 Proton-induced lipid intermixing between vesicles consisting of PA (30 mol%), PE (55 mol%), phosphatidylethanol-*N*-lactobionamide (10 mol%), and fluorophore (5 mol%) as a function of final pH. Adjustment of pH was made either in the absence of lectin (open symbols) or 1 min after the addition of *Ricinus communis* agglutinin I at 60 µg/ml (filled symbols). Q_3 = [(initial flourescence) - (flourescence 3 min after acidification)]/initial fluorescence. (Reproduced from Ref. 43.)

a glycophospholipid (phosphatidylethanol-*N*-lactobionamide) was included
in mixed phospholipid vesicles and *R. communis* agglutinin I was used
to bring the vesicles in contact (43). Vesicles composed of PA, PE, and
glycolipid showed significant lipid intermixing (determined by resonance
energy transfer) (44) upon acidification to a final pH as high as 6.5 after
being brought in contact by the lectin (Fig. 3). In the absence of lectin
no lipid intermixing occurred. Later experiments have shown that the results
are virtually identical irrespective of whether *R. communis* agglutinin or
peanut agglutinin is used as a lectin, which argues against a role for
direct interaction of the lectin protein with phospholipids. In addition,
both lectins are without effect on glycolipid-free vesicles and the effects
on glycolipid-containing vesicles are eliminated by the soluble ligand methyl-
β-D-galactoside. The kinetics of lipid intermixing upon acidification was
very rapid compared to that induced by calcium and reached its maximum
at pH 6 in about 20 s. It was followed by a rather extensive leakage from
the vesicles after a short lag period (43). The lipid intermixing observed
in this case appears to reflect complete fusion of the vesicles since it is
accompanied by intermixing of vesicle contents (Fig. 4). The latter is
somewhat less extensive than the former, in part because of the leakage
induced by acidification.

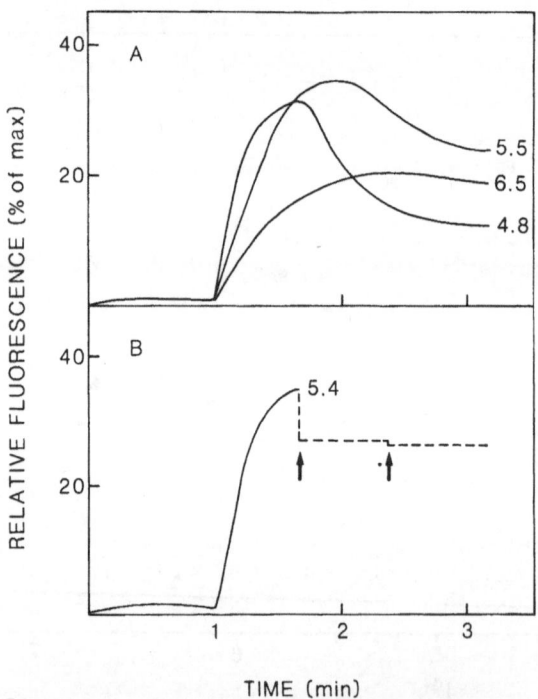

FIGURE 4 Proton-induced fusion of vesicles consisting of PE/PA/phospha-
tidylethanol-*N*-lactobionamide (60:30:10), assessed by the terbium/dipicolinic-
acid assay. The addition of peanut agglutinin (60 μg/ml, at pH 7.4) was followed 1 min later by acidification to the pH value indicated; in (B) pH was
returned to the original value (first arrow) and lectin-mediated agglutination
was reversed by the addition of methyl-β-D-galactoside (second arrow).

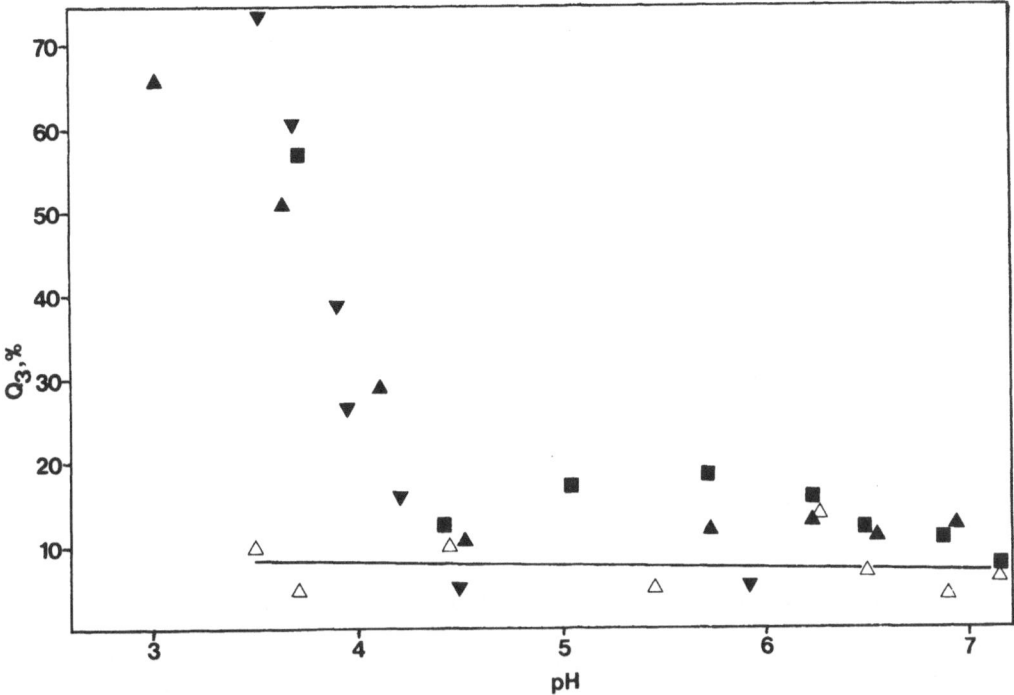

FIGURE 5 Proton-induced lipid intermixing between vesicles consisting of
PS (30 mol%), PE (55 mol%), phosphatidylethanol-*N*-lactobionamide (10 mol%),
and fluorophore (5 mol%), as a function of final pH. Q_3 is defined as in
Figure 3. Adjustment of pH was made either in the absence of lectin (open
symbols) or 1 min after the addition of *Ricinus communis* agglutinin I at
60 μg/ml (filled symbols). (Reproduced from Ref. 43.)

The phospholipid specificity noted for lectin-dependent, proton-induced
fusion shows certain similarities to that already demonstrated for Ca^{2+}-
induced fusion (1,9,10) with two notable exceptions, namely phosphatidylethanol (45) and PI. These phospholipids are virtually resistant to Ca^{2+}-
induced, but susceptible to H^+-induced, fusion. Vesicles containing glycolipid and either PS, or PS in mixture with PE, show significant lipid
intermixing only when the pH drops below 4.5 (Fig. 5). Also, lipid intermixing as well as the ensuing release of vesicle content is strictly dependent
on lectin-mediated intermembrane contact. Results from gel chromatography
corroborate the findings on proton-induced lipid and content intermixing
in the case of PS-containing vesicles (Fig. 6C,D) and show a threshold
pH of approximately 4.5 also for the growth in vesicle size. In contrast,
any significant growth in size of PA-containing vesicles after reversal
of pH and lectin agglutination (Fig. 6A,B) could not be detected, despite
concomitant lipid and contents intermixing. The reason for this is not
known, but similar findings have been reported for the proton-induced
fusion among PA-containing vesicles brought in contact by polyhistidine
(46) and for glycolipid/phosphatidylethanol vesicles interacting via a lectin.

The results discussed above indicate that PA is the anionic phospholipid, of those investigated, that is most sensitive to H^+-induced fusion.

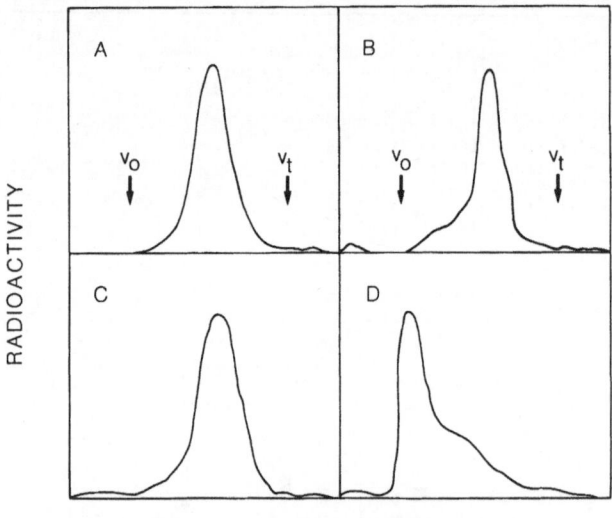

FRACTION NUMBER

FIGURE 6 Gel chromatography on Sephacryl S-1000 of sonicated vesicles consisting of PE/PA/phosphatidylethanol-*N*-lactobionamide (60:30:10) (A and B), or PS/glycophospholipid (90:10) (C and D). (A) Untreated vesicles; (B) after treatment with *Ricinus communis* agglutinin I and acidification to pH 5.3, followed by neutralization and reversal of agglutination as in Figure 4; (C) after treatment as in (B), except at pH 5.5; (D) after treatment as in (B) except at pH 4.0.

PE has a permissive effect on H^+-induced fusion, while PC is inhibitory. The similarity in phospholipid specificity between proton- and Ca^{2+}-induced fusion suggests that a dehydration of phospholipid headgroups is critical to the triggering of fusion in both cases. PA exists predominantly as a mixture of divalent and monovalent forms at pH 7.4 and its protonation should increase upon even moderate acidification (47). This would be expected to reduce the hydrated size of the PA headgroup (24). PS is known to have its serine carboxyl group protonated at about pH 4.5 (48,49), i.e., the same pH at which proton-induced fusion becomes apparent. Also, protonation should result in a reduction of headgroup polarity.

VI. COMPARISON WITH INTERMEMBRANE CONTACT DUE TO PEPTIDE-PHOSPHOLIPID INTERACTION

It is of interest to compare the effects on membrane fusion of lectin-mediated intervesicle binding with those resulting from direct interaction of various peptides and proteins with vesicle phospholipids.

The intervesicle contact induced between anionic lipid vesicles by cationic peptides, especially oligomers and polymers of L-lysine, has been extensively investigated (50-55). Peptides containing five or more lysine residues can induce rapid and extensive lipid intermixing among vesicles composed of PA and PE at neutral pH without causing complete fusion, i.e., intermixing of vesicle contents (Fig. 7). Polylysine and polyhistidine

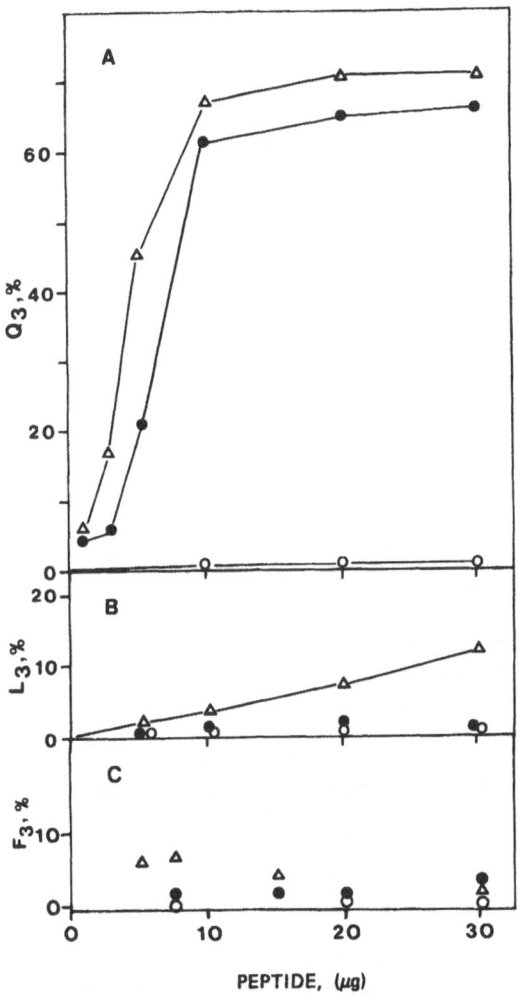

FIGURE 7 Lipid intermixing, vesicle leakage, and intermixing of vesicle
contents induced by lysine peptides differing in size. Lipid intermixing
among vesicles containing 30 mol% PA and 55 or 60 mol% PE was monitored
by fluorescence resonance energy transfer (A); leakage by the release
of encapsulated calcein (B); and vesicle content intermixing by the formation
of fluorescent terbium/dipicolinic-acid complex (C). Results shown were
recorded 3 min after the addition of lysine peptide and are expressed
as percent of the maximal value, determined by calibration. Symbols denote
trilysine (o), pentalysine (●), and polylysine (2-5 kDa) (△). (Reproduced
from Ref. 52.)

(46,53), as well as pentalysine, polyarginine, and polyornithine (Bondeson,
J., and Sundler, R., in preparation), can aggregate PS-containing vesicles
and promote fusion induced by acidification. Except for reports of polylysine-
induced fusion among sonicated cardiolipin-containing vesicles (50,51) and
asolectin vesicles (53), basic poly(amino acid)s have not been found capable
of inducing complete membrane fusion at neutral pH (46,55), and our recent

results support this conclusion. Nanomolar concentrations of polylysine
have been found to lower the Ca^{2+}-threshold for fusion between large
unilamellar vesicles containing PC and cardiolipin (55) by inducing intervesicle contact, while higher concentrations of polylysine inhibit fusion,
presumably by blocking Ca^{2+}-binding sites. The former effect shows a
certain similarity to the effects of lectin-mediated (see Section IV) and
synexin-mediated (56) intermembrane contact, although the lipid composition
of vesicles was different in the latter studies.

In the case of proton-induced fusion, the characteristics of the process
differ depending on whether intervesicle contact is established with lectins
(43) or basic peptides (46,53). The pH dependence for proton-induced
fusion among PS-containing vesicles shows that quite extensive fusion
is induced at pH 5-6 after treatment with basic oligopeptides or polypeptides
(46; Bondeson, J., and Sundler, R., in preparation), while a significantly
higher proton concentration is required after lectin-mediated contact (Fig. 5).
This suggests differences in the mechanism for the induction of fusion.
It has been proposed (46,53) that the histidine groups of polyhistidine
may themselves trigger fusion by becoming cationic upon protonation. However, the pH dependence appears to be similar whether polymers of histidine
or lysine are utilized (Bondeson, J., and Sundler, R., in preparation).
Other findings indicate a very tight interaction between the lysine oligomers
and polymers and anionic phospholipid headgroups (54,57), leading to at
least partial headgroup dehydration (54). The basic peptides may thus
not only induce intervesicle contact, by electrostatically linking the negatively charged vesicles, but also reduce their surface hydration and thereby
facilitate Ca^{2+}-induced fusion and increase the sensitivity toward proton-
induced fusion above that achieved with lectin-mediated contact.

Several different amphiphilic proteins, such as melittin (58), insulin
(59), colicin (60), and cardiotoxin (61), have been reported to induce
intervesicle contact, with or without true fusion. Here, a combination of
electrostatic interactions and penetration of hydrophobic peptide segments
into the acyl chain region appears to be involved. This type of interaction
thereby differs from both lectin-mediated bridging of vesicles and the
intervesicle contact and modulation of the headgroup layer induced by
polar, basic peptides. In the case of melittin, this interaction is accompanied
by considerable vesicle lysis that may precede—rather than follow—vesicle
lipid intermixing. In our hands melittin causes extensive lipid intermixing
and vesicle leakage, but neither intermixing of vesicle contents nor growth
in vesicle size. Clathrin, the major coat protein, has been shown to interact with both liposomes containing PC (62) and PS (63) and to mediate
fusion at, or below, pH 5. This protein is known to undergo a conformational changes upon acidification, and it is likely that an increased exposure
of hydrophobic regions is involved in both clathrin-vesicle interaction
and the triggering of fusion.

VII. CONCLUDING REMARKS

Intermembrane contact is necessary, but in itself insufficient, for membrane
fusion. The incorporation of glycolipids differing in size of the carbohydrate
group in model membranes has made it possible to manipulate the distance
between contacting membranes and further define the role of this parameter
in membrane fusion. Results from studies on anionic lipid vesicles and
their mutual interaction, mediated by lectins or basic peptides, provide

further support for the concept of qualitative differences among intermembrane contacts, even if those leading to departure from the bilayer
state are not considered. For simplicity, the types of intermembrane adhesion
discerned can be divided in two groups; (1) contact without any major
perturbation of the hydrated and stabilizing headgroup layer of the contacting membranes (e.g., that mediated by lectin-glycolipid interaction), and
(2) contact involving electrostatic interactions that lead to varying degrees
of headgroup dehydration [e.g., that induced by basic poly(amino acid)s,
Mg^{2+} ions, protons, or Ca^{2+} ions]. The latter type of contact may be
accompanied by complete fusion (fusion/fission), by an apparently incomplete
fusion (fusion without fission), or by no fusion but a clearly increased
sensitivity to other fusogens. In this context, protons have been found
to be quite inefficient in inducing intermembrane contact when compared
to their fusogenic activity and to the di- and polyvalent agents. This
means that phospholipid headgroup protonation, as well as the binding
of Ca^{2+} ions in certain cases, may render noncontacting membranes competent for fusion, although this ability is not expressed. In such instances
the control of fusion would be transferred to agents that can mediate intermembrane contact.

ACKNOWLEDGMENTS

Studies carried out in the authors' laboratory received financial support
from the Swedish Medical Research Council (Project 5410), the Albert
Påhlsson Foundation, and the Medical Faculty, University of Lund. Assistance in preparation of the manuscript by Ms. Gesa Johnson and Ms. Brigitta
Jönsson is gratefully acknowledged.

REFERENCES

1. Nir, S., Bentz, J., Wilschut, J., and Düzgüneş, N. (1983). Aggregation and fusion of phospholipid vesicles. *Prog. Surface Sci.* 13:1-124.
2. Portis, A., Newton, C., Pangborn, W., and Papahadjopoulos, D. (1979).
   Studies on the mechanism of membrane fusion: Evidence for an intermembrane Ca^{2+}-phospholipid complex, synergism with Mg^{2+}, and
   inhibition by spectrin. *Biochemistry* 18:780-790.
3. Sundler, R., and Papahadjopoulos, D. (1981). Control of membrane
   fusion by phospholipid head groups. I. Phosphatidate/phosphatidylinositol
   specificity. *Biochim. Biophys. Acta* 649:743-750.
4. Düzgüneş, N., Wilschut, J., Fraley, R., and Papahadjopoulos, D.
   (1981). Studies on the mechanism of membrane fusion. Role of head
   group composition in calcium- and magnesium-induced fusion of mixed
   phospholipid vesicles. *Biochim. Biophys. Acta* 642:182-195.
5. Sundler, R., Düzgüneş, N., and Papahadjopoulos, D. (1981). Control
   of membrane fusion by phospholipid head groups. II. The role of phosphatidylethanolamine in mixtures with phosphatidate and phosphatidylinositol. *Biochim. Biophys. Acta* 649:751-758.
6. Papahadjopoulos, D., Vail, W. Y., Newton, C., Nir, S., Jacobson, K.,
   Poste, G., and Lazo, R. (1977). Studies on membrane fusion. III.
   The role of calcium-induced phase changes. *Biochim. Biophys. Acta*
   465:579-598.

7. Wilschut, J., Düzgüneş, N., and Papahadjopoulos, D. (1981).
   Calcium/magnesium specificity in membrane fusion: Kinetics of aggregation and fusion of phosphatidylserine vesicles and the role of bilayer
   curvature. *Biochemistry* 20:3126-3133.

8. Hoekstra, D. (1982). Role of lipid phase separations and membrane
   hydration in phospholipid vesicle fusion. *Biochemistry* 21:2833-2840.

9. Sundler, R. (1984). Role of phospholipid head group structure and
   polarity in the control of membrane fusion. In: *Biomembranes*, Vol. 12
   M. Kates and L. A. Manson, eds.). Plenum Press, New York, pp.
   563-583.

10. Düzgüneş, N., Wilschut, J., and Papahadjopoulos, D. (1985). Control
    of membrane fusion by divalent cations, phospholipid head groups
    and proteins. In: *Physical Methods on Biological Membranes and Their
    Model Systems* (F. Conti, W. E. Blumberg, J. de Gier, and F. Pocchiari,
    eds.). Plenum Press, New York, pp. 193-218.

11. Curatolo, W., Yau, A. O., Small, D. M., and Sears, B. (1978). Lectin-
    induced agglutination of phospholipid/glycolipid vesicles. *Biochemistry*
    17:5740-5744.

12. Sundler, R., and Wijkander, J. (1983). Protein-mediated intermembrane
    contact specifically enhances Ca^{2+}-induced fusion of phosphatidate-
    containing membranes. *Biochim. Biophys. Acta* 730:391-394.

13. Düzgünes, N., Hoekstra, D., Hong, K., and Papahadjopoulos, D.
    (1984). Lectins facilitate calcium-induced fusion of phospholipid vesicles
    containing glycosphingolipids. *FEBS Lett.* 173:80-84.

14. Hoekstra, D., Düzgüneş, N., and Wilschut, J. (1985). Agglutination
    and fusion of globoside GL-4 containing phospholipid vesicles mediated
    by lectins and calcium ions. *Biochemistry* 24:565-572.

15. Hoekstra, D., and Düzgüneş, N. (1986). *Ricinus Communis* agglutinin-
    mediated agglutination and fusion of glycolipid-containing phospholipid
    vesicles: Effect of carbohydrate head group size, calcium ions and
    spermine. *Biochemistry* 25:1321-1330.

16. Düzünes, N., and Hoekstra, D. (1986). Agglutination and fusion
    of glycolipid-phospholipid vesicles mediated by lectins and calcium
    ions. *Studia Biophysica* 111:5-10.

17. De Kroon, A. I. P. M., van Hoogevest, P., Geurts van Kessel,
    W. S. M., and de Kruijff, B. (1985). Influence of glycophorin incor-
    poration on Ca^{2+}-induced fusion of phosphatidylserine vesicles.
    Biochemistry 24:6382-6389.

18. Surolia, A., Bachhawat, B. K., and Podder, S. K. (1975). Interaction
    between lectin from *Ricinus Communis* and liposomes containing
    gangliosides. *Nature* 257:802-804.

19. Grant, C. W. M., and Peters, M. W. (1984). Lectin-membrane inter-
    actions. Information from model systems. *Biochim. Biophys. Acta* 779:
    403-422.

20. Slama, J. S., and Rando, R. R. (1980). Lectin-mediated aggregation
    of liposomes containing glycolipids with variable hydrophilic spacer
    arms. *Biochemistry* 19:4595-4600.

21. Orr, G. A., Rando, R. R., and Bangerter, F. W. (1979). Synthetic
    glycolipids and the lectin-mediated aggregation of liposomes. *J. Biol.
    Chem.* 254:4721-4725.

22. Hampton, R. Y., Holz, R. W., and Goldstein, I. J. (1980). Phospho-
    lipid, glycolipid, and ion dependencies of concanavalin A- and *Ricinus
    Communis* agglutinin I-induced agglutination of lipid vesicles. *J. Biol.
    Chem.* 255:6766-6771.

23. Sundler, R. (1982). Agglutination of glycolipid-phospholipid vesicles by concanavalin A. Evidence for steric modulation of lectin binding by phospholipid head groups. *FEBS Lett.* 141:11-13.

24. Sundler, R. (1984). Studies on the effective size of phospholipid head groups in bilayer vesicles using lectin-glycolipid interaction as a steric probe. *Biochim. Biophys. Acta* 771:59-67.

25. Irvine, R. F., Hemington, N., and Dawson, R. M. C. (1979). The calcium-dependent phosphatidylinositol phosphodiesterase of rat brain. Mechanisms of suppression and stimulation. *Eur. J. Biochem.* 99:525-530.

26. Hofmann, S. L., and Majerus, P. W. (1982). Modulation of phosphatidylinositol-specific phospholipase C activity by phospholipid interactions, diglycerides, and calcium ions. *J. Biol. Chem.* 257:14359-14364.

27. Lundberg, G. A., Jergil, B., and Sundler, R. (1986). Phosphatidyl-inositol 4-phosphate kinase from rat brain. Activation by polyamines and inhibition by phosphatidylinositol 4,5-bisphosphate. *Eur. J. Biochem.* 161:257-262.

28. Van der Bosch, J., and McConnell, H. M. (1975). Fusion of dipalmitoyl-phosphatidylcholine vesicle membranes induced by concanavalin A. *Proc. Natl. Acad. Sci. USA* 72:4409-4413.

29. Tycko, B., and Maxfield, F. R. (1982). Rapid acidification of endocytic vesicles containing alpha-2-macro-globulin. *Cell* 28:643-651.

30. Ohkuma, S., and Poole, B. (1978). Fluorescence probe measurement of the intralysosomal pH in living cells and the perturbation of pH by various agents. *Proc. Natl. Acad. Sci. USA* 75:3327-3331.

31. Anderson, R. G.W., and Pathak, R. K. (1985). Vesicles and cisternae in the *trans* Golgi apparatus of human fibroblasts are acidic compartments. *Cell* 40:635-643.

32. Brown, M. S., Anderson, R. G. W., and Goldstein, J. L. (1983). Recycling receptors: The round-trip itinerary of migrant membrane proteins. *Cell* 32:663-667.

33. Tartakoff, A. M. (1983). Perturbation of vesicular traffic with the carboxylic ionophore monensin. *Cell* 32:1026-1028.

34. White, J., Kielian, M., and Helenius, A. (1983). Membrane fusion of enveloped animal viruses. *Q. Rev. Biophys.* 16:151-195.

35. Hope, M. J., Walker, D. C., and Cullis, P. R. (1983). Ca^{2+} and pH-induced fusion of small unilamellar vesicles consisting of phosphatidylethanolamine and negatively charged phospholipids: A freeze fracture study. *Biochem. Biophys. Res. Commun.* 110:15-22.

36. Ellens, H., Bentz, J., and Szoka, F. C. (1985). H^+- and Ca^{2+}-induced fusion and destabilization of liposomes. *Biochemistry* 24:3099-3106.

37. Yatvin, M. B., Kreutz, W., Horwitz, B. A., and Shinitzky, M. (1980). pH-sensitive liposomes: Possible clinical implications. *Science* 210:1253-1255.

38. Connor, J., Yatvin, M. B., and Huang, L. (1984). pH-sensitive liposomes: Acid-induced liposome fusion. *Proc. Natl. Acad. Sci. USA* 81:1715-1718.

39. Düzgüneş, N., Staubinger, R. M., Baldwin, P. A., Friend, D. S., and Papahadjopoulos, D. (1985). Proton-induced fusion of oleic acid-phosphatidylethanolamine liposomes. *Biochemistry* 24:3091-3098.

40. Ellens, H., Bentz, J., and Szoka, F. C. (1984). pH-induced destabilization of phosphatidylethanolamine-containing liposomes: Role of bilayer contact. *Biochemistry* 23:1532-1538.

41. Bentz, J., Ellens, H., Lai, M-Z., and Szoka, F. C. (1985). On the correlation between H_{II} phase and the contact-induced destabilization of phosphatidylethanolamine-containing membranes. *Proc. Natl. Acad. Sci. USA* 82:5742–5745.

42. Nayar, R., and Schroit, A. J. (1985). Generation of pH-sensitive liposomes: Use of large unilamellar vesicles containing *N*-succinyldioleoylphosphatidylethanolamine. *Biochemistry* 24:5967–5971.

43. Bondeson, J., Wijkander, J., and Sundler, R. (1984). Proton-induced membrane fusion. Role of phospholipid composition and protein-mediated intermembrane contact. *Biochim. Biophys. Acta* 777:21–27.

44. Struck, D. K., Hoekstra, D., and Pagano, R. E. (1981). Use of resonance energy transfer to monitor membrane fusion. *Biochemistry* 20:4093–4099.

45. Bondeson, J., and Sundler, R. (1987). Phosphatidylethanol counteracts calcium-induced membrane fusion but promotes proton-induced fusion. *Biochim. Biophys. Acta* 899:258–264.

46. Wang, C-Y., and Huang, L. (1984). Polyhistidine mediates an acid-dependent fusion of negatively charged liposomes. *Biochemistry* 23:4409–4416.

47. Eibl, H. and Blume, A. (1979). The influence of charge on phosphatidic acid bilayer membranes. *Biochim. Biophys. Acta* 553:476–488.

48. MacDonald, R. C., Simon, S. A., and Baer, E. (1976). Ionic influences on the phase transition of dipalmitoylphosphatidylserine. *Biochemistry* 15:885–891.

49. Tokutomi, S., Ohki, K., and Ohnishi, S-I. (1980). Proton-induced phase separation in phosphatidylserine/phosphatidylcholine membranes. *Biochim. Biophys. Acta* 596:192–200.

50. Gad, A. E., Silver, B. L., and Eytan, G. D. (1982). Polycation-induced fusion of negatively-charged vesicles. *Biochim. Biophys. Acta* 690:124–132.

51. Gad, A. E. (1983). Cationic polypeptide-induced fusion of acidic liposomes. *Biochim. Biophys. Acta* 728:377–382.

52. Bondeson, J., and Sundler, R. (1985). Lysine peptides induce lipid intermixing but not fusion between phosphatidic acid-containing vesicles. *FEBS Lett.* 190:283–287.

53. Uster, P. S., and Deamer, D. W. (1985). pH-dependent fusion of liposomes using titrable polycations. *Biochemistry* 24:1–8.

54. Carrier, D., and Pézolet, M. (1986). Investigation of polylysine-dipalmitoylphosphatidylglycerol interactions in model membranes. *Biochemistry* 25:4167–4174.

55. Gad, A. E., Bental, M., Elyashiv, G., and Weinberg, H. (1985). Promotion and inhibition of vesicle fusion by polylysine. *Biochemistry* 24:6277–6282.

56. Hong, K., Düzgüneş, N., and Papahadjopoulos, D. (1981). Role of synexin in membrane fusion. *J. Biol. Chem.* 256:3641–3644.

57. Hartman, W., and Galla, H-J. (1978). Binding of polylysine to charged bilayer membranes. Molecular organization of a lipid-peptide complex. *Biochim. Biophys. Acta* 509:474–490.

58. Eytan, G. D., and Almary, T. (1983). Melittin-induced fusion of acidic liposomes. *FEBS Lett.* 156:29–32.

59. Farias, R. N., Vinals, A. L., and Morero, R. D. (1985). Insulin-mediated fusion of negatively charged phospholipid vesicles at low pH. *Biochem. Biophys. Res. Commun.* 128:68–74.

60. Pattus, F., Cavard, D., Crozel, V., Baty, D., Adrian, M., and Lazdunski, C. (1985). pH-dependent membrane fusion is promoted by various colicins. *EMBO J.* 4:2469-2474.

61. Batenburg, A. M., Bougis, P. E., Rochat, H., Verkleij, A. J., and de Kruijff, B. (1985). Penetration of a cardiotoxin into cardiolipin model membranes and its implications on lipid organization. *Biochemistry* 24:7101-7110.

62. Blumenthal, R., Henkart, M., and Steer, C. J. (1983). Clathrin-induced pH-dependent fusion of phosphatidylcholine vesicles. *J. Biol. Chem.* 258:3409-3415.

63. Hong, K., Yoshimura, T., and Papahadjopoulos, D. (1985). Interaction of clathrin with liposomes: pH-dependent fusion of phospholipid membranes induced by clathrin. *FEBS Lett.* 191:17-23.

64. Wilschut, J., and Papahadjopoulos, D. (1979). Ca^{2+}-induced fusion of phospholipid vesicles monitored by mixing of aqueous contents. *Nature* 281:690-692.

III
FUSOGENIC PROPERTIES OF VIRUSES

13

Membrane Fusion and the Infectious Entry of Viruses into Cells
An Overview

JOHN LENARD

University of Medicine and Dentistry of New Jersey—Robert Wood Johnson Medical School at Rutgers, Piscataway, New Jersey

I. INTRODUCTION

Every virus has a distinct structure, defined by specific interactions between a limited number of viral gene products (nucleic acids and proteins). Some viruses also possess lipids, which are not viral products but are taken from the host cell at the time of exit or budding (reviewed in Refs. 5-7). The genome then enters the nucleus, or whichever subcellular site is required by its replication program, but the initial entry into the cell itself, i.e., into the cytoplasm, seems to occur independently of subsequent events. Viruses with similar replication strategies may thus have quite different modes of entry, while similar uncoating processes may be shared by quite different viruses.

Every virus that contains lipid seems to have it arranged in a membrane-like bilayer surrounding the genome (reviewed in Refs. 1,2). It seems self-evident that this must fuse with a cellular membrane, at some point and in some way, in order to accomplish infectious entry of the viral genome into the cell. Experimental confirmation of this postulate has been obtained for several types of commonly studied virus (as shown in Fig. 1), and no exceptions are known. There remain, however, a much larger number of lipid-containing viruses which, although fusion is generally assumed, have not been critically tested (2).

Since the need for the viral genome to cross the cell membrane is not limited to lipid-containing viruses, but is common to all viruses, fusion proteins may well resemble membrane-active proteins that perform the same function for nonlipid viruses (e.g., reoviruses, see Ref. 8). The fusion proteins might ultimately be recognized as part of a larger class of viral uncoating proteins, all acting to destabilize cell membranes by a limited number of common molecular mechanisms (cf. Fig. 1).

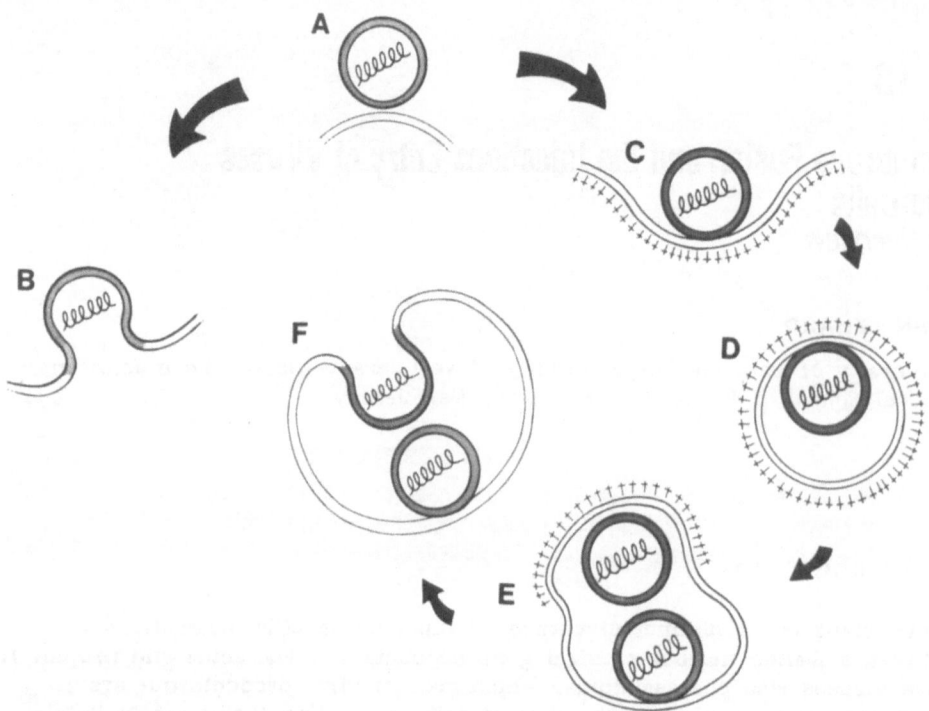

FIGURE 1 Uncoating of an enveloped virus, i.e., introduction of the
genome into the cytoplasm of the cell to be infected, requires initial attachment (A) of the virion onto the cell surface. Fusion may then occur directly
at the plasma membrane (B), or inside acidified endosomes (F) after internalization by endocytosis via the coated vesicle pathway (C-E). Interaction
of certain viral protein coats with cell membranes may be depicted in this
same schematic way.

II. FUSING VIRUSES AND FUSION ASSAYS

It is not possible to demonstrate directly that a particular fusion process
causes infection of a cell. This is because many viral particles need to
be present if a cell culture is to be infected with any efficiency, even
though a single particle suffices to infect a single cell. The infecting particle cannot be experimentally distinguished from other particles which may
be nonproductively engaged. Further, readily measurable consequences
of infection do not appear for at least several hours after initiation, i.e.,
after the fusion event. The argument that fusion is required to initiate
viral infection is therefore made somewhat indirectly, by correlating fusion
activity of the virus with infectivity under the widest possible range of
conditions. For this purpose, several methods to measure viral fusion
activity have been developed (see also Chapter 14).

The measurement of fusion between adjacent cells in a monolayer
(*cell-cell fusion*) is perhaps the most widely used, since the basic observation is readily made during routine microscopic examination of infected

cell cultures. Two types of cell-cell fusion have been distinguished (9). "Fusion from without" occurs very soon after addition of large amounts of virus to the cells. Noninfectious virions effectively induce fusion from without. The fusion activity is therefore not a property of infected cells, but is intrinsic to the viral particle. "Fusion from within" is induced by much less virus, and it occurs only many hours later than fusion from without. Infection of the cell is required, since irradiated virions are inactive, and viral protein must be produced. Fusion from within is now known to arise from active viral fusion proteins, newly produced by the infected cell, incorporated into the cell membrane.

Hemolysis, or the leakage of hemoglobin from erythrocytes after fusion with viruses, provides another simple assay of viral fusion. Its usefulness is actually surprising, since on the one hand fusion is intrinsically a non-leaky process, and on the other erythrocyte hemolysis can be induced under many conditions (e.g., hypotonicity) that do not promote fusion. Under certain careful conditions of preparation, viruses can, in fact, be prepared that, though fusible and infectious, are not hemolytic (10-12). These are readily converted to hemolytic viruses by freeze-thawing or brief sonication, however, and most conventionally prepared fusing viruses are hemolytic. Virus-induced hemolysis requires *virus-erythrocyte fusion* as a necessary prerequisite, and so hemolysis has remained a convenient measure of viral fusion.

Several procedures have been employed to study *virus-liposome fusion*. These have been particularly useful in delineating the precise proteins and lipids required for viral fusion reactions. Electron microscopy has been successfully employed to detect virus-liposome fusion products (13-16). A more quantitative assay was developed by trapping a nuclease or protease in the target liposomes. Fusion of suitably labeled Semliki Forest virus (SFV) (16) or influenza virus (17) with such liposomes could then be determined from the degradation of normally inaccessible internal viral components. Fusion has recently been measured directly by one of several lipid dilution techniques, in which fluorescent lipid molecules are incorporated either into the target liposomes (18-20) or into virions (21). Fusion of such a labeled bilayer with an unlabeled one, followed by rapid lateral diffusion, dilutes the concentration of fluorescent lipid and thus increases the mean distance between fluorescent molecules. This relieves self-quenching between like molecules or resonance energy transfer between unlike ones. Spin-labeled lipids have also been successfully used in a dilution assay of membrane fusion (22).

Fusion of enveloped viruses with planar bilayers has also been reported (23). As with hemolysis, however, detection of *virus-planar bilayer fusion* is indirect, relying for detection on the conductance increase arising from leakage by the fused viral membrane (23).

Paradoxically, *virus-cell fusion*, the reaction that actually initiates infection, has been the most difficult to measure accurately. Although electron micrographs of viruses fusing with cells abound in the literature, quantitation by this method is generally unsatisfactory (5). Hence, it is not surprising that the various reports of virus-cell fusion using electron microscopy engendered considerable disagreement about its location and frequency, and other techniques had to be brought to bear before further progress could be made. Nonetheless, current ideas about the role of virus-cell fusion in initiating infection are supported by electron micrographs showing that large numbers of virions fuse with particular cell membranes under predicted experimental conditions (24-28). The resonance energy transfer and fluorescence dequenching techniques using fluorescent lipid analogs, discussed above in connection with virus-liposome fusion, have been successfully extended to the study of virus-cell fusion (21) and of fusion between reconstituted viral envelopes and intact cells or cell membranes (see below).

Virus-induced fusion reactions have been extensively studied using these and other techniques for only a few different types of viruses: SFV, influenza virus, Sendai virus, and vesicular stomatitis virus (VSV), and, far less extensively, certain retroviruses. Their behavior will be considered in greater detail below.

III. THE PARADIGM OF THE VIRAL FUSION PROTEIN

For SFV, influenza, Sendai, and VSV there is strong evidence that the mediators of viral fusion are viral membrane glycoproteins (Table 1). These are encoded by the viral genome, possess signal sequences, cross the endoplasmic reticulum membrane during synthesis, proceed normally through the Golgi to the plasma membrane, and remain anchored in the bilayer of the budded viral particle by hydrophobic transmembrane sequences (reviewed in Refs. 1,2). Their external domains (about 90% of the total protein) act by destabilizing the target (cell) membrane so as to facilitate the production of a metastable fusion intermediate, by some nonenzymatic, calcium independent means (review in Ref. 29). No additional proteins, either viral or cellular, are required for fusion activity by these four viruses.

The lack of a requirement for cell proteins has been demonstrated by studies of virus-liposome fusion. All four viruses fuse with liposomes (13-17,22,30,31) and/or with planar lipid bilayers (23) devoid of protein. These studies have also revealed pronounced differences between the four viruses with regard to the required lipid composition of the target membrane. SFV fusion requires cholesterol (16,32). The other three viruses show preferences, but no absolute requirements, for particular lipids (13,17,22, 23,30).

The most convincing evidence that a single viral glycoprotein carries the fusing activity has been obtained when the cloned glycoprotein gene was expressed in animal cells. The gene product was shown to be synthesized normally and to appear on the plasma membrane, presumably after normal processing through the endoplasmic reticulum and Golgi apparatus. Cell-cell fusion was demonstrated between cells expressing this single viral gene product (33-35).

Additional evidence was obtained by studying the fusion activity of liposomes containing isolated viral membrane glycoproteins ("virosomes"). These have been shown to fuse with liposomes and to possess hemolytic and cell fusing activity (36-46). Activity is generally no more than 25% that of intact virions, suggesting that the arrangement of glycoproteins on the viral surface, and/or the rigidity of the underlying structure, influences fusion activity. Virosome activity has provided strong evidence for the identity of the fusing proteins of SFV (43) and VSV (36), since each has only a single type of glycoprotein structure on its surface. However, the existence of two different, and often incompletely separated, glycoproteins in influenza and Sendai virus have led to some confusion about whether one or both proteins are required for activity. The fusion activity of cloned HA protein has definitively answered this question for

TABLE 1 Glycoproteins of Some Enveloped Viruses

Virus	Family	Glycoprotein[a]	Function	Structure/sequence (Ref.)
Influenza	Orthomyxoviridae	$[(HA_1(44)+HA_2(30)]_3$	Fusion Attachment	75 76
		$[NA(60)]_4$	Receptor destruction Other (?)	77 78
Sendai	Paramyxoviridae	$[F_1(52)+F_2(11)]_n$	Fusion	79,80
		$[HN(67)]_n$	Attachment Receptor destruction Fusion (?)	81,80
Semliki Forest virus (SFV)	Togaviridae	$[E_1(51)+E_2(52)+E_3(11)]$	Fusion Attachment (?)	82
Vesicular stomatitis virus (VSV)	Rhabdoviridae	$[G(61)]_1$	Fusion Attachment (?)	83
Murine mammary tumor virus (MMTV)	Retroviridae	$[gp52+gp36]_n$	Fusion (?) Attachment	84

[a]Products derived by proteolysis from a common precursor are linked by a "+"; approximate molecular weight in kD shown in parentheses; degree of self-association on viral surface is indicated by subscript outside brackets (n = not known).

influenza (35), but not yet (as of March 1986) for Sendai. Virosomes containing only the Sendai F protein, and lacking detectable viral attachment protein HN, have been reported to be fusogenic when suitable provisions are made for viral attachment to the target membrane (47). Perhaps even more persuasive, a Sendai mutant lacking >95% of its HN could nonetheless infect Hep2 cells and induce cell-cell fusion after binding to the cell's asialoglycoprotein receptor (48) and behaved identically with wild-type virus in several fusion reactions (49). Others find, however, that both F and HN must be present and active in viral envelopes or virosomes for fusion to occur (44-46,50).

Based on these findings, and on the similarities in primary amino acid sequence between the influenza HA protein and the Sendai F protein (reviewed in Ref. 29), an independently acting single viral gene product that mediates fusion has become the paradigm for the uncoating reactions of all lipid-containing viruses. It seems entirely possible, however, that fusion reactions of other viruses, not yet fully characterized, may require several viral proteins to act in concert, or that some viral fusion proteins may need to act in conjunction with one or more specific cell membrane protein(s). Some of the retroviruses, which possess highly specific cell "receptor" requirements (51,52), and for which only cell-cell fusion has so far been demonstrated, might even require a cellular surface protein for fusion. The many lipid-containing viruses whose entry mechanisms remain uncharacterized may reveal a variety of different protein and lipid requirements for fusion.

IV. pH DEPENDENCE OF VIRAL FUSION

The pH dependence of viral fusion has proven to be a critical determinant of the route by which a virus infects a cell (reviewed in Refs. 29,53,54; see Fig. 1). SFV, VSV, and influenza virus undergo fusion only at mildly acidic pH, generally below 6.0, with maximal activity around pH 5.0. Sendai differs from these, exhibiting substantial fusion activity over the pH range 5-9, but it is most active above pH 7 (47,55). The pH dependence of fusion of each virus is qualitatively similar regardless of the assay used—whether cell-cell fusion, virus-erythrocyte fusion, virus-liposome fusion, or virus-cell fusion—although assay-related differences are also seen (53,56,57). As expected from the paradigm of the independent viral fusion protein, the pH dependence of fusion is a property conferred by the HA gene of influenza (56-58).* Of these four viruses, therefore, only Sendai possesses fusion activity at the neutral pH characteristic of tissue culture medium or extracellular fluid. It alone is capable of fusing with the plasma membrane under normal physiological conditions, and persuasive evidence now exists that this is, in fact, the route by which it uncoats its genome to initiate infection (7,62).

For SFV, VSV, and influenza virus, on the other hand, a different site of uncoating must exist. Early thinking centered on the lysosome, which was known to be highly acidic, and which provided the ultimate

---

*Reports from two laboratories have claimed that influenza possesses fusion activity at neutral pH (31,59,60), in apparent contradiction to the findings cited in the body of the text. This discrepancy may be explained by a recent finding, which showed that heat-induced changes in HA conformation rendered the protein fusogenic at neutral pH (61).

destination for many molecules and particles that were actively brought
into the cell by the process of adsorptive endocytosis. Lysosomes were
also known, however, to contain a rich array of proteases and degradative
enzymes at levels amply sufficient to rapidly inactivate the viral fusion
protein. Further investigation of the process of adsorptive endocytosis
revealed, however that material brought into the cell resided initially in
prelysosomal compartments (termed endosomes, receptosomes, or CURL),
and that these compartments, though devoid of degradative enzymes, were
practically as acidic as lysosomes (63; reviewed in Refs. 53,54). The uncoating of SFV by fusion with the endosomal membrane after internalization
by absorptive endocytosis has now been thoroughly documented (25,64),
and it seems virtually certain that VSV and influenza virus initiate infection
in the same way. An important part of this argument has been the demonstration that so-called lysosomotropic amines, which raise endosomal and
lysosomal pH above the point at which viral fusion can occur, effectively
inhibit viral infection, provided they are present very early in infection
(reviewed in Ref. 53). They are ineffective if added even 30 min after
the virus and the cells are warmed together, and the inhibition is reversible
upon removing the amines, as predicted from their postulated mechanism
of action.

In contrast to the four viruses discussed above, the retroviruses
present a more complex picture. Several of the murine leukemia viruses
initiate cell-cell fusion only at pH's above 6.4 (65), implying that they
are inactive at endosomal pH, and therefore that they must penetrate the
cell by fusion with the plasma membrane. Of the seven strains examined,
however, one showed appreciable fusion activity at pH 5, suggesting that
no general statement may be applicable. Mouse mammary tumor viruses,
on the other hand, mediate cell-cell fusion only after treatment at low pH
(66), and infection was inhibited by lysosomotropic amines (67).

Many speculations have been made regarding the relative selective
advantages of entry by fusion with the plasma membrane versus internal
membranes such as endosomes. Fusion with the plasma membrane leaves
viral antigens on the cell surface, where they are subject to immune surveillance (7). On the other hand, fusion at neutral pH permits the viral
infection to spread by cell-cell fusion (fusion from within) without the
formation of new viral particles, and without the need to reinitiate infection
from the outside. The selective advantage of one or the other may change
in complex ways as strains develop over relatively short periods of time.
It should not be surprising, therefore, to find closely related viruses
that differ in the pH dependence of fusion, a change that can probably
be made by changing only a very small number of amino acids in a viral
fusion protein (e.g., Ref. 56). It has recently been suggested from electron microscope evidence that Western equine encephalitis virus, an alpha-
virus in the same family as SFV, may fuse with the plasma membrane of
the cell to be infected (68). Although this is not at all a definitive report,
the discovery of such differences, even between very closely related
viruses, may perhaps be anticipated in the future.

V. RECEPTORS AND FUSION

The term receptor has been loosely used to describe the cell surface site
to which viruses attach en route to a productive infection. These sites
should not be confused, however, with true cell surface receptors, which

bind specific ligands for purposes useful to the cell (or to the organism of which it is part), and which participate actively in the function of the ligand, e.g., by phosphorylating intracellular proteins. The paradigm of the independent viral fusion protein precludes direct involvement in the fusion process by any cell surface viral receptor. Indeed, a Sendai virus mutant lacking the viral HN binding protein (through which Sendai usually attaches to cell surface sialic acid) was shown to initiate infection and cell-cell fusion by binding to a completely unfamiliar surface protein, the asialoglycoprotein receptor on Hep-2 cells (48). In another report, hemolysis and cell-cell fusion by Sendai occurred after binding of the virus to anti-Sendai antibodies covalently attached to the surface of neuraminidase-treated erythrocytes (45). In a similar vein, VSV, which is normally unable to initiate infection through the apical membranes of MDCK cells (69), acquired this ability if influenza HA molecules (with their affinity for terminal sialic acid groups) were present to bind the VSV to the apical surface (70). In these cases, the function of these arti-ficial cell surface receptor molecules was apparently restricted to assuring the attachment of the virus particle to a suitable region of the cell surface.

The cell surface attachment site is, therefore, a critical determinant of the ability of the virus to enter the cell. Several requirements for productive initial attachment of the virion to the cell surface seem evident.

1. The binding must permit access to the membrane bilayer under the appropriate pH conditions for fusion. A virus that attaches, for exam-ple, to the tip of a rigid protein extending 100 Å from the membrane may be unable to fuse.

2. The segment of bilayer to which the virus has access must have a lipid composition compatible with the requirements of the viral fusion protein. Plasma membranes consist of hundreds of different lipid species, and studies of even simple mixtures of synthetic lipids have made it abundantly clear that lateral phase segregation into areas of different composition is a ubiquitous property of mixed bilayers. Certain proteins may also nucleate regions of selected lipid composition immediately surround-ing them. Very little is known about the nature of local variations within lipid systems as complex as plasma membranes.

3. Those viruses that fuse only at low pH must bind to a cell surface site that enters endosomes, and must remain bound after internalization, at the pH required for fusion. Conversely, viruses that fuse at neutral pH should infect more efficiently by binding to sites that are not internal-ized.

4. The appropriate sites must be accessible to the external viral particles. In cell culture, where no obvious anatomical restrictions exist, the commonly observed enhancement of infectivity by polycations may indi-cate that a general electrostatic barrier exists that limits the close approach of negatively charged virions to negatively charged plasma membranes. Infection by viruses with strict receptor requirements such as the retro-viruses, as well as by viruses apparently lacking such requirements, such as VSV and SFV, are similarly enhanced by polycations (36,71,72). Other treatments that reduce the net negative charge on the cell surface, such as treatment with protease of neuraminidase (73), are similarly effective in enhancing infection or fusion by VSV. Hemolysis by SFV and VSV is enhanced by similar treatments (36,55,74), suggesting similar mechanisms of productive binding to erythrocytes and cultured cells. Although the detailed charge distribution on the surface of any plasma membrane is completely unknown, these findings suggest that electrostatic repulsions

limit access to otherwise appropriate cell surface attachment sites. Interestingly, the activities of influenza and Sendai virus, which lack sialic acid and are therefore much less negatively charged, are not appreciably enhanced by polycations (55).

It may be expected that in the next few years the general applicability of the viral fusion protein paradigm will be more thoroughly tested as more viruses are carefully examined. The role of specific cellular receptor or binding proteins in determining the efficiency of virus-cell fusion is another promising area for future progress. The unique role of these cell attachment sites will be completely understood only in the context of a detailed molecular knowledge of the fusion mechanism underlying viral infection.

Note added: Final revision of this paper was completed in May 1986. Its ideas are based on literature published before that time, and, with one exception (49), no later references are included.

REFERENCES

1. Lenard, J. (1978). Virus envelopes and plasma membranes. *Annu. Rev. Biophys. Bioeng.* 7:139-165.
2. Lenard, J. and Compans, R. W. (1974). The membrane structure of lipid-containing viruses. *Biochim. Biophys. Acta* 344:51-94.
3. Patzer, E. J., Wagner, R. R., and Dubovi, E. J. (1979). Viral Membranes: model systems for studying biological membranes. *CRC Crit. Rev. Biochem.* 6:165-217.
4. Simons, K. and Garoff, H. (1980). The budding mechanism of enveloped animal viruses. *J. Gen. Virol.* 50:1-21.
5. Dales, S. (1973). Early events in cell-animal virus interactions. *Microbiol. Rev.* 37:103-135.
6. Dimmock, N. J.(1982). Initial stages in infection with animal viruses. *J. Gen. Virol.* 59:1-22.
7. Fan, D. P. and Sefton, B. M. (1978). The entry into host cells of Sindbis virus, vesicular stomatitis virus and Sendai virus. *Cell* 15:985-992.
8. Borsa, J., Morash, B. D., Sargent, M. D., Copps, T. P., Lievaart, P. A., and Szekely, J. G. (1979). The modes of entry of reovirus particles into L cells. *J. Gen. Virol.* 45:161-170.
9. Bratt, M. A. and Gallaher, W. R. (1969). Preliminary analysis of the requirements for fusion from within and fusion from without by Newcastle disease virus. *Proc. Natl. Acad. Sci. USA* 64:536-543.
10. Homma, M., Shimizu, K., Shimizu, Y. K., and Ishida, N. (1976). On the study of Sendai virus hemolysis. I. Complete Sendai virus lacking in hemolytic activity. *Virology* 71:41-47.
11. Shimizu, Y. K., Shimizu, K., Ishida, N., and Homma, M. (1976). On the study of Sendai virus hemolysis. II. Morphological study of envelope fusion and hemolysis. *Virology* 71:48-60.
12. Vaananen, P. and Kaariainen, L. (19). Haemolysis by two alphaviruses: Semliki Forest and Sindbis virus. *J. Gen. Virol.* 43:593-601.
13. Haywood, A. M. (1974). Fusion of Sendai virus with model membranes. *J. Mol. Biol.* 87:625-628.
14. Haywood, A. M. (1975). "Phagocytosis" of Sendai virus by model membranes. *J. Gen. Virol.* 29:63-68.

15. Haywood, A. M. and Boyer, B. P. (1981). Initiation of fusion and disassembly of Sendai virus membranes into liposomes. *Biochim. Biophys. Acta* 646:31-35.

16. White, J. and Helenius, A. (1980). pH-dependent fusion between the Semliki Forest virus membrane and liposomes. *Proc. Natl. Acad. Sci. USA* 77:3273-3277.

17. White, J., Kartenbeck, J., and Helenius, A. (1982). Membrane fusion activity of influenza virus. *EMBO J.* 1:217-222.

18. Gibson, S., Jung, C. Y., Takahashi, M., and Lenard, J. (1986). Radiation inactivation analysis of influenza virus reveals different target sizes for fusion, leakage and neuraminidase activities. *Biochemistry* 25:6264-6268.

19. Stegmann, T., Hoekstra, D., Scherphof, G., and Wilschut, J. (1985). Kinetics of pH-dependent fusion between influenza virus and liposomes. *Biochemistry* 24:3107-3113.

20. Struck, D. K., Hoekstra, D., and Pagano, R. E. (1981). Use of resonance energy transfer to monitor membrane fusion. *Biochemistry* 20:4093-4099.

21. Hoekstra, D., Klappe, K., deBoer, T., and Wilschut, J. (1985). Characterization of the fusogenic properties of Sendai virus: Kinetics of fusion with erythrocyte membranes. *Biochemistry* 24:4739-4745.

22. Maeda, T., Kawasaki, K., and Ohnishi, S. (1981). Interaction of influenza virus hemagglutinin with target membrane is a key step in virus-induced hemolysis and fusion at pH 5.2. *Proc. Natl. Acad. Sci. USA* 78:4133-4137.

23. Young, J.D.-E., Young, G. P. H., Cohn, Z. A., and Lenard, J. (1983). Interaction of enveloped viruses with planar bilayer membranes: Observation on Sendai, influenza, vesicular stomatitis and Semliki Forest viruses. *Virology* 128:186-194.

24. Bachi, T., Deas, J. E., and Howe, C. (1977). Virus-erythrocyte membrane interaction. In: *Virus-Infection and the Cell Surface* (G. Poste and G. L. Nicholson, eds.). Cell Surface Reviews, Vol. 2. Elsevier, North Holland, Amsterdam, pp. 83-127.

25. Helenius, A. (1984). Semliki Forest virus penetration from endosomes: A morphological study. *Biol. Cell* 51:181-185.

26. Matlin, K., Reggio, H., Helenius, A., and Simons, K. (1981). The infective entry of influenza virus into MDCK cells. *J. Cell Biol.* 91:601-613.

27. Matlin, K., Reggio, H., Simons, K., and Helenius, A. (1982). The pathway of vesicular stomatitis virus leading to infection. *J. Mol. Biol.* 156:609-631.

28. White, J., Kartenbeck, J., and Helenius, A.(1980). Fusion of Semliki Forest virus with the plasma membrane can be induced by low pH. *J. Cell Biol.* 87:264-272.

29. White, J., Kielian, M., and Helenius, A. (1983). Membrane fusion proteins of enveloped animal viruses. *Q. Rev. Biophys.* 16:151-195.

30. Haywood, A. M. and Boyer, B. P. (1984). Effect of lipid composition upon fusion of liposomes with Sendai virus membranes. *Biochemistry* 23:4161-4166.

31. Haywood, A. M. and Boyer, B. P. (1985). Fusion of influenza virus with liposomes at pH 7.5. *Proc. Natl. Acad. Sci. USA* 82:4611-4615.

32. Kielian, M. and Helenius, A. (1984). Role of cholesterol in fusion of Semliki Forest virus with membranes. *J. Virol.* 52:281-283.

33. Florkiewicz, R. Z. and Rose, J. K. (1984). A cell line expressing vesicular stomatitis virus glycoprotein fuses at low pH. *Science* 225: 721-723.

34. Kondor-Koch, C., Burke, B., and Garoff, H. (1983). Expression of Semliki Forest virus proteins from cloned complementary DNA. I. The fusion activity of the spike glycoprotein. *J. Cell Biol.* 97:644-651.

35. White, J., Helenius, A., and Gething, M. J. (1982). Haemagglutinin of influenza virus expressed from a cloned gene promotes membrane fusion. *Nature* 300:658-659.

36. Bailey, C. A., Miller, D. K., and Lenard, J. (1984). Effects of DEAE-dextran on infection and hemolysis by VSV. Evidence that nonspecific electrostatic interactions mediate effective binding of VSV to cells. *Virology* 133:111-118.

37. Chejanovsky, N. and Loyter, A. (1985). Fusion between Sendai virus envelopes and biological membranes. The use of fluorescent probes for quantitative estimation of virus-membrane fusion. *J. Biol. Chem.* 260:7911-7918.

38. Citovsky, V. and Loyter, A. (1985). Fusion of Sendai virions or reconstituted Sendai virus envelopes with liposomes or erythrocyte membranes lacking virus receptors. *J. Biol. Chem.* 260:12072-12077.

39. Harmsen, M. C., Wilschut, J., Scherphof, G., Hulstaert, C., and Hoekstra, D. (1985). Reconstitution and fusogenic properties of Sendai virus envelopes. *Eur. J. Biochem.* 149:591-599.

40. Hsu, M.-C., Scheid, A., and Choppin, P. (1979). Reconstruction of membranes with individual paramyxovirus glycoproteins and phospholipids in cholate solution. *Virology* 95:476-491.

41. Inoue, J., Nojima, S., and Inoue, K. (1985). The activity of membranes reconstituted from HVJ envelope protein and lipids to induce hemolysis and fusion between liposomes and erythrocytes. *Biochim. Biophys. Acta* 816:321-331.

42. Kruse, C. A., Wisnieski, B. J., and Popjak, G. (1984). Characterization of a glycoprotein fusogen isolated from Sendai virus. *Biochim. Biophys. Acta* 797:40-50.

43. Marsh, M., Bolzau, E., and Helenius, A. (1983). Interactions of Semliki Forest virus spike glycoprotein rosettes and vesicles with cultured cells. *J. Cell Biol.* 96:455-461.

44. Nakanishi, M., Uchida, T., Kim, J., and Okada, Y. (1982). Glycoproteins of Sendai virus (HVJ) have a critical ratio for fusion between virus envelopes and cell membranes. *Exp. Cell Res.* 142:95-101.

45. Nussbaum, O., Zakai, N., and Loyter, A. (1984). Membrane-bound antiviral antibodies as receptors for Sendai virions in receptor-depleted erythrocytes. *Virology* 138:185-196.

46. Ozawa, M., Asano, A., and Okada, Y. (1979). Biological activities of glycoproteins of HVJ (Sendai virus) studied by reconstitution of hybrid envelope and by concanavalin A-mediated binding: A new function of HANA protein and structural requirement for F protein hemolysis. *Virology* 99:197-202.

47. Hsu, M.-C., Scheid, A., and Choppin, P. (1982). Enhancement of membrane-fusing activity of Sendai virus by exposure of the virus to basic pH is correlated with a conformational change in the fusion protein. *Proc. Natl. Acad. Sci. USA* 79:5862-5866.

48. Markwell, M. A., Portner, A., and Schwartz, A. L. (1985). An alternative route of infection for viruses: Entry by means of the asialoglyco-

protein receptor of a Sendai virus mutant lacking its attachment protein. *Proc. Natl. Acad. Sci. USA* 82:978-982.

49. Gibson, S., Bundo-Morita, K., Portner, A., and Lenard, J. (1988). Fusion of a Sendai mutant deficient in HN protein (ts 271) with cardiolipin liposomes. *Virology* 163:226-229.

50. Miura, N., Uchida, T., Kim, J., and Okada, Y. (1982). HVJ (Sendai virus)-induced envelope fusion and cell fusion blocked by anti-HVJ protein antibody that does not inhibit hemagglutination activity of HVJ. *Exp. Cell Res.* 141:409-420.

51. Bishayee, S. M., Strand, M., and August, J. T. (1978). Cellular membrane receptors for virus envelope glycoprotein: Properties of the binding reaction and influence of different reagents on the substrate and the receptors. *Arch. Biochem. Biophys.* 189:161-171.

52. Delarco, J., and Todaro, G. J. (1976). Membrane receptors for murine leukemia viruses: Characterization using the purified viral envelope glycoprotein, gp71. *Cell* 8:365-371.

53. Lenard, J. and Miller, D. K. (1983). Entry of enveloped viruses into cells. In: *Receptor-Mediated Endocytosis, Receptors and Recognition*, Series B, Vol. 15 (P. Cuatrecasas and T. Roth, eds.). Chapman & Hall, London, pp. 121-138.

54. Marsh, M. (1984). The entry of enveloped viruses into cells by endocytosis. *Biochem. J.* 218:1-10.

55. Lenard, J. and Miller, D. K. (1981). pH-dependent hemolysis by influenza, Semliki Forest virus and Sendai virus. *Virology* 110:479-482.

56. Doms, R. W., Gething, M.-J., Henneberry, J., White, J., and Helenius, A. (1986). Variant influenza virus hemagglutinin that induces fusion at elevated pH. *J. Virol.* 57:603-613.

57. Gething, M.-J., Doms, R. W., York, D., and White, J. (1986). Studies on the mechanism of membrane fusion: Site-specific mutagenesis of the hemagglutinin of influenza virus. *J. Cell Biol.* 102:11-23.

58. Lenard, J., Bailey, C., and Miller, D. (1982). pH dependence of influenza virus-induced hemolysis is determined by the haemagglutinin gene. *J. Gen. Virol.* 62:353-355.

59. Huang, R. T. C., Rott, R., Wahn, K., Klenk, H.-D., and Kohama, T. (1980). The function of the neuraminidase in membrane fusion induced by myxoviruses. *Virology* 107:313-319.

60. Huang, R. T. C., Wahn, K., Klenk, H.-D., and Rott, R. (1980). Fusion between cell membranes and liposomes containing the glycoproteins of influenza virus. *Virology* 104:294-302.

61. Ruigrok, R. W. H., Martin, S. R., Wharton, S. A., Skehel, J. J., Bayley, P. M., and Wiley, D. C. (1986). Conformational changes in the hemagglutinin of influenza virus which accompany heat-induced fusion of virus with liposomes. *Virology* 155:484-497.

62. Nagai, Y., Hamaguchi, M., Toyoda, T., and Yoshida, T. (1983). The uncoating of paramyxoviruses may not require a low pH mediated step. *Virology* 130:263-268.

63. Tycko, B. and Maxfield, F. R. (1982). Rapid acidification of endocytic vesicles containing alpha-2-macroglobulin. *Cell* 28:643-651.

64. Marsh, M., Bolzau, E., and Helenius, A. (1983). Penetration of Semliki Forest virus from acidic prelysosomal organelles. *Cell* 32:931-940.

65. Portis, J. L., McAtee, F. J., and Evans, L. H. (1985). Infectious entry of murine retroviruses into mouse cells: Evidence of a post-adsorption step inhibited by acidic pH. *J. Virol.* 55:806-812.

66. Redmond, S., Peters, G., and Dickson, C. (1984). Mouse mammary tumous virus can mediate cell fusion at reduced pH. *Virology* 133:393-402.

67. Andersen, K. B. and Nexo, B. A. (1983). Entry of murine retroviruses into mouse fibroblasts. *Virology* 125:85-98.

68. Houk, E. J., Kramer, L. D., Hardy, J. L., and Chiles, R. E. (1985). Western equine encephalomyelitis virus: In vivo infection and morphogenesis in mosquito mesenteronal epithelial cells. *Virus Res.* 2:123-138.

69. Fuller, S., Von Bonsdorff, C.-H., and Simons, K. (1984). Vesicular stomatitis virus infects and matures only through the basolateral surface of the polarized epithelial cell line, MDCK. *Cell* 38:65-77.

70. Fuller, S., Von Bonsdorff, C.-H., Simons, K. (1985). Cell surface hemagglutinin can mediate infection by other animal viruses. *EMBO J.* 4:2475-2485.

71. Miyamoto, K. and Gilden, R. V. (1971). Electron microscopic studies of tumor viruses. I. Entry of murine leukemia virus into mouse embryo fibroblasts. *J. Virol.* 7:395-406.

72. Notter, M. F. D., Leary, J. F., and Baldusi, P. C. (1982). Absorption of Rous sarcoma virus to genetically susceptible and resistant chicken cells studied by laser flow cytometry. *J. Virol.* 41:958-964.

73. Griffin, J. A., Basak, S., and Compans, R. W. (1983). Effects of hexose starvation and the role of sialic acid in influenza virus release. *Virology* 125:324-334.

74. Mifune, K., Ohuchi, M., and Mannin, K. (1982). Hemolysis and cell fusion by rhabdoviruses. *FEBS Lett.* 137:293-297.

75. Wilson, I. A., Skehel, J. J., and Wiley, D. C. (1981). Structure of the haemagglutinin membrane glycoprotein of influenza virus at 3 Å resolution. *Nature* 289:366-373.

76. Porter, A. G., Barber, C., Carey, N. H., Hallwell, R. A., Threlfall, G., and Emtage, J. S. (1979). Complete nucleotide sequence of an influenza virus haemagglutinin gene from cloned DNA. *Nature* 282:471-477.

77. Varghese, J. N., Laver, W. G., Colman, P. M. (1983). Structure of the influenza virus glycoprotein antigen neuraminidase at 2.9 Å resolution. *Nature* 303:35-40.

78. Fields, S., Winter, G., and Brownlee, G. G. (1981). Structures of the neuraminidase gene in human influenza virus A/PR/8/34. *Nature* 290:213-217.

79. Blumberg, B. M., Giorgi, C., Rose, K., and Kolakofsky, D. (1985). Sequence determination of the Sendai virus fusion protein gene. *J. Gen. Virol.* 66:317-333.

80. Shioda, T., Iwasaki, K., and Shibuta, H. (1986). Determination of the complete nucleotide sequence of the Sendai virus genome RNA and the predicted amino acid sequences of the F, HN and L proteins. *Nucleic Acids Res.* 14:1545-1563.

81. Blumberg, B. M., Giorgi, C., Roux, L., Rajn, R., Dowling, P., Chollet, A., and Kolakofsky, D. (1985). Sequence determination of the Sendai virus HN gene and its comparison to the influenza virus glycoprotein. *Cell* 41:269-278.

82. Garoff, H., Frischauf, A.-M., Simons, K., Lehrach, H., and Delius, H. (1980). Nucleotide sequence of cDNA coding for Semliki Forest Virus membrane glycoprotein. *Nature* 288:236-241.

83. Rose, J. K., and Gallione, C. J. (1981). Nucleotide sequence of the mRNAs encoding the vesicular stomatitis virus G and M proteins

determined from cDNA clones containing the complete coding regions. *J. Virol.* 39:519-528.

84. Redmond, S. M. S. and Dickson, C. (1983). Sequence and expression of the mouse mammary tumour virus *env* gene. *Embo. J.* 2:125-131.

85. Chejanovsky, N., Eytan, G. D., and Loyter, A. (1984). Fusion between Sendai virus envelopes and biological membranes as monitored by energy transfer methods. *FEBS Lett.* 174:304-309.

86. Lonberg-Holm, K. and Philipson, L. (1974). Early interaction between animal virus and cells. *Monogr. Virol.* 9:1-149.

87. Ozawa, M. and Asano, A. (1981). The preparation of cell fusion inducing proteoliposomes from purified glycoproteins of HVJ (Sendai virus) and chemically defined lipids. *J. Biol. Chem.* 256:5954-5956.

88. Uchida, T., Kim, J., Yamaizumi, M., Miyake, Y., and Okada, Y. (1979). Reconstitution of lipid vesicles associated with HVJ (Sendai virus) spikes. *J. Cell Biol.* 80:10-20.

89. Vainstein, A., Hershkovitz, M., Israel, S., Rabin, S., and Loyter, A. (1984). A new method for reconstitution of highly fusogenic Sendai virus envelopes. *Biochim. Biophys. Acta* 773:181-188.

14

Fusion of Enveloped Viruses
From Microscopic Observation to Kinetic Simulation

DICK HOEKSTRA

University of Groningen, Groningen, The Netherlands

I. INTRODUCTION

As exemplified by numerous chapters in this volume, many biological
processes involve membrane fusion. Although the physiological results
mediated by these fusion events are largely understood, the molecular
mechanism(s) by which the fusion reaction itself occurs is less clear.
Yet, substantial insight has been gained, particularly from studies involving
the use of artificial model membranes, such as liposomes. In many respects,
liposomes provide a useful tool to define *fundamental* features of a membrane
fusion reaction; i.e., studies involving these systems readily allow one
to recognize and appreciate certain trends regarding parameters affecting
intermembrane interactions, as occurs prior to and during membrane fusion.
Thus, the results of such studies should be considered as a guide rather
than facts concerning the molecular details underlying the mechanism of
fusion of the much more complex biological membranes. The complexity
of biological membranes, on the one hand, and, on the other, the realization
that insight into the mechanism of biological membrane fusion is only to
be obtained when studying fusion in a biological membrane system per
se has initiated numerous studies using relatively simple biological membranes
such as those derived from erythrocytes ("ghosts") or viruses. The latter,
in particular, have become a very attractive and popular model, as membrane-
bounded or enveloped viruses, which exploit their membrane fusion capacity
to deliver their genomes into host cells for replication, usually have a
relatively simple membrane lipid and protein composition (1,2). The lipids
are derived from the host cell plasma membrane on which they grow, while
some viruses may contain as few as one or two (virus-specific) membrane
glycoproteins.

Having selected and defined the experimental system, the next, obvious
question is: how does one determine whether or not fusion occurs? In
this regard, many biochemical and biophysical techniques have been employed. However, based on work involving fusion of liposomes, it has
become apparent that quantitative assays which allow a continuous monitoring

of the fusion reaction are most advantageous, as they provide the possibility
to distinguish whether certain physicochemical alterations of fusing membranes
result from or actually cause fusion (cf. Chapters 4 and 5).

In this chapter, the various techniques that are frequently used to
register or monitor the fusion of enveloped viruses are described. In particular, the application of fluorescence assays is emphasized as it allows one
to monitor the membrane merging process in a sensitive and, above all,
continuous manner. Rather than presenting a detailed overview of our
work on virus-target membrane fusion, a brief account is given of some
early events occurring during virus-membrane interaction. These initial
stages of the interaction process, although still fairly ill defined, are of
obvious importance to clarify the mechanism of virus fusion activity. In
conjunction with experimental approaches based on fluorescence, a theoretical
model has been developed, simulating the overall fusion reaction. With
this model the separate rate constants of aggregation and fusion can be
determined, the significance of which is discussed.

II. METHODS TO DETECT VIRUS FUSION ACTIVITY

The procedures used to determine viral fusion activity can roughly be
divided into two categories. One type of approach is to use indirect techniques such as hemolysis (3,4), infectivity (5,6), or cell-cell fusion (7,8).
Alternatively, direct methods are also available and are commonly based
on the use of radioisotopes (5,9), electron spin probes (10,11) or, more
recently, fluorescent labels (12-18). The most distinct difference between
these types of approaches is that indirect methods rely on events that
are secondary to the actual virus-membrane fusion event. By contrast,
the direct procedures involve the use of "fusion-reporting" molecules,
which are located in the viral membrane, or, when liposomes are used,
they can be incorporated in the target membrane as well. Upon fusion,
the environment of the probe changes, which can be revealed by chemical
or physical procedures. Clearly, each assay has its advantages and disadvantages. These and the various elements involved in the assays are
summarized below.

A. Membrane Insertion of Viral Antigens and Release of
Intracellular Compounds

The various stages that can be discerned in the interaction between viruses
and cells depend on the incubation temperature. At temperatures below
approximately 10°C, virions avidly bind to cell surface receptors but fusion
between viral and target membrane does not occur (Fig. 1a). When added
at 37°C, the viral particles attach to the cell surface, which is rapidly
followed either by fusion with the plasma membrane or by internalization
via receptor-mediated endocytosis followed by fusion with the endosomal
membrane "from within." Viruses, such as Sendai and Newcastle disease
virus, which belong to the family of paramyxoviruses, but also measles
and mumps virions, fuse directly with the plasma membrane at neutral pH.
However, most other viruses (cf. Chapter 13) enter the cell via endocytosis
and fusion is triggered at a mildly acidic pH, generated in the endosomal
compartment (cf. Chapter 15). The ability of a virus to fuse with a cellular
membrane can be elegantly demonstrated by electron microscopy (19).
Upon fusion, the viral antigens diffuse laterally from the side of initial

interaction and randomize with the target membrane components. As shown
in Figure 1, this randomization can be visualized, using ferritin-labeled
viral antibodies. Similarly, electron microscopic studies also provide a
possibility to examine, at the molecular level, alterations of the target
membrane that *might* be related to fusion, most notably the occurrence
of a redistribution of intramembrane particles (7,8). In particular, erythrocytes, being devoid of intracellular membranes, serve as a valuable biological
model to investigate the molecular properties of the mechanism of virus-
membrane interactions (3,10,11,19-24). In contrast to nucleated cells,
red blood cells are experimentally much more readily accessible. In addition,
the erythrocyte membrane is affected by the interaction with viruses in
such a manner that hemoglobin is released during such an interaction (3,4;
see Fig. 1c). Therefore, the release of hemoglobin is frequently used
as a measure of viral fusion activity. With paramyxoviruses optimal release
is seen in the neutral pH range, whereas orthomyxoviruses and togaviruses—
both fusing at mild acidic pH—induce the release in the mildly acidic pH
range.

  Nucleated cells also display permeability changes in their plasma membranes when interacting, at appropriate conditions, with viruses, although
leakage is usually restricted to ions such as K^+ and Na^+ and low-molecular-
weight compounds (16,25). Obviously, monitoring the release of these
compounds to register fusion is much more tedious than the release of
hemoglobin from erythrocytes. However, the release of intracellular molecules
depends on (1) the susceptibility of nucleated cells for the virus (26),
(2) the amount of virus added to nucleated cells or erythrocytes, and
(3) the viral membrane structure. The latter parameter is affected by
the age of the virus (27) and also depends on whether the viruses have
been stored frozen, i.e., have undergone a freeze-thawing cycle before
use (28). For example, Sendai virus, grown in embryonated eggs and
harvested 24 h after infection, does fuse with erythrocytes without the
induction of significant hemolysis. Hemolysis is observed when such virions
undergo a freeze and thawing cycle or when viruses are used that have
been harvested 72 h after infection (cf. Chapter 16). These observations
signify that a lack of hemolytic activity per se does not necessarily imply
a lack of fusion activity. Furthermore, it is thought that hemoglobin, ions,
and metabolites escape from erythrocytes or nucleated cells *after* integration
of the "damaged" viral membrane into the target membrane (29,30). A
distinct difference in the permeability properties of "early" vs. "late"
harvest viruses could be demonstrated by showing that penetration of
a negative staining dye, uranylacetate, occurs into the latter but not the
former particles (30). The evidence thus obtained reveals the following
sequence of events: After attachment the virus fuses with the target
membrane, and subsequently, ions or metabolites (in case of nucleated
cells) or hemoglobin (in case of erythrocytes) may be released, depending
on the virus's history, into the extracellular environment. Evidently, the
process of release is *secondary* to the actual fusion step.

B. Virus-Induced Cell-Cell Fusion

A typical manifestation of viruses, when incubated with cultured cells
or erythrocytes at 4°C, represents their ability to cause cell-cell agglutination. When the temperature is subsequently raised to 37°C, virus-induced
cell-cell fusion takes place (Fig. 2) provided that the incubation pH corresponds to that necessary for expression of the viral fusion activity. This

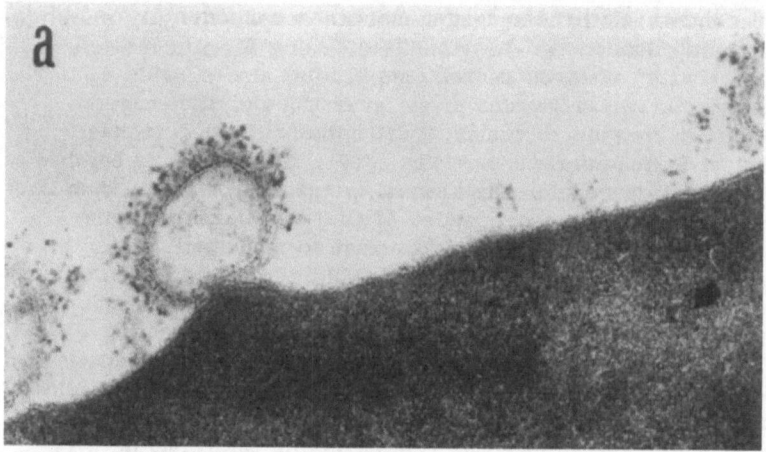

FIGURE 1 Interaction of Sendai virus with human erythrocytes, visualized
by electron microscopy. (a) Bound Sendai virions are seen after an incubation of the virus and erythrocytes at 2-4°C for 30 min. The sample was
stained with ferritin-labeled antiviral antibody. After the temperature was
raised to 37°C for 10 min (b), ferritin-labeled antibodies can be seen that
react with viral antigens, after the latters' insertion into the erythrocyte
membrane as a result of fusion (arrow). Another virus is still bound to
the cell surface. After another 30 min viral antigens are dispersed over
the entire surface (c), reflecting the randomization of cell membrane and
viral membrane components. The occurrence of hemolysis is indicated by
the reduced electron density of the cytoplasm in b (cf. to a) and the
complete absence of hemoglobin in c. The electron-dense material still seen
in c represents remnants of viral nucleoprotein, attached to the primary
site of virus-cell interaction. (From Ref. 30, with permission.)

type of fusion, also called "fusion from without," is triggered by a relatively
large exogenous viral dose, without the need for viral replication. Cell
fusion can also be induced "from within" and is initiated at a later stage
of the virus replicative cycle following infection of cells with a relatively
low dose of virus (31). In essence, the occurrence of cell-cell fusion,
commonly determined by light microscopic techniques, merely reflects a
certain property of the virus rather than providing substantial insight
into the mechanism of virus fusion per se. The formation of polykaryons
becomes apparent only as a result of osmotically driven cell swelling (7),
a secondary process taking place after an initial local membrane fusion
event of apposed cell membranes (see also Chapter 7). Hence, the kinetics
of both events will differ, which is essential to bear in mind when attempting
to relate certain membrane alterations as being the cause rather than the
result of membrane fusion (see below).

   In this regard it is interesting to note that nonhemolytic virus preparations, although capable of inducing membrane fusion, do not seem to induce
extensive cell fusion. However, this is only misleading, as cytoplasmic
connections between adjacent cells can be seen but, because of the lack
of osmotic expansion of these sites, light microscopy leads thus to an
erroneous interpretation. When such treated cells are subsequently exposed
to hypotonic media, polykaryon formation becomes readily apparent (31,32).

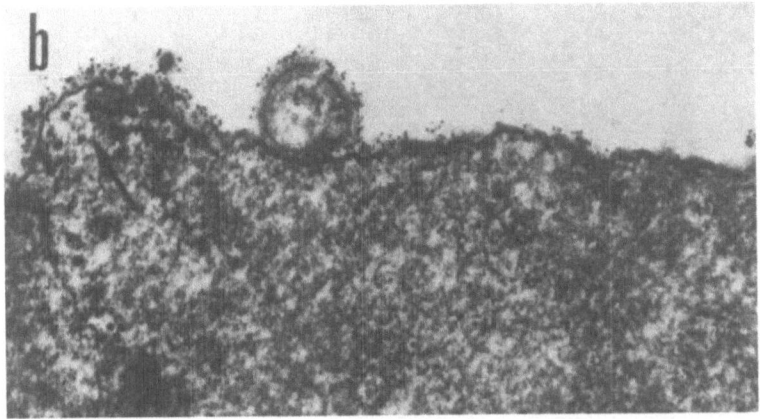

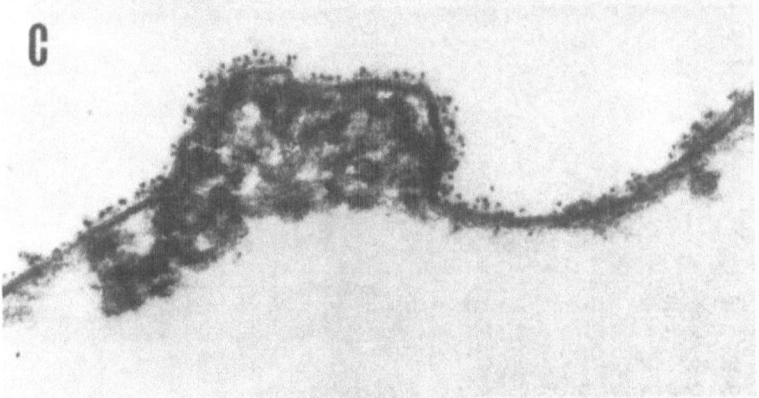

C. Virus Fusion and Infectivity

Obviously, enveloped viruses fuse with plasma or endosomal membranes
to introduce their nucleocapsid into the cytoplasm for replication, an
essential step in viral survival. Hence viral infection of the cells indicates
that membrane fusion must have taken place. As such, infectivity may serve
as a useful parameter to characterize virus-membrane fusion in a qualitative
manner, for example to determine the route of entry (5,33). Yet, assaying
infectivity is not necessarily a direct reflection of the fusion between viral
and cellular membranes per se. Significant differences in virus replication
in cells at different stages of the cell cycle and between actively replicating
and nonreplicating cells have been described for a wide range of host
cells in vitro (34).

The procedures described thus far all reflect a means to determine
in an indirect and qualitative manner, i.e., according to an "all-or-nothing"
approach, whether or not fusion between a virus and a cellular membrane
has occurred. To put it differently, they do not provide an accurate
insight into such relevant questions as numbers of virus particles that
bind to the cell surface vs. those that actually fuse and the kinetics of
these processes in relation to the kinetics of endocytosis, the latter thought
to mediate the entry of many families of enveloped viruses. Answers to
such questions are pertinent in order to obtain insight at the molecular

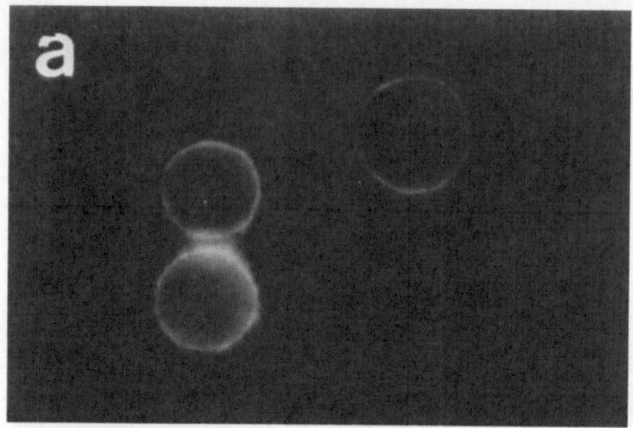

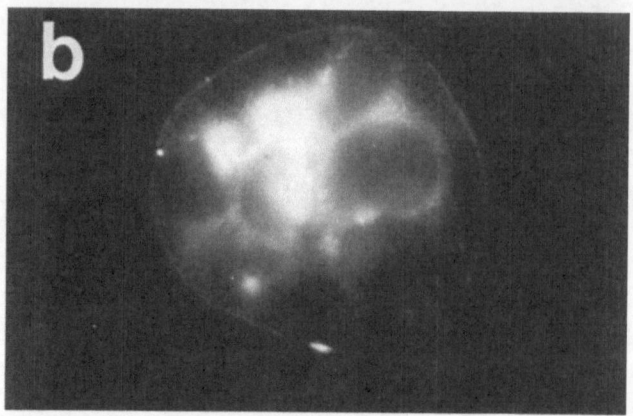

FIGURE 2   Virus-induced cell-cell fusion. Chinese hamster V79 fibroblasts
were incubated with Sendai virus on ice. After 5 min the incubation was
continued at 37°C. After another 3 min the sample was diluted 10-fold
and subsequently incubated for 7 min at 37°C. Prior to addition of virus,
the plasma membrane was labeled with the fluorescent lipid analog C_6-NBD-
PC (a). After fusion (b) the cells were examined in a fluorescence microscope. Note the presence of at least six nuclei in the fused cell. For details
see Ref. 68.

level of the mechanism of fusion. This desire has been recognized in recent
years, and numerous studies have been published in which the kinetics
of the early interactions between viruses and target membranes have been
investigated using either radioactive markers, electron spin probes, or
fluorophores.

D.   Use of Radiolabels to Monitor Virus-Cell Fusion

Virions can be readily labeled with radioactive markers by growing them
in embryonated eggs or cultured cells in the presence of radioactive amino
acids or nucleic acids, such as ^{35}S-methionine or ^3H-uridine (5,9). Alternatively, it is also possible to label the particles exogenously with ^{125}I,
which results primarily in the labeling of the viral membrane glycoproteins

and internal proteins (35). Parameters such as the fraction of total virions that binds to the cells and the kinetics of this binding process can thus be determined by carrying out incubations at low temperature and removal of nonbound particles by washing the cells. Raising the temperature to 37°C will initiate the fusion reaction. It is then necessary to distinguish, as a function of time, between the number of viruses that remain bound to the cell surface and those that actually fuse or, depending on the virus family, those that are internalized by endocytosis. This distinction can be made by treating the cells after the appropriate incubation time with proteolytic enzymes, such as proteinase K (cf. Fig. 3), as reported, for example, for Semliki Forest virus (SFV) (5,36), or by a rather harsh treatment with the reducing agent dithiothreitol (DTT), as reported for Sendai virus (37). Alternatively, it has also been shown that in systems in which Sendai virus was incubated with liposomes as target membranes, bound viruses can be distinguished from fused viruses by fractionation on sucrose gradients (see Chapter 16). Furthermore, as viral binding is usually mediated by attachment to cell surface sialic acid residues, neuraminidase treatment may also be effective in removing bound, as opposed to fused, viruses (38,39). Yet, the exact nature of viral binding per se is not entirely clear. Removal of nonbound viruses, after allowing the binding to take place at 4°C, does not necessarily imply that the remaining fraction is irreversibly bound. Subsequent incubation at 37°C may also lead to a partial release of bound virions which may (39) or may not (40)

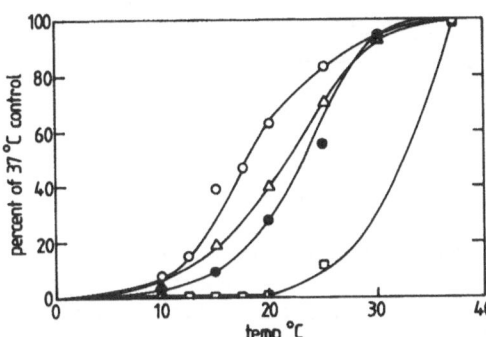

FIGURE 3 Internalization, uncoating, infection, and degradation of Semliki Forest virus (SFV) in BHK-21 cells as a function of temperature. ^{35}S-methionine-labeled SFV were incubated with BHK cells for 1 h at the indicated temperatures. The amount of internalized virus (○) was determined by measuring the fraction of virus that could not be removed from the cell surface by proteinase K treatment. Degradation (□) in the lysosomal compartment was estimated from the fraction of TCA-soluble radioactivity. Uncoating (△) was assayed using ^{3}H-uridine SFV. After interaction with virus, the cells were lysed and treated with RNase, which causes degradation of the intracellularly released viral nucleocapsid. Infection (●) was measured by ^{3}H-uridine incorporation. For experimental details see Ref. 5. The results show that at temperatures at which the delivery of viruses to the secondary lysosomes is blocked (below 20°C, □), internalization (○), uncoating (△), and infection (●) can take place, indicating that SFV nucleocapsids are released into the cytoplasm from a prelysosomal compartment. This compartment has been identified as the endosomes. (From Ref. 5, with permission.)

be due to viral neuraminidase activity. The thus remaining bound fraction
may not be equally susceptible to proteolytic enzymes or exogenous neuraminidase, and it is therefore not clear whether this entire fraction can
be effectively removed. If not, fusion would be overestimated. In this
context, it is pertinent to note that in a study in which the interaction
of liposomes with cells infected with influenza virus and expressing the
viral hemagglutinin at the cell surface was studied, a bound fraction of
liposomes could be identified that had not yet fused but neither could
be removed by enzymatic treatments (39).

When studying the interaction of nucleic acid-labeled virions with cells
in culture, a possibility is provided to unambiguously demonstrate the
intracellular delivery of the nucleocapsid. In addition, by comparing the
kinetics of delivery vs. those of endocytosis at temperature-controlled
conditions, it has been elegantly demonstrated (Fig. 3) that the penetration
of SFV takes place at the level of the endosomes rather than the lysosomes
(5).

E.   Application of Electron Spin Probes

The introduction of the nucleocapsid into the intracellular environment
and hence fusion can also be revealed by loading virions with the electron
spin probe tempocholine (41). Incorporation is accomplished by simply
incubating the virus with the probe for several hours, and it was suggested
that it associates preferably with the nucleocapsid. At the probe concentration applied, the electron spin resonance (ESR) spectrum shows an exchange-
broadened signal. When the labeled viruses fuse with the plasma membrane,
the broadened signal is converted to a sharp signal with a concomitant
increase in peak height, due to the release and dilution of the tempocholine
from the preloaded virus into the cytoplasm. Spin-label electron spin resonance techniques have also been used to monitor the mixing of viral and
cell membrane lipids, an obvious consequence of membrane fusion. These
procedures (10,42) entail the insertion of spin-labeled phospholipids into
the viral membrane at relatively high probe concentrations (10-20 mol%
with respect to total lipid). As a result an ESR spectrum is obtained that
is characterized by strong spin-spin exchange interactions, similarly as
observed for tempocholine-labeled virions. The ESR peak height increases
upon fusion of the labeled virus with nonlabeled target membranes, and
from the net increase in peak height the relative contribution of fusion
can be assessed. With lower spin probe concentrations, an alternative
procedure can be followed which seems to be more sensitive and, therefore,
reflects more accurately the kinetics and molecular characteristics of virus-
target membrane fusion. This approach relies on differences in the ESR
spectrum as a function of bilayer fluidity (11). The membrane fluidity
of erythrocytes and viruses is different, the viral membrane being more
rigid (43). Upon fusion between the labeled virus and an erythrocyte,
the probe will thus sense a different environment and its spectral characteristics, i.e., before vs. after fusion, will change. These changes can
be quantitatively related to the fractions of spin-labeled virions that bind
and fuse with the erythrocyte membrane.

One of the major problems in the quantitation of membrane fusion by
lipid dilution is that intermembrane transfer of individual probe molecules
may occur by a "nonfusion" mechanism involving either a transfer via
a collision-mediated mechanism or a spontaneous transfer of free monomers
through the aqueous phase. In this regard, the proper choice of spin-

labeled lipid is important. Spin labels, such as the nitroxide probe, should preferably be linked to the glycerol backbone of the phospholipid derivative by a relatively long carbon spacer, as the shorter the spacer, the more readily the probe may transfer between labeled and nonlabeled membranes, analogous to monomeric transfer seen for fluorescently labeled phospholipid (44) or fatty-acid derivatives (45). If such precautions are taken, a substantial contribution of lipid exchange to the fusion event can be rigorously excluded on the time scale of fusion, and reliable kinetic and quantitative insight in virus-target membrane fusion can be obtained. Yet, electron spin resonance techniques cannot be considered routine laboratory procedures while the analysis of the data can be quite cumbersome and sometimes difficult to interpret.

III. APPLICATION OF FLUORESCENCE ASSAYS

Although ESR techniques provide the possibility to determine the early stages of viral fusion, a convenient and continuous monitoring of the fusion reaction is not possible. Fluorescence assays appear to provide an attractive alternative in this respect, as may be inferred from the rapid advancement in the development of numerous such assays, allowing the measurement of fusion in both artificial and biological membrane systems (Fig. 4). These assays offer a number of advantages over other techniques discussed thus far. Among others, these advantages include (1) the sensitivity, (2) the ability of most of these assays to monitor continuously the kinetics of fusion,

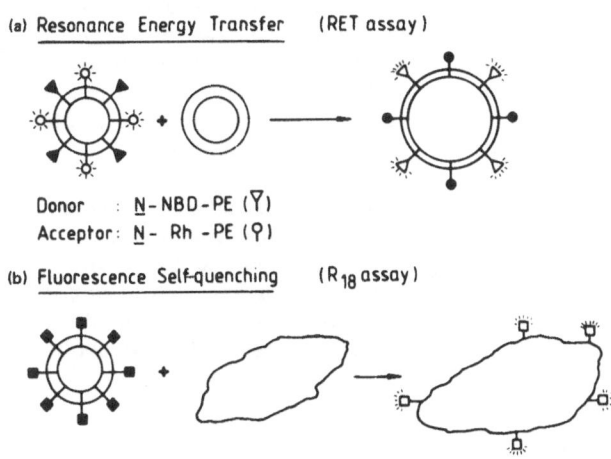

(a) Resonance Energy Transfer   (RET assay)

Donor : $\underline{N}$-NBD-PE (Y)
Acceptor: $\underline{N}$- Rh -PE (9)

(b) Fluorescence Self-quenching   (R_{18} assay)

Probe: Octadecyl rhodamine B chloride (R_{18})

FIGURE 4 Schematic representation of the principles of fluorescence assays used to monitor the fusion of viruses with artificial and biological membranes. (a) Liposomes are labeled with fluorescent donor and acceptor lipid analogs. Upon fusion with (nonlabeled) virions the surface density of the probes will decrease, causing an increase in donor fluorescence, which is monitored as a measure of fusion. (b) A virus is labeled with a self-quenching concentration of R_{18}. Upon fusion the self-quenching is relieved and the increase in fluorescence is monitored as a measure of fusion. For details see Refs. 12 and 51.

(3) the relative ease by which quantitative data can be obtained, (4) the possibility of visualizing the interaction of viruses with cells by fluorescence microscopy and hence the ability to determine the intracellular sites to which molecules move during or after fusion, and (5) the convenient accessibility of fluorescence techniques. However, when lipid mixing is monitored, using fluorescently labeled lipid derivatives, these techniques share the same disadvantage as noted above for spin-labeled lipids, i.e., the possibility of lipid exchange, which requires inclusion of rigorous control experiments.

The application of fluorescent probes in studies concerning virus-target membrane interaction offers various possibilities. Fusion can be followed, for example, by means of fluorescent labeling of the viral proteins (17,46) or by monitoring the mixing of membrane lipids after insertion of fluorescently tagged lipid derivatives into the native viral membrane (12,16; cf. Fig. 4) or in their reconstituted envelopes (14,15,47). With artificial membranes the fluorescent lipids can also be inserted into the target membranes (13,48; cf. Fig. 4), but successful application of this approach will very much depend on the efficiency by which the virus fuses, which, in turn, may depend on the target membrane composition (49). Finally, in systems consisting of reconstituted viral envelopes and liposomes it is also possible to monitor the fusion by following the mixing of aqueous contents, similarly as in pure liposomal systems (cf. Chapter 17).

A. Fluorescently Tagged Viral Proteins

When Sendai virus is incubated with fluorescein-isothiocyanate, fluorescently labeled virions are obtained in which more than 80% of the fluorophore appears to be associated with the binding (HN), fusion (F), and matrix (M) protein (17). The biological activities of the virus such as hemagglutination, cell-cell fusion, and hemolysis are not affected as a result of the labeling. The fraction of viruses bound or fused at a variety of conditions can then be determined by application of fluorescence recovery after photobleaching (FRAP). This technique relies on the fact that bleaching of bound fluorescently labeled virions does not result in the recovery of fluorescence, whereas recovery does take place when fusion has occurred, because of the lateral mobility of fluorescently labeled viral membrane proteins in the target membrane. This procedure allows one to determine the distribution of viral envelope-cell fusion per cell, relative to the entire cell population, but also the fusion susceptibility over the surface of single cells (18). Besides fluorescent labeling of proteins, fluorescent lipid probes can also be applied, using this technique.

B. Use of Exogenous Fluorescent Lipid Analogs

Fluorescent lipid(-like) probes have become popular tools in monitoring membrane fusion in both artificial and biological systems, including systems involving the fusion of viral membranes. The methods (Fig. 4) are based either on the principle of resonance energy transfer (50-52) or on the relief of fluorescence self-quenching (12,15). The approach of resonance energy transfer relies on the interactions between two different fluorophores, i.e., fluorescently labeled phospholipids or other lipophilic molecules, provided the emission band of one fluorophore, the energy donor, overlaps with the excitation band of the second, the energy acceptor, and that the two probes exist in close physical proximity. When these conditions are met, the energy from a photon absorbed by the energy donor can be transferred to the energy acceptor in a radiationless manner which will then fluoresce as though it had been excited directly. Fusion is commonly monitored by following the relief of energy transfer, which is revealed as an increase in donor fluorescence upon merging of labeled membranes containing both the donor and acceptor fluorophore, with unlabeled ones.

As donor and acceptor molecules, the use of two nonexchangeable, fluorescently tagged phospholipids, N-(7-nitro-2,1,3-benzoxadiazol-4-yl)phosphatidylethanolamine (N-NBD-PE) and N-(lissamine rhodamine B sulfonyl)phosphatidylethanolamine (N-Rh-PE), respectively, has become quite popular. The application of this approach has been most profitable in studies involving the fusion of reconstituted viral envelopes (14,15), as the probes are readily incorporated into the viral membrane during reconstitution whereas direct incorporation in native viral membranes is difficult to achieve. However, fusion of native virions can be studied with this experimental approach by using liposomes, labeled with both fluorophores, as the target membranes (13,48,49). To investigate the fusion of intact viruses with biological membranes a method has been developed based on the relief of fluorescence self-quenching (12,23). In this assay, a fluorescent lipidlike probe, octadecyl rhodamine B chloride (R_{18}), is inserted into the viral membrane by addition of an ethanolic solution of the fluorophore to the viral suspension. At a sufficiently high concentration (ca. 3-4 mol% with respect to total lipid) efficient self-quenching of the fluorescence occurs. Fusion of the labeled virus with either artificial or biological target membranes will result in dilution of the probe and a concomitant relief of self-quenching (cf. Fig. 5). The increase in fluorescence thus observed is taken as a measure of fusion. Once inserted into the

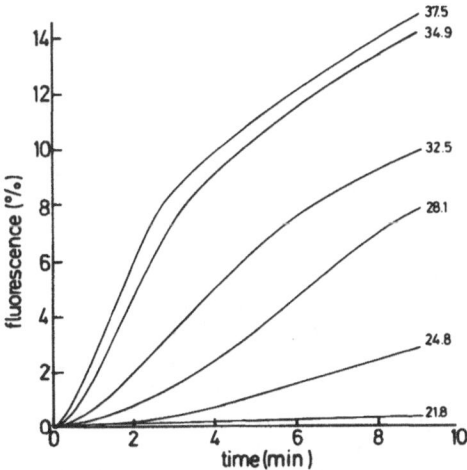

FIGURE 5 Fusion of Sendai virus with erythrocyte membranes as a function of temperature. R_{18}-labeled Sendai virus and erythrocyte ghosts were equilibrated separately at the indicated temperatures (°C). At time zero the samples were mixed in a cuvette and the relief of R_{18} fluorescence self-quenching was monitored continuously at the specified temperatures. Note that the time span of the lag phase decreases with increasing temperature. For experimental details see Ref. 23.

viral membrane the probe does not dissociate from the membranes by either
spontaneous transfer of free monomers through the aqueous phase or a
collision-mediated transfer process. By contrast, the C_{18} derivate of fluorescein redistributes spontaneously (12) when labeled and nonlabeled membranes
are mixed. This observation indicates that the choice of probe is critical
but also that interactions between the rhodamine headgroup of R_{18} and
either viral phospholipids and/or proteins may contribute to its nonexchangeable behavior after membrane insertion. This may also explain why, in
contrast to previous assumptions, the probe does not become randomly
distributed in the lateral plane of the viral bilayer upon exogenous addition.
On the other hand, the self-quenching increases linearly with concentration
(up to 9 mol%) when incorporated into liposomal bilayers during vesicle
preparation (12).

A nonrandom distribution of other fluorescent probes, when incorporated
into viral membranes, has also been reported. Exogenous insertion of the
phospholipid analog pyrene-sulfonylphosphatidylethanolamine into Sendai
virus membranes similarly results in an unexpected high degree of self-
quenching at relatively low concentrations (16). This also appears to be
the case when N-NBD-PE is incorporated into reconstituted membranes
containing the M-protein of vesicular stomatitus virus (53). The origin
of a nonrandom insertion could therefore be related to a preferential association of the probes with distinct domains in the viral membrane. When
virions are labeled with a relatively low concentration of R_{18} (ca. 1 mol%,
i.e., ca. 2 mol% with respect to the outer leaflet lipids), a partial relief
of fluorescence quenching is observed upon treatment with reagents that
modify the protein structure, such as DTT and trypsin (K. Klappe and
D. Hoekstra, unpublished observations). In phosphatidylcholine vesicles,
this amount of probe (2 mol%) results in ca. 20% fluorescence quenching.
These observations suggest that viral proteins may affect the lateral distribution of the probe. Moreover, such interactions could also explain,
at least in part, the tight interaction displayed by R_{18} after its insertion
into biological membranes, as opposed to the membrane-association behavior
of the fluorescein-labeled C_{18} chain analog (see above).

In addition to an interaction with proteins, the mechanism of insertion
may as well contribute to an artificial enhancement of the fluorescence
self-quenching. Exogenous labeling of phospholipid vesicles causes a higher
self-quenching than occurs when the fluorophore is incorporated during
vesicle preparation, indicating that the nonrandom distribution of the
fluorescent probes is not exclusively a result of a specific association
with proteins. It turns out, however, that the interactions giving rise
to an enhanced degree of self-quenching are not irreversible. Upon fusion
of R_{18}-labeled virions with both artificial and biological membranes, the
probe instantaneously and completely redistributes, resulting in a random
distribution in the newly formed, i.e. fused, membranes. For example,
when R_{18}-labeled Sendai virus (average diameter 1500 Å) and nonlabeled
liposomes (diameter 1000 Å) are mixed at a lipid ratio of 1:1, complete
merging should lead to a net increase of fluorescence of ca. 40%. Similarly,
when they are mixed at a 1:4 ratio (virus/liposomes), a final fluorescence
level of 78% is anticipated upon complete randomization of the viral and
liposomal membrane components. At certain conditions, as will be discussed
below, such efficient fusion between virions and liposomes can occur and
the level of probe dilution seen experimentally is entirely consistent with
the theoretical predictions (49,54). Moreover, preparation of mock fusion
products by extracting the virus/liposome mixtures with octylglucoside

and reconstitution of the membranes by slow dialysis revealed that the calculated probe density in the mock products and the relief of self-quenching observed experimentally were entirely consistent with the density-related self-quenching seen when constructing a calibration curve with R_{18}-labeled liposomes. Hence, within the fusion product, the distribution of the probe is random in contrast to its distribution within the viral membrane after labeling.

That the kinetics of this redistribution upon fusion are extremely fast can be inferred from experiments in which fusion between virions and liposomes has been compared, using different assays. Thus, with the resonance energy transfer assay, in which the fusion reporting probes are located in the liposomal target membrane, the kinetics of fusion are virtually identical to those seen for the R_{18} assay (12,24). Furthermore, a comparison with nonrelated assays, based on the use of monitoring fusion with isotopically labeled viral proteins (55; see also Chapter 16) or an analysis of the fusion reaction by sucrose density gradient centrifugation (49) further confirms the profitable use of this assay. In addition, a direct comparison of fusion between Sendai virus and erythrocyte membranes with three different assays that rely on relief of self-quenching (R_{18} assay), lateral mobility of viral components (FRAP assay), or chemical removal of nonfused virions (DDT treatment) demonstrated that all three assays gave virtually identical results (18). In summary, although certain aspects of the R_{18} assay are not entirely clear, it turns out that in practice the assay has proved to be a convenient and reliable tool for monitoring membrane fusion, in particular for monitoring the fusion of viruses with biological membranes. The relevance of being able to follow fusion in a quantitative and continuous manner is illustrated in the following section. Rather than presenting a detailed overview (see other chapters in this volume), some aspects specifically related to the use of kinetic, fluorescence assays in studies concerning virus-target membrane fusion are emphasized.

IV. INTERACTION OF VIRUSES WITH ARTIFICIAL AND BIOLOGICAL MEMBRANES

Membrane fusion entails a two-step reaction. The overall reaction involves initial binding of apposed membranes followed by merging of the bilayers in a subsequent reaction. With enveloped viruses, it has been well established that both processes are accomplished by one or more proteins. In case of a Sendai virus, a paramyxovirus that introduces its nucleocapsid into the cytoplasm of a cell by fusion with the plasma membrane at neutral pH, two proteins are involved in the overall reaction. One type of protein, HN, mediates the binding of the virus to sialic acid-containing glycoproteins and/or glycolipids at the cell surface, while the other, F, subsequently triggers the actual fusion reaction (56). In contrast to Sendai virus, influenza virus, an orthomyxovirus that enters the cell by endocytosis followed by fusion from within at mild acidic pH in the endosomal compartment, displays the binding and fusion capacity by virtue of only one protein, the hemagglutinin (HA). Both activities are located, however, at different positions of the HA molecule, which consists of two disulfide-linked subunits, HA_1 and HA_2, the former expressing the binding activity while the fusion activity is contained in the HA_2 polypeptide (57,58). Fusion between viruses and target membranes takes place at 37°C, irrespective of whether the virions are preattached at low temperature. A preincubation in the cold

(4°C) enhances the initial rate of fusion owing to the fact that a majority
of the virus particles is bound at the onset of the temperature increase
(23,24). Typically, prior attachment of Sendai virus to erythrocyte ghosts
results in a 1.5- to 2-fold increase in the initial fusion rate, as determined
with the R_{18} assay. However, within 10-15 min after the onset of the incubation at 37°C, the levels of fluorescence intensity are almost the same, i.e.,
irrespective of the preincubation period in the cold, and the final extents
of fusion, as established after a 20-h incubation period, are indistinguishable in both cases (40).

The overall shape of the fluorescence tracings (cf. Fig. 5) at these
different conditions is very similar; i.e., initially a gradual increase in
fluorescence is seen, followed by a relatively rapid development of fluorescence, which, in turn, slows down after 4-5 min. The persistent presence,
in both cases, of a lag phase at the early onset of the fusion reaction
is quite intriguing and demonstrates in particular the validity of applying
sensitive fluorescence assays. A priori, one would anticipate that a lag
phase, seen when virus and ghosts are mixed at 37°C (i.e., without a
preattachment at 4°C), is related to the time-dependent process of viral
adhesion to the target membrane. Apparently, this is not the case, since
this phase is still clearly discernible after preattachment, although the
time span has shortened (10-15 vs. 20-30 s). Influenza virus, capable
of fusing with erythrocyte ghosts at mild acidic pH, displays a similar
lag phase when added directly to ghosts at 37°C (24). After a lag period
of ca. 6 s the fluorescence increases rapidly, leveling off after ca. 1 min.
Preattachment results in an increase in the initial fusion rate, similarly
as observed for Sendai virus. In this case, however, a lag phase is no
longer apparent; i.e., fusion starts immediately after the temperature
is raised to 37°C. Presumably, this distinction in initial events relies on
a difference in the kinetics by which the viral proteins bring about and/or
undergo certain alterations necessary to trigger the fusion reaction (31,58;
see Chapter 15).

As noted above, in case of influenza only one protein is involved in
both binding and fusion, whereas Sendai virus requires two separate
proteins. It appears that with influenza virus, the molecular changes in
HA occur on a millisecond scale or faster. With Sendai virus, the molecular
events that convey the fusion activity to the virus seem more complex
and are likely related to the fact that two different proteins are involved,
each carrying out its own specific function. The binding capacity of Sendai
virus to biological target membranes is an exclusive property of HN; i.e.,
the function of the F protein is rendered useless upon removal of HN.
This implies that only by virtue of HN does a direct molecular interaction
between the target membrane and F become possible. The constraint thus
imposed on a joint concerted action of both proteins is apparent since
the HN molecules are located at the focal points of membrane contact and,
hence, membrane fusion. In principle, therefore, these molecules could
pose as a sterical and/or physical barrier for the F protein and in that
manner act as regulators of the membrane fusion reaction. Indicative for
such a regulatory role are experiments in which the fusion of Sendai virus
was examined as a function of temperature. At 4°C, when the virus binds
as avidly as at 37°C, no fusion takes place (40). Yet, as discussed elsewhere (23), there is no reason to assume that the F protein is in a fusion-
inactive state at those conditions. However, fusion becomes apparent only
at a threshold temperature of ca. 20°C and the initial fusion rate rapidly
increases when the temperature is raised to 37°C (Fig. 5).

Around 20°C the onset of the fusion reaction is delayed, but the lag phase becomes increasingly less with increasing temperature, ranging between ca. 8 min at 23°C and 30 s at 37°C. Interestingly, the rapid increase in the kinetics of fusion in the temperature range between 23°C and 37°C overlaps remarkably well with a substantial increase in the rotational mobility of the viral proteins, occurring within the same temperature range (59). Moreover, upon labeling the viral proteins with eosinyl-5-maleimide and determining, as a function of temperature, the extent of fluorescence self-quenching, an abrupt relief of quenching becomes apparent around 22°C (D. Hoekstra and K. Klappe, unpublished observation). This change in self-quenching would reflect a change in the mutual distances between viral membrane proteins, indicating that they become more dispersed above 22°C. As a result, their motional freedom will be enhanced. Taken together, these observations indicate that when the viral proteins are packed closely together (perhaps existing as aggregated complexes in the plane of the bilayer), the molecular rearrangements between or within the viral proteins are hampered such that protein modulations necessary to trigger the fusion reaction are prevented. Apparently, at such conditions, the binding function of Sendai virus is not impaired; i.e., irrespective of their state of inter-molecular packing, the function of HN can be expressed. Being located near the focal sites of fusion, the binding protein could thus by virtue of steric interference refrain the F protein from interaction with the target membrane and thereby impede the fusion reaction.

The profound effect of mobility constraints on the ability of viral membrane proteins to trigger fusion (59,60) may also explain the distinct differences observed between the initial kinetics of fusion of Sendai virus and their reconstituted envelopes. With the nonionic detergent octylglucoside, the viral lipids and membrane glycoprotein can be extracted and reconstituted, using a slow dialysis procedure (14). The reconstituted viral particles thus obtained are very similar to those of the native virions with respect to appearance and membrane composition. Furthermore, equal amounts of the viral proteins, HN and F, are introduced in the plasma membranes of erythrocytes (14) and various murine cell lines (unpublished observation), when the cells are incubated with either the reconstituted virions or the native virus. Yet, the initial fusion rate of the reconstituted envelopes is notably faster (ca. 1.5- to 2-fold) than that of the native virions, while the lag phase commonly seen for the intact virus is conspicuously lacking upon fusion of the envelopes (14). Also in this case, it is tempting to explain the differences in the initial rates in terms of a potential correlation between the fusion activity and the rotational mobility of the viral proteins; i.e., the fusion activity is facilitated with increasing protein mobility. Since both HN and F are transmembrane proteins (61-63), it seems possible that their mobility in the intact virus is restricted by the matrix, M, protein, located just underneath the viral membrane. In the reconstituted envelope, this protein is no longer present and hence an enhanced mobility and therefore a facilitation of fusion could be the result.

An analysis of the effect of (cell-attached) viral particle density versus the rate of fusion particularly benefitted from the ability to determine in a sensitive manner the initial fusion rates by means of fluorescence assays. It could thus be demonstrated that the initial fusion rate of the virus becomes relatively suppressed when large amounts of virions are attached to the membrane surface (40). This suggests that adjacent particles affect one another such that fusion becomes impaired. At the conditions of these experiments, the viral binding per cell was not saturated, implying that perhaps the number of fusogenic sites at a cell surface is limited. Indeed, using erythrocyte ghosts it was found that ca. 1200 Sendai viruses bind per ghost whereas ca. 150 virions can maximally fuse, despite the fact that all virions are fusion-active (40,64). The exact number of attached virions that eventually fuse is dependent on the input multiplicity (number of virus particles added per ghost or cell) and on the cell type. In conjunction with work of others (37,65) these results may imply that specific fusion sites exist on a target membrane surface, but further work is required to substantiate this possibility.

Fluorescence assays are particularly helpful in analyzing the interaction of viruses with liposomes, as it is rather difficult and tedious in this case to distinguish unequivocally between fused and nonfused particles, using assays that rely on the application of radioactive markers to detect fusion (66). Work carried out with these systems has shown that pure negatively charged bilayers, consisting of phosphatidylserine (PS) or cardiolipin (CL), are particularly amenable to fuse with various viruses (13,48,49; Fig. 6). However, with such vesicles as target membranes, the mechanism of fusion appears to deviate from that seen at conditions thought to resemble physiological conditions, i.e., with ghosts and cultured cells as target membranes. For example, in the case of fusion of influenza virus, there is no strict

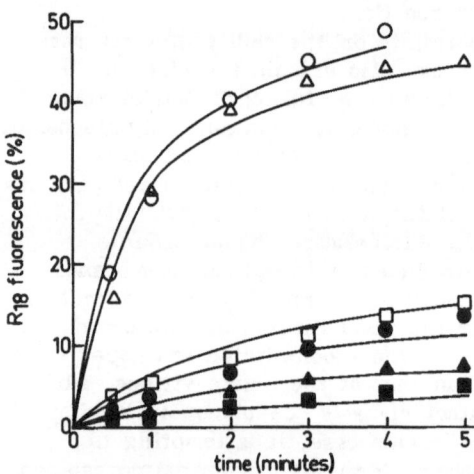

FIGURE 6 Sendai virus-liposome fusion: pH dependence and kinetic simulation. The virus was suspended in a medium of pH 7.4 (closed symbols) or 5.0 (open symbols). Fusion (at 37°C), initiated by injecting the liposomes into the medium, was continuously monitored by following R_{18} fluorescence increase. The drawn lines are the experimentally obtained fluorescence tracings, while the calculated values, obtained by kinetic simulation, are given by data points. The vesicle compositions were CL (circles); CL/DOPC, 1:1 (triangles); and PS (squares). Note that with these liposomes Sendai virus displays fusion activity at low pH, which is not observed with biological membranes. By kinetic simulation it could be shown that the liposomal composition and pH modulated the fusion reaction itself as the rate constant of fusion varied from 1 to 0.005 s^{-1}, whereas the rate constant of aggregation varied only from 1.8×10^8 to 0.7×10^8 $M^{-1} \cdot s^{-1}$. For further details see Ref. 54.

dependence on the low pH-induced conformational change in the HA molecule (24), an essential element in the fusion of these viruses at biological conditions (58; Chapter 15). With Sendai virus, the typical lag phase, seen when the virus fuses with biological target membranes, is lacking; i.e., the virus fuses immediately upon mixing with the vesicles. Furthermore, the fusion of Sendai virus with erythrocyte membranes is strongly inhibited at mild acidic pH (23). Yet, with CL and PS vesicles a strong enhancement in fusion activity is seen at those conditions (Fig. 6; Ref. 47). However, with vesicles consisting of zwitterionic phospholipids, cholesterol, and gangliosides, the latter acting as viral receptors, the lag phase shows up again, and based on a number of other, virus-specific, criteria, it turns out that with such vesicles as target membranes, the fusogenic behavior of both Sendai (49) and influenza virus (24) most closely resembles that seen with biological targets.

V. APPLICATION OF A MASS ACTION KINETIC MODEL TO ANALYZE VIRUS-MEMBRANE FUSION

In essence, many of the fusion assays, including the fluorescence assays, follow the *overall* fusion reaction, which, as noted above, involves an initial binding step of adjacent membranes and the actual fusion reaction per se. To obtain a more detailed picture of membrane merging it would therefore be desirable to analyze the fusion reaction in terms of these distinct steps. Such a possibility is provided by the application of a theoretical model (54,64,67), based on a mass action kinetic model (see Chapter 5), in conjunction with the fluorescence fusion assays. This mathematical model views the overall fusion reaction as a sequence of a second-order process of membrane adhesion, followed by the first-order fusion reaction itself. By varying the concentrations of interacting particles, the rate constants for aggregation and fusion are obtained. Thus, at high particle concentrations, the fusion reaction per se determines the overall rate, whereas at low particle concentrations the initial binding step will be essentially rate-limiting. The relevance of determining these rate constants is given by the fact that the aggregation rate constant is a reflection of the balance of forces that mediates close apposition (see also Chapter 2). The fusion rate constant provides a direct measure of the significance of parameters, such as pH and temperature, affecting the molecular events involved in the fusogenic destabilization of the apposed bilayers.

It was observed that fusion between virions, such as Sendai virus or influenza virus, and liposomes resulted in the formation of fusion products that consisted of a single virus and several liposomes (54,67). This conclusion was established by comparing the final extents of probe dilution, i.e., fluorescence increase, with calculated values obtained by applying the various theoretical models, as described in Chapter 5. These experimental observations and theoretical predictions were further substantiated by showing that addition of nonlabeled liposomes to labeled virus-liposome fusion products resulted in a further increase in fluorescence intensity to a final level consistent with the level as predicted for the final virus/liposome ratio (49). On the other hand, no increase in intensity was observed upon addition of virus particles, indicating that virions did not fuse with the fusion products. It will be of interest to examine the onset of fusion between a virus and fusion products consisting of many ($\geqslant 10$) liposomes. Such experiments coupled with the theoretical analysis should provide a convenient procedure for examining the effect of viral glycoproteins on the rate constants of adhesion and fusion.

A direct consequence of the fact that virus/liposome fusion products consist of a single virus and several liposomes is the outcome that a certain fraction of the virus fraction will not fuse, unless the liposome/virus ratio is large. The fraction of fully active virions remaining unfused is described by $\exp\text{-}L_0/V_0$, in which L_0 and V_0 are the molar concentrations of liposomes and viruses, respectively. It can thus be predicted that for a 1:1 population ca. 33% of the viruses remain unfused, which agrees well with the experimentally determined value of 39% (67).

The kinetic model can adequately simulate and predict the outcome of fusion of virus particles with liposomes at a variety of conditions (Fig. 6). In the case of Sendai virus, it was observed experimentally that these particles are particularly prone to fusion with acidic phospholipid vesicles at low pH, compared to fusion at neutral pH (49). The kinetic analysis revealed (54) that the rate constant of fusion exhibited the largest variation as a function of pH and liposome composition, varying from 1 to 0.005 s^{-1}, whereas the rate constant of aggregation varied from 1.8×10^8 to 0.7×10^8 $M^{-1} \cdot s^{-1}$. Hence, pH affects the molecular factors that are directly involved in fusion, rather than those related to aggregation. Additional experimental work revealed that at low pH, the fusion mechanism likely involves a direct participation of HN, besides F, in the fusion reaction, presumably related to the physical and geometrical properties of the vesicle bilayer per se (49).

The model has also been successfully applied to studies involving the interaction between Sendai virus and erythrocyte ghosts (Fig. 7). The

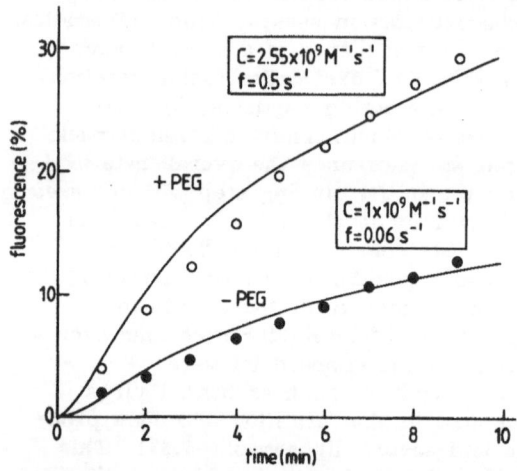

FIGURE 7   Kinetic simulation of virus-erythrocyte membrane fusion. Fusion between R_{18}-labeled Sendai virus and erythrocyte ghosts was monitored continuously in the presence and absence of 4% (w/v) poly(ethylene glycol) (PEG) at 37°C (drawn lines). Calculated values obtained by simulating fusion with the kinetic model are given by data points. The rate constants of aggregation (C) and fusion (f) are indicated. As can be seen, the dehydrating agent PEG drastically enhances the overall kinetics of fusion. From the kinetic analysis it becomes evident that the increase can be attributed to an increase of the fusion reaction per se (as the fusion rate constant in the presence of PEG increases by about an order of magnitude) rather than to a substantial modulation of the binding step between virus and target membrane.

fusion rate constants were relatively small ($0.04-0.07$ s^{-1}), but the rate constants of adhesion were very high, close to those expected in diffusion-controlled processes (64). The picture that emerges from these studies is that of a highly dynamical process in which virus particles adhere and dissociate continuously during their interaction with target membranes. The simultaneous occurrence of membrane fusion further adds to the complexity of the overall interaction process. Needless to say that convenient, accurate, and easy-to-apply techniques will be important tools to unravel these complex events. The application of theoretical procedures to kinetically analyze the various steps in the overall process, in conjunction with fluorescence assays as described here, will no doubt offer the possibilities to distinguish and characterize the various parameters affecting the fusion process (cf. Fig. 7) and, hence, to improve our insight into the mechanism of fusion of enveloped viruses.

ACKNOWLEDGMENTS

I thank the collaborators who contributed to the work in this paper, in particular Karin Klappe, Shlomo Nir, Jan Wilschut, Toon Stegmann, and Hettie Hoff. The secretarial assistance of Rinske Kuperus and Lineke Klap is much appreciated.

Part of the original work cited was carried out under the auspices of the Netherlands Foundation of Chemical Research (SON). Financial support from the Netherlands Organization for Scientific Research (NWO) and NATO (Research Grant 151.81) is acknowledged.

REFERENCES

1. Blough, H. A. and Tiffany, J. M. (1973). Lipids in viruses. *Adv. Lipid Res.* 11:267-339.
2. Lenard, J., and Millar, D. K. (1983). Entry of enveloped viruses into cells. In: *Receptor-Mediated Endocytosis* (P. Cuatrecasas and T. F. Roth, eds.). Chapman and Hall, London, pp. 121-138.
3. Bachi, T., Deas, J. E., and Howe, C. (1977). Virus-erythrocyte membrane interaction. In: *Virus Infection and the Cell Surface* (G. Poste and G. L. Nicholson, eds.), *Cell Surf. Rev.* 2, Elsevier-North Holland, Amsterdam, 83-127.
4. Maeda, T., and Ohnishi, S-I. (1980). Activation of influenza virus by acidic media causes hemolysis and fusion of erythrocytes. *FEBS Lett.* 122:283-287.
5. Marsh, M., Bolzau, E., and Helenius, A. (1983). Penetration of Semliki Forest virus from acidic prelysosomal vacuoles. *Cell* 32:931-940.
6. Miller, D. K., and Lenard, J. (1980). Inhibition of vesicular stomatitus virus infection by spike glycoprotein. Evidence for an intracellular, G protein-requiring step. *J. Cell Biol.* 84:430-437.
7. Knutton, S., and Pasternak, C. A. (1979). The mechanism of cell-cell fusion. *Trends Biochim. Sci.* 4:220-223.
8. Spear, P. G. (1987). Virus-induced cell fusion. In: *Cell Fusion* (A. E. Sowers, ed.). Plenum Press, New York, pp. 3-32.
9. Haywood, A. M., and Boyer, B. P. (1982). Sendai virus membrane fusion: Time course and effect of temperature, pH, calcium and receptor concentration. *Biochemistry* 24:6041-6046.

10. Maeda, T., Asano, A., Ohki, K., Okada, Y., and Ohnishi, S. (1975). A spin-label study on fusion of red blood cells induced by hemaggluti-nating virus of Japan. *Biochemistry* 14:3736-3741.

11. Lyles, D. S., and Landsberger, F. R. (1979). Kinetics of Sendai virus envelope fusion with erythrocyte membranes and virus-induced hemolysis. *Biochemistry* 18:5088-5095.

12. Hoekstra, D., De Boer, T., Klappe, K., and Wilschut, J. (1984). Fluorescence method for measuring the kinetics of fusion between bio-logical membranes. *Biochemistry* 23:5675-5681.

13. Stegmann, T., Hoekstra, D., Scherphof, G., and Wilschut, J. (1985). Kinetics of pH-dependent fusion between influenza virus and liposomes. *Biochemistry* 24:3107-3113.

14. Harmsen, M. C., Wilschut, J., Scherphof, G., Hulstaert, C., and Hoekstra, D. (1985). Reconstitution and fusogenic properties of Sendai virus envelopes. *Eur. J. Biochem.* 149:591-599.

15. Chejanovsky, N., and Loyter, A. (1985). Fusion between Sendai virus envelopes and biological membranes. The use for quantitative estimation of virus-membrane fusion. *J. Biol. Chem.* 260:7911-7918.

16. Micklem, K. J., Nyaruwe, A., and Pasternak, C. A. (1985). Permea-bility changes resulting from virus-cell fusion: temperature dependence of the contributing processes. *Mol. Cell. Biochem.* 66:163-173.

17. Henis, Y. I., and Jenkins, T. M. (1983). Detection of Sendai virus fusion with human erythrocytes by fluorescence photobleaching recovery. *FEBS Lett.* 151:134-138.

18. Aroeti, B., and Henis, Y. I. (1986). Fluorescence photobleaching recovery as a method to quantitate viral envelope-cell fusion: Application to study fusion of Sendai virus envelopes with cells. *Biochemistry* 25:4588-4596.

19. Pinto da Silva, P., Shimizu, K., and Parkison, C. (1980). Fusion of human erythrocytes induced by Sendai virus: Freeze fracture aspects. *J. Cell Sci.* 43:419-432.

20. Knutton, S. (1980). Studies of membrane fusion. VI. Mechanism of the membrane fusion and cell swelling stages of Sendai virus-mediated cell fusion. *J. Cell Sci.* 43:103-118.

21. Sekiguchi, K., and Asano, A. (1978). Participation of spectrin in Sendai virus-induced fusion of human erythrocyte ghosts. *Proc. Natl. Acad. Sci. USA* 75:1740-1744.

22. Lalazar, A., and Loyter, A. (1979). Involvement of spectrin in membrane fusion: Induction of fusion in human erythrocyte ghosts by proteolytic enzymes and its inhibition by antispectrin antibody. *Proc. Natl. Acad. Sci. USA* 76:318-322.

23. Hoekstra, D., Klappe, K., De Boer, T., and Wilschut, J. (1985). Characterization of the fusogenic properties of Sendai virus: Kinetics of fusion with erythrocyte membranes. *Biochemistry* 24:4739-4745.

24. Stegmann, T., Hoekstra, D., Scherphof, G., and Wilschut, J. (1986). Fusion activity of influenza virus. A comparison between biological and artificial target membrane vesicles. *J. Biol. Chem.* 261:10966-10969.

25. Pasternak, C. (1984). Virally-mediated changes in cellular permeability. In: *Membrane Processes: Molecular Biological Aspects and Medical Appli-cations* (G. Benga, H. Baum, and F. Kummerow, eds.). Springer, New York, pp. 140-166.

26. Poste, G., and Pasternak, C. A. (1978). Mechanisms of virus-induced cell fusion. *Cell Surf. Rev.* 5:306-349.

27. Homma, M., Shimizu, Y. K., and Ishida, N. (1976). On the study of Sendai virus hemolysis. I. Complete Sendai virus lacking in hemolytic activity. *Virology* 71:41-47.

28. Young, J. D.-E., Young, G. P. H., Cohn, Z. A., and Lenard, J. (1983). Interaction of enveloped viruses with planar bilayer membranes: Observation on Sendai, influenza, vesicular stomatitus and Semliki Forest viruses. *Virology* 128:186-194.

29. Patel, K., and Pasternak, C. A. (1985). Permeability changes elicited by influenza and Sendai viruses: Separation of fusion and leakage by pH-jump experiments. *J. Gen. Virol.* 66:767-775.

30. Hosaka, Y., and Shimizu, K. (1977). Cell fusion by Sendai virus. *Cell Surf. Rev.* 2:129-155.

31. White, J., Kielian, M., and Helenius, A. (1983). Membrane fusion proteins of enveloped animal viruses. *Q. Rev. Biophys.* 16:151-195.

32. Knutton, S., and Bachi, T. (1980). The role of cell swelling and hemolysis in Sendai virus-induced cell fusion and in the diffusion of incorporated viral antigens. *J. Cell Sci.* 42:153-167.

33. Helenius, A., Kartenbeck, J., Simons, K., and Fries, E. (1980). On the entry of Semliki Forest virus into BHK-21 cells. *J. Cell Biol.* 84:404-420.

34. Smith, H. (1977). Host and tissue specificities in virus infections of animals. *Cell Surf. Rev.* 2:1-46.

35. Wolf, K., Kahan, I., Nir, S., and Loyter, A. (1980). The interaction between Sendai virus and cell membranes. *Exp. Cell Res.* 130:361-369.

36. Fries, E., and Helenius, A. (1979). Binding of Semliki Forest virus and its isolated glycoprotein to cells. *Eur. J. Biochem.* 97:213-220.

37. Chejanovsky, N., Beigel, M., and Loyter, A. (1984). Attachment of Sendai virus particles to cell membranes: Dissociation of adsorbed virus particles with dithiothreitol. *J. Virol.* 49:1009-1013.

38. Markwell, M. A. K., Svennerholm, L., and Paulson, J. C. (1981). Specific gangliosides function as host receptors for Sendai virus. *Proc. Natl. Acad. Sci. USA* 78:5406-5410.

39. Van Meer, G., and Simons, K. (1983). An efficient method for introducing defined lipids into the plasma membrane of mammalian cells. *J. Cell Biol.* 97:1365-1374.

40. Hoekstra, D., and Klappe, K. (1986). Sendai virus-erythrocyte membrane interaction: Quantitative and kinetic analysis of viral binding, dissociation and fusion. *J. Virol.* 58:87-95.

41. Maeda, T., Kuroda, K., Toyama, S., and Ohnishi, S.-I. (1981). Interaction of hemagglutinating virus of Japan with erythrocytes as studied by release of a spin probe from virus. *Biochemistry* 20:5340-5345.

42. Umeda, M., Nojima, S., and Inoue, K. (1985). Effect of lipid composition on HVJ mediated fusion of glycophorin liposomes to erythrocytes. *J. Biochem.* 97:1301-1310.

43. Abidi, T. F., and Yeagle, P. L. (1984). Surface properties of Sendai virus envelope. *Biochim. Biophys. Acta* 775:419-425.

44. Nichols, J. W., and Pagano, R. E. (1983). Resonance energy transfer assay of protein-mediated lipid transfer between vesicles. *J. Biol. Chem.* 258:5368-5371.

45. Storch, J., and Kleinfeld, A. M. (1986). Transfer of long-chain fluorescent free fatty acids between unilamellar vesicles. *Biochemistry* 25:1717-1726.

46. Henis, Y. I., Gutman, O., and Loyter, A. (1985). Sendai virus envelope glycoproteins become laterally mobile on the surface of human erythrocytes following fusion. *Exp. Cell Res.* 160:514-526.

47. Amselem, S., Barenholz, Y., Loyter, A., Nir, S., and Lichtenberg, D. (1986). Fusion of Sendai virus with negatively charged liposomes as studied by pyrene-labeled phospholipid liposomes. *Biochim. Biophys. Acta* 860:301-313.

48. Eidelman, O., Schlegel, R., Tralka, T. S., and Blumenthal, R. (1984). pH-dependent fusion induced by vesicular stomatitus virus glycoprotein reconstituted into phospholipid vesicles. *J. Biol. Chem.* 259:4622-4628.

49. Klappe, K., Wilschut, J., Nir, S., and Hoekstra, D. (1986). Parameters affecting fusion between Sendai virus and liposomes. Role of viral proteins, liposome composition and pH. *Biochemistry* 25:8252-8260.

50. Stryer, L. (1978). Fluorescence energy transfer as a spectroscopic ruler. *Annu. Rev. Biochem.* 47:819-832.

51. Struck, D. K., Hoekstra, D., and Pagano, R. E. (1981). Use of resonance energy transfer to monitor membrane fusion. *Biochemistry* 20:4093-4099.

52. Hoekstra, D. (1982). Role of lipid phase separations and membrane hydration in phospholipid vesicle fusion. *Biochemistry* 21:2833-2840.

53. Wiener, J. R., Pal, R., Barenholz, Y., and Wagner, R. R. (1985). Effect of the vesicular stomatitus virus matrix protein on the lateral organization of lipid bilayers containing phosphatidylglycerol: use of fluorescent phospholipid analogues. *Biochemistry* 24:7651-7658.

54. Nir, S., Klappe, K., and Hoekstra, D. (1986). Mass action analysis of kinetics and extent of fusion between Sendai virus and phospholipid vesicles. *Biochemistry* 25:8261-8266.

55. Haywood, A. M. and Boyer, B. P. (1984). Effect of lipid composition upon fusion of liposomes with Sendai virus membranes. *Biochemistry* 23:4161-4166.

56. Scheid, A., Graves, M. C., Silver, S. M., and Choppin, P. W. (1978). Studies on the structure and function of paramyxovirus glycoproteins. In: *Negative Strand Viruses and the Host Cell* (B. W. J. Mahy and R. D. Barry, eds.). Academic Press, New York, pp. 181-193.

57. Wilson, I. A., Skehel, J. J., and Wiley, D. C. (1981). The hemagglutinin membrane glycoprotein of influenza virus: structure at 3A resolution. *Nature* 289:366-373.

58. Doms, R. W., Helenius, A., and White, J. (1985). Membrane fusion activity of the influenza hemagglutinin. The low pH-induced conformational change. *J. Biol. Chem.* 260:2973-2981.

59. Lee, P. M., Cherry, R. J., and Bachi, T. (1983). Correlation of rotational mobility and flexibility of Sendai virus spike glycoproteins with fusion activity. *Virology* 128:65-76.

60. Junankar, P. R., and Cherry, R. J. (1986). Temperature and pH dependence of the haemolytic activity of influenza virus and the rotational mobility of the spike glycoproteins. *Biochim. Biophys. Acta* 854:198-206.

61. Lyles, D. S. (1979). Glycoproteins of Sendai virus are transmembrane proteins. *Proc. Natl. Acad. Sci. USA* 76:5621-5625.

62. Blumberg, B. M., Giorgi, C., Rose, K., and Kolakofsky, D. (1985). Sequence determination of the Sendai virus fusion protein gene. *J. Gen. Virol.* 66:317-331.

63. Blumberg, B., Giorgi, C., Roux, L., Raju, R., Dowling, P., Chollet, A., and Kolakofsky, D. (1985). Sequence determination of the Sendai

virus HN gene and its comparison to the influenza virus glycoproteins. *Cell* 41:269-278.

64. Nir, S., Klappe, K., and Hoekstra, D. (1986). Kinetics and extent of fusion between Sendai virus and erythrocyte ghosts: Application of a mass action kinetic model. *Biochemistry* 25:2155-2161.

65. Richardson, C. D., and Choppin, P. W. (1983). Oligopeptides that specifically inhibit membrane fusion by paramyxoviruses: Studies on the site of action. *Virology* 131:518-532.

66. Haywood, A. M., and Boyer, B. P. (1985). Fusion of influenza virus membranes with liposomes at pH 7.5. *Proc. Natl. Acad. Sci. USA* 82: 4611-4615.

67. Nir, S., Stegmann, T., and Wilschut, J. (1986). Fusion of influenza virus with cardiolipin liposomes at low pH: Mass action analysis of kinetics and extent. *Biochemistry* 25:257-266.

68. Hoekstra, D. (1983). Topographical distribution of a membrane-inserted fluorescent phospholipid analogue during cell fusion. *Exp. Cell Res.* 144:482-488.

15

Influenza Virus Hemagglutinin and Membrane Fusion

ROBERT W. DOMS,* JUDY WHITE,† FRANCOIS BOULAY,‡ and
ARI HELENIUS

Yale University School of Medicine, New Haven, Connecticut

I. INTRODUCTION

Enveloped animal viruses penetrate into their host cells by membrane fusion
(see also Chapter 13). The fusion reaction, which occurs between the
viral envelope and a cellular membrane, is mediated by specific, virally
encoded proteins (1,2). Several fusion proteins have been identified by
biochemical and genetic means, and some have been isolated from virus
particles or infected cells. A few have been cloned, sequenced, and analyzed
in great detail.

The structure and properties of the viral fusion factors are presently
the focus of considerable interest because by mediating an important step
in the viral replication cycle, they constitute a potential target for antiviral
strategies. They also provide the most experimentally accessible models
for the study of protein-mediated membrane fusion, and they are increasingly
employed as fusogens in the bulk delivery of macromolecules into cells (3),
cell-cell fusion, and implantation of lipids into cellular membranes (4).
For an extensive discussion of these applications of membrane fusion techniques, refer to the last part of this book.

Although the fusion proteins of different viruses have common features,
they are distinct proteins with quite varied properties. Sequence homologies
between fusion proteins of different virus families are rare, and studies
on their fusion mechanisms suggest principal differences in their mode
of action (5). They are, however, all integral membrane glycoproteins
and often contain fatty acyl groups. The single transmembrane region is
usually located close to the C-terminus with the bulk of the mass external
to the viral membrane forming spikelike projections 5-20 nm in length.

Present affiliations:
*National Institute of Allergy and Infectious Diseases, National Institutes
of Health, Bethesda, Maryland.
†University of California, San Francisco, California.
‡Centre National de la Recherche Scientifique, Grenoble, France.

They are frequently oligomeric and are often initially synthesized as precursor forms which are activated to their mature fusion active state by posttranslational proteolytic cleavages (1,2). As each virus particle contains several hundred copies of the fusion proteins, they constitute a major component of the viral membrane.

The general mechanisms of virus entry and fusion have recently been reviewed (1,6). In this chapter we focus on the best-characterized of the viral fusion proteins, the hemagglutinin(HA) of influenza virus. During the last few years a large number of studies have addressed the functional and structural properties of this interesting, well-characterized protein. For a more complete background in the biology of influenza virus, refer to recent reviews (7-10).

II. THE CELL BIOLOGY OF VIRUS ENTRY

A. The Structure of Influenza Virus

Influenza belongs to a family of viruses called the myxo- or orthomyxoviridae. There are three classes, termed A, B, and C, differing slightly in composition and general properties (8). The virus particles are either spherical or rod-shaped, and they display considerable size heterogenity (Fig. 1). The virus aquires its lipid bilayer membrane by a budding reaction at the host cell plasma membrane (7). The membrane of influenza viruses contains two types of integral membrane spike glycoproteins, the hemagglutinin (HA) and neuraminidase (NA). Contained within the particle is the genome and accessory proteins. The genome is in the form of eight negative sense RNA strands associated with multiple copies of the nucleo- and matrix proteins and a set of RNA polymerases (7,8). Because of the segmented nature of the genome, the recombination rate in cells infected with multiple virus strains is extremely high. This contributes to the rapid generation of antigenic diversity and thereby to the pathogenic potential of the virus (11-13).

HA and NA mediate the early interactions between the virus and host cell. They are exceptionally well characterized at the structural and biochemical levels. Both have large ectodomains which form the spike projections visible in Figure 1. The ectodomains can be released in water-soluble forms by digestion with proteases and have been crystallized and their three-dimensional structures determined to 2.8-3.0 Å resolution (14,15). HA has a molecular weight of 76 kD and occurs as a homotrimer, while NA, with a molecular weight of 60 kD, is a homotetramer. The HA and NA sequences from numerous strains of influenza have been determined, and their antigenic structures have been mapped in detail (11,14,16).

B. The Mechanism of Entry

As the first step in entry the influenza virus binds to sialic acid residues on plasma membrane glycoproteins of a host cell via sialic acid binding sites present in each HA (17,18). The bound virus may either be released by the action of the viral NA (and proceed to bind to another site or another cell), or it may be internalized by endocytosis and delivered to endosomes (19-23). Endosomes are prelysosomal vacuoles which function in the endocytic pathway as sites for molecular sorting, recycling, and processing of incoming receptors, ligands, and fluid components in the cell (24; see also Chapter 19). They have an acid, internal pH (pH 5-6.5) generated by membrane-associated, ATP-driven proton pumps (25). The

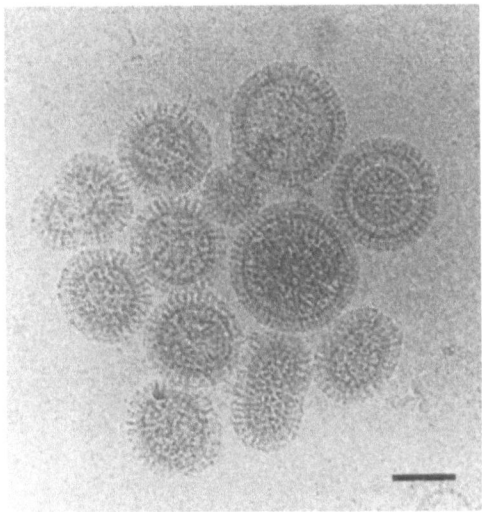

FIGURE 1   Frozen-hydrated images of influenza virus at pH 7.0 prepared by the technique of Lepault et al. (80). Bar represents 100 nm.

low pH is of crucial importance in triggering virus penetration; it induces an irreversible conformational change in the HA, which, as discussed in detail below, mediates the fusion between the viral and endosomal membranes (1,22,22a,23,23a). As a result, the nucleocapsid gains entry to the cytoplasm and the viral genome and polymerase enzymes are transported into the nucleus where RNA replication begins (8).

The endocytic entry pathway is not unique to influenza virus. The majority of enveloped animal viruses studied thus far are internalized by endocytosis into their host cells (1,26-31). The E1/E2/E3 heterotrimer of Semliki Forest virus (SFV) (5), the homotrimeric G glycoprotein of vesicular stomatitis virus (VSV) (32), and G1 of La Cosse virus (33) have been shown to underto acid-induced conformational changes which lead to activation of membrane fusion. In no case are the data concerning the fusion reaction, however, as extensive as for influenza HA.

III.   PROPERTIES OF INFLUENZA VIRUS FUSION

Efficient fusion between influenza virus and host cell membranes occurs only at pH values lower than about 6.0 (34,36). Acid-induced fusion is also observed with a variety of other membranes such as liposomes (4,37-39), red blood cell membranes (34,37,39), and black lipid membranes (40). Results obtained with a variety of target membranes have generally been in good agreement. The most salient findings are as follows:

1. Efficient fusion occurs only at low pH (Fig. 2 and Refs. 4,34,36, 37,39). The pH threshold varies with the strain of influenza virus, with the midpoint ranging from approximately pH 5.2 to pH 6.0 (41). Under ideal conditions up to 95% of the virus particles in a given preparation may fuse (37). When a virus/liposome mixture is acidified, viruses begin to fuse immediately. Fusion of the virus is arrested when the pH is elevated and resumes upon reacidification (39). Low pH is thus required during the actual fusion event—it is not just needed for preactivation of the virus.

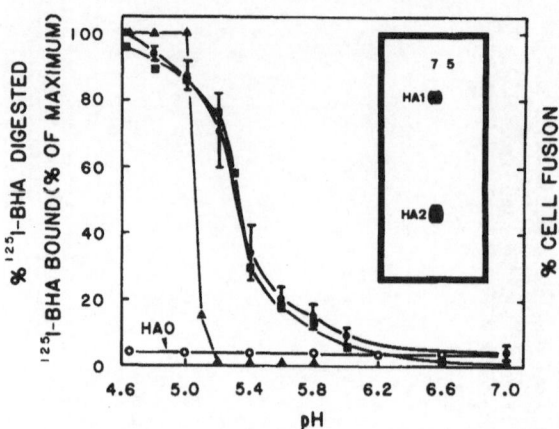

FIGURE 2   pH dependence of the conformational change and membrane
fusion. The pH dependence of membrane fusion for the X-31 strain was
determined by fusion from within (57). The extent of fusion both before
(○) and after (▲) trypsin activation of the HA$_0$ precursor is shown.
^{125}I-BHA was incubated with liposomes at the indicated pH for 15 min,
37°C. The samples were neutralized and the extent of liposome binding
(■) and conversion to the proteinase K-sensitive form determined (●).

2. Provided that the virus particles and target membranes are intact,
the fusion reaction does not result in the transient rupture of either
participating membrane; i.e., the reaction is not "leaky" (37,40). "Leakiness" is observed, however, when influenza fuses with red blood cells,
a reaction that leads to hemolysis (34,36).

3. Fusion is relatively rapid. The T$_{1/2}$ at 37°C is approximately 15-30 s
(37,39). Using a mass action kinetic model, Nir et al. (38) have dissected
the fusion reaction into two discrete steps: a rate limiting step, corresponding to binding, and a first-order step, involving the fusion reaction itself
(see also Chapter 5).

4. Fusion is relatively independent of the target membrane's lipid
composition (37), though it is somewhat more efficient when phosphatidylethanolamine (PE) is present (4,37). This is in marked contrast to Semliki
Forest virus, which displays an absolute requirement for cholesterol (42),
and to Sendai virus, which also displays a cholesterol requirement (43).
Inclusion of protein or sialic acid receptors (in the form of glycoproteins
or gangliosides) is not a prerequisite for fusion (37-39).

5. Fusion occurs over a broad temperature range, with activity
observed at 4°C (37). The kinetics of the fusion reaction are slowed
as temperature is decreased. Fusion, albeit highly leaky, also occurs
with liposomes below their gel to liquid-crystalline phase transition, suggesting that the lipids of the target membrane need not be in a fluid state (44).

6. Fusion is independent of the presence or absence of divalent cations,
and it occurs over a wide range of ionic conditions (37).

Whereas fusion between influenza virus and biological membranes has
not been observed at neutral pH, recent studies have indicated that it
can occur with certain types of artificial vesicles (39,45). Stegmann et al.
(39) found that influenza could fuse with liposomes possessing a high
negative charge at neutral pH. The pH dependence, efficiency, and kinetics

of virus fusion with the cardiolipin vesicles were found to be markedly different from those of fusion with liposomes containing a more biologically relevant lipid composition (39). Fusion with cardiolipin vesicles was, for instance, not inactivated by pretreatment of the virus with acid pH, a treatment that abolishes fusion between the virus and biological membranes presumably owing to aggregation of virus particles and HA spikes (46). It was therefore concluded that a different mechanism of fusion was operating in the two instances and that fusion with negatively charged vesicles at neutral pH is nonspecific and nonphysiological (39). This may reflect a general susceptibility of negatively charged membranes to fusion (47).

Another report of influenza fusion at neutral pH used ganglioside-containing liposomes as target membranes. It relied on binding assays and negative stain electron microscopy to analyze fusion (45). As neither of these assays demonstrated mixing of membrane components or fusion quantitatively, the results are difficult to interpret.

IV. HEMAGGLUTININ, THE FUSION FACTOR

The hemagglutinin (HA) is the factor responsible for influenza virus-induced membrane fusion at acid pH (1). It was first noted by Klenk et al. (48) that the cleavage of the precursor, called HA_O, to the mature HA is absolutely required for viral infectivity (Fig. 3). Subsequent studies showed that it was required for acid-induced fusion activity of the virus (49,52). This implicated the HA as a factor involved in the reaction, but it did not eliminate the possible involvement of other proteins. The final

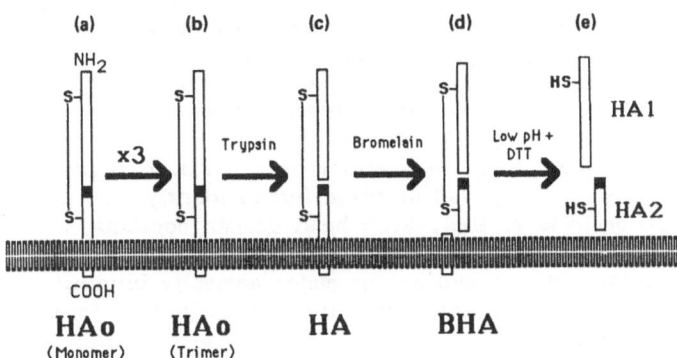

FIGURE 3 Different forms of the influenza hemagglutinin. Monomeric HA_O (a) forms a homotrimer in the endoplasmic reticulum (b) (83,83a). HA_O is converted to HA (c) by a posttranslational proteolytic cleavage or by trypsin digestion. The resulting HA_1 and HA_2 subunits are held together by noncovalent interactions as well as an interchain disulfide bond. The sequence of the highly conserved, uncharged amino terminus of HA_2 (black portion in the schematic) is shown. Bromelain digestion releases the molecule in water soluble form. Acid treatment of BHA (d) followed by reduction of the disulfide bond releases HA_1 in monomeric form from BHA_2 (e).

proof that HA is, indeed, necessary and sufficient for fusion activity was obtained by expressing HA in tissue culture cells without the presence of other viral proteins (49). Cells that expressed the HA in its mature cleaved form fused with each other when exposed to low pH. Fusion studies with reconstituted vesicles containing isolated HA have confirmed this conclusion (51,52). Thus, HA is capable of catalyzing fusion in the absence of other viral proteins provided it is anchored in one of the fusing membranes (49).

Several reports have suggested that the NA spike glycoprotein may be involved in the fusion reaction (53,54), but none of these have been compelling. It is possible, however, that the NA may affect the pH dependence and/or efficiency of fusion in some way.

V. ACID-INDUCED CONFORMATIONAL CHANGES IN THE HEMAGGLUTININ

A variety of biochemical, genetic, morphological, and immunological techniques have been employed to study the effects of low pH on the HA molecule. The results show that HA undergoes a major, irreversible conformational change approximately at the pH of fusion. To understand the nature and implications of this alteration, the structure of HA in its neutral pH form needs to be discussed in some detail.

A. The Structure of HA

As mentioned above, the crystal structure of the ectodomain fragment of HA is known. Bromelain digestion of intact viruses or a lysate of infected cells releases nearly the entire N-terminal ectodomain of HA (95% of the mass) in water-soluble form. The HA_2 chain is thereby cleaved close to the point where it emerges from the viral membrane (Fig. 3, Ref. 55). The resulting soluble fragment, termed BHA, has been crystallized and its three-dimensional structure determined to 3Å resolution (Ref. 15 and shown schematically in Fig. 4).

The BHA is a trimeric protein with two distinct parts: a head region and a fibrous stem. The head is composed of three independently folded globular domains made up entirely of HA_1. Each head domain contains an eight-stranded, antiparallel β-sheet structure with associated reverse turns. The sialic acid binding site as well as the major antibody binding sites, termed the loop, hinge, and tip/interface, are located in this domain (indicated in Fig. 4, Ref. 16).

The head domain rests on the stem, which is composed of HA_2 and the very C-terminal and N-terminal sequences of HA_1. The C-terminus of HA_1 is located close to the viral membrane, 2.2 nm from the N-terminus of HA_2, to which it is originally linked in the HAO precursor. The most notable feature of the stem domain is the complex of three 7.6-nm-long α-helices (one from each HA_2 subunit) which form a triple-stranded coiled coil in the center of the stem. It is stabilized by electrostatic, hydrophobic, and numerous other noncovalent interactions. Biochemical studies have shown that the HA trimer is, indeed, very stable. It fails to dissociate completely even after treatment with sodium dodecyl sulfate (56). Tucked between the long α-helices near the base of the molecule are the uncharged amino termini of the HA_2 subunits. Their position is stabilized by noncovalent interactions with HA_1 and with residues on the same and adjoining

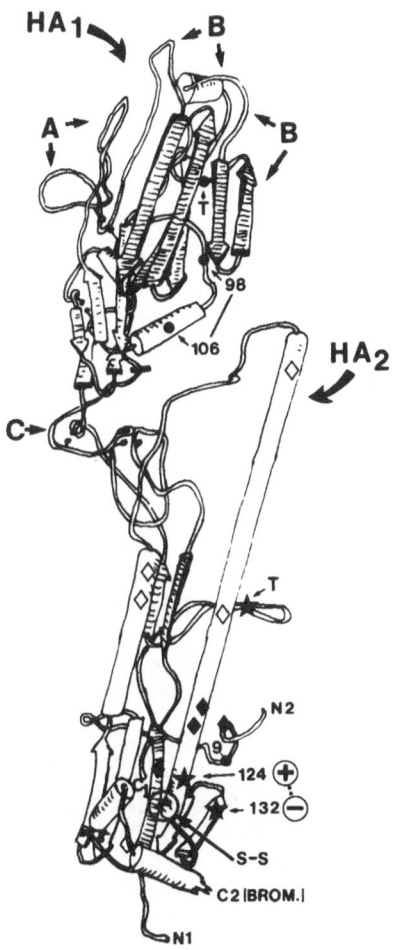

FIGURE 4 The three-dimensional structure of HA. A schematic representation of a BHA monomer is shown (see Ref. 15). Three HA subunits interact
primarily along the long α-helix in HA_2 to form the trimeric HA structure.
The amino and carboxy termini of both the HA_1 and HA_2 subunits are
indicated. (The C-terminus of HA_2 is the site at which bromelain cleaves
HA). Also shown is the hinge epitope (C), loop epitope (A), tip/interface
epitope (B), the sequence recognized by antipeptide monoclonal "C" (HA_1
residues 98-106), the interchain disulfide bond (S—S), and the trypsin
cleavage sites exposed after acid treatment of the molecule (residues 27
and 224 of HA_1, indicated by "T"). Several of the amino acid substitutions
identified in variant HA's which mediate fusion at a higher pH than the
wt (69) are indicated. The substitutions fall into two categories: those
which affect subunit:subunit interactions (◇) and those which affect the
position of the amino terminus of HA_2 (◆). As an example of the former
category, aspartic acid 132 of HA_2 forms a salt bridge with arginine 124
of the adjacent HA_2 subunit (indicated by stars at the base of the molecule).
When aspartic acid 132 is replaced with asparagine, the salt bridge can no
longer be formed and the protein catalyzes fusion 0.3 pH units higher
than the wild type.

HA$_2$ subunits. The single HA$_1$-HA$_2$ interchain disulfide bond is also located in the stem region, near the base of the molecule.

B. The Effects of Acid Treatment on the Structure and Properties of HA

1. The Acquisition of Amphiphilic Properties

Many of the initial studies designed to examine the acid-induced conformational change in HA have used BHA as a model system because of its solubility in water and the ease of isolation and labeling with radioactive isotopes (57,58). Although incable of inducing fusion, it undergoes many of the changes that are observed in the intact HA. For instance, when BHA is exposed to mildly acidic pH (e.g., pH 5.0 for 1 min), it acquires amphiphilic properties as judged by the following criteria:

1. It aggregates in aqueous solution (58). Aggregation is prevented by the inclusion of nonionic detergent, and it is due to elements located primarily in the BHA$_2$ subunit.

2. It interacts with nonionic detergents, like Triton X-100 or X-114 (57,58). When HA$_1$ and BHA$_2$ subunits are separated from one another by reduction of the single interchain disulfide bridge (59), BHA$_2$ is found to be the subunit that partitions into the Triton phase while HA$_1$ remains water soluble (57). When assayed by direct binding experiments, approximately 10 Triton X-100 molecules bind to each acid BHA monomer (60). The results suggest that BHA$_2$ possesses a hydrophobic detergent binding site.

3. It binds to liposomes (57,58). The pH dependence of BHA-liposome binding displays a midpoint of approximately pH 5.4 (Fig. 2 and Ref. 57). Binding is rapid ($T_{1/2}$ of approximately 10 sec), independent of temperature and divalent cations, and occurs with liposomes of varied composition (57). The interaction is hydrophobic, but it can be reversed by high pH (57), suggesting a less firm attachment than observed for transmembrane proteins. Recent studies using a photoactivatable phosphatidylcholine derivative indicate that only the BHA$_2$ chain is hydrophobically associated with the lipid bilayer (61,62,62a). This is in agreement with the finding that HA$_1$ can be easily removed if the interchain disulfide bond is reduced (59).

These observations show that the solubility properties of BHA change dramatically and irreversibly upon exposure to acid. A previously unexposed hydrophobic moiety in BHA$_2$ is exposed, mediating the interaction with detergents and lipids. The acquisition of a hydrophobic surface group in BHA and its attachment to liposomes is contingent on the activating cleavage (57). When a water-soluble, trimeric form of HA$_0$ is prepared by expressing a truncated HA gene lacking the transmembrane and cytoplasmic domains (63), the resulting anchor minus HA$_0$ remains water soluble and does not bind to liposomes after acid treatment (64). The N-terminal peptide of HA$_2$, a sequence often referred to as the "fusion sequence," is probably involved in the exposed moiety, as shown below.

In addition to the acquisition of amphiphilic properties, the alteration in conformation is detectable by several other biochemical techniques. Whereas the neutral BHA and HA are remarkably resistant to a variety of proteases, the acid forms are susceptible to digestion with trypsin, proteinase K, and other proteases (57,58). While complete degradation is often observed, limiting digestion with trypsin has defined two distinct cleavage sites located in HA$_1$, one at residue 224 close to the domain inter-

SITE-SPECIFIC MUTANTS OF THE HA2 AMINO TERMINUS

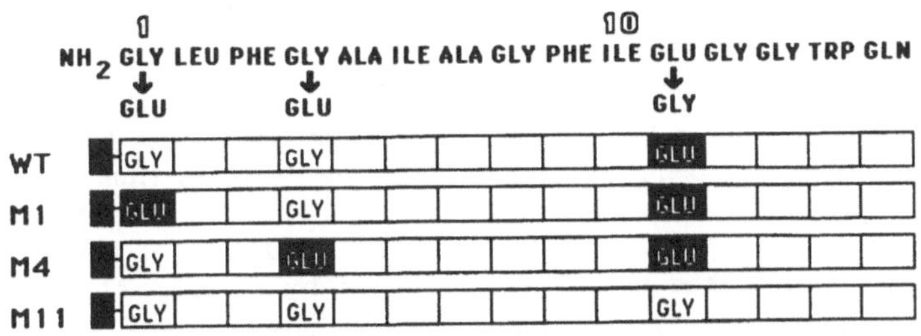

PHENOTYPES OF SITE-SPECIFIC HA MUTANTS

	FUSION				BHA		
	RBC-CELL		CELL-CELL		CONVERSION		LIPID BINDING
	pH	%	pH	%	pH	T1/2	T1/2
WT	5.3	90	5.3	90	5.3	3˙	3˙
M1	–	0	–	0	5.0	6˙	16˙
M4	5.7	50	5.6	50	5.5	1˙	1˙
M11	5.3	90	4.6	<2	5.3	3˙	3˙

FIGURE 5 The site-directed mutants M1, M4, and M11 of the Japan strain of influenza are shown. Charged residues are indicated by black boxes. The threshold pH of red blood cell:cell fusion or cell:cell fusion is indicated, as is the percentage of all cells undergoing fusion. The ability of BHA obtained from the mutant and wild-type proteins to convert to the acid conformation (determined by trypsin digestion) and bind to liposomes was also determined. The pH for half-maximal conversion and liposome binding was identical in all cases.

face at the top of the molecule, and the other at residue 27 which, in the neutral form, is tucked between adjoining HA_2 α-helices midway up the stem (Fig. 4, Ref. 58). The pH dependence with which HA and BHA convert to their protease-sensitive forms displays a midpoint of pH 5.4 for the X-31 strain (Fig. 2, Ref. 57). Conversion is rapid ($T_{1/2}$ = 10 sec), irreversible, and largely independent of temperature and ionic conditions. HA and BHA convert to the acid form at nearly identical pH (57).

Other differences between the neutral and acid conformations have been described: (1) Circular dichroism measurements detect an irreversible change in BHA which has been interpreted as resulting from a shift in domain positions without major secondary structure alterations (58). (2) The interchain disulfide bridge located at the base of the molecule (Fig. 5) becomes accessible to reducing agents. Reduction thus leads to dissociation of monomeric HA_1 subunits from HA_2 (59).

Although not identical in its responses to acid, BHA is generally found to be sufficiently similar to HA to serve as a useful model with which to examine the acid-induced conformational changes occurring in the ectodomain of intact HA. It is important to note that many properties of BHA and HA remain unchanged after acid treatment. The molecules retain their sialic

acid binding sites (58) as well as most of their antibody binding sites
(65-68). Thus, the alteration apparently does not lead to gross denaturation
of the molecule.

2. Fusion Variants and Mutants

Two genetic approaches have been taken to analyze HA-induced membrane
fusion. The first involved isolation and analysis of variants that catalyze
fusion at a pH more basic than the wild type (69-71). The second was
the expression and characterization of site-specific mutants of HA (72).
Results from both approaches have been reviewed by Gething et al. (73).

Fusion variants. Fusion variants of influenza virus arise spontaneously.
They can be isolated from virus stocks by plaque purification (70) or
specifically selected for by growth of virus in the presence of lysosomotropic
weak bases such as amantadine (69). Such bases elevate the pH of endosomes
(74), thus giving a variant with a higher fusion pH an advantage over
the wild type. In all cases tested, single or double amino acid substitutions
have been found, usually in the HA_2 chain (69,70,75).

The substitutions fall into two general categories. The first are residues
that are involved in interactions between adjacent HA subunits in the stem,
particularly between the long HA_2 α-helices. Their locations indicate that
interactions along nearly the entire length of the stem are important in
stabilizing the trimer (several indicated by ◇ in Fig. 4, Ref. 69). The
second class of variants possess single amino acid changes which affect
residues in or around the HA_2 N-terminus (indicated by ◆ in Fig. 4,
Refs. 69-71). By changing the environment around the amino terminus,
these substitutions apparently allow it to become exposed at higher pH.
The results underscore the importance of this hydrophobic region in membrane fusion.

Site-specific mutagenesis of HA. Many observations have indirectly
suggested that the hydrophobic N-terminus of HA_2 is involved with fusion,
and it is widely referred to as the "fusion peptide." To test the validity
of this hypothesis, the effect of site-specific mutations in the N-terminal
region of HA_2 has been analyzed (Fig. 5 and Ref. 72). Mutant genes of
HA were placed in SV40 vectors and the HA expressed in CV-1 cells (63).
Fusion activity was assayed by red blood cell fusion and by fusion of
infected CV-1 cells to one another (72). In addition, mutant BHA was
generated by bromelain digestion of infected cell lysates and its ability
to undergo the conformational change and bind to liposomes determined.

The first mutant protein, called M1 (Fig. 5), contained a glutamic
acid in place of glycine at position 1 of HA_2, thus introducing a negative
charge on the amino terminal residue. The protein failed to catalyze membrane fusion under any conditions. The corresponding BHA could, however,
undergo a conformational change at acid pH which resulted in protease
sensitivity as well as liposome binding. Interestingly, these changes occurred
at a lower pH and with markedly slower kinetics than in the wild-type
protein.

The second mutant, M4, contained a glutamic acid residue in place
of glycine at position 4. M4 fused red cells and CV-1 cells together at
a pH 0.3-0.4 units higher than the wild type, though only half as efficiently. The pH shift was accurately reflected by M4 BHA. Whether the
decreased fusion efficiency was due to the substitution itself, or, more
simply, to interruption of the hydrophobic stretch is not yet known.

Finally, the highly conserved glutamic acid residue at position 11 was replaced with a glycine. The resulting protein, M11, was almost indistinguishable from the wild-type protein with regard to fusion activity, liposome binding, or the conformational change. That fusion and the conformational change occurred with the same pH dependence as wild type is consistent with the position of glu 11, whose side chain is not involved in any stabilizing interactions. The only significant difference was that M11 did not fuse CV1 cells efficiently, even though its fusion activity as assayed by red cell fusion was like that of the wild type. The reason for the discrepancy between fusion assays is not clear. It could be explained by the more intimate binding between HA and the sialic-acid-rich red blood cells.

The site-specific mutants have provided the most direct evidence so far that the N-terminal peptide of HA_2 is, indeed, involved in fusion. The fact that all of the site-specific mutants were capable of interacting with liposomes in a manner similar to the wild-type HA indicated, moreover, that it is possible for HA_2 to attach hydrophobically to the target bilayer without causing fusion. This suggests that hydrophobic attachment of HA_2 to the target membrane via the amino terminal peptide may be required for fusion but that it is not sufficient by itself. Stegmann et al., using liposomal model systems, have recently made a similar suggestion (75a).

3. Quaternary Structure of the Low-pH Form

The location of the amino acid substitutions in the fusion variants and the exposure of trypsin cleavage sites indicated that changes occur along the entire length of the HA ectodomain after acid treatment. Indeed, recent studies have shown that BHA dissociates into dimers and monomers upon acid treatment (56,76). Intact HA, however, remains trimeric at acid pH, though some dissociation occurs with very dilute solutions (56). HA also remains resistant to the dissociative effects of SDS, indicating that stable subunit:subunit interactions are still present after the acid-induced alteration. BHA, by contrast, is extremely sensitive to dissociation by SDS (56).

4. Morphology of Acid-Treated HA

At neutral pH, the HA spikes appear as well-ordered rectangular projections approximately 13.5 nm in length as seen by negative stain electron microscopy (Fig. 6A and Ref. 57,77-79). Following acid treatment, individual spikes are more difficult to distinguish (52,57,77,78 and Fig. 6B). A wirey, disordered layer extending approximately 14 nm from the membrane is usually observed. Examination of unstained virus particles by cryoelectron microscopy (80) supports this interpretation; the disordered appearance is thus not an artifact of negative staining.

Two approaches have been taken to visualize individual virus spikes by negative stain electron microscopy. Ruigrok et al. (77,81) examined isolated BHA and HA, as well as reconstituted liposomes containing HA. They observed that HA becomes thinner and longer after acid treatment and concluded that trimeric contacts in the head domain are broken and the HA_2 stem region is elongated. On the basis of the morphology, they suggested that the amino terminus of HA_2 (the "fusion peptide") may interact with the virus membrane during fusion rather than with the target membrane.

We have taken a somewhat different approach to the morphology of acid-treated HA. By partially digesting virus particles with bromelain, it was possible to obtain particles that displayed decreased spike densities

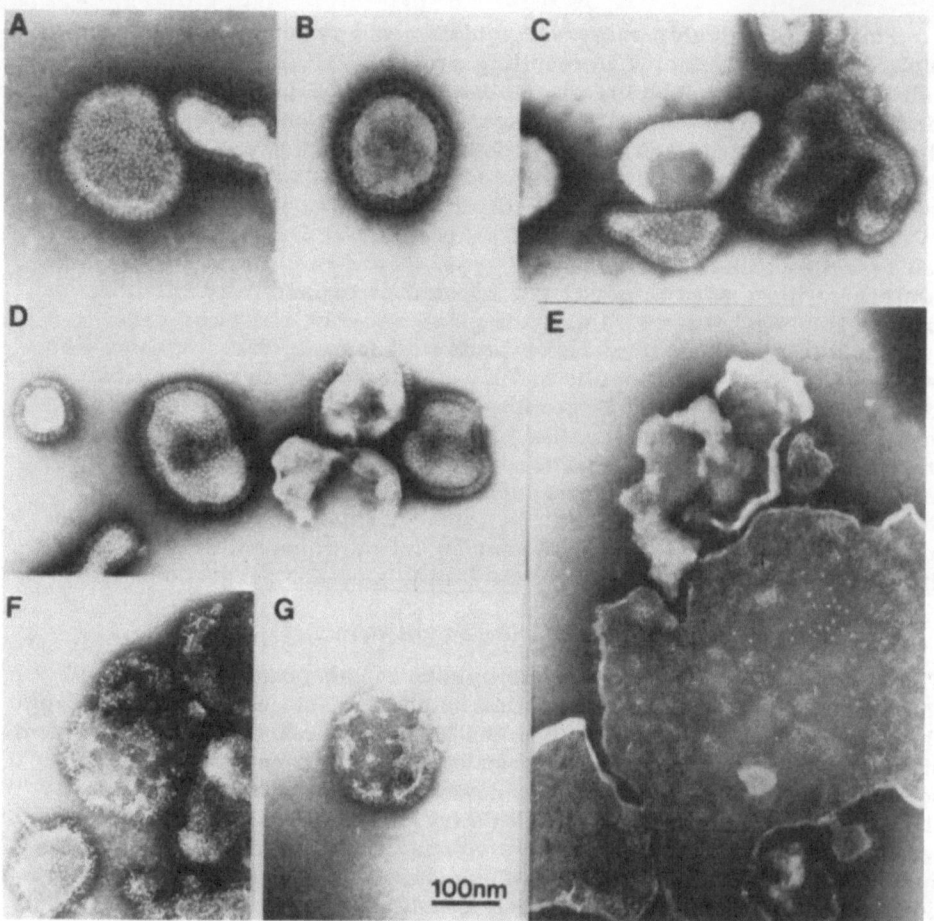

FIGURE 6 Negative-stain electron microscopy of influenza at: pH 7.0
(A), pH 5.0 (B), pH 7.0 after partial removal of HA spikes by bromelain
(C and D), and acid treatment of the particles in C and D for 15 s at
pH 5.0 (E,F,G). Note the aggregates on the surface of these viruses.
Bar represents 100 nm.

(Fig. 6C and D and Ref. 56). Upon acid treatment for a short time, star-
shaped aggregates could be observed on the viral surface (Fig. 6E, F,
and G). Thin, 9-10-nm-long arms, six to nine in number, radiated from
a central nexus. We interpret these images to be higher-order structures
of trimers which were associating via their transmembrane and/or HA_2
domains, the "arms" representing individual HA_1 subunits. It was clear
that each rosette contained more than one HA trimer. Whether such multimeric structures are relevant for membrane fusion remains to be determined.

5. Changes in Antigenic Structure

Crystallization of the low pH form of BHA has been impeded by its hydrophobic character. In the absence of crystal structure information, the
structural analysis of the acid conformation has relied on indirect tech-

niques. One of the most powerful approaches has been to use poly- and monoclonal antibodies which recognize known regions of the protein.

Loss of epitopes after acid treatment. As indicated above, antibodies to influenza are directed almost exclusively to three major antibody binding sites on the neutral form of HA (A, B, and C in Figure 4; Refs. 11-13,16). These have been extensively analyzed and their locations on the surface of the molecule are well known. Following acid treatment, a number of antibodies against the tip/interface site, particularly those which bind close to the trimer interface, no longer recognize HA, indicating that a structural alteration occurs in this region (65,67,82). The binding of antibodies against the loop and hinge epitopes, by contrast, is usually not affected by acid treatment (68).

Exposure of new epitopes after acid treatment. If acid-treated virus is used for immunization and selection, monoclonal antibodies specific to the acid conformation can be generated (57,65,66,82,83). Though the exact locations of the epitopes are not known, it is clear that such antibodies recognize sites both in HA_1 and HA_2 (66,67,83). The pH dependence with which these epitopes appear is the same as that for acquisition of protease sensitivity and liposome binding. BHA and HA display the same changes in antigenic structure.

Antipeptide antibodies. More recent studies have employed antibodies directed against defined peptides of the HA to highlight the regions of the molecule that change in response to low pH (64,84). The reactivity of antisera directed against peptides located on the surface of the molecule (e.g., the amino terminus of HA_1 and a peptide located in the loop binding site) was unaffected by pretreatment of the HA at acid pH (84). In contrast, the reactivities of antisera directed against peptides located in regions of subunit:subunit interactions were markedly enhanced by briefly exposing the HA to mildly acidic pH. The most striking example was a monoclonal antibody against a nine-amino-acid determinant that is totally buried in the globular head region interface (residues 98-106 of HA_1, Fig. 4). This antibody displayed virtually no reactivity (<0.1%) with neutral pH HA trimers, yet precipitated >90% of HA trimers that had been pretreated at pH 4.8. The two peptides that showed the next greatest increase in reactivity with their corresponding antisera upon low pH treatment were the C-terminus of HA_1 and residues 1-15 of the fusion peptide.

Taken together these results corroborate the previous interpretation that the conformational change at 37°C involves partial dissociation of subunit contacts without gross denaturation of the individual protein domains. Further, by examining the kinetics with which each epitope becomes exposed, it should be possible to establish an "unfolding pathway" for the transition of HA to its fusion-active state (84).

C. The Effects of Acid Treatment on the Structure and Properties of HA_0

HA_0, the intracellular precursor to HA, is converted to HA by a proteolytic cleavage which occurs at or close to the time of insertion into the plasma membrane. A large number of other viral spike glycoproteins also undergo activating cleavages (85,86), such as the F glycoprotein of Sendai virus (85). While the inability of HA_0 to catalyze membrane fusion has been well documented (49,52), its ability to undergo an acid-induced conformational

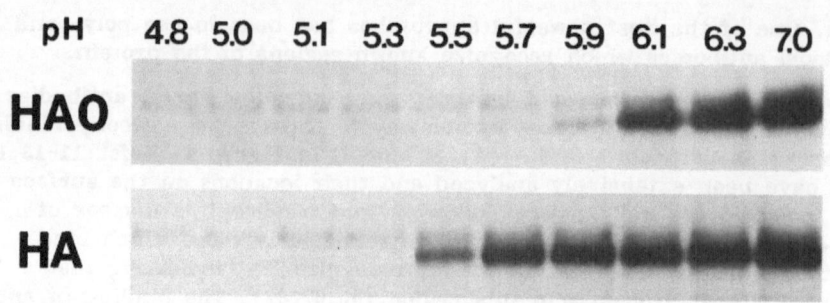

FIGURE 7 pH dependence with which HA and HA$_O$ become trypsin sensitive. The isolated proteins were incubated at the indicated pH for 30 min at 37°C, reneutralized, digested with trypsin, and visualized after SDS poly-acrylamide gel electrophoresis and autoradiography.

change has been in dispute. Daniels et al. (65) and Ruigrok et al. (78) have reported that HA$_O$ does not undergo the conformational change, whereas Jackson et al. (67) and Bächi et al. (87) obtained the opposite result based on antigenic changes.

When this question was reexamined in recent studies, it was found that HA$_O$ undergoes an irreversible conformational change quite similar to that seen for HA (64). The change is detected most efficiently above 32°C by antibodies specific to the acid and neutral conformations, respectively, and by trypsin sensitivity (Fig. 7 and Table 1). There are, however, significant differences between the responses of HA and HA$_O$ to acid pH: the changes are much slower and not as cooperative as in HA.

It is apparent that the activating cleavage of HA$_O$ to HA has several structural and functional consequences. First, it results in a conformational change in the neutral form of the molecule (see Ref. 15). Second, the activated molecule changes in such a way that it can respond to acid pH more rapidly and in a more cooperative fashion. Third, the cleavage allows the conformational change to occur so that a hydrophobic moiety is exposed. Without this moiety, which presumably includes the HA$_2$ N-terminus, the HA cannot interact with the target membrane hydrophobically and cause fusion. One role of the HA$_O$ precursor is probably to allow transport of the fusion factor to the plasma membrane of the infected cell without premature fusion activity.

D. The Conformational Change Occurs During Virus Entry

The studies described above have, for the most part, examined the effects of acidification on the structures of purified HA$_O$, HA, and BHA. It is apparent, however, that the conformational change also occurs in intact virions and during virus entry into cells. Bächi et al. (87) provided the first direct confirmation that HA undergoes an acid-induced conformational change following endocytosis. Antibodies specific to the acid conformation of HA were found to react with viruses in intracellular vacuoles. Reactivity was inhibited if cells were preincubated with lysosomotropic agents which elevate endosomal pH. We have made similar observations with conformation specific antibodies to the X-31 strain of influenza. By using immunoelectron microscopy, we found that antibodies specific to the neutral conformation of HA reacted with virus particles as long as they were present on the

TABLE 1 Responses of HA and HA$_O$ to Acid pH

Antibody	Subunit of HA	Epitope	HA		HA$_O$	
			pH	$T_{1/2}$	pH	$T_{1/2}$
N1	HA$_1$	TIP	5.5	0:11	6.1	1:30
N2	HA$_1$	TIP	5.5	0:10	6.0	—
A1	HA$_2$	—	5.5	<0:10	5.5	5:00
A2	HA$_1$	—	5.5	—	5.3	5:00
C	HA$_1$	98–106	5.5	—	6.0	—

Metabolically labeled HA$_O$ was isolated in Triton X-100 from CV-1 cells infected with the X-31 strain of influenza virus. HA was derived from the HAO by mild trypsin digestion. The proteins were incubated with buffers of various pH at 37°C for 30 min, reneutralized, and immuno-precipitated with the indicated monoclonal antibodies or digested with trypsin. The values reflect the pH at which 50% of the molecules were precipitated or digested. To determine the kinetics of conversion, the proteins were incubated at pH 5.0 for various periods of time (expressed in minutes) prior to reneutralization and processed as shown. Antibody C was raised against a synthetic peptide corresponding to the indicated residues of HA$_1$ (64).

cell surface but not with HA which had been delivered to intracellular vacuoles (Doms and Helenius, unpublished results). The vacuolar viruses stained with the acid-specific anti-HA antibodies.

VI. THE MECHANISM OF FUSION

While it has been possible to determine some of the irreversible changes in HA and some of the general properties of the fusion reaction, the underlying mechanism remains unclear. The difficulty in interpreting the data in the form of a coherent mechanism is, in part, due to the relative lack of data on the lipid-protein interactions and the changes in lipid configuration during fusion. We cannot expect to understand in detail how the fusion reaction takes place without additional information about the lipid bilayer perturbations involved. Unfortunately, sensitive methodology to study short-lived, highly localized changes in bilayer structure is not available.

The best we can do, at present, is to consider models that may be used as a basis for further experimentation. Three alternative mechanisms have been proposed (1,77,88). The emphasis in our treatment is on the one that we favor. For details about the others the reader is referred to the papers by Ruigrok and co-workers (77) and Landsberger and Sehgal (88).

As a starting point, our model assumes that HA, in its acidic conformation, directly participates in the fusion reaction, that fusion occurs without changes in lipid or protein covalent bonds, and that the reaction occurs in a focal point of limited size. On the basis of studies with BHA (57,71), we think that the hydrophobic moiety exposed in the ectodomain of HA interacts with the target membrane. The interaction is hydrophobic and therefore relatively nonspecific. Since proteins are not required in the target membrane for fusion, and since the interaction can be detected by photoactivatable lipid analogs (61,62,62a), it is likely that the interaction is directly with the lipid bilayer. In addition to the hydrophobic N-termini of the three HA_2 subunits, the interaction may also involve elements of the stem such as the central α-helices in HA_2. To explain why BHA is extractable with base from liposomal membranes (57), we suggest that the hydrophobic moiety penetrates only into one of the bilayer leaflets.

As a result of the interaction HA becomes, at least transiently, integrated into both the viral and target membranes. This dual association may help to bring the membrane surfaces into close proximity and thus overcome the hydration force imposed by surface-associated water molecules (see Chapters 3 and 4 and Refs. 89,90). The hydration force constitutes the main barrier to membrane fusion. In view of what is known about its magnitude, it seems plausible that the primary role of HA, and other fusion proteins, may simply be to perturb the hydration layers and allow close approach of the two membranes. Once the hydration force has been locally overcome, fusion of the lipid bilayers may occur spontaneously. This has been suggested previously in model lipid membrane studies using Ca^{2+} and other agents that can bring about close association and fusion of bilayers with which they interact (90).

The possibility that HA induces a local perturbation in the bilayer configuration which allows the lipids in the two membranes to adopt new geometrical alignments and fuse more easily should also be considered. Such a possibility is suggested by the site-specific mutant M1 (71). It is apparently able to associate with the target membrane, but unable to cause fusion, demonstrating that attachment alone is not sufficient for fusion. A perturbation could be caused by the insertion of polar groups into the hydrophobic part of the bilayer (these could involve the top parts of the coiled coils of the HA_2 subunits, which form the center of the stem), or by changing lipid packing geometry by inserting a proteinaceous moiety not easily accommodated by a bilayer.

Since fusion is nonleaky to ions and proteins, we assume that it proceeds without membrane rupture and resealing. We suggest that throughout all stages of the reaction, a double membrane barrier exists between the fusing compartments and the bulk solution. This means that fusion occurs in two steps; the two outermost bilayer leaflets must fuse followed by the two innermost. It may be that the function of the viral protein is to induce the first half of this reaction. The second step may occur spontaneously.

Although consistent with many of the results described above, the model described is not without problems. The most troubling is the location of the N-termini of the HA_2 subunits. As seen in the neutral pH structure, the fusion peptides are much closer to the viral than to the target bilayer. Between the two are the HA_1 domains. Clearly, a major change must occur in order for the amino termini to interact with the target membrane. As discussed above, a major structural alteration occurs at acid pH which we have interpreted to include a partial dissociation and unfolding of the trimeric structure along the central axis. This may allow the HA_2 amino termini access to the target bilayer. Whether this is needed to overcome the hydration force and bring the viral and target membranes into close proximity remains to be seen.

Does the fusion reaction require cooperativity between several HA molecules? A cooperative mechanism is suggested by our finding that about 80% of the spikes in a virus particle must be in the acidic form before fusion activity is detected (57) and the observation that a certain density of HA on the cell surface is necessary before cell:cell fusion is observed (3). Rosettelike structures are seen by electron microscopy in acid-treated viruses where spike density has been reduced (Fig. 6, Ref. 56). These may arise from several HA trimers whose HA_2 chains form a central nexus from which the HA_1 subunits radiate outward. We have speculated that fusion may occur in the center of such complexes. A very hydrophobic central core could be generated in which the HA_1 chains would not impose a steric barrier against bilayer interaction. In addition, the resulting high-protein density could drastically affect the configuration of lipids, which is a prerequisite for fusion of the bilayers.

HA-induced fusion also requires cooperativity at the level of the individual trimer. Recent studies using hybrid HA trimers have shown that acid-activation of one subunit raises the pH threshold required for activating the remaining two (91).

In contrast to the model described above, where the new hydrophobic moieties interact with the target membrane, Ruigrok et al. (77) have suggested that the moiety may interact with the viral membrane and modify it in such a way that it becomes prone to fusion. It is not clear how the virus in this case would interact with the target membrane and how the hydration force would be overcome. Landsberger and Sehgal (88) have suggested that HA interacts with phospholipid molecules in a fashion similar to phospholipid binding proteins, and by sequestering lipid sufficiently to destabilize the target membrane and induce fusion.

VII. CONCLUSIONS AND FUTURE DIRECTIONS

The influenza HA offers the best hope of understanding a biological membrane fusion system on the molecular level. A large amount of information is already available. However, as illustrated by the differences between the proposed mechanistic models, we have not yet reached a point where protein-mediated fusion can be understood. It will now be important to obtain more data on the role of the lipids and lipid-protein interactions. Promising results in this direction have recently been obtained using photo-activatable lipid analogs (61,62,62a). Further studies with well-characterized photoactivatable probes attached to defined positions in membrane lipids should make it possible to obtain information on the depth of protein penetration into the target membrane. Biophysical techniques, such as electron-spin resonance (ESR) and fluorescence spectroscopy, may shed light on the nature of the subsequent membrane perturbation. By using reconstituted HA virosomes (88) with different protein to lipid ratios, the number and density of spikes required for fusion can be determined.

The three-dimensional structure of acidic HA at atomic resolution should also be pursued. It may now be possible to genetically engineer a BHA_0 molecule that remains water soluble after acid treatment (cf. Refs. 63,64). Such a molecule could conceivably be crystallized and its three-dimensional structure may approximate that of acid-treated HA.

Finally, it is crucial that studies continue on other fusion proteins. Although different in their properties, they may share some basic functional principles. Viral proteins with fusion activity are numerous and relatively easily accessible. Cellular fusion proteins are being intensively searched for in numerous laboratories. It will be interesting to find out to what degree they share properties with HA.

REFERENCES

1. White, J., Kielian, M., and Helenius, A. (1983). Membrane fusion proteins of enveloped animal viruses. *Q. Rev. Biophys.* 16:151-195.
2. Stegmann, T., Doms, R. W., and Helenius, A. (1989). Protein-mediated membrane fusion. *Annu. Rev. Biophys. Biophys. Chem.* 18:187-211.
3. Doxsey, S. J., Sambrook, J., Helenius, A., and White, J. (1985). An efficient method for introducing macromolecules into living cells. *J. Cell Biol.* 100:704-714.
4. van Meer, G., Davoust, J., and Simons, K. (1985). Parameters affecting low-pH mediated fusion of liposomes with the plasma membrane of cells infected with influenza virus. *Biochemistry* 24:3593-3602.
5. Kielian, M. and Helenius, A. (1985). pH-induced alterations in the fusogenic spike protein of Semliki Forest virus. *J. Cell Biol.* 98:139-145.
6. Kielian, M. and Helenius, A. (1986). Entry of alpha viruses. In: *The Togaviridae and Flaviviridae* (S. Schlesinger and M. J. Schlesinger, eds.), Plenum Press, New York, pp. 91-119.
7. Wiley, D. C. (1985). Viral membranes. In: *Virology* (B. N. Fields et al., eds.). Raven Press, New York, pp. 45-68.
8. Lamb, R. A., and Choppin, P. W. (1983). The structure and replication of influenza virus. *Annu. Rev. Biochem.* 52:467-506.
9. Wiley, D. C. and Skehel, J. J. (1987). The structure and function of the hemagglutinin membrane glycoprotein of influenza virus. *Annu. Rev. Biochem.* 56:365-394.
10. Doms, R. W., Stegmann, T., and Helenius, A. (1989). Penetration of influenza virus into host cells. In: *Concepts in Viral Pathogenesis III* (A. L. Notkins and M. B. A. Oldstone, eds.). Springer Verlag, New York, pp. 114-120.
11. Webster, R. G., Laver, W. G., Air, G. M., and Schild, G. C. (1982). Molecular mechanisms of variation of influenza viruses. *Nature (London)* 296:115-121.
12. Palese, P., and Young, J. F. (1982). Variation of influenza A, B, and C viruses. *Science* 215:1468-1473.
13. van Rompuy, L., Min Jou, W., Verhoeyen, M., Huylebroeck, D., and Fiers, W. (1983). Molecular variation of influenza surface antigens. *Trends Biochem. Sci.* 8:414-417.
14. Varghese, J. N., Laver, W. G., and Colman, P. M. (1983). Structure of the influenza virus glycoprotein antigen neuraminidase at 2.9A resolution. *Nature (London)* 303:35-40.
15. Wilson, I. A., Skehel, J. J., and Wiley, D. C. (1981). Structure of the haemagglutinin membrane glycoprotein of influenza virus at 3A resolution *Nature (London)* 289:366-373.
16. Wiley, D. C., Wilson, I. A., and Skehel, J. J. (1981). Structural identification of the antibody sites of Hong Kong influenza haemagglutinin and their involvement in antigenic variation. *Nature (London)* 289:373-378.
17. Rogers, G. N., Paulson, J. C., Daniels, R. S., Skehel, J. J., and Wiley, D. C. (1983). Single amino acid substitutions in influenza

haemagglutinin change receptor binding specificity. *Nature (London)* 304:76–78.

18. Weiss, N., Brown, S. H., Cusack, S., Paulson, J. C., Skehel, J. J., and Wiley, D. C. (1988). Structure of the influenza virus haemagglutinin complexed with its receptor, sialic acid. *Nature* 333:426–431.

19. Matlin, K. S., Reggio, H., Helenius, A., and Simons, K. (1981). Infectious entry pathway of influenza virus in a canine kidney cell line. *J. Cell Biol.* 91:601–613.

20. Dourmashkin, R. R., and Tyrrell, D. A. J. (1974). Electron microscopic observations on the entry of influenza virus into susceptible cells. *J. Gen. Virol.* 24:129–141.

21. Patterson, S., Oxford, J. S., and Dourmashkin, R. R. (1979). Studies on the mechanism of influenza virus entry into cells. *J. Gen. Virol.* 43:223–229.

22. Yoshimura, A., and Ohnishi, S-I. (1984). Uncoating of influenza virus in endosomes. *J. Virol.* 51:497–504.

22a. Stegmann, T., Morselt, H. W. M., Scholma, J., and Wilschut, J. (1987). Fusion of influenza virus in an intracellular acidic compartment measured by fluorescence dequenching. *Biochim. Biophys. Acta* 904: 165–170.

23. Yoshimura, A., Kuroda, K., Yamashina, S., Maeda, T., and Ohnishi, S-I. (1982). Infectious cell entry mechanism of influenza virus. *J. Virol.* 43:284–293.

23a. Nussbaum, O., and Loyter, A. (1987). Quantitative determination of virus-membrane fusion events. Fusion of influenza virions with plasma membranes and membranes of endocytic vesicles in living cultures cells. *FEBS Lett.* 221:61–67.

24. Helenius, A., Mellman, I., Wall, D., and Hubbard, A. (1983). Endosomes. *Trends Biochem. Sci.* 8:245–250.

25. Mellman, I., Fuchs, R., and Helenius, A. (1986). Acidification of the endocytic and exocytic pathways. *Annu. Rev. Biochem.* 55:663–700.

26. Gollins, S. W., and Porterfield, J. S. (1986). pH-dependent fusion between the Flavivirus West Nile and liposomal model membranes. *J. Gen. Virol.* 67:157–166.

27. Pazmino, N. H., Yuhas, J. M., and Tennant, R. W. (1974). Inhibition of murine RNA tumor virus replication and oncogenesis by chloroquine. *Int. J. Cancer* 14:165–217.

28. Rhim, J. S., Lane, W. T., and Huebner, R. J. (1972). Amantadine hydrochloride: inhibitory effect on murine parcoma virus infection in cell cultures. *Proc. Soc. Exp. Biol. Med.* 139:1258–1260.

29. Wallbank, A. M., Matler, R. E., and Klinikowski, N. G. (1966). 1-Adamatanamine hydrochloride: Inhibition of Rous and Esh sarcoma viruses in cell culture. *Science* 152:1258–1260.

30. Matlin, K., Reggio, H., Simons, K., and Helenius, A. (1982). The pathway of vesicular stomatitis entry into MDCK-cells. *J. Cell Biol.* 91:601–613.

31. Gonzalez-Scarano, F., Pobjecky, N., and Nathanson, N. (1984). LaCrosse bunyavirus can mediate pH-dependent fusion from without. *Virology* 132:222–225.

32. Doms, R. W., Keller, D. S., Helenius, A., and Balch, W. E. (1987). Role for adenosine triphosphate in regulating the assembly and transport of vesicular stomatitis virus G protein trimers. *J. Cell Biol.* 105:1957–1969.

33. Gonzalez-Scarano, F. (1985). La Crosse virus G1 glycoprotein undergoes a conformational change at the pH of fusion. *Virology* 140:209-216

34. Maeda, T., and Ohnishi, S. (1980). Activation of influenza virus by acidic media causes hemolysis and fusion of erythrocytes. *FEBS Lett.* 122:282-287.

35. White, J., Matlin, K., and Helenius, A. (1981). Cell fusion by Semliki Forest, influenza and vesicular stomatitis viruses. *J. Cell Biol.* 89:674-679.

36. Huang, R. T. C., Rott, R., and Klenk, H-D. (1981). Influenza viruses cause hemolysis and fusion of cells. *Virology* 110:243-247.

37. White, J., Kartenbeck, J., and Helenius, A. (1982). Membrane fusion activity of influenza virus. *EMBO J.* 1:217-222

38. Nir, S., Stegmann, T., and Wilschut, J. (1986). Fusion of influenza virus with cardiolipin liposomes at low pH: Mass action analysis of kinetics and extent. *Biochemistry* 25:257-266.

39. Stegmann, T., Hoekstra, D., Scherphof, G. L., and Wilschut, J. (1986). Fusion activity of influenza virus. A comparison between artificial and biological target membrane vesicles. *J. Biol. Chem.* 261:10966-10969.

40. Young, J. D-E., Young, G. P. H., Cohn, Z. A., and Lenard, J. (1983). Interaction of enveloped viruses with planar bilayer membranes: Observations on Sendai, influenza, vesicular stomatitis, and Semliki Forest viruses. *Virology* 128:186-194.

41. Beyer, W. E. P., Ruigrok, R. W. H., van Driel, H., and Masurel, N. (1986). Influenza virus strains with a fusion threshold of pH 5.5 or lower are inhibited by amantadine. *Arch. Virol.* 90:173-181.

42. White, J., and Helenius, A. (1980). pH-dependent fusion between the Semliki Forest virus membrane and liposomes. *Proc. Natl. Acad. Sci. USA* 77:3273-3277.

43. Mooney, J. J., Dalrymple, J. M., Alving, C. R., and Russell, P. K. (1975). Interaction of Sindbis virus with liposomal model membranes. *J. Virol.* 15:225-231.

44. Maeda, T., Kawasaki, K., and Ohnishi, S-I. (1981). Interaction of influenza virus hemagglutinin with target membrane lipids is a key step in virus-induced hemolysis and fusion at pH 5.2. *Proc. Natl. Acad. Sci. USA* 78:4133-4137.

45. Haywood, A. M., and Boyer, B. P. (1985). Fusion of influenza virus membranes with liposomes at pH 7.5. *Proc. Natl. Acad. Sci. USA* 82:4611-4615.

46. Junankar, P. R., and Cherry, R. J. (1986). Temperature and pH dependence of the haemolytic activity of influenza virus and of the rotational mobility of the spike glycoproteins. *Biochim. Biophys. Acta* 854:198-206.

47. Lampe, P. D., and Nelsestuen, G. L. (1982). Myelin basic protein-enhanced fusion of membranes. *Biochim. Biophys. Acta* 693:320-325.

48. Klenk, H-D., Rott, R., Orlich, M., and Blodorn, J. (1975). Activation of influenza A viruses by trypsin treatment. *Virology* 68:426-439.

49. White, J., Helenius, A., and Gething, M-J. (1982). Haemagglutinin of influenza virus expressed from a cloned gene promotes membrane fusion. *Nature (London)* 300:658-659.

50. Deleted in proof.

51. Nussbaum, O., Lapidot, M., and Loyter, A. (1987). Reconstitution of functional influenza virus envelopes and fusion with membranes and liposomes lacking Virus receptors. *J. Virol.* 61:2245-2252.

52. Stegmann, T., Morselt, H. W. M., Booy, F. P., van Breemen, J. F. L., Scherphof, G., and Wilschut, J. (1987). Functional reconstitution of influenza virus envelopes. *EMBO J.* 6:2651-2659.

53. Huang, R. T. C., Dietsch, E., and Rott, R. (1985). Further studies on the role of neuraminidase and the mechanism of low pH dependence in influenza virus-induced membrane fusion. *J. Gen. Virol.* 66:295-301.

54. Huang, R. T. C., Rott, R., Wahn, K., Klenk, H-D, and Kohama, T. (1980). The function of the neuraminidase in membrane fusion induced by myxoviruses. *Virology* 107:313-319.

55. Brand, C. M., and Skehel, J. J. (1972). Crystalline antigen from the influenza virus envelope. *Nature (London) New Biol.* 238:145-147.

56. Doms, R. W., and Helenius, A. (1986). The quaternary structure of the influenza virus hemagglutinin after acid treatment. *J. Virol.* 60:833-839.

57. Doms, R. W., Helenius, A. H., and White, J. (1985). Membrane fusion activity of the influenza virus hemagglutinin. The low pH induced conformational change. *J. Biol. Chem.* 260:2973-2981.

58. Skehel, J. J., Bayley, P. M., Brown, E. B., Martin, S. R., Waterfield, M. D., White, J. M., Wilson, I. A., and Wiley, D. C. (1982). Changes in the conformation of influenza hemagglutinin at the pH optimum of virus mediated membrane fusion. *Proc. Natl. Acad. Sci. USA* 79:968-972.

59. Graves, P. N., Schulman, J. L., Young, J. F., and Palese, P. (1983). Preparation of influenza virus subviral particles lacking the HA_1 subunit of HA: unmasking of cross-reactive HA_2 determinants. *Virology* 126:106-116.

60. Doms, R. W., and Helenius, A. (1988). Properties of a viral fusion protein. In: *Molecular Mechanisms of Membrane Fusion* (S. Ohki et al., eds.). Plenum Press, New York, pp. 385-398.

61. Boulay, F., Doms, R. W., and Helenius, A. (1986). Photolabeling of the influenza hemagglutinin with a hydrophobic probe. In: *Positive Strand RNA Viruses* (M. A. Brinton and R. R. Ruickert, eds.). Alan R. Liss, New York, pp. 103-112.

62. Harter, C., Bachi, T., Semenza, G., and Brunner, J. (1988). Hydrophobic photolabeling identifies BHA2 as the subunit mediating the interaction of bromelain-solubilized influenza virus hemagglutinin with liposomes at low pH. *Biochemistry* 27:1856-1864.

62a. Harter, C., James, P., Bächi, T., Semenza, G., and Brunner, J. (1989). Hydrophobic binding of the membranes occurs through the fusion peptide. *J. Biol. Chem.* 264:6459-6464.

63. Gething, M-J., and Sambrook, J. (1982). Construction of influenza hemagglutinin genes that code for intracellular and secreted forms of the protein. *Nature (London)* 300:598-603.

64. Boulay, F., Doms, R. W., Wilson, I., and Helenius, A. (1987). The influenza hemagglutinin precursor as an acid sensitive probe of the biosynthetic pathway. *EMBO J.* 6:2643-2650.

65. Daniels, R. S., Douglas, A. R., Skehel, J. J., and Wiley, D. C. (1983). Analyses of the antigenicity of influenza hemagglutinin at the pH optimum for virus-mediated membrane fusion. *J. Gen. Virol.* 64:1657-1662.

66. Webster, R. G., Brown, L. E., and Jackson, D. C. (1983). Changes in the antigenicity of the hemagglutinin molecule of H3 influenza virus at acidic pH. *Virology* 126:587-599.

67. Jackson, D. C., Murray, J. M. Anders, E. Margot, White, D. O., Webster, R. G., and Brown, L. E. (1983). Expression of a unique antigenic determinant of influenza virus hemagglutinin at pH 5. In:

The Origin of the Pandemic Influenza Viruses (W. G. Laver, ed.). Elsevier, New York.

68. Jackson, D. C., and Nestorowicz, A. (1985). Antigenic determinants of influenza virus hemagglutinin. XI. Conformational changes detected by monoclonal antibodies. *Virology* 145:72-83.

69. Daniels, R. S., Downie, J. C., Hay, A. J., Knossow, M., Skehel, J. J., Wang, M. L., and Wiley, D. C. (1985). Fusion mutants of the influenza virus hemagglutinin glycoprotein. *Cell* 40:431-439.

70. Doms, R. W., Gething, M-J., Henneberry, J., White, J., and Helenius, A. (1986). Variant influenza virus hemagglutinin that induces fusion at elevated pH. *J. Virol.* 57:603-613.

71. Rott, R., Orlich, M., Klenk, H-D., Wang, M. L., Skehel, J. J., and Wiley, D. C. (1984). Studies on the adaptation of influenza viruses to MDCK cells. *EMBO J.* 3:3329-3332.

72. Gething, M-J., Doms, R. W., York, D., and White, J. (1986). Studies on the mechanism of membrane fusion: Site-specific mutagenesis of the hemagglutinin of influenza virus. *J. Cell Biol.* 102:11-23.

73. Gething, M-J., Doms, R. W., White, J., and Helenius, A. (1986). Studies on the mechanism of membrane fusion. In: *Protein Engineering: Applications in Science, Medicine, and Industry* (M. Inouye and R. Sarma, eds.). Academic Press, New York.

74. Gonzalez-Noriega, A., Grubb, J. H., Talkad, V., and Sly, W. S. (1980). Chloroquine inhibits lysosomal enzyme pinocytosis and enhances lysosomal enzyme secretion by impairing receptor recycling. *J. Cell Biol.* 85:839-852.

75. Daniels, R. S., Jefferies, S., Yates, P., Schild, G. C., Rogers, G. N., Paulson, J. C., Wharton, S. A., Douglas, A. R., Skehel, J. J., and Wiley, D. C. (1987). The receptor-binding and membrane fusion properties of influenza virus variants selected using anti-hemagglutinin monoclonal antibodies. *EMBO J.* 6:1459-1465.

75a. Stegmann, T., Booy, I. P., and Wilschut, J. (1987). Effects of low pH on influenza virus. Activation and inactivation of the membrane fusion capacity of the hemagglutinin. *J. Biol. Chem.* 262:17744-17749.

76. Nestorowicz, A., Laver, G, and Jackson, D. C. (1985). Antigenic determinants of influenza virus hemagglutinin. X. A comparison of the physical and antigenic properties of monomeric and trimeric forms. *J. Gen. Virol.* 66:1687-1695.

77. Ruigrok, R. W. H., Wrigley, N. G., Calder, L. J., Cusak, S., Wharton, S. A., Brown, E. B., and Skehel, J. J. (1986). Electron microscopy of the low pH structure of influenza virus hemagglutinin. *EMBO J.* 5:41-49.

78. Ruigrok, R. W. H., Cremers, A. F. M., Beyer, W. E. P., and de Ronde-Verloop, F. M. (1984). Changes in the morphology of influenza particles induced at low pH. *Arch. Virol.* 82:181-194.

79. Hewat, E. A., Cusack, S., Ruigrok, R. W. H., and Verwey, C. (1984). Low resolution structure of the influenza C glycoprotein determined by electron microscopy. *J. Mol. Biol.* 175:175-193.

80. Lepault, J., Booy, F. P., and Dubochet, J. (1983). Electron microscopy of frozen biological suspensions. *J. Micros.* 129:89-102.

81. Ruigrok, R. W., Aitken, A., Calder, L. J., Martin, S. R., Skehel, J. J., Wharton, S. A., Weis, W., and Wiley, D. C. (1988). Studies on the structure of the influenza virus haemagglutinin at the pH of membrane fusion. *J. Gen. Virol.* 69:1847-1857.

82. Yewdell, J. W., Gerhard, W., and Bächi, T. (1983). Monoclonal anti-hemagglutinin antibodies detect irreversible antigenic alterations that

coincide with the acid activation of influenza virus A/PR/8/34-mediated hemolysis. *J. Virol.* 48:239–248.

83. Copeland, C. S., Doms, R. W., Bolzau, E. M., Webster, R. G., and Helenius, A. (1986). Assembly of influenza hemagglutinin trimers and its role in intracellular transport. *J. Cell Biol.* 103:1179–1192.

84. White, J., and Wilson, I. A. (1987). Anti-peptide antibodies detect steps in a protein conformational change: low-pH activation of the influenza virus hemagglutinin. *J. Cell Biol.* 105:2887–2897.

85. Scheid, A., and Choppin, P. (1974). Identification of biological activities of paramyxovirus glycoproteins. Activation of cell fusion, hemolysis and infectivity by proteolytic cleavage of an inactive precursor protein of Sendai virus. *Virology* 57:475–490.

86. Nagai, Y., and Klenk, H-D. (1977). Activation of precursors to both glycoproteins of Newcastle disease virus by proteolytic cleavage. *Virology* 77:125–134.

87. Bächi, T., Gerhard, W., and Yewdell, J. W. (1985). Monoclonal antibodies detect different forms of influenza virus during viral penetration and biosynthesis. *J. Virol.* 55:307–313.

88. Landsberger, F. R., and Sehgal, P. B. (1986). Protein-mediated fusion of viral and cellular membranes. In: *Virus Attachment and Entry Into Cells* (R. L. Crowell and K. Longberg-Holm, eds.). ASM Press, pp. 66–71.

89. Rand, R. (1981). Interacting phospholipid bilayers: measured forces and induced structural changes. *Annu. Rev. Biophys. Bioeng.* 10:277–314.

90. Wilschut, J. and Hoekstra, D. (1984). Membrane fusion: from liposomes to biological membranes. *Trends Biochem. Sci.* 9:479–483.

91. Boulay, F., Doms, R. W., Webster, R. G., and Helenius, A. (1988). Post-translational oligomerization and cooperative acid activation of mixed influenza hemagglutinin trimers. *J. Cell Biol.* 106:629–639.

16

Fusion of Viruses with Phospholipid Vesicles at Neutral pH

ANNE M. HAYWOOD

University of Rochester Medical Center, Rochester, New York

I. INTRODUCTION

Many viruses are enveloped, i.e., have a membrane surrounding the nucleocapsid. The membranes of some of these viruses contain only a few proteins and so are much simpler than most biological membranes. Viral membranes interact with the host cell membranes to effect viral entry and exit. Therefore, they are excellent membranes for investigating membrane-membrane interactions. To use viral membranes in this way it is necessary that the complexities of the host cell membrane do not obscure the functions of the viral membrane. For this reason liposomes are often used in place of host cells as relatively simple target membrane vesicles.

For the virus-liposome system to be used to probe biological interactions it is important both that biological conditions be simulated and that any interaction such as membrane fusion not be considered an isolated phenomenon but as part of the continuum of viral entry. Liposomes made of lipid compositions similar to those of host cell membranes and temperatures and ionic conditions reflecting those found in vivo are most likely to give data that reflect the biological situation. In vivo membrane fusion is preceded by the binding of virus to receptors and followed by viral disassembly. The nature and kinetics of the binding can be expected to influence fusion. Viral disassembly causes fusion to be more easily identified by electron microscopy.

Viral membrane fusion in vivo and viral structure are briefly reviewed before considering fusion of viral membranes with phospholipid vesicles as a model for biological membrane fusion. Sendai virus is used as the prototype.

II. HISTORY AND BIOLOGICAL ASPECTS OF MEMBRANE FUSION DIRECTED BY VIRUSES

Pathologists have recognized for nearly 60 years that the presence of syncytia can be used as a diagnostic criterion for infections by paramyxoviruses and

by herpesviruses. The presence of synctia was first noted for measles
infections even before it was known that measles was caused by a virus
(1,2) and is still used in diagnosis of herpes infections. Formation of syncytia
is used as an assay for human immunodeficiency virus. The membranes
of the infected cells fuse with membranes of adjacent cells but only after
there has been synthesis of viral proteins, including the viral membrane
proteins that are incorporated into the plasma membrane of the infected
cell. This has been dubbed "fusion from within" (3). Whether a virus
can cause fusion from within is also determined by whether infection results
in cytolysis. Only if the cell survives the infection will polykaryocytes
be seen. Vesicular stomatitis virus is an example of a virus that usually
is cytolytic, but can cause fusion from within at neutral pH if certain
strains or mutants are used or if protein synthesis is decreased after
the eclipse period of viral replication (4).

Sendai virus, a mouse paramyxovirus, was isolated in the mid-1950s
(5). Shortly thereafter it was realized that cell-cell fusion can be caused
by the addition of very high multiplicities of Sendai virus to cells even
under conditions that do not allow viral replication (6). This has been
dubbed "fusion from without." Until it was discovered that polyethylene
glycol could cause fusion, Sendai virus was utilized to form cell hybrids
to study mammalian genetics (7). Virus-induced cell fusion has been reviewed (8).

By 1950 it was recognized that paramyxoviruses can cause hemolysis.
Hemolysis was early recognized to be enhanced by treatments that partially
damage the viral structure, such as repeated freezing and thawing, mild
sonication, or changes in osmotic pressure (9). Hemolysis has been used
as an assay for the capacity of Sendai virus to fuse. Erythrocytes are,
however, not a host cell and it must be remembered that they, like liposomes, are model membranes for studying viral entry.

In 1968 Morgan and Howe (10) demonstrated by electron microscopy
that Sendai viruses can fuse their membranes with cellular membranes,
thereby releasing their nucleocapsids into the cytoplasm. They proposed
that this constitutes a method of viral entry.

In the early 1970s it was recognized that specific gangliosides could
serve as Sendai virus receptors (11,12) and that ganglioside-containing
liposomes could both envelop virus as in the initial steps of endocytosis
(13) and fuse with viral membranes (14). This made it evident that the
viral proteins can direct membrane fusion without participation of the host
cell proteins, so that the Sendai virus membrane could be used for studying
membrane interactions and particularly membrane fusion (15).

III. THE STRUCTURE OF SENDAI VIRUS

Sendai virus [also called hemagglutinating virus of Japan (HVJ)] is a
member of the Paramyxoviridae (16). The virion contains six proteins,
including five viral-coded proteins and host-derived actin. It is likely
that these proteins are multifunctional and that only part of their activities
are currently understood. The virus particles as usually isolated are pleomorphic and mostly 150-250 nm in diameter, although particles up to 600
nm in diameter can be seen (17).

A. Sendai Virus Membrane Components

The membrane contains only three proteins (18). Two of the membrane proteins, the HN and the F proteins, are glycoproteins and can be seen as spikes that extend about 120 Å from the viral surface (17).

1. The HN Protein

The HN protein (also called HANA) has neuraminidase (sialidase) activity and binds to host cell receptors. Binding to erythrocytes results in hemagglutination, which is often used as a biological assay to measure the amount of virus present. The polypeptide has a molecular weight of 67,000 (19) and is found in the virus mainly as disulfide-linked dimers and tetramers (20). The HN protein and the F proteins are transmembrane proteins, at least after fusion has occurred (21). The HN protein is anchored in the membrane at its N-terminal end (22,23). Studies with monoclonal antibodies suggest the exterior portion of the protein has four functional sites (24,25). Monoclonal antibodies to the HN protein which inhibit hemolysis but do not block hemagglutination or neuraminidase activity have been identified (24,26). Therefore, it is likely the HN participates in the fusion process by more than merely promoting binding. Another paramyxovirus, mumps, causes extensive syncytial formation (fusion from within) only when a variant lacking neuraminidase activity is used for infection (27).

2. The F Protein

The F protein is synthesized as a precursor F_0 protein and converted by proteolytic cleavage to the active F protein which consists of the disulfide-linked polypeptide subunits, F_1 and F_2. The active F protein is necessary for the virus to be able to fuse and to be infective (28,29). It is thought to be anchored in the membrane by a hydrophobic region near the C-terminal end and to have 42 amino acids on the cytoplasmic side of the membrane (31,31). Activation of radioactive photoactivable hydrophobic reagents in Sendai virus gave a pattern of F-protein labeling that is consistent with the protein being anchored in the bilayer near the C-terminal end and also with an interaction of the more distal section of the protein with the bilayer (32). This raised the question whether the F protein acts on the viral or host membrane. Monoclonal antibodies to the F protein can be divided into two groups, only one of which inhibits hemolysis (24). It has been suggested the F protein also recognizes a receptor (33,34), but that this receptor is not on erythrocytes (33). The ability of liposomes reconstituted with HN and F proteins to fuse with cells varies with the ratio of the F and HN protein (35). This suggests that an interaction between the F and HN proteins may be important for fusion.

3. The M Protein

The M protein, the third membrane protein, is small (M.W. 38,500 for the Harris strain) and highly basic (36). When the M protein is isolated, it forms filaments and sheets (37,38). The virus contains host-derived actin (39), and the M protein associates with this actin (40). The M protein is thought to form a matrix on the inner side of the bilayer and to participate in assembly by organizing the glycoproteins and binding the nucleocapsid (38,41-43).

4. Membrane Lipids

Whether the lipid composition of the viral membrane is nearly the same
as that of the host plasma membrane from which the virus membrane is
derived or whether the virus proteins modulate the lipid composition is
controversial. In studies with MK cells and BHK21 cells, the lipid composition of a paramyxovirus membrane was similar to that of the host membrane
(44), but this was not so in studies that used egg chorioallantoic or chick
embryo cells as host (45,46). This leaves unresolved whether there are
specific viral protein-lipid interactions.

B. Viral Internal Components

The interior of the virus contains the RNA, and the nucleoprotein (NP),
L and P proteins. The nucleocapsid is a ribonucleoprotein (RNP) composed
of a single-stranded, minus-strand RNA (M.W. 5×10^6) which is associated
with many NP subunits to make a flexible rod with helical symmetry. Minus-
strand RNA does not function as a messenger and so must be copied into
plus-strand RNA before viral proteins can be made. For this reason the
virion contains transcriptase (RNA polymerase) (47-50), methyltransferase
(51), and polyadenylase (52) activities. The L and P proteins contribute
to the transcriptase activity (49).

C. Young ("Early Harvest") and Aged ("Late Harvest") Sendai Virus

The structure of the virus and its ability to cause hemolysis vary with
the age of the virus. As discussed in the section on disassembly, these
differences probably determine whether the viruses disassemble after fusion
with liposomes. When the chorioallantoic cavity of embryonated eggs is
injected with virus, the virus infects some of the chorioallantoic cells and
produces virus in about a day (20-24 h). Some of the virus resulting
from the first round of infection in turn infects those cells that were not
previously infected, and these produce more virus. To obtain the maximum
amount of virus, it is customary to harvest the virus from the embryonated
egg at 48 or 72 h, which gives aged ("late harvest") virus. The aged
virus contains a mixture of virus that was formed during the last round
of replication and of virus which has aged in the incubated egg for several
days. When virus is harvested at 24 h to give young ("early harvest")
virus, its structure is different from that of aged (late harvest) virus
(53,54). A detailed electron microscopic study (55) showed young virions
have a fairly uniform size and are rod-shaped, with the nucleocapsids
strands regularly folded along the long axis of the virions. Aged virus
is very pleomorphic and includes large, irregularly shaped virions which
are presumably the older virus. These virions have randomly folded nucleocapsid strands. On freeze-fracture the young virus has no intramembrane
particles (IMP); whereas, many of the late harvest virus have IMP on
the E but not the P faces. The distribution of IMP is different on different
virions. When young virions were treated with cross-linker, a NP-M hetero-
dimer was formed, and the amount of this dimer that could be formed decreased as the virions were aged (20). It has been suggested that the
M protein acts as a "viroskeleton" and that aging of the virus disrupts
the interactions with the M protein (55). Aged virions fuse with each other
to make large virions when incubated at 37°C, whereas young virions are

resistant to fusion with each other even in the presence of polyethylene glycol (56). Therefore, the "viroskeleton" also appears to have an effect on the ease with which the virion fuses.

D. Virus-Associated Enzyme Activities

In addition to the transcriptase, methyltransferase, and polyadenylase activities, other enzyme activities also are associated with the virion. These include protease (57,58), protein kinase (59,60), and protein phosphatase activities (59). Which protein(s) contribute these enzyme activities is not defined for all the activities, and in some cases the activity may result from host proteins that are associated with or included in the virus. As might be expected since the virus contains protein kinase and protein phosphatase activities, the viral proteins exist in several forms that differ in their degree of phosphorylation (61,62). Adenosine diphosphatase and adenosine triphosphatase activities have been reported associated with several enveloped viruses including Sendai virus, and it was proposed that this enzymatic activity is derived from the host and in some way is related to the mode of release of enveloped viruses (63). These enzymes resemble what was later defined as nucleoside diphosphate phosphohydrolase, which has been described in vesicular stomatitis virus (64). Interestingly, the presence of an ATP diphosphohydrolase has been associated with the release of microvesicles in several nonviral systems (65).

Despite the relatively few proteins the virus contains, it has many activities, and studies with monoclonal antibodies suggest there are more functions of the known proteins to be uncovered. Some of these activities participate in fusion and disassembly and some, such as the transcriptase, probably function after fusion and disassembly. Use of liposomes in place of the host cell should make it possible to investigate these activities without the complication of host proteins. Further, the presence of these activities in liposomes after fusion can be used to monitor whether fusion and disassembly with liposomes reproduce the events occurring after entry into cells.

IV. REQUIREMENTS FOR MEMBRANE FUSION

The requirements for biological membrane fusion include (1) specific binding, (2) close approach, (3) destabilization of the bilayer, (4) mixing of the membrane components and restabilization of the bilayer. These requirements involve most of the steps in entry, which include binding of virus to its receptors, penetration by membrane fusion and/or by endocytosis, and disassembly (uncoating) of the virus with activation of the viral enzymes and the initiation of plus-strand RNA synthesis.

In the following discussion, three concepts are developed:

1. Binding and close approach determine much of the fusion process.
2. The deformation of the membrane that results from binding and enveloping the virus is involved in fusion.
3. Whether viral disassembly follows fusion depends on the condition of the virus. Totally intact virus often is not disassembled after fusion with liposomes. This suggests that disassembly is an active process that involves either the viral enzymatic activities, host components, or both.

A. Binding

Viral receptors participate in the binding of virus particles to cell membranes. They are one of the factors that determine the specificity of the tissues to be infected. Viral binding to receptors may well initiate changes in both the cell and virus. In incorporating receptors into liposomes, both the structural details of the receptor and the optimum concentration must be considered.

1. The Receptors for Myxo- and Paramyxoviruses Contain Sialic Acid

The receptors for both the myxoviruses (influenza) and paramyxoviruses (Sendai virus) have long been known to contain sialic acid (substituted neuraminic acid) as shown by the fact that treatment of cells with sialidase (neuraminidase) prevents these viruses from binding to and infecting the cells (66). The viral sialidases also destroy the virus receptors and were originally called receptor-destroying enzymes (RDE). Since the surface of the virus is covered with HN glycoprotein spikes, the binding is multivalent. Sialoglycoproteins have multiple sialic acid residues and can inhibit hemagglutination (67). They were therefore initially assumed to be the viral receptors.

Gangliosides subsequently were shown also to be able to act as Sendai virus receptors when incorporated in liposomes so that they are arrayed over the liposome surface and thus are capable of multivalent binding with virus (11,12). This was demonstrated by showing that during centrifugation at 8800 × *g* for 10 min Sendai virus cosediments with large ganglioside-containing liposomes but not with liposomes with no net charge or with liposomes containing other negatively charged lipids. The negatively charged virus also cosediments with positively charged stearylamine-containing liposomes. Ganglioside-containing but not stearylamine-containing liposomes can compete with the erythrocyte receptors and cause inhibition of hemagglutination.

Sialoglycoproteins, such as glycophorin, when incorporated in liposomes also bind Sendai virus (68,69). Liposomes containing glycophorin fuse with Sendai virus (68,70).

The relative importance of gangliosides and sialoglycoproteins as receptors in vivo is not known. Sialidase treatment removes the receptor activity of cells and their susceptibility to Sendai virus infection. Susceptibility can be restored by the addition of exogenous receptor gangliosides which become incorporated into the cell membranes (71). Therefore, it seems likely that gangliosides play some role in viral infection in vivo.

Not only viruses but many substances [hormones (72), lectins (73), and toxins (74)] can bind both gangliosides and sialoglycoproteins. In some cases it has been postulated that these substances interact with complexes containing both gangliosides and sialoglycoproteins. In the case of the thyrotropin receptor, the glycoprotein is apparently important in high-affinity recognition but the ganglioside is vital for all the steps that result in coupling to the adenylate cyclase system. The fact that these other molecules use sialoglycoconjugates as receptors raises the interesting question whether the virus uses a receptor that provides a physiological function and whether the cellular changes that result from viral binding facilitate infection. Similarly, data discussed below suggest that binding of viral glycoproteins to receptors results in changes in the virus. Whether gangliosides are able to duplicate completely the effects of the viral receptors in the cell is unknown.

2. *Different Myxo- and Paramyxoviruses Recognize Different Receptors*

The sialic acid residues of gangliosides have different linkages to the other sugars present in the gangliosides. The other sugars also vary. Since different myxo- and paramyxoviruses have different receptors, they can be expected to bind to different gangliosides and sialoglycoproteins. Initial studies using the ganglioside G_{M1} and di- and trisialogangliosides isolated from bovine brain indicated that Sendai viruses does not bind to G_{M1} and led to the suggestion that it binds to gangliosides with a terminal *N*-acetyl neuraminic acid (12). Holmgren et al. (75) examined the ability of Sendai virus to bind to individual brain gangliosides and confirmed and extended the receptor specificity. They found that the ganglioside G_{D1a} but not G_{D1b} or G_{M1} binds Sendai virus. Ganglioside G_{D1a} contains a terminal *N*-acetylneuraminic acid in an α2-3-ketosidic linkage to the adjacent galactose, which in turn is linked to an *N*-acetylgalactosamine (NeuAcα2-3Galβ1-3GalNAc). Gangliosides containing a terminal NeuAcα2-8NeuAcα2-3Galβ1-3GalNAc, such as G_{Q1b} and G_{P1c}, bind Sendai virus even more efficiently.

Umeda and co-workers (76), using the formation of a ternary complex of liposome, virus, and erythrocytes as an assay, found that on the basis of mol sialic acid per mol of phospholipid, human erythrocyte sialosylparagloboside is as efficient as glycophorin and considerably more efficient than G_{D1a} and G_{T1b} as a receptor for Sendai virus. Sialosylparagloboside has the terminal sequence NeuAcα2-3Galβ1-4GlcNAc.

Different strains of influenza virus recognize different sialic acid linkages (2-3 vs. 2-6), and the binding is also influenced by which sugars in addition to sialic acid are present (77). If gangliosides are to be used when studying fusion of influenza virus membranes with liposomes, these differences in receptors have to be taken into account.

When gangliosides are used as receptors in studies of liposome fusion with myxo- and paramyxoviruses, it is also important to consider the amount of gangliosides used, since the high concentrations sometimes used result in decreased extents of fusion (78).

3. *Modulation of Receptor Binding by Carbohydrate Polymers*

The adhesion of Sendai virus to liposomes containing receptor gangliosides can be increased by the presence of other carbohydrate polymers. While it is possible to cosediment virus and liposomes containing ganglioside receptors at low speeds (e.g., 8800 × *g*), centrifugation at high speeds (300,000 × *g*) removes virus from liposomes containing ganglioside G_{D1a} (79). In the presence of ficoll or dextran, however, the adhesion of virus to liposomes containing receptor ganglioside (G_{D1a}) at 0-4°C is greatly increased, so that the bound virus is not released by centrifugation at 300,000 × *g*. Ficoll and dextran increase adhesion of virus only to liposomes containing gangliosides that have receptor activity and not to other liposomes including liposomes that contain gangliosides that have no receptor activity (80).

Cells, of course, have polysaccharides in their extracellular matrix and glycocalyx, so it is quite possible that polysaccharides play a part in the binding of viruses to cells. Consistent with this also is the demonstration of WuDunn and Spear (81) that the initial interaction of herpes

simplex virus with cells is binding to heparan sulfate, which is present
on the surface of many vertebrate cells. They propose this interaction
is only the first in a cascade of interactions leading to viral membrane
fusion.

Enhanced binding to receptors in the presence of other macromolecules
may be a general phenomenon, since in the presence of dextran and of
serum albumin lectins show increased high affinity binding to liposomes
containing a receptor glycoprotein (82,83). Similar increases in binding
to cells were not reported. The use of liposomes has highlighted an interaction that would be difficult to demonstrate in vivo where the glycocalyx
and extracellular matrix are already present.

4. For Most Viruses the Receptors Are Not Known

More is known about the receptors for the orthomyxo- and paramyxoviruses
than about the receptors for other viruses. In the late 1980s the use of monoclonal antibodies and of gene transfer of receptors to nonsusceptible cells
made it possible to start identifying the proteins involved in binding of some
other virus groups (84). In studies on the binding of viruses to liposomes
two general phenomena, which may represent nonspecific binding, have
been observed. One is that viruses often bind when the pH is lowered.
and the other is that they often bind and/or fuse to liposomes containing
negatively charged lipids.

The first observation that low pH affects the interaction of virus with
membranes was that of Mooney et al. (85), who noted that, just as Sindbis
virus hemagglutinates goose and sheep erythrocytes with a pH optimum
of 6.0, similarly Sindbis virus binds to liposomes prepared from sheep
erythrocyte lipids with an optimum at pH 6.0. This binding was performed
at 37°C and may have represented both binding and fusion. The binding
of Semliki Forest virus and of its isolated spike glycoproteins to different
cells is also dependent on pH, and the suggestion was made that virus
can bind to specific glycoproteins at neutral pH and to plasma membrane
lipids at low pH (86). Influenza virus hemagglutinin released by bromelain
(BHA) binds to liposomes of diverse compositions at pH 5 and only to
liposomes containing gangliosides at pH 7.0 (87).

Sendai virus and influenza virus membranes fuse with liposomes that
contain any of a variety of negatively charged lipids (88,89), so they
must bind to such liposomes. Thus, there is considerable evidence that
viruses can fuse with liposomes that do not contain specific receptors.
The question is whether receptors are necessary if membrane fusion of
Sendai virus with liposomes is to model in vivo entry accurately. The
parameters affecting fusion of Sendai virus with liposomes that contain
cardiolipin (CL) or phosphatidylserine (PS) vary from those that contain
phosphatidylcholine (PC), phosphatidylethanolamine (PE), cholesterol,
and GD_{1a} (89). These parameters are also changed when fusion occurs
at low pH, and it was suggested that fusion with liposomes can be achieved
in a manner that, at least in part, is not of physiological significance.

5. Rearrangements in Sendai Virus Structure
After Binding to Cells at 37°C

When Sendai viruses is allowed to bind to erythrocytes or to HeLa or
Lettree cells in the cold and the temperature is raised to 37°C, major
rearrangements occur in the viral membrane (17,90,91). During these
rearrangements the envelopes of many of the bound virions become con-

voluted with numerous infoldings of the membrane. In freeze-fracture replicas these virus particles show linear ridges about 30 nm wide and up to 0.5 µm long on the E faces and a complementary arrangement of linear grooves on the P faces. These linear ridges are always smooth and devoid of IMP. The 14-nm IMP present before binding and warming are replaced by IMP that are about 9 nm in diameter. The virus particles lose their clearly defined spikes. An active F protein is necessary for these rearrangements, and it was suggested that the function of the F spike is to cause the structural reorganization of the envelope components. Those virus particles that underwent these structural changes all appeared to fuse with the cell membrane (91). Other virus particles remained spherical and did not fuse. Although Pinto da Silva et al. (92) failed to observe these structural changes, this is probably related to the fact they did not see any virus directly participating in the fusion event. Volsky and Loyter (93) did observe these changes in viruses fusing with erythrocytes and interpreted the ridges as filaments of viral M protein.

These rearrangements only occur at higher temperatures and only when the virus is bound to a cell (90,91), which suggests that they are mediated by the viral membrane proteins as a consequence of binding to receptors. If this is so, liposomes should contain receptors if viruses are to fuse with liposomes by the same mechanism that they do with cells.

B. "Endocytosis" of Virus and Stabilization of Binding

Since the binding of paramyxo- and orthomyxoviruses to sialoglycoconjugates in cells or liposomes occurs at 0-4°C, and the viral neuraminidase is active only at higher temperatures, the binding of these viruses is usually studied at 0-4°C. Viral penetration, on the other hand, occurs only at higher temperatures and is usually studied at 37°C.

When Sendai viruses are added to liposomes containing gangliosides and the temperature is raised to 37°, the viruses are enveloped by the liposomes, as shown in Figure 1. If liposomes do not contain gangliosides or other lipids with a net charge, they do not envelop the virus. Therefore,

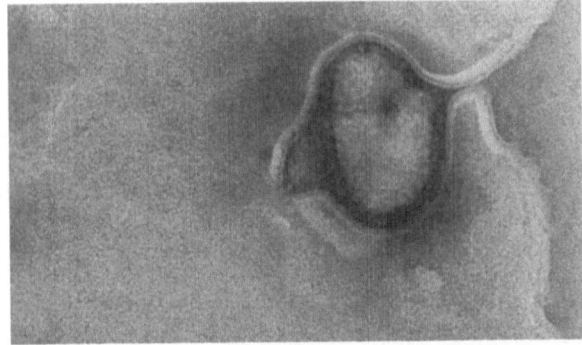

FIGURE 1 Envelopment of Sendai virus by a ganglioside-containing liposome. Liposomes were made from 1.14 µmol of PC, 0.63 µmol of PE, and brain gangliosides containing 0.2 µmol sialic acid in 0.2 ml PBS. Sendai virus was added for 55 min at 0°C and then incubated for 6 min at 37°C. The samples were stained with 2% ammonium molybdate, pH 7. Magnification: 124,050×.

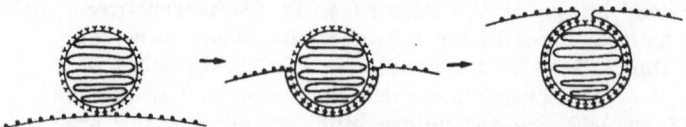

FIGURE 2 Deformation of a liposome and envelopment of Sendai virus
as a result of multivalent binding of viral proteins to gangliosides.
Υ , Viral HN protein. ● , Ganglioside receptor.

envelopment was proposed to be a consequence of binding of the virus
to liposomes (13). For two spherical bodies to accomplish multivalent binding,
deformation of at least one surface is necessary (94). Since the virus
is a relatively rigid particle, multivalent binding is associated with the
deformation of the cell or liposomes and results in "endocytosis" of the
virus (13). This is diagrammed in Figure 2. A similar mechanism was later
demonstrated to be involved in the engulfment of particles by macrophages
(95).

The kinetics of binding and membrane fusion of Sendai virus with
liposomes that contain only PE and ganglioside G_{D1a} have been studied
under conditions where the virus was present in considerable excess (78,96).
The association between virus and liposomes follows consecutive irreversible
three-step kinetics. The first step is very rapid and appears to constitute
binding that is mostly reversible by centrifugation. The second step involves
binding that is not reversed by centrifugation but is by dithiothreitol
(DTT). DTT alters the viral HN protein. This step was termed binding
stabilization, is first order, and has an activation energy between 16 and
23 kcal/mol when the temperature is above the transition temperature of
the liposomal lipids. This step is decreased in both rate and extent if
the F glycoprotein is not present. The last step is fusion, which was defined as a DTT-irreversible association of virus with liposome. This is a
zero-order process and also requires the F glycoprotein. Above the transition temperature the activation energy of the fusion step is negligibly
small, which suggests activation of the fusion step may be entropy controlled. The rate constants for the binding stabilization are smaller than
the fusion rate constants, so that at physiological temperatures the rate-
limiting step relates to viral binding. It was suggested that the binding
stabilization may result from lateral movement of gangliosides and ligands
to match each other in the contact area or from changes in the orientation
of the oligosaccharide chains of the gangliosides (78). Another possibility
is that this step represents the envelopment of the virus which results
in a multivalent and therefore more stable binding. Yet another possibility
is that after the virus undergoes structural rearrangements, a different
kind of binding occurs. In any case, it is clear that the binding must
play a major role in the fusion process.

After viruses have bound to liposomes or cells, the virus still has
to solve the problem of close approach of the bilayers and how to discretely
destabilize the bilayers before the fusion event can occur.

C. Close Approach of the Bilayers

For membrane fusion to occur the bilayers have to approach each other
to within roughly 10 Å. Three factors inhibit this close approach.

1. Membrane Proteins

Sendai virus glycoproteins extend 120 Å beyond the bilayer (17). Even
the gangliosides extend as much as 25 Å beyond the bilayer. The membrane
reorganization that follows binding of Sendai virus to erythrocytes at 37°C
included a loss of clearly defined spikes and a rearrangement of IMP (17,
90,91). Therefore, it seems likely that the virus glycoproteins are rearranged
during the reorganization of the membrane so close approach is possible.

2. Charge Repulsion

Biological membranes carry a net negative charge and close approach is
inhibited by electrostatic repulsion resulting from negative charges on
two surfaces (97). At pH 7.4 in PBS Sendai virus has a ζ potential of
-17 mV. Similarly, liposomes containing brain gangliosides, 0.125 mol sialic
acid per mol of phosphatidylcholine (PC), have a ζ potential of -18 mV
in PBS at neutral pH (11). Cells also have ζ potentials in this range.
Therefore, considerable electrostatic forces preventing the close approach
between the virus and either a cell or a ganglioside-containing liposome
are to be expected. The repulsive force (V_R) between two surfaces involves
the expression (98)

$$V_R = \frac{D}{4} \frac{r_1 r_2}{r_1 + r_2} [(\psi_1 + \psi_2)^2 \ln(1 + e^{-kH}) + (\psi_1 + \psi_2)^2 \ln(1 - e^{kH})]$$

where ψ_1, ψ_2 are the surface potentials of the particles; r_1, r_2 are the
radii of curvature of the particles; H is the distance of closest approach
of the particles; D is the dielectric constant; and k is the inverse Debye-
Hückel length.

Since the virus is outside the liposome or cell in buffer or extracellular
fluid, it cannot change the ionic strength. Similarly, it is difficult to change
the surface potentials. When only one virus infects a cell, any decrease
in surface potential due to cleavage of sialic acid by viral sialidase is probably compensated for by rearrangement of surface charges. Thus the only
variable involved in this expression that the virus can easily change is
the radius of curvature. Poste (99) has performed theoretical calculations
for the electrostatic repulsion between two cells with radii of 12 µm and
estimated that for two such cells to achieve a separation of 10 Å a kinetic
energy of 1100-1200 kT would be required. He calculated that if the cells
extended microvilli with radii of 0.6 µm, only 6 kT would be required
for approach within 10 Å. These calculations give an idea of how formation
of a region with a very small radius of curvature in a liposome or cell
could considerably facilitate close approach of a virus. Any postulated
mechanism of membrane fusion has to include a means to reduce the electrostatic force between the membranes, and a reduction of radius of curvature
seems a likely mechanism.

3. Water

One more barrier to close approach of the bilayers is water. Abidi and
Yeagle (100) have found by ^{31}P NMR spectroscopy that the surface of
Sendai virus gives a spectrum similar to that given by lipids that are
dehydrated. If the F protein is removed by trypsin, the spectrum expected
for a hydrated bilayer is obtained. These observations raise the possibility
that the F glycoprotein participates in removing water.

D. Destabilization of Bilayers

Once the bilayers are close to each other, the bilayer must be destabilized
so that the lipids in the bilayers can mix. The destabilization must be
very discrete since the infected cell will not survive if it becomes too
leaky. This means the region of destabilization of the bilayer is probably
local and circumscribed.

V. CHARACTERISTICS OF FUSION OF SENDAI VIRUS

The fact that Sendai virus can fuse with liposomes indicates that host
proteins are not required for fusion. Fusion of Sendai virus with liposomes,
like fusion with cells, requires an active F_1 protein (101). Experiments
were undertaken to determine if the characteristics of fusion in liposomes
are similar to those in cells. Such a similarity implies that the viral proteins
direct fusion relatively independently of the host cell proteins and that
fusion with liposomes is indeed accurately modeling in vivo fusion.

A. Time, Temperature, and pH Requirements

The time, temperature, and pH requirements of fusion of Sendai virus
membranes with liposomes composed of PC/PE/Chol/G_{D1a} in the mol ratios
0.7/0.3/0.66/0.03 have been determined (79). This composition was chosen
since it contains receptor and some of the major lipids present in the plasma
membrane of host cells. The amount of fusion was assayed by measuring
the amount of radiolabeled Sendai virus protein incorporated into liposomes.
The free virus and the virus that were bound but not fused were separated
from the liposome by centrifugation in a discontinuous sucrose gradient.
The radioactivity found in the liposomal fraction was therefore due solely
to viral proteins incorporated into the liposomes. This assay requires controls to make certain the bound virus is removed whenever the basic conditions are changed and, while very satisfactory for use with Sendai virus,
has to be modified substantially for use with influenza virus.

1. Time Course

The time course of fusion was followed and found to be consistent with
the time course of viral entry. The extent of fusion increases most rapidly
over the first 15 min but continues to increase more slowly for at least
the 6.5 h tested. There is some further increase up to 18 h, but, depending on the virus preparation, a maximum of 20-30% of the viruses fuses.
For infection of cell cultures by paramyxoviruses and myxoviruses (influenza virus) it is customary to adsorb the virus to cells for 1 h to give
it sufficient time to enter. Therefore, the slow time course of fusion is
consistent with the time course of viral entry. Fusion of virus with liposomes
at neutral pH is much slower and much less complete than fusion of liposomes
with other liposomes induced by Ca^{2+} or polycations or than fusion of
viral membranes with liposomes at low pH. Thus, in comparison to the
rate and extent of fusion that can be achieved with other systems, viral
fusion at neutral pH is inefficient. The important fact, however, is that
the course of viral fusion at neutral pH is similar in both rate and extent
to viral entry.

2. Temperature Dependence

When Sendai virus is incubated with liposomes at different temperatures for 2 h, fusion becomes detectable as the temperature is raised to 20°C and the extent increases up to 43°C. Above 43°C the extent of fusion decreases, presumably as a result of denaturation of the viral proteins. This is consistent with the observation that the hemagglutinating activity of the Z strain of Sendai virus is stable at 40°C but not at 45°C (5).

3. Absence of Divalent Cation Requirement

Virus membrane fusion does not require the presence of Ca^{2+} and can occur even in the presence of EDTA.

4. Effect of pH

The extent of fusion, when measured as a function of pH, showed a broad maximum between pH 7.5 and 8. These conditions are also consistent with the conditions found for fusion with cells. With liposomes it is possible to measure the effect of pH extremes. The extent of fusion increases as the pH goes toward 11.0. This could be related to titrating either an amine group on the PE in the liposome or an amine group of the virus. As the pH drops from 7.5 to 6.0 the extent of fusion decreases, but at pH 4.0 there is a dramatic increase in the extent of fusion which considerably exceeds that obtained at physiological pH. The isoelectric point of aggregates of Sendai virus is 4.0 as measured in a microelectrophoresis apparatus in the laboratory of A.D. Bangham (Haywood, unpublished results), so the abrupt rise in the amount of fusion could be related to neutralization of the negatively charged groups on the virus.

B. The Role of the Liposomal Lipid Composition

Fusion of Sendai virus with liposomes provides an opportunity to determine how the lipid composition of the target membrane influences its fusion with a biological membrane. The lipid composition of liposomes was varied to test various hypotheses of membrane fusion that involve lipid composition. It was possible to eliminate most such hypotheses (88). The extent of fusion was measured by the transfer of viral proteins to liposomes. Sendai virus does not fuse with liposomes composed of lipids with no net charge but fuses with liposomes containing PC and any of a range of negatively charged lipids including phosphatidylinositol (PI), phosphatidic acid (PA), phosphatidylglycerol (PG), dicetyl phosphate, and ganglioside G_{M1}, which does not have receptor activity. For fusion to occur, however, these negatively charged lipids have to be present in about 10-fold the mole fraction that is necessary for ganglioside G_{D1a}, which does have receptor activity. The lipids that have a negative charge also have an oxygen available to form hydrogen bonds, so the negative charge does not have to be the critical property for binding and fusion. Gangliosides have both a carboxyl group and ample other residues capable of hydrogen bonding.

The role of cholesterol in fusion of viral membranes with liposomes at both neutral and low pH is controversial. In the above studies (88) cholesterol did not have a significant effect except when the ganglioside concentration was suboptimal. Then cholesterol decreased the extent of fusion. In another study (102) it was noted that cholesterol did not appreciably affect the rate of transfer of spin-labeled PC from virus to liposomes containing glycophorin as receptor. Yet another study (103), however, claimed that cholesterol is required for fusion of liposomes with Sendai virus, but this study had many technical problems, including the use of ganglioside concentrations that were so high they could be expected to inhibit fusion, and the fusion assays were problematic, as illustrated by the observation that significant amounts of fusion were supposed to be occurring at 4°C. When Sendai virus-induced lysis of liposomes was measured with glycophorin-containing liposomes, cholesterol was needed (104). On the other hand, when ganglioside-containing liposomes were used, cholesterol suppressed the leakage rate induced by Sendai virus (105). It seems likely that the controversy about cholesterol in these studies as well as those on other viruses probably relates to the differences in total lipid compositions used.

Another method of measuring membrane fusion is to incorporate octadecylrhodamine B chloride (R_{18}) into the virion at concentrations that result in self-quenching of fluorescence and then measure the relief of self-quenching that occurs during fusion with another membrane (106). This assay can also be used to microscopically identify liposomes or erythrocyte ghosts that have fused with virus, since the fluorescence is spread through the fused membrane (Haywood, unpublished results). When, however, R_{18}-labeled Sendai virus are fused with chicken erythrocytes that are nucleated, the fluorescence is very bright in the region of nucleus (Haywood, unpublished results). Care has to be taken in interpreting the presence of R_{18} in interior organelles, since lipid analogs are transported from the plasma membrane to interior membranes (107). To be able to conclude that R_{18} fluorescence dequenching identifies the site of membrane fusion, it is necessary to view the dequenching event rather than the fluorescent labeling at some time after fusion.

In general, the results from quantitative experiments in which R_{18} fluorescence dequenching was used as an assay and experiments in which protein transfer to liposome was used are in agreement. One set of investigators, however, using R_{18}-labeled Sendai virus, found fluorescence dequenching when virus is incubated with liposomes containing only phosphatidylcholine (PC) and cholesterol (108). The source of the different results obtained by the two methods is not clear. R_{18} contains two positively charged amino groups, so it reduces the net negative charge of the virus. This might change the interaction of the virus with liposomes composed of PC and cholesterol.

While the different studies on the role of liposomal composition in fusion with viral membranes were performed differently and gave somewhat different results, no clear indication of a lipid requirement emerged. Since virus can fuse with liposomes of a wide range of compositions, many possible mechanisms of fusion that depend on specific lipid compositions can be eliminated. Since fusion can occur in the absence of gangliosides, fusion as a result of the proclivity of gangliosides to make micelles can be eliminated. Similarly, there is no requirement for lipids that can undergo a lamellar to hexagonal phase transition, even though PE to some degree potentiates fusion. Fusion occurs in the presence of EDTA, so an anhydrous complex with Ca^{2+} or a Ca^{2+}-induced lateral phase transition is not involved. Fusion also occurs with liposomes composed only of PS, so a packing defect between two different lipids is not involved. However, it is not possible to eliminate the possibility that the virus binds to some of the lipids and thereby causes a lateral phase separation. It seems most likely, however, that the viral proteins rather than a lateral phase separation of phospholipids determine whether fusion occurs.

Preliminary experiments (Haywood and Boyer, unpublished results) indicate that the lipid composition may play a role in the distribution of viral membrane proteins in liposomes after fusion. When liposomes contain PS, it is commonly noted that smaller pieces of membrane and filamentous forms bud off from the fused liposome. If cardiolipin (CL) is included in the liposomes, this tendency of the fused liposomes to break off pieces becomes very marked.

While virus can fuse with liposomes of a wide range of compositions, including those that do not contain receptors, it is not clear whether the same mechanism of fusion is operative in all cases. For Sendai virus to fuse with liposomes without gangliosides, negatively charged lipids have to be present in 10-fold the molar ratio that is required for gangliosides. This could imply that other lipids can inefficiently fulfill the same function or that virus can be made to fuse by a mechanism that only partially represents the in vivo mechanism. Sendai virus fusion with liposomes composed of just CL or PS does not show the same dependence on intact viral proteins as does fusion with cells (89,109). DTT, which reduces the disulfide bonds in the viral glycoproteins, decreases the amount of fusion with erythrocytes and with liposomes containing G_{D1a}, but has less effect on fusion when CL or PS liposomes are the target membranes.

The fusion of Sendai virus membranes that occurs at pH 4.0 differs from fusion at neutral pH. Fusion at pH 4.0 is not only much greater in extent (79) but also in rate (89). Fusion with CL liposomes occurs at low pH with virus that has been treated with trypsin and therefore does not involve an active F protein (89). Similarly, liposomes that have been reconstituted to contain a fluorescent probe and HN but not F protein show fluorescence dequenching when incubated with PS liposomes at pH 4.0 (109). Further, this fluorescence dequenching at pH 4.0 is very rapid and reaches 80-85% with concentrations of liposomes lipid as low as 20 μM, which is over 100-fold less lipid than is required to demonstrate fusion at neutral pH. Since fusion at low pH is much more rapid than at neutral pH and can occur in the absence of the F protein, and since it is very unlikely that there is a cellular compartment with pH at 4 or lower, the highly efficient fusion of Sendai virus membranes with liposomes containing negatively charged lipids at low pH is likely to represent a laboratory rather than a biological phenomenon.

Many questions remain to be resolved. Is the mechanism of fusion of viruses with liposomes at neutral pH the same as the mechanism of fusion of viruses with cells except when the membrane composition is very different from the in vivo composition as is the case, for instance, with pure PS or CL liposomes? Are receptors necessary for fusion and/or disassembly to occur by a biological mechanism?

VI. POSTULATED MECHANISMS OF SENDAI VIRUS MEMBRANE FUSION

At present there is no agreement on the mechanism of Sendai virus membrane fusion. Several hypotheses are currently being considered.

A. Fusion Requires Formation of a Very Curved Region of Membrane

When Sendai virus was incubated with liposomes for only a few minutes, electron microscopy revealed that the virus membranes fuse with the leading edge of a region of the liposome that is enveloping the virus (110). The radius of curvature of this leading edge is very small. With further incubation the fused viruses are released from the region of envelopment, probably as a result of the action of the viral sialidase on the receptors. Therefore, *all* the fusion events are seen to occur with the leading edge of a vacuole only if fusion is observed during the first few minutes of incubation. Once this pattern became evident for liposomes, it was possible to find examples in the micrographs of viruses fusing with cells (10). The F-protein-dependent changes in the viral surface described by Knutton (91) result in convolutions of the viral surface, so that the virus surface also has regions with small radii of curvature.

Fusion at the leading edge of a developing endocytic vacuole and the changes in the viral surface are diagrammed in Figure 3. This presents a simple solution to many of the requirements for fusion. Reduction of the radius of curvature of the liposome decreases the term $r_1 r_2 / r_1 + r_2$ in the equation for charge repulsion. Not only will the charge repulsion between the virus and the liposome or cell be reduced at this very curved region, but also this region can be expected to be stressed since it is at the juncture between the liposomal surface tension and the binding of the virus. Curvature increases the surface free energy. Figure 4 shows the packing of the lipids at a very curved region. The phospholipids in the outer monolayer tend to become separated. Thus, only the specific region of the membrane that is capable of close approach to the virus is destabilized.

This mechanism is consistent with the kinetic analysis of Tsao and Huang (96). The rate-limiting steps in this mechanism would be expected to be the binding and envelopment of virus. Once the curved leading edge of a developing vacuole was formed, the problems of approach and destabilization should be solved, and the fusion event would consist only of the mixing of the bilayers.

The energetics of membrane deformation and adhesion have been described by Evans and Parsegian (111). As is intuitively evident, "rigid" bodies are not easily deformed and so have limited ability to adhere, whereas "flaccid" bodies are easily deformed and can form large contact areas with little work. In the case of viruses and liposomes or cells, the virus is relatively rigid, whereas liposomes and cells are usually deformable and so a large contact area can be made if the liposome or cell envelopes the virus. The rigidity and hence the deformability of the liposomes, however, depends on the characteristics of the liposomes. The work of membrane deformation includes changes in membrane area, extension of the membrane without any change in area (surface shear), and bending of the membrane to produce the membrane curvature. Among the characteristics of the liposome that have a bearing on the work of deformation are the surface area to volume ratio and the membrane elastic and viscous properties.

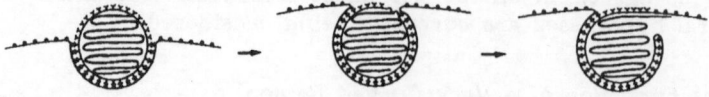

FIGURE 3 Fusion of Sendai virus at the leading edge of a region of membrane enveloping virus. Υ, Viral HN protein. ●, Ganglioside receptor.

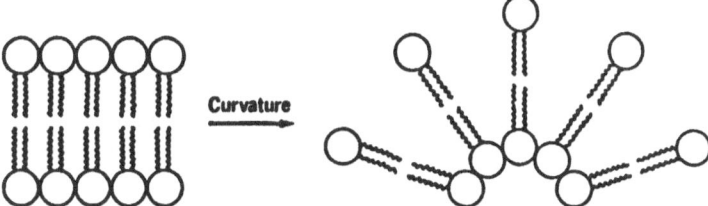

FIGURE 4   Separation of phospholipids in a bilayer as a result of curvature.

The surface area to volume relates to liposome size as well as osmotic activity. The *excess* surface area, i.e., that area which is in excess of the amount needed to enclose the volume, is important for envelopment of virus. This excess surface area should limit the number of viruses that can fuse with a cell or liposome. This, of course, is true only if the virus itself does not contribute *excess* surface area when it fuses. The virus membrane, however, encloses RNA and proteins and is closely associated with the viral content through the M protein, so it seems unlikely that it has significant surface area in excess of the volume of viral contents. Since more viruses can bind than can be enveloped, the number of viruses that bind should also be greater than the number that can fuse. This result was found with Sendai virus binding and fusion with erythrocytes (112), although the other interpretations of this result given by the authors are equally possible. Liposomes that are in the same size range as the virus would not be expected to have enough excess surface area to completely envelope the virus, but should nevertheless produce a curved leading edge after partial envelopment of the virus. Small liposomes have been observed to fuse with virus after only partial envelopment of the liposomes (Haywood and Boyer, unpublished results). This also means that one virus would be unlikely to fuse with more than one small liposome, but that several small liposomes should be able to partially envelop one virus and fuse with it. This also appears to be the case (89). With SUV the membrane thickness is no longer negligible in relation to the diameter of the vesicle, and above the transition temperature the bending stiffness is greatly increased over that of larger vesicles (111). Therefore, SUV should not be expected to fuse well, and preliminary data indicate this is so (88). When the lipids are below their transition temperature, the shear rigidity of larger vesicles will be considerably greater than when the lipids are in the liquid state, so the rate of binding and hence fusion should be decreased, and this also is the case (96).

Membrane expansion has been postulated to play a role in membrane fusion (113). Osmotic swelling drives Sendai virus-mediated cell-cell fusion (114). Cell swelling should assist fusion at the leading edge of an endocytic vacuole, since expansion of the host membrane would increase stress on the leading edge and contribute to separation of the phospholipids in the outer monolayer and resulting exposure of hydrophobic domains. Ohki (115) has suggested that increased hydrophobicity or increased free energy of the membrane surface can explain lipid membrane fusion caused by a range of fusion agents. Certainly this applies to fusion at the leading edge of a developing endocytic vesicle or fusion driven by osmotic swelling. Another effect of cell swelling is to alter the cytoskeleton, which is also likely to be important in membrane fusion and disassembly.

B.   Insertion of the Viral F-Protein Terminus into the Target Bilayer

One hypothesis for Sendai virus membrane fusion postulates insertion of the N-terminal end (the distal end) of the F protein into the target membrane, thereby destabilizing it. This has been proposed on the basis of the following findings: (1) There is a long hydrophobic amino acid sequence at the NH_2 terminus of the F_1 polypeptide that is generated by the proteolytic cleavage that activates the protein (116). (2) The NH_2-terminal sequence is highly conserved among paramyxoviruses (117). (3) Oligopeptides with sequences resembling this NH_2-terminal sequence specifically inhibit membrane fusion (117). (4) The activation cleavage of the F protein involves a conformational change that exposes a hydrophobic region on the protein (118). (5) When the hydrophobic photoaffinity label TID {3-(trifluoromethyl)-3-($m[^{125}I]$iodophenyl)diazirine} was incorporated into cardiolipin or phosphatidylserine liposomes and Sendai virus was added and irradiated, the F protein was preferentially labeled during the initial stages of fusion (119). The fact that the F protein extends 120 Å from the bilayer poses problems with this mechanism, since perturbation of the host membrane would occur at a considerable distance from the viral membrane. This mechanism does not solve the problem of charge repulsion. Insertion of the N-terminal region into the host membrane on the basis of the hydrophobicity alone would imply that all the F proteins at the area of the contact would become inserted, and this might make the membrane quite leaky.

As described above, the F protein is necessary for the structural rearrangements in the virus that occur prior to fusion (90), participates in binding stabilization (96), and may help remove water (100), so the requirement for the F protein can be accounted for without postulating that it directly perturbs the host bilayer.

Inherent in most studies on membrane fusion has been the assumption that there is a specific process that destabilizes the bilayer. The idea that the F protein might actively drive fusion by inserting itself into the host membrane was derived with this assumption. Another possibility is that the key process for membrane fusion is removal of the barriers to close approach. The idea that the F protein participates in fusion by rearranging the viral surface and removing water is consistent with this possibility.

VII.   FUSION OF INFLUENZA VIRUS MEMBRANES

There have been reports of fusion of influenza virus with the plasma membrane of cells since the 1960s. These were for the most part electron microscopic studies, and with the exception of the report of Fidgen and Tisdale (120), the micrographs of influenza virus fusion were never as convincing as the micrographs of Sendai virus fusion. Influenza virus does not usually induce cell-cell fusion at neutral pH. Influenza virus is a negative-strand RNA virus and in many ways is similar to the paramyxoviruses. Its membrane also has an M protein and two glycoproteins, but the hemagglutination and fusion functions are associated with one glycoprotein, the hemagglutinin (HA), and the neuraminidase activity with the second glycoprotein (NA). Influenza virus differs from Sendai virus in that the different strains have isoelectric points between pH 5 and 6 rather than pH 4.0, and as discussed below, it does not disassemble as readily. The general similarity, however, makes it seem probable that influenza virus enters cells by the same mechanism as does Sendai virus.

A. Fusion of Influenza Virus Membranes with Liposomes at Neutral pH

Fusion of egg grown influenza virus X-31 with G_{D1a}-containing liposomes at pH 7.4 was demonstrated both by electron microscopy and by measuring the transfer of viral protein to liposomes (121). Influenza binds more tightly to negatively charged liposomes than does Sendai virus, so to remove viruses that are bound but not fused by centrifugation, it is necessary to add fetuin, a sialoglycoprotein, prior to centrifugation. It is very important when studying fusion of influenza at neutral pH to make sure the virus is not aggregated. Aggregated virus can be disaggregated by a very brief sonication (121). Aggregation appears to occur when purified virus is frozen and thawed and can be prevented by the addition of bovine serum albumin prior to freezing (122).

1. Time Course

The time course of fusion of influenza virus membranes at 37°C and at neutral pH is consistent with the rate of viral entry (122). Thus, 6% of the virus has fused by 2 min, 11% of the virus has fused after 5 min, and 23% after 1 h. This is also generally consistent with the rate of Sendai virus membrane fusion at neutral pH.

2. Temperature Dependence

The amount of fusion of influenza virus at neutral pH was measured after incubation for 1 h at different temperatures (122). The extent of fusion increases semilogarithmically with temperature from 15°C to 50°C. Thus, there is considerably more fusion at 50°C than at 37°C, so incubation at 50°C provides a method for increasing the efficiency of influenza virus membrane fusion at neutral pH. Consistent with the high rate of fusion at 50°C is the fact that influenza virus HA has been reported to be very resistant to heat (5,123).

3. Effect of Liposome Composition

When liposomes contain only lipids that have no net charge, influenza virus X-31 does not fuse at neutral pH. Addition of ganglioside G_{D1a} to the liposome composition results in fusion which reaches a maximum level when the G_{D1a} is present at 3 mol%. The addition of 5 mol% PS to liposomes results in some fusion but less than occurs when G_{D1a} is present.

B. Fusion of Influenza Virus with Erythrocyte Ghosts

When R_{18}-labeled influenza virus PR8 is incubated with human erythrocyte ghosts at neutral pH, the R_{18} becomes distributed through the surface of some of the ghosts to give an appearance similar to that visible when R_{18}-labeled Sendai virus is incubated with ghosts (Haywood, unpublished results).

C. Fusion of Influenza Virus Membranes at Low pH

Influenza viruses fuse with liposomes (124,125) and cause hemolysis (126, 127) and cell-cell fusion (128) at low pH. The pH required for fusion varies between 5 and 6, depending on the virus strain. The characteristics of fusion at low pH are different from those at neutral pH but are similar to the characteristics of Sendai virus membrane fusion at low pH.

At low pH influenza virus membrane fusion is very rapid (usually mainly complete by 2 min) and involves up to 90% of the viruses (124,125). Significant levels of fusion can be seen when the liposome concentration is quite low, e.g., 25 µM lipid phosphorus (129). Such low concentrations are convenient because they avoid problems with light scattering when fluorescence assays are used. The very rapid rate of fusion at low pH is unlike the rate of viral entry, but if fusion does proceed at low pH in vivo, it would occur in an endocytic vacuole and the binding and envelopment of virus would be the rate-limiting step.

Unlike fusion at neutral pH, fusion at low pH can occur with liposomes that contain no net charge (121,124,125). A requirement for receptors is apparently bypassed at low pH.

Because fusion at low pH does not require receptors, is much faster and more complete than fusion at neutral pH, and can be detected with low concentrations of liposomes, experimental conditions that result in fusion at low pH may fail to yield significant fusion at neutral pH.

When the published isoelectric points of different influenza viruses are compared to the pH at which maximum binding and low pH fusion occur, there appears to be a correlation. A maximum for hemagglutination was found at pH 4.1 for Sendai virus and at pH 5.35 for influenza A/PR8 (130). The isoelectric point of influenza WSN was reported to be 6.0-6.5 (131), and that of influenza A/PR8 was reported to be 5.3 (132). Influenza WSN and A/PR8 cause hemolysis with maxima around pH 6.0 (126) and pH 5.2 (124), respectively. Therefore, for Sendai virus and at least two strains of influenza, the low pH at which fusion occurs is close to or the same as the isoelectric point. While more viruses need to be investigated to determine whether there is a relationship between the isoelectric point and the low pH at which fusion occurs rapidly and efficiently, such a relationship would provide a mechanism for dealing with the problem of charge repulsion between the virus and liposome or cell. The change in pH, of course, would also affect the target membrane, so that the nature of the target membrane could introduce some variation in this relationship. If the effect of low pH is to titrate the charge on the virus, alterations of the viral glycoproteins would be predicted to change the pH at which virus can fuse. Influenza virus HA mutants which vary in the pH at which they fuse have been described (133). Further, if the effect of low pH is to titrate the charge on the virus, fusion would be expected to follow rapidly after the addition of acid, which is the case.

At the time it was initially noted that efficient fusion occurs at low pH, it was suggested that viruses such as influenza, vesicular stomatitis virus (VSV), and the togaviruses could not fuse at neutral pH and that a low pH step is required for viral entry. Since coated endosomes have an internal pH between 5 and 6 (134), it was postulated that fusion occurs in the endosomes. A conformational change in the influenza HA is induced at pH values which correspond to those that maximize membrane fusion (135). This low pH-induced change has been postulated to be necessary for membrane fusion. The fact that mutants of the influenza HA fuse at different pH's has been used as support for this hypothesis (133). Basic ("lysosomotropic") drugs, such as the phenothiazines, and NH_4^+ in very high doses inhibit viral entry. This inhibition has been interpreted to result from an increase of the pH within the endosomes (136,137). These drugs, however, have many pharmacological actions (138-142) and have been postulated to inhibit viral replication by a range of mechanisms (143-149).

The conformational change in the viral HA protein that occurs at low pH is postulated to reposition a hydrophobic peptide on the HA protein so that it can insert itself in the host bilayer. This peptide is similar to the fusion peptide on the Sendai F protein. In a very careful study, Harter et al. (150) examined the labeling of BHA, the bromelain-solubilized portion of the HA protein, after it was added to liposomes containing PC plus [^{125}I]TID or [^{3}H]PTPC, a carbene-generating, membrane-directed reagent which they synthesized. The HA1 region was labeled at both pH 5 and 7 when [^{125}I]TID was used but only at pH 5 when [^{3}H]PTPC was used. Many other proteins, including water-soluble proteins with biological roles that seem unrelated to membrane fusion, were also labeled in similar experiments, so BHA is not specific in its association with the bilayer.

The hypothesis that fusion of the membranes of viruses such as influenza and VSV requires a low pH-induced change in a viral protein was based mainly on the idea these viruses are unable to direct fusion at neutral pH (4). In the case of Sindbis virus-induced cell-cell fusion at low pH, the low pH step has to be followed by a return to neutral pH before fusion occurs (151).

Fusion of Sendai virus and fusion of influenza virus membranes have similar characteristics at neutral pH. The characteristics of fusion at low pH are different from those at neutral pH, but again the characteristics are similar for the two viruses. Therefore, it seems likely that the two viruses fuse by similar mechanisms but that the mechanisms are different at neutral and low pH. In the case of Sendai virus, fusion at low pH does not require an F protein and occurs at a pH so low that it is unlikely to occur in the cell. Therefore, fusion of Sendai virus membranes at low pH appears to be a laboratory rather than a biological phenomenon. Since the fusion of influenza virus membranes occurs at a low pH that is in the range that occurs in the endosome, the possibility cannot be eliminated that fusion of influenza virus membranes occurs through a low pH step in the endosome. However, it seems likely that fusion of influenza virus membranes at low pH is similar to fusion of Sendai virus membranes at low pH. While fusion at low pH is appealing because of its speed and efficiency, it must be remembered that the rate and efficiency of viral entry parallel the rate and efficiency of fusion at neutral pH.

VIII. FUSION OF HUMAN IMMUNODEFICIENCY VIRUS AND HERPES SIMPLEX VIRUS AT NEUTRAL pH

In addition to the paramyxoviruses, human immunodeficiency virus (HIV) and herpes simplex virus (HSV) are established to fuse their membranes at neutral pH. HIV appears in endocytic vacuoles (152), but this is true for all viruses, including Sendai virus, when added at high multiplicities, since many virus particles will be treated as just foreign bodies. Low pH is not required for HIV entry. Stein et al. (153) demonstrated that when the endosomal compartments are neutralized with lysosomotropic agents, HIV nucleocapsid entry was not inhibited as shown by the presence of nonintegrated HIV DNA in the cells. McClure et al. (154) also demonstrated that weak bases do not inhibit HIV entry. They commented that fusion at neutral pH does not have to occur at the plasma membrane but can also occur within an endocytic organelle.

Herpes simplex virus (HSV) appears to fuse with the plasma membrane at neutral pH. Consistent with this was the observation that HSV membrane Fc receptors are in the cell membrane shortly after infection in the absence of protein synthesis (155). Antibodies to the HSV glycoprotein D that permit adsorption but block viral infection prevent release of naked nucleocapsids into the cytoplasm near the plasma membrane (156). Ammonium chloride does not appear to inhibit HSV entry into Vero cells, which is consistent with entry at neutral pH (157). In fact, there is a mild acidic pH inhibition of HSV penetration (158).

IX. DISASSEMBLY (UNCOATING) OF SENDAI VIRUS

Disassembly (uncoating) of the virus follows membrane fusion. This is often considered as a part of fusion, but can be identified as a separate step (159). Viral disassembly can be defined as all the series of rearrangements of viral components that lead to the distribution of viral membrane proteins in the host membrane, release of the viral internal proteins inside the cell, activation of the viral enzymes, and formation of the RNA polymerase complex that synthesizes messenger RNA.

Since disassembly has often been considered a passive process that automatically followed fusion (160) and since the techniques to study it as a separate stage in viral replication have been available only in recent years, very little is known about it. Preliminary data indicate that disassembly does not occur after fusion of myxoviruses or paramyxoviruses with liposomes if the virus is completely undamaged and intact.

Microscopically, fusion is usually identified by the changes in the virus structure that are a result of disassembly. If the entry process stopped after formation of the fusion bridge, which without disassembly should be very narrow, it would be difficult to identify fusion by electron microscopy. Only after disassembly occurs is there widening of the fusion bridge, migration of the viral proteins into the host or liposomal membrane, gradual loss of the contrours of the virus, and at least partial uncoiling of the ribonucleoprotein. Therefore, the identification of some virus groups, such as the paramyxoviruses, as viruses that fuse with the plasma membrane may really be a reflection not of their ability to fuse, but of their ability to disassemble.

The gross changes in viral structure that cause the viral outlines to merge with that of the cell constitute the beginning of viral disassembly. Caldwell and Lyles showed that Sendai virus M protein and nucleocapsid remain attached to the cytoplasmic surface of fused erythrocytes (161). They suggested that the initial transcription events could occur in association with the membrane or that another step in disassembly is necessary to release nucleocapsids into the cytoplasm.

During disassembly the associations between viral components must be released. These were formed during virus assembly, which has been reviewed (162). During assembly of the paramyxoviruses and myxoviruses the viral glycoproteins become incorporated into the host cell membrane in a random pattern. These proteins then aggregate in patches. Since the M protein is also present in these patches, it is thought that the patching results from an association of the M protein with the glycoproteins. The M protein also associates with the nucleocapsid and thereby organizes all the viral components in one region. The virion is formed by the budding off of this region of the host plasma membrane containing viral proteins associated with the nucleocapsid. To some degree disassembly may be the reverse of assembly. It seems possible that if disassembly does not occur after fusion, the process could be reversible and the fused virus might bud off again.

A. Fusion of Young (Early Harvest) vs. Aged (Late Harvest) Sendai Virus with Erythrocytes

Aged (late harvest) Sendai virus has been used in most studies on membrane fusion. Whether virus causes hemolysis or disassembles after fusion with erythrocyte membranes depends on whether young (early harvest) or aged (late harvest) virus is used. As described in the section on virus structure, the virion changes greatly if incubated in eggs for 48-72 h and then harvested (late harvest) rather than being harvested at 24 h (early harvest). When young virus fuses with red cells, hemolysis does not result, whereas when aged virus fuses with erythrocytes, hemolysis follows. Changes in the structure of the virus can also be brought about by repeated freezing and thawing or sonication (53,54). These procedures cause the virions to become permeable to negative stain and to change from nonhemolytic to hemolytic (53).

The question whether nonhemolytic young (early harvest) virus can disassemble after fusion with a red cell was addressed by Knutton and Bächi (163). Using immunofluorescence, they showed that diffusion of antigens (glycoproteins) from fused young Sendai virus membranes is restricted in the plane of erythrocyte membranes unless the erythrocytes are put in a hypotonic medium. They postulated that the resultant swelling perturbs the structural organization of the erythrocyte membrane and concluded that an erythrocyte membrane perturbation appears to be a prerequisite for the lateral diffusion of viral elements. A possibility they could not address using erythrocytes is that a perturbation in the viral structure is also a prerequisite for the lateral diffusion of viral elements. With liposomes it is possible to investigate the role of the viral structure in restricting diffusion of viral membrane proteins.

B. Fusion of Young (Early Harvest) vs. Aged (Late Harvest) Sendai Virus with Liposomes

Two patterns of disassembly can be seen after fusion of Sendai virus with liposomes. These are diagrammed in Figure 5. In both patterns the viral HN protein must be released from its receptor, which is a possible function of the viral sialidase.

In one pattern complete morphological disassembly occurs rapidly. Release of the virus from the receptors proceeds concurrently. This is diagrammed in Figure 5a and the micrographs representing these steps have previously been published (101). Figure 6 shows an example of a liposome where viral membrane fusion has been followed by viral disassembly. The viral glycoprotein spikes have become distributed in the liposomal membrane and the RNP is inside the liposome and is partially uncoiled. The viral RNP stays associated with the membrane. These experiments, like those of Caldwell and Lyles (161), led to the question whether mRNA

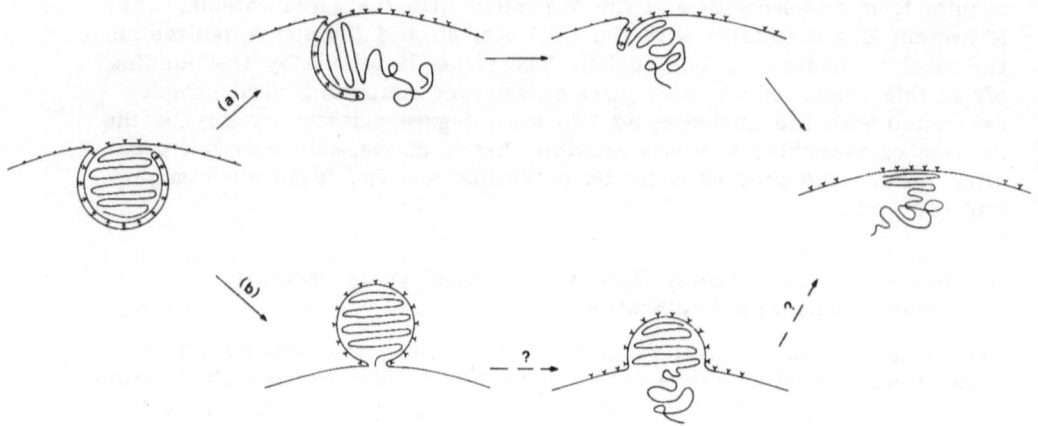

FIGURE 5 Sendai virus disassembly after fusion with liposomes.

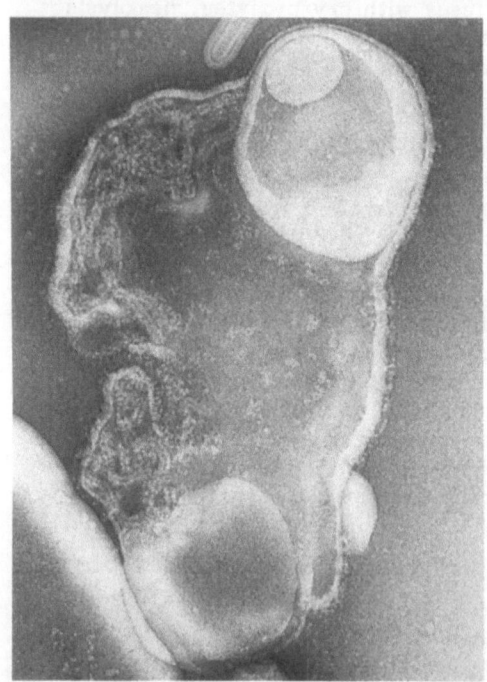

FIGURE 6 Fusion and disassembly of aged ("late harvest") Sendai virus.
Liposomes were made from 0.7 µmol of PC, 0.3 µmol of PE, 0.66 µmol of
cholesterol, and brain gangliosides containing 0.1 µmol sialic acid. Sendai
virus was added at 0°C for 1 h and then incubated at 37°C for 2 h. The
grids were stained with 2% ammonium molybdate, pH 7.1. Magnification:
95,345×.

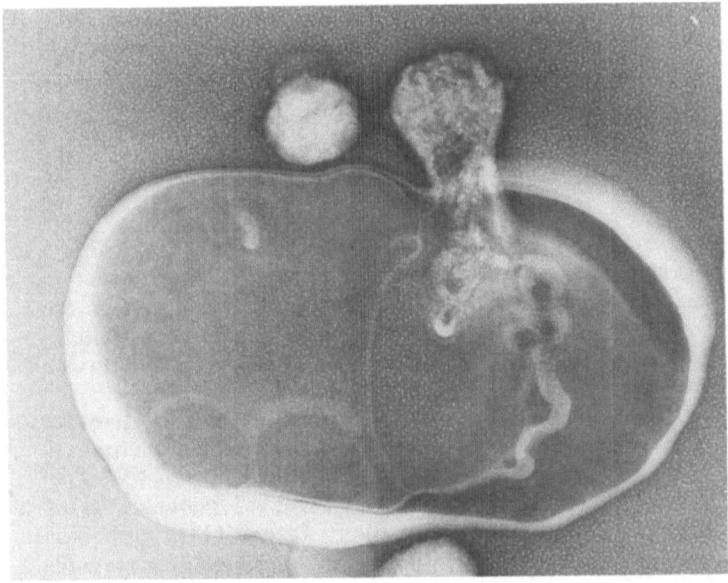

FIGURE 7   Fusion of young ("early harvest") Sendai virus. Liposomes were made from 0.7 μmol of PC, 0.3 μmol of PE, 0.66 μmol of cholesterol, and 0.03 μmol of G_{D1a}. "Early harvest" Sendai virus, which had been frozen once, was added and incubated at 40°C for 2 h. The sample was stained with 2% ammonium molybdate pH 7.1. Magnification: 95,345×.

synthesis might be in association with the membrane (14), and whether there are more steps involved in disassembly and activation of the RNA polymerase. When disassembly occurs immediately following fusion with liposomes, it is likely the interactions between the viral components were already disrupted before the fusion event.

In the second pattern, diagrammed in Figure 5b, the virus binding to receptors is released so the virus is freed from the developing endocytic vesicle but the viral components stay associated with each other and the viral membrane proteins do not dissociate into the liposomal membrane. The viral RNP stays associated with the virion structure and is not released into the interior of the liposome. Figure 7 illustrates such a fusion event where only a little disassembly has occurred despite incubation for 1 h at 37°C. This micrograph was the result of a preliminary experiment using young virus, which had been frozen once. Further work is needed to establish whether young virus are uniformly unable to disassemble after fusion with liposomes.

It would seem to be to the virus's advantage in vivo if the virus does not just randomly dump its contents into the cell but rather if disassembly is a carefully regulated event that results in deposition of the nucleocapsid at an intracellular locus where degradative enzymes will be minimized and host factors necessary for replication will be maximized. Regulation of disassembly might be a function of the viral enzymes such as the protein kinase, which after fusion could utilize components of the host cytoplasm such as ATP. In a natural infection viruses are passed from cell to cell and are not incubated for 1 or more days. Therefore, the virus in a natural infection virus should be like young virus. When virus is made more permeable to negative stain and able to produce hemolysis by repeated freezing and thawing, it also becomes less infective (53). This suggests nonhemolytic virus may be required to initiate infection, even though hemolytic virus can fuse its membrane.

X. DISASSEMBLY OF INFLUENZA VIRUS

Differences in young and aged virus have not been described for influenza, and the virions in a preparation of influenza, even when harvested after 72 h, are generally fairly homogeneous and impermeable to negative stain. Therefore, it is to be expected that when influenza virus fuses with liposomes at neutral pH, disassembly will not occur, so fusion will be hard to identify microscopically.

Whenever fusion of influenza membranes at neutral pH has been reported to be detectable by electron microscopy, there have been experimental conditions that should change the viral structure. Thus, Fidgen and Tisdale's procedure included fusion of virus with erythrocyte ghosts at neutral pH on grids which were then washed with distilled water (120). This would have resulted in osmotic swelling, which would have perturbed the structural organization of the viral membranes. When influenza virus fusion and disassembly with liposomes at neutral pH were demonstrated by micrographs, the viruses had been previously briefly sonicated in order to disaggregate them (121). [Brief sonication has been reported to enhance hemolytic activity at low pH (124).] When influenza virus has been neither frozen nor sonicated, it is hard to demonstrate fusion at neutral pH by microscopy, as illustrated in Figure 8. Some of the viral glycoproteins appear to have formed a collar in the region where the virus and liposomes are joined (white arrow). One small cluster of viral glycoproteins (black arrows) can be seen in the liposome membrane at a distance from the virus, which makes it seem likely fusion has occurred.

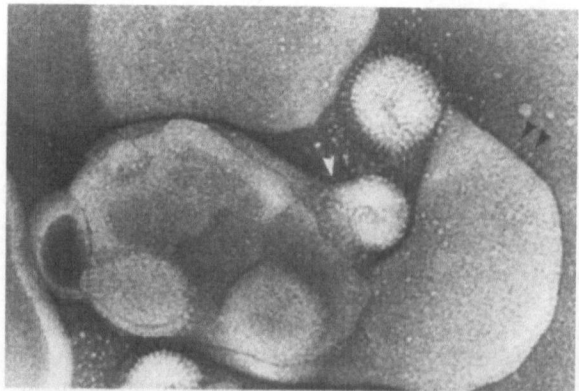

FIGURE 8 Probable fusion of influenza virus membrane without disassembly. Liposomes were made from 0.7 μmol of PC, 0.3 μmol of PE, 0.66 μmol of cholesterol, and 0.05 μmol of G_{D1a} in PBS, pH 7.5. Influenza virus X-31 was added and incubated at 37°C for 2 h. Magnification: 128,400×. White arrow indicates juncture of virus and liposome. Black arrows indicate several viral glycoprotein spikes in liposomal membrane.

At low pH influenza virus structure becomes markedly changed. The virion becomes permeable to negative stain and a loosening of the general structure is quite evident by electron microscopy (87,121,124). This may well account for why influenza virus membrane fusion is easily demonstrated by microscopy at low pH both with liposomes and with cells.

XI. HOW DO VIRUSES ENTER CELLS?

Viral entry has long been thought to occur either by endocytosis or by direct fusion of viral and plasma membranes (164-167). For viruses that replicate at specific sites in the cell, entry by an endocytic vacuole would have the advantage that the virus could be delivered to its replication site. Because electron microscopic studies require the use of a very high ratio of virus to cells and because endocytosis of virus can be the response of a cell to a foreign body, the presence of virus in an endocytic vacuole does not necessarily indicate this virus is involved in initiating an infection. If a virus that has undergone endocytosis is to initiate infection, the viral nucleoprotein must be released from both the viral and the vacuolar membranes and enter the cytoplasm. Several studies have been designed to search for virus that is fusing with endocytic vacuoles (168,169) and have failed to reveal a convincing micrograph of such an event. To the author's knowledge no clear instance of such fusion has ever been demonstrated.

Microscopic studies have revealed that influenza virus enters both coated and uncoated vesicles (168,169). Influenza virus uptake into vesicles (viropexis) has been shown to occur in the presence of inhibitors of glycolysis, oxidative phosphorylation, membrane Na^+/K^+ transport, and microfilament and microtubule function (170). A mechanism for uptake into uncoated vesicles similar to that for envelopment of Sendai virus by liposomes has been proposed (169,170). If fusion does occur with vesicular membranes, it could be with the membrane of an uncoated vesicle at neutral pH. Regions of endocytic vesicles containing virus have been described as budding (169), and such budding also results in envelopment of the virus by a section of the vesicles with the formation of a very curved region of the vesicular membrane.

The fact that the virus does not have to disassemble immediately after fusion also raises a completely new possibility. The virus membrane could fuse with the cell plasma membrane during the course of endocytosis. If disassembly does not occur, endocytosis would be completed. The result would be an endocytic vesicle that had one or more fusion bridges, as diagrammed in Figure 9. The contents of the virion, therefore, would have one or more portals through the vesicular membrane to the cellular contents. The virus then would be released from the vesicles not upon the initiation of fusion, but upon the initiation of disassembly. This mechanism might allow very specific trafficking of the vesicles, since disassembly

FIGURE 9 Fusion of the viral membrane followed by endocytosis of the virus.

might be triggered only when the vesicles reached a specific environment in the cell.

XII. MEDICAL IMPLICATIONS

The medical implications of the interactions of the viral membranes with their host cells involve both immunological considerations and the potential for development of antiviral agents. Sendai virus, a paramyxovirus type 1, is a respiratory virus of mice and produces disease in infant mice that is very similar to the disease produced by the human paramyxovirus type 1 in human infants. Therefore, advances made with the liposome system can easily be tested in mice for their clinical relevance.

The viral proteins incorporated into the host membrane during fusion of the viral membrane with the host membrane at entry and during assembly of the virus at the completion of the replicative cycle stimulate the cellular immune system to respond to the infected cell. The presentation of these antigens will determine what components of the immune system respond, which in turn has a bearing on whether the immune response is helpful or contributes to the pathogenesis of disease. The disassembly process in conjunction with the properties of the host membrane will determine this presentation.

The development of antiviral agents should be greatly facilitated by finding steps in viral replication that are directed by the viral proteins rather than use the host synthetic apparatus. Since most of the virus-coded proteins are included in the virion and many are involved in entry, the steps of viral entry, especially disassembly, are likely targets for antiviral agents. Supporting this idea that it should be possible to find antiviral agents that inhibit viral entry is the fact that amantadine and its congener, rimantadine (at concentrations too low to affect endosomal pH) can inhibit entry of influenza, type A (171). These drugs are currently being used clinically, although how they inhibit entry is not known. It should be possible to determine this mechanism if the steps in entry are understood.

ACKNOWLEDGMENTS

Part of the investigations from the author's laboratory were supported by grants PCM 78-08931 and PCM 82-05896 from the National Science Foundation and by grant AI-15540 from the National Institutes of Health. Some of the unpublished results were obtained at the Institut für Immunologie und Virologie der Universität Zürich while the author was supported by Senior International Fellowship TWO1147 from the Fogarty International Center of the National Institutes of Health.

REFERENCES

1. Warthin, A. S. (1931). Occurrence of numerous large giant cells in the tonsils and pharyngeal mucosa in the prodromal stage of measles. *Arch. Pathol.* 11:864-874.
2. Finkeldey, W. (1931). Über Riesenzellbefunde in den Gaumenmandeln zugleich ein Beitrag zur histopathologie der Mandelveränderungen in Maserninkubationsstadium. *Virchow's Arch. Pathol. Anat. Physiol.* 281:323-329.

3. Bratt, M. A., and Gallaher, W. R. (1969). Preliminary analysis of the requirements for fusion from within and fusion from without by Newcastle disease virus. *Proc. Natl. Acad. Sci. USA* 64:536-540.

4. Storey, D. G., and Kang, C. Y. (1985). Vesicular stomatitis virus-infected cells fuse when the intracellular pool of functional M protein is reduced in the presence of G protein. *J. Virol.* 53:374-383.

5. Fukai, K., and Suzuki, T. (1955). On the characteristics of a newly-isolated hemagglutinating virus from mice. *Med. J. Osaka Univ.* 6:1-15.

6. Okada, Y. (1958). The fusion of Ehrlich's tumor cells caused by HVJ virus in vitro. *Biken's J.* 1:103-110.

7. Harris, H., and Watkins, J. F. (1965). Hybrid cells derived from mouse and man: Artificial heterokaryons of mammalian cells derived from different species. *Nature* 205:640-646.

8. Spear, P. G. (1987). Virus-induced cell fusion. In: *Cell Fusion* (A. E. Sowers, ed.). Plenum Press, New York, pp. 3-32.

9. Hosaka, Y. (1958). On the hemolytic activity of HVJ. I. An analysis of the hemolytic reaction. *Biken's J.* 1:70-89.

10. Morgan, C., and Howe, C. (1968). Structure and development of viruses as observed in the electron microscope. IX. Entry of parainfluenza I (Sendai) virus. *J. Virol.* 2:1122-1132.

11. Haywood, A. M. (1974). Characteristics of Sendai virus receptors in a model membrane. *J. Mol. Biol.* 83:427-436.

12. Haywood, A. M. (1975). Model membranes and Sendai virus: Surface-surface interactions. In: *Negative Strand Viruses*, Vol. 2 (R. D. Barry and B. W. J. Mahy, eds.). Academic Press, London, pp. 923-928.

13. Haywood, A. M. (1975). "Phagocytosis" of Sendai virus by model membranes. *J. Gen. Virol.* 29:63-68.

14. Haywood, A. M. (1974). Fusion of Sendai viruses with model membranes. *J. Mol. Biol.* 87:625-628.

15. Haywood, A. M. (1978). Interactions of liposomes with viruses. *Ann. N.Y. Acad. Sci.* 308:275-280.

16. Kingsbury, D. W., Bratt, M. A., Choppin, P. W., Hanson, R. P., Hosaka, Y., ter Meulen, V., Norrby, E., Plowright, W., Rott, R., and Wunner, W. H. (1978). Paramyxoviridae. *Intervirology* 10:137-152.

17. Knutton, S. (1977). Studies of membrane fusion. II. Fusion of human erythrocytes by Sendai virus. *J. Cell Sci.* 28:189-210.

18. Compans, R. W., and Klenk, H-D. (1979). Viral membranes. In: *Comprehensive Virology*, Vol. 23 (H. Fraenkel-Conrat, and R. R. Wagner, eds.). Plenum Press, New York, pp. 293-407.

19. Shimizu, K., Shimizu, Y. K., Kohama, T., and Ishida, N. (1974). Isolation and characterization of two distinct types of HVJ (Sendai virus) spikes. *Virology* 62:90-101.

20. Markwell, M. K., and Fox, C. F. (1980). Protein-protein interactions within paramyxoviruses identified by native disulfide bonding or reversible chemical cross-linking. *J. Virol.* 33:152-166.

21. Lyles, D. S. (1979). Glycoproteins of Sendai virus are transmembrane proteins. *Proc. Natl. Acad. Sci. USA* 76:5621-5625.

22. Hiebert, S. W., Paterson, R. G., and Lamb, R. A. (1985). Hemagglutinin-neuraminidase protein of the paramyxovirus simian virus 5: Nucleotide sequence of the mRNA predicts an N-terminal membrane anchor. *J. Virol.* 54:1-6.

23. Blumberg, B., Giorgi, C., Roux, L., Raju, R., Dowling, P., Chollet, A., and Kolakofsky, D. (1985). Sequence determination of the Sendai virus HN gene and its comparison to the influenza virus glycoproteins. *Cell* 41:269-278.

24. Örvell, C., and Grandien, M. (1982). The effects of monoclonal anti-
    bodies on biologic activities of structural proteins of Sendai virus.
    J. Immunol. 129:2779-2787.

25. Yewdell, J., and Gerhard, W. (1982). Delineation of four antigenic
    sites on a paramyxovirus glycoprotein via which monoclonal antibodies
    mediate distinct antiviral activities. *J. Immunol.* 128:2670-2675.

26. Miura, N., Uchida, T., and Okada, Y. (1982). HVJ (Sendai virus)-
    induced envelope fusion and cell fusion are blocked by monoclonal
    anti-HN protein antibody that does not inhibit hemagglutination activity
    of HVJ. *Exp. Cell Res.* 141:409-420.

27. Waxham, M. N. and Wolinsky, J. S. (1986). A fusing mumps virus
    variant selected from a nonfusing parent with the neuraminidase inhibitor
    2-deoxy-2,3-dehydro-N-acetylneuraminic acid. *Virology* 151:286-295.

28. Homma, M., and Ohuchi, M. (1973). Trypsin action on the growth
    of Sendai virus in tissue culture cells. III. Structural differences
    of Sendai viruses grown in eggs and tissue culture cells. *J. Virol.*
    12:1457-1465.

29. Scheid, A., and Choppin, P. W. (1974). Identification of biological
    activities of paramyxovirus glycoproteins. Activation of cell fusion,
    hemolysis and infectivity by proteolytic cleavage of an inactive pre-
    cursor protein of Sendai virus. *Virology* 57:475-490.

30. Blumberg, B. M., Giorgi, C., Rose, K., and Kolakofsky, D. (1985).
    Sequence determination of the Sendai virus fusion protein gene. *J. Gen.
    Virol.* 66:317-331.

31. Hsu, M-C., and Choppin, P. W. (1984). Analysis of Sendai virus
    mRNAs with cDNA clones of viral genes and sequences of biologically
    important regions of the fusion protein. *Proc. Natl. Acad. Sci. USA*
    81:7732-7736.

32. Moscufo, N., Gallina, A., Schiavo, G., Montecucco, C., and Tomasi, M.
    (1987). Multiple lipid interactions of the Sendai virus fusogenic protein.
    J. Biol. Chem. 262:11490-11496.

33. Peterhans, E., Baechi, T., and Yewdell, J. (1983). Evidence for
    different receptor sites in mouse spleen cells for the Sendai virus
    hemagglutinin-neuraminidase (HN) and fusion (F) glycoproteins.
    Virology 128:366-376.

34. Polos, P. G. and Gallaher, W. R. (1982). Independent effects of phyto-
    hemagglutinin on cell fusion and attachment by Newcastle disease virus.
    Virology 119:268-275.

35. Nakanishi, M., Uchida, T., Kim, J., and Okada, Y. (1982). Glyco-
    proteins of Sendai virus (HVJ) have a critical ratio for fusion between
    virus envelopes and cell membranes. *Exp. Cell Res.* 142:95-101.

36. Blumberg, B. M., Rose, K., Simona, M. G., Roux, L., Giorgi, C.,
    and Kolakofsky, D. (1984). Analysis of the Sendai virus M gene and
    protein. *J. Virol.* 52:656-663.

37. Hewitt, J. A., and Nermut, M. V. (1977). A morphological study of
    the M-protein of Sendai virus. *J. Gen. Virol.* 34:127-136.

38. Heggeness, M. H., Smith, P. R., and Choppin, P. W. (1982). In
    vitro assembly of the nonglycosylated membrane protein (M) of Sendai
    virus. *Proc. Natl. Acad. Sci. USA* 79:6232-6236.

39. Wang, E., Wolf, B. A., Lamb, R. A., Choppin, P. W., and Goldberg,
    A. R. (1976). The presence of actin in enveloped viruses. In: *Cell
    Motility*, Book B (R. Goldman, T. Pollard, and J. Rosenbaum, eds.).
    Cold Spring Harbor Press, Cold Spring Harbor, NY, pp. 589-599.

40. Giuffre, R. M., Tovell, D. R., Kay, C. M., and Tyrrell, D. L. J. (1982). Evidence for an interaction between the membrane protein of a paramyxovirus and actin. *J. Virol.* 42:963-968.

41. Shimizu, K., and Ishida, N. (1975). The smallest protein of Sendai virus: Its candidate function of binding nucleocapsid to envelope. *Virology* 67:427-437.

42. Yoshida, T., Nagai, Y., Yoshii, S., Maeno, K., Matsumoto, T., and Hoshino, M. (1976). Membrane (M) protein of HVJ (Sendai virus): Its role in virus assembly. *Virology* 71:143-161.

43. Büechi, M., and Bächi, T. (1982). Microscopy of internal structures of Sendai virus associated with the cytoplasmic surface of host membranes. *Virology* 120:349-359.

44. Klenk, H. D., and Choppin, P. W. (1969). Lipids of plasma membranes of monkey and hamster kidney cells and of parainfluenza virions grown in these cells. *Virology* 38:255-268.

45. Blough, H. A., and Lawson, D. E. M. (1968). The lipids of paramyxoviruses: A comparative study of Sendai and Newcastle disease virus. *Virology* 36:286-292.

46. Israel, A., Audubert, F., and Semmel, M. (1975). Phospholipids in Newcastle disease virus infected cells. *Biochim. Biophys. Acta* 375: 224-235.

47. Stone, H. O., Portner, A., and Kingsbury, D. W. (1971). Ribonucleic acid transcriptases in Sendai virions and infected cells. *J. Virol.* 8: 174-180.

48. Robinson, W. S. (1971). Ribonucleic acid polymerase activity in Sendai virions and nucleocapsid. *J. Virol.* 8:81-86.

49. Hamaguchi, M., Yoshida, T., Nishikawa, K., Naruse, H., and Nagai, Y. (1983). Transcriptive complex of Newcastle disease virus. I. Both L and P proteins are required to constitute an active complex. *Virology* 128:105-117.

50. Leppert, M., Rittenhouse, L., Perrault, J., Summers, D. F., and Kolakofsky, D. (1979). Plus and minus strand leader RNAs in negative strand virus-infected cells. *Cell* 18:735-747.

51. Colonno, R. J., and Stone, H. O. (1975). Methylation of messenger RNA of Newcastle disease virus in vitro by a virion-associated enzyme. *Proc. Natl. Acad. Sci. USA* 72:2611-2615.

52. Weiss, S. R., and Bratt, M. A. (1974). Polyadenylate sequences on NDV mRNA synthesized in vivo and in vitro. *J. Virol.* 13:1220.

53. Homma, M., Shimizu, K., Shimizu, Y. K., and Ishida, N. (1976). On the study of Sendai virus hemolysis. I. Complete Sendai virus lacking in hemolytic activity. *Virology* 71:41-47.

54. Shimizu, Y. K., Shimizu, K., Ishida, N., and Homma, M. (1976). On the study of Sendai virus hemolysis. II. Morphological study of envelope fusion and hemolysis. *Virology* 71:48-60.

55. Kim, J., Hama, K., Miyake, Y., and Okada, Y. (1979). Transformation of intramembrane particles of HVJ (Sendai virus) envelopes from an invisible to visible form on aging of virions. *Virology* 95:523-535.

56. Kim, J. and Okada, Y. (1987). Difference in capacities for virion-to-virion fusion of young and aged HVJ (Sendai virus): A model of membrane fusion. *J. Membr. Biol.* 97:241-249.

57. Israel, S., Ginsberg, D., Laster, Y., Zakai, N., Milner, Y., and Loyter, A. (1983). A possible involvement of virus-associated protease in the fusion of Sendai virus envelopes with human erythrocytes. *Biochim. Biophys. Acta* 732:337-346.

58. Sugawara, K-E., Tashiro, M., and Homma, M. (1982). Intermolecular association of HANA glycoprotein of Sendai virus in relation to the expression of biological activities. *Virology* 117:444-455.

59. Lamb, R. A. (1975). The phosphorylation of Sendai virus proteins by a virus particle-associated protein kinase. *J. Gen. Virol.* 26:249-263.

60. Roux, L., and Kolakofsky, D. (1974). Protein kinase associated with Sendai virions. *J. Virol.* 13:545-547.

61. Lamb, R. A., and Choppin, P. W. (1977). The synthesis of Sendai virus polypeptides in infected cells. III. Phosphorylation of polypeptides. *Virology* 81:382-397.

62. Hsu, C-H., Morgan, E. M., and Kingsbury, D. W. (1982). Site-specific phosphorylation regulates the transcriptive activity of vesicular stomatitis virus NS protein. *J. Virol.* 43:104-112.

63. Neurath, A. R., and Sokol, F. (1963). Association of myxoviruses with an adenosine diphosphatase/adenosine triphosphatase as revealed by chromatography on DEAE-cellulose and by density gradient centrifugation. *Z. Naturforsch.* 18b:1050-1052.

64. Testa, D., and Banerjee, A. K. (1979). Nucleoside diphosphate kinase activity in purified cores of vesicular stomatitis virus. *J. Biol. Chem.* 254:9075-9079.

65. Beaudoin, A. R., Vachereau, A., Grondin, G., St-Jean, P., Rosenberg, M. D., and Strobel, R. (1986). Microvesicular secretion, a mode of cell secretion associated with the presence of an ATP-diphosphohydrolase. *FEBS Lett.* 103:1-2.

66. Hirst, G. K. (1959). Viral-host relation. In: *Viral and Rickettsial Infections of Man* (T. M. Rivers and F. L. Horsfall, eds.). Lippincott, Philadelphia, pp. 96-144.

67. Gottschalk, A., Belyavin, G., and Biddle, F. (1972). Glycoproteins as influenza virus haemagglutinin inhibitors and as cellular virus receptors. In: *Glycoproteins; Their Composition, Structure and Function* (A. Gottschalk, ed.). Elsevier, New York, pp. 1082-1096.

68. MacDonald, R. I., and MacDonald, R. C. (1975). Assembly of phospholipid vesicles bearing sialoglycoprotein from erythrocyte membrane. *J. Biol. Chem.* 250:9206-9214.

69. Oku, N., Nojima, S., and Inoue, K. (1981). Studies on the interaction of Sendai virus with liposomal membranes. Sendai virus-induced agglutination of liposomes containing glycophorin. *Biochim. Biophys. Acta* 646:36-42.

70. Oku, N., Inoue, K., Nojima, S., Sekiya, T., and Nozawa, Y. (1982). Electron microscopic study on the interaction of Sendai virus with liposomes containing glycophorin. *Biochim. Biophys. Acta* 691:91-96.

71. Markwell, M. A. K., Svennerholm, L., and Paulson, J. C. (1981). Specific gangliosides function as host cell receptors for Sendai virus. *Proc. Natl. Acad. Sci. USA* 78:5406-5410.

72. Kohn, L. D., Aloj, S. M., Tombaccini, D., Rotella, C. M., Toccafondi, R., Marcocci, C., Corda, D., and Grollman, E. F. (1985). The thyrotropin receptor. In: *Biochemical Actions of Hormones 12* (G. Litwack, ed.). Academic Press, New York, pp. 457-512.

73. Surolia, A., Bachhawat, B. K., and Podder, S. K. (1975). Interaction between lectin from Ricinus communis and liposomes containing gangliosides. *Nature* 257:802-804.

74. Fishman, P. H., Moss, J., Richards, R. L., Brady, R. O., and Alving, C. R. (1979). Liposomes as model membranes for ligand-receptor inter-

actions: Studies with choleragen and glycolipids. *Biochemistry* 18: 2562-2567.

75. Holmgren, J., Svennerholm, L., Elwing, H., Fredman, P., and Strannegård, Ö. (1980). Sendai virus receptor: Proposed recognition structure based on binding to plastic-adsorbed gangliosides. *Proc. Natl. Acad. Sci. USA* 77:1947-1950.

76. Umeda, M., Nojima, S., and Inoue, K. (1984). Activity of human erythrocyte gangliosides as a receptor to HVJ. *Virology* 133:172-182.

77. Rogers, G. N., and Paulson, J. C. (1983). Receptor determinants of human and animal influenza virus isolates: Differences in receptor specificity of the H3 hemagglutinin based on species of origin. *Virology* 127:361-373.

78. Tsao, Y-s. (1985). Interactions of Sendai virus with liposome target membrane. Doctoral thesis, The University of Tennessee, Knoxville.

79. Haywood, A. M. and Boyer, B. P. (1982). Sendai virus membrane fusion: time course and effect of temperature, pH, calcium, and receptor concentration. *Biochemistry* 21:6041-6046.

80. Haywood, A. M., and Boyer, B. P. (1986). Ficoll and dextran enhance adhesion of Sendai virus to liposomes containing receptor (ganglioside (G_{D1a}). *Biochemistry* 25:3925-3929.

81. WuDunn, D., and Spear, P. G. (1989). Initial interaction of herpes simplex virus with cells is binding to heparan sulfate. *J. Virol.* 63:52-58.

82. Ketis, N. V., and Grant, C. W. M. (1982). Co-operative binding of concanavalin A to a glycoprotein in lipid bilayers. *Biochim. Biophys. Acta* 689:194-202.

83. Grant, C. W. M., and Peters, M. W. (1984). Lectin-membrane interactions. Information from model systems. *Biochim. Biophys. Acta* 779: 403-422.

84. White, J. M. and Littman, D. R. (1989). Viral receptors of the immunoglobulin superfamily. *Cell* 56:725-728.

85. Mooney, J. J., Dalrymple, J. M., Alving, C. R., and Russell, P. K. (1975). Interaction of Sindbis virus with liposomal model membranes. *J. Virol.* 15:225-231.

86. Fries, E., and Helenius, A. (1979). Binding of Semliki Forest virus and its spike glycoproteins to cells. *Eur. J. Biochem.* 97:213-220.

87. Doms, R. W., Helenius, A., and White, J. (1985). Membrane fusion activity of the influenza virus hemagglutinin. *J. Biol. Chem.* 260:2973-2980.

88. Haywood, A. M., and Boyer, B. P. (1984). Effect of lipid composition upon fusion of liposomes with Sendai virus membranes. *Biochemistry* 23:4161-4166.

89. Klappe, K., Wilschut, J., Nir, S., and Hoekstra, D. (1986). Parameters affecting fusion between Sendai virus and liposomes. Role of viral proteins, liposome composition and pH. *Biochemistry* 25:8252-8260.

90. Knutton, S. (1976). Changes in viral envelope structure preceding infection. *Nature* 264:672-673.

91. Knutton, S. (1978). The mechanism of virus-induced cell fusion. *Micron* 9:133-154.

92. Pinto da Silva, P., Shimizu, K., and Parkison, C. (1980). Fusion of human erythrocytes induced by Sendai virus: Freeze-fracture aspects. *J. Cell Sci.* 43:419-432.

93. Volsky, D. J., and Loyter, A. (1978). Role of Ca^{++} in virus-induced membrane fusion. Ca^{++} accumulation and ultrastructural changes induced by Sendai virus in chicken erythrocytes. *J. Cell Biol.* 78:465-479.

94. Bongrand, P., and Bell, G. I. (1984). Cell-cell adhesion: Parameters and possible mechanisms. In: *Cell Surface Dynamics, Concepts and Models*, Vol. 3 (A. S. Perelson, C. DeLisi, and F. W. Wiegel, eds.). Marcel Dekker, New York, pp. 459-493.

95. Griffin, F. M. Jr., Griffin, J. A. Leider, J. E., and Silverstein, S. C. (1975). Studies on the mechanism of phagocytosis. I. Requirements for circumferential attachment of particle-bound ligands to specific receptors on the macrophage plasma membrane. *J. Exp. Med.* 142:1263-1282.

96. Tsao, Y-S., and Huang, L. (1986). Kinetic studies of Sendai virus-target membrane interactions: Independent analysis of binding and fusion. *Biochemistry* 25:3971-3976.

97. Gingell, D., and Ginsberg, L. (1978). Problems in the physical interpretation of membrane interaction and fusion. In: *Membrane Fusion* (G. Poste and G. L. Nicolson, eds.). North-Holland, New York, pp. 791-833.

98. Weiss, L., and Harlos, J. P. (1972). *Short-Term Interactions Between Cell Surfaces*. Pergamon Press, New York, pp. 355-405.

99. Poste, G. (1970). Cell surface changes in virus-induced cell fusion. 1. The importance of microvilli. *Microbios* 2:227-239.

100. Abidi, T. F., and Yeagle, P. L. (1984). Surface properties of Sendai virus envelope. *Biochim Biophys. Acta* 775:419-425.

101. Haywood, A. M., and Boyer, B. P. (1981). Initiation of fusion and disassembly of Sendai virus membranes into liposomes. *Biochim. Biophys. Acta* 646:31-35.

102. Umeda, M., Nojima, S., and Inoue, K. (1985). Effect of lipid composition on HVJ-mediated fusion of glycophorin liposomes to erythrocytes. *J. Biochem.* 97:1301-1310.

103. Hsu, M-C., Scheid, A., and Choppin, P. W. (1983). Fusion of Sendai virus with liposomes: Dependence on the viral fusion protein (F) and the lipid composition of liposomes. *Virology* 126:361-369.

104. Kundrot, C. E., Spangler, E. A., Kendall, D. A., MacDonald, R. C., and MacDonald, R. I. (1983). Sendai virus-mediated lysis of liposomes requires cholesterol. *Proc. Natl. Acad. Sci. USA* 80:1608-1612.

105. Tsao, Y-S., and Huang, L. (1985). Sendai virus induced leakage of liposomes containing gangliosides. *Biochemistry* 24:1092-1098.

106. Hoekstra, D., de Boert, T., Klappe, K., and Wilschut, J. (1984). Fluorescence method for measuring the kinetics of fusion between biological membranes. *Biochemistry* 23:5675-5681.

107. Pagano, R. E. and Longmuir, K. J. (1983). Intracellular translocation and metabolism of fluorescent lipid analogues in cultured mammalian cells. *TIBS* 8:157-161.

108. Citovsky, V., Blumenthal, R., and Loyter, A. (1985). Fusion of Sendai virions with phosphatidyl-cholesterol liposomes reflects the viral activity required for fusion with biological membranes. *FEBS Lett.* 193:135-140.

109. Chejanovsky, N., Zakai, N., Amselem, S., Barenholz, Y., and Loyter, A. (1986). Membrane vesicles containing the Sendai virus binding glycoprotein, but not the viral fusion protein, fuse with phosphatidylserine liposomes at low pH. *Biochemistry* 25:4810-4817.

110. Haywood, A. M. (1983). Virus infection of liposomes (or virology for the microveterinarian). In: *Liposome Letters* (A. D. Bangham, ed.). Academic Press, London, pp. 277-287.

111. Evans, E. A., and Parsegian, V. A. (1983). Energetics of membrane deformation and adhesion in cell and vesicles aggregation. *Ann. NY Acad. Sci.* 416:13-33.

112. Nir, S., Klappe, K., and Hoekstra, D. (1986). Kinetics and extent of fusion between Sendai virus and erythrocyte ghosts: Application of a mass action kinetic model. *Biochemistry* 25:2155-2161.

113. Pollard, H. B., Pazoles, C. J., Creutz, C. E., and Zinder, O. (1979). The chromaffin granule and possible mechanisms of exocytosis. *Int. Rev. Cytol.* 58:160-198.

114. Knutton, S. (1980). Studies of membrane fusion. VI. Mechanism of the membrane fusion and cell swelling stages of Sendai virus-mediated cell fusion. *J. Cell Sci.* 43:103-118.

115. Ohki, S. (1984). Effects of divalent cations, temperature, osmotic pressure gradient, and vesicle curvature on phosphatidylserine vesicle fusion. *J. Membr. Biol.* 77:265-275.

116. Gething, M.-J., White, J., and Waterfield, M. (1978). Purification of the fusion protein of Sendai virus: analysis of the NH_2-terminal sequence generated during precursor activation. *Proc. Natl. Acad. Sci. USA* 75:2737-2740.

117. Richardson, C. D., Scheid, A., and Choppin, P. W. (1980). Specific inhibition of paramyxovirus and myxovirus replication by oligopeptides with amino acid sequences similar to those at the N termini of the F_1 or HA_2 viral polypeptides. *Virology* 105:205-222.

118. Hsu, M-C., Scheid, A., and Choppin, P. W. (1981). Activation of the Sendai virus fusion protein (F) involves a conformational change with exposure of a new hydrophobic region. *J. Biol. Chem.* 256:3557-3563.

119. Novick, S. L. and Hoekstra, D. (1988). Membrane penetration of Sendai virus glycoproteins during the early stages of fusion with liposomes as determined by hydrophobic photoaffinity labeling. *Proc. Natl. Acad. Sci. USA* 85:7433-7437.

120. Fidgen, K. J., and Tisdale, M. (1981). An "on grid" electron microscopic method for studying the interaction and fusion of influenza A virus with human erythrocyte membranes. *J. Virol. Methods* 3:271-276.

121. Haywood, A. M., and Boyer, B. P. (1985). Fusion of influenza virus membranes with liposomes at pH 7.5. *Proc. Natl. Acad. Sci. USA* 82:4611-4615.

122. Haywood, A. M., and Boyer, B. P. (1986). Time and temperature dependence of influenza virus membrane fusion at neutral pH. *J. Gen. Virol.* 67:2813-2817.

123. Howe, C., Lee, L. T., and Rose, H. M. (1961). Collocalia mucoid: A substrate for myxovirus neuraminidase. *Arch. Biochem. Biophys.* 95:512-520.

124. Maeda, T., Kawasaki, K., and Ohnishi, S.-I. (1981). Interaction of influenza virus hemagglutinin with target membrane lipids is a key step in virus-induced hemolysis and fusion at pH 5.2. *Proc. Natl. Acad. Sci. USA* 78:4133-4137.

125. White, J., Helenius, A., and Kartenbeck, J. (1982). Membrane fusion activity of influenza virus. *EMBO J.* 1:217-222.

126. Lenard, J., and Miller, D. K. (1981). pH dependent hemolysis by influenza, Semliki Forest virus, and Sendai virus. *Virology* 110:479-482.

127. Sato, S. B., Kawasaki, K., and Ohnishi, S.-I. (1983). Hemolytic activity of influenza virus hemagglutinin glycoproteins activated in mildly acidic environments. *Proc. Natl. Acad. Sci. USA* 80:3153-3157.

128. White, J., Matlin, K., and Helenius, A. (1981). Cell fusion by Semliki forest, influenza, and vesicular stomatitis viruses. *J. Cell Biol.* 89: 674–679.

129. Nir, S., Stegmann, T., and Wilschut, J. (1986). Fusion of influenza virus with cardiolipin liposomes at low pH: Mass action analysis of kinetics and extent. *Biochemistry* 25:257–266.

130. Tischer, I. (1962). Untersuchungen über die pH-Abhängigkeit der Haemagglutination und die Beziehung zwischen Oberflächenladung und Receptoraffinität einiger Myxoviren. *Arch. Ges. Virusforsch.* 12:89–107.

131. Lakshmi, M. V., and Schulze, I. T. (1978). Effects of sialylation of influenza virions on their interactions with host cells and erythrocytes. *Virology* 88:314–324.

132. Miller, G. L., Lauffer, M. A., and Stanley, W. M. (1944). Electrophoretic studies on PR8 influenza virus. *J. Exp. Med.* 8:549–559.

133. Daniels, R. S., Downie, J. C., Hay, A. J., Knossow, M., Skehel, J. J., Wang, M. L., and Wiley, D. C. (1985). Fusion mutants of the influenza virus hemagglutinin glycoprotein. *Cell* 40:431–439.

134. Tycko, B., Keith, D. H., and Maxfield, F. R. (1983). Rapid acidification of endocytic vesicles containing asialoglycoprotein in cells of a human hepatoma line. *J. Cell Biol.* 97:1762–1776.

135. Skehel, J. J., Bayley, P. M., Brown, E. B., Martin, S. R., Waterfield, M. D., White, J. M., Wilson, I. A., and Wiley, D. C. (1982). Changes in the conformation of influenza virus hemagglutinin at the pH optimum of virus-mediated membrane fusion. *Proc. Natl. Acad. Sci. USA* 79:968–972.

136. Helenius, A., Kartenbeck, J., Simons, K., and Fries, E. (1980). On the entry of Semliki Forest virus into BHK-21 cells. *J. Cell Biol.* 84:404–420.

137. Miller, D. K., and Lenard, J. (1981). Antihistaminics, local anesthetics, and other amines as antiviral agents. *Proc. Natl. Acad. Sci. USA* 78:3605–3609.

138. Bar-Sagi, D., and Prives, J. (1983). Trifluoperazine, a calmodulin antagonist, inhibits muscle cell fusion. *J. Cell Biol.* 97:1375–1380.

139. Horwitz, S. B., Chia, G. H., Harracksingh, C., Orlow, S., Pifko-Hirst, S., Schneck, J., Sorbara, L., Speaker, M., Wilk, E. W., and Rosen, O. M. (1981). Trifluoperazine inhibits phagocytosis in a macrophagelike cultured cell line. *J. Cell Biol.* 91:798–802.

140. Low, P. S., Lloyd, D. H., Stein, T. M., and Rogers, J. A. III (1979). Calcium displacement by local anesthetics. *J. Biol. Chem.* 254:4119–4125.

141. Ros, M., and Glushko, V. (1982). Inhibition of erythrocyte membrane-associated protein kinase by substituted phenothiazines. *Biophys. J.* 37:140a.

142. Volpi, M., Sha'afi, R. I., Epstein, P. M., Andrenyak, D. M., and Feinstein, M. B. (1981). Local anesthetics, mepacrine, and propranolol are antagonists of calmodulin. *Proc. Natl. Acad. Sci. USA* 78:795–799.

143. Bukrinskaya, A. G., Vorkunova, N. K., Kornilayeva, G. V., Narmanbetova, R. A., and Vorkunova, G. K. (1982). Influenza virus uncoating in infected cells and effect of rimantidine. *J. Gen. Virol.* 60:49–59.

144. Durand, D. P., Chalgren, S. D., and Franke, V. (1970). Effect of chloroquine on myxovirus replication. *Antimicrobial Agents and Chemotherapy* (G. L. Mabby, ed.). American Society of Microbiology, Bethesda, MD, pp. 105–108.

145. Fletcher, R. D., Hirschfield, J. E., and Forbes, M. (1965). A common mode of antiviral action for ammonium ions and various amines. *Nature* 207:664–665.

146. Haywood, A. M., Simons, R. L., and Douglas, R. G., Jr. (1981). Inhibition of influenza virus plaque formation by drugs that inhibit osmotic lysis of erythrocytes. *Abstracts of Twenty-First Interscience Conference of Antimicrobial Agents and Chemotherapy*. American Society of Microbiology, Washington, DC, p. 339.

147. Poste, G., and Reeve, P. (1972). Inhibition of virus-induced fusion by local anaesthetics and phenothiazine tranquillizers. *J. Gen. Virol.* 16:21–28.

148. Shimizu, Y., Yamamoto, S., Homma, M., and Ishida, N. (1972). Effect of chloroquine on the growth of animal viruses. *Arch. Ges. Virusforsch.* 36:93–104.

149. Richardson, C. D., and Vance, D. E. (1978). The effect of colchicine and dibucaine on the morphogenesis of Semliki Forest virus. *J. Biol. Chem.* 253:4584–4589.

150. Harter, C., Bächi, T., Semenza, G., and Brunner, J. (1988). Hydrophobic photolabeling identified BHA2 as the subunit mediating the interaction of bromelain-solubilized influenza virus hemagglutinin with liposomes at low pH. *Biochemistry* 27:1856–1864.

151. Edwards, J., and Brown, D. T. (1986). Sindbis virus-mediated cell fusion from without is a two-step event. *J. Gen. Virol.* 67:377–380.

152. Pauza, C. D. and Price, T. M. (1988). Human immunodeficiency virus infection of T cells and monocytes process via receptor-mediated endocytosis. *J. Cell Biol.* 107:959–968.

153. Stein, B. S., Gowda, S. D., Lifson, J. D., Penhallow, R. C., Bensch, K. G., and Engleman, E. G. (1987). pH-independent HIV entry into CD4-positive T cells via virus envelope fusion to the plasma membrane. *Cell* 49:659–668.

154. McClure, M. O., Marsh, M., and Weiss, R. A. (1988). Human immunodeficiency virus infection of CD4-bearing cells occurs by a pH-independent mechanism. *EMBO J.* 7:513–518.

155. Para, M. F., Baucke, R. B., and Spear, P. G. (1980). Immunoglobulin G(Fc)-binding receptors on virions of herpes simplex virus type 1 and transfer of these receptors to the cell surface by infection. *J. Virol.* 34:512–520.

156. Fuller, A. O. and Spear, P. G. (1987). Antiglycoprotein D antibodies that permit adsorption but block infection of herpes simplex virus 1 prevent virion-cell fusion at the cell surface. *Proc. Natl. Acad. Sci. USA* 84:5454–5458.

157. Koyama, A. H. and Uchida, T. (1987). The mode of entry of herpes simplex virus type I into Vero cells. *Microbiol. Immunol.* 31:123–130.

158. Rosenthal, K. S., Killius, J., Hodnichak, C. M., Venetta, T. M., Gyurgyik, L., and Janiga, K. (1989). Mild acidic pH inhibition of the major pathway of HSV entry into Hep-2 cells. *J. Gen. Virol.* 70:857–867.

159. Haywood, A. M., and Boyer, B. P. (1982). Fusion and disassembly of Sendai virus membranes with liposomes. *Biophys. J.* 37:128–129.

160. Dimmock, N. J. (1982). Initial stages in infection with animal viruses. *J. Gen. Virol.* 59:1–22.

161. Caldwell, S. E. and Lyles, D. S. (1981). Interaction of Sendai virus proteins with the cytoplasmic surface of erythrocyte membranes following viral envelope fusion. *J. Biol. Chem.* 256:4838–4842.

162. Dubois-Dalcq, M., Holmes, K. V., and Rentier, B. (1984). *Assembly of Enveloped RNA Viruses.* Springer-Verlag, Vienna, New York, pp. 44-82.

163. Knutton, S., and Bächi, T. (1980). The role of cell swelling and haemolysis in Sendai virus-induced cell fusion and in the diffusion of incorporated viral antigens. *J. Cell Sci.* 42:153-167.

164. Dales, S. (1973). Early events in cell-animal virus interactions. *Bacteriol. Rev.* 37:103-135.

165. Bukrinskaya, A. G. (1982). Penetration of viral genetic material into host cell. *Adv. Virus Res.* 27:141-204.

166. Dimmock, N.J. (1982). Initial stages in infection with animal viruses. *J. Gen. Virol.* 59:1-22.

167. Kohn, A. (1985). Membrane effects of cytopathogenic viruses. *Prog. Med. Virol.* 31:109-167.

168. Dourmashkin, R. R., and Tyrrell, D. A. J. (1974). Electron microscopic observations on the entry of influenza virus into susceptible cells. *J. Gen. Virol.* 24:129-141.

169. Matlin, K. S., Reggio, H., Helenius, A., and Simons, K. (1981). Infectious entry pathway of influenza virus in a canine kidney cell line. *J. Cell Biol.* 91:601-613.

170. Patterson, S., Oxford, J. S., and Dourmashkin, R. R. (1979). Studies on the mechanism of influenza virus entry into cells. *J. Gen. Virol.* 43:223-229.

171. Hoffman, C. E. (1980). Structure, activity and mode of action of amantadine HCl and related compounds. *Antibiotics Chemother.* 27:233-250.

17

The Role of Envelope Glycoproteins in the Fusion of Sendai Virus with Liposomes

ABRAHAM LOYTER and VITALY CITOVSKY

The Hebrew University of Jerusalem, Jerusalem, Israel

I. INTRODUCTION

Infection of cells by enveloped viruses belonging to the paramyxovirus group, such as Sendai virus and Newcastle disease virus, occurs by fusion of the viral envelope with the cell plasma membrane (1,2). Binding of Sendai virus to specific plasma membrane receptors, sialic-acid-containing glycolipids or glycoproteins, is mediated by the viral hemagglutinin/neuraminidase (HN) glycoprotein, while virus-membrane fusion is promoted by the viral fusion (F) glycoprotein (1,2). These two viral envelope glycoproteins are sufficient to promote the overall process of virus-membrane fusion, as can be inferred from experiments using reconstituted Sendai virus envelopes (RSVE). It has been well established that lipid bilayers into which the HN and F glycoproteins are incorporated are as fusogenic as intact Sendai virions (3,4).

Until recently, it was assumed that the only function of the HN glycoprotein is to mediate binding of the virus to cell receptors. However, findings from several laboratories suggest that the HN polypeptide may also play an active role in the fusion step itself. It has been shown that anti-HN monoclonal antibodies inhibit the fusogenic activity of Sendai virus, without interfering with its capacity to bind to recipient cells (5). Moreover, membrane vesicles containing only the viral F glycoprotein, fail to fuse with cell plasma membranes even after binding mediated by other ligands (6). Only envelopes containing, in addition to the F glycoprotein, also the HN polypeptide fuse readily with living cells (6).

Infection of cultured cells by Sendai virus is accompanied by an increase in the membrane permeability to molecules of low and high molecular weight (7). Similarly, virus-induced hemolysis has been shown to reflect a process of virus-membrane fusion and has been used as a semiquantitative method to estimate the fusogenic activity of Sendai virus (8). Various inhibitors that block the virus's ability to infect cells also inhibit its hemolytic activity (9). Sendai virions can be rendered noninfectious as well as nonhemolytic by treatment with trypsin, phenylmethylsulfonylfluoride (PSMF) or dithiothreitol

(DTT). Trypsinization and treatment with PMSF significantly inhibit the viral fusion protein, whereas DTT irreversibly reduces the disulfide bonds of the viral HN glycoprotein (9,10). Trypsinized or PMSF-treated virions are able to bind to, but not to fuse with or lyse, living cells, while DTT-treated virions neither bind to nor fuse with biological membranes. Also, viruses bearing the uncleaved fusion protein, namely F_0, are neither fusogenic nor hemolytic (1,2).

Recently, it has been shown that energy transfer or fluorescence dequenching methods can be used to follow, on a quantitative basis, fusion processes between enveloped virions and biological membranes (11,12). Incubation of fluorescently labeled virions with either human erythrocytes or living cells results in fluorescence dequenching. Using these methods, it has been demonstrated that incubation of trypsin-, PMSF-, or DTT-treated Sendai virus with living cells does not result in fluorescence increase (12-14).

All the above features and properties make Sendai virus an excellent candidate to study questions related to the mechanism of virus-membrane interaction and fusion, as well as elucidation of the molecular mechanism of membrane fusion in general. However, in spite of extensive efforts made during the last few years to study the various steps involved in the interaction between Sendai virus or its isolated envelope glycoproteins and biological membranes or liposomes, the following questions still remain unanswered. (1) What is the function of membrane components that serve as virus receptors? Are they only passive binding sites or do they also play an active role in the virus-membrane fusion process? (2) What is the detailed mechanism by which the viral glycoproteins promote fusion between the lipid bilayer of the viral envelope and the cell plasma membrane?

The interaction between enveloped viruses and liposomes may serve as an excellent tool and model system for the elucidation of some aspects relating to the above questions. In this chapter, we give a brief review of the various approaches used and the results obtained from studies on the interaction of Sendai virus with liposomes of various compositions. In addition, we present a summary of our recent work on the subject.

II. VIRUS-LIPOSOME FUSION: A BRIEF SUMMARY OF THE EXISTING DATA

Haywood (15) was the first to study the interaction between Sendai virions and lipid vesicles (see also Chapter 16). Using electron microscopy techniques, she demonstrated that Sendai virions bind to and fuse with liposomes composed of neutral lipids such as phosphatidylcholine (PC), sphingomyelin (SM), phosphatidylethanolamine (PE), or cholesterol-bearing sialoglycolipids. Since electron microscopy was the only method used to follow virus-liposome fusion, no quantitative data could be obtained. Later, Haywood and Boyer (16) studied fusion between Sendai virus and liposomes by the use of discontinuous sucrose gradient which allowed the separation of virus-liposome fusion products from unfused virions. Employing ^{35}S-labeled virions and liposomes composed of neutral lipids (PC, PE, and cholesterol), it was shown, confirming previous observations, that the presence of virus receptors such as the ganglioside G_{D1a} is obligatory to allow virus-liposome fusion (16). Fusion was temperature and pH dependent, being maximal at 37-40°C and above pH 7.0. In a more recent study (17), the effect of lipid composition on the process of virus-liposome fusion was investigated, using the same sucrose gradient method. Fusion was observed with liposomes composed of only PC, provided that ganglioside (G_{D1a}) molecules were present. Cholesterol was not necessary, and at low concentrations of gangliosides, its presence decreased the degree of virus-liposome fusion (17). Furthermore, specific virus receptors were not required when negatively charged phospholipids such as phosphatidic acid (PA), dicetylphosphate (DCP), or others were incorporated into the liposome bilayer.

Using freeze-fracture techniques, Oku et al. (18) showed that fusion between Sendai virions and liposomes (composed of PC, DCP, and cholesterol) depends on the presence of a sialic-acid-containing receptor, the human erythrocyte glycophorin. Oku et al. (19) also used virus-induced increase in the permeability of loaded liposomes as a method to follow virus-liposome fusion processes. This approach is based on the assumption that the viral hemolytic activity is identical to and reflects the viral fusogenic ability. Release of methylumbelliferyl-phosphate, enclosed within the liposome, following incubation with Sendai virus, was found to be absolutely dependent on the presence of a specific receptor (glycophorin) in the liposome membrane (19). Neither trypsinized virions nor virus containing the uncleaved fusion protein (F_0) was able to induce lysis of loaded liposomes. However, no attempts were made to study the effect of cholesterol or the ability of the virus to induce lysis in liposomes lacking the negatively charged lipid DCP (19). The liposomes employed by Oku and colleagues (19) contained higher amounts of DCP than those used by Haywood and Boyer (17). In spite of this difference, Oku et al. (19) found that lysis was dependent on the presence of sialic-acid-containing components such as glycophorin.

Virus-induced lysis of loaded liposomes as a method to study the viral fusogenic properties has also been used by Kundrot et al. (20). It was clearly shown that virus-induced release of calcein from liposomes composed of neutral lipids (PC and cholesterol) was depedent on the presence of cholesterol and of virus receptors, namely glycophorin or gangliosides. The virus failed to lyse liposomes composed of only PC, even in the presence of relatively high concentrations of virus receptors.

Dependency of the virus-liposome fusion processes on the presence of cholesterol has also been observed by Hsu et al. (21). Like Haywood and Boyer (17), these authors used sucrose gradients to separate free virions from those fused with liposomes. Fusion was also studied by estimating the digestion of the viral nucleoproteins following fusion with liposomes loaded with trypsin. Using the sucrose gradient method, Hsu et al. (21) observed that fusion of Sendai virions can occur with liposomes composed of PC and cholesterol and lacking virus receptors. Evidently, these results are in contradiction with those obtained by Haywood and Boyer (17), who showed dependency on the presence of gangliosides in liposomes composed of neutral phospholipids. The results of Hsu et al. (21) also seem to be in contrast with those obtained by Oku et al. (19) and Kundrot et al. (20), although the latter two groups have used virus-induced lysis to follow virus-liposome fusion. Evidently, no clear picture emerges from these studies, particularly regarding the need for specific virus receptors in virus-liposome fusion, while such receptors are absolutely required for fusion of Sendai virus with biological membranes (1,2).

For a better understanding of the fusion process between Sendai virus and liposomes and in order to support the view that this process indeed reflects the viral activity needed for its infectivity, the following questions should be studied and answered.

1. What is the function of virus receptors in the process of virus-liposome fusion?
2. Can negatively charged phospholipids replace virus receptors in the fusion process?
3. Is cholesterol necessary for fusion of the virus with liposomes composed of neutral lipids?

III. INTERACTION OF SENDAI VIRUS WITH NEGATIVELY CHARGED LIPOSOMES

A. Intact Virus

1. Virus-Induced Lysis of Loaded Liposomes

Fusion processes occurring between enveloped viruses, such as Sendai or influenza virus, and living cells lead to an increase in the cell membranes' permeability (1,2). Sendai virus has been shown to induce hemolysis at pH values above 6.5-7.0 (8), while the hemolytic activity of influenza virus becomes manifest at pH 5.2, a value at which the viral fusion protein is activated (22, see also Chapter 15). Based on these observations, virus-induced release of liposome contents has been used to follow virus-liposome fusion processes (19,20). Experiments in our laboratory, confirming previous observations, have shown that Sendai virus can induce the release of contents from negatively charged, loaded liposomes (Table 1; Ref. 23). No virus receptors are needed in the liposome bilayer for the lysis of such liposomes. Sendai virus induces a relatively high degree of lysis, as revealed by the release of the self-quenched carboxyfluorescein (CF), of liposomes composed of only phosphatidylserine (PS) or of PC/DCP (Table 1; Ref. 23) and lacking any sialoglycoproteins or sialoglycolipids. These observations support previous results by Haywood and Boyer (17), claiming that no

TABLE 1 Induction of Lysis in Negatively Charged Liposomes by Sendai Virus

Sendai virus treated with:	Hemolysis (%)	CF release (%)		
		PC/DCP	PS	CL
None	84	64	67	93
DTT	0	64	78	96
PMSF	9	63	61	94
Trypsin	5	8	13	25

Sendai virus (10 µg of protein) was incubated for 30 min at 37°C with 2 µg of large unilamellar liposomes, containing 80 mM carboxyfluorescein (CF), prepared by octylglucoside removed with Bio-beads SM-2 (30) and composed of PC/DCP (molar ratio 3:1), PS, or cardiolipin (CL). CF fluorescence was measured before and after the incubation, and the fluorescence obtained in the presence of 0.1% (v/v) Triton X-100 was considered to represent 100% release. Treatment of the virus was DTT (3 mM), PMSF (7 mM), or trypsin (60 µg/mg of viral protein) was carried out at 37°C for 30 min (9).

virus receptors are needed for fusion between Sendai virus and negatively charged liposomes.

In order to study whether the lytic activity observed indeed reflects the activity necessary for the virus' infectivity and fusogenic ability, we have investigated the effect of several inhibitors known to inactivate the viral biological functions. PMSF- or DTT-treated Sendai virus was shown to be noninfectious and nonfusogenic and failed to lyse red blood cells (Table 1; Refs. 9,10). Such nonhemolytic virions are able to lyse negatively charged liposomes to the same degree as non treated virions (23). Only trypsinized virions fail to lyse such negatively charged liposomes, indicating that viral envelope glycoproteins are involved in the release of contents from the liposomes (Table 1). Essentially the same results have been obtained with liposomes composed of other negatively charged phospholipids, such as phosphatidylinositol (PI) or phosphatidylglycerol (PG) (23). Liposomes composed of cardiolipin (CL) show the highest sensitivity to the viral lytic activity. Even trypsinized virions cause a 25% lysis upon incubation with CF-loaded liposomes (Table 1).

The presence of cholesterol greatly reduces the susceptibility of the negatively charged liposomes to the viral lytic activity (Fig. 1; Ref. 23). The incorporation of cholesterol has been shown to increase their stability and to reduce the leakage of their contents (24).

2. Intermixing of Viral and Liposomal Lipids

Fusion between enveloped viruses and biological membranes can be determined by following intermixing of the viral envelope lipids with lipids of the recipient membrane (11-14). In our work, we have used two methods for monitoring intermixing of Sendai virus and liposomal lipids. First, fluorescent molecules, such as octadecylrhodamine B (R_{18}) or N-(4-nitrobenzo-2-oxa-1,3-diazole)phosphatidylethanolamine (N-NBD-PE), have been inserted into envelopes of intact Sendai virions or reconstituted virus envelopes, respectively (12-14; see also Chapter 14). Fusion of fluorescently labeled viral envelopes with human erythrocytes or living cultured cells results in fluorescence dequenching (12-14). Second, the fluorescent

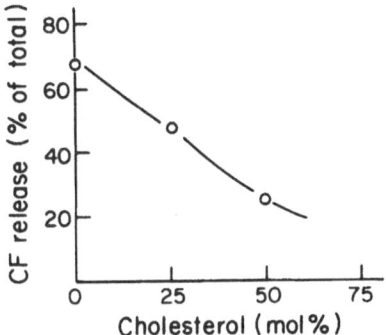

FIGURE 1 Lysis of liposomes composed of negatively-charged phospholipids: effect of cholesterol. Liposomes containing CF and composed of PC/DCP [molar ratio (1:0.3)] plus the indicated amounts of cholesterol were prepared as described in Table 1. Virus-induced lysis of the loaded liposomes was determined as described in Table 1.

molecule pyrenyldodecanoylphosphatidylcholine has been incorporated into the liposome bilayer (25). Fusion between unlabeled intact virions and fluorescently labeled liposomes result in an increase of the fluorescence degree due to dilution of fluorescent probes. Both approaches have clearly demonstrated that incubation of Sendai virions with negatively charged liposomes results in virus-liposome fusion. For such fusion to occur, no specific virus receptors, i.e., sialoylycoproteins or sialoglycolipids, are required (Table 2; Refs. 14,25). Maximum fusion is observed at pH values below 5.0, while at pH 7.0 or above, a much lower degree of virus-liposome fusion is obtained (25).

Similar to processes of virus-induced lysis of negatively charged liposomes, virus-liposome fusion as revealed by the fluorescence dequenching method can also be observed with nonhemolytic, PMSF- or DTT-treated virions (Table 2; Ref. 14). Only treatment with trypsin significantly reduces the viral ability to fuse with negatively charged liposomes (14). Results similar to those obtained with fluorescently labeled virions (Table 2) have been obtained by incubation of fluorescently labeled liposomes and unlabeled intact Sendai virions (25).

B. Reconstituted Sendai Virus Envelopes (RSVE)

1. *RSVE-Induced Lysis of Loaded Liposomes*

Intact Sendai virions can be solubilized by either nonionic or ionic detergents (3). Removal of the detergent from a mixture containing the viral glycoproteins and phospholipids results in the formation of fusogenic, reconstituted viral envelopes (3,26; see also Chapter 18). Such envelopes

TABLE 2 Fusion of Sendai Virus with Negatively Charged Phospholipid Vesicles: Intermixing of Viral and Liposomal Lipids

Sendai virus treated with:	R_{18} DQ (%)	
	PC/DCP	PS
None	46	29
DTT	44	25
PMSF	46	28
Trypsin	6	9

Sendai virus (2 μg of viral protein), labeled with octadecyl rhodamine (R_{18}) to a surface density of about 3 mol% relative to viral phospholipid (14), was incubated with 10 μg of PC/DCP (molar ratio 3:1) or PS liposomes (prepared as described in Table 1, except that the CF was omitted) for 30 min at 37°C. The extent of fluorescence dequenching (DQ) was determined from the fluorescence intensity before and after the incubation, taking the intensity in the presence of 0.1% (v/v) Triton X-100 as the 100% value (14). Treatment of the (labeled) virus with DTT, PMSF, or trypsin was carried out as described in Table 1.

TABLE 3 Lysis of Negatively Charged Liposomes Is Promoted by the HN Glycoprotein of Sendai Virus

System	Hemolysis (%)	CF release from PS liposomes (%)
RSVE	86	48
Trypsinized RSVE	0	4
PMSF-treated RSVE	7	48
Reconstituted F vesicles	5	2
Reconstituted HN vesicles	0	43
Trypsinized reconstituted HN vesicles	0	9
Reconstituted (F + HN) vesicles	78	43

Reconstituted Sendai virus envelopes (RSVE), or reconstituted vesicles bearing the viral F, HN, or both F and HN glycoproteins, were prepared as described in Ref. 30. Incubation with CF-containing PS liposomes, determination of CF release, and treatment of the reconstituted viral envelopes with PMSF or trypsin were carried out as described in Table 1.

contain only the two viral glycoproteins HN and F. We have used Triton X-100 to obtain fusogenic RSVE (3,26). However, other detergents, such as octylglucoside, have also been used for solubilization of Sendai virus (27) or other enveloped viruses (2).

RSVE, obtained by solubilization of intact virions with Triton X-100, are hemolytic (see Table 3) as well as fusogenic (14,26). Similar to intact virions, RSVE also induce lysis of loaded, negatively charged liposomes lacking virus receptors (Table 3; Refs. 23,28). Treatment of RSVE with PMSF or DTT completely blocks the hemolytic activity, indicating that this activity is promoted by the viral glycoproteins and not by the residual amount of detergent left in the reconstituted viral envelopes (9,26,28). This view is further confirmed by the results showing that reconstituted vesicles containing only the HN or the F glycoprotein are not hemolytic (Table 3; Refs. 28,29). Hemolysis could only be induced by vesicles containing both glycoproteins.

From our experiments (Table 3; Ref. 28), it is clear that neither PMSF nor DTT blocks the ability of RSVE to lyse negatively charged liposomes. But, the lytic activity of RSVE, similar to that of the intact virus, can be inhibited by treatment with trypsin (Table 3; Refs. 23,28), indicating once again that intact viral glycoproteins are required for the expression of this activity. Interestingly, vesicles bearing only the viral HN glycoprotein, although not hemolytic, are capable of lysing negatively charged liposomes (Table 3; Ref. 28). This is not observed with vesicles bearing only the F glycoprotein (Table 3). Therefore, the lytic and fusogenic activities observed with intact virions toward negatively charged liposomes may be due to the presence of the viral HN glycoprotein and not necessarily to the activity of the viral fusion (F) polypeptide. Lysis of, and fusion with, biological membranes and living cells is promoted only by viral envelopes bearing both the F and the HN glycoproteins (3).

TABLE 4 Reconstituted Viral Envelopes Bearing the HN Glycoprotein
Can Fuse with Negatively Charged Liposomes

System	N-NBD-PE DQ (%)	
	PC	PS
RSVE	18	60
PMSF-treated RSVE	N.D.	60
Pronase-treated RSVE	9	15
Reconstituted F vesicles	3	11
Reconstituted HN vesicles	10	60
PMSF-treated reconstituted HN vesicles	N.D.	65
Reconstituted (F + HN) vesicles	N.D.	58

RSVE and reconstituted vesicles bearing the viral F, HN, or both F
and HN glycoproteins, labeled with N-NBD-PE to a surface density of
6 mol% relative to viral phospholipid (12,30), were incubated for 20
min at 37°C with 10 µg of either PC or PS liposomes (prepared as
described in Tables 1 and 2). Under these conditions the degree of
fluorescence quenching is proportional to the surface density of the
probe (43). The extent of fluorescence dequenching (DQ) was determined from the fluorescence intensity before and after the incubation,
taking the intensity in the presence of 0.1% (v/v) Ammonyx-LO as the
100% value. Treatment of the reconstituted viral envelopes with PMSF
was carried out as described in Table 1, treatment with pronase
(0.5 mg/ml) as described in Ref. 31.

2. Intermixing of RSVE and Liposomal Lipids

Fluorescence dequenching methods have also been used to study interaction
and fusion between RSVE and negatively charged liposomes (12,30,31).
Fluorescently labeled RSVE can be prepared either by the addition of
N-NBD-PE molecules to the reconstitution system of the viral envelopes
(12) or by insertion of R_{18} in already formed RSVE (14). In both systems,
the addition of the fluorescent molecules to the viral envelopes results
in fluorescence self-quenching. Fluorescence dequenching is observed
upon fusion of fluorescent RSVE with biological membranes, living cultured
cells, or lipid vesicles (12-14,30,31; see also Chapter 32).

Neither virus receptors nor cholesterol molecules are needed to allow
fusion of RSVE with negatively charged liposomes (Table 4; Refs. 14,25,31).
Incorporation of cholesterol into negatively charged liposomes does not
affect their ability to fuse with RSVE. Nonhemolytic RSVE, treated with
PMSF or DTT, readily fuse with liposomes composed of PS or of PC/DCP
(Table 4; Ref. 31). A high degree of fluorescence dequenching is also
observed upon incubation of vesicles containing only the viral HN glycoprotein and negatively charged liposomes (31). No such increase is seen
with vesicles containing only the fusion (F) polypeptide of Sendai virus
(Fig. 2). It thus appears that fusion of RSVE with negatively charged
liposomes displays the same features as observed for virus-induced lysis
of the same liposomes.

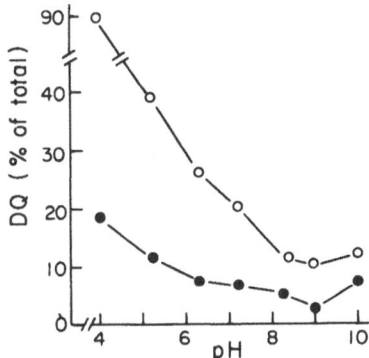

FIGURE 2 Fusion of reconstituted envelopes bearing Sendai virus glyco-proteins with PS liposomes: pH curves. PS liposomes and *N*-NBD-PE-labeled, reconstituted vesicles, bearing either F (●-●) or HN (○-○) glycoproteins, were prepared as described in Tables 2 and 4, respectively. To induce fluorescence dequenching, the reconstituted F or HN vesicles (1 μg) were incubated with 10 μg of PS liposomes in 150 mM of NaCl, buffered with 10 mM of glycine HCl (for pH 3.0-6.0), Tris HCl (for pH 7.0-8.0), and glycine NaOH (for pH 9.0-12.0). All other experimental conditions were as described in Table 4.

TABLE 5   RSVE Obtained by Solubilization of Sendai Virus with Octylglucoside Can Fuse with Negatively Charged Liposomes But Not With Erythrocyte Membranes

Sendai virus solubilized with:	R_{18} DQ (%)	
	PC/DCP/Chol	HEG
Triton X-100		
RSVE	54	60
Trypsinized RSVE	7	3
Octylglucoside		
RSVE	49	7
Trypsinized	7	8

RSVE were prepared from viral envelopes, solubilized in either Triton X-100 or octylglucoside (0.5 ml of a 4% detergent solution per 10 mg total viral protein), as described in Refs. 14 and 32, and labeled with R_{18}, as described in Table 2. Fluorescence dequenching (DQ) was performed following the incubation of the RSVE (2 μg) with either 10 μg PC/DCP/cholesterol (PC/DCP/Chol) liposomes (molar ratio 1.0:0.3:0.5) or 200 μg human erythrocyte ghosts (HEG), prepared as in Ref. 42, and was determined as described in Table 2. Treatment with trypsin was done as in Table 1.

Fusion of RSVE with negatively charged liposomes is almost as efficient as fusion with living cells, such as hepatoma tissue culture cells (HTC) (14), or biological membranes, such as human erythrocyte ghosts (HEG) (Table 5; Refs. 12-14). However, experiments in our laboratory have shown that only RSVE, obtained after solubilization of the virus with Triton X-100, have the capacity to effectively fuse with HEG or HTC. RSVE obtained by solubilization with octylglucoside (27) fuse with negatively charged liposomes, but neither with HEG nor with HTC (Table 5; Ref. 14). Trypsinization affects the fusogenic properties of either RSVE preparation, showing that also for fusion of RSVE with negatively charged liposomes intact viral glycoproteins are needed.

Maximum fusion between RSVE or HN vesicles and negatively charged phospholipids is obtained at low pH (Fig. 2; Ref. 31), whereas the fusogenic activity of the virus toward living cells is maximally expressed at pH values between 6.5 and 9.5 (3,12). Hence, also regarding the pH optimum, a clear discrepancy appears between the fusogenic properties of the virus toward negatively charged liposomes and plasma membranes of living cells.

IV. INTERACTION OF SENDAI VIRUS WITH LIPOSOMES COMPOSED OF NEUTRAL LIPIDS

A. Intact Virus

1. Virus-Induced Lysis of Loaded Liposomes

Liposomes composed of only neutral lipids, as opposed to those composed of negatively charged phospholipids, are not susceptible to the lytic activity of Sendai virus. Virus-induced lysis of such liposomes is observed only after the incorporation of virus receptors, such as sialoglycolipids, into the lipid vesicle membrane (Table 6; Refs. 20,28). Release of the liposome content is most pronounced with liposomes bearing both sialoglycolipids and sialoglycoproteins, as observed first by Kundrot et al. (20) and confirmed by experiments in our laboratory (Table 6; Ref. 28). Virus-induced lysis is observed also with liposomes composed of neutral lipids (PC and cholesterol) and gangliosides (28), while liposomes bearing only glycophorin do readily bind to Sendai virus, but they are not lysed (Table 6; Ref. 28). Nonhemolytic Sendai virus, treated with PMSF, DTT, or trypsin, fails to induce lysis in PC/cholesterol liposomes bearing gangliosides as virus receptors (Table 6; Ref. 28). Furthermore, induction of lysis is absolutely dependent on the presence of cholesterol, the incorporation of which into the liposome bilayer gradually increases the degree of lysis observed (Fig. 3; Table 6). Thus, it appears that virus-induced lysis of PC/cholesterol liposomes, unlike that of negatively charged liposomes, is characterized by the following features: (1) absolute dependency on virus receptors, (2) inhibition by PMSF or DTT treatment of the virions, and (3) requirement for the presence of cholesterol in the liposome bilayer.

2. Intermixing of Viral and Liposomal Lipids

Incubation of fluorescently labeled (R_{18}) Sendai virus with PC/cholesterol liposomes results in fluorescence dequenching (Table 7; Ref. 14). No specific virus receptors are needed: the same increase in fluorescence is observed upon incubation of the virus with liposomes containing or lacking virus receptors (Table 7; Ref. 14). Increase in fluorescence is completely

TABLE 6 Virus-Induced Lysis of Liposomes Composed of Neutral Lipids: Requirement for Virus Receptors

System	CF release (%)				
	PC	PC/Chol	PC/Chol/gangl	PC/Chol/GP	PC/Chol/gangl/GP
Experiment 1					
Sendai virus treated with					
None	6	4	18	5	40
DTT	N.D.	3	4	4	3
PMSF	N.D.	0	2	0	2
Trypsin	N.D.	2	5	0	7
Experiment 2					
Untreated virus and neuraminidase-treated liposomes	N.D.	N.D.	6	2	4

CF-containing liposomes were prepared as described in Table 1 and consisted of PC, PC/cholesterol (molar ratio 1.0:0.5), PC/cholesterol/gangliosides (molar ratio 1.0:0.5:0.3), PC/cholesterol/glycophorin (molar ratio 1.0:0.5:0.001), or PC/cholesterol/gangliosides/glycophorin (molar ratio 1.0:0.5:0.3:0.001). The gangliosides (gangl) were from egg yolk, the glycophorin (GP) from human erythrocytes (44). Sendai virus was treated with DTT, PMSF, or trypsin as in Table 1, and the liposomes were treated with neuraminidase (Experiment 2), as described in Ref. 28. CF release from the liposomes upon incubation with the virus (30 μg of protein) was determined as described in Table 1.

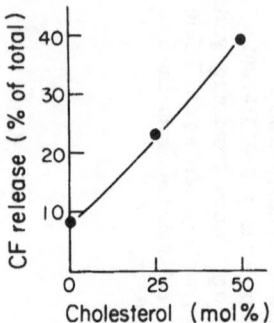

FIGURE 3 Lysis of neutral liposomes bearing virus receptors by Sendai virions: requirement for cholesterol. CF-containing liposomes, composed of PC/gangliosides/glycophorin (molar ratio 1:0.3:0.001) and the indicated amounts of cholesterol were prepared and lysed by Sendai virus as described in Table 6.

dependent on the presence of cholesterol and is not induced by nonhemolytic virions (Table 7). Fusion of Sendai virus with PC/cholesterol liposomes lacking any virus receptors has already been reported, based on a completely different assay system (21).

Incorporation of SM or PE into the liposome bilayer does not change its susceptibility to fusion with Sendai virions. Fusion with such liposomes still requires the presence of cholesterol and does not result in release of the liposome content. Thus, it seems that under certain conditions, fusion of Sendai virus can be a nonleaky process. It should be noted, however, that similar to the incorporation of sialoglycolipids, the insertion of negatively charged phospholipids also renders the PC/cholesterol liposomes susceptible to the viral lytic activity (28; see also Fig. 1).

B. Reconstituted Sendai Virus Envelopes (RSVE)

1. *RSVE-Induced Lysis of Loaded Liposomes*

RSVE behave exactly as intact Sendai virions regarding their interaction with PC/cholesterol liposomes. No CF release is observed following incubation of RSVE with loaded liposomes composed of only PC and cholesterol, while a high degree of lysis is obtained when the liposomes also contain gangliosides and glycophorin (Table 8; Ref. 28). RSVE-induced lysis is not observed with PMSF-, DTT-, or trypsin-treated RSVE (Table 8; Ref. 28). Induction of lysis is observed neither with vesicles containing only the viral F glycoprotein nor with those bearing only the HN glycoprotein; similar to RSVE-induced hemolysis, lysis of the liposomes only occurs with vesicles containing both the HN and F glycoproteins within the same membrane (Table 8; Ref. 28).

As already mentioned and shown before (26,30), CF as well as other molecules can be enclosed within the RSVE, if added during the reconstitution process. Being impermeable, the RSVE retain most of the trapped CF (30). Incubation of RSVE with empty PC/cholesterol liposomes containing gangliosides and glycophorin does not result in release of the RSVE content (Table 8; Ref. 30). These results further support the possibility raised

TABLE 7 Fusion of Sendai Virus with Liposomes Composed of
Neutral Lipids

Sendai virus treated with:	R_{18} DQ (%)		
	PC	PC/Chol	PC/Chol/gangl/GP
None	6	48	40
DTT	N.D.	9	5
PMSF	N.D.	9	1
Trypsin	N.D.	9	2

Sendai virus was labeled with R_{18} as described in Table 2
and labeled virus was treated with DTT, PMSF, or trypsin as
described in Table 1. Liposomes of the indicated compositions
were obtained as described in Table 6, except that CF was
omitted during the preparation. Fluorescence dequenching
(DQ) was induced by incubation (45 min at 37°C) of R_{18}-
labeled Sendai virions (2 μg) with liposome preparations (20 μg)
and its degree was estimated as described in Table 2.

TABLE 8 Release of Liposome but Not of Viral Envelope Content
upon Interaction of RSVE with Lipid Vesicles Bearing Virus Receptors

System	CF release (%)	
	PC/Chol	PC/Chol/gangl/GP
Experiment 1		
RSVE	0	40
PMSF-treated RSVE	2	2
Reconstituted F vesicles	0	0
Reconstituted HN vesicles	0	3
Reconstituted (F + HN) vesicles	1	36
Experiment 2		
CF-RSVE	0	2

CF-containing (Experiment 1), or empty (Experiment 2), liposomes of
the compositions indicated were prepared as described in Tables 1 and
6, RSVE and reconstituted vesicles, bearing viral glycoproteins, were
prepared as described in Table 3, and CF-containing RSVE as in Ref. 25.
Treatment with PMSF and determination of CF release were done as indicated in Table 1, except that 30 μg of viral protein was used in Experiment 1 and 5 μg in Experiment 2.

before (30) that fusion of RSVE, as well as of intact virions, with liposomes results in the formation of nonleaky vesicles which do not release their contents. Therefore, the lysis observed upon incubation of intact virions or RSVE and loaded liposomes must occur before the fusion process is completed.

In summary, induction of lysis by RSVE in loaded PC/cholesterol liposomes containing specific viral receptors shows the same characteristics as lysis by intact virions, but differs markedly from virus-induced lysis of negatively charged liposomes.

2. Intermixing of RSVE and Liposomal Contents

Fusion of RSVE and liposomes composed of PC/cholesterol can be demonstrated by experiments monitoring the mixing of their contents. Incubation of terbium-loaded RSVE with liposomes containing the anion of dipicolinic acid (DPA) results in an increase of the fluorescence intensity, indicating the formation of the Tb/DPA complex (30). Formation of such a complex can only result from the fusion of the loaded viral envelopes with the liposomes (32; see also Chapter 4). The Tb/DPA complex formation is not observed with PMSF-, DTT-, or trypsin-treated RSVE (30).

3. Intermixing of RSVE and Liposomal Lipids

Fusion of RSVE with liposomes composed of PC/cholesterol, and lacking any virus receptors, has also been demonstrated by experiments using fluorescence dequenching methods. Essentially the same degree of fluorescence dequenching is observed upon incubation of fluorescently labeled RSVE with PC/cholesterol liposomes, irrespective of whether these liposomes contain specific viral receptors or not (30). No fluorescence dequenching is observed with liposomes composed of only PC, showing the absolute requirement for cholesterol to allow fusion of RSVE and liposomes of the above composition. Fusion is not observed with PMSF-, DTT-, or trypsin-treated RSVE (30).

DTT treatment of Sendai virions, under the experimental conditions used, affects the viral HN glycoprotein, but not the F polypeptide (33). The requirement for a nonreduced HN glycoprotein for fusion with PC/cholesterol liposomes supports the view mentioned before (5,6,29) that the viral HN glycoprotein, besides being the viral binding protein, also plays an active role in the fusion step itself. This view is further confirmed by experiments using vesicles bearing the isolated viral glycoproteins. No fluorescence dequenching is observed by incubating vesicles containing either the viral F or HN glycoprotein and liposomes of neutral lipids, bearing or lacking virus receptors. Also, incubation of such liposomes with a mixture of the viral HN and F vesicles does not result in any increase of fluorescence dequenching, showing that to allow fusion, the viral glycoproteins must be present within the same membrane (30). The only possible explanation for these observations is that the viral HN is needed together with the F glycoprotein for fusion with liposomes lacking any virus receptors. These experiments clearly show the active involvement of the viral HN glycoprotein in the fusion event.

V. DISCUSSION AND CONCLUSIONS

Studies of the last few years have clearly shown that enveloped animal viruses are able to fuse with phospholipid vesicles. Virus-liposome fusion has been demonstrated for Sendai (3,17,21), influenza (34,35), Semliki Forest (36), herpes simplex (37), and vesicular stomatitis viruses (38). It appears that a common denominator characterizing such virus-liposome fusion processes is that fusion is independent of the phospholipid composition and does not require the presence of specific receptors which may mediate binding of the virions to the liposomes. This is in contrast to the interaction between animal viruses and living cells, which, in most cases, is dependent on the association between a virus-binding protein and a membrane receptor (1,2). These discrepant observations, as well as other studies, raise the question whether the fusion of enveloped animal viruses with liposomes indeed reflects in its detailed mechanisms the viral biological activities necessary for penetration and infectivity.

In the present chapter, we have mainly summarized our studies regarding the interaction and fusion of Sendai virus with liposomes composed of negatively charged or neutral lipids. Due to the following features, studies with Sendai virions may be more advantageous than studies with other enveloped virions. First, fusion of Sendai virus with biological membranes occurs at neutral and high pH values. These conditions may minimize membrane fusion processes that might be promoted by electrostatic interaction between the positively charged viral envelope glycoproteins and negatively charged membrane groups. Such processes are more pronounced at low pH values (31). Second, various inhibitors render Sendai virus nonfusogenic as well as noninfective. The use of specifically inactivated Sendai virus is important for studies in which the ability of the virus to fuse with biological membranes is compared with its ability to fuse with phospholipid vesicles. Third, the availability of fusogenic, reconstituted viral envelopes allows studies with isolated viral components.

We believe that the information gained from studies on the interaction between Sendai virus and liposomes might be of substantial help in the elucidation of the molecular mechanisms of virus-membrane fusion and cellular penetration. The following conclusions can be drawn regarding the interaction of Sendai virions and liposomes. First, Sendai virus effectively fuses with and lyses liposomes composed of negatively charged phospholipids. No specific virus receptor, i.e., sialic-acid-containing components, is needed. However, fusion with negatively charged liposomes does not seem to be promoted by the viral biological activity needed for its penetration and infectivity. This statement is based primarily on the following observations. (1) Nonfusogenic, noninfective Sendai virions efficiently fuse with and lyse negatively charged liposomes. (2) Vesicles containing only the viral HN glycoprotein, but not the fusion (F) polypeptide, readily fuse with and lyse negatively charged liposomes, but do not fuse with biological membranes. (3) Maximum virus-liposome fusion is obtained at acid pH values. Furthermore, it has been well established that in most plasma membranes of animal cells, the negatively charged phospholipids are located on the inner side of the membrane not facing the invading virions (39). Therefore, unless a special mechanism is suggested, in most plasma membranes the negatively charged phospholipids are not available to interact with the viral envelope components. Also, fusion of Sendai virus with liposomes composed of neutral lipids shows the same features that characterize fusion with biological membranes, thus reflecting the viral fusogenic activity necessary for infectivity. This conclusion is based on the use of various inhibitors as well as vesicles containing isolated viral components. Fusion of Sendai virus with liposomes composed of neutral lipids possesses the following features. (1) The presence of cholesterol within the liposome bilayer if obligatory to allow virus-liposome fusion. (2) No virus receptors are needed for virus-liposome fusion, but fusion with liposomes lacking virus receptors is a nonleaky process: Sendai virus fails to induce lysis of liposomes composed of neutral lipids lacking virus receptors. (3) The incorporation of gangliosides (in the presence or absence of sialoglyco-proteins) renders liposomes composed of neutral lipids susceptible to the viral lytic activity. (4) Fusion with PC/cholesterol liposomes, lacking or bearing virus receptors, occurs only with viral envelopes containing both the viral HN and F polypeptides.

The observation that cholesterol is needed to allow fusion with liposomes composed of neutral lipids raises the possibility suggested before (40) that cholesterol constitutes the putative membrane receptor for the viral fusion glycoprotein. Fusion of Semliki Forest virus has also been shown to be absolutely dependent on the presence of cholesterol in the recipient liposomes (36). It is conceivable that the hydrophobic moiety of the viral fusion polypeptide, which is needed for promoting the fusion process, interacts specifically with membrane cholesterol molecules (40). Perhaps this property is shared by all enveloped viruses.

From our studies, a new and active function may be attributed to Sendai virus membrane receptors, especially to sialoglycolipids. Association between Sendai virus and sialoglycolipids incorporated in liposomes induces an increase in the liposomal permeability. Both the viral HN and F glyco-proteins are required to induce lysis in animal cells as well as in loaded liposomes. A virus-induced increase in the permeability of biological membranes may promote osmotic swelling and, consequently, unmask the membrane lipids, events that have been suggested to be required for virus-membrane as well as cell-cell fusion processes (41). If virus-induced lysis is required to allow fusion of the virus with biological membranes, it is expected that lysis will precede the actual fusion process. Indeed, studies using the interaction between Sendai virus and liposomes indicate that virus-induced lysis of liposomes precedes their fusion with the lipid bilayer rather than being a consequence of it. The possibility that this property, namely induction of lysis following attachment to the appropriate receptor, is shared by all enveloped animal viruses cannot be excluded. Seeking a membrane component that will render liposomes susceptible to the viral lytic activity may be an efficient method for the isolation and characterization of functional membrane receptors for enveloped viruses.

ACKNOWLEDGMENTS

This work was supported by a grant from the National Council for Research and Development, Jerusalem, Israel, and from the Gesellschaft für Strahlung and Umweltforschung, Munich, Federal Republic of Germany.

REFERENCES

1. Poste, G., and Pasternak, C. A. (1978). Virus-induced cell fusion. In *Cell Surface Reviews*, Vol. 5 (G. Poste and G. L. Nicolson, eds.). Elsevier/North Holland, Biomedical Press, Amsterdam, pp. 305-367.
2. White, J., Kielian, M., and Helenius, A. (1983). Membrane fusion proteins of enveloped animal viruses. *Q. Rev. Biophys.* 16:151-195.
3. Loyter, A., and Volsky, D. J. (1982). Reconstituted Sendai virus envelopes as carriers for the introduction of biological material into

animal cells. In: *Membrane Reconstitution* (G. Poste and G. L. Nicolson, eds.). pp. 215-266, Elsevier/North Holland Biomedical Press, Amsterdam.

4. Ozawa, M., Asano, A., and Okada, Y. (1979). Biological activities of glycoproteins of HVJ (Sendai virus) studied by reconstitution of hybrid envelopes and by Concanavalin-A-mediated binding: A new function of HANA protein and structural requirements for F protein hemolysis. *Virology* 99:197-202.

5. Miura, N., Uchida, T., and Okada, Y. (1982). HVJ (Sendai virus)-induced envelope fusion and cell fusion are blocked by monoclonal anti-HN protein antibody that does not inhibit hemagglutination activity of HVJ. *Exp. Cell Res.* 141:409-420.

6. Gitman, A. G., Kahane, I., and Loyter, A. (1985). Use of virus-attached antibodies or insulin molecules to mediate fusion between Sendai virus envelopes and neuraminidase-treated cells. *Biochemistry* 24:2762-2768.

7. Bashford, C. L., Micklem, K. J., and Pasternak, C. A. (1985). Sequential onset of permeability changes in mouse ascites cells induced by Sendai virus. *Biochim. Biophys. Acta* 814:247-255.

8. Maeda, Y., Kim, J., Koseki, I., Mekada, E., Shiokawa, Y., and Okada, Y. (1977). Modification of cell membranes with viral envelopes during fusion of cells with HVJ (Sendai virus). *Exp. Cell Res.* 108:95-106.

9. Israel, S., Ginsberg, D., Laster, Y., Zakai, N., Milner, Y., and Loyter, A. (1983). A possible involvement of virus-associated protease in the fusion of Sendai virus envelopes with human erythrocytes. *Biochim. Biophys. Acta* 732:337-346.

10. Ozawa, M., Asano, A., and Okada, Y. (1979). The presence and cleavage of interpeptide disulfide bonds in viral glycoproteins. *J. Biochem. (Tokyo)* 86:1361-1369.

11. Hoekstra, D., De Boer, T., Klappe, K., and Wilschut, J. (1984). Fluorescence method for measuring the kinetics of fusion between biological membranes. *Biochemistry* 23:5675-5681.

12. Chejanovsky, N., and Loyter, A. (1985). Fusion between Sendai virus envelopes and biological membranes. *J. Biol. Chem.* 260:7911-7918.

13. Chejanovsky, N., Henis, Y. I., and Loyter, A. (1986). Fusion of fluorescently labeled Sendai virus envelopes with living cultured cells as monitored by fluorescence dequenching. *Exp. Cell Res.*, in press.

14. Citovsky, V., Blumenthal, R., and Loyter, A. (1985). Fusion of Sendai virions with phosphatidylcholine-cholesterol liposomes reflects the viral activity required for fusion with biological membranes. *FEBS Lett.* 193:135-140.

15. Haywood, A. M. (1974). Characteristics of Sendai virus receptors in a model membrane. *J. Mol. Biol.* 83:427-436.

16. Haywood, A. M., and Boyer, B. P. (1982). Sendai virus membrane fusion: Time course and effect of temperature, pH, calcium and receptor concentration. *Biochemistry* 21:6041-6046.

17. Haywood, A. M., and Boyer, B. P. (1984). Effect of lipid composition upon fusion of liposomes with Sendai virus membranes. *Biochemistry* 23:4161-4166.

18. Oku, N., Inoue, K., Nojima, S., Sekiya, T., and Nosawa, Y. (1982). Electron microscopic studies on the interaction of Sendai virus with liposomes containing glycophorin. *Biochim. Biophys. Acta* 691:91-96.

19. Oku, N., Nojima, S., and Inoue, K. (1982). Studies on the interaction of HVJ (Sendai virus) with liposomal membranes induced permeability increase of liposomes containing glycophorin. *Virology* 116:419-427.

20. Kundrot, C. E., Spangler, E. A., Kendall, D. A., MacDonald, R. C., and MacDonald, R. I. (1983). Sendai virus-mediated lysis of liposomes requires cholesterol. *Proc. Natl. Acad. Sci. USA* 80:1608-1612.

21. Hsu, M., Scheid, A., and Choppin, P. W. (1983). Fusion of Sendai virus with liposomes: Dependence on the viral fusion protein (F) and the lipid composition of liposomes. *Virology* 126:361-369.

22. Huang, R. T. C., Rott, R., and Klenk, H-D. (1981). Influenza viruses cause hemolysis and fusion of cells. *Virology* 110:243-247.

23. Amselem, S., Loyter, A., Lichtenberg, D., and Barenholtz, Y. (1985). The interaction of Sendai virus with negatively charged liposomes: Virus-induced lysis of carboxyfluorescein-loaded small unilamellar vesicles. *Biochim. Biophys. Acta* 820:1-10.

24. Papahadjopoulos, D., Nir, S., and Ohki, S. (1971). Permeability properties of phospholipid membranes: Effect of cholesterol and temperature. *Biochim. Biophys. Acta* 266:571-583.

25. Amselem, S., Barenholtz, Y., Loyter, A., Nir, S., and Lichtenberg, D. (1986). Fusion of Sendai virus with negatively charged liposomes as studied by pyrene-labeled phospholipid liposomes. *Biochim. Biophys. Acta*, in press.

26. Vainstein, A., Hershkovitz, M., Israel, S., Rabin, S., and Loyter, A. (1984). A new method for reconstitution of highly fusogenic Sendai virus envelopes. *Biochim. Biophys. Acta* 773:181-188.

27. Harmsen, M. C., Wilschut, J., Scherphof, G., Hulstaert, C., and Hoekstra, D. Reconstitution and fusogenic properties of Sendai virus envelopes. *Eur. J. Biochem.* 149:591-599.

28. Citovsky, V., Zakai, N., and Loyter, A. (1986). Fusion of Sendai virus with liposomes: Specific requirement for liposome-associated sialoglycolipids, but not sialoglycoproteins, to allow lysis of phospholipid vesicles by reconstituted viral envelopes. *Exp. Cell Res.*, in press.

29. Citovsky, V., Yanai, P., and Loyter, A. (19). The use of circular dichroism to study conformational changes induced in Sendai virus envelope glycoproteins. *J. Biol. Chem.* 261:2235-2239.

30. Citovsky, V., and Loyter, A. (1985). Fusion of Sendai virions or reconstituted Sendai virus envelopes with liposomes or erythrocyte membranes lacking virus receptors. *J. Biol. Chem.* 260:12072-12077.

31. Chejanovsky, N., Amselem, S., Barenholtz, Y., and Loyter, A. (1986). Membrane vesicles containing the Sendai virus binding protein (HN) but not the viral fusion protein (F) fuse with phosphatidylserine liposomes at low pH. *Biochemistry*, in press.

32. Wilschut, J., and Papahadjopoulos, D. (1979). Ca^{2+}-induced fusion of phospholipid vesicles, monitored by mixing of aqueous contents. *Nature* 281:690-692.

33. Tomasi, M., and Loyter, A. (1981). Selective extraction of biologically active F glycoprotein from dithiothreitol-reduced Sendai virus particles. *FEBS Lett.* 131:381-385.

34. Kawasaki, K., Sato, S. B., and Ohnishi, S-I. (1983). Membrane fusion activity of reconstituted vesicles of Influenza virus hemagglutinin glycoproteins. *Biochim. Biophys. Acta* 733:286-290.

35. Stegmann, T., Hoekstra, D., Scherphof, G., and Wilschut, J. (1985). Kinetics of pH-dependent fusion between Influenza virus and liposomes. *Biochemistry* 24:3107-3113.

36. White, J., and Helenius, A. (1980). pH-dependent fusion between the Semliki Forest virus membrane and liposomes. *Proc. Natl. Acad. Sci. USA* 77:3273-3277.

37. Citovsky, V., Loyter, A., and Katz, E. (1986). Fusogenic properties of herpes simplex virus followed by fluorescence dequenching, in preparation.

38. Eidelman, O., Schlegel, R., Tralka, T. S., and Blumenthal, R. (1984). pH-dependent fusion induced by vesicular stomatitis virus glycoprotein reconstituted into phospholipid vesicles. *J. Biol. Chem.* 259:4622-4628.

39. Verkleij, A. J., Zwaal, R. F. A., Roelofsen, B., Comfurius, P., Kastelijn, D., and Van Deenen, L. L. M. (1973). The asymmetric distribution of phospholipids in the human cell membrane: A combined study using phospholipases and freeze-etch electron microscopy. *Biochim. Biophys. Acta* 323:178-193.

40. Asano, K., and Asano, A. (1985). Why is a specific amino acid sequence of F glycoprotein required for the membrane fusion reaction between envelope of HVJ (Sendai virus) and target cell membranes? *Biochem. Int. (Tokyo)* 10:115-122.

41. Lucy, J. A., and Ahkong, Q. F. (1986). An osmotic model for the fusion of biological membranes. *FEBS Lett.* 199:1-11.

42. Fairbanks, G., Steck, T. L., and Wallach, D. M. E. (1971). Electrophoretic analysis of the major polypeptides of the human erythrocyte membranes. *Biochemistry* 10:2606-2617.

43. Schroit, A. J., and Pagano, R. E. (1981). Capping of a phospholipid analog in the plasma membrane of lymphocytes. *Cell* 23:105-112.

44. Hamaguchi, H., and Clefe, H. (1972). Solubilization and comparative analysis of mammalian erythrocyte membrane glycoproteins. *Biochem. Biophys. Res. Commun.* 47:459-463.

18

Functional Reconstitution of Viral Envelopes

ANNE WALTER,* OFER EIDELMAN,† MICHEL OLLIVON,‡
and ROBERT BLUMENTHAL

*National Cancer Institute, National Institutes of Health, Bethesda,
Maryland*

I. INTRODUCTION

Enveloped viruses inject their genetic content into the cytoplasm of the
host cell after fusion between the viral and either the cell plasma or
endosomal membranes (1-4). The viral envelope includes components to
recognize the target cell, move to an appropriate fusion site, bring the two
membranes to close apposition, and finally, promote membrane fusion. In
this, the viral fusion event is probably similar to other biological fusion processes, such as exocytosis and secretion, fertilization, and parasite invasion.

The viral envelope is densely loaded with a small variety of glycoproteins, making it the only known example of a source of components
required for fusion in a relatively pure form. These fusion-inducing proteins
have a single transmembrane segment, with the majority of the polypeptide
facing outward, and are organized in clusters that appear as spikes in
electron micrographs (5,6). Although many of these proteins are well characterized biochemically, how fusion is induced is not yet understood. To
study the molecular mechanisms of fusion, it is necessary to manipulate
each component involved. Reconstitution of the viral fusion proteins into
virosomes offers several advantages over the intact virus for these sorts
of studies. One advantage is the relatively clean preparation with minimal
or no contamination from the viral internal contents. Other advantages
are associated with the ability to manipulate the composition of the virosome
and especially to introduce labels or to chemically modify the lipid and/or
the protein. These advantages include making it feasible to: (1) study
viral fusion using fluorescent, electron spin resonance (ESR), or radiolabeled probes, (2) study protein-protein interactions, and (3) study
lipid-protein interactions.

Present affiliations:
*Wright State University, Dayton, Ohio.
†The Hebrew University of Jerusalem, Jerusalem, Israel.
‡Centre Nationale de la Recherche Scientifique, Thiais, France.

Although it is possible to label intact virions with fluorescent or electron-spin resonance (ESR) probes and monitor transfer of those probes to cells or liposomes as a result of fusion, the repertoire of fusion assays in that case is limited only to outer monolayer mixing (7-12). Reconstitution enables one to use the plethora of fusion assays developed for vesicle-vesicle fusion, i.e., lipid mixing, contents mixing, or outer vs. inner monolayer mixing (13-15). Incorporation of fluorescent or radioactive tracers in the reconstituted membrane allows determination of the routes and fates of the fusing vesicles (e.g., fusion with the plasma membrane or an endocytic vesicle). Furthermore, it is possible to separate fusion from binding events, and to monitor fusion against a high background of bound, but nonfusing vesicles.

Protein-protein interactions in the context of viral fusion may be interspecific, involving discrete roles for two or more types of proteins, or intraspecific, involving cooperativity among identical peptides in the membrane. When two or more types of glycoproteins are found in the viral envelope, it is only possible to describe their individual functions and how they work in concert by preparing virosomes with a defined composition. For example, the discrete roles of the Sendai HN and F proteins have been delineated by reconstituting the two proteins together and separately (16-19). Since the proteins are densely packed in the intact virion envelope membrane and also tend to self-associate when reconstituted in virosomes (see, e.g., Refs. 17,20,21), it is reasonable to hypothesize that some sort of cooperativity between protein units is important for fusion. Indeed, studies on fusion of virosomes containing the G protein of vesicular stomatitis virus (VSV) indicate that viral proteins appear to remain clustered together even after the fusion event (20). In some cases, as with influenza virus, three peptide units are organized into a "spike" (6); it is hypothesized that a concerted action of the peptides forming the spike is central to the induction of fusion (see also Chapter 15). By regulating the protein density or altering the interaction among the subunits, it is possible to determine whether either sort of cooperativity is essential for fusion.

To elucidate the catastrophic rearrangement that occurs when the two sets of membrane phospholipids coalesce requires observation of the transient events. This can be achieved by specific labeling of each of the constituents. To date, reconstitution and specific labeling of the viral fusion proteins has helped to define interactions with the virosomal membrane (22-24). Interactions with the target membrane and the events during fusion will require an even greater ability to define and label the constituents of the reconstituted virosome.

Viral fusion proteins have been used for cellular delivery of substances that were incorporated in lipid vesicles for both practical and research applications (see refs. listed in Table 1; see also Chapter 32). Reconstitution is an indispensable step in the formation of any delivery vesicle based on the use of a fusion protein, as a means to incorporate the material to be delivered and the specific targeting components. Reconstituted viral coat proteins devoid of the viral replicating machinery also may provide an excellent nonvirulent vaccine (25-27).

Thus, developing the ability to reconstitute viral envelope proteins that express biological function will contribute toward understanding at the molecular level how enveloped viruses merge with cell or endosomal membranes, and toward developing medical and biotechnological applications. In this chapter we outline how reconstitution of viral protein can be done, give some idea of how to improve the existing approaches, and review examples of viral reconstitution. Finally, we attempt to delineate the more promising directions for future research.

TABLE 1ᵃ Viral Spike Glycoproteins Reconstituted into Lipid Vesicles

Virus/detergent method	Application	Assay	Reference
Sendai/Triton-X-100 Dialysis	Reconstitution methodology	Hemolysis, EM, cell fusion	44
	Implantation of red cell band 3	Expression of anion transport in Band-3 negative cells	74
			71-73
	Implantation of theta antigen	Susceptibility to complement lysis	83
	Implantation of the EBV-receptor	Infection by EBV	75,76
	Mechanism of fusion	Hemolysis, cell-cell fusion	96
	Roles of HN and F proteins	Hemolysis	19
	Mechanism of viral penetration	Infection by EBV	101
	Implantation of H-2a antigens	Detection of active antigens	84
	Implantation of cromolyn-binding protein	Restoration of Ca²⁺ influx and degranulation	102
	Injection of SV40-DNA into cells	Expression of injected DNA	89,91
	Microinjection of EBV- DNA	Induction of EBV nuclear antigen	85
	Mechanism of fusion	Hemolysis, ESR	57
	Mechanism of fusion	Hemolysis, RET	63
Sendai/Triton X-100 SM-2 BioBeads	Reconstitution methodology	Hemolysis, EM, cell fusion	43
	Reconstitution methodology	No fusion assay	97
b	Mechanism of fusion	Hemolysis	87,103

(continued)

(Table 1, continued)

Virus/detergent method	Application	Assay	Reference
	Microinjection of Ricin A and SV40 DNA	Cell killing(Ricin A) Expression of SV40 nuclear T- antigen	88
	Mechanism of fusion	RET lipid mixing, hemolysis	8,16,100, 104-106
Sendai/octylglucoside			
Dialysis	Implantation of surface antigens	Lysis of target cells by anti-Sendai CTL	77
c	Identify binding proteins	No fusion assay	78,79
d	Membrane protein degradation	Transfer of CF and HRP	98
		Immunofluorescence, protease digestion	45
Slow dialysis	Mechanism of fusion	Lipid mixing by RET	55
Rotary dialysis	Reconstitution methodology	No fusion assay	62
Sendai/NP-40			
Dialysis	Delivery of diphteria toxin A chain	EM, cell killing	90
Chromatography and sucrose gradient	Roles of HN and F proteins	Cell killing	18
Sendai/Tween 20			
Gelchromatography[e]	Cell fusion	Hemolysis	52,107
Sendai/DOC			
Dialysis	Implantation of surface antigens	Recognition and lysis of target cells by anti-Sendai CTL	80-82
c	Mechanism of fusion	Hemolysis	17,99

Influenza/octylglucoside			
Dialysis	Lipid-protein interactions	No fusion assay	23
VSV/octylglucoside			
Dialysis	Mechanism of reconstitution	No fusion assay	51
	Lipid-protein interactions	No fusion assay	24,108,109
	Antibody production	No fusion assay	110
Slow dialysis	Mechanism of fusion	Lipid mixing by RET, EM with liposome targets	20
VSV/DOC			
Dialysis	Triggering of a secondary CTL response	No fusion assay	111
VSV/lyso-PC			
Adsorption to BSA	Protein-lipid interactions	Hemagglutination, no fusion assay	22
Herpes/octylglucoside			
Dialysis	Mechanism of fusion	EM	95
Rabies/octylglucoside			
Dialysis	Vaccination, formation of immunosomes[f]	No fusion assay	60
Semliki/octylglucoside			
Dialysis	Mechanism of reconstitution	No fusion assay	31,54
c	Induction of endocytosis	Hemolysis, irreversible ligand binding	50
Sindbis/Triton (insertion into cholate equilibration vesicles)			
Dilution	Reconstitution technique	Octadecylrhodamine dilution	21
Rubella/Tween 80-ether			
Dialysis	Vaccination	Aggregation, no fusion assay	27
Measles/octylglucoside			
Dialysis[g]	Vaccination	Hemagglutination no fusion assay	26

(continued)

(Table 1, continued)

CF, carboxyfluorescein; CTL, cytotoxic T lymphocytes; DOC, deoxycholate; EBV, Epstein-Barr virus; EM, electron microscopy; ESR, electron spin resonance; RET, fluorescence resonance energy transfer; HRP, horseradish peroxidase; PC, phosphatidylcholine; VSV, vesicular stomatitis virus

[a]This table is complete through May 1986, the original deadline for submission of the manuscript.

[b]The virosomes were targetted by covalently attached insulin or antibodies.

[c]Viral membrane proteins were isolated with Triton X-100 and transfered into cholate or octylglucoside.

[d]Poly(ethyleneglycol) was needed to induce virosome-cell fusion.

[e]Also named Emasol; procedure was done in a bicarbonate buffer pH 10.

[f]Immunosomes are formed by adding protein/detergent to preformed liposomes.

[g]Viral membrane proteins were isolated with NP-40 and transferred into octylglucoside.

II. THEORY AND PRACTICE OF RECONSTITUTION

The considerations for planning and carrying out the reconstitution of viral proteins into vesicles are very similar to those for reconstituting any intrinsic membrane protein. Strategies for extracting, handling, and reconstituting membrane proteins have been extensively reviewed (28-34). In this section we describe, in brief, the reconstitution process, outline the basic procedures for reconstitution, and develop criteria for evaluating the final product. Structurally, viral spike glycoproteins differ from membrane receptor and transport proteins in that the major portion extends into the aqueous phase, with only a small hydrophobic tail in the membrane. This property may affect the reconstitution process simply because both the hydrophilic and hydrophobic regions must be protected from irreversible denaturation throughout the procedure.

To be useful, both for practical and for research purposes, reconstituted viral fusion proteins must be able to induce fusion between membranes in a biologically relevant manner. This requires that the fusion proteins be anchored and oriented in the reconstituted membrane in a fashion similar to the original viral envelope. In many cases, one or more auxiliary proteins are required for binding or for activation of the protein directly responsible for fusion. Evaluating the reconstituted virosomes obviously depends on having an appropriate target membrane, either host cells or vesicles, with the appropriate receptors and lipid composition. As will be seen below, it is also possible to observe false positives, meaning fusion that is not biologically relevant, either due to the choice of target membranes or as secondary effects deriving from the reconstitution procedure. This occurrence of nonspecific fusion is sometimes useful when fusion is needed as a tool and occasionally when one wishes to study protein action. However, these nonbiological activities, rather than shedding light on biological fusion, often only lead to increasing "confusion."

A. Reconstitution: A Molecular View

Reconstitution, when applied to membrane proteins, means that the protein in question has been extracted from its native membrane with the aid of detergent and placed in a newly formed one by subsequent detergent removal. The solubilized protein is purified (with or without its original lipids) and resituated, alone or in combination with other proteins, in a bilayer composed of native and/or exogenous lipid. It is this ability to recombine the protein in question with other proteins, lipids, and tracers that makes reconstitution so appealing.

Detergents solubilize membranes by intercalating among the phospholipids until the mole fraction of detergent is so high that the mixture no longer forms a bilayer (see e.g., Refs. 30,35-37,39). Evidence for this, which differs in detail among detergents, derives from measurements made as a function of detergent concentration. These include observing turbidity of aqueous lipid/detergent mixtures, electron-microscopic (EM) examination of the particles, detection of the distribution of lipid in different environments by nuclear magnetic resonance (NMR), examination of energy transfer between fluorescent lipid probes, and measurement of partition coefficients of detergents and of relative fluorescence anisotropy of hydrophobic probes. The detergent partitions into the membrane until some open vesicles are observed, maintaining the lamellar state apparently stabilized by detergent on the edges (38-40). At higher detergent concentration, cylindrical, then spheroidal lipid- and/or protein-saturated detergent micelles appear. At somewhat higher detergent concentration, the equilibrium state of this four-part mixture is pure detergent, lipid/detergent, protein/detergent, and probably lipid/protein/detergent micelles in excess water (Fig. 1) (35,36,39,41). This state is commonly referred to as solubilization of the membrane components. At much higher detergent concentrations, the mixed micelles contain only one lipid or protein molecule in excess detergent so that the membrane components are fully dispersed (see e.g., Ref. 41). Ideally, the structure of the protein/detergent micelle will be such that the membrane-spanning regions of the protein are protected from the aqueous environment, without disrupting the structure of the extramembraneous portions.

The solubilized proteins of interest are purified to the extent desired. Exogenous components (lipids, protein, lipid or aqueous labels, and material to be encapsulated) can be added to the reconstitution mixture, and reconstitution is achieved by reversing the solubilization process to form virosomes. The molecular rearrangements that occur during membrane formation, as detergent is removed, may follow the same pathway as the solubilization event. When the detergent concentrations, limiting the solubilities of lipid and protein, are similar, the detergent/lipid and detergent/protein mixed micelles coalesce into three-component micelles and then cylindrical micelles that appear to cooperatively form closed vesicles (*Biophys. J.* 56:669, 1989). If the rate of detergent removal is slow enough, the observed intermediates will be the metastable structures shown in Figure 1 (39). By increasing the rate of detergent removal, equilibria may not be reached at each stage. This will affect the type of vesicles obtained and has sometimes been utilized in systems where the three-component structures were not stable.

The concentration of detergent needed for each phase will be a function of many factors, including the particular lipids, protein, and their concentrations. These transitions can be manipulated by extrinsic factors such as temperature, ionic strength, choice of salt, or presence of sugars, to increase the chance that the protein will be successfully reconstituted. For instance, the choice of temperature will change the miscibility of protein and lipids since hydrophobic interactions will be weaker at lower temperature whereas ionic ones will be greater, and vice versa (e.g., Ref. 102). Ionic interactions can be masked or inhibited by salt concentrations, and glycoprotein aggregation is often prevented in the presence of sugars. Thus, in theory, by adjusting these external parameters, the desired lipid and protein phases can be obtained.

B. Solubilization and Virosome Reconstitution: Some Practical Notes

1. *Solubilization*

The first decision to be made when planning a reconstitution protocol is the means of solubilizing the viral membrane proteins. Since this first step affects the entire reconstitution procedure, it is worth ensuring that an optimal choice is made. Criteria for a good detergent are the ability to solubilize and to protect the protein from denaturation while being sufficiently water soluble to be removed efficiently. Successful solubilization of the membrane protein implies formation of protein/detergent micelles or three-component protein/lipid/detergent micelles (Fig. 1) without disrupting and solubilizing other viral components.

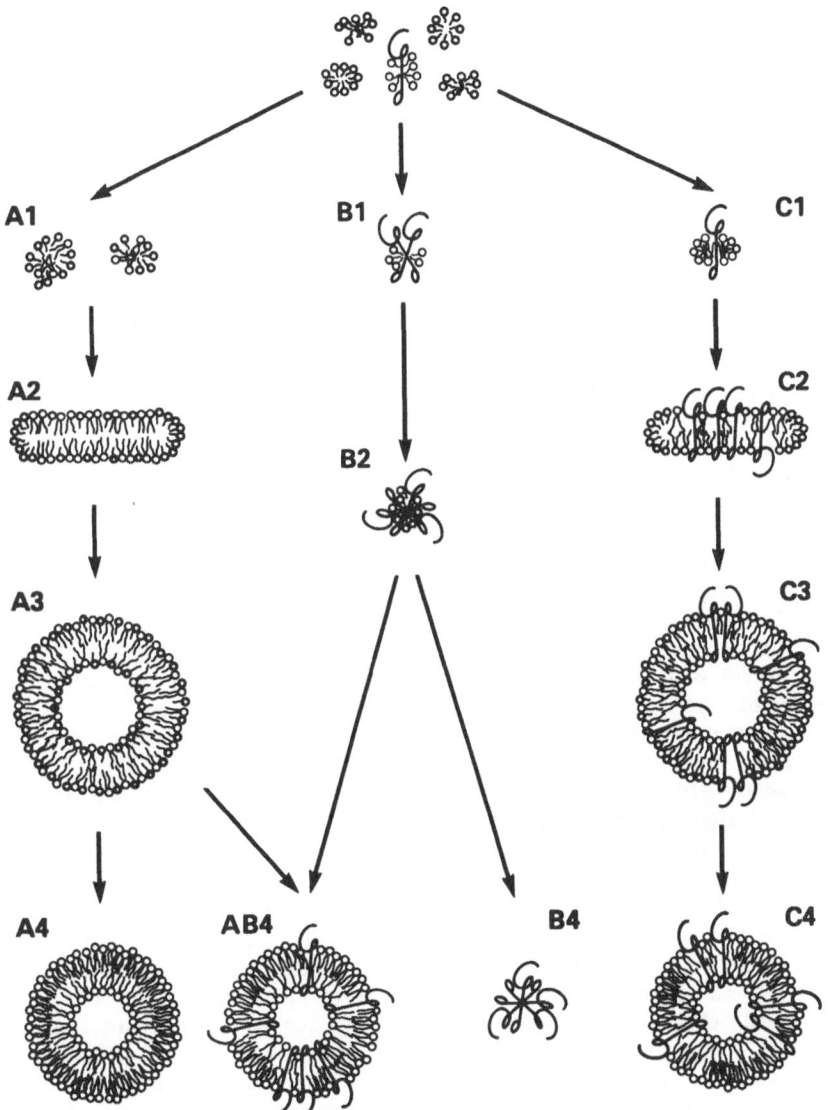

FIGURE 1 Proposed pathways for lipid and protein reconstitution from detergent solutions into vesicles. Detergent (ρ), solubilized lipids ($\mathfrak{Q}$), and proteins ($\mathcal{G}$) can follow three pathways as the detergent concentration is lowered toward zero. First the lipids and proteins begin to aggregate separately (A1, B1) or together (C1) into lipid, protein, or lipid/protein saturated mixed detergent micelles. At lower detergent concentrations, the lipids form detergent-saturated lamellar disks (A2, C2) that may contain protein (C2). The lamellar phases close (A3, C3) to detergent-swollen vesicles, which as the detergent concentration approaches zero shrink to lipid (A4), or protein-containing lipid (C4) vesicles. The protein/detergent micelles (B1) can self-aggregate (B2) and ultimately form pure protein aggregates or rosettes (B4) when the detergent level nears zero. However, the protein/detergent micelles (B2) may also interact with detergent-swollen vesicles (A3), which may result in proper insertion of the glycoproteins (AB4). For a current model, see Vinson et al., *Biophys. J.* 56:669, 1989.

To be useful, the detergent must protect the hydrophobic region from the aqueous environment without irreversibly denaturing the water-soluble regions of the protein. Clearly, if detergents that are the most efficient solubilizers also denature or otherwise affect protein function (see Section II.C), then it is necessary to accept a less efficient detergent for an effective final product. Efficient detergents with a relatively high critical micellar concentration (cmc) that generally do not affect soluble protein function (42) include octylglucoside, CHAPS, CHAPSO, and Zwittergents 310 and 312. These sorts of detergents may not work well with very hydrophobic membrane proteins, but might be good for viral membrane proteins that are predominantly extramembranous and thus have large regions that may be structurally similar to soluble proteins. However, Triton X-100 and $C_{12}E_8$, detergents with a relatively low cmc, have been used effectively for several functional viral reconstitutions (43,44) where the milder octylglucoside appeared to be ineffective for a functional reconstitution (45).

It is feasible to minimize the amount of detergent required for solubilization. Rivnay and Metzger (46) carefully titrated their system to find the minimum detergent concentration required for solubilization as a function of the total lipid used. They observed that the effective cmc (i.e., the concentration needed for solubilization extrapolated to zero added lipid and protein) in the presence of phospholipid and biological membranes is lower than the pure detergent cmc as predicted (see, for example, Refs. 39,47). Rivnay and Metzger's empirical observation was that the solubilization of rat basophil leukemia cell membranes containing the IgE receptor occurred when the excess detergent above the effective cmc is between 1.5 and 2 times the phospholipid concentration. Paternostre et al. (49a) used Triton X-100 at an effective ratio of 5.5 to reconstitute the vesicular stomatitis glycoprotein. However, Mimms et al. (48) and Ueno et al. (49) noted that a large excess of detergent is needed to form vesicles of uniform size and to minimize nonvesicular structures. Thus, subsequent addition of detergent might be necessary.

Given reasonable, nondenaturing solubilization by several detergents, the detergent of choice would be one with a relatively high cmc, since these are the easiest to remove. It is possible to change the detergent used to solubilize the protein with one that is easier to remove. For example, octylglucoside or cholate can replace Triton X-100 by separating the protein from the first detergent on a sucrose gradient (17,30) or affinity column (26) containing the new detergent. Some of the viral spike glycoproteins form reversible aggregates, known as rosettes, when detergent is removed, and these can be resolubilized by subsequent addition of a second detergent (50). The caveat is that the proteins may not remain functional after self-aggregation and solubilization.

To dissolve the viral envelope, the virus is transferred to a buffer with the desired detergent and incubated at conditions determined to be optimal for the particular virus. Viral membrane proteins have been successfully solubilized under conditions ranging from occasional gentle swirling on ice in the presence of octylglucoside (51) to violent shaking at room temperature in the presence of Triton X-100 (43). Often protein denaturation and/or solubilization of unwanted viral components can be avoided by keeping the detergent levels low, near the cmc. Similarly, varying the incubation time, temperature (which affects the cmc of the detergent differently from lipid or protein aggregation), ionic strength, and degree of mechanical agitation may dramatically improve the preparation.

2. Isolation

Once the protein is solubilized, it must be separated from the remaining viral components. With some protocols, the nonenvelope components maintain their association in the nucleocapsid under conditions that effectively dissolve the envelope proteins. In this case, the nucleocapsid can be pelleted by a low-speed centrifugation leaving the solubilized membrane proteins and lipids in the supernatant. Since the viral membrane proteins are glycoproteins, they can also be isolated on lectin affinity columns (see, e.g., Refs. 25,26). The purity of the separation can be checked, for example, by polyacrylamide gel electrophoresis. More extensive purification can be performed in the presence of detergent by gel filtration chromatography, or affinity chromatography using lectins, specific antibodies, or other ligands with differential affinity for the membrane proteins (e.g., 18, 66a,b,98). Separating the viral lipids from the viral protein is readily accomplished by flotation on a sucrose density gradient which includes the requisite concentration of detergent (51). (This is a good time to alter the detergent as described above, if necessary.)

After the protein is purified, exogenous lipid can be added to the isolated protein/detergent micelles. Depending on the purpose of the reconstitution, lipids with specific headgroups, acyl chains, or particular mixtures of lipids may be used. Moreover, trace amounts of radiolabeled lipids, lipids with attached fluorophores, spin labels, or photoreactive groups are added at this stage of the protocol. The lipids may be added as lipid/detergent micelles or may be solubilized from a thin dry film with the protein/detergent mixture. The advantage of the latter protocol is that dilution is minimized, but care must be taken to have enough excess detergent to fully solubilize the lipids. The amount needed can be calculated from the lipid/detergent ratio at solubilization or determined empirically by stepwise detergent addition until the mixture is clarified.

3. Vesicle Formation

Once the lipid, protein, and detergent have had time to form structures, presumably two- and three-component micelles, the detergent is removed to form the virosomes. How this is achieved is determined by the type of detergent and by the peculiarities of the given protein/lipid system. Possible variations are infinite, but the most important parameters seem to be the rate of detergent removal and the temperature (e.g., 49a,102a). The more hydrophylic detergents, such as octylglucoside and cholate, can be removed from the micelles fairly readily by dilution with excess buffer. In practice, this means elution over a gel filtration column (48,52), dialysis (20), or rapid dilution followed by dialysis (28-34). The more hydrophobic detergents, with a lower cmc, can be dialyzed but not very efficiently unless a "sink," such as a hydrophobic resin, is added to the buffer (44).

During dialysis or other modes of detergent removal, the detergent/lipid and detergent/protein micelles eventually must come together to form the virosomes. If at any stage during detergent removal, the lipids and proteins are in different micelles, they might remain separated and reconstitution might fail. For example, it was observed that the G protein from VSV aggregates at different octylglucoside concentrations than the lipids (39,53) (see Fig. 1, B1). Furthermore, unlike many of the transport proteins that can be reconstituted with one or two molecules per vesicle, the viral glycoproteins seem to "prefer" to be at a density comparable to that found in the intact virus (20,44,54). This feature, which probably is useful for the formation of the viral coat, makes it difficult to study such questions as whether this density is a necessary requirement for function or to examine a single functional unit. If the proteins aggregate before the formation of a lipid lamellar phase, one must try to develop a protocol to form vesicles with the protein incorporated at the desired density oriented with the extracellular domain facing outward. One strategy might be to adjust parameters (e.g., temperature, ionic strength, detergents) until quasistable three-component micelles can be formed during equilibrium dialysis. Indeed, the strategy, called slow dialysis, which permits the protein/lipid/detergent mixture to come to equilibrium at several intermediate states in the process of vesicle formation, appears to work for the G protein of VSV (20) and the Sendai virus spike proteins (55). Since holding proteins for long periods in detergent also tends to denature them, octylglucoside dialysis may be substituted by dilution at a rate just slow enough to allow putative intermediates to form (hours vs. days for dialysis) at a temperature and ionic strength where lipids and proteins do coalesce. The detergent removal is finished rapidly with SM-2 Bio-Beads. By combining these techniques, functional VSV virosomes have been formed (A. Walter et al., unpublished results).

Alternatively, an effective strategy might be to *very rapidly* remove the detergent leaving the protein to be immediately "solubilized" by the lipid. Reconstitution of rhodopsin, a very hydrophobic protein that spans the membrane at least seven times, is an example that works in this way (36). Unfortunately, the viral spike glycoproteins tend to aggregate and form rosettes if the detergent is removed rapidly (31,51).

Due to the low cmc of the more hydrophobic detergents, such as Triton X100 and $C_{12}E_8$, monomer concentrations are very low and consequently detergent removal by dialysis is exceedingly slow. Triton removal from proteins or lipids is facilitated by the addition of hydrophobic resins such as SM-2 Bio-Beads to the mixture (49,49a,56). The use of these resins has been adapted to reconstitute the Sendai viral envelope proteins (16,43,44) and the vesicular stomatitis virus coat protein (49a). The hydrophobic resin is placed in an excess amount outside the dialysis bag (44,57) or added directly to the aqueous protein/lipid/detergent dispersion (43). The adsorption is facilitated by agitation (43) or repeated passage over a small column (49). Lipids (and protein) also bind to the hydrophobic resins, and there must be enough contact time or repeated passage over the same hydrophobic columns to allow the lipid to reequilibrate with the vesicles as they form. This happens because the free energy of a lipid in a vesicle is much lower than the free energy of the phospholipid associated with the resin. Moreover, the removal of detergent from the inner monolayer is slow, and thus, unless the time is sufficient, significant detergent can remain in the inner monolayer since stable bilayer structures exist at significant detergent-to-lipid ratios (≈1:1 for octyl glucoside).

Another approach to reconstitution is to add viral proteins to preformed vesicles and have them insert at detergent levels just sufficient to swell but not disrupt the bilayers (Fig. 1, AB4). This has worked with the asialoglycoprotein receptor (59) and Semliki Forest virus glycoproteins (31). Insertion of the VSV (51) and of rabies virus (60) glycoprotein from rosettes to preformed vesicles has been reported, but it is not clear that the membrane-spanning region was properly inserted [nor has this been confirmed for any of the other reconstitution protocols (61); see also

Section II.C.2]. Triton-solubilized monomers of the Sindbis virus glycoprotein added to preformed vesicles above the gel to liquid crystalline
phase transition temperature also resulted in virosomes of defined size
with properly oriented proteins (21). The latter approach is thought to
increase the likelihood that the large extracellular domain of the protein
would be oriented to the outside.

Also, by varying the protein-to-lipid ratio during incubation and subsequently selecting virosomes of a given density by density gradient
centrifugation, vesicles with a given protein density were obtained (21).

Other ways to induce protein insertion are mild sonication and freeze-
thaw cycles of vesicle/protein mixtures. However, both approaches tend
to denature proteins. In one case, where vesicles with large encapsulation
volumes were desired, a protocol for forming vesicles from Ca^{2+}-precipitated
phosphatidylserine (PS) designed for vesicle preparation in the absence
of protein was adapted to accommodate the Sendai glycoproteins by lowering
the rate of Ca^{2+} removal (62).

C. Evaluation of Reconstituted Virosomes

1. Assays for Virosome Fusion

The criterion for an effective reconstitution of a viral fusion protein is
the ability of the reconstituted virosome to fuse in a manner analogous
to that of the intact virus. Virosomes with properly oriented spikes (e.g.,
seen in EM) that yet did not fuse have been observed (45; see also Section III). Conversely, several proteins are known to cause fusion among
some targets [e.g., cardiolipin (CL) or PS vesicles] that are unlikely
mimics of natural target membranes (15,63,64). In any new virosome preparation, the specific protein requirement for fusion can be established with
straightforward controls such as protein-free liposomes and virosomes in
which the viral protein has been compromised by proteolysis, high temperature, or inactivated otherwise.

To establish that the virosome fusion activity is akin to the biological
one, it is necessary to have several hallmarks for a particular viral fusion.
These might include requirements for a specific receptor or specific lipids
in the target membrane, or knowledge of specific inhibitors or a particular
pH dependence for the process. Unfortunately, in most cases, little is
known about such specific characteristics. Among the best examples demonstrating similarity to native function is the inhibition of Sendai virosome
fusion by specific inhibitors, analogous to that seen with the native virus
(8). "Specific" receptors for viral fusion, such as gangliosides for Sendai
virus, have been implicated in a number of cases, but few of them have
been identified. These receptors might be important only for binding,
a prerequisite that may have nothing to do with the actual molecular rearrangements required for the fusion process. Recently, it has become
feasible to compare the extent and rate of virosome lipid mixing directly
with that of the intact virus using the same assay. Rhodamine-labelled
probes [octadecylrhodamine and rhodamine-labeled phosphatidylethanolamine
(*N*-Rh-PE)] have been incorporated into intact virus membranes from ethanol
solutions (7-11) and spin-labelled phosphatidylcholine (PC) introduced
into viral envelopes using the PC-specific lipid exchange protein (12).

There are numerous assays to test for virosome fusion, physiological
ones developed to study virus activity and those that measure mixing either
between the virosome and target membranes or of the two aqueous contents.
Physiological assays of viral fusion include measurement of polykaryon

formation by virosome or virus-induced fusion between cells (50,65) and
of hemolysis. The underlying assumption for the latter measurement is
that hemolysis is commensurate with fusion. Fusion of viruses with the
plasma membrane seems to be accompanied by immediate and profound
changes in the passive permeability of the target membrane (66,67). It
has been argued that those permeability changes are not a necessary concomitant of fusion, but could be an artifact if the virion membranes are
damaged due to aging, freeze-thawing, or sonication (55,68-70). An alternative interpretation of leaky viral fusion is that some portion of the viral
spike glycoprotein destabilizes the target membrane and that this destabilization is, in fact, under certain circumstances part of the fusion phenomenon.
Since the hemolysis assay does not always report fusion (16) and is often
neither an indicator nor a necessary concomitant of biological fusion, it
cannot be used as a reliable fusion assay.

Studies of the mechanism of fusion utilize biophysical techniques, mainly
based on fluorescence or ESR, to assess successful reconstitution leading
to virosomes that are able to fuse with target membranes. Fluorescent
probes for both lipid- and contents-mixing assays have been used extensively for vesicle-vesicle fusion and have recently been reviewed in depth
(13,14; see also Chapter 4). The fluorescence assays have the advantage
of being able to follow the fusion process continuously and they are con-
figured so that very small extents of fusion can be detected. Other probes
for membrane mixing include the introduction of ESR probes or of a unique
membrane protein into the target, such as the anion transport protein
(71-74), a receptor such as the Epstein-Barr virus receptor (75,76), or
cell-surface antigens (77-85). Membrane mixing may also be followed by
fluorescence recovery after photobleaching (FRAP), which permits measuring
the extent of fusion to single cells and regional differences on the cell
membrane but not the kinetics of the fusion event (86). Contents-mixing
assays include the fluorescent assays which require a fluorescent compound
and either a quenching or enhancing agent to be encapsulated in the virosome
and target vesicles such as the ANTS/DPX pair (13). Both contents mixing
and lipid mixing may be observed and quantified by standard fluorescence
measurements or digitally processed videomicroscopy (49a). Contents mixing
can also be observed by encapsulation in the virosome of anything that
will be uniquely detected in the final product, such as a toxin that will
affect cell activity or DNA that can be discerned phenotypically (87-91).

2. Chemical and Physical Characterization

In order to gather quantitative information about the function of virosomes,
it is essential to characterize them physically and chemically (28-34). Vesicle
sizes and size distribution should be defined and, if possible, controlled,
since the rate of fusion can be a function of vesicle diameter (e.g., Ref. 92).
Vesicle size and size distributions can be determined by EM, light scattering, or gel filtration on Sephacryl, Sephadex, or HPLC columns (93).
The advantage of the latter is that vesicles are separated by size, allowing
selection of a more homogeneous preparation, if necessary. One problem
with column separation is to find elution conditions that minimize virosome
sticking to the matrix. The number of lamellae per vesicle can be observed
by EM or calculated from the ratio of lipid surface area to contents volume,
in conjunction with an independently determined average diameter.

The absence of residual detergent in the final product can be ensured
either directly, by measuring the amount of residual detergent (e.g., with
a radiolabel or chemically), or indirectly, by appropriate controls for the

specificity of membrane fusion, as suggested above. Total detergent removal is critical when fusion function is being reconstituted since detergents do destabilize membranes and can induce lipid exchange (which would report as fusion in lipid-mixing assays; see, e.g., Ref. 39) or hemolysis, which is sometimes considered a hallmark for viral function (see Section II.C.1).

Knowledge of the actual protein-to-lipid ratio, the orientation of the proteins, and protein distribution among vesicles in the final product is necessary for quantitation and for discerning molecular mechanisms for fusion. The first parameter can be determined chemically or, if possible, by following radiolabeled protein and lipid. The orientation can be determined from the percentage of protein susceptible to protease cleavage (50). This is particularly clear-cut for viral proteins that have large extraviral domains that, when oriented properly, are susceptible to specific proteases leaving a small peptide so that the percentage facing outward can be readily discerned on polyacrylamide gels. Whether the virosome protein actually spans the membrane, as in the intact virus, has not been discerned for any reconstitution protocol. Another glycoprotein with a single membrane-spanning region is the H-2K^k protein, one of the class I major histocompatibility complex molecules. That this glycoprotein is inserted as a hairpin rather than across the lipid bilayer when reconstituted by deoxycholate dialysis was demonstrated by labeling both the N and C terminal ends with a fluorescent probe prior to reconstitution and measuring the accessibility of the probe to Co^{2+}, an impermeant quenching agent (61).

An equally important parameter is the distribution of protein among the vesicles and subsequently ensuring that the preparation is free from protein-free vesicles or unincorporated protein in the form of rosettes (Fig. 1, A4, B4) since rosettes of viral spike glycoproteins have been reported to induce fusion among lipid vesicles or cells (57,66,94). The latter can be achieved by gel filtration chromatography, by sucrose density gradient separation, or by visualizing viral spike glycoproteins by negative stain electron microscopy. The protein-to-lipid ratio can be estimated and the vesicles separated into subpopulations by determining the lipid and protein distribution on sucrose density gradients (21,50,51). The number of proteins per vesicle is determined for each fraction from tracers and/or density calculations.

When several viral envelope proteins are being reconstituted simultaneously, their individual distributions among the vesicles may affect the fusion results. There is no a priori reason to believe they will be uniformly dispersed among the vesicles, especially since proteins may have different solubilities in detergents and thus are likely to "come out of solution" at different stages during the detergent removal. This was observed, for example, when Johnson et al. (95) discovered a subpopulation of virosomes that was deficient in one of the four herpes simplex glycoproteins they were coreconstituting and, therefore, did not bind to cells. Thus, if there is a functionally recalcitrant subpopulation of virosomes, one must suspect that perhaps one of the necessary components is missing from those virosomes.

III. VIRAL SPIKE GLYCOPROTEINS RECONSTITUTED INTO LIPID VESICLES: APPLICATIONS

The list of applications for which viral spike glycoproteins were incorporated into lipid vesicles (Table 1, see also Chapter 32) indicates that the specific objectives of reconstitution have been sixfold:

1. Study of the mechanism of reconstitution and development of specific reconstitution protocols for viral proteins
2. Study of lipid-protein interactions
3. Induction of an immune response (antibody production or cytotoxic activity)
4. Insertion of membrane receptors or transport proteins into cell membranes
5. Intracellular delivery of drugs, toxins, or genetic material
6. Study of the mechanisms of fusion induced by viral proteins

The first three objectives do not require fusion properties of the resulting virosomes and therefore will not be discussed in the context of this book. Objectives 4 and 5 are related to biotechnological and clinical applications of virosomes. The criterion for successful reconstitution in those cases is success; that is, the function to be transferred is expressed in the target cell. As shown in Table 1, both cell membrane components and encapsulated materials have been delivered to cells with the aid of reconstituted viral proteins.

Realization of the technological goals of viral-protein-induced implantation and delivery clearly demonstrates that viral proteins can be reconstituted in a way to induce fusion between virosomes and cells. However, using these sorts of virosomes to help describe biologically relevant fusion is more difficult because the virosome behavior first must be shown to be analogous to that of the intact virus. Reconstitutions can be successful by physicochemical criteria, yet the resulting virosomes do not always induce fusion. A case in point is the study by Hare and Huston (45), who constructed hybrid virosomes containing the red cell band 3 and the Sendai HN and F glycoproteins by dialysis of octylglucoside. The resulting virosomes bound to Hepatoma cells but did not fuse. Treatment of cell-associated virosomes with poly(ethylene glycol) (PEG) resulted in fusion with about 5% transfer efficiency, which provides 3×10^6 band 3 molecules per cell, as shown by fluid phase marker transfer, immunofluorescence, and protease digestion. The failure of the octylglucoside-derived virosomes to fuse (without PEG) was attributed to inactivation of the fusion proteins in octylglucoside solutions. Although one report indicates that octylglucoside is not a good detergent for functional reconstitution of Sendai virus envelopes (8), the effect of octylglucoside is not clearly implicated since others have formed fusogenic Sendai virosomes using octylglucoside (55, 77-79). The activity of the proteins in the study of Hare and Huston (45) was not checked independently since those authors could not form virosomes by Triton X-100, a detergent known to conserve Sendai virus function.

The incomplete removal of Triton X-100 from protein/lipid complexes is a prevalent worry in the reconstitution literature. For instance, Harmsen et al. (55) noted that fluorescence quenching of NBD in Triton X-100-dialyzed Sendai virosomes containing N-NBD-PE and N-Rh-PE could not be solely due to fluorescence resonance energy transfer (RET), and that the efficiency of RET was affected by residual membrane-associated detergent. Vainstein et al. (43) measured residual Triton in Sendai virosomes using ^{3}H-labeled Triton X-100 and found 42 moles of Triton per 100 moles of phospholipid remained if the detergent was removed by Spectrapor dialysis tubing. Addition of SM-2 BioBeads to the dialysis buffer significantly improves removal of Triton X-100. For instance, Ozawa and Asano (96) measured only 0.1-0.2 mol of residual Triton X-100 per mol of glycoproteins, and if we assume 100 lipids per protein in the Sendai virosomes, this number would be 0.1-0.2 mol of Triton X-100 per 100 moles of lipid. Direct addition of SM-2 BioBeads instead of dialysis yielded a ratio of 2-3 moles of Triton per 100 moles of phospholipid (43). One critique that has been leveled at the Triton X-100 Sendai virosomes is that fusion or hemolysis could be induced by residual Triton X-100. This seems unlikely, because both fusion and hemolysis were blocked by agents (DTT and PMSF) that affect protein function (16). Those agents do not inhibit binding of the virion to the cell surface. Moreover, Sendai virosomes prepared from Triton X-100 are nonleaky to small molecules (43,97), indicating that any residual Triton does not affect the permeability properties. It cannot, however, be ruled out that the Triton plays a synergistic role in viral-protein-induced fusion.

Virosomes that mimic at least some characteristics of the native virus have been prepared and manipulated to address some of the issues relevant to the fusion process. In particular, by selecting the viral envelope glyco-proteins that are reconstituted, the individual proteins required for herpes simplex and Sendai virus binding to cells have been identified (16,17,95, 98,99). The Sendai F protein is needed to induce fusion, probably in cooperation with the HN protein (16,17,99). In fact, there is evidence that optimal fusion occurs when the HN and F proteins are present together in a ratio of 1:2 in Sendai virosomes (18). Of interest now, for Sendai and other enveloped viruses with two or more membrane proteins, is to describe how these work in concert.

Viral envelope proteins reconstituted with fluorescently labeled lipids have also been used to measure the kinetics and/or the extents of fusion (20,21,49a,55,100). A fluorescent assay to measure contents mixing between two types of vesicles has been used to compare Sendai virosome- and virus-induced fusion (16). Nucleated cells have been shown to have regions that are recalcitrant to fusion by measuring the mobility of *N*-NBD-PE, appearing in cell membranes from labeled Sendai virosomes, using fluorescence photobleaching recovery (86).

The message of Table 1 is that the experimental and practical possibilities for utilizing reconstituted viral fusion proteins, as outlined in the Introduction, are just beginning to be realized. The results already obtained indicate that attaining the technological and basic science goals is feasible. Use of reconstituted viral proteins alone, or in conjunction with other targeting molecules, has succeeded as a means for delivering both membrane proteins and vesicle contents to intact cells. Virosomes and systems for studying the mechanism of fusion are being developed, and these experimental model systems are being shown to be a realistic means for approaching questions pertaining to viral fusion. The future of this research is in the, as yet largely unexplored, molecular aspects of fusion.

ACKNOWLEDGMENTS

This work was supported in part by NIH grant #HL-40158 (O.E.).

REFERENCES

1. Blumenthal, R. (1987). Membrane fusion. *Current Topics Membranes Transport* 29:203-254.
2. Bukrinskaya, A. G. (1982). Penetration of viral genetic material into host cells. *Adv. Virus Res.* 27:141-200.

3. Howe, C., Coward, J. E., and Fenger, T. W. (1980). Viral invasion: Morphological, biochemical and biophysical aspects. In: *Comprehensive Virology*, Vol. 16 (H. Fraenkel-Conrat and R. R. Wagner, eds.). Plenum Press, New York, pp. 1–71.

4. White, J., Kielian, M., and Helenius, A. (1983). Membrane fusion proteins of enveloped animal viruses. *Q. Rev. Biophys.* 16:151–195.

5. Rose, J. K., and Gallione, C. J. (1981). Nucleotide sequences of the mRNA's encoding the Vesicular Stomatitis virus G and M proteins determined from c DNA clones containing the complete coding regions. *J. Virol.* 39:519–528.

6. Wilson, I. A., Skehel, J. J., and Wiley, D. C. (1981). Structure of the haemagglutinin membrane glycoprotein of influenza virus at 3 A resolution. *Nature* 289:366–373.

7. Blumenthal, R., Bali-Puri, A., Walter, A., Covell, D., and Eidelman, O. (1986). pH-dependent fusion of Vesicular stomatitis virus with Vero cells: Measurement of kinetics and extent by fluorescence dequenching. *J. Biol. Chem.* 262:13614–13619.

8. Citovsky, V., Blumenthal, R., and Loyter, A. (1985). Fusion of Sendai virions with phosphatidylcholine-cholesterol liposomes reflects the viral activity required for fusion with biological membranes. *FEBS Lett.* 193:135–140.

9. Hoekstra, D., de Boer, T., Klappe, K., and Wilschut, J. (1984). Fluorescence method for measuring the kinetics of fusion between biological membranes. *Biochemistry* 23:5675–5681.

10. Lorge, P., Cabiaux, V., Long, L., and Ruysschaert, J. M. (1986). Fusion of Newcastle disease virus with liposomes: Role of the lipid composition of liposomes. *Biochim. Biophys. Acta* 858:312–316.

11. Nir, S., Klappe, K., and Hoekstra, D. (1986). Kinetics and extent of fusion between Sendai virus and erythrocyte ghosts: Application of mass action kinetic model. *Biochemistry* 25:2155–2166.

12. Yamada, S., and Ohnishi, S. (1986). Vesicular stomatitis virus binds and fuses with phospholipid domain in target cell membranes. *Biochemistry* 25:3703–3708.

13. Düzgünes, N., and Bentz, J. (1988). Fluorescence assays for membrane fusion. In: *Spectroscopic Membrane Probes*, Vol. 1 (L. M. Loew, ed.). CRC Press, Boca Raton, FL, Chapter 6, pp. 117–159.

14. Morris, S. J., Bradley, D., Gibson, C. C., Smith, P. D., and Blumenthal, R. (1988). Use of membrane-associated fluorescence probes to monitor fusion of bilayer vesicles: Application to rapid kinetics using pyrene excimer/monomer fluorescence. In: *Spectroscopic Membrane Probes*, Vol. 1 (L. M. Loew, ed.). CRC Press, Boca Raton, FL, Chapter 7, pp. 161–191.

15. Walter, A., Steer, C. J., and Blumenthal, R. (1986). Polylysine induces pH-dependent fusion of acidic phospholipid vesicles: A model for polycation-induced fusion. *Biophys. Biochim. Acta* 861:319–330.

16. Citovsky, V., and Loyter, A. (1985). Fusion of Sendai virions or reconstituted Sendai virus envelopes with liposomes or erythrocyte membranes lacking virus receptors. *J. Biol. Chem.* 260:12072–12077.

17. Hsu, M. C., Scheid, A., and Choppin, P. W. (1979). Reconstitution of membranes with individual paramyxovirus glycoproteins and phospholipid in cholate solution. *Virology* 95:476–491.

18. Nakanishi, M., Uchida, T., Kim, J., and Okada, Y. (1982). Glycoproteins of Sendai virus (HVJ) have a critical ratio for fusion between virus envelopes and cell membranes. *Exp. Cell Res.* 142:95–101.

19. Ozawa, M., Asano, A., and Okada, Y. (1979). Biological activities of glycoproteins of HVJ (Sendai virus) studied by reconstitution of hybrid envelope and by concanavalin A-mediated binding: A new function of HANA protein and structural requirement for F protein in hemolysis. *Virology* 99:197–202.

20. Eidelman, O., Schlegel, R., Tralka, T. S., and Blumenthal, R. (1984). pH-dependent fusion induced by vesicular stomatitis virus glycoprotein reconstituted into phospholipid vesicles. *J. Biol. Chem.* 259:4622–4628.

21. Scheule, R. K. (1986). Novel preparation of functional Sindbis virosomes. *Biochemistry* 25:4223–4232.

22. Altstiel, L. D., and Landsberger, F. R. (1981). Lipid-protein interactions between the surface glycoprotein of vesicular stomatitis virus and the lipid bilayer. *Virology* 115:1–9.

23. Lyles, D. S., McKinnon, K. P., and Parce, J. W. (1985). Labeling of the cytoplasmic domain of the influenza virus hemagglutinin with fluorescein reveals sites of interaction with membrane lipid bilayers. *Biochemistry* 24:8121–8128.

24. Petri, W. H., Jr., Pal, R., Barenholz, Y., and Wagner, R. R. (1981). Fluorescence anisotropy of a fatty acid covalently linked in vivo to the glycoprotein of vesicular stomatitis virus. *J. Biol. Chem.* 256:2625–2627.

25. Casali, P., Sissons, J. G., Buchmeier, M. J., and Oldstone, M. B. (1981). In vitro generation of human cytotoxic lymphocytes by virus. Viral glycoproteins induce nonspecific cell-mediated cytotoxicity without release of interferon. *J. Exp. Med.* 154:840–855.

26. Casali, P., Sissons, J. G., Fujinami, R. S., and Oldstone, M. B. (1981). Purification of measles virus glycoproteins and their integration into artificial lipid membranes. *J. Gen. Virol.* 54:161–171.

27. Trudel, M., Ravaoarinoro, M., and Payment, P. (1980). Reconstitution of rubella hemaglutinin on liposomes. *Can. J. Microbiol.* 26:899–904.

28. Cabantchik, Z. I., and Darmon, A. (1985). Reconstitution of membrane transport systems. In: *Structure and Properties of Cell Membranes*, Vol. 3 (G. Benga, ed.). CRC Press, Boca Raton, FL.

29. Darszon, A. (1983). Strategies in the reassembly of membrane proteins into lipid bilayer systems and their functional assay. *J. Bioenerg. Biomembr.* 15:321–334.

30. Helenius, A., and Simons, K. (1975). Solubilization of membranes by detergents. *Biochim. Biophys. Acta* 415:29–79.

31. Helenius, A., Sarvas, M., and Simons, K. (1981). Asymmetric and symmetric membrane reconstitution by detergent elimination. Studies with Semliki-Forest-virus spike glycoprotein and penicillinase from the membrane of *Bacillus licheniformis. Eur. J. Biochem.* 116:27–35.

32. Jones, O. T., Earnest, J. P., and McNamee, M. G. (1987). Solubilization and reconstitution of membrane proteins. In: *Biological Membranes: A Practical Approach* (J. B. C. Findley and W. H. Evans, eds.). IRL Press, London.

33. Klausner, R. D., van Renswoude, J., Blumenthal, R., and Rivnay, B. (1984). Reconstitution of membrane receptors. In: *Molecular and Chemical Characterization of Membrane Receptors* (J. C. Venter and L. C. Harrison, eds.). Alan R. Liss, New York, pp. 209–239.

34. Levitzki, A. (1985). Reconstitution of membrane receptor systems. *Biochim. Biophys. Acta* 822:127–153.

35. Helenius, A., McCaslin, D. R., Fries, E., and Tanford, C. (1979). Properties of detergents. *Methods Enzymol.* 56:734–749.

36. Jackson, M. L., and Litman, B. J. (1985). Rhodopsin-egg phosphatidyl-
    choline reconstitution by an octyl glucoside dilution procedure. *Biochim.
    Biophys. Acta* 812:369-376.
37. Tanford, C., and Reynolds, J. (1975). Characterization of membrane
    proteins in detergent solutions. *Biochim. Biophys. Acta* 417:133-170.
38. Fromherz, P., and Ruppel, D. (1985). Lipid vesicle formation: The
    transition from open discs to closed shells. *FEBS Lett.* 179:155-159.
39. Ollivon, M., Eidelman, O., Blumenthal, R., and Walter, A. (1988).
    Micelle-vesicle transitions of egg phosphatidylcholine and octylglucoside
    in water. *Biochemistry* 27:1695-1703.
40. Schurtenberger, P., Mazer, N., and Kanzig, W. (1985). Micelle to
    vesicle transition in aqueous solutions of bile salt and lecithin. *J. Phys.
    Chem.* 89:1042-1049.
41. Eidelman, O., Blumenthal, R., and Walter, A. (1988). Composition
    of octyl glucoside-phosphatidylcholine mixed miceles. *Biochemistry*
    27:2839-2846.
42. Womack, M. D., Kendall, D. A., and MacDonald, R. C. (1983). Deter-
    gent effects on enzyme activity and solubilization of lipid bilayer
    membranes. *Biochim. Biophys. Acta* 733:210-215.
43. Vainstein, A., Hershkovitz, M., Israel, S., Rabin, S., and Loyter, A.
    (1984). A new method for reconstitution of highly fusogenic Sendai
    virus envelopes. *Biochim. Biophys. Acta* 773:181-188.
44. Volsky, D. J., and Loyter, A. (1978). An efficient method for re-
    assembly of fusogenic Sendai virus envelopes after solubilization of
    intact virions with Triton X-100. *FEBS Lett.* 92:190-194.
45. Hare, J. F., and Huston, M. (1985). Virosome-mediated implantation
    of red cell band 3 into the plasma membrane of cultured hepatoma
    cells. *Exp. Cell Res.* 161:317-330.
46. Rivnay, B., and Metzger, H. (1982). Reconstitution of the receptor
    for immunoglobin E into liposomes. *J. Biol. Chem.* 257:12800-12808.
47. Tanford, C. (1980). *The Hydrophobic Effect: Formation of Micelles
    and Biological Membranes*, 2nd ed., Wiley, New York.
48. Mimms, L. T., Zampighi, G., Nozaki, Y., Tanford, C., and Reynolds,
    J. A. (1981). Phospholipid vesicle formation and transmembrane protein
    incorporation using octyl glucoside. *Biochemistry* 20:833-840.
49. Ueno, M., Tanford, C., and Reynolds, J. A. (1984). Phospholipid
    vesicle formation using nonionic detergents with low monomer solubility.
    Kinetic factors determine vesicle size and permeability. *Biochemistry*
    23:3070-3076.
49a. Paternostre, M. T., Lowy, R. J., and Blumenthal, R. (1989). pH-
    dependent fusion of reconstituted vesicular stomatitis virus envelopes
    with vero cells. *FEBS Lett.* 243:251-258.
50. Marsh, M., Balzau, E., White, J., and Helenius, A. (1983). Inter-
    actions of Semliki Forest virus spike glycoprotein rosettes and vesicles
    with cultured cells. *J. Cell Biol.* 96:455-461.
51. Petri, W. H., Jr., and Wagner, R. R. (1979). Reconstitution into
    liposomes of the glycoprotein of vesicular stomatitis virus by detergent
    dialysis. *J. Biol. Chem.* 254:4313-4316.
52. Shimizu, K., Hosaka, Y., and Shimizu, Y. K. (1972). Solubilization
    of envelopes of HVJ (Sendai virus) with alkali-emasol treatment and
    reassembly of envelope particles with removal of the detergent.
    J. Virol. 9:842-850.
53. Blumenthal, R., Ollivon, M., Margolis, D., and Eidelman, O. (1985).
    Dissolution and reconstitution of VSV G protein into egg PC vesicles:

Studies using energy transfer between lipid probes. *Biophys. J.* 47: 256a.

54. Helenius, A., Fries, E., and Kartenbeck, J. (1977). Reconstitution of Semliki forest virus membrane. *J. Cell Biol.* 75:866-880.

55. Harmsen, M. C., Wilschut, J., Scherphof, G., Hulstaert, C., and Hoekstra, D. (1985). Reconstitution and fusogenic properties of Sendai virus envelopes. *Eur. J. Biochem.* 149:591-599.

56. Holloway, P. W. (1973). A simple procedure for the removal of Triton X 100 from protein samples. *Anal. Biochem.* 53:304-308.

57. Inoue, J-I., Nojima, S., and Inoue, K. (1985). The activity of membranes reconstituted from HVJ envelope proteins and lipids to induce hemolysis and fusion between liposomes and erythrocytes. *Biochim. Biophys. Acta* 816:321-331.

58. Goni, F. M., Urbaneja, M-A., Arrondo, J. L. R., Alonso, A., and Dunani, A. A. (1986). The interaction of phosphatidylcholine bilayers with Triton X-100. *Eur. J. Biochem.* 160:659-665.

59. Klausner, R. D., Bridges, K., Tsunoo, H., Blumenthal, R., Weinstein, J. M., and Ashwell, G. (1980). Reconstitution of the hepatic asialoglyco-protein with phospholipid vesicles. *Proc. Natl. Acad. Sci. USA* 77:5087-5091.

60. Perrin, P., Sureau, P., and Thibodeau, L. (1985). Structural and immunogenic characteristics of rabies immunosomes. *Dev. Biol. Stand.* 60:483-491.

61. Cardoza, J. D., Kleinfeld, A. M., Stallcup, K. C., and Mescher, M. F. (1984). Hairpin configuration of h-2K^k in liposomes formed by detergent dialysis. *Biochemistry* 23:4401-4409.

62. Gould-Fogerite, S., and Mannino, R. J. Rotary dialysis: Its application to the preparation of large liposomes and large proteoliposomes (protein-lipid vesicles) with high encapsulation efficiency and efficient reconstitution of membrane proteins. *Anal. Biochem.* 148:15-25.

63. Amselem, S., Loyter, A., Lichtenberg, D., and Barenholz, Y. (1985). The interaction of Sendai virus with negatively charged liposomes: Virus-induced lysis of carboxyfluorescein-loaded small unilamellar vesicles. *Biochim. Biophys. Acta* 820:1-10.

64. Walter, A., Margolis, D., Mohan, R., and Blumenthal, R. (1986). Apocytochrome c induces pH-dependent vesicle fusion. *Membr. Biochem.* 6:217-237.

65. Toister, Z., and Loyter, A. (1973). The mechanism of cell fusion II. Formation of chicken erythrocyte polykaryons. *J. Biol. Chem.* 248: 422-432.

66. Kruse, C. A., Wisnieski, B. J., and Popjak, G. (1984). Characterization of a glycoprotein fusagen isolated from Sendai virus. *Biochim. Biophys. Acta* 797:40-50.

66a. Lambert, D. M., and Pons, M. W. (1983). Respiratory syncytial virus glycoproteins. *Virology* 130:204-214.

66b. Lapidot, M., Nussbaum, O., and Loyter, A. (1987). Fusion of membrane vesicles bearing only the influenza hemagglutinin with erythrocytes, living cultured cells, and liposomes. *J. Biol. Chem.* 262:13736-13741.

67. Pasternak, C. A., and Micklem, K. J. (1974). Permeability changes during cell fusion. *J. Membrane Biol.* 14:293-303.

68. Homma, M., Shimizu, K., Shimizu, Y. K., and Ishida, N. (1976). On the study of Sendai virus hemolysis I. Complete Sendai virus lacking hemolytic activity. *Virology* 71:41-47.

69. Vaananen, P., and Kaariainen, L. (1980). Fusion and haemolysis of erythrocytes caused by three togaviruses: Semliki Forest, Sindbis and rubella. *J. Gen. Virol.* 46:467-475.

70. Young, J.D-E., Young, G. P. H., Cohn, Z. A., and Lenard, J. (1983). Interaction of enveloped viruses with planar bilayer membranes, observations on Sendai, influenza, vesicular stomatitis and Semliki Forest viruses. *Virology* 128:186-194.

71. Beigel, M., and Loyter, A. (1983). Fusion-mediated implantation of band 3 into living cells. A new system to study degradation of membrane proteins. *Exp. Cell Res.* 148:95-103.

72. Beigel, M., Volsky, D. J., Ginsburg, H., Cabantchik, Z. I., and Loyter, A. (1980). Functional incorporation of the human erythrocyte chloride exchange system into plasma membranes of Friend erythroleukemic cells by Sendai virus-induced cell fusion. *Exp. Cell Res.* 126:448-453.

73. Cabantchik, Z. I., Volsky, D. J., Ginsburg, H., and Loyter, A. (1980). Reconstitution of the erythrocyte anion transport system: in vitro and in vivo approaches. *Ann. NY Acad. Sci.* 341:444-454.

74. Volsky, D. J., Cabantchik, Z. I., Beigel, M., and Loyter, A. (1979). Implantation of the isolated human erythrocyte anion channel into plasma membranes of Friend erythroleukemic cells by use of Sendai virus envelopes. *Proc. Natl. Acad. Sci. USA* 76:5440-5444.

75. Tsukuda, K., Volsky, D. J., Shapiro, I. M., and Klein, G. (1982). Epstein-Barr virus (EBV) receptor implantation onto human B lymphocytes changes immunoglobulin secretion patterns induced by EBV infection. *Eur. J. Immunol.* 12:87-90.

76. Volsky, D. J., Shapiro, I. M., and Klein, G. (1980). Transfer of Epstein-Barr virus receptors to receptor-negative cells permits virus penetration and antigen expression. *Proc. Natl. Acad. Sci. USA* 77: 5453-5457.

77. Abidi, T. F., and Flanagan, T. D. (1984). Cell-mediated cytotoxicity against targets bearing Sendai virus glycoproteins in the absence of viral infection. *J. Virol.* 50:380-386.

78. Al-Ahdal, M. N., Nakamura, I., and Flanagan, T. D. (1985). Cytotoxic T-lymphocyte reactivity with individual Sendai virus glycoproteins. *J. Virol.* 53:53-57.

79. Al-Ahdal, M. N., Abidi, T. F., and Flanagan, T. D. (1986). The interaction of Sendai virus glycoprotein-bearing recombinant vesicles with cell surfaces. *Biochim. Biophys. Acta* 854:157-168.

80. Hale, A. H., Lyles, D. S., and Fan, D. P. (1980). Elicitation of anti Sendai virus cytotoxic T lymphocytes by viral and H-2 antigens incorporated into the same lipid bilayer by membrane fusion and by reconstitution into liposomes. *J. Immunol.* 124:724-731.

81. Hale, A. H., Lyles, D. S., Paulus, L. K., and Ruebush, M. J. (1980). Minimal molecular requirements for reactivity of tumor cells with T cells. *J. Immunol.* 124:2063-2070.

82. Hale, A. H., Ruebush, M. J., Lyles, D. S., and Harris, D. T. (1980). Antigen-liposome modification of target cells as a method to alter their susceptibility to lysis by cytotoxic T lymphocytes. *Proc. Natl. Acad. Sci. USA* 77:6105-6108.

83. Prujansky-Jakobovits, A., Volsky, D. J., Loyter, A., and Sharon, N. (1980). Alteration of lymphocyte surface properties by insertion of foreign functional components of plasma membrane. *Proc. Natl. Acad. Sci. USA* 77:7247-7251.

84. Volsky, D. J., Ahrlung-Richter, L., Dalianis, T., and Klein, G. (1981). Implantation of mouse histocompatibility antigens into membranes of cultured tumor cells. *Eur. J. Immunol.* 11:341-344.

85. Volsky, D. J., Gross, T., Sinangil, F., Kuszynski, C., Bartzatt, R., Dambaugh, T., and Kieff, E. (1984). Expression of Epstein-Barr virus (EBV) DNA and cloned DNA fragments in human lymphocytes following Sendai virus envelope-mediated gene transfer. *Proc. Natl. Acad. Sci. USA* 81:5926-5930.

86. Aroeti, B., and Henis, Y. I. (1986). Fluorescence photobleaching recovery as a method to quantitate viral envelope-cell fusion: Application to study fusion of Sendai virus envelopes with cells. *Biochemistry* 25:4588-4596.

87. Gitman, A. G., Graessmann, A., and Loyter, A. (1985). Targeting of loaded Sendai virus envelopes by covalently attached insulin molecules to virus receptor-depleted cells: fusion-mediated microinjection of ricin A and simian virus 40 DNA. *Proc. Natl. Acad. Sci. USA* 82:7309-7313.

88. Gitman, A. G., Kahane, I., and Loyter, A. (1985). Use of virus-attached antibodies or insulin molecules to mediate fusion between Sendai virus envelopes and neuraminidase-treated cells. *Biochemistry* 24:2762-2768.

89. Loyter, A., Vainstein, A., Graessmann, M., and Graessmann, A. (1983). Fusion-mediated injection of SV40-DNA. Introduction of SV40-DNA into tissue culture cells by the use of DNA-loaded reconstituted Sendai virus envelopes. *Exp. Cell Res.* 143:415-425.

90. Uchida, T., Kim, J., Yamaizumi, M., Miyake, Y., and Okada, Y. (1979). Reconstitution of lipid vesicles associated with HVJ (Sendai virus) spikes. Purification and some properties of vesicles containing nontoxic fragment A of diphtheria toxin. *J. Cell Biol.* 80:10-20.

91. Vainstein, A., Razin, A., Graessmann, A., and Loyter, A. (1983). Fusogenic reconstituted Sendai virus envelopes as a vehicle for introducing DNA into viable mammalian cells. *Methods Enzymol.* 101:492-512.

92. Ohki, S. (1984). Effects of divalent cations, temperature, osmotic pressure gradient, and vesicle curvature on phosphatidylserine vesicle fusion. *J. Membrane Biol.* 77:265-275.

93. Ollivon, M., Walter, A., and Blumenthal, R. (1986). Sizing and separation of liposomes, biological vesicles and viruses by high-performance liquid chromatography. *Anal. Biochem.* 152:262-274.

94. Wharton, S. A., Skehel, J. J., and Wiley, D. C. (1986). Studies of influenza haemoglutinin-mediated membrane fusion. *Virology* 149:27-35.

95. Johnson, D. C., Wittels, M., and Spear, P. G. (1984). Binding to cells of virosomes containing Herpes Simplex virus type 1 glycoproteins and evidence for fusion. *J. Virol.* 52:238-247.

96. Ozawa, M., and Asano, A. (1981). The preparation of cell fusion-inducing proteoliposomes from purified glycoproteins of HVJ (Sendai virus) and chemically defined lipids. *J. Biol. Chem.* 256:5954-5956.

97. Bartzatt, R. L., and Volsky, D. J. (1984). Determination of the internal volume of reconstituted Sendai virus envelopes by quenching of calcein fluorescence. *Biosci. Rep.* 4:551-557.

98. Peterhans, E., Baechi, T., and Yewdell, J. (1983). Evidence for different receptor sites in mouse spleen cells for the Sendai virus hemagglutinin-neuraminidase (HN) and fusion (F) glycoproteins. *Virology* 128:366-376.

99. Scheid, A., Hsu, M., and Choppin, P. W. (1980). Role of paramyxovirus glycoproteins in the interactions between viral and cell membranes. *Soc. Gen. Physiol. Ser.* 34:119-130.

100. Chejanovsky, N., and Loyter, A. (1985). Fusion between Sendai
     virus envelopes and biological membranes. The use of fluorescent
     probes for quantitative estimation of virus-membrane fusion. *J. Biol.
     Chem.* 260:7911-7918.

101. Shapiro, I. M., Klein, G., and Volsky, D. J. (1981). Epstein-Barr
     virus co-reconstituted with Sendai virus envelopes infects Epstein-Barr
     virus-receptor negative cells. *Biochim. Biophys. Acta* 676:19-24.

102. Mazurek, N., Bashkin, P., Loyter, A., and Pecht, I. (1983). Restora-
     tion of Ca2+ influx and degranulation capacity of variant RBL-2H3
     cells upon implantation of isolated cromolyn binding protein. *Proc.
     Natl. Acad. Sci. USA* 80:6024-6018.

102a. Miguel, M. G., Eidelman, O., Ollivon, M., and Walter, A. (1989).
     Temperature dependence of the vesicle-micelle transition of egg phos-
     phatidylcholine and octylglucoside. *Biochemistry* 28:8921-8928.

103. Gitman, A. G., and Loyter, A. (1984). Construction of fusogenic
     vesicles bearing specific antibodies. Targeting of reconstituted Sendai
     virus envelopes toward neuraminidase-treated human erythrocytes.
     J. Biol. Chem. 259:9813-9320.

104. Chejanovsky, N., Eytan, G. D., and Loyter, A. (1984). Fusion be-
     tween Sendai virus envelopes and biological membranes as monitored
     by energy transfer methods. *FEBS Lett.* 174:304-309.

105. Citovsky, V., Yanai, P., and Loyter, A. (1986). The use of circular
     dichroism to study conformational changes induced in Sendai virus
     envelope glycoproteins. A correlation with the viral fusogenic activity.
     J. Biol. Chem. 261:2235-2239.

106. Henis, Y. I., Gutman, O., and Loyter, A. (1985). Sendai virus en-
     velope glycoproteins become laterally mobile on the surface of human
     erythrocytes following fusion. *Exp. Cell. Res.* 160:514-526.

107. Ohuchi, M., and Homma, M. (1976). Trypsin action on the growth
     of Sendai virus in tissue culture cells. IV. Evidence for activation
     of Sendai virus by cleavage of a glycoprotein. *J. Virol.* 18:1147-1150.

108. Petri, W. H., Jr., Estep, T. N., Pal, R., Thompson, T. E., Biltonen,
     R. L., and Wagner, R. R. (1980). Thermotropic behavior of di-
     palmitoylphosphatidylcholine vesicles reconstituted with the glycoprotein
     of vesicular stomatitis virus. *Biochemistry* 19:3088-3091.

109. Petri, W. H., Jr., Pal, R., Barenholz, Y., and Wagner, R. R. (1981).
     Fluorescence studies of dipalmitoylphosphatidylcholine vesicles recon-
     stituted with the glycoprotein of vesicular stomatitis virus. *Biochemistry*
     20:2796-2800.

110. Miller, D. K., Feuer, B. I., Vanderoef, R., and Lenard, J. (1980).
     Reconstituted G protein-lipid vesicles from vesicular stomatitis virus
     and their inhibition of VSV infection. *J. Cell Biol.* 84:421-429.

111. Loh, D., Ross, A. H., Hale, A. H., Baltimore, D., and Eisen, H. N.
     (1979). Synthetic phospholipid vesicles containing a purified viral
     antigen and cell membrane proteins stimulate the development of cyto-
     toxic T lymphocytes. *J. Exp. Med.* 150:1067-1074.

IV
INTRACELLULAR MEMBRANE TRAFFIC

19

Membrane Cycling Through the Endocytotic and Exocytotic Pathways
An Overview

MARK MARSH

Institute of Cancer Research, London, England

PAUL QUINN

University College, London, England

I. INTRODUCTION

Eukaryotic cells are organized into specialized membrane-bound compartments or organelles, with distinct functions and biochemical compositions. In the endocytotic and exocytotic pathways, which account for 60-70% of the total cell membrane (Table 1), proteins and glycoproteins are transported between compartments by highly regulated membrane fusion and fission reactions. The routes of transport, the organelles involved, and the biochemical modifications that occur during transport have been studied in detail (1-5). From these and other studies it is clear that membrane traffic is extensive and must be carefully regulated. Membrane must be cycled to maintain the balance between individual compartments, and components must be sorted to preserve the biochemical composition of the various organelles. Here we provide an overview of transport and recycling in the endocytotic and exocytotic pathways. We have not attempted to present a comprehensive discussion of the literature and, where appropriate, recent reviews are cited. Moreover, specific aspects of transport and recycling are mentioned only superficially and left for other authors in this volume to cover in more detail.

II. THE ENDOCYTOTIC PATHWAY

Endocytosis is the process by which cells internalize media and ligands from the extracellular milieu via invaginations of the plasma membrane to form endocytotic vesicles. Different mechanisms of endocytosis have been described in various cell types (6). Some cells, macrophages for example, internalize large particles (diameter >200 nm), such as bacteria and yeast, by a receptor-dependent process termed phagocytosis. Although, in general, only specialized cells carry out phagocytosis, virtually all cells constitutively internalize fluid and small particles (<200 nm diameter).

TABLE 1 Summary of Membrane Surface Areas and Volumes of
BHK Organelles

	Absolute surface (μm^2)	Absolute volume (μm^3)
Cell		1400 ± 220
Cytoplasm		1050 ± 210
Plasma membrane	2,200 ± 470	
Coated pits	(35 ± 6)[a]	
ER (including nuclear envelope)	5,870 ± 990	101 ± 18
Golgi stack	1,960 ± 540	20 ± 5
Endosomes (total)	430 ± 105	7 ± 2
Lysosomes and prelysosomes	370 ± 11	37 ± 11
Mitochondria		67 ± 15
Outer membrane	1,080 ± 280	
Inner membrane	3,950 ± 711	
Total membrane	15,860 ± 3,209	

[a]This value is included in the plasma membrane total.
Source: Griffiths et al. (1989) (26a).

This pinocytotic uptake can be either receptor-dependent (receptor-mediated endocytosis) or independent (fluid-phase endocytosis).

Aspects of pinocytosis have been quantitated for several cell types. Where appropriate, some of these values are given to emphasize both the extent of membrane transport and the variability between different cells.

A. Receptor-Mediated Endocytosis

Many of the interactions between a cell and its environment are mediated through specific high-affinity receptors expressed on the cell surface. Often these interactions result in the internalization and intracellular processing of both the ligand and the receptor. Receptor-mediated endocytosis (R-ME) and adsorptive endocytosis (uptake through the opportunistic use of cell surface receptors) are processes whereby a wide variety of important macromolecules, small particles, and their receptors are taken up and concentrated by the cell. The list of ligands internalized by R-ME is extensive and includes nutrient carriers such as low-density lipoprotein (LDL) and transferrin (Tf), serum proteins, growth factors and polypeptide hormones, lymphokines and other immune modulators, immunoglobulins, immune complexes, and a number of viral pathogens and polypeptide toxins (2,7-9).

Internalization occurs through specialized domains of the plasma membrane called coated pits. These regions have a morphologically visible coat on the cytoplasmic aspect of the membrane composed of a protein complex termed clathrin (10). Coated pits continually invaginate and pinch off, enclosing a vesicle (average diameter 100 nm) within a closed clathrin basket—an endocytotic coated vesicle (7,11). In baby hamster kidney (BHK) cells 1-2% of the cell surface is coated and each cell has between 1500 and 3000 coated pits. As approximately 1500 coated vesicles are internalized each minute (12,13), the average lifetime of a coated pit on the cell surface is between 1-2 min. Following invagination, the clathrin is released from the vesicle (11,14) by means of an ATP-dependent uncoating enzyme (15), and the uncoated vesicles fuse with organelles of the endosome compartment.

Many of the receptors involved in R-ME, including the LDL-R and the galactose-binding asialoglycoprotein receptor (ASGP-R), are preferentially localized in coated pits (2,16-18). These receptors, all of which are transmembrane proteins, initially appear on the cell surface individually but are rapidly relocated into the pit regions (7,11,19). Other receptors, including those for epidermal growth factor (EGF-R) and insulin (I-R), appear not to cluster in coated pits and are distributed more randomly on the cell surface. These receptors, however, relocate into coated pits, on binding their ligands (18). Receptors have been broadly classified into two groups (20). Class 1 receptors primarily effect information transfer. They direct changes in cell metabolism and behavior, require ligand to induce internalization, are frequently down-regulated, and generally do not cluster into coated pits. Class 2 receptors primarily mediate internalization and continuously recycle in the presence or absence of ligand. They affect cell metabolism only indirectly, are not down-regulated by ligand, and are found to cluster in coated pits.

The mechanisms by which constitutively internalized receptors interact with coated pits are now being elucidated (20a). Comparison of the primary sequences of many receptors has failed to reveal common structural features that might suggest how clustering occurs. For the LDL-R, two mutations have been found where the receptors are able to bind LDL but fail to localize into coated pits and, in turn, fail to internalize efficiently. These point mutations give rise, in one instance, to a receptor in which the cytoplasmic domain is truncated close to the membrane, and in the other, to a change at position 807 from tyrosine to cysteine (2,21,22). This tyrosine is believed to be critical for the interaction with the adaptor components of the clathrin coat (20a,22a).

The fact that coated pits can preferentially internalize certain plasma membrane components has given rise to the idea that they act as molecular filters regulating the composition of internalized membrane (16,17). Apart from the distinction between receptors that cluster into coated pits and those that do not, coated pits apparently express little specificity for particular receptors. Consequently, different receptors and ligands can internalize through the same coated pits and coated vesicles (8,16,18,23).

Cells appear to have some capacity to modulate endocytosis through coated pits and vesicles. Endocytosis, together with other membrane transport processes, is inhibited during mitosis (24). On the other hand, specific growth factors can transiently increase fluid-phase uptake (25). This increase, for PC12 cells exposed to nerve growth factor and EGF, is due to an increase in the number of endocytotic coated vesicles (26). For a more detailed discussion of the properties of coated vesicles see Ref. 20a.

B. Fluid Phase Endocytosis

The uptake of ligands through specific receptors is accompanied by continual internalization of the surrounding media (6). The internalization of fluid can be measured using soluble markers, such as horseradish peroxidase (HRP) or ^{3}H-sucrose, which are neither bound to the cell surface nor degraded by the cell. The uptake of such markers is proportional to their extracellular concentration and is inefficient compared to R-ME. However, the volume of fluid internalized can be considerable. Mouse peritoneal macrophages internalize the equivalent of 25% of their cell volume each hour (6). BHK cells internalize 0.3 μl/10^7 cells/h (12), or 1.5% of the cell volume (Table 1). The uptake in BHK cells can be accounted for by approximately 1500 coated vesicles leaving the cell surface each minute. The similarity between this estimate and the number of vesicles involved in R-ME suggests that most, if not all, BHK cell fluid phase pinocytosis occurs through coated vesicles and that both receptor-mediated and fluid phase endocytosis are properties of the same vesicular transporters (12,13,26a).

C. The Fate of Internalized Membrane

1. Receptor-Mediated Endocytosis

Following internalization ligand/receptor complexes are delivered to acidic prelysosomal organelles termed endosomes [also referred to as receptosomes, pinosomes, intermediate vesicles, or CURL (compartment for uncoupling receptor from ligand)] and subsequently redistributed to different destinations within the cell (Fig. 1). This delivery involves fusion of uncoated endocytotic vesicles either with each other or with organelles of the endosome compartment. The fusion reaction is specific and requires both ATP and components of the cytosol (26b,c). Biochemical studies have demonstrated that the internalized complexes are sorted by one of four principal alternatives (2,5,8).

1. Receptors such as the LDL-R and the ASGP-R bind their ligands in a pH-dependent manner. Within the acidic, endocytotic environment the complexes dissociate and the unoccupied receptors recycle to the cell surface, while ligands pass to the lysosomes and are degraded (7,27).

2. For a second group of receptors, including EGF-R and the murine macrophage receptor for the Fc domain of $IgG_{1/2b}$ (Fc-R), the ligands do not dissociate from the receptor at endosomal pH's and both the ligands and receptor pass to the lysosomes (28). Lysosomal degradation leads to a net loss of receptors from the cell surface and, consequently, influences the cell's responsiveness to ligand. The signal directing the Fc-R to the lysosomes is specified in part by ligand-induced cross-linking of receptors. Monovalent ligands (e.g., Fab fragments of an antibody to the receptor) internalize and recycle together with the receptor, while multivalent ligands (soluble immune complexes or antireceptor Fab coupled to colloidal gold) deliver the receptor and ligands to the lysosomes (29).

3. As with the previous group, Tf and its receptor (Tf-R) do not dissociate at low pH. The low pH of endosomes reverses the pH-dependent association of iron with Tf, but the apo-Tf (iron-free Tf) remains receptor-bound and recycles to the cell surface with the receptor. On returning to neutral pH, apo-Tf dissociates from the Tf-R and is released from the cell surface (30,31).

4. Finally, in polarized epithelial cells receptors and ligand internalized at one plasma membrane domain are delivered to a different domain of the cell surface by transcytosis. IgA and its receptor, for example, are internalized at the basolateral surface, transported through the cell, and released (following proteolytic cleavage of the receptor) at the apical domain (32). Similarly, maternal IgG is transported across the gut epithelium in neonatal

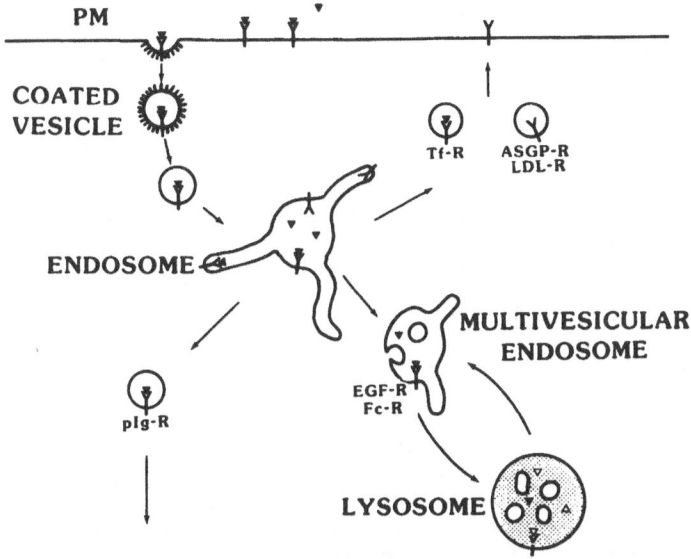

FIGURE 1 Schematic representation of the pathways involved in receptor uptake and recycling. PM = plasma membrane.

rat. However, in this case, transport is from the apical to the basolateral membrane. Ligand-receptor binding is specialized to occur in the acid conditions of the gut and is reversed at neutral pH following translocation to the basolateral domain (33).

2. Fluid Phase Endocytosis

Internalized fluid markers follow the intracellular pathways specified above nonselectively and in proportion to the relative volume of the vesicular traffic. In macrophages, BHK, and Madin-Darby canine kidney (MDCK) cells, the intracellular accumulation of internalized fluid phase markers is curvilinear; an initial rapid accumulation is followed, after 40-60 min, by a slower linear accumulation (12,34-36). The first phase corresponds most closely to the absolute rate of uptake. As the intracellular compartments are filled, an increasing fraction of the marker is recycled to the cell surface and released, and then the rate of accumulation declines. The second phase reflects the situation at steady state when uptake is balanced by accumulation and recycling.

Recycling can be measured directly. When labeled cells are returned to culture in marker-free medium, marker is rapidly released and can be assayed either in the medium or as a decrease in cell-associated activity. For macrophages the $T_{1/2}$ of the rapid recycling phase is 6 min and the fraction of internalized marker recycled is 4/5 of the volume internalized. The remaining 20% is retained by the cell or recycles only very slowly (34,36). The fact that markers are released rapidly after uptake suggests that ligand recycles from an early endocytotic compartment, while accumulation occurs in late endosomes or lysosomes.

In epithelia, fluid phase markers are transported across the cells. A quantitative estimate for fluid phase transcytosis has been made for MDCK cells (35). In monolayers, grown on filters, approximately 3×10^{-8}

nl/cell/min is transported simultaneously in each direction. Although the
vesicles involved in transport are derived from the endosomes and have
not been characterized, the volume transported is equivalent to about
100 coated vesicles passing in either direction each minute.

3. Membrane Recycling

The area of membrane internalized by 1500 coated vesicles/cell/min corresponds to 75 μm^2. In a BHK cell this would result in the internalization
of the entire plasma membrane every 30-60 min (6,12,37). The biosynthetic
capacity of the cell cannot provide replacement membrane and, therefore,
recycling must occur (6). Recycling is reflected in the reutilization of
specific receptors (e.g., LDL-R, ASGP-R, Tf-R), the reflux of internalized
fluid phase markers, and the reappearance of internalized antibodies bound
to specific cell surface components (29,38,39).

The bulk recycling of internalized membrane has been demonstrated
directly by following modified cell surface components. The oligosaccharide
chains on cell surface glycoproteins and glycolipids can be labeled by
the enzymatic addition of monosaccharides, e.g., galactose (37), or by
$NaB[^3H]_4$ reduction after $NaIO_4$ oxidation (40). On reincubation at 37°C,
the labeled membrane is internalized and equilibrates with the intracellular
pools. The intracellular galactose label becomes resistant to removal by
β-galactosidase and can be distinguished from cell surface label. In
Dictyostelium amoebae and macrophage-like $P388D_1$ cells, the steady-state
distribution of the intracellular and cell surface pools is 1:2 and 1:4,
respectively (37,41). If, after incubation at 37°C, the labeled cells are
treated with β-galactosidase and returned to culture, the internal label
will recycle to the cell surface and become accessible to further treatment
with β-galactosidase. An area equivalent to the entire cell surface is turned
over every 45 and 21 min, respectively, and most of the internalized membrane (90%) is recycled to the cell surface within minutes (37,40,41).

The composition of the internal membrane at steady state appears,
by one dimensional SDS-gel electrophoresis, to be qualitatively similar
to the initial plasma membrane pool. Endocytotic membranes have also been
characterized by radioiodination using internalized lactoperoxidase (LPO).
In J774 macrophages the pattern of polypeptides labeled, following fluid-
phase endocytosis of LPO, is qualitatively similar to the pattern generated
by labeling the cell surface (42). However, in experiments where conjugated
LPO is internalized by R-ME, receptors, such as the Tf-R, that are constitutively recycled are present in higher concentrations than in the plasma
membrane (43). The picture that emerges suggests that most membrane
components are internalized and recycled as part of bulk membrane flow
but that certain receptors are present at higher concentrations in endosomal
membranes as a consequence of their preferential internalization in coated
vesicles. Direct characterization of the endosomal membrane composition
will require the isolation of these organelles (43a).

D. Endosomes

1. Morphology

Light microscope studies in cultured cells show that fluorescent ligands
collect into either spherical or elongated vesicles distributed in the periphery
of the cell and adjacent to the plasma membrane. These peripheral endosomes
undergo saltatory movement for approximately 10 min and are then frequently

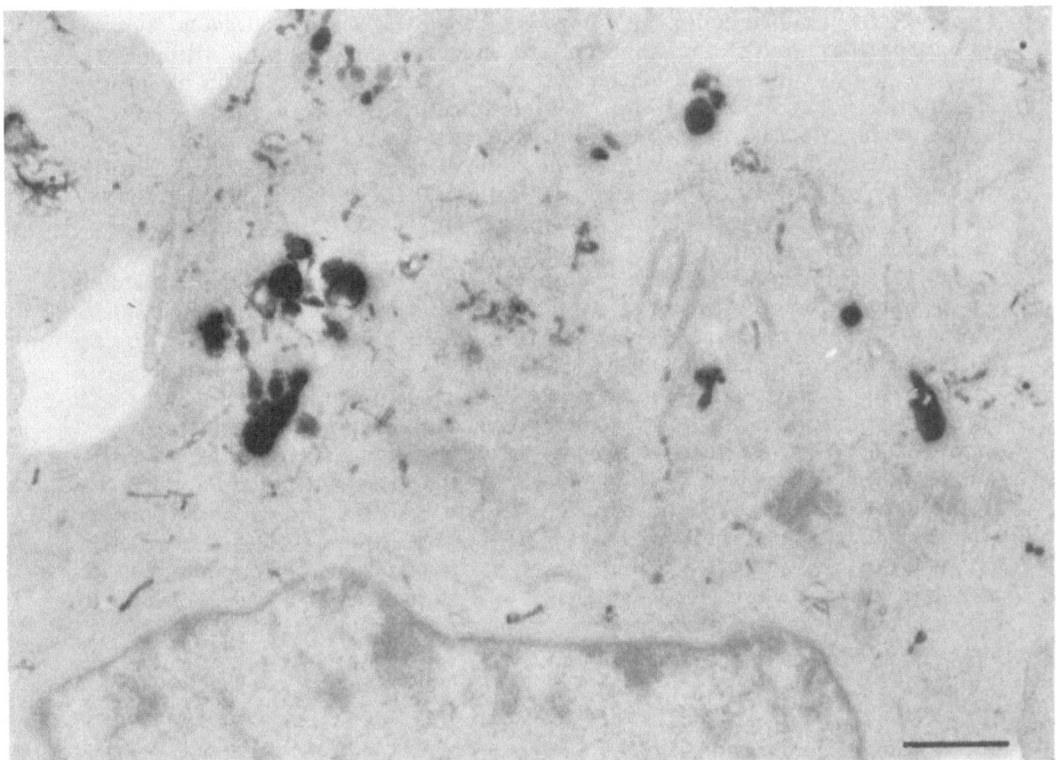

FIGURE 2 Semithin section of a BHK cell labeled to display the endosomes.
BHK-21 cells were labeled with horseradish peroxidase for 15 min at 37°C,
conditions in which the label is confined to the endosome compartment
(see Section II.D). The cells were washed, fixed, reacted with diamino-
benzidine, and processed for electron microscopy. Semithin (0.25 μm)
sections were cut and viewed without staining. The HRP/DAB reaction
product is seen in complex organelles consisting of a central vacular portion
and a number of associated tubules. Bar is equivalent to 1 μm.

relocated into the perinuclear region. The relocation appears to be mediated
by microtubules (44,45). In cells treated with microtubule disrupting agents,
endosomes fail to undergo relocation and the subsequent delivery of internalized ligands to the lysosome compartment is slowed (8,44).

Within 1-2 min of internalization, R-ME and fluid-phase markers appear,
by electron microscopy, in electron-lucent vacuoles, small vesicles and
tubules in the periphery of the cell, adjacent to the plasma membrane.
After 5-15 min the ligands appear in the perinuclear endosomes, close
to Golgi cisternae and lysosomes. These perinuclear endosomes consist
of vacuoles, small vesicles, tubules, and multivesicular bodies (8,46-50).
The endosomal vacuoles are usually 0.3-1 μm in diameter, the vesicles
and tubules 50-100 nm in diameter, and the tubules up to 4 μm in length.
Semithin (0.25 μm) sections of HRP-labeled cells (Fig. 2), and reconstruc-
tions from serial thin sections, indicate that the endosome compartment
in BHK cells is composed of organelles consisting of vesicles and cisternae,
from which a number of tubules project. These tubules may occasionally
connect between the endosomal vesicles, but the majority of endosomes

appear to be independent (51). Stereological measurements indicate that fluid phase ligands fill the whole of the endosome compartment within 5 min and occupy up to 2.5% of the cell volume. In BHK cells, the endosomes account for about 4% of the total cell membrane and have a surface area corresponding to 20% of the plasma membrane (26a).

Essentially similar observations have been made in intact tissues. During the uptake of ASGP into liver, coated vesicles deliver internalized ligands from the sinusoidal membrane to peripheral endosomes, which appear as an elaborate system of tubules and vesicles underlying the cell surface (23,50,52). Subsequently the ligands move to perinuclear endosomes which are frequently larger and often contain internal vesicles (multivesicular bodies). ASGP is uncoupled from the ASGP-R in endosomes (CURL). Using antibodies and colloidal gold conjugates, the receptors and ligands can be localized in cryosections. The unoccupied receptors are found in endosomal tubules, while the free ligand remains within the vesicular endosome (46). In sections of hepatocytes labeled to identify both ASGP-R and the transcytotic polymeric IgA receptor, both receptors are found in the same endosomes but in mutually exclusive tubules (23) and subsequently in distinct vesicles in the Golgi/lysosome region (54). These data suggest that the endosomal tubules may be involved in sorting and recycling. Although low pH is required to dissociate ligands from receptors, it remains unclear how membrane proteins are physically segregated into distinct domains within the endosome.

2. Cell Fractionation

Cell fractionation experiments have shown endosomes to have a lower buoyant density than lysosomes (1.02–1.05 g.cm^{-3} vs. 1.1–1.5 g.cm^{-1}), allowing them to be distinguished by centrifugation in various density gradient media (55–58). Furthermore, density differences can be used to distinguish early from late endosomes (52,59). Thus, it has been shown that fluid phase markers and ligands, such as ASGP, pass through a low-density endosome compartment en route to the higher-density lysosomes.

Density gradient fractionation has confirmed that ASGP dissociates from the ASGP-R before the complex reaches the lysosomes, and that only the ligand is found in the heavier-density lysosome fractions (27). The movement into receptor-negative vesicles is believed to occur prior to the transfer into lysosomes, such that ligand-positive/receptor-negative endosomes can be distinguished from ligand-positive/receptor-positive endosomes (59).

Affinity-purified antibodies specific for ASGP-R have been used to isolate membrane vesicles containing the ASGP-R. The antibody can be used to distinguish vesicles that contain both the ligand and receptor from vesicles that contain only the ligand. After 2.5 min of internalization over 98% of the internalized radiolabeled ASGP can be precipitated. During the next 12 min up to 30% of the internalized label transfers to endosomal vesicles which will not precipitate with anti ASGP-R antibodies. Subsequently the ligand appears in the lysosomes. At 16°C ligand is not transferred to the lysosomes (60), but remains in receptor-positive, immunoprecipitable vesicles. Morphologically, the receptor-positive organelles correspond to the vesiculotubular structures, while the receptor-negative structures are multivesicular. The multivesicular endosomes, devoid of receptor, represent the final prelysosomal endosomal compartment (61).

3. Acidification

Weak bases (NH_4Cl) as well as carboxylic ionophores (monensin and nigericin) raise the pH of acidic organelles (62). As a consequence, in cells treated with NH_4Cl, iron remains bound to Tf, and internalized LDL and ASGP, for example, fail to dissociate from their receptors and fail to move to the lysosomes (27). The fusion of endosomes with lysosomes is not itself inhibited by NH_4Cl. Missorting is a secondary effect of the inability to dissociate ligand from receptor (29). In the presence of weak bases receptors which are constitutively internalized accumulate intracellularly and do not recycle to the cell surface (63,64). Together these data indicate that endosomes are acidic and that low pH is important for function.

Direct evidence for the acidic nature of endosomes has come from several sources. Fluorescein-labeled ligands, used to follow endocytosis morphologically (see Section II.D.1), were initially used as intracellular pH probes in intact macrophages. The emission intensity of fluorescein changes as a function of pH (62). Thus internalized FITC-labeled alpha-2-macroglobulin, EGF, or Tf have been found to enter a low-pH compartment (about 5.5) prior to entry into lysosomes (65-67).

A second source has been the studies on the entry of fusogenic enveloped viruses (see also the preceding section of the book). Viruses such as Semliki Forest virus (SFV) undergo membrane fusion at pH 6.0 as part of their strategy for entering target cells (14,68). The fusion occurs rapidly after R-ME of the virions, indicating that the viruses encounter a low pH early after uptake in the endosome. In addition, a mutant of SFV (FUS-1), which undergoes fusion at pH 5.5, also fuses within endosomes (69). Kinetic experiments show that the wild-type virus encounters pH 6.0 with a $T_{1/2}$ of 15 min, whereas FUS-1 encounters pH 5.5 with a $T_{1/2}$ of 45 min (70). It is not clear whether the viruses pass sequentially through different endosomal compartments of decreasing pH, or whether they are transported within an organelle that undergoes progressive acidification (70). Finally, the acidic nature of endosomes has been demonstrated by electron microscopy. The reagent 3-(2,4-dinitroanilino)-3'-amino-N-methyl-dipropylamine (DAMP) is weakly basic and concentrates in acidic compartments where it can be detected by immunocytochemistry. In cells incubated with DAMP, endosomes and lysosomes are stained (71).

The mechanism of endosome acidification has now been characterized using permeabilized cells and isolated organelles. Endosomes are acidified by a proton ATPase with properties similar to those described in other vacuolar organelles, including lysosomes, *trans* Golgi network, coated vesicles, and secretory granules (see Ref. 72). These pumps are not electrogenic, and proton translocation proceeds without direct molecular coupling with other anions or cations. The membrane potential derived from H^+-translocation must therefore be dissipated by the influx of external anions or efflux of alkali cations. The vacuolar proton pumps can be distinguished from other proton ATPases by their sensitivity to various inhibitors. Endosomal proton translocation is not affected by inhibitors of the mitochondrial F_1F_0 ATPase (oligomycin, efrapeptin) or by inhibitors of plasma-membrane-type ATPases (vanadate, ouabain). However, acidification is blocked by the alkylating reagent N-ethylmaleimide (NEM) and the nucleotide analog 4-chloro-7-nitrobenzo-2-oxa-1,3-diazole (NBD-Cl) (55,72,73).

4. Transfer of Ligands to Lysosomes

The lysosomes are the terminal degradative compartment of the endocytotic pathway. They are located primarily in the perinuclear region of the cell

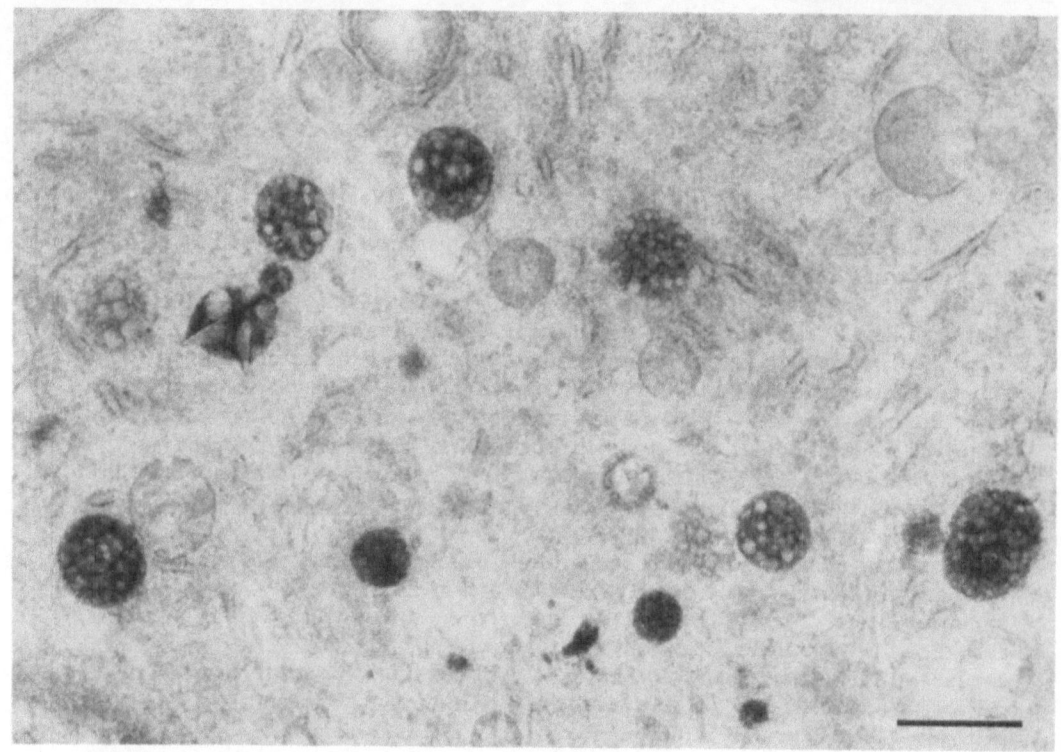

FIGURE 3   Semithin section of a BHK cell labeled to display the lysosomes.
BHK-21 cells were labeled as described for Figure 2 and then chased with
HRP-free medium for 2 h at 37°C before fixation. In contrast to Figure 2,
the label is seen in discrete, spherical organelles which lack the distinctive
tubular structures characteristic of the endosomes. Bar is equivalent to
0.5 μm.

and account for 3% of the cell volume and 2% of the cell membrane (Table 1;
Ref. 6). They contain a wide range of acid hydrolases and are acidified
to a pH of 4.8 by a mechanism similar to, but distinguishable from, that
found in endosomes (62,73,74). Fluid phase markers and receptor-mediated
ligands are first delivered to lysosomes within 10-20 min of uptake from
the cell surface, and delivery from endosomes to lysosomes is inhibited
at temperatures between 16°C and 20°C (56,60).

Morphologically, the lysosomes appear as discrete oval or spherical
vacuoles, frequently multivesicular, and largely depleted of the prominent
tubules characteristic of the endosomes (Fig. 3). Biochemical and cell
fractionation data indicate that the acid hydrolases probably transit through
endosomes en route to the lysosome (75). The mannose-6-phosphate receptors
(MPR) that transport the hydrolases from the Golgi to the endocytotic
pathway can be found in endosomes but not in lysosomes (76,77). The
hydrolase/MPR interaction is pH-sensitive and, like the LDL-R, the MPR
can release its ligand and recycle. Although it is unclear from where this
recycling occurs, it appears that the receptor does not enter the lysosomes
(76).

Two models have been proposed for the mechanism by which ligands pass from the endosome to the lysosome (47). Either ligands are transferred from a stable endosome compartment to lysosomes via shuttle vesicles or, alternatively, forming endosomes mature into lysosomes. The current evidence can be used in support of either model. Endosomal contents are clearly relocated within the cell. There is a progressive decrease in the endosomal pH, a loss of endosomal tubular structures, and a gradual acquisition of acid hydrolases and lysosomal membrane components. Whether by maturation or an alternative mechanism, the transit of membrane through the endocytotic pathway clearly involves numerous tightly controlled fusion and fission reactions. The number of lysosomes and the volume of the compartment remain constant. Thus lysosomes must go through a cycle of fusion with incoming endosomes, followed by removal of membrane through invagination and degradation (condensation) or recycling. The evidence indicates that a combination of the two may occur.

III. THE EXOCYTOTIC PATHWAY

The synthesis and intracellular transport of secretory proteins, proteoglycans, lysosomal hydrolases, and integral plasma membrane proteins are mediated by the endoplasmic reticulum (ER), the Golgi complex, and, for proteins stored before secretion, secretory granules (4). Collectively these organelles have been termed the exocytotic pathway (Fig. 4).

A. The Endoplasmic Reticulum

In brief, nascent polypeptides are synthesized on membrane-bound polysomes and cotranslationally inserted into the membrane of the rough ER (rER) (79). Proteins lacking membrane anchors, such as secretory proteins or lysosomal hydrolases, pass through the membrane and are released into the rER lumen, while integral membrane proteins remain anchored by one or more hydrophobic membrane-spanning domains (80). Processing of the newly synthesized polypeptides begins during, or soon after, translocation. The signal sequences, required for membrane translocation, are usually N-terminal and are cleaved during translation (80); high-mannose oligosaccharides are transferred to appropriate asparagine residues (N-linked) contained within the recognition sequence Asn-X-Ser/Thr [X ≠ PRO (3,81)]; and multimeric proteins are assembled. Following synthesis, those proteins which are not residents of the ER are transported to their correct functional locations. The transport routes are well documented (82), but the mechanisms involved are poorly understood.

Proteins destined for the cell surface are first exported from the ER to the Golgi complex. The residence time in the ER appears to vary between different proteins. In hepatocytes, for example, secretory proteins transported to the plasma membrane via the constitutive pathway (see below) are secreted at different rates (83-85), and in *Dictyostelium* different lysosomal hydrolases are delivered to lysosomes at different rates (86). Transit through the Golgi complex and delivery to the cell surface, or lysosomes, appear equally rapid for all proteins, so the variability is assumed to be in the rates at which proteins exit the ER. This suggests that transport is regulated or receptor-mediated (85,86). Using viral membrane proteins, major histocompatibility (MHC) antigens, or the LDL-R, attempts

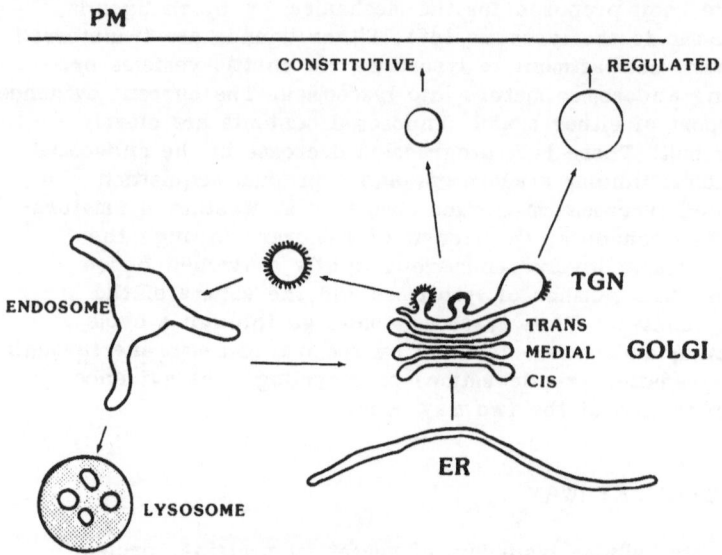

FIGURE 4 Schematic representation of the pathways involved in the transport of newly synthesized proteins. PM = plasma membrane; ER = endoplasmic reticulum; TGN = *trans* Golgi network. A single constitutive pathway is shown from the TGN to the PM, but it is not yet clear that PM proteins and constitutive secretory products are transported in the same vesicles.

have been made to identify specific signals involved in transport or sorting. These studies, although largely unsuccessful (87,88), have indicated that mutations which prevent newly synthesized proteins from folding correctly also inhibit their subsequent transport. In one well-studied example, the influenza virus hemagglutinin (HA), the protein is a trimer of identical transmembrane polypeptides (89). These are assembled in the ER and can be identified by antibodies that distinguish between trimeric HA and the monomeric subunits (88,90). The trimerization occurs within 3–12 min of synthesis, before the acquisition of endoglycosidase H resistance [Endo H removes *N*-linked oligosaccharides, but not *O*-linked or complex oligosaccharides, and can, therefore, be used to distinguish ER from Golgi forms of a glycoprotein (91)] and prior to transport from the ER. However, some of the HA fails to fold correctly and, although trimerized, does not become resistant to Endo H and fails to leave the ER. Thus oligomerization per se is not sufficient for transport and the protein must also adopt the correct conformation. A mutant HA that is unable to fold correctly is found associated with a 70-kD protein (BIP) which is also known to bind IgG heavy chain in the ER of β cells (92). BIP is related to the gp70 heat shock protein and both are believed to play a role in protein folding (93). Correct folding appears to be necessary for the transport of a number of other membrane and secretory proteins, including the glycoprotein of vesicular stomatitis virus, immunoglobulins, and MHC antigens (92,94–96), and appears to function as a quality control mechanism to prevent export of aberrant proteins (97). The observed rates of transit to the cell surface would then be governed by the time required for different proteins to adopt their correct tertiary and quaternary structures. The conformational constraints are apparently quite subtle, as closely related membrane proteins, such as the H-2K^k and H-2D^k MHC antigens, can be transported at different rates within the same cells (98).

Proteins are also concentrated between the ER and Golgi (99). In BHK cells infected with SFV, the newly synthesized spike proteins are concentrated by a factor of 9 to a concentration of 860 copies per μm^2 in Golgi membranes. There is little evidence for further concentration during transport through the subcompartments of the Golgi complex (100).

Morphological studies have failed to demonstrate any physical continuity between the ER and the Golgi complex, and it is widely assumed that transport between them occurs via shuttle vesicles. The vesicles are believed to originate at specializations of the ER termed transitional elements. These domains are distinguished by the absence of ribosomes and appear morphologically as budding vesicle profiles closely opposed to the first, or *cis*, Golgi cisterna (4,101). Currently, transitional elements are the best candidates for the site of selection and concentration of components for transport from the ER.

B. The Golgi Apparatus

Following export from the ER, proteins are transported to the Golgi, where they undergo various posttranslational modifications including oligosaccharide processing, sulfation, and phosphorylation (1,3). The Golgi comprises a stack of closely apposed, but physically separate, flattened cisternae. The number of cisternae varies widely (between 3 and 20) both in the same cell and between different cell types. However, as yet only three functional compartments have been identified: (1) the *cis* compartment containing mannosidase I, *N*-acetylglucosaminylphosphotransferase and *N*-acetylglucosamine-1-phosphodiester-*N*-acetylglucosaminidase, (2) the medial compartment containing mannosidase II and *N*-acetylglucosaminyltransferase, and (3) the *trans* compartment containing the terminal transferases. The *cis* and *trans* compartments were defined by fractionation on analytical gradients (102) and subsequently, the medial compartment was defined using the effects of monensin treatment on membrane protein transport (103,104). The subdivisions of the Golgi complex have been confirmed by immunocytochemical studies using antibodies to compartment-specific enzymes (105,106). Nevertheless, variations in the number of cisternae in a Golgi stack make it difficult to correlate the biochemically defined compartments with individual cisternae. It remains unclear whether more than three functional compartments exist. Membrane and secretory proteins transported from the ER pass vectorially through the Golgi stacks, progressing from the *cis* through the medial to the *trans* face (99,107).

C. The *Trans* Golgi Network (TGN)

From the Golgi, proteins are sorted for transport to their final destinations. Morphological experiments show that viral membrane proteins destined for different plasma membrane domains, and lysosomal hydrolases, transit the Golgi together (108-111). Sorting appears to occur in a tubuloreticular compartment associated with the *trans* side of the Golgi, termed the *trans* Golgi network (TGN) (77,108).

From the TGN proteins can follow five possible pathways (Fig. 4; 108). These lead to (1) the plasma membrane, which in polarized cells forms two domains (apical and basolateral), (2) lysosomes, and (3) the secretory vesicles, which must also be considered as at least two routes depending on whether secretion is regulated or constitutive (77,112). Constitutive secretion may well occur via the same pathway used for plasma membrane proteins, but this remains to be formally demonstrated. Transport from the TGN is presumed to occur via vesicular carriers. Morphological studies have identified three distinct classes of budding profiles associated with the TGN (77,108). The first of these is a coated structure, which reacts with anticlathrin antibodies, contains the MPR, and is presumed to give rise to coated vesicles that carry lysosomal hydrolases to the endocytotic pathway (75). A second class of vesicles are also coated but do not label for clathrin and do not contain the MPR. In cells that exhibit both constitutive and regulated secretion, a third type of vesicle has been identified. These vesicles frequently have limited regions of clathrin coating, contain proteins destined for regulated secretion, and are believed to be the precursor of the secretion granules (77,112,113). The mechanism by which proteins segregate to different domains of the TGN is unclear. However, the presence of clathrin and non-clathrin-containing coat structures suggests that proteins may be selected by direct or indirect interactions with coat components in a manner analogous to R-ME (see Section II.A).

As with the sorting in the endocytotic pathway, the TGN appears to be acidic. Tubuloreticular elements associated with the Golgi, which probably represent TGN, are labeled by the weak bases DAMP and primaquine (114,115), and an H^+-ATPase, with properties similar to the endosomal H^+-ATPase, has been demonstrated in isolated Golgi membranes (116). In a manner similar to that operating in endosomes, low pH in the TGN may regulate the association and dissociation of receptors and ligands, but on the basis of the known pH dependence of MPR binding to lysosomal hydrolases, the TGN itself probably does not acidify below pH 6.0.

The endosome compartment and the TGN appear as functional equivalents in the endocytotic and exocytotic pathways, respectively. The similarity is more than just functional. The organelles resemble each other morphologically and, as a consequence of their physical proximity, are frequently difficult to distinguish. Both organelles appear to be acidified by similar H^+-ATPases (71,72,114), to use the low pH to facilitate sorting, to be affected by weak bases and carboxylic ionophores (63,64,117-119), and to be inhibited in their transport function at temperatures between 16°C and 20°C (56,60,108,120). Despite these similarities, endosomes and TGN appear to be distinct organelles. Overlap of the endocytotic and exocytotic pathways is not extensive and is difficult to demonstrate biochemically. Furthermore, in BHK cells, infected with VSV to label the exocytotic pathway and labeled with HRP in the endocytotic pathway, the endosomes and TGN, though often physically close, are independent structures (77,108).

IV. RECYCLING

The endocytotic and exocytotic pathways continually carry membrane to and from the cell surface. As discussed, internalization through the endocytotic pathway is considerable and membrane must be recycled (6). In the exocytotic pathway the extent of recycling is less clear. The ER is the site of synthesis of both lipids and proteins and must therefore be a net exporter of both. Although the amount of membrane exported is still unknown, some recycling to the ER is believed to occur. In comparison, regulated secretion results in substantial amounts of membrane being delivered to the cell surface, and much of this is believed to recycle to intracellular organelles (1,121,122).

Recycling through the endocytotic pathway has been studied in some detail. Specific receptors, such as the LDL-R, can cycle 100 times or more before degradation (63), while most other membrane components appear to internalize and recycle, together with fluid phase components, as part of the basal plasma membrane turnover (see Sections II.B and II.C). Considerable efforts have been made to understand the routes of this recycling.

Morphologically, internalized membrane, ligands, and fluid markers are seen to pass from the cell surface to the endosomes where membrane is retrieved prior to fusion of endosomes with lysosomes. Retrieval appears to occur through endosomal tubules which are presumed to pinch off from the endosomes (123). However, the route by which these components are returned to the cell surface remains unclear. Following uptake, specific receptors such as EGF-R and Tf-R, or the galactose-binding lectin ricin, are found in membrane systems closely apposed to components of the Golgi stack. Consequently the Golgi or TGN have been proposed to function in endocytic sorting and recycling (8,124,125). However, given the morphological similarity and physical proximity of some endosomal components with elements of the TGN and Golgi cisternae, it is impossible, without independent markers, to determine the precise nature of the labeled compartments.

In professional secretory cells (e.g., plasma cells), fluid phase tracers and Fab fragments of anti-Tf-R antibodies, can be relocated to Golgi cisternae following uptake from the cell surface (1,126). In order to define and quantify the recycling routes biochemically, cell surface glycoproteins have been modified so that they become substrates for enzymes specific to the Golgi processing pathway. One approach has been to remove terminal sialic acid residues from cell surface glycoproteins using neuraminidase. Following reculture, the cells can be assayed to determine whether or not specific components are internalized and reappear on the cell surface in a resialated form (127,128). The Tf-R, which cycles every 10-20 min (129,130), is resialated with a $T_{1/2}$ of 2-3 h, suggesting that 16% of the Tf-R passes through the sialyl transferase containing *trans* Golgi or TGN during each cycle. Other experiments have used deoxynorijimicin to prepare cells containing functional cell surface Tf-R carrying high mannose N-linked oligosaccharides. When the inhibitor is removed, the Tf-R oligosaccharides are processed slowly ($T_{1/2}$ = 6 h) to the complex type (128). The implication is that the internalized receptor can pass through a compartment containing mannosidase 1 (i.e., *cis* Golgi) and that this route is taken by approximately 1-2% of the receptors each cycle (102).

These data show that internalized plasma membrane proteins can recycle through compartments of the Golgi. Although the reprocessing reactions may be inefficient, it appears that most of the Tf-R recycles without passing through the Golgi compartments. Nevertheless, although the proportion of the total receptor recycling through the early Golgi is small, it would be considerably more than the newly synthesized receptor in transit through the Golgi (128). Whether or not these routes reflect normal pathways of cycling is unclear. The components that have been studies are modified and may be earmarked for repair or, alternatively, contain the address information to direct them to the sites where oligosaccharides are modified. Furthermore, the interpretation of these experiments is based on the current knowledge of the distribution of the relevant processing enzymes (1,3), and as the intracellular locations of many of these have only been tentatively assigned, the precise routes of recycling remain open to question. In BHK cells we have failed to find internalized HRP in Golgi cisternae (51). Similarly, membrane proteins and ligand/receptor complexes internalized at one plasma membrane domain of epithelial cells do not appear to pass through either the Golgi or the lysosomes during transcytosis to the other plasma membrane domain (23,33,131).

The membranes of secretory granules contain unique polypeptides required for the accumulation, concentration, and processing of the granule content. Specific antibodies against some of these polypeptides have enabled the fate of the granule membrane proteins to be traced after secretion. Following fusion with the plasma membrane and release of content, the membrane antigens are rapidly removed from the cell surface ($T_{1/2}$ = 10 min, Ref. 121) through coated pits and vesicles to endosomes and multivesicular bodies (122). A portion of the membrane proteins of chromaffin granules appear to be reutilized, as indicated by immunogold labeling, in newly formed secretory vesicles. From the current view of secretory vesicle formation it would seem that these polypeptides must recycle via the TGN, or earlier stations of the Golgi, in order to reenter the granules.

V. MECHANISMS OF SORTING

To ensure that newly synthesized glycoproteins and polypeptides are delivered to their correct functional sites, and to prevent membrane traffic randomizing otherwise biochemically distinct organelles, the contents and membrane of the endocytotic and exocytotic pathways are continually monitored and sorted. The processes involved in this sorting function with considerable efficiency. For example, during the R-ME of Tf on the basolateral surface of MDCK cells, the degree of missorting to the apical domain is less than 0.2% (132). The receptor is recycled efficiently to the correct domain, and the distribution between the apical and basolateral membranes is maintained. The mechanisms by which sorting occurs and the address information contained within individual polypeptides are, with the exception of the ER and lysosomal recognition markers, unknown.

Sorting involves the selective transfer of components from one compartment to another. The transfer potentially involves either (1) the movement of nonselected membrane, and subsequent selective recycling, or (2) the movement of a selected set of polypeptides. First, in the exocytotic pathway resident ER proteins are excluded from the Golgi (3,133). A C-terminal sequence of four amino acids (KDEL) either prevents their exit from the ER or causes efficient recycling back from the next compartment (133a). Second, proteins destined for the plasma membrane move through the Golgi stack while the resident cisternal proteins are retained in restricted subcompartments, and third, coated vesicles mediate the preferential uptake of specific proteins from the cell surface.

Transport between compartments is presumed to occur via vesicular carriers (4,113,134). The vesicles provide a means to both physically separate the components for transport from those to be retained and maintain the donor and acceptor compartments as distinct entities. Presumably a single class of vesicles is sufficient for transport between any two compartments. Thus transport from the ER, between cisternae of the Golgi, and from the plasma membrane may occur via a single class of vesicles in each instance. The vesicles in any one of these sets will have to carry a targetting address for delivery to the *cis* Golgi, the next Golgi cisternae, or the endosomes, respectively. However, when components are sorted in one site (e.g., the TGN) and targetted to different destinations, different classes of transport vesicle will be required. These will presumably carry different sets of address labels and different components. The endocytotic and exocytotic pathways both have points at which sorting to multiple destinations can occur. In the TGN it appears that there are different types of vesicles to mediate transport (see Section III.C), while in the endosome it appears that receptors can be sorted into different membrane domains (see Section II.D).

A.   Soluble Proteins

The cell needs to sort both membrane-bound and soluble proteins. For example, newly synthesized lysosomal hydrolases are separated from secretory proteins destined for the cell surface. For the hydrolases sorting is effected by converting the soluble proteins into membrane-bound proteins through the use of specific receptors: The *N*-linked oligosaccharides of the hydrolases are tagged with M-6-P groups which enable the polypeptides to bind to the MPR in the Golgi. The receptor/ligand complexes are sorted and delivered to the endocytotic pathway where the ligands dissociate and the receptors recycle (75,76,135). The mechanism is analogous to R-ME, where soluble ligands are sorted from the extracellular media. In cells with two secretory pathways, such a mechanism may also account for the sorting of secretory proteins into either the constitutive (e.g., laminin and immunoglobulin) or the regulated (e.g., insulin and adrenocorticotrophic hormone [ACTH]) pathways (112). For discussions of the possible mechanisms involved in stimulus-response coupling during regulated secretion, refer to Chapters 23 and 24.

A carrier must allow proteins to bind prior to sorting, and dissociate afterward. Many ligand-receptor interactions are influenced by pH. In the endocytotic pathway, changes in the pH of different compartments enable receptors to first bind and subsequently release specific ligands. Many of the well-characterized ligands (e.g., LDL, ASGP, lysosomal hydrolases) bind at neutral pH and dissociate in acid conditions. In contrast, maternal immunoglobulin in the neonatal rat gut binds to the IgG receptors at acid pH and dissociates in neutral conditions (33). Changes in pH also appear to be important in transport through the exocytotic pathway. In AtT 20 cells, newly synthesized secretory proteins are sorted for either regulated or constitutive secretion (117). When these cells are treated with a weak base (chloroquine), ACTH, for example, is diverted from the regulated to the constitutive pathway, suggesting that a pH-dependent carrier mechanism is involved in sorting. Studies using the DAMP reagent in human fibroblasts have demonstrated that vesicles and cisternae associated with the *trans* Golgi, and containing a constitutively secreted protein, fibronectin, are acidic (114).

B.   Membrane Proteins

While carrier proteins are proposed to mediate the sorting of soluble proteins, there is little information concerning the mechanisms by which these carrier proteins, and other membrane proteins, are sorted. Clathrin-coated membranes involved in R-ME are currently the best-studied situation in which membrane proteins are selectively sorted for transport. Specific receptors, such as the LDL-R, are localized in clathrin-coated pits (11,18),

while others, such as the EGF-R and MHC class 1 antigen and Thy-1,
are excluded (see Section II.A; 18,19). The LDL-R is concentrated at
least 100-fold in coated pits on human fibroblasts (11). Conversely, Thy-1
has a coated pit concentration 100-fold less than that of the surrounding
membrane (16,17). This ability to both concentrate at least 100-fold and
purify 10^4-fold would be sufficient in itself for the sorting required during
vesicular transport.

Although clathrin coats appear to play a role in sorting and transport
at the plasma membrane and at the TGN, there is little evidence that they
have a sorting function elsewhere. The coat structures associated with
transitional elements of the ER and Golgi cisternae do not label with anticlathrin antibodies (77,113), and the vesicles believed to mediate transport
between Golgi cisternae in vitro have a morphological coat which does not
react with anticlathrin antibodies (136). Furthermore, yeast mutants which
lack the gene for clathrin, although slow growing, are viable and able
to correctly sort and transport newly synthesized proteins (137). Potentially
other coat proteins could be involved in sorting.

C.   Role of Aggregation and Conformation in Sorting

Unlike the signal sequences which direct nascent membrane proteins to
the ER, mitochondria, or nucleus, the addresses by which exocytotic and
endocytotic membrane proteins are targetted to different functional sites
remain unclear. Genetic attempts are beginning to identify relevant domains
and sequences involved in targetting (87). For the LDL-R and the polymeric
immunoglobulin receptor (22,138) it is apparent that sequences in the cytoplasmic domain of the polypeptides are important for the endocytotic properties of the receptor. However, for several different systems it has become
apparent that there are specific conformational requirements for transport
of proteins along different pathways, and that modifications to the requisite
conformation inhibit transport (see Section III.A).

Receptor cross-linking is involved in sorting of internalized receptors
and ligands in the endocytotic pathway. As discussed in Section II, following internalization the macrophage Fc-R is sorted to either the recycling
pathway or the lysosomal pathway, depending on the ability of the bound
ligands to aggregate or cross-link the receptors (29). Cross-linking of
the LDL-R or Tf-R with antireceptor antibodies also redirects the receptors
from the recycling pathway to the lysosomes and causes down-regulation
of the LDL-R, MPR, and Tf-R (135,139,140). In each of these cases cross-
linking is mediated by multivalent ligands which do not dissociate from
the receptors at the pH's encountered in the endosomes.

Ligand-induced aggregation may also be important for the internalization
of receptors from the cell surface. Early morphological descriptions of
R-ME suggested that receptors "cluster" into coated pits, that ligands
induce the clustering of randomly distributed receptors, and that receptor
cross-linking is important for association with coated pits (8,141). Higher-
resolution studies have further shown that the Tf-R, on A431 cells, emerges
monodisperse but rapidly aggregates into clusters associated with coated
pits (19). In addition enveloped viruses, which clearly make multiple interactions with cell surface components, are internalized through coated pits
and coated vesicles. When SFV envelope proteins are formed into oligomers
of different valency, the efficiency of internalization reflects their potential
to cross-link receptors (142). For certain cell surface components, cross-
linking with specific antibodies does not lead to their internalization (16,17),
and the relevance of cross-linking for endocytosis remains unclear.

D. Role of Low pH

It is now apparent that sorting is influenced by pH. The role of pH in receptor-ligand interaction and the sorting of soluble proteins has been discussed (see Sections II.D.3, III.C). Acidification provides one mechanism by which ligands, such as LDL, ASGP, lysosomal hydrolases, and iron/ transferrin, can be dissociated from their receptors. Agents that elevate the pH in acidic organelles inhibit dissociation and severely affect receptor-mediated endocytosis and receptor recycling (63,64). Furthermore, these reagents clearly influence the sorting of soluble proteins in the exocytotic pathway (see Section VA). However, the effects of weak bases and iono-phores are not confined to the dissociation of ligands from receptors. In cells treated with monensin, for example, LDL-R's are redistributed from the surface to an intracellular site, presumed to be the endosome (63). Similarly, the transport of newly synthesized membrane proteins in transit from the ER to the cell surface is inhibited (74,103,104).

The mechanisms by which the transport of membrane proteins is inhibited are unclear, but may again involve a perturbation of a required tertiary or quaternary structure. For the EGF-R and ASGP-R, the receptors undergo conformational changes at pH values corresponding to those at which ligands dissociate (143). Thus, in cells treated with weak bases, the internalized receptors may fail to undergo a conformational change required to direct them to appropriate sorting pathways.

VI. CONCLUSIONS

We have outlined the pathways of membrane transport involved in exocytosis and endocytosis. Even though considerable information already exists concerning these pathways and the organelles involved in transport, there is still much to be learned. Many of the vesicular intermediates are ill defined, and the compositions and properties of the organelles involved in transport are poorly characterized. The specific nature of the recognition reactions between transport vesicles and their target membranes still awaits characterization in molecular terms, as do the sorting signals associated with specific proteins. Furthermore, virtually nothing is currently known about the molecular mechanisms involved in the membrane fusion and fission events that make vesicular transport possible. Much of this information will only be derived by isolation of homogeneous populations of the relevant organelles, biochemical analysis of these preparations and the reconstitution of their functions in vitro. The following papers in this section describe a number of the current attempts to understand some of these events in detail.

REFERENCES

1. Farquhar, M. G. (1985). Progress in unravelling pathways of Golgi traffic. *Annu. Rev. Cell Biol.* 1:447-488.
2. Goldstein, J. L., Brown, M. S., Anderson, R. G. W., Russell, D. W., and Schneider, W. J. (1985). Receptor-mediated endocytosis: Concepts emerging from the LDL receptor system. *Annu. Rev. Cell Biol.* 1:1-39.
3. Kornfeld, R., and Kornfeld, S. (1985). Assembly of asparagine-linked oligosaccharides. *Annu. Rev. Biochem.* 54:631-664.

4. Palade, G. E. (1975). Intracellular aspects of the process of protein secretion. *Science* 189:347-358.

5. Simons, K., and Fuller, S. D. (1985). Cell surface polarity in epithelia. *Annu. Rev. Cell Biol.* 1:243-288.

6. Steinman, R. M., Brodie, S. E., and Cohn, Z. A. (1976). Membrane flow during pinocytosis. *J. Cell Biol.* 68:665-687.

7. Brown, M. S., and Goldstein, J. L. (1979). Receptor-mediated endocytosis: insights from the lipoprotein receptor system. *Proc. Natl. Acad. Sci. USA* 76:3330-3337.

8. Pastan, I., and Willingham, M. C. (1985). The pathway of endocytosis. In: *Endocytosis* (I. Pastan and M. C. Willingham, eds.). Plenum Press, New York.

9. Steinman, R. M., Mellman, I. S., Muller, W. A., and Cohn, Z. A. (1983). Endocytosis and the recycling of plasma membrane. *J. Cell Biol.* 96:1-27.

10. Pearse, B. M. F., and Bretscher, M. (1981). Membrane recycling by coated vesicles. *Annu. Rev. Biochem.* 50:85-101.

11. Anderson, R. G. W., Brown, M. S., and Goldstein, J. L. (1977). Role of the coated endocytotic vesicle in the uptake of receptor-bound low density lipoprotein in human fibroblasts. *Cell* 10:351-364.

12. Marsh, M., and Helenius, A. (1980). Adsorptive endocytosis of Semliki Forest virus. *J. Mol. Biol.* 142:439-454.

13. Marsh, M., Matlin, K., Simons, K., Reggio, H., White, J., Kartenbeck, J., and Helenius, A. (1981). Are lysosomes a site of enveloped-virus penetration? *Cold Spring Harbor Symp. Quant. Biol.* 146:835-843.

14. Helenius, A., Kartenbeck, J., Simons, K., and Fries, E. (1980). On the entry of Semliki Forest virus into BHK-21 cells. *J. Cell Biol.* 84:404-420.

15. Rothman, J. E., and Schmid, S. L. (1986). Enzymatic recycling of clathrin from coated vesicles. *Cell* 46:5-9.

16. Bretscher, M. S., and Pearse, B. M. F. (1984). Coated pits in action. *Cell* 38:3-4.

17. Bretscher, M., Thompson, J., and Pearse, B. (1980). Coated pits act as molecular filters. *Proc. Natl. Acad. Sci. USA* 77:4156-4159.

18. Hopkins, C. R. (1985). Coated pits and their role in membrane receptor internalization. In: *Molecular Mechanisms of Transmembrane Signalling* (Cohen and Houslay, eds.). Elsevier, New York, pp. 337-357.

19. Hopkins, C. R. (1985). The appearance and internalization of transferrin receptors at the margins of spreading human tumour cells. *Cell* 40:199-208.

20. Kaplan, J. (1981). Polypeptide-binding membrane receptors: Analysis and classification. *Science* 212:14-20.

20a. Glickman, J. N., Conibear, E., and Pearse, B. M. F. (1989). Specificity of binding of clathrin adaptors to signals on the mannose-6-phosphate/insulin-like growth factor II receptor. *EMBO J.* 8:1041-1047.

21. Davis, C. G., Lehrman, M. A., Russell, D. W., Anderson, R. G. W., Brown, M. S., and Goldstein, J. L. (1986). The J. D. mutation in familial hypercholesterolemia: Amino acid substitution in cytoplasmic domain impedes internalization of LDL receptors. *Cell* 45:15-24.

22. Lehrman, L. A., Goldstein, J. L., Brown, M. S., Russell, D. W., and Schneider, W. J. (1985). Internalization-defective LDL receptors produced by genes with nonsense and frameshift mutations that truncate the cytoplasmic domain. *Cell* 41:735-743.

22a. Lazarovits, J., and Roth, M. (1988). A single amino acid change in the cytoplasmic domain allows the influenza virus hemagglutinin to be endocytosed through coated pits. *Cell* 53:743–752.

23. Geuze, H. J., Slot, J. W., Strous, G. J. A. M., Peppard, J., von Figura, K., Hasilik, A., and Schwartz, A. L. (1984). Intracellular receptor sorting during endocytosis comparative immunoelectron microscopy of multiple receptors in rat liver. *Cell* 37:195–204.

24. Berlin, R. D., Oliver, J. M., and Walter, R. J. (1978). Surface functions during mitosis. 1. Phagocytosis, pinocytosis and mobility of surface bound Con A. *Cell* 15:327–341.

25. Haigler, H. T., McKanna, J. A., and Cohen, S. (1979). Direct visualization of the binding and internalization of a ferritin conjugate of epidermal growth factor in human carcinoma cells A-431. *J. Cell Biol.* 81:382–395.

26. Connolly, J. L., Green, S. A., and Greene, L. A. (1984). Comparison of rapid changes in surface morphology and coated pit formation of PC12 cells in response to nerve growth factor, epiderman growth factor, and dibutyryl cyclic AMP. *J. Cell Biol.* 98:457–465.

26a. Griffiths, G., Back, R., and Marsh, M. (1989). A quantitative analysis of the endocytotic pathway in BHK cells. *J. Cell Biol.* 109:2703–2720.

26b. Braell, W. A. (1987). Fusion between endocytotic vesicles in a cell free system. *Proc. Natl. Acad. Sci. USA* 84:1137–1141.

26c. Warren, G., Woodman, P., Pypaert, M., and Smythe, E. (1988). Cell-free assays and the mechanism of receptor mediated endocytosis. *TIBS* 13:462–465.

27. Harford, J., Bridges, K., Ashwell, G., and Klausner, R. D. (1983). Intracellular dissociation of receptor-bound asialoglycoproteins in cultured hepatocytes. *J. Biol. Chem.* 258:3191–3197.

28. Beguinot, L., Lyall, R. M., Willingham, M. C., and Pastan, I. (1984). Down regulation of the epidermal growth factor receptor in KB cells is due to receptor internalisation and subsequent degradation in lysosomes. *Proc. Natl. Acad. Sci. USA* 81:2384–2388.

29. Ukkonen, P., Lewis, V., Marsh, M., Helenius, A., and Mellman, I. (1986). Transport of macrophage Fc receptors and Fc receptor-bound ligands to lysosomes. *J. Exp. Med.* 163:952–971.

30. Dautry-Varsat, A., Ciechanover, A., and Lodish, H. F. (1983). pH and the recycling of transferrin during receptor-mediated endocytosis. *Proc. Natl. Acad. Sci. USA* 80:2258–2262.

31. Klausner, R. D., Ashwell, G., van Renswoude, J., Harford, J. B., and Bridges, K. R. (1983). Binding of apotransferrin to K562 cells: Explanation of the transferrin cycle. *Proc. Natl. Acad. Sci. USA* 80: 2263–2266.

32. Solari, R., and Kraehenbuhl, J. P. (1984). Biosynthesis of the IgA antibody receptor: A model for the trans-epithelial sorting of a membrane glycoprotein. *Cell* 36:61–71.

33. Abrahamson, D. R., and Rodewald, R. (1981). Evidence for the sorting of endocytotic vesicular contents during the receptor-mediated transport of IgG across the newborn rat intestine. *J. Cell Biol.* 91:270–280.

34. Besterman, J. M., Airhart, T. A., Woodsworth, R. C., and Low, R. B. (1981). Exocytosis of pinocytosed fluid in cultured cells: kinetic evidence for rapid turnover and compartmentation. *J. Cell Biol.* 91:716–727.

35. von Bonsdorff, C. H., Fuller, S. D., and Simons, K. (1985). Apical and basolateral endocytosis in Madin-Darby canine kidney (MDCK) cells grown on nitrocellulose filters. *EMBO J.* 4:2781–2792.

36. Swanson, J. A., Yirinec, B. C., and Silverstein, S. C. (1985). Phorbol esters and horseradish peroxidase stimulate pinocytosis and redirect the flow of pinocytosed fluid in macrophages. *J. Cell Biol.* 100:851-859.

37. Thilo, L., and Vogel, G. (1980). Kinetics of membrane internalization and recycling during pinocytosis in Dictyostelium discoidium. *Proc. Natl. Acad. Sci. USA* 77:1015-1019.

38. Ciechanover, A., Schwartz, A., Dautry-Varsat, A., and Lodish, H. F. (1983). Kinetics of internalization and recycling of transferrin and the transferrin receptor in a human hepatoma cell. *J. Biol. Chem.* 258:9681-9689.

39. Louvard, D. (1980). Apical membrane aminopeptidase appears at site of cell-cell contact in cultures kidney epithelial cells. *Proc. Natl. Acad. Sci. USA* 77:4132-4136.

40. Fishman, J. B., and Cook, J. S. (1986). The sequential transfer of internalized, cell surface sialoglycoconjugates through the lysosomes and Golgi complex in HeLa cells. *J. Biol. Chem.* 261:11896-11905.

41. Burgert, H. G., and Thilo, L. Internalization and recycling of plasma membrane glycoconjugates during pinocytosis in the macrophage cell line P388D$_1$. *Exp. Cell Res.* 144:127-142.

42. Mellman, I. S., Steinman, T. S., Unkeless, J. C., and Cohn, Z. A. (1980). Selective iodination and polypeptide composition of pinocytic vesicles. *J. Cell Biol.* 86:712-722.

43. Watts, C. (1984). In situ ^{125}I-labelling of endosome proteins with lactoperoxidase conjugates. *EMBO J.* 3:1965-1970.

43a. Schmid, S. L., Fuchs, R., Male, P., and Mellman, I. (1988). Two distinct sub-populations of endosomes involved in membrane recycling and transport to lysosomes. *Cell* 52:73-83.

44. Herman, B., and Albertini, D. F. (1984). A time lapse video image intensification analysis of cytoplasmic organelle movements during endosome translocation. *J. Cell Biol.* 98:565-576.

45. Schnapp, B. T., Vale, R. D., Sheetz, M. P., and Reese, T. S. (1985). Single microtubules from squid axoplasm support bidirectional movement of organelles. *Cell* 40:455-462.

46. Geuze, H. J., Slot, J. W., Strous, G. J. A. M., Lodish, H. F., and Schwartz, A. L. (1983). Intracellular site of asialoglycoprotein receptor-ligand uncoupling: Double-label immunoelectron microscopy during receptor-mediated endocytosis. *Cell* 32:277-287.

47. Helenius, A., Mellman, I., Wall, D., and Hubbard, A. (1983). Endosomes. *Trends Biochem. Sci.* 8:245-250.

48. Hopkins, C. R. (1983). Intracellular routing of transferrin and transferrin receptors in epidermoid carcinoma A431 cells. *Cell* 35:321-330.

49. Miller, K., Beardmore, J., Kanety, H., Schlessinger, J., and Hopkins, C. R. (1986). Localization of the epidermal growth factor (EGF) receptor within the endosomes of EGF-stimulated epidermoid carcinoma (A431) cells. *J. Cell Biol.* 102:500-509.

50. Wall, D. A., and Hubbard, A. L. (1981). Galactose-specific recognition system of mammalian liver: Receptor redistribution on the hepatocyte cell surface. *J. Cell Biol.* 90:687-695.

51. Marsh, M., Griffiths, G., Dean, G. E., Mellman, I., and Helenius, A. (1986). Three dimensional structure of endosomes in BHK-21 cells. *Proc. Natl. Acad. Sci. USA* 83:2899-2903.

52. Wall, D. A., and Hubbard, A. L. (1985). Receptor-mediated endocytosis of asialoglycoproteins by rat liver hepatocytes: biochemical characterization of the endosomal compartments. *J. Cell Biol.* 101:2104-2112.

53. Duncan, R., and Pratten, M. K. (1977). Membrane economics in endo-
    cytotic systems. *J. Theor. Biol.* 66:727-735.
54. Hoppe, C. A., Connolly, T. P., and Hubbard, A. L. (1985). Trans-
    cellular transport of polymeric IgA in the rat hepatocytes: biochemical
    and morphological characterization of the transport pathway. *J. Cell
    Biol.* 101:2113-2123.
55. Galloway, C. J., Dean, G. E., Marsh, M., Rudnick, G., and Mellman, I.
    (1983). Acidification of macrophage and fibroblast endocytic vesicles
    in vitro. *Proc. Natl. Acad. Sci. USA* 80:3334-3338.
56. Marsh, M., Bolzau, E., and Helenius, A. (1983). Penetration of Semliki
    Forest virus from acidic prelysosomal vacuoles. *Cell* 32:931-940.
57. Merion, M., and Sly, W. S. (1983). The role of intermediate vesicles
    in the adsorptive endocytosis and transport of ligand to lysosomes
    by human fibroblasts. *J. Cell Biol.* 96:644-650.
58. Tolleshaug, H., Berg, T., Frölich, W., and Norum, K. R. (1979).
    Intracellular localization and degradation of asialoetuin in isolated
    rat hepatocytes. *Biochem. Biophys. Acta* 585:71-84.
59. Baenziger, J. U., and Fiete, D. (1986). Separation of two populations
    of endocytic vesicles involved in receptor-ligand sorting in rat hepato-
    cytes. *J. Biol. Chem.* 261:7445-7454.
60. Dunn, W. A., Hubbard, A. L., and Aronson, N. W. (1980). Low
    temperature selectively inhibits fusion between pinocytic vesicles and
    lysosomes during heterophagy of ^{125}I-asialofetuin by the perfused
    rat liver. *J. Biol. Chem.* 255:5971-5978.
61. Mueller, S. C. and Hubbard, A. L. (1986). Receptor-mediated endo-
    cytosis of asialoglycoproteins by rat hepatocytes: Receptor-positive
    and receptor-negative endosomes. *J. Cell Biol.* 102:932-942.
62. Ohkuma, S., and Poole, B. (1978). Fluorescence probe measurement
    of the intralysosomal pH in living cells and the perturbation of pH
    by various agents. *Proc. Natl. Acad. Sci. USA* 75:3327-3331.
63. Basu, M., Goldstein, J. L., Anderson, R. G. W., and Brown, M. S.
    (1981). Monensin interrupts the recycling of low density lipoprotein
    receptors in human fibroblasts. *Cell* 24:493-502.
64. Gonzalez-Noriega, A., Grubb, J. H., Talkad, V., and Sly, W. S.
    (1980). Chloroquine inhibits lysosomal enzyme pinocytosis and enhances
    lysosomal enzyme secretion by impairing receptor recycling. *J. Cell
    Biol.* 85:839-852.
65. van Renswoude, J., Bridges, K. R., Harford, J. B., and Klausner,
    R. D. (1982). Receptor-mediated endocytosis of transferrin and the
    uptake of Fe in K562 cells: Identification of a nonlysosomal acidic
    compartment. *Proc. Natl. Acad. Sci. USA* 79:6186-6190.
66. Tycko, B., Keith, C. H., and Maxfield, F. R. (1983). Rapid acidifica-
    tion of endocytic vesicles containing asialoglycoprotein in cells of a
    human hepatoma line. *J. Cell Biol.* 97:1762-1776.
67. Tycko, B., and Maxfield, F. R. (1982). Rapid acidification of endocytic
    vesicles containing alpha$_2$ macroglobulin. *Cell* 28:643-651.
68. Marsh, M. (1984). The entry of enveloped viruses into cells by endo-
    cytosis. *Biochem. J.* 218:1-10.
69. Kielian, M., Keranen, S., Kääriäinen, L., and Helenius, A. (1984).
    Membrane fusion mutants of Semliki Forest virus. *J. Cell Biol.* 98:139-
    145.
70. Kielian, M. C., Marsh, M., and Helenius, A. (1986). Kinetics of
    endosome acidification detected by mutant and wild type Semliki Forest
    virus. *EMBO J.* 5:3103-3109.

71. Anderson, R. G. W., Falck, J. R., Goldstein, J. L., and Brown, M. S. (1984). Visualisation of acidic organelles in intact cells by electron microscopy. *Proc. Natl. Acad. Sci. USA* 81:4838-4842.

72. Mellman, I., Fuchs, R., and Helenius, A. (1986). Acidification of the endocytotic and exocytotic pathways. *Annu. Rev. Biochem.* 55:663-700.

73. Merion, M., Schlessinger, J., Brooks, R. M., Moehring, J. M., Moehring, J., and Sly, W. S. (1983). Defective acidification of endosomes in Chinese himster ovary cell mutants "cross-resistant" to toxins and viruses. *Proc. Natl. Acad. Sci. USA* 80:5315-5319.

74. Robbins, A. R., Oliver, C., Bateman, J. L., Krag, S. S., Galloway, C. J., and Mellman, I. S. (1984). A single mutation in Chinese hamster ovary cells impairs both Golgi and endosome functions. *J. Cell Biol.* 99:1296-1308.

75. von Figura, K., and Hasilik, A. (1986). Lysosomal enzymes and their receptors. *Annu. Rev. Biochem.* 55:167-193.

76. Brown, W. J., Goodhouse, J., and Farquhar, M. G. (1986). Mannose-6-phosphate receptors for lysosomal enzymes cycle between the Golgi complex and endosomes. *J. Cell Biol.* 103:1235-1247.

77. Griffiths, G., and Simons, K. (1986). The trans Golgi network: Sorting at the exit site of the Golgi complex. *Science* 234:438-443.

78. Muller, W. A., Steinman, R. M., and Cohn, Z. A. (1980). The membrane proteins of the vacuolar system: II. Bidirectional flow between secondary lysosomes and plasma membrane. *J. Cell Biol.* 86:304-314.

79. Blobel, G., and Dobberstein, B. (1975). Transfer of proteins across membranes. II. Reconstitution of functional rough microsomes from heterologous components. *J. Cell Biol.* 67:852-862.

80. Blobel, G., Walter, P., Chang, C. N., Goldman, B. M., Erickson, A. M., and Lingappa, V. R. (1979). In: *Secretory Mechanisms* (C. R. Hopkins and C. J. Duncan, eds.). Cambridge University Press, Cambridge, England, pp. 9-36.

81. Hubbard, S. C., and Ivatt, R. J. (1981). Synthesis and processing of asparagine-linked oligosaccharides. *Annu. Rev. Biochem.* 50:555-583.

82. Palade, G. E. (1982). Problems in intracellular membrane traffic. *CIBA Found. Symp.* 92:1-14.

83. Fries, E., Gustafson, L., and Peterson, P. A. (1984). Four secretory proteins synthesized by hepatocytes are transported from endoplasmic reticulum to Golgi complex at different rates. *EMBO J.* 3:147-152.

84. Ledford, B. E., and Davis, D. F. (1983). Kinetics of serum protein secretion by cultured hepatoma cells: Evidence for multiple secretory pathways. *J. Biol. Chem.* 258:3304-3308.

85. Lodish, H. F., Kong, N., Snider, M., and Strous, G. J. A. M. (1983). Hepatoma secretory proteins migrate from rough endoplasmic reticulum to Golgi at characteristic rates. *Nature* 304:80-83.

86. Cardelli, J. A., Golumbeski, G. S., and Dimond, R. I. (1986). Lysosomal enzymes in Dictyostelium discoideum are transported to lysosomes at distinctly different rates. *J. Cell Biol.* 102:1264-1270.

87. Garoff, H. (1985). Using recombinant DNA techniques to study protein targeting in the eukaryotic cell. *Annu. Rev. Cell Biol.* 1:403-445.

88. Gething, M-J., McCammon, K., and Sambrook, J. (1986). Expression of wild-type and mutant forms of influenza hemagglutinin: The role of folding in intracellular transport. *Cell* 46:939-950.

89. Wilson, I. A., Skehel, J. J., and Wiley, D. C. (1981). Structure of the haemagglutinin membrane glycoprotein of influenza virus at 3Å resolution. *Nature* 289:366-373.

90. Copeland, C. S., Doms, R. W., Bolzau, E. M., Webster, R. G., and Helenius, A. (1986). Assembly of influenza hemagglutinin trimers and its role in intracellular transport. *J. Cell Biol.* 103:1179-1191.

91. Tarentino, A. L., and Maley, F. (1974). Purification and properties of an endo-B-acetylglucosaminidase from *S. griseus*. *Mol. Cell Biol.* 4:688-694.

92. Bole, D. G., Hendershot, L. M., and Kearney, J. F. (1986). Post translational association of immunoglobulin heavy chain binding protein with nascent heavy chains in non-secreting and secreting hybridomas. *J. Cell Biol.* 102:1558-1566.

93. Pelham, H. R. B. (1986). Speculations on the functions of the major heat shock and glucose regulated proteins. *Cell* 46:959-961.

94. Kreis, T. E., and Lodish, H. F. (1986). Oligomerization is essential for transport of vesicular stomatitis viral glycoprotein to the cell surface. *Cell* 46:929-937.

95. Owen, M. J., Kissonerglus, A. M., and Lodish, H. F. (1980). Biosynthesis of HLA-A and HLA-B antigens in vivo. *J. Biol. Chem.* 255:9678-9684.

96. Sege, K., Rask, L, and Peterson, P. A. (1981). Role of B2-microglobulin in the intracellular processing of HLA antigens. *Biochemistry* 20:4523-4530.

97. Rose, J. K., and Doms, R. W. (1988). Regulation of protein export from the ER. *Annu. Rev. Cell Biol.* 4:254-288.

98. Williams, D. B., Swiedler, S. J., and Hart, G. W. (1985). Intracellular transport of membrane glycoproteins: Two closely related histocompatibility antigens differ in their rates of transit to the cell surface. *J. Cell Biol.* 101:725-734.

99. Quinn, P., Griffiths, G., and Warren, G. (1984). Density of newly synthesized plasma membrane proteins in intracellular membranes. *J. Cell Biol.* 98:2142-2147.

100. Griffiths, G., Brands, R., Burke, B., Louvard, D., and Warren, G. (1982). Viral membrane proteins acquire galactose in *trans* Golgi cisternae during intracellular transport. *J. Cell Biol.* 95:781-792.

101. Saraste, J., and Kuisman, E. (1984). Pre- and post-Golgi vacuoles operate in the transport of Semliki Forest virus membrane glycoproteins to the cell surface. *Cell* 38:535-549.

102. Dunphy, W. G., Fries, E., Urbani, L. J., and Rothman, J. E. (1981). Early and late functions associated with the Golgi complex reside in different compartments. *Proc. Natl. Acad. Sci. USA* 78:7453-7457.

103. Griffiths, G., Quinn, P., and Warren, G. (1983). Dissection of the Golgi complex. I. *J. Cell Biol.* 96:835-850.

104. Quinn, P., Griffiths, G., and Warren, G. (1983). Dissection of the Golgi complex. II. *J. Cell Biol.* 96:851-856.

105. Dunphy, W. G., Brands, R., and Rothman, J. E. (1985). Attachment of terminal N-acetylglucosamine to asparagine linked oligosaccharides occurs in central cisternae of the Golgi stack. *Cell* 40:463-472.

106. Roth, J., Lentze, M. J., and Berger, E. G. (1985). Immunocytochemical demonstration of ecto-galactosyl transferase in adsorptive intestinal cells. *J. Cell Biol.* 100:118-125.

107. Bergman, J. E., Tokuyasu, K. T., and Singer, S. J. (1981). Passage of an integral membrane protein, the vesicular stomatitis virus glycoprotein, through the Golgi apparatus en route to the plasma membrane. *Proc. Natl. Acad. Sci. USA* 78:1746-1750.

108. Griffiths, G., Pfeiffer, S., Simons, K., and Matlin, K. (1985). Exit of newly synthesised membrane proteins from the trans cisterna of the Golgi complex to the plasma membrane. *J. Cell Biol.* 101:949-964.

109. Matlin, K. S., and Simons, K. (1984). Sorting of a plasma membrane glycoprotein occurs before it reaches the cell surface in cultured epithelial cells. *J. Cell Biol.* 99:2131-2139.

110. Pfeiffer, S., Fuller, S. D., and Simons, K. (1985). Intracellular sorting and basolateral appearance of the G-protein of vesicular stomatitus virus in Madin-Darby canine kidney cells. *J. Cell Biol.* 101:470-476.

111. Rodriguez-Boulin, E. J., and Pendergast, M. (1980). Polarised distribution of viral envelope proteins in the plasma membrane of infected epithelial cells. *Cell* 20:45-54.

112. Kelly, R. B. (1985). Pathways of protein secretion in eukaryotes. *Science* 230:25-32.

113. Orci, L., Halban, P., Amherdt, M., Ravazzola, M., Vassalli, J. D., and Perrellet, A. (1984). A clathrin coated, Golgi related compartment of the insulin secreting cell accumulates proinsulin in the presence of monensin. *Cell* 39:39-47.

114. Anderson, R. G. W., and Pathak, R. K. (1985). Vesicles and cisternae in the *trans* Golgi apparatus of human fibroblasts are acidic compartments. *Cell* 40:635-643.

115. Schwartz, A. L., Strous, G. J. A. M., Slot, J. W., and Geuze, H. J. (1985). Immuno-electron micrographic localisation of acidic intracellular compartments in hepatoma cells. *EMBO J.* 4:899-940.

116. Glickman, J., Croen, K., Kelly, S., and Al Awquati, Q. (1983). Golgi membranes contain an electrogenic H^+ pump in parallel to a chloride conductance. *J. Cell Biol.* 97:1303-1308.

117. Moore, H. P., Gumbiner, B., and Kelly, R. B. (1983). Chloroquine diverts ACTH from a regulated to a constitutive secretory pathway in AtT-20 cells. *Nature* 302:434-436.

118. Oda, K., and Ikehara, Y. (1985). Weakly basic amines inhibit the proteolytic conversion of proalbumin to serum albumin in cultured rat hepatocytes. *Eur. J. Biochem.* 152:605-609.

119. Vladutin, G. D., and Rattazzi, H. C. (1980). The effect of monensin on hexosaminidase transport in normal and T cell fibroblasts. *Biochem. J.* 192:813-820.

120. Matlin, K. S., and Simons, K. (1983). Reduced temperature prevents transfer of a membrane glycoprotein to the cell surface but does not prevent terminal glycosylation. *Cell* 34:233-243.

121. Bonifacino, J. S., Perez, P., Klausner, R. D., and Sandoval, I. V. (1986). Study of the transit of an integral membrane protein from secretory granules through the plasma membrane of secreting rat basophilic leukaemia cells using a specific monoclonal antibody. *J. Cell Biol.* 102:516-522.

122. Patzak, A., and Winkler, H. (1986). Exocytotic exposure and recycling of membrane antigens of chromaffin granules: Ultrastructural evaluation after immunolabelling. *J. Cell Biol.* 102:510-515.

123. Yamashiro, D. J., Tycho, B., Fluss, S., and Maxfield, F. R. (1984). Segregation of transferrin to a mildly acidic (pH 6.5) para-Golgi compartment in the recycling pathway. *Cell* 37:789-800.

124. Van Deurs, B., Tønnessen, T. I., Peterson, O. W., Sandvig, K., and Olsnes, S. (1986). Routing of internalized ricin conjugates to the Golgi complex. *J. Cell Biol.* 102:37-47.

125. Gonatos, N. K., Steiber, W. F., Hickey, S. H., and Gonatos, J. O. (1984). Endosomes and Golgi vesicles in adsorptive and fluid phase endocytosis. *J. Cell Biol.* 99:909-917.

126. Woods, J. W., Doriaux, M., and Farquhar, M. G. (1986). Transferrin receptors recycle to cis and middle as well as *trans* Golgi cisternae in Ig-secreting myeloma cells. *J. Cell Biol.* 103:277-286.

127. Snider, M. D., and Rogers, O. C. (1985). Intracellular movement of cell surface receptors after endocytosis: resialylation of asialotransferrin receptor in human erythroleukemia cells. *J. Cell Biol.* 100:826-834.

128. Snider, M. D., and Rogers, O. (1986). Membrane traffic in animal cells: Cellular glycoproteins return to the site of Golgi manosidase 1. *J. Cell Biol.* 103:265-275.

129. Bleil, J. D., and Bretscher, M. S. (1982). Transferrin receptor and its recycling in HeLa cells. *EMBO J.* 1:351-355.

130. Watts, C. (1985). Rapid endocytosis of the transferrin receptor in the absence of bound transferrin. *J. Cell Biol.* 100:633-637.

131. Pesonen, M., and Simons, K. (1984). Transcytosis of the G-protein of vesicular stomatitis virus after implantation into the apical plasma membrane of Madin-Darby canine kidney cells. (1) Involvement of endosomes and lysosomes. *J. Cell Biol.* 99:796-802.

132. Fuller, S. D., and Simons, K. (1986). Transferrin receptor polarity and recycling accuracy in "tight" and "leaky" strains of Madin-Darby canine kidney cells. *J. Cell Biol.* 103:1767-1781.

133. Brands, R., Snider, M. D., Hino, Y., Park, S. S., Gilboin, H. V., and Rothman, J. E. (1985). The retention of membrane proteins by the endoplasmic reticulum. *J. Cell Biol.* 101:1724-1732.

133a. Munro, S., and Pelham, H. R. B. (1987). A c-terminal signal prevents secretion of luminal ER proteins. *Cell* 48:899-907.

134. Balch, W. E., Dunphy, W. G., Braell, W. A., and Rothman, J. E. (1984). Reconstitution of the transport of protein between successive compartments of the Golgi measured by the coupled incorporation of *N*-acetyl glucosamine. *Cell* 39:405-416.

135. von Figura, K., Gieselman, V., and Hasilik, A. (1984). Antibody to mannose-6-phosphate specific receptor induces receptor deficiency in human fibroblasts. *EMBO J.* 3:1281-1282.

136. Orci, L., Glick, B. S., and Rothman, J. E. (1986). A new type of coated vesicular carrier that appears not to contain clathrin: Its possible role in protein transport within the Golgi stack. *Cell* 46:171-184.

137. Payne, G. S., and Schekman, M. (1985). A test of clathrin function in protein secretion and cell growth. *Science* 230:1009-1014.

138. Mostov, K. E., de Bruyn Kops, A., and Deitcher, D. L. (1986). Deletion of the cytoplasmic domain of the polymeric immunoglobulin receptor prevents basolateral localization and endocytosis. *Cell* 47:359-364.

139. Anderson, R. G. W., Brown, M. S., Beisiegel, U., and Goldstein, J. L. (1982). Surface distribution and recycling of the low density lipoprotein receptor as visualized with anti-receptor antibodies. *J. Cell Biol.* 93:523-531.

140. Ekblom, P., Thesleff, I., Lehto, V-P., and Virtanen, I. (1983). Distribution of the transferrin receptor in normal human fibroblasts and fibrocarcinoma cells. *Int. J. Cancer* 31:111-120.

141. Maxfield, F. R., Schlessinger, Y., Schechter, I., Pastan, I., and Willingham, M. C. (1983). Collection of insulin, EGF and alpha$_2$ macroglobin in the same patches on the surface of cultured fibroblasts and common internalisation. *Cell* 14:805-810.

142. Marsh, M., Bolzau, E., White, J., and Helenius, A. (1983). Interactions of Semliki Forest virus spike glycoprotein rosettes and vesicles with cultured cells. *J. Cell Biol.* 96:455-461.

143. Di Paola, M., and Maxfield, F. R. (1984). Conformational changes in the receptors for epidermal growth factor and asialoglycoproteins induced by the mildly acidic pH found in endocytic vesicles. *J. Biol. Chem.* 259:9163-9171.

144. Griffiths, G., Warren, G., Quinn, P., Mathieu-Costello, O., and Hoppeler, H. (1984). Density of newly synthesised plasma membrane protein in intracellular membranes. I. Stereological studies. *J. Cell Biol.* 98:2133-2141.

20

Cell-Free Systems for Studying the Pathway of Receptor-Mediated Endocytosis

JOHN DAVEY* and GRAHAM BARRY WARREN†
University of Dundee, Dundee, Scotland

I. INTRODUCTION

Endocytosis is the process by which cells internalize extracellular molecules by enclosing them within inward foldings of the plasma membrane that seal to form intracellular vesicles (1). Some molecules are selectively concentrated for uptake by first binding to specific receptors at the plasma membrane, a process known as receptor-mediated endocytosis. Probably the best-studied example of such a molecule is low-density lipoprotein (LDL), which delivers cholesterol to cells (2) and can be used to illustrate the steps on the endocytic pathway (Fig. 1). LDL binds to specific, high-affinity receptors located in regions of the plasma membrane called coated pits, indentations of the cell surface possessing a characteristic cytoplasmic coat. The pit invaginates and pinches off to form a coated vesicle. These uncoat, the coat subunits apparently returning to the plasma membrane and the uncoated, or partially uncoated, vesicle fusing with a prelysosomal compartment called the endosome. The acidic environment within this compartment causes the LDL to dissociate from its receptor, and while the receptor is recycled back to the plasma membrane, the LDL is delivered to lysosomes where degradation and the release of cholesterol occurs.

Our understanding of this process at the molecular level is incomplete; although a few of the proteins involved have been characterized in some detail, there must be many others that remain to be identified and characterized. As examples, consider the various receptors at the plasma membrane. Some, such as the LDL receptor, are concentrated in coated pits in the absence of their ligand, while others, such as the EGF receptor, only concentrate there after binding their ligand. Several of these receptors have now been cloned and sequenced. Recent work (3) suggests that their

Present affiliations:
*University of Birmingham, Birmingham, England.
†Imperial Cancer Research Fund, London, England.

449

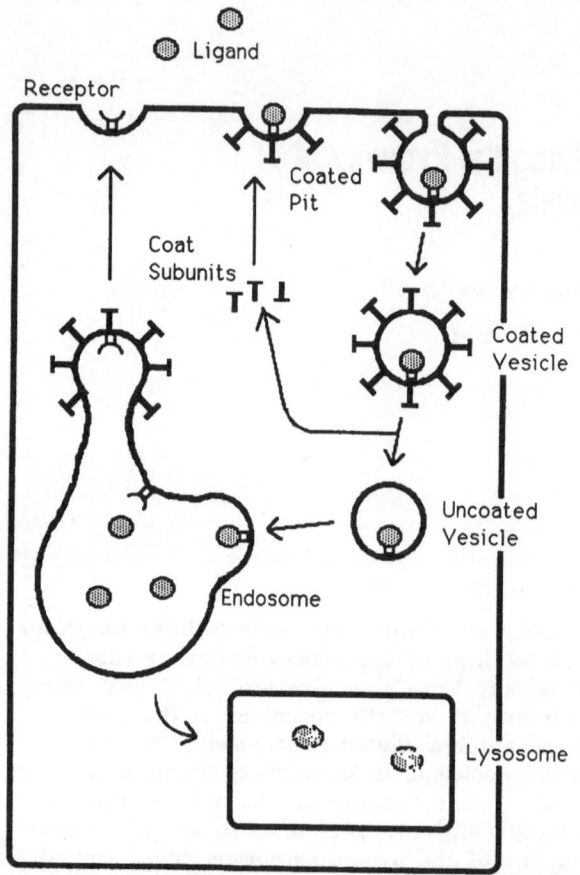

FIGURE 1   The pathway of receptor-mediated endocytosis.

cytoplasmic tails are important in their interaction with the coated pit.
In addition to concentrating the receptors, the coated pit must bud and
seal to form a coated vesicle. It is not known which, if any, of the presently characterized components of the coat structure are responsible for
these various steps. Once the vesicles have been uncoated, they must
recognize and fuse with the next compartment on the pathway. This fusion
must be specific and rigorously controlled.

   Analysis of the endocytic mechanism requires the development of cell-
free assays that faithfully reconstitute the steps on the pathway as they
occur in the intact cell. By effectively removing the plasma membrane,
it becomes possible to manipulate the environment in the immediate neighbor
hood of the step being studied, and this should lead to identification,
purification, and characterization of the factors controlling each step.
Rothman and his colleagues have used this approach to purify the enzyme
responsible for the uncoating of coated vesicles (4), while the recent reassembly of coated vesicles in a cell-free system (5) could give molecular
insights into the functions associated with the coat structure.

   We have previously described a cell-free assay to monitor the fusion
of uncoated endocytic vesicles with endosomes (6), and rather than merely
reproduce those results here, what we have done is describe the results

in the context of the thoughts that guided the development of the assay. Our aim is to encourage more people to develop cell-free assays for studying other parts of the endocytotic pathway and for studying vesicle fusion in general.

II. DESIGNING THE ASSAY

Good design is the key to a successful assay. Careful planning at this stage saves both time and money later. The basic principle involves taking two substances that interact (such as an enzyme and its substrate or an antibody and its antigen) and so arrange conditions that one is present in one compartment and the other in the next compartment on the endocytotic pathway. Since the two substances can only interact once the two compartments have fused, the interaction is an indirect measure of the fusion event. In choosing which two substances to use, the following should be considered:

1. The uptake pathway for both substances must be well characterized so that the substances can be precisely positioned.
2. Both substances must use the same pathway if all the steps are to be eventually accessible.
3. The interaction must be quick to measure so that the assay will be useful for purifying the proteins involved in the fusion event.
4. The substances used should preferably offer the potential of a morphological study to complement the biochemical one.
5. Preparation of the assay components should be quick and inexpensive.

Our present assay satisfies all but one of these points. We chose to use enveloped viruses because the uptake of several of them had been characterized extensively in studies that led to the elucidation of their infection pathway (see Ref. 7). The common pathway leading to infection is illustrated schematically in Figure 2. The pathway for virus uptake is the same as that for LDL up to the endosome. The low pH in this compartment converts the viral spike proteins to potent fusogens which cause a rapid and irreversible fusion of the viral envelope membrane with the endosomal membrane, thereby expelling the viral genome into the cytoplasm and initiating the process of infection.

The interaction we chose was that between an enzyme and its substrate. The enzyme is provided by fowl plague virus (FPV), an animal virus that has two kinds of glycoproteins inserted into its lipid envelope. The major species is hemagglutinin, which is involved in attaching the virus to the target cell, while the other is neuraminidase, the role of which is unclear. The substrate is provided by Semliki Forest virus (SFV), another enveloped animal virus. Glycoproteins inserted into the lipid envelope of this virus possess terminal sialic acid residues and these can be radioactively labeled by growing the virus in the presence of ^{3}H-N-acetylmannosamine, a specific precursor of sialic acid (8). Interaction between the two viruses allows the neuraminidase of the FPV to remove the tritium-labeled sialic acid residues (^{3}H-SA) from the SFV. Since the released ^{3}H-SA is TCA-soluble while the ^{3}H-SA bound to the SFV is TCA-precipitable, the interaction can be followed very simply by the appearance of ^{3}H-SA in the TCA-soluble fraction.

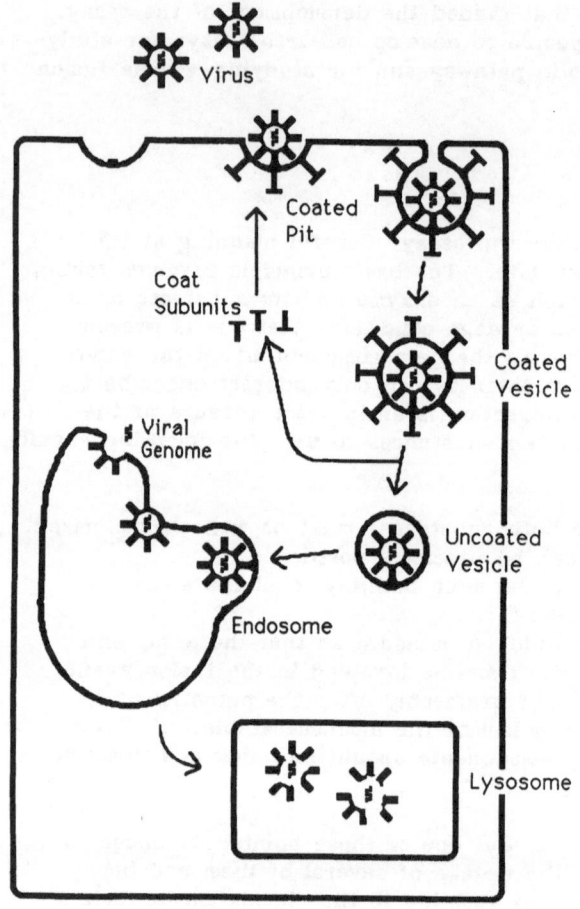

FIGURE 2 The pathway leading to virus infection.

To measure the fusion between endocytotic vesicles and endosomes
we put the enzyme into the endocytotic vesicles and the substrate into
the endosomes. This arrangement was chosen since the delivery of a single
enzyme into the compartment containing the substrate will convert all of
that substrate to product, thus giving a built-in amplification and increasing
the sensitivity of the assay.

Two sets of membranes were prepared: donor membranes containing
the enzyme in the endocytotic vesicles and acceptor membranes containing
the substrate in endosomes. To put the FPV into encytotic vesicles we
exploited observations made by Matlin et al. (9) showing that a 15-min
incubation at 37°C causes the majority of cell-bound FPV to become internalized as far as endocytotic vesicles. After quenching at 4°C to stop further
transport the cells were homogenized under isoosmotic conditions and a
postnuclear supernatant prepared from the homogenate was stored in aliquots
in liquid nitrogen. SFV possessing ^{3}H-SA (^{3}H-SFV) was transferred quantitatively to endosomes by incubating cells for 2 h at 20°C, which stops
transfer from endosomes to lysosomes (10), and then removing remaining

surface SFV with protease. Postnuclear supernatants were prepared as above. Mixing of the donor and acceptor membranes allows fusion between endocytotic vesicles containing FPV and endosomes vesicles containing ³H-SFV causing the release of ³H-SA. A schematic outline of the assay is presented in Figure 3.

The separation of the assay into two sets of membranes serves two functions. First, it ensures that a truly cell-free event is being measured since fusion can only occur if the plasma membrane of both sets of cells has been ruptured. Second, it eliminates the need to perform prior purification of the vesicles and the endosomes since the two sets of membranes complement each other.

DONOR MEMBRANES

1. BIND FPV TO CELLS AT 4°C
2. INTERNALIZE AT 37°C FOR 15 MIN
3. QUENCH TO 4°C

ACCEPTOR MEMBRANES

1. BIND ³H-SIALIC ACID-SFV TO CELLS AT 4°C
2. INTERNALIZE AT 20°C FOR 2 H
3. REMOVE SURFACE SFV USING PROTEASE AT 4°C

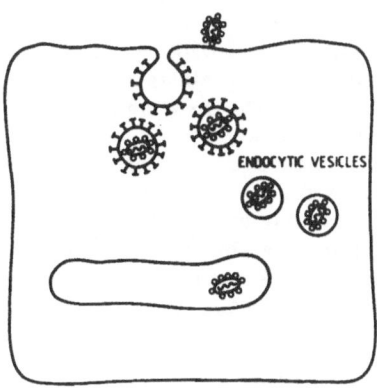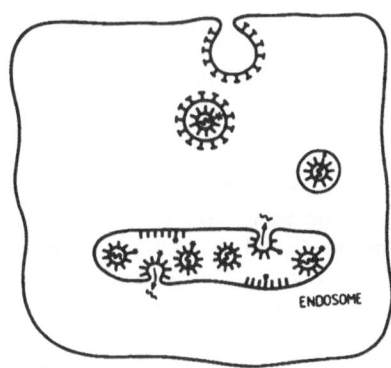

4. PREPARE POST NUCLEAR SUPERNATANTS

5. MIX DONOR AND ACCEPTOR MEMBRANES IN THE PRESENCE OF AN ENERGY SOURCE

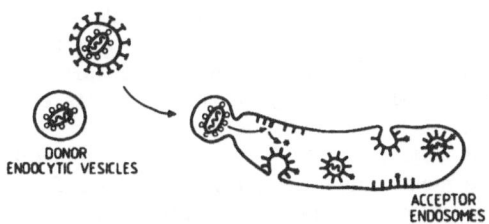

6. MONITOR RELEASE OF ³H-SIALIC ACID (⁕)

FIGURE 3 Outline of the cell-free assay to measure an endocytotic fusion event.

III. INITIAL EXPERIMENTS

For the standard assay, acceptor membranes were mixed with either standard donor membranes or mock donor membranes both in the presence and
in the absence of ATP. Mock donor membranes were prepared in the same
way as standard donor membranes, except that no FPV was added to the
cells, and were used to estimate the release of ^{3}H-SA by cellular enzymes.
Since we felt it was reasonable to assume that the fusion would require
energy, incubations were done in the presence and absence of ATP. The
presence of ATP was maintained by including a regenerating system of
creatine phosphate and creatine phospholinase, while a hexokinase/glucose-
based depleting system ensured its absence. From early experiments it
became clear that the homogenization procedure ruptured many vesicles
and released their viruses. Since the ^{3}H-SA produced by the interaction
of these free viruses masked the release of the ^{3}H-SA resulting from
vesicle fusion, it was necessary for us to inhibit any cleavage reactions
occurring outside sealed vesicles. The compound 2,3-dehydro-2-deoxy-N-
acetylneuraminic acid (DDNA) is a sialic acid analog which competitively
inhibits neuraminidases from a variety of sources (11), including FPV.
It is a small, acidic molecule (M_r = 290), which would be expected to
diffuse easily throughout the incubation mixture but not across membranes.
A typical example of the results obtained when sufficient DDNA was included
to inhibit all of the neuraminidase activity outside sealed vesicles is shown
in Figure 4A. Subtracting the release of ^{3}H-SA due to the FPV neuraminidase
seen with the mock donor from that seen with the standard donor produces
the third pair of columns in Figure 4A. Since the cleavage reaction itself
is not influenced by ATP, the simplest explanation for the greater release
seen in the presence of ATP than in its absence is that the ATP allows
the vesicles containing the viruses to fuse. Support for this suggestion
was the finding that the addition of detergent (Triton X-100) to the incubation mixture abolished the ATP-dependent release (Fig. 4A). The detergent
presumably lyses the vesicles and allows the DDNA to gain access to,
and inhibit, the cleavage reaction.

The above discussion involves subtraction of the releases seen with
the mock-treated donor membranes from those seen with the standard donor.
Such a subtraction assumes not only that the two donor preparations are
identical except for the presence or absence of the FPV, but also that
the various cleavage reactions are additive and there is no competition
between the endogenous neuraminidases and the FPV neuraminidase for
the ^{3}H-SFV. Whereas differences between the two donor preparations were
minimized by producing both on the same day from identical sets of cells,
the second assumption was more troublesome. Without knowing the nature
of the endogenous release it was not possible to know whether it competed
with the FPV, and we decided therefore to try to minimize this endogenous
release. Without knowing its cause it was not possible to know how to
inhibit it, and purely empirical means had to be used. Several modifications
to the method of preparing the membranes were tried, and we found that
the most successful single change with respect to emphasizing the release
due to fusion was to increase the level of homogenization. The results
of these changes are shown in Figure 4B. The endogenous release was
greatly reduced, while the release due to the FPV was emphasized. In
the early experiment (Fig. 4A) the release due to the fusion of vesicles
containing FPV represented only 20% of the total release seen with the
standard donor in the presence of ATP, but in the later experiment (Fig. 4B)
it represented 75% of the total release.

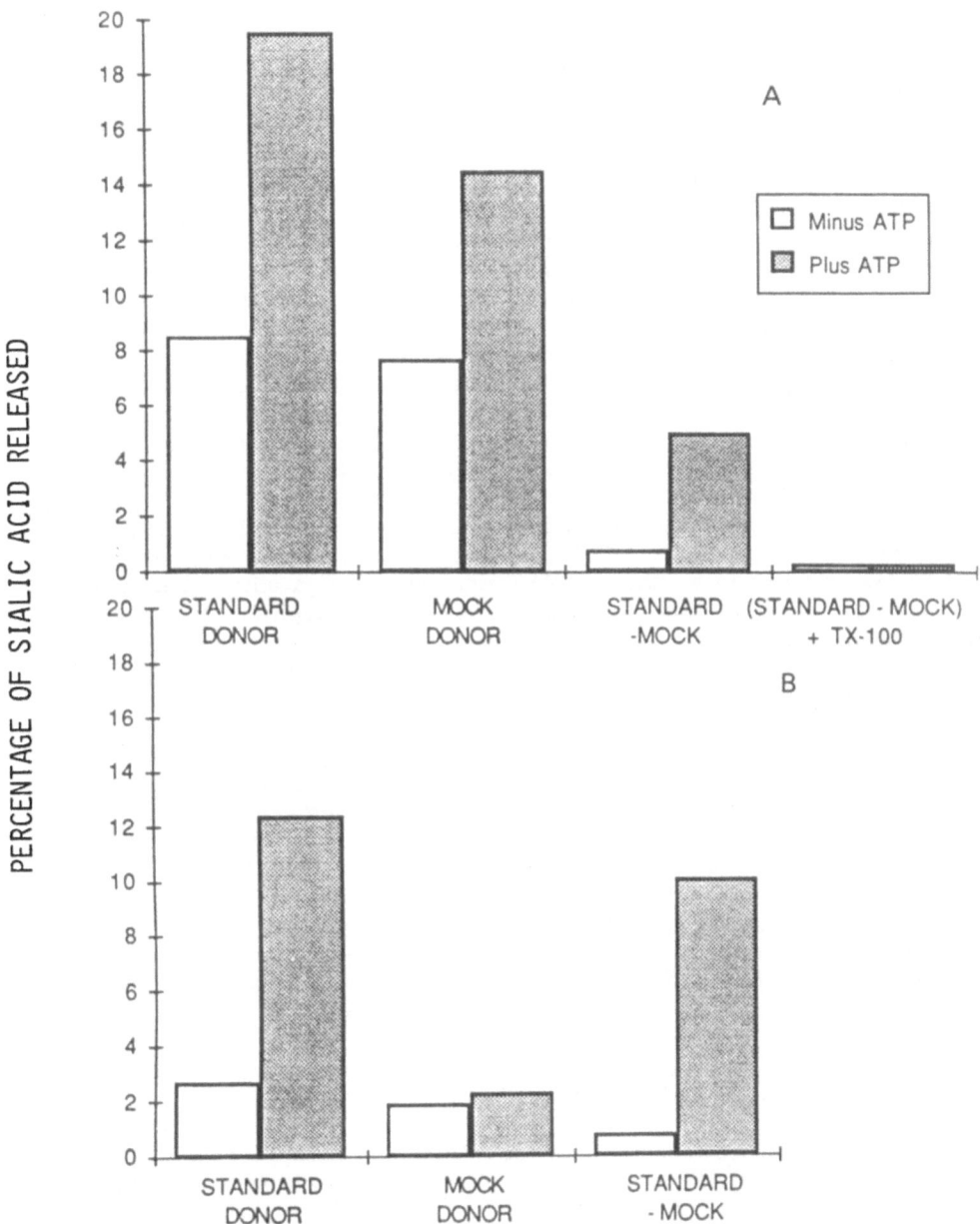

FIGURE 4 Typical results of the standard assay. Standard acceptor membranes were incubated with either standard donor membranes or mock-treated
donor membranes in the presence of an ATP-depleting system or an ATP-
regenerating system. The third pair of columns shows the differences
in the releases observed with the different donor membranes and represents
the release due to the FPV. All incubations contained DDNA and were
for 2 h at 37°C. The membrane preparations used in A were typical of
those produced early in the development of the assay, while those in B
were homogenized to a greater level to inhibit the endogenous releases.
The final pair of columns in A shows the result of performing the assay
in the presence of 0.1% (w/v) Triton X-100.

IV. PHYSIOLOGICAL RELEVANCE

Even though the conditions in the cell-free system were chosen to approximate those in the intact cell, it was still necessary to establish the physiological relevance of the fusion event being studied. A common criticism leveled at this type of assay is that the fusion is nonspecific and that the vesicles in the mixture merely aggregate and fuse. It is therefore necessary to provide evidence to the contrary. For us this involved preparing donor and acceptor membranes from cells in which the viruses had been positioned at different locations along the endocytotic pathway. In addition to the standard conditions, membranes were prepared from cells possessing surface-bound viruses and from cells in which the ciruses had been internalized into lysosomes. Figure 5 shows the results of incubating the different donor and acceptor membranes in the fusion assay. Acceptor membranes prepared from cells with surface-bound ^{3}H-SFV gave low release irrespective of the location of the FPV, while with the standard acceptor membranes the FPV was only capable of releasing ^{3}H-SA when it had been internalized in the standard way.

Acceptor membranes prepared from cells in which the ^{3}H-SA-SFV had been internalized for 2 h at 37°C were not assayed because msot of the ^{3}H-SA was already in a TCA-soluble form, even though chloroquine was used to try to inhibit the lysosomal hydrolases. Although the lysosomal degradation of FPV was almost completely inhibited by chloroquine, it seems that SFV, and particularly its sialic acid residues, are especially sensitive to lysosomal hydrolases.

The conclusion that only specific vesicles containing the viruses can fuse in the assay is only valid if the donor and acceptor membranes other than the standard preparations contain viruses within sealed vesicles. It is conceivable, for example, that lysosomes containing FPV could act as donors but that they are especially sensitive to lysis and were all broken during the homogenization. The DDNA would then inhibit the FPV neuraminidase. We found, however, that in each of the donor and acceptor preparations used in Figure 5, between 15 and 20% of the viruses were present within sealed vesicles. Overall, the data show that only sealed vesicles derived from certain positions along the endocytotic pathway can participate in the fusion event.

V. PARAMETERS OF THE ASSAY

Two of the properties to be determined with a new assay are its efficiency and time dependence. To calculate the efficiency of some assays, including ours, it is first necessary to determine how much of the sample is capable of participating in the assay. With our system, ^{3}H-SFV not present within sealed vesicles at the start of the incubation cannot participate in the fusion reaction and cannot therefore be considered in calculations concerning efficiency.

The percentage of the total ^{3}H-SFV that was present within sealed vesicles in a standard acceptor preparation was determined using bacterial neuraminidase in the presence and absence of detergent and was about 16-17%. Since nearly 12% of the total ^{3}H-SA in this same acceptor preparation was released owing to the fusion of vesicles containing viruses, this means that over 70% of the ^{3}H-SA-SFV present within sealed vesicles at

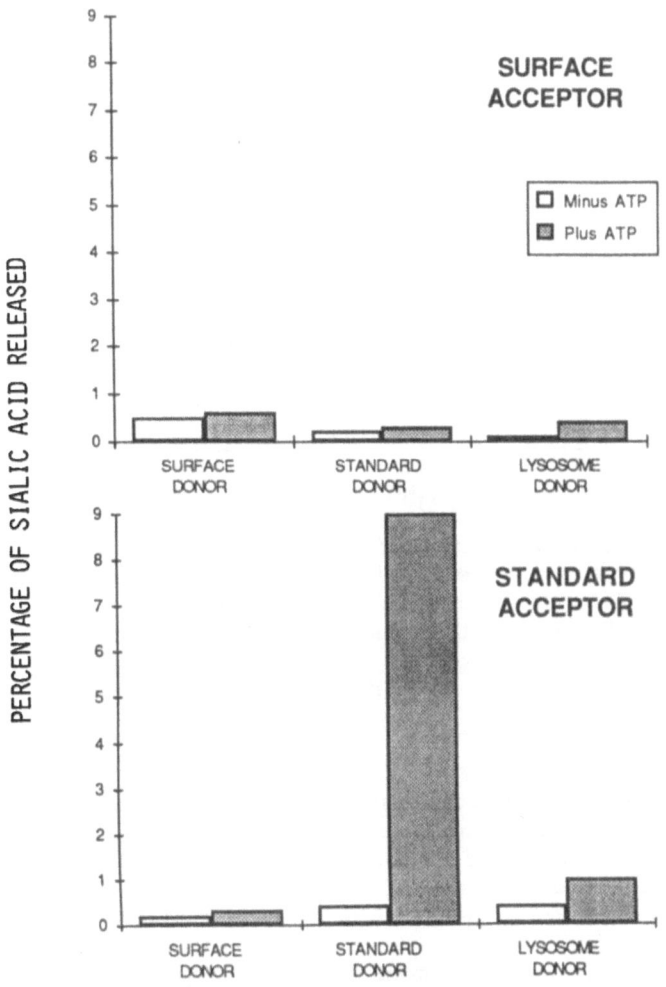

FIGURE 5 Only certain closed vesicles fuse. In addition to the standard
internalization conditions membranes were prepared from cells with surface-
bound viruses and from cells in which the viruses had been internalized
into lysosomes. Mock-treated donor membranes were used to estimate the
release not involving FPV, and only the release due to the FPV neuraminidase is shown. Incubations were for 2 h at 37°C and contained DDNA.
The membranes were prepared and assayed in 0.1 mM chloroquine to try
to inhibit lysosomal hydrolases.

the start of the assay was cleaved as a result of those vesicles fusing
with vesicles containing FPV. The assay is therefore very efficient.

   The time dependence of the release of ^{3}H-SA resulting from the fusion
of vesicles containing FPV with vesicles containing ^{3}H-SA-SFV is shown
in Figure 6. The release is almost complete by 2 h and has a half-time
of about 35 min. This is slower than the fusion in the intact cell and
illustrates two important points regarding cell-free assays. First, the conversion from a cellular to a cell-free system usually involves a dilution

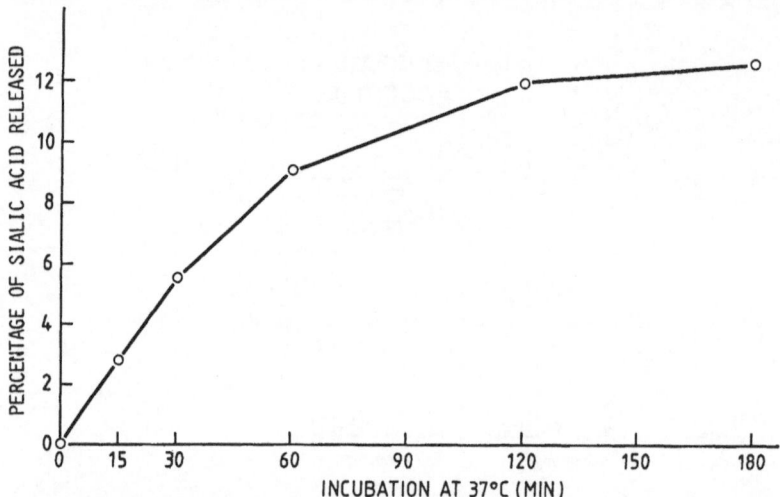

FIGURE 6 Time-dependence of the release of ^{3}H-SA following the fusion of vesicles containing FPV and vesicles containing ^{3}H-SFV. Standard donor and acceptor membranes were incubated at 37°C with DDNA and either an ATP-depleting or an ATP-regenerating system. Mock-treated donor membranes were used to estimate the release not involving FPV. The time course shows the release of H-SA by a reaction requiring both FPV and ATP.

of the components and this might decrease the rate of reaction, and second, assays often measure events that are secondary to the event of interest. In our assay we monitor the release of ^{3}H-SA as a consequence of fusion, we do not measure the fusion event itself. It is possible that fusion is more rapid and that there is then a delay before the neuraminidase cleaves the ^{3}H-SA from the ^{3}H-SFV.

It would be preferable if biochemical studies could be supported by morphological studies, with either the light or electron microscope being used to visually monitor the fusion event. Ideally, samples taken at the start of the incubation would show the substances in different vesicles while samples taken at the end would show vesicles containing both substances. For this approach it must be possible to identify both substances A and B, and in this respect our use of FPV and SFV is advantageous since both are distinguishable by electron microscopy.

VI. USE OF THE ASSAY

The fusion of two vesicles is likely to be controlled at two levels, first at the level of the initial recognition process between the vesicles and then at the level of the fusion event itself. Identification of the factors that control these events presents the most interesting use of cell-free assays. Preliminary data, presented in Figures 7 and 8, suggest that the fusion event in our assay requires the presence of both cytosolic factors and proteins present on the cytoplasmic face of the acceptor vesicles, both trypsinized membranes and membranes without cytosol being incapable of fusing. A possible approach to identify the membrane-bound proteins

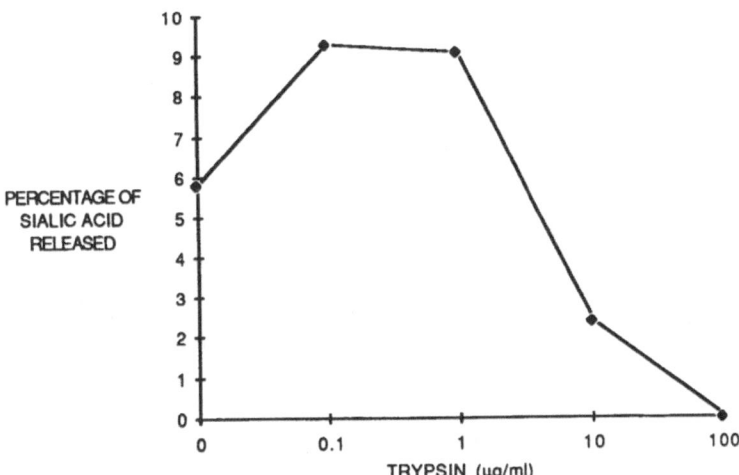

FIGURE 7 Factors on the cytoplasmic face of acceptor vesicles are required for fusion. Standard acceptor membranes were incubated with trypsin at 4°C for 1 h. The trypsin was inhibited and the membranes used in a standard cell-free fusion assay. Only the release due to a reaction requiring both FPV and ATP is shown.

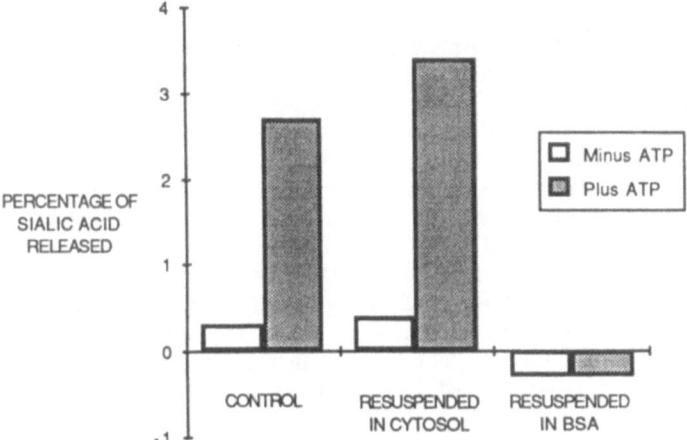

FIGURE 8 Cytosolic factors are required for fusion. The membrane fractions of standard donor and acceptor preparations were prepared on a sucrose gradient and mixed in the presence of cytosol or buffer containing BSA (cytosol is defined as the supernatant following centrifugation of mock-treated acceptor membranes at 100,000 g for 1 h). The control is a standard assay performed with the same preparations but unfractionated. Only the release due to the FPV neuraminidase is shown.

is the production of antibodies against the membrane proteins and the use of these antibodies to inhibit fusion. The intention is that a specific antibody would bind to the protein involved in controlling fusion and would impair the activity of that protein. A more productive approach, however, is likely to be the reactivation of an inactivated system. For example, since we know that cytosolic factors are required for fusion, one approach would now be to fractionate the cytosol and determine the ability of each fraction to support the fusion of the partially purified vesicles. Central to such an approach is the speed at which the assay can be performed. Assays that rely either on visualization of the fusion event by light microscopy and especially by electron microscopy or that rely on the use of gradients to separate the product from the starting material are not amenable to the analysis of a large number of samples, as would be generated, for example, by the fractionation of cytosol. Our assay can determine whether or not fractions are active within a couple of hours of obtaining them, and this is the sort of speed required if one is to isolate active factors.

Another consideration with regard to the analysis of numerous samples is the speed and cost at which the assay components can be produced. Our present assay is poor in this respect since the membrane preparations are both expensive and time consuming. This does, however, illustrate the point that no assay is perfect and even when a system is working there will always be changes that could be made; the important thing is to decide which changes should be made and when.

VII. CONCLUDING REMARKS

Cell-free systems represent the best approach to understanding the molecular basis of many cellular events. Here we have tried to use our own experiences in outlining some of the factors to be considered when developing such systems. If we were to make a final comment it would be that many cellular events seem to proceed quite readily in the test tube and that most work is concerned with removing irrelevant or unwanted reactions so that one can focus on a single event. It is only then that the factors controlling that event can be analyzed.

We wish to point out that this review was written over 4 years ago. This assay was then the only available cell-free system for studying such events along the endocytotic pathway. It has since been improved and many new assays have been developed. The reader is referred to Warren et al. (12) for a more recent treatment of this subject.

REFERENCES

1. Silverstein, S. C., Steinman, R. M., and Cohn, Z. A. (1977). Endocytosis. Annu. Rev. Biochem. 46:669-722.
2. Brown, M. S., Anderson, R. G. W., and Goldstein, J. L. (1983). Recycling receptors: The round-trip itinerary of migrant membrane proteins. Cell 32:663-667.
3. Lehrman, M. A., Goldstein, J. L., Brown, M. S., Russell, D. W., and Schneider, W. J. (1985). Internalization-defective LDL receptors produced by genes with nonsense and frameshift mutations that truncate the cytoplasmic domain. Cell 41:735-743.

4. Schlossman, D. M., Schmid, S. L., Braell, W. A., and Rothman, J. E. (1984). An enzyme that removes clathrin coats: Purification of an uncoating ATPase. *J. Cell Biol.* 99:723-733.

5. Pearse, B. M. F. (1985). Assembly of the mannose-6-phosphate receptor into reconstituted clathrin coats. *EMBO J.* 4:2457-2460.

6. Davey, J., Hurtley, S. M., and Warren, G. (1985). Reconstitution of an endocytotic fusion event in a cell-free system. *Cell* 43:643-652.

7. Marsh, M. (1984). The entry of enveloped viruses into cells by endocytosis. *Biochem. J.* 218:1-10.

8. Monaco, F., and Robbins, J. (1973). Incorporation of N-acetylmannosamine and N-acetylglucosamine into thryoglobulin in rat thyroid *in vitro*. *J. Biol. Chem.* 248:2072-2077.

9. Matlin, K. S., Reggio, H., Helenius, A., and Simons, K. (1981). Infectious entry pathway of influenza virus in a canine kidney cell line. *J. Cell Biol.* 91:602-613.

10. Marsh, M., Bolzau, E., and Helenius, A. (1983). Penetration of Semliki Forest virus from acidic prelysosomal vacuoles. *Cell* 32:931-940.

11. Meindl, P., Bodo, G., Palese, P., Schulman, J., and Tuppy, H. (1974). Inhibition of neuraminidase activity by derivatives of 2,3-dehydro-2-deoxy-N-acetylneuraminic acid. *Virology* 58:457-463.

12. Warren, G., Woodman, P., Pypaert, M., and Smythe, E. (1988). Cell-free assays and the mechanism of receptor-mediated endocytosis. *Trends Biochem. Sci.* 13:462-465.

21

Specific Conditions for Fusion of Membranes of Nuclear Envelopes, Endoplasmic Reticulum, and Golgi Apparatus from Vertebrate Cells

JACQUES M. PAIEMENT and J. J. M. BERGERON

University of Montreal and McGill University,
Montreal, Quebec, Canada

I. INTRODUCTION

Membrane fusion has been implicated in dynamic exchange processes occurring among intracellular compartments of the secretory system and endocytic apparatus of all eukaryotic cells. Fusion is necessary for the transfer of secretory proteins from the rough endoplasmic reticulum (RER) through the Golgi apparatus and secretory vesicles to the plasma membrane (1). It is also involved in lysosome-plasma membrane interactions via endosomes (2,3) and has been implicated in the reassembly of nuclei, ER, and Golgi apparatus within postmitotic cells (4-6). Studies of cytoplasmic membrane fusion have in the past been largely carried out on the exocytotic process (secretory granule membrane-plasma membrane fusion) (7,8). Only recently have suitable experimental systems been made available for the study of fusion occurring at the level of other cytoplasmic membranes (9-17). The work to be described here pertains mainly to the demonstration of fusion within RER, Golgi, and nuclear envelope membranes in vitro. Among our objectives is to examine to what extent such fusions are associated with membrane functions. Furthermore, since the factors (specific nucleotides and cations) that control fusion events at the level of these membranes have been shown to differ from those implicated in other experimental systems, we are particularly interested in knowing why this is so. Finally, in order to place the membrane fusion events in perspective to what is actually occurring in the living cell, a model system, i.e., the microinjection of fragmented organelles into *Xenopus laevis* oocytes and embryos, is presently being used to demonstrate and characterize fusion of cytoplasmic membranes in vivo.

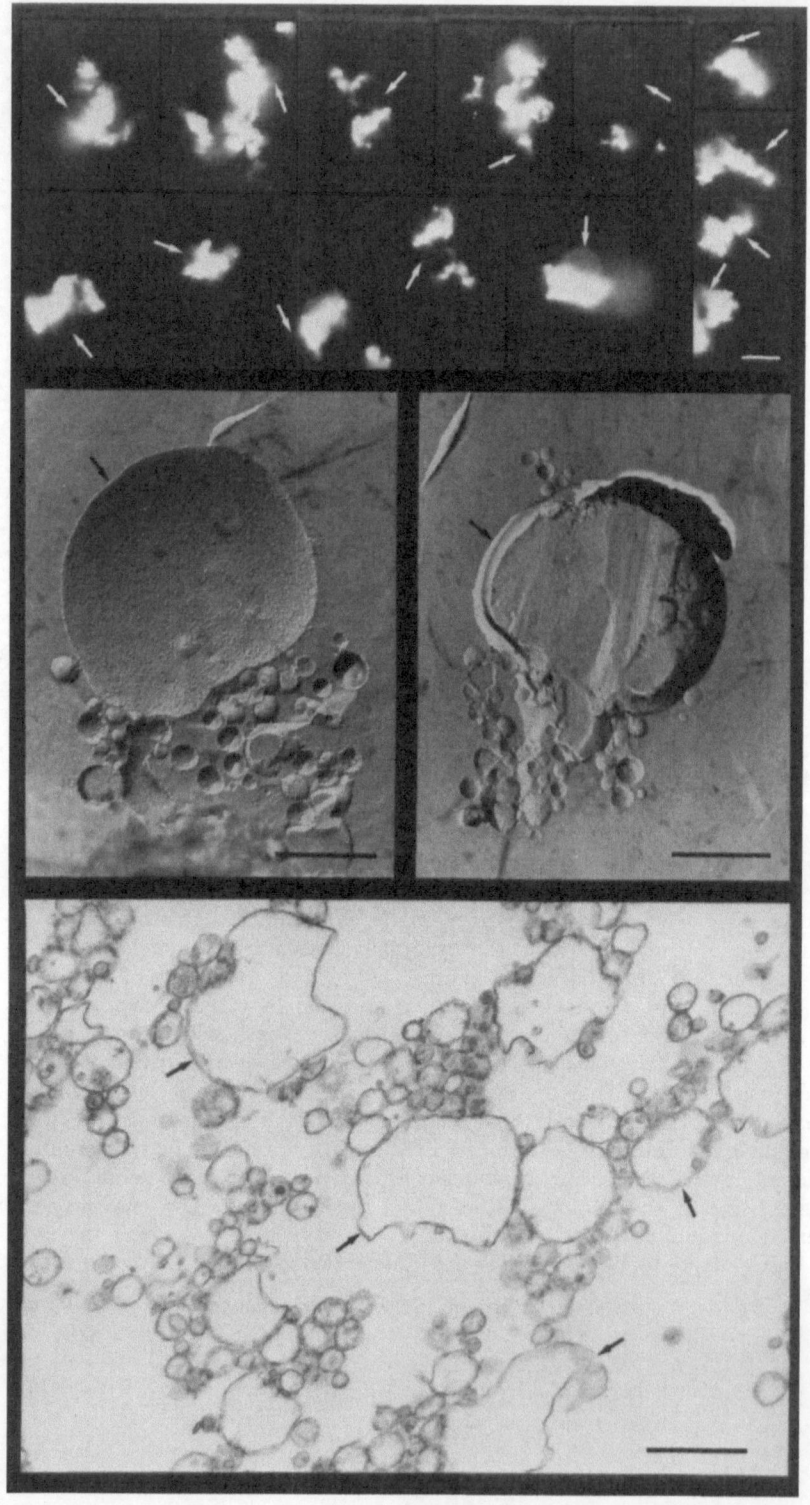

II. FUSION OF ROUGH ER IS PROMOTED IN VITRO BY A SINGLE NUCLEOTIDE AND SPECIFIC DIVALENT CATIONS

A. Initial Description and Definition

In 1980 it was observed that conditions which promoted posttranslational core glycosylation in RER membranes in vitro also promoted membrane coalescence (18). Coalescence was defined as the formation of large membrane-bounded elements by fusion of the membranes of small RER vesicles. Formation of large fusion products was demonstrated by thin-section electron microscopy of fixed and embedded microsomes, by freeze-fracture electron microscopy of rapidly frozen and cleaved membranes, as well as by light microscopy of unfixed membranes using fluorescent lipid probes (Fig. 1). Fusion has also been defined quantitatively as an increase in membrane surface area as evaluated by morphometric measurement (19). Therefore, the criteria for fusion as used in the initial discovery and throughout this chapter are all morphological.

B. Factors Effecting Fusion of RER Membranes

The initial discovery (18) was made on ER membranes that had been rendered competent to incorporate nucleotide sugars (i.e., UDP-*N*-acetylglucosamine or GDP-mannose) into endogenous dolichol glycolipid intermediates and protein acceptors in vitro (20-22). These detailed biochemical observations were themselves iconoclastic, as up to that time it had been accepted that core glycosylation occurred only cotranslationally (23). This is now known not to be the case; indeed, even proteins can be translocated posttranslationally (24,25). The main requirements for the glycosylation of endogenous constituents of rough microsomes from rat liver were the prior removal of the ribosomes from the membranes and subsequent incubation with the nucleotide sugars in the presence of GTP (20-22). Fusion of rough microsomes

---

FIGURE 1 Definition of ER fusion in vitro. Upper panel: Evidence for fusion by fluorescence microscopy. Stripped rough microsomes (100-300 μg protein) were preincubated for 15 min at 10°C in the presence of 1 μg dihexylcarbocyanine iodide (DiOC$_6$) and subsequently incubated for 120 min at 37°C with GTP and Mg^{2+}. They were then examined without further treatment. The fluorescence micrographs show clumps of microsomes at the periphery of which can be seen large fluorescent membrane fusion products (arrows). All micrographs were taken at the same magnification. Bar = 10 μm.

Middle panel: Evidence for fusion by freeze-fracture electron microscopy. Stripped rough microsomes were incubated 60 min at 37°C with GTP and Mg^{2+}. They were then rapidly frozen and treated for freeze-fracture analysis. The concave cytoplasmic fracture face of a membrane fusion product dominates the picture on the left (arrow) and a cross-fractured membrane fusion product dominates the picture on the right (arrow). The protoplasmic and luminal fracture faces of membranes belonging to unfused microsomes are aggregated at the periphery and below the large fusion products. Left photomicrograph from Ref. 18. Bar = 0.5 μm.

Lower panel: Evidence for fusion by thin-section electron microscopy. Stripped rough microsomes were incubated 60 min at 37°C with GTP and Mg^{2+}. They were then fixed and processed for electron microscopy by routine procedures. The membranes of the large fusion products (arrows) are closely apposed to those of small unfused vesicles. Bar = 0.5 μm.

TABLE 1   Comparison of Cation and Nucleotide Specificities for Membrane Fusion in RER Membranes and Golgi Membranes

Cation or nucleotide	Amount of fusion[b]	
	RER[a]	Golgi
Mg^{2+}	++	-
Mn^{2+}	+++	+++
Ca^{2+}	+	-
Zn^{2+}	-	-
Cu^{2+}	-	-
Cd^{2+}	-	N.D.[c]
ATP	-	+++
CTP	-	+++
GTP	+++	+++
ITP	-	+++
UTP	-	+++

[a]Stripped RER membranes were incubated in medium containing 0.5 mM GTP plus 2 mM of the test cation or in the presence of medium containing 7.5 mM Mg^{2+} and 2 mM Mn^{2+} plus 1 mM of the test nucleotide. Incubations were done at 37°C for 60 min. Golgi membranes were incubated in medium containing 2 mM ATP plus 20 mM of the test cation or in the presence of medium containing 20 mM Mn^{2+} plus 2 mM of the test nucleotide. Golgi incubations were carried out at 37°C for 60 min.
[b]Degree of fusion was assayed as previously suggested by Creutz (37) and involved semiquantitative determination of the amount of formation of large vesicles as seen in the electron microscope: +++, extensive fusion, many large vesicles evident; ++, intermediate fusion, large vesicles distinct but smaller; +, limited fusion, very few vesicles two or three times the size of the average population of vesicles; -, no fusion, all vesicles the same size as unincubated vesicles.
[c]N.D. not determined.

exhibits similar requirements. It requires prior removal of membrane-bound ribosomes and the presence of physiological concentrations of GTP and Mg^{2+} in the incubation medium. It will occur irrespective of what non-denaturing protocols are used to strip off the ribosomes provided GTP and the appropriate cation is present (18). Only GTP is maximally effective in stimulating fusion at concentrations down to at least 10 µM. This is well within estimates of physiological concentrations of GTP (26,27); other nucleotides, including the nonhydrolyzable analogs of GTP, i.e., GppCp and GppNp (18), are ineffective. Either Mg^{2+} and/or Mn^{2+} (at 100 µM) must be present in the incubation medium with GTP in order for fusion to occur. In the case of Mg^{2+} this is well within the physiological range (28,29). Other cations including Ca^{2+} are ineffective in stimulating fusion (Table 1).

The actions of GTP and divalent cations can be dissociated from each other as follows: The incubation of stripped rough microsomes in buffered isotonic sucrose at 37°C leads to irreversible inactivation and membranes are unable to respond by fusion in a subsequent incubation with GTP and Mg^{2+}. The presence of GTP in the preincubation medium prevents inactivation but the membranes neither aggregate nor fuse. However, in a subsequent postincubation in the presence of Mg^{2+} membranes fuse (Fig. 2). Thus, nucleotide and cation effects can be uncoupled and such factors act in a coordinated manner.

Fusion among stripped RER occurs in a time- and temperature-dependent manner. Morphometric measurements (19) have demonstrated that fusion is linear with time until 30 min and reaches a plateau after 60 min. Furthermore, incubations at different temperatures (10–37°C) reveal that stripped RER fuse only at temperatures above 17–20°C.

The contribution of both lipid and protein in RER membrane fusion has been studied. Of the various phospholipids so far examined phosphatidylglycerol (PG) was found to form lipid bilayers and to fuse with stripped RER membranes in a GTP-dependent fashion. This occurred in the presence of Mg^{2+} and is consistent with reports that PG is readily stabilized by the presence of Mg^{2+} (30,31). Plant phosphatidylinositol, in contrast, did not form recognizable lipid bilayers under similar incubation conditions. This lipid, however, was found to be a potent inhibitor of GTP/Mg^{2+}-stimulated fusion, a fact that may be related to steric effects of the large size of the headgroup of this molecule following incorporation into the microsomes (32,33). Finally, treatment with phospholipase C inhibited GTP/Mg^{2+} fusion of stripped RER.

Controlled proteolytic digestion experiments have been carried out on stripped RER. They showed that membrane fusion could be inhibited

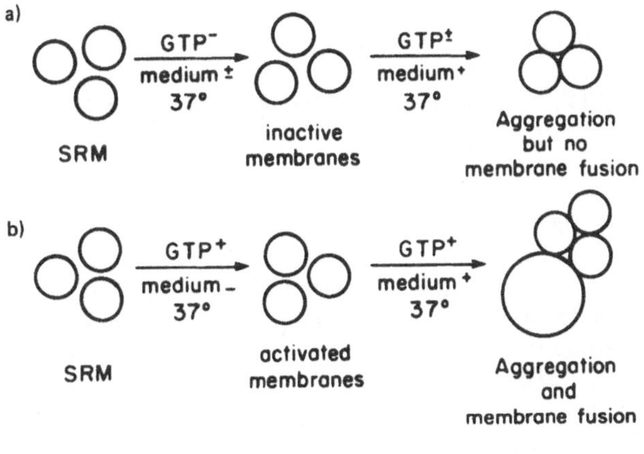

FIGURE 2 GTP by itself does not lead to membrane fusion but does protect stripped ER from inactivation. (a) When stripped RER (SRM) are incubated either in buffered sucrose or with additional Mg^{2+} (medium) at 37°C the membranes are rapidly (≈10 min) inactivated irreversibly. Subsequent incubation with GTP in the presence of Mg^{2+} (medium) does not lead to membrane fusion. (b) However, the addition of buffered GTP to stripped ER prevents inactivation. Such "activated" membranes are not aggregated. Addition of Mg^{2+} (medium) leads to aggregation and fusion. (From Ref. 43.)

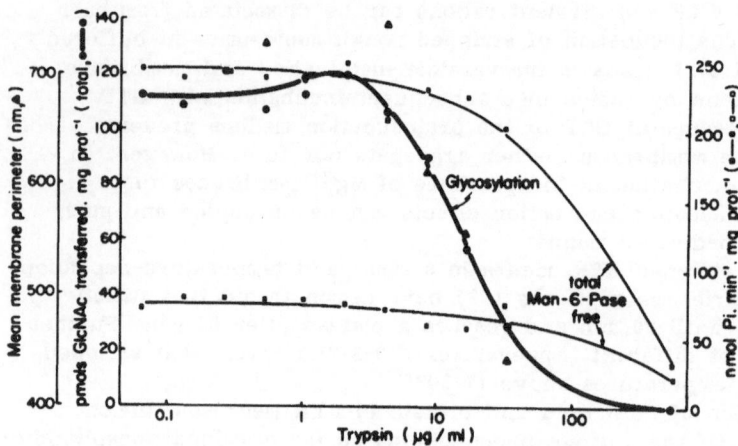

FIGURE 3   Controlled proteolytic digestion of stripped RER. Membranes
were incubated at 10°C with varying concentrations of trypsin in buffered
sucrose. The ratio of membrane protein to trypsin varied from 876:1 to
13:1. Following trypsin treatment, stripped RER was incubated at 37°C
with a 10-fold concentration of soybean trypsin inhibitor as well as GTP
and Mg^{2+}. After 60 min of incubation, samples were fixed, processed for
transmission electron microscopy, and fusion (▲) quantified by morphometry
(19). Glycosylation was quantified by evaluating the incorporation of
[^{3}H]GlcNAc from nucleotide sugar precursor into endogenous dolichol lipid
and glycopeptide acceptors (●—●). Samples were also evaluated for membrane leakiness induced by trypsin by determination of the latency of
mannose-6-phosphatase activity (defined as the difference between total
(+ detergent) and free (- detergent) mannose-6-phosphatase activity.
These results show that trypsin induced loss of membrane fusion and glycosylation was coincident with 50% inhibition occurring at 15 µg/ml trypsin.
In contrast, latency of mannose-6-phosphatase activity was maintained
until at least 50 µg/ml of trypsin. Therefore, a cytosolically oriented pep-
tide(s) may be responsible for GTP/Mg^{2+}-induced fusion and glycosylation.
(From Ref. 43.)

by mild trypsinization (membrane-protein to trypsin-protein ratios of 100
to 1, at 10°C for 30 min). The loss of membrane fusion occurred coincident
with the loss of the ability to core glycosylate in vitro (Fig. 3). Control
experiments showed that at these trypsin concentrations spatial continuity
of stripped microsomes was maintained as small molecules such as mannose-
6-phosphate were still impermeant (Fig. 3). Therefore, a cytosolically
exposed protein(s) of ER membranes was concluded to be linked causally
to fusion. A number of GTP-binding proteins have recently been identified
associated with purified rough microsomes and could be potential candidates
in the initiation of the fusion process (33a,b,c).

C. Membrane Specificity

Under the specified conditions for RER fusion described above, RER membranes do not fuse in vitro with mitochondria, lysosomes, plasma membranes,
or Golgi apparatus. The sole exception is nuclear envelopes; however, this

will be dealt with in detail below, as well as the lack of fusion with Golgi elements. GTP/Mg^{2+} promotes fusion of RER from species as disparate as canine pancreas and amphibian oocytes, attesting to the universality of the phenomenon.

D. Leakiness of Fusion

Intactness of isolated RER membranes is often evaluated by the latency of mannose-6-phosphatase (34), an intraluminal enzyme. The ratio of the degree of hydrolysis of this enzyme in the absence or presence of detergent is considered a quantitative measure of ER membrane intactness. Neither GTP nor divalent cations alter latency of this enzyme activity; however, the combination of both renders the membranes leaky (Fig. 4). Similarly, nucleoside diphosphatase activity has also been used to evaluate membrane leakiness (35). By EM cytochemistry this leakiness can be localized exclusively to fused membranes (Fig. 5). As fused membranes are also the site of GTP/Mg^{2+}-enhanced core glycosylation (19), then enhanced glycosylation and loss of latency is most simply explained as increased leakiness of ER membranes induced by fusion. Several other in vitro model fusion systems also show enhanced leakiness (36-41). In the specific case of the ER fusion shown here, this may suggest that the in vivo equivalent is either absent or controlled by as yet undescribed cytosolic factors.

E. Localization of a Fusion Domain in RER Membranes

Controlled proteolytic digestion experiments reveal that a cytosolically oriented peptide(s) factor is required for RER membrane fusion (see above). The restricted location of this factor beneath membrane-bound polyribosomes

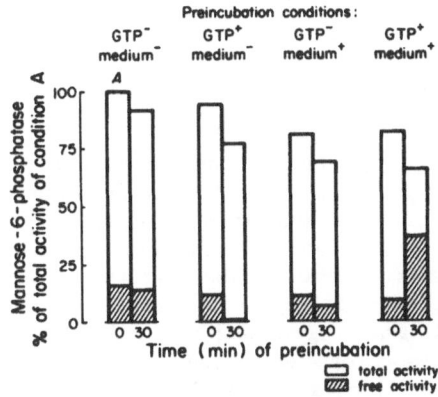

FIGURE 4 GTP/Mg^{2+} together lead to loss of latency of stripped RER to mannose-6-phosphate but neither GTP nor Mg^{2+} by itself affects the intactness of the membranes. Stripped RER were incubated for 0 or 30 min in the presence or absence of buffered GTP with or without Mg^{2+} (medium). Mannose-6-phosphatase activity was assayed in the presence (total activity) or absence (free activity) of detergent. The proportion of free activity increases markedly during the 30-min incubation with GTP/Mg^{2+} indicating a loss of membrane impermeability to mannose-6-phosphate. (From Ref. 43).

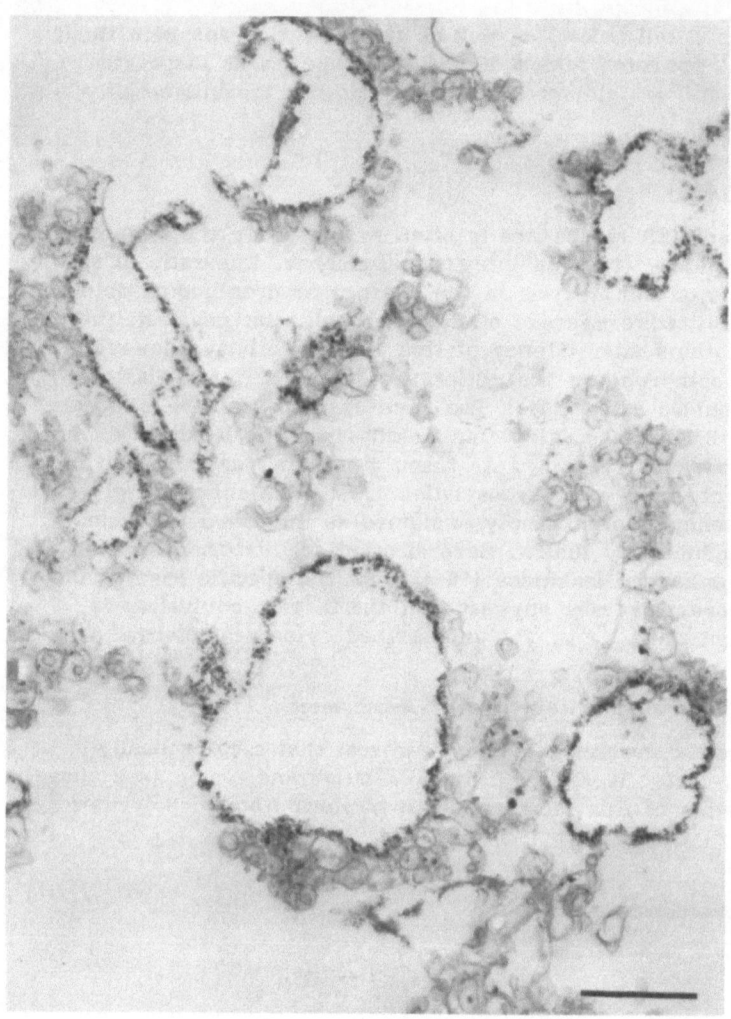

FIGURE 5 Evaluation of substrate accessibility to an intraluminal enzyme
(nucleoside diphosphatase) of stripped RER using electron microscope cytochemistry. Stripped rough microsomes are shown following incubation in
the presence of GTP and Mg^{2+} for 60 min at 37°C and subsequent processing for nucleoside diphosphatase cytochemistry. After incubation, membranes
were fixed briefly (30 min in 1% glutaraldehyde) and then incubated for
120 min at 37°C in the presence of 4 mM thiamine pyrophosphate and lead
nitrate. They were then washed and processed for electron microscopy.
Lead phosphate reaction product is particularly evident along the lumenal
border of the membranes of large fusion products. Scant amount of reaction
product is seen in smaller unfused microsomes which are aggregated at the
periphery of the fusion product. Bar = 0.5 μm. (From Ref. 43.)

is furthermore indicated by ribosome capping experiments. Several years
previously Ojakian and colleagues showed that exposure of rough microsomes
to low concentrations of ribonuclease led to the aggregation of polysomes
to a restricted cluster on the surface of the microsome (42). We have confirmed this observation and have calculated by morphometry that similarly
treated membranes have as much as 50% of their surface area devoid of
ribosomal particles (i.e., containing large, bare membrane patches). Such
microsomes when subsequently incubated in the presence of GTP and Mg^{2+}
are unable to fuse despite the increased surface area available for membrane
interaction. Therefore, a cytosolically oriented protein probably buried
beneath the ribosome in vivo is concluded to be causally linked to the
fusion phenomenon described here (43).

F. Postulated Mechanisms of Action of GTP and Mg^{2+} in RER Membrane Fusion in Vitro

The removal of ribosomes from ER membranes leads to the exposure of
a membrane constituent that is modified in the presence of GTP. This
modified constituent can then interact with specific cations to lead to
distinct morphological and biochemical changes in ER membranes. Since
the nonhydrolyzable analogues of GTP are ineffective in bringing about
similar changes, it is suggested that GTP may act through a phosphorylation
mechanism. Either proteins or phospholipids could serve as targets of a

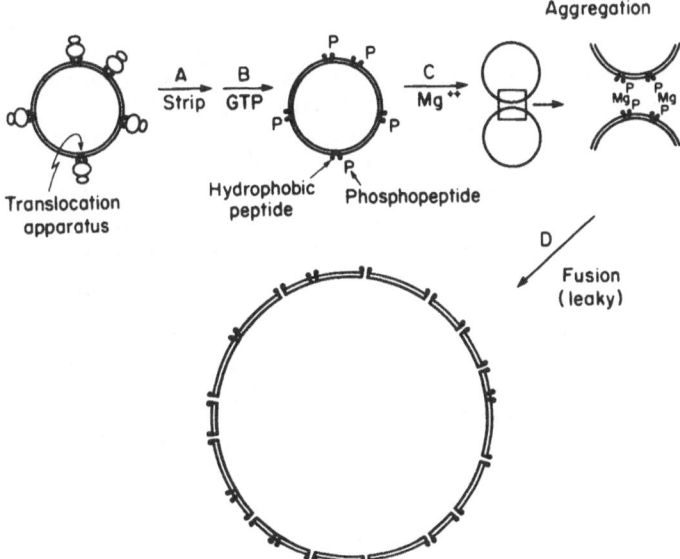

FIGURE 6 Implication of protein phosphorylation in the fusion of RER
in vitro: Step A: Removal of ribosomes by any of the available nondenaturing
techniques leading to exposure of the hypothetical translocation apparatus
originally beneath each ribosome. Step B: GTP-dependent phosphorylation
of a constituent of the translocation apparatus leading to the exposure of
a hydrophobic peptide. Membranes are unaggregated. Step C: Aggregation
of membranes by Mg^{2+} or Mn^{2+} leading to the functional interaction of
contiguous hydrophobic peptides leading (step D) to membrane coalescence,
i.e., fusion.

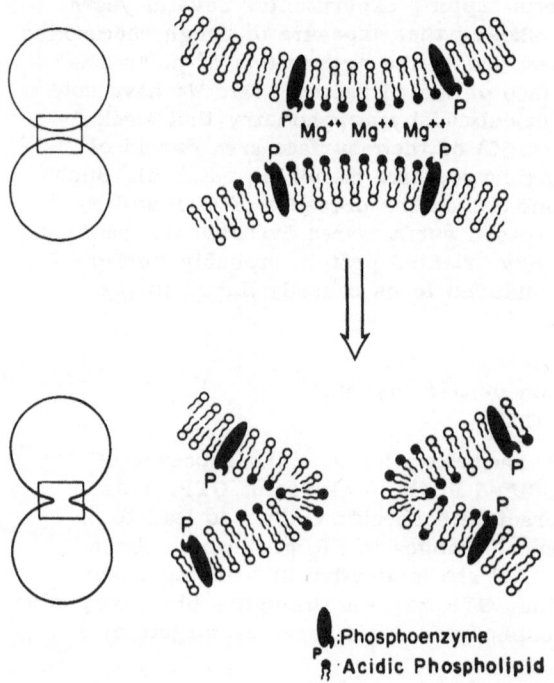

Phosphoenzyme
Acidic Phospholipid

FIGURE 7 Implication of phospholipid synthesis in the fusion of RER membranes in vitro. A protein kinase requiring GTP phosphorylates the enzyme CTP: phosphatidate cytidyltransferase (45). This enzyme stimulates the formation of CDP-diacylglycerol, which is a required intermediate in the synthesis of several acidic phospholipids including phosphatidylglycerol. The latter accumulate in a membrane microdomain and complex with Mg^{2+}. A similar domain of complexed acidic phospholipids forms in the outer bilayer of an adjacent vesicle, creating a fusion competent region. Fusion follows, leading to bilayer reorganization. Flipflop of phospholipids is speculated to occur during such fusion events.

GTP-specific phosphorylation. For example (Fig. 6), the phosphorylation of a specific membrane protein could result in the exposure of a hydrophobic sequence in the peptide, thereby allowing it to perturb adjacent lipid bilayers of aggregated, stripped rough microsomes and thus enabling membrane coalescence, in a similar fashion to that proposed by Helenius (44) for the fusion of viral proteins with endosomal membranes. In this model the cations promote membrane aggregation. An alternative possibility (Fig. 7) would involve a phosphoenzyme that is a step in an enzymatic reaction (i.e., involved in phospholipid synthesis) which could modify the lipid environment of the membranes near ribosome-binding sites and in the region of membrane aggregation. In this model GTP activates formation of acidic phospholipids (45). Such phospholipids are expected to complex with Mg^{2+} and in zones of membrane apposition become involved in fusion (46,47). Finally, there is a possibility that GTP does not induce fusion via a phosphorylation mechanism but causes a conformational change in an RER-specific G protein which then interacts with a specific effector leading to membrane fusion, in analogy with that proposed for exocytosis (47a).

III. NUCLEI AND NUCLEAR ENVELOPE FUSION

Biochemical studies have long emphasized the qualitative biochemical similarity between the nuclear envelope membrane and the membrane of the RER (48). Indeed, morphological studies have often documented continuity between the ribosome-studded outer nuclear membrane and the membrane of the rough ER (49,50). It therefore stood to reason that the GTP divalent cation-induced fusion phenomenon be assessed in isolated nuclei.

During in vitro incubations under ER fusion conditions, nuclei aggregate, making contact at multiple sites along the outer nuclear membranes. Fusion occurs. It proceeds along the surface of nuclei between regions of the nuclear pores in such a way that the resulting fusion products (contact zones between fused nuclear pairs) are easily recognized in the electron microscope. Zones of fusion are characteristically delimited by two continuous outer membranes and contain confluent perinuclear spaces. Bridges of ribosome-studded outer nuclear membranes span the contact zone and are continuous with the inner membranes of adjacent nuclei. Aggregated nuclei with similar membrane organizations have been described previously in fused gametes (51-54). The formation of fused nuclei is summarized in diagrammatic form in Figure 8. As well as having been described in detail using thin sections, this phenomenon has been demonstrated by freeze-fracture electron microscopy and a quantitative assay has been devised to determine kinetic, nucleotide, and temperature dependence as well as cell and species specificity (55,56). The quantitative assay

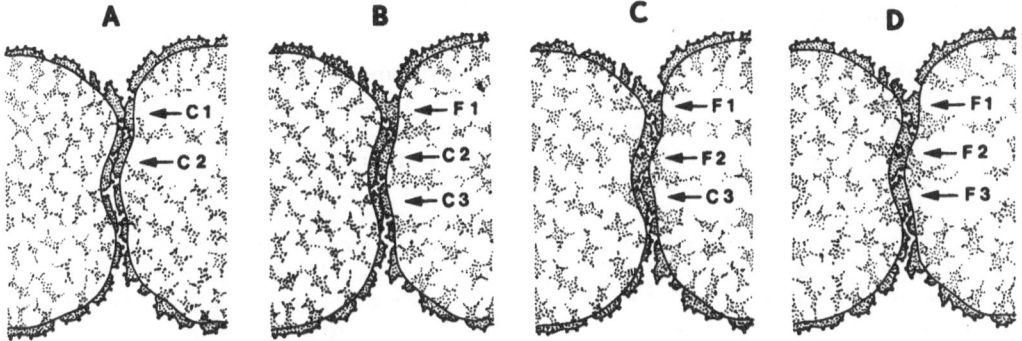

FIGURE 8 Diagrammatic representation of outer nuclear membrane fusion. In A, nuclei make contact at several sites (C_1 and C_2) along their respective outer membranes, at regions slightly distant from nuclear pores. In B, fusion occurs at one site (F_1) leading to the formation of continuous outer membranes and confluent perinuclear spaces. In B, a third contact site is formed (C_3). In C, the fusion process extends between nuclear pores to incorporate a second fusion site (F_2) and leads to the formation of a bridge of ribosome-studded outer membrane, which spans the confluent nuclear space and is continuous with the inner membranes of adjacent nuclei at the level of the pores. In D, fusion occurs at a third site (F_3) and leads to the formation of a second bridge of ribosome-studded outer membrane and continuous outer membranes at the lower limit of the contact site. Similar fusion products have been described in situ between the pronuclei of gametes (51-54) and are thought to represent nuclear fusion intermediates en route to the formation of the zygote nucleus.

consists of counting closely aggregated nuclear pairs and determining the percent number of pairs showing the morphological properties of fused nuclear pairs described above. Using this assay, fusion was shown to occur at GTP concentrations as low as 10 µM (55). Kinetic analysis revealed it to occur in a linear fashion for up to 60 min and temperature studies showed that it occured only above 20°C (56). The tissue and species specificity of this fusion process was also examined. It was found that GTP and specific cations (Mg^{2+} and Mn^{2+}) can promote fusion between homologous nuclear membranes from various sources as well as between heterologous nuclear membranes following mixing of different types of nuclei (56). Therefore, the capacity of outer nuclear membranes to fuse in the presence of GTP may be a feature common to all eukaryotic cells.

Many of the properties of this fusion process are similar to those shown for RER membrane fusion; however, a major difference is noted. Whereas the removal of at least 2/3 of the ribosomal RNA from rough microsomes is a prerequisite for fusion, this is not the case for nuclei. Isolated nuclei fuse with their normal complement of ribosomes. However, isolated nuclei have been shown to be damaged, i.e., they have lost latency to mannose-6-phosphatase (57). Therefore, potential GTP-sensitive sites normally sequestered beneath ribosomes in nuclear envelope membranes may have become exposed during the rigors of nuclear isolation. Alternatively, the frequency of binding of ribosomes to outer membranes in vivo may be lower than that to ER membranes and consequently some GTP-sensitive sites remain exposed following cell fractionation. These sites would then be activated by GTP in vitro.

IV. GOLGI APPARATUS MEMBRANES

The phenomenon of nucleotide-dependent, manganese-stimulated fusion of Golgi apparatus in vitro was uncovered as a logical consequence of prior work on ER fusion. As ER fusion had been observed during in vitro core glycosylation reactions, similar studies were designed to evaluate Golgi membrane structure during in vitro terminal glycosylation. Initial attention was focused on the transfer of galactose (Gal) to endogenous acceptors within isolated Golgi fractions. Optimal conditions for transfer (Table 2) revealed a surprising requirement for relatively high concentrations of Mn^{2+} as well as a broad nucleotide specificity. Purified galactosyl transferase has been reported, by contrast, to require only low levels of Mn^{2+} and to have no requirement for ATP (58). When galactosylation was carried out in purified Golgi membranes, membrane coalescence was observed (Fig. 9b). Fused Golgi membranes were the site of transferred galactose and fusion was regulated by ATP and Mn^{2+} (59). No other cation could substitute for Mn^{2+}, although any nucleotide could substitute for ATP (43; Table 1).

The initial hypothesis was modeled on that used to correlate ER fusion and core glycosylation, namely, that fusion may be causally linked to the transfer of nucleotide sugars as an intermediate across Golgi membranes. A thorough search for possible intermediates such as dolichol-like galacto lipids (59) proved negative (60), and therefore a tentative model (61) involving such lipids as possible intermediates during fusion-related translocation of sugar across the bilayer was discarded.

Observations were extended to sialic acid (NeuAc) transfer across the Golgi membranes. Optimal conditions for incorporation of sialic acid

TABLE 2 Optimal Conditions for Sugar Transfer to Endogenous Acceptors in Intact Golgi Membranes

Nucleotide sugar donor	pH	$[Mn^{2+}]$ (mM)	[ATP] (mM)	Km (μM)	V_{max} (nmol · 10 min^{-1} · mg/prot^{-1})
UDP-GlcNAc	6.5	15	2	0.7	0.6
UDP-Gal	6.2-6.5	20	2	0.8	1.5
CMP-NeuAc	5.8-6.5	0-2	0	0.4	0.3

Intact Golgi fractions were isolated by a one-step protocol and incubated with the corresponding radiolabeled nucleotide sugar in the absence of any detergent for 10 min at 37°C as described (60).

into endogenous glycoproteins from the nucleotide sugar donor CMP-sialic acid showed marked differences to Gal transfer (Table 2). No Mn^{2+} requirement was observed. Thus, sialic acid transfer was nearly identical whether Golgi membrane structure was perturbed (fusion) or not. Furthermore, the preferred site of uptake of sialic acid into endogenous acceptors was not fused membranes but rather a network which resisted fusion (Fig. 9d) on the *trans* pole of the Golgi apparatus (60). N-acetylglucosamine (GlcNAc) transfer was also assessed. Relatively high concentrations of Mn^{2+} were optimal for in vitro transfer from the parent nucleotide sugar (Table 2). However, significantly less incorporation of N-acetylglucosamine as opposed to galactose into fused membranes was observed (60).

It is now well documented that specific nucleotide sugar transporters operate to channel nucleotide sugars and nucleoside monophosphates across Golgi membranes (62). There is, therefore, no a priori need to invoke fusion as relevant to specific nucleotide sugar transport across the Golgi membrane bilayer.

Our studies can perhaps be best explained in light of current models of Golgi apparatus structure and function. Cytochemical and biochemical work has long emphasized the compartmental nature of the Golgi apparatus (63). Within the limits of sensitivity of current immunocytochemical and subcellular fractionation protocols, sugar transferases for N-acetylglucosaminyl, galactosyl, and sialyl transferases also appear to have been differentially localized (64-76b). Our studies therefore point to at least part of the GlcNAc and galactose transferring compartments (medial to *trans*) as participating in fusion, with minor contribution for the compartment for sialylation.

An alternative explanation for our findings has been advanced by Braell et al. (77). These investigators postulated a nonspecific fusion between Golgi *cis* and medial saccules containing acceptors with Golgi *trans* saccules containing transferase enzymes. Our data do not support this interpretation. Radiolabeled acceptors have been evaluated by sodium dodecyl sulfate polyacrylamide gel electrophoresis. Approximate differences of 2000 in M_r were observed for labeled GlcNAc, Gal, or sialic acid acceptors (Fig. 10). However, addition of unlabeled sugars (e.g., unlabeled UDP-Gal or CMP-sialic acid) did not alter the distinctive profiles for acceptors. Therefore, we conclude that cell-free transport of the acceptors from the site of one transferase to the next was not operative under our conditions. Furthermore, as near-maximal transfer of sialic acid was observed under

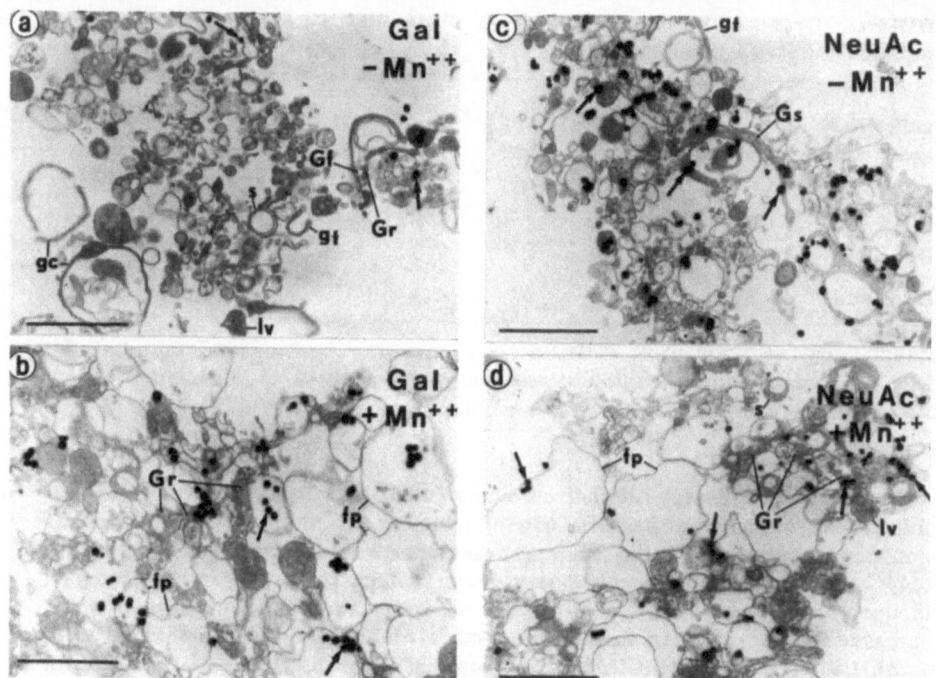

FIGURE 9   [³H]galactose transferred to endogenous acceptors is localized
to fused Golgi membranes. Labeled sialic acid ([³H]NeuAc) is, however,
transferred to endogenous acceptors in nonfused membranes. Purified Golgi
fractions were incubated with labeled nucleotide sugars either in the
absence (−Mn²⁺) or presence (+Mn²⁺) of 30 mM MnCl₂. After incubation,
the fractions were fixed and processed for electron microscope radioautography. Silver grains indicating sites of sugar incorporation are
observed over fused membranes for galactose uptake 9b but mainly unfused
membranes for NeuAc transfer 9d. Silver grains (arrows) indicate the
sites of incorporation of labeled sugar. fp, fusion products; Gf, Golgi
fenestrations; Gr, tubular reticulum of Golgi complexes; Gs, Golgi saccules;
gc, Golgi cisternae; gt, Golgi tubules; lv, lipoprotein-filled vesicles;
s, signet ring structures. Bar = 1 μm. (From Ref. 60.)

nonfusogenic conditions, the hypothesis of Braell et al. (77) is unsupported
by currently available data.

     What, then, is the physiological relevance of the nucleotide-dependent
Mn²⁺-induced fusion of a subset of Golgi membranes? Mn²⁺ levels in situ
are several orders of magnitude less than those required for fusion (29).
One must therefore postulate other cytosolic factors that would carry out
the Mn²⁺ function observed here. No candidates have thus far been identified, and possible scenarios such as biogenesis or reconstitution of Golgi
membranes must remain at present highly speculative.

V.   MEMBRANE INTERACTION SPECIFICITY

In an attempt to understand how cytoplasmic membranes recognize and
interact with each other, we have carried out a series of recombinant

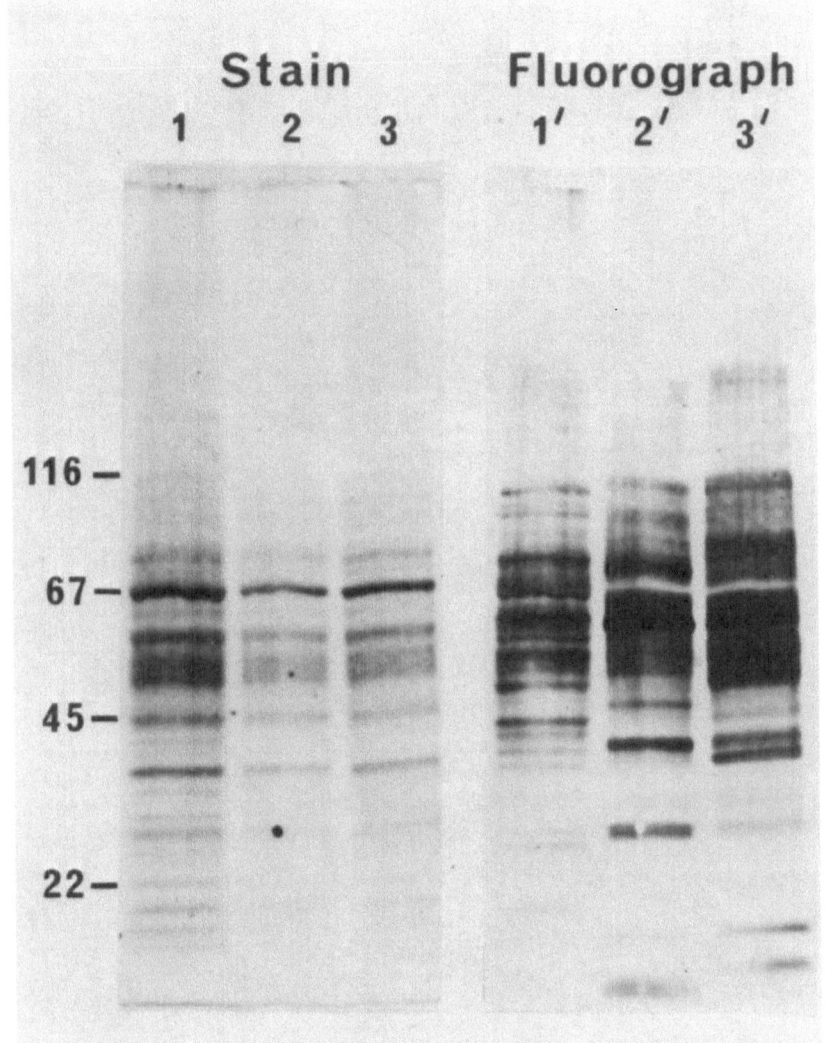

FIGURE 10 Endogenous acceptors in Golgi apparatus for sugar elongation by labeled GlcNAc, Gal, or NeuAc are of distinct, but related molecular weights. Fluorography of labeled proteins from intact Golgi fraction after incubation in the presence of UDP-GlcNAc (lanes 1, 1'), UDP-Gal (lanes 2, 2'), and CMP-NeuAc (lanes 3, 3'). After incubation, approximately equal amounts of radioactivity were electrophoresed. Molecular weight markers are indicated and expressed in kilodaltons. Several endogenous acceptors are radiolabeled and are displaced upward by 2000 in molecular weight in the order GlcNAc (lane 1'), Gal (lane 2'), and NeuAc (lane 3'). Addition of unlabeled sugars (e.g., unlabeled UDP-Gal to the incubation of lane 1' or CMP-NeuAc to the incubation of lane 2') did not alter appreciably the distinctive profiles for labeled GlcNAc or Gal acceptors. (From Ref. 60.)

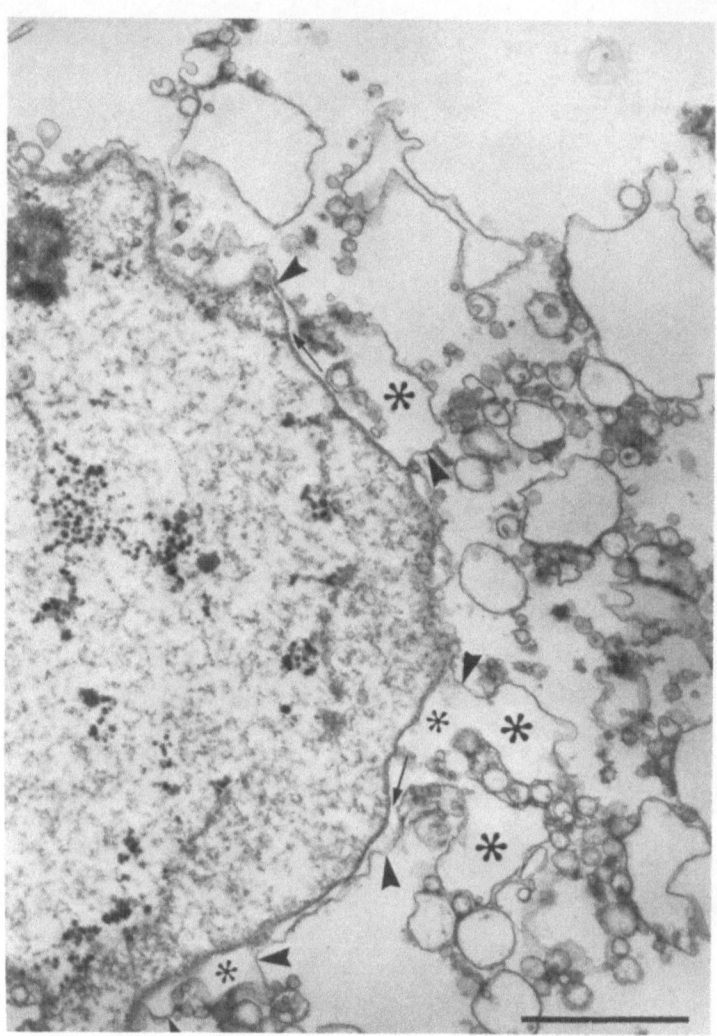

FIGURE 11   Formation of hybrid membrane products after mixing stripped
RER membranes with nuclear envelope membranes. Stripped rough microsomes
were incubated with purified nuclei for 120 min at 37°C in the presence
of 0.5 mM GTP, 7.5 mM Mg^{2+}, and 2 mM Mn^{2+}. This led to the fusion
of the outer membrane of the nucleus with microsomes to form large irregular
membrane extensions. Arrowheads denote points of deflection of the hybrid
membrane products and the arrows indicate confluency between the lumina
(*) of such large extensions with that of the perinuclear space. Bar = 1 μm.
(From Ref. 78.)

experiments. It was found that when nuclei were mixed with stripped rough microsomes in the presence of GTP and cations this led to the formation of hybrid membrane products; i.e., ER membranes fused with outer nuclear membranes (Fig. 11). From the morphometric data of Table 1 in Paiement (78) we can calculate that the amount of ER membrane added per nucleus is 6.96 μm of membrane per 2 h. This value divided by the average length of the microsomal membrane (0.525 μm) (19) gives a total number of fusions per nucleus for 2 h of 13.26, or 6.63 fusions/nucleus/h. Therefore, the process of fusion between ER membranes and outer nuclear membranes is slow and may reflect early saturation of GTP-sensitive sites along the outer nuclear membranes during incubation. The conditions that lead to ER-nuclear envelope membrane fusion are ineffective in stimulating fusion between mitochondrial or between Golgi membranes and do not promote fusion between either of these types of membranes and nuclear envelope membranes. Other cellular membranes that do not respond to GTP and cations include smooth microsomal membranes, lysosomal membranes, and plasma membranes.

Recombinant studies using ER and Golgi membranes have been carried out to determine whether these two types of membranes can fuse together and form hybrid membrane products (43). A systematic evaluation of the nucleotide and divalent cation requirements for homologous membrane fusion revealed that ER membranes and Golgi membranes could be induced to fuse by the same factors, i.e., the presence of Mn^{2+} and GTP (Table 1). Any differences were mainly quantitative. With this information, we attempted to fuse stripped RER with Golgi membranes. Despite exhaustive study, no evidence for ER-Golgi hybrids was found. Thus, using RER with ribosomal remnants as markers and liver Golgi components with intraluminal lipoprotein particles as markers, at no time were hybrid vesicles observed surrounded by ribosomal remnants and containing in their lumina lipoprotein particles. Clumps of fused vesicles were either of ER origin or of Golgi origin. Furthermore, using sialic acid-labeled glycoproteins as a Golgi marker, by EM radioautography silver grains (sialo-labeled glycopeptides) were not observed over fused ER vesicles. Finally, in a purely biochemical study, radiolabeled vesicular stomatitis "G" glycoprotein was inserted into ER vesicles by cell-free translation. Mixing of these ER vesicles with purified Golgi apparatus under fusion conditions (GTP/Mn^{2+}) did not lead to transfer of core glycosylated "G" to the Golgi membranes as evaluated by sensitivity to endoglycosidase H (43). These studies therefore demonstrate the remarkably exquisite specificity of the in vitro fusions documented herein. We conclude that the specificity for interaction between ER and Golgi membranes is determined by very select microdomains or subcompartments which have yet to be defined. Specific vesicular carriers may be involved in the translocation of material between the ER and Golgi (1) and such transport has recently begun to be characterized with cell-free systems (78a,b,c,e).

VI. COMPARISON TO OTHER IN VITRO MODELS OF FUSION

A. Nucleotide Requirements for Cytoplasmic Membrane Fusion

ATP has been implicated in fusion processes at the level of plasma membranes (95) and Golgi membranes (15,59), as well as at the level of endocytotic membranes (16). In the studies of Rothman and colleagues (15), vesiculation is now known to be the ATP-sensitive step (78d). cAMP was found to promote phagolysosome fusion (12) and GTP to activate fusion among outer nuclear envelope membranes and rough endoplasmic reticulum membranes (18,55). GTP-dependent membrane fusion was also demonstrated using crude liver microsomes and the presence of polyethyleneglycol (78f). Recently GTP has been shown to be involved in ER-to-Golgi transport (78e) and has been implicated in vesicular membrane traffic between Golgi saccules (78d). The exact mechanism of action of these nucleotides is unknown. In two cases the nucleotide has been shown to work coordinately with cations to stimulate fusion (43,78e). It may be that other cytoplasmic membrane fusion processes are under similar dual control.

B. Cation Requirements for Cytoplasmic Membrane Fusion

Ca^{2+} plays an important role in the fusion between secretory vesicles and the plasma membrane (79,80) among secretory vesicles themselves (9,10, 81,82), between coated vesicles and lysosomes (13), and between intermediate vesicles and *cis* Golgi (78e). Golgi membrane fusion was found to be Mn^{2+} specific, this cation as well as Mg^{2+} promoted RER membrane fusion. Ca^{2+} was ineffective in stimulating either RER membrane fusion or Golgi membrane fusion (43, Table 1). We cannot at present explain why in our hands Ca^{2+} did not promote Golgi vesicle fusion (10). It therefore appears that at the level of specific cytoplasmic membranes fusion is stimulated by different cations. This differential requirement with respect to cations may be a consequence of the varying lipid and protein compositions of the membrane systems involved.

The absolute dependence on Mn^{2+} for Golgi membrane fusion (59,60; Table 1) is exceptional and may involve a subcompartment of this organelle. Both Mn^{2+} and Ca^{2+} were previously reported to be able to stimulate fusion among Golgi and plasma membrane elements (11).

C. Requirements for Fusion of Artificial Membranes

RER membrane fusion has been shown to have a specific requirement for either Mg^{2+} or Mn^{2+} (Table 1). The Mg^{2+} dependence of RER membrane fusion is similar to the previously demonstrated requirement for this same cation in phosphatidylserine vesicle fusion (41,83,84), in phosphatidate vesicle membrane fusion (33,85), in phosphatidylglycerol vesicle fusion (86), in the fusion of phosphatidylserine/phosphatidylethanolamine vesicles (87,88), in the fusion of phosphatidate/phosphatidylcholine vesicles (89), and in the fusion of phosphatidylethanolamine/phosphatidylinositol membranes (90). However, unlike the case of fusion of phospholipid vesicles, where it was observed that Ca^{2+} could substitute for Mg^{2+}, this has been found not to be the case for RER membrane fusion. It is nevertheless noteworthy that Mg^{2+}-induced fusion of PS/PE vesicles was much more extensive and rapid at temperatures above 20°C (88). GTP-dependent, Mg^{2+}-stimulated RER membrane fusion does not occur at temperatures below 20°C but occurs equally well at temperatures between 20°C and 37°C. Nuclear membrane fusion exhibits the same temperature dependence as stripped RER membrane fusion (55).

VII. PHYSIOLOGICAL SIGNIFICANCE

Cytoplasmic vesicles assist in the vectorial transport of a variety of macromolecules and metabolites (1,2). The traffic of vesicular carriers requires at least two separate types of membrane fusion events; one helps to separate vesicles as they bud out from a larger membrane organelle system, while another allows the fusion of different kinds of organelle membranes with one another. A key factor of all in vivo fusion phenomena in interphase cells is the conservation of the surface of donor and acceptor compartments (91,92). This is distinct to the in vitro fusion we define in our studies where the organelle, i.e., ER and Golgi apparatus membranes, undergo large increases in their surface area.

Nuclear envelope, ER, and Golgi membrane fusion has been reported mainly in early postmitotic cells. Such fusions lead to large increases in the surface area of the various membrane components (4-6). Recently Burke and Gerace (17) and Lohka and Masui (14) have described in vitro model systems to study nuclear envelope formation. A role for GTP and Mg^{2+} in the coalescence of nuclear and ER membranes was not ruled out in these systems. For example, the membranes employed could have been "primed" in vivo by this nucleotide before homogenization. Such "primed" membranes (see Fig. 2) could subsequently undergo fusion in vitro in the presence of Mg^{2+}, leading to nuclear envelope formation. A sequence of steps implicating both ER and nuclear envelope membranes in nuclear envelope formation has previously been described (78) and should be reconsidered in light of new findings suggesting that GTP and Mg^{2+} work coordinately to induce membrane fusion at the level of RER membranes (43).

The cryptic location (ribosome sequestered) of a GTP-sensitive factor [protein(s)] of ER has led to the consideration that the GTP/divalent cation-dependent fusion phenomenon described here may be an artifactual expression of a hydrophobic peptide(s) involved in the transport of nascent chains through the ER membranes. Connolly and Gilmore have recently documented a requirement for GTP, and its nonhydrolyzable analogs, in the binding of polyribosomes to stripped RER (92a). Although the 41-kDa G protein of canine pancreatic ER may not be related to protein translocation (33b), other putative G proteins could be involved (33c). GTP-specific events may also be relevant to RER structure and may contribute to the maintenance and reformation of the 3-D organization of this membrane system in both interphase and early postmitotic cells (18).

With respect to nuclear membrane fusion, this process may be relevant to the fusion of pronuclei occurring during karyogamy of gametes (51-54). Although GTP is found in all cell types and would be expected to be involved in this process, it does not stimulate nuclear membrane fusion in syncytial cells which contain numerous nuclei (i.e., in muscle cells and syncytiotrophoblasts of the placenta). Cytoskeletal elements may prevent nuclear apposition and fusion in such cell types.

VIII. FUSION OF CYTOPLASMIC MEMBRANES IN VIVO

The evidence that intracellular membranes fuse in vivo comes mainly from electron microscopy of cells in steady state (92). This system is not amenable to obvious experimentation. Hence, a model system is needed whereby intracellular membranes can be easily manipulated and their behavior documented with time. Currently the frog oocyte and embryo are being explored for their potential usefulness for studies of intracellular membrane fusion. As it was previously shown that GTP-dependent fusion was a property of ER and nuclear membranes obtained from numerous tissues and species (56,78), it was predicted that following the microinjection of rat liver

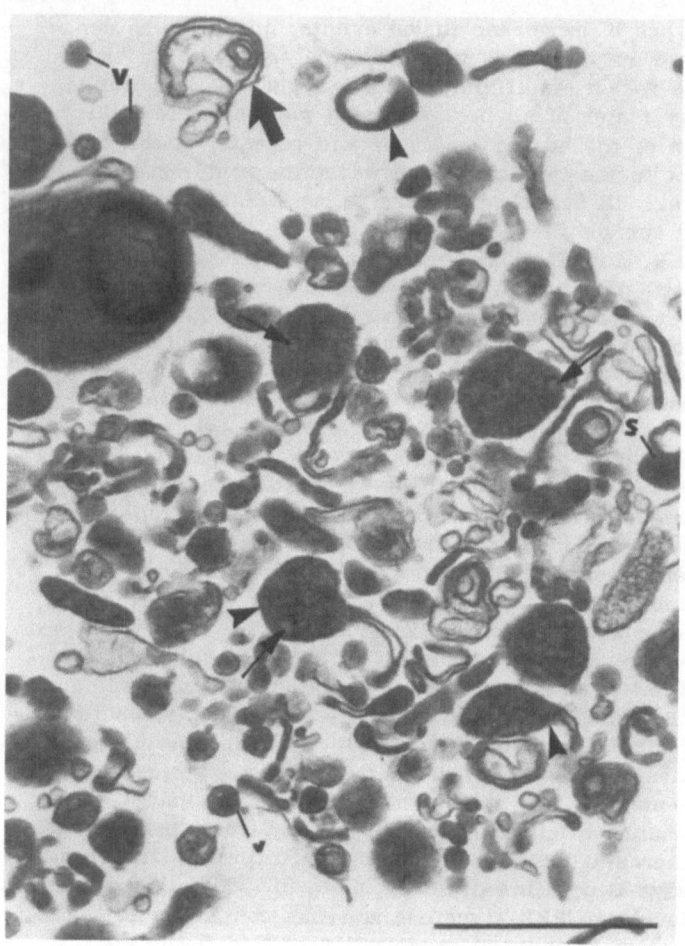

FIGURE 12   The appearance of dispersed Golgi fragments (Golgi intermediate fraction, Ref. 60) fixed in glutaraldehyde and postfixed in reduced
osmium immediately after preparation. The fraction contains a heterogeneous
population of elements including vesicles (v), double membrane structures
some without content (large arrow), and others terminating in bulbous
ends filled with lipoprotein particles (arrowheads). Lipoprotein particles
measuring 20-80 nm in diameter are easily distinguished within many of
these elements (small arrows) and represent the luminal marker for rat
liver Golgi apparatus. Bar = 0.5 μm.

microsomes into frog oocytes, such microsomes should fuse and integrate
into the general pool of cytoplasmic membranes of the host cell.

    Microinjection experiments using rat liver RER microsomes and disrupted
rat liver Golgi membranes have shown the feasibility of employing such an
approach (93,93a,b,c,96). Microinjected organelles are localized within the
oocyte cytoplasm by the simultaneous injection of an oil droplet together
with the membrane fractions, and by using morphological and cytochemical
markers characteristic of rat liver ER and Golgi fractions to identify elements
of the injected organelles. A time-dependent reorganization of organellar
structures has been documented implicating membrane fusion events (93,
93a,b,c,96). Results obtained with Golgi derivatives are particularly informa-

tive because such derivatives contain a hepatic marker (lipoprotein particles) which has the advantage of being within membrane-limited fragments. Following microinjection of completely dispersed rat liver Golgi microsomes (Fig. 12) they become decorated with membrane-coated regions (Fig. 13). Lipoprotein particles are often found as integral components of Golgi stacks (Fig. 14) and are observed within elements in direct continuity with fenestrated Golgi saccules (Fig. 15). The results can be interpreted in several ways either of which implicates membrane fusion processes. The lipoprotein particles were either transported to and incorporated within preexisting host Golgi apparatus or the injected fragments coalesced among themselves

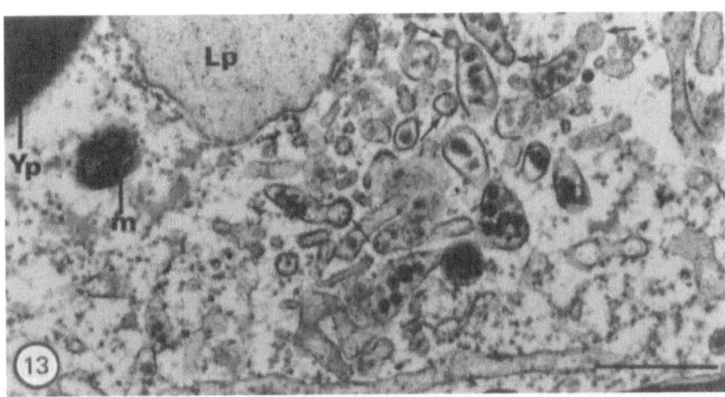

FIGURE 13 Electron micrograph showing the cytoplasmic region of a host oocyte 120 min after microinjection with Golgi fragments. Numerous vesicles containing lipoprotein particles are seen in the middle of the field. Coated regions can be seen at the periphery of some of these vesicles (arrows). Lp, lipid droplet; (m), mitochondrion; (Yp), yolk platelet. Bar = 0.5 μm. (From Ref. 93.)

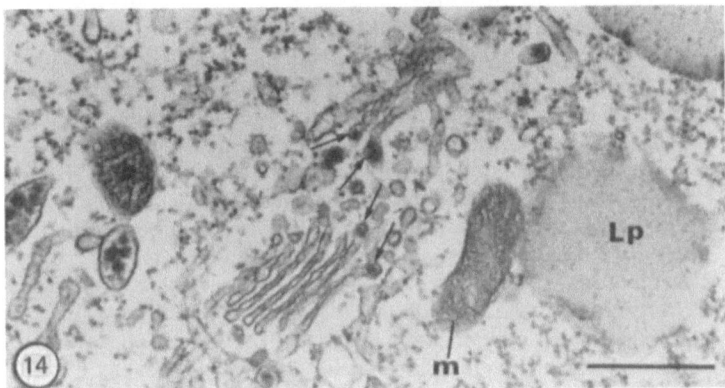

FIGURE 14 Oocyte cytoplasm near the site of microinjection 120 min after injection. Two Golgi complexes can be seen. The bulbous ends of several of the stacked saccules contain lipoprotein particles (arrows). (m), mitochondrion; (Lp), lipid droplet. Bar = 0.5 μm. (From Ref. 93.)

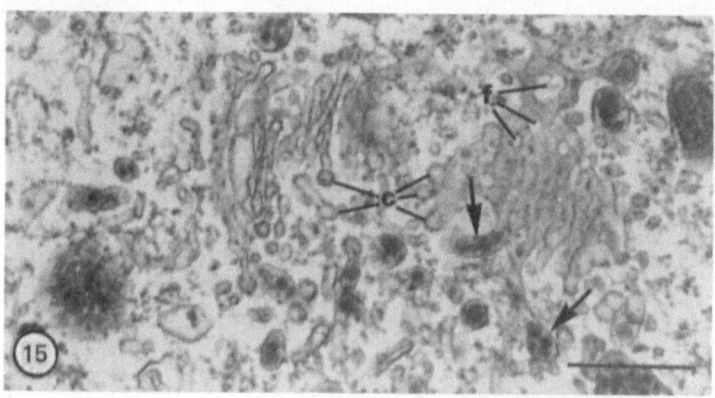

FIGURE 15   Oocyte cytoplasm 120 min after microinjection of Golgi fragments. Stacked golgi saccules can be seen cut tangentially as well as a
Golgi saccule cut obliquely and seen *en face*. Lipoprotein particles can
be seen within tubules and vesicles at the periphery of these elements
as well as in the dilate rim of the obliquely section saccule (arrows).
(c), coated regions; (f), fenestrations. Bar = 0.5 μm. (From Ref. 93.)

and/or with host membranes to reconstitute Golgi apparatus. These studies
have recently been confirmed by using radiolabeled Golgi fragments and
electron microscope radioautography (93c). Therefore, microinjection of
cellular subfractions into live cells may be useful to study membrane fusion
in vivo and to determine its potential role in organelle assembly and functional organization.

IX. CONCLUSIONS

The factors affecting fusion among three types of intracellular membranes—
nuclear envelope, rough endoplasmic reticulum, and Golgi membranes—have
been examined. The cation and nucleotide requirements for the fusion
of these membranes have been shown to be unique. Recombinant studies
using mixtures of cytoplasmic membranes and mixtures of membranes from
different cell types demonstrated membrane specificity and cell specificity.
Correlated morphological and biochemical studies indicated coincident structural and molecular changes. The data lead to the following conclusions:
The capacity for nuclear envelopes and rough endoplasmic reticulum membranes to fuse in the presence of physiological concentrations of GTP and
Mg^{2+} and Mn^{2+} is a property common to all eukaryotic cells. Recombinant
experiments point to the existence of membrane recognition factors at the
surface of different organelles. Such factors would, in part, account for
known membrane interaction specificity. Finally, biochemical changes occurring coincident with certain fusion processes point to the possible relevance
of membrane-perturbing events in molecular translocation processes.

Future work must aim at an understanding of the mechanism of the
in vitro phenomena described herein and at defining in vivo models that
will permit experimental control of the behavior (fusion) of cytoplasmic
membranes within intact cells. Although interpretation of results obtained
using natural membranes is made difficult by the molecular complexity of
such membranes, their use as models to study fusion is justified by the

fact that results obtained with them are more likely to be representative of what is happening in the living cell. Furthermore, such membrane systems permit the analysis of certain questions that could not otherwise be examined, e.g., the question of how interaction between different membrane types is controlled. Technology similar to that applied to purely artificial membrane vesicles (94) when applied to studies of the interactions of natural membranes will likely prove informative and relevant. Recombinant studies employing natural membranes and liposomes with specified lipid compositions may help to single out the relevant lipidic molecular species controlling the fusion phenomena described here. With respect to the role of proteins in fusion, it becomes imperative to determine their exact contribution in membrane recognition and membrane coalescence. Ultimately it would be hoped to be able to reconstitute such factors into bilayer membranes and demonstrate specific binding and/or fusion activity. Finally, experimental models are required to study cytoplasmic membrane fusion in vivo. Such model systems are required (1) to provide proof of fusion in vivo, (2) to define the factors involved, (3) to quantitate the events, and (4) to demonstrate directly a role in cell physiology.

ACKNOWLEDGMENTS

This work was supported by grants from the Medical Research Council of Canada, and the Fonds de la recherche en santé du Quebéc. Appreciation is extended to Line Roy and Micheline Fortin for their assistance in setting up and carrying out morphometric fusion assays and to Marjory Jolicoeur and Jeanne Fournel for their help with correlated biochemical assays. We thank Chantal Joseph for typing the manuscript.

REFERENCES

1. Palade, G. (1975). Intracellular aspects of the process of protein synthesis. *Science* 189:347-358.
2. De Duve, C. and Wattiaux, R. (1966). Functions of lysosomes. *Annu. Rev. Physiol.* 28:435-492.
3. Bergeron, J. J. M., Cruz, J., Khan, M. N., and Posner, B. I. (1985). Uptake of insuline and other ligands into receptor-rich endocytic components of target cells: The endosomal apparatus. *Annu. Rev. Physiol.* 47:383-403.
4. Porter, K. R., and Machado, R. D. (1960). Studies on the endoplasmic reticulum. IV. Its form and distribution during mitosis in cells of onion root tip. *J. Biophys. Biochem. Cytol.* 7:167-183.
5. Franke, W. W. (1977). Structure and function of nuclear membranes. *Biochem. Soc. Symp.* 43:125-135.
6. Warren, G. (1985). Membrane traffic and organelle division. *Trends Biochem. Sci.* 10:439-443.
7. Poste, G., and Allison, A. C. (1973). Membrane fusion. *Biochim. Biophys. Acta* 300:421-465.
8. Düzgüneş, N. (1985). Membrane fusion. In: *Subcellular Biochemistry*, Vol. 11 (D. B. Roodyn, ed.). Plenum Press, New York, pp. 195-286.
9. Gratzl, M., and Dahl, G. (1976). Ca^{2+}-induced fusion of Golgi-derived secretory vesicles isolated from rat liver. *FEBS Lett.* 62:142-145.

10. Quinn, P. S., and Judah, J. D. (1978). Calcium-dependent Golgi-vesicle fusion and cathepsin B in the conversion of proalbumin into albumin in rat liver. *Biochem. J.* 172:301-309.

11. Baydoun, E. A. H., and Northcote, D. H. (1980). Measurement and characteristics of fusion of isolated membrane fractions from maize root tips. *J. Cell Sci.* 45:169-186.

12. Oates, P. J., and Touster, O. (1980). In vitro fusion of *Acanthamoeba* phagolysosomes. III. Evidence that cyclic nucleotides and vacuole sub-populations respectively control the rate and the extent of vacuole fusion in *Acanthamoeba* homogenates. *J. Cell Biol.* 85:804-810.

13. Altstiel, L., and Branton, D.(1983). Fusion of coated vesicles with lysosomes: Measurement with a fluorescence assay. *Cell* 32:921-929.

14. Lohka, M. J., and Masui, Y. (1984). Roles of cytosol and cytoplasmic particules in nuclear envelope assembly and pronuclear formation in cell-free preparations from amphibian eggs. *J. Cell Biol.* 98:1222-1230.

15. Balch, W. E., Glick, B. S., and Rothman, J. E. (1984). Sequential intermediates in the pathway of intercompartmental transport in a cell-free system. *Cell* 39:525-536.

16. Davey, J., Hurtley, S. M., and Warren, G. (1985). Reconstitution of an endocytotic fusion event in a cell-free system. *Cell* 43:643-652.

17. Burke, B., and Gerace, L. (1986). A cell free system to study re-assembly of the nuclear envelope at the end of mitosis. *Cell* 49:639-652.

18. Paiement, J., Beaufay, H., and Godelaine, D. (1980). Coalescence of microsomal vesicles from rat liver: A phenomenon occurring in parallel with enhancement of the glycosylation activity during incubation of stripped rough microsomes with GTP. *J. Cell Biol.* 86:29-37.

19. Paiement, J., and Bergeron, J. J. M. (1983). Localization of GTP-stimulated core glycosylation to fused microsomes. *J. Cell Biol.* 96:1791-1796.

20. Godelaine, D., Beaufay, H., and Wibo, M. (1977). Incorporation of N-acetylglucosamine into endogenous acceptors of rough microsomes from rat liver: Stimulation by GTP after treatment with pyrophosphate. *Proc. Natl. Acad. Sci. USA* 74:1095-1099.

21. Godelaine, D., Beaufay, H., Wibo, M., and Amar-Costesec, A. (1979). The dolichol pathway of protein glycosylation in rat liver. Stimulation by GTP of the incorporation of N-acetylglucosamine in endogenous lipids and proteins of rough microsomes treated with pyrophosphate. *Eur. J. Biochem.* 96:17-26.

22. Godelaine, D., Beaufay, H., and Wibo, M. (1979). The dolichol path-way of protein glycosylation in rat liver. Incorporation of mannose into endogenous lipids and proteins of rough microsomes. *Eur. J. Biochem.* 96:27-34.

23. Rothman, J. E., and Lodish, H. F. (1977). Synchronized transmembrane insertion and glycosylation of a nascent membrane protein. *Nature* 269:775-780.

24. Mueckler, M., and Lodish, H. F. (1986). The human glucose transporter can insert posttranslationally into microsomes. *Cell* 44:629-637.

25. Waters, M. G., and Blobel, G. (1986). Secretory protein translocation in a yeast cell-free system can occur posttranslationally and requires ATP hydrolysis. *J. Cell Biol.* 102:1543-1550.

26. Clifford, A. J., Ruimallo, J. A., Baliga, B. S., Munro, H. W., and Brown, P. R. (1972). Liver nucleotide metabolism in relation to amino acid supply. *Biochim. Biophys. Acta* 277:443-458.

27. Van den Berghe, G., Bronfman, M., Vanneste, R., and Hers, H. G. (1977). The mechanism of adenosine triphosphate depletion in the liver after load of fructose. A kinetic study of liver adenylate deaminase. *Biochem. J.* 162:602-609.

28. Corkey, B. E., Duzynski, J., Rich, T. L., Matschinsky, B., and Williamson, R. (1986). Regulation of free and bound magnesium in rat hepatocytes and isolated mitochondria. *J. Biol. Chem.* 261:2567-2574.

29. Schramm, V. L. (1982). Metabolic regulation: Could Mn^{2+} be involved? *Trends Biochem. Sci.* 7:369-371.

30. Tocanne, J. F., Ververgaert, P. H. J. T., Verkleij, A. J., and van Deenen, L. L. M. (1974). A monolayer and freeze-etching study of charged phospholipids: I. Effects of ions and pH on the ionic properties of phosphatidylglycerol and lysylphosphatidylglycerol. *Chem. Phys. Lipids* 12:201-219.

31. Ververgaert, P. H. J. T., de Kruyff, B., Verkleij, B., Tocanne, J. F., and van Deenen, L. L. M. (1975). Calorimetric and freeze-etch study of the influence of Mg^{++} on the thermotropic behavior of phosphatidylglycerol. *Chem. Phys. Lipids* 14:97-101.

32. Düzgüneş, N., Wilschut, J., Fraley, R., and Papahadjopoulos, D. (1981). Studies on the mechanism of membrane fusion: Role of headgroup composition in calcium- and magnesium-induced fusion of mixed phospholipid vesicles. *Biochim. Biophys. Acta* 642:182-195.

33. Sundler, R., and Papahadjopoulos, D. (1981). Control of membrane fusion by phospholipid headgroups. I. Phosphatidate/phosphatidylinositol specificity. *Biochim. Biophys. Acta* 649:743-750.

33a. Lanoix, J., Roy, L., and Paiement, J. (1989). Detection of GTP-binding proteins in purified derivatives of rough endoplasmic reticulum. *Biochem. J.* 262:497-503.

33b. Audigier, Y., Nigam, S. K., and Blobel, G. (1988). Identification of a G protein in rough endoplasmic reticulum of canine pancreas. *J. Biol. Chem.* 263:16352-16357.

33c. Robinson, A., and Austen, B. (1987). GTP-dependent ADP-ribosylation of a 22 kDa protein in the endoplasmic reticulum membrane. *FEBS Lett.* 218:63-67.

34. Arion, W. J., Ballas, L. M., Lange, A. J., and Wallis, B. K. (1976). Microsomal membrane permeability and the hepatic glucose-6-phosphatase system: Interactions of the system with D-mannose-6-phosphate and D-mannose. *J. Biol. Chem.* 251:4901-4907.

35. Godelaine, D., Beaufay, H., Wibo, M., and Ravoet, A. M. (1983). Alteration of membrane barrier in stripped rough microsomes from rat liver on incubation with GTP: Its relevance to the stimulation by this nucleotide of the dolichol pathway for protein glycosylation. *J. Cell Biol.* 97:340-350.

36. Aldwinkle, T. J., Ahkong, Q. F., Bangham, A. D., Fisher, D., and Lucy, J. A. (1982). Effects of poly(ethylene glycol) on liposomes and erythrocytes. Permeability changes and membrane fusion. *Biochim. Biophys. Acta* 689:548-560.

37. Creutz, C. E. (1981). *Cis*-unsaturated fatty acids induce the fusion of chromaffin granules aggregated by synexin. *J. Cell Biol.* 91:247-256.

38. Gad, A. E., Silver, B. L., and Eytan, G. D. (1982). Polycation-induced fusion of negatively-charged vesicles. *Biochim. Biophys. Acta* 690:124-132.

39. Hunt, G. R. A., and Jones, I. C. (1983). A ^{1}H-NMR investigation of the effect of ethanol and general anaesthetics on ion channels and

membrane fusion using unilamellar phospholipid membranes. *Biochim. Biophys. Acta* 736:1-6.

40. Messineo, F. C., Rathier, M., Favreau, C., Watras, J., and Takenaka, H. (1984). Mechanism of fatty acid effects on sarcoplasmic reticulum. III. The effect of palmitic and oleic acids on sarcoplasmic reticulum function. A model for fatty acid membrane interactions. *J. Biol. Chem.* 259:1336-1343.

41. Wilschut, J., Düzgüneş, N., Hong, K., Hoekstra, D., and Papahadjopoulos, D. (1983). Retention of aqueous contents during divalent cation-induced fusion of phospholipid vesicles. *Biochim. Biophys. Acta* 734: 309-318.

42. Ojakian, G. K., Kreibich, G., and Sabatini, D. D. (1977). Mobility of ribosomes bound to microsomal membranes. A freeze-etch and thin section electron microscope study of the structure and fluidity of the rough endoplasmic reticulum. *J. Cell Biol.* 72:530-551.

43. Paiement, J., Rindress, D., Smith, C. E., Poliquin, L., and Bergeron, J. J. M. (1987). Properties of a GTP sensitive microdomain in rough microsomes. *Biochim. Biophys. Acta* 898:6-22.

44. Helenius, A., Marsh, M., and White, J. (1980). The entry of viruses into animal cells. *Trends Biochem. Sci.* 5:104-106.

45. Sribney, M., Dove, J. L., and Lyman, E. M. (1977). Studies on the synthesis of CDP-Diacylglycerol: Stimulation by GTP and inhibition by ATP and fluoride. *Biochem. Biophys. Res. Commun.* 79:749-755.

46. Bearer, E. L., and Friend, D. S. (1982). Modifications of anionic-lipid domains preceding membrane fusion in Guinea pig sperm. *J. Cell Biol.* 92:604-615.

47. Papahadjopoulos, D. (1978). Calcium-induced phase changes and fusion in natural and model membranes. In: *Membrane Fusion* (G. Poste and G. L. Nicolson, eds.). Elsevier/North-Holland Biomedical Press, Amsterdam, pp. 765-790.

47a. Goud, B., Salminen, A., Walworth, N. C., and Novick, P. J. (1988). A GTP-binding protein required for secretion rapidly associates with secretory vesicles and the plasma membrane in yeast. *Cell* 53:753-768.

48. Franke, W. W., Scheer, U., Krohne, G., and Jarasch, E. D. (1981). The nuclear envelope and the architecture of the nuclear periphery. *J. Cell Biol.* 91:39-50s.

49. Watson, M. L. (1955). The nuclear envelope: Its structure and relation to cytoplasmic membranes. *J. Biophys. Biochem. Cytol.* 1:257-270.

50. Fawcett, D. W. (1965). In: *Intracellular Membranous Structure* (S. Seno and E. V. Cowdry, eds.). Japan Society for Cell Biology, Okayana, pp. 15-40.

51. Jensen, W. A. (1964). Observations on the fusion of nuclei in plants. *J. Cell Biol.* 23:669-672.

52. Longo, F. J., and Anderson, E. (1968). The fine structure of pronuclear development and fusion in the sea urchin, *Arbacia punctulata*. *J. Cell Biol.* 39:339-368.

53. Urban, P. (1969). The fine structure of pronuclear fusion in the coenocytic marine alga *Bryopsis hypnoides Lamouroux*. *J. Cell Biol.* 42:606-611.

54. Franke, W. W. (1974). Structure, biochemistry and functions of the nuclear envelope. In: *International Review of Cytology*, suppl. 4 (G. H. Bourne and J. F. Danielli, eds.). Academic Press, London, pp. 71-236.

55. Paiement, J. (1981). GTP-dependent fusion of outer nuclear membranes in vitro. *Exp. Cell Res.* 134:93-102.

56. Paiement, J. (1984). GTP stimulates fusion between homologous and heterologous nuclear membranes. *Biochim. Biophys. Acta* 777:274-282.

57. Arion, W. J., Schulz, L. O., Lange, A. J., Telford, J. N., and Walls, H. E. (1983). The characteristics of liver glucose-6-phosphatase in the envelope of isolated nuclei and microsomes are identical. *J. Biol. Chem.* 258:12661-12669.

58. Powell, J. T., and Brew, K. (1976). A comparison of the interactions of galactosyltransferase with a glycoprotein substrate (Ovalbumin) and with α-lactalbumine. *J. Biol. Chem.* 251:3653-3663.

59. Paiement, J., Rachubinski, R. A., Ng Ying Kin, N. M. K., Sikstrom, R. A., and Bergeron, J. J. M. (1982). Membrane fusion and glycosylation in the rat hepatic Golgi apparatus. *J. Cell Biol.* 92:147-154.

60. Bergeron, J. J. M., Paiement, J., Khan, M. N., and Smith, C. E. (1985). Terminal glycosylation in rat hepatic Golgi fractions: Heterogeneous locations for sialic acid and galactose acceptor and their transferases. *Biochim. Biophys. Acta* 821:393-403.

61. Bergeron, J. J. M., Paiement, J., Rachubinski, R., Ng Ying Kin, N. M. K., and Sikstrom, R. (1981). Membrane fusion and the mechanism of terminal glycosylation within the Golgi apparatus of rat liver hepatocytes. *Biophys. J.* 37:121-122.

62. Sommers, L. W., and Hirshberg, C. B. (1982). Transport of sugar nucleotides into rat liver Golgi. A new Golgi marker activity. *J. Biol. Chem.* 257:10811-10817.

63. Farquhar, M. G. (1985). Progress in unraveling pathways of Golgi traffic. *Annu. Rev. Cell Biol.* 1:447-488.

64. Dunphy, W. G., and Rothman, J. E. (1983). Compartmentation of asparagine-linked oligosaccharide processing in the Golgi apparatus. *J. Cell Biol.* 97:270-275.

65. Dunphy, D. G., Fries, E., Urbani, L. J., and Rothman, J. E. (1981). Early and late functions associated with the Golgi apparatus reside in distinct compartments. *Proc. Natl. Acad. Sci. USA* 78:7453-7457.

66. Elhammer, A., and Kornfeld, S. (1984). Two enzymes involved in the synthesis of O-linked oligosaccharides are localized on membranes of different densities in mouse lymphoma BW5147 cells. *J. Cell Biol.* 98:327-331.

67. Goldberg, D. E., and Kornfeld, S. (1983). Evidence for extensive subcellular organization of asparagine-linked oligosaccharide processing and lysosomal enzyme phosphorylation. *J. Biol. Chem.* 258:3159-3165.

68. Griffiths, G., Brands, R., Burke, B., Louvard, D., and Warren, G. (1982). Viral membrane proteins acquire galactose in Trans Golgi cisternae during intracellular transport. *J. Cell Biol.* 95:781-792.

69. Roth, J. (1983). Application of lectin-gold complexes for electron microscopic localization of glycoconjugates on thin section. *J. Histochem. Cytochem.* 31:987-999.

70. Sato, A., and Spicer, S. S. (1982). Untrastructural visualization of galactose in the glycoprotein of gastric surface cells with a peanut lectin conjugate. *Histochem. J.* 14:125-138.

71. Sato, A., and Spicer, S. (1982). Ultrastructural visualization of galactosyl residues in various alimentary epithelial cells with the peanut lectin-horseradish peroxidase procedure. *Histochemistry* 73:607-624.

72. Kramer, M. F. and Geuze, J. J. (1980). Comparison of various methods to localize a source of radioactivity in ultrastructural autoradiographs.

The site of [^{3}H]-galactose incorporation in surface mucous cells of the rat stomach. *J. Histochem. Cytochem.* 28:381-387.

73. Bennet, G., and O'Shaughnessy, D. (1981). The site of incorporation of sialic acid residues into glycoproteins and the subsequent fates of these molecules in various rat and mouse cell types as shown by radioautography after injection of [^{3}H]N-acetylmannosamine I. Observations in hepatocytes. *J. Cell Biol.* 88:1-15.

74. Dunphy, W. G., Brands, R., and Rothman, J. E. (1985). Attachment of terminal N-acetylglucosamine to asparagine-linked oligosaccharides occurs in central cisternae of the Golgi stack. *Cell* 40:463-472.

75. Roth, J., and Berger, E. G. (1982). Immunocytochemical localization of galactosyltransferase in Hela cells: Codistribution of thiamine pyrophosphatase in *trans*-Golgi cisternae. *J. Cell Biol.* 92:223-229.

76. Slot, J. W., and Geuze, J. J. (1983). Immunoelectron microscopic exploration of the Golgi complex. *J. Histochem. Cytochem.* 31:1049-1056.

76a. Yvan, L., Barriocanal, J. G., Bonifacino, J. S., and Sandoval, I. V. (1987). Two integral membrane proteins located in the *cis*-middle and *trans*-part of the Golgi system acquire sialylated N-linked carbohydrates and display different turnovers and sensitivity to cAMP-dependent phosphorylation. *J. Cell Biol.* 105:215-227.

76b. Gonatas, J. O., Mezitis, S. G. E., Sticher, A., Fleisher, B., and Gonatas, N. K. (1989). MG-160: A novel sialoglycoprotein of the medial cisternae of the Golgi apparatus. *J. Biol. Chem.* 264:646-653.

77. Braell, W. A., Balch, W. E., Dobbertin, D. C., and Rothman, J. E. (1984). The glycoprotein that is transported between successive compartments of the Golgi in a cell-free system resides in stacks of cisternae. *Cell* 39:511-524.

78. Paiement, J. (1984). Physiological concentrations of GTP stimulate fusion of the endoplasmic reticulum and the nuclear envelope. *Exp. Cell Res.* 151:354-366.

78a. Beckers, C. J. M., Keller, D. S., and Balch, W. E. (1987). Semi-intact cells permeable to macromolecules: Use in reconstitution of protein transport from the endoplasmic reticulum to the Golgi complex. *Cell* 50:523-534.

78b. Lodish, H. F., Kong, N., Hirani, S., and Rasmussen, J. (1987). A vesicular intermediate in transport of hepatoma secretory proteins from the rough endoplasmic reticulum to the Golgi complex. *J. Cell Biol.* 104:221-230.

78c. Paulik, M., Nowack, D. D., and Morré, D. J. (1988). Isolation of a vesicular intermediate in the cell-free transfer of membrane from transitional elements of the endoplasmic reticulum to Golgi apparatus cisternae of rat liver. *J. Biol. Chem.* 263:17738-17748.

78d. Orci, L., Malhotra, V., Amherdt, M., Serafini, T., and Rothman, J. (1989). Dissection of a single round of vesicular transport: Sequential intermediates for intercisternal movement in the Golgi stack. *Cell* 56:357-368.

78e. Beckers, C. J. M., and Balch, W. E. (1989). Calcium and GTP: Essential components in vesicular trafficking between the endoplasmic reticulum and Golgi apparatus. *J. Cell Biol.* 108:1245-1256.

78f. Comerford, J. G., and Dawson, A. P. (1988). The mechanism of action of GTP on Ca^{++} efflux from rat liver microsomal vesicles. *Biochem. J.* 249:89-93.

79. Crabb, J. H., and Jackson, R. C. (1985). In vitro reconstitution of exocytosis from plasma membrane and isolated secretory vesicles. *J. Cell Biol.* 101:2263-2273.

80. Zimmerberg, J., Sardet, C., and Epel, D. (1985). Exocytosis of sea urchin egg cortical vesicles in vitro is retarded by hyperosmotic sucrose: Kinetics of fusion monitored by quantitative light-scattering microscopy. *J. Cell Biol.* 101:2398-2410.

81. Dahl, G., and Gratzl, M. (1976). Calcium-induced fusion of isolated secretory vesicles from the islet of langerhans. *Cytobiologie* 12:344-355.

82. Judah, J. D., and Quinn, P. S. (1978). Calcium ion-dependent vesicle fusion in the conversion of proalbumin to albumin. *Nature* 271:384-385.

83. Papahadjopoulos, D., Vail, W. J., Newton, C., Nir, S., Jacobson, K., Poste, G., and Lazo, R. (1977). Studies on membrane fusion III. The role of calcium-induced phase changes. *Biochim. Biophys. Acta* 465: 579-598.

84. Wilschut, J., Düzgüneş, N., and Papahadjopoulos, D. (1981). Calcium/magnesium specificity in membrane fusion: Kinetics of aggregation and fusion of phosphatidylserine vesicles and the role of bilayer curvature. *Biochemistry* 20:3126-3133.

85. Ohki, S., and Ohshima, H. (1985). Divalent cation-induced phosphatidic acid membrane fusion. Effect of ion binding and membrane surface tension. *Biochim. Biophys. Acta* 812:147-154.

86. Papahadjopoulos, D., Vail, W. J., Panghorn, W. A., and Poste, G. (1976). Studies on membrane fusion II. Induction of fusion in pure phospholipid membranes by calcium ions and other divalent metals. *Biochim. Biophys. Acta* 448:265-283.

87. Hoekstra, D., and Martin, O. C. (1982). Transbilayer redistribution of phosphatidylethanolamine during fusion of phospholipid vesicles. Dependence of fusion rate, lipid phase separation and formation of nonbilayer structures. *Biochemistry* 21:6097-6103.

88. Düzgüneş, N., Paiement, J., Freeman, K. B., Lopez, N. G., Wilschut, J., and Papahadjopoulos, D. (1984). Modulation of membrane fusion by ionotropic and thermotropic phase transitions. *Biochemistry* 23:3486-3494.

89. Liao, M. J., and Prestegard, J. H. (1980). Ion specificity in fusion of phosphatidic acid-phosphatidylcholine mixed lipid vesicles. *Biochim. Biophys. Acta* 601:453-461.

90. Sundler, R., Düzgünes, N., and Papahadjopoulos, D. (1981). Control of membrane fusion by phospholipid head groups II. The role of phosphotidylethanolamine in mixtures with phosphatidate and phosphatidylinositol. *Biochim. Biophys. Acta* 649:751-758.

91. Meldolesi, J., Borgese, N., De Camilli, P., and Ceccarelli, B. (1978). Cytoplasmic membranes and the secretory process. In: *Membrane Fusion* (G. Poste and G. L. Nicolson, eds.). North-Holland, Amsterdam, pp. 509-627.

92. Morré, D. J., Kartenbeck, J., and Franke, W. W. (1979). Membrane flow and interconversions among endomembranes. *Biochim. Biophys. Acta* 559:71-152.

92a. Connolly, T., and Gilmore, R. (1986). Formation of a functional ribosome-membrane junction during translocation requires the participation of a GTP-binding protein. *J. Cell Biol.* 103:2253-2261.

93. Paiement, J. (1986). Morphology of endoplasmic reticulum and Golgi elements following microinjection of rat liver microsomes into *Xenopus laevis* oocyte cytoplasm. *Exp. Cell Res.* 166:510-518.

93a. Paiement, J., Kan, F. W. K., Lanoix, J., and Blain, M. (1988). Cytochemical analysis of the reconstitution of endoplasmic reticulum after microinjection of rat liver microsomes into *Xenopus* oocytes. *J. Histochem. Cytochem.* 36:1263–1273.

93b. Dominguez, J. M., and Paiement, J. (1989). Reconstitution of endoplasmic reticulum in rapidly dividing cells of early *Xenopus* embryos. *Am. J. Anat.* 186:99–113.

93c. Paiement, J., Jolicoeur, M., Fazel, A., and Bergeron, J. J. M. (1989). Reconstitution of the Golgi apparatus after microinjection of rat liver Golgi fragments into *Xenopus* oocytes. *J. Cell Biol.* 108:1257–1269.

94. Wilschut, J., and Hoekstra, D. (1984). Membrane fusion: From liposomes to biological membranes. *Trends Biochem. Sci.* 9:479–483.

95. Vilmart-Seuwen, J., Kersken, H., Stürzl, R., and Plattner, H. (1986). ATP keeps exocytosis sites in a primed state but is not required from membrane fusion: An analysis with Paramecium cells in vivo and in vitro. *J. Cell Biol.* 103:1279–1288.

96. Paiement, J., Dominguez, J. M., McLeese, J., Bernier, J., Roy, L., and Bergeron, M. (1990). Morphogenesis of endoplasmic reticulum in *Xenopus* oocytes after microinjection of rat liver smooth microsomes. *Am. J. Anat.* 187:183–192.

22

Reconstitution of Transport Between Early Compartments of the Secretory Pathway Using Cell-Free Systems

WILLIAM E. BALCH

Scripps Clinic and Research Foundation, La Jolla, California

I. OVERVIEW

To understand the molecular basis for sorting of membrane and secreted protein through the organelles of the secretory pathway of eukaryotic cells it is essential to elucidate the biochemical basis of the transport processes controlling delivery between compartments. The secretory pathway has been explored intensively in vivo since the discovery by Palade (1) of the central role of the endoplasmic reticulum (ER) and the Golgi apparatus in the targetting of proteins to secretory granules, lysosomes, and the cell surface. These lines of investigation have been the subject of numerous reviews (2-12). In contrast, until recently little attention was directed toward dissecting the biochemistry of the secretory pathway through reconstitution of transport between specific compartments using cell-free systems.

The focus of this chapter is twofold: to present some of the basic principles used in development of novel assays to study specific segments of the secretory pathway in vitro, and to provide an overview of the success this approach has had in beginning to define the enzymatic mechanisms underlying export of protein from the ER and transport between compartments of the Golgi apparatus.

II. COMPARTMENTALIZATION OF THE SECRETORY PATHWAY IN VIVO

Two general functions distinguish the biochemical properties of organelles of the secretory pathway. They are observed in the specialization of each compartment to carry out a selected class of posttranslational modifications and the capacity of these compartments to process and transport a broad spectrum of proteins through their membrane-delimited boundaries. The ER is a single, extensive reticular network occupying the entire cytoplasm of the eukaryotic cell and is the common site of synthesis of all proteins found in subsequent compartments of the exocytotic pathway (2). The biochemical basis for cotranslational insertion (3), attachment of asparagine-linked core oligosaccharides to the nascent chain (13), and subsequent ER-associated posttranslational modifications, including proteolysis of the signal peptide (14) and processing of N-linked oligosaccharides (15-17), has been the subject of investigation in vivo and in vitro. The Golgi apparatus is a stack of functionally distinct compartments. On the basis of morphological, biochemical, cytochemical, and immunocytochemical studies, the current "consensus" Golgi structure contains at least three compartments (Fig. 1). These compartments, the *cis*, *medial* and *trans* Golgi, house a variety of processing enzymes which modify proteins in transit through each compartment (see Ref. 6 for a recent review). The best studied are the enzymes that sequentially process the high-mannose, asparagine-linked core oligosaccharides acquired in the ER to the complex oligosaccharide structure containing additional N-acetylglucosamine, galactose, and sialic acid residues prior to exit from the *trans* Golgi (16,17). This oligosaccharide-processing pathway and the compartmental organization of the enzymes involved are presented in Figure 1.

A genetic analysis of the secretory pathway in yeast has established that a common biochemical machinery regulates transport of all protein between subcellular compartments of the secretory pathway (18). Both morphological and biochemical studies implicate a role for intermediate carrier vesicles in vivo between the ER and the Golgi (19-21), between Golgi subcompartments (1,22-26), and between the Golgi and the cell surface (27). Functional studies on the transport of a viral glycoprotein between early compartments of the secretory pathway have provided direct evidence that transport in vivo is dissociative (25,26). The synthesis, posttranslational processing, and transfer of the G protein of vesicular stomatitis virus (VSV) between cell compartments have been studied extensively in vivo and are indistinguishable from the pathway followed by normal cellular membrane-bound glycoproteins (28). During synthesis and insertion into the ER the G protein acquires two high-mannose core oligosaccharides (29). Subsequent transport between the *cis*, *medial*, and *trans* compartments can be evaluated on the basis of the processing of the G-protein oligosaccharides (Fig. 1, Ref. 30). A mutant cell line [Chinese hamster ovary (CHO), clone 158] deficient in a Golgi-associated processing enzyme found in the *medial* compartment [N-acetylglucosamine transferase I (Tr I, Fig. 1)] was infected with VSV to populate the secretory pathway with G protein. This cell line was fused to an uninfected wild-type cell line containing the (missing) processing enzyme. Using this novel approach, it was shown that the G protein was transferred with high efficiency from the *cis* Golgi compartment of the mutant cell line to the *medial* Golgi compartment of the wild-type cell line through their common cytoplasm in vivo (25,26). Such transfer between compartments of the exocytotic pathway could occur only if transport in vivo proceeds via a mechanism involving carrier vesicles.

Further analysis using this cell/cell fusion assay revealed that inter-Golgi transfer in vivo is vectorial. The G protein could be transported between compartments only in the *cis*-to-*trans* direction (26). These studies provided the first in vivo evidence that functional boundaries regulate specific transfer of protein between secretory compartments. Additional evidence for these functional boundaries was provided by the discovery that transport between the ER and Golgi subcompartments requires a

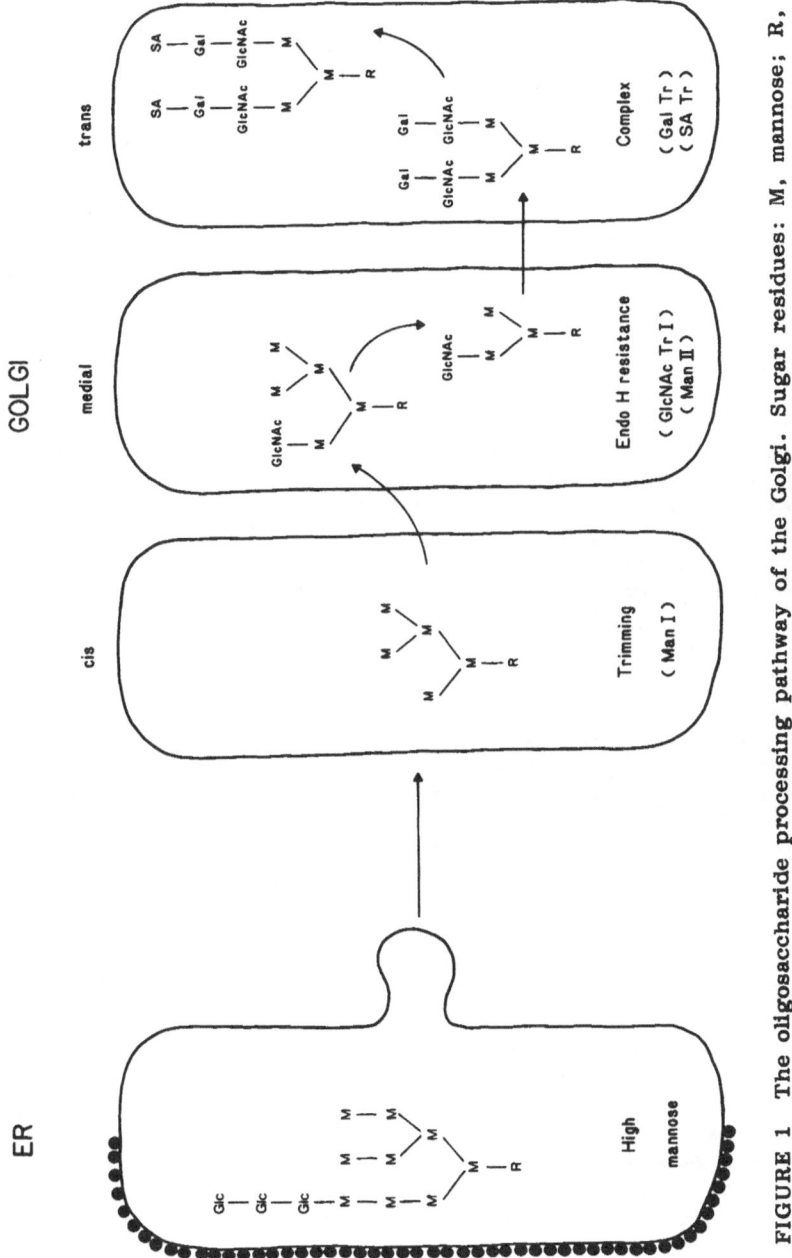

FIGURE 1 The oligosaccharide processing pathway of the Golgi. Sugar residues: M, mannose; R, polypeptide chain; Glc, glucose; GlcNAc, N-acetylglucosamine; Gal, galactose; SA, sialic acid. Enzymes: man I, mannosidase I; GlcNAc Tr I, N-acetylglucosamine transferase I; Man II, mannosidase II; Gal Tr, galactosyl transferase; SA Tr, sialyl transferase.

remarkably high threshold of cellular ATP (31). Transport between compartments was blocked when the cellular ATP pool was reduced by only 50% (31). Transport between all compartments was found to occur via a two-step process in which an initial ATP-sensitive export (budding) step was followed by a second ATP-insensitive delivery (fusion) step (31). From these and other studies (18,32), it is apparent that transfer in vivo between compartments through vesicular intermediates is a highly regulated, energy-dependent process.

III. TRANSPORT OF PROTEIN IN CELL-FREE SYSTEMS

Measurement of transport between compartments in vitro is based on the concept of an organelle complementation assay introduced by Fries and Rothman (33). A cell homogenate containing the "donor" compartment is prepared from one cell line and incubated in vitro with an "acceptor" compartment prepared separately using a different cell line. The donor compartment houses the protein to be transported prior to incubation in vitro; the acceptor compartment contains a processing activity which modifies the newly transferred protein during incubation of the two compartments in vitro to "score" delivery. Through the judicious choice of donor/acceptor pairs, movement of a marker protein between specific subcellular compartments can be measured directly using this transport-coupled processing assay. The exocytotic pathway is particularly amenable to this approach. A large variety of modifications accompany transport of many proteins to the cell surface. These include glycosylation (17), sulfation (34), phosphorylation (35,36), proteolytic cleavage (37), and addition of fatty acids (38). These modifications occur in specific subcellular compartments and serve as landmarks with which to delienate segments of the transport pathway for biochemical studies.

Several criteria are used to correlate the events observed in vitro with those occurring in vivo. It is necessary to establish the cellular localization of the transported protein prior to incubation (the donor compartment) and to define the cellular localization of the processing activity marking the acceptor compartment. This information identifies the segment of the total cellular pathway being studied in vitro. Since the events occurring in vitro are a consequence of transfer between sealed membrane-bound compartments in vivo, transport should be sensitive to detergents, and the transported protein should be found in sealed compartments before and after incubation. In addition, authentic transport in vitro should occur under physiological conditions with the efficiency, specificity, and kinetics approaching those observed in vivo; it should also measure a vectorial process dependent on the distinct donor and acceptor membranes. Application of these criteria during the development of new assays will greatly facilitate interpretation of results and help to ensure that the membrane fusion events observed in vitro are relevant to the living cell.

IV. INTER-GOLGI TRANSPORT IN VITRO

A. Inter-Golgi Assay

Successful reconstitution of transport of protein between Golgi compartments in vitro by Fries and Rothman (33) relies on two important observations: knowledge of the cellular distribution of the processing intermediates of

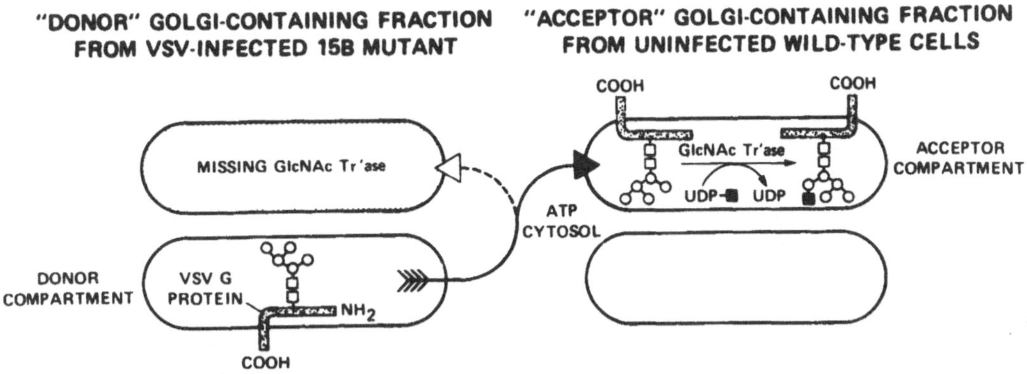

"DONOR" GOLGI-CONTAINING FRACTION FROM VSV-INFECTED 15B MUTANT

"ACCEPTOR" GOLGI-CONTAINING FRACTION FROM UNINFECTED WILD-TYPE CELLS

FIGURE 2 Assay for transport of protein between successive Golgi compartments in vitro. (Reprinted, with permission, from Ref. 43.)

the asparagine–linked oligosaccharides of the VSV G protein acquired during transport to the cell surface (Fig. 1), and the availability of mutant Chinese hamster ovary (CHO) cell lines which are defective in selected processing steps. An outline of the inter-Golgi transport assay is illustrated in Figure 2. A donor compartment is prepared from the VSV-infected mutant cell line clone 15B. Clone 15B is missing the *medial* Golgi enzyme N-acetylglucosamine transferase I (Tr I), which processes the $man_5GlcNAc_2$ oligosaccharide form of G protein to the $GlcNAc_1man_5GlcNAc_2$ form, containing an additional N-acetylglucosamine residue (GlcNAc) (Fig. 1) (39). The activity of Tr I is critical for further processing to the complex structure since the enzymes involved in these subsequent steps function sequentially. In clone 15B, the G protein is transported normally to the cell surface in the $man_5GlcNAc_2$- oligosaccharide form, since processing by Tr I, not transport, is the only known defect in this cell line (40).

A donor Golgi fraction prepared from the VSV-infected 15B cell line is mixed in vitro with an acceptor Golgi fraction prepared from the wild-type cell line containing the missing Tr I activity (Fig. 2). If the G protein is transported from the mutant donor compartment to the wild-type acceptor compartment, transfer in vitro can then be detected by one of several methods. An early approach was to follow the oligosaccharide processing of [^{35}S]methionine-labeled G protein by conversion of the G-protein oligosaccharides to a form resistant to the enzyme endoglycosidase H (endo H) (41). Endo H resistance is conferred by delivery of the G protein to the *medial* Golgi compartment housing the sequentially acting enzymes Tr I and mannosidase II in the wild-type acceptor membranes (Figs. 1 and 2). Since endo-H-sensitive and endo-H-resistant forms are of different molecular weight after treatment with the enzyme, the extent of G-protein transport can be quantitated by SDS/gel electrophoresis (33).

These early studies provided two central pieces of information about transport in vitro. First, the G protein was found to only transiently occupy a compartment in the clone 15B cells in vivo which was active in vitro (42). Kinetic studies revealed that this compartment was not the ER; rather, the G protein had to be chased from the ER to the *cis* Golgi compartment for efficient transfer in vitro to the *medial* acceptor compartment conferring endo H resistance. Further chase of G protein in vivo into the *medial* and *trans* Golgi compartments in clone 15B, prior to preparation of the donor Golgi, resulted in a striking loss of activity in vitro (26,42). These results suggest that the assay is specific for a single stage of Golgi transport; movement of G protein from a *cis* to a *medial* Golgi compartment in vitro. Second, transfer from the *cis* Golgi compartment to the *medial* Golgi compartment is very efficient (42). Nearly 100% of the G protein present in the active *cis* Golgi donor compartment was transported in vitro during a 60-min incubation. Furthermore, transport in vitro required the G protein to be present in a sealed membrane-bound compartment before and after incubation for appearance of the processed endo-H-resistant structure (33). These early results define explicitly the high level of specificity of the membrane fusion events being measured in this assay.

The next step in understanding the biochemical properties of inter-Golgi transport in vitro was a consequence of development of a rapid, filter-binding assay illustrated in Figure 2. Instead of measuring the conversion of G protein to the endo-H-resistant structure using [^{35}S]methionine-labeled G protein, Golgi fractions were prepared from unlabeled infected 15B cells and incubated in the presence of the acceptor wild-type Golgi fraction and UDP-[^{3}H]GlcNAc (uridine diphosphate-[^{3}H]N-acetylglucosamine) (43). UDP-[^{3}H]GlcNAc is the activated sugar nucleotide substrate for Tr I which adds GlcNAc to the G protein in the *medial* Golgi compartment (Fig. 1 and 2). If the G protein is transported from the *cis* Golgi donor compartment to the acceptor compartment containing Tr I in vitro, it will be modified to the form containing [^{3}H]GlcNAc (Fig. 2). Immunoprecipitation of G protein after incubation using a monospecific anti-G-protein antibody and collection on a filter for scintillation counting provides a rapid, sensitive, highly reliable quantitative measure of the extent of G-protein transport.

B.   Properties of Inter-Golgi Transport In Vitro

Initial studies established that in addition to the donor and acceptor Golgi compartments, both ATP and a high-speed supernatant fraction (cytosol) prepared from cell homogenates were required for transport of the G protein in vitro (33). In order to determine the relationships between the requirement for donor and acceptor membrane fractions, and the cytosol and ATP dependency, a kinetic analysis revealed that transport could be dissected into a series of intermediate steps shown in Figure 3 (44).

The first step in transport is formation of the *primed donor intermediate* (step 1 in Fig. 3). Priming requires both ATP and cytosol in the incubation, occurs rapidly (within just a few minutes), and is independent of the acceptor Golgi compartment. In addition, formation of the primed donor intermediate is sensitive to treatment of the donor membrane with trypsin and *N*-ethylmaleimide (NEM), a sulfhydryl-alkylating reagent. These results suggest that a number of protein components must coordinate an initial step in transport.

What is the primed donor? Morphological characterization of the primed donor provides a surprising and tantalizing glimpse of an isolated first step in compartmental transfer which may be occurring in the living cell. When unincubated and primed Golgi fractions were compared using electron microscopy, the quiescent cisternal elements of the Golgi prepared from cell homogenates were transformed into a network of tubules and budding vesicles with electron-dense coats (Fig. 4) similar to the steady-state morphology of the Golgi observed in vivo (1,7,22). These results provided the first biochemical evidence that transport between Golgi compartments may involve formation of carrier vesicles.

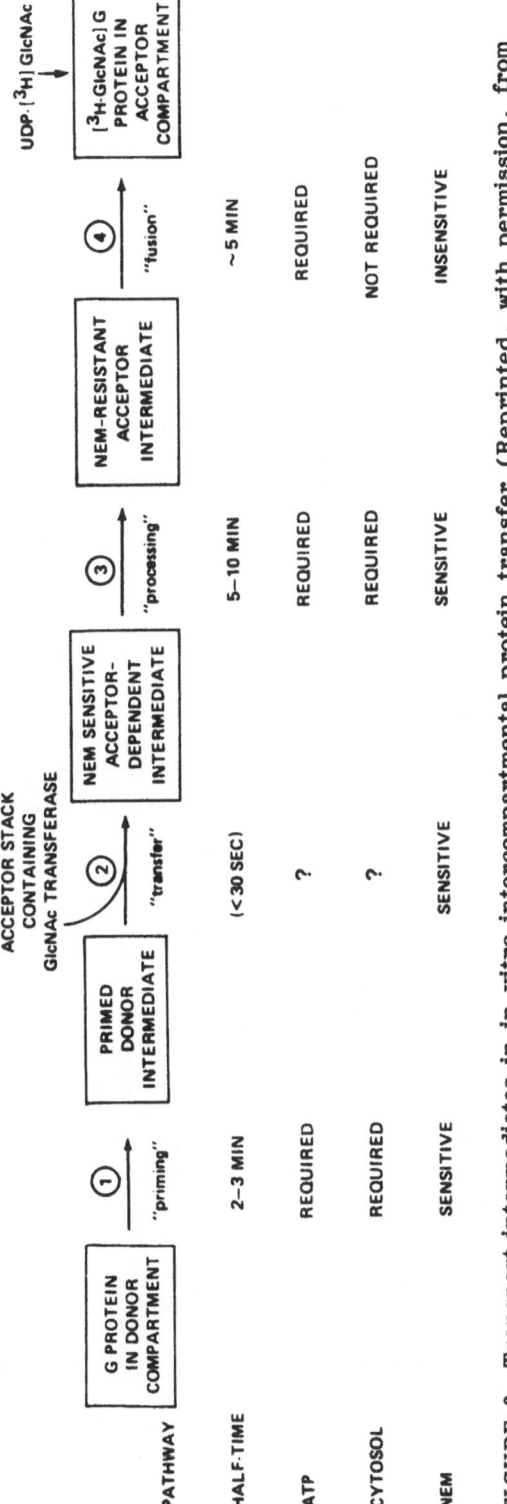

FIGURE 3 Transport intermediates in in vitro intercompartmental protein transfer (Reprinted, with permission, from Ref. 44.)

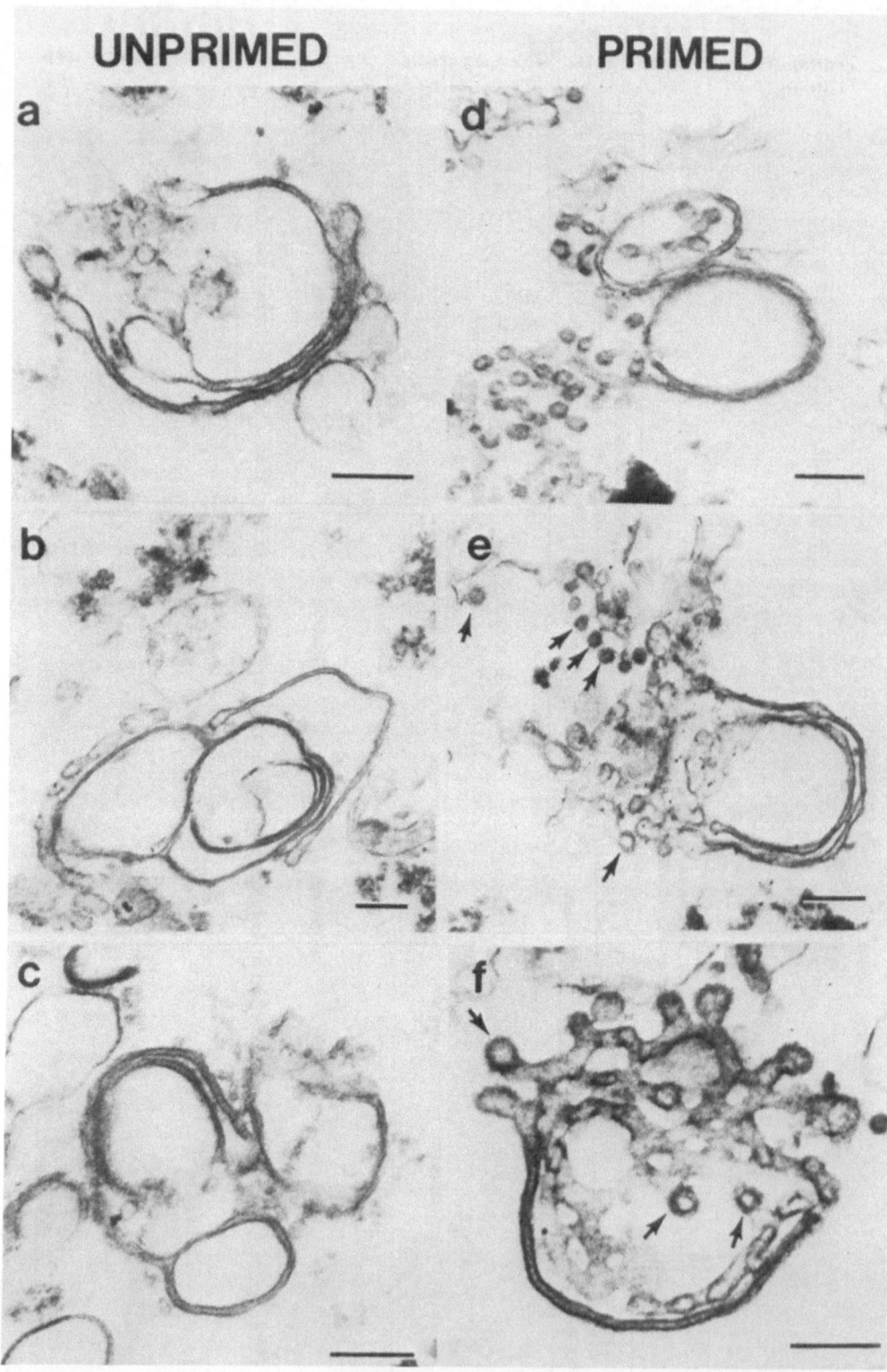

UNPRIMED   **PRIMED**

a

d

b

e

c

f

FIGURE 4 Morphology of the Golgi stacks of the donor membrane before and after priming. Views of representative Golgi stacks in unprimed (a–c) and primed (d–f) conor fractions at high magnification (ranging from 37,500× to 61,000×). Bars: 0.25 µm. The shape and features of apparently budding vesicles are frequently suggestive of electron-dense cytoplasmic coats (arrows in d and e). (Reprinted, with permission, from Ref. 23.)

Subsequent studies on the morphology of the primed donor examined the distribution of G protein in the golgi membrane using immunoelectron microscopy (24). G protein was distributed randomly between the budding vesicles and the remainder of the Golgi cisternae membranes. The implication of this finding is that entry of G protein into a budding site is stochastic: the number of G-protein molecules transported in an export vesicle is proportional to the density of G in the membrane. This interpretation was supported by the observation that the vesicles containing the G protein did not contain a clathrin coat. Clathrin is an important component in the formation of coated pits at the cell surface during receptor-mediated endocytosis (45) and provides for the electron-dense, cagelike (coat) structure encompassing the budding vesicle (46). Receptor/ligand complexes are selectively concentrated in coated pits during endocytosis (45,73; see also Chapter 19). In vivo studies have established that clathrin coats do form on Golgi cisternae, but are found exclusively on the *trans* Golgi compartment(s) (47). However, when G protein is accumulated in the *trans* Golgi by prolonged incubation at 20°C in vivo, it cannot be detected in clathrin-containing coated structures (47). The electron-dense coat delimiting the budding, G-protein-containing, vesicles formed in vitro was not reactive with anticlathrin antibody, suggesting that a new, previously unrecognized protein or protein complex may catalyze formation of carrier vesicles that transport G protein through the Golgi. This interpretation is consistent with the recent genetic evidence in which the clathrin gene was deleted from yeast without adverse effect on either the growth or kinetics of transport of protein through the yeast secretory pathway (48). Since budding vesicles have not been observed when Golgi compartments were incubated in the presence of ATP but in the absence of cytosol, soluble components in the cytoplasm must play a key role in budding. Their role in priming of both ATP and the NEM/trypsin-sensitive components found on the donor membrane remains to be elucidated.

The next intermediate step observed in vitro required the addition of the acceptor Golgi compartment (step 2 in Fig. 3). The formation of this intermediate from the primed donor is rapid, is characterized by its sensitivity to NEM, but can be distinguished from the primed donor by its resistance to dilution of the assay mixture. This intermediate is referred to as the *NEM-sensitive, acceptor-dependent intermediate* (step 2 in Fig. 3). Dilution resistance is measured by the change in the cytosol requirement for subsequent delivery of G protein to the processing site of mannosidase I in the acceptor Golgi compartment. In this case, the requirement for rate-limiting concentrations of early-acting cytosolic components for formation of the primed donor is replaced by a requirement for a new cytosolic component (factor B) which is found in a relative excess to the early-acting components (49). The implication of these results is that this step defines the formation of a new intermediate *prefusion complex* (Fig. 5): a "loose" association of a transport vesicle with its target acceptor Golgi element. Transport activity is now independent of a diffusion step which would be expected to be only important for transport from the primed donor to the acceptor, and it requires an additional component present in the cytosol for completion of delivery.

While the initial interaction of the transport vesicle with the acceptor Golgi to form the NEM-sensitive prefusion is rapid, the subsequent step(s) leading to vesicle-acceptor fusion and delivery of G protein to the processing site of mannosidase I is slow. The next intermediate following the initial interaction of a transport vesicle with the acceptor Golgi compartment is

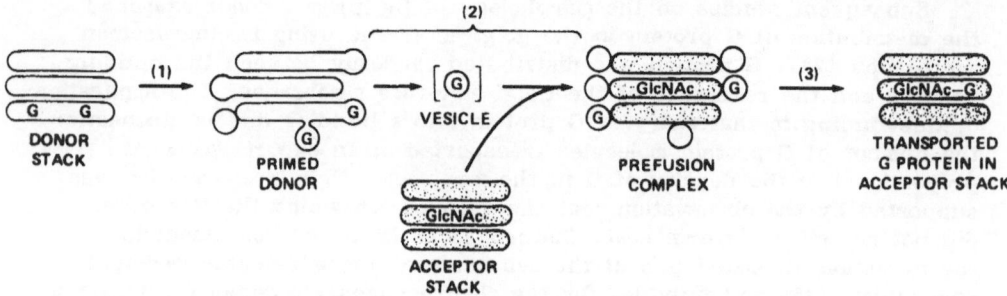

FIGURE 5   Working hypothesis to relate the kinetic intermediates to stages in a vesicle-mediated transport process. (Reprinted, with permission, From Ref. 23.)

the formation of the *NEM-resistant intermediate* (step 3 in Fig. 3). Consumption of this intermediate is insensitive to NEM and independent of cytosol (44). At present, little is understood about the series of events that culminate in membrane-membrane fusion. This slow step in transport might be interpreted as the period in which vesicle recognition of the target membrane is completed and the membrane fusion events are activated ("processing" in Fig. 3). Since addition of GlcNAc to G protein delivered to the interior of the acceptor compartment is not rate-limiting (23), slow diffusion of G protein in the membrane of the acceptor compartment after fusion to the site of processing by Man I is unlikely. Despite differences in sensitivity to NEM and cytosolic factors during intermediate steps in transport, a consistent feature throughout transport (with the possible exception of step 2 in Fig. 3, in which the vesicle diffuses to the target membrane) is the requirement for energy in the form of ATP. A working model summarizing these results is shown in Figure 5.

Fractionation of the cytosol has revealed at least four soluble components that are required for transport of the G protein in vitro (49). The role of each of these components in transport is just starting to be unraveled by the use of stage-specific functional assays (49). One component, factor B, appears to have a role in vesicle-acceptor fusion, as discussed above. Recently, it has been discovered that palmitoyl-CoA play a role in the final steps of transport (50). The metabolism of palmitoyl-CoA in vitro is regulated by an NEM-sensitive component which is associated with the Golgi membranes. Although the role for this cofactor remains to be established, it has been suggested that it is important in the events leading to vesicle-acceptor membrane fusion.

Perhaps one of the most surprising observations stemming from these in vitro studies is the evolutionary conservation of this metabolic pathway (51). All mammalian Golgi fractions tested, including rat liver, bovine brain, and fish kidney, can act as acceptors in vitro. The soluble factors were found in an even greater range of species. Cytosol fractions obtained from plants as well as animals and the lower eukaryotes substitute for the requirement of the CHO cytosol. Conservation of the essential aspects of this pathway of posttranslational transport and processing of protein suggests that this pathway is evolutionarily very ancient.

V. EXPORT OF PROTEIN FROM THE ER IN VITRO

A. Use of the G Protein of VSV as a Model to Study Export from the ER

An important stage in the delivery of protein to the cell surface is the selective export of proteins from their site of synthesis in the cellular endoplasmic reticulum. While the synthesis, insertion, and core glycosylation have been studied in vitro by a number of laboratories (11-13), little is known of the subsequent biochemical events controlling export from this organelle and delivery to the Golgi. Isolation of yeast temperature-sensitive mutants which are pleiotropically defective in export of both membrane-bound and secreted protein from the ER has established that a common biochemical basis regulates protein transfer from this organelle (4,18). In addition, comparative studies defining differences in the kinetics of transport of many different membrane-bound and secreted proteins (53-55), the altered secretion of genetically engineered mutant proteins (56-58), and the requirement for cofactors (59) and quaternary interactions (60-63) for export suggest that the signals present in the primary sequence of a protein provide important information facilitating an initial step in the interaction with this common transport machinery.

We have used the G protein to analyze the initial ER export step. Transport of G protein from the ER to the *cis* Golgi compartment in vivo results in the oligosaccharides of the G protein being trimmed from the high mannose forms ($man_{8-9}GlcNAc_2$) to the $man_5GlcNAc_2$ species as a result of the early Golgi oligosaccharide-processing enzyme mannosidase I (Man I). As illustrated previously (Figs. 1 and 2), trimming of the G protein can be followed directly in the clone 15B cell line missing Tr I. In addition, export of the G protein from the ER in vivo can be readily distinguished from later transport steps between Golgi compartments through use of the temperature-sensitive VSV mutant strain ts045 in which the G protein is found exclusively in the ER when cells are incubated at the restrictive temperature during infection (64-66). Shift of cells from the restrictive to the permissive temperature in vivo results in the synchronous release of G protein to the Golgi as shown morphologically (28) and bio-chemically through trimming to the $man_5GlcNAc_2$ form (31). Transport of ts045 G protein between the ER and the Golgi in vivo occurs in two biochemically distinct steps: an early, temperature-sensitive, ATP-dependent (budding) step, followed by a temperature- and ATP-insensitive step (fusion) providing for delivery of G protein to the *cis* Golgi compartment containing Man I (31).

B. An In Vitro Assay for ER Export

Early studies established that the conditions used to achieve inter-Golgi transport in vitro were not sufficient to support transport of G from the ER to the *medial* Golgi compartment (26,42). Apparently, during homogenization of cells, the donor ER or acceptor Golgi functions (or both) may be inactivated. Alternatively, incubation conditions necessary to achieve transport in vitro between the ER and Golgi are different from those required for transport between Golgi compartments.

To explore these various possibilities we have designed an in vitro assay to specifically focus on this initial stage (67). As shown in Figure 6, the donor/acceptor pairs are different than those found in the inter-Golgi

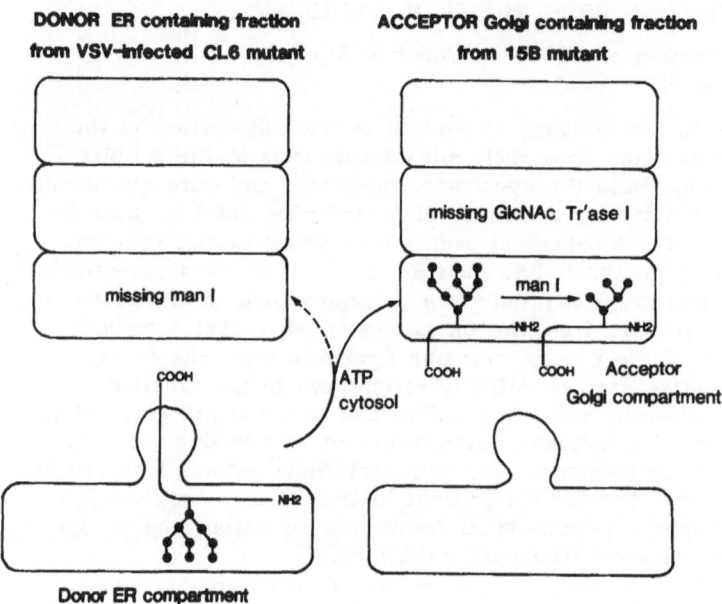

DONOR ER containing fraction
from VSV-infected CL6 mutant

ACCEPTOR Golgi containing fraction
from 15B mutant

missing man I

missing GlcNAc Tr'ase I

man I

Acceptor
Golgi compartment

ATP
cytosol

Donor ER compartment

FIGURE 6   Assay for transport of G protein between the endoplasmic reticulum and the *cis* Golgi compartment (see Ref. 67).

assay. To measure export from the ER, a donor ER fraction is prepared from the mouse L-cell mutant clone 6. Clone 6 is defective in processing oligosaccharides beyond the $man_8GlcNAc_2$ form, suggesting that it is missing the *cis* Golgi processing enzyme mannosidase I (68,69). In VSV-infected clone 6 cells, the G protein is transported to the cell surface in the $man_{8-9}GlcNAc_2$ oligosaccharide form because processing, not transport, is the only known defect in this cell (69). For preparation of the donor ER, clone 6 is infected with the ts045 mutant virus strain and the G protein is labeled with [^{35}S]methionine at the restrictive temperature to ensure that all labeled G protein is found in the ER prior to homogenization. This donor ER homogenate is mixed with an acceptor Golgi homogenate prepared from the 15B cell line missing Tr I, but containing mannosidase I. If G protein is trimmed in vitro from the $man_{8-9}GlcNAc_2$ ER form to the $man_5GlcNAc_2$ form, then G must be transferred from the ER prepared from the clone 6 cells to the *cis* Golgi trimming site containing mannosidase I found in the 15B homogenate. The trimming of G protein can be readily detected by the increased electrophoretic mobility of the $man_5GlcNAc_2$ oligosaccharide form of G protein during SDS/gel electrophoresis, or by analysis of the oligosaccharide structure using high-performance liquid chromatography (HPLC) (31).

Incubation in vitro of donor ER membranes from clone 6 and acceptor Golgi membranes from clone 15B results in transport of G protein between sealed membrane compartments with the kinetics and efficiency observed in vivo (67). However, one important difference is required to reconstitute export from the ER in vitro when compared to inter-Golgi transport. Efficient transfer is observed only when the donor and acceptor membranes are incubated in the presence of a homogenate prepared from mitotic cells. The rationale for this result stems from the observation that during mitosis the nuclear envelope, ER, and Golgi are selectively disassembled in CHO cells in preparation for cell division (69). Incubation of mitotic homogenates in vitro results in reversal of the process, leading to reassembly of the intact nuclei (70). A similar result occurs in the ER export assay. In addition to transport of G protein, both the nuclear envelope and the ER compartments are found to reassemble (67; W. E. Balch, unpublished results). These results suggest that the efficiency of transport observed in vitro for this stage is coupled to functional reassembly of organelles. This result seems particularly suited for reconstitution of ER export, since the ER is a single organelle and is extensively vesiculated during homogenization of interphase cell populations. In contrast, the Golgi is a compact structure which remains relatively intact during cell homogenization.

While our understanding of export of protein from the ER is in its infancy, early studies show that transfer is dependent on ATP, a cytosol fraction, and contains NEM- and trypsin-sensitive components on the donor compartment (67). The relationship of these requirements to those observed during inter-Golgi transfer in vitro or for the proposed role of organelle reassembly in transport efficiency remains to be established.

VI. PERSPECTIVES

The central thesis of this in vitro approach to the study of interorganelle sorting of protein is that it will be difficult to understand the molecular basis of protein traffic along the cell secretory pathway until we understand the enzymological rudiments of the transport process itself. Knowledge of the biochemical principles of budding and fusion of carrier vesicles will then provide a mechanism to understand putative sorting signals and the sorting machinery that operates in the cell.

While morphological studies have provided strong evidence that transport in vivo between all compartments is vesicular (1), little progress has been made to date in distinguishing between the two models for protein flow into carrier vesicles at early stages of the secretory pathway. On one hand, constitutive transport of protein from the ER through the Golgi and to the cell surface may be a consequence of a bulk flow mechanism (5). In this interpretation, transfer of proteins is largely stochastic, principally reflecting their capacity for diffusion into forming carrier vesicles. Proteins not destined for transport are retained in organelles by a mechanism independent of the transport machinery. In the opposite model, transport is selective; specific export signals present on a protein may trigger interaction with transport-specific receptors at each stage of compartmental transfer (71,72). Of course, it is not unreasonable to assume that both mechanisms may function in concert. Clathrin-coated vesicles, which are known to play an important role in selective transport of protein during receptor-mediated endocytosis (45), are found only in the *trans* Golgi region where sorting of protein to secretory granules (5,22) (and possibly to the lysosome) occurs. Since the G protein does not occupy these coated structures during constitutive transport to the cell surface in vivo (5,47) or in vitro (24), these results suggest that two pathways may be operating in parallel. Without the ability to distinguish between these two simple models for a single protein at each stage of transport, interpretation of the significance of putative protein-sorting signals and associated machinery will be difficult at best.

What is the cytosolic machinery catalyzing transport? The obvious paradigm offered by the extensive studies on the formation of coated vesicles during receptor-mediated endocytosis is no longer tenable for study of export during early stages of transport of protein from the ER through the Golgi. This conclusion is a consequence of both genetic (18) and the present biochemical and morphological evidence provided by study of the inter-Golgi transport of the G protein in vitro (24). The role of clathrin as a molecular filter (73) carrying out only a limited role in the formation of carrier vesicles seems logical. The entire biochemical machinery, both membrane-bound and cytosolic, triggering formation of vesicles along the exocytotic pathway remains to be resolved. This problem can only be effectively addressed using sensitive in vitro systems.

In addition to the problem of budding carrier vesicles, we also lack a reasonable working model for the exquisite regulation of membrane fusion events occurring in vivo. While the fusogenic activity of viruses (74) and a variety of chemicals have provided insight into the capacity and possible mechanisms by which biological membranes can fuse, none of these sytems are compatible with our current view of the specificity and physiological conditions that prevail in the cytoplasm of the eukaryotic cell. The biochemistry of these events, based on biologically relevant transport assays such as those described here, should yield new insight into this problem. Nature's solution to the problem of specificity of fusion (and budding) is likely to have both common and unique components for each stage of the pathway.

Finally, we have the problem of defining the relationships between protein flow and lipid flow along the compartments of the secretory pathway. Secretory compartments are differentiated on the basis of not only their protein composition, but also their lipid composition (6-8). The mechanisms that lead to these differences, the role (if any) of such differences in protein transport, and the overall balance of membrane-lipid flow and protein transfer in rapidly growing or stationary secretory cells remain to be determined. Since these differences may be stage-specific, our understanding may again require the study of these and related problems using cell-free assays which focus on lipid transfer.

The problem of interorganelle membrane sorting is obviously a rich one with many exciting discoveries to be made through study of reconstituted systems in vitro.

REFERENCES

1. Palade, G. E. (1975). Intracellular aspects of the processing of protein synthesis. *Science* 189:347-358.
2. Wickner, W. T., and Lodish, H. F. (1985). Multiple mechanisms of insertion into and across membranes. *Science* 230:400-407.
3. Walters, P., Gilmore, R., and Bloebel, G. (1984). Protein translocation across the ER. *Cell* 38:5-8.
4. Schekman, R., and Novick, P. (1982). The secretory process and yeast cell-surface assembly. In *The Molecular Biology of the Yeast Saccharomyces: Metabolism and Gene Expression* (J. Strathern, E. Jones, and J. Broach, eds.). Cold Spring Harbor Laboratory, Cold Spring Harbor, NY, pp. 361-393.
5. Kelly, R. P. (1985). Pathways of protein secretion in eukaryotes. *Science* 230:25-32.

6. Dunphy, W. G., and Rothman, J. E. (1985). Compartmental organization of the Golgi stack. *Cell* 42:13-21.

7. Farquhar, M. G., and Palade, G. E. (1981). The Golgi apparatus (complex)-(1954-1981)-from artifact to center stage. *J. Cell Biol.* 91: 77s-106s.

8. Farquhar, M. G. (1985). Progress in unraveling pathways of Golgi traffic. *Annu. Rev. Cell Biol.* 1:447-488.

9. Garoff, H. (1985). Using recombinant DNA techniques to study protein targeting in the eucaryotic cell. *Annu. Rev. Cell Biol.* 1:403-445.

10. Schekman, R. (1985). Protein localization and membrane traffic in yeast. *Annu. Rev. Cell Biol.* 1:115-143.

11. Rothman, J. E., and Lenard, J. (1984). Membrane traffic in animal cells. *Trends Biochem. Sci.* 100:176-178.

12. Tartakoff, A. M. (1982). Simplifying the complex Golgi. *Trends Biochem. Sci.* 7:174-176.

13. Robbins, P., Hubbard, S. C., Turco, S. J., and Wirth, D. F. (1977). Proposal for a common oligosaccharide intermediate in synthesis of membrane glycoproteins. *Cell* 12:893-900.

14. Sabatini, D. D., Kreibich, G., Morimoto, T., and Adesnik, M. (1982). Mechanisms for the incorporation of proteins in membranes and organelles. *J. Cell Biol.* 92:1-22.

15. Bischoff, J., Liscum, L, and Kornfeld, R. (1986). The use of 1-deoxymannojirimycin to evaluate the role of various α-mannosidases in oligosaccharide processing in intact cells. *J. Biol. Chem.* 261:4766-4774.

16. Hubbard, S. C., and Ivatt, R. J. (1981). Synthesis and processing of asparagine-linked oligosaccharides. *Annu. Rev. Biochem.* 50:555-583.

17. Kornfeld, R., and Kornfeld, S. (1985). Assembly of asparagine-linked oligosaccharides. *Annu. Rev. Biochem.* 54:631-664.

18. Novick, P., Ferro, S., and Schekman, R. (1981). Order of events in the yeast secretory pathway. *Cell* 25:461-469.

19. Saraste, J., and Hedman, K. (1983). Intracellular vesicles involved in the transport of Semliki Forest virus membrane proteins to the cell surface. *EMBO J.* 2:2001-2006.

20. Saraste, J., and Kuismanen, E. (1984). Pre- and post-Golgi vacuoles operate in the transport of Semliki Forest virus membrane glycoproteins to the cell surface. *Cell* 38:535-549.

21. Saraste, J., Palade, G. E., and Farquhar, M. G. (1986). Temperature-sensitive steps in the transport of secretory proteins through the Golgi complex in exocrine pancreatic cells. *Proc. Natl. Acad. Sci. USA*, in press.

22. Strous, G. J. A. M., Willemsen, R., van Kerkhof, P., Slot, J. W., Geuze, H. J., and Lodish, H. F. (1983). Vesicular Stomatitis virus glycoprotein, albumin, and transferrin are transported to the cell surface via the same Golgi vesicles. *J. Cell Biol.* 97:1815-1822.

23. Balch, W. E., Glick, B. S., and Rothman, J. E. (1984). Sequential intermediates in the pathway of intercompartmental transport in a cell-free system. *Cell* 39:525-536.

24. Orci, L., Glick, B. S., and Rothman, J. E. (1986). A new type of coated vesicular carrier that appears not to contain clathrin: Its possible role in protein transport within the Golgi stack. *Cell* 46:171-184.

25. Rothman, J. E., Urbani, L. J., and Brands, R. (1984). Transport of protein between cytoplasmic membranes of fused cells: Correspondence to processes reconstituted in a cell-free system. *J. Cell Biol.* 99:248-259.

26. Rothman, J. E., Miller, R. L., and Urbani, L. J. (1984). Intercompartmental transport in the Golgi complex is a dissociative process: Facile transfer of membrane protein between two Golgi populations. *J. Cell Biol.* 99:260-271

27. Arnheiter, H., Dubois-Dalcq, M., and Lazzarini, R. A. (1984). Direct visualization of protein transport and processing in the living cell by microinjection of specific antibodies. *Cell* 39:99-109.

28. Bergmann, J. E., Tokuyasu, K. T., and Singer, S. J. (1981). Passage of an integral membrane protein, the vesicular stomatitis virus glycoprotein, through the Golgi apparatus en route to the plasma membrane. *Proc. Natl. Acad. Sci. USA* 76:4350-4354.

29. Li, I., Tabas, I., and Kornfeld, S. (1978). The synthesis of complex oligosaccharides. I. Structure of the lipid-linked oligosaccharide precursor of the complex-type oligosaccharides of the vesicular stomatitis virus G protein. *J. Biol. Chem.* 253:7762-7770.

30. Kornfeld, S., Li, E., and Tabas, I. (1978). The synthesis of complex oligosaccharides. II. Characterization of the processing intermediates of the synthesis of the complex oligosaccharide units of vesicular stomatitis virus G protein. *J. Biol. Chem.* 253:7771-7778.

31. Balch, W. E., Elliott, M. M., and Keller, D. S. (1986). ATP-coupled transport of vesicular stomatitis virus G protein between the endoplasmic reticulum and the Golgi. *J. Biol. Chem.* 261: in press.

32. Jamieson, J., and Palade, G. (1968). Intracellular transport of secretory proteins in the pancreatic endocrine cell. IV. Metabolic requirements. *J. Cell Biol.* 39:589-603.

33. Fries, E., and Rothman, J. E. (1980). Transport of vesicular stomatitis virus glycoprotein in a cell-free extract. *Proc. Natl. Acad. Sci. USA* 77:3870-3874.

34. Young, R. W. (1973). The role of the Golgi complex in sulfate metabolism. *J. Cell Biol.* 57:175-189.

35. Kaplan, A., Achord, D. T., and Sly, W. W. (1977). *Proc. Natl. Acad. Sci. USA* 74:2026.

36. Tabas, I., and Kornfeld, S. (1980). Biosynthetic intermediates of α-glucuronidase contain high mannose oliosaccharides with blocked phosphate residues. *J. Biol. Chem.* 225:6633-6640.

37. Gumbiner, B., and Kelly, R. (1981). *Proc. Natl. Acad. Sci. USA* 78:318-322.

38. Schmidt, M. F. G., and Schlesinger, M. J. (1980). Relation of fatty acid attachment to the translation and maturation of vesicular stomatitis and Sindbis virus membrane glycoproteins. *J. Biol. Chem.* 3334-3339.

39. Gottlieb, C., Baenziger, J., and Kornfeld, S. (1975). Deficient uridine diphosphate-*N*-acetylglucosamine: Glycoprotein *N*-acetylglucosaminetransferase activity in a clone of Chinese hamster ovary cells with altered surface glycoproteins. *J. Biol. Chem.* 250:3303-3309.

40. Schlesinger, S., Gottlieb, C., Feil, P., Gelb, N., and Kornfeld, S. (1976). Growth of enveloped RNA viruses in a line of Chinese hamster ovary cells with deficient *N*-acetyl-glucosaminyltransferase activity. *J. Virol.* 17:239-246.

41. Tarentino, A. L., Plummer, T. H., and Maley, F. (1974). The release of intact oligosaccharides from specific glycoproteins by endo-β-*N*-acetylglucosaminidase H. *J. Biol. Chem.* 249:818-824.

42. Rothman, J. E., and Fries, E. (1981). Transport of newly synthesized vesicular stomatitis viral glycoprotein to purified Golgi membranes. *J. Cell Biol.* 89:162-168.

43. Balch, W. E., Dunphy, W. G., Braell, W. A., and Rothman, J. E. (1984). Reconstitution of the transport of protein between successive compartments of the Golgi measured by the coupled incorporation of *N*-acetylglucosamine. *Cell* 39:405-416.

44. Wattenberg, B. W., Balch, W. E., and Rothman, J. E. (1986). A novel prefusion complex formed during protein transport between Golgi cisternae in a cell-free system. *J. Biol. Chem.* 261:2202-2207.

45. Goldstein, J. L., Anderson, R. G. W., and Brown, M. S. (1979). Coated pits, coated vesicles, and receptor-mediated endocytosis. *Nature* 279:679-684.

46. Pearse, B. M. F., and Bretscher, M. S. (1981). Membrane recycling by coated vesicles. *Annu. Rev. Biochem.* 50:85.

47. Griffiths, G., Pfeiffer, S., Simons, K., and Matlin, K. (1985). Exit of newly synthesized membrane proteins from the trans cisterna of the Golgi complex to the plasma membrane. *J. Cell Biol.* 101:949-964.

48. Payne, G. S., and Schekman, R. (1985). A test of clathrin function in protein secretion and cell growth. *Science* 230:1009-1014.

49. Wattenberg, B. W., and Rothman, J. E. (1986). Multiple cytostolic components promote intra-Golgi protein transport. Resolution of a protein acting at a late stage, prior to membrane fusion. *J. Biol. Chem.* 261:2208-2221.

50. Glick, B. S., and Rothman, J. E. A possible role for fatty acyl coenzyme A in facilitating intracellular protein transport. *Nature*, in press.

51. Paquet, M. R., Pfeffer, S., Burczak, J. D., Glick, B. S., and Rothman, J. E. (1986). Components responsible for transport between successive Golgi cisternae are highly conserved in evolution. *J. Biol. Chem.* 261:4367-4370.

52. Strous, G. J. A. M., and Lodish, H. F. (1980). Intracellular transport of secretory and membrane proteins in hepatoma cells infected by vesicular stomatitis virus. *Cell* 22:709-717.

53. Ledford, B. E., and Davis, D. F. (1983). Kinetics of serum protein secretion by cultured hepatoma cells: Evidence for multiple secretory pathways. *J. Biol. Chem.* 258:3304-3308.

54. Fries, E., Gustafsson, L., and Peterson, P. A. (1984). Four secretory proteins synthesized by hepatocytes are transported from the endoplasmic reticulum to Golgi complex at different rates. *EMBO J.* 3:147-152.

55. Yeo, K-T., Parent, J. B., Yeo, T-K., and Olden, K. (1985). Variability in transport rates of secretory glycoproteins through the endoplasmic reticulum and Golgi in human hepatoma cells. *J. Biol. Chem.* 260:7896-7902.

56. Rose, J. K., and Bergmann, J. E. (1982). Expression from cloned cDNA of cell-surface secreted forms of the glycoprotein of vesicular stomatitis virus G protein. *Cell* 30:753-762.

57. Rose, J. K., and Bergmann, J. E. (1983). Altered cytoplasmic domains affect intracellular transport of vesicular stomatitis virus G protein. *Cell* 34:513-524.

58. Emr, S. D., Schauer, I., Hansen, W., Esmon, P., and Schekman, R. (1984). Invertase B-galactosidase hybrid proteins fail to be transported from the endoplasmic reticulum in Saccharomyces cerevisiae. *Mol. Cell. Biol.* 4:2347-2355.

59. Rask, L., Valtersson, C., Anundi, H., Kvist, S., Eriksson, U., Dallner, G., and Peterson, P. A. (1983). Subcellular localization in

normal and vitamin A-deficient rat liver of vitamin A serum transport proteins, albumin, ceruloplasmin and class I major histocompatibility antigens. *Exp. Cell Res.* 143:91-102.

60. Owen, M. J., Kissonerghis, A. M., and Lodish, H. F. (1980). Biosynthesis of HLA-A and HLA-B antigens in vivo. *J. Biol. Chem.* 255: 9678-9684.

61. Sege, K., Rask, L., and Peterson, P. A. (1981). Role of α-2-microglobulin in the intracellular processing of HLA antigens. *Biochemistry* 20:4523-4530.

62. Kvist, S., Wiman, K., Claesson, L., Peterson, P. A., and Dobberstein, B. (1982). Membrane insertion and oligomeric assembly HLA-DR histocompatibility antigens. *Cell* 29:61-69.

63. Mains, P. E., Sibley, C. H. (1982). Control of IgM synthesis in the murine pre-B cell line, 70Z/3. *J. Immunol.* 128:1664-1670.

64. Lafay, F. (1974). Envelope proteins of vesicular stomatitis virus: Effect of temperature-sensitive mutations in complementation groups III and V. *J. Virol.* 14:1220-1228.

65. Knipe, D. M., Baltimore, D., and Lodish, H. F. (1977). Maturation of viral proteins in cells infected with temperature-sensitive mutants of vesicular stomatitis virus. *J. Virol.* 21:1149-1158.

66. Zilberstein, A., Snider, M. D., Porter, M., and Lodish, H. F. (1980). Mutants of vesicular stomatitis virus blocked at different stages in maturation of the viral glycoprotein. *Cell* 21:417-427.

67. Balch, W. E., Hermanowsky, A. L., and Keller, D. S. (1986). Reconstitution of transport of vesicular stomatitis virus G protein from the endoplasmic reticulum to the Golgi using a cell-free assay. *Cell,* in press.

68. Gottlieb, C., and Kornfeld, S. (1976). Clone 6. *J. Biol. Chem.* 251: 7761-7768.

69. Robbins, E., and Gonatas, K. (1964). The ultrastructure of a mammalian cell during the mitotic cycle. *J. Cell Biol.* 21:429-463.

70. Burke, B., and Gerace, L. (1986). A cell free system to study reassembly of the nuclear envelope at the end of mitosis. *Cell* 44:639-652.

71. Fitting, T., and Kabat, D. (1982). Evidence for a glycoprotein "signal" involved in transport between subcellular organelles: Two membrane glycoproteins encoded by murine leukemia virus reach the cell surface at different rates. *J. Biol. Chem.* 257:14011-14017.

72. Lodish, H. F., Kong, N., Snider, M., and Strous, G. J. A. M. (1983). Hepatoma secretory proteins migrate from rough endoplasmic reticulum to Golgi at characteristic rates. *Nature* 304:80-83.

73. Bretscher, M. S., Thomson, J. J., and Pearse, B. M. F. (1980). Coated pits act as molecular filters. *Proc. Natl. Acad. Sci. USA* 77: 4156-4159.

74. White, J., Kielian, M., and Helenius, A. (1983). Membrane fusion proteins of enveloped animal viruses. *Q. Rev. Biophys.* 16:151-195.

23

Cytoplasmic Determinants of Exocytotic Membrane Fusion

PETER I. LELKES

University of Wisconsin Medical School, Milwaukee Clinical Campus, Milwaukee, Wisconsin

HARVEY B. POLLARD

National Institute of Diabetes, Digestive, and Kidney Diseases, National Institutes of Health, Bethesda, Maryland

I. INTRODUCTION

Fusion of the intracellular storage organelles with the cytoplasmic face of the plasma membranes is the ultimate event in the stimulus–secretion coupling in most secretory cells. This process, termed exocytosis, is the basis for the release of a variety of hormones, neurotransmitters, and other bioactive compounds (1). The fusion process has traditionally comprised three distinct steps: (1) close approach of the fusing partners, (2) coalescence of the membranes, in particular of the lipid bilayers, and (3) merging of the included volumes. In the case of exocytosis, these volumes are the vesicle interior and the extracellular medium. In secreting cells the fusion process can only be detected as a consequence of step 3. Thus, we infer from the secretion event that step 2, membrane coalescence, has just occurred. Of course, step 1, close approach, must have preceded step 2, but at a time that is very difficult to specify. Therefore, our index of fusion in the exocytotic process is really an index of recent merging of membrane bilayers prior to and concomitant with release of the granule contents. Membrane fusion has been studied in numerous model systems which are described in great detail elsewhere in this book. Besides leading to important methodological advances on how to assay membrane fusion in these respective model systems, all these experimental approaches have generated valuable insights into some of the potential requirements for fusion, such as the role of calcium and pH, of osmotic forces, and of cytosolic and/or membrane-bound "fusion" proteins (2).

However, a central question is how far we can extrapolate from these results, obtained in the simplified and controlled environment of a test tube, to real events occurring in the highly complex cytosolic environment. Furthermore, are those parameters that seem to govern membrane fusion in vitro indeed part of the requirements of the fusion process in vitro, or are they simply instructive artifacts (3)?

In the "real world" of cells, stimulated secretion, viz. exocytotic membrane fusion, occurs under tight spatial and temporal regulation in order to prevent the uncontrolled release of, for example, neurotransmitters or hormones. The nature and localization of intracellular factors governing membrane contact and fusion in situ and the mechanisms of their activation following stimulation of the cells remain largely elusive. Cellular stimulation generally occurs via specific receptors and triggers a host of membrane-associated events, such as depolarization of the plasma membrane, opening of (voltage-dependent) ion channels, and transmembrane movement of ions and charges. Activation of most of the secretory cells is accompanied by a significant increase in the level of ionized intracellular calcium, $[Ca^{2+}]_i$. This, in turn, might trigger a variety of subsequent cellular events, such as activation of enzymatic events (kinases, lipases, proteases, etc.) (4). In some systems, such as basophils and mast cells, degranulation (another expression for exocytotic release of granule contents) is believed to be largely triggered by the stimulus-mediated mobilization of intracellular calcium.

However, under certain experimental conditions and in some systems (5) exocytosis can occur even in the absence of extracellular calcium and/or without raising $[Ca^{2+}]_i$. And in yet another set of secretory cells, stimulus-secretion coupling is accompanied by a transient drop in the intracellular calcium levels (6).

This variability in the calcium response accompanying stimulus-secretion coupling must therefore raise doubts about its pivotal role for membrane fusion. For example, does the initial modulation in the intracellular calcium level provide the necessary and sufficient signal to trigger exocytosis, as previously suggested (7)? Or does this first event merely serve to generate a number of other messengers, such as inositolphosphatides, protons, and other metabolic products, which then in turn could trigger exocytosis? Does the fusion of biological membranes in situ require calcium at all, or is it rather calcium-independent and controlled by one of the above-mentioned agents? Or have we, till now, in our search for the "fusion factor(s)" essentially missed the proper directions and should we rather be looking for other specific intracellular (cytoplasmic or membrane-bound) determinants which are quiescent in resting cells and would be activated upon stimulation of the cells?

In this chapter, we survey a number of such intracellular factors that are potentially of relevance for initiating, mediating, or facilitating membrane contact and fusion. We wish to emphasize that this chapter is not meant as an exhaustive review. Instead, our goal is to get some general ideas across that we feel are important for our current understanding of exocytotic membrane fusion and that also will be indicative of where future work may probably be directed. Based on our personal experiences, we draw our examples mainly from work done on exocytosis in endocrine organs, such as adrenal chromaffin cells. However, we also attempt to correlate those findings with similar or divergent results in other systems, such as neutrophils, mast cells, platelets, parathyroid cells, or sperm-egg fusion.

II. INTRACELLULAR CALCIUM

Measurement of fusion in biological systems is fundamentally hard to do in a quantitative manner. But as indicated above, secretion is one process that is readily accessible. And yet in secretion, as also mentioned above, we are again confronted with the indirect character of the signals. Except for the final fusion event, various modifying agents may only be affecting the various efficiencies of the signal transduction apparatus. For the present purposes, we assume that nature has not evolved more than one basic mechanism for fusing membranes, and that all the differences we see in secretory competence are only variations in the expression of various control mechanisms. Our main problem remains, however, to distinguish what is "control" and what is "fusion."

An excellent example of this dilemma is the role of calcium in the fusion process. The involvement of calcium in the exocytosis process has been central to our thinking about exocytosis for many years (4,7). In many secretory systems, stimulation of secretion is accompanied by an elevation in intracellular calcium concentration (8,9). Furthermore, calcium in micromolar quantities can provoke release of vesicular contents from cells permeabilized by either high voltage or detergents (10,11). However, in more recent studies, the infallibility of this causal relationship between calcium and fusion has become difficult to sustain.

This seems certainly to be the case for fusion during endocytosis and/or membrane retrieval following exocytosis. It appears that some of these processes, and hence presumably the respective membrane fusion events per se, are essentially independent of extracellular calcium (12-14). For example, endocytotic membrane retrieval does not seem to require elevation of the free intracellular calcium levels, but will be inhibited by drugs that interfere with energy metabolism, microfilament organization, or phospholipase A_2 activity, viz., inhibitors of other putative mediators of membrane fusion (see below). In some cases, elevations of the intracellular calcium levels by secretagogues or by calcium ionophores even seem to inhibit endocytosis (15).

Furthermore, numerous other membrane fusion events occur at low calcium concentrations (~100 nM) within the cell, for example in the Golgi or in the lysosomal apparatus, or in the interactions between the rough endoplasmic reticulum and the plasma membranes (16). Even in the case of exocytosis, per se, from intact secretory cells, basal levels of non-stimulated exocytotic secretion do not require an elevation of intracellular calcium levels. These "basal" secretory events involve exocytotic fusion events very much like the evoked events and occur continuously, albeit at a low rate, at calcium levels that are in the order of 50-150 nM, depending on the cell type (9). It is generally accepted, however, that calcium is a key component in cellular signaling during "stimulus-secretion coupling" and probably *the* ubiquitous messenger for receptor-mediated exocytotic hormone and neurotransmitter release (17).

One of the best-characterized examples of calcium-dependent exocytosis is catecholamine release from adrenal medullary chromaffin cells (1,4). However, even in this cell a close evaluation of the available information leaves ambiguous the predicted causal relationship between the intracellular calcium concentration and the rate of exocytotic membrane fusion. For example, simply raising $[Ca^{2+}]_i$ by ionophores or via calcium-containing liposomal vectors, is not sufficient to stimulate chromaffin cell secretion (18,19). Similar results were observed in neutrophils (20). On the other hand, following stimulation of the cells by depolarization with high potassium, catecholamine release rapidly ceases in spite of continued high levels of intracellular calcium. In the case of other secretagogues such as the sodium-channel agonist veratridine, the rate of calcium influx is rather slow and is initially proportionate to the rate of catecholamine secretion; however, the release process is sustained rather than pulsatile even though the cells will eventually reach maximal levels of $[Ca^{2+}]_i$ (1). Similar findings as to the discrepancy between $[Ca^{2+}]_i$ and the secretory competence of the respective cells have also been reported for neutrophils (20) and PC12 cells (21), pancreatic acini (22), platelets (23), and pituitary cells (24).

The situation with neutrophils is illustrative of the confusing calcium situation. Upon stimulation of human neutrophils with the chemoattractant peptide f-methionyl-leucyl-phenylalanine (FMLP), maximal levels of $[Ca^{2+}]_i$ are attained at agonist concentrations approximately 100 times below the dose required to elicit exocytosis (25). In a medically relevant aspect of neutrophil function, Cabrini and DeTongi report that in patients suffering from cystic fibrosis neutrophils have increased resting levels of free cytosolic calcium and decreased competence to secrete (26). More compelling information on the lack of causality between intracellular calcium levels and secretion in neutrophils comes from studies with certain secretagogues, such as diacylglycerol, phorbol esters, or cAMP, which can stimulate exocytotic secretion without a measurable rise in the levels of free intracellular calcium (5,23,27). Clearly, all these results do not fit with a simple calcium-directed membrane fusion hypothesis. Another example for calcium-independent membrane fusion is the case of degranulation elicited by ATP.

In certain cell types, such as mast cells (29) or parathyroid cells (30), addition of nonhydrolysable GTP analogs, such as GTP-γ-S seems to be a sufficient intracellular stimulus to elicit slow, but clearly calcium-independent, secretion. This result might suggest that at least in some cells, activation of a GTP-binding protein might trigger an as yet unknown calcium-independent fusogen. Indeed, inactivation of the GTP-binding protein by pertussis toxin inhibits degranulation, e.g., in neutrophils (31). However, the role for guanine-binding proteins as "calcium-substitutes" in providing alternate sources for fusion-promoting compounds might be restricted to certain cell types only, since the effects of GTP and pertussis toxin in permeabilized chromaffin cells are marginal, at best (Ref. 32; P. I. Lelkes and H. B. Pollard, unpublished results).

The lack of a requirement for elevated calcium in the fusion process is even more dramatically emphasized in a number of cell systems where secretion is actually turned on by a decrease in intracellular calcium concentration. In fact, raising intracellular calcium will actually turn off the secretory mechanism in parathyroid cells (6). However, Brown and his colleagues found that under certain experimental conditions the secretion of parathyroid hormone (PTH) is inhibited by ouabain and low extracellular potassium, without a concomitant increase in the cytosolic calcium concentration (33). In permeabilized parathyroid cells, on the other hand, secretion of PTH is, strangely enough, not dependent on the free calcium concentration (28). These confusing and seemingly contradictory results must thus serve as further caveats for those who may wish to draw compelling conclusions from permeabilized-cell models regarding the action of calcium in the secretion process.

Therefore, although changes in intracellular calcium are pivotal in many cases for intracellular signal transduction, it is clear that there is no direct, compelling evidence linking calcium and the membrane fusion process in cellular systems.

III. ROLE OF PROTONS IN FUSION PROCESSES

A. H⁺ in Secretory Processes

Stimulation of secretory processes in many cell types is accompanied by transient changes in the intracellular pH. For example, in the chromaffin cells, addition of secretagogues such as acetylcholine or nicotine or elevated K^+ results in a transient acidification of the cytoplasm of ~0.1 pH units. The time course of this acidification parallels that of secretion (L. M. Rosario and H. B. Pollard, unpublished results). While 0.1 pH unit could appear to be rather paltry, recent studies by Tsien and Poenie indicate that cells can support quite disparate pH values in different regions of the same cell (34). Therefore, a change of 0.1 pH unit, averaged over an entire cell population, might be rather profound when considered from a microscopic viewpoint.

Another example for cells exhibiting rapid changes in internal pH upon stimulation are the human neutrophils: Addition of the chemotactic factor FMLP induces the release of lysosomal enzymes and coincidentally lowers the internal pH by about 0.2 pH unit (35). It remains to be established whether blockade of acidification also blocks secretion. Interestingly, the neutrophil is one of those cells where changes in intracellular calcium concentration alone have been found to be insufficient to elicit the secretion process (20). Similarly, CO_2-induced cytoplasmic acidification is the first cellular signal leading to the exocytotic insertion of H^+-pumps in the turtle-bladder luminal membrane (36).

The interrelationship between pH and calcium in situ was studied by Abercombie and Hart using isolated cytoplasm from *Myxicola* axons (37). Microinjection of HCl, sufficient to lower the average pH by 1 unit, resulted in the average increase of the cytoplasmic calcium by 300 µM. However, raising the pCa by 2.2 units did not significantly affect the resting pH. These data thus indicate that, at least in the model studied, a rapid stimulus-induced acidification might result in a subsequent increase in $[Ca^{2+}]_i$, sufficient to trigger those events between a transient acidification and a rapid increase in $[Ca^{2+}]_i$. This seems to be a general phenomenon accompanying cellular stimulation (38). Thus, acidification could be one of the earliest stages in intracellular signal transduction, following binding of the various agonists to their respective receptors. The subsequent activation of a Na^+/H^+ antiporter and the resulting alkalinization would then appear to be prolonged cellular responses related to later phases in the stimulus-secretion coupling process. However, in other systems, such as in platelets, the initial, rapid acidification has not been observed, and hence platelet activation has been equated with the slower phase of Na^+/H^+ exchange (39).

At present, we cannot distinguish whether calcium or protons represent themselves or each other during the signal transduction process.

B. Changes in pH and Their Relation to Fusion, In Vitro

The best known examples of pH-induced fusion are found in the virus literature. As described in Chapter 15, certain proteins are rendered fusogenic when encountering a low-pH environment. Proton-induced fusion, possibly relevant to the virus systems, has also been described using liposomes of specific composition (40,41). The underlying mechanism appears to involve changes in the structure of the lipid phase. For example, at low pH some of these lipids undergo changes to hexagonal phases (see Chapter 2), resulting in leaky hemifusion. While the latter may not necessarily be a biologically relevant fusion process, it nonetheless points out that localized point defects, possibly induced by local pH decrements, might cause otherwise juxtaposed membranes to fuse.

In a system more related to exocytosis, chromaffin granules can be fused with acidic liposomes (42,43), by simply mixing them together (Fig. 1).

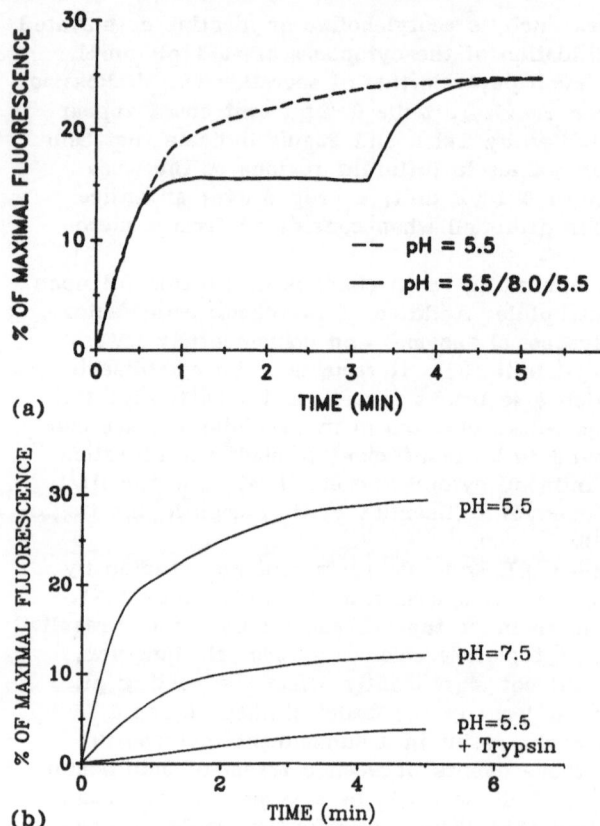

(a)

(b)

FIGURE 1 (a) pH-induced fusion of chromaffin granule ghosts with large
unilamellar liposomes composed of cardiolipin. Total lipid concentration
was 20 μM. The fusion was assayed using a fluorescence resonance energy
transfer assay. In the control experiment, liposomes containing 0.5 mol%
each of the fluorescent phospholipids *N*-NBD-PE and *N*-Rh-PE, were
incubated in a buffer containing 150 mM KCl, 5 mM EGTA, 10 mM Tris-
acetate with a fivefold excess (molar ratio lipid/lipid) of unlabeled chromaffin
granule ghosts at pH 5.5 (upper curve). Rapidly raising the pH of the
medium to pH 8.0, by injecting NaOH, resulted in the virtual arrest of
the fusion reaction, while subsequent reversal of the pH (by injecting HCl),
allowed the fusion reaction to resume (lower curve) (P. I. Lelkes, unpublished results). (b) pH dependence and trypsin sensitivity of the fusion
of intact chromaffin granules with acidic liposomes. Large unilamellar liposomes
composed of PS/PE/octadecylrhodamine at a molar ratio of 7:3:0.1 were
prepared by reverse-phase evaporation. The liposomes were incubated
with a fivefold (molar ratio lipid/lipid) excess of intact chromaffin granule
ghosts. Total lipid concentration was 100 μM. The relief of the concentration-
dependent self-quenching of the fluorescent probe was taken as an indication
for the occurrence of fusion between the liposomes and the granules. For
studying the protease sensitivity of the reaction, aliquots of the granules
were incubated for 15 min with 1 mg/ml trypsin (followed by the addition
of 1 mg/ml trypsin inhibitor). The granules were then washed and resuspended and the fusion experiments were carried out as described in
Refs. 42 and 43.

This system does not require calcium, but proceeds to fuse in a sustained manner at pH 5.5. Raising the pH to 8.0 attenuates the fusion process, but subsequently lowering the pH back to 5.5 allows fusion to resume. The fusion capacity of this system seems to depend on a protein component of the granule membrane, since trypsin treatment impairs fusion competence irreversibility.

In contrast to this model system, chromaffin granules cannot be fused to one another purely by the addition of calcium or protons. Instead, a specific protein, synexin, is required, which at moderately acidic pH values (pH 6.0-6.5) will cause the granules to aggregate in a calcium-dependent fashion (44). Subsequent fusion is induced by addition of micromolar concentrations of arachidonic acid. Interestingly, the fusion step, per se, is entirely independent of calcium, since it proceeds equally well in the presence of EGTA (45).

Thus, we conclude that in either of the synexin-dependent systems, or in the synexin-independent granule-liposome fusion system, fusion per se is promoted in a pH-dependent manner, irrespective of the calcium requirement.

C. What Do Protons Do To Promote Fusion?

Without recourse to specific models, it is obvious that changing the pH could titrate charged groups on amino acid side chains, on phospholipids, or on complex lipids. Indeed, exposure of chromaffin granule ghosts to membrane impermeant carboxyl-group modifiers renders the membrane resistant to fusion by synexin or to fusion to acidic liposomes (P. I. Lelkes and H. B. Pollard, unpublished observations). Furthermore, while the pH in a homogeneous solution is uniform, the pH at surfaces is nonuniform. For example, the concentration of a cation at a surface, c_s, depends on the surface potential, $\Delta\psi$, in the following way,

$$c_s = c_b \cdot e^{-\Delta\psi/kT}$$

where c_b is the concentration in the bulk phase, and kT has its usual values.

The preferential movement of H^+ across surfaces formed by phospholipid head groups has also been demonstrated experimentally using a monolayer on a Langmuir-type trough (46). At the very minimum, this means that movement of H^+ across a membrane could generate local low ("nonphysiological") pH values. The exact pH value depends on the volume considered, but even at the level of the light microscope, the local bulk phase pH value can vary by as much as 0.4 pH unit (47).

IV. THE ROLE OF ATP

It has been one of the more firmly established dogmas that exocytosis is an energy-consuming process which necessarily requires the hydrolysis of ATP. This view has seemingly been confirmed in numerous electrically or chemically permeabilized cell systems where calcium-dependent release is found to occur only in the presence of MgATP. However, as stressed by Knight and Baker (10), so far no one has really been able to pinpoint the actual site of action for ATP. Recently, the notion of an absolute requirement for ATP has been challenged by Gratzl and co-workers (48).

These authors observed calcium-induced catecholamine secretion to occur from alpha-toxin-treated PC12 cells in the absence of ATP. Similarly, Rojas and co-workers have demonstrated calcium-induced ATP release from electrically as well as from chemically permeabilized chromaffin cells in the absence of exogenous ATP (49). Both these results could simply mean that in these cells the ATP pools might not have been completely depleted. Indeed, the original observations were performed on permeable cells, allowed to incubate for a period to "deplete" the endogenous ATP. However, recent data on secretion from saponin-permeabilized platelets clearly illustrate that thrombin can induce exocytotic serotonin release in spite of extensive ATP depletion (50). Plattner and co-workers, studying exocytosis in *Paramecium*, observed that the energy-requiring processes do not occur during the secretory events proper, but that they are rather related to some subsequent phase. Indeed, using an in vitro cell-free model of *Paramecium* cortices, ATP was found to inhibit exocytosis (51; see also Chapter 25). Thus, in situ, ATP might actually have to be removed from the fusion site during the secretory activity. In line with these findings, Brooks and Brooks reported that thiophosphorylation irreversibly inhibits calcium-induced catecholamine secretion from chemically skinned chromaffin cells (52). Therefore, it remains conceivable that in intact cells, the role of ATP is limited to a housekeeping function which is not involved in the exocytotic fusion step, rather than in energy-consuming processes which are relevant for retaining or regaining fusion competence.

V. TRANSMEMBRANE ELECTRICAL POTENTIAL

In many secretory cells, one of the earliest consequences of receptor-mediated or nonspecific stimulation is a depolarization of the plasma membrane which results in the opening of voltage-sensitive sodium and calcium channels (53). This, in turn, is believed to be the major source of an increase in $[Na^+]_i$, and in $[Ca^{2+}]_i$, which then will trigger further steps of intracellular signal transduction, as discussed above. However, there appears to be no direct correlation between the transmembrane electrical potential and membrane fusion. For example, in chromaffin cells addition of excess potassium in the absence of extracellular calcium will depolarize the cells; however, practically no catecholamine secretion will occur (4,54). These data seem to be in line with more recent findings that the triggering of degranulation in mast cells and basophils by IgE does not seem to involve depolarization and/or to require the opening of voltage-sensitive ion channels (55). Moreover, potassium-induced depolarization was found to suppress exocytosis in rat basophilic leukemia cells (56). Thus, secretion, viz., membrane fusion, and depolarization can clearly be decoupled in many cell types.

However, fertilization is one example of membrane fusion that might be controlled by changes in membrane potential. In some, but not all, species, sperm-egg fusion appears to be directly regulated by the egg transmembrane electrical potential (for a recent review, see Ref. 57). Upon fertilization, a distinctive shift in the transmembrane potential is observed, which is believed to be inhibitory for further fusion events between the egg and the sperm membranes. These potential shifts, which, depending on the species, can be either depolarizing or hyperpolarizing, represent an effective means of blocking polyspermy.

The mechanism by which the transmembrane electrical potential regulates the fusion competence of the participating membranes is at present unknown. However, it has been speculated that a putative fusion protein on the sperm membrane will interact differentially with its receptor on the egg membrane, depending on the potential-sensitive disposition of proteins on the latter (58). The disposition of membrane-bound proteins, viz., vertical displacement, the availability of binding sites, the exposure of cryptic sites, etc., can be regulated indirectly via changes in the fluidity of the lipid bilayer; such changes were shown to be dependent on the transmembrane electrical potential (59,60).

Alternatively, changes in the transmembrane potential can directly modulate the orientation of proteins in the membranes or at their surfaces (61). All of these changes might conceivably change the fusion competence in these systems, for example, by modulating protein phosphorylation (62). Indeed, Prives and Shinitzky have reported that the fusion of myoblasts is preceded by an increase in the microviscosity of the plasma membranes (63).

It might be tempting to speculate that a similar mechanism of membrane fusion might also be the basis for "compound" exocytosis. In many secretory systems, intracellular storage granules exist in close proximity to each other. During their saltatory movement in the cytoplasm, they frequently collide with each other, as can easily be observed in the light microscope. Under resting conditions, the granules will seldom fuse with each other. And yet, when the cells are stimulated, the membranes of those granules that have fused with the plasma membrane appear to be recognized as "fusogenic" since new granules will fuse to such a granule "prefused" with the plasma membrane (1,44).

In the case of adrenal chromaffin cells, the electrochemical properties of granules and their membranes have been well characterized (64). Under resting conditions, chromaffin granules in situ have a slightly positive potential of about +10 mV with respect to the cytoplasm. However, it seems plausible to assume that upon exocytotic fusion, the granule membrane will acquire the potential of the plasma membrane. This would hyperpolarize the granule membrane. In terms of the possible potential dependence of membrane fusion, this might mean that upon hyperpolarization granule membranes could expose otherwise cryptic receptors for cytosolic or membrane-bound proteins that would mediate granule-granule contact and fusion. Considering the possible involvement of synexin in compound exocytosis (see below), the fact that synexin-induced charge displacement within membranes is voltage-sensitive (65) might at least suggest that such charge displacement processes could also occur in vivo.

Since various secretagogues induce different degrees of depolarization of the plasma membrane (54), the magnitude of the hyperpolarization of the granule membrane should also depend on the stimulus. This in part could explain the differences in efficiency and frequency in occurrence of compound exocytosis observed with various secretagogues.

Furthermore, it is tempting to hypothesize that such a voltage-dependent mechanism might also be involved in regulating some of the intracellular fusion events in receptor-mediated endocytosis, such as the fusion between receptosomes and endosomes. The mechanism(s) governing these steps are, at present, essentially unknown.

VI. OSMOTIC FORCES

Based on chemiosmotic considerations, H. B. Pollard and co-workers suggested that exocytotic membrane fusion might be driven by osmotic forces acting across the chromaffin granule membrane (66-68). According to this hypothesis,

experimental conditions causing vesicle lysis should sustain or cause exocytosis, while other conditions, causing vesicle stabilization and/or shrinkage, would impair membrane fusion. To date, this hypothesis has neither
been fully proved nor completely disproved, but it certainly has stimulated
many interesting experiments (69).

The inhibitory effect of hyperosmolar solutions on exocytotic hormone
release from intact cells has been observed in a number of systems, such
as in platelets (68), parathyroid cells (70), PC 12 cells (71), and also
in chromaffin cells (1,72,73). Detailed studies, however, have revealed
that at least in the chromaffin cell, the bulk of the osmotic effect on high
K^+-induced release residues in alterations of the calcium channel activity:
the impairment of calcium influx by hypertonic solutions can fully account
for the inhibition of catecholamine release (E. Heldman and H. B. Pollard,
unpublished observations). However, in the case of secretion induced
by nicotinic agonists, the inhibitory effect appears to reside at a step
distal to calcium entry (69; E. Heldman and H. B. Pollard, unpublished
observations).

In some model systems of exocytosis, using either permeabilized cells
(1,10,74,108) or cell-free cortical membrane patches from sea urchin eggs
(75), it has been confirmed that hyperosmotic media do indeed interfere
with exocytotic membrane fusion. However, vesicle shrinkage does not
seem to be a major factor in impairing exocytosis, since severely shrunken
granules still can undergo membrane fusion in all systems investigated
(74,76,110).

As to the mechanism of how osmotic forces might impair exocytosis,
Holz has speculated that the increase in internal ionic strength achieved
via reducing the free water space might inhibit other intracellular components required for mediating the events leading up to the fusion step
(69). On the other hand, Baker and Knight have concluded that the
inability of chloride ions to support exocytosis in electrically permeabilized
chromaffin cells contradicts the involvement of chemiosmotic forces, at
least as defined by Cl^--driven lysis of chromaffin granules in the presence
of ATP and Ca^{2+} (77). Thus osmotic forces might have a variety of cellular
effects, both on the level of the plasma membrane, as well as intracellularly,
not necessarily related to the fusion step.

The importance of membrane swelling induced by osmotic forces had
been clearly demonstrated for the fusion of lipid vesicles with planar bilayer membranes [for reviews, see Ref. 78 and Chapter 8]. Similarly,
Ahkong and Lucy observed osmotic swelling of erythrocytes during electrically induced fusion (79) and proposed a general osmotic model for the
fusion of biological membranes (80). To correlate these results and to
test these hypotheses in an in vivo system, two groups have independently
studied exocytosis from giant mast cells of the beige mouse (81,82). Using
sophisticated techniques to simultaneously measure membrane capacitance
and video-enhanced optical parameters, both groups conclude that the
exocytotic fusion event as defined by membrane capacitance changes clearly
precedes the osmotic swelling of the granules. However, the conclusion
is based on the explicit assumption that the observed morphological changes
are those associated with granule swelling and not with changes in core
structure following fusion.

The role of osmotic forces in exocytotic membrane fusion is at present
challenged, at least in the mast cells of the beige mouse. Future work
will have to clarify whether this challenge is valid, and if so, whether
it can be generalized to other kinds of biological fusion events.

VII. INTRACELLULAR PROTEINS

In the preceding sections, we dealt mainly with small molecules such as protons, calcium, ATP, etc., which either directly or indirectly might control exocytotic fusion. In this section, we discuss the possible regulatory role of some larger polypeptides and proteins. Some of these proteins, such as protein kinase C or synexin, turn out to translocate from a cytoplasmic to a membrane-bound state, concomitant with or as a consequence of cell stimulation.

In considering the involvement of intracellular proteins in the fusion process, we can distinguish two distinct classes of such proteins. One of these classes must include proteins of known functions, such as kinases, lipases, proteases, and calcium-binding proteins like calmodulin. In this class, we assume that the proteins, in addition to their known functions, might themselves participate in some way in fusion or might generate products that would be involved in the fusion process. In the second class, we have identified several proteins of basically unknown function, which have been directly implicated in membrane contact and fusion events. Such proteins include synexin and other members of the gene family, such as endoxin II, lipocortin I, calpactin I (p 36), protein II and calelectrin 67 kD.

A. Protein Kinases

Protein kinase C is a ubiquitous protein that has been implicated in mediation of both receptor-linked processes (83,84) and stimulus–secretion coupling (7). Examples of the latter process include the stimulation of neutrophils and of parathyroid cells, degranulation of basophils, activation of platelets, stimulation of insulin release from pancreatic beta cells, and the release of neurotransmitters. In all these cells, exocytotic secretory events can be induced by phorbol esters, which are believed to mimic the effect of diacylglycerol, a specific activator of protein kinase C (85). For example, in human neutrophils, there is good evidence that activation of protein kinase C is a sufficient stimulus to elicit the entire spectrum of chemotactic responses, including degranulation at vanishingly low concentrations of free cytosolic calcium (86), without actually raising the level of intracellular calcium. However, in several other systems, such as the chromaffin cells, phorbol esters will hardly affect secretion by themselves. Yet, they will potentiate suboptimal secretion evoked by calcium ionophores, such as A23187, or by potassium, provided the cells were preincubated in a low-calcium medium (87a,b). Indeed, in some secretory systems, such as platelets (88), PC12 cells (21), or neutrophils (89), activation of protein kinase C by phorbol ester was shown to synergistically augment the effects of other less potent secretagogues, such as calcium ionophores. Similarly, in digitonin-permeabilized chromaffin cells, activation of protein kinase C did not cause secretion by itself, but was found to lower the calcium requirement for catecholamine release (90). Thus, activation of protein kinase C might indeed mediate exocytotic hormone release. However, the question remains, is protein kinase C, itself, involved in the fusion process, or is it rather one of its phosphorylated substrates that exerts control over fusion event?

Under resting conditions, the majority of protein kinase C is in soluble form; however, upon stimulation of the cells, the protein becomes membrane-associated in a calcium- and lipid-dependent fashion. Indeed, Creutz and his co-workers identified protein kinase C as one of the "chromobindins"

that binds to chromaffin granule membranes in a calcium-dependent manner
(91). In stimulated chromaffin cells, protein kinase C is found in association
both with the cytoplasmic face of the plasma membranes as well as with
that of granule membranes. This is important, since protein kinase C is
enzymatically active only in a membrane-bound or a physically equivalent
state. Thus, there are some circumstances in which both secretion occurs
and protein kinase C can and does bind to membranes in the cell. It is
presently conventional to expect that only the "activated" protein kinase C,
in its membrane-bound form, will phosphorylate granule membrane and/or
plasma membrane proteins. However, it may be instructive that protein
kinase C also phosphorylates at least one of the other chromobindins (92).
This advances the possibility that a soluble protein product of protein
kinase C action might also be involved in protein kinase C action during
the fusion process. However, whether protein kinase C will also induce
membrane fusion, in the sense that it cross-links and/or coalesces adjacent
membranes, remains to be tested. Nonetheless, while the possible fusogenic
role of the protein kinase C molecule is purely speculative, there is over-
whelming evidence for a number of other cellular events (e.g., protein
synthesis, channel activation, stimulus-secretion-synthesis coupling, etc.)
being accompanied by phosphorylation.

An increasingly large number of different proteins are currently being
identified in the various secretory systems that are phosphorylated upon
cell stimulation. Yet, the roles of most of these phosphoproteins in intracellular signal transduction or, more specifically, in the fusion process
remain unsubstantiated. In chromaffin cells, for instance, stimulus-induced
^{32}P incorporation into several proteins has been repeatedly observed, but
one of these proteins, for instance, is tyrosine hydroxylase, a key enzyme
in catecholamine synthesis (93). At least one of the "chromobindins" is
a protein kinase substrate which is phosphorylated upon stimulation of
of the cells (94). Furthermore, a number of intrinsic chromaffin granule
membrane proteins are phosphorylated in a Ca^{2+}/calmodulin- and/or c-AMP-
dependent fashion. However, as for protein kinase C substrates, their
relevance for exocytosis also remains undetermined (95).

So far, only reversible phosphorylation events have been considered
as having possibly regulatory functions in membrane fusion. However,
(calcium-dependent) dephosphorylation might be also potentially relevant.
After all, to reversibly phosphorylate a site implies it must be dephosphorylated to prepare it for the next activation step. Indeed, Tallant and
Wallace have identified a Ca^{2+}/calmodulin-dependent phosphatase in human
platelets and suggest a role for Ca^{2+}-dependent dephosphorylation in
platelet activation (96). In some endocrine cells, such as cultured thyroid
cells (97) and chromaffin cells (98), a stimulus-induced dephosphorylation
of some smaller proteins with molecular weights of approximately 20 kD
has been reported. However, no attempt has been made so far to correlate
these phosphoproteins to the fusion event proper.

The notion that dephosphorylation and not phosphorylation might play
a key element in exocytosis has recently been substantiated by two groups
studying exocytosis and membrane fusion in unicellular organisms such as
Paramecium (99,100). Both groups have identified a 65-kD protein, which
upon stimulation is rapidly dephosphorylated in a Ca^{2+}/calmodulin-dependent
fashion and then gradually becomes rephosphorylated, as fusion competence
of these cells is regained. Thus, at least in *Paramecium*, a major role of
the secretagogue indeed seems to include the stimulation of phosphoprotein
dephosphorylation concomitant with a depression of protein kinase activity.

These data appear to provide the first hard-core evidence for a causal connection between dephosphorylation and exocytosis.

B. Phospholipases and Their Products

Two distinct phospholipases, phospholipase C and phospholipase A_2, and a diacylglycerol lipase have been implicated in regulating membrane fusion events in a variety of cells. Phospholipase C cleaves off water-soluble species, inositolphosphate (IP), inositoldiphosphate (IP_2), and inositoltriphosphate (IP_3), from the respective inositol phospholipids, leaving a water-insoluble diacylglycerol moiety (85). This diacylglycerol molecule may be involved in activating protein kinase C, as described in the previous section or, alternatively, it might itself induce the release of calcium from the endoplasmatic reticulum. The soluble product of phospholipase C action, IP_3, may mobilize calcium from endogenous stores in the endoplasmic reticulum (83). While the plasma membrane phospholipase C enzyme has generally been implicated in the generation of calcium-mobilizing IP_3, another cytosolic phospholipase C activity has also been detected that binds in vitro to chromaffin granules in a calcium-dependent manner (101). The possible direct roles of either of these proteins in the fusion process have not been considered or investigated.

Phospholipase A_2 activity has been implicated as a key component in membrane fusion mechanism(s), most probably since both of its reaction products, lysophospholipids as well as arachidonic acid, are putative fusogens (102). However, an alternative source of arachidonic acid is the enzyme diacylglycerol lipase. The latter enzyme uses the product of the membrane-bound phospholipase C, diacylglycerol, as a substrate to generate arachidonic acid. Interestingly, in chromaffin cells, a diacylglycerolipase is localized on the plasma membranes, but it is calcium-insensitive (103).

Phospholipase A_2 has been detected in chromaffin granule membranes, but only in vanishingly small quantities in plasma membrane (103). In vitro, it is activated by relatively high concentrations of calcium (~1 mM). Lysophospholipids, the products of phospholipase A_2 activity, which destabilize lipid bilayers, have been invoked as fusogens for erythrocyte membranes (104). However, since the chromaffin granule membrane is exceptionally rich in lysolipids (64), the relevance of this finding for exocytotic fusion remains obscure.

In many cells, such as mast cells, platelets, neutrophils, and chromaffin cells, the coupling of stimulation to secretion is accompanied by the liberation of arachidonic acid and/or its metabolites (8,105,106). Conversely, addition of exogenous arachidonic acid promotes catecholamine secretion from digitonin-permeabilized chromaffin cells (107). In the latter case, a close relationship was established between arachidonic acid activity and secretion. The data, however, seemed somewhat confusing, because, while arachidonic acid action in permeabilized cells requires divalent cations, magnesium seems as potent as calcium. Similarly, exogenous arachidonic acid was shown to cause insulin release from pancreatic beta cells (108). Furthermore, a metabolite of arachidonic acid, prostaglandin E_1, promotes the fusion of myoblasts (109). These data therefore indicate that arachidonic acid, or its metabolites, might directly act on membrane fusion.

By itself, at quite high concentrations (~30 µM) arachidonic acid uncouples mitochondria and is a general membrane perturbant, inducing massive leakage from liposomes (P. I. Lelkes, unpublished observations). However, at lower concentrations, (ca. 5 µM) it accelerates calcium-driven

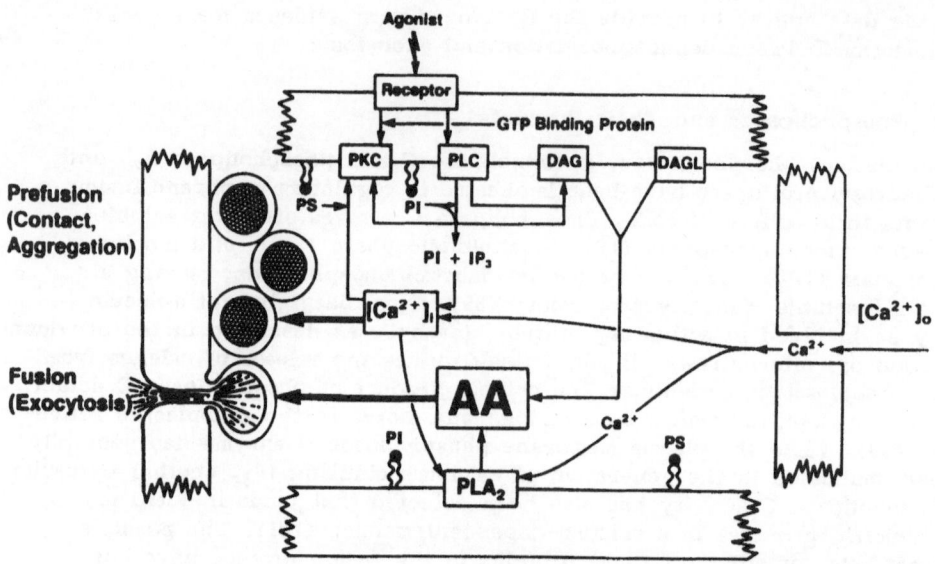

FIGURE 2 Schematic depiction of the possible cellular mechanisms of calcium-dependent events that can produce arachidonic acid. According to this scheme, calcium would be needed for the intracellular signal transduction leading up to and including plasma membrane-granule and granule-granule contact. Arachidonic acid, in turn, might regulate membrane fusion in a calcium-independent fashion. PKC, Protein kinease C; PLC, phospholipase C; DAG, diacylglycerol; DAGL, diacylglycerol lipase; PS, phosphatidylserine; PI, phosphatidylinositol; PLA$_2$, phospholipase A$_2$.

polymerization of synexin molecules (110) and it promotes fusion of chromaffin granules previously aggregated by the calcium-binding protein synexin (45) in a Ca^{2+}-independent fashion. However, the entire spectrum of fusion-promoting properties is not unique to arachidonate but it is shared by many *cis*-unsaturated fatty acids. And yet, arachidonic acid is the most abundant of all naturally occurring *cis*-unsaturated fatty acids, and hence its fusogenic competence might be of physiological relevance.

Garcia-Gil and Siraganian found that the source and the amount of arachidonic acid released upon stimulation of rat basophils depended on the type of stimulus applied (111). Thus, IgE-mediated degranulation induced the release of arachidonic acid predominantly from phosphatidylinositol (PI) and phosphatidylserine (PS), while stimulation of the cells by calcium ionophores caused a doubling of the amount of arachidonic acid release predominantly from phosphatidylethanolamine (PE) and phosphatidylcholine (PC). These data suggest that, depending on the stimulus provided, the cells can activate various pathways of arachidonic acid production with different degrees of efficiency. Thus, we hypothesize that the liberation of arachidonic acid by diverse pathways could be a central goal in the cellular signal transduction scheme(s) leading to exocytosis. The existing data at least suggest that arachidonic acid or its breakdown products might then be utilized to fuse membranes previously in contact. A scheme for the putative role of arachidonic acid as a central agent in modulating membrane fusion is presented in Figure 2.

C. Proteinases

By analogy to protease-dependent examples of viral fusion, the possible involvement of proteolytic enzyme activity has also been investigated in terms of the exocytotic process. Indeed, a proteolytic enzyme, probably a serine esterase, seems to be one of the first intracellular components to be activated upon mast cell stimulation (for a summary see Ref. 112). Lucy observed the activation of intracellular proteolytic activity during chemically induced erythrocyte fusion and hypothesized that endogenous proteinases might also modulate exocytotic membrane fusion (113). In terms of specific mechanisms, Lucy speculated that hydrophobic sequences, cleaved from cellular polypeptides, might induce membrane fusion in situ (114).

Recently, Strittmatter and his co-workers suggested that metalloendo-protease(s) (MEPase), located at the cytoplasmic face of the plasma membranes, might directly modulate membrane fusion (115). They based this hypothesis on the fact that a number of "specific" inhibitors and substrates of MEPase activity were also found to inhibit mast cell degranulation, exocytotic secretion from chromaffin cells, and neurotransmitter release at the frog muscular junction. In addition, the same compounds also inhibited myoblast fusion as well as fertilization in the sea urchin (116). While the majority of metalloendoprotease activity in both myoblasts and chromaffin cells is found in association with the plasma membranes, Mundy and Stritt-matter presented evidence that it is not the plasma-membrane-bound form of MEPase activity, but rather the cytoplasmic isozyme(s), that is inhibited by the specific MEPase antagonists (117,118). These results raise doubts about the actual involvement of a metalloprotease activity in the fusion step proper, in particular in myoblasts.

Recently, we showed that in adrenal medullary chromaffin cells, inhibitors of metalloprotease activity block catecholamine secretion, not by preventing exocytotic fusion of the granules with the plasma membranes, but rather by inhibiting secretagogue-stimulated rise in intracellular calcium levels (119). Furthermore, "specific" metalloendoprotease antagonists by themselves caused an accelerated calcium efflux from intact cells. In addition, direct exposure of the granules to genuine metalloendoproteases activity, either in situ in detergent-permeabilized cells or in the test tube, did not enhance their fusion competence. Therefore, we might conclude that the oligopeptide inhibitors and hence presumably the metalloendoprotease activity per se are not mediating membrane fusion. Rather, MEPase activity, or a factor affected by MEPase inhibitors, may be required to regulate intracellular calcium homeostasis. Thus, it is not surprising that in detergent-permeabilized cells, where these mechanisms are short-circuited, catecholamine secretion stimulated by micromolar calcium is unaffected by metalloendoprotease antagonists.

In summary, the appealing idea of soluble and/or membrane-bound proteases as actual mediators of membrane fusion has not yet been substantiated by hard experimental evidence. Rather, proteolytic activity, an abundant feature occurring at the plasma membrane, in the cytoplasm and/or in intracellular organelles, seems to be related to regulation of signal transduction processes.

D. Calmodulin

The ubiquitous calcium-regulatory protein calmodulin has been implicated in virtually every cytosolic event that might be mediated by calcium. Its consistent presence in a wide variety of secretory cells has been shown by immunocytochemical and by biochemical techniques (120). Calmodulin was originally thought to be a soluble, cytosolic protein; however, in digitonin-permeabilized chromaffin cells, calmodulin could also be visualized by light microscopy in colocalization with a typical granule marker protein, dopamine-β-hydroxylase (98). However, while the total amount of calmodulin in these cells remained unchanged, its subcellular distribution was altered from a soluble to a preferentially membrane-bound form, depending on the intracellular calcium levels (121).

Because it is so ubiquitous, it has been quite difficult to envision a specific role for calmodulin, or for the many "calmodulin-binding proteins" in regulating the fusion event. Previous studies, employing "selective" pharmacological calmodulin antagonists, such as the phenothiazine drugs, have revealed that some of these compounds interfere rather nonspecifically with a number of cellular functions. Trifluoroperazine, at doses affecting calmodulin, is a good local anesthetic and blocks synexin. Promethazine, at similar doses, blocks secretion and blocks synexin, but is a calmodulin blocker and local anesthetic at only 100-fold greater concentrations (1). These latter drug effects thus exclude calmodulin as the site of phenothiazine action in secretory cells.

However, recently several additional lines of evidence have been presented that might suggest a direct involvement of calmodulin and/or some calmodulin-binding proteins in exocytotic secretion. Using a cell-free model system for exocytosis, Steinhardt and Alderton demonstrated that antibodies raised against calmodulin could inhibit calcium-induced degranulation of cortical surfaces of sea urchin eggs (122). One caveat here is the high antibody concentration and prolonged incubation time needed to observe the effect. In chromaffin cells, Kenigsberg and Trifaro have shown that microinjection of monospecific anticalmodulin antibodies (and of their F_{AB}-fragments) resulted in partial inhibition of acetylcholine- or potassium-induced catecholamine secretion (123). Similarly, Plattner and his collaborators have shown that anticalmodulin antibodies will inhibit synchronous exocytosis in *Paramecium* cells (124; see also Chapter 25). The latter authors have also established a causal link between calmodulin, the dephosphorylation of a 65-kD protein, and exocytosis in *Paramecium*.

In the chromaffin system, calmodulin-binding sites were shown to exist both on the granule membranes as well as on the cytoplasmic face of the plasma membranes (121). One of the membranous receptors for calmodulin in chromaffin cells has been identified as a 65-kD protein (125). This protein was immunologically cross-reactive with granule surface antigens of identical molecular weight found in synaptic vesicles and in storage vesicles from anterior and posterior pituitary cells, platelets, and pancreatic β cells. Yet to be determined is whether these proteins are related to the 65-kD protein, previously described by Gilligan and Satir in *Paramecium* and in *Tetrahymena* (99), and whether this 65-kD calmodulin receptor in chromaffin cells is also dephosphorylated upon cell stimulation. However, it remains tempting to speculate that calmodulin might be involved in the regulating the attachment of storage organelles to the fusion site at the plasma membrane and that triggering of exocytosis might require a concerted action of a protein kinase and/or a phosphatase.

VIII. STRUCTURAL PROTEINS: THE CYTOSKELETON

The cytoplasm of eukaryotic cells contains a number of intracellular organelles embedded in an essentially saturated protein solution. This notion may be of particular importance when trying to mimic the fusion of biological membrane vesicles in a test tube, and when attempting to extrapolate from such in vitro fusion experiments to what might be going on in situ. Besides a variety of "soluble" proteins, the cytosol contains three main classes of cytoskeletal proteins (microfilaments, intermediate filaments, and microtubules) and a large number of as yet uncharacterized structural proteins binding to and/or forming the cytoskeleton. Cytoskeletal architecture is not static, but rather a highly dynamic complex of interacting and reacting proteins. Their assembly and disassembly is tightly controlled by calcium and specific cytoskeleton-binding proteins. So also is their binding to intracellular membranes. This cytoskeletal complex can constitute a sizable part of the cytoplasmic proteins (up to 40% or more). This highly dynamic, interconnected network is believed to regulate intracellular organelle movement, close approach to the plasma membrane, and possibly even membrane fusion.

Although there is sufficient evidence for the involvement of the cytoskeleton in intracellular signal transduction mechanisms, we must conclude that there is no direct proof, as yet, linking cytoskeletal proteins to membrane fusion proper. However, there are four essential properties of cytoskeletal proteins which might pertain to their involvement in exocytosis. These are (1) their contractility, (2) their viscoelastic properties, (3) their ability to bind to lipid membranes, and (4) their interactions with specific membrane proteins.

A. Cytoskeletal Proteins Interact with Membranes

Cytoskeletal proteins have been demonstrated in many secretory cells including chromaffin cells (98,121,126). Furthermore, it was shown that these proteins bind to biological membrane surfaces (127-129). For example, in the chromaffin cell system, α-actinin and α-fodrin have been demonstrated to be present on the chromaffin granule membrane, while myosin and ankyrin (the spectrin "receptor" in erythrocytes) could not be detected (130). Actin and fodrin were also found in close association with the cytoplasmic face of the plasma membranes (126,131). However, under stringent conditions of isolation, actin, α-fodrin, and α-actinin are absent from granule membranes but actin is present on plasma membranes (132).

Intracellular storage granules from a variety of secretory cells bind tubulin. However, the functional role of this association with microtubules is not understood. In many systems, association between microtubular and cytoplasmic organelles has been related to intracellular movement (133). However, such organelle movement can also occur along actin cables (134). In spite of these divergent data, however, the presence of microtubule-associated proteins in neuronal and secretory cells and their putative interactions with secretory granules suggest some functional role for membrane-microtubule interactions in exocytosis.

Interestingly, all three classes of cytoskeletal proteins [microfilaments (135,136), microtubules (137,138), and intermediate filaments (139,140)] and also some of the actin-binding proteins bind to phospholipid membrane vesicles. Provided that membrane fusion involves modulation in the lipid bilayer stability, one can assume that such binding to lipids might represent another, as yet unexplored mechanism, by which cytoskeletal proteins might mediate membrane fusion.

Among the actin-binding proteins, vinculin contains covalently bound lipid and specifically binds to lipid membranes containing acidic phospholipids (141,142). Similarly, band 4.1, a peripheral erythrocyte-membrane anchor protein for spectrin, preferentially interacts with PS-containing lipid membranes (143). Interestingly, band 4.1 was determined to be a spectrin-binding protein immunologically related to synapsin I, a synaptic vesicle protein (144). Synapsin I, on the other hand, was found to be functionally related to the family of the calpactins, binding and cross-linking vesicles and cytoskeletal proteins (see below).

By contrast, α-actinin specifically binds to artificial membranes containing glycerides and fatty acids (145). Furthermore, in the presence of diacylglycerol and palmitic acid (lipid products liberated during PI turnover), supramolecular complexes between α-actinin and actin are formed in vitro, reminiscent of microfilament bundles in vivo. The latter results suggest, but do not prove, a possible interrelationship between the lipid composition of the biological membranes and the state of assembly of the adjacent microfilaments (146).

Most of the association of the actomyosin system with biological membranes is believed to occur via "specific" receptors or membrane-associated cytoskeletal binding proteins. Some of these binding proteins might be truly membrane-associated. However, as outlined above, some of the cytoskeleton-binding proteins, like the calpactins or some of the chromobindins, might also be of cytosolic origin and will bind to the cytoskeletal proteins and/or to the membranes only in the presence of calcium. In this respect, it is remarkable that the cytoskeletal association of vinculin, one of the proteins believed to form the linkage between actin and membrane surfaces, has been shown to be dependent on the thrombin-induced activation of platelets (147). Some authors have also found evidence linking the α-actinin/actin complex to the diacylglycerol-mediated signal transduction process in activated platelets (148). In chromaffin cells, Aunis, Trifaro, and their colleagues identified some calcium-activated actin-binding proteins related to gelsolin and brevin (149). These and other such proteins (e.g., fodrin) could serve as a calcium-dependent regulatory mechanism to dynamically control the gel-sol state of the cytoskeleton in secretory cells in general (121,130).

In search for a functional role of cytoskeletal proteins in exocytosis, Fowler and Pollard (127,128) have characterized the interactions between actin and chromaffin granule membranes. They demonstrated that the viscosity of a chromaffin granule membrane solution containing F-actin is dramatically increased in the presence of low calcium concentrations (<0.1 μM), indicating a strong cross-linking effect of the granules. At higher calcium concentrations (>0.1 μM), occurring for example when the cells are stimulated, the viscosity of the actin-granule solution is strongly decreased because of the dissociation of the actin filaments from the granule membranes. These findings have been confirmed and extended by Aunis and Perrin, who suggest that the actin-binding and cross-linking properties of the granule membranes might be related to spectrinlike molecules (fodrin) residing on the granule membranes (129,130). Similar results have been obtained for the interactions of actin with isolated plasma membranes from chromaffin cells. However, as shown by Aunis and associates, fodrin is apparently not associated with chromaffin granules in resting cells. Furthermore, fodrin is absent from EDTA-washed granule membranes, which are nonetheless easily cross-linked by F-actin (126).

B. Cytoskeletal Contractility

Based on the well-known contractile properties of muscle, it is plausible to assume that cytoskeletal proteins dynamically interact with secretory organelles in a calcium-dependent fashion and direct organelle movement toward the site of exocytosis. Indeed, directed movement of intracellular organelles on actin or tubulin cables in vitro has been documented by video microscopy (133,134). Thus, cytoskeletal proteins could be viewed as "railroad tracks" regulating intracellular trafficking and providing for the transport of secretory organelles to preformed fusion sites on the plasma membrane. The intracellular movement of secretory granules could therefore occur either as a direct consequence of the contractility of the cytoskeletal proteins and/or indirectly via calcium-activated proteins that concomitantly bind to membranes and to cytoskeletal proteins.

However, under resting conditions, cytoskeletal filaments could actually obstruct biological membrane fusion by virtue of their mere presence. For example, the microfilamentous web close to the cytoplasmic face of the plasma membrane can be as much as 1000 nm thick. Therefore, it has been postulated that membrane fusion during exocytosis requires the physical removal of cytoskeletal proteins from the fusion site (19,44). Indeed, in the prefusion state in many systems, e.g., *Paramecium* (150), *Tetrahymena* (151), *Limulus* amebocytes (152), or in "active zones" of presynaptic nerve terminals (153), secretory granules are found in close apposition to plasma membranes. The two membranes frequently appear to be separated only by a very narrow gap (~100 Å), which is often filled with filamentous, cytoplasmic material. To date, no one has identified this material, but it is reasonable to assume that it might include cytoskeletal proteins (154). Thus, cytoskeletal proteins might be regulatory in that they prevent the random occurrence of "spontaneous," basal secretion.

Consistently, effective lysosomal secretion from neutrophils requires destabilization of F-actin with cytochalasins (155). Conversely, Nielsen and Johansen have observed a decrease in histamine secretion upon chemical induction of an increase in the number of microfilaments by treating mast cells with dimethylsulfoxide (156). In chromaffin cells, stimulation of exocytosis was found to be accompanied by a grossly visual reorganization of the submembranal web of actin filaments (157) (P. I. Lelkes, unpublished observations) and of an actin-binding protein, α-fodrin (158). In addition, exocytosis is accompanied by a translocation of membrane-bound, filamentous, F-actin into the cytosol, as shown biochemically in chromaffin cells as well as in other secretory cells (157,159). Also, destabilization of F-actin by altering the cytosolic F-G equilibrium with actin-specific drugs and proteins enhances exocytotic fusion from intact cells (19,160). Conversely, reinforcement of the actomyosin system will result in a decreased secretory response in digitonin-permeabilized chromaffin cells (146). In line with these results, Aunis and his co-workers have demonstrated that antibodies to α-fodrin will specifically block a significant portion of catecholamine release from permeabilized chromaffin cells (161), possibly by preventing the stimulus-induced reorganization of α-fodrin.

C. Regulation of the Function of Membrane Proteins

The functional role of cytoskeletal proteins in mediating exocytosis might extend beyond the mere structural aspects mentioned above. Friedman and co-workers have presented evidence for a direct correlation between the state of assembly of microfilaments and the regulation of transmembranal ion fluxes (19,160,162). In particular, they showed that the enhancement of the microfilament system by heavy meromyosin caused concomitant increases in calcium and sodium influxes, mediated by voltage-dependent calcium channels and the amiloride-sensitive Na^+/H^+ exchanger, respectively.

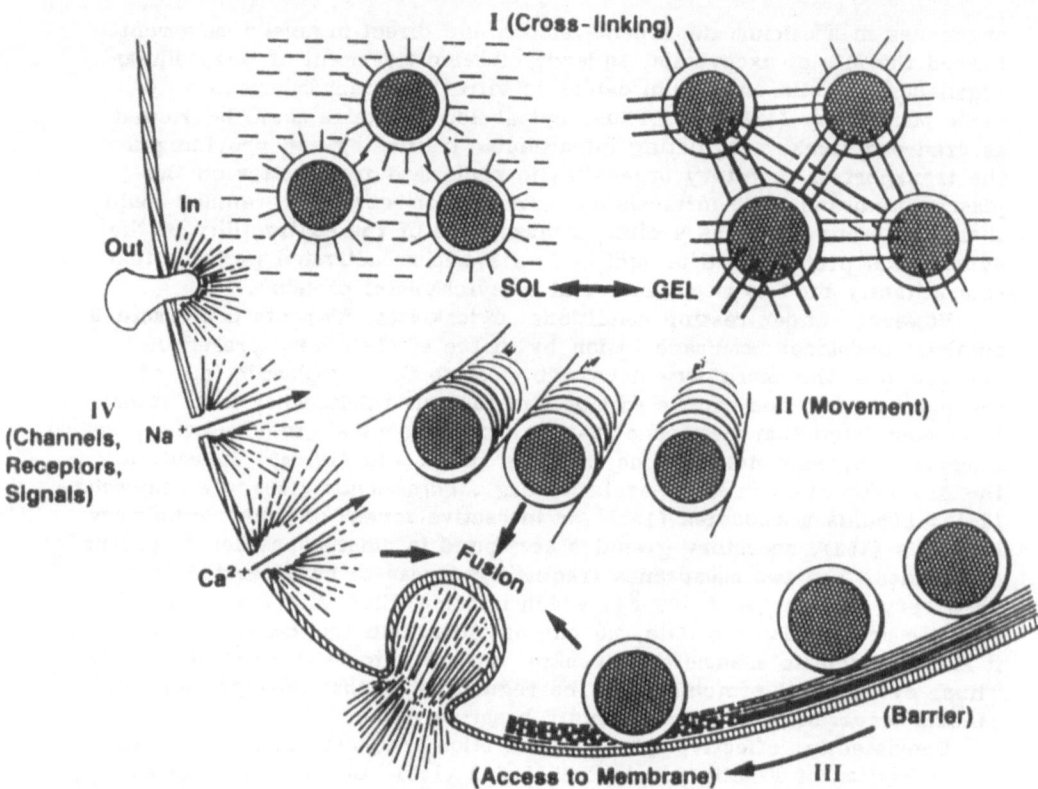

FIGURE 3 A schematic impression of the possible modes of cytoskeletal involvement in regulating membrane contact and fusion. For details, see text.

These latter findings are consistent with previous reports on the regulatory role of the cytoskeleton for receptor function and mobility also in other secretory systems (163).

In conclusion, cytoskeletal proteins have been invoked to participate in stimulus-secretion coupling (130,164,165). Figure 3 is a graphic summary of the different modes by which, as we suggest, the cytoskeleton could modulate exocytosis, in particular, the fusion step.

IX. "FUSION PROTEINS"

The proteins described here have well-defined functions and were of interest to us because they potentially might either regulate or even participate in the fusion event. However, little direct evidence links these proteins to the act of fusion. By contrast, there exists a separate class of proteins present in secreting cells that actually might promote membrane contact and cause fusion. These proteins were discovered by investigators trying to devise in vitro model systems for exocytosis, using biological membranes and cytosolic extracts from cells. These proteins can be divided into two distinct categories: those which seem to be intrinsic membrane proteins, and those which are cytosolic and become recruited to the membrane in response to specific signals.

A. Membrane-Bound Proteins

Intrinsic membrane proteins in secretory granules have been implicated in in vitro fusion processes between granules and other membranes. These model systems include fusion of isolated plasma membranes, granule-granule fusion, fusion of granules with plasma membranes, and fusion of granules with defined liposomes. (For a review of earlier studies, see Ref. 166.) The major evidence for involvement of vesicle membrane proteins was the sensitivity of the reaction to treatment of the membranes with proteases such as trypsin, thermolysin, chymotrypsin, and others. Using electron-microscopic techniques, Gratzl and his co-workers found that a small fraction (up to 10%) of the biological membrane vesicles aggregate and fuse in the presence of micromolar concentrations of calcium. This calcium requirement is about 2-3 orders of magnitude less than that required to fuse liposomes prepared from phospholipid extracts of the same membrane preparations. Clearly, these data indicate that the fusion of biological membranes and of liposomes are different, at least as far as their requirements for metal ions is concerned.

In order to study the cytoplasmic requirements for biological membrane fusion in vitro, several groups have attempted to reconstitute cell-free models of exocytosis comprising isolated plasma membrane vesicles and storage granules. For example, Lelkes and co-workers found only very limited interaction (fusion?) between chromaffin cell membrane vesicles and chromaffin granules (167) perhaps due to the fact that the isolated plasma membranes spontaneously form right-side-out vesicles. This finding can probably be generalized to all kinds of plasma membrane vesicle preparations and might serve as a caveat for using this type of in vitro model system for exocytosis.

A more promising approach to study exocytosis in vitro using the same ingredients would be to invert the orientation of the plasma membranes. To this end, Scott and co-workers attached isolated chromaffin cell plasma membrane vesicles to polylysine-coated polyacrylamide beads (168). Following osmotic lysis, the membrane patches remaining on the beads were used to measure calcium-dependent binding of synexin to the cytoplasmic face of the plasma membranes. Preliminary data indicate that this system might eventually be suitable for studying (synexin-mediated) interactions between chromaffin granules and the inverted plasma membrane patches (J. H. Scott and H. B. Pollard, unpublished observations).

Another approach to obtain access to the inner face of the plasma membrane has been the use of cortical cell-surface complexes derived from sea urchin eggs. In this preparation, the eggs are 'glued' to a polylysine-coated coverslip and mechanically dislodged to yield a "plasma membrane lawn" containing inside-out membrane patches and cortical granules which remain in close apposition to the membranes. Exocytosis can then be triggered merely by the addition of micromolar calcium (169,170).

In studying cytosolic parameters that might affect the exocytotic fusion in this system, Whitaker and Baker found that trifluoroperazine, an inhibitor of calmodulin as well as of synexin activity, could inhibit calcium-triggered exocytosis (169). By contrast, numerous anticytoskeletal drugs and antibodies against actin were ineffective in blocking the fusion event.

Their conclusions, however, that the cytoskeleton is not involved in the fusion step contradict other findings in intact and also in permeabilized cells (see above). Jackson et al. demonstrated that in order for fusion to occur, intact S-S bonds were required on the surface of either the plasma membrane lawn or the cortical granules (171). Furthermore, calcium-induced exocytosis can be retarded by hyperosmotic solutions (75), a finding reminiscent of similar data on the role of osmotic forces in exocytosis from intact and also from permeabilized chromaffin cells (see above). Finally, Sasaki identified a heat- and protease-sensitive protein with a molecular weight of about 100 kD in cortical homogenates from eggs of several species of sea urchins that modulated the calcium-sensitivity of the in vitro fusion event (172). The latter experiment indicates that the system of isolated cortical complexes as such might be a valid, yet incomplete, cell-free model for granule-membrane interactions, since essential cytosolic parameters, such as soluble proteins, might be lost during preparation. Nevertheless, this model might be uniquely suitable for the systematic study of cytosolic and membranous components that can trigger and/or modulate exocytotic membrane fusion.

In order to overcome the above-mentioned complications with the orientation of plasma membrane vesicles, several groups are currently studying the interactions of granule membranes with liposomes (42,43). In this simplified cell-free model for exocytosis, the lipid composition of the liposomes can be matched to that of the cytoplasmic face of the plasma membranes, which is rich in acidic phospholipids. In the case of fusion of the chromaffin granules with acidic liposomes, the process is essentially calcium-independent, but pH-dependent (see above). This suggests the involvement of titrable groups, probably carboxylic residues on the membrane proteins, in the fusion process. This observation is analogous to similar findings on the fusion of bacterial membranes with acidic liposomes (173). The pH dependence of this type of fusion is also reminiscent of certain aspects of synexin-mediated granule-granule fusion, as described below.

Another interesting process of low-pH-induced fusion is mediated by clathrin, the coat protein of endocytotic residues. Clathrin does not seem to be an integral membrane protein, since it is readily purified as a soluble protein, forming a three-armed trimer, a triskelion. However, in situ, it appears to be very tightly bound to intracellular membrane structures, as inferred from the bristle-coated, punctate fluorescence appearance of cells microinjected with anticlathrin antibodies (174). Purified clathrin was found, upon lowering the pH below 6.5, to form stable complexes with neutral PC liposomes (175). At pH 6.0, these liposomes fused. However, the ability of clathrin to fuse *neutral* liposomes at low pH in the absence of calcium appears to be unique, not shared by any of the control proteins investigated, including synexin. This finding certainly warrants further studies for the role of clathrin in membrane fusion in vivo. At present, we can only speculate that clathrin might be intimately involved in the initial fusion events during endocytosis, either at the plasma membrane or at some subsequent fusion steps of the coated pits with the Golgi apparatus.

Some years ago, Meyer and Burger used a chromaffin-granule-membrane affinity column to isolate three different proteins from the plasma membranes of bovine adrenal chromaffin cells, on the basis of the affinity of these proteins for chromaffin granule membranes (176). The major polypeptide thus isolated (about 80% of the total binding material) had a molecular weight of about 50 kD. The authors suggested that this protein might function as plasma-membrane-located receptor for chromaffin granules during exocytosis. Antibodies raised against this protein, when microinjected into PC12 cells, were shown to partially block potassium-evoked catecholamine secretion, suggesting that this 50-kD protein might be directly involved in exocytosis. Although at present no further information is available, the existence of this or similar proteins is encouraging to those interested in the quest of membrane-bound fusion proteins.

B. Cytosolic Proteins (Synexins)

The prototype for proteins in this category is synexin, which aggregates (177) and, with the help of arachidonic acid, fuses chromaffin granules to one another (45). This property has been exploited to identify and purify other functionally similar proteins, some of which have been found to be members of the same gene family. Attempts have been made to categorize these proteins according to either physicochemical or biochemical properties. For example, Geisow and collaborators found that many of these proteins contain a consensus sequence (178), somewhat characteristic of a 16-amino-acid portion of lipocortin, repeated in lipocortin 1 (4 times), p36 (4 times), endonexin II (4 times), and calelectrin 67K (2 times). More recently, synexin has been shown to have an analogous structure with the fourfold repeat, but with a unique, lengthy hydrophobic N-terminal leader sequence (178a).

So far, the only known functional properties common to these proteins is that they bind to membranes in a calcium-dependent fashion. In addition, some of these proteins, such as calelectrins (179), synexins, calpactins (180), and calcimedins (181), are calcium-binding proteins themselves. Furthermore, they have a preference for binding to acidic phospholipids, rather than neutral ones. [For some tentative summaries on the various proteins, see Refs. 1, 182, and 183.]

Yet another class of functionally related proteins has recently been described by Lee and Pollard (cited in Ref. 1). A new, synexinlike protein with a molecular weight of approximately 38 kD, called ν-synexin, aggregates chromaffin granules. However, this aggregation occurs only at vanishingly low calcium concentrations and is markedly inhibited at calcium levels above 1 μM.

In considering other approaches to making sense of this apparent welter of proteins, new data regarding the protein calpactin may hint at a more functional distinction between these proteins. Calpactin is a complex of two 36-kD heavy chains and two 11-kD light chains, located in the actin-rich region just below the plasma membrane of many cells (184). The protein binds to PS in a calcium-dependent manner, but also bundles F-actin filaments at high calcium concentrations (~1 mM). Therefore, it has been proposed that calpactin and perhaps other proteins, sharing the consensus lipocortin sequence, may just adsorb to the membrane and bind it to otherwise free F-actin molecules. In this sense, calpactin might be yet another cytoskeleton-anchoring protein, of cytosolic origin, but which also binds membranes.

An alternative description of these proteins has been offered by Smith and Dedman, who have proposed to consider all of these proteins in the category of calcimedins (185). That is, they bind to hydrophobic supports such as phenothiazine-Sepharose in a calcium-dependent manner. Indeed, some of them will do this, although it is difficult to reconcile this view with the fact that they prefer specific interactions with charged phospholipid headgroups.

By contrast, synexin seems different in many ways from both calpactins and calcimedins. Functionally, synexin does not bundle F-actin, nor does it interact with actin at all, as assayed by falling ball viscometry. Also, synexin binds to phenothiazine-Sepharose columns, but does so in a calcium-independent manner (H. B. Pollard, unpublished observations). In fact, synexin does bind to artificial and biological membranes, like the other proteins, but it also does much more. Uniquely, synexin truly fuses chromaffin granule membranes in vitro, and the mechanism is currently being elucidated (186). Synexin *aggregates* intact granules in a calcium-dependent manner, and these aggregates can be induced to fuse by addition of small concentrations (5 μM) of arachidonic acid. The aggregation step is absolutely pH dependent, while the arachidonic acid fusion step is independent of external calcium (45).

Synexin-induced *fusion* of chromaffin granule *ghosts* is only qualitatively similar, but nonetheless instructive as to the mechanisms of synexin action. In a series of papers, Stutzin and co-workers (186-188) have shown that chromaffin granule ghosts fuse directly upon addition of synexin. This fusion process is primarily dependent on [H^+] and only partially (~40%) modulated by calcium. In this system, membrane mixing, measured by the dilution of fluorescent lipid probes, is slightly faster than volume mixing, measured by dilution of an aqueous marker entrapped in the membrane vesicles. Similar to the cases for granule-liposome fusion, the pH dependence for this fusion event seems to reside in membrane carboxyl groups rather than in the synexin molecule.

The mechanism by which synexin fuses membranes has also been studied by membrane capacitance techniques using PS bilayers constructed at the tip of a patch pipet (65,189,190). These data suggest that synexin inserts into the core of the bilayer in a calcium- and pH-dependent manner. A variety of independent physicochemical studies seem consistent with the concept that synexin assumes a more hydrophobic conformation when exposed to calcium and the proper acidic (pH 6.0-6.5) environment. In fact, once inserted into the membrane, as defined by capacitance criteria, synexin becomes capable of sustaining a conductance state selective for calcium (65,191). This is conventionally interpreted as a "channel" and provides further, substantive support for the concept that synexin indeed enters the hydrophobic interior of the bilayer.

Based on these data, we have recently formulated a hypothesis to explain how synexin might drive and direct the fusion process between biological membranes. Upon suitable ionic activation by both protons and calcium, synexin assumes a hydrophobic conformation. This allows the protein to recognize the specific membrane and to penetrate into the substance of the bilayer. If synexin simultaneously enters two juxtaposed membranes, the protein can form a hydrophobic bridge between the two bilayers. This facilitates the movement of surface phospholipids across the bridge and results in membrane mixing, as described experimentally above. The joining of the contralateral monolayers can occur by a model we have described elsewhere (190,191) and would accompany volume mixing. We anticipate that this model, which is depicted schematically in Figure 4, might prove to be quite general.

Our own contribution to the classification of cytosolic proteins with fusogenic potential is shown in Figure 5. In considering these proteins

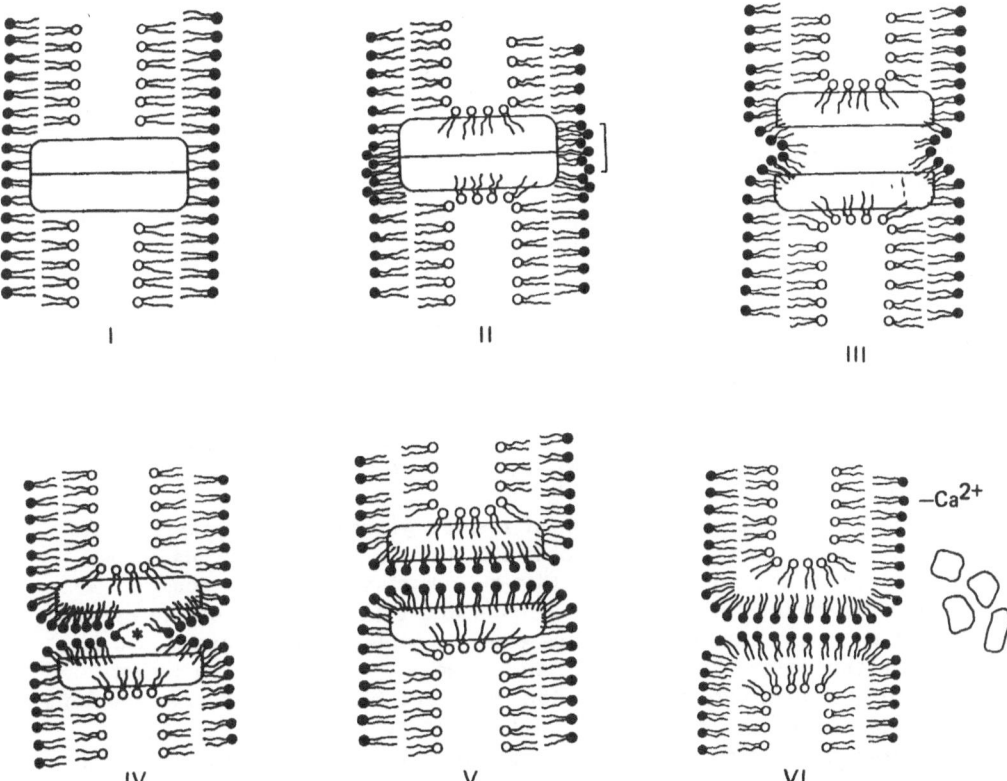

FIGURE 4   The "hydrophobic bridge hypothesis" for synexin-induced membrane contact and fusion (for details, see Refs. 190, 191). (I) Synexin polymerizes in the presence of calcium and the resulting hydrophobic structures bind to acidic phospholipids. Subsequently, the synexin polymer penetrates into the bilayer of each membrane, causing them to adhere (stage I). Membrane adhesion is, of course, a necessary preliminary step to eventual fusing. (II) Phospholipids on the *cis* monolayers of the adherent membranes cross the synexin bridge toward each other (stage II). The nonpolar fatty acid tails remain oriented toward either the interior of the bilayer or the hydrophobic bridge formed by polymerized synexin. (III) The synexin polymer bridge, immersed in a nonpolar environment, may undergo dissociation (stage III). (IV) The *trans* monolayer phospholipids continue to intrude into the nonpolar space between synexin molecules until the fatty acid chains of the leading phospholipids make contact (stage IV). The asterisk represents the fact that the fatty acid chains prefer to interact with the substance of the hydrophobic bridge. (V) The trans-monolayers fold in and the lipids might reorient across the synexin bridge. (VI) The *cis* monolayers and the *trans* monolayers of the two membranes are now respectively continuous, thus generating a genuinely fused structure (stage V). The previously separated volumes are now continuous. Bits of the hydrophobic bridge material, however, may be left in the substance of the fused bilayers.

in the context of membrane fusion, we realized that they might all be considered as arrested at various stages of a reaction mechanism that ultimately
will lead to fusion. Thus, some cytoplasmic proteins will merely bind to
membranes; we call them B-type (Fig. 5a). Certainly, some of the chromobindins (e.g., calmodulin), which bind to chromaffin granules but do not
aggregate them, might be included in this class. Other proteins come pre-
bound to the membranes; we call them MB-type (Fig. 5b). Although they
are less well characterized, we might consider them as intrinsic membrane
proteins, such as the ones described by Meyer and Burger (176) or by
Bental et al. (43). Another example for an MB-type protein might perhaps
be clathrin, which in situ is found only (?) in membrane-associated form
(174). Finally, a third set of proteins appears to be able to bind to both
cytoskeleton and membranes; we call these BCM-type proteins (Fig. 5c).
A number of the classical cytoskeletal-binding proteins, such as fodrin,
vinculin, and caldesmin (182), can be accommodated in this class. Furthermore, recent evidence suggests that some of the calpactins, such as calpactin II and synapsin I, are also members of this family of BCM-proteins
(84,85).

Going further in the direction of fusion, we realized that there exists
a class of proteins which, when bound to the membranes, will also *cross-
link* (x) them; we call these the BX-type proteins (Fig. 5d). Some of
the chromobindins (e.g., the calelectrins and other members of the synexin
family) may certainly belong to this class of proteins. By contrast, so far
we have encountered only one protein, synexin, which binds to target
membranes and *inserts* itself into the membrane bilayers, while *cross-linking*
(x) them at the same time. We call this an IX-type protein (Fig. 5e).
This class of proteins is unique, in that it provides a feasible mechanism
for directly proceeding from membrane binding to *cross-linking* to fusion.
Of all the chromobindins, calelectrins, calcimedins, and so forth, so far
only synexin has fulfilled the criteria of an IX-type protein. Studies similar
to those currently carried out in our laboratory might, however, reveal
other such IX-type proteins that mediate membrane contact and fusion.

We feel that the mnemonics implicit in our nomenclature, as well as
the scheme (Fig. 6), will facilitate remembering and classifying these
putative fusion proteins, especially since their number is certainly bound
to increase over the next few years.

X. ADDENDUM

Since the completion of our chapter, some important new results have been
obtained regarding the structure of synexin and its relationship to some
other calcium-dependent, membrane-binding proteins. The successful cloning
and sequencing of human synexin at NIH (192,178a) has shown synexin to
contain the ca. 36-Kd tetrad repeat characteristic of lipocortin I, calpactin I
heavy chain (AKA p36), endonexin II, protein II, and calelectrin 67K.
The homology among these proteins in this domain is approximately 50%.
However, the synexin molecule also contains a unique 167-amino-acid *N*-
terminal segment which is highly hydrophobic. This hydrophobic segment,
as well as the extensive hydrophobic character of the tetrad repeat, lends

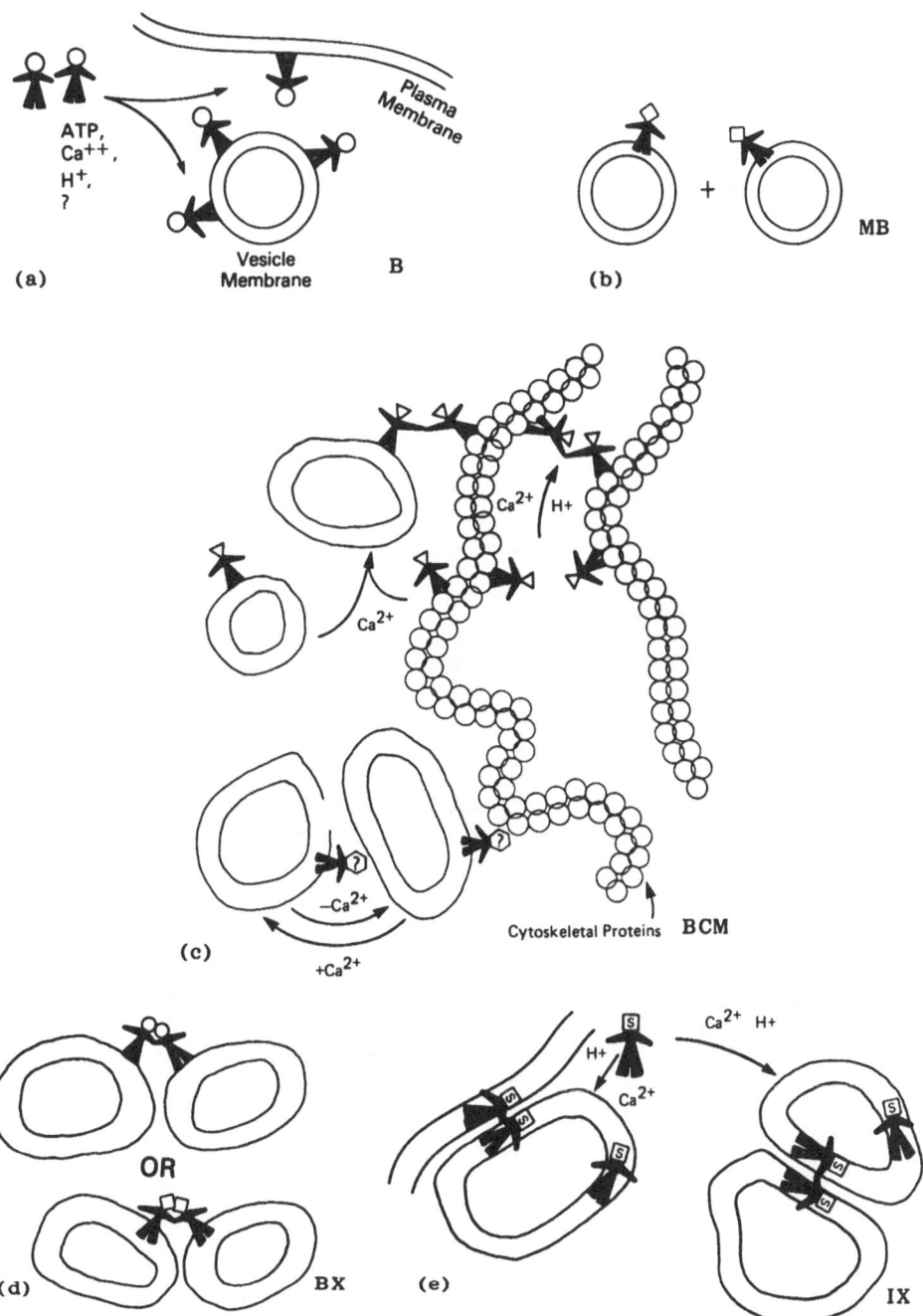

FIGURE 5 Schematic impressions of the typology of the different classes of potentially fusogenic proteins. B, cytoplasmic proteins that *bind* to membranes; MB, intrinsic/membrane-associated proteins that are pre-*bound* to the *membranes*; BCM, proteins that *bind* both to *cytoskeleton* and to *membranes*; BX, proteins that *bind* and *cross*(x)-link membranes; IX, proteins that *insert* into membranes and *cross*-link them. For details, see text.

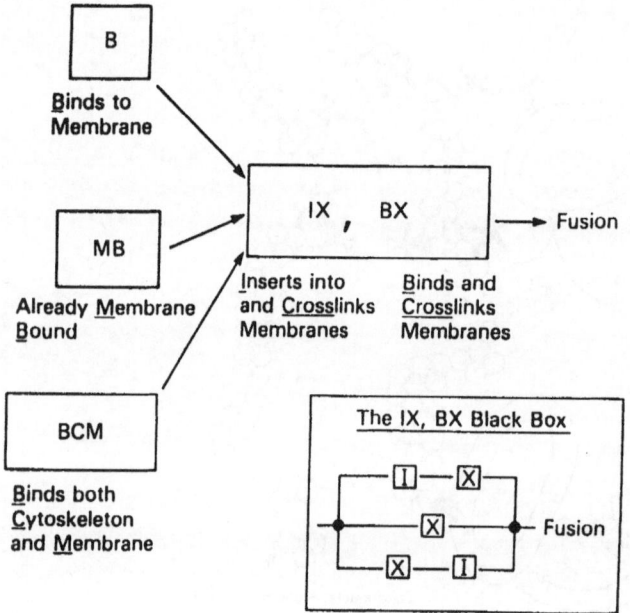

FIGURE 6   Flow diagram of the various pathways and mechanisms that will lead up to membrane fusion mediated by cytosolic proteins. The terminology is the same as used for Figure 5. For details, see text.

```
         S: MSYPGYPPTGYPPFPGYPPAGQESSFPPSGQYPYPSGFPPMGGGAYPQVPSSGYPGAGGYP
            APGGYPAPGGYPGAPQPGGAPSYPGVPPGQGFGVPPGGAGFSGYPQPPSQSYGGGPAQVP
                                          *
                 LPGGFPGGQMPSQYPGGQPTYPSQPATVTQVTQGTIRPAANFDAIR
         E2:                             MAQVLRGTVIDFPGFDERA
AMINO    L1: MAMVSEFLKQAWFIENEEQEYVQIVKSSKGGPGSAVSPYPTFNPSS
TERMINAL C1:               MSTVHEILCKLSLEGDHSTPPSAYGSVKAYINFDAER
DOMAIN   P2:                             AAKGGVKAASGFNAAE
         c:                              MAKPAQGAKYRGSIHDFPGFDPNQ

             *       *    * *    *                ***      **      *
         S: DAEILRKAMKGFGTDEQAIVDVVANRSNDQRQKIKAAFKTSYGKDLIKDLKSELSGNMEELIIALFMPPTYY
         E2:    T      L     ES LTLLTS    A    E S     LF R  LD     T KF K  V  MK SRL
FIRST    L1: VAA H  IMVK V  AD I LLITK N A   Q    LQET  P LET  KA T HL  VV    LKT AQF
REPEAT   C1: INIET I TK V  VD  NLIT     A    D AF YQRRTK E ASA    A   HL TV  G LKT AQ
         P2:  QT      L    D  ISLY  TA   E RT Y STI R  LD       F QV  GMMT TVL
         c1:    A YT      S KE  L IITS    R    EVCQSY SL     A    Y  T KF R   VG MR  A C
         c5:   KA      L    DT I IITH    V    Q RQT  SHF R  MT       I  DLAR   G M   AH

             *       * **  * ** *  **       *        *     ***           *
         S: DAWSLRKAMQGAGTQERVLIEILCTRINQEIREIVRCYQSEFGRDLEKDIRSDTSGHFERLLVSMCQGNRDENQSINHQMAQE
         E2: YE KH LK     N K  T  IAS  PE L A KQV EE Y SS  D VVG      YYQ M  VLL A    PDAG DEAQVEQ
SECOND   L1: DE  A KL  D DT     AS    K  D N V RE LK   A  T      D RNA L LAK D S  DFGV EDL  DS
REPEAT   C1: SE KAS K L  D DS    I S     LQ  N V KEMYKT       I      D RK M ALAK R A  DG VIDYELIDQ
         P2: VQE R      D GC      S  PE  R NQT  LQY S D      FM Q V  LSA G   GNYLDDALVRQ
         c2: KEIKD IS I  D KC     AS    EQMHQL AA KDAYE    A IG      QKM  VLL  T E  DDVVSEDLV Q
         c6: KQ K  E    D KA      A    A  A NEA KEDYHKS  DALS          R I I LAT H E  GGENLD AREDAQVAA

                                *      *                            * *
         S: DAQRLYQAGEGRLGTDESCFNMILATRSFPQLRATMEAYSRMANRDLLSSVSREFSGYVESGLKTILQCALNRPAFF
         E2:    A F     LKW    EK IT FG    VSH   KVFDK MTISGFQIEETID  T  NL QL LAVVKSIRSI  YL
THIRD    L1: RA  E     R K    VNV T T  Y      RVFQK TKYSKH MNKVLDL LK DI KC TA VK  TSK A
REPEAT   C1: RD  D  VK K      VPKWIS MTE   V H QKVFDR KSYSPY M E IRK VK DL NAFLNLV  IQ K LY
         P2:  D E     KKW     VK LTV CS NRNH  HVFDE K ISQK IEQ IKS T   SF DA LA VKCMR KS Y
         c3: V D E    LKW     AQ IY  GN KQH   LVFDE LKTTGKPMKA IRG L  DF KLMLAVVK IRST EY
         c7: EILEIADTPS DKTSL  TR MT   C  TY H  RVFQEFIK T Y VEHTIKK M  D RDAFVA V SVK K L

                  *      * *** **   * *  * **** *            *  *      * *  *   *
         S: AERLYYAMKGAGTDDSTLVRIVVTRSEIDLVQIKQMFAQMYQKTLGTMIAGDTSGDYRRLLLAIVGQ
         E2:    T        H I VM S     FN RKE RKNFATS YS  K        KKA  LLC EDD
FOURTH   L1: K HQ     V  RHKA I M S    MND  AFYQK  GIS CQA LDE K      EKI V LC GN
REPEAT   C1: D  DS    K  R KV I M S    V MLK RSE KRK G S YYY QQ  K    QKA  YLC GDD
         P2:    KS    L  N  I VM S A   MMD RAN KRL G S YSF K         KV   ILC GDD
         c4:    FK    L  R N  I M S   L MLD REI RTK E S YS  KN    E KKT  KLS GDD
         c8: DK KS       EK  T M S     LN RRE IEK D S HQA E         FLKA  LC GED
```

FIGURE 7 Comparison of human synexin with related proteins. (From Ref. 178a.)

strong structural support to the hydrophobic bridge hypothesis for membrane fusion, described above. A detailed comparison of the human synexin sequence with the homologous proteins, endonexin II (#2), lipocortin 1 (L1), calpactin heavy chain (C1), protein II (P2), and calelectrin 67K (c) is shown in Figure 7.

REFERENCES

1. Pollard, H. B., Ornberg, R., Levine, M., Kelner, K., Morita, K., Levine, Forsberg, E., Brocklehurst, K. W., Duong, L., Lelkes, P. I., Heldmann, E., and Youdin, M. (1985). Hormone secretion by exocytosis with emphasis on information from the chromaffin cell system. In:

Vitamins and Hormones, Vol. 42 (G. D. Aurbach, editor-in-chief). Academic Press, Orlando, FL, pp. 109-196.

2. Blumenthal, R. (1985). Membrane fusion. *Curr. Topics Membrane Transp.* 22:172-251.

3. Rand, R. P., and Parsegian, V. A. (1986). Mimickry and mechanism in phospholipid models of membrane fusion. *Annu. Rev. Physiol.* 48: 201-212.

4. Douglas, W. W. (1968). Stimulus-secretion coupling: The concept and clues from chromaffin and other cells. *Br. J. Pharmacol.* 34:453-474.

5. Rink, T. J., Sanchez, A., and Hallam, T. J. (1983). Diacylclycerol and phorbol ester stimulate secretion without raising cytoplasmic free calcium in human platelets. *Nature* 305:317-318.

6. Nemeth, E. F., Wallace, J., and Scarpa, A. (1986). Stimulus-secretion coupling in bovine parathyroid cells. Dissociation between secretion and net changes in cytosolic Ca^{2+}. *J. Biol. Chem.* 261:2668-2674.

7. Rubin, D. P. (1982). *Calcium and Cellular Secretion*. Plenum Press, New York, and London.

8. Cantin, M. (1984). *Cell Biology of the Secretory Process*. Karger, Basel.

9. Tsien, R. Y., Pozzan, T., and Rink, T. J. (1984). Measuring and manipulating cytosolic Ca^{2+} with trapped indicators. *Trends Biochem. Sci.* 9:263-266.

10. Knight, D. E., and Baker, P. F. (1982). Calcium-dependence of catecholamine release from bovine adrenal medullary cells after exposure to intense electric fields. *J. Membr. Biol.* 68:107-140.

11. Brooks, J. C., and Treml, S. (1983). Catecholamine secretion from chemically skinned cultured chromaffin cells. *J. Neurochem.* 40:468-473.

12. Truneh, A., Mishal, Z., and Leserman, L. D. (1984). A calmodulin antagonist increases the apparent rate of endocytosis of liposomes bound to MHC molecules via monoclonal antibodies. *Exp. Cell Res.* 155:50-63.

13. Furuichi, K., Ra, C., Isersky, C., and Riverea, J. (1986). Comparative evaluation of the effect of pharmacological agents on endocytosis and coendocytosis of IgE by rat basophilic leukemia cells. *Immunology* 58:105-110.

14. Grafenstein, H. von, Roberts, C. S., and Baker, P. F. (1986). Kinetic analysis of the triggered exocytosis/endocytosis secretory cycle in cultured bovine adrenal medullary cells. *J. Cell Biol.* 103:2343-2352.

15. Korc, M., Matrisian, L. M., and Magun, B. E. (1984). Cytosolic calcium regulates epidermal growth factor endocytosis in rat pancreas and cultured fibroblasts. *Proc. Natl. Acad. Sci. USA* 81:461-465.

16. Palade, G. E. (1982). Problems in intracellular membrane traffic. In: *Membrane Recycling* (D. Evered and G. M. Collins, eds.). *Ciba Founda. Symp.*, Bath, U.K., Pitman Press, 92:1-14.

17. Rasmussen, H. (1986). The calcium messenger system. *N. Engl. J. Med.* 314:1094-1101, 1164-1170.

18. Suchard, S. J., Lattanzio, F. A. Jr., Rubin, R. W., and Pressman, B. C. (1982). Stimulation of catecholamine secretion from cultured chromaffin cells by an ionophore-mediated rise in intracellular sodium. *J. Cell Biol.* 94:531-539.

19. Friedman, J. E., Lelkes, P. I., Rosenheck, K., and Oplatka, A. (1986). Control of stimulus-secretion coupling in adrenal medullary chromaffin cells by microfilament specific macromolecules. *J. Biol. Chem.* 261:5745-5750.

20. Pozzan, T., Lew, D. P., Wollheim, C. B., and Tsien, R. Y. (1983). Is cytosolic ionized calcium regulating neutrophil activation? *Science* 221:1413-1415.

21. Pozzan, T., Gatti, G., Dozio, N., Vicentini, L. M., and Meldolesi, J. (1984). Ca^{2+}-dependent and -independent release of neurotransmitters from PC 12 cells: a role for protein kinase C activation? *J. Cell Biol.* 99:628-638.

22. Ochs, D. L., Korenbrot, J. I., and Williams, J. A. (1985). Relation between free cytosolic calcium and amylase release by pancreatic acini. *Am. J. Physiol.* 249:6389-6398.

23. Rink, T. J. (1986). Stimulus-secretion coupling mechanisms in human platelets. *Agents Actions* 20 (suppl):147-169.

24. Guild, S., Itoh, Y., Kebabian, J. W., Luini, A., and Reisine, T. (1986). Forskolin enhances basal and potassium-evoked hormone release from normal and malignant pituitary tissue: the role of calcium. *Endocrinology* 118:268-279.

25. Korchak, H. M., Vienne, K., Rutherford, L. E., Wilkenfeld, C., Finkelstein, M. C., and Weismann, G. (1984). Stimulus-response coupling in the human neutrophil. II. Temporal analysis of changes in the cytosolic calcium and calcium efflux. *J. Biol. Chem.* 259:4076-4082.

26. Cabrini, G., and DeTogni, P. (1985). Increased cytosolic calcium in cystic fibrosis neutrophils effect on stimulus-secretion coupling. *Life Sci.* 26:11561-11567.

27. Sha'afi, R. I., Shecyk, J., Yassin, R., Molski, T. F., Volpi, M., Naccache, P. M., White, J. R., Feinstein, M. B., and Becker, E. L. (1986). Is a rise in intracellular concentration of free calcium necessary or sufficient for stimulated cytoskeletal-associated actin? *J. Cell Biol.* 102:1459-1463.

28. Muff, R., and Fischer, J. A. (1986). Parathyroid hormone secretion does not respond to changes of free calcium in electropermeabilized bovine parathyroid cells but is stimulated with phorbol ester and cyclic AMP. P. *Biophys. Res. Commun.* 139:1233-1238.

29. Fernandez, J. M., Neher, E., and Gomperts, B. D. (1984). Capacitance measurements reveal stepwise fusion events in degranulating mast cells. *Nature* 312:453-455.

30. Oetting, M., LeBoff, M., Swiston, L., Preston, J., and Brown, E. (1986). Guanine nucleotides are potent secretagogues in permeabilized parathyroid cells. *FEBS Lett.* 208:99-104.

31. Becker, E. L., Kermode, J. C., Naccache, P. H., Yassin, R., Munoz, J. J., Marsh, M. L., Huang, C. K., and Sha'afi, R. I. (1986). Pertussis toxin as a probe of neutrophil activation. *Fed. Proc.* 45:2151-2155.

32. Bittner, M. A., Holz, R. W., and Neubig, R. R. (1986). Guanine nucleotide effects on catecholamine secretion from digitonin-permeabilized adrenal chromaffin cells. *J. Biol. Chem.* 261:10182-10188.

33. Brown, E. M., Watson, E. J., Thatcher, J. G., Koletsky, R., Dawson-Hughes, B. F., Posillico, J. T., and Shoback, D. M. (1987). Ouabain and low extracellular potassium inhibit PTH secretion from bovine parathyroid cells by a mechanism that does not involve increases in the cytosolic calcium concentration. *Metabolism* 36:36-42.

34. Tsien, R. Y. and Poenie, M. (1987). Measurement and manipulation of cytosolic free calcium with high spatial and temporal resolution. *Biophys. J.* 51:389a.

35. Yuli, I., and Oplatka, A. (1987). Cytosolic acidification as an early transductory signal of human neutrophil chemotaxis. *Science* 235:340-342.

36. Cannon, C., van Adelsberg, J., Kelly, S., and Al-Awqati, Q. (1985). Carbon-dioxide induced exocytotic insertion of H^+ pumps in turtle-bladder luminal membrane: role of cell pH and calcium. *Nature* 314:443-446.

37. Abercrombie, R. R., and Hart, C. E. (1986). Calcium and proton buffering and diffusion in isolated cytoplasm from Myxicola axons. *Am. J. Physiol.* 250:C391-405.

38. Ives, H. E., and Daniel, T. O. (1987). Interrelationship between growth factor-induced pH changes and intracellular Ca^{2+}. *Proc. Natl. Acad. Sci. USA* 84:1950-1954.

39. Siffert, W., and Akkerman, J. W. N. (1987). Activation of sodium-proton exchange is a prerequisite for Ca^{2+}-mobilization in human platelets. *Nature* 325:456-458.

40. Connor, J., and Huang, L. (1987). Acid-induced fusion of liposomes. In: *Cell Fusion* (A. E. Sowers, ed.). Plenum Press, New York, pp. 285-299.

41. Szoka, F. (1987). Lipid vesicles: Model system to study membrane-membrane destabilization and fusion. In: *Cell Fusion* (A. E. Sowers, ed.). Plenum Press, New York, pp. 209-240.

42. Bental, M., Lelkes, P. I., Scholma, J., Hoekstra, D., and Wilschut, J. (1984). Ca^{++}-independent, protein-mediated fusion of chromaffin granule ghosts with liposomes. *Biochim. Biophys. Acta* 774:296-300.

43. Bental, M., Lelkes, P. I., and Wilschut, J. (1990). Protein mediated Ca^{2+}-independent fusion of liposomes with bovine adrenal chromaffin granules. Characterization of the system and implications for the mechanism of exocytotic secretion, manuscript in preparation.

44. Pollard, H. B., Creutz, C. E., Fowler, V., Scott, J., and Pazoles, C. J. (1982). Calcium-dependent regulation of chromaffin granule movement, membrane contact, and fusion during exocytosis. *Cold Spring Harb. Symp. Quant. Biol.* 46:819-834.

45. Creutz, C. (1981). Cis-unsaturated fatty acids induce the fusion of chromaffin granules aggregated by synexin. *J. Cell Biol.* 91:247-256.

46. Pratts, M., Teissie, J., and Tocanne, J. F. (1986). Lateral proton conduction at lipid-water interfaces and its implications for the chemiosmotic coupling hypothesis. *Nature* 322:756-758.

47. Heiple, J. M., and Taylor, D. L. (1980). Intracellular pH in single motile cells. *J. Cell Biol.* 86:885-890.

48. Ahnert-Hilger, G., Bhakdi, S., and Gratzl, M. (1985). Minimal requirements for exocytosis. A study using PC12 cells permeabilized with staphyloccal alpha-toxin. *J. Biol. Chem.* 260:12730-12734.

49. Rojas, E., Santos, R. M., Stutzin, A., and Pollard, H. B. (1986). Optical detection of ATP release from stimulated endocrine cells. A universal marker of exocytotic secretion of hormones. In: *Ionic channels in cells and model systems* (R. Latorre, ed.). Plenum Press, New York and London, pp. 163-178.

50. Ruggiero, M., Zimmerman, T. P., and Lapetina, E. G. (1985). ATP depletion in human platelets caused by permeabilization with saponin does not prevent serotonin secretion induced by collagen. *Biochem. Biophys. Res. Commun.* 131:620-627.

51. Vilmart-Seuwen, J., Kersken, H., Sturzl, R., and Plattner, H. (1986). ATP keeps exocytosis sites in a primed state but is not required for

membrane fusion: an analysis with *Paramecium* cells in vivo and in vitro. *J. Cell Biol.* 103:1279-1288.

52. Brooks, J. C., and Brooks, M. (1987). Protein thiophosphorylation associated with secretory inhibition in permeabilized chromaffin cells. *Life Sci.* 37:1869-1875.

53. Kilpatrick, D. L. (1984). Ion channels and membrane potential in stimulus-secretion coupling in adrenal paraneurons. *Can. J. Physiol. Pharmacol.* 62:477-483.

54. Friedman, J. E., Lelkes, P. I., Levis, E., Rosenheck, K., Schneeweiss, F., and Schneider, A. S. (1985). Membrane potential and catecholamine secretion by bovine adrenal chromaffin cells: Use of tetraphenylphosphonium distribution and carbocyanine dye fluorescence. *J. Neurochem.* 44:1391-1402.

55. Sagi-Eisenberg, R., and Pecht, I. (1984). Resolution of cellular compartments involved in membrane potential changes accompanying IgE-mediated degranulation of rat basophilic leukemia cells. *EMBO J.* 3:497-500.

56. Mohr, F. C., and Fewtrell, C. (1987). Depolarization of rat basophilic leukemia cells inhibits calcium uptake and exocytosis. *J. Cell Biol.* 104:783-792.

57. Jaffee, L. A., and Cross, N. L. (1986). Electrical regulation of sperm egg fusion. *Annu. Rev. Physiol.* 48:191-200.

58. Monroy, A. (1985). Processes controlling sperm-egg fusion. *Eur. J. Biochem.* 152:51-56.

59. Lelkes, P. I. (1979). Potential-dependent rigidity changes in lipid membrane vesicles. *Biochem. Biophys. Res. Commun.* 90:656-662.

60. Shinitzky, M., and Henkart, P. (1979). Fluidity of cell membranes—Current concepts and views. *Int. Rev. Cytol.* 60:121-147.

61. Blumenthal, R., Kempf, C., Van Renswoude, J., Weinstein, J. N., and Klausner, R. (1983). Voltage-dependent orientation of membrane proteins. *J. Cell Biochem.* 22:55-67.

62. Hershkowitz, M., Heron, D., Samuel, D., and Shinitzky, M. (1982). The modulation of protein phosphorylation and receptor binding in synaptic membranes by changes in lipid fluidity: Implications for ageing. *Prog. Brain Res.* 56:419-434.

63. Prives, J., and Shinitzky, M. (1977). Increased membrane fluidity precedes fusion of muscle cells. *Nature* 268:761-763.

64. Winkler, H., and Carmichael, S. W. (1983). The chromaffin granule. In: *The Secretory Process, Vol. 1. The Secretory Granule* (A. M. Poisner and J. M. Trifaro, eds.). Elsevier, Amsterdam, pp. 3-79.

65. Pollard, H. B., and Rojas, E. (1987). Calcium-activated synexin forms highly selective, voltage gated calcium channels in phosphatidylserine bilayer membranes. *Proc. Natl. Acad. Sci. USA* 85:2974-2978.

66. Pollard, H. B., Pazoles, C. J., Creutz, C. E., and Zinder, O. (1979). The chromaffin granule and possible mechanisms of exocytosis. *Int. Rev. Cytol.* 58:1549-176.

67. Pollard, H. B., Pazoles, C. J., Creutz, C. E., Scott, J. M., Zinder, O., and Hotchkiss, A. (1984). An osmotic mechanism for exocytosis from dissociated chromaffin cells. *J. Biol. Chem.* 259:1114-1121.

68. Pollard, H. B., Tack-Goldman, K., Pazoles, C. J., Creutz, C. E., and Shulman, N. R. (1977). Evidence for control of serotonin secretion from human platelets by hyroxyl ions and osmotic lysis. *Proc. Natl. Acad. Sci. USA* 74:5295-5299.

69. Holz, R. W. (1986). The role of osmotic forces in exocytosis from adrenal chromaffin cells. *Annu. Rev. Physiol.* 48:175–189.

70. Brown, W. M., Pazoles, C. J., Creutz, C. E., Aurbach, G., and Pollard, H. B. (1978). Role of anions in parathyroid hormone release from dispersed bovine parathyroid cells. *Proc. Natl. Acad. Sci. USA* 75:876–880.

71. Englert, D. F., and Pearlman, R. L. (1981). Permanent anions are not required for norepinephrine secretion from pheochromocytoma cells. *Biochim. Biophys. Acta* 674:136–143.

72. Hampton, R. Y., and Holz, R. W. (1983). The effects of osmolality on the stability and function of cultured chromaffin cells and the role of osmotic forces in eoxytosis. *J. Cell Biol.* 96:1082–1088.

73. Ladona, M. G., Bader, M. F., And Aunis, D. (1987). Influence of hypertonic solutions on catecholamine release from intact and permeabilized cultured chromaffin cells. *Biochim. Biophys. Acta* 927:18–25.

74. Holz, R. W., and Senter, R. A. (1986). Effects of osmolality and ionic strength on secretion from adrenal chromaffin cells permeabilized with digitonin. *J. Neurochem.* 46:1835–1842.

75. Zimmerberg, J., and Whitaker, M. (1985). Exocytosis of sea urchin egg cortical vesicles in vitro is retarded by hyperosmotic sucrose: kinetics of fusion monitored by quantitative light-scattering microscopy. *J. Cell Biol.* 101:2398–2410.

76. Zimmerberg, J., and Whitaker, M. (1985). Irreversible swelling of secretory granules during exocytosis caused by calcium. *Nature* 315:581–584.

77. Baker, P. F., and Knight, D. E. (1984). Chemiosmotic hypothesis of exocytosis: A critique. *Biosci. Rep.* 4:285–298.

78. Finkelstein, A., Zimmerberg, J., and Cohen, F. S. (1986). Osmotic swelling of vesicles; Its role in the fusion of planar phospholipid bilayer membranes and its possible role in exocytosis. *Annu. Rev. Physiol.* 48:163–174.

79. Ahkong, Q. F., and Lucy, J. A. (1986). Osmotic forces in artificially induced cell fusion. *Biochim. Biophys. Acta* 858:206–216.

80. Lucy, J. A., and Ahkong, Q. F. (1986). An osmotic model for the fusion of biological membranes. *FEBS Lett.* 199:1–11.

81. Breckenridge, L. J., and Almers, W. (1987). Final steps in exocytosis observed in a cell with giant secretory granules. *Proc. Natl. Acad. Sci. USA* 84:1945–1949.

82. Zimmerberg, J., Curran, M., Cohen, F. S., and Brodwick, M. (1987). Simultaneous electrical and optical measurements show that membrane fusion precedes secretory granule swelling during exocytosis of beige mouse mast cells. *Proc. Natl. Acad. Sci. USA* 84:1585–1589.

83. Nishizuka, Y. (1984). The role of protein kinase C in cell surface signal transduction and tumor promotion. *Nature* 308:693–698.

84. Sibley, D. R., Benovic, J. L., Caron, M. G., and Lefkowitz, R. J. (1987). Regulation of transmembrane signalling by receptor phosphorylation. *Cell* 48:913–922.

85. Berridge, M. J. (1984). Inositol triphosphate and diacylglycerol as second messengers. *Biochem J.* 220:345–360.

86. DiVirgilio, F., Lew, D. P., and Pozzan, T. (1984). Protein kinase activation of physiological processes in human neutrophils at vanishingly small cytosolic Ca^{2+} levels. *Nature* 310:691–693.

87a. Morita, K., Brocklehurst, K. W., Tomares, S. M., and Pollard, H. B. (1985). The phorbol ester TPA enhances A23187—but not carbachol and high K$^+$-induced catecholamine secretion from cultured bovine adrenal chromaffin cells. *Biochem. Biophys. Res. Commun.* 129:511-516.

87b. Brocklehurst, K. W., Morita, K., and Pollard, H. B. (1985). Characterization of protein kinase C and its role in catecholamine secretion from bovine adrenal-medullary cells. *Biochem. J.* 228:35-42.

88. Yamanishi, J., Takai, Y., Kaibuchi, K., Sano, K., Castagna, M., and Nishizuka, Y. (1983). Synergistic functions of phorbol ester and calcium in serotonin release from human platelets. *Biochem. Biophys. Res. Commun.* 112:778-786.

89. Dale, M. M., and Penfield, A. (1984). Synergism between phorbol ester and A23187 in superoxide production by neutrophils. *FEBS Lett.* 175:170-172.

90. Brocklehurst, K. W., and Pollard, H. B. (1985). Enhancement of Ca^{2+}-induced catecholamine release by the phorbol ester cTPA in digitonin-permeabilized cultured bovine adrenal chromaffin cells. *FEBS Lett.* 183:107-110.

91. Creutz, C. E., Dowling, L. G., Sando, J. J., Villar-Palasi, C., Whipple, J. H., and Zaks, W. J. (1983). Characterization of the chromobindins: soluble proteins that bind to the chromaffin granule membrane in the presence of Ca^{2+}. *J. Biol. Chem.* 258:14664-14674.

92. Summers, T. A., and Creutz, C. E. (1985). Phosphorylation of a chromaffin granule-binding protein by protein kinase C. *J. Cell Biol.* 260:2437-2443.

93. Haycock, J. W., Meligeni, J. A., Bennet, W. F., and Waymire, J. C. (1982). Phosphorylation and activation of tyrosine hydroxilase mediate the acetylcholine-induced increase in catecholamine biosynthesis in adrenal chromaffin cells. *J. Biol. Chem.* 257:12641-12648.

94. Michener, M. L., Dawson, W. B., and Creutz, C. E. (1986). Phosphorylation of chromaffin granule binding-protein in stimulated chromaffin cells. *J. Biol. Chem.* 261:6548-6555.

95. Geisow, M. J., and Burgoyne, R. D. (1987). An integrated approach to secretion. Phosphorylation and Ca^{2+}-dependent binding of proteins associated with chromaffin granules. *Ann. NY Acad. Sci.* 493:563-576.

96. Tallant, E. A., and Wallace, R. W. (1985). Characterization of a calmodulin-dependent protein phosphatase from human platelets. *J. Biol. Chem.* 260:7744-7751.

97. Ikeda, M., Deery, W. J., Nielsen, T. B., Ferdows, M. S., and Field, J. B. (1986). Dephosphorylation of 19 K and 21 K polypeptides in response to thyroid-stimulating hormone in cultured thyroid cells. *Endocrinology* 119:591-599.

98. Trifaro, J. M., and Kenigsberg, R. L. (1987). Chromaffin cell calmodulin. In: *Stimulus-Secretion Coupling in Chromaffin Cells* (K. Rosenheck and P. I. Lelkes, eds.). CRC Press, Boca Raton, FL, pp. 125-153.

99. Gilligan, D. M., and Satir, B. H. (1982). Protein phosphorylation/dephosphorylation and stimulus-secretion coupling in wild type and mutant Paramecium. *J. Biol. Chem.* 257:13903-13906.

100. Zieseniss, E., and Plattner, H. (1985). Synchronous exocytosis in *Paramecium* cells involves very rapid (less than or equal to 1 s), reversible dephosphorylation of a 65-kD phosphoprotein in exocytosis-competent strains. *J. Cell Biol.* 101:2028-2035.

101. Creutz, C. E., Dowling, L. B., Kyger, E. M., and Franson, C. (1985). Phosphatidylinositol-specific phospholipase C activity of chromaffin granule-binding proteins. *J. Biol. Chem.* 260:7171-7173.

102. Izumi, F., Yanagihara, N., Wada, A., Toyohira, Y., and Kobayashi, H. (1986). Lysis of chromaffin granules by phospholipase A_2-treated plasma membranes: A cell-free model for exocytosis in adrenal medulla. *FEBS Lett.* 196:349-352

103. Zahler, P., Reist, M., Pilarska, M., and Rosenheck, K. (1986). Phospholipase C and diacylglycerol lipase activities associated with plasma membranes of chromaffin cells from bovine adrenal medulla. *Biochim. Biophys. Acta* 877:372-379.

104. Lucy, J. A. (1970). The fusion of biological membranes. *Nature* 227: 814-817.

105. Frye, R. A. and Holz, R. W. (1984). The relationship between arachidonic acid release and catecholamine secretion from cultured bovine adrenal chromaffin cells. *J. Neurochem.* 43:146-150.

106. Hotchkiss, A., Pollard, H. B., Scott, J., and Axelrod, J. (1981). Release of arachidonic acid from adrenal chromaffin cell cultures during secretion of epinephrine. *Fed. Proc.* 40:256.

107. Frye, R. A., and Holz, R. W. (1985). Arachidonic acid release and catecholamine secretion from digitonin-treated chromaffin cells: effects of micromolar calcium, phorbol ester, and protein alkylating agents. *J. Neurochem.* 44:265-273.

108. Metz, S. A., Draznin, B., Sussman, K. E., and Leitner, J. W. (1987). Unmasking of arachidonate-induced insulin release by removal of extracellular calcium. Arachidonic acid mobilized cellular calcium in rat islet of Langerhans. *Biochem. Biophys. Res. Commun.* 142:251-258.

109. Entwistle, A., Curtis, D. H., and Zalin, R. J. (1986). Myoblast fusion is regulated by a prostanoid of the one series independently of a rise in cyclic AMP. *J. Cell Biol.* 103:857-866.

110. Sterner, D. C., Zaks, W. J., and Creutz, C. E. (1985). Stimulation of the Ca^{2+}-dependent polymerization of synexin by *cis*-unsaturated fatty acids. *Biochem. Biophys. Res. Commun.* 132:505-512.

111. Garcia-Gil, M., and Siraganian, R. P. (1986). Source of the arachidonic acid released on stimulation of rat basophilic leukemia cells. *J. Immunol.* 136:3825-3828.

112. Metzger, H., Alcaraz, G., Hohman, R., Kinet, J-P., Pribluda, V., and Quarto, R. (1986). The receptor with high affinity for immunoglobulin E. *Annu. Rev. Immunol.* 4:419-470.

113. Lucy, J. A. (1984). Fusogenic mechanisms. In: *Cell Fusion*, Ciba foundation symposium, Vol. 103 (D. Evered and J. Whelan, eds.). Pitman, London, pp. 28-44.

114. Lucy, J. A. (1984). Do hydrophobic sequences cleaved from cellular polypeptides induce membrane fusion reactions in vivo? *FEBS Lett.* 166:223-231.

115. Strittmatter, W. J., Couch, C. B., and Mundy, D. I. (1985). Role of proteins in the fusion of biological membranes. In: *Membrane Fluidity in Biology*, Vol. IV (R. C. Aloia and J. Boggs, eds.). Academic Press, New York, pp. 259-291.

116. Strittmatter, W. J., Couch, C. B., and Mundy, D. I. (1985). Role of metalloendoprotease in the fusion of biological membranes. In: *Cell Fusion* (A. E. Sowers, ed.). Plenum Press, New York, pp. 99-121.

117. Mundy, D. I., and Strittmatter, W. J. (1986). Characterization of adrenal chromaffin metalloendoproteases and their involvement in exocytosis. *J. Cell Biol.* 103:459a.

118. Couch, C. B., and Strittmatter, W. J. (1984). Specific blockers of myoblast fusion inhibit a soluble and not a membrane-associated metalloendoprotease in myoblasts. *J. Biol. Chem.* 259:5396-5399.

119. Lelkes, P. I., and Pollard, H. B. (1987). Inhibitors of metalloendoprotease activity inhibit exocytosis from bovine adrenal medullary chromaffin cells by modulating intracellular calcium homeostasis. *J. Biol. Chem.* 262:15496-15505.

120. Egsmose, C., Bock, E., Mollgard, T., and Thorn, N. A. (1985). Immunocytochemical demonstration of calmodulin in cells secreting by exocytosis. *Experientia* 41:1340-1342.

121. Trifaro, J. M., and Fournier, S. (1987). Calmodulin and the secretory vesicle. *Ann. NY Acad. Sci.* 493:417-434.

122. Steinhardt, R. A., and Alderton, J. M. (1982). Calmodulin confers calcium sensitivity on secretory exocytosis. *Nature* 295:154-155.

123. Kenigsberg, R., and Trifaro, J. M. (1985). Microinjection of calmodulin antibodies into cultured chromaffin cells blocks catecholamine release in response to stimulation. *Neuroscience* 14:325-347.

124. Plattner, H., Lumpert, C. J., Gras, U., Vilmart-Seuwen, J., Stecher, B., Hohne, B., Momayezi, M., Pape, R., and Kersken, H. (1987). Enzymatic regulation of membrane fusion during synchronous exocytosis in Paramecium cells. In: *Molecular Mechanisms of Membrane Fusion* (S. Ohki, ed.), pp. 477-494.

125. Bader, M. F., Hikita, T., and Trifaro, J. M. (1985). Calcium-dependent calmodulin binding to chromaffin granule membranes: presence of a 65-kilodalton calmodulin binding protein. *J. Neurochem.* 44:526-539.

126. Aunis, D., Guerold, B., Bader, M. F., and Cielski-Treska, J. (1980). Immunocytochemical and biochemical demonstration of contractile proteins in chromaffin cells in culture. *Neuroscience* 5:2261-2277.

127. Fowler, V. M., and Pollard, H. B. (1982). Chromaffin granule membrane-F-actin interactions are calcium sensitive. *Nature* 295:336-339.

128. Fowler, V. M., and Pollard, H. B. (1982). In vitro reconstitution of chromaffin granule-cytoskeleton interactions: Ionic factors influencing the association of F-actin with purified chromaffin granule membranes. *J. Cell Biochem.* 18:295-311.

129. Aunis, D., and Perrin, D. (1984). Chromaffin granule membranes-F-actin interaction and spectrin-like protein of subcellular granules: a possible relationship. *J. Neurochem.* 42:1558-1569.

130. Aunis, D., Perrin, D., and Langley, D. K. (1987). Cytoskeletal proteins and chromaffin cell activity. In: *Stimulus-Secretion Coupling in Chromaffin Cells* (K. Rosenheck and P. I. Lelkes, eds.). CRC Press, Boca Raton, FL, pp. 155-175.

131. Lelkes, P. I., Naquira, D., Friedman, J. E., Rosenheck, K., and Schneider, A. A. (1982). Plasma membrane vesicles from bovine adrenal chromaffin cells: Characterization and fusion with chromaffin granules. In: *Synthesis, Storage and Secretion of Adrenal Catecholamines* (F. Izumi, ed.). Pergamon Press, Oxford, New York, pp. 143-150.

132. Zinder, O., Hoffman, P. G., Bonner, W. M., and Pollard, H. B. (1978). Comparison of chemical properties of purified plasma membranes and secretory vesicle membranes from bovine adrenal medulla. *Cell Tissue Res.* 188:153-170.

133. Sheetz, M. P., Vale, R., Schnapp, B., Schroer, T., and Reese, T. (1987). Movements of vesicles on microtubules. *Ann. NY Acad. Sci.* 493:409-416.

134. Kachar, B. (1985). Direct visualization of organelle movement along actin filaments dissociated from characean algae. *Science* 227:1355-1357.

135. Ostlund, R. E., Leung, J. T., and Kipnes, D. M. (1977). Muscle actin filaments bind pituitary secretory granules in vitro. *J. Cell Biol.* 73:78-87.

136. Utsumi, K., Okimasu, E., Morimoto, Y. M., Nishihara, Y., and Miyara, M. (1982). Selective interaction of cytoskeletal proteins with liposomes. *FEBS Lett.* 141:176-180.

137. Caron, J. M., and Berlin, R. D. (1979). Interaction of microtubule proteins with phospholipid vesicles. *J. Cell Biol.* 81:665-671.

138. Klausner, R. D., Kumar, N., Weinstein, J. N., Blumenthal, R., and Flavin, M. (1981). Interaction of tubulin with phospholipid vesicles. I. Association with vesicles at the phase transition. *J. Biol. Chem.* 256:5879-5885.

139. Perides, G., Scherbarth, A., Kuhn, S., and Traub, P. (1986). An electron microscopic study of the interaction in vitro of vimentin intermediate filaments with vesicles prepared from Ehrlich ascites tumor cell lipids. *Eur. J. Cell Biol.* 41:313-325.

140. Traub, P., Perides, G., Schimmel, H., and Scherbarth, A. (1986). Interaction in vitro of nonepithelial intermediate proteins with total cellular lipids, individual phospholipids, and a phospholipid mixture. *J. Biol. Chem.* 261:10558-10568.

141. Burn, P., and Burger, M. M. (1987). The cytoskeletal protein vinculin contains transformation-sensitive covalently bound lipid. *Science* 235:476-479.

142. Niggli, V., Dimitrov, D. P., Brunner, J., and Burger, M. M. (1986). Interaction of the cytoskeletal component vinculin with bilayer structures analyzed with a photoactivatable phospholipid. *J. Biol. Chem.* 2261:6912-6918.

143. Sato, S. B., and Ohnishi, S. (1983). Interaction of a peripheral protein of the erythrocyte membrane, band 4.1, with phosphatidylserine-containing liposomes and erythrocyte inside-out vesicles. *Eur. J. Biochem.* 130:19-25.

144. Baines, A. J., and Bennet, V. (1986). Synapsin I is a microtubule-bundling protein. *Nature* 319:145-147.

145. Meyer, R. K., Schindler, H., and Burger, M. M. (1982). Alpha-actinin interacts specifically with model membranes containing glycerides and fatty acids. *Proc. Natl. Acad. Sci. USA* 79:4280-4284.

146. Lelkes, P. I., Friedman, J. E., Rosenheck, K., and Oplatka, A. (1986). Destabilization of actin filaments as a requirement for the secretion of catecholamines from permeabilized chromaffin cells. *FEBS Lett.* 208:375-383.

147. Asyee, G. M., Sturk, A., and Muszbek, L. (1987). Association of vinculin to the platelet cytoskeleton during thrombin-induced aggregation. *Exp. Cell Res.* 168:358-364.

148. Burn, P., Rotman, A., Meyer, R. K., and Burger, M. M. (1985). Diacylglycerol in large alpha actinin/actin complexes and in the cytoskeleton of activated platelets. *Nature* 314:469-472.

149. Bader, M. F., Trifaro, J. M., Langley, O. K., Thierse, D., and Aunis, D. (1986). Secretory cell actin-binding proteins: identification of a gelsolin-like protein in chromaffin cells. *J. Cell Biol.* 1102:636-646.

150. Plattner, H., Matt, H., Kersken, H., Haacke, B., and Sturzl, R. (1984). Synchronous exocytosis in *Paramecium* cells. I. A novel approach. *Exp. Cell Res.* 151:6-13.

151. Satir, B. H., Reichman, M., and Orias, E. (1986). Conjugation rescue of an exocytosis-competent membrane microdomain in Terahymena thermophila mutants. *Proc. Natl. Acad. Sci. USA* 90:40-54.

152. Ornberg, R. L., and Reese, T. S. (1981). Beginnings of exocytosis in *Limulus* amoebocytes. *J. Cell Biol.* 90:40-54.

153. Kelly, R. B., and Hooper, J. E. (1982). Cholinergic vesicles. In: *The Secretory Granule. The Secretory Process*, Vol. I (A. M. Poisner and J. M. Trifaro, eds.). Elsevier, Amsterdam, pp. 81-118.

154. Kersken, H., Momayezi, M., Braun, C., and Plattner, H. (1986). Filamentous actin in Paramecium cells: functional and structural changes correlated with phalloidin affinity labeling in vivo. *J. Histochem. Cytochem.* 34:455-465.

155. Smolen, J. E., Korchak, H. M., and Weissmann, B. (1984). Stimulus-secretion coupling in human polymorphonuclear leukocytes. In: *Cell Biology of the Secretory Process* (M. Cantin, ed.). Karger, Basel, pp. 517-545.

156. Nielsen, E. H. and Johnasen, T. (1986). Effects of dimethylsulfoxide (DMSO), nocodazole, and taxol on mast cell histamine secretion. *Acta Pharmacol. Toxicol. (Copenh.)* 59:214-219.

157. Cheek, T. R., and Burgoyne, R. D. (1986). Nicotine-evoked disassembly of cortical actin filaments in adrenal chromaffin cells. *FEBS Lett.* 207:110-114.

158. Perrin, D., and Aunis, D. (1985). Reorganization of alpha-fodrin induced by stimulation in secretory cells. *Nature* 315:589-592.

159. Oplatka, A., Friedman, J. E., and Rosenheck, K. (1985). Reorganization of the cytoskeleton in transformed and in tumor cells may be intimately coupled with changes in ion transport through the plasma membrane as well as with subsequent steps which affect proliferation and invasiveness. In: *Cell Membranes and Cancer* (T. Gleotti et al., eds.). Elsevier, Amsterdam, pp. 117-123.

160. Friedman, J. E., Lelkes, P. I., Rosenheck, K., and Oplatka, A. (1980). The possible implication of membrane-associated actin in stimulus-secretion coupling in adrenal chromaffin cells. *Biochem. Biophys. Res. Commun.* 96:1717-1723.

161. Perrin, D., Langley, O. K., and Aunis, D. (1987). Anti-a-fodrin inhibits secretion from permeabilized chromaffin cells. *Nature* 326:498-501.

162. Harish, G. E., Levi, R., Rosenheck, K., and Oplatka, A. (1984). Possible involvement of actin and myosin in Ca^{2+} transport through the plasma membrane of chromaffin cells. *Biochem. Biophys. Res. Commun.* 119:652-656.

163. Robertson, D., Holowka, D., and Baird, B. (1985). Cross-linking of immunoglobulin E-receptor complexes induces their interaction with cytoskeleton of rat basophilic leukemia cells. *J. Immunol.* 136:4565-4572.

164. Hall, P. F. (1984). The role of the cytoskeleton in hormone action. *Can. J. Biochem. Cell Biol.* 62:653-665.

165. Howell, S. L., and Tyhurst, M. (1986). The cytoskeleton and insulin secretion. *Diabetes Metab. Rev.* 2:107-123.

166. Gratzl, M., Schudt, C., Eckerdt, R., and Dahl, G. (1980). Fusion of isolated biological membranes; a tool to investigate basic processes of exocytosis and cell-cell fusion. In: *Membrane Structure and Function*, Vol. 3 (E. E. Bittar, ed.). Wiley Interscience, New York, pp. 59-92.

167. Lelkes, P. I., Lavie, E., Naquira, D., Schneeweiss, F., Schneider, A. S., and Rosenheck, K. (1980). Acetylcholine induced in vitro fusion between cell membrane vesicles and chromaffin granules from the bovine adrenal medulla. *FEBS Lett.* 115:129-133.

168. Scott, J. H., Creutz, C. E., Pollard, H. B., and Ornberg, R. (1985). Synexin binds in a calcium-dependent fashion to oriented chromaffin cell plasma membranes. *FEBS Lett.* 180:17-23.

169. Whitaker, M. J., and Baker, P. F. (1983). Calcium dependent exocytosis in an in vitro secretory granule plasma membrane preparation from sea urchin eggs and the effects of some inhibitors of cytoskeletal function. *Proc. R. Soc. London* 218:397-413.

170. Crabb, J. H., and Jackson, R. C. (1985). In vitro reconstitution of exocytosis from plasma membranes and isolated secretory vesicles. *J. Cell Biol.* 101:2263-2273.

171. Jackson, R. C., Ward, K. K., and Haggerty, J. G. (1985). Mild proteolytic digestion restores exocytotic activity to *N*-ethylmaleimide-inactivated cell surface complexes from sea urchin eggs. *J. Cell Biol.* 101:8-11.

172. Sasaki, H. (1984). Modulation of calcium sensitivity by a specific cortical protein during sea urchin egg cortical vesicle exocytosis. *Dev. Biol.* 1101:125-135.

173. Driessen, A. J. M., Hoekstra, D., Scherphof, G., Kalicharan, R. D., and Wilschut, J. (1985). Low pH-induced fusion of liposomes with membrane vesicles derived from Bacillus subtilis. *J. Biol. Chem.* 260: 10880-10887.

174. Pastan, I., and Willingham, M. (1985). The pathway of endocytosis. In: *Endocytosis* (I. Pastan and M. Willingham, eds.). Plenum Press, New York, pp. 1-44.

175. Blumenthal, R., Henkart, P., and Steer, C. J. (1983). Clathrin-induced pH-dependent fusion of phosphatidylcholine vesicles. *J. Biol. Chem.* 258:3409-3415.

176. Meyer, D. I., and Burger, M. M. (1979). Isolation of a protein from the plasma membrane of adrenal medulla which binds to secretory vesicles. *J. Biol. Chem.* 254:9854-9859.

177. Creutz, C. E., Pazoles, C. J., and Pollard, H. B. (1978). Identification and purification of an adrenal medullary protein (synexin) that causes calcium-dependent aggregation of isolated chromaffin granules. *J. Biol. Chem.* 253:2858-2866.

178a. Burns, A. L., Magendzo, K., Shirvan, A. K., Srivastava, M., Rojas, E., Alijani, M. R., and Pollard, H. B. (1989). Calcium channel activity of purified human synexin and structure of the human synexin gene. *Proc. Natl. Acad. Sci. USA* 86:3798-3802.

178a. Burns, A. L., Magendzo, K., Rojas, E., Cheung, B., Srivastava, A., Johnson-Seaton, D., Alijani, M., and Pollard, H. B. (1988). Calcium channel and membrane fusion: properties of human synexin and cloning of the human synexin gene. *Proc. Natl. Acad. Sci. USA*, in press.

179. Sudhoff, T. C., Ebbecke, M., Walker, J. H., Fritsche, U., and Boustead, C. (1984). Isolation of mammalian calelectrins: A new class of ubiquitous Ca^{2+}-regulated proteins. *Biochemistry* 13:1103-1109.

180. Glenney, J. (1986). Phospholipid-dependent Ca^{2+}-binding by the 36-kDa tyrosine substrate (calpactin) and its 33 kDa core. *J. Biol. Chem.* 261:7247-7252.

181. Moore, P. B., Kraus-Friedmann, N., and Dedman, J. R. (1984). Unique calcium-dependent hydrophobic binding proteins: possible

independent mediators of intracellular calcium distinct from calmodulin. *J. Cell Sci.* 72:121-133.

182. Geisow, M. J., and Walker, J. H. (1986). New proteins involved in cell regulation by Ca^{2+} and phospholipids. *TIBS* 11:420-423.

183. Creutz, C. E., Zaks, W. J., Hamman, H. C., Crane, S., Martin, W. H., Gould, K. L., Oddie, K. M., and Parsons, S. J. (1987). Identification of chromaffin granule binding proteins. Relationship of the chromobindins to calelectrin, synhibin, and the tyrosine kinase substrates p35 and p36. *J. Biol. Chem.* 262:1860-1868.

184. Glenney, J. R., Jr., Tack, B., and Powell, M. A. (1987). Calpactins: two distinct Ca^{++}-regulated phospholipid- and actin-binding proteins isolated from lung and placenta. *J. Cell Biol.* 104:503-511.

185. Smith, V. A., and Dedman, J. R. (1986). An immunological comparison of several novel calcium-binding proteins. *J. Biol. Chem.* 261:15815-15818.

186. Stutzin, A. (1986). A fluroescence assay for monitoring and analyzing fusion of biological membrane vesicles in vitro. *FEBS Lett.* 197:274-280.

187. Nir, S., Stutzin, A., and Pollard, H. B. (1987). Effect of synexin on aggregation and fusion of chromaffin granule ghosts at pH 6. *Biophys. Biochim. Acta* 903:309-318.

188. Stutzin, A., Cabantchik, Z. I., Lelkes, P. I., and Pollard, H. B. (1987). Synexin-mediated fusion of bovine chromaffin granule ghosts. Effect of pH. *Biochim. Biophys. Acta* 905:205-212.

189. Rojas, E., and Pollard, H. B. (1987). Membrane capacity measurements suggest a calcium-dependent insertion of synexin into phosphatidyl-serine bilayers. *FEBS Lett.* 217:25-31.

190. Pollard, H. B., Rojas, E., and Burns, A. L. (1987). Synexin and chromaffin granule membrane fusion. A new "hydrophobic bridge" hypothesis for driving and directing of the fusion process. *Ann. NY Acad. Sci.* 493:524-541.

191. Pollard, H. B., Rojas, E., Burns, A. L., and Parra, C. (1987). Synexin, calcium and the hydrophobic bridge hypothesis for membrane fusion. In: *Molecular Mechanisms of Membrane Fusion* (S. Ohki, et al., eds.). Plenum Press, New York, pp. 341-355.

192. Pollard, H. B., Burns, A. L., and Rojas, E. (1988). A molecular basis for synexin driven calcium-dependent membrane fusion. *J. Exp. Biol.* 139:267-286.

24

Permeabilized Cells
An Approach to the Study of Exocytosis

MANFRED GRATZL

University of Ulm, Ulm, Federal Republic of Germany

I. INTRODUCTION

Biological membranes are highly asymmetrical structures that separate cells or subcellular compartments differing greatly in their composition. Well-known cellular activities, such as secretion by exocytosis, uptake of extracellular material by phagocytosis, formation of the multinucleated myotubes during development of striated skeletal muscle, and fertilization, require direct (open) communication between cells or subcellular compartments. This is achieved by the process of membrane fusion. At least two types of membrane fusion can be distinguished: in one type, extracellular membrane surfaces interact with each other (e.g., cell-cell fusion), while in the other, interactions occur between membrane surfaces facing the cytoplasm (e.g., exocytosis).

Exocytotic membrane fusion is difficult to analyze because the interacting membrane surfaces are not accessible from the outside of the cell. Moreover, the complexity of the processes in the chain of events between stimulation of a cell and release of secretory product makes it difficult to study or manipulate adequately membrane fusion in intact cells.

Within the past few years, procedures have been developed that allow exocytotic membrane fusion to be investigated by permeabilization of the plasma membrane of secretory cells. While they leave the exocytotic machinery intact, they allow for modifying the cytoplasmic composition as desired and determining exocytotic output as a function of various manipulations. In this way, the properties of exocytotic membrane fusion can be recorded and compared with the fusion properties of more simple model systems or with the secretory process as exhibited by intact cells.

II. PERMEABILIZATION TECHNIQUES

Three different techniques have been used to permeabilize cells: application of physical force (high-voltage discharges), of detergents, and of pore-forming proteins.

When cells in suspension are briefly exposed to electric fields, their membranes become permeable to solutes. The size of the membrane-bound structure permeabilized depends on the strength of the electric field applied. Hence, one can permeabilize the cell membrane, permitting access to the cytosol, without risking damage of intracellular organelles. This technique has been initially applied to bovine adrenal medullary cells (1-3) and subsequently also to other secretory cells. After this treatment, the adrenal medullary cells behave as if their cell membrane contains pores allowing passage for at least 1 h of substances of up to a molecular weight of about 1000.

Detergents such as digitonin and saponin have also been applied to chromaffin cells (4-9). Within a narrow range of concentration and time, these substances are suitable for permeabilizing the cell membrane without causing leakage of substances from secretory vesicles. Since the size of the pores created is not uniform, i.e., the holes obtained with saponin vary between 0.1 μm and 1 μm (10), intact chromaffin secretory vesicles, which have a diameter of about 0.25 μm, can escape. A careful examination of digitonin-permeabilized chromaffin cells shows that compared to alpha toxin poration (see below) significant changes in the ultrastructure and in the secretory behavior occur with digitonin (10a). The cells also lose proteins essential for exocytosis (10b). During measurement of Ca^{2+}-induced release of chromogranin A and noradrenalin from digitonin-permeabilized chromaffin cells in primary cultures, it has been shown that part of the secretory product can be sedimented by centrifugation (9), which indicates that the secretory product remains membrane bound. Thus, the release observed from digitonin-treated cells does not occur solely by exocytosis. A further drawback of the use of detergents as permeabilizing agents is their possible influence on the exocytotic process itself, a conclusion based on the observed inhibition by detergents of catecholamine release from electrically permeabilized cells (2).

Natural proteins produced from T lymphobytes (11) or bacteria (12,13) or derived from the complement complexes (14) insert in target membranes and lead to the formation of "stabilized" pores because every hole is surrounded by a protein ring (14a).

Two of these proteins, the staphylococcal α-toxin and streptolysin O from streptococci, were valuable for probing the exocytotic process. They attack both the cell membrane of PC 12 cells (a pheochromocytoma cell line from rat) as well as of bovine adrenal chromaffin cells in culture (9,15,16,16a-f).

Unlike the pores created by electrical discharges or detergents, all α-toxin pores are the same size (see Fig. 1). In target membranes, the water-soluble toxin monomers (molecular weight 34 kD) form ring-structured hexamers surrounding a pore with a diameter of 2-3 nm (12). These structures do not permit the passage of myoglobin (17 kD) or Dextran 4 (4 kD) (12,13). Consequently, the toxin monomers cannot enter the cells. Also, the cytoplasmic enzyme lactate dehydrogenase is not released from chromaffin cells in primary cultures or from PC 12 cells under this treatment. However, the rapid efflux of $^{86}Rb^{+}$ from the cells demonstrates complete accessibility of the cytosol for small molecules (9,15,16). Also, the rapid equilibration of cellular ATP or Ca^{2+} or externally added substances like inositol-1,4,5-trisphosphate, different vanadate species, and so forth, indicates an effective poration of the plasma membrane for molecules up to a molecular mass of about 1 kD (16a,16g). Since catecholamines or the protein chromogranin A was not released from the secretory vesicles, α-toxin's attack is strictly confined to the plasma membrane. Therefore, the technically simple permea-

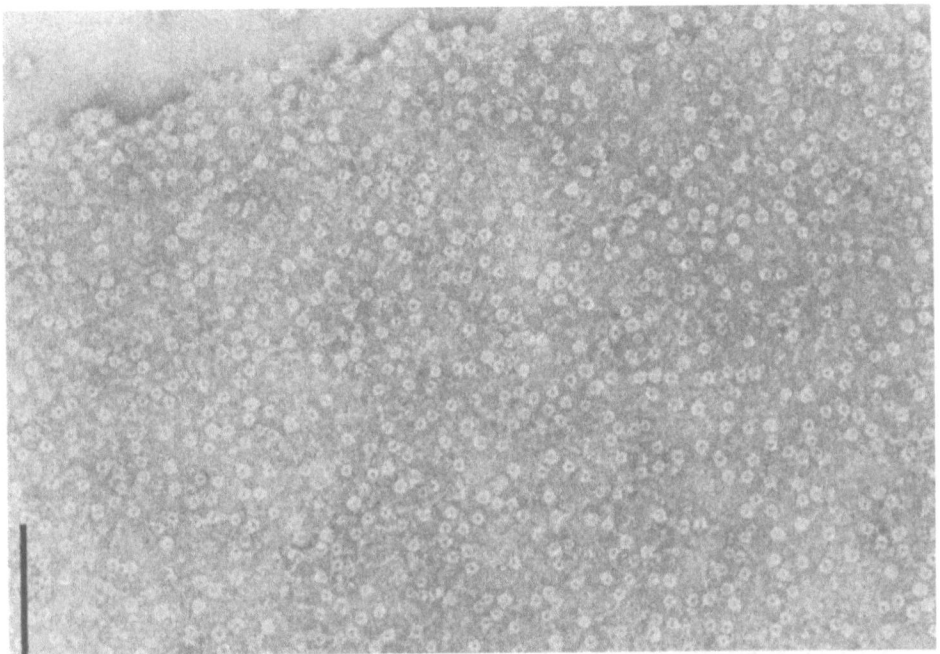

FIGURE 1 Electron micrograph of a negatively stained fragment of rabbit erythrocyte membrane, lysed with *Staphylococcus aureus* α-toxin (20 µg toxin/10^8 cells). Numerous ring-shaped toxin polymers are seen over the membrane. The hexameric form of the toxin has an outer diameter of 10 nm and exhibits a central stain deposit that indicates an inner diameter of about 2.5 nm. Sodium silicotungstate staining. Scale bar indicates 100 nm. (Courtesy of J. Tranum-Jensen, Institute of Anatomy, Department C, University of Copenhagen, Denmark.)

bilization by α-toxin represents an ideal technique for studying exocytotic membrane fusion.

Streptolysin O (SLO) from beta-hemolytic streptococci produces large pores in target membranes (14a). Catecholamine-secreting cells permeabilized with SLO retain an intact exocytotic machinery and antibodies to intracellular constituents, as well as tetanus toxin and botulinum A toxin, can be introduced directly into these cells (16d-f).

In addition to the three procedures described here, another closely related technique is the use of hemolytic Sendai virus as a permeabilizing agent (17). However, the size of the holes created with this procedure cannot be precisely defined. Therefore, this technique must be considered inferior to permeabilization with α-toxin or streptolysin O.

III. RELEASE STUDIES WITH PERMEABILIZED CHROMAFFIN CELLS

A. Ca^{2+} Requirement

The concept of stimulus-secretion coupling was developed around 25 years ago (18). This coupling concept describes calcium as having a crucial role in the regulation of secretion by exocytosis. Injection of Ca^{2+} into mast

cells (19) and nerve (20) provided direct evidence for the role of intracellular Ca^{2+} as a trigger substance.

The Ca^{2+} concentration is precisely controlled within secretory cells
by systems present in the cell membrane and in subcellular structures.
In resting cells, the free Ca^{2+} concentration is close to 10^{-7} M but increases
upon receptor activation and/or depolarization owing to an influx of Ca^{2+}
from the extracellular space which contains high (mM) concentrations of
Ca^{2+} (cf. Refs. 21-23). Receptor activation and linked processes can be
circumvented by ionophores which have often been used to facilitate the
Ca^{2+} influx. A complete exchange of the intracellular fluid as well as its
exact control (e.g., by buffering substances) can be achieved if pores
sufficiently large in size to permit exchange of small or even high-molecular
weight substances can be created in the cell membrane.

Compared to intact cells, permeabilized cells require much less Ca^{2+}
for the release of secretory product. An example is given for pheochromocytoma cells (PC 12) (Fig. 2). Intact cells require mM concentrations of
Ca^{2+} but α-toxin permeabilized cells respond already to μM Ca^{2+}.

Exocytosis in intact cells is the complete transfer of small as well as
large molecules from a intracellular vesicular compartment to the extracellular
space, a process morphologically characterized by fusion of secretory vesicles with the cell membrane. Therefore, a parallel release of small and
large secretory products (catecholamines as well as dopamine-B-hydroxylase
and chromogranins) but not of cytoplasmic lactate dehydrogenase was used
as a biochemical parameter for secretion by exocytosis from chromaffin
cells (cf. Refs. 24-26). Cell preparations permeabilized by high-voltage
discharges or by α-toxin or streptolysin O meet these criteria, suggesting
that release occurs by exocytosis (1-3,9,15,16). Proof for exocytotic release
from permeabilized catecholamine-secreting cells was also obtained in a
study of the catecholamine metabolism in these cells (16c). It was shown
that the enzymes oxidizing catecholamines, which are present in the cytoplasm, cannot come into contact with the membrane-bound catecholamines
indicating that the cytoplasm is avoided during the release of catecholamines
in α-toxin permeabilized cells (16c). As an example, the Ca^{2+}-dependent
release of the soluble contents of the storage vesicles, but not of the
cytosolic marker enzyme, from α-toxin permeabilized adrenal medullary
chromaffin cells in tissue culture is shown in Figure 3.

The range of Ca^{2+} concentrations required for exocytosis in permeabilized cells certainly is in accordance with the measurements of Ca^{2+} within
intact cells. Adrenal medullary chromaffin cells or pheochromocytoma cells
(PC 12) contain about 0.1 μM free Ca^{2+} within the cytoplasm (27,28).
Stimulation increases this value by a factor of roughly 3 or more. The
measurement of free Ca^{2+} concentrations within cells certainly is not an
easy procedure. One can assume that the true Ca^{2+} concentration in these
cells is slightly higher than estimated for the following two reasons. First,
Ca^{2+} determination using indicator substances only gives the average concentration within a given cell and not the concentration close to the plasma
membrane where, owing to its influx from outside, the largest increase
could be envisaged. Second, these substances, in order to provide a good
signal, must be present within the secretory cells in fairly high amounts,
and the indicator, being a Ca^{2+} chelator itself, would thus report a smaller
Ca^{2+} concentration than actually exists. In addition to the difficulty of
adjusting free Ca^{2+} concentrations to the low values used in studies with
permeabilized cells, the possible influence on subcellular Ca^{2+} pools by
the permeabilization procedure or by the media used may also have con-

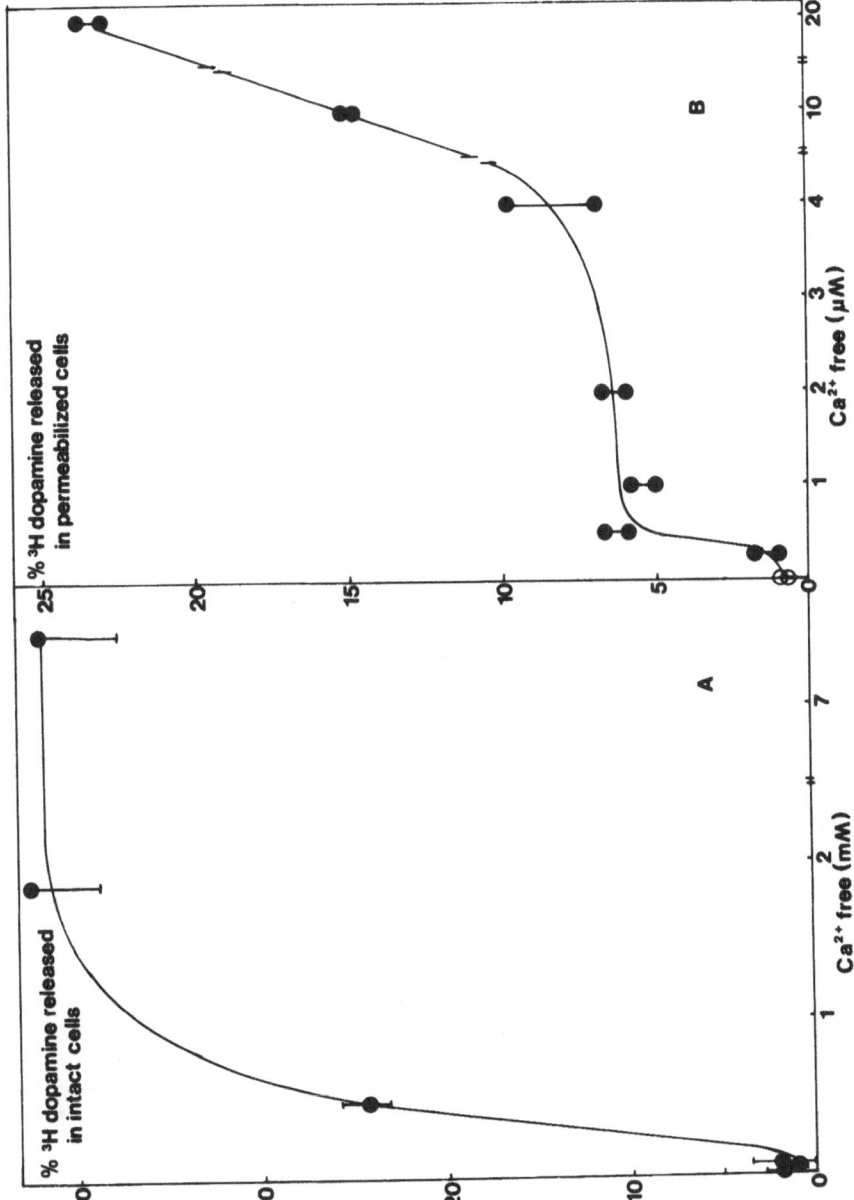

FIGURE 2 Ca²⁺ requirement of dopamine release by intact (A) and by α-toxin permeabilized PC12 cells. (From Ref. 15, by permission).

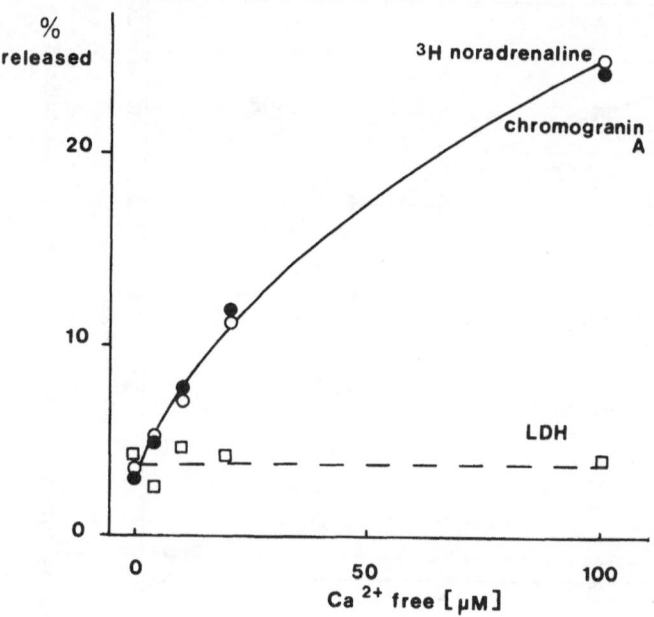

FIGURE 3 Ca^{2+} dependence of noradrenalin and chromogranin release
from α-toxin-permeabilized chromaffin cells. Ca^{2+} triggers parallel secretion
of noradrenalin and chromogranin A but has no effect on lactate dehydrogenase from the cells. (From Ref. 9, by permission.)

tributed to the reported differences in the Ca^{2+} sensitivity of catecholamine
release by permeabilized chromaffin cells. Nonetheless, there is no doubt
that about 1000 times less Ca^{2+} needs to be added to elicit hormone release
from permeabilized chromaffin cells than from intact cells.

B.   Role of ATP

Freshly isolated adrenal medullary cells permeabilized electrically require
ATP to be stimulated successfully with μM concentrations of Ca^{2+} (1-3).
If the same cells are kept in tissue culture and permeabilized with detergents, the effect of ATP is inconsistent (4-6). PC 12 cells permeabilized
with α-toxin do not require ATP at all (16). These differences in ATP
requirement may reflect cell-type-specific properties of the release process
or the different permeabilization procedures used.

It is likely that the lack of ATP sensitivity of permeabilized PC 12
cells is inherent to this pheochromocytoma cell line. While chromaffin cells
in primary culture exhibit decreased secretion when ATP production by
glycolysis and oxidative phosphorylation is blocked (29,30), PC 12 cells
do not respond to such a treatment (31). Thus, the properties of secretion
by intact cells are in accordance with findings obtained with permeabilized
cells. The reason for the different behavior of PC 12 and chromaffin cells
is not clear. However, a fact worth considering is that, in PC 12 cells,
the preferential arrangement of secretory vesicles is near the cell membrane
(32), whereas in chromaffin cells the secretory vesicles are distributed
throughout the cytoplasm (cf. Ref. 33). In other words, movement of

secretory vesicles toward the cell membrane seems to be essential in chromaffin cells but not in PC 12 cells. In conjunction with this, the sole requirement of organelle movement for ATP is of considerable interest (34,35). Another possibility could be that a phosphorylation step is essential to prime for exocytosis. Along this line, studies have been carried out using intact chromaffin cells, permeabilized cells, or subcellular fractions (36-39). Finally, an effect of ATP-dependent proton translocation across the secretory vesicle membrane has been considered (40).

Despite the efforts carried out, the mode of ATP action in exocytosis is still unknown. It is noteworthy that chromaffin cells kept in tissue culture, similar to PC 12 cells after permeabilization with α-toxin, were not dependent on ATP. However, after a washout period the ATP requirement of catecholamine release by the chromaffin cells could be clearly demonstrated (Fig. 4). Under these precisely controlled conditions ATP could not be substituted by any of the nucleotides tested (9). A parallel finding has been reported with freshly isolated chromaffin cells (3) This clear-cut difference in the ATP requirement between chromaffin cells and PC 12 cells may help to find the ATP-dependent step in exocytotic secretion.

C. Effect of Neurotoxins

The clostidial neurotoxins tetanus toxin and botulinum A toxin belong to the most poisonous substances known (20a). By means of the large pores generated by SLO, neurotoxins can be introduced into chromaffin cells (16d-f). It has been observed that tetanus toxin and botulinum A toxin following separation of their disulfide linked-chains are able to block Ca^{2+}-induced catecholamine release. The heavy chain of the toxins has no effect, but the light chains block exocytosis (16d-f). Thus, a new tool, the light chains of the clostridial neurotoxins for the analysis of the components involved in exocytosis, has been detected (16d-f). Furthermore, appropriate permeabilized cells should permit identification of the target of tetanus toxin within the cells, as well as detection of its reactive domain within the light chain, and consequently provide an approach to elucidate the mechanism of the action of these neurotoxins at the molecular level.

D. Modulation of Catecholamine Release

All observations reported agree that the amount or nature of monovalent cations does not modify the characteristics of secretion by permeabilized cells. From the divalent cations tested Mg^{2+} could not be substituted for Ca^{2+} and high concentrations (>mM) and Mg^{2+} inhibit the release elicited by μM Ca^{2+} (1-3,6,16). The reported ability of Sr^{2+}, Cd^{2+}, Mg^{2+}, and Ba^{2+} to replace or block the effect of Ca^{2+} (6) must be interpreted with caution because the experiments were carried out with very high Ca^{2+} concentrations (mM). Moreover, no chelators were present during these experiments, i.e., it is not known how much Ca^{2+} was still present or whether a redistribution of ions or an actual influence on the Ca^{2+}-sensitive target was the cause. Thus, further experiments are required to characterize the effect of divalent cations in order to obtain a solid basis for the comparison of secretion by intact secretory cells with such permeabilized cell preparations.

Freshly isolated chromaffin cells permeabilized electrically demonstrate a large inhibition of Ca^{2+}-induced catecholamine release by chloride and

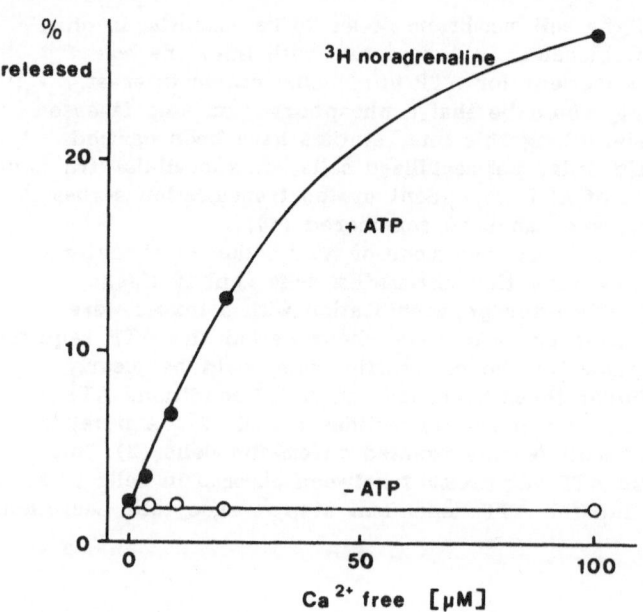

FIGURE 4   Effect of ATP on Ca^{2+}-evoked catecholamine release from α-toxin-permeabilized chromaffin cells. (From Ref. 9, by permission.)

other anions (3), but chromaffin cells kept in primary cultures and permeabilized with digitonin or saponin (4,6) are not affected by the nature of the anions present. This difference is not yet understood. Recently, it has been observed that changes in the medium composition with respect to anions likewise do not modify dopamine release from α-toxin-permeabilized PC 12 cells in culture (16).

Permeabilized cells appear to be an almost ideal tool to elucidate the role of intracellular messengers. That is why the participation of the protein kinase C system in secretion by exocytosis also has been analyzed by using such preparations. Phorbolesters, which can substitute the natural diacylglycerol as activator of protein kinase C, apparently decrease the requirement for Ca^{2+} of catecholamine release in freshly isolated, electrically permeabilized cells (41), as well as in digitonin-permeabilized chromaffin cells in primary culture (42). The latter, however, is less sensitive to this drug. Surprisingly 1-oleyl-2-acetyl-glycerol (OAG), an activator of purified kinase C, does not show a detectable effect in electrically permeabilized cells (43). In permeabilized rat pheochromocytoma cells (PC 12) activation of protein kinase C by the diacylglycerol analog OAG or the phorbolester TPA ameliorates Ca^{2+} induced exocytosis (16c,43a). Using α-toxin-permeabilized PC 12 cells, it could also be shown that the activation by TPA is dependent on the presence of Mg^{2+}/ATP (16c), which clearly indicates the involvement of an active protein kinase C system. Exposure of these cells to GTP-γ-S inhibits Ca^{2+}-dependent release in bovine cells but stimulates it in chicken cells. As the inhibitory action on bovine cells persisted in the presence of phorbolester, it was concluded that a GTP binding protein (G protein) may inhibit the action of protein kinase C (43). A Ca^{2+}-independent stimulatory effect of GTP analogs was reported for bovine chromaffin cells in primary culture (43b), whereas

Ca^{2+}-triggered exocytosis in α-toxin-permeabilized pheochromocytoma cells (PC 12) was inhibited by $GTP_\gamma S$ (16c). This effect could be abolished by pertussis toxin but not by cholera toxin, indicating that exocytosis by PC 12 cells can be modulated by a pertussis toxin-sensitive G protein (16c). Further analysis of the G proteins present in chromaffin cells seems to be necessary to fully understand the role of these proteins in intracellular signal transduction. In this context it is noteworthy that low- and high-molecular-mass GTP binding proteins have been detected at a strategic location, namely the secretory vesicle membrane in chromaffin cells (43c,d).

Application of drugs interacting with calmodulin in intact chromaffin cells (44,45) and injection of anticalmodulin antibodies into chromaffin cells using erythrocytes as a vehicle (46) suggested the possibility of calmodulin controlling the Ca^{2+} sensitivity of the release process. The reported data using permeabilized cells do not support this view, at least as far as trifluoperazine is concerned. Neither electrically nor detergent-permeabilized cells responded to low concentrations of this drug; 10 µM trifluoperazine caused the Ca^{2+}-independent catecholamine release to increase, suggesting a damage of secretory vesicles (3,16d,47). Since introduction of an antibody against calmodulin into SLO-permeabilized PC 12 cells did not affect the Ca^{2+}-dependent secretory response, an involvement of calmodulin in the final steps of exocytosis is less likely (16d).

The different techniques to permeabilize the cell membrane of chromaffin cells allow for introducing small and large molecular substances into the intracellular environment which would not penetrate otherwise. The examples given above demonstrate the usefulness of permeabilized cell preparations to determine the molecular requirements for exocytosis.

IV. COMPARISON WITH OTHER PERMEABILIZED SECRETORY CELLS

A. Endocrine and Exocrine Pancreas

Much of our knowledge on the process of secretion by exocytosis stems from numerous investigations of the chromaffin cell. It was therefore feasible to use mainly chromaffin cells as an object for permeabilization studies.

However, such studies have also been conducted on other endocrine cells: The pancreatic islet cells produce several key hormones in carbohydrate metabolism, including insulin, glucagon, and somatostatin. Knowledge of the release process of these hormones is of particular importance in understanding diabetes and other endocrine disorders. The high-voltage technique has been used to determine the molecular requirements for exocytosis in islet cells (48-51).

It is not surprising that also in these cells µM Ca^{2+} was sufficient to release insulin after permeabilization. The ATP requirement has been studied in more detail (48). Cells that had received an ATP washout treatment were found to be dependent on this nucleotide. On the other hand, cells that were not washed out did not require the addition of ATP. This is similar to the situation seen with α-toxin-permeabilized chromaffin cells in primary culture, which also exhibit ATP dependence only after a washout period (9).

Elevation of the glucose level in the extracellular fluid in the presence of Ca^{2+} is the natural stimulus for insulin secretion from intact pancreatic

B cells. Neither glucose nor glucose-6-phosphate modified insulin release
from permeabilized cells, but phosphoenolpyruvate stimulated the release.
This suggests that the latter substance may act as an intracellular modulator
of insulin release. Further experiments along this line may be helpful to
settle a long-lasting controversy on whether glucose itself or some of its
metabolites may act as the principal stimulus of insulin release (cf. Ref. 52).

Inhibition of insulin release from pancreatic islets has been reported
for phalloidin (50), which stabilizes F-actin, cytochalasin B (49), which
alters microfilament function, and vinblastin (49), which impairs microtubule
function. By contrast, none of these substances affected catecholamine
release from electrically permeabilized chromaffin cells (3).

Insulin secretion has also been investigated using digitonin-permeabilized
islets of Langerhans (53,54). Both forskolin and phorbolesters, in addition
to Ca^{2+}, have been observed to stimulate insulin release in this preparation
(53). Concerning phorbolesters, a parallel finding was recently noted using
electrically permeabilized islets; thus a physiological role for protein kinase
C cascade in insulin secretion can be envisioned (51).

Also, exocrine pancreatic acinar cells which produce digestive enzymes
were rendered permeable by intense electric fields (55). Without Ca^{2+}
present, less than 0.5% of the cellular content of amylase was released
but about 4% was released upon addition of 10 μM free Ca^{2+}. Cyclic nucleotides did not affect the exocytotic machinery within these cells. However,
similar to the behavior of pancreatic B cells and chromaffin cells, amylase
release was increased by phorbolesters.

B. Platelets and Mast Cells

Secretion from and aggregation of platelets are important processes cooperat-
ing in hemostasis. Both can be elicited by thrombin, a protein with a central
role in the regulation of blood clotting. This protein induces serotonin
release from intact platelets even in the absence of extracellular Ca^{2+}.
It also causes phosphatidyl inositol breakdown leading to diacylglycerol
formation, activation of protein kinase C, and phosphorylation of proteins
with molecular masses of 40 kD and 20 kD, the latter being light-chain
platelet myosin (cf. Ref. 56). When permeabilized by high-voltage discharges, platelets in the presence of ATP and μM concentrations of Ca^{2+}
release serotonin from the storage vesicles and acid hydrolases from the
lysosomes. This release is increased by further addition of thrombin (57-61).
In permeabilized platelets with low (0.1 μM) concentration of Ca^{2+}, thrombin
also increases diacylglycerol formation, presumably by hydrolysis of phos-
phatidylinositols (61) and phosphorylation of proteins with a molecular
weight of about 40 kD and 20 kD. Diacylglycerol itself causes enhanced
phosphorylation of these two proteins (59). The results obtained so far
are consistent with the view that secretion from platelets as studied with
permeabilized cells is dependent on μM Ca^{2+} and members of the
phosphatidylinositol/protein kinase C cascade are probably potent modulators.
The relation of protein phosphorylation to the secretory response is not
yet known (cf. Chapter 25). However, further studies may help to provide
some insight into the nature of mechanisms involved in this ATP-dependent
step of exocytosis.

Mast cells permeabilized with Sendai virus release histamin upon addition
of μM Ca^{2+}. This requires no ATP and a nonhydrolyzable ATP derivative
had no effect (17). However, when permeabilized with streptolysin O, rat
mast cells that have been pretreated with metabolic inhibitors secrete histamine
provided that Ca^{2+} and a nucleoside triphosphate are present (59a).

C. Sea Urchin Eggs

Within seconds after fusion of sperm and egg, the membranes of cortical vesicles fuse with the cell membrane. This example of massive exocytosis has been studied with intact eggs, with a preparation containing cortical vesicles plus cell membrane ("isolated cortices"), as well as with permeabilized eggs, with the aim of obtaining information on the mechanism of exocytosis in these cells.

The properties of the co-called cortical reaction in intact eggs are very similar to the secretion by exocytosis observed in other cells. It is associated with a rise in intracellular free Ca^{2+} (62,63), can also be activated by the ionophor A23187 (64), as well as by injection of Ca^{2+} (65). As a measure of secretion, enzymes present in the cortical vesicles are determined (66,67).

A preparation of cortical vesicles attached to the inner surface of the plasma membrane (isolated cortices) can be obtained from sea urchin eggs adhering to a coated surface, after shearing away the upper part of the cell (68). Addition of Ca^{2+}, Ba^{2+}, and Sr^{2+}, but not of Mg^{2+}, elicits discharge. Treatment with drugs affecting microtubules or microfilaments, with cyclic nucleotides, or with ATP neither enhanced nor inhibited this process (69). Experiments carried out with isolated cortices as well as with eggs permeabilized by high-voltage discharges have shown that a supply of ATP is not necessary in these preparations (70). However, the sensitivity of exocytosis to Ca^{2+} is increased by ATP and the fraction of vesicles reacting falls in the absence of this nucleotide. Finally, in a reconstituted system, consisting of purified cortical vesicles and plasma membrane, it was shown that ATP was not necessary for exocytosis (71). In vitro exocytosis in isolated egg cortices is inhibited by elevated Mg^{2+} concentrations. This inhibition can be overcome by increasing the free Ca^{2+} concentration (72). Many other conceivable modulators of the Ca^{2+}-dependent cortical reaction did not show any effect. Trifluoperazine inhibited Ca^{2+}-dependent exocytosis, but because of the increased Ca^{2+}-independent release observed at slightly higher concentrations of this drug, the authors were not confident in attributing its inhibitory effects to an action on calmodulin rather than to its detergent-like activity (72). Another group using an antibody to calmodulin, which inhibits exocytosis in isolated cortices, concluded that calmodulin or other similar Ca^{2+}-binding modulatory proteins may be involved in exocytosis (73).

V. CONCLUSION

Investigation of exocytotic membrane fusion in permeabilized cell preparations has shown that there is an absolute requirement for Ca^+ in the µM range. The target (receptor) of this cation, i.e., the protein and/or the lipid it interacts with, remains to be identified.

The specificity of Ca^{2+} binding to its receptor is to be compared with the ionic requirements of the secretory vesicle/plasma membrane interaction. While there is agreement on the fact that Mg^{2+} in high (mM) concentrations inhibits the effect elicited by µM Ca^{2+} and that Mg^{2+} cannot replace Ca^{2+}, it has not yet been completely worked out whether other divalent or trivalent cations can act like Ca^{2+} does.

Another unsolved problem is the question of exactly how ATP acts in exocytotic secretion. It must be borne in mind that some endocrine cells can do without ATP. Also, in the sea urchin egg there is no direct influence of ATP. Furthermore, isolated secretory vesicle fusion requires no ATP added (74-76). Thus, it may be concluded that the exocytotic membrane fusion catalyzed by Ca^{2+} has to be differentiated from other ATP-requiring processes which may precede the fusion step. In particular, ATP may assure the movement of the secretory vesicles toward the cell membrane or the docking of the secretory vesicle to the cell membrane prior to membrane fusion.

Modulation of exocytosis via the protein kinase C system is certainly an ATP-dependent step. Within the intact cell this pathway is affected by receptor-coupled, GTP-binding proteins at the level of the plasma membrane. Modulation of exocytosis via intracellular GTP binding proteins is indicated by the effects of GTP analogs on Ca^{2+}-triggered exocytosis in permeabilized cells and first reports on the intracellular distribution of these proteins within secretory cells. However, the nature and the role of these proteins in exocytotic membrane fusion remain to be elucidated.

Recent reports on the inhibitory effect of the clostridial neurotoxins on exocytotic release of catecholamines from permeabilized chromaffin cells indicate that a novel tool has been detected to define intracellular molecules participating in exocytotic membrane fusion. Further work concerning the intracellular action of tetanus toxin and botulinum A toxin will certainly allow us to define the mechanism of action of these highly poisonous neurotoxins.

The processes of exocytotic membrane fusion in different secretory cells probably have the same principal basic mechanisms. Permeabilizing cell preparations appears to be an ideal method for following these mechanisms.

ACKNOWLEDGMENTS

The work by the author and his colleagues referred to in this chapter was supported by grants from the Deutsche Forschungsgemeinschaft (Gr 681) and the Forschungsschwerpunkt No. 24 of the State of Baden-Württemberg. I thank Mrs. B. Lauman for editing and Mrs. B. Mader for preparing the manuscript.

REFERENCES

1. Baker, P. F., and Knight, D. E. (1978). Calcium-dependent exocytosis in bovine adrenal medullary cells with leaky plasma membranes. *Nature* 276:620-622.

2. Baker, P. F., and Knight, D. E. (1981). Calcium control of exocytosis and endocytosis in bovine adrenal medullary cells. *Phil. Trans. R. Soc. Lond. B* 296:83-103.

3. Knight, D. E., and Baker, P. F. (1982). Calcium-dependence of catecholamine release from bovine adrenal medullary cells after exposure to intense electric fields. *J. Membrane Biol.* 68:107-140.

4. Brooks, J. C., and Treml, S. (1983). Catecholamine secretion in chemically skinned cultured chromaffin cells. *J. Neurochem.* 40:468-473.

5. Dunn, L. A., and Holz, R. W. (1983). Catecholamine secretion from digitonin-treated adrenal medullary chromaffin cells. *J. Biol. Chem.* 258:4989-4993.

6. Wilson, S. P., and Kirshner, N. (1983). Calcium-evoked secretion from digitonin-permeabilized adrenal medullary chromaffin cells. *J. Biol. Chem.* 258:4994-5000.

7. Holz, R. W., and Senter, R. A. (1985). Plasma membrane and chromaffin granule characteristics in digitonin-treated chromaffin cells. *J. Neurochem.* 45:1548-1557.

8. Morita, K., Levine, M., Heldman, E., and Pollard, H. B. (1985). Ascorbic acid and catecholamine release from digitonin-treated chromaffin cells. *J. Biol. Chem.* 260:15112-15116.

9. Bader, M. F., Thierse, D., Aunis, D., Ahnert-Hilger, G., and Gratzl, M. (1986). Characterization of hormone and protein release from α-toxin permeabilized chromaffin cells in primary culture. *J. Biol. Chem.* 261:5777-5783.

10. Brooks, J. C., and Carmichael, S. W. (1983). Scanning electron microscopy of chemically skinned bovine adrenal medullary chromaffin cells. *Mikroskopie* 40:347-356.

10a. Grant, N. J., Aunis, D., and Bader, M. F. (1987). Morphology and secretory activity of digitonin- and alpha-toxin-permeabilized chromaffin cells. *Neuroscience* 23:1143-1155.

10b. Sarafian, T., Aunis, D., and Bader, M-F. (1987). Loss of proteins from digitonin-permeabilized adrenal chromaffin cells essential for exocytosis. *J. Biol. Chem.* 262:16671-16676.

11. Masson, D., and Tschopp, J. (1985). Isolation of a lytic, pore-forming protein (Perforin) from cytolytic T-lymphocytes. *J. Biol. Chem.* 260:9069-9072.

12. Füssle, R., Bhakdi, S., Sziegoleit, A., Tranum-Jensen, J., Kranz, T., and Wellensiek, H. J. (1981). On the mechanism of membrane damage by *Staphylococcus aureus* α-toxin. *J. Cell Biol.* 91:83-94.

13. Bhakdi, S., and Tranum-Jensen, J. (1984). Mechanism of complement cytolysis and the concept of channel forming proteins. *Phil. Trans. R. Soc. Lond. B* 306:311-324.

14. Tranum-Jensen, J., and Bhakdi, S. (1981). Freeze-fracture analysis of the membrane lesion of human complement. *J. Cell Biol.* 97:618-626.

14a. Bhakdi, S., and Tranum-Jensen, J. (1987). Damage to mammalian cells by proteins that form transmembrane pores. *Rev. Physiol. Biochem. Pharmacol.* 107:147-223.

15. Ahnert-Hilger, G., Bhakdi, S., and Gratzl, M. (1985). α-toxin permeabilized rat pheochromocytoma cells: A new approach to investigate stimulus-secretion coupling. *Neurosci. Lett.* 58:107-110.

16. Ahnert-Hilger, G., Bhakdi, S., and Gratzl, M. (1985). Minimal requirements for exocytosis: A study using PC12 cells permeabilized with staphylococcal α-toxin. *J. Biol. Chem.* 260:12730-12734.

16a. Lind, I., Ahnert-Hilger, G., Fuchs, G., and Gratzl, M. (1987). Purification of alpha-toxin from *Staphylococcus aureus* and application to cell permeabilization. *Anal. Biochem.* 164:84-89.

16b. Ahnert-Hilger, G., and Gratzl, M. (1987). Further characterization of dopamine release by permeabilized PC12 cells. *J. Neurochem.* 49:764-770.

16c. Ahnert-Hilger, G., Bräutigam, M., and Gratzl, M. (1987). Ca^{2+}-stimulated catecholamine release from alpha-toxin permeabilized PC12 cells: Biochemical evidence for exocytosis and its modulation by protein kinase C and G-proteins. *Biochemistry* 26:7842-7848.

16d. Ahnert-Hilger, G., Bader, M-F., Bhakdi, S., and Gratzl, M. (1989). Introduction of macromolecules into chromaffin cells by permeabilization with streptolysin O: Inhibitory effect of tetanus toxin on catecholamine secretion. *J. Neurochem.* 52:1751–1758.

16e. Ahnert-Hilger, G., Weller, U., Dauzenroth, M. E., Habermann, E., and Gratzl, M. (1989). The tetanus toxin light chain inhibits exocytosis. *FEBS Lett.* 242:245–248.

16f. Stecher, B., Weller, U., Habermann, E., Gratzl, M., and Ahnert-Hilger, G. (1989). The light chain but not the heavy chain of botulinum A toxin inhibits exocytosis from permeabilized adrenal chromaffin cells. *FEBS Lett.* 255:391–394.

16g. Föhr, K. J., Scott, J., Ahnert-Hilger, G., and Gratzl, M. (1989). Inhibition of inositol trisphosphate induced calcium release from permeabilized endocrine cells by decavanadate. *Biochem. J.* 262:83–89.

17. Gomperts, B. D., Baldwin, J. M., and Micklem, K. J. (1983). Rat mast cells permeabilized with sendai virus secrete histamine in response to Ca^{2+} buffered in the micromolar range. *Biochem. J.* 210:737–745.

18. Douglas, W. W., and Rubin, R. P. (1961). The role of calcium in the secretory response of the adrenal medulla to acetylcholine. *J. Physiol. (London)* 159:40–57.

19. Kanno, T., Cochrane, D. E., and Douglas, W. W. (1973). Exocytosis (secretory granule extrusion) induced by injection of calcium into mast cells. *Can. J. Physiol. Pharmacol.* 51:1001–1004.

20. Miledi, R. (1973). Transmitter release induced by injection of calcium ions into nerve terminals. *Proc. R. Soc. Ser. B* 183:421–425.

20a. Habermann, E., and Dreyer, F. (1986). Clostridial neurotoxins: Handling and action at the cellular and molecular level. *Curr. Top. Microbiol. Immunol.* 129:93–179.

21. Borle, A. B. (1981). Control, modulations and regulation of cell calcium. *Rev. Physiol. Biochem. Pharmacol.* 90:13–153.

22. Rubin, R. P. (1982). *Calcium and Cellular Secretion*, 2nd ed. Plenum Press, New York, London.

23. Rasmussen, H., and Barrett, P. Q. (1984). Calcium messenger system: An integrated view. *Physiol. Rev.* 64:938–984.

24. Kirshner, N., Sage, H. I., Smith, W. J., and Kirshner, A. G. (1966). Release of catecholamines and specific protein from adrenal glands. *Science* 154:529–531.

25. Viveros, O. H., Arqueros, L., Connet, R. J., and Kirshner, N. (1968). Release of catecholamines and dopamine β-oxidase from the adrenal medulla. *Life Sci.* 7:609–618.

26. Schneider, F. H., Smith, A. D., and Winkler, H. (1967). Secretion from the adrenal medulla: Biochemical evidence for exocytosis. *Br. J. Pharmacol. Chemother.* 31:94–104.

27. Knight, D. E., and Kesteven, N. T. (1983). Evoked transient intracellular free Ca^{2+} changes and secretion in isolated adrenal medullary cells. *Proc. R. Soc. Lond. B* 218:177–199.

28. Meldolesi, J., Huttner, W. B., Tsien, R. Y., and Pozzan, T. (1984). Free cytoplasmic Ca^{2+} and neurotransmitter release: Studies on PC12 cells and synaptosomes exposed to α-latrotoxin. *Proc. Natl. Acad. Sci. USA* 81:620–624.

29. Rubin, R. P. (1969). The metabolic requirements for catecholamine release from the adrenal medulla. *J. Physiol. (London)* 202:197–209.

30. Kirshner, N., and Smith, W. J. (1969). Metabolic requirements for secretion from the adrenal medulla. *Life Sci.* 8:799–803.

31. Reynolds, E. E., Melega, W. P., and Howard, B. D. (1982). Adenosine 5′-triphosphate independent secretion from PC12 phaeochromocytoma cells. *Biochemistry* 21:4795–4799.

32. Watanabe, D., Torda, M., and Meldolesi, J. (1983). The effect of α-latrotoxin on the neurosecretory PC12 cell line: Electron microscopy and cytotoxicity studies. *Neuroscience* 10:1011-1024.

33. Bader, M. F., Ciesielski-Treska, J., Thiersé, D., Hesketh, J. E., and Aunis, D. (1981). Immunocytochemical study of microtubules in chromaffin cells in culture and evidence that tubulin is not an integral protein of the chromaffin granule membrane. *J. Neurochem.* 37:917-933.

34. Allen, R. D., Weiss, D. G., Hayden, J. H., Brown, D. T., Fujiwake, H., and Simpson, M. (1985). Gliding movement of and bi-directional transport along microtubules from squid axoplasm: Evidence for an active role of microtubules in cytoplasmic transport. *J. Cell Biol.* 100:1736-1752.

35. Vale, R. D., Schnapp, B. I., Reese, T. S., and Sheetz, N. P. (1985). Organelle, bead, and microtubule translocations promoted by soluble factors from the squid giant axon. *Cell* 40:559-569.

36. Schulman, H., and Greengard, P. (1978). Ca^{2+}-dependent protein phosphorylation system in membranes from various tissues, and its activation by "calcium-dependent-regulator." *Proc. Natl. Acad. Sci. USA* 75:5432-5436.

37. Amy, C. M., and Kirshner, N. (1981). Phosphorylation of adrenal medulla cell proteins in conjunction with stimulation of catecholamine secretion. *J. Neurochem.* 36:847-854.

38. Treiman, M., Weber, W., and Gratzl, M. (1983). 3',5'-Cyclic adenosine monophosphat- and Ca^{2+}-calmodulin-dependent endogenous protein phorphorylation activity in membranes of the bovine secretory vesicles: Identification of two phosphorylated compounds as tyrosin hydroxylase and protein kinase regulatory subunit type II. *J. Neurochem.* 40:661-669.

39. Niggli, V., Knight, D. E., Baker, P. F., Vigny, A., and Henry, J-P. (1984). Tyrosine hydroxylase in "leaky" adrenal medullary cells: Evidence for in situ phosphorylation by separate Ca^{2+} and cyclic AMP-dependent systems. *J. Neurochem.* 43:646-658.

40. Knight, D. E., and Baker, P. F. (1985). The chromaffin granule proton pump and calcium dependent exocytosis in bovine adrenal medullary cells. *J. Membrane Biol.* 83:147-156.

41. Knight, D. E., and Baker, P. F. (1983). The phorbol ester TPA increases the affinity of exocytosis for calcium in "leaky" adrenal medullary cells. *FEBS Lett.* 160:98-100.

42. Brocklehurst, K. W., and Pollard, H. B. (1985). Enhancement of Ca^{2+}-induced catecholamine release by the phorbol ester TPA in digitonin-permeabilized cultured bovine adrenal chromaffin cells. *FEBS Lett.* 183:107-110.

43. Knight, D. E., and Baker, P. F. (1985). Guanine nucleotides and Ca^{2+}-dependent exocytosis. Studies on two adrenal cell preparations. *FEBS Lett.* 189:345-349.

43a. Peppers, S. C., and Holz, R. W. (1986). Catecholamine secretion from digitonin-treated PC12 cells. Effects of Ca^{2+}, ATP, and protein kinase C activators. *J. Biol. Chem.* 261:14665-14670.

43b. Bittner, M. A., Holz, R. W., and Neubig, R. R. (1986). Guanine nucleotide effects on catecholamine secretion from digitonin-permeabilized adrenal chromaffin cells. *J. Biol. Chem.* 261:10182-10188.

43c. Toutant, M., Aunis, D., Bochaert, J., Homburger, V., and Rouot, B. (1987). Presence of three pertussis toxin substrates and $G_0\alpha$ immunoreactivity in both plasma and granule membranes of chromaffin cells. *FEBS Lett.* 215:339-344.

43d. Burgoyne, R. D., and Morgan, A. (1989). Low molecular mass GTP-binding proteins of adrenal chromaffin cells are present on the secretory granule. *FEBS Lett.* 245:122–126.

44. Kenigsberg, R. L., Coté, A., and Trifaró, J. M. (1982). Trifluoperazine, a calmodulin inhibitor, blocks secretion in cultured chromaffin cells at a step distal from calcium entry. *Neuroscience* 7:2277–2286.

45. Brooks, J. C., and Treml, S. (1983). Effect of trifluoperazine on catecholamine secretion by isolated bovine adrenal medullary chromaffin cells. *Biochem. Pharmacol.* 32:371–373.

46. Kenigsberg, R. L., and Trifaro, J. M. (1984). Microinjection of calmodulin antibodies into cultured chromaffin cells blocks catecholamine release in response to stimulation. *Neuroscience* 14:335–347.

47. Brooks, J. C., and Treml, S. (1984). Effect of trifluoperazine and calmodulin on catecholamine secretion by saponin-skinned cultured chromaffin cells. *Life Sci.* 34:669–674.

48. Pace, C. P., Tarvin, J. T., Neighbors, A. S., Pirkle, J. A., and Greider, M. H. (1980). Use of a high voltage technique to determine the molecular requirements for exocytosis in islet cells. *Diabetes* 29: 911–918.

49. Yaseen, M. A., Pedley, K. C., and Howell, S. L. (1982). Regulation of insulin secretion from islets of Langerhans rendered permeable by electric discharge. *Biochem. J.* 206:81–87.

50. Stutchfield, J., and Howell, S. L. (1984). The effect of phalloidin on insulin secretion from islets of Langerhans isolated from rat pancreas. *FEBS Lett.* 175:393–396.

51. Jones, P. M., Stutchfield, J., and Howell, S. L. (1985). Effects of Ca^{2+} and a phorbolester on insulin secretion from islets of Langerhans permeabilized by high-voltage discharge. *FEBS Lett.* 191:102–106.

52. Hedeskov, C. J. (1980). Mechanism of glucose-induced insulin secretion. *Physiol. Rev.* 60:442–509.

53. Tamagawa, T., Niki, H., and Niki, A. (1985). Insulin release independent of a rise in cytosolic free Ca^{2+} by forskolin and phorbolester. *FEBS Lett.* 183:430–432.

54. Colca, J. R., Wolf, B. A., Commens, P. G., and McDaniel, M. L. (1985). Protein phosphorylation in permeabilized pancreatic islet cells. *Biochem. J.* 288:529–536.

55. Knight, D. E., and Koh, E. (1984). Ca^{2+} and cyclic nucleotide dependence of amylase release from isolated rat pancreatic acinar cells rendered permeable by intense electric fields. *Cell Calcium* 5:401–418.

56. Nishizuka, Y. (1983). Calcium phospholipid turnover and transmitter signalling. *Phil. Trans. R. Soc. London B* 302:101–112.

57. Knight, D. E., Hallam, T. J., and Scrutton, M. C. (1982). Agonist selectivity and second messenger concentration in Ca^{2+}-mediated secretion. *Nature* 296:256–257.

58. Knight, D. E., and Scrutton, M. C. (1980). Direct evidence for a role for Ca^{2+} in amine storage granule secretion by human platelets. *Thromb. Res.* 20:437–446.

59. Knight, D. E., and Scrutton, M. C. (1984). Cyclic nucleotides control a system which regulates Ca^{2+} sensitivity of platelet secretion. *Nature* 309:66–68.

59a. Howell, T. W., Cockcroft, S., and Gomperts, B. D. (1987). Essential synergy between Ca^{2+} and guanine nucleotides in exocytotic secretion from permeabilized rat mast cells. *J. Cell Biol.* 105:191–197.

60. Knight, D. E., Niggli, V., and Scrutton, M. C. (1984). Thrombin and activators of protein kinase C modulate secretory responses of permeabilized human platelets induced by Ca^{2+}. *Eur. J. Biochem.* 143: 437–446.

61. Haslam, R. J., and Davidson, M. M. L. (1984). Potentiation by thrombin of the secretion of serotonin from permeabilized platelets equilibrated with Ca^{2+} buffers. *Biochem. J.* 222:351–361.

62. Ridgway, E. B., Gilkey, J. C., and Jaffe, L. F. (1977). Free calcium increases explosively in activating medaka eggs. *Proc. Natl. Acad. Sci. USA* 74:623–627.

63. Steinhardt, R., Zucker, R., and Schatten, G. (1977). Intracellular calcium release at fertilization in the sea urchin egg. *Dev. Biol.* 58: 185–196.

64. Steinhardt, R. A., and Epel, D. (1974). Activation of sea urchin eggs by a calcium ionophore. *Proc. Natl. Acad. Sci. USA* 71:1915–1919.

65. Hollinger, T. G., and Schuetz, A. W. (1976). "Cleavage" and cortical granule breakdown in *Rana pipiens* oocytes induced by direct micro-injection of calcium. *J. Cell Biol.* 71:395–401.

66. Schuel, H., Wilson, W. L., Chen, K., and Lorand, L. (1973). A trypsin-like protease localized in cortical granules isolated from unfertilized sea urchin eggs by zonal centrifugation. Role of enzyme in fertilization. *Dev. Biol.* 34:175–186.

67. Vacquier, V. D., Epel, D., and Douglas, L. A. (1972). Sea urchin eggs release protease activity at fertilization. *Nature* 237:34–36.

68. Vacquier, V. D. (1976). Isolated cortical granules: A model system for studying membrane fusion and calcium-mediated exocytosis. *J. Supramolec. Struct.* 5:27–35.

69. Vacquier, V. D. (1975). The isolation of intact cortical granules from sea urchin eggs: Calcium ions trigger granule discharge. *Dev. Biol.* 43:62–74.

70. Baker, P. F., and Whitaker, M. J. (1978). Influence of ATP and calcium on the cortical reaction in sea urchin eggs. *Nature* 276:513–515.

71. Crabb, J. H., and Jackson, R. C. (1985). In vitro reconstitution of exocytosis from plasma membrane and isolated secretory vesicles. *J. Cell Biol.* 101:2263–2273.

72. Whitaker, M. J., and Baker, P. F. (1984). Calcium dependent exocytosis in an in vitro secretory granule plasma membrane preparation from sea urchin eggs and the effects of some inhibitors of cytoskeletal function. *Proc. R. Soc. Lond. B* 218:397–413.

73. Steinhardt, R. A., and Alderton, J. M. (1982). Calmodulin confers calcium sensitivity on secretory exocytosis. *Nature* 295:154–155.

74. Gratzl, M., Dahl, G., Russell, J. T., and Thorn, N. A. (1977). Fusion of neurohypophyseal membranes in vitro. *Biochim. Biophys. Acta* 470: 45–57.

75. Ekerdt, R., Dahl, G., and Gratzl, M. (1981). Membrane fusion of secretory vesicles and liposomes: Two different types of fusion. *Biochim. Biophys. Acta* 646:10–22.

76. Gratzl, M., Schudt, C., Ekerdt, R., and Dahl, G. (1980). Fusion of isolated biological membranes. A tool to investigate basic processes of exocytosis and cell-cell fusion. In *Membrane Structure and Function* (E. E. Bittar, ed.). Wiley, New York, pp. 59–92.

25

Ultrastructural Aspects of Exocytosis

HELMUT PLATTNER

University of Konstanz, Konstanz, Federal Republic of Germany

I. INTRODUCTION

Evidence is accumulating that the organization of membranes at sites of exocytotic membrane fusion might be principally similar in quite different systems. This includes the presence of both membrane-integrated proteins [assumed to correspond mainly to the membrane-intercalated particles (MIP) seen after freeze-fracturing] and membrane-associated proteins. Among others, the presence of Ca^{2+}-dependent enzymes [(phospho-) protein phosphatase/kinase] or of Ca^{2+}-sensitivity conferring constituents (including calmodulin) must be taken into consideration. Some of these elements might contribute to "membrane-connecting materials," while the molecular equivalent of MIP is not yet known. Altogether a probable regulatory function of these proteins for exocytotic membrane fusion is quite likely. This is supported by the fact, that, at least in *Paramecium* cells, the absence of these structural elements from fusogenic sites in different mutations or in their phenocopies always entails the incapability to perform exocytotic membrane fusion. However, the occurrence of highly regular MIP aggregates at fusion sites, as they occur in *Paramecium*, is the exception, rather than the rule. Thus, it remains to be seen whether the existence of pre-formed exocytotic fusion sites is a general phenomenon. In all cases analyzed with reliable ultrastructural methods, exocytotic membrane fusion operates without previous restructuring of the membranes involved; it is always a focal event (with an "instability focus" of a diameter in the 10-nm range) and it operates without the formation of a lipidic diaphragm. Wherever preparative artifacts had been reliably excluded, no exceptions from these rules have been observed. This holds even for widely different systems, such as protozoa, nerve terminals, oocytes, mast cells, adrenal chromaffin cells, and so forth. For further structural analyses it will be challenging to identify in situ the different structural components occurring at sites involved in exocytosis. References for all these aspects will be given throughout the text.

II. WHAT ARE THE PROBLEMS?

Understanding the structural mechanisms that allow different cells to secrete widely different secretory products by exocytosis depends largely on the actual state of the art of specimen preparation. Therefore, when one compares, for example, the first (1) and the last (2) published micrographs of exocytotic events in the adrenal medulla, the striking difference is not only a matter of aesthetics, but also of information content.

Evidently fast-freezing techniques (cryofixation), estimated to "fix" dynamic processes with a time resolution in the msec range (3) are in many ways superior to chemical fixations. Cryofixation has obtained particular interest in this field of research, since in combination with freeze fracturing, it also allows for an analysis of ultrastructural details within membranes. These include MIP that are considered mostly as (hetero-)oligomeric membrane-integrated proteins (for review, see Ref. 4) and occasionally also as inverted lipid micelles (5,6). As we shall see, both these structures might be of some interest for an understanding of membrane fusion.

One should also keep in mind the mechanisms underlying chemical fixation techniques. In my opinion (7), the destruction of membrane-associated proteins (that are now assumed to occur on exocytosis sites; see below) by osmiumtetroxide fixatives led some time ago to the assumption that they would be cleared away from sites of undergoing exocytosis. Many investigators had assumed that MIP, too, have to "leave the scene" to allow exocytotic membrane fusion to go on. This postulate has also been assumed to represent most likely an artifact (7-10).

Another problem is the fact that exocytosis is a spatially and temporally restricted event and, thus, difficult to catch. If "caught" it might be difficult to differentiate between the actual occurrence of exocytosis and other phenomena accompanying this event. Most secretory systems keep reacting to a stimulus over a long time period and at many sites diffusely scattered over their surface. Only a few systems react synchronously at defined sites (see Refs. 7,11). In order to obtain a picture of general validity, I shall compare such widely different systems, but concentrate on those that appear most favorable for a thorough analysis of ultrastructural aspects. These are systems with regularly structured preformed exocytosis sites, such as the *Paramecium* cell (11,12). But can a ciliated protozoon teach us a lesson on membrane fusion in higher eukaryotes? As we shall see, step by step very similar features are revealed in widely different systems, at least in principle. I shall try to indicate to what extent such generalizations can be made at this time.

The molecular events that lead to the fusion of two membranes were—and to a certain extent still are—poorly understood with regard to biochemical and biophysical aspects. This provoked the development of a bewildering variety of hypotheses on membrane fusion, as reviewed before (7). Even now it remains questionable whether we are dealing with the overlap or cooperation of possibly widely different factors. Ever since the phenomenon of Ca^{2+}-mediated stimulus-secretion coupling was detected (see 13,14), it was consistently attempted to pinpoint the function of calcium for membrane fusion. Even at present, the increase of the intracellular free Ca^{2+} concentration is the only common denominator for ideas about exocytosis (14), though this also has been challenged in the years since submission of this chapter. Let us, therefore, have a closer look at such Ca^{2+}-mediated phenomena and their possible ultrastructural consequences.

III. POSSIBLE ROLES OF CA²⁺ AND THEIR ULTRASTRUCTURAL IMPLICATIONS

Table 1 lists a variety of effects of Ca^{2+} with possible relevance for exocytotic membrane fusion. This includes enzymatic and nonenzymatic effects, partly on lipids and partly on proteins. (Only occasional references are given in the text, since many of these aspects are covered in other chapters of this book.)

Such aspects comprise different categories. With regard to nonenzymatic effects, it is difficult to judge any ultrastructural consequences of local charge screening (aspect 1) and of Ca^{2+} mobilization by formation of phosphatidylinositol breakdown products (aspect 6a). Aspect 2, critically discussed by Papahadjopoulos' group (15) on the basis of their studies with synthetic model systems, would entail the elimination of protein MIP from fusion sites. A plethora of material has been presented before in the literature to show that this model can be used to explain exocytotic membrane fusion. As reviewed some time ago (7) this became untenable when freeze-fracture analyses had made use of fast cryofixation techniques [nerve terminals (8); mast cells (10); oocytes (9); *Limulus* amebocytes (16); adrenal chromaffin cells (2); *Paramecium* cells (17)]. The formation of lipidic particles (aspect 3), as propagated by Verkleij's group (5,6), is an interesting possibility, though less defined transitional fusion stages also appear possible (15,18). Evidence is accumulating that distinct membrane-associated proteins have to be enriched at a certain site for fusion to occur (aspect 4). Synexin had been identified first (19), but recently calmodulin and calmodulin-activated as well as other Ca^{2+}-binding proteins are a matter of discussion as well (see below). Current concepts assuming the presence of proteins at fusogenic sites include also aspects 6b and 9-12 in Table 1. In my opinion, three hypotheses can now be dismissed (5, 7, and 8 in Table 1). Aspect 5 would entail gross cell damage. With regard to aspect 7, which was proposed by Puskin and Kochwa (20), there are now thoroughly analyzed examples showing that the arrangement of microfilaments to be postulated is just not realized in these cells [*Paramecium* (21)] and that fusion occurs even after a total block of contractile forces [*Paramecium* (22,23); oocytes (24)]. All these papers also present evidence that microtubules cannot be directly involved in exocytotic membrane fusion either.

The important conclusion from Table 1 is that the presence of proteins, integral and membrane-associated, can be expected at fusion sites (except for cytoskeletal elements of the microfilament and microtubule type). A detailed account of this postulate will be presented throughout this article.

IV. TIME COURSE OF EXOCYTOSIS IN DIFFERENT SYSTEMS

Most secretory systems can be stimulated to release secretory materials for many minutes (e.g., different endocrine and exocrine gland cells, see Refs. 14,25). For a typical time course in response to different trigger agents I refer to activation curves presented in Refs. 14 and 25 or for adrenal chromaffin cells by Schneider and his co-workers (26). This type of cell might be called "slow systems," when one refers to the overall reaction (however, see below for individual events).

A few other cell types release their contents within the range of 1 min or a fraction thereof. All these cells do not display any long-range intracellular organelle movements. Thrombocytes display tubular extensions

TABLE 1 Fusogenic Effects Proposed for Ca^{2+} in the Literature and Ultrastructural Aspects to be Expected as Possible Consequences

Functional aspects	Ultrastructural consequences expected
Nonenzymatic effects	
Effects on lipids	
1. Charge screening (phosphatidylserin)	Unknown
2. Liquid-solid state phase transition and "cis-complex" formation	Elimination of MIP (proteins) from fusion zone
3. Phase transition involving inverted micelle formation (hexagonal phase HII)	Formation of a lipidic MIP precisely at fusion site
Effects involving proteins	
4. "Activation" of Ca^{2+} binding proteins (synexin, etc.)	Some proteins would have to be present at the fusion zone
Enzymatic effects	
Effects on lipids	
5. Formation of lysophospholipids	See 3; hypothesis abandoned particularly because membrane leakage would result
6. Phosphatidylinositolturnover	
(a) Ca^{2+} mobilization (phosphatidic acid)	Unknown
(b) Protein kinase C activation (diacylglycerol)	See 4
Effects involving proteins	
7. Actomyosin activation—tensile forces } Intense contact formation	Actomyosin elements would have to be arranged vertically to cell membrane or around fusing vesicles
8. Limited microtubule depolymerization }	Microtubules would have to lead to fusion sites and be present during fusion
9. Ca^{2+}-ATPase—removal of stabilizing ATP	See 4
10. Ca^{2+}-protease (or metalloprotease) activation	See 4
11. Protein kinase activation	See 4
12. Phosphoprotein dephosphorylation	See 4 (occurrence of proteins at fusion site)

of their surface membrane into which secretory granules release their contents (27,28). In oocytes (or egg cells), cortical granules are so closely apposed to the cell surface that they just have to fuse with the cell membrane (29,30). Mast cells release their contents by "compound exocytosis," i.e., secretory granules that have already fused with the cell membrane allow for the fusion of further granules deeper inside the cell (see, for example, Ref. 31). It appears that a quick reaction by exocytosis is required for these systems to be successful in starting blood coagulation, in avoiding polyspermy, or in inducing an inflammatory response.

Let us turn now to the "fastest systems." In response to appropriate triggers, the ciliated protozoon *Paramecium tetraurelia* can discharge the majority (~1000) of its extrusive organelles ("trichocysts") within ~1 sec (32,33). One can, thus, easily calculate that *Paramecium* performs exocytosis 10^4 times more synchronously than an average gland cell, e.g., the chromaffin cell. Table 2 summarizes events occurring during this 1 sec and afterward, when the store of trichocysts is reestablished. The essential point is that a *Paramecium* cell has ~95% of its trichocysts permanently attached at the cell membrane. These can undergo exocytosis immediately when the cell senses an appropriate trigger. We found polyamino compounds, particularly aminoethyldextran (AED) with 1 aminogroup per 1 kD dextran ($M_r = 40,000$), to be the most appropriate trigger agents (32,33). My

TABLE 2 Time Periods Required for Individual Steps During and After Exocytosis in *Paramecium* Cells

	Individual event	All events in one cell
Membrane fusion	1 msec[a]	< 1 s
Exocytotic opening,	0.05 sec[b]	< 1 s[d]
extrusion of contents	1 msec[c]	< 1 s[d]
Membrane resealing	?	< 1 s[d]
65-kD protein dephosphorylation		< 1 s[e]
65-kD protein rephosphorylation		~20 s[e]
Ghost removal	~1 min[f]	~10 min[f]
Plug formation	min[g]	9 h[g]
Docking of trichocysts	1 min[h]	9 h[i]
Adaptation stage of docking site (rosette assembly)	5 min[i]	9 h[i]

(Membrane fusion through Membrane resealing rows bracketed as: 1 s[b])

[a]See Section VIII.
[b]Data from Deitmer and Plattner (unpublished observations).
[c]Data from Ref. 37.
[d]Inferred from [b].
[e]Data from Ref. 38.
[f]Data from Refs. 22,39,40.
[g]Data from Refs. 39,40.
[h]Data from Ref. 41.
[i]Data from Ref. 40.

group exploits this quick response to analyze aspects that largely depend on synchrony.

One should expect nerve cells to be the fastest exocytosis systems. Yet they release only a small portion of their secretory contents upon one electric stimulation. Individual exocytotic events might be fast also in other systems (see below), yet the reaction appears sluggish because secretory activity is not synchronous. These aspects are highly relevant for an ultrastructural analysis of exocytotic membrane fusion.

A reliable judgment of the actual time required can be achieved almost only ultrastructurally by fast-freezing techniques (to be discussed in detail later) or by electrophysiological measurements. "Usual" gland cells display a lag-period of 1 sec before they react to a trigger (34). Surface capacitance changes indicating the temporal insertion and retrieval of membrane packages of individual secretory vesicles revealed rapid fluctuations in chromaffin cells (35), though on/off phenomena might originate from different individual organelles. As we shall see, this makes it difficult to analyze ultrastructural events during exocytosis performance in such "slow" cells. In *Paramecium* capacitance changes last 50 msec for individual events and all together ≤ 1 sec for synchronous AED-mediated exocytosis (Deitmer and Plattner, unpublished observations). These values are believed to reflect the time period required for an exocytotic event or for all events in one cell (synchrony), respectively. Therefore, ultrastructural changes could easily be followed in these systems (17,22). In nerve terminals an exocytotic opening stays open only for a msec period (8,36); in this case electrical triggering could also be combined with fast freezing and ultrastructural follow-up procedures like freeze fracturing and freeze substitution (8,36).

The occurrence of relatively synchronous exocytosis in (a) mast cells and (b) oocytes has also allowed for an unequivocal ultrastructural analysis of membrane changes in the course of secretory activity [mast cells (10); oocytes (9)].

In conclusion, exocytotic membrane fusion probably is a very rapid process in all the different systems, but the degree of synchrony certainly is widely different. The consequence would be that even for "slow systems" one would require fast-freezing methods to catch membrane fusion. If such phenomena can be recognized after chemical fixation, they are probably induced by the fixative action (Ref. 7 for this aspect). As discussed later, a direct correlation of ultrastructural changes with exocytosis performance can be achieved only with synchronous systems, whereas statistical evaluations have to be applied for "slow systems" such as chromaffin cells. Some other aspects pertinent to this chapter are discussed in Section VIII.

V. PREFORMED EXOCYTOSIS SITES AND THEIR ORGANIZATION

Couteaux [42] was the first to recognize the occurrence of preformed exocytosis sites in the form of the "*zone actives*" in motor endplates of the frog. These sites display an amorphous to fibrillar mass of electron-dense materials closely apposed to the cell membrane. Freeze-fracture studies (8,36) revealed double rows of MIP [particles, speculatively interpreted as Ca^{2+}-channels (43)]. Exocytotic fusion sites, thus, display ultrastructural elements that can be tentatively interpreted as membrane-associated and -integrated proteins. In addition, in nerve terminals, microtubules emanate frequently from sites of potential vesicle discharge (44) and, thus, could possibly contribute to establish defined exocytosis sites.

By now, the occurrence of preformed exocytosis sites has been most thoroughly analyzed in *Tetrahymena* (45) and *Paramecium* cells (11,12,46-48). Particularly the *Paramecium* system allowed for a clear dissociation of individual events (that lead to exocytosis) in space as well as in time (Tables 2 and 3). In most other systems such clear distinctions would not be possible. Figure 1 presents the freeze-fracture appearance of a "trichocyst" docking site on the cell membrane of a *Paramecium* cell. Within the cell membrane the fusogenic zone is surrounded by a double "ring" of MIP. The "ring" encircles a "rosette" of ~9 (on average) slightly larger MIP (Fig. 1a). The actual trichocyst tip, frequently bulging out in a nipple form (46), is devoid of MIP. Three MIP rings surround a slightly lower portion of the trichocyst ("annulus"; Fig. 1b,c). The upper trichocyst membrane is flanked also by a "collar" (Figs. 1, 2), including an amorphous "collar matrix" and "collar fibers" [a parallel dense packing is seen that gives this membrane portion a crenated appearance on freeze-fracture replicas (Fig. 10)]. Of particular interest is the occurrence of an amorphous to fibrillar electron dense mass of "connecting material" between the trichocyst tip and the cell membrane (Fig. 2) (21,49). All these structures are sensitive to proteolytic enzymes (50,51).

What can be said about the molecular constituents of a trichocyst docking site? There is no experimental evidence that "rosette" MIP represent Ca^{2+}-channels (52,53). We have recently found that exocytotic membrane fusion is inhibited by ATP (54). Fusion also involves the rapid (≤ 1 sec) dephosphorylation of a phosphoprotein of an $M_r = 65$ kD (38,55). From this one can conclude that exocytotic sites must be endowed with an enzymatic machinery that might form part of the functional equivalent of the ultrastructural elements just described. We could also ascertain the presence of calmodulin at potential exocytosis sites and its probable participation in exocytosis induction [because of antibody inhibition (56)]. Furthermore, in agreement with the effect of ATP just mentioned, we have obtained evidence for the occurrence of a Ca^{2+}-dependent ATPase and/or phosphatase activity; this activity had been localized some time ago by cytochemical methods precisely at sites of trichocyst release (57). All these data suggest that preformed exocytosis sites in *Paramecium* dispose of efficiently arranged, Ca^{2+}-dependent proteins.

Additional evidence along these lines comes from work with mutant *Paramecium* strains, carried out by the groups of Beisson and Satir and by my own group (33,38,46-48,52,55,58). Some of these mutants do not synthesize trichocysts (strain "trichless," *tl*); others do not convey trichocysts to the cell membrane (*tam*); still others are not capable of performing exocytotic membrane fusion ("nondischarge" mutations, *nd*).

In *nd* strains, isolated and characterized by Beisson and her coworkers, potential exocytosis sites are characterized by the occurrence of a "ring" without a "rosette" (46-48) and without "connecting materials" (48,58). Since the combined cytochemical Ca^{2+}-dependent phosphatase activity and the phosphoprotein dephosphorylation step (that occurs in response to AED-mediated synchronous exocytosis in normally secreting strains) do not occur in fusion inhibited *nd* strains (38,58), this further specifies some possible functional aspects of the ultrastructural details observed in normally secreting *Paramecium* strains.

Is it possible that microtubules or microfilaments would specify predetermined exocytosis sites? Microtubules are not attached at preformed

TABLE 3 Summary of the Data on the Organization of Trichocyst Docking Sites in Relation to Exocytosis Capacity in Wild-Type and Mutant Strains

Site	Differentiation	Strains							
		wt	tl	tam8	tam38	nd6	nd7	nd9-28	nd9-18
Unoccupied sites	Parenthesis	+	+	+	+	+	+	+	+
	Plug	+	+	+	+	+	+	+	+
Occupied sites	Ring	+	/	/	/	+	+	+	+
	Rosette	+	/	/	/	–	–	–	+
	Connecting material collar (fibers + matrix)	+	/	/	/	+	+	+	+
	Exocytosis capacity	+	–	–	–	–	–	–	+
	65-kD phosphoprotein dephosphorylation capacity	+	–	Not done	–	–	–	–	+

+, Present; –, absent; /, not relevant for a given strain.
Source: Data mainly from Pouphile et al. (48); data on phosphoproteins are from Zieseniss and Plattner (38).

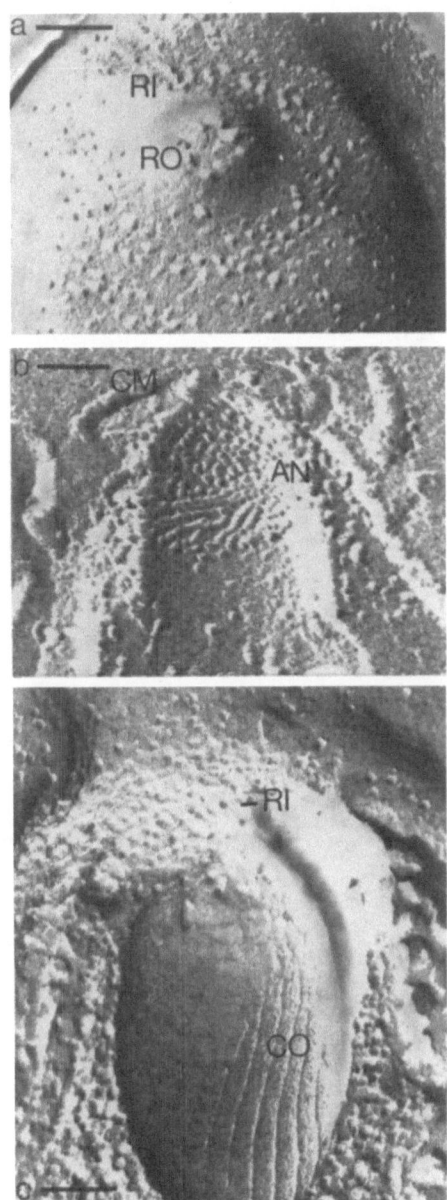

FIGURE 1 Freeze-fracture of a trichocyst docking site in *Paramecium tetraurelia*. (a) RI = ring, RO = rosette contained in the cell membrane (CM). (b, c) Trichocyst tip with AN = annulus on an outer membrane leaflet (b) and CO = collar on an inner membrane leaflet (i.e., fracture half, c) of a trichocyst membrane. For further explanations see text. Unpublished micrographs. Bars = 0.1 μm.

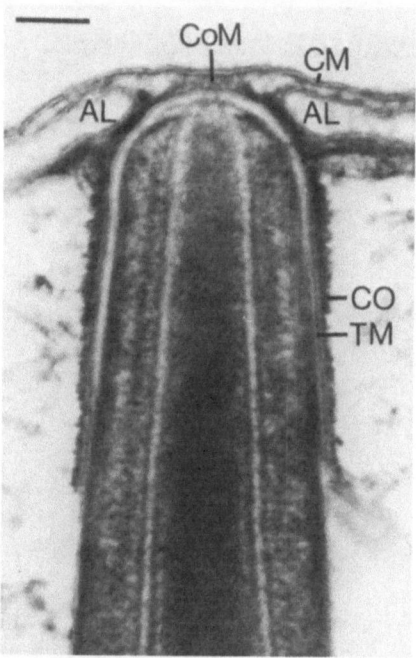

FIGURE 2 Ultrathin median section through a trichocyst docking site. CM = cell membrane with underlying alveolar cavities (AL). TM = trichocyst membrane with apposed CO = collar. CoM = connecting material between CM and TM. Tannic acid was present in the fixative to show CoM. Unpublished micrograph. Bar = 0.1 μm.

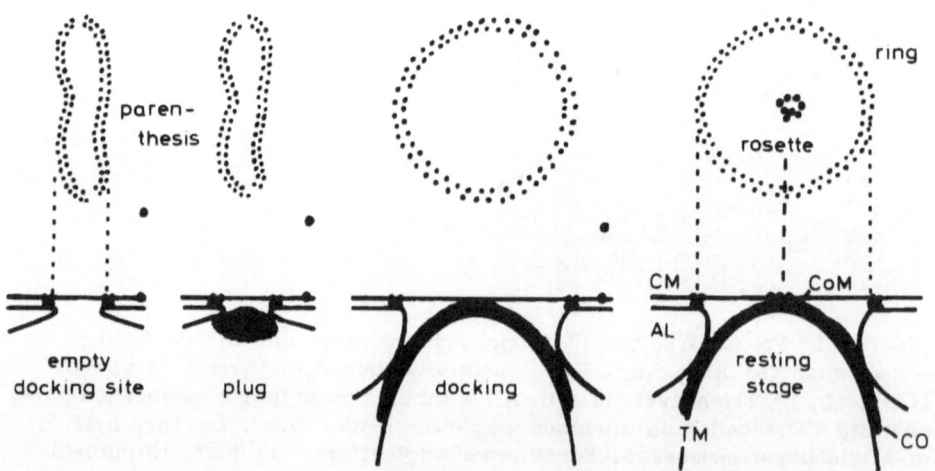

FIGURE 3 Biogenesis of a preformed exocytosis site in *Paramecium tetraurelia*. For symbols, see Figures 1 and 2; for explanations see text. [Reproduced from Pape and Plattner (40), with permission of the publisher.]

exocytosis sites (21,22). F-actin bundles, though present in the cortex of *Paramecium* cells according to immuno- and affinity labeling (23,59), surround trichocyst tips only at a certain distance below the fusogenic zone. Since they are oriented in a direction parallel to the cell surface, they could hardly contribute to the precise determination of fusion sites or to the fusion mechanism. In conclusion, neither microtubules nor microfilaments are thought to specify exocytotic fusion sites in *Paramecium*. Both these systems might, however, represent a kind of long-range signal for the docking of trichocysts to the cell membrane (21). The specificity of membrane interaction must be definitely controlled by other means.

To understand how this might occur, it was helpful to analyze the biogenesis of preformed exocytosis sites in *Paramecium* (Fig. 3, derived from Refs. 39,40). Normally there are hardly any empty docking sites. These are short-lived states formed after trichocyst release and clearing of ghosts. Within fractions of an hour a "plug" is accumulated, which consists of a mass of amorphous to fibrillar electron-dense proteins (21,51). When a trichocyst is docked, the "plug" material is distributed over the trichocyst tip. While empty sites in freeze fracturing are devoid of "rosettes" and contain "parenthesis" instead of "ring" stages (Fig. 3), parentheses are transformed to "rings"; with a delay of ~5 min, "rosette" MIP are assembled (40). Only at this stage is the system fully reactive to exocytosis triggering (40). All these situations may be considered as phenocopies of *tl* and *tam* ("parenthesis" with "plug") and of *nd* ("ring" without "rosette") mutations whose ultrastructural features were described in Refs. 46-48.

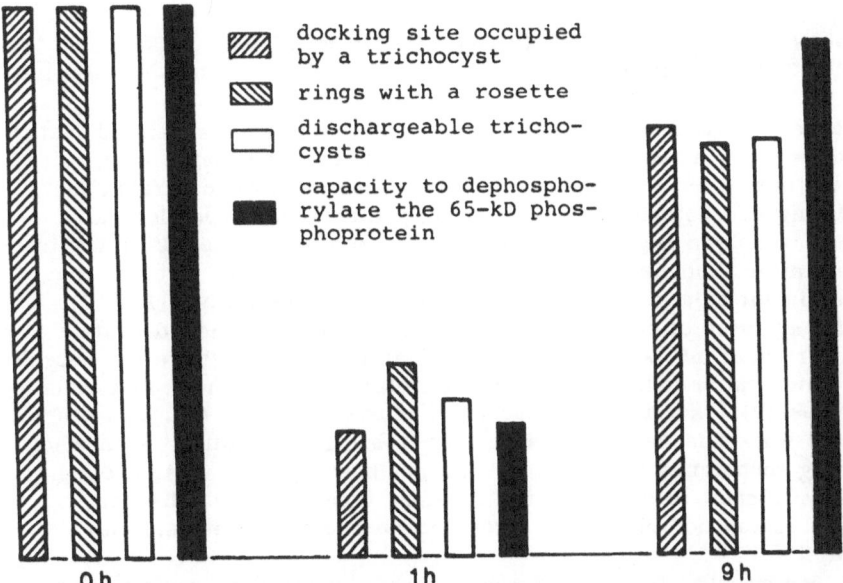

FIGURE 4 Coincidence of ultrastructural and functional parameters pertinent to exocytosis capacity in *Paramecium tetraurelia* cells. Left: Normal cells. Middle: Cells depleted of most of their trichocysts 1 h before. Right: The same, but 9 h later, when the trichocyst set has largely been regenerated. For further explanations, see text. [According to Zieseniss and Plattner (38).]

An extensive analysis of mutations has revealed that "connecting material" (see above) might, at least in part, be carried along by trichocysts as they dock at the cell membrane (48) rather than being all together part of the "plug" [as assumed before (51)]. The requirement of a cytosolic component for establishing exocytosis competence had also been inferred from microinjection experiments in which *nd9-28°C* cells regained their capacity for exocytotic membrane fusion, when supplemented with wild-type cytosol (47,60). Some collar components might be derived from the "plug."

As can be derived from Table 3, the capability of exocytosis performance in different *Paramecium* mutants is coupled with the coincidental occurrence of "rosettes," "connecting materials," and the capacity to dephosphorylate a phosphoprotein of M_r = 65 kD. The last phenomenon could be caught only when cells were immediately fixed, either by using picric acid as a trigger agent (55) or by inactivating cells at different intervals during synchronous AED-induced exocytosis (38). This coincidence also applies to different stages occurring after trichocyst depletion by AED (Fig. 4). We could show this by applying a second AED trigger at different time points over a period of up to 8 h (sufficient to reinstall an almost complete set of trichocysts (38).

Except for these two examples, the neuromuscular junction and the protozoan cells, one can only speculate as to the occurrence of predetermined exocytosis sites in other systems. Only in part can this predetermination reside in a particular arrangement of microtubules, since it is now generally agreed that membrane proteins incorporated into different sections of a cell membrane must be sorted according to a program inherent to the particular components.

VI. FUSION SITES IN SYSTEMS DEVOID OF HIGH-GRADE STRUCTURAL ORDER

I have never claimed that regular MIP aggregates would be a general feature of exocytosis sites in many different systems. When such claims were raised, they turned out to have different reasons. It has been demonstrated that MIP clustering can be caused (a) by keeping the temperature close to the thermal transition region of lipids, (b) by the occurrence of endocytosis immediately after exocytosis, or (c) by the presence of MIP clusters that were independent of any membrane fusion (for a discussion see Ref. 7).

Careful analyses of different exocytotic systems have then definitely shown that regular MIP aggregates (of the kind described above for neuromuscular junctions or ciliated protozoa) are clearly absent from mast cells (10), oocytes (9), exocrine pancreas (61), chromaffin cells (2), and others. Most of this work had been done with fast-freezing techniques. Even when fast-freezing techniques (see Ref. 3) had not been applied (61), cooling rates were sufficient to ascertain that, before fusion occurs, MIP are neither aggregated nor dispersed, provided no chemical pretreatments were used.

As in *Paramecium*, the contact zone between chromaffin granules and the cell membrane in adrenal medullary cells also displays electron-dense materials (Fig. 5, from Ref. 62). This might contain not only synexin (19) and calmodulin (shown to affect catecholamine release by Ref. 63), but possibly many other components. Some of them might be identical with the cytosolic components shown to bind reversibly to chromaffin granules before they are transported to the cell membrane (64).

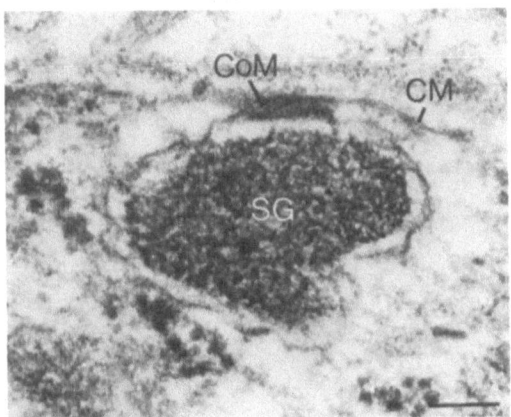

FIGURE 5 Ultrathin section showing the peripheral region of a chromaffin cell. CM = cell membrane, SG = secretory granule. Note the presence of CoM = connecting material between both membranes. [Reproduced from Aunis et al. (62), with permission of authors and publishers.] Bar = 0.1 µm.

The presence of a similar kind of membrane "connecting material" in other cell types was reviewed before (7). It was later discovered in some more cell types, including oocytes (29) and thrombocytes [after freeze substitution (28)]. It is still denied, however, for mast cells, though it remains to be seen how the point contacts between membranes, as shown in figures by Lawson (65) and Uvnäs (31), can be maintained without the support by such materials.

The ultrastructure of exocytosis performance in chromaffin cells (freshly isolated from bovine adrenal medullae) during carbachol-triggered catecholamine secretion is documented in Figures 6 and 7. All these micrographs were obtained by fast-freezing technology. The following details can be discerned (2). (a) No MIP clustering occurs, but MIP dispersal from fusion sites is not recognizable either (Fig. 6a). (b) Fusion starts on a small dimple on top of a little funnellike depression. The dimple originally has a diameter of 10 nm. InFigure 6b it can be unequivocally identified as a fusion stage as it increases to 20 nm. (c) As soon as it expands, etching techniques (sublimation of ice) show that the dimple represents an open connection to the outside medium (Fig. 6c, etc.). This implies that a diaphragm, assumed for some time to be always formed in the course of exocytotic membrane fusion (for a review of this aspect, see Ref. 7), is not formed. (d) The transition between the two membranes undergoing fusion is abrupt; this can also be seen in lateral view (Fig. 7) and it can account for the craterlike appearance in cross-fractures (Fig. 6c-e). (e) The exocytotic opening expands until it attains a size comparable to a chromaffin granule (Fig. 6f).

It is difficult to ascertain that these ultrastructural details are really due to exocytosis, rather than, for instance, to endocytosis. However, since the catecholamine output and the number of exocytotic openings (of different classes of diameters; Fig. 8) increase in concert, this led us to the interpretation of our micrographs as just described (2). Furthermore, these stages were observed within a short stimulation period during which endocytosis is negligibly small (see below).

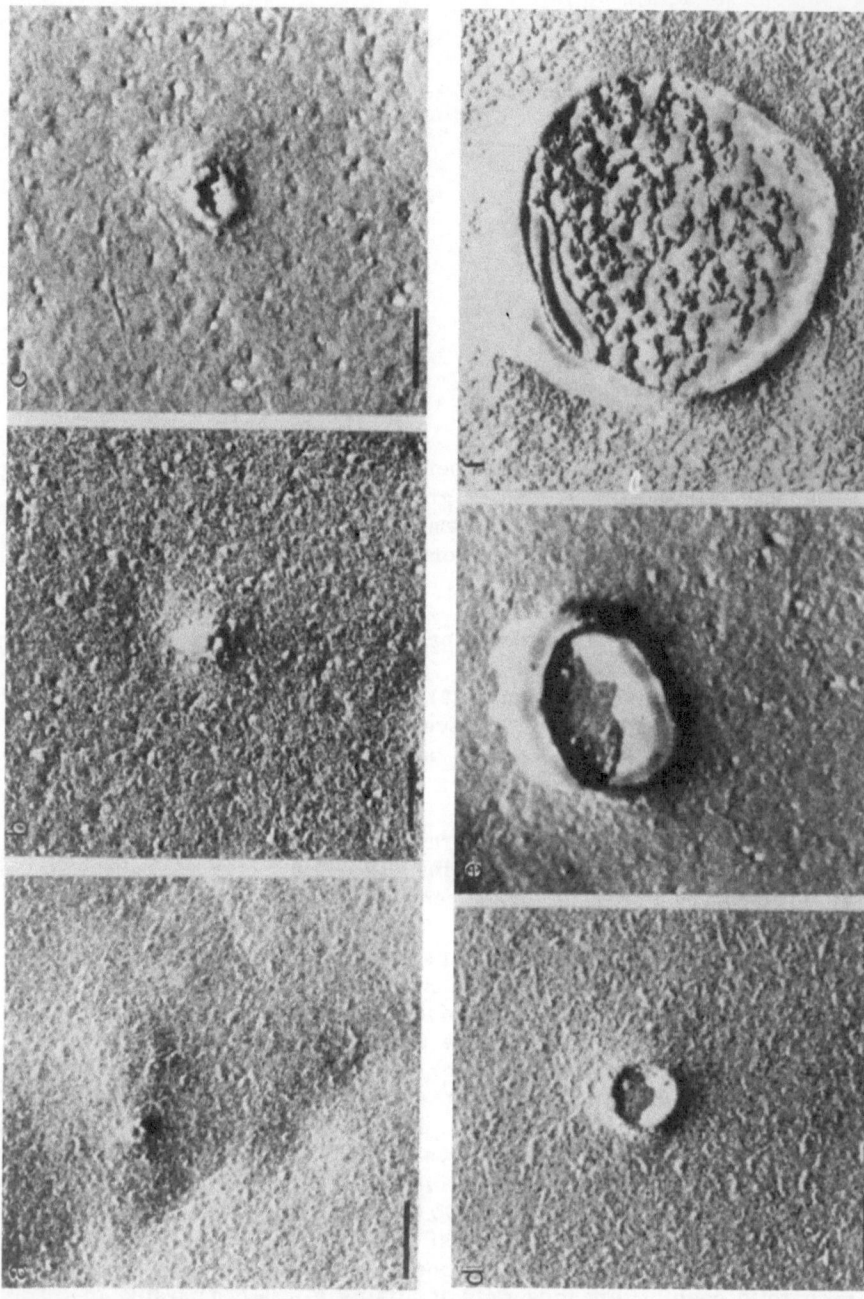

FIGURE 6 Freeze-fractured cell membrane of a chromaffin cell during carbachol-triggered exocytosis. For explanation, see text. [Reproduced from Schmidt et al. (2), with permission of the publishers.] Bars = 0.1 µm.

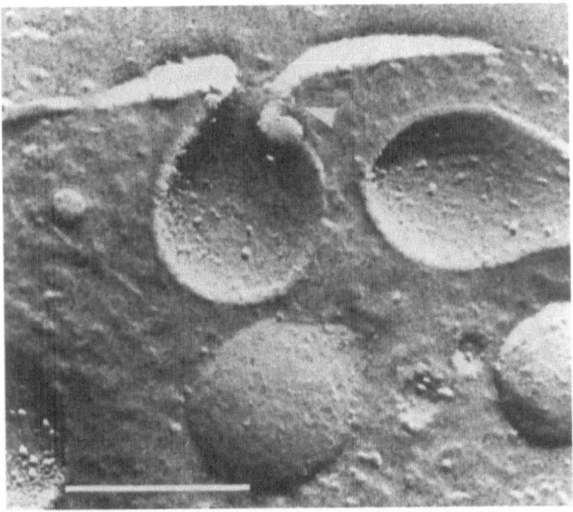

FIGURE 7 Cross-fractured stimulated chromaffin cell with an exocytotic profile (arrowhead). [Reproduced from Schmidt et al. (2), with permission of the publishers.] Bar = 0.1 μm.

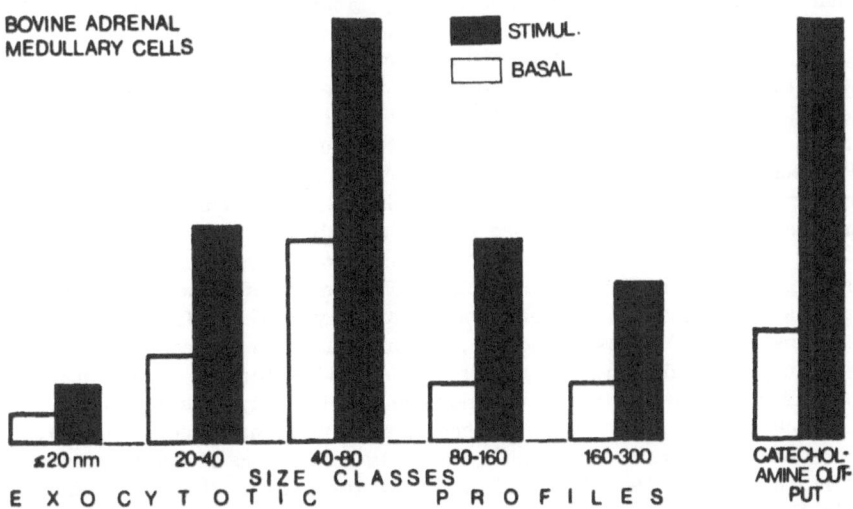

FIGURE 8 Correlation of functional and ultrastructural (freeze fracture) data obtained from stimulated chromaffin cells. [According to Schmidt et al. (2).]

There is hardly any description of a dimplelike structure in other
systems. As to the rest of our observations, quite similar phenomena were
reported for membrane fusion in quick-frozen mast cells (10), oocytes
(9), and *Limulus* amebocytes (16). Aspects in agreement with our studies
(2) are that membranes undergoing fusion show no internal restructuring.
A depression by which the cell membrane makes a focal contact with the
secretory granule was also observed in quite different systems [mast cells
(65), amebocytes (16), chromaffin cells (2)]. As mentioned, in *Paramecium*
it is a nipplelike structure on top of the trichocyst that makes that intimate
contact (46). It seems that this focal membrane contact is a permanent
aspect (which then could represent a preformed fusion site) rather than
a transient feature occurring only when exocytosis is soon to take place.
One could philosophize that the small curvature of this membrane portion
might facilitate membrane fusion or that the ignition site for membrane
fusion should be kept small, thus allowing for a better control. The "connecting materials" observed at these sites might restrict the fusogenic
zone. [Beyond this it might contain components to "ignite" fusion (see
above) and to restrict exchange of membrane materials during fusion (see
below)]. Another feature in common to different systems is the presence
of loosely scattered MIP and of peripheral proteins (see above) close to
the site of potential exocytosis, though in a form less conspicuous than
in *Paramecium*.

Recently, peripheral and particularly integral proteins have been recognized for their capacity to induce membrane fusion (5,66-69). If so, MIP
that remain quite close to fusion sites during exocytosis could play such
a role. This could then be modulated by a changing degree of hydrophobicity, e.g., by partial proteolysis (67) or by reversible (de-)phos-
phorylation (38), etc. The same could apply to membrane-associated proteins.
One is still looking for an equivalent of viral fusion proteins, for which
White et al. (70) have recently argued that they might have been "con-
fiscated" from eukaryotic cells during evolution. It also requires more
detailed correlated structure-function analyses (e.g., by means of affinity-
and immunolabeling) to see how close Ca^{2+} channels or receptor proteins
have to stay to the actual fusion site for fusion to occur.

It is tempting to consider the dimple in Figure 6a as a local instability
focus or as an inverted lipid micelle, postulated by Verkleij (5,6) to be
a fusion intermediate. However, so far, no one has ever experimentally
identified the physicochemical nature of this minute structure and there
is no consensus yet on the actual occurrence of precisely this intermediate
(15,18). Since the dimple in Figure 6 seems to expand as fusion goes on,
one may reasonably assume, at least, that it is some form of lipid arrangement. It is questionable whether one can ever get hold of the real fusion
process in the electron microscope. There are estimates that the time required for actual membrane fusion in pure lipid systems is in the μsec
range (71,72). This would by far surpass detection limits [msec; (3)]
even by the fastest ultrastructural methods. Nevertheless, the frequent
presence of this dimple structure on top of the funnellike contact zones
would be an argument in favor of its possible function as a latent fusogenic
stage. Clearly, more detailed work is required on this important aspect,
since we are not sure enough about the lifespan of the ultrastructural
stages just described, though it does occur in vivo (2).

VII. EXOCYTOSIS PERFORMANCE IN THE REGULARLY STRUCTURED SYSTEM OF THE *PARAMECIUM* CELL

Micrographs presented in Figure 9 were obtained from cells that were triggered by AED and immediately frozen (17). All these stages may occur side by side in one cell within a short period of time. When no chemical fixation is used, "ring" and "rosette" MIP adhere consistently to the PF-(upper row) or to the EF-face (lower row), respectively (Fig. 9a,a'). The smallest seizable opening was—due to less favorable conditions for cryofixation—only several 10 nm in diameter (b,b'); it forms within the "rosette" and—because it can be "etched"—it does not contain a diaphragm, just as in other systems (see above). In Figure 9 (c,c') the opening expands further, while "rosette" MIP are laterally dispersed, probably beyond the "ring." In d,d', trichocyst contents are seen to come out. Clearly, the fusion zone remains within the "ring," even when the exo-cytosis opening reaches a maximal diameter (e,e') corresponding to the thickness of a trichocyst. One can then recognize the partly granular (from "annuli") and partly crenated (from "collar") texture of the upper trichocyst membrane. Resealing in Figure 9 (f,f') entails the collapse of the "ring" to a "parenthesis," with the intermediate of an "oval ring" [(22); not shown here].

VIII. MAINTENANCE OF MEMBRANE SPECIFICITY

Figure 10 summarizes, based on Figures 9 and 11, how the specific structure (and inherently the specific composition) of the two membranes underoing fusion is maintained in a *Paramecium* cell. The occurrence of endogenous ultrastructural tags makes it very attractive for such analyses. Figure 9 has already revealed that "ring" MIP persist in a "parenthesis" form [with an unchanged number of MIP of ~70 (22)]. Figure 11a now shows that normally the entire trichocyst membrane, including the membrane specializations contained in its uppermost membrane region ("annulus" and "collar"), becomes completely detached after the discharge of a tricho-cyst. In the uppermost position of a ghost membrane one can recognize only some glycokalyx elements, stainable by heavy metals; these components probably have spread from the cell membrane by lateral diffusion (22). The scheme of Figure 10 takes this into account; it also illustrates the very rare (0.3% of all cases analyzed) occurrence of "false" incorporation of trichocyst membrane constituents into the resealed cell membrane (22).

The electron-dense materials surrounding the trichocyst tip probably exert a stabilizing function and impede the exchange of more membrane materials. One might assume that "connecting material," which normally seems to tie "rosette" MIP to the fusogenic site, would also be dispersed over the cell membrane during exocytosis, since it is no longer observed afterward. In conclusion, these membrane-associated materials might con-tribute to maintain membrane specificity in *Paramecium* cells.

Similar mechanisms, including restriction of material exchange during fusion by apposed proteins, could occur also in other systems, since in principle, these contain very similar structural components at their exo-cytosis sites. Whereas the retrieval of trichocyst ghost membranes does not operate via "coated" (clathrin) pits or vesicles (22), this molecular filter mechanism is widely distributed among other exocytotic systems (73), where it may contribute to the maintenance of membrane specificity.

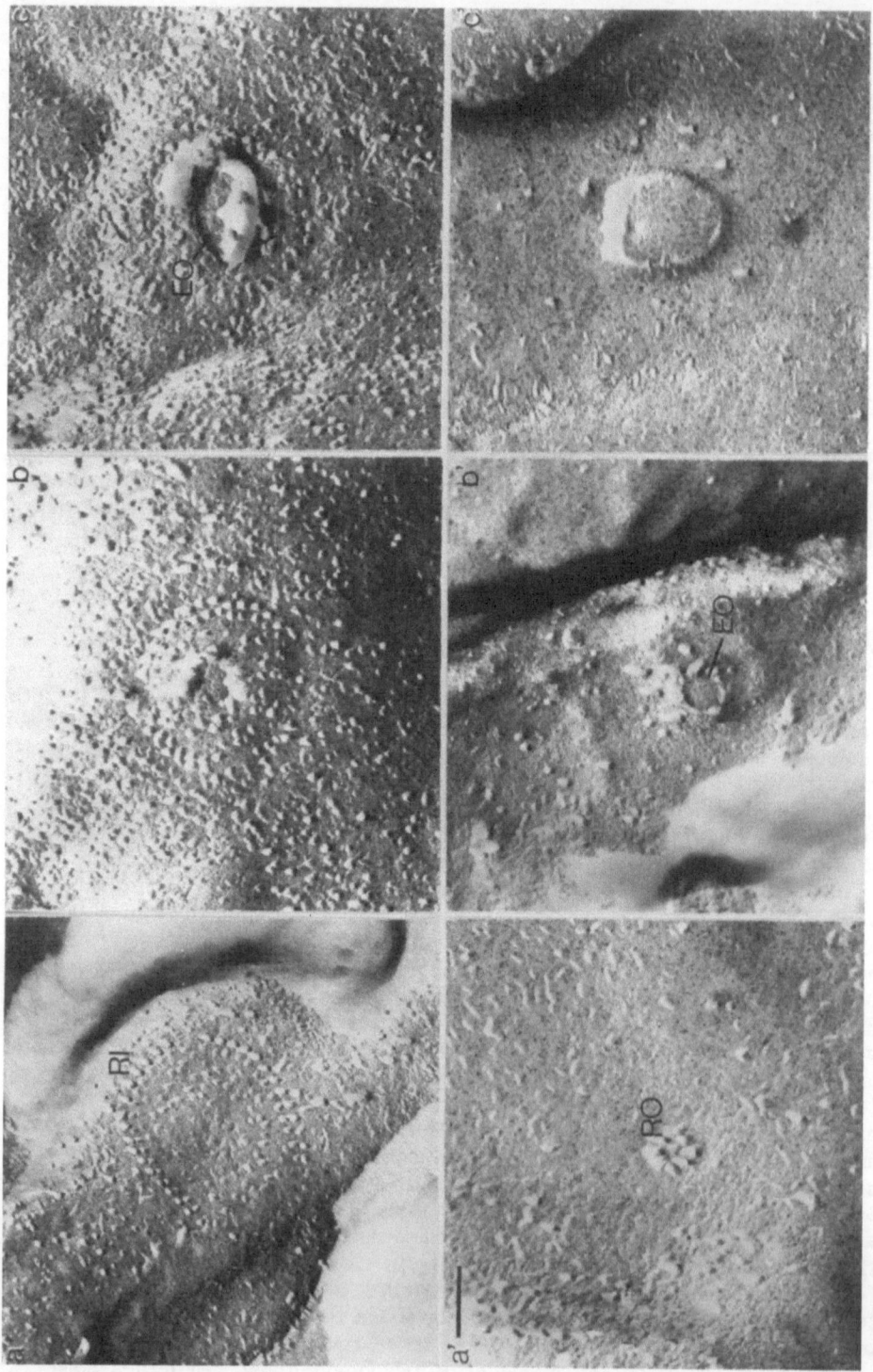

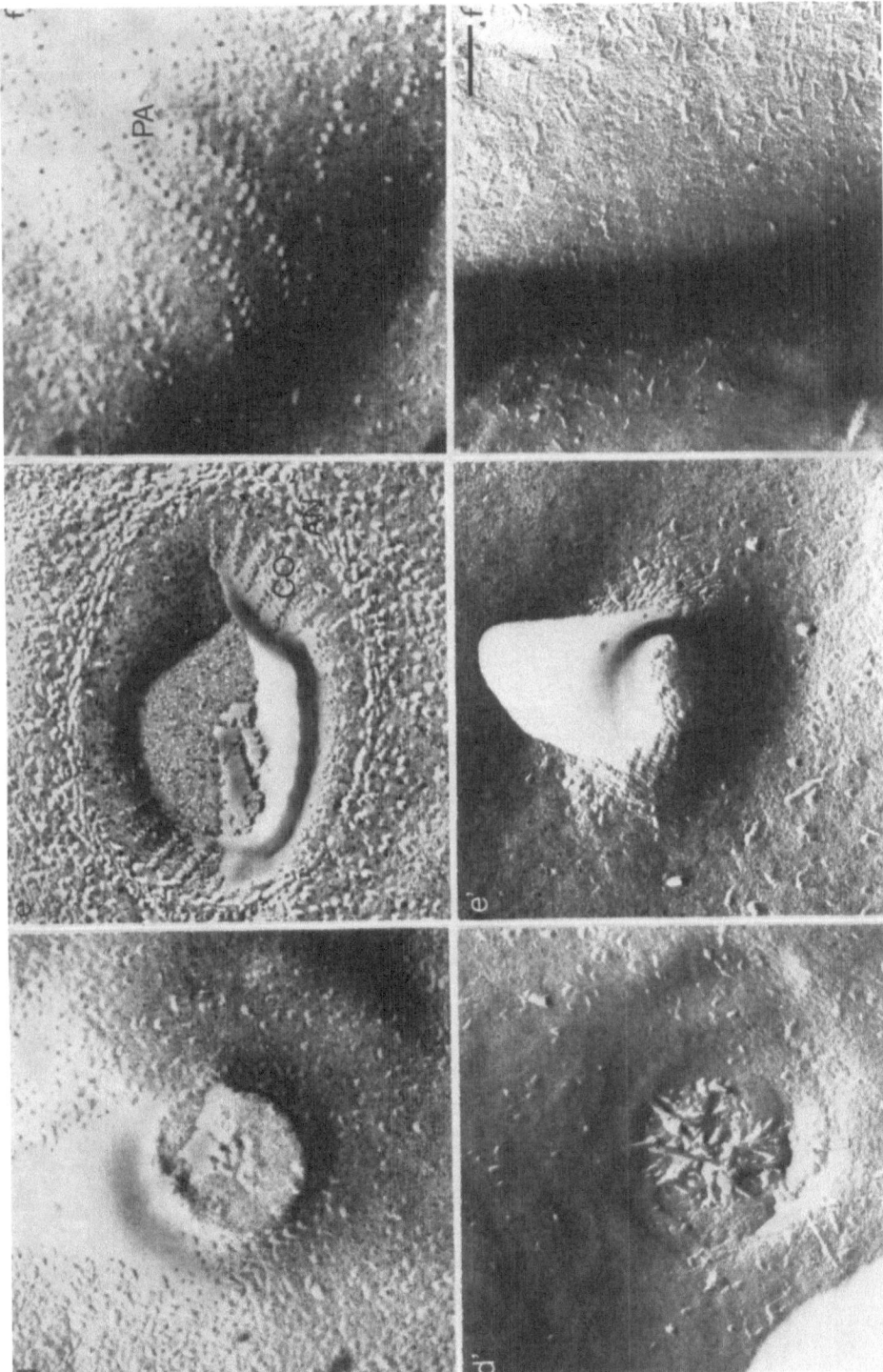

FIGURE 9 Ultrastructural changes observed during stimulated exocytosis of trichocysts in *Paramecium tetraurelia*. RI = ring; RO = rosette; EO = exocytotic opening; PA = parenthesis. For further explanations, see Figure 10 and text. [Reproduced from Olbricht et al. (17), with permission of the publishers.] Bars = 0.1 μm.

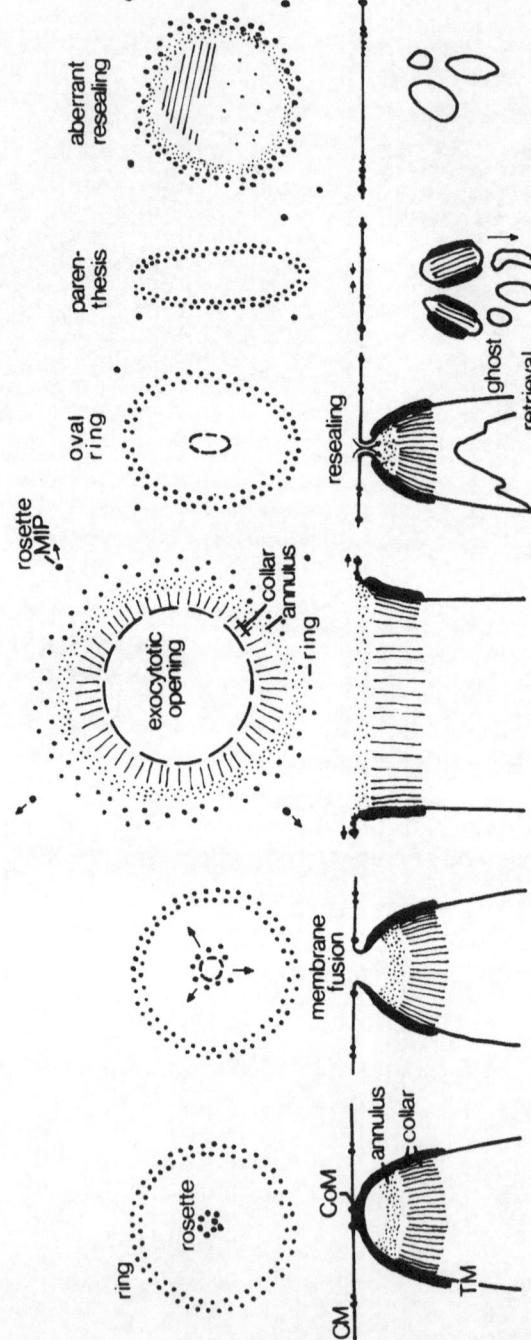

FIGURE 10 Scheme of trichocyst exocytosis derived from Figures 9 and 11. For symbols, see Figures 1 and 2; arrows indicate direction of movement. For further explanations see text. [Reproduced from Plattner et al. (22), with permission of the publishers.]

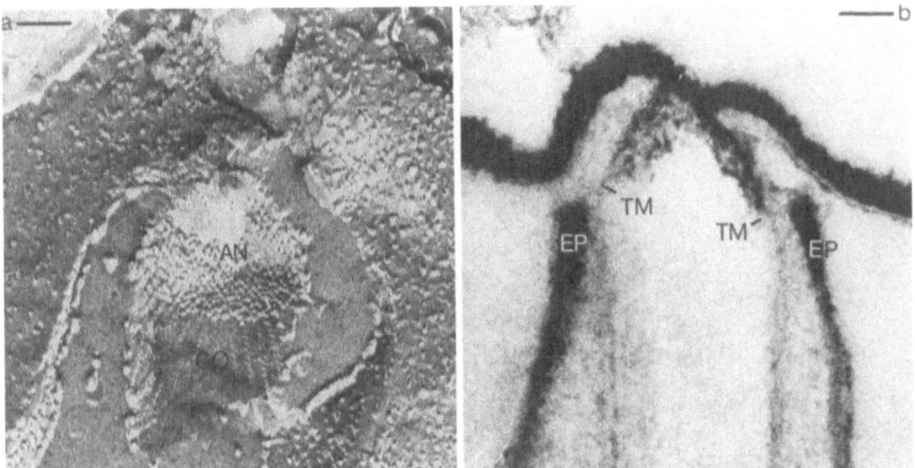

FIGURE 11 Ghosts detached from the cell membrane after trichocyst exocytosis in *Paramecium tetraurelia*. AN = annulus; CO = collar; TM = trichocyst membrane. (a) Freeze-fracture replica. (b) Ultrathin section stained
for (glyco-)proteins with phosphotungstic acid. Note the complete detachment
of TM in (a) and the additional shift of some stained materials from the
cell membrane down to TM (this region being unstained without previous
exocytosis) in (b). EP = epiplasm located outside trichocyst. [Reproduced
from Plattner et al. (22), with permission of the publishers.] Bars = 0.1 μm.

Another factor relevant for maintaining membrane specificity is the
short time period during which an exocytotic opening stays open (see
Section IV). For *Paramecium* we assume an opening time of only 50 msec
(Deitmer and Plattner, unpublished results). For the transmitter release
in frog neuromuscular junctions one had previously postulated a total
integration of the synaptic vesicle membrane into the cell membrane, followed
by a delayed recovery via coated pits (74). A recent reinvestigation demonstrates an immediate retrieval with only little delay (36). For adrenal medullary cells, antibody labeling (of lumenal membrane components of chromaffin
granules exposed during stimulated secretion) indicated long opening times
(see, for example, Ref. 75). Yet, according to more recent analyses the
actual opening time might be much shorter (76), though still in the range
of minutes.

Only in some exceptional cases the cell does not retrieve the excess
of membranes inserted into its surface by exocytosis. The classical—and
perhaps quite unique—example, where this occurs under physiological conditions, is the cortical granule reaction in the fertilized egg (30).

IX. COMMON FEATURES OF EXOCYTOTIC
MEMBRANE FUSION

The rather regularly structured system of the *Paramecium* cell and its
mutant strains has taught us important aspects of structure-function
correlation of exocytotic membrane fusion. It is now time for a generalization.

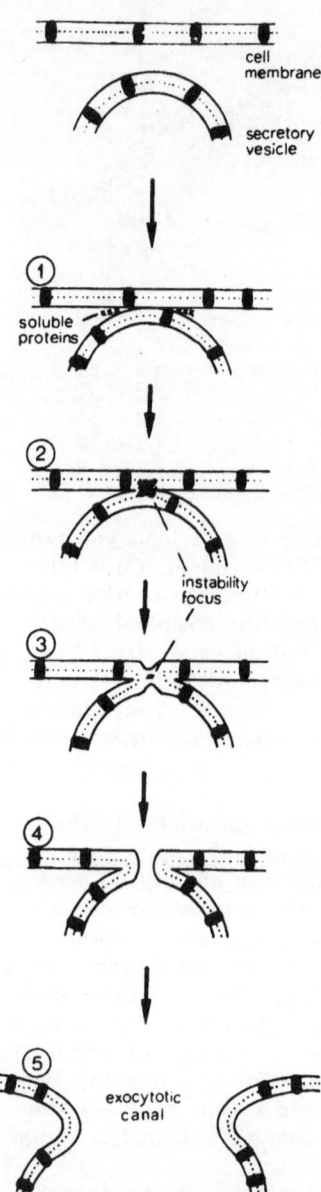

FIGURE 12 Scheme of exocytotic membrane fusion ("focal fusion"), according to Plattner (7). Black dots indicate membrane-integrated proteins (MIP) visualized by freeze fracturing (dotted line). For further explanations, see text. [Reproduced from Plattner (7), with permission of the publishers.]

Based on a comparison of the different exocytotic systems analyzed so far by "advanced" ultrastructural preparation techniques (including fast freezing), the general scheme for exocytotic membrane fusion presented in Figure 12 was developed. (1) MIP (bona fide membrane-integrated proteins) and adsorbed proteins remain closely associated with actual fusion sites, though regular MIP aggregates (as they occur in protozoa and in neuromuscular junctions) are not required for fusion to take place. (2) Fusion is focal and it is followed by the extension of the exocytotic opening about to the size of the secretory granule. (3) No MIP-bare zone and no diaphragm are formed during fusion. (4) The specificity of the membranes involved in exocytosis is maintained—if at all, as in most cases—by totally different mechanisms.

NOTE ADDED IN PROOF

Since this manuscript was submitted in Spring 1986, many new data have accumulated, as summarized from the point of view of structure-function correlation in the following reference: Plattner, H. (1989). Regulation of membrane fusion during exocytosis. *Int. Rev. Cytol.* 119:197-286.

ACKNOWLEDGMENTS

I gratefully acknowledge the cooperation of my students and co-workers who have helped to develop some of the concepts presented here. I thank Ms. D. Bliestle for photographic work, Ms. E. Rehn for secretarial help, as well as the "SFB 156" for financial support that allowed for some of the work included here.

REFERENCES

1. DeRobertis, E., and Vazferreira, A. (1957). Electron microscope study of the excretion of catechol-containing droplets in the adrenal medulla. *Exp. Cell Res.* 12:568-574.
2. Schmidt, W., Patzak, A., Lingg, G., Winkler, H., and Plattner, H. (1983). Membrane events in adrenal chromaffin cells during exocytosis: A freeze-etching analysis after rapid cryofixation. *Eur. J. Cell Biol.* 32:31-37.
3. Plattner, H., and Bachmann, L. (1982). Cryofixation: A tool in biological ultrastructural research. *Int. Rev. Cytol.* 79:237-304.
4. Zingsheim, H. P., and Plattner, H. (1976). Electron microscopic methods in membrane biology. In *Methods in Membrane Biology*, Vol. 7 (E. D. Korn, ed.). Plenum Press, New York, pp. 1-146.
5. DeKruiff, B., Cullis, P. R., Verkleij, A. J., Hope, M. J., Van Echteld, C. J. A., and Taraschi, T. F. (1985). Lipid polymorphism and membrane function. In *The Enzymes of Biological Membranes, Vol. 1. Membrane Structure and Dynamics* (A. N. Martonosi, ed.). Plenum Press, New York, London, pp. 131-204.
6. Verkleij, A. J. (1984). Lipidic intramembranous particles. *Biochim. Biophys. Acta* 779:43-63.
7. Plattner, H. (1981). Membrane behaviour during exocytosis. *Cell Biol. Int. Rep.* 5:435-459.

8. Heuser, J. E., Reese, T. S., Dennis, M. J., Jan, Y., Jan, L., and Evans, L. (1979). Synaptic vesicle exocytosis captured by quick freezing and correlated with quantal transmitter release. *J. Cell Biol.* 81:275-300.

9. Chandler, D. E., and Heuser, J. (1979). Membrane fusion during secretion. Cortical granule exocytosis in sea urchin eggs as studied by quick-freezing and freeze-fracture. *J. Cell Biol.* 83:91-108.

10. Chandler, D. E., and Heuser, J. E. (1980). Arrest of membrane fusion events in mast cells by quick-freezing. *J. Cell Biol.* 86:666-674.

11. Plattner, H. (1987). Synchronous exocytosis in *Paramecium* cells. In *Cell Fusion* (A. E. Sowers, ed.). Plenum-Press, New York, pp. 69-98.

12. Plattner, H., Miller, F., and Bachmann, L. (1973). Membrane specializations in the form of regular membrane-to-membrane attachment sites in *Paramecium*. A correlated freeze-etching and ultrathin-sectioning analysis. *J. Cell Sci.* 13:687-719.

13. Douglas, W. W. (1974). Involvement of calcium in exocytosis and the exocytosis-vesiculation sequence. *Biochem. Soc. Symp.* 39:1-28.

14. Rubin, R. P. (1984). Stimulus-secretion coupling. In *Cell Biology of the Secretory Process* (M. Cantin, ed.). Karger, Basel, München, Paris, pp. 52-72.

15. Düzgünes, N., Paiement, J., Freeman, K. B., Lopez, N. G., Wilschut, J., and Papahadjopoulos, D. (1984). Modulation of membrane fusion by ionotropic and thermotropic phase transitions. *Biochemistry* 23:3486-3494.

16. Ornberg, R. L., and Reese, T. S. (1981). Beginning of exocytosis captured by rapid-freezing of *Limulus* amebocytes. *J. Cell Biol.* 90:40-54.

17. Olbricht, K., Plattner, H., and Matt, H. (1984). Synchronous exocytosis in *Paramecium* cells. II. Intramembranous changes analyzed by freeze-fracturing. *Exp. Cell Res.* 151:14-20.

18. Bentz, J., Ellens, H., Lai, M. Z., and Szoka, F. C. (1985). On the correlation between H_{II} phase and the contact-induced destabilization of phosphatidylethanolamine-containing membranes. *Proc. Natl. Acad. Sci. USA* 82:5742-5745.

19. Pollard, H. B., Creutz, C. E., Fowler, V., Scott, J., and Pazoles, C. J. (1981). Calcium-dependent regulation of chromaffin granule movement, membrane contact, and fusion during exocytosis. *Cold Spring Harbor Symp. Quant. Biol.* 46:819-834.

20. Puskin, S., and Kochwa, S. (1974). Regulation of neurotransmitter release by a complex of actin with relaxing protein isolated from rat brain synaptosomes. *J. Biol. Chem.* 249:7711-7714.

21. Plattner, H., Westphal, C., and Tiggemann, R. (1982). Cytoskeleton-secretory vesicle interactions during the docking of secretory vesicles at the cell membrane in *Paramecium tetraurelia* cells. *J. Cell Biol.* 92:368-377.

22. Plattner, H., Pape, R., Haacke, B., Olbricht, K., Westphal, C., and Kersken, H. (1985). Synchronous exocytosis in *Paramecium* cells. VI. Ultrastructural analysis of membrane resealing and retrieval. *J. Cell Sci.* 77:1-17.

23. Kersken, H., Vilmart-Seuwen, J., Momayezi, M., and Plattner, H. (1986). Filamentous actin in *Paramecium* cells: Mapping by phalloidin affinity labeling in vivo and in vitro. *J. Histochem. Cytochem.* 34:443-454.

24. Whitaker, M. J., and Baker, P. F. (1983). Calcium-dependent exo-

cytosis in an in vitro secretory granule plasma membrane preparation from sea urchin eggs and the effects of some inhibitors of cytoskeletal function. *Proc. R. Soc. Lond. Ser. B.-Biol. Sci.* 218:397–413.

25. Campbell, A. K. (1983). *Intracellular Calcium. Its Universal Role as Regulator*. Chichester, New York, Brisbane, Toronto, Singapore.

26. Schneider, A. S., Herz, R., and Rosenheck, K. (1977). Stimulus-secretion coupling in chromaffin cells isolated from bovine adrenal medulla. *Proc. Natl. Acad. Sci. USA* 74:5036–5040.

27. White, J. G. (1984). The secretory process in platelets. In *Cell Biology of the Secretory Process* (M. Cantin, ed.). Karger, Basel, München, Paris, pp. 546–569.

28. Morgenstern, E., Edelmann, L., Reimers, H. J., Miyashita, C., and Haurand, M. (1985). Fibrinogen distribution on surfaces and in organelles of ADP-stimulated human blood platelets. *Eur. J. Cell Biol.* 38:292–300.

29. Guraya, S. S. (1982). Recent progress in the structure, origin, composition, and function of cortical granules in animal egg. *Int. Rev. Cytol.* 78:257–360.

30. Epel, D., and Vacquier, V. D. (1978). Membrane fusion events during invertebrate fertilization. In *Membrane Fusion* (G. Poste and G. L. Nicolson, eds.). North-Holland, Amsterdam, New York, Oxford, pp. 1–63.

31. Uvnäs, B. (1982). Mast cell granules. In *The Secretory Granule* (A. M. Poisner and J. M. Trifaró, eds.). Elsevier, Amsterdam, New York, Oxford, pp. 357–384.

32. Plattner, H., Matt, H., Kersken, H., Haacke, B., and Stürzl, R. (1984). Synchronous exocytosis in Paramecium cells. I. A novel approach. *Exp. Cell Res.* 151:6–13.

33. Plattner, H., Stürzl, R., and Matt, H. (1985). Synchronous exocytosis in *Paramecium* cells. IV. Polyamino compounds as potent trigger agents for repeatable trigger-redocking cycles. *Eur. J. Cell Biol.* 36:32–37.

34. Petersen, O. H. (1980). Exploring the exocrine-pancreas by microelectrodes—Lessons from the first decade of pancreatic electrophysiology. In *Biology of Normal and Cancerous Exocrine Pancreatic Cells* (A. Ribet, L. Pradayrol, and C. Susini, eds.). Elsevier North-Holland, Amsterdam, pp. 65–82.

35. Neher, E., and Marty, A. (1982). Discrete changes of cell membrane capacitance observed under conditions of enhanced secretion in bovine adrenal chromaffin cells. *Proc. Natl. Acad. Sci. USA* 79:6712–6716.

36. Torri-Tarelli, F., Grohovaz, F., Fesce, R., and Ceccarelli, B. (1985). Temporal coincidence between synaptic vesicle fusion and quantal secretion of acetylcholine. *J. Cell Biol.* 101:1386–1399.

37. Matt, H., Bilinski, M., and Plattner, H. (1978). Adenosinetriphosphate, calcium and temperature requirements for the final steps of exocytosis in Paramecium cells. *J. Cell Sci.* 32:67–86.

38. Zieseniss, E., and Plattner, H. (1985). Synchronous exocytosis in *Paramecium* cells involves very rapid (≤ 1 s), reversible dephosphorylation of a 65 kD-phosphoprotein in exocytosis-competent strains. *J. Cell Biol.* 101:2028–2035.

39. Haacke, B., and Plattner, H. (1984). Synchronous exocytosis in *Paramecium* cells. III. Rearrangement of membranes and membrane-associated structural elements after exocytosis performance. *Exp. Cell Res.* 151:21–28.

40. Pape, R., and Plattner, H. (1985). Synchronous exocytosis in Para-
    mecium cells. V. Ultrastructural adaptation phenomena during re-
    insertion of secretory organelles. *Eur. J. Cell Biol.* 36:38-47.

41. Aufderheide, K. J. (1977). Saltatory motility of uninserted trichocysts
    and mitochondria in *Paramecium tetraurelia*. *Science* 198:299-300.

42. Couteaux, R. and Pécot-Dechavassine, M. (1974). Les zones specialisées
    des membranes présynaptiques, *C. R. Acad. Sci. (Paris) Ser. D.*
    278:291-293.

43. Pumplin, D. W., Reese, T. S., and Llinás, R. (1981). Are the pre-
    synaptic membrane particles the calcium channels? *Proc. Natl. Acad.
    Sci USA* 78:7210-7213.

44. Lacy, P. E. (1975). Endocrine secretory mechanisms. A review. *Am. J.
    Pathol.* 79:170-187.

45. Satir, B., Schooley, C., and Satir, P. (1973). Membrane fusion in
    a model system. Mucocyst secretion in *Tetrahymena*. *J. Cell Biol.* 56:
    153-176.

46. Beisson, J., Lefort-Tran, M., Pouphile, M., Rossignol, M., and
    Satir, B. (1976). Genetic analysis of membrane differentiation in
    Paramecium. Freeze-fracture study of the trichocyst cycle in wild-type
    and mutant strains. *J. Cell Biol.* 69:126-143.

47. Lefort-Tran, M., Aufderheide, K., Pouphile, M., Rossignol, M. and
    Beisson, J. (1981). Control of exocytotic process: Cytological and
    physiological studies of trichocyst mutants in *Paramecium tetraurelia*.
    J. Cell Biol. 88:301-311.

48. Pouphile, M., Lefort-Tran, M., Plattner, H., Rossignol, M., and
    Beisson, J. (1986). Genetic dissection of the morphogenesis and dynamics
    of exocytosis sites in *Paramecium*. *Biol. Cell* 56:151-162.

49. Westphal, C., and Plattner, H. (1981). Ultrastructural analysis of
    the cell membrane-secretory organelle interaction zone in *Paramecium
    tetraurelia* cells. I. In situ characterization by electron "staining"
    and enzymatic digestion. *Biol. Cell* 42:125-140.

50. Vilmart, J. and Plattner, H. (1983). Membrane integrated proteins
    at preformed exocytosis sites. *J. Histochem. Cytochem.* 31:626-632.

51. Westphal, C., and Plattner, H. (1981). Ultrastructural analysis of
    the cell membrane-secretory organelle interaction zone in *Paramecium
    tetraurelia* cells. II. Changes involved in the attachment of secretory
    organelles. *Biol. Cell* 42:141-146.

52. Matt, H., Plattner, H. Reichel, K., Lefort-Tran, M., and Beisson, J.
    (1980). Genetic dissection of the final exocytosis steps in *Paramecium
    tetraurelia* cells: Trigger analyses. *J. Cell Sci.* 46:41-60.

53. Gilligan, D. M., and Satir, B. H. (1983). Stimulation and inhibition
    of secretion in *Paramecium*: Role of divalent cations. *J. Cell Biol.*
    97:224-234.

54. Vilmart-Seuwen, J., Kersken, H., Stürzl, R., and Plattner, H. (1986).
    ATP keeps exocytosis sites in a primed state but is not required for
    membrane fusion: An analysis with *Paramecium* cells in vivo and in
    vitro. *J. Cell Biol.* 103:1279-1288.

55. Gilligan, D. M., and Satir, B. H. (1982). Protein phosphorylation
    and stimulus-secretion coupling in wild type and mutant *Paramecium*.
    J. Biol. Chem. 257:13903-13906.

56. Momayezi, M., Kersken, H., Gras, U., Vilmart-Seuwen, J., and
    Plattner, H. (1986). Calmodulin in *Paramecium tetraurelia*: Localization
    from the in vivo to the ultrastructural level. *J. Histochem. Cytochem.*
    34:1621-1638.

57. Plattner, H., Reichel, K., and Matt, H. (1977). Bivalent-cation-

stimulated ATPase activity at preformed exocytosis sites in *Paramecium* coincides with membrane-intercalated particle aggregates. *Nature* 267: 702-704.

58. Plattner, H., Reichel, K., Matt, H., Beisson, J. Lefort-Tran, M., and Pouphile, M. (1980). Genetic dissection of the final exocytosis steps in *Paramecium tetraurelia* cells: Cytochemical determination of Ca^{2+}-ATPase activity over preformed exocytosis sites. *J. Cell Sci.* 46:17-40.

59. Kersken, H., Momayezi, M., Braun, C., and Plattner, H. (1986). Filamentous actin in *Paramecium* cells: Functional and structural changes correlated with phalloidin affinity labeling in vivo. *J. Histochem. Cytochem.* 34:455-465.

60. Beisson, J., Cohen, J., Lefort-Tran, M., Pouphile, M., and Rossignol, M. (1980). Control of membrane fusion in exocytosis. Physiological studies on a *Paramecium* mutant blocked in the final step of the trichocyst extrusion process. *J. Cell Biol.* 85:213-227.

61. Tanaka, Y., DeCamilli, P., and Meldolesi, J. (1980). Membrane interactions between secretion granules and plasmalemma in three exocrine glands. *J. Cell Biol.* 84:438-453.

62. Aunis, D., Hesketh, J. E., and Devilliers, G. (1979). Freeze-fracture study of the chromaffin cell during exocytosis: Evidence for connections between the plasma membrane and secretory granules and for movements of plasma membrane-associated particles. *Cell Tissue Res.* 197:433-441.

63. Kenigsberg, R. L., and Trifaró, J. M. (1985). Microinjection of calmodulin antibodies into cultured chromaffin cells blocks catecholamine release in response to stimulation. *Neuroscience* 14:335-347.

64. Burgoyne, R. D., and Geisow, M. J. (1982). Phosphoproteins of the adrenal chromaffin granule membrane. *J. Neurochem.* 39:1387-1396.

65. Lawson, D. (1980). Rat peritoneal mast cells: A model system for studying membrane fusion. In *Membrane-Membrane Interactions* (N. B. Gilula, ed.). Raven Press, New York, pp. 27-44.

66. Batenburg, A. M., Bougis, P. E., Rochat, H., Verkleij, A. J., and DeKruiff, B. (1985). The penetration of a cardiotoxin into cardiolipin model membranes and its implications on lipid organizations. *Biochemistry* 24:7101-7110.

67. Lucy, J. A. (1985). Calcium ions, enzymes, and cell fusion. In *The Enzymes of Biological Membranes*, Vol. 1. *Membrane Structure and Dynamics* (N. Martonosi, ed.). Plenum Press, New York, London, pp. 371-391.

68. Young, T. M., and Young, J. D. (1984). Protein-mediated intermembrane contact facilitates fusion of lipid vesicles with planar bilayers. *Biochim. Biophys. Acta* 775:441-445.

69. Zimmerberg, J., Cohen, F. S., and Finkelstein, A. (1980). Micromolar Ca^{2+} stimulates fusion of lipid vesicles with planar bilayers containing a calcium-binding protein. *Science* 210:906-908.

70. White, J., Kielian, M., and Helenius, A. (1983). Membrane fusion proteins of enveloped animal viruses. *Q. Rev. Biophys.* 16:151-195.

71. Dimitrov, D. S., and Jain, R. K. (1984). Membrane stability. *Biochim. Biophys. Acta* 779:437-468.

72. Morris, S. J., Gibson, C. C., Smith, P. D., Greif, P. C., Stirk, C. W., Bradley, D., Haynes, D. H., and Blumenthal, R. (1985). Rapid kinetics of Ca^{2+}-induced fusion of phosphatidylserine/phosphatidylethanolamine vesicles. *J. Biol. Chem.* 260:4122-4127.

73. Evered, D., and Collins, G. M. (eds.) (1982). *Membrane Recycling*, Ciba Foundation Symposium 92, Ciba Foundation. Pitman Books, London.

74. Heuser, J. E., and Reese, T. S. (1981). Structural changes after transmitter release at the frog neuromuscular junction. *J. Cell Biol.* 88:564–580.
75. Dowd, D. J., Edwards, C., Englert, D., Mazukiewicz, J. E., and Ye, H. Z. (1983). Immunofluorescent evidence for exocytosis and internalization of secretory granule membrane in isolated chromaffin cells. *Neuroscience* 10:1025–1033.
76. Geisow, M. J., Childs, J., and Burgoyne, R. D. (1985). Cholinergic stimulation of chromaffin cells induces rapid coating of the plasma membrane. *Eur. J. Cell Biol.* 38:51–56.

V
CELL-CELL FUSION

26

Fusion of Myoblasts

KAREN A. KNUDSEN

Lankenau Medical Research Center, Philadelphia, Pennsylvania

I. INTRODUCTION

Early cytological observations of developing muscle led to two major theories explaining the generation of the syncytia formed during embryogenesis or the regeneration of muscle. The first theory postulated nuclear division without concomitant cytoplasmic division, while the second postulated fusion of mononucleate cells to form multinucleate cells. In the late 1950s and early 1960s, with the development of muscle cell culture, support began to grow for the theory of cell fusion. Today it is accepted that during the development of skeletal muscle, single muscle cells fuse to form multinucleate cells that synthesize muscle-specific proteins and assemble myofibrils.

The fusion of myoblasts does not result simply from a rapid intermingling of protein-free regions of the lipid bilayers of cells in close contact. Rather, myoblast fusion is a complex, multistep process involving an ordered sequence of events including: (1) expression of the fusion-competent phenotype (differentiation), (2) specific myoblast-myoblast adhesion (specific adhesion), and (3) fusion of lipid bilayers (membrane union).

In this chapter I will synthesize a picture of myoblast fusion consistent with existing published data. The literature was reviewed several years ago by Bischoff (1) and more recently by Wakelam (2). Therefore, instead of exhaustively reviewing the literature again, I will recount some of the major discoveries and incorporate them into an integrated view of myoblast fusion. To begin, it is helpful to recall the microscopic accounts of myogenesis.

II. MICROSCOPIC OBSERVATIONS

Observation of myoblast fusion in vitro at the level of the light microscope can be summarized as follows: myogenic cells isolated from muscle tissue adhere to the matrix, proliferate, assume a bipolar morphology, align, and fuse to form multinucleate myotubes with the nuclei frequently initially clustered in a centralized site. With cultures of chick embryo myoblasts, fusion begins 35–40 h after plating of the cells, has a maximal rate at 48–52 h, and is essentially complete after 60–70 h. Rodent cell lines fuse over a period of days only after confluence has been reached and mitogenic stimulus present in serum is reduced. One observation that has proved extremely useful to investigators studying myogenesis is that myotube formation is dependent on the presence of Ca^{2+} in the culture medium (3). In low Ca^{2+} medium chick cells proliferate and differentiate into fusion-competent myoblasts but do not fuse until the Ca^{2+} concentration is raised to about 1 mM. Thus, culturing cells in low Ca^{2+} permits accumulation of a relatively synchronized population of cells poised to fuse upon the addition of Ca^{2+}.

In the early stages of myoblast fusion, the formation of cytoplasmic contiguity can only be established with certainty using ultrastructural techniques. Fine-structural analysis of fusing myoblasts, using serial sectioning and tilting stage analysis to eliminate ambiguity between actual fusion of the plasma membranes and apparent cytoplasmic continuity due to oblique sectioning, showed extensive membrane interdigitation and what appeared to be a single, enlarging cytoplasmic bridge (4). In addition, transmission electron microscopy showed electron-opaque material localized at discrete points of cell–cell contact in a fusing rat myoblast line (102). These observations suggested that fusion occurs at a specialized site(s) on the myoblast surface membrane. Cinemicrographic studies of myogenesis supported this hypothesis by demonstrating that fusion tended to follow end-to-end cell contact rather than side-to-side contact (5).

Correlating electrical coupling with ultrastructural studies, Rash and Fambrough (6) found that closely apposed, but unfused, cells sometimes displayed weak electrical coupling. Such cells were observed by electron microscopy to have gap junction-like structures. The authors reasoned that cell coupling via these close junctions was salient to myoblast fusion because it was rare and occurred at the expected time of fusion and because junction remnants persisted during the early postfusion period. High-efficiency coupling was also observed and was accompanied by the formation of a cytoplasmic bridge between cells. No depolarization occurred during the formation of the cytoplasmic bridge, indicating that membrane integrity was not lost during the fusion process.

Studies on developing rat muscle revealed that the gap junctions also existed in vivo and appeared to form prior to multinucleate cell formation (7). Rash and Staehelin (7) speculated that the gap junctions mediated initial events leading to myoblast fusion. Gap junctions and ionic coupling did not appear to be sufficient for myoblast fusion, however, since Kalderon et al. (8) demonstrated that both phenomena occurred under conditions that blocked myotube formation.

Using ultrastructural, freeze-fracture techniques, Kalderon and Gilula (9) showed that fusion of chick myoblasts took place at particle-free plasma membrane domains that sometimes had intracellular unilamellar, particle-free vesicles associated with them. The lipid-like, cytoplasmic vesicles were absent in cultures treated with agents that prevented myotube formation, such as 5-bromodeoxyuridine (BudR), cycloheximide, or phospholipase C. The authors suggested that the vesicles "initiate the generation of particle-depleted membrane domains, both being essential components in the fusion process." On the other hand, Fumagalli et al. (10) did not observe either unilamellar vesicles or particle-free areas in preparations of fusing rat myoblasts, suggesting that vesicles and particle-free domains may not be a universal requirement for myoblast fusion.

The general picture that emerges from the morphological observations of many investigators is that myotube formation follows from a coordinated sequence of events. Myogenic cells proliferate, differentiate, migrate, and align, and then begin the fusion process.

III. KINETICS OF FUSION

When studied at the level of the light microscope, myotubes form over a period of hours or days. However, since myoblast fusion is a multistep process and since the cells are not synchronized perfectly, many events, in addition to lipid bilayer union, are occurring during the period required to form recognizable myotubes. In primary chick cultures, even if the cells have already differentiated and are fusion-competent, the myoblasts still must migrate, adhere to each other, and only then, fuse. As discussed above, initial fusion of myoblasts cannot be unambiguously identified by light microscopy and is only recognizable some time after actual cytoplasmic continuity has been established.

To measure more precisely the actual event of cytoplasmic bridging between two myoblasts, Rash and Fambrough (6) monitored the electrical coupling of individual myoblasts observed by the light microscope to be in close contact. As described above, high-efficiency coupling was correlated with cytoplasmic contiguity and occurred after about 20 min of monitoring cells. Sometimes the cells displayed low-efficiency coupling prior to acquiring high-efficiency coupling. In either case, the switch to high-efficiency coupling was relatively fast, taking about a minute.

To study the kinetics of fusion in an entire population of myoblasts, Neff et al. (11) cultured myoblasts as aggregates in suspension, thus eliminating the need for cell migration. Chick embryo myoblasts were cultured for 48-52 h in low Ca^{2+} medium to generate a synchronized, fusion-competent (but unfused) population of myoblasts. Fusion was initiated by the addition of Ca^{2+}, and the onset of membrane contiguity was assayed using the transfer of a fluorescent lipid dye. Significant membrane contiguity was observed within 20-30 min. Distinguishable multinucleate cell morphology at the level of the light microscope was not apparent for at least another hour, however.

The above studies suggest that an initial, relatively slow (20-30 min) process occurs between fusion-competent myoblasts in close contact. Subsequently, a more rapid (seconds-1 min) process leads to the union of lipid bilayers and the formation of cytoplasmic continuity.

IV. INHIBITORS OF MYOBLAST FUSION

To begin to describe myoblast fusion at a molecular level, investigators have been experimentally manipulating myotube formation by altering culture conditions and by adding various agents to muscle cells for more than a decade. Because myotube formation is a complex process, it is not surprising that an impressive list of inhibitors of myotube formation has been generated. Generally, the exact mechanism of action of the inhibitors in myogenesis is still a matter of speculation. However, taken together, the results of inhibitor studies support the microscopic observations in indicating that myoblast fusion is a multistep process. In addition, the inhibitor studies implicate roles in myogenesis for a variety of molecules, including, for example, glycoproteins, proteins, proteases, cytoskeletal elements, and lipids.

For the sake of simplicity, the effectors of myotube formation can be grouped according to the way they inhibit the process (summarized in Table 1). For example, agents or culture conditions that affect the differentiation of precursor cells into fusion-competent myoblasts also affect myotube formation. Examples of these include frequent changes of culture medium (12-14) and treatment of cells with BudR (15), dimethylsulfoxide (DMSO) (16), tumor promoters (17), FGF and TGF-β growth factors (17a,b), and antibodies (CSAT) that bind to the integrin glycoprotein complex involved in myoblast-matrix adhesion (17c). Expression of the muscle phenotype, including fusion competency, appears to involve the products of cellular oncogenes (17a,d-f) and genes (e.g., MyoD1) that regulate the coordinate expression of muscle-specific molecules (17g-j).

Myotube formation can also be inhibited by preventing the specific adhesion of fusion-competent myoblasts. Examples of inhibitors in this category include EDTA, energy poisons, cycloheximide, trypsin, inhibitors of cholesterol synthesis, tunicamycin (18,19), and antibodies to the surface of fusion-competent myoblasts (19). In addition, lectins have been reported (20) to inhibit myotube formation and may affect specific adhesion by binding to adhesion glycoproteins on the cell surface.

Still other agents or cell culture conditions block myotube formation by preventing membrane union. Many of these appear to perturb lipids, including phospholipases (21-23), lysophosphatidylcholine (24), exogenous cholesterol (25), lysomotropic amines (26), sodium butyrate (27,28) and manipulation of endogenous lipids (29,30). Membrane union, but not specific adhesion (18,19), is also inhibited by the following known inhibitors of

TABLE 1 Inhibitors of Myotube Formation Can Be Grouped According to the Step of Myogenesis They Appear to Affect

Differentiation	Specific adhesion	Membrane union
BudR	EGTA	EGTA
TPA	Cycloheximide	20 mM Mg^{2+}
DMSO	Tunicamycin	Cytochalasin B
CSAT	25-OH-cholesterol	Colchicine
FGF	Lectins	Elaidate
TGF-β	Energy poisons	1,10-phenanthroline
PG inhibirors	Temp. <30°C	Trifluoperazine
	PI-PLC	Sodium butyrate
		Chloroquine
		Phospholipase C
		TPA
		PG inhibitors

myotube formation: the cytoskeletal-disrupting agents cytochalasin B and colchicine (31,32); trifluoperazine (33), an antagonist of the Ca^{2+}-binding protein calmodulin; and inhibitors of metalloendoproteases (e.g., 1,10-phenanthroline) (34,35). In addition, myotube formation was reported to be inhibited by agents that block prostaglandin synthesis (36). Although the mechanism by which the inhibitors of prostaglandin synthesis act is unclear, it was speculated that the fusion process is regulated by a prostanoid (37). Finally, tumor promoters, in addition to affecting the differentiation step in myogenesis, were reported to block fusion of fusion-competent myoblasts (17), probably by affecting membrane union, since TPA has little effect on the aggregation of fusion-competent myoblasts (Dank and Knudsen, unpublished observations).

At this point, I will focus primarily on the events subsequent to myoblast differentiation, i.e., the adhesion and fusion of fusion-competent myoblasts, and on the molecules that have been purported to be involved in these events.

V. A CLOSER LOOK AT ADHESIVE INTERACTION AMONG MYOBLASTS

The initial event in the process of myoblast fusion is a special adhesive interaction between fusion-competent cells. Using either the light or electron microscope, a close interaction of myoblasts is observed to precede the formation of multinucleate cells. Mixing fusion-competent myoblasts in suspension, Knudsen and Horwitz (38) demonstrated that a specific, Ca^{2+}-dependent adhesion of myoblasts preceded fusion. This cell-cell interaction appeared unique, since fusion-competent myoblasts agglutinated by lectins (18) or grown as aggregates in low Ca^{2+} medium (11) did not fuse.

The specific, Ca^{2+}-dependent adhesion of myoblasts appeared to be an integral part of myoblast fusion. The strength of the interaction increased with time as the myoblast aggregates became increasingly resistant to dispersion until, after 1-2 h, the cells formed distinguishable multinucleate cells (myoballs) (38). In addition, the initial, Ca^{2+}-dependent myoblast interaction had optima for Ca^{2+} concentration, pH, temperature, and culture age that closely paralleled those for myotube formation (38). Finally, certain inhibitors of myoblast fusion, such as EGTA (18) and tunicamycin (19) (Table 1), appeared to exert their action by preventing Ca^{2+}-dependent myoblast adhesion. To date, no agent has been found to inhibit Ca^{2+}-dependent myoblast aggregation without also blocking myoblast fusion. However, while required for myoblast fusion, the Ca^{2+}-dependent interaction of myoblasts is not sufficient, since many inhibitors of myotube formation have no effect on the Ca^{2+}-dependent myoblast adhesion, but, rather, appear to prevent membrane union (18,19) (Table 1).

My laboratory has shown that the Ca^{2+}-dependent adhesion of fusion-competent myoblasts appears to be mediated by surface membrane glycoproteins (19). Furthermore, it appears that some adhesion glycoproteins have a Ca^{2+}-sensitive conformation (19). If fusion-competent myoblasts were treated with trypsin in the absence of Ca^{2+}, the cells did not aggregate when Ca^{2+} was added. On the other hand, if the cells were exposed to Ca^{2+} for 5-10 min before being treated with trypsin, they continued to display the ability to aggregate in the presence of Ca^{2+}.

Gibralter and Turner (39) confirmed the existence of the specific, Ca^{2+}-dependent adhesion of chick myoblasts and also noted the existence of a Ca^{2+}-independent interaction. In addition, Pizzey et al. (39a) observed both a Ca^{2+}-independent and a Ca^{2+}-dependent interaction of mouse myoblasts and noted that the Ca^{2+}-dependent adhesion was destroyed by treating the cells with trypsin in the absence of Ca^{2+} but was protected from the action of trypsin if Ca^{2+} was present. The existence of both Ca^{2+}-dependent and Ca^{2+}-independent cell-cell interactions has been well documented in other cell types (40-42, reviewed in Ref. 43).

Two molecules known to mediate cell-cell adhesion of other cell types are expressed by skeletal muscle. One is the Ca^{2+}-independent neural cell adhesion molecule (NCAM), a member of the immunoglobulin supergene family (reviewed in Refs. 43a,b). NCAM is expressed by developing skeletal muscle and has been shown to have a muscle-specific extracellular domain (43c,d) and to mediate nerve-muscle cell interaction (43e,f). Like nerve NCAM, muscle NCAM is expressed as isoforms with different molecular weights, variable amounts of polysialic acid, and both hydrophobic amino acid and glycosyl-phosphotidylinositol linkages to the cell surface membrane (43h-j). Concomitant with fusion during myogenesis, there is a striking shift in the expression of NCAM isoforms (43i,j), involving a decrease in a higher-molecular-weight polypeptide with a transmembrane anchor and an increase in a lower-molecular-weight polypeptide with a lipid anchor. In chick cultures this shift parallels the expression of the fusion-competent phenotype and the formation of myotubes but occurs even in the absence of cell fusion. In the postnatal rat, NCAM is down-regulated and concentrated in synaptic regions (43g).

The second adhesion molecule that has been reported to be expressed by skeletal muscle is N-cadherin (43k,l), originally identified in mouse and chicken brain cells. N-cadherin is a member of a family of related Ca^{2+}-dependent adhesion molecules involved in morphogenesis (reviewed in Ref. 43m). Cadherins, like NCAM, appear to mediate cell-cell adhesion by a homotypic interaction, involving the binding of cadherin molecules on one cell to cadherin molecules on an adjacent cell. N-cadherin is distinguished from NCAM by being (1) the product of a different gene, (2) immunologically unrelated to NCAM, (3) functional on cells only in the presence of Ca^{2+}, (4) protected from the action of trypsin by the presence of Ca^{2+}. N-cadherin has an intracellular domain that interacts with the cytoskeleton.

Recently we have explored the role of NCAM and N-cadherin in the interaction of fusion-competent myoblasts using specific antibodies to the two adhesion molecules. We found that antibodies to NCAM inhibited Ca^{2+}-independent adhesion, suggesting that NCAM mediates the adhesion of myoblasts observed in the absence of Ca^{2+} (103). In addition we noted that inositol-specific phospholipase C inhibited myoblast aggregation (43n) and released the lipid-linked isoform of NCAM. Anti-NCAM also affected Ca^{2+}-dependent adhesion, suggesting that NCAM may somehow affect the Ca^{2+}-dependent adhesion system. In contrast, antibodies to N-cadherin inhibited Ca^{2+}-dependent myoblast aggregation but had no significant effect on Ca^{2+}-independent myoblast interaction (104). When added together to cells expressing both NCAM and N-cadherin, the antibodies to NCAM and N-cadherin inhibited aggregation occurring in both the absence and presence of Ca^{2+} (Fig. 1).

In summary, a specific adhesion step involving both Ca^{2+}-independent and Ca^{2+}-dependent mechanisms precedes lipid bilayer union and appears to be an integral part of myoblast fusion. Specific adhesion appears to be necessary, but not sufficient, for cell fusion and is likely mediated

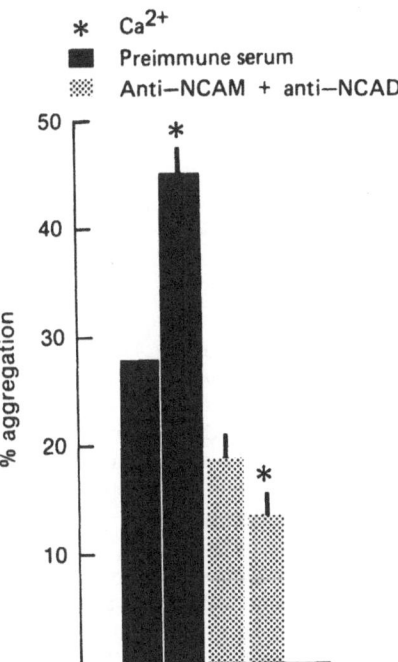

FIGURE 1 Inhibition of the aggregation of fusion-competent chick embryo myoblasts by antibodies to *N*-cadherin and NCAM. Chick myoblasts were prepared from the pectoral muscle of 11-day embryos. The cells were pre-plated twice to enrich for myoblasts, cultured for 52-h in low Ca^{2+} medium (Ca^{2+}-free-DMEM containing 10% FCS and 5% EE), and harvested by pipetting. The cells were resuspended at a concentration of 4×10^5 cells/ml in Hank's balanced salt solution containing 15 mM Hepes and 0.5% glucose and mixed for 18 min at 37°C in the presence of 1 mM EGTA or 1.5 mM Ca^{2+} (*). Ten minutes prior to the addition of EGTA or Ca^{2+}, a combination of pre-immune serum (1:30) and preimmune IgG (75 µg/ml) or anti-*N*-cadherin (anti-NCAD) serum (1:30) and anti-NCAM IgG (75 µg/ml) was added to the cells. Anti-*N*-cadherin was a gift from Dr. Masatoshi Takeichi (Kyoto University, Kyoto, Japan), and anti-NCAM IgG and normal rabbit IgG were gifts from Drs. Michiko Watanabe and Urs Rutishauser (Case Western Reserve University School of Medicine, Cleveland, OH).

by cell-surface glycoproteins including NCAM and *N*-cadherin. We postulate that the specific interaction of fusion-competent myoblasts in the presence of Ca^{2+} triggers subsequent intracellular and membrane events leading to lipid bilayer union and multinucleate cell formation.

VI. CELL-SURFACE PROTEINS AND MYOBLAST FUSION

The identify of the cell-surface molecules involved in both myoblast adhesion and membrane union has interested a number of investigators who have addressed the question using several different approaches. One approach employed the radiolabeling of surface proteins, either by ^{125}I or by

metabolic incorporation of radiolabeled sugars into glycoproteins, followed
by SDS-polyacrylamide gel electrophoretic analysis of membrane preparations
of cells at different stages of myogenesis or treated with known inhibitors
of myotube formation. Using this approach, Cates and Holland (44)
reported that metabolic labeling of fusing chick cells with radioactive
fucose revealed an increased synthesis and accumulation of a 70,000-Dalton
glycoprotein concomitant with the onset of fusion. This change was blocked
by treating the cells with the fusion-inhibiting agent BudR. Labeling pre-,
mid-, and postfusion chick cultures with ^{125}I, Moss et al. (45) concluded
there were both quantitative and qualitative changes during myogenesis
and emphasized two bands of 245,000-Daltons found in prefusion cells that
appeared as a single band during and after fusion.

Analyzing rat muscle cells, surface labeled with ^{125}I, Hynes et al.
(46) reported no detectable qualitative changes pre- and postfusion, but
noted an increase in the level of fibronectin subsequent to fusion. Gardner
and Fambrough (47) found no regulation of fibronectin synthesis in fusing
chick cultures, however. In contrast to Hynes et al. (46), Pauw and David
(48) observed alterations in the ^{125}I-labeling pattern of rat myoblasts
during alignment and fusion, and emphasized a 66,000-Dalton band labeled
during alignment of cells, but not during fusion. Rosenberg et al. (49),
studying plasma membranes isolated from rat myoblasts metabolically labeled
with radioactive sugars, concluded that the predominant change in surface
glycoproteins coincident with fusion was a reduction in the expression
of a 105,000-Dalton band and an increase in a 90,000-Dalton one. This
change was prevented by BudR and phenanthroline, both inhibitors of
myotube formation. Finally, Walsh and Phillips (50) ^{125}I-labeled a clonal
mouse myoblast cell line and noted five proteins of 205,000, 160,000, 70,000,
64,000, and 53,000 Daltons, respectively, as being unique to myoblasts,
while bands of 150,000, 140,000, 54,000, and 36,000 Daltons were found
only in myotubes.

Lectins have also been used to probe for cell-surface changes during
myotube formation. Generally, the approach has been to use radiolabeled
lectins to detect glycoproteins isolated from surface membranes and separated
by SDS-polyacrylamide gel electrophoresis. In this way, Walsh and Phillips
(50) used ^{125}I-concanavalin A (Con A) and ^{125}I-wheat germ agglutinin
(WGA) to detect glycoproteins isolated from rat myoblasts and myotubes.
Con A identified seven glycoproteins found only in myoblasts and in newly
formed myotubes, while WGA identified two glycoproteins not expressed
in mature myotubes. Using a similar approach, Senechal et al. (51) concluded that major changes in glycosylated molecules occurred during myotube formation. Three Con A-binding bands were unique to myotubes.
and four were unique to myoblasts, while the expression of four increased
and another four decreased in myotubes as compared to myoblasts.
Similarly, Holland et al. (52) found developmentally regulated changes
in glycoproteins from rat myoblasts as they fused into myotubes. Coincident with fusion was a decrease in lectin labeling of bands of 136,000
and 49,000 Daltons and a shift in a broad 115,000-Dalton band to 107,000.
These changes were inhibited by growing the cells in BudR.

A third approach to studying surface membrane molecules involved
in myoblast fusion has utilized nonfusing mutants. Cates et al. (53) isolated
two classes of Con A-resistant, nonfusing mutants from the L6 rat myoblast
cell line. One type (RII) was unable to transfer mannose to its lipid-linked
form, and thus, Con A binding to glycoproteins isolated from these cells
was reduced 80% when compared to the wild type. Analysis of a second

type of mutant (RI), which had an unknown biochemical lesion, revealed only one major glycoprotein of 46,000 Daltons that did not bind Con A to the same extent as the wild type. Somatic cell hybridization between types RI and RII restored both fusion and the expression of the 46,000-Dalton glycoprotein. Thus, the authors speculated that this protein may paly some role in myoblast fusion.

Finally, antibodies have been used to probe for cell-surface molecules involved in myoblast fusion. Friedlander and Fischman (54) demonstrated the existence of developmental stage-specific cell-surface antigens, but did not characterize these biochemically. More recently, monoclonal antibodies were used to probe for developmental changes. Grove et al. (55) and Lee and Kaufman (56) reported quantitative changes in cell-surface determinants during myoblast differentiation, but none of these molecules were shown to be involved in fusion.

It is difficult to draw any sort of conclusion from the literature, except that the expression of surface molecules changes during myogenesis. The data have been collected from different species (rat, mouse, chicken) and even, in the case of chick cultures, a mixture of cell types (myoblasts and fibroblasts). There is no consensus at this time concerning surface molecules likely involved in myoblast fusion. Clearly, this area remains a challenge for investigators studying myogenesis.

Because myoblast fusion is a multistep process and is difficult to study in its entirety, my laboratory has begun to dissect the molecular mechanism of myoblast fusion by first focusing on the initial, specific adhesion of fusion-competent myoblasts and the cell-surface molecules involved. Using an immunological approach described by Gerisch (57) and later used by other investigators studying cell-cell and cell-substratum adhesive interactions (reviewed in Ref. 43), we initially generated a heterogeneous, polyclonal antiserum against fusion-competent chick embryo myoblasts (19). This antiserum (anti-myo) inhibited the Ca^{2+}-dependent adhesion of myoblasts (and myotube formation), presumably through the binding of specific antibodies to adhesion-related antigens. The effect of anti-myo was blocked by a glycoprotein fraction isolated by sequential lectin affinity chromatography (WGA and *Lens culinaris* lectin) from 11-day chick embryo pectoral muscle. More recently we have shown that antibodies specific to NCAM and *N*-cadherin perturb myoblast adhesion and fusion (see Section V).

VII. THE ROLE OF LIPIDS IN MYOBLAST FUSION

It is likely that lipids are directly involved in myoblast fusion. Myotube formation is inhibited by treating cells in ways that disturb the lipid composition, such as treating cells with phospholipase C (PLC) (21-23), which degrades phospholipids and, in addition, stimulates phospholipid metabolism (58) and by altering the endogenous membrane lipid composition (25,29,59). Inositol-specific PLC may also act by releasing lipid-linked proteins from the surface of myoblasts.

Whereas the lipid-to-protein ratio in the plasma membranes of muscle cells was reported to be unusually high, no gross differences were observed when comparing the lipid composition of prefusion myoblasts, fusing myoblasts, and myotubes (60). However, as compared to that of fibroblasts, the surface of chick embryo myoblasts was found to be enriched in phosphatidylethanolamine and phosphatidylserine (61). The two- to threefold increase in these potentially fusogenic lipids on the external leaflet of the myoblast surface membrane was found throughout the prefusion period and was speculated to contribute to the ability of fusion-competent myoblasts to fuse (61).

The ganglioside GD_{1a} was reported to increase almost threefold just prior to fusion in a rat myogenic cell line and be absent in mutants unable to fuse (62). In contrast, the accumulation of GD_{1a} was reduced, and GD_3 increased, in fusing quail myoblasts (63). Leskawa et al. (63a) noted that in fusion of L6 rat muscle cells, the biosynthesis of glycosphingolipids was tightly regulated. In avian cells, a generalized increase in glycolipid content at the time of cell recognition and aggregation was observed (64). The increase occurred even when fusion was blocked by treating cells with EGTA or BudR.

Coincident with the initiation of fusion by adding Ca^{2+} to EGTA-treated chick embryo myoblast cultures, Wakelam and Pette (65) noted a rapid (within 15 min) breakdown of phosphatidylinositol, with a concomitant increase in 1,2-diacylglycerol. No breakdown of phosphatidylinositol was detected in fusion-incompetent cells or in cells treated with sodium butyrate, an inhibitor of myotube formation. Wakelam (66) also measured a rapid, Ca^{2+}-induced breakdown in phosphatidylinositol phosphate and phosphatidyl-inositol 4,5-bisphosphate with an increase in synthesis of 1,2-diacylglycerol and phosphatidic acid. Other phospholipids were not affected. The poly-phosphoinositide breakdown was also stimulated by the addition of Sr^{2+}, which Schudt et al. previously reported (67) as the only cation other than Ca^{2+} able to promote myoblast fusion. In contrast, high concentrations of Mg^{2+}, which block cell fusion (67), inhibited the polyphosphoinositide breakdown (66). Thus, Wakelam (66) postulated that the breakdown of inositol phospholipids and the synthesis of phosphatidic acid and diacyl-glycerol were salient to myoblast fusion.

A general increase in diacylglycerol and/or phosphatidic acid did not appear sufficient to stimulate myoblast fusion, however, since their levels were raised by agents that inhibited, rather than promoted, myoblast fusion. Grove and Schimmel (68) noted that the fusion-blocking tumor promoter 12-O-tetradecanoylphorbol 13-acetate (TPA) caused a rapid (15-30 Min) twofold increase in the 1,2-diacylglycerol (DAG) level in differentiated chick embryo myoblasts. The stimulation of the DAG level was insensitive to Ca^{2+} and appeared to derive from breakdown of acetylcholine (69). TPA also stimulated synthesis of phosphatidic acid, phosphatidylinositol, and phosphatidylcholine (69). The authors speculated that the TPA stimu-lated a membrane-associated phospholipase C, which hydrolyzed phos-phatidylcholine and produced 1,2-diacylglycerol, which was then converted to phosphatidic acid, phosphatidylinositol, and back to phosphatidylcholine. The effects of TPA likely derive from an initial stimulation of protein kinase C (reviewed in Ref. 69a).

The availability of cholesterol was shown to modulate the fusion of myoblasts. van der Bosch et al. (25) observed that cholesterol added exo-genously to the medium of chick embryo myoblast cultures inhibited myo-tube formation. Cornell et al. (70) demonstrated that cholesterol synthesis was necessary for myoblast fusion. When cultures of chick embryo myoblasts were grown in lipid-depleted medium and treated with an inhibitor of cholesterol synthesis, 25-hydroxycholesterol, myotube formation was inhibited. Lowrey and Horwitz (71) concluded that 25-hydoxycholesterol somehow blocked the specific adhesion of fusion-competent myoblasts, since Ca^{2+}-dependent myoblast aggregation was inhibited 90% within 2 h after the inhibitor was added to myoblast cultures. No alterations in bulk protein, DNA, RNA, or phospholipid synthesis, or in the cholesterol content of the plasma membrane, resulted from inhibiting cholesterol synthesis (70). In addition, the inhibitor did not affect the synthesis of two developmentally regulated proteins—creatinine phosphokinase and acetylcholine receptor—or decrease the synthesis or activity of lactate dehydrogenase as rapidly, or to the same extent, as it affected myoblast aggregation (71). The authors concluded "that the level of cholesterol synthesis influences myoblast fusion in a selective manner and does not arise from a general inhibition of either membrane protein incorporation or synthesis or from an inhibition of the expression of differentiation-dependent proteins."

In summary, lipids appear to play several roles in myoblast fusion. Certainly, a particular lipid composition appears essential for lipid bilayer union, since altering endogenous lipid composition can affect myoblast fusion. In addition, as will be discussed later, the breakdown of phosphatidylinositol 4,5-bisphosphate may be an important signal in stimulating myoblast fusion. Finally, synthesis of cholesterol is important to the phenotypic expression of specific myoblast adhesion.

VIII. THE ROLE OF CALCIUM IN MYOBLAST FUSION

Myotube formation is a Ca^{2+}-dependent process (3). Sr^{2+} is the only other divalent cation reported to partially promote myoblast fusion, while Mg^{2+}, at high concentrations, actually blocks myotube formation (67). Calcium appears to play multiple roles in myoblast fusion. It stimulates an initial, specific adhesion of myoblasts (38). In addition, Ca^{2+} appears to protect the cell surface adhesion glycoproteins from the action of exogenously added trypsin, suggesting that the cation may induce a specific conformation in molecules involved in mediating adhesion (19).

In addition to being essential for myoblast adhesion, Ca^{2+} appears to be required for membrane union (18,19,72). David et al. (72) noted that when the Ca^{2+} concentration was lowered after the myoblasts had aligned on culture dishes, myotube formation was inhibited. In addition, we observed that if Ca^{2+} was removed following myoblast aggregation in suspension, the cells remained as aggregates but did not fuse into multinucleate myoballs (19), suggesting that Ca^{2+} acts in at least two different ways to promote myoblast fusion.

Several other lines of evidence suggest Ca^{2+} has multiple sites of action. First, Knudsen and Horwitz (18) observed that, while 20 mM Mg^{2+} promoted initial myoblast aggregation, the apparent strength of the interaction did not increase and the cells failed to fuse, even if 1 mM Ca^{2+} was present. Thus, it appeared that Mg^{2+} could substitute for Ca^{2+} in stimulating myoblast adhesion, but not in promoting membrane union. Second, trifluoperazine, an antagonist of the Ca^{2+}-binding protein calmodulin, was shown to inhibit myotube formation (33). Trifluoperazine had little effect on myoblast aggregation (19), however, suggesting that calmodulin (along with Ca^{2+}) was involved in membrane union but not specific adhesion.

The entry of Ca^{2+} into the cell appears to be required for myoblast fusion. Initially, using the ionophore A23187 to increase cytoplasmic Ca^{2+}, Schudt and Pette (73) concluded that an increase in cytoplasmic Ca^{2+} was not necessary to stimulate fusion, but, rather, that Ca^{2+} acted at the cell surface. More recent results provide a different picture. David et al. (72) and Schollmeyer (74) demonstrated that the addition of A23187 induced precocious fusion in chick embryo and rat myoblasts, respectively. David et al. (72) also reported a measurable increase in net Ca^{2+} influx just prior to fusion. Furthermore, David et al. (72) demonstrated that a Ca^{2+} channel blocker, D600, inhibited cell fusion. These observations led David et al. (72) to propose that an increase in intracellular Ca^{2+} triggered membrane union of myoblasts in close contact.

In short, Ca^{2+} plays multiple key roles in myogenesis. While not required for the differentiation events resulting in fusion-competent myoblasts, Ca^{2+} (about 1 mM) is essential for both Ca^{2+}-dependent myoblast interaction and membrane union. Ca^{2+} stimulates adhesion of fusion-competent myoblasts, perhaps by inducing a special conformation in adhesion glycoproteins. Furthermore, intracellular Ca^{2+} from extracellular and/or intracellular pools, working in concert with calmodulin, appears essential for membrane union. Since both Ca^{2+} (75) and calmodulin (76) are known to play regulatory roles in intracellular events, an increase in intracellular Ca^{2+} concentration may trigger the molecular events that lead to membrane union.

IX. THE ROLE OF cAMP AND PROTEIN KINASES IN MYOBLAST FUSION

Wahrmann et al. (77) noted that addition of cyclic AMP (cAMP), dibutyryl cyclic AMP (db cAMP), or theophylline to a rat cell line 24 h after plating reversibly inhibited myotube formation. Similarly, Zalin (78) noted a delay in the onset of fusion when adding db cAMP to chick embryo myoblast cultures 24 h after plating. High levels of cAMP in young cultures appeared to disturb the expression of the fusion-competent phenotype.

When Zalin and Montague (79) measured endogenous cAMP levels in chick embryo myoblast cultures, they found that the basal intracellular level increased 10- to 15-fold for approximately 1 h, 5-6 h before the onset of myoblast fusion. Adenylate cyclase activity and cAMP-sensitive protein kinase activity also increased during this time, leading the authors to speculate that a transient rise in cAMP concentration was involved in the control of the expression of myoblast differentiation. When prostaglandin E_1 (PGE_1) was added exogenously 10 h prior to the onset of myoblast fusion, it induced a rise in the intracellular cAMP level and also induced precocious formation of myotubes (80,81). David and Higginbotham (81) noted that an increase in intracellular cAMP at this time was not sufficient to stimulate fusion, however, since the increase was observed even when myotube formation was inhibited by low Ca^{2+} medium, again suggesting that cAMP was involved in expression of the fusion-competent phenotype.

More recently, Entwistle et al. (37) reported that an increase in cAMP (induced by the addition of 10^{-8} M PGE_1 or PGE_2) at the onset of fusion blocked myotube formation. The authors suggested that, while an experimentally induced increase in cAMP levels 10 h prior to the onset of myotubule formation stimulated precocious fusion, an increase in cAMP in differentiated cultures beginning to fuse antagonized a process involved in myotube formation.

Schützle et al. (82) observed a gradual rise in cAMP following the onset of fusion in chick embryo cultures, but did not observe a rise in cAMP levels prior to myoblast fusion. They further noted that cells grown in low Ca^{2+} medium exhibited only a small rise in cAMP, but that, following the addition of Ca^{2+}, an increase in cAMP levels occurred concomitant with fusion. These authors postulated that the large increase in cAMP was not salient to the fusion process itself but, rather, was a consequence of the stimulation of myoblast fusion.

Thus, it appears that cAMP plays multiple roles in myogenesis. An early, transient rise in the intracellular cAMP level prior to the onset of fusion may be involved in signaling the expression of the fusion-competent phenotype. An increase in the cAMP level at the onset of fusion does not appear to be a fusion-promoting signal, however, since an experimentally induced increase in cAMP at this time inhibited myotube formation. On the other hand, a sustained rise in the cAMP level subsequent to myoblast fusion may signal altered protein synthesis in the myotubes.

A role for protein kinase C in myoblast fusion is implicated by the observations of several laboratories that tumor promotors known to stimulate protein kinase C activity inhibit the fusion of myoblasts expressing the fusion-competent phenotype (17,18; Dank and Knudsen, unpublished observations). In contrast to cAMP-dependent protein kinase activity, the total calcium-, phospholipid-dependent protein kinase C activity was observed by Adamo et al. (82a) to decline as embryonic chick muscle cells fused. These authors also noted that a shift in the subcellular distribution of protein kinase C from a particulate to a soluble fraction occurred as the myoblasts formed myotubes and that TPA stimulated the phosphorylation of several proteins in myotubes that were spontaneously phosphorylated in myoblasts. On the other hand, Toutant and Sobel (82b) noted that TPA stimulated the phosphorylation of certain cytoplasmic proteins in myoblasts but did not stimulate phosphorylation of the same proteins in myotubes. Finally, Lognonne and Wahrmann (82c) reported that Ca^{2+} induced the phosphorylation of a 48-kD protein present at the surface of fusing rat myoblasts.

Farzaneh et al. (82d) observed that TPA could overcome the delay in myoblast fusion induced by inhibitors of prostanoid biosynthesis and, thus, suggested that protein kinase C activation may be involved in mediating the activity of PGE_1 during myoblast differentiation and fusion (see Section X). However, they did not show the effect of TPA on fusion-competent cells in the absence of prostanoid synthesis inhibitors.

The above data, taken together with the studies on inositol phospholipid turnover (see Section VII) suggest that protein kinase C may play a role in myoblast fusion. This role is perhaps complex and may involve both stimulation and down-regulation of cell fusion.

X. THE ROLE OF PROSTAGLANDINS IN MYOBLAST FUSION

Prostaglandins have been postulated to play a role in myogenesis. Zalin (36) reported that physiological concentrations of PGE_1, added to chick embryo muscle cultures prior to the onset of fusion, provoked a burst of myoblast fusion 5 h later. In addition, both Zalin (36) and David and Higginbotham (81) noted that indomethacin, an inhibitor of cyclooxygenase and, thus, prostaglandin synthesis, blocked fusion when added to 24-h chick embryo cultures. The effect of indomethacin was reversed by PGE (36). Zalin (36) speculated that prostaglandin synthesis was required for generation of the transient increase in the intracellular level of cAMP observed prior to the onset of myoblast fusion (see Section IX).

David and Higginbotham (81) also observed that the addition of PGE_1 (10^{-6} M) to chick embryo cultures prior to the onset of fusion stimulated an increase in net Ca^{2+} influx in chick embryo myoblast cultures, in addition to increasing intracellular cAMP and stimulating precocious fusion. The PGE_1-stimulated Ca^{2+} influx appeared to follow or be independent of the increase in cAMP, however. Inhibitors of prostaglandin synthesis blocked the increase in Ca^{2+} influx observed in control cells, indicating that the cation influx was influenced by PGE_1.

Entwistle et al. (37) observed that indomethacin blocked myoblast fusion without affecting cell proliferation, alignment, or the expression of either acetylcholine receptors or creatine phosphokinase. However, the concentrations of PGE_1 (10^{-9} M) required to reverse the indomethacin block of myotube formation had no effect on the cAMP level, suggesting that the prostaglandin regulated fusion through a cAMP-independent mechanism. The authors noted that eicosatrienoate and linoleate, but not PGE_2 or arachidonate, reversed the indomethacin effect, suggesting the fusion-regulating prostanoid was of the one series. Zalin (81a) also noted that inhibitors of prostanoid synthesis depressed prostaglandin E levels and differentiation in human myoblast cultures.

In contrast to the above observations, Schützle et al. (82) found that indomethacin addition had no effect on myotube formation in chick embryo cultures. However, these authors noted that indomethacin inhibited the rise in cAMP accompanying fusion in control cultures and, in addition, resulted in lowered creatine kinase levels. Based on these results, Schützle et al. concluded that prostaglandins played no primary role in myoblast fusion but, rather "that prostaglandin synthesis is a consequence of the stimulation of the stimulation of myoblast fusion and that via cyclic AMP it stimulates protein synthesis." Similarly, Steiner et al. (83) observed no inhibition in the fusion of cloned mouse myoblasts treated with aspirin or indomethacin, both inhibitors of cyclooxygenase activity. In contrast, they noted that inhibitors of the lipoxygenase pathway inhibited myoblast fusion. Thus, they concluded that leukotrienes or hydroxyeicosatetraeonic acids, rather than prostaglandins, play an active role in myoblast fusion.

The apparently contradictory results described above likely emanate from differences in when the assay measuring myotube formation was terminated and scored. For instance, my laboratory (Goretsky and Knudsen, unpublished observations) observed a 20-30% inhibition of myoblast fusion by indomethacin if myotube formation was scored 4-20 h after the addition of Ca^{2+} to initiate fusion, but only a 5% inhibition if myotube formation was scored 54 h after Ca^{2+} addition, even though fresh indomethacin was added every 24 h. The activity of indomethacin in 24-h, cell-conditioned medium was confirmed by the ability of the medium to inhibit platelet secretion by more than 85%. In other words, inhibiting prostaglandin synthesis delays myoblast fusion (81b), rather than blocking it.

Entwistle et al. (81b) have suggested that myoblast fusion may be initiated by the depolarization of a high resting membrane potential in fusion-competent myoblasts leading to a concomitant rise in intracellular Ca^{2+}. In support of this hypothesis, they observed that the delay in myoblast fusion due to inhibitors of prostanoid synthesis was overcome by adding PGE_1, raising extracellular potassium, or adding the acetylcholine receptor agonist carbachol. The effect of each reagent was inhibited by Ca^{2+} channel blockers. The same authors further noted that antagonists of the nicotinic acetylcholine receptor delayed spontaneous myoblast fusion (81c).

The work of Hausman and Berggrun (81d) has shown that myoblasts express a developmentally regulated binding of PGE_1 that peaks prior to myoblast interaction. These authors have proposed a sequence of events during myogenesis involving synthesis and release of PG, transient increased binding of PG, myoblast-myoblast adhesion, and membrane fusion.

Thus, prostaglandins may participate in inducing the fusion-competent phenotype, perhaps by causing a rise in intracellular Ca^{2+}. They may also affect the rate of lipid bilayer union.

XI. PROTEASES AND MYOBLAST FUSION

Some years ago Couch and Strittmatter (34) noted that inhibitors of metalloendoproteases blocked myotube formation. The inhibitors were effective when added to fusion-competent rat myoblasts shortly (2 h) before initiation of fusion by addition of Ca^{2+} to cells grown in low Ca^{2+} medium. This was in contrast to serine and thiol protease inhibitors, which did not affect myotube formation, and a carboxyprotease inhibitor, which only transiently delayed fusion. These results suggested a specific role for metalloendoproteases at the time of fusion. Further work by Couch and Strittmatter (35) indicated that the activity of a soluble metalloendoprotease with a pI of 4.8 was salient to myoblast fusion. The authors speculated that the enzyme was perhaps involved in: (1) posttranslational processing of plasma membrane proteins essential for fusion, (2) initiating a cascade of events leading to fusion, or (3) generating a fusogenic peptide that induced membrane union.

Similarly, we found that the metalloendoprotease inhibitor 1,10-phenanthroline blocked the fusion of chick embryo myoblasts (19). However, the inhibitor had no effect on the initial, Ca^{2+}-dependent aggregation of fusion-competent myoblasts (19), suggesting that the enzyme was not involved in specific adhesion of myoblasts. Instead, the metalloendoprotease appeared to be involved in membrane union, since 1,10-phenanthroline prevented the fusion of myoblasts aggregated in the presence of Ca^{2+} (19).

Rosenberg et al. (49) presented evidence that inhibitors of metalloendoproteases altered the expression of the plasma membrane glycoproteins of rat myoblasts. They noted that the inhibitors prevented a fusion-correlated reduction in the expression of a 105,000-Dalton glycoprotein and the appearance of a 90,000-Dalton glycoprotein (both of unknown function), suggesting the metalloendoprotease may play a role in the surface remodeling that accompanies myoblast fusion. Similarly, Baldwin and Kayalar (83a) showed that metalloendoprotease inhibitors blocked both fusion and biochemical differentiation of L_6 rat myoblasts.

In addition to the soluble metalloendoprotease, other proteases may be involved in myoblast fusion. Kaur and Sanwal (84) observed an increase in a Ca^{2+}-activated neutral proteinase about the time of fusion. In addition, Schollmeyer (74) demonstrated redistribution of a Ca^{2+}-activated protease, concomitant with an influx of Ca^{2+} and the initiation of rat myoblast fusion. The Ca^{2+}-activated protease redistributed from a dispersed, random pattern in proliferating myoblasts to a predominantly peripheral pattern in prefusion myoblasts. At the same time fibronectin disappeared from the surface of fusing myoblasts, leading Schollmeyer (74) to speculate that an influx of Ca^{2+} activated the protease and promoted rearrangement of the membrane to accommodate fusion.

In summary, the activity of a metalloendoprotease appears to be important to the membrane union step of myoblast fusion. Perhaps, along with Ca^{2+}-activated proteases, it is involved in a remodeling of the plasma membrane necessary for lipid bilayer union. For example, in a model system, the fusion of rat erythrocytes by a membrane-mobility agent unable to fuse human erythrocytes was correlated with rat erythrocyte membrane protein degradation by Ca^{2+}-activated proteases (85). Such protease activity was missing in human erythrocytes.

XII. MEMBRANE REMODELING AND MYOBLAST FUSION

There is ample evidence from studies employing surface radiolabeling, lectins, and antibodies to indicate that the surface of the myoblast is remodeled during fusion (see Section VI; also Refs. 74,86-90). Furthermore, an increase in membrane fluidity and the movement of molecules embedded in the membrane accompanies myoblast fusion (91-95b). Herman and Fernandez (96) correlated the changes in topography and lateral translational mobility of concanavalin A (Con A) receptors on the surface of chick embryo muscle cells with altered membrane fluidity during fusion and concluded that changes in membrane fluidity modulated receptor mobility. The increase in membrane fluidity appears to be salient to myoblast fusion, since exposure of myoblasts to fatty acids that tend to decrease fluidity was observed to delay or block fusion, while exposure to those that increase fluidity enhanced fusion (29,30,59). Changes in membrane electrical properties (96a) and membrane order as measured by EPR (96b) also accompany myoblast differentiation.

To study surface membrane events during myoblast fusion in an isolated system, Schudt et al. (97) prepared plasma membrane vesicles from cultured fusion-competent myoblasts. The authors noted that the vesicles fused in the presence of Ca^{2+}. The amount of vesicle fusion, like myotube formation, was proportional to the temperature and dependent on culture age (98). Pore formation, as detected using freeze-fracture techniques and electron microscopy, was preceded by vesicle adhesion and by an aggregation of intramembranous particles (IMP) in the area of contact between two vesicles (97).

Both major alterations in the surface membrane and an increase in membrane fluidity appear to be salient to myoblast fusion. Rearrangement of surface membrane proteins may be initiated prior to and during specific myoblast adhesion, while biochemical modification and further surface remodeling likely continues until lipid bilayer union is completed.

XIII. THE CYTOSKELETON AND MYOBLAST FUSION

A role for the cytoskeleton in myoblast fusion has been implicated by the results of Holtzer et al. (32), who observed that cytoskeleton disrupting agents, such as cytochalasin B and colcemid, inhibit myotube formation. Knudsen and Horwitz (18) noted that these agents had little effect on the initial, Ca^{2+}-dependent aggregation of fusion-competent myoblasts, however. Rather, the inhibitors prevented fusion of aggregated myoblasts, suggesting that the cytoskeleton participated in membrane union. Fulton et al. (99) showed that extensive reorganization of the cytoskeleton occurred during myoblast fusion. It may be that the cytoskeleton plays an active role in remodeling the myoblast surface membrane so that union of the lipid bilayers can proceed.

XIV. SUMMARY

Although complete understanding of myoblast fusion is still a future goal, several conclusions can be drawn at this time. Ignoring the events involved in myoblast differentiation and focusing only on those events effecting myoblast fusion, we can conclude that fusion consists of at least two essential steps—specific myoblast adhesion and membrane union. The adhesion step includes both a Ca^{2+}-independent mechanism and a Ca^{2+}-dependent process involving cell surface glycoproteins having a Ca^{2+}-sensitive conformation. The molecules mediating myoblast adhesion appear to include both the Ca^{2+}-independent adhesion molecule, NCAM, and the Ca^{2+}-dependent adhesion molecule, N-cadherin. Other additional adhesion molecules may also be involved. Specific adhesion precedes membrane union by about 20 min, suggesting that further molecular events are required before lipid bilayer union can occur. Therefore, it seems reasonable to postulate that myoblast fusion is analogous to a receptor-mediated process in which the specific adhesion of myoblasts triggers subsequent events that lead to membrane union.

Membrane union requires Ca^{2+}, a particular lipid bilayer composition, and the participation of calmodulin, the cytoskeleton, and a metalloendoprotease. Myoblast fusion is accompanied by the expression of new surface membrane (glyco)proteins, the modification of existing ones, a rearrangement of surface molecules, and an increase in surface membrane fluidity. These dynamic molecular alterations probably accompany membrane union and may be initiated by specific myoblast adhesion.

Which molecules are responsible for coupling specific myoblast adhesion and membrane union is a matter of speculation. Cyclic AMP is a well-known second messenger and appears to affect protein synthesis throughout myogenesis, but does not appear to act directly in myoblast fusion. On the other hand, an increase in intracellular Ca^{2+} is a possible candidate for the signal that initiates membrane union following myoblast adhesion, since fusion is Ca^{2+}-dependent and since Ca^{2+} appears to enter the cell during myotube formation. An increase in intracellular Ca^{2+} may also result from hydrolysis of phosphatidylinositol 4,5-bisphosphate, since the breakdown product, inositol trisphosphate, is known to mobilize intracellular Ca^{2+} (reviewed in Ref. 100).

As a stimulus to our thinking, and hopefully that of others, we have designed the following model for myoblast fusion (depicted diagrammatically in Fig. 2). In our model, myoblast fusion is initiated by the interaction of Ca^{2+}-independent and Ca^{2+}-dependent adhesion molecules on the surface of myoblasts expressing the fusion-competent phenotype. Adhesion may stimulate the entry of Ca^{2+} into the myoblast. It also may trigger the breakdown of inositol phospholipids via phospholipase C (PLC), which may be coupled to adhesion through a GTP-binding protein (G). The breakdown of inositol phospholipids may increase membrane fluidity. In addition, the hydrolysis of 4,5-bisphosphate (PIP_2) generates inositol trisphosphate (IP_3), which is known to mobilize Ca^{2+} from internal stores and may raise (or sustain) the intracellular Ca^{2+} level. Calcium, perhaps in concert with calmodulin, may activate Ca^{2+}-sensitive proteins, including proteases and, perhaps, cytoskeletal components, which, in turn, may act to remodel the myoblast surface and promote membrane union. In addition, diacylglycerol (DAG), generated from the hydrolysis of PIP_2, may activate protein kinase C, a Ca^{2+}- and phospholipid-dependent enzyme

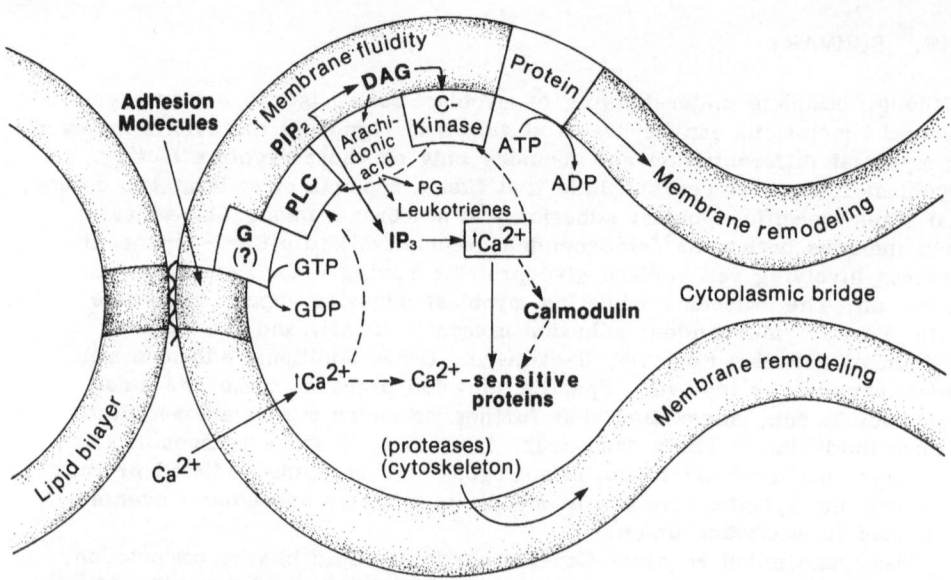

FIGURE 2   Schematic model postulating events in myoblast fusion. G, GTP-binding protein; PLC, phospholipase C; PIP_2, phosphatidylinositol 4,5-bisphosphate; DAG, 1,2-diacylglycerol; IP_3, inositol triphosphate; C-kinase, protein kinase C; PG, prostaglandins.

(reviewed in Ref. 101). Newly phosphorylated proteins may alter cellular functions and, perhaps, participate in remodeling the myoblast surface to promote lipid bilayer union (or to dampen it since TPA, which stimulates protein kinase C, can inhibit myoblast fusion). As arachidonic acid is released from DAG, prostaglandins and leukotrienes may be synthesized. While prostaglandins are probably not involved directly in evoking myoblast fusion, they may affect the rate of fusion and may also stimulate adenyl cyclase, thus increasing the cAMP level, stimulating cAMP-dependent protein kinases activity and protein phosphorylation, and, thereby, altering protein synthesis. A sustained increase in cAMP may also signal changes that dampen or inhibit further myoblast fusion since an experimentally induced increase in cAMP at the onset of fusion inhibits myotube formation.

Although the field is still in its infancy and many challenges face investigators studying myogenesis, a picture of myoblast fusion is emerging. Hopefully, this chapter will stimulate imaginative contemplation and creative experimentation that will lead to an increased understanding of myoblast fusion at the molecular level.

ACKNOWLEDGMENT

I express my appreciation to Ms. Sena Smith for technical assistance and to Drs. Alan F. Horwitz, Nicole Neff, and Margaret Wheelock for their helpful comments on this manuscript.

Work from our laboratory has been supported by grants from the Muscular Dystrophy Association, the National Science Foundation (DCB-8615890), and NIH (AR3745).

REFERENCES

1. Bischoff, R. (1978). Myoblast fusion. In: *Membrane Fusion* (G. Poste and G. L. Nicolson, eds.). Elsevier/North-Holland, New York, pp. 127–179.

2. Wakelam, M. J. O. (1985). The fusion of myoblasts. *Biochem. J.* 228: 1–12.

3. Shainberg, A., Yagil, G., and Yaffe, D. (1969). Control of myogenesis in vitro by Ca^{2+} concentration in nutritional medium. *Exp. Cell. Res.* 58:163–167.

4. Lipton, B. H., and Konigsberg, I. R. (1972). A fine-structural analysis of the fusion of myogenic cells. *J. Cell Biol.* 53:348–364.

5. Fear, J. (1977). Observations on the fusion of chick embryo myoblasts in culture. *J. Anat.* 124:437–444.

6. Rash, J. E., and Fambrough, D. (1973). Ultrastructural and electrophysiological correlates of cell coupling and cytoplasmic fusion during myogenesis in vitro. *Dev. Biol.* 30:166–186.

7. Rash, J. E., and Staehelin, L. A. (1974). Freeze-cleave demonstration of gap junctions between skeletal myogenic cells in vivo. *Dev. Biol.* 36:455–461.

8. Kalderon, N., Epstein, M. L., and Gilula, N. B. (1977). Cell-to-cell communication and myogenesis. *J. Cell Biol.* 75:788–806.

9. Kalderon, N., and Gilula, N. B. (1979). Membrane events involved in myoblast fusion. *J. Cell Biol.* 81:411–425.

10. Fumagalli, G., Brigonzi, A., Tachikawa, T., and Clementi, F. (1981). Rat myoblast fusion: Morphological study of membrane apposition, fusion, and fission during controlled myogenesis in vitro. *J. Ultrastruct. Res.* 75:112–125.

11. Neff, N., Decker, C., and Horwitz, A. (1984). The kinetics of myoblast fusion. *Exp. Cell Res.* 153:25–31.

12. Doering, J. L., and Fischman, D. A. (1977). A fusion-promoting macromolecular factor in muscle conditioned medium. *Exp. Cell Res.* 105:437–443.

13. Delain, D., Wahrmann, J. P., and Gros, F. (1981). Influence of external diffusible factors on myogenesis of the cells of the line L6. *Exp. Cell Res.* 131:217–224.

14. Linkhart, T. A., Clegg, C. H., and Hauschka, S. D. (1981). Myogenic differentiation in permanent clonal mouse myoblast cell lines: Regulation by macromolecular growth factors in the culture medium. *Dev. Biol.* 86:19–30.

15. Bischoff, R., and Holtzer, H. (1970). Inhibition of myoblast fusion after one round of DNA synthesis in 5-bromodeoxyuridine. *J. Cell Biol.* 44:134–150.

16. Blau, H. M., and Epstein, C. J. (1979). Manipulation of myogenesis in vitro: Reversible inhibition by DMSO. *Cell* 17:95–108.

17. Cohen, R., Pacifici, M., Rubinstein, N., Biehl, J., and Holtzer, H. (1977). Effect of a tumour promoter on myogenesis. *Nature* 266:538–540.

17a. Olson, E. N. (1988). Regulation of myogenic differentiation by polypeptide growth factors and oncogenes. *Cancer Bull.* 40:304–309.

17b. Allen, R. E. and Boxhorn, L. K. (1987). Inhibition of skeletal muscle satellite cell differentiation by transforming growth factor-beta. *J. Cell Physiol.* 133:567–572.

17c. Menko, A. S. and Boettiger, D. (1987). Occupation of the extracellular matrix receptor, integrin, is a control point for myogenic differentiation. *Cell* 51:51–57.

17d. Gossett, L. A., Zhang, W., and Olson, E. N. (1988). Dexamethasone-dependent inhibition of differentiation of C2 myoblasts bearing steroid-inducible N-ras oncogenes. *J. Cell Biol.* 106:2127-2137.

17e. Leibovitch, M-P., Leibovitch, S. A., Hillion, J., Guillier, M., Schmitz, A., and Harel, J. (1987). Possible role of c-fos, c-N-ras and c-mos proto-oncogenes in muscular development. *Exp. Cell Res.* 170:80-92.

17f. Denis, N., Blanc, S., Leibovitch, M. P., Nicolaiew, N., Dautry, F., Raymondjean, M., Kruh, J., and Kitzis, A. (1987). c-myc oncogene expression inhibits the initiation of myogenic differentiation. *Exp. Cell Res.* 172:212-217.

17g. Tapscott, S. J., Davis, R. L., Thayer, M. J., Cheng, P-F., Weintraub, H., and Lassar, A. B. (1988). MyoD1: A nuclear phosphoprotein requiring a myc homology region to convert fibroblasts to myoblasts. *Science* 242:405-411.

17h. Lassar, A. B., Paterson, B. M., and Weintraub, H. (1986). Transfection of a DNA locus that mediates the conversion of 10T1/2 fibroblasts to myoblasts. *Cell* 47:649-656.

17i. Braun, T., Buschhausen-Denker, G., Bober, E., Tannich, E., and Arnold, H. H. (1989). A novel human muscle factor related to but distinct from myoD1 induces myogenic conversion in 10T1/2 fibroblasts. *EMBO J.* 8:701-709.

17j. Wrigt, W. E., Sassoon, D. A., and Lin, V. K. (1989). Myogenin, a factor regulating myogenesis, has a domain homologous to myoD. *Cell* 56:607-617.

18. Knudsen, K. A., and Horwitz, A. F. (1978). Differential inhibition of myoblast fusion. *Dev. Biol.* 66:294-307.

19. Knudsen, K. A. (1985). The calcium-dependent myoblast adhesion that precedes cell fusion is mediated by glycoproteins. *J. Cell Biol.* 101:891-897.

20. Den, H., Malinzak, D. A., Keating, H. J., and Rosenberg, A. (1975). Influence of concanavalin A, wheat germ agglutinin, and soybean agglutinin on the fusion of myoblasts in vitro. *J. Cell Biol.* 67:826-834.

21. Nameroff, M., Trotter, J. A., Keller, J. M., and Munar, E. (1973). Inhibition of cellular differentiation by phospholipase C. I. Effects of the enzyme on myogenesis and chondrogenesis in vitro. *J. Cell Biol.* 58:107-118.

22. Nameroff, M., and Munar, E. (1976). Inhibition of cellular differentiation by phospholipase C. II. Separation of fusion and recognition among myogenic cells. *Dev. Biol.* 49:288-293.

23. Schudt, C., and Pette, D. (1976). Influence of monosaccharides, medium factors and enzymatic modification of fusion of myoblasts in vitro. *Cytobiologie* 12:74-84.

24. Reporter, M., and Norris, G. (1973). Reversible effects of lysolecithin on fusion of cultured rat muscle cells. *Differentiation* 1:83-95.

25. van der Bosch, J., Schudt, C., and Pette, D. (1973). Influence of temperature, cholesterol, dipalmitoyllecithin and Ca^{2+} on the rate of muscle cell fusion. *Exp. Cell Res.* 82:433-438.

26. Kent, C. (1982). Inhibition of myoblast fusion by lysosomotropic amines. *Dev. Biol.* 90:91-98.

27. Fiszman, M. Y., Montarras, D., Wright, W., and Gros, F. (1980). Expression of myogenic differentiation and myotube formation by chick embryo myoblasts in the presence of sodium butyrate. *Exp. Cell Res.* 126:31-37.

28. Minty, C., Montarras, D., Fiszman, M. Y., and Gros, F. (1981). Butyrate-treated chick embryo myoblasts synthesize new proteins. *Exp. Cell Res.* 133:63-72.

29. Horwitz, A., Wight, A., Ludwig, P., and Cornell, R. (1978). Inter-related lipid alterations and their influence on the proliferation and fusion of cultured myogenic cells. *J. Cell Biol.* 77:334-357.

30. Horwitz, A., Wight, A., and Knudsen, K. (1979). A role for lipid in myoblast fusion. *Biochem. Biophys. Res. Commun.* 86:514-521.

31. Sanger, J. W. (1974). The use of cytochalasin B to distinguish myoblasts from fibroblasts in cultures of developing chick striated muscle. *Proc. Natl. Acad. Sci. USA* 71:3621-3625.

32. Holtzer, H., Croop, J., Dienstman, S., Ishikawa, H., and Somlyo, A. P. (1975). Effects of cytochalasin B and colcemide on myogenic cultures. *Proc. Natl. Acad. Sci. USA* 72:513-517.

33. Bar-Sagi, D., and Prives, J. (1983). Trifluoperazine, a calmodulin antagonist, inhibits muscle cell fusion. *J. Cell Biol.* 97:1375-1380.

34. Couch, C. B., and Strittmatter, W. J. (1983). Rat myoblast fusion requires metalloendoprotease activity. *Cell* 32:257-265.

35. Couch, C. B., and Strittmatter, W. J. (1984). Specific blockers of myoblast fusion inhibit a soluble and not the membrane-associated metalloendoprotease in myoblasts. *J. Biol. Chem.* 259:5396-5399.

36. Zalin, R. J. (1977). Prostaglandins and myoblast fusion. *Dev. Biol.* 59:241-248.

37. Entwistle, A., Curtis, D. H., and Zalin, R. J. (1986). Myoblast fusion is regulated by a prostanoid of the one series independently of a rise in cyclic AMP. *J. Cell Biol.* 103:857-866.

38. Knudsen, K. A., and Horwitz, A. F. (1977) Tandem events in myoblast fusion. *Dev. Biol.* 58:328-338.

39. Gibralter, D., and Turner, D. C. (1985). Dual adhesion systems of chick myoblasts. *Dev. Biol.* 112:292-307.

39a. Pizzey, J. A., Jones, G. E., and Walsh, F. S. (1988). Requirements for the Ca^{2+}-independent component in the initial intercellular adhesion of C2 myoblasts. *J. Cell Biol.* 107:2307-2317.

40. Takaeichi, M. (1977). Functional correlation between cell adhesive properties and some cell surface proteins. *J. Cell Biol.* 75:464-474.

41. Brackenbury, R., Rutishauser, U., and Edelman, G. M. (1981). Distinct calcium-independent and calcium-dependent adhesion systems of chicken embryo cells. *Proc. Natl. Acad. Sci. USA* 78:387-391.

42. Magnani, J. L., Thomas, W. A., and Steinberg, M. S. (1981). Two distinct adhesion mechanisms in embryonic neural retina cells. I. A kinetic analysis. *Dev. Biol.* 81:96-105.

43. Damsky, C. H., Knudsen, K. A., and Buck, C. A. (1984). Integral membrane glycoproteins in cell-cell and cell-substratum adhesion. In: *The Biology of Glycoproteins* (R. J. Ivatt, ed.). Plenum Press, New York, pp. 1-64.

43a. Edelman, G. M. (1986). Cell adhesion molecules in the regulation of animal form and tissue pattern. *Annu. Rev. Cell Biol.* 2:81-116.

43b. Cunningham, B. A., Hemperly, J. J., Murray, B. A., Prediger, E. A., Brackenbury, R., and Edelman, G. M. (1987). Neural cell adhesion molecule: Structure, immunoglobulin-like domains, cell surface modulation, and alternative RNA splicing. *Science* 236:799-806.

43c. Dickson, G., Gower, H. J., Barton, C. H., Prentice, H. M., Elsom, V. L., Moore, S. E., Cox, R. D., Quinn, C., Putt, W., and Walsh, F. S. (1987). Human muscle neural cell adhesion molecule (N-CAM):

Identification of a muscle-specific sequence in the extracellular domain. *Cell* 50:1119-1130.

43d. Prediger, E. A., Hoffman, S., Edelman, G. M., and Cunningham, B. A. (1988). Four exons encode a 93-base-pair insert in three neural cell adhesion molecule mRNAs specific for chicken heart and skeletal muscle. *Proc. Natl. Acad. Sci. USA* 85:9616-9620.

43e. Grumet, M., Rutishauser, U., and Edelman, G. M. (1982). Neural cell adhesion molecule is on embryonic muscle cells and mediates adhesion to nerve cells in vitro. *Nature (London)* 295:693-695.

43f. Rutishauser, U., Grumet, M., and Edelman, G. (1983). Neural cell adhesion molecule mediates initial interactions between spinal cord neurons and muscle cells in culture. *J. Cell Biol.* 97:145-152.

43g. Covault, J., and Sanes, J. R. (1986). Distribution of N-CAM in synaptic and extrasynaptic portions of developing and adult skeletal muscle. *J. Cell Biol.* 102:716-730.

43h. Barton, C. H., Dickson, G., Gower, H. J., Rowett, L. H., Putt, W., Elsom, V., Moore, S. E., Goridis, C., and Walsh, F. S. (1988). Complete sequence and in vitro expression of a tissue-specific phosphatidylinositol-linked N-CAM isoform from skeletal muscle. *Development* 104:165-173.

43i. Covault, J., Merlie, J. P., Goridis, C., and Sanes, J. R. (1986). Molecular forms of N-CAM and its RNA in developing and denervated skeletal muscle. *J. Cell Biol.* 102:731-739.

43j. Moore, S. E., Thompson, J., Kirkness, V., Dickson, J. G., and Walsh, F. S. (1987). Skeletal muscle neural cell adhesion molecule (N-CAM): Changes in protein and mRNA species during myogenesis of muscle cell lines. *J. Cell Biol.* 105:1377-1386.

43k. Hatta, K., and Takeichi, M. (1986). Expression of N-cadherin adhesion molecules associated with early morphogenetic events in chick development. *Nature* 320:447-449.

43l. Hatta, K., Takagi, S., Fujisawa, H., and Takeichi, M. (1987). Spatial and temporal expression pattern of N-cadherin cell adhesion molecules correlated with morphogenetic processes of chicken embryos. *Dev. Biol.* 120:215-227.

43m. Takeichi, M. (1988). The cadherins: Cell-cell adhesion molecules controlling animal morphogenesis. *Development* 102:639-655.

43n. Knudsen, K. A., Smith, L., and McElwee, S. (1989). Involvement of cell surface phosphatidylinositol-anchored glycoproteins in cell-cell adhesion of chick embryo myoblasts. *J. Cell Biol.* 109:1779-1786.

44. Cates, G. A., and Holland, P. C. (1978). Biosynthesis of plasma-membrane proteins during myogenesis of skeletal muscle in vitro. *Biochem. J.* 174:873-881.

45. Moss, M., Norris, J. S., Peck, E. J., Jr., and Schwartz, R. J. (1978). Alterations in iodinated cell surface proteins during myogenesis. *Exp. Cell Res.* 113:445-450.

46. Hynes, R. O., Martin, G. S., Shearer, M., Critchley, D. R., and Epstein, C. J. (1976). Viral transformation of rat myoblasts: Effects on fusion and surface properties. *Dev. Biol.* 48:35-46.

47. Gardner, J. M., and Fambrough, D. M. (1983). Fibronectin expression during myogenesis. *J. Cell Biol.* 96:474-485.

48. Pauw, P. G., and David, J. D. (1979). Alterations in surface proteins during myogenesis of a rat myoblast cell line. *Dev. Biol.* 70:27-38.

49. Rosenberg, J., Szabo, A., Rheuark, D., and Kayalar, C. (1985). Correlation between fusion and the developmental regulation of membrane glycoproteins in L_6 myoblasts. *Proc. Natl. Acad. Sci USA* 82:8409-8413.

50. Walsh, F. S., and Phillips, E. (1981). Specific changes in cellular glycoproteins and surface proteins during myogenesis in clonal muscle cells. *Dev. Biol.* 81:229-237.

51. Senechal, H., Schapira, G., and Wahrmann, J. P. (1982). Changes in plasma membrane glycoproteins during differentiation of an established myoblast cell line and a non-fusing α-amanitin-resistant mutant. *Exp. Cell Res.* 138:355-365.

52. Holland, P. C., Pena, S. D. J., and Guerin, C. W. (1984). Developmental regulation of neuraminidase-sensitive lectin-binding glycoproteins during myogenesis of rat L_6 myoblasts. *Biochem. J.* 218:465-473.

53. Cates, G. A., Brickenden, A. M., and Sanwal, B. D. (1984). Possible involvement of a cell surface glycoprotein in the differentiation of skeletal myoblasts. *J. Biol. Chem.* 259:2446-2650.

54. Friedlander, M., and Fischman, D. A. (1979). Immunological studies of the embryonic muscle cell surface. Antiserum to the prefusion myoblast. *J. Cell Biol.* 81:193-214.

55. Grove, B. K., Schwartz, G., and Stockdale, F. E. (1981). Quantitation of changes in cell surface determinants during skeletal muscle cell differentiation using monospecific antibody. *J. Supramol. Struct. Cell. Biochem.* 17:147-152.

56. Lee, H. U., and Kaufman, S. J. (1981). Use of monoclonal antibodies in the analysis of myoblast development. *Dev. Biol.* 81:81-95.

57. Gerisch, G. (1980). Univalent antibody fragments as tools for the analysis of cell interactions in *Dictyostelium*. *Curr. Top. Dev. Biol.* 14:243-270.

58. Kent, C. (1979). Stimulation of phospholipid metabolism in embryonic muscle cells treated with phospholipase C. *Proc. Natl. Acad. Sci. USA* 76:4474-4478.

59. Nakornchai, S., Falconer, A. R., Fisher, D., Goodall, A. H., Hallinan, T., and Lucy, J. A. (1981). Effects of retinol, fatty acids and glycerol monooleate on the fusion of chick embryo myoblasts in vitro. *Biochim. Biophys. Acta* 643:152-160.

60. Kent, C., Schimmel, S. D., and Vagelos, P. R. (1974). Lipid composition of plasma membranes from developing chick muscle cells in culture. *Biochim. Biophys. Acta* 360:312-321.

61. Sessions, A., and Horwitz, A. F. (1981). Myoblast aminophospholipid asymmetry differs from that of fibroblasts. *FEBS Lett.* 134:75-78.

62. Whatley, R., Ng, S. K-C., Rogers, J., McMurray, W. C., and Sanwal, B. D. (1976). Developmental changes in gangliosides during myogenesis of a rat myoblast cell line and its drug resistant variants. *Biochem. Biophys. Res. Commun.* 70:180-185.

63. Dubois, C., Hauttecoeur, B., Coulon-Morelec, M-J., Montarras, D., Rampini, C., and Fiszman, M. Y. (1984). Changes in ganglioside metabolism during in vitro differentiation of quail embryo myoblasts. *Dev. Biol.* 105:509-517.

63a. Leskawa, K. C., Erwin, R. E., Buse, P. E., and Hogan, E. L. (1988). Glycosphingolipid biosynthesis during myogenesis of rat L6 cells in vitro. *Mol. Cell. Biochem.* 83:47-54.

64. McEvoy, F. A., and Ellis, D. E. (1977). Glycolipids and myoblast differentiation. *Biochem. Soc. Trans.* 5:1719-1721.

65. Wakelam, M. J. O., and Pette, D. (1982). The breakdown of phosphatidylinositol in myoblasts stimulated to fuse by the addition of Ca^{2+}. *Biochem. J.* 202:723-729.

66. Wakelam, M. J. O. (1983). Inositol phospholipid metabolism and myo-
    blast fusion. *Biochem. J.* 214:77-82.
67. Schudt, C., van der Bosch, J., and Pette, D. (1973). Inhibition of
    muscle cell fusion in vitro by Mg^{2+} and K^+ ions. *FEBS Lett.* 32:296-298.
68. Grove, R. I., and Schimmel, S. D. (1981) Generation of 1,2-
    diacylglycerol in plasma membranes of phorbol ester-treated myoblasts.
    Biochem. Biophys. Res. Commun. 102:158-164.
69. Grove, R. I., and Schimmel, S. D. (1982). Effects of 12-O-
    tetradecanoylphorbol 13-acetate on glycerolipid metabolism in cultured
    myoblasts. *Biochem. Biophys. Acta* 711:272-280.
69a. Nishizuka, Y. (1986). Studies and perspectives of protein kinase C.
    Science 233:305-311.
70. Cornell, R. B., Nissley, S. M., and Horwitz, A. F. (1980). Cholesterol
    availability modulates myoblast fusion. *J. Cell Biol.* 86:820-824.
71. Lowrey, C. H., and Horwitz, A. F. (1982). Effect of inhibitors of
    cholesterol synthesis on muscle differentiation. *Biochim. Biophys. Acta*
    712:430-432.
72. David, J. D., See, W. M., and Higginbotham, C-A. (1981). Fusion
    of chick embryo skeletal myoblasts: Role of calcium influx preceding
    membrane union. *Dev. Biol.* 82:297-307.
73. Schudt, C., and Pette, D. (1975). Influence of the ionophore A 23
    187 on myogenic cell fusion. *FEBS Lett.* 59:36-38.
74. Schollmeyer, J. E. (1986). Role of Ca^{2+} and Ca^{2+}-activated protease
    in myoblast fusion. *Exp. Cell Res.* 162:411-422.
75. Rasmussen, H. (1970). Cell communication, calcium ion, and cyclic
    adenosine monophosphate. *Science* 170:404-412.
76. Klee, C. B., Crouch, T. H., and Richman, P. G. (1980). Calmodulin.
    Annu. Rev. Biochem. 49:489-515.
77. Wahrmann, J. P., Winand, R., and Luzzati, D. (1973). Effect of cyclic
    AMP on growth and morphological differentiation of an established
    myogenic cell line. *Nat. New Biol.* 245:112-113.
78. Zalin, R. J. (1973). The relationship of the level of cyclic AMP to
    differentiation in primary cultures of chick muscle cells. *Exp. Cell
    Res.* 78:152-158.
79. Zalin, R. J., and Montague, W. (1974). Changes in adenylate cyclase,
    cyclic AMP, and protein kinase levels in chick myoblasts, and their
    relationship to differentiation. *Cell* 2:103-108.
80. Zalin, R. J., and Leaver, R. (1975). The effect of a transient increase
    in intracellular cyclic AMP upon muscle cell fusion. *FEBS Lett.* 53:33-
    36.
81. David, J. D., and Higginbotham, C.-A. (1981). Fusion of chick embryo
    skeletal myoblasts: interactions of prostaglandin E_1, adenosine 3':5'
    monophosphate, and calcium flux. *Dev. Biol.* 82:308-316.
81a. Zalin, R. J. (1987). The role of hormones and prostanoids in the
    in vitro proliferation and differentiation of human myoblasts. *Exp.
    Cell Res.* 172:265-281.
81b. Entwistle, A., Zalin, R. J., Bevan, S., and Warner, A. E. (1988).
    The control of chick myoblast fusion by ion channels operated by
    prostaglandins and acetylcholine. *J. Cell Biol.* 106:1693-1702.
81c. Entwistle, A., Zalin, R. J., Warner, A. E., and Bevan, S. (1988).
    A role for acetylcholine receptors in the fusion of chick myoblasts.
    J. Cell Biol. 106:1703-1712.
81d. Hausman, R. E., and Berggrun, D. A. (1987). Prostaglandin binding
    does not require direct cell-cell contact during chick myogenesis in
    vitro. *Exp. Cell Res.* 168:457-462.

82. Schützle, U. B., Wakelam, M. J. O., and Pette, D. (1984). Prostaglandins and cyclic AMP stimulate creatine kinase synthesis but not fusion in cultured embryonic chick muscle cells. *Biochim. Biophys. Acta* 805:204-210.

82a. Adamo, S., Caporale, C., Nervi, C., Ceci, R., and Molinaro, M. (1989). Activity and regulation of calcium-, phospholipid-dependent protein kinase in differentiating chick myogenic cells. *J. Cell Biol.* 108:153-158.

82b. Toutant, M., and Sobel, A. (1987). Protein phosphorylation in response to the tumor promoter TPA is dependent on the state of differentiation of muscle cells. *Dev. Biol.* 124:370-378.

82c. Lognonne, J. L., and Wahrmann, J. P. (1986). Spontaneous myoblast fusion is mediated by cell surface Ca^{2+}-dependent protein kinase(s). *Exp. Cell Res.* 166:340-356.

82d. Farzaneh, F., Entwistle, A., and Zalin, R. J. (1989). Protein kinase C mediates the hormonally regulated plasma membrane fusion of avian embryonic skeletal muscle. *Exp. Cell Res.* 181:298-304.

83. Steiner, S., Manley, G., and Adams, T. (1984). Effect of inhibitors of the lipoxygenase pathway on mouse myoblast fusion. *Exp. Cell Res.* 155:289-293.

83a. Baldwin, E., and Kayalar, C. (1986). Metalloendoprotease inhibitors that block fusion also prevent biochemical differentiation in L_6 myoblasts. *Proc. Natl. Acad. Sci. USA* 83:8029-8033.

84. Kaur, H., and Sanwal, B. D. (1981). Regulation of the activity of a calcium-activated neutral protease during differentiation of skeletal myoblasts. *Can. J. Biochem.* 59:743-747.

85. Kosower, N. S., Glaser, T., and Kosower, E. M. (1983). Membrane-mobility agent-promoted fusion of erythrocytes: Fusibility is correlated with attack by calcium-activated cytoplasmic proteases on membrane proteins. *Proc. Natl. Acad. Sci. USA* 80:7542-7546.

86. Furcht, L. T., Wendelschafer-Crabb, G., and Woodbridge, P. A. (1977). Cell surface changes accompanying myoblast differentiation. *J. Supramol. Struct.* 7:307-322.

87. Kaufman, S. J., and Foster, R. F. (1985). Remodeling of the myoblast membrane accompanies development. *Dev. Biol.* 110:1-14.

88. Kaufman, S. J., Foster, R. F., Haye, K. R., and Faiman, L. E. (1985). Expression of a developmentally regulated antigen on the surface of skeletal and cardiac muscle cells. *J. Cell Biol.* 100:1977-1987.

89. Haye, K. R., Foster, R. F., Goff, J. P., and Kaufman, S. J. (1986). Endocytosis of α_2-macroglobulin is developmentally regulated during myogenesis. *Dev. Biol.* 114:470-474.

90. Wakshull, E., Bayne, E. K., Chiquet, M., and Fambrough, D. M. (1983). Characterization of a plasma membrane glycoprotein common to myoblasts, skeletal muscle satellite cells, and glia. *Dev. Biol.* 100: 464-477.

91. Prives, J., and Shinitzky, M. (1977). Increased membrane fluidity precedes fusion of muscle cells. *Nature* 268:761-763.

92. Herman, B. A., and Fernandez, S. M. (1978). Changes in membrane dynamics associated with myogenic fusion. *J. Cell. Physiol.* 94:253-264.

93. Sandra, A., Leon, M. A., and Przybylski, R. J. (1977). Suppression of myoblast fusion by concanavalin A: Possible involvement of membrane fluidity. *J. Cell Sci.* 28:251-272.

94. Elson, H. F., and Yguerabide, J. (1979). Membrane dynamics of differentiating cultured embryonic chick skeletal muscle cells by fluorescence microscopy techniques. *J. Supramol. Struct.* 12:47-61.

95. Weidekamm, E., Schudt, C., and Brdiczka, B. (1976). Physical properties of muscle cell membranes during fusion. A fluorescence polarization study with the ionophore A23187. *Biochim. Biophys. Acta* 443:169-180.

95a. Kawasaki, Y., Wakayama, N., and Seto-Ohshima, A. (1988). Lateral motion of fluorescent molecules embedded into cell membranes of clonal myogenic cells, L6, changes upon cell maturation. *FEBS Lett.* 231: 321-326.

95b. Sawyer, J. T., and Akeson, R. A. (1986). Differential redistribution of lectin receptor classes on clonal rat myotubes and myoblasts. *J. Cell Sci.* 83:181-196.

96. Herman, B. A., and Fernandez, S. M. (1982). Dynamics and topographical distribution of surface glycoproteins during myoblast fusion: A resonance energy transfer study. *Biochemistry* 21:3275-3283.

96a. Bonincontro, A., Cametti, C., Hausman, R. E., Indovina, P. L., and Santini, M. T. (1987). Changes in myoblast membrane electrical properties during cell-cell adhesion and fusion in vitro. *Biochim. Biophys. Acta* 903:89-95.

96b. Santini, M. T., Indovina, P. L., and Hausman, R. E. (1987). Changes in myoblast membrane order during differentiation as measured by EPR. *Biochim. Biophys. Acta* 896:19-25.

97. Schudt, C., Dahl, G., and Gratzl, M. (1976). Calcium-induced fusion of plasma membranes isolated from myoblasts grown in culture. *Cytobiologie* 13:211-223.

98. Dahl, G., Schudt, C., and Gratzl, M. (1978). Fusion of isolated myoblast plasma membranes. An approach to the mechanism. *Biochim. Biophys. Acta* 514:105-116.

99. Fulton, A. B., Prives, J. Farmer, S. R., and Penman, S. (1981). Developmental reorganization of the skeletal framework and its surface lamina in fusing muscle cells. *J. Cell Biol.* 91:103-112.

100. Berridge, M. J., and Irvine, R. F. (1984). Inositol trisphosphate, a novel second messenger in cellular signal transduction. *Nature* 312: 315-321.

101. Nishizuka, Y. (1984). The role of protein kinase C in cell surface signal transduction and tumour promotion. *Nature* 308:693-698.

102. Engel, L. C., Egar, M. W., and Przybylski, R. J. (1985). Morphological characterization of actively fusing L_6 myoblasts. *Eur. J. Cell Biol.* 39:360-365.

103. Knudsen, K. A., McElwee, S. A., and Smith, L. (1990). A role for the neural cell adhesion molecule, NCAM, in myoblast interaction during myogenesis. *Dev. Biol.* 138:159-168.

104. Knudsen, K. A., Myers, L., and McElwee, S. A. (1990). A role for the Ca^{2+}-dependent adhesion molecule, N-cadherin, in myoblast interaction during myogenesis. *Exp. Cell Res.* 188 (in press).

27

Fusion of Sperm and Egg Plasma Membranes During Fertilization

CHARLES G. GLABE

University of California—Irvine, Irvine, California

KEELUNG HONG

University of California—San Francisco, San Francisco, California

VICTOR D. VACQUIER

University of California—San Diego, La Jolla, California

I. INTRODUCTION

Spern are one of few cell types that fuse with the plasma membrane of another cell under physiological conditions. One of the functions of the sperm is to deliver its haploid nucleus to the egg cytoplasm. In this sense, the sperm is like a virus with a very large genome that is propelled by a flagellum. Using the electron microscope, the Colwins discovered that the union of the two gametes during fertilization involved the fusion of sperm and egg plasma membranes (1). Although the ultrastructure of gamete fusion during fertilization has been well studied, our understanding of its molecular basis has increased very little. In this chapter we focus on recent reports that bear on the molecular mechanism of gamete plasma membrane fusion. For a more detailed analysis of membrane fusion during fertilization the reader is directed to several excellent reviews (2-6).

Experimental observations with model membrane systems in vitro suggest that two requirements must be met for the fusion of two adjacent phospholipid bilayers: (1) The outer leaflets of adjacent bilayers must closely approach each other to a distance of less than approximately 13 Å. (2) The phospholipids in the outer leaflets must undergo a change in organization that results in their coalescence. The topology of the bilayer dictates that a transient intermediate stage must exist where the outer leaflets are continuous while the inner leaflets remain separate.

The egg plasma membrane is typically covered with glycocalyx or extracellular envelope (Figs. 1 and 2). The specific recognition and adhesion of spern to the egg surface facilitates fertilization of the egg by homologous sperm. In terms of sperm-egg fusion, some portion of the sperm plasma membrane (such as that covering the acrosomal process) must penetrate the egg glycocalyx and also any material coating the acrosomal membrane before the sperm and egg plasma membranes are able to closely approach each other. These structures would also tend

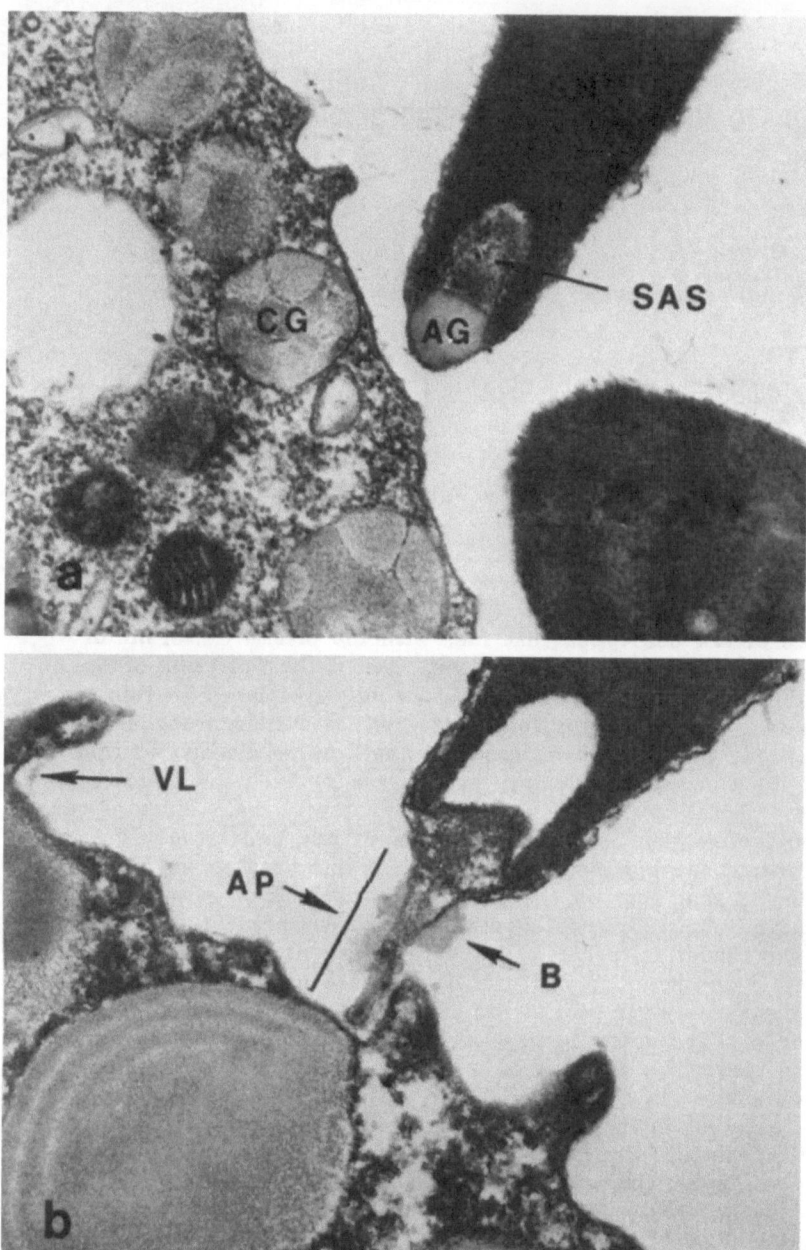

FIGURE 1 Electron micrograph of sperm-egg interaction during sea urchin fertilization. (a) The anterior region of the sperm is shown approaching the egg surface. The acrosome granule (AG) is located at the apex of the sperm overlying an indentation of the sperm nucleus (SN) known as the subacrosomal space (SAS). Cortical granules (CG) are located beneath the egg plasma membrane (50,000×). (b) The anterior region of an acrosome-reacted sperm is shown adhering to the egg vitelline layer (VL) covering the egg plasma membrane. The acrosomal granule has undergone exocytosis and an acrosomal process (AP) is formed by the polymerization of actin filaments in the subacrosomal space. The acrosomal process is coated with bindin, the persistent contents of the acrosomal granule, which mediates the adhesion of the acrosomal process to the vitelline layer.

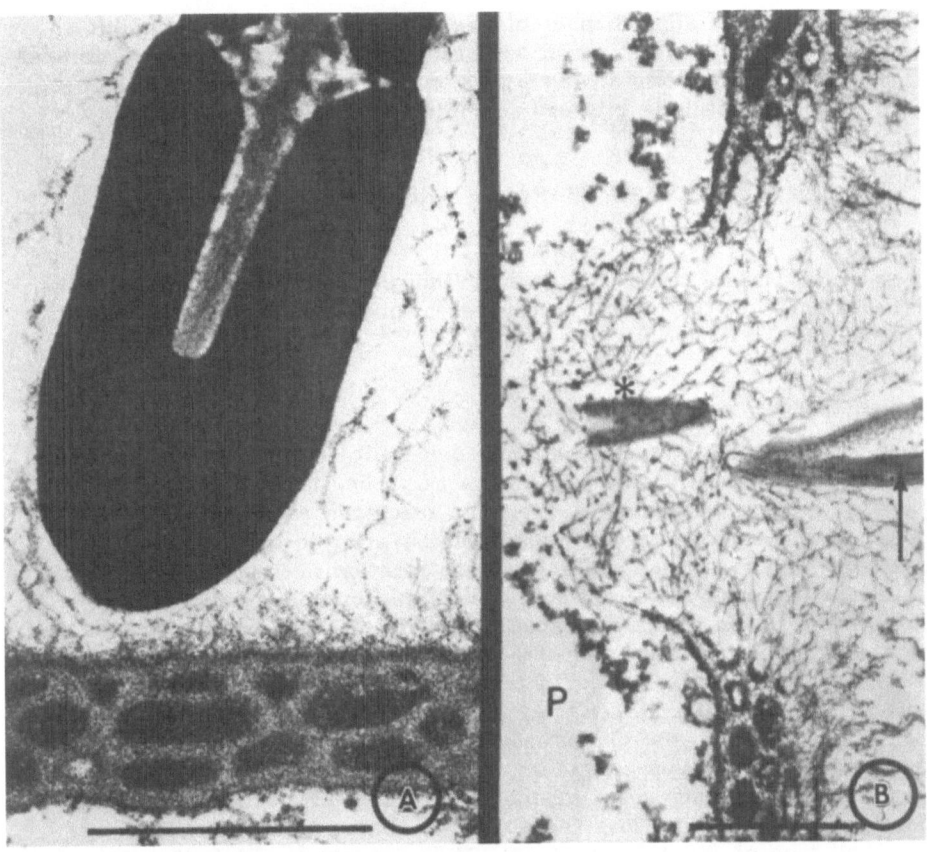

FIGURE 2 Electron micrograph of sperm-egg interaction in the abalone,
Haliotus rufescens. (A) The anterior end of the large intact acrosomal granule
binds to fibrous material on the outer surface of the vitelline layer. The
acrosomal granule contains a darker ovoid material (D) at its distal end
and a lighter material (L) at the proximal end. The partially preformed
acrosomal process is located in an indentation of the acrosomal granule
and the sperm nucleus and appears to couple the granule to the nucleus.
(B) The anterior end of the sperm is shown dissolving a hole in the vitelline
layer. The sperm is coming from the right. The acrosome granule has
undergone exocytosis, releasing the lysin from the granule. The acrosomal
process (arrow) is covered by the former membrane of the acrosomal
granule. The membrane is coated with a striated, paracrystalline material.
A mass of vitelline layer fibers surrounds the acrosomal process. The
bar in both panels is 1 µm.

to prevent the close apposition of plasma membranes and hence inhibit membrane fusion (7). As in model membrane systems, the plasma membranes may also have to overcome strong hydration and weaker ionic forces that tend to repel approaching bilayers (cf. Chapter 7).

II. THE GAMETE PLASMA MEMBRANES ARE SPECIALIZED FOR FUSION

Because most plasma membranes do not readily fuse with one another when cells make contact, the plasma membranes of one or both gametes must be specialized for fusion.

A. The Sperm Plasma Membrane

The sperm membrane is specialized for fusion with other plasma membranes. Fusion of the sperm with the egg plasma membrane is restricted to specific regions of the spermatozoon (1,8,9). The fusogenic regions are often exposed or modified by the acrosome reaction prior to or during sperm-egg contact. In sea urchins, the acrosome reaction is induced by a fucose sulfate-rich glycoconjugate of the jelly coat surrounding the egg. The acrosome reaction in sea urchin sperm consists of the exocytosis of the acrosomal granule and the extension of the acrosomal process resulting from the polymerization of actin in the subsacrosomal space (10) (Fig. 1). In sea urchins and many other marine animals, the site of plasma membrane fusion is restricted to the plasma membrane covering the acrosomal process of the sperm. This fusion-competent region of the sperm is the former inner surface of the acrosomal granule membrane. After the acrosome reaction, the sperm rapidly lose the ability to adhere to and fertilize eggs with a half-time of about 20 s (11,12). It is not yet clear whether this rapid loss of fertilizability results from a loss in the ability of acrosome-reacted sperm to adhere to the egg surface "receptors" or is due to a limited lifetime of the fusion-competent state. In addition to fusing with the egg plasma membrane, the acrosomal membrane is capable of fusing with other plasma membranes, such as the sperm plasma membrane overlying the flagellum (Glabe, unpublished observations), mitochondrion and acrosomal process of adjacent sperm (13), the plasma membrane of blastomeres, blastula cells, and adult somatic cells (14).

In mammalian sperm, the region of the sperm plasma membrane that participates in fusion is restricted to the equatorial segment and post-acrosomal segment of acrosome-reacted sperm (9). The membrane in this region also appears to be specialized for fusion. After the acrosome reaction, intramembrane particle-free regions accumulate on the P face of the membrane in the postacrosomal region of guinea pig sperm (15). Phospholipid vesicles bind to and fuse with the guinea pig sperm plasma membrane at this specialized postacrosomal region (15). *Chlamydomonas* gametes also display particle-free areas on regions of the plasma membrane which are specialized for membrane fusion (16).

B. The Egg Plasma Membrane

The egg is also specialized for ensuring successful fusion of the plasma membrane of sperm. Many types of eggs have specific cell surface receptors for sperm which serve to increase the probability of successful fusion by

sperm contacting the egg surface (17,18). The receptors appear to fall
into two general classes: receptors located on the glycocalyx immediately
adjacent to the egg plasma membrane and receptors in the extracellular
coat surrounding the egg, known as the zona pellucida or vitelline envelope.
Sea urchin sperm bind to a glycocalyx known as the vitelline layer (Fig. 1).
Mammalian oocytes appear to have both types of receptors since sperm
can adhere both to the zona pellucida and to the glycocalyx overlying
the egg plasma membrane (9,19).

Many eggs are also adapted to ensure fusion with only one sperm.
Inmost types of eggs, fusion of the egg plasma membrane with more than
one sperm results in abnormal cleavage and ultimately developmental arrest.
Several different mechanisms have evolved to prevent polyspermic fertilization (for reviews see Refs. 4,6,20). In eggs of many different species,
a rapid, partial, and temporary block to polyspermy protects the egg from
penetration by supernumerary sperm until a more permanent block is established. This so-called "rapid" block to polyspermy is mediated by a change
in the membrane potential of the egg (6,21). The sperm initiates a depolarization in the plasma membrane of these types of eggs. In sea urchins,
anuran amphibians, and the Echiuroid worm, *Urechis caupo*, the membrane
potential becomes positive after rapid depolarization from a resting potential
of -40 to -70 mV. This positive potential is maintained for a length of
time that depends on the type of egg. During the period of time when
the potential is positive, the egg plasma membrane is refractory to fusion
with the supernumerary sperm. If inseminated, voltage-clamped eggs are
given a brief pulse (28-40 msec) to -60 mV, the eggs subsequently fertilize
(22). This demonstrates that the membrane of adherent sperm is capable
of rapidly fusing with the egg plasma membrane if the positive voltage
is not maintained. If the membrane potential is clamped at positive levels
in these types of eggs, the eggs remain unfertilized after addition of sperm
although the sperm bind normally to the egg surface. If the potential is
clamped at -30 mV, the eggs become polyspermic (6,21,23).

Experiments with cross-species fertilization indicate that this ability
to modulate fusion by the membrane potential of the egg resides in the
sperm. Sperm of *Urechis* efficiently fertilize eggs of the sea urchin *Strongylocentrotus purpuratus* and induce a normal activation potential in the egg. In
Urechis eggs, a more positive potential is required to prevent fertilization
than in the sea urchin *S. purpuratus* (50 mV for *Urechis* vs. 20 mV for
S. purpuratus). Voltage-clamping *S. purpuratus* eggs at 20 mV inhibited
fertilization by sea urchin sperm, but had no effect on fertilization of
the eggs by *Urechis* sperm. However, raising the voltage-clamped potential
to 50 mV inhibited fertilization by *Urechis* sperm (24). A similar situation
exists with *Xenopus* gametes and gametes of the salamander *Notophthalmus*.
Fertilization of salamander eggs is not regulated by the membrane potential
of the egg, and fertilization in the salamander is normally polyspermic
(25). Salamander sperm can fertilize *Xenopus* eggs. When the *Xenopus*
eggs were voltage-clamped at positive values, fertilization proceeded at
normal values with salamander sperm while fertilization with homologous
Xenopus sperm was inhibited (26).

The region of the egg plasma membrane that is capable of fusion with
the sperm plasma membrane is restricted in some eggs. In many different
species, the egg surface is covered with microvilli. In sea urchins (2),
mammals (9), and polychaete worms (27) fusion with the sperm membrane
is commonly observed to occur at the tips of the microvilli. In most amphibians, sperm only fuse with the animal pole of the egg. In the frog,

Discoglossus, the site of sperm fusion is restricted to a differentiated region of the membrane known as the animal dimple. Fucose-containing glycoconjugates are localized to this region of fusion-competent membrane (28).

III. SPECIFIC MOLECULES ARE LOCALIZED ON THE FUSOGENIC REGIONS OF THE SPERM PLASMA MEMBRANE

In many marine animals, the membrane of the acrosomal process is coated with the exposed contents of the acrosomal granule (2,29). These include lysin proteins (2), which are necessary for the penetration of the sperm through the extracellular envelopes, and adhesive proteins (bindins), which mediate the adhesion of sperm to the egg surface (30,31). It is widely presumed that acrosomal enzymes hydrolyze a path for sperm to penetrate the egg envelopes since the acrosome of mammalian sperm contains the protease acrosin (32) and a trypsinlike protease activity has been localized to the acrosomal process of sea urchin sperm (33). Although enzyme activities have been reported and extracts of sperm from marine invertebrates are capable of dissolving the egg envelopes (34), there has been no demonstration of an enzyme-substrate relationship of a sperm enzyme with the egg investments. The acrosomal granule of the snail, *Tegula*, and the abalone, *Haliotis*, contains a protein lysin that dissolves the vitelline envelope by a nonenzymatic, stoichiometric mechanism (Fig. 2) (35,36). The abalone lysin does not cleave macromolecular components of the vitelline envelope and it does not produce reducing sugar, peptides, lysophosphatides, or free sulfhydryl groups (36).

Acrosome granules from sea urchin sperm contain the adhesive protein bindin (30,31,37). Isolated acrosome granules species specifically agglutinate eggs (30,38,39), and monospecific antibody to the 24,000-molecular-weight bindin polypeptide localizes bindin to the acrosomal process of acrosome-reacted sperm (29). Acrosomal adhesive proteins have also been isolated from oyster sperm (40) and *Urechis caupo* (41). The complementary receptors for bindin on the vitelline layer of *S. purpuratus* eggs are proteoglycan-like glycoconjugates containing sulfated fucose heteropolysaccharide chains (17,18,31). Similar glycoconjugates have been described for other type of eggs (42). Sulfated fucans from the egg surface and algal analogs bind selectively to bindin particles (17). Other sulfated polysaccharides such as glycosaminoglycans bind only weakly in comparison. The negatively charged sulfate esters are critical for the binding of fucans to bindin. Removal of the sulfate groups by solvolysis results in the inactivation of fucan binding (17,43), whereas chemical resulfation of desulfated fucans restores binding activity (43). A domain of the bindin polypeptide containing clusters of basic amino acids may be the active site for polysaccharide binding. This domain can be specifically crosslinked to photoactivatible derivatives of fucoidin, and specific modification of basic amino acids abolishes fucan-binding activity (43).

The fusogenic region of mammalian sperm is also differentiated with respect to the types of molecules that are restricted to this site. The plasma membrane of mammalian sperm shows a great diversity of antigenic and morphological specializations, which are restricted to particular domains of the surface (44-46). In a recent report, six different monoclonal antibodies were produced that react with antigens restricted to the equatorial segment of the sperm head (47). Two of the six antibodies displayed

significant fertilization inhibitory activity. The antibodies also inhibited
the fertilization of zona-free eggs, suggesting that the antibodies inhibit
sperm-egg plasma membrane fusion rather than penetration of the zona.

A similar antibody, PH-30, has been reported in guinea pig sperm
(48). The PH-30 antigen is restricted to the posterior head region of the
sperm, the site where sperm-egg plasma fusion occurs. PH-30 antibody
precipitates two surface iodinated polypeptides of 60,000 and 44,000 molecular weight from detergent lysates of sperm. Preincubation of acrosome-reacted
sperm with saturating levels of PH-30 antibody inhibits the fertilization
of zona-free guinea pig eggs by 75% and also decreases the average number
of sperm that fuse per egg by 75% (48).

IV. ACROSOMAL PROTEINS INDUCE THE FUSION OF PHOSPHOLIPID VESICLES IN VITRO

We recently found that bindin from sea urchin sperm and lysin from abalone
sperm interact with phospholipid vesicles and induce their fusion in vitro.
Bindin associates with phospholipid vesicles (49) and induces their aggregation. An unusual feature of this interaction is that it requires phospholipids
in a gel or crystalline state (Fig. 3). This stringent requirement for gel-
phase phospholipids is observed with both neutral and acidic polar headgroups. The association of bindin with the bilayer is reversible. Aggregated
vesicles disaggregate as the temperature is raised above the phase transition
temperature, and sucrose density gradient centrifugation reveals that the
bindin polypeptide is not associated with fluid-phase vesicles (49). These
observations and the finding that vesicle-associated bindin and free bindin
are equally susceptible to proteolysis suggest that bindin associates peripherally rather than integrating within the bilayer. Bindin also associates
with vesicles that contain both gel-phase and fluid-phase domains such
as vesicles formed of the binary mixture of brain phosphatidylserine (PS)
and dipalmitoylphosphatidylcholine (DPPC). When bindin is added to these
vesicles, they aggregate and fuse (50). After incubation with bindin,
these mixed-phase vesicles are much larger (Fig. 4): The average diameter
of the vesicles after incubation is 190 ± 109 nm, compared to an average
diameter of 39 ± 20 nm for vesicles incubated in the absence of bindin.

A resonance energy transfer assay (51) was used to characterize the
bindin-induced fusion of vesicles in further detail. In this assay, two
fluorescent probes are incorporated into the membrane of mixed-phase
DPPC:PS (1:1) vesicles at a density where efficient energy transfer occurs
between the probes. The efficiency of energy transfer is proportional
to the concentration of the fluorescence energy acceptor in the bilayer.
The fluorescent vesicles are mixed with an excess of unlabeled target vesicles to quantify fusion. After bindin addition, there is a significant decrease
in the efficiency of energy transfer compared to controls incubated in the
absence of bindin (50), indicating that the bilayers of the fluorescent
probe vesicles have coalesced with the nonfluorescent target vesicles.
Bindin induces the fusion of vesicles in the absence of calcium, but the
rate of fusion is higher in the presence of calcium (Fig. 5). The rate
of bindin-induced vesicle fusion is also increased by low pH (pH 5-6)
and the presence of phosphatidylethanolamine (PE) in the vesicles (Glabe,
unpublished observations), conditions that also increase the spontaneous
fusion rate of the vesicles.

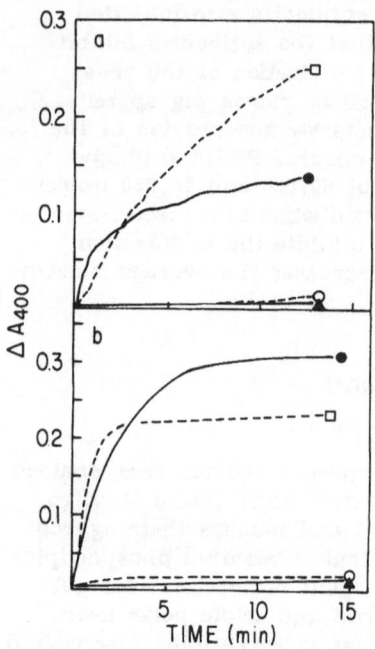

FIGURE 3 Bindin-induced aggregation of different types of phospholipid
vesicles at 20°C or 26°C. Fifty micrograms of bindin was added to a suspension of 100 µl vesicles in 0.7 ml 0.54 M NaCl, 20 mM Tris, pH 8.0 (TBS).
(a) ▲, Dioleoyl-, ▲, Dilauroyl-, ▲, Dilinoleoylphosphatidylcholine at 20°C.
●, Dimyristoylphosphatidylcholine, 20°C. ○, Dimyristoylphosphatidylcholine,
26°C. □, Dipalmitoylphosphatidylcholine, 26°C. (b) Bindin-induced aggregation of dimyristoylphosphatidylglycerol (DMPG) and dipalmitoylphosphatidylglycerol (DPPG) vesicles at 20°C and 26°C. ▲, DMPG vesicles in the
absence of bindin, 20°C. ●, DMPG vesicles in the presence of bindin
at 20°C. ○, DMPG vesicles in the presence of bindin at 26°C. □, DPPG
vesicles in the presence of bindin at 26°C.

    The ability of bindin to induce the fusion of some types of phospho-
lipid vesicles in vitro may be related to the fusagenic activity of the sperm
acrosome process, but this remains to be established. One of the obvious
problems that must be addressed is the fact that the fusion rate of sperm
in vivo is much faster than the observed rate of bindin-induced vesicle
fusion in vitro. This suggests that pure lipid vesicles are a poor model
for fusion of sperm and egg plasma membranes. It is clear that even in
the presence of calcium, the rate of vesicle fusion is much slower than
the rate of bindin-induced vesicle aggregation (Figs. 3 and 5). Thus,
bindin may promote vesicle fusion in vitro by bringing the bilayers into
close apposition, rather than by catalyzing the coalescence of the outer
leaflets of adjacent vesicles.
    A similar type of fusion enhancement by aggregation has been observed
for the soybean agglutinin (SBA)-induced fusion of vesicles containing
the glycolipid globoside (52). Preaggregated vesicles fuse upon Ca^{2+} addition at a rate six- to eightfold higher than the rate for vesicles incubated
with Ca^{2+} or SBA alone. Further studies on the fusion activity of bindin
in a system reconstituted from sperm and egg surface components, such

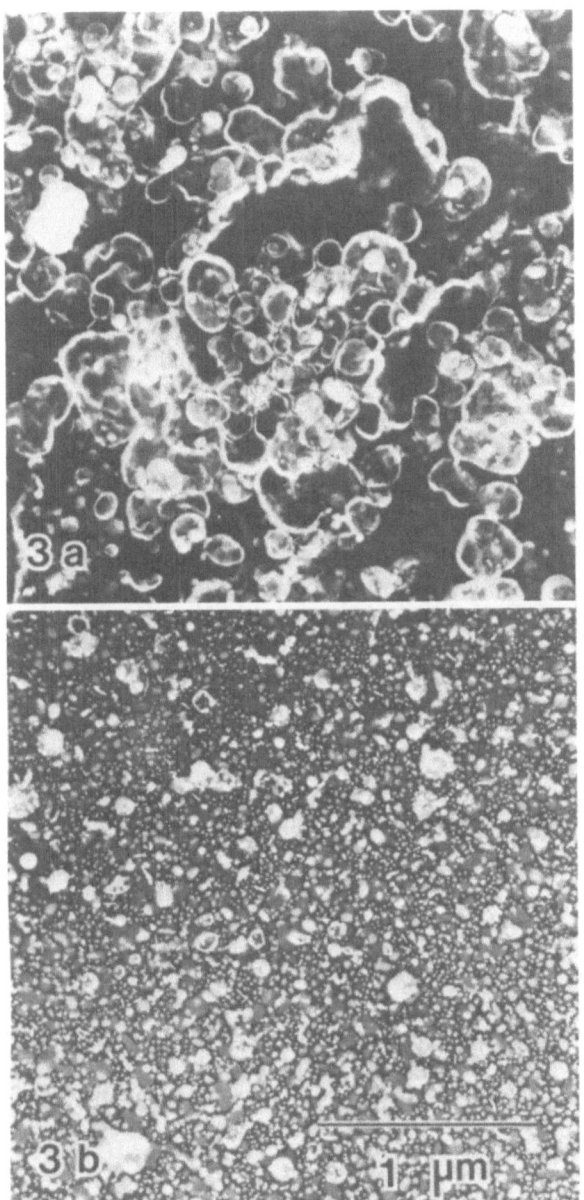

FIGURE 4 Electron micrographs of negatively stained DPPC:PS vesicles before and after incubation with bindin. Vesicle samples containing 0.1 ml of DPPC:PS vesicles in 0.7 ml TBS were applied to a carbon and parlodion-coated electron microscope grid and stained with potassium phosphotungstate. (a) Vesicles incubated with 50 µg of bindin for 16 h. (b) Control vesicles incubated in the absence of bindin. Both micrographs ×29,000.

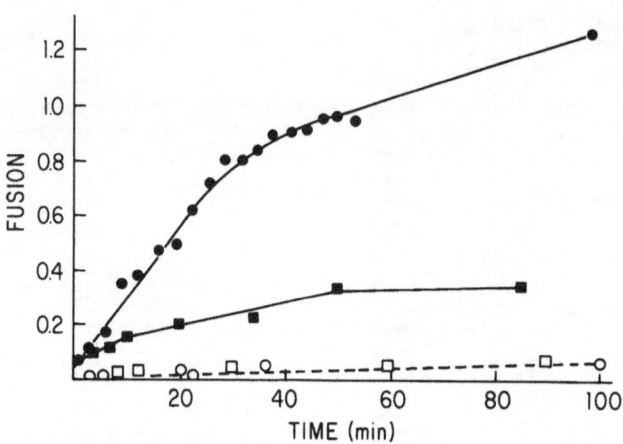

FIGURE 5   Kinetics of bindin-induced vesicle fusion. The efficiency of
energy transfer was determined over a large range of fluorescent to nonfluorescent lipid ratios. The fusion index unit is defined as the decrease
in energy transfer efficiency which corresponds to a dilution of the probe
vesicles by 1/2; that is, the fusion of one fluorescent vesicle with one
nonfluorescent vesicle. ●, Vesicles after bindin addition in the presence
of 2 mM Ca^{2+}. ■, Vesicles after bindin addition in the absence of calcium.
○, Vesicles in the absence of bindin and in the presence of 2 mM Ca^{2+}.
□, Vesicles in the absence of bindin and Ca^{2+}.

as the suflated fucose glycoconjugate receptors for bindin, may help clarify
the basis for the discrepancy of the fusion rate in vitro and in vivo.
       The acrosomal lysin from abalone sperm also interacts with phospholipids
and induces vesicle fusion in vitro (53). Lysin binds nonspecifically to
phospholipid vesicles and causes the release of liposome-encapsulated carboxyfluorescein. This lack of specificity for the headgroups of the phospholipids suggests that lysin interacts hydrophobically with the bilayer.
Although lysin binds in a nonspecific fashion, only vesicles containing
phosphatidylserine were aggregated upon association with lysin (Fig. 6).
Addition of lysin induced the fusion of phosphatidylserine-containing
vesicles, but not the fusion of phosphatidylcholine vesicles. The rate of
fusion was higher in vesicles that also contained phosphatidylcholine.
Addition of Ca^{2+} to vesicles pretreated with lysin enhanced fusion (Fig. 7a).
This is the same type of effect observed for the fusion of soybean agglutinin
aggregated vesicles mentioned above. Pretreatment of the vesicles with
calcium inhibited lysin-induced vesicle fusion (Fig. 7b). Addition of EDTA
to the calcium-pretreated vesicles also stimulated the rate and extent of
fusion. This stimulation of fusion of aggregated vesicles by both addition
and chelation of calcium may be due to destabilization of phospholipid
packing at phase boundaries by a calcium-dependent phase transition while
adjacent bilayers are in close apposition.
       The sequence of abalone lysin has been determined (54). Analysis
of the secondary structure of lysin predicted by the Chou-Fassman algorithm
suggests that the amino-terminal half of the polypeptide may be arranged
in four α-helical segments. The four helical segments are amphipathic,
with hydrophobic residues on one side of the helix and basic amino acid
residues on the other face. This amphipathic organization is typical of

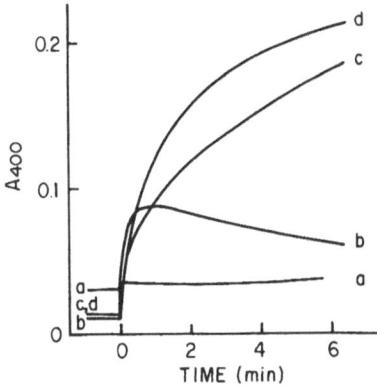

FIGURE 6 Time course of the lysin-induced turbidity change of liposomes composed of: (a) PC; (b) PS/PC(2:1); (c) PS/PE(2:1); (d) PS/PC/cholesterol (1:2:1). Lysin (18 μg) was added to vesicles at time 0.

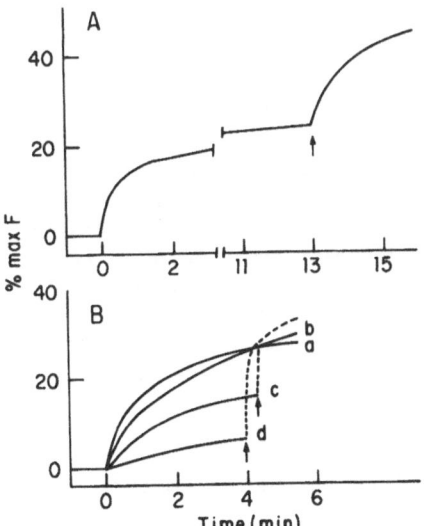

FIGURE 7 Liposome fusion induced by lysin. Labeled liposomes, PS/PC (1:2), were mixed with unlabeled liposomes in a 1:9 molar ratio. Lysin was added to liposomes of 50 M total phospholipid in 1 ml at 25°C, pH 7.4, time = 0. (a) Labeled PS/PC (1:2) liposomes were mixed with unlabeled PS/PC (1:2) liposomes. Ca^{2+} (2 mM) enhanced fusion, as indicated by the arrow. (b) Fusion induced by lysin was inhibited by the presence of Ca^{2+}. Liposomes were preincubated with (a) 0.1, (b) 0.25, (c) 0.5, (d) 1.0 mM Ca^{2+} before the addition of lysin (18 μg). The inhibition was reversed (broken lines) by the addition of EDTA (pH 7.4) at twice the Ca^{2+} concentration. EDTA was added where the arrows indicate.

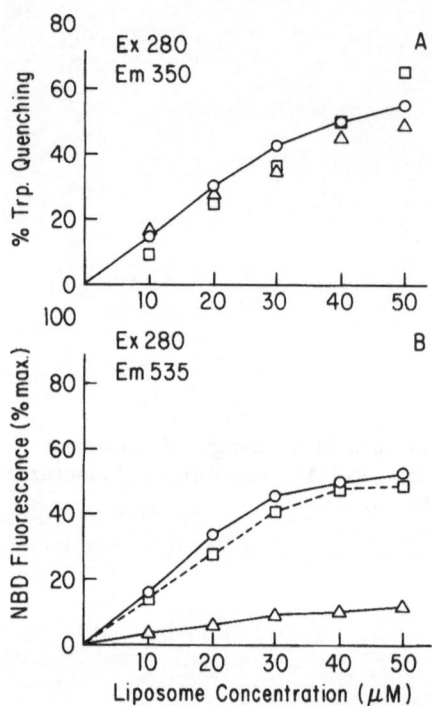

FIGURE 8 Titration of lysin (25 µg) by acceptor liposomes (*N*–NBD–PE labeled, 2 mol%). ○ , PS/PC (1:2); □ , PS/PE (1:2); △ , PC. (a) The quenching of tryptophan fluorescence (ex 280 nm, em 350 nm) as a function of liposome concentration at each addition. Quenching was calculated as the percent decrease of fluorescence after the addition of liposomes as compared to the lysin fluorescence in the absence of liposomes. (b) NBD fluorescence (ex 280 nm, em 535 nm) as a function of liposome concentration. The fluorescence of 50 µM NBD-labeled liposomes excited at 340 nM was used to define 100% fluorescence.

the plasma apolipoproteins and is believed critical for the phospholipid-binding properties of the apolipoproteins (55,56).

The interaction of lysin with phospholipid bilayers was further characterized by the quenching of lysin tryptophan fluorescence and resonance energy transfer between tryptophan and a fluorescent phospholipid acceptor, *N*-NBD-PE (2 mol%) (Hong, unpublished data). The critical distance for efficient energy transfer ranges from 2.0 to 5.0 nm, lengths comparable to the diameter of most proteins and the thickness of the bilayer. When liposomes are added to lysin, a quenching of tryptophan fluorescence (ex 280 nm, em 350 nm) is observed (Fig. 8a). The binding of lysin to vesicles, as measured by the quenching of tryptophan fluorescence, shows no significant differences in the three types of liposomes tested. This result confirms previous observations based on the lysin-induced release of carboxyfluorescein (53). Efficient resonance energy transfer between tryptophan and NBD was observed to occur only for PS-containing liposomes (Fig. 8b). Lysin contains four tryptophan residues, three of which are located in predicted α-helical regions (54). We have proposed that these α-helical regions are the focal point for the interaction of lysin with liposomes (53). Since half of the *N*-NBD-PE should be contained on the inner leaflet of the bilayer, the theoretical maximal acceptor population is 50% of the total *N*-NBD-PE in the membrane if the tryptophan-containing domains do not penetrate the bilayer. We found that the maximal transfer efficiency is 50% (compared to the emission of NBD when excited at 340 nm), which suggests that resonance energy transfer across the bilayer is not significant. The efficiency of energy transfer as a function of liposome concentration is slightly higher for PS/PC (1:2) liposomes than PS/PE (1:2) liposomes (Fig. 8b). This may indicate a slightly larger penetration of lysin into PS/PC bilayer and would thus help explain the fact that the rate of lysin-induced fusion of PS/PC vesicles is higher than of PS/PE liposomes. It is clear that the interaction of the α-helical regions of lysin with neutral PC liposomes is different from PS-containing liposomes. For PC liposomes, the NBD fluorescence is about one-quarter of that of PS-containing liposomes, perhaps indicating that on the average only one of the tryptophan residues is close enough to the membrane surface to allow efficient energy transfer.

Sea urchin sperm bindin and abalone lysin are both localized to the region of the sperm plasma membrane that is the fusion-competent region. Thus, bindin and lysin are in the correct location at the appropriate time to participate in membrane fusion. If these acrosomal proteins are instrumental for the fusion of sperm and egg plasma membranes, then they would both play two roles in fertilization: In addition to a putative role in plasma membrane fusion, bindin mediates the adhesion of sperm to the egg surface and lysin plays a role in the penetration of the vitelline envelope. The influenza virus hemagglutinin also performs dual roles during viral infection (57). The influenza virus hemagglutinin functions in the adhesion of the virus to the cell surface by binding sialic acid residues of glycolipids and glycoproteins and also induces the fusion of membranes by interacting with phospholipid bilayers. However, in contrast to our understanding of the role of viral fusion proteins in infection, the hypothesis that acrosomal proteins paly a critical role in the fusion of gamete plasma membranes during fertilization awaits further verification.

V. HOW DOES THE MEMBRANE POTENTIAL OF THE EGG REGULATE SPERM FUSION?

Several possible mechanisms have been proposed to account for the regulation of sperm-egg plasma membrane fusion by the membrane potential of the egg (2,4,6). One hypothesis is that sperm "receptors" in the egg plasma membrane change conformation to an inactive state or become inaccessible upon depolarization of the membrane. These receptors would presumably be distinct from the sperm [receptors" for sperm adhesion on the vitelline layer, since sperm adhesion is not affected by the membrane potential of the egg. An alternative hypothesis is that the interaction of fusagenic components of the sperm plasma membrane, analogous to viral "fusion proteins," with the egg plasma membrane is regulated by membrane potential. Membrane potential can regulate the translocation of proteins across the bilayer and influence the asymmetrical orientation of membrane proteins (58). Voltage-dependent conformational changes are believed to be important for the voltage-dependent gating of ion channels in electrically excitable cells (59). The latter alternative seems more likely since the studies of cross-species fertilization previously discussed show that the

TABLE 1 Sequence Homology Between Bindin[a] and *Electrophorus* Sodium Channel[b]

Bindin:	VAL ARG GLU GLN VAL LEU SER ALA MET GLN GLU GLU
	GLU GLU GLU GLU
Na channel:	PRO SER GLU GLN ASP PRO LEU ALA LYS GLU GLU GLU
	GLU GLU GLU GLU
Bindin:	GLU GLU ASP ALA ALA THR GLY ALA
Na channel:	GLU GLU PRO GLU GLU LEU GLU SER

[a]Vacquier, V. D., and Walsh, K. A., unpublished observations.
[b]Noda et al. (61).

voltage dependence of cross-fertilization follows the species of sperm. This hypothesis requires that there be some mechanism whereby the putative sperm fusagenic components can sense the membrane potential of the egg. In this regard, it is intriguing that the bindin polypeptide shares a rather striking sequence homology with the *Electrophorus* sodium channel in a highly negatively charged polypeptide domain (Table 1). In both bindin and the sodium channel, a segment of eight and nine contiguous glutamic acid residues is present. This sequence is not an artifact of protein sequencing, since the nucleic acid sequence of bindin confirms the presence of the eight glutamic acid residues (60). Noda and co-workers speculate that this anionic domain and three other acidic domains in cooperation with four repeated basic domains may constitute the voltage gating mechanism of the sodium channel (61). Similarly, this negatively charged domain in bindin may be free to move or change conformation in response to membrane potential across the membrane. If bindin is involved in gamete plasma membrane fusion, then its fusagenic activity may be regulated by the membrane potential of the egg. We are currently investigating whether membrane potential influences the rate of bindin-induced vesicle fusion in vitro.

VI. DOES MEMBRANE FUSION ACTIVATE THE EGG?

There is considerable evidence that, in addition to delivering the sperm genome into the egg cytoplasm, the sperm initiates the activation of the unfertilized egg. Sperm can activate the egg without the nucleus of the sperm being incorporated into the cytoplasm (62-66). Mitosis fails to occur under these circumstances because of the absence of the sperm centriole. It is likely that fusion of the sperm plasma membrane with the egg plasma membrane occurs under these conditions: Fluorescent label from fluorescein isothiocyanate (FITC)-labeled sperm is transferred to the egg by these nonincorporated sperm (66). The fluorescent label is mostly in low-molecular-weight hydrophobic species (phospholipids?) and labeled proteins make up only 10% of the ^{125}I FITC label (67). After sperm-egg fusion, the fluorescence segregates into two compartments: a large restricted region of the embryo surface and a small cytoplasmic locus near the point of sperm entry (68). The process of fusion alone does not appear to be sufficient to activate the egg, since sea urchin eggs fertilized by mussel sperm fail to activate after sperm-egg fusion (64,65) and eggs are not activated by fusion with one another (69).

It is not yet known how the sperm activates the egg. There is considerable evidence implicating the influx of Na^+ along with a corresponding efflux of H^+ (resulting in an alkalinization of the cytoplasm) as well as a release of free Ca^{2+} from intracellular stores) as second messengers in egg activation. Perhaps fusion of the sperm and egg plasma membrane supplies membrane components, such as ion channels, or cytoplasmic components required for activation of the egg components which ultimately initiate development. Recently, evidence has been presented to suggest that guanyl nucleotide binding proteins (G proteins) may be an intermediate in the transduction of the sperm activation signal (70). The G protein is thought to stimulate the production of inositol-1,4,5-triphosphate (Ip_3), which elevates the level of intracellular free calcium, which in turn causes exocytosis.

An alternative hypothesis is that the sperm-activating signal is transduced across the bilayer prior to membrane fusion. This intriguing possibility is supported by the recent report that the isolated adhesive protein from the acrosomal granules of *Urechis* sperm activates eggs upon binding to the egg surface (71). Additional experimental observations support the idea that activation of the egg (as measured by the sperm-induced increase in the plasma membrane conductance) precedes membrane fusion. Membrane fusion was observed only in the eggs fixed 4-8 s after the initiation of the conductance increase as determined by electron microscopy (72) or the transfer of the fluorescent dye Hoechst 33342 from the egg cytoplasm to the sperm nucleus (73). Other experimental results argue that membrane fusion is coincident with the conductance increase: the fusion of the sperm and egg plasma membrane reflected in an increase in the capacitance of a small patch of the egg plasma membrane due to an increase in membrane area (74). The change in membrane conductance was observed to be entirely coincident with a corresponding change in capacitance.

If the sperm does indeed activate the egg prior to the fusion of the sperm and egg plasma membranes, then this mechanism presents an apparent paradox: If the sperm activates the egg (and thus initiates membrane depolarization) prior to membrane fusion, how can the fertilizing sperm fuse with the egg plasma membrane when the supernumerary sperm are prevented from fusion by the depolarization of the membrane? This suggests that the activating sperm may have reached an intermediate stage in the fusion process that is not sensitive to the membrane potential.

ACKNOWLEDGMENTS

This work was supported by NIH grants HD 21379 to C.G.G., GM28117 to K.H. and HD 12986 to V.D.V.

REFERENCES

1. Colwin, A. L., and Colwin, L. H. (1964). Role of the gamete membranes in fertilization. In: *Cellular Membranes in Development* (M. Locke, ed.). Academic Press, New York, pp. 233-279.

2. Epel, D., and Vacquier, V. D. (1978). Membrane fusion events during invertebrate fertilization. In: *Membrane Fusion* (G. Poste and G. L. Nicolson, eds.). Elsevier/North-Holland Biomedical Press, Amsterdam, pp. 1-63.

3. Shapiro, B. M. (1984). Molecular aspects of sperm-egg fusion. In: *Cell Fusion*. Ciba Foundation symposium 103. Pitman, London, pp. 87-99.

4. Gould-Somero, M., and Jaffe, L. A. (1984). Control of cell fusion at fertilization by membrane potential. In: *Cell Fusion: Gene Transfer and Transformation* (R. F. Beers and E. G. Bassett, eds.). Raven Press, New York.

5. Monroy, A. (1985). Processes controlling sperm-egg fusion. *Eur. J. Biochem.* 152:51-56.

6. Jaffe, L. A., and Cross, N. L. (1986). Electrical regulation of sperm-egg fusion. *Annu. Rev. Physiol.* 48:191-200.

7. Maroudas, N. G. (1975). Polymer exclusion, cell adhesion and membrane fusion. *Nature* 254:695-696.

8. Longo, F. J. (1973). Fertilization: A comparative ultrastructural review. *Biol. Reprod.* 9:149-215.

9. Yanagimachi, R. (1978). Sperm egg association in mammals. *Curr. Topics Dev. Biol.* 12:83-105.

10. Tilney, L. G., Keihart, D. P., Sardet, C., and Tilney, M. (1978). Polymerization of actin IV. Role of Ca++ and H+ in the assembly of actin and in membrane fusion in the acrosomal reaction of echinoderm sperm. *J. Cell Biol.* 77:536-550.

11. Vacquier, V. D. (1979). The fertilizing capacity of sea urchin sperm decreases rapidly after induction of the acrosome reaction. *Dev. Growth Differ.* 21:61-69.

12. Kinsey, W. H., Rubin, J. A., and Lennarz, W. A. (1980). Studies on the specificity of sperm binding in echinoderm fertilization. *Dev. Biol.* 74:245-250.

13. Collins, F. (1976). A reevaluation of the fertilizin hypothesis of sperm agglutination and the description of a novel form of sperm aggregation. *Dev. Biol.* 49:381-394.

14. Kato, K. H., Iwaikawa, Y., and Sugiyama, M. (1983). Fusion of spermatozoa with embryonic cells and somatic cells in the sea urchin. *Dev. Growth Diff.* 25:571-583.

15. Friend, D. S., Orci, L., Perrelet, A., and Yanagimachi, R. (1977). Membrane particle changes attending the acrosome reaction in guinea pig spermatozoa. *J. Cell Biol.* 74:561-577.

16. Weiss, R. L., Goodenough, D. A., and Goodenough, U. W. (1977). Membrane differentiations at sites specialized for cell fusion. *J. Cell Biol.* 72:144-160.

17. Glabe, C. G., Grabel, L. B., Vacquier, V. D., and Rosen, S. D. (1982). Carbohydrate specificity of sea urchin sperm bindin: A cell surface lectin mediating sperm-egg adhesion. *J. Cell. Biol.* 94:123-128.

18. Rossignol, D. P., Earles, B. J., Decker, G. L., and Lennarz, W. J. (1984). Characterization of the sperm receptor on the surface of eggs of *Strongylocentrotus purpuratus*. *Dev. Biol.* 104:308-321.

19. Talbot, P., and Chacon, R. S. (1982). Ultrastructural observations on binding and membrane fusion between human sperm and zona pellucida free hamster oocytes. *Fertil. Steril.* 37:240-248

20. Nuccitelli, R., and Grey, R. D. (1984). Controversy over the fast, partial, temporary block to polyspermy in sea urchins: A reevaluation. *Dev. Biol.* 103:1-17.

21. Jaffe, L. (1976). Fast block to polyspermy in sea urchins is electrically mediated. *Nature* 261:68-71.

22. Shen, S. S., and Steinhardt, R. A. (1984). Time and voltage windows for reversing the electrical block to fertilization. *Proc. Natl. Acad. Sci. USA* 81:1436-1439.

23. Lynn, J. W., and Chambers, E. L. (1984). Voltage clamp studies of fertilization in sea urchin eggs. I. Effect of clamped membrane potential on sperm entry, activation and development. *Dev. Biol.* 102: 98-109.

24. Jaffe, L. A., Gould-Somero, M., and Holland, L. Z. (1982). Studies of the mechanism of the electrical polyspeprmy block using voltage clamp during cross species fertilization. *J. Cell Biol.* 92:616-621.

25. Charbonneau, M., Moreau, M., Picheral, B., Vilain, J. P., and Guerrier, P. (1983). Fertilization of amphibian eggs: A comparison of electrical responses between anurans and urodeles. *Dev. Biol.* 98: 304-318.

26. Jaffe, L. A., Cross, N. L., and Picheral, B. (1983). Studies of the voltage dependent polyspermy block using cross species fertilization of amphibians. *Dev. Biol.* 98:319-326.

27. Sato, M., and Osanai, K. (1983). Sperm reception by an egg microvillus in the polychaete, *Tylorynchus heterochaetus. J. Exp. Zool.* 227:459-469.

28. Denis-Donini, S., and Campanella, C. (1977). Ultrastructural and lectin binding changes during the formation of the animal dimple in oocytes of *Discoglossus pictus* (Anura). *Dev. Biol.* 61:140-152.

29. Moy, G. W., and Vacquier, V. D. (1979). Immunoperoxidase localization of bindin during the adhesion of sperm to sea urchin eggs. *Curr. Topics Dev. Biol.* 13 (Pt. 1):31-44.

30. Vacquier, V. D., and Moy, G. W. (1977). Isolation of bindin: The protein responsible for adhesion of sperm to sea urchin eggs. *Proc. Natl. Acad. Sci. USA* 74:2456-2460.

31. Glabe, C. G., Lennarz, W. J., and Vacquier, V. D. (1982). Sperm surface components involved in sea urchin fertilization. In: *Cellular Recognition* (W. A. Frazier and L. Glaser, eds.). Alan R. Liss, New York, pp. 821-832.

32. McRorie, R. A., and Williams, W. L. (1974). Biochemistry of mammalian fertilization. *Annu. Rev. Biochem.* 43:777-801.

33. Green, J. D., and Summers, R. G. (1980). Ultrastructural demonstration of a trypsin-like protease in acrosomes of sea urchin sperm. *Science* 209:398-400.

34. Dan, J. C. (1967). Acrosome reaction and lysins. In: *Mammalian Cell Membranes*, Vol. 1 (C. B. Metz and A. Monroy, eds.). Academic Press, New York, pp. 237-293.

35. Haino-Fukushima, K. (1974). Studies on the egg membrane lysin of *Tegula pfeifferi*: The reaction mechanism of the lysin. *Biochim. Biophys. Acta* 352:179-191.

36. Lewis, C. A., Talbot, C. F., and Vacquier, V. D. (1982). A protein form abalone sperm dissolves the egg vitelline layer by a nonenzymatic mechanism. *Dev. Biol.* 92:227-239.

37. Vacquier, V. D., and Moy, G. W. (1978). Macromolecules mediating sperm egg recognition and adhesion during sea urchin fertilization. In *Cell Reproduction* (E. R. Dirksen, D. Prescott, and C. F. Fox, eds.). (ICN-UCLA Symposia on Molecular and Cellular Biology), Vol. 12, Academic Press, New York, pp. 379-389.

38. Glabe, C. G., and Vacquier, V. D. (1977). Species specific agglutination of eggs by bindin isolated from sea urchin sperm. *Nature* 267:836-838.

39. Glabe, C. G., and Lennarz, W. J. (1979). Species specific sperm adhesion in sea urchins. A quantitative investigation of bindin mediated egg agglutination. *J. Cell Biol.* 83:595-604.

40. Brandriff, B., Moy, G. W., and Vacquier, V. D. (1978). Isolation of bindin from sperm of the oyster *Crassostria gigas*. *Gamete Res.* 1:89-99.

41. Stefano-Hornedo, J. L. (1981). Estudios sobre una fraccion acrosomal de *Urechis caupo*. Ph.D. thesis, Instituto Polytecnico National, Mexico, DF.

42. Kinsey, W. H., and Lennarz, W. J. (1981). Isolation of a glycopeptide fraction from the surface of the sea urchin egg that inhibits sperm-egg binding and fertilization. *J. Cell Biol.* 91:325-331.

43. DeAngelis, P. L., and Glabe, C. G. (1985). Biochemical characterization of the interaction of sulfated fucans and the sperm adhesive protein, bindin. *J. Cell Biol.* 101:264 (abstr.).

44. Bearer, E. L., and Friend, D. S. (1980). Anionic lipid domains: Correlation with functional topography in a mammalian cell membrane. *Proc. Natl. Acad. Sci. USA* 77:6601-6605.

45. Myles, D. G., Primakoff, P., and Bellve, A. R. (1981). Surface domains of guinea pig sperm defined with monoclonal antibodies. *Cell* 23:433-439.

46. Eddy, E. M., and Koehler, J. K. (1982). Restricted domains of the sperm surface. *Scan. Electron. Microsc.*, Part 3:1313-1323.

47. Saling, P. M., Irons, G., and Waibel, R. (1985). Mouse sperm antigens that participate in fertilization. I. Inhibition of sperm fusion with the egg plasma membrane using monoclonal antibodies. *Biol. Reprod.* 33:515-526.

48. Primakoff, P., Hyatt, H., and Tredick-Kline, J. (1987). Identification of a sperm surface protein with a potential role in sperm-egg membrane fusion. *J. Cell Biol.* 104:141-149.

49. Glabe, C. G. (1985). Interaction of the sperm adhesive protein, bindin, with phospholipid vesicles. I. Specific association of bindin with gel-phase phospholipid vesicles. *J. Cell Biol.* 100:794-799.

50. Glabe, C. G. (1985). Interaction of the sperm adhesive protein, bindin, with phospholipid vesicles. II. Bindin induces the fusion of mixed phase vesicles that contain phosphatidylcholine and phosphatidylserine in vitro. *J. Cell Biol.* 100:800-806.

51. Struck, D. K., Hoekstra, D., and Pagano, R. E. (1981). Use of resonance energy transfer to monitor membrane fusion. *Biochemistry* 20:4093-4099.

52. Hoekstra, D., Duzgunes, N., and Wilschut, J. (1985). Agglutination and fusion of globoside GL-4 containing phospholipid vesicles mediated by lectins and calcium ions. *Biochemistry* 24:565-572.

53. Hong, K., and Vacquier, V. D. (1986). Fusion of liposomes induced by a cationic protein from the acrosome granule of abalone spermatozoa. *Biochemistry* 25:543-549.

54. Fridberger, A., Sundelin, J., Vacquier, V. D., and Peterson, P. A. (1985). Amino acid sequence of an egg lysin protein from abalone spermatozoa that solubilizes the vitelline layer. *J. Biol. Chem.* 260:9092-9099.

55. Segrest, J. P., Jackson, R. L., Morrisett, J. D., and Gotto, A. M.

55. Segrest, J. P., Jackson, R. L., Morrisett, J. D., and Gotto, A. M. (1974). A molecular theory of lipid protein interactions in the plasma lipoproteins. *FEBS Lett.* 38:247–253.

56. Sparrow, J. T., and Gotto, A. M. (1982). Apolipoprotein lipid interactions: Studies with synthetic polypeptides. *CRC Crit. Rev. Biochem.* 13:87–107.

57. White, J., Kielan, M., and Helenius, A. (1983). Membrane fusion proteins of enveloped animal viruses. *Q. Rev. Biophys.* 16:151–195.

58. Weinstein, J. N., Blumenthal, R., van-Renswoude, J., Kempf, C., and Klausner, R. D. (1982). Charge clusters and the orientation of membrane proteins. *J. Membr. Biol.* 66:203–212.

59. Catterall, W. A. (1984). The molecular basis of neuronal excitability. *Science* 223:653–661.

60. Gao, B., Klein, L. E., Britten, R. J., and Davidson, E. H. (1986). Sequence of mRNA coding for bindin, a species specific sea urchin sperm protein required for fertilization. *Proc. Natl. Acad. Sci. USA* 83:8634–8638.

61. Noda, M. N., Shimizu, S., Tanabe, T., Takai, T., Kayano, T., Ikeda, T., Takahasi, H., Nakayama, H., Kanaoka, Y., Minamino, N., Hayashidal, H., and Numa, S. (1984). Primary structure of electrophorus electricus sodium channel deduced from cDNA sequence. *Nature* 312:121–127.

62. Hiramoto, Y. (1962). An analysis of the mechanism of fertilization by means of enucleation of sea urchin eggs. *Exp. Cell Res.* 28:323–334.

63. Gould-Somero, M., Holland, L., and Paul, M. (1977). Cytochalasin B inhibits sperm penetration into eggs of *Urechis caupo* (Echiura). *Dev. Biol.* 58:11–22.

64. Longo, F. J. (1977). An ultrastructural study of cross fertilization (Arbacia sperm X Mytilus eggs). *J. Cell Biol.* 73:14–26.

65. Osani, K., and Kyozuka, K. (1984). Cross fertilization between sea urchin eggs and mussel spermatozoa: Sperm entry without egg activation. *Zool. Sci.* 1:245–254.

66. Gundersen, G. G., Gabel, C. A., and Shapiro, B. M. (1982). An intermediate state of fertilization involved in an internalization of sperm components. *Dev. Biol.* 93:59–72.

67. Gundersen, G. G., and Shapiro, B. M. (1984). Sperm surface proteins persist after fertilization. *J. Cell Biol.* 99:1343–1353.

68. Gundersen, G. G., Medill, L., and Shapiro, B. M. (1986). Sperm surface proteins are incorporated into the egg membrane and cytoplasm after fertilization. *Dev. Biol.* 113:207–217.

69. Bennett, J., and Mazia, D. (1981). Interspecific fusion of sea urchin eggs. Surface events and cytoplasmic mixing. *Exp. Cell Res.* 131:197–207.

70. Turner, P. R., Jaffe, L. A., and Fein, A. (1986). Regulation of cortical vesicle exocytosis in sea urchin eggs by inositol 1,4,5-triphosphate and GTP-binding protein. *J. Cell Biol.* 102:70–76.

71. Gould, M., Stephano, J. L., and Holland, L. Z. (1986). Isolation of protein from Urechis sperm acrosomal granules that binds sperm to eggs and initiates development. *Dev. Biol.* 117:306–318.

72. Longo, F. J., Lynn, J. W., McCulloh, D. H., and Chambers, E. L. (1986). Correlative ultrastructural and electrophysiological studies of sperm-egg interactions of the sea urchin, *Lytechinus pictus*. *Dev. Biol.* 118:155–166.

73. Hinkley, R. E., Wright, B. D., and Lynn, J. W. (1986). Rapid visual detection of sperm-egg fusion using the DNA-specific fluorchrome Hoechst 33342. *Dev. Biol.* 118:148-154.
74. McCulloh, D. H., and Chambers, E. L. (1986). Fusion and "unfusion" of sperm and eggs are voltage dependent in the sea urchin Lytechinus variegatus. *J. Cell Biol.* 103:236a (abstr.).

VI
APPLICATIONS OF MEMBRANE FUSION TECHNIQUES

28

Experimental and Technological Applications of Cell-Cell Fusion
An Overview

DAVID S. ROOS

University of Pennsylvania, Philadelphia, Pennsylvania

Scientific investigation of cell fusion phenomena has been characterized by a mutually productive interplay between technological development and biological application (1-4). This brief review outlines some of the major advances that have led to our current situation, where cell-cell fusion* is commonly used for a variety of purposes in genetics, biology, and biotechnology. Recent investigations of the biochemical components of cell membranes that play a role in regulating cell fusion are also discussed. Results of these studies suggest a variety of new approaches to the controlled application of fusion technology.

I. THE DEVELOPMENT OF CELL FUSION TECHNOLOGY

Early observations of cell fusion date from the 19th century, when pathologists noted the prevalence of multinucleated cells in various human diseases (7-11). As techniques were developed for cell and tissue culture, multinucleated cells were also observed in vitro (12,13), particularly in association with viral infections (14-17), but cell-cell fusion remained essentially a morphological curiosity until the publication of a report by Barski et al. describing hybridization between distinct cell types in mixed culture (18).

Because different cells exhibit markedly different characteristics in culture, the ability to hybridize two distinguishable cell types offers many exciting prospects for studying the control of gene expression. Ephrussi

---

*The development of vesicle fusion techniques has provided a useful adjunct to cell-cell fusion which is often better suited to the delivery of specific substances to target cells. As many of these techniques are described elsewhere in this book, I have focused this discussion on applications requiring fusion between cells. More detailed discussion of the technical aspects of cell-cell fusion can be found in recent books (5,6,98) and in Chapters 29 and 30.

and co-workers rapidly seized on the potential of this approach, confirming the hybridization of cell types karyologically (19), and embarking on a study of antigen expression in hybrid mouse cells (20). Unfortunately, the range of possible experiments utilizing hybrid cells was severely circumscribed by two technological difficulties. First, the frequency of spontaneous hybridization is very low for most cultured cells, and second, the emergence of hybrids from a mixed culture requires the unlikely good fortune that the fused cells exhibit sufficient vigor to outgrow the two parental cell types.

The first major step in improving hybridization frequencies followed from the observations of Okada et al. (21,22) that Sendai virus [also called hemagglutinating virus of Japan (HVJ)] is capable of forming syncytia in homogeneous cell cultures. Harris and Watkins demonstrated that high titers of UV-inactivated Sendai virus can also be used to form heterokaryons between different cell types, even across species borders (23,24). This discovery improved the ability to form cell hybrids manyfold, and virus-induced fusion techniques remain in use to this day (although they have in recent years been supplanted by chemical fusogens for most applications).

The isolation of rare hybrids from a large excess of unfused parental cell types requires some form of selectable phenotype. Appropriate culture media had been developed by Szybalski et al. for the selection of revertants from mutant cells deficient in the salvage pathways for nucleotide metabolism (25), and Littlefield used this "HAT" system to isolate hybrid cells in which each parent's enzymatic deficiency could be complemented by wild-type genes from the other fusion partner (26), effectively adapting the powerful tools of bacterial genetics to somatic animal cells (27). Naturally occuring phenotypes, such as the slow growth rates of one parent, were also developed as selectable markers usable in conjunction with the HAT selection scheme (28), minimizing the need for preexisting mutants in both partner cell types. Although other hybrid selection schemes have been developed, including nongenetic techniques (29,30), the HAT system is still widely used.

II. APPLICATIONS OF CELL-CELL FUSION: GENE EXPRESSION AND GENETIC MAPPING

The combination of improved fusion techniques and schemes for the isolation of hybrids dramatically increased the spectrum of feasible experiments. From the earliest days, the malignant phenotype has received a great deal of attention in cell hybrids. The original hybrids isolated by Barski et al. (18) were derived from cells that differed in their tumorigenicity, and the majority of the resulting hybrids remained highly tumorigenic (31). In different systems, however, malignancy is suppressed in hybrid cells, and the subject remains a matter of controversy (32–35). Studies on gene expression and interaction in hybrid cells have sought to address many important questions. Are specific mutant phenotypes dominant, co-dominant, or recessive in hybrid cells? How are these findings affected by gene dosage? How are the differentiated functions of cells from different tissues regulated in hybrids between them? How do DNA replication and mitosis occur when two genomes are juxtaposed in the same cell, and how do mitochondria, ribosomes, and other organelles interact? Reports on hybrid cells proliferated exponentially in the ensuing years, examining the fate of numerous enzymes, structural proteins, nucleic acids, sub-cellular organelles, and other distinctive characteristics of different species or tissue types. Some of these early studies are reviewed in Refs. 36 and 37.

Adding to the utility of hybridization experiments for genetic analysis of was the important discovery of chromosomal loss in interspecies hybrids (38). Most human-mouse cell hybrids, for example, preferentially lose human chromosomes when maintained in continuous culture. This observation allowed the correlation of mutations or phenotypes with specific chromosomes. Weiss and Green showed that a single human chromosome [later shown to be chromosome 17 (39)] was universally maintained in hybrids with mouse cells deficient in thymidine kinase (TK) activity, effectively mapping the human TK gene (38). Both selective retention and loss of specific chromosomes have been used to map specific loci in somatic cell hybrids (40), including even a gene that promotes cell fusion (41)!

Improvements in fusion technology also spawned an industry beyond the direct application of hybrid cells to scientific inquiry, by facilitating the production of cells expressing a combination of desirable traits derived from different sources. Joining the secretion of a single, homogeneous antibody species characteristic of terminally differentiated B lymphocytes with the capacity of myeloma cell lines to proliferate indefinitely in culture, Köhler and Milstein were able to create cell lines producing monoclonal antibodies suitable for immunological and biochemical work (42; see also Chapter 30).

III. MODERN TECHNIQUES FOR THE INDUCTION OF FUSION

Although the use of enveloped viruses to promote cell fusion is much more broadly applicable than waiting for spontaneous hybrids to arise, serious limitations still apply to this technique. Some cell types are not readily susceptible to virus-induced fusion, leaving certain cells (including many human cells) largely outside the realm of hybridization studies. Where viral fusion is effective, preparation and purification of the large quantities of virus necessary for fusion "from without" is time consuming,* and the introduction of viral lipids, proteins, and nucleic acids may complicate the analysis of results.

One approach to the resolution of these problems has been to artificially construct vesicles that mimic the relevant properties of fusogenic viruses. The purification of viral fusion proteins and characterization of biochemical events required for fusion to occur (43-47) has allowed the development of a variety of vesicle fusion systems, usually incorporating the Sendai virus F-protein (48,49). These vesicles have not proved particularly useful for intercellular fusion, but hold considerable promise as delivery systems, especially for the targetting of drugs, nucleic acids, and so forth to specific cells, discussed in Chapter 32.

---

*Enveloped viruses are typically characterized by a highly restricted host range, requiring that most fusion reactions be carried out by incubation with high titers of UV-inactivated virions (typically 10^2-10^3 particles/cell). Fusion "from without" is usually used even for cells that can be directly infected with fusogenic viruses, as it is generally not desirable to produce hybrids infected with virus.

The greatest contribution to the improvement of fusion protocols over
the past decade has been the development of chemical fusion techniques.
A wide variety of chemicals are known to be active in promoting cell hybridization (50,51). By far the most widely used fusogen remains poly(ethylene
glycol) (PEG), first reported to promote the hybridization of plant protoplasts by Kao and Michayluk in 1974 (52).* It was soon discovered that
PEG is capable of fusing animal cells as well (62,63), and techniques for
the efficient hybridization of cells in mixed culture were established (56,
64-66). Although the precise biophysical mechanism of PEG-induced fusion
remains unclear, it seems likely to involve the perturbation of membrane
hydration (67; see also Chapter 11). Aside from its ease of preparation
and use, PEG obviates the problems of introducing extraneous biological
material into the system and is widely applicable, promoting fusion of cells
as diverse as animals, plants, fungi, protozoa, and bacteria (68,69). PEG
is also capable of causing fusion between lipid vesicles and vesicle-cell
fusion.

PEG has been successfully applied to the production of a wide range
of cell hybrids, and these studies have further expanded our understanding
of the control of gene expression. Hybrid cells have been used to study
the expression of tissue-specific proteins characteristic of liver (70,71),
muscle (72), and cells of the erythroid (73,74) and lymphoid (75,76)
lineages. PEG has also become the method of choice in the production of
lymphocyte hybridomas (57,58), and it is now possible to prepare monoclonal antibody-secreting hybridomas readily, in sufficient quantity for
experimental, diagnostic, and therapeutic use (see Chapter 30).

Improved fusion techniques have been helpful in genetic analyses as
well. Whereas early mapping studies required constant surveillance of large
numbers of hybrids retaining various combinations of chromosomes, microcell fusion has simplified the production and maintenance of cell lines bearing
a reduced complement of chromosomes (77). Nonnuclear traits can also be
mapped by hybridization with enucleated cells or other subcellular fractions
(78). For the mapping of cloned genes, screening is carried out by filter
hybridization with DNA isolated from well-characterized panels of somatic
cell hybrids (79). The development of recombinant DNA technology has
greatly extended the usefulness of hybrid cells in genetic mapping, expanding the range of characteristics that can be mapped from the phenotypic
level down to the basic nucleic sequence.

A. Suspension Fusion: Practical Aspects

Current fusion techniques allow the production of hybrid cells to meet
most requirements, but certain difficulties remain. Fusion of suspension
cultures has always been inefficient compared with that in monolayer culture
(64), and it is particularly problematic for chemical fusogens, which do
not agglutinate cells as effectively as do fusogenic viruses. Even the fusion
of cells in monolayer cultures, while usually sufficiently effective to generate

---

*There has been some discussion in the literature of whether the fusion-
promoting activity of PEG solutions is caused by poly(ethylene glycol)
itself or by some contaminating factor (53-55). Although there is considerable variation in the fusogenicity of different commercial preparations of
PEG (56-58), it now appears clear that purified PEG is indeed capable
of inducing fusion (59-61). Commercial preparations may contain a number
of contaminants, however, both stimulatory and inhibitory to the fusion
process.

selectable hybrids, is not efficient enough to allow careful control of the dose in experiments involving the delivery of drugs or other compounds to target cells.

The need for better suspension fusion techniques is particularly acute for hybridoma technology, as most lymphoid cells must be grown in suspension culture. Current technology has made the production of murine B-cell hybridomas a routine matter (57,75,80), but improved techniques would be helpful for samples that have proved relatively resistant to hybridization, such as human lymphocytes (81) or where the stimulating immunogen is weak. Considerable advances have been made in the isolation of infrequent hybrids (30), but even where hybrids can be produced readily, it is desirable to maximize the efficiency of the fusion reaction. It is likely that certain subpopulations of cells are more prone to hybridization than others and suboptimal fusion techniques may therefore skew the spectrum of hybrids formed (57,82).

Several investigators have tried to improve suspension fusion technology by increasing the frequency of intercellular contact. One approach has been to generate an artificial cell monolayer, either by centrifuging cells onto a coverslip (83,84) or by cocultivation with an attached cell line (85). Cells can also be artificially agglutinated through the use of lectins (86,87). It would be interesting to explore the possibility of using lectins (or other compounds) to transiently attach cells to monolayers for the duration of the fusogenic treatment, followed by cell release with glycosidases (or other appropriate agents).

Two factors are commonly involved in the failure of a fusion reaction to produce hybrid cells: poor fusion yield and high toxicity of the fusogen. The association of cell toxicity with PEG (and other fusogenic agents) under conditions that promote fusion is probably not coincidental. A fine line separates the perturbation of membranes necessary for fusion to occur from the disruption of the membrane altogether (67,88). Increasing the severity of the fusion conditions (usually by increasing the length of treatment or the concentration of the fusogen) invariably increases both fusion and cell death. Yet the thresholds for unacceptable toxicity and useful hybridization frequencies are not identical and vary with respect to each other depending on the specific treatment. The goal is to determine conditions under which acceptable hybridization occurs with minimal toxicity, and to maximize the size of this window.

The best guide in solving problems where the generation of hybrids appears difficult is to examine the culture shortly after fusion. When fusion is rare, it is not surprising to find low frequencies of hybridization.* High levels of toxicity, characterized by large numbers of dead cells and membrane fragments, also depress the fusion yield, both directly (by killing cells that might otherwise have hybridized) and indirectly (by releasing toxic compounds into the medium). Toxicity can be minimized by recognizing that PEG-treated membranes are very unstable and treating cells gently (64,90). In general, we have found that PEG concentrations of ~40-45% are most appropriate for suspension hybridizations (slightly higher concentrations are usually more effective for monolayer fusion where the PEG solution can be removed more efficiently). Resuspended cell pellets should be exposed to PEG for no longer than 30-90 s, followed by gentle dilution with warm medium. Small PEG molecules are usually toxic to cells below

---

*Note, however, that it is important to distinguish between fusion and hybridization. Conditions that produce the most extensive fusion are often not optimal for the production of binucleate hybrids (64,89).

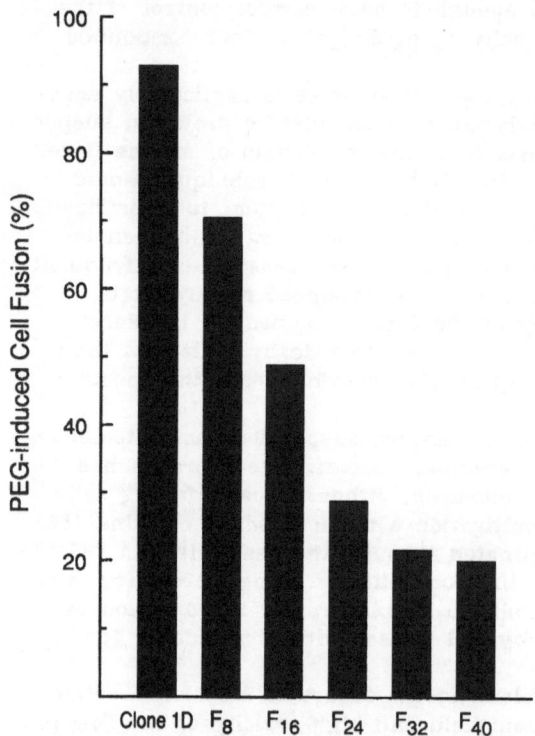

FIGURE 1   Fusion response of PEG-resistant cell lines. A graded series
of increasingly PEG-resistant cell lines has been isolated from a highly
fusible murine fibroblast parent LM TK-Clone 1D (94). Cell lines designated
F_1, F_2, F_3, ..., F_{40} were generated by repeated cycles of PEG treatment
followed by outgrowth of the remaining unfused cells; subscripts indicate
the number of cycles of selection used to create each cell line. Fusion
was carried out by treating parallel cultures of the various cell lines with
50% PEG 1000 (J. T. Baker Chemical Co., Phillipsburg, NJ) for 1 min.
Percent fusion was scored as the number of nuclei that remained in mononucleate cells following treatment. (Reprinted, with permission, from Refs.
88 and 95.)

fusogenic concentrations, while very long chain polymers are less effective
at promoting fusion and are awkward to use because of their high viscosity
(56,75). Best results are usually obtained with PEG of molecular weight 1000-
4000. Certain commercial preparations of PEG are more effective than others
in promoting fusion (56-58), presumably because of contaminating factors
and variable amounts of short polymer PEG. Under some conditions, the
inclusion of additional compounds, such as dimethylsulfoxide, may serve
to stabilize membranes exposed to PEG, decreasing toxicity without appreciably affecting fusion response (57,91).

B.   Future Directions

Because fusion depends on both cells in the reaction, it is often possible
to control the frequency of hybridization by choosing an appropriate partner, and by varying the ratio of one cell type to the other (82,89,92).

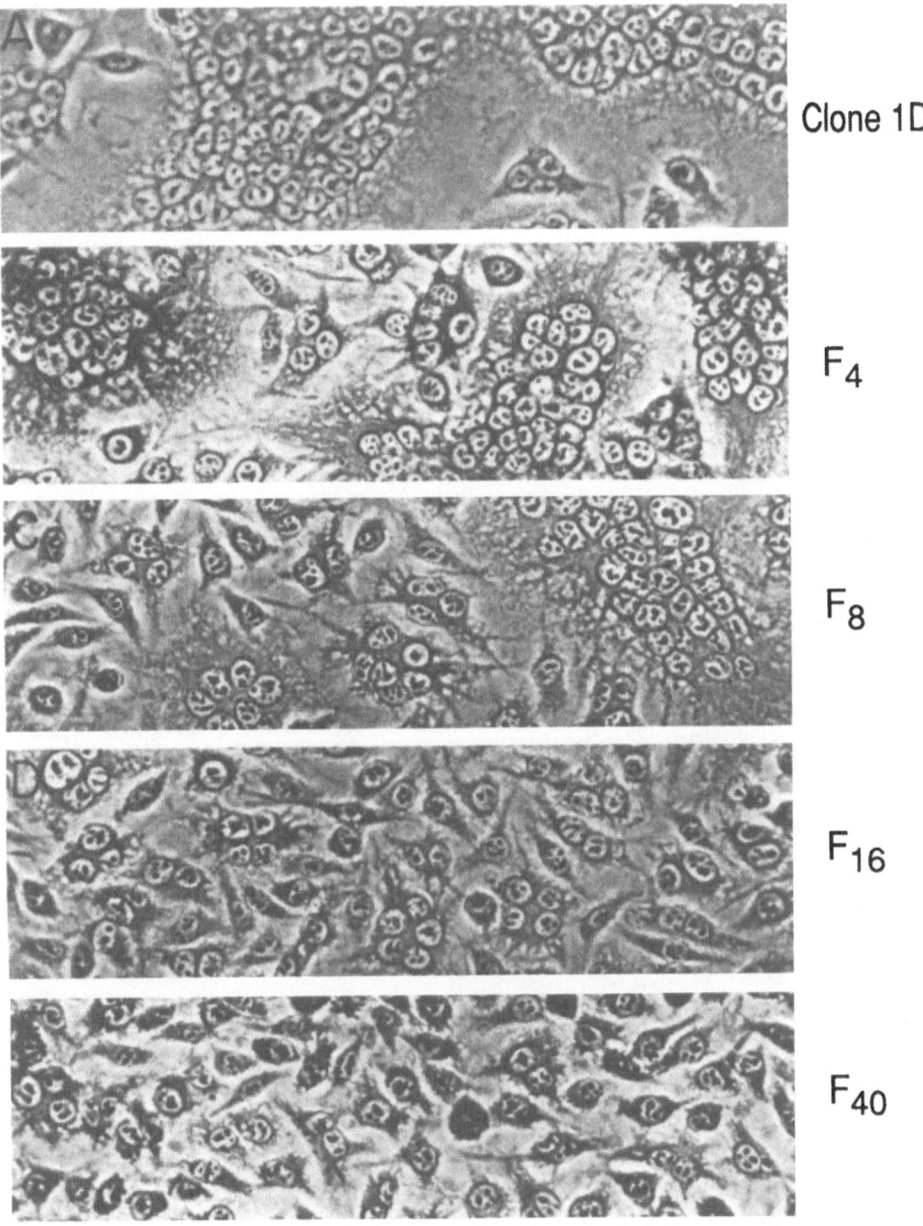

Clone 1D

F_4

F_8

F_{16}

F_{40}

We have obtained good results by mixing a 10-fold excess of one relatively PEG-resistant cell line with a minority of highly fusible partners. A variety of myeloma cell lines which differ in their susceptibility to PEG-induced fusion are currently available for murine hybridoma production (93). In work originally designed to elucidate the mechanisms of PEG-induced cell fusion, we have previously isolated a series of cell lines that are increasingly resistant to the fusogenic effects of PEG (94) (Fig. 1). The parental TK-mouse L-cell fibroblast cell line (designated Clone 1D) is highly susceptible to PEG-induced fusion. F_{40} mutants, isolated from the Clone 1D line by the failure to fuse through 40 cycles of PEG treatment, are highly resistant

Clone 1D cells   F₄₀ cells

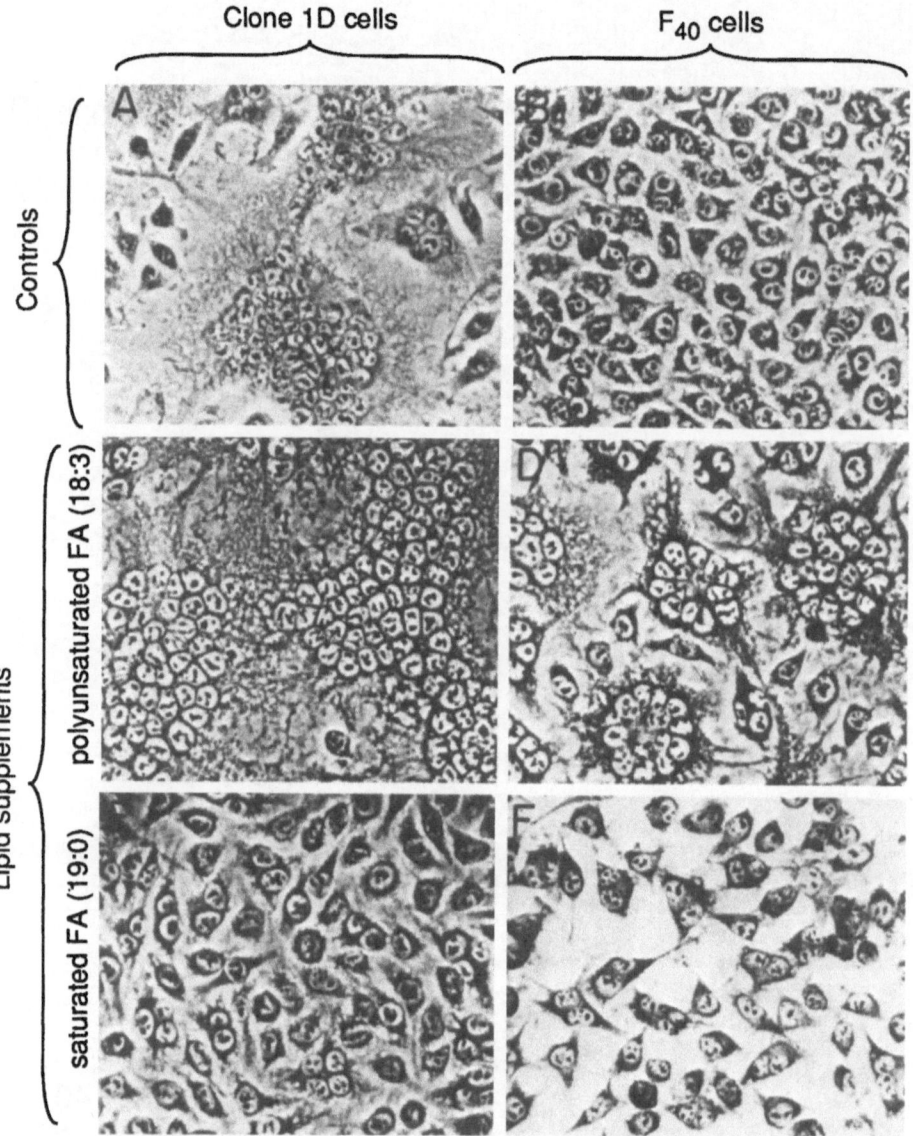

FIGURE 2   Control of PEG-induced fusion by lipid supplements. Parallel
cultures of parental cells (Clone 1D; panels A, C, E) and highly PEG-
resistant mutants (F_{40}; panels B, D, F) were grown for 18 h in complete
medium supplemented with 10^{-4} M linolenic acid (18:3; panels C and D)
or nonadecanoic acid (19:0; panels E and F) conjugated to delipidated
serum albumin. (In the presence of 10% fetal bovine serum, the added
lipids constitute ~60% of the available fatty acids in the medium.) Control
samples were supplemented with delipidated BSA only (panels A and B).
All cultures were then treated with PEG, incubated in fresh medium (without added lipid) for 1.5 h, fixed in methanol, and stained for microscopy.
Polyunsaturated fatty acids strongly stimulate fusion in F_{40} cells, effectively
complementing the mutant phenotype (compare panel D with control cells in
panel B), while saturated fatty acids render Clone 1D cells resistant to
fusion (compare panels E and A), mimicking the fusion response of untreated F_{40} cells. (Reprinted, with permission, from Ref. 96.)

to fusion and form hybrids ~100-fold less frequently than do Cline 1D cells. Similar mutants have also been isolated from a variety of other cell lines, including lymphoid cells grown in suspension (82). Using such techniques, it should be possible to develop a battery of myeloma cell lines that are highly effective as fusion partners (easy to culture at high density, rapidly growing, capable of supporting high levels of immunoglobin production defined by the introduced genetic material, stable in long-term culture, etc.) but differ in their susceptibility to PEG-induced fusion. In order to immortalize a particular cell type, it would only be necessary to select an appropriate immortalizing partner from this panel.

The fusion-resistant mutant cell lines described above are characterized by several alterations in lipid metabolism (95), and altered fusion response appears to be mediated by changes in acyl-chain saturation (96,97). By manipulating the culture medium in which the cells are grown, it is possible to control the susceptibility of these cells to PEG-induced fusion, as shown in Figure 2. PEG-resistant cells can be made fusible through the incorporation of polyunsaturated fatty acids into membrane lipids, and fusible cells are rendered PEG-resistant by increasing acyl-chain saturation. Figure 3 illustrates that beginning with a panel of cell lines that differ

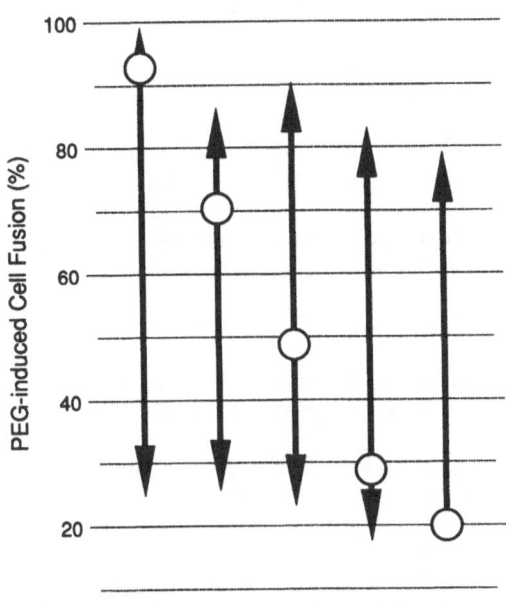

FIGURE 3 Control of fusion response in cultures intermediate in susceptibility to PEG-induced fusion. A panel of cell lines ranging in sensitivity to PEG-induced fusion between the highly fusible Clone 1D line and the highly fusion-resistant F_{40} mutants were divided into triplicate culture dishes and treated with PEG after 18 h in the presence of either normal medium (O), or medium supplemented with 100 µM polyunsaturated fatty acid (↑) or saturated fatty acid (↓). Individual cultures can be manipulated virtually at will to produce any degree of fusion desired, regardless of the original fusion response characteristic of these cell lines. Data presented are the compiled averages from more than 20 separate experiments.

greatly in their inherent sensitivity to PEG-induced fusion, fusion response can be controlled over nearly the entire range defined by the Clone 1D and F_{40} cell lines. Virtually any degree of fusion desired can be achieved, solely by the manipulation of lipid composition. Similar control can also be exerted over virus-induced fusion through the modification of cellular membrane lipids (60,99). Although the role these lipids play in regulating the fusion reaction is not yet understood (97), the ability to control fusion in a rapid, transient manner has obvious value for technological applications (82). We anticipate that further generalization of these findings will provide great flexibility in the design of fusion protocols.

REFERENCES

1. Harris, H. (1970). *Cell Fusion*. Harvard University Press, Cambridge.
2. Ephrussi, B. (1972). *Hybridization of Somatic Cells*. Princeton University Press, Princeton, NY.
3. Davidson, R. L. (1977). Genetics of cultured mammalian cells, as studied by somatic cell hybridization. *Natl. Cancer Inst. Monogr.* 48: 21-30.
4. Littlefield, J. W. (1987). The early history of mammalian somatic cell fusion. In: *Cell Fusion* (A. E. Sowers, ed.). Plenum Press, New York, pp. 421-426.
5. Sowers, A. E., ed. (1987). *Cell Fusion*. Plenum Press, New York.
6. Gottesman, M. M., ed. (1987). Molecular genetics of mammalian cells, *Methods Enzymol.* 151.
7. Müller, J. (1838). *Über den feineren Bau die Formen der krankhaften Geschwülste*, Berlin, quoted by Faber, K. (1892). *J. Pathol. Bacteriol.* 1:549-559.
8. Robin, C. (1849). Sur l'existence de deux espèces nouvelles d'éléments anatomiques qui se trouvent dans le canal médullaire des os. *C. R. Séanc. Soc. Biol.*, pp. 149-155.
9. Virchow, R. (1858). Reizung und Reizbarkeit. *Virchow's Arch. Path. Anat. Physiol.* 14:1-63.
10. Langerhans, T. (1868). Über Riesenzellen mit wandständigen Kernen in Tuberkeln und die fibröse Form des Tuberkels. *Virchow's Arch. Path. Anat. Physiol.* 42:382-404.
11. Krauss, E. (1884). Beiträge zur Riesenzellenbildung in epithelialen Geweben. *Virchow's Arch. Path. Anat. Physiol.* 95:249-272.
12. Lambert, R. A. (1912). The production of foreign body giant cells in vitro. *J. Exp. Med.* 15:510-515.
13. Lewis, W. H. (1927). The formation of giant cells in tissue cultures and their similarity to those in tuberculous lesions. *Am. Rev. Tuberc. Pulm. Dis.* 15:616-619.
14. Enders, J. F., and Peebles, T. C. (1954). Propagation in tissue cultures of cytopathic agents from patients with measles. *Proc. Soc. Exp. Biol. Med.* 86:277-286.
15. Henle, G., Deinhardt, F., and Girardi, A. (1954). Cytolytic effects of mumps virus in tissue cultures of epithelial cells. *Proc. Soc. Exp. Biol. Med.* 87:386-393.
16. Chanock, R. M. (1956). Association of a new type of cytopathogenic myxovirus with infantile croup. *J. Exp. Med.* 104:555-576.
17. Marston, R. Q. (1958). Cytopathogenic effects of haemadsorption virus type I. *Proc. Soc. Exp. Biol. Med.* 98:853-856.

18. Barski, G., Sorieul, S., and Cornefert, F. (1960). Production dans des cultures in vitro de deux souches cellulaires en association de cellules de caractére "hybride." *C. R. Acad. Sci. (Paris)* 251:1825-1827.

19. Sorieul, S., and Ephrussi, B. (1961). Karyological demonstration of hybridization of mammalian cells in vitro. *Nature* 160:653-654.

20. Spencer, R., Hauschka, T., Amos, D., and Ephrussi, B. (1964). Codominance of isoantigens in somatic hybrids of murine cells grown in vitro. *J. Natl. Cancer Inst.* 33:893-903.

21. Okada, Y., Suzuki, T., and Hosaka, Y. (1957). Interaction between influenza virus and Ehrlich's tumor cells III. Fusion phenomenon of Ehrlich's tumor cells by the action of H.V.J. Z strain. *Med. J. Osaka Univ.* 7:709-717.

22. Okada, Y. (1958). Analysis of giant polynuclear cell formation caused by H.V.J. virus from Ehrlich's ascites tumor cells. *Exp. Cell Res.* 26:98-107.

23. Harris, H., and Watkins, J. F. (1965). Hybrid cells derived from mouse and man: artificial heterokaryons of mammalian cells from different species. *Nature* 205:640-646.

24. Harris, H., Watkins, J. F., Ford, C. E., and Schoefl, G. I. (1966). Artificial heterokaryons of animal cells from different species. *J. Cell Sci.* 1:1-30.

25. Szybalski, W., Szybalska, E. H., and Ragni, G. (1962). Genetic studies with human cell lines. *Natl. Cancer Inst. Monogr.* 7:75-89.

26. Littlefield, J. W. (1964). Selection of hybrids from matings of fibroblasts in vitro and their presumed recombinants. *Science* 145:709-710.

27. Lederberg, J. (1958). Genetic approaches to somatic cell variation. *J. Cell Comp. Physiol.* 1:383-392.

28. Davidson, R. L., and Ephrussi, B. (1965). A selective system for the isolation of hybrids between L cells and normal cells. *Nature* 205:1170.

29. Wright, W. E. (1984). Toxin-antitoxin selection for isolating somatic cell fusion products between any cell types. *Proc. Natl. Acad. Sci. USA* 81:7822-7826.

30. Kamarck, M. E. (1987). Fluorescence-activated cell sorting of hybrid and transfected cells. *Methods Enzymol.* 151:150-165.

31. Barski, G., and Cornefert, F. (1962). Characteristics of 'hybrid'-type clonal cell lines obtained from mixed cultures in vitro. *J. Natl. Cancer Inst.* 28:801-821.

32. Croce, C. M. (1980). Cancer genes in cell hybrids. *Biochim. Biophys. Acta* 605:411-430.

33. Stanbridge, E. J., Der, C. J., Doersen, C-J., Nishimi, R. Y., Peehl, D. M., Weissman, B. E., and Wilkinson, J. E. (1982). Human cell hybrids: analysis of transformation and tumorigenicity. *Science* 215:252-259.

34. Harris, H. (1985). Suppression of malignancy in hybrid cells: The mechanism. *J. Cell Sci.* 79:83-94.

35. Lagarde, A. E., and Kerbel, R. S. (1985). Somatic cell hybridization in vivo and in vitro in relation to the metastatic phenotype. *Biochim. Biophys. Acta* 823:81-110.

36. Davidson, R. L. (1974). Gene expression in somatic cell hybrids. *Annu. Rev. Genet.* 8:195-218.

37. Ringertz, N. R., and Savage, R. E. (1976). *Cell Hybrids*. Academic Press, New York.

38. Weiss, M., and Green, H. (1967). Human-mouse hybrid cell lines containing partial complements of human chromosomes and functioning human genes. *Proc. Natl. Acad. Sci. USA* 58:1104-1111.

39. Miller, O., Allerdice, P., and Miller, D. (1971). Human thymidine kinase gene locus: assignment to chromosome 17 in a hybrid of man and mouse cells. *Science* 173:244-245.

40. Ruddle, F. H., and Creagan. (1975). Parasexual approaches to the genetics of man. *Annu. Rev. Genet.* 9:407-486.

41. Wright, C. E., and Shows, T. B. (1979). Genetics of cell fusion: human chromosome 10 assignment of a gene (FUSE) that promotes polykaryocyte formation. *Somatic Cell Genet.* 5:503-517.

42. Köhler, G., and Milstein, C. (1975). Continuous culture of fused cells secreting antibody of predefined specificity. *Nature* 256:495-497.

43. Homma, M., and Ohuchi, M. (1973). Trypsin action on the growth of Sendai virus in tissue culture cells III. Structural differences of Sendai viruses grown in eggs and tissue culture cells. *J. Virol.* 12:1457-1465.

44. Scheid, A., and Choppin, P. W. (1974). Identification of the biological activities of paramyxovirus glycoproteins. Activation of cell fusion, hemolysis and infectivity of an inactive precursor protein of Sendai virus. *Virology* 57:475-490.

45. Scheid, A., Hsu, M-C., and Choppin, P. W. (1980). Role of paramyxovirus glycoproteins in the interactions between viral and cell membranes. In: *Membrane-Membrane Interactions* (N. B. Gilula, ed.). Raven Press, New York, pp. 119-130.

46. White, J., Kielian, M., and Helenius, A. (1983). Membrane fusion proteins of enveloped animal viruses. *Q. Rev. Biophys.* 16:151-195.

47. Spear, P. G. (1987). Virus-induced cell fusion. In: *Cell Fusion* (A. E. Sowers, ed.). Plenum Press, New York, pp. 3-32.

48. Hsu, M-C., Scheid, A., and Choppin, P. W. (1979). Reconstitution of membranes with individual paramyxovirus glycoproteins and phospholipid in cholate solution. *Virology* 95:476-491.

49. Volsky, D. J., and Loyter, A. (1979). An efficient method for reassembly of Sendai virus envelopes after solubilization of intact virions with Triton X-100. *FEBS Lett.* 92:190-194.

50. Lucy, J. A. (1978). Mechanisms of chemically induced cell fusion. In: *Membrane Fusion* (G. Poste and G. L. Nicholson, eds.). Elsevier/North-Holland, Amsterdam, pp. 267-304.

51. Klebe, M. J., and Mancuso, M. G. (1981). Chemicals which promote cell hybridization. *Somatic Cell Genet.* 7:473-488.

52. Kao, K. N., and Michayluk, M. R. (1974). A method for high-frequency intergenic fusion of plant protoplasts. *Planta* 115:355-367.

53. Honda, K., Maeda, Y., Sasakawa, S., Ohno, H., and Tsuchida, E. (1981). Activities of cell fusion and lysis of the hybrid type of chemical fusogens I. Structure and function of the promoter of cell fusion. *Biochem. Biophys. Res. Commun.* 101:165-171.

54. Honda, K., Maeda, Y., Sasakawa, S., Ohno, H., and Tsuchida, E. (1981). The components contained in polyethylene glycol of commercial grade (PEG-6000) as a cell fusogen. *Biochim. Biophys. Acta* 640:82-90.

55. Wojcieszyn, J. W., Schlegel, R. A., Lumley-Sapanski, K., and Jacobson, K. A. (1983). Studies on the mechanism of polyethylene glycol-mediated cell fusion using fluorescent membrane and cytoplasmic probes. *J. Cell Biol.* 96:151-159.

56. Davidson, R. L., O'Malley, K. A., and Wheeler, T. B. (1976). Polyethylene glycol molecular weight and concentration. *Somatic Cell Genet.* 2:271–280.

57. Fazekas de St. Groth, S., and Scheidegger, D. (1980). Production of monoclonal antibodies: strategy and tactics. *J. Immunol. Methods* 35:1–21.

58. Lane, R. D., Crissman, R. S., and Lachman, M. F. (1984). Comparison of polyethylene glycols as fusogens for producing lymphocyte-myeloma hybrids. *J. Immunol. Methods* 72:71–76.

59. Smith, C. L., Ahkong, Q. F., Fisher, D., and Lucy, J. A. (1982). Is purified poly(ethylene glycol) able to induce cell fusion? *Biochim. Biophys. Acta* 692:109–114.

60. Roos, D. S. (1985). Membrane fusion lipid composition, and tumorigenicity of cultured cells. Ph.D. thesis, The Rockefeller University, New York.

61. Hui, S. W., Isac, T., Boni, L. T., and Sen, A. (1985). Action of polyethylene glycol on the fusion of human erythrocyte membranes. *Membr. Biol.* 84:137–146.

62. Ahkong, Q. F., Fisher, D., Tampion, W., and Lucy, J. A. (1975). Mechanisms of cell fusion. *Nature* 253:194–195.

63. Pontecorvo, G. (1975). Production of mammalian somatic cell hybrids by means of polyethylene glycol treatment. *Somatic Cell Genet.* 4:397–400.

64. Davidson, R. L., and Gerald, P. S. (1976). Improved techniques for the induction of mammalian somatic cell hybridization by polyethylene glycol. *Somatic Cell Genet.* 2:165–176.

65. Maul, G. G., Steplewski, Z., Weibel, J., and Koprowski, H. (1976). Time sequence and morphological evaluations of cells fused by polyethylene glycol 6000. *In Vitro* 12:787–796.

66. Pontecorvo, G., Riddle, P. N., and Hales, A. (1977). Time and mode of fusion of human fibroblasts treated with polyethylene glycol (PEG). *Nature* 265:257–258.

67. Boni, L. T., and Hui, S. W. (1987). The mechanism of polyethylene glycol-induced fusion in model membranes. In: *Cell Fusion* (A. E. Sowers, eds.). Plenum Press, New York, pp. 301–330.

68. Ahkong, Q. F., Howell, J. I., Lucy, J. A., Safwat, F., Davey, M. R., and Cocking, E. C. (1975). Fusion of hen erythrocytes with yeast protoplasts induced by polyethylene glycol. *Nature* 255:66–67.

69. Hirachi, Y., Kurono, M., and Kotani, S. (1979). Polyethylene glycol-induced fusion of L-forms of *Staphylococcus aureus*. *Biken J.* 22:25–29.

70. Rankin, J. K., and Darlington, G. J. (1979). Expression of human hepatic genes in mouse hepatoma-human amniocyte hybrids. *Somatic Cell Genet.* 5:1–10.

71. Petit, C., Levilliers, J., Ott, M-O., and Weiss, M. C. (1986). Tissue-specific expression of the rat albumin gene: genetic control of its extinction in microcell hybrids. *Proc. Natl. Acad. Sci. USA* 83:2561–2565.

72. Blau, H. M., Pavlath, G. K., Hardeman, E. C., Chiu, P-P., Silberstein, L., Webster, S. G., Miller, S. C., and Webster, C. (1985). Plasticity of the differentiated state. *Science* 230:758–766.

73. Papayannopoulou, T., Brice, M., and Stamatoyannopoulos, G. (1986). Analysis of human hemoglobin switching in MEL x human fetal erythroid cell hybrids. *Cell* 46:469–476.

74. Baron, M. H., and Maniatis, T. (1986). Rapid reprogramming of globin gene expression in transient heterokaryons. *Cell* 46:591–602.

75. Hämmerling, G. J., Hämmerling, U., and Kearney, J. F., eds. (1981). *Monoclonal Antibodies and T-Cell Hybridomas: Perspectives and Technical Advances.* Elsevier/North-Holland, Amsterdam.

76. Uchida, T., Ju, S-T., Fay, A., Liu, Y-N., and Dorf, M. E. (1985). Functional analysis of macrophage hybridomas I. Production and initial characterization. *J. Immunol.* 134:772–778.

77. Saxon, P. J., and Stanbridge, E. J. (1987). Transfer and selective retention of single specific human chromosomes via microcell-mediated chromosome transfer. *Methods Enzymol.* 151:313–325.

78. Shay, J. W. (1987). Cell enucleation, cybrids, reconstituted cells, and nuclear hybrids. *Methods Enzymol.* 151:221–237.

79. D'Eustachio, P., and Ruddle, F. H. (1983). Somatic cell genetics and gene families. *Science* 220:919–924.

80. Oi, V. T., and Herzenberg, L. A. (1980). Immunoglobin-producing hybrid cell lines. In: *Selected Methods in Cellular Immunology* (B. B. Mishell and S. M. Shigii, eds.). W. H. Freeman, San Francisco, pp. 351–372.

81. Olsson, L., and Kaplan, H. S. (1983). Human-human monoclonal antibody-producing hybridomas: technical aspects. *Methods Enzymol.* 92:3–16.

82. Roos, D. S., Davidson, R. L., and Choppin, P. W. (1987). Control of cell fusion in polyethylene glycol-resistant cell mutants: applications to fusion technology. In: *Cell Fusion* (A. E. Sowers, ed.). Plenum Press, New York, pp. 123–144.

83. O'Malley, K. A., and Davidson, R. L. (1977). A new dimension in suspension fusion techniques with polyethylene glycol. *Somatic Cell Genet.* 3:441–448.

84. Anders, G. J. P. A., Wierda, J., Nienhaus, A. J., and Idenburg, V. J. S. (1978). Time and cell systems as variables in fusion experiments with polyethylene glycol. *Hum. Genet.* 42:319–322.

85. Brahe, C., and Serra, A. (1981). A simple method for fusing human lymphocytes with rodent cells in monolayer by polyethylene glycol. *Somatic Cell Genet.* 7:109–115.

86. Mercer, W. E., and Schlegel, R. A. (1979). Phytohemagglutinin enhancement of cell fusion reduces polyethylene glycol cytotoxicity. *Exp. Cell Res.* 120:417–421.

87. Szoka, F., Magnusson, K-E., Wojcieszyn, J., Hou, Y., Derzko, Z., and Jacobson, K. (1981). Use of lectins and polyethylene glycol for fusion of glycolipid-containing liposomes with eukaryotic cells. *Proc. Natl. Acad. Sci. USA* 78:1685–1689.

88. Roos, D. S., Robinson, J. M., and Davidson, R. L. (1983). Cell fusion and intramembrane particle distribution in polyethylene glycol-resistant cells. *J. Cell Biol.* 97:909–917.

89. Röhme, D., and Thorburn, D. (1981). Quantitative cell fusion: Derivation and application of theoretical models. *Somatic Cell Genet.* 7:43–57.

90. Lane, R. D. (1985). A short duration polyethylene glycol fusion technique for increasing production of monoclonal antibody-secreting hybridomas. *J. Immunol. Methods* 81:223–228.

91. Norwood, T. H., Zeigler, C. J., and Martin, G. M. (1976). Dimethyl sulfoxide enhances polyethylene glycol-mediated cell fusion. *Somatic Cell Genet.* 2:263–270.

92. Davidson, R. L., and Ephrussi, B. (1970). Factors affecting the defective mating rate of mammalian cells. *Exp. Cell Res.* 61:222-226.
93. Kozbor, D., Dexter, D., and Roder, J. C. (1983). A comparative analysis of the phenotypic characteristics of available fusion partners for the construction of human hybridomas. *Hybridoma* 2:7-16.
94. Roos, D. S., and Davidson, R. L. (1980). Isolation of mouse cell lines resistant to the fusion-inducing effect of polyethylene glycol. *Somatic Cell Genet.* 6:381-390.
95. Roos, D. S., and Choppin, P. W. (1985a). Biochemical studies on cell fusion I. Lipid composition of fusion-resistant cells. *J. Cell Biol.* 101:1578-1590.
96. Roos, D. S., and Choppin, P. W. (1985b). Biochemical studies on cell fusion II. Control of fusion response by lipid alteration. *J. Cell Biol.* 101:1591-1598.
97. Roos, D. S. (1988). Control of cell membrane fusion by lipid composition. In: *Molecular Mechanisms of Membrane Fusion* (S. Ohki, D. Doyle, T. D. Flanagan, S.-W. Hui, and E. Mayhew, eds.). Plenum Press, New York, pp. 273-288.
98. Ohki, S., Doyle, D., Flanagan, T. D., Hui, S.-W., and Mayhew, E., eds. (1988). *Molecular Mechanisms of Membrane Fusion.* Plenum Press, New York.
99. Roos, D. S., Duchala, C. S., Stephensen, C. B., Holmes, K. V., and Choppin, P. W. (1990). Control of virus-induced cell fusion by host cell lipid composition. *Virology* 175:345-357.

29

Electrofusion and Electropermeabilization in Genetic Engineering

U. ZIMMERMANN

University of Würzburg, Würzburg, Federal Republic of Germany

I. INTRODUCTION

The use of electric-field pulses of high intensity and very short duration for cell-to-cell fusion and gene transfer provides an efficient, promising method for the somatic hybridization and genetic engineering of organisms. The applications of cell electrofusion and of cell membrane electropermeabilization (electric-field-induced transfer of normally membrane-impermeable substances and particles) are already numerous. In addition, further basic research in the field of interactions of electric fields with biological systems will certainly provide more information and, in turn, more applications. The field-pulse techniques and their applications have been reviewed several times from different viewpoints (1-13). Therefore, this chapter presents a discussion of recent applications, developments, improvements, and modifications of electrofusion and electropermeabilization techniques in order to facilitate the application of these physical techniques by biologists, biochemists, and immunologists. The aims are twofold. On one hand, it is hoped that researchers who are not so familiar with physical techniques will develop more confidence in their ability to recognize and apply this new physical approach in their research. On the other hand, the chapter is aimed at stimulating further research into the characterization and standardization of cellular parameters that control the hybrid yield in electrofusion and electro-gene transfer. The author believes that although the application of field-pulse technology to research and industrial problems is very broad, the knowledge about many cellular parameters involved in electrofusion and electropermeabilization is still very poor.

II. PRINCIPLES OF ELECTROFUSION AND ELECTROPERMEABILIZATION

Both electrofusion and electropermeabilization are based on the reversible increase in membrane permeability in response to an electrical breakdown

of the membrane in the presence of an external electric field (1-16). Therefore, the same apparatus can be used for both applications. The difference
between the two applications is that electrofusion requires membrane contact
between at least two cells, whereas electropermeabilization is performed
on freely suspended cells at a relatively low density. The establishment
of membrane contact often needs further equipment and a different medium
composition compared to electropermeabilization. An interesting exception
is seen in cell suspensions of high density where application of the field
pulse leading to membrane breakdown is sufficient to give fusion due to
the very small average distance between the cells. This was the approach
by which electrofusion of cells was demonstrated for the first time (17; for
a detailed discussion, see also Ref. 3).

Because of the relevance of electrical membrane breakdown in both
applications, I will discuss first some important features of this phenomenon
in respect to electrofusion and electro-gene transfer.

Electrical breakdown of lipid or lipid-protein membranes occurs if the
cells are exposed to a field of sufficient strength (18-20). Owing to charge
separation across the membrane capacitor a potential difference is built
up in response to the external field pulse. The magnitude of the generated
potential depends on the setup used for the application of the external
field. If two electrodes are used, one on each side of the membrane (e.g.,
when using intra- and extracellular electrodes in giant algal cells or eggs
or in squid axon or when performing breakdown in planar lipid bilayer
membranes or membrane sheets), the generated potential difference is
directly proportional to the applied field.

However, when field pulses are injected via two external electrodes
into a suspension of small particles surrounded by a closed membrane
(cells, vesicles, ets., see Figs. 1 and 2), the generated membrane potential
V_g is given by the integrated Laplace equation (21,22):

$$V_g = 1.5 \cdot a \cdot E \cdot \cos \theta \tag{1}$$

where V_g is the generated potential difference across the membrane, a
the radius of the cell, E the field strength, and θ the angle between
a given membrane site and the field vector. This equation holds for spherical cells; for ellipsoidal cells a similar, but more complicated, relationship
is obtained (see Ref. 22). Because of the radius dependence of the generated membrane potential difference, much higher voltages can be built
up across membranes of vesicles or cells than across membrane sheets
when a given field strength is applied to a suspension of these materials
(for a detailed discussion, see Refs. 7 and 12). Furthermore, the angular
dependence of the generated membrane potential in spherical cells (or
particles surrounded by a thin shell) indicates that the generated membrane
potential is not uniformly distributed over the membrane surface as is
the case when, for example, intra- and extracellular electrodes are used.
In a spherical (or ellipsoidal) cell the generated membrane potential always
has its maximum value at membrane sites oriented in the field direction
and its minimum value (zero) at membrane sites located perpendicular to
the field direction.

Therefore, when the critical field strength is applied, the breakdown
voltage will only be reached at those parts of the membrane oriented in
the field direction. Breakdown in other membrane areas only occur (depending on the angle θ) if supracritical field strengths are applied. The spatial
permeabilization of the membrane as a function of the field strength was

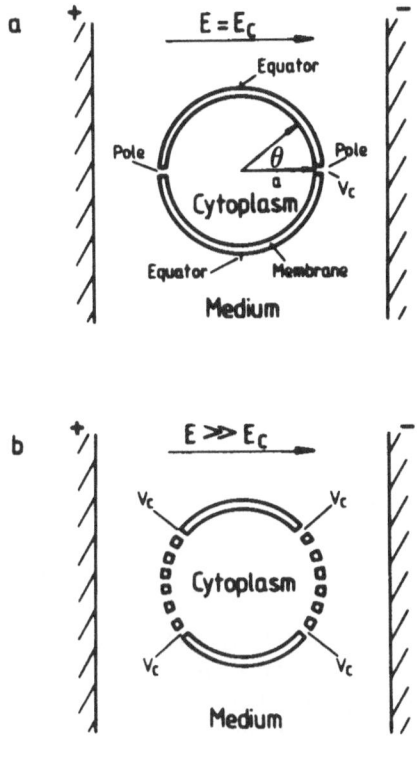

$$V_C = 1.5 \cdot a \cdot E_C \cdot \cos\theta$$

FIGURE 1 Schematic diagram of a cell exposed to high electrical field
strengths. The membrane potential V_g, which is built up across the membrane in response to the external field, is at its highest value at membrane
sites oriented in field direction (poles) and progressively decreases toward
the equator (perpendicular to the field direction). For membrane sites
in perpendicular orientation to the field direction, the potential V_g is zero.
The breakdown voltage V_C is therefore first reached in field direction,
$E = E_C$. It is only reached in membrane sites oriented at a certain angle
θ to the field direction if supracritical field strengths ($E > E_C$) are applied.
Breakdown of the membrane is indicated by the formation of transmembrane
pores.

recently demonstrated for freely (and dielectrophoretically aligned) plant
protoplasts by means of chemotactic bacteria and solute uptake (22a).

The breakdown voltate of most of the membranes is of the order of
1 V (3,12). The critical field strength for breakdown is therefore in the
range of 1-25 kV/cm, depending on the radius of the cell [see Eq. (1)],
on the temperature, and on the exposure time of the cells to the external
electric field (1-16). When calculating the required field strength for breakdown of the membrane and, in turn, for electropermeabilization, it must
be remembered that Eq. (1) holds only for the stationary state. This means
that it is assumed that the generated potential difference has reached its

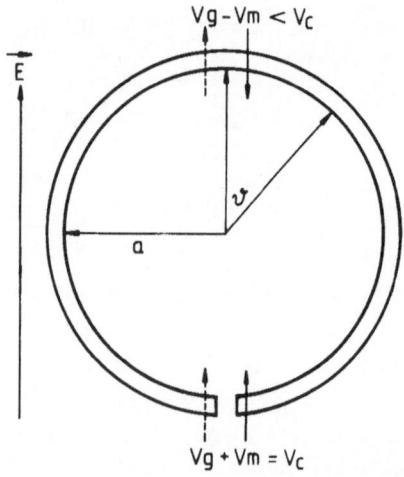

FIGURE 2 Schematic diagram of a cell exposed to a uniform electrical field (E) between two large electrodes. a, radius of the protoplast; V'_m, resting transmembrane potential; V_g, generated (superimposed) membrane potential; θ, angle between a given membrane site and the field direction.

maximum value. Because of the resistivities of the intra- and extracellular solutions as well as of the membrane resistance and of the capacitance, the charge separation across the membrane does not occur instantaneously. As always when dealing with a relaxation process, such as membrane charging, there will be a time lag while the system approaches a new equilibrium after field application. A relaxation process can be described by the relaxation time τ, which is the time required to reach 63.2% of the final steady-state potential difference [according to Eq. (1)]. Because of the asymptotic character of the curve of a relaxation process, about five relaxation times are needed before the voltage difference equilibrates.

For a spherical cell suspended in a solution into which a field pulse is injected, the relaxation time is given by (for derivation of the equation and for the assumptions made, see Refs. 12,22):

$$\frac{1}{\tau} = \frac{1}{R_m C_m} + \frac{1}{C_m \cdot a(\rho_i + 0.5\rho_e)} \tag{2}$$

where R_m is the specific membrane resistance (Ωcm^2), C_m the specific membrane capacitance ($\mu\text{F/cm}^2$), ρ_i the internal specific resistance (Ωcm), and ρ_e the external specific resistance (Ωcm). It follows from Eq. (2) that the relaxation time depends on the radius of the cells. The final value of the potential difference defined by Eq. (1) is therefore reached in much shorter times for small cells than for larger ones (e.g., eggs). The relaxation time also depends on the external resistivity and therefore on the electrolyte concentration. Electropermeabilization is generally performed in solutions containing sufficient amounts of electrolytes, whereas the conventional electrofusion of cells (for definition, see below) is normally carried out in weakly conductive solution (because of heat development in more conductive solutions). Therefore, longer pulse lengths are required for fusion than for gene transfer in order to reach the same degree of permeabilization of the membrane. These two consequences of Eq. (2) have to be taken into account when the optimum pulse length is selected for a particular experiment, because the exposure time is a critical parameter in electrofusion and electro-gene transfer.

Electrical breakdown and the associated changes in membrane resistance and permeability are only harmless to the suspended cells if the pulse does not exceed a certain value. Otherwise, resealing of the membranes with the concomitant return of the original resistance and impermeability is not observed. When the field pulse is too long, the reversible breakdown becomes irreversible, leading to deterioration of the membranes and of the cells. There are several mechanisms and processes for the change from reversible to irreversible breakdown, which are discussed at some length elsewhere (3,12). As a rule, the following guidelines can be used for different experimental limits. For electropermeabilization performed in physiological solutions (very high electrolyte content, about 150 mM univalent salts) the duration of the field pulse should not exceed 1 to 5 μs (see, e.g., Fig. 13 in Ref. 1). Exceptions are red blood cells if they are not nucleated; in this case pulse lengths of 40 μs can be applied without damage to the cells. In gene-transfer experiments which are performed in conductive solutions containing about 30 mM univalent salts (12) pulse lengths up to 10-15 μs can be applied without reduction in the yield of transformants due to damage of the cells. Exceptions are protoplasts of some yeast strains, which exhibit a high breakdown voltage (2 V compared to the normal value of 1 V). For electrofusion much higher pulse lengths have to be applied if microscopic cells are used because of the very low content of electrolyte in the external solution. Under these conditions the pulse lengths are normally in the range of 20-50 μs. If larger cells or eggs are used, the pulse length has to be extended correspondingly for each of the above-mentioned experimental conditions; e.g., fusion of eggs may need pulse lengths of the order of some milliseconds (9,12).

III. MECHANISM OF REVERSIBLE ELECTRICAL BREAKDOWN: USE OF TERMS

The term "reversible electrical breakdown" was introduced (14,23) in order to distinguish it from irreversible changes to the membrane and the cell in response to field pulses of excessively high intensity and/or duration. The term refers to the primary event within the membrane, which then leads to secondary changes in the membrane properties and subsequently—owing to osmotic processes—to changes in the cell interior (14,15). The term derived from solid-state physics reveals nothing about the molecular mechanism leading to electrical breakdown. A number of models have been proposed recently (24-33). One postulates that electrical breakdown occurs because of local electromechanical compression and tension of the membrane resulting in local instabilities within the membrane once the breakdown voltage is exceeded. The theoretical analysis of models based on this assumption, particularly the rigorous approach of Dimitrov (33), show that the pulse-length and pressure dependence of the breakdown voltage

and the influence of surfactants on the magnitude of the breakdown voltage
can be explained in terms of electromechanical instabilities in the membrane
in the presence of an electric field. The finding of Deuticke and co-workers
(34) that a considerable mobility of lipid molecules between the two asymmetrical leaflets of the bilayer structure in red blood cells membranes can
occur when critical field strengths are applied demonstrates that the electromechanical concept is too simple and that the structural changes are considerably more complex.

Most of the authors assume that breakdown leads to the formation
of transmembrane pores, but the origin of such pores is controversial.
Some authors assume that the electromechanically induced instabilities lead
to pore formation, others that existing defects in the membrane can increase
in size under the influence of the electric field. Evidence for the formation
of pores comes from electrical breakdown experiments on planar lipid bilayer
membranes. Benz et al. (35,36) showed that reversible electrical breakdown
in lipid membranes made up of oxidized cholesterol leads to a change in
the resistance of the lipid bilayer by more than eight orders of magnitude
in the presence of 1 M KCl solution on both sides. Such dramatic changes
in the membrane resistance can apparently only be explained by assuming
the presence of pores filled with conductive solutions. Similar changes
in the membrane resistance, once reversible breakdown has occurred,
have also been observed in lipid bilayer membranes of different composition,
stabilized by uranyl ions (37). Such marked membrane resistance changes,
however, have not as yet been observed in cell membranes (38,39). In
a lipid bilayer membrane the resealing times of the field-generated pertuba-
tions are very short (about 2-20 µs depending on temperature, Ref. 40)
compared with the long-lived defects in biological membranes, which can
persist for 10 min and more at low temperature (3,15).

On the basis of permeation experiments Schwister and Deuticke (41)
proposed that stable, long-lived pores of diameters less than 2 nm are
generated by the field pulses. These pores may arise from induced mismatches at lipid-protein boundaries. Such mismatches would be consistent
with an enhanced transbilayer mobility of erythrocyte phospholipids following
field application (34). We therefore have to conclude that conformational
changes of proteins or nonuniform aggregation of protein and lipid molecules
as a result of breakdown (3) gives rise to these long-lived permeabilization
stages of the membrane. The tangential components of the electric field
in the membrane may also be involved in this aggregation (2). In addition,
rearrangements of the membrane components should occur very easily because of the high diffusion coefficients of the components in a membrane
exhibiting a large number of defects (12).

In this context, evidence that organelles and whole cells can be electrically translocated across the membrane of a host cell under appropriate
field conditions (4,8,12) also seems to rule out the formation of stable
pores.

Convincing evidence against the formation of pores just after the application of the breakdown pulse was recently reported by Stenger and Hui
(42). These authors studied the sequence of events during electrofusion
of human erythrocytes by rapid-quench freeze-fracture electron microscopy.
The red blood cells were pretreated with pronase (3,12,43) and closely
positioned by dielectrophoresis. The authors observed the emergence of
transient point defects attributed to intercellular contact 2 s after the
breakdown pulse. After 10 s, most of these transient attachment sites

vanish, but a few fused gaps continue to develop into cytoplasmic bridges. According to Stenger and Hui (42), their data suggest that there is an important morphological distinction between the type, or stage, of membrane defect that results in electrofusion and that which manifests itself as a long-lived "pore" measurable by leakage experiments.

Although the pore model is certainly a very lucid conception, the molecular interpretation of electrical breakdown in terms of pores is not absolutely proven. It now seems that the stage after electropermeabilization (or electrofusion) has to be defined before the relevance of pores can be discussed.

The term electrical breakdown means that transmembrane defects occur, which can be considered as pores, channels, irregular leaks, or orientational defects (44). The term does not say anything about the molecular mechanism or the structural changes.

The term "electroporation," introduced by Neumann et al. (45) very recently instead of the term "reversible electrical breakdown," is appealing, but may not be totally accurate under the circumstances. In fact, it may well be misleading since, as mentioned above, other permeability changes in the membrane structure induced by the field pulse are not taken into account. This could lead to completely erroneous interpretations of the events involved in electrofusion and electro-gene transfer and hence to erroneous experimental concepts. Therefore, I would regard the use of the term "electroporation" as inappropriate, at least for the time being. The terms "electropermeabilization" and "electroinjection," on the other hand, are quite acceptable, if we place more emphasis on the consequences of reversible electrical breakdown. The increased permeability state of the membrane is, after all, the significant event as far as biological applications are concerned. Like the term "reversible electrical breakdown," these terms give no indication of the type of molecular changes induced in the membrane by the critical field pulse.

IV. ELECTROPERMEABILIZATION

Gene transfer and cell fusion are complementary methods in genetic engineering. The field-pulse technique can be used for both applications. Zimmermann and co-workers (16,20,23,46) were able to demonstrate more than one decade ago that electropermeabilization of the cell membrane would allow membrane-impermeable substances (of low molecular weight up to macromolecules) to be entrapped in cells without destroying their integrity and function. Most of these experiments were made by using the high-voltage discharge technique (1). The discharge chamber consisted of two flat, parallel platinum electrodes, 1 cm apart, mounted in a rectangular well in a plexiglass chamber. The cell suspension was added into the well up to the upper edge of the electrodes. One electrode was grounded, whereas the other one was connected via a switch (spark gap) to a capacitor. The capacitor was loaded by a high-voltage source to several kV and then discharged through the cell suspension. Using this experimental setup it was possible to incorporate dyes (1,15), drugs (16), proteins (16,23), enzymes (46), latex particles (47,48), and DNA (3,20,49) into various host cells. The work done on this field until 1983 is reviewed elsewhere (16).

There have been other studies of electrically induced DNA transfer and the production of stable transformants. Neumann et al. (45) reported on electrically induced gene transfer in mouse lyoma cells, Potter et al. (50) injected genes into pre-B lymphocytes by electrical means, Schnettler and Zimmerman (3), Karube et al. (51), and Salek et al. (52) obtained stable transformants when they incorporated different plasmids in yeast (*Saccharomyces cerevisiae*) protoplasts after application of a field pulse of appropriate intensity and duration.

Electric-field-mediated DNA transformation in plant protoplasts was reported by Langridge et al. (53) and by Fromm et al. (54), leading either to transient gene expression or to stable transformants. Although most of these authors claim that this technique has substantial benefits over current methods, the yield of stable transformants in some of these experiments was rather low and a relatively high DNA (plasmid) concentration (10-50 µg/ml) was needed to obtain sufficient numbers of clones. The main reason for these negative results is based on inappropriate conditions during the application of the breakdown pulse and the subsequent resealing process. The physical limits outlined above, the temperature profile during the whole process, and medium composition play a crucial role in successful electro-gene transfer. For a critical discussion of this subject matter the reader is referred to a recent review article (12).

Shilito et al. (55) tried to increase the efficiency of electrically induced direct gene transfer in plant protoplasts by a combined treatment consisting of a breakdown pulse, poly(ethylene glycol) (PEG) and a heat shock (5 min at 45°C). Using 10-30 µg/ml plasmid DNA and 50 µg carrier DNA, they obtained about 300 stable transformants per 3.5×10^5 treated protoplasts (corresponding to about 90 clones per 10^6 cells and per 1 µg plasmid DNA). Even though it seems to work in protoplasts of *Nicotiana tabacum*, this modified gene-transfer technique is very empirical. It also lacks one of the main advantages of the electropermeabilization procedure: i.e., the clear-cut understanding and control of the process.

For mouse L cells and for murine myeloma cells Stopper et al. (56, 57,57a) worked out a protocol for electro-gene transfer, which yields stable transformants in the order of about 10^{-3} (58). The characteristic features of this protocol are: dispase pretreatment of the cells, field application in a weakly buffered isotonic medium containing 30 mM KCl and the corresponding amount of inositol at low temperature (4°C), several consecutive pulses of 10 kV/cm strength and 5-µs duration at an interval of 1-2 min, low plasmid concentration (1 µg/ml), and performance of the resealing process in a special medium.

The electrolyte concentration of the medium in which the pulse application is carried out seems to be important. This is probably because membrane breakdown occurs asymmetrically only in one hemisphere of the cell (59,60). Asymmetrical breakdown occurs due to the superposition of the external field with the intrinsic field within the membrane, which depends on the ionic strength of the external solution (12). The intrinsic field is only parallel to the external field vector in one hemisphere; in the other one it is antiparallel (Fig. 1b). Therefore, if the intrinsic field is sufficiently strong, breakdown field strength will be reached first in the membrane of the hemisphere in which the two field vectors are in parallel.

Asymmetrical breakdown must also play a role in conventional electrofusion; however, no investigations of this effect have been performed so far (for further details, see Ref. 22a).

Another important point of the gene-transfection protocol of Stopper et al. (56-58) also seems to be the much longer time interval of 1 min between subsequent pulses. The resealing of lipid-protein membranes at 4°C is indeed slow (15), but the lipid domains within the membrane may reseal completely at this temperature (see above); thus a field can be built up across the membrane again. Due to Brownian movement of the cells during the time interval, the next breakdown will take place in different parts of the membrane, leading to a higher probability that DNA molecules can enter the cell.

Using the protocol of Stopper et al. (56,57,57a), Ruppert (61) succeeded for the first time in cotransfection of mouse L cells. Stable transformants were obtained, where two electrically incorporated genes could be expressed. Recently, Däumler and Zimmermann (58a) showed that electrotransfection of mouse L cells and macrophages in *strongly* hypoosmolar solutions gave very high yields of stable transformants, which significantly exceeded the clone number obtained under isoosmolar conditions. The cells survived these extremely low osmolarities for 1 h without any apparent deterioration of cellular or membrane functions. Highest yields were obtained in buffered 75-mOsmol solutions containing 30 mM KCl and an appropriate amount of inositol provided that the strength of the breakdown pulse was matched to the dramatic increase in cell volume at low osmolarity. The absolute clone number depended on the postincubation time in the hypoosmolar solution after application of a single breakdown pulse at 4°C. The absolute number of stable transformants was maximum when the postincubation was restricted to 2 min. Toward longer incubation times the absolute number decreased; even the relative clone number was similar. This was because of a corresponding decrease in the number of viable cells.

Finally, I would like to point out that there are many other applications of electropermeabilized cells. An excellent review article covering these applications was written by Knight and Scrutton (62). Here, I will discuss only the use of electropermeabilization in electrofusion, for which the establishment of closed membrane contact should be considered first.

V. ELECTROFUSION

A. Membrane Contact

In the conventional electrofusion technique membrane contact is established between at least two freely suspended cells by the application of an inhomogeneous alternating electric field of a frequency between 10 kHz and 2 MHz and of a peak amplitude of 1-100 V (depending on the volume). In the inhomogeneous field the cells will move to give formation of pearl chains (Fig. 3), which lie within the electrode gap or emanate from the electrodes, depending on the divergence of the field and of the ratio of the relative dielectric constants of the cells to that of the medium. This process, termed dielectrophoresis, had been discussed in detail several times (1-15,63,64, 64a). An interesting modification of dielectrophoresis in cell fusion may be dielectrophoretic levitation (65). With this technique it is possible to align the cells in special field geometries distant from the electrodes using gravity to compensate for the dielectrophoretic forces. Because of heat development, dielectrophoresis must be carried out in low-conductivity media (isotonic sugar solutions). The use of such media is not expected to be damaging to cells any more than it is in the well-established technique of cellular (free-flow) electrophoresis (66). This assumption is supported

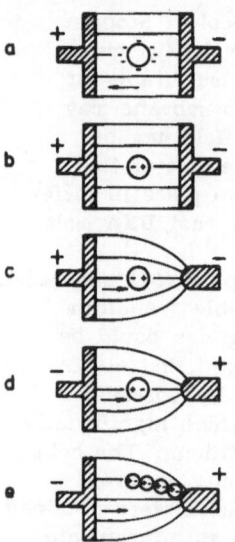

FIGURE 3 Diagrammatic representation of electro- and dielectrophoresis.
Because of the net charge on the outer surface of the membrane, cells
or charged particles migrate in the direction of an electrode (electrophoresis)
in a homogeneous electrical field (plate capacitor) (a). The direction of
migration depends on the sign of the charge and on the direction of the
external voltage. In general, cells have a negative surface charge. In
addition, a dipole is induced in a dielectric particle or in a cell (regardless
of whether these particles carry a net charge or not) (b). Since the electrical field intensity is equal on both sides of the particle or cell (b),
this induced charge does not contribute to movement. Uncharged particles
are thus unable to migrate in a homogeneous field. In an inhomogeneous
electrical field, on the other hand, uncharged particles are also able to
migrate because the electrical field exerts a net force on the electrically
induced dipole (c); the field intensity is not equal on both sides of the
particle, resulting in a net force acting on the particle. This effect is
known as dielectrophoresis. In contrast to the direction of migration during
electrophoresis, the direction of the dielectrophoretic migration of the
particles and cells is not reversed when the external electrical voltage
between the two electrodes is reversed (d). The cells adhere to each other
and form chains at the electrodes along the electrical field lines, because
the dipoles induced in the cells attract each other (e).

by many papers in the literature (see, e.g., Ref. 67). In particular, the
membrane permeability of red blood cells to various ions after breakdown
has been shown to increase at higher ionic strengths (68).

   Use of dielectrophoresis in combination with a breakdown pulse (so-
called conventional electrofusion) leads to the production of large numbers
of viable hybrids. This method has been used by most workers so far
(see below).

   However, for special purposes, alternative methods that allow the
establishment of membrane contact in conductive solutions may be useful.
For example, Kramer et al. (70,71) were able to produce magnetically coated
erythrocytes by the use of Fe_3O_4 particles. By means of two perpendicular

magnetic fields the coated cells could be attracted into a very small volume. The force directed toward the center of this region was sufficient (the fields had strong gradients) to give contact between the cells. Application of two field pulses gave fusion. The possible incorporation of Fe_3O_4 particles into the fusion products may be used to advantage in a subsequent magnetic isolation of fused cells, allowing, for example, the isolation of hybrids formed from commercially important yeast cells (which possess no genetic markers usable for selection).

In a sonic field (71,72), forces are exerted on cells in a somewhat analogous manner to those exerted during dielectrophoresis. The force exerted in the sonic field depends on the density difference between particles and medium, whereas in dielectrophoresis it is the difference in dielectric constants that is important. However, sound wavelengths may be used that are much smaller than the fusion chamber. This permits not only the production of pearl chains (in a purely propagating wave), but also the concentration of cells at standing-wave-pressure maxima. The effect is so powerful that cells are often completely absent from other regions of the standing-wave pattern. Vienken et al. (72) were able to use 1 MHz ultrasound (1 mm wavelength) to establish contact between red blood cells, myeloma cells, or myeloma and lymphocyte cells (Fig. 4). The standing-wave maxima of 0.5 mm periodicity were developed between two electrodes. Application of breakdown pulses led to fusion. By variation of the cell density, preferential formation of two-cell hybrids was possible.

Recently, we have shown that a brief application of an alternating electric field (about 1 s dielectrophoresis) after sonic concentration and prior to pulse application results in very high yields (demonstrated for hybridoma cells). In addition, it is quite conceivable that the use of low-conductivity media favors high yields of hybrids because of the lesser decrease in membrane resistance once the breakdown pulse has been applied to the contact zone of the two cells. This will shield the cell interior against the field, which is still there for some time because of the very rapid event of breakdown (69). Owing to the low electrolyte concentration in the fusion media, the pulse length must be longer than that which is optimal for transfection (see the discussion above).

Empirically it was found that sufficient dielectrophoretic force has to be applied to the cells to cause slight deformation of the cells in the contact zone within the pearl chain. Otherwise, no fusion is observed after the application of the breakdown pulse. The question was open how close the membranes have to approach each other at this stage. By using rapid-quench freeze-fracture electron microscopy, Stenger and Hui (42) showed that dielectrophoresis leading to subsequent fusion enabled adjacent human red blood cells pretreated with pronase to approach each other within 15 nm. An aqueous boundary separating the two adjacent cells was still detectable. The distance of 15 nm represents the distance at which the attractive dielectrophoretic and Van der Waals forces are balanced by the electrostatic repulsion between the pronase-treated cells. The mutual approach of two cells within the 15 nm observed due to dielectrophoresis would presumably invoke an exponentially increasing repulsion by electrostatic and hydration forces. According to these results, Stenger and Hui (42) suggest that repulsive forces successfully oppose direct lipid bilayer contact between adjacent cell membranes prior to the breakdown-pulse application. The pulse caused a brief disruption of the aqueous boundary leading to the formation of defects and discontinuous areas.

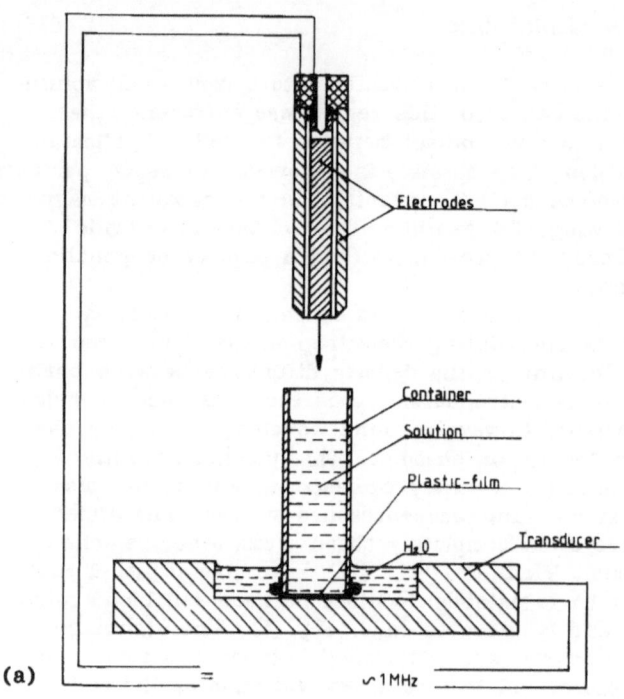

FIGURE 4 (a) Setup for electroacoustic fusion. For further details, see Ref. 73. (b) To illustrate the banding of red blood cells following sound irradiation, a Camlab-Microslide is filled with a suspension of erythrocytes and inserted into the container of the electroacoustic fusion setup. Two seconds after the beginning of irradiation, cells are no longer suspended homogeneously, but show a banding over the whole length of the microslide due to the sound aggregation forces.

(a)

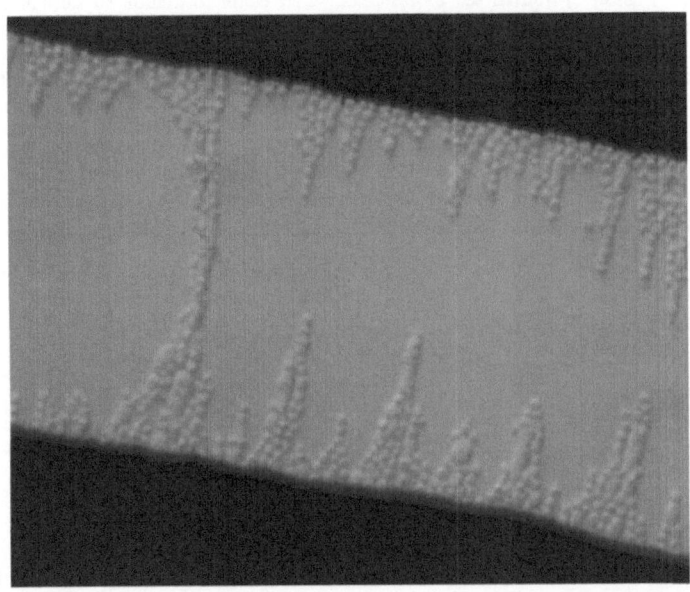

(b)

FIGURE 5   (a) In principle, the rotational fusion chamber consists of a cylindrical casing, which may be made to act as a centrifuge by spinning it around its axis of symmetry. In practice, the floor of the chamber consists of a Plexiglas carrier plate onto which electrodes have been previously vacuum-evaporated. Of the various electrode configurations, (a) shows a configuration that has proved suitable for the mass production of electrically fused yeast hybrids and hybridoma cells. The electrodes cover the circular area of the carrier disk, with electrode pairs (two electrodes running in parallel at a distance of 125 microns) radiating out from the center. The thickness of the vacuum-evaporated electrodes is about 1 micron. The circular electrode plate (or disk) is surrounded by a lower-lying, annular channel into which the fused cells can be centrifuged after the fusion process has been completed. (b) Fusion of yeast cells in the rotational fusion chamber after application of the breakdown pulse (see Refs. 90, 91 for experimental conditions). The chamber can be sealed with a lid, allowing the entire fusion process to be completed under sterile conditions.

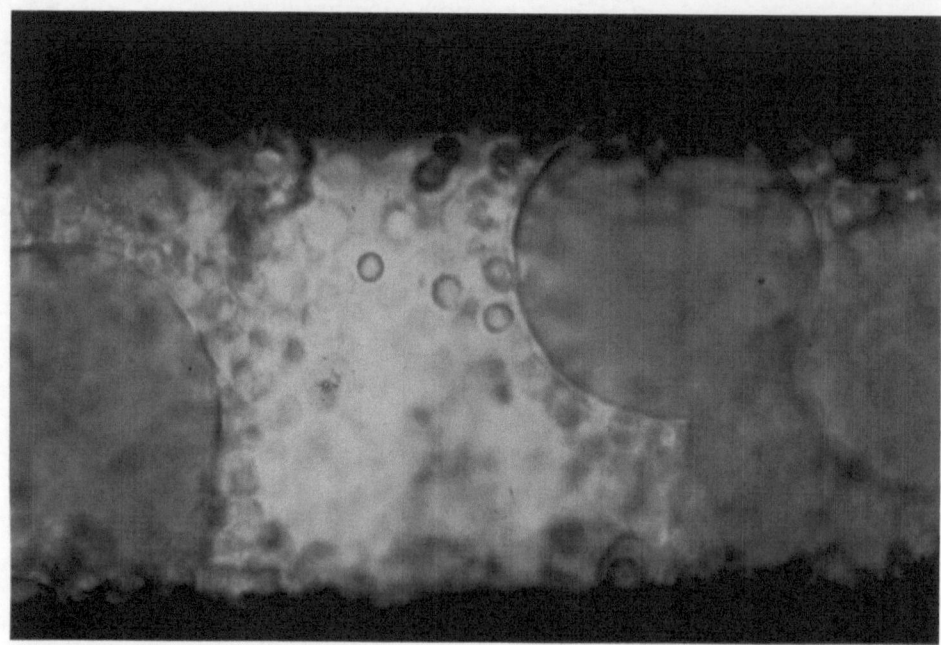

FIGURE 6   Fusion of human red blood cells. Cell chain formation of human erythrocytes was achieved at high suspension densities (400 V/cm, 1 MHz, 0.3 mol mannitol solution). Fusion was induced by three subsequent electrical field pulses of 6 kV/cm strength and 5 microseconds duration. The giant cells are surrounded by erythrocytes that were not exposed to the electrical field.

Electron microscopic studies also gave support to the assumption of Zimmermann (3) that regions of the membrane free of integral proteins appeared to be the sites where fusion is initiated. Pretreatment of cells with digestive enzymes (which is known to facilitate fusion), together with the effects of the alternating field, may favor the emergence of such areas in the membrane contact zone (3,12).

In the electromechanical method, cells are pulsed either in solution of high suspension density (3,11) or at low suspension density and then brought in close contact by centrifugation. Large quantities of cells can be handled without the addition of fusogenic chemicals. The disadvantage is that the fusion process cannot be observed (e.g., by light microscopy); in this respect this particular method is similar to chemical or viral induction of fusion. There are other possibilities based on the application of physical forces to bring cells into close contact (71). Chemicals, such as PEG, have also been used in combination with a field pulse to establish the required cell contact. Lo et al. (73) were able to induce proximity between B lymphocytes and myeloma cells by synthesizing a selective cross-bridge between the two. In this technique, immunogen-avidin complexes are used to bridge the B cells bearing surface immunoglobulin to biotinylated myeloma cells. Bridged rosettes are then fused selectively by pulse application. This method, which is useful in hybridoma production, has been criticized for several reasons (71,72,74). Wojchowski and Sytkowski (74) published a new modification of the avidin-mediated electrofusion which seems to be very effective. This method is discussed below.

B. Electrofusion in Specific Systems

The many reports published in the last years on electrofusion demonstrate that this technique can be applied to all living cells, liposomes, and planar lipid bilayer membranes, and combinations of them (Figs. 2,5-7) (Figs. 5a,b and 6, see color plate). The universality of the electrofusion technique results from the fact that both electrical breakdown and the physical procedures leading to the required membrane contact reflect characteristic features of lipid-protein and lipid bilayer membranes. Membranes of dead cells do not exhibit the electrical breakdown phenomenon (e.g., see Ref. 75). An increasing number of publications show that the advantages of this new technique over the conventional ones are recognized by many authors. These are:

a. Prediction of the fusion parameters for fusion of new cell combinations provided that the relevant physical parameters given by Eqs. (1) and (2) are known. Equipment to measure these parameters is available (1-13).

b. The fusion process can be observed under the light microscope; the experienced experimenter can tell at once whether a fusion was successful or not.

c. The fusion process is nearly synchronous because breakdown occurs in all field-exposed cells at the same time; thus studies directed at revealing single steps in the fusion process and changes in membrane structure can be performed. This is very important when large amounts of fusing cells are required (e.g., for electron microscopy).

d. The fusion technique permits the fusion of only a few cells (cell pairs) or thousands or millions of cells, depending on the experimental conditions and the procedure used for the establishment of membrane contact. Special chambers (e.g., see Fig. 4) are available (12).

e. The fusion conditions are gentle because only a low percentage of the total membrane area of a cell is subjected to breakdown; in consequence the yield of hybrids is very high provided that the correct medium and fusion parameters have been selected.

Results obtained on electrofusion up to the middle of 1985 have been extensively reviewed (1-15). In this review article I shall therefore restrict myself only to a brief summary of these data in order to discuss in more detail the most recent findings.

1. Artificial Lipid Bilayer Systems

Electrofusion of planar black lipid membranes consisting of different lipids has so far only been reported by Melykian et al. (76). The two planar membranes were brought into proximity by the application of a small hydrostatic pressure difference. If symmetrically charged or uncharged membranes were used, fusion could be obtained only by external field pulses. On the other hand, if charged membranes were pushed together by the hydrostatic pressure difference, spontaneous fusion could occur, indicating that under some circumstances the intrinsic field is high enough to fuse the membranes once contact has been established (77). Electrofusion of liposomes made up of natural or polymerizable lipids was reported by Büschl et al. (78). In these experiments large liposomes and the standard electrofusion technique were used. Liposomes without surface charge exhibited negative dielectrophoresis (movement and pearl chain formation distant from the electrodes) in the kHz frequency range when the experiments were performed in distilled water, whereas either positively or negatively charged liposomes showed positive dielectrophoresis (movement to the electrodes). Fusion between liposomes and cells is also possible provided the tonicity of the medium is suitable.

2. Bacteria

Shivarova et al. (79) described fusion of protoplasts of *Bacillus thuringiensis* by combined chemical (PEG) and field fusion. One strain, containing the plasmid pUB 110, was resistant to kanamycin; the second was sensitive to kanamycin and produced a brown pigment. The authors reported that colonies were obtained in selective media being kanamycin-resistant and able to produce the pigment. The first electrofusion of bacteria (strains

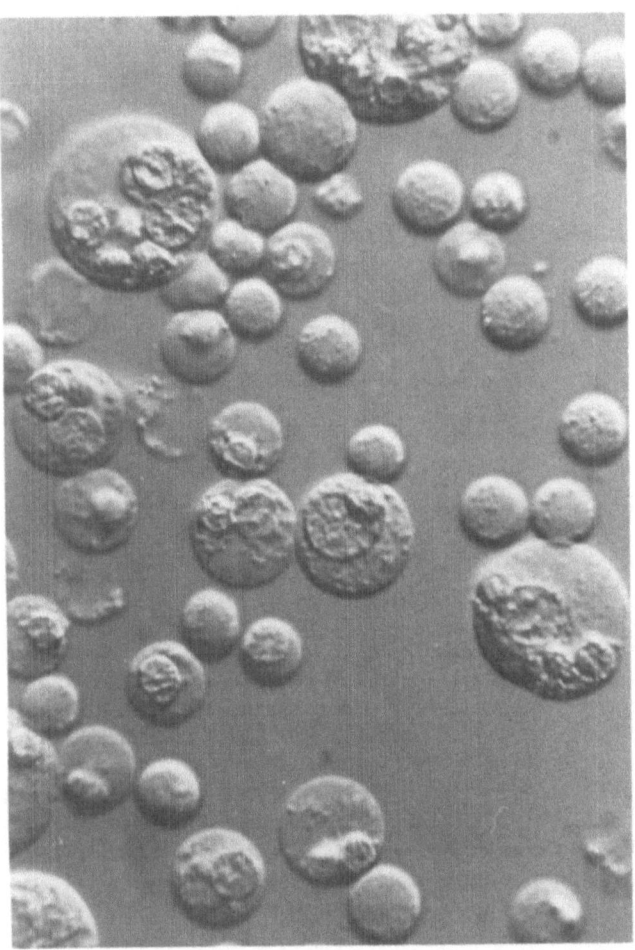

FIGURE 7 Interference phase-contrast micrograph of electrically fused
murine erythroleukaemic cells (Friend cells). Cells were fused using the
following protocol. Cells were suspended in an isotonic 0.3 M mannitol
solution supplemented with 1 mg/ml Pronase (Serva GmbH, Heidelberg).
Then the cells were collected dielectrophoretically into "strings of pearls"
between the electrodes (frequency of the a.c. field, 1 MHz; field strength,
300 V cm^{-1}). Cells were fused by injecting a d.c. field pulse of 2 kV
cm^{-1} and 40 μs duration. Rounding-up of the cells is initiated by adding
phosphate-buffered saline and is completed after 5 min. It is evident
that the nuclei of some cells were in the process of fusion. This may
lead to an exchange of genetic material.

with different genetic markers of *Escherichia coli* and of *Salmonella typhi-murium*) was reported by Ruthe and Adler (80), using giant sheroplasts. They obtained many colonies exhibiting the genetic markers of the parental fusion partners, also when electrofusion was performed between cells of *E. coli* and *S. typhimurium*.

3. Fungi

Electrofusion of fungal protoplasts has been demonstrated by Schnettler and Zimmermann (quoted in Ref. 12) using fusion conditions and media almost identical to those developed for the fusion of yeast protoplasts (81). Electrofusion of different wall-less (slime) mutants of *Neurospora crassa* was reported by Fikus et al. (82). The authors pointed out that this tech-nique is at present the only one that permits formation of diploid slime variants.

4. Yeast

Weber et al. (83) fused different genetically marked strains of the yeast *Saccharomyces cerevisiae* by combined treatment of the cells with PEG and field pulses. Halfmann et al. (84,85) were the first to report that protoplasts of yeast mutants of the same species can be fused with high yields by means of the standard electrofusion technique (Fig. 4). Using appropriate genetic markers and respiratory deficient mutants, they could show by using selection media that fusion leads to both plasmogamy and karyogamy. Karyogamy occurred only when haploid partners were subjected to electrofusion (85). The yield of hybrids was always some magnitudes higher when electrofusion was used rather than PEG-induced fusion.

Halfmann et al. (86) were also the first to demonstrate that it is possible to fuse yeast partners of different size. They electrofused dielectrophoretic-ally aligned cells of a haploid strain of *S. cerevisiae* with cells of a poly-ploid strain (baker's yeast). The hybrids obtained after transfer to selection medium were heterokaryons, which gave rise to spontaneous parental-type segregants on nutrionally complete media. Stable hybrids could be obtained after several passages on selection media.

Emeis and Zimmermann (87) were successful in intergeneric electro-fusion between cells of different genera. These authors have also shown that mutagenic treatment and subsequent electrofusion of cells of *Candida*, which exhibit an interesting pectinase activity, lead to hybrids of 10-fold pectinase activity.

Schnettler and Zimmermann (81) considerably improved the electro-fusion technique as applied to yeast cells by using new fusion chambers, so that very large numbers of yeast hybrids of different species can now be produced. They were also able to show that vector DNA in yeast cells can be transferred into other cells by means of electrofusion. Broda et al. (88) investigated variations in hybrid yield, using the improved protocol for electrofusion of yeast protoplasts (81). They showed that the yield of hybrids under optimal electrofusion conditions depends only on the 'quality' of the protoplasts. Electronic measurements of the size distribu-tions of the parental fusion partners showed that considerable variations in the modal volume, shape, and skewness of the respective cell population occurred despite apparent constant culture and protoplast preparation conditions. Broda et al. (88) formulated a simple, quantitative guideline for prediction of hybrid yield on the basis of the respective size distribution. They also showed, that a 24-h preincubation of yeast protoplasts prior to electrofusion in special media at 29°C can improve the size distribution of the fusion partners with respect to hybrid frequency (elimination of skewness, enlargement of the modal volume, etc.). This work is an important step in the establishment of reproducible and predictable electrofusion conditions. It simultaneously showed that much more effort has to be made in the future to improve the conditions of cell culture and protoplast preparations.

5. Plant Protoplasts

The enormous potential for plant breeding inherent in electrofusion of plant protoplasts has resulted in the publication of many papers (89-108). In addition, the peculiar membrane properties of plant protoplasts makes them very easy to fuse, which is not always the case with mammalian cells. However, one factor that is frequently overlooked is that the system parameters determine the viability and regenerative capacity of the fusion products (e.g., excessive pulse duration despite the relatively small cell radius, see Ref. 89). Most of this work is reviewed elsewhere (1-15).

Some special applications of protoplast fusion were reported recently. Gaff et al. (105) electrically fused protoplasts from desiccation-tolerant grasses with those from desiccation-sensitive grasses. Saga et al. (99) used protoplasts isolated from marine algae for electrofusion. Salhani et al. (106) also demonstrated that interkingdom electrofusion between a plant protoplast and a mammalian cell is possible. Properties of both parents were expressed in the fusion product. Hampp et al. (104) fused electrically evacuolated plant protoplasts. This work may provide a new way of achieving better regeneration of fused hybrids of different origin for plant-breeding purposes because of the lack of the vacuole; owing to simultaneous breakdown of the tonoplast membrane, the vacuole content might interfere with the viability of the fused products. In this context, the field strength required for electrofusion of evacuolated protoplasts is half that required for fusion of normal protoplasts (if the different radii are taken into account). These results can only be explained if the two membranes, tonoplast and plasmalemma, are arranged in series from an electrical point of view.

Proof that electrically fused plant protoplasts are able to regenerate to the callus stage was first provided by Koop et al. (102), who fused two mesophyll protoplasts of *N. tabacum* by electrical means. Recently, Koop and Schweiger (103) also reported regeneration of plants after electrofusion of defined pairs of tobacco leaf protoplasts at remarkably high frequencies (40%). For regeneration, fusion products were grown individually in a microculture system. The results demonstrated that electrofusion of selected pairs of protoplasts can be used for somatic hybridization of higher plants and, in combination with individual-cell-culture techniques, for regeneration of plants with good yields.

Tempelaar and Jones (97,98) also succeeded in the regeneration of calli and shoots from electrofused suspension cultures with mesophyll protoplasts. Kohn et al. (107) also reported somatic hybridization of electrofused protoplasts isolated from suspension cultures of nitrate-reductase-deficient mutant cell lines of *N. tabacum*. These authors concluded that somatic hybrids derived from electrofused plant protoplasts were viable and capable of regeneration into calli and exhibiting organogenetic responses by shoot differentiation.

The rapid growth of the number of publications in the field of electrofusion of plant protoplasts seems to indicate the great potential of this technique in plant breeding, and we can expect further developments in this area in the near future.

6. Animal Cells

Human red blood cells have been fused by several electrofusion techniques: by application of the standard electrofusion technique, by electroacoustic and electromagnetic fusion methods, and by injection of breakdown pulses into dense suspensions of cells within a discharge chamber.

Electrofusion of human red blood cells (Fig. 5) is best carried out after pretreatment of the cells with digestive enzymes (43,109). Although fusion without pretreatment is possible, it leads to small yields of fused products. Ahkong and Lucy (110) showed that human red blood cells incubated in hypotonic erythritol solutions fuse more readily after application of the field pulse than less swollen cells in 200-400 mM erythritol. The authors concluded that the extent of fusion induced by the breakdown pulse is governed by the combination of electrocompressive and osmotic forces. It is known from other work (111,112) that the value of the breakdown voltage decreases with increasing pressures (turgor pressure = pressure gradients and absolute hydrostatic pressure). Therefore, the results may be interpreted in terms of a larger perturbation of the membrane at low tonicity compared to isotonic or hypertonic conditions, because the same field strength was used in both sets of experiments.

Nucleated animal cells, such as Friend cells (113; Fig. 6), Ehrlich ascites tumor cells (114), lymphoblasts (115), lymphoma cells (116,117), and 3T3 cells (118), have been fused electrically. At present, pretreatment of such cells with proteolytic enzymes is not required; a brief pretreatment can, however, increase the fusion rate. Ohno-Shosaku and Okada (115-117) investigated in detail facilitation of electrofusion of animal cells by proteolytic enzymes and by phospholipases. These and other authors (81) showed that a small amount of calcium ions in the solution is a prerequisite for electrofusion of such cells and that part of the facilitation of electrofusion by proteolytic enzymes is due to contamination of the enzymes with calcium.

Podesta et al. (119) reported steroid hormone and cyclic AMP production in adrenal-Leydig-cell hybrids generated by electrofusion. Leydig cells were prepared by collagenase digestion of adult rat testis and fused with rat adrenocortical cells. Fusion was only achieved between these cells after nine pulses of 15 μs duration. Normally (for fusion of other cells mentioned above), a sequence of only two to three pulses is required using pulse lengths of 5-20 μs, the required field strengths being about 1.5-3 times higher than that calculated from Eq. (1) for membrane regions oriented in field direction.

Tsong, Teissie, and co-workers (120,121) described electric-field-pulse-mediated fusion of 3T3-C2 fibroblasts in monolayer cultures, using natural cell-to-cell contact. Using the same technique, Finaz et al. (122) fused a coculture of ID and CH rodent cell lines. Hybrids were obtained in selection medium with a frequency of 1×10, which was 100 times higher than with PEG. Natural contact was also used when clustered cells of the mould *Dictyostelium* were subjected to breakdown pulses (123).

Much of the cell fusion work carried out so far has used lymphocytes and myeloma cells. This is because of the great potential of the hybridoma cells produced by this procedure, which I will discuss in a separate section.

7. Hybridoma

The production of hybridoma cells that secrete monoclonal antibodies (124) is based on the fusion of myeloma cells and spleen cells from appropriately immunized animals, usually mice or rats (see Chapter 30). Hybridoma cells can be produced either by standard electrofusion or by electroacoustic fusion. Using a special electrohydraulic procedure, Vienken and Zimmermann (125) obtained high yields of pronase-pretreated mouse myeloma/lymphocyte pairs by dielectrophoresis. Field-pulse application resulted in fusion and formation of hybridoma cells. Production of human hybridoma cells by standard electrofusion was reported at the same time by Bischoff et al. (126). In both reports electrofusion was performed in isotonic sugar (e.g., mannitol) solutions which did not contain any further ingredients.

Improvement of the fusion medium and the resealing conditions led recently to a dramatic increase in the yield of hybridomas from electrofusion. Vienken and Zimmermann (127) found that as in the case of electrofusion of yeast cells (81), small concentrations of divalent cations (0.1 mM calcium ions and 0.5 mM magnesium ions) as well as buffer substances played an important role. These authors also showed that the sugar used for maintenance of isotonicity considerably influences the viability of the hybridoma cells. When mannitol was replaced by inositol, a high frequency of electrofusion of lymphocytes with myeloma cells was obtained. Owing to the presence of divalent ions, it was not necessary to pretreat the cells by proteolytic enzymes. In the experiments of these authors it was, however, necessary that the fused products be treated afterward with special resealing media. In the meantime, Schmitt et al. (128) have shown that the use of resealing media is not required if fusion is performed in media as used in the electrofusion of yeast cells (isotonic concentrations of sorbitol, 0.1 mM calcium, and 0.5 mM magnesium ions). If suitable field conditions are used, high yields of mouse hybridoma cells can be obtained (128) which secrete monoclonal antibodies. This was shown by immunization of mice with IDNP hemocyanin or by in vitro stimulation of lymphocytes using lipopolysaccharide (LPS). These authors also showed by electronic determination of the size distributions of lymphocytes stimulated in vivo or in vitro and of myeloma cells that culture conditions of the myeloma cells as well as the time of stimulation of the lymphocytes and of harvest of the myeloma cells considerably affect the yield of hybridoma cells. This finding is similar to that mentioned above for electrofusion of yeast cells.

Karsten et al. (129) and Brown et al. (130) were also able to produce hybridoma cells capable of secreting monoclonal antibodies, using the standard electrofusion technique and sugars described by Vienken et al. (127) and Bischoff et al. (126). In both studies the cells were pretreated with pronase or dispase. Whereas Karsten et al. (129) reported a 10-fold increase in yield when using electrofusion isntead of PEG-induced fusion, Brown et al. (130) found only a slight increase in yield under electrofusion conditions compared to PEG-induced fusion. Possible reasons for the result of Brown et al. (130) may have been the pretreatment of the cells with digestive enzymes, which always leads to less viable hybrids, or the antigen used for immunization. Electronic size distribution measurements have shown (Zimmermann et al., unpublished data) that immunization does not always lead to stimulated lymphocytes of favorable size for electrofusion with myeloma cells. Under these conditions, only those myeloma cells which are smaller should be used. The size can be controlled, as mentioned above, by culture conditions and time of harvest. In the meantime many reports

on the production of antibody-secreting hybridoma cells have been published
(130a-130p). As in the case of electrotransfection (see above), considerable
progress in the field of hybridoma production could recently be achieved
by electrofusion of myeloma cells with stimulated lymphocytes in *strongly*
hypoosmolar solutions containing sorbitol, divalent cations, and albumin
(130n,o). The number of viable hybrids was higher than any value for
chemically or electrically mediated fusion reported in the literature. Optimum
clone numbers were obtained in fusion media of 75 mOsmol osmolarity.
Similar results were obtained for fusion of myeloma cells with the murine
hybridoma cell line G8. Due to the dramatic increase in volume, the field
strength of the breakdown pulse (leading to fusion of the dielectrophoretically aligned cells) has to be reduced, as predicted by theory (22). The
efficiency of hypoosmolar electrofusion allowed the use of very few cells
(130n,o). Most important, in contrast to isoosmolar conditions, the fusion
and hybrid yield in hypoosmolar electrofusion was reproducible over long
periods of time and less dependent on variations between cultures.

Further improvement of hybridoma electrofusion is expected if it becomes
possible to preselect secreting lymphocytes prior to fusion. A promising
step in this direction was recently described by Arnold and Zimmermann
(131,132). They showed that secreting lymphocytes can be distinguished
from nonsecreting ones by the electrorotation technique.

Another interesting approach of combined preselection and electric-
field-pulse-mediated fusion of myeloma cells with lymphocytes was recently
reported by Lo et al. (73), as discussed above. A considerable improvement
to the avidin-biotin method of Lo et al. (73) was published recently by
Wojchowski and Sytkowski (74). In the first step of their approach, biotin
was attached both to immunogen and to the surface of murine myeloma
cells using N-hydroxysuccinimidobiotin. Biotinylated immunogen was then
incubated with splenocytes derived from immunized mice and allowed to
bind to surface immunoglobulins of B cells. Using streptavidin, biotinylated
myeloma cells then were bridged to those B cells bearing biotinylated
immunogen. Selective fusion of bridged cells was performed by breakdown-
pulse application. A high frequency of hybridomas was obtained, all of
which secreted high titers of antibodies to a selected immunogen, keyhole
limpet hemocyanin. This simplified avidin-mediated electrofusion technique
seems to be very powerful.

8. Blastomeres and Eggs

Berg (133) described electrofusion of mouse blastomeres containing between
two and eight cells, using attached microelectrodes. Blastomeres of two-cell
mouse embryos with an intact *zona pellucida* were also fused electrically
by Kubiak and Tarkowski (134) using standard electrofusion conditions
and media of different conductivity. These authors tested the viability
of the fusion products and demonstrated the development of the fused
two-cell blastomeres into tetraploid blastocytes, which, however, died after
implantation. Nuclear transplantation in sheep embryos using the conventional electrofusion technique was recently reported by Williadsen (135).
Blastomeres from eight- and 16-cell embryos were fused with enucleated
or nucleated halves of unfertilized eggs. Williadsen (135) showed that fully
viable embryos may be obtained by embedding the fusion products in agar,
followed by culture of the reconstituted embryos in the ligated oviducts
of ewes in diestrus. The yield of fused blastomere-egg fragments was
considerably higher when electrofusion was performed than when inactivated

Sendai virus was used. In the Sendai-virus-induced fusion the whole process of fusion took up to 4 h, whereas fusion was completed within 1 h when a field pulse was applied. Richter et al. (136) fused denuded sea urchin eggs. The fused eggs had intact nuclei and could be fertilized, but underwent abortive cleavage. Savage and Grey (137) were successful in electrofusion of *Xenopus* eggs. For fusion of eggs, pulse lengths of longer duration were required than with other cells because of the large radius of these systems.

C. Electrofusion in Membrane Research

Electrofusion has great potential not only in somatic hybridization, but also in membrane research. Since the fusion process is nearly synchronous for all cells exposed to the pulse, the individual intermediate stages can be studied by electron microscopy, as mentioned above (42,138). The possibility of observation of the fusion between two pairs of cells under the light microscope permits studies of movement of membrane components by fluorescent labeling (139,140) or by electrical means (141). Diffusion coefficients can be calculated from such measurements. Verhoek-Köhler et al. (92) showed, by studying electrofusion of cell pairs and multiples, that the rate of rounding-up of mesophyll protoplasts of *Avena sativa* is affected by the energy charge of the cells (which could be influenced by various metabolic inhibitors). The authors also demonstrated in various fused multiples of plant protoplasts that the ATP/ADP ratio did not change. This indicates that leakage of intracellular solutes during the fusion process is very small. Hampp et al. (104) found evidence, during the fusion of evacuolated plant protoplasts, that various physical parameters such as viscosity and surface tension may also be involved in the kinetics of electrofusion. Salhani et al. (142) investigated the hydraulic conductivity of fused plant protoplasts and found that the hydraulic conductivity of the fused cells was identical to that of the individual cells. This indicates that the original membrane resistance, at least to water transport, had been restored. Similar results were obtained for electrofused red blood and ghost cells with the interesting exception that long-lived fusogenic states of several minutes seem to be established in the membrane of the fused product (143).

Strong support for this assumption is also given by the early observation that giant red blood cells can be formed by electrofusion of thousands of cells (1–15,109; Fig. 6). Formation of giant cells, either from red blood cells (109) or from nucleated cells (113), has been performed by subjecting high cell densities to dielectrophoresis before application of several pulses of supracritical field strength (Figs. 6,7). This procedure leads to breakdown not only in the contact zone between adjacent cells in a chain, but also in other parts of the membrane due to the angle dependence of the generated membrane potential difference. In consequence three-dimensional fusion occurs between cells within a chain and between adjacent chains. Under these conditions it took several minutes before the giant cells (diameter several 100 μm) rounded up. It was still possible to obtain further fusion between such giant cells afterward if they were simply pushed together. This clearly indicates that some fusogenic state of the membrane persisted.

Giant fused cells are interesting systems in membrane research because microelectrodes can be inserted for direct measurement of the electrical properties of the membrane. This opens the way to direct studies on the action of membrane-active substances, including pharmaceuticals.

VI. CONCLUSIONS

The current state of the art is that the field-pulse techniques for electro-permeabilization and cell-cell fusion are not only capable of competing with the conventional chemical methods, but they may open new perspectives for biotechnology. The theoretical and experimental work in this field is very promising, but needs to be extended. This necessarily involves an improvement of the present methods and a better understanding and control of the biophysical parameters involved in electrofusion and electrotransfection.

The fundamental experiments in this field, i.e., reversible electrical breakdown and electric-field-mediated entrapment of molecules in cells, go back more than one decade, and they are only now gaining increasing relevance. One reason for this is undoubtedly the fact that the introduction of physical techniques into biology, biotechnology, and medical technology, which all have a strong biochemical bias, still meets with substantial difficulties. The present trend in this field to develop protocols for electrofusion and electrotransfection by combination of chemicals with electric-field-pulse application seems to support this view. I believe that it is only a question of time and effort before scientists become aware that the use of physical forces alone may well lead to success. The vectorial character of physical forces, which in principle always permits prediction and control of the induced effects, makes physical techniques per se superior to those using chemicals. The difficulty with this particular technique is that at present we know too little about the interactions of electric fields with living material.

ACKNOWLEDGMENT

This work has been supported by grants of the BMFT (DFVLR) and the DFG (SFB 176). I am greatly indebted to Dr. W. M. Arnold and Dr. K-H. Büchner for critically reading and discussing the manuscript.

REFERENCES

1. Zimmermann, U., Scheurich, P., Pilwat, G., and Benz, R. (1981). Cells with manipulated functions: New perspectives for cell biology, medicine and technology. *Angewandte Chemie, Int. Ed. Engl.* 20:325-344.
2. Zimmermann, U., and Vienken, J. (1982). Electric field-induced cell-to-cell fusion. *J. Membrane Biol.* 67:165-182.
3. Zimmermann, U. (1982). Electric field-mediated fusion and related electrical phenomena. *Biochim. Biophys. Acta* 694:227-277.
4. Zimmermann, U. (1983). Electrofusion of cells: Principles and industrial potential. *Trends Biotechnol.* 1:149-155.
5. Zimmermann, U., and Vienken, J. (1984). Electric field-mediated cell-to-cell fusion. In: *Cell Fusion, Gene Transfer and Transformation* (R. F. Beers and E. G. Bassett, eds.). Raven Press, New York, pp. 171-187.
6. Zimmerman, U., Vienken, J., Pilwat, G., and Arnold, W. M. (1984). Electrofusion of cells: Principles and potential for the future. In: *Cell Fusion*, Ciba Foundation symposium 103 (D. Evered and J. Whelan, eds.). Pitman, London, pp. 60-73.

7. Zimmermann, U., Büchner, K. H., and Arnold, W. M. (1984). Electro-fusion of cells: Recent development and relevance for evolution. In: *Charge and Field Effects in Biosystems* (M. J. Allen and P. N. R. Usherwood, eds.). Abacus, UK, pp. 293-317.

8. Zimmermann, U., Vienken, J., and Pilwat, G. (1984). Electrofusion of cells. In: *Investigative Microtechniques in Medicine and Biology*, Vol. 1 (J. Chayen and L. Bitensky, eds.). Marcel Dekker, New York, pp. 89-168.

9. Arnold, W. M., and Zimmermann, U. (1984). Electric field-induced fusion and rotation of cells. In: *Biological Membranes*, Vol. 5 (D. Chapman, ed.). Academic Press, London, pp. 389-454.

10. Zimmerman, U., and Vienken, J. (1984). Electrofusion of cells. In: *Hybridoma Technology in Agricultural and Veterinary Research* (N. J. Stern and H. R. Gamble, eds.). Rowman & Allanheld, Totowa, NJ, pp. 173-200.

11. Zimmermann, U., Vienken, J., Halfmann, J., and Emeis, C. C. (1985). Electrofusion: A novel hybridization technique. In: *Advances in Biotechnological Processes*, Vol. 4 (A. Mizrahi and A. L. van Wezel, eds.). Alan R. Liss, New York, pp. 79-151.

12. Zimmermann, U. (1986). Electrical breakdown, electropermeabilization and electrofusion. *Rev. Physiol. Biochem. Pharmacol.* 105:175-256.

13. Zimmermann, U. (1987). Electrofusion of cells. In: *Methods of Hybridoma Formation* (A. H. Bartal and Y. Hirshaut, eds.). Humana Press, New York, pp. 97-150.

14. Zimmermann, U., Pilwat, G., Beckers, F., and Riemann, F. (1976). Effects of external electric fields on cell membranes. *Bioelectrochem. Bioenerg.* 3:58-83.

15. Zimmermann, U., Vienken, J., and Pilwat, G. (1980). Development of drug carrier systems: Electrical field-induced effects in cell membranes. *Bioelectrochem. Bioenerg.* 7:553-574.

16. Zimmerman, U. (1983). Cellular drug-carrier systems and their possible targeting. In: *Target Drgus* (E. Goldberg, ed.). Wiley, New York, pp. 153-200.

17. Zimmermann, U., and Pilwat, G. (1978). The relevance of electric field induced changes in the membrane structure to basic membrane research and clinical therapeutics and diagnostics. In: *Abstract of the Sixth International Biophysics Congress*, Kyoto, Japan, IV-19-(H), p. 140.

18. Zimmermann, U., Schulz, J., and Pilwat, G. (1973). Transcellular ion flow in *Escherichia coli B* and electrical sizing of bacteria. *Biophys. J.* 13:1005-1013.

19. Zimmermann, U., Pilwat, G., and Riemann, F. (1974). Reversible dielectric breakdown of cell membranes by electrostatic fields. *Z. Naturforsch.* 29c:304-310.

20. Zimmermann, U., Pilwat, G., and Riemann, F. (1984). Dielectric breakdown of cell membranes. *Biophys. J.* 14:881-889.

21. Schwan, P. (1983). Biophysics of the interaction electromagnetic energy with cells and membranes. In: *Biological Effects and Dosimetry of Static and Electromagnetic Fields* (M. Grandolfo, S. M. Michaelson, and A. Rindi, eds.). Plenum Press, New York, pp. 213-231.

22. Jeltsch, E., and Zimmermann, U. (1979). Particles in a homogeneous electrical field: a model for the electrical breakdown of living cells in a Coulter counter. *Bioelectrochem. Bioenerg.* 76:349-384.

22a. Mehrle, W., Hampp, R., and Zimmermann, U. (1988). Electric pulse induced membrane permeabilisation. Spatial orientation and kinetics of solute efflux in freely suspended and dielectrophoretically aligned plant mesophyll protoplasts. *Biochim. Biophys. Acta* 978:267-275.

23. Zimmermann, U., Pilwat, G., and Riemann, F. Dielectric breakdown of cell membranes. In: *Membrane Transport in Plants* (U. Zimmermann and J. Dainty, eds.). Springer Verlag, Berlin, Heidelberg, New York, pp. 146-153.

24. Coster, H. G. L., and Zimmermann, U. (1975). The mechanism of electrical breakdown in the membranes of Valonia utricularis. *J. Membrane Biol.* 22:73-90.

25. Coster, H. G. L., Steudle, E., and Zimmermann, U. (1976). Turgor pressure sensing in plant cell membranes. *Plant Physiol.* 58:636-643.

26. Zimmermann, U., Buchner, K. H., and Benz, R. (1982). Transport properties of mobile charges in algal membranes, influence of pH and turgor pressure. *J. Membrane Biol.* 67:183-197.

27. Abidor, I. G., Arakelyau, V., Chernomordik, L., Chizmadzhev, Y., Pastushenko, V., and Tarasevich, M. (1979). Electric breakdown of bilayer membranes. The main experimental facts and their qualitative discussion. *Bioelectrochem. Bioenerg.* 6:37-52.

28. Petrov, A. G., Mitov, M. D., and Derzhauski, A. I. (1980). Edge energy and pore stability in bilayer lipid membranes. In: *Advances in Liquid Crystal Res. Appl.* (L. Bata, ed.). Pergamon Press, Oxford, pp. 695-737.

29. Sugar, I. P. (1983). Effect of mechanical and electrical pressure on the phase transition properties and stability of phospholipid bilayers. A pore model of electrical breakdown. In: *Physical Chemistry of Transmembrane Ion Motions* (G. Spach, ed.). Elsevier, Amsterdam, pp. 21-28.

30. Dimitrov, D. S., and Jain, R. K. (1984). Membrane stability. *Biochim. Biophys. Acta* 779:437-468.

31. Weaver, J. C., Mintzer, R. A., Ling, H., and Sloan, St. R. (1986). Conduction criteria for transient aqueous pores and reversible electrical breakdown in bilayer membranes. *Bioelectrochem. Bioenerg.* 15:229-241.

32. Powell, K., Derrick, E., and Weaver, J. C. (1986). Quantitative theory of reversible electrical breakdown in bilayer membranes. *Bioelectrochem. Bioenerg.* 15:243-245.

33. Dimitrov, D. S. (1984). Electric field-induced breakdown of lipid bilayers and cell membranes: A thin viscoelastic field model. *J. Membrane Biol.* 78:53-60.

34. Dressler, V., Schwister, K., Haest, C. W. M., and Deuticke, B. (1983). Dielectric breakdown of the erythrocyte membrane enhances transbilayer mobility of phospholipids. *Biochim. Biophys. Acta* 732:304-307.

35. Benz, R., Beckers, F., and Zimmermann, U. (1979). Reversible electrical breakdown of lipid bilayer membranes. A charge pulse relaxation study. *J. Membrane Biol.* 48:181-204.

36. Benz, R., and Zimmermann, U. (1980). Relaxation studies on cell membranes and lipid bilayers in the high electric field range. *Bioelectrochem. Bioenerg.* 7:723-739.

37. Chernomodik, L. V., Sukharev, S. I., Abidor, I. G., and Chizmadchev, Y. A. (1983). *Biochim. Biophys. Acta* 736:202-213.

38. Coster, H. G. L., and Zimmerman, U. (1975). Dielectric breakdown in the membranes of Valonia utricularis: The role of energy dissipation. *Biochim. Biophys. Acta* 382:410-418.

39. Zimmermann, U., Groves, M., Schnabl, H., and Pilwat, G. (1980). Development of a new Coulter counter system: Measurement of the volume, internal conductivity and dielectric breakdown voltage of a single guard cell protoplast of *Vicia faba*. *J. Membrane Biol.* 52:37.

40. Benz, R., and Zimmermann, U. (1981). The resealing process of lipid bilayers after reversible electrical breakdown. *Biochim. Biophys. Acta* 640:169-178.

41. Schwister, K., and Deuticke, B. (1985). Formation and properties of aqueous leaks induced in human erythrocytes by electrical breakdown. *Biochim. Biophys. Acta* 816:332-348.

42. Stenger, D. A., and Hui, S. W. (1985). Kinetics of ultrastructural changes during electrically-induced fusion of human erythrocytes. *J. Membrane Biol.*, in press.

43. Zimmermann, U., Pilwat, G., and Richter, H. P. (1981). Electric field-stimulated fusion: Increased field stability of cells induced by pronase. *Naturwissenschaften* 68:577-578.

44. Sackmann, E. (1984). Physical basis of trigger processes and membrane structures. In: *Biological Membranes* (D. Chapman, ed.). Academic Press, London, pp. 105-144.

45. Neumann, E., Schaeffer-Ridder, M., Wang, Y., and Hofschneider, P. (1982). Gene transfer into mouse lyoma cells by electroporation in high electric fields. *EMBO J.* 1:841-845.

46. Zimmermann, U., Riemann, F., and Pilwat, G. Enzyme loading of electrically homogeneous human red blood cell ghosts prepared by electrical breakdown. *Biochim. Biophys. Acta* 436:460-474.

47. Vienken, J., Jeltsch, E., and Zimmermann, U. (1978). Penetration and entrapment of large particles in erythrocytes by electrical breakdown techniques. *Cytobiologie* 17:182-196.

48. Schüssler, W., and Ruhenstroth-Bauer, G. (1984). Stomatocytosis of latex particles by rat erythrocytes by the electrical breakdown technique. *Blut* 49:213-217.

49. Auer, D., Brandner, G., and Bodemer, W. (1976). Dielectric breakdown of the red blood cell membrane and uptake of SV40 DNA and mammalian RNA. *Naturwissenschaften* 63:391.

50. Potter, H., Weir, L., and Leder, P. (1984). Enhancer-dependent expression of human k immunoglobulin genes introduced into mouse pre-B lymphocytes by electroporation. *Proc. Natl. Acad. Sci. USA* 81:7161-7165.

51. Karube, I., Tamiya, E., and Matsouka, H. (1985). Transformation of Saccharomyces cerevisae spheroplasts by high electric pulse. *FEBS Lett.* 182:90-94.

52. Salek, A., Schnettler, R., and Zimmermann, U. (1990). Transmission of killer activity into laboratory and industrial strains of *Saccharomyces cerevisiae* by electroinjection. *FEMS Lett.*, in press.

53. Langridge, W. H. R., Li, B. J., and Szalay, A. A. (1985). Electric field mediated DNA transformation in plant protoplasts. In: *Biotechnology in Plant Science, Relevance to Agriculture in the Eighties, Program and Abstracts for an International Symposium*. Cornell University, Ithaca, NY, p. 25.

54. Fromm, M. E., Taylor, L. P., and Walbot, V. (1986). Stable transformation of maize after gene transfer by electroporation. *Nature* 319: 791-793.

55. Shilito, R. D., Saul, M. W., Paszkowski, J., Muller, M., and Potrykus, I. (1985). High efficiency direct gene transfer to plants. *Biotechnology* 3:10-15.

56. Stopper, H., Zimmermann, U., and Wecker, E. (1985). High yields of DNA transfer into mouse L-cells by electropermeabilization. *Z. Naturforsch.* 40c:929-932.

57. Zimmerman, U., and Stopper, H. (1985). Elektrofusion und Elektropermeabilisierung von Zellen. *Pharmatechnol. Biotechnol.* 3:26-36.

57a. Stopper, H., Zimmermann, U., and Neil, G. A. (1988). Increased efficiency of transfection of murine hybridoma cells with DNA by electropermeabilization. *J. Immunol. Meth.* 109:145-151.

58. Stopper, H., Jones, H., and Zimmermann, U. (1987). Improved DNA-transfection by electropermeabilization. *Biochim. Biophys. Acta* 900:38-44.

58a. Däumler, R., and Zimmermann, U. (1989). High yields of stable transformants by hypo-osmolar plasmid electroinjection. *J. Immunol. Meth.* 122:205-210.

59. Farkas, D. L., Korenstein, R., and Malkin, S. (1980). Electroselection in the photosynthetic membrane: Polarized luminescence induced by an external electric field. *FEBS Lett.* 120:236-242.

60. Mehrle, W., Zimmerman, U., and Hampp, R. (1985). Evidence for asymmetrical uptake of fluorescent dyes through electropermeabilized membranes of *Avena* mesophyll protoplasts. *FEBS Lett.* 185:89-94.

61. Ruppert, A. (1986). Aufbau eines Systems zur Komplementation mit Influenza-Proteinen in Eukaryonten Zellen. Diplomarbeit, Universität GieBen.

62. Knight, D. E., and Scrutton, C. (1986). Gaining access to the cytosol: the technique and some applications of electropermeabilization. *Biochem. J.* 234:497-506.

63. Takashima, S., and Schwan, P. (1985). Alignment of microscopic particles in electric fields and its biological implications. *Biophys. J.* 47:513-518.

64. Sauer, F. A. (1985). Interaction-forces between microscopic particles in an external electromagnetic field. In: *Interactions Between Electromagnetic Fields and Cells* (A. Chiabrera, C. Nicolini, and H. P. Schwan, eds.). Plenum Press, New York, pp. 181-202.

64a. Mehrle, W., Hampp, R., Zimmermann, U., and Schwan, H. P. (1988). Mapping of the field distribution around dielectrophoretically aligned cells by means of small particles as field probes. *Biochim. Biophys. Acta* 939:561-568.

65. Jones, T. B., and Kraybill, J. P. (1986). Active feedback-controlled dielectrophoretic levitation. *J. Appl. Physiol.* 60:1247-1252.

66. Hannig, K. (1982). New aspects in preparative and analytical continuous free-flow cell electrophoresis. *Electrophoresis* 3:235-243.

67. Förster, E., and Emeis, C. C. (1985). Quantitative studies on the viability of yeast protoplasts following dielectrophoresis. *FEMS Microbiol. Lett.* 26:65-69.

68. Serpersu, E. H., Kinosita, K., and Tsong, T. Y. (1985). Reversible and irreversible modification of erythrocyte membrane permeability by electric field. *Biochim. Biophys. Acta* 812:779-785.

69. Benz, R., and Zimmermann, U. (1980). Pulse length dependence of the electrical breakdown in lipid bilayer membranes. *Biochim. Biophys. Acta* 597:637-642.

70. Kramer, I., Vienken, K., Vienken, J., and Zimmermann, U. (1984). Magneto-electrofusion of human erythrocytes. *Biochim. Biophys. Acta* 772:407-410.

71. Arnold, W. M., and Zimmermann, U. (1986). Cell fusion and measurement of membrane properties using electric fields and other forces. *Biochem. Soc. Trans.* 14:246-249.

72. Vienken, J., Zimmermann, U., Zenner, H. P., Coakley, W. T., and Gould, R. K. (1985). Electro-acoustic fusion of erythrocytes and myeloma cells. *Biochim. Biophys. Acta* 820:259-264.

73. Lo, M. M. S., Tsong, T. Y., Conrad, M. K., Strittmatter, S. M., and Hester, L. D. (1984). Monoclonal antibody production by receptor-mediated electrically induced cell fusion. *Nature* 310:792-794.

74. Wojchowski, D. M., and Sytkowski, A. J. (1986). Hybridoma production by simplified avidin-mediated electrofusion. *J. Immunol. Meth.* 90:173-177.

75. Coster, H. G. L., and Zimmermann, U. (1975). Direct demonstration of electrical breakdown in the membranes of Valonia utricularis. *Z. Naturforsch.* 30c:77-79.

76. Melykian, G. B., Abidor, I. G., Chernomodik, L. V., and Chailakhyan, L. M. (1983). Electrostimulated fusion and fission of bilayer lipid membranes. *Biochim. Biophys. Acta* 730:395-398.

77. Puchikova, T. V., Putvinskii, A. V., Vladimirov, Y. A., and Parnev, O. M. (1982). Electrical breakdown of membranes of phospholipid vesicles by the diffusion potential. *Biophysics* 26:268-274.

78. Büschl, R., Ringsdorf, H., and Zimmermann, U. (1982). Electric-field-induced fusion of large liposomes from natural and polymerizable lipids. *FEBS Lett.* 150:38-42.

79. Shivarova, N., Grigorova, R., Förster, W., Jacob, H-E., and Berg, H. (1983). Microbiological implications of electric field effects. Part VIII. Fusion of Bacillus thuringiensis protoplasts by high electric field pulses. *Bioelectrochem. Bioenerg.* 11:181-185.

80. Ruthe, H. J., and Adler, J. (1985). Fusion of bacterial spheroplasts by electric fields. *Biochim. Biophys. Acta* 819:105-113.

81. Schnettler, R., and Zimmermann, U. (1985). Influence of media composition on the yield of electrofused yeast hybrids. *FEMS Microbiol. Lett.* 27:195-198.

82. Fikus, M., Grzesiuk, E., Marszalek, P., Rozycki, S., and Zielinski, J. (1985). Electrofusion of Neurospora crassa slime cells. *FEMS Microbiol. Lett.* 27:123-127.

83. Weber, H., Förster, W., Jakob, H-E., and Berg, H. (1981). Enhancement of yeast protoplast fusion by electric field effects. In: *Current Developments in Yeast Research* (G. G. Stewart and I. Russel, eds.), Pergamon Press, Toronto, New York, pp. 219-224.

84. Halfmann, H. J., Röcken, W., Emeis, C. C., and Zimmermann, U. (1982). Transfer of mitochondrial function into a cytoplasmic respiratory-deficient mutant of Saccharomyces yeast by electrofusion. *Curr. Gen.* 6:25-28.

85. Halfmann, H. J., Emeis, C. C., and Zimmermann, U. (1983). Electro-fusion of haploid *Saccharomyces* yeast cells of identical mating type. *Arch. Microbiol.* 134:1-4.

86. Halfmann, H. J., Emeis, C. C., and Zimmermann, U. (1983). Electro-fusion and genetic analysis of fusion products of haploid and polyploid *Saccharomyces* yeast cells. *FEMS Microbiol. Lett.* 20:13-16.

87. Emeis, C. C., and Zimmermann, U. (1985). Neue Entwicklungen auf dem Gebiet der Elektrofusion. In: *Tierische Zellkulturen*, 2. BMFT-Statusseminar, Jülich, BMFT, ed., pp. 81-95.

88. Broda, H. G., Schnettler, R., and Zimmermann, U. (1987). Parameters controlling yeast hybrid yield in electrofusion: Relevance of pre-incubation and the slowness of the size distributions of bath fusion partners. *Biochim. Biophys. Acta* 899:25-34.

89. Senda, M., Takeda, J., Abe, S., and Nakamura, T. (1979). Induction of cell fusion of plant protoplasts by electrical stimulation. *Plant Cell Physiol.* 20:1441-1443.

90. Zimmermann, U., and Scheurich, P. (1981). High frequency fusion of plant protoplasts by electric fields. *Planta* 151:26-32.

91. Vienken, J., Ganser, R., Hampp, R., and Zimmermann, U. (1981). Electric field induced fusion of isolated vacuoles and protoplasts of different developmental and metabolic provenience. *Physiol. Plant.* 53:64-70.

92. Verhoek-Köhler, B., Hampp, R., Ziegler, H., and Zimmermann, U. (1983). Electrofusion of mesophyll protoplasts of Avena sativa. *Planta* 158:199-204.

93. Vienken, J., Zimmermann, U., Ganser, R., and Hampp, R. (1983). Vesicle formation during electrofusion of mesophyll protoplasts of Kalanchoe daigremontiana. *Planta* 157:331-335.

94. Jacob, H.-E., Siegemund, F., and Bauer, E. (1984). Fusion von pflanzlichen Protoplasten durch elektrischen Feldpuls nach Dielektrophorese. *Biol. Zentralbl.* 103:77-82.

95. Zachrisson, A., and Bornmann, Ch. H. (1984). Application of electric field fusion in plant tissue culture. *Physiol. Plant.* 61:314-320.

96. Watts, J. W., and King, J. M. (1984). A simple method for large-scale electrofusion and culture of plant protoplasts. *Bioscience Rep.* 4:335-342.

97. Tempelaar, M. J., and Jones, M. G. K. (1985). Fusion characteristics of plant protoplasts in electric fields. *Planta* 165:205-216.

98. Tempelaar, M. J., and Jones, M. G. K. (1985). Directed electrofusion between protoplasts with different responses in a mass fusion system. *Plant Cell Rep.* 4:92-95.

99. Saga, N., Polne-Fuller, M., and Gibor, A. (1984). Protoplasts from seaweeds: Production and fusion. In: *Proceedings of a Workshop on the Present Status and Future Directions for Biotechnology Based on Algal Biomass Production.* University of Colorado Press, Boulder.

100. Bates, G. W. (1985). Electrical fusion for optimal formation of protoplast heterokaryons in Nicotiana. *Planta* 165:217-224.

101. Bates, G. W., and Hasenkampf, C. A. (1985). Culture of plant somatic hybrids following electrical fusion. *Theor. Appl. Genet.* 70:227-233.

102. Koop, H. U., Dirk, J., Wolff, D., and Schweiger, H. G. (1983). Somatic hybridization of two selected single cells. *Cell Biol. Int. Rep.* 7:1123-1128.

103. Koop, H. U., and Schweiger, H. G. (1986). Regeneration of plants after electrofusion of selected pairs of protoplasts. *Eur. J. Cell Biol.* 39:46-49.

104. Hampp, R., Steingräber, M., Mehrle, W., and Zimmermann, U. (1985). Electric field-induced fusion of evacuolated mesophyll protoplasts of oat. *Naturwissenschaften* 72:91-92.

105. Gaff, D. F., Ziegler, H., and Zimmerman, U. (1985). Electrofusion of protoplasts from desiccation tolerant grass species with desiccation sensitive grass protoplasts. *J. Plant Physiol.* 120:375-380.

106. Salhani, N., Vienken, J., Zimmerman, U., Ward, M., Davery, M. R., Clothier, R. H., Balls, M., Cocking, E. C., and Lucy, J. A. (1985). Haemoglobin synthesis and cell wall regeneration by electric field induced interkingdom heterokaryons. *Protoplasma* 126:30-35.

107. Kohn, H., Schieder, R., and Schieder, O. (1985). Somatic hybrids in tobacco mediated by electrofusion. *Plant Sci.* 38:121-128.

108. Morikawa, H., Sugino, K., Hayashi, Y., Takeda, J., Senda, M., Hiai, A., and Yamada, Y. (1986). Interspecific plant hybridization by electrofusion in Nicotiana. *Biotechnology*, 57-60.

109. Scheurich, P., and Zimmermann, U. (1981). Giant human erythrocytes by electric field-induced cell-to-cell fusion. *Naturwissenschaften* 68: 45-46.

110. Ahkong, Q. F., and Lucy, J. A. (1986). Osmotic swelling in artificially-induced cell fusion. *Biochim. Biophys. Acta* 858:206-216.

111. Zimmermann, U., Beckers, F., and Coster, H. G. L. (1977). The effect of pressure on the electrical breakdown in the membranes of Valonia utricularis. *Biochim. Biophys. Acta* 464:399-416.

112. Zimmermann, U., Pilwat, G., Pequeux, A., and Gilles, R. (1980). Electromechanical properties of human erythrocyte membranes. *J. Membrane Biol.* 54:103.

113. Pilwat, G., Richter, H. P., and Zimmerman, U. (1981). Giant culture cells by electric field-induced fusion. *FEBS Lett.* 133:169-174.

114. Herzog, R., Müller-Wellensick, A., and Voelter, W. (1986). Fusion tierischer Tumorzellen im elektrischen Feld. *Chemiker-Zeitung* 110: 11-16.

115. Ohno-Shosaku, T., Hama-Inaba, H., and Okada, Y. (1984). Somatic hybridization between human and mouse lymphoblast cells produced by an electric pulse-induced fusion technique. *Cell Struc. Funct.* 9:193-196.

116. Ohno-Shosaku, T., and Okada, Y. (1984). Facilitation of electrofusion of mouse lymphoma cells by the proteolytic action of proteases. *Biochem. Biophys. Res. Commun.* 120:138-143.

117. Ohno-Shosaku, T., and Okada, Y. (1985). Electric pulse-induced fusion of mouse lymphoma cells: roles of divalent cations and membrane lipid domains. *J. Membrane Biol.* 85:269-280.

118. Zimmermann, U., Pilwat, G., and Pohl, H. A. (1982). Electric field-mediated cell fusion. *J. Biol. Phys.* 10:43-50.

119. Podesta, E. J., Solano, A. R., Vedia, L. M., Paladini, A., Sanchez, M. L., and Torres, H. N. (1984). Production of steroid hormone and cyclic AMP in hybrids of adrenal and Leydig cells generated by electrofusion. *Eur. J. Biochem.* 145:329-332.

120. Teissie, J., Knutson, V. P., Tsong, T. Y., and Lane, M. D. (1982). Electric pulse-induced fusion of 3T3 cells in monolayer culture. *Science* 216:537-538.

121. Blangero, C., and Teissie, J. (1983). Homokaryon production by electrofusion: A convenient way to produce a large number of viable mammalian fused cells. *Biochem. Biophys. Res. Commun.* 114:663-669.

122. Finaz, C., Lefevre, A., and Teissie, J. (1984). A new, highly efficient technique for generating somatic cell hybrids. *Exp. Cell Res.* 150: 477-482.

123. Neumann, E., Gerisch, G., and Opatz, K. (1980). Cell fusion induced by high electric impulses applied to Dictyostelium. *Naturwissenschaften* 67:414-415.

124. Goding, J. (1980). Antibody production by hybridomas. *J. Immunol. Meth.* 39:285-308.

125. Vienken, J., and Zimmermann, U. (1982). Electric field-induced fusion: Electro-hydraulic procedure for production of heterokaryon cells in high yield. *FEBS Lett.* 137:11-13.

126. Bischoff, R., Eisert, R. M., Schedel, I., Vienken, J., and Zimmermann, U. (1982). Human hybridoma cells produced by electrofusion. *FEBS Lett.* 147:64-68.

127. Vienken, J., and Zimmermann, U. (1985). An improved electrofusion technique for production of mouse hybridoma cells. *FEBS Lett.* 180: 278-280.

128. Schmitt, J. J., Zimmermann, U., and Neil, G. A. (1989). Efficient generation of stable antibody forming hybridoma cells by electrofusion. *Hybridoma* 8(1):107-115.

129. Karsten, U., Papsdorf, G., Roloff, G., Stolley, P., Abel, H., Walther, I., and Weiss, H. (1985). Monoclonal anti-cytokeratin antibody from a hybridoma clone generated by electrofusion. *Cancer Clin. Oncol.* 21:733-740.

130. Brown, S. M., Ahkong, Q. F., Sage, A. D., and Lucy, J. A. (1986). *Biochem. Soc. Trans.* 14:298.

130a. Abel, H., Stolley, P., Dreyer, G., Walther, I., Seidel, B., Handschack, W., Schwarz, K., Platzer, C., Gröbel, C., Karsten, U., and Micheel, B. (1987). Hybridoma technique by means of electrofusion—a concise experiment report. *Stud. Biophys.* 119:67-68.

130b. Gathuru, J. K., Miyoshi, I., and Naiki, M. (1987). Electrically stimulated fusion of human, UC 729-6 cell lin potentially useful for generating antibody-secreting human-human hybridomas. *Jpn. J. Vet. Res.* 35:109-113.

130c. Glassy, M. C. (1988). Creating hybridomas by electrofusion. *Nature* 333:579-580.

130d. Gravekamp, C., Bol, S. J. L., Hagemeijer, A., and Bolhuis, R. L. H. (1985). Production of human T-cell hybridomas by electrofusion. In: *Human Hybridomas and Monoclonal Antibodies* (E. G. Engelmann, S. K. Foung, and J. Larrick, eds.). pp. 323-339.

130e. Gravekamp, C., Santoli, D., Vreugdenhil, R., Collard, J. G., and Golhuis, R. L. H. (1987). Efforts to produce human cytotoxic T-cell hybridomas by electrofusion and PEG fusion. *Hybridoma* 6(2):121-133.

130f. Mangoldt, D., Schumann, I., and Stelzner, A. (1987). Hybridoma production by electrofusion. *Allerg. Immunol.* 33:63-64.

130g. O'Hare, M. J., and Edwards, P. A. W. (1987). Human monoclonal antibodies and the LICR-Lon-HMy2 system. In: *Immunology Series Human Hybridomas: Diagnostic and Therapeutic Applications* (A. J. Strelkavskas, ed.). Marcel Dekker, New York, pp. 47-64.

130h. Ohnishi, K., Chiba, J., Goto, Y., and Tokumaga, T. (1987). Improvement in the basic technology of electrofusion for generation of antibodyproducing hybridomas. *J. Immunol. Meth.* 100:181-189.

130i. Pratt, M., Mikhalev, A., and Glassy, M. C. (1987). The generation of Jg-secreting UC 729-6 derived human hybridomas by electrofusion. *Hybridoma* 6(5):469-477.

130j. Stenger, D. A., Kubiniec, R. T., Purucker, W. J., Liang, H., and Hui, S. W. (1988). Optimization of electrofusion parameters for efficient production of murine hybridomas. *Hybridoma* 7(5):505-518.

130k. Suzuki, M., Tamiya, E., Matsuoka, H., Sugi, M., and Karube, I. (1986). Electrical stimulation of hybridoma cells producing monoclonal antibody to camp. *Biochim. Biophys. Acta* 889:149-155.

130l. Tsoneva, I., Panova, I., Doinov, P., Dimitrov, D. S., and Strahilov, D. (1988). Hybridoma production by electrofusion: Monoclonal antibodies against the HC antigene of salmonella. *Stud. Biophys.* 126(1):31-35.

130m. Yeandle, S., Sieckman, D., Yeh, T. Y., Hardegan, N., and Hofmann, G. (1987). Hybridoma production by electrofusion as assayed by flow microfluorimetry. *Fed. Proc.* 43(3):674.

130n. Schmitt, J. J., Zimmermann, U., and Gessner, P. (1989). Electrofusion of osmotically treated cells. High and reproducible yields of
hybridoma cells. *Naturwissenschaften* 76:122-123.

130o. Schmitt, J. J., and Zimmermann, U. (1989). Enhanced hybridoma
production by electrofusion in strongly hypo-osmolar solutions. *Biochim.
Biophys. Acta* 983:42-50.

130p. Perkins, S., Zimmermann, U., Gessner, P., and Foung, S. K. H.
(1989). Formation of hybridomas secreting human monoclonal antibodies
with mouse-human fusion partners. In: *Electromanipulation in Hybridoma
Technology* (C. Borrebaeck and I. Hagen, eds.). Stockton Press,
pp. 47-70.

131. Arnold, W. M. and Zimmermann, U. (1982). Rotating-field-induced
rotation and measurement of the membrane capacitance of single mesophyll cells of Avena sativa. *Z. Naturforsch.* 37c:908-915.

132. Arnold, W. M., and Zimmermann, U. (1983). German patent application,
official designation P3325 843 O, recorded July 18.

133. Berg, H. (1982). Fusion of blastomeres and blastocytes of mouse
embryos. *Bioelectrom. Bioenerg.* 9:223-228.

134. Kubiak, J. Z., and Tarkowski, A. K. (1985). Electrofusion of mouse
blastomeres. *Exp. Cell Res.* 157:561-566.

135. Williadsen, S. M. (1986). Nuclear transplantation in sheep embryos.
Nature 320:63-65.

136. Richter, H. P., Scheurich, P., and Zimmermann, U. (1981). Electric
field-induced fusion of sea urchin eggs. *Dev. Growth. Diff.* 23:479-
486.

137. Savage, J. S., and Grey, R. D., Poster, *Am. Soc. Cell Biol.*, San
Antonio, TX.

138. Sowers, A. E. (1983). Fusion of mitochondrial inner membranes by
electric fields produces inside-out vesicles. *Biochim. Biophys. Acta*
735:426-428.

139. Sowers, A. (1984). Characterization of electric field-induced fusion
in erythrocyte ghost membranes. *J. Cell Biol.* 99:1989-1996.

140. Sowers, A. (1985). Movement of a fluorescent lipid label from a labeled
erythrocyte membrane to an unlabeled erythrocyte membrane following
electric field-induced fusion.

141. Donath, E. and Arndt, R. (1984). Electric field-induced fusion of
enzyme-treated human red cells: Kinetics of intermembrane protein
exchange. *Gen. Physiol. Biophys.* 3:239-249.

142. Salhani, N., Schnabl, H., Kuppers, G., and Zimmermann, U. (1982).
The hydraulic conductivity as a criterion for the membrane integrity
of protoplasts fused by an electric field pulse. *Planta* 155:140-145.

143. Sowers, A. (1986). A long-lived fusogenic state is induced in erythrocyte ghosts by electric pulses. *J. Cell. Biol.* 102:1358-1362.

30

The Production of Monoclonal Antibodies

LOU de LEIJ and T. HAUW THE

University Hospital Groningen, Groningen, The Netherlands

I. INTRODUCTION

The in vitro production of somatic cell hybrids by fusion of different types of cells is a commonly used technique for the study of various biological issues. This chapter deals with a special application of cell hybridization technology, namely, the production of monoclonal antibodies, also called *hybridoma technology*. Basically this technology aims to combine the specific antibody-producing capacity of·immune lymphocytes with the in vitro growth ability of established myeloma cell lines. As a result, continuously growing hybrid cell lines (hybridomas), secreting specific antibodies, are generated.

Since its first description in 1975 (1), this particular type of cell hybridization has become very popular and has led to the development of numerous hybridomas, producing equally high numbers of different monoclonal antibodies. Monoclonal antibodies have already proven to be extremely useful for both fundamental and applied research applications; in addition, their use for industrial and therapeutic purposes is growingly acknowledged. As a corollary of this, many review articles and books have been devoted to the production and application of monoclonal antibodies (see Refs. 2-4). The aim of the present chapter is not to repeat these reviews, but rather to describe the biological and physical principles under-lying the production of monoclonal antibodies. From these principles, new research trends in hybridoma technology are indicated. In addition, as an illustration of the power of hybridoma technology, the prospective appli-cation of monoclonal antibodies in clinical tumor immunology is discussed.

II. THE IMMUNOLOGICAL BASIS OF MONOCLONAL ANTIBODY PRODUCTION

A. Introduction

Each hybridoma is characterized by the unique specificity of its produced (monoclonal) antibodies; i.e., such antibodies will match with only one

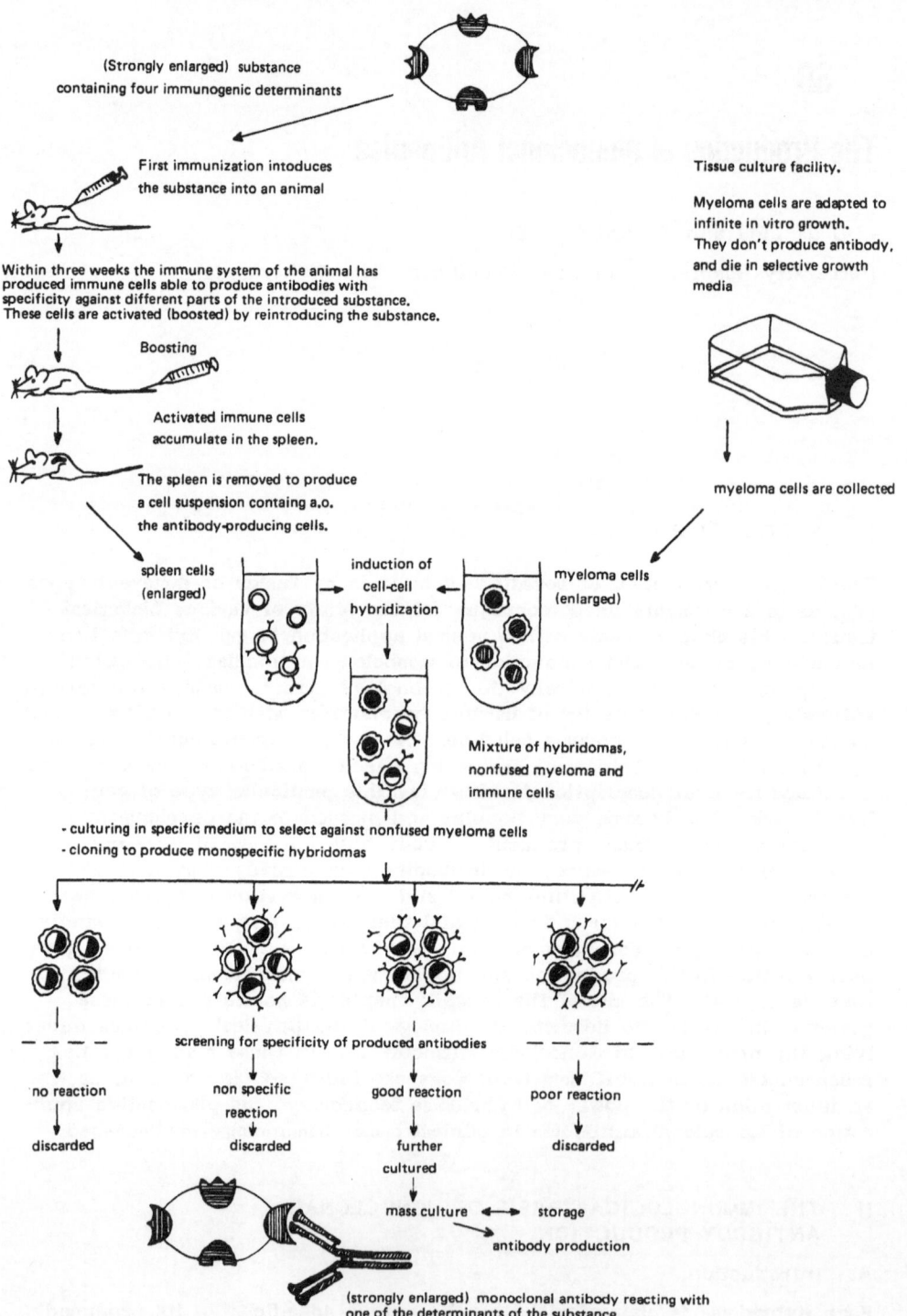

FIGURE 1 Schematic presentation of the construction of a new hybridoma according to standard technology.

epitope on a particular target substance and bind to this with a given, fixed constant of binding (K_a).

Figure 1 summarizes the different steps in the preparation of a new hybridoma. First, a suitable animal (e.g., a mouse) is injected with the substance against which the future monoclonal antibodies must be directed. This substance is now called "antigen." The injection evokes a first immunological reaction in the animal ("priming"). Then, after a rest period of minimally 3 weeks, the generated immune reactivity is boosted by reinjecting the antigen. Further boosting is possible by repeating this scheme of resting and antigen injection. The timing of the last antigen boost is important, since it has been shown that an optimal amount of suitable antibody-producing cells is present in the spleen of the animal only if an intravenously applied booster is given 3-4 days before sacrifying the animal (5). The spleen is removed and the isolated spleen cells can be used in the hybridization experiment. After fusion, the obtained hybridomas are biochemically preselected (6), cloned, cultured, and assessed for specificity of produced antibodies. Only the hybridomas producing relevant antibodies are selected and propagated further.

As stated above, the spleen of the immunized animal contains, along with many irrelevant cells, also the immune lymphocytes harboring the information on the specificity of eventually produced monoclonal antibodies. This information, which is in fact the most important aspect of the future hybridoma, is acquired in the animal as a result of immunological programming and selection of immune lymphocytes. To provide a better understanding of this process, we will first present a survey of the recent advances in knowledge about the biological events underlying the generation of the immune response. Subsequently, new in vitro procedures, aimed at substitution of the in vivo immunization protocols, are discussed.

B. The Generation of Antibody Specificity

The immune lymphocytes responsible for antibody production in an immunized animal are the B lymphocytes. In 1959, Burnet postulated that the diversity of antibody specificities normally present in the serum of an immunized animal had a clonal cellular basis; i.e., this diversity should be based on the presence of a large number of different B lymphocytes, each having the capacity to synthesize only one specific type of antibody molecule (7). This theory has been proven to be correct. Moreover, the immune system synthesizes not only a diverse array of antibody specificities as such, but, rather responds specifically to a given antigenic stimulus by synthesizing the matching specific antibodies. It appears that almost any applied antigen can evoke such a specific response (8). Given the cellular basis of antibody specificity, this implies the existence of an almost unlimited number of different B lymphocytes with preexisting specificities toward all kinds of possible antigenic stimuli. This assumption has turned out to be essentially correct and only recently the molecular-biological principles underlying this phenomenon are being unraveled.

It has been shown that all precursor B cells, present in the bone marrow, have common chromosome regions encoding for antibody molecules (germ-line genes). When these precursor cells mature into (pre)-B lymphocytes, a gene rearrangement process is started and this will proceed in each B lymphocyte in a strictly cell-specific way. During this process, a large number of germ-line antibody gene fragments are (partly) deleted and/or combined (9) to form, in an unpredictable and variable way (10), the actual genes coding for an antibody molecule. In this way B cells are generated which contain an almost infinite number of possible gene arrangements. Every day large amounts of B cells are produced from bone marrow precursor cells, and, of course, not all of these B cells contain genes coding for relevant antibody specificities. Which B cells will survive, multiply, and further mature is largely dependent on the relevancy of their antigen specificity. This, in turn, is guided by the actual supply of antigen; i.e., only the B cells coding for antigen-reactive antibodies are selected. After maturation, further somatic gene diversification events (11) occur in the surviving, antigen-specific B cells ("memory cells"). This increases the capacity of the immune system to react and further adapt to a renewed application of the same antigenic stimulus ("boosting"), adding in this way to a further fine-tuning of the antibody response.

C.   Regulation of Antibody Production

The mere interaction of a specific B cell with antigen is normally not sufficient to trigger antibody production. The reason for this lies in the fact that the immune system has elaborated a delicate way to discriminate "self" antigens from "nonself" antigens, only the latter being the target of immune reactions. To this end, the action of B cells is controlled by another type of immune cell, the "regulator" T cells. Regulator T cells are the first immune cells responding to a newly applied antigen. This response is triggered, however, only if the antigen is present in a very specific way, i.e., if present on so-called "antigen-presenting cells" ("APC's"). The main property of these APC's is their capability to aspecifically entrap antigen, to partially degrade this ("processing"), and to present the "processed" antigen on its surface in conjunction with class II molecules of the major histocompatibility complex (MHC). The exact structure of MHC molecules is highly individual-specific, and these molecules serve in the immune reaction as a built-in "hallmark" for the recognition of "self." Regulator T cells have a specific receptor (the T-cell receptor), which recognizes these "self" class II molecules, and if a foreign "processed" antigen is associated with such a molecule, this complex is recognized as "nonself." As a result the regulator T cell will become activated. Different regulator T cells have an ability to recognize different kinds of "processed" antigens.

The cellular events preceding an antibody response can be described as follows (12). First, regulator T cells with specificity for "processed" parts of the antigen are activated as indicated above. At the same time, other parts of the (native) antigen bind to the surface of matching, specific B cells. This binding is caused by the fact that (mature) B cells bear on their surface a specialized type of antibody, which has an antigen-recognition specificity identical to that of eventually secreted antibody molecules. After specific binding to the B cell, the antigen is internalized and similarly "processed" and presented as in the APC's. The activated specific regulator T cells recognize the similarly "processed" antigen on the B cell, and only then will the T cells deliver "help" signals to this B cell. The "help" signals finally activate the B cell. Activation is followed by multiplication and differentiation. A major portion of the activated cells differentiates into antibody-producing cells. Some of the daughter B (and T) cells differentiate into "memory" cells. These cells give the immune system the opportunity to respond better to a similar antigenic exposure in the future ("booster").

In summary, B and T cells respond differently to antigen. B cells recognize a native part of the antigen, whereas T cells supply the "help" necessary for a subsequent activation of these B cells only after recognition of "processed" antigens. These "processed" antigens are always presented in the context of "self" class II molecules of the major histocompatibility complex. In this way, T cells can recognize an antigen as nonself and, as a result, can guide the immune response toward such antigens.

D.   In Vitro Immunization

Not all antigens can be used for in vivo immunization. For instance, weakly immunogenic or toxic compounds can be expected to cause problems. This is especially true for an intended production of human-derived monoclonal antibodies, since the deliberate application of all kinds of immunizing substances to humans is impossible for ethical reasons. Nevertheless, the special production of human-derived monoclonal antibodies could be important. The main reason for this is that human-derived antibodies are nonimmunogenic in humans and, in addition, are apparently more suited to fit in with other (human) immune functions. These properties are desirable if the produced monoclonal antibodies are intended to be used in vivo (e.g., for antitumor or antivirus therapy). To provide an alternative for in vivo immunization, extensive research has been carried out to develop suitable in vitro immunization protocols (13).

As indicated above, a large number of different B cells exist in an unprimed animal or human. Each specific B cell is present, however, at very low frequency. Therefore, the aim of an in vitro immunization protocol is, essentially, to multiply the B cells with the desired specificity. For such a purpose, both specific antigen and T regulator "help" are necessary.

Recently, the molecular vectors executing this "help" have been investigated in detail (14,15). It appears that T-lymphocyte-derived "help" involves both direct T-cell/B-cell interactions and the (indirect) action of a set of secreted regulator molecules (lymphokines). Lymphokines relevant to B-cell activation, proliferation, and maturation have been identified and isolated. These are B-cell growth factor (BCGF), interleukin-1 (IL-1), interleukin-2 (IL-2), B-cell differentiation factor (BCDF), and γ-interferon. Thus, to provide "help" in an in vitro immunization protocol, cell-cell interactions must be mimicked, e.g., by the addition of lectines, while the lymphokines relevant to B-cell activation and proliferation must be supplied as well. Mishel and Dutton were the first to demonstrate that, in vitro, primary, antigen-specific stimulation of mouse spleen cells is possible (16). In vitro immunization of mouse-derived B lymphocytes is now a more or less established technique. In the human system, difficulties still exist, however, and only a few reports have claimed positive results (17). Apparently, the developed, successful procedures for the mouse system cannot simply be transferred to human lymphocytes, and more studies have to be done in order to develop a suitable technique for this purpose.

The advantages of in vitro immunization can be summarized as follows: First, the procedure can be carried out under standardized conditions,

which makes it possible to monitor all different steps in the immunization
process. Second, there is a better opportunity to make antibodies against
toxic substances. This is especially important for the production of human-
derived monoclonal antibodies. Third, antibodies against poorly immunogenic
or nonimmunogenic antigens (e.g., self or autoantigens) can be generated
more easily. This is so because "help" is provided irrespective of the recognition of the antigen as nonself; in addition, the normally occurring suppressor
mechanisms, prohibiting a response against autoantigens in vivo, are absent.
From a theoretical point of view, it can be anticipated that immunization
with auto- or alloantigens (human-derived lymphocytes immunized with
human tumor cells) will trigger a different repertoire of antibody specificities
as compared to immunization with xenoantigen (mouse lymphocytes immunized
with human-derived tumor cells). Fourth, it has been shown that extremely
low amounts of antigen are needed to activate B cells in vitro.

In conclusion, in vitro immunization has clear advantages over in vivo
immunization, and it can be anticipated, therefore, that this technique
will become increasingly important for hybridoma technology in the near
future. This applies especially to the production of human-derived antibodies,
although a good protocol for such a purpose has still to be developed.

III. FUSION PROCEDURES

A. Introduction

It was noted as early as 1838 that multinucleated cells were present in
tumor biopsies (18), and while this observation was extended subsequently
to other in vivo situations, one of the main speculations about their nature
was that they originated from somatic cell hybridization. After the introduction of tissue culture techniques, a detailed description of the formation
of multinucleated cells became possible, demonstrating that, indeed, somatic
cell hybridization was involved. Subsequently, it was shown, that binucleated
cells, when entering mitosis, form one nuclear spindle, resulting after
cell division in the formation of two mononucleated daughter cells, each
containing a double number of chromosomes.

While the above studies were done on "spontaneous" fusion events,
deliberate induction of somatic cell hybridization became possible, initially
through the use of fusion factors derived from specific viruses. Most of
this work has been done with the fusion factors present in Newcastle disease and Sendai viruses. When it became apparent that the fusion activity
could be dissociated from viral infectivity (19), "inactivated" viruses or
isolated virus proteins were used as fusogens. In addition to the use of
viruses, other reliable fusion-inducing agents and procedures have been
developed since then. Of these, poly(ethylene glycol) and electrofusion
are relevant for hybridoma formation and will be discussed below (see
also Chapters 28 and 29, respectively).

A major impetus for cell-hybridization research has been the isolation
of enzyme-deficient parental cell lines, allowing hybridization protocols
in which the nonfused parental cells can be biochemically eliminated after
fusion (6). This is especially useful if the percentage of formed hybrids
is relatively low, as is the case with most currently employed fusion techniques. In the case of the production of hybridomas, one parental cell
type, namely the B lymphocyte, is unable to grow in normal tissue culture
medium. However, the myeloma cells do so quite well and will quickly over-
grow the (few) newly formed hybridomas, if the fusion mixture is not
incubated in a selective (HAT) culture medium (see also Chapter 28).

B. Fusion Induction by Poly(ethylene Glycol)

Although the first fusions of lymphoid cells by Koehler and Milstein were performed with inactivated Sendai virus as a fusogen (1), poly(ethylene glycol) (PEG) clearly is the most widely employed fusogen in the production of hybridomas now. The reason for this is that PEG induces much higher fusion frequencies than Sendai virus. The low fusion frequency obtained with Sendai virus is probably due to the lack of suitable receptors for Sendai virus on lymphoid cells. PEG was originally employed to fuse plant protoplasts (20), and Pontocorvo (21) was the first to apply this fusogen to mammalian cells. Despite its general application, little is known about the exact mechanisms underlying PEG-induced cell fusion (22; see also Chapters 11 and 28). When antioxidants and/or polymerization agents (like α-tocopherol and other phenolic derivates) are thoroughly removed from commercially available PEG, its fusogenic capacity is drastically lowered (23). Ultrapure PEG still induces the membranes of different cells to form close contacts, with lipid probes spreading from one cell membrane to the other (23), but the additives present in commercial PEG apparently are necessary to induce the cells to fuse. Much effort has been put into the identification of commercially available PEG preparations that are most effective in inducing fusion (24), and which molecular weight and concentration of PEG is optimal (25). In agreement with others, we found that PEG 4000 (Merck 9727) in a concentration of 50% (w/v) is a good choice. A critical parameter in inducing fusion proved to be the pH of the fusion medium. We found a pH of 8.0 to result in optimal hybridoma formation (26). The same pH has been reported to be optimal for the induction of somatic cell hybrids using either Sendai virus or lysolecithin as fusogens (27). In addition to a better induction of cell fusion, a pH of 8.0 appears to stimulate the formation of the nuclear envelope and prevents premature chromosome condensation in the newly formed hybrids (28).

Although the percentage of hybrid formation is definitely higher than that obtained with Sendai virus, PEG has still as a major drawback that the obtained fusion frequencies are rather low (10^{-3} to 10^{-4}), which is probably due to its toxicity. This may not pose a problem, when sufficient B cells are available, as is the case when a complete spleen (containing about 10^9 cells in the mouse system) is used as a source of antibody-producing cells. However, in the case of in vitro immunization protocols, this could become a limiting factor. Therefore, alternative procedures for fusion induction, like the one discussed below, have to be exploited further.

C. Electrofusion

Although electrofusion can be considered an established technique for the production of cell hybrids in plant protoplast systems (29), its application for the production of hybridomas is still in its infancy. The reason for this probably stems from the fact that specifically adapted fusion procedures and media had to be developed, which have only recently been successfully applied (30). Since the principles of electrofusion are described in detail in the preceding chapter of this book, just the general features will be given here. In addition, a special application of electrofusion for the specific construction of hybridomas will be discussed.

Before different cells can be fused by electrofusion, close membrane contacts between the cells must be established. Fusion is induced then by applying a single or, if necessary, a few, direct current pulses of high intensity (kV/cm range). These pulses induce the transient formation of pores in the cell membranes. If there is a close contact between the cells, these short-lived pores can merge, causing membrane fusion at the site of contact.

The formation of close cell-cell contacts can be accomplished either by electrical or by (immuno)chemical techniques. In the case of the former technique, this is done by the application of a nonuniform alternating field or by the application of a sequence of direct current pulses of low intensity. In this way dipoles are induced in the cells (dielectrophoresis), which will result in the formation of chains of cells (pearl-chains) oriented parallel to the field lines (31). In the case of establishing cell-cell contact by nonelectrical procedures, aspecific cross-linking of the cells by the application of lectines such as concanavalin A is possible. Another approach is to pretreat the fusion partners with pronase, which also induces aspecific cell aggregate formation. Hybridization induced by electrofusion of these latter aggregates have met, however, with only limited success.

A more sophisticated procedure for the establishment of specific cell-cell contacts has been described by Lo and co-workers (32). This procedure is based on two phenomena. First, as indicated in Section IIC, B cells contain on their surface an antibody species which has an antigen-binding specificity identical to that of the eventually produced antibodies. Second, the protein avidin has a very strong natural affinity for the vitamin biotin. Taking advantage of these phenomena, the procedure is as follows. The antigen against which a monoclonal antibody has to be prepared is conjugated to avidin. Then, this avidin-antigen complex is added to a spleen cell suspension prepared from an immunized animal. Only the B cells containing the right antigen recognition ability will bind the antigen-avidin complex. Separately, a myeloma cell suspension is treated to covalently attach biotin to the surface of the myeloma cells. When these two cell populations are mixed, the biotin-labeled myeloma cells will bind to the avidin-labeled B cells. In this way only B cell/myeloma pairs are formed between a myeloma, on the one hand, and a B cell with the desired antigen specificity, on the other. Then, fusion is induced by an electrical pulse. This approach has the following advantages. First, antibody specificity is preselected by the specific establishment of cell-cell contacts, diminishing the number of hybridomas to be screened after the fusion. Second, the tightness and stringency of the cell-cell adhesion may be controlled in the process by the amount of antigen added, by varying the incubation time, or by the addition of competing, cross-reacting antigens. Third, since spleens of unboosted animals can be used, only small amounts of antigen are necessary in the whole procedure, particularly if this fusion procedure were to be combined with the in vitro immunization protocol, described in the previous section.

In conclusion, electrofusion appears a well-suited technique for fusion induction in hybridoma technology. The fusion percentage may be much higher than that obtained with other fusion techniques, while, in addition, preferential induction of hybridoma formation appears to open new avenues to a more directed approach of hybridoma construction.

IV. MONOCLONAL ANTIBODIES IN TUMOR IMMUNOLOGY

A. Introduction

Monoclonal antibodies are used in clinical tumor immunology for both ex vivo and, more recently, in vivo purposes. Although this last item, i.e., the administration of monoclonal antibodies to patients, is still in an early experimental phase, its prospects are promising. In this section, only in vitro applications of monoclonal antibodies, including their use in serum assays and in oncological pathology, are dealt with. The latter point will be illustrated with immunohistopathological observations in lung cancer.

B. Monoclonal Antibodies in Serum Assays

Monoclonal-antibody-based serum assays have been developed to monitor tumor progression. At the moment this application is still limited to the replacement of already existing assays measuring rather aspecific tumor-associated markers, such as the carcinoembryonic antigen (33). The further development of monoclonal antibodies, directed against tumor markers, will make such an application feasible for specific monitoring of the clinical course of different tumors. An attractive application of monoclonal antibodies in such serum assays is the "sandwich assay" (Fig. 2). In this assay, one monoclonal antibody specifically concentrates a tumor marker from a serum sample onto a solid phase ("catching"), while, subsequently, this marker is visualized and quantitated with the aid of a (radio- or enzyme-labeled) second monoclonal antibody. This second monoclonal antibody should be directed to an epitope on the tumor marker, different from the epitope recognized by the catching monoclonal antibody. Because serum is easy to sample and because such assays can provide objective quantitative data, it can be anticipated that specific serum assays will become important tools for the future monitoring and managing of cancer treatment.

C. Monoclonal Antibodies in Immunohistopathology

A growing number of monoclonal antibodies are used as routine "staining tools" in pathology (34). At the moment this application is used as a supplement to normal tumor diagnosis, which is still based on morphological criteria. Only in tumor cases showing poor morphology, monoclonal-antibody-based immunohistopathology is the method of choice to establish a differential tumor diagnosis. It can be anticipated, however, that the use of monoclonal antibodies for pathological purposes will become more important in the near future. First, a growing number of monoclonal antibodies detecting tumor-associated antigens are being isolated and characterized. Among these, some also react with paraffin-embedded tissues and apparently identify tumors in a way matching with the currently employed histopathological classifications. Second, not all antigenic features of a tumor need to parallel morphologically based subdivisions. Therefore, it can be anticipated that new subclassification of cancer might emerge from the application of monoclonal-antibody-based immunohistopathology. The clinical relevance of such new classifications should be established, however, only by careful clinical studies and comparison with the established morphological criteria. This point is discussed more extensively below.

D. Immunohistopathology of Lung Cancer

Lung cancer is the most common cause of cancer death in Western societies and its incidence in the Third World is rapidly increasing. A number of different primary lung cancers can be discriminated on histopathological criteria. Of these small cell lung cancer (SCLC) appears to be the most malignant (35), since it is characterized by both a high growth rate and

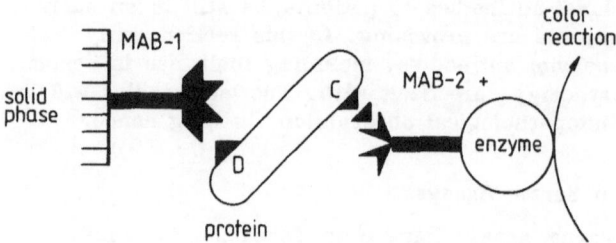

FIGURE 2 Monoclonal-antibody-based sandwich assay. The first monoclonal
antibody (MAB-1) is attached to a solid phase. After application of the
sample, this antibody catches and concentrates the antigen from the sample.
The second antibody (MAB-2) detects and quantitates the caught antigen
by means of a covalently attached tracer molecule. This tracer can be
a radiolabeled substance in the case of a radioimmuno-assay (RIA) or an
enzyme in the case of an enzyme-linked immuno-sorbent assay (ELISA).
Adapted from Ref. 33.

a propensity to early and extensive metastasis. However, unlike the other
lung cancers (commonly called non-SCLC), SCLC is initially highly
sensitive to chemo- or radiotherapy, resulting in almost complete tumor
regressions in the majority of patients. Despite these encouraging initial
results, treatment-resistant recurrences develop in almost all cases after
a relatively short time, and the overall 5-year survival is disappointingly
low as a result. Because of its different clinical course (almost always
excluding surgical intervention as a treatment) and sensitivity to chemotherapy, the differential diagnosis of SCLC vs. non-SCLC is very important
for judging which treatment modality should be chosen. In the following,
the possible use of monoclonal antibodies for such a purpose is illustrated
and discussed.

In order to prepare monoclonal antibodies that could be helpful for
establishing a differential diagnosis in lung cancer, we have immunized
mice with SCLC-derived cell lines. After the subsequent establishment
and testing of a large number of hybridomas, it was concluded that the
produced monoclonal antibodies define differentiation-related, rather than
tumor-specific, antigens on SCLC. It turned out that the majority of monoclonal antibodies produced detected two kinds of SCLC-associated antigens
(see Table 1): antigens also present on normal epithelial structures and
antigens also expressed by various neural and/or (neuro)endocrine tissues.
This result matches with the suggestion (36) that SCLC represents a malignant counterpart of the normal pulmonary neuroendocrine cells, which
are epithelial cells with a proposed paracrine function. When applied to
a large panel of lung cancers (37), it turned out that all histologically
diagnosed lung carcinomas, comprising both non-SCLC (71 cases) as well
as SCLC (43 cases), expressed the epithelial cell marker defined by the
monoclonal antibody MOC-31, whereas the noncarcinomatous cancers present
in the lung (e.g., mesothelioma, lymphoma) were unreactive with this
antibody. MOC-31 can be considered, therefore, as a suitable "general"
marker for lung carcinoma.

With respect to the monoclonal antibodies directed against neuroendo-
drine antigens, it turned out that histologically diagnosed non-SCLC was

TABLE 1 Specificity of Anti–SCLC Monoclonal Antibodies

Antigen class[a]	Monoclonal antibodies prepared against SCLC	Reactions in normal and malignant lung tissues: possible implications
Epithelium-associated	MOC-31, and many others	Normal tissue: present on all lung epithelia
		Malignant tissue: present in both SCLC and non-SCLC
		Indicative for an endodermal derivation of all lung cancers including SCLC
Neuroendocrine-associated	MOC-1 MOC-1g1	Normal tissue: only present on normal pulmonary neuroendocrine cells
		Malignant tissue: only present in SCLC
		Indicative for a relationship between normal pulmonary neuroendocrine cells and SCLC
	MOC-21 MOC-32 MOC-51 MOC-52	Normal tissue: not present on normal adult lung epithelia, present only in very early (1st trimester) lung tissue
		Malignant tissue: only present in SCLC
		Indicative for a reexpression of fetal antigens in SCLC

[a]Judged from the occurrence of the antigens on a panel of normal and malignant tissues.

generally unreactive, whereas SCLC (and carcinoid, which is a separate, but rarely occurring, endocrine lung tumor) was reactive. These results indicate a general match between morphologically based diagnosis, on the one hand, and the specific presence of neuroendocrine-associated antigens on SCLC and not on non-SCLC, on the other. This is in agreement with the previously established notion (38), that SCLC (and carcinoid) is an "endocrine" tumor, whereas non-SCLC is not. However, a small group of staining results did not fit in with the histologically based diagnosis.

First, some (5-10%) of the assessed non-SCLC cases (37) showed a focal or even complete reactivity with some of the monoclonal antibodies directed against neuroendocrine related antigens. This means that these non-SCLC tumors have at least some "endocrine" features. Whether this finding has a clinical meaning remains to be established. The critical question in this respect is: does the correlation nonresponsiveness to chemotherapy and no early metastasis (true in the majority of, but sure not in all, non-SCLC cases) correspond to the morphologically based classification

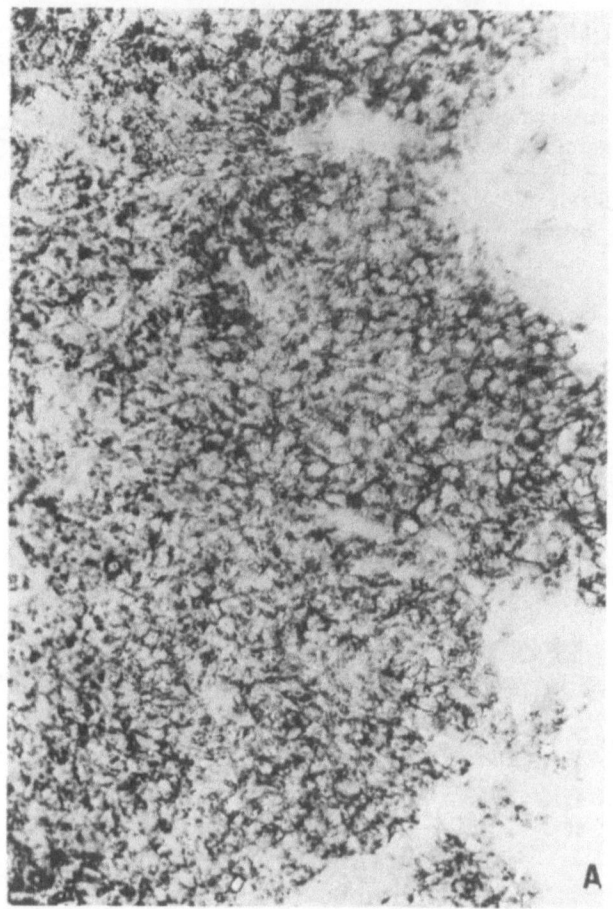

FIGURE 3 Bronchoscopically procured malignant lung tissue biopsies stained
with a peroxidase staining technique, based on the use of a monoclonal
antibody (MOC-21). The biopsies were taken from the same patient at
different time points, i.e., at the beginning of disease when the tumor
showed a good reaction to (chemo)therapy (A) and at the end-stage, when
the tumor was changed to become refractory to further treatment (B).
A positive reaction can be appreciated as a dark (reddish-brown) color.
The sections are weakly counterstained with hematoxylin.

"non-SCLC" or should this classification be replaced by the immunohisto-
logically based classification "nonendocrine"? Second, some of the morphologically typed SCLC cases showed no or only a limited reactivity with
the monoclonal antibodies directed against neuroendocrine antigens. Whether
these cases represent a clinically relevant subgroup should also be established in carefully conducted clinical studies. In fact, we have recently
obtained strong indications that treatment-resistant SCLC expresses less
neuroendocrine antigens than treatment-sensitive SCLC (39; Fig. 3), which
is in agreement with the suggestion that the "endocrine" nature of lung
tumors is the best indication for sensitivity to chemotherapy.

In summary, pathological practice is based now on a careful description of the morphology of the cells and tissues present in a biopsy. The correlation between aberrant biopsy morphology and clinical diagnosis has been established in a long-standing tradition. Monoclonal-antibody-based immunohistopathology adds an extra dimension to this morphological tradition, since it characterizes a biopsy by the visualization of the different antigenic features present. It can be anticipated that some of these antigenic features may be clinically relevant. Our results with lung cancer provide an indication that this is indeed the case. Thus, assessing of antigens relevant to prognosis of responsiveness to (chemo)therapy may well become an important extension of pathological practice in the near future.

V. SUMMARY AND CONCLUSIONS

In this chapter, the immunological and physical principles underlying hybridoma technology have been described. These principles can provide insight into newly emerging lines of hybridoma research. In addition, as

an example of one of the uses of monoclonal antibodies, its impact on oncological pathology is briefly indicated.

A hybridoma obtains the genetic information, coding for the specificity
of produced monoclonal antibodies, from the immune lymphocytes present
in an immunized animal. Recent knowledge on the process of in vivo "programming" of immune lymphocytes during immunization can be used in
in vitro immunization protocols. This may provide new, controllable, ways
to obtain specific immune lymphocytes in the near future.

New trends in the induction of hybridoma formation are indicated.
Especially, the preselection of immune lymphocytes on the guide of their
antigen-binding properties, followed by a specific hybridization induction
by means of electrofusion, provides a new avenue for a more specific construction of hybridomas.

Preliminary results obtained with immunohistopathology in lung cancer
appear impressive, and if such results can be extended to other cancers
or diseases, it can be anticipated that immunohistopathology will revolutionize
the tools available to the pathologist in the near future.

REFERENCES

1. Koehler, G., and Milstein, C. (1975). Continuous cultures of fused
   cells secreting antibody of predefined specificity. *Nature* 256:495-497.
2. Langone, J. J., and Van Vunakis, H. (1986). Immunochemical techniques. Part I: Hybridoma technology and monoclonal antibodies. In:
   Methods in Enzymology, Vol. 121 (S. P. Colowick and N. O. Kaplan,
   eds.). Academic Press, Orlando, FL.
3. Milstein, C. (1986). Overview: Monoclonal antibodies. In: *Handbook
   of Experimental Immunology. Vol. 4: Application of Immunological Methods
   in Biomedical Sciences* (D. M. Weir, ed.). Blackwell, Oxford, pp.
   107.1-107.12.
4. DePinho, R. A., Feldman, L. B., and Scharff, M. D. (1986). Tailor-
   made monoclonal antibodies. *Ann. Intern. Med.* 104:225-233.
5. Staehli, C., Staehelin, Th., and Miggiano, V. (1983). Spleen cell
   analysis and optimal immunization for high-frequency production of
   specific hybridomas. In: *Methods in Enzymology*, Vol. 92 (S. P.
   Colowick and N. O. Kaplan, eds.). Academic Press, Orlando, FL,
   pp. 26-36.
6. Littlefield, S. W. (1964). Selection of hybrids from matings of fibroblast in vitro and their presumed recombinants. *Science* 145:709-710.
7. Burnet, F. M. (1959). *The Clonal Selection Theory of Acquired
   Immunity.* Vanderbilt and Cambridge University Presses, Nashville,
   TN.
8. Landsteiner, K. (1936). *The Specificity of Serological Reactions.*
   Charles C. Thomas, Springfield, IL.
9. Brack, C., Hirama, M., Lenhard-Schuller, R., and Tonegawa, S.
   (1978). A complete immunoglobulin gene is created by somatic recombination. *Cell* 15:1-14.
10. Tonegawa, S. (1983). Somatic generation of antibody diversity. *Nature*
    302:575-581.
11. Manser, T., Wysocki, L. J., Gridley, T., Near, R. I., and Gefter,
    M. L. (1985). The molecular evolution of the immune response. *Immunol.
    Today* 6:94-100.

12. Lanzavecchia, A. (1985). Antigen-specific interaction between T and B cells. *Nature* 314:537-539.

13. Reading, C. L. (1986). In vitro immunization for the production of antigen specific lymphocyte hybridomas. In: *Methods in Enzymology*, Vol. 121 (S. P. Colowick and N. O. Kaplan, eds.). Academic Press, Orlando, FL, pp. 18-27.

14. Borrebaeck, C. A. K. (1986). In vitro immunization for production of murine and human monoclonal antibodies: present status. *TIBTECH* 4:147-153.

15. Kinashi, T., Harada, N., Severinson, E., Tanabe, T., Sideras, P., Konishi, M., et al. (1986). Cloning of complementary DNA encoding T-cell replacing factor and identity with B-cell growth factor II. *Nature* 324:70-73.

16. Mishel, R. I., and Dutton, R. W. (1967). Immunization of dissociated spleen cell cultures from normal mice. *J. Exp. Med.* 126:413-442.

17. Ho, M. K., Rand, N., Murray, J., Kato, K., and Rabin, H. (1985). In vitro immunization of human lymphocytes. I. Production of human monoclonal antibodies against bombesin and tetanus toxoid. *J. Immunol.* 135:3831-3838.

18. Mueller, J. (1838). Ueber den feineren bau und die formen der krankheften gaschwuelste. Quoted by K. Faber, *J. Path. Bact.* 1:349-358 (1893).

19. Kohn, A. (1965). Polykaryocytosis induced by Newcastle disease virus in monocultures of animal cells. *Virology* 26:228-245.

20. Kao, K. N., and Michiayluk, M. R. (1974). Method for high frequency intergeneric fusion of plant protoplasts. *Planta* 115:355-367.

21. Pontocorvo, G. (1975). Production of mammalian somatic cell hybrids by means of polyethylene glycol treatment. *Somat. Cell Genet.* 1:397-400.

22. Westerwoudt, R. J. (1986). Factors affecting production of monoclonal antibodies. In: *Methods in Enzymology*, Vol. 121 (S. P. Colowick and N. O. Kaplan, eds.). Academic Press, Orlando, FL, pp. 3-18.

23. Honda, K., Maeda, Y., Sasakawa, J., Ohno, H., and Tsuchida, E. (1981). Activities of cell fusion and lysis of the hybrid type of chemical fusogens. (I). Structure and function of the promotor of cell fusion. *Biochem. Biophys. Res. Commun.* 100:442-448.

24. Wojcieszyn, J. W., Schlegel, R. A., Lumley-Sapanski, K., and Jacobson, K. A. (1983). Studies on the mechanism of poly(ethylene glycol)-mediated cell fusion using fluorescent membrane and cytoplasmic probes. *J. Cell Biol.* 96:151-159.

25. Ochiai, K., and Ikeda, T. (1985). Effect of polyethylene glycol molecular weight and concentration on hybridization of mouse myeloma and spleen cells. *IRCS Med. Sci.* 13:322-323.

26. de Leij, L., Poppema, S., and The, T. H. (1983). Cryopreservation of newly formed hybridomas. *J. Immunol. Meth.* 62:69-72.

27. Croce, C. M. Koprowski, ., and Eagle, H. (1972). Effect of environmental pH on the efficiency of cellular hybridization. *Proc. Natl. Acad. Sci. USA* 69:1953-1956.

28. Obara, Y., Chai, S., Weinfeld, H., and Sandberg, A. A. (1974). Prophasing of interphase nuclei and induction of nuclear envelopes around metaphase chromosomes in HeLa and Chinese hamster homo- and heterokaryons. *J. Cell Biol.* 62:104-113.

29. Tempelaar, M. J., and Jones, M. K. (1985). Fusion characteristics of plant protoplasts in electric fields. *Planta* 165:205-216.

30. Vienken, J., and Zimmermann, U. (1985). An improved electrofusion technique for production of mouse hybridoma cells. *FEBS Lett.* 182: 278-280.

31. Zimmermann, U. (1982). Electric field-mediated fusion and related electrical phenomena. *Biochim. Biophys. Acta* 694:227-277.

32. Lo, M. M. S., Tsong, T. Y., Konrad, M. K., Strittmatter, S. M., Hester, L. D., and Snyder, S. H. (1984). Monoclonal antibody production by receptor-mediated electrically induced cell fusion. *Nature* 310: 792-794.

33. Hedin, A., Carlsson, L., Berglund, A., and Hammarstroem, S. (1983). A monoclonal antibody-enzyme immunoassay for serum carcinoembryonic antigen with increased specificity for carcinomas. *Proc. Natl. Acad. Sci. USA* 80:3470-3474.

34. Ruiter, D., Fleuren, G. J., and Warnaar, S., eds. (1987). *Application of Monoclonal Antibodies in Tumor Pathology*. Martinus Nijhoff, The Hague.

35. Iannuzzi, M. C., and Scoggin, C. H. (1986). State of art: Small cell lung cancer. *Am. Rev. Respir. Dis.* 134:593-608.

36. Bensch, K. G., Corrin, B., Pariente, R., and Spencer, H. (1968). Oat cell carcinoma of the lung: Its origin and relationship to bronchial carcinoid. *Cancer* 22:1163-1172.

37. de Leij, L., Broers, J., Ramaekers, F., Berendsen, H., and Wagenaar, S. (1987). Monoclonal antibodies in clinical and experimental pathology of lung cancer. In: *Application of Monoclonal Antibodies in Tumor Pathology* (D. Ruiter, G. J. Fleuren, and S. Warnaar, eds.). Martinus Nijhoff, The Hague, pp. 191-210.

38. Becker, K. L., and Gazdar, A. F., eds. (1984). *The Endocrine Lung in Health and Disease*. Saunders, Philadelphia.

39. Berendsen, H. H., de Leij, L., Postmus, P. E., Poppema, S., Elema, J. D., Sluiter, H. J., and The, T. H. (1987). Small cell lung cancer. Tumor cell phenotype detected by monoclonal antibodies and response to chemotherapy in small cell lung cancer. *Chest* 21:11s-12s.

31

pH-Sensitive Liposomes
Introduction of Foreign Substances into Cells

NEJAT DÜZGÜNEŞ, ROBERT M. STRAUBINGER,*
PATRICIA A. BALDWIN,† and DEMETRIOS PAPAHADJOPOULOS
University of California—San Francisco, San Francisco, California

I. INTRODUCTION

The use of liposomes for drug delivery in vivo (1-7) and for the introduction of foreign molecules into cells in vitro (8-14) is under active investigation. The concept of pH-sensitive liposomes for drug delivery was introduced by Yatvin et al. (15), who proposed that such liposomes could be used for the release of encapsulated drugs in tissues, such as primary tumors, metastases, and sites of infection, where the pH is thought to be below the physiological pH. Here we review the recent development of pH-sensitive liposomes for the introduction of fluorescent compounds and macromolecules into the cytoplasm.

Certain types of liposomes are internalized by cells via phagocytosis or endocytosis (16-19). Recent studies with CV-1 cells, an epithelioid cell line derived from African green monkey kidney cells, have revealed that negatively charged liposomes are endocytosed in coated pits and subsequently sequestered in endosomes (20,21). Earlier studies had indicated that the pH of the lumen of endosomes is mildly acidic (22,23). That liposomes indeed encounter a low-pH compartment after endocytosis was revealed by experiments comparing the intracytoplasmic delivery of fluorescent compounds that have differential permeabilities through membranes at acidic pH (20).

Several lipid-enveloped viruses, such as influenza, vesicular stomatitis virus, and Semliki Forest virus, are thought to microinject their genome into the cytoplasm of the host cell by fusing with the membrane of endosomes in which they are internalized, after acidification of the endosome interior (24-27,27a; see also Chapter 19). In view of these findings, the question then arises: Can we design liposomes that will destabilize or fuse with the endosome membrane at mildly acidic pH, thereby releasing their aqueous

---

Present affiliations:
*State University of New York—Buffalo, Buffalo, New York.
†California Biotechnology, Inc., Mountain View, California.

contents into the cytoplasm? The design of liposomes that destabilize and fuse with one another at low pH is the first step toward this goal.

II. MEMBRANE FUSION INDUCED BY MILDLY ACIDIC pH

Liposomes of various compositions can be induced to fuse at a pH below neutral (Table 1). The thresholds vary from pH 2 for phosphatidylserine (PS) large unilamellar vesicles (LUV) to near pH 7 for small unilamellar vesicles (SUV) consisting of (egg) phosphatidylethanolamine (PE) and palmitoylhomocysteine (PHC) (28,29). Although mildly acidic pH by itself is sufficient to cause fusion of liposomes of certain compositions, such as PE, a mixture of oleic acid (OA) and TPE [phosphatidylethanolamine obtained by transphosphatidylation of egg phosphatidylcholine (PC)], or a mixture of cholesterylhemisuccinate (CHEMS) and TPE (28,30–32), auxiliary mediators such as polyhistidine or lectins are necessary for the fusion of liposomes of other compositions (e.g., PS/PE, asolectin, PS/PC) (33,34).

In Table 1 the term "fusion" has been used in the widest sense. Many of the studies listed have utilized only lipid mixing assays or electron microscopy. As discussed in detail previously (35), mixing of membrane components does not necessarily imply the coalescence of aqueous contents. In the context of proton-dependent delivery of liposome contents into cells, coalescence of the aqueous compartments would likely be required for cytoplasmic delivery. Growth of vesicles to larger structures may also proceed via pathways that do not lead to mixing of aqueous contents. For example, CHEMS/TPE liposomes undergo lipid mixing at low pH, as measured by the dilution of N-NBD-PE and N-Rh-PE from labeled to unlabeled liposomes. However, their aqueous contents do not intermix, although some leakage of contents does occur (36,28). Similarly, transformation of membranes from the bilayer to the hexagonal (H_{II}) phase at low pH does not imply that mixing of aqueous contents has occurred (30).

Different mechanisms may be operative in proton-induced membrane fusion in the various systems listed in Table 1. Wang and Huang (33), who studied polyhistidine-mediated fusion, propose that the poly(amino acid), which is positively charged at acidic pH, binds to negatively charged lipids. Fusion then proceeds via lateral phase separation of the negatively charged lipids from the zwitterionic lipids. Uster and Deamer (34) have suggested that polyhistidine causes vesicle aggregation even at pH 7.4, possibly owing to hydrophobic interactions between the imidazole group and the membrane, and that at lower pH the polyhistidine binding can influence acyl chain packing and lead to phase separation. Partial protonation of the headgroups of acidic phospholipids at mildly acidic pH is thought to reduce the polarity and hydration of the membrane surface and lead to fusion, as long as intermembrane contact is achieved by an independent means, such as with lectins in the case of vesicles containing glycolipids (37). Studies on clathrin-mediated fusion of LUV at reduced pH indicate that protein binding and a conformational change leading to the exposure of hydrophobic domains of the protein are involved in this process (81). While protein binding may mediate the close approach of the membranes, the conformational change is thought to cause the insertion of the hydrophobic domain into the lipid bilayer and perturbation of lipid packing.

The pH dependence of the fusion of OA/TPE liposomes is shown in Figure 1. The substitution of PS for OA or of PC for TPE abolishes the pH sensitivity of fusion and destabilization (32,38,39). In the case of

TABLE 1 Proton-Induced Fusion of Phospholipid Vesicles[a]

Vesicle composition and type	Mediator	Threshold pH[b]	Reference
PS (SUV)	—	3.5	Papahadjopoulos et al. (71)
PS (LUV)	—	2	Ellens et al. (28)
PS/PE, PA/PE, PG/PE, PS/PC, CL/PC (SUV)	Polyhistidine	≤6.5 (5)	Wang and Huang (33)
PS/PE/PC/chol, asolectin (SUV, LUV)	Polyhistidine	≤6.5 (6.0)	Uster and Deamer (34)
PE, DOPE, TPE, DOPE-Me, DOPE/DOPC (LUV)	—	4.5	Ellens et al. (30,72,85)
PE/PA/PELBA (LUV)	Ricinus communis agglutinin	≤6.5 (3.8)	Bondeson et al. (37)
PE/PS/PELBA (LUV)		<4.5 (3.5)	Bondeson et al. (37)
PA/PC (SUV)	Insulin	<4.5 (3.6–3.9)	Farias et al. (73)
PE/PHC (SUV)	—	<7 (4.8)	Connor et al. (29)
PC (SUV)	Albumin	<4	Schenkmann et al. (74)
PC (SUV)	Albumin fragments	<4	Garcia et al. (75)
PS/SPE,PA/SPE (SUV)	—	3	Hope et al. (76)

(continued)

(Table 1, continued)

Vesicle composition and type	Mediator	Threshold pH[b]	Reference
CHEMS/TPE (LUV)	—	<5	Ellens et al. (28)
			Lai et al. (31)
			Bentz et al. (77)
DOPC (SUV)	Clathrin	6	Blumenthal et al. (78)
PS/PC (LUV)	Clathrin[c]	5	Hong et al. (79)
PS, PS/PC (LUV)	Clathrin	<5	Maezawa et al. (81)
OA/TPE (extruded MLV)	—	<6.5	Düzgüneş et al. (32)
Asolectin, *E. coli* phospholipids (SUV)	Colicins	<6.5 (4.2)	Pattus et al. (80)
PC (SUV)	GALA[c]	5	Parente et al. (82)
PC/PS (SUV), CAIPEI/PC (SUV)	CAIPEI	5	Oku et al. (83)

OAPA/DEPE (LUV)	—	5.5 (4.5)	Leventis et al. (65a)
SCD/TPE (extruded MLV)	—	6 (4.5)	Epand et al. (84)
DPPC (SUV)	Diptheria toxin	4.3	Cabiaux et al. (86)
Asolectin (SUV)	Tetanus toxin	4.3	Cabiaux et al. (87)

[a]Abbreviations: ANTS, 1-aminonaphthalene-3,6,8-trisulfonic acid; CAIPEI, cetylacetyl(imidazol-4-ylmethyl) polyethylenimine; CHEMS, cholesterylhemisuccinate; DEPE, dielaidoylphosphatidylethanolamine; DEPS, dielaidoylphosphatidylserine; DPA, dipicolinic acid; DMPE, dimyristoylphosphatidylethanolamine; DOPC, dioleoylphosphatidylcholine; DOPE, dioleoylphosphatidylethanolamine; DOPE-Me, monomethyl DOPE; DOPS, dioleoylphosphatidylserine; DPPC, dipalmitoylphosphatidylcholine; DPX, p-xylylene-bis-pyridinium bromide; GALA, a 30-amino acid polypeptide with the repeat unit Glu-Ala-Leu-Ala; LUV, large unilamellar vesicles (approx. 100 nm diameter); MLV, multilamellar vesicles; NBD, N-(7-nitro-2,1,3-benzoxadiazol-4-yl) phosphatidylethanolamine; OA, oleic acid; OAPA, N-oleoyl-2-aminopalmitic acid; PA, phosphatidate (phosphatidic acid); PC, (egg) phosphatidylcholine; PE, (egg) phosphatidylethanolamine; PELBA, phosphatidylethanol-N-lactobionamide; PHC, palmitoyl homocysteine; PS, (bovine brain) phosphatidylserine; SCD, 2,3-seco-5α-cholestan-2,3-dioic acid; SOPE, succinyl phosphatidylethanolamine; SPE, soybean phosphatidylethanolamine; SUV, small unilamellar vesicles (approx. 30 nm diameter); TPE, phosphatidylethanolamine synthesized by transphosphatidylation of egg phosphatidylcholine.
[b]The pH at which the maximal effect is observed is given in parentheses.
[c]These mediators are ineffective in inducing the fusion of PC (LUV).

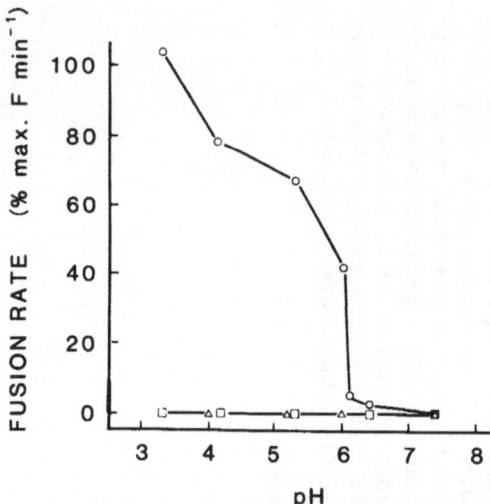

FIGURE 1   pH dependence of the initial rate of fusion of liposomes composed
of OA/TPE (circles), OA/PC (squares), and PS/PE (triangles) (all 3:7
molar ratio). Fusion was measured by a fluorescent probe dilution assay,
using a 1:9 ratio of labeled to unlabeled liposomes. Reproduced, with permission, from Düzgünes et al. (32).

OA/TPE liposomes, low pH induces the destabilization of the membrane
before the intermixing of aqueous contents, suggesting that the points
of destabilization provide nucleation sites for fusion (32). OA may function
as a titratable negative charge which initially provides stability to the
vesicles composed predominantly of PE. Thus, neutralization of OA would
lead to fusion of the PE component of the membrane, since liposomes composed of pure PE (prepared at pH 9) aggregate at neutral pH (40) and
fuse at pH 5 (30,41). The presence of PE in liposomes has been shown
to enhance membrane fusion in a number of systems (42,43). OA may also
serve as a fusogen, as demonstrated in studies with erythrocytes and
chromaffin granules (44,45). The observation that TPE liposomes containing
arachidonic, elaidic, or stearic acids fuse at low pH with different kinetics
suggests that the role of the fatty acids is not only to provide a titratable
surface charge, but also to affect the phase behavior of the PE (46,47).
     The susceptibility of phosphatidic acid (PA)- or fatty-acid-containing
membranes to low-pH-induced fusion (32,37) raises the possibility that
the formation of these lipids during the stimulation of secretory cells (48,49)
modulates fusion events accompanying intracellular membrane traffic, as
in the budding of vesicles from the compartment of uncoupling of receptor
and ligand (CURL, Refs. 50-52).
     Biological membrane vesicles can also be induced to fuse with liposomes
at low pH. Fusion of asolectin (soybean lipids) vesicles (SUV) with inner
mitochondrial membranes at pH 6.5 has been utilized as a method to change
the protein/lipid ratio in the latter membranes (53; see also Chapter 3).
Such asolectin-enriched mitochondrial inner membranes were fused with
each other (optimally at pH 6.5 in the presence of Ca^{2+}) to produce membranes large enough for lateral diffusion or microelectrode measurements
(54). Fusion of asolectin SUV with rye protoplasts requires substantially

lower pH, in the range 4-5 (54a). Similarly, LUV containing cardiolipin (CL) or PS fuse at pH below 6 with bacterial inner membranes (55; see also Chapter 3).

Other pH-sensitive liposomes have been developed which release their contents but do not appear to undergo fusion. Liposomes composed of dioleoylphosphatidylethanolamine (DOPE) and succinyl-DOPE (SOPE) (7:3) become destabilized below pH 5, although pure SOPE liposomes release their contents at pH 7.4 (56). Liposomes containing palmitoyl histidine, in addition to dipalmitoyl PC and diheptadecanoyl PC (16:10:74), respond to pH 6 and the presence of serum by releasing encapsulated calcein (57). Maximal destabilization at acidic pH is obtained when both the PC and N-acylamino acid have similar chain lengths in the range C-16 and C-17 (57a). The release of 6-carboxyfluorescein from PC vesicles in the presence of 0.03% poly(α-ethylacrylic acid) in the aqueous medium is also achieved when the pH is lowered to 6.5 (57b).

III. DELIVERY OF FOREIGN MOLECULES INTO THE CYTOPLASM MEDIATED BY pH-SENSITIVE LIPOSOMES

Since certain types of liposomes can undergo fusion in the pH range found in the endocytotic pathway, it is possible that they can also destabilize or fuse with the endosome membrane, after they are internalized. Figure 2 shows a schematic view of the intracellular fate of pH-sensitive and pH-insensitive liposomes and their contents. Liposomes composed of acidic phospholipids are internalized by the coated pit pathway after adhering to the plasma membrane (20). The nature of the cell surface molecules to which liposomes bind is not known. The contents of coated vesicles ultimately proceed to other organelles with a variety of morphologies, including endosomes. The lumen of endocytic vesicles is acidified soon after their formation.

If pH-insensitive liposomes contain a weak acid such as carboxyfluorescein (CF), to which the liposome and endosome membranes are permeable at acidic pH, intracytoplasmic delivery of the encapsulated contents will be achieved. Studies with chloroquine and NH_4Cl indicate that when the endosomal pH is raised, molecules such as CF are not delivered into the cytoplasm (20). In contrast, a highly charged molecule such as calcein, which is not membrane-permeant in the pH range encountered in the endocytotic pathway (pH 4.6-5.5; 59,60), remains sequestered in vesicles of the endocytotic pathway. Thus, pH-insensitive fluorescent markers must be used to ascertain the role of pH-sensitive liposomes in mediating cytoplasmic delivery of the markers.

The molecules delivered to the cytoplasm by pH-sensitive liposomes, the types of liposomes, and the cell lines used for these studies are summarized in Table 2. We have utilized liposomes composed of OA and TPE, which exhibit low-pH-mediated fusion and destabilization (32,38), for the introduction of fluorescent probes and macromolecules to the cytoplasm of CV-1 cells (39,58). Liposomes containing calcein give rise to a diffuse cytoplasmic fluorescence in a high proportion of the cells in culture (Fig. 3). A number of manipulations to the system reduce considerably the delivery of calcein into the cytoplasm. These include azide and deoxyglucose, metabolic inhibitors that inhibit endocytosis, weakly basic amines such as chloroquine and NH_4Cl, which raise the endosomal pH, and the proton-carrying ionophore monensin. However, the failure of lysosomotropic amines to completely inhibit cytoplasmic delivery suggests that pH-sensitive liposomes

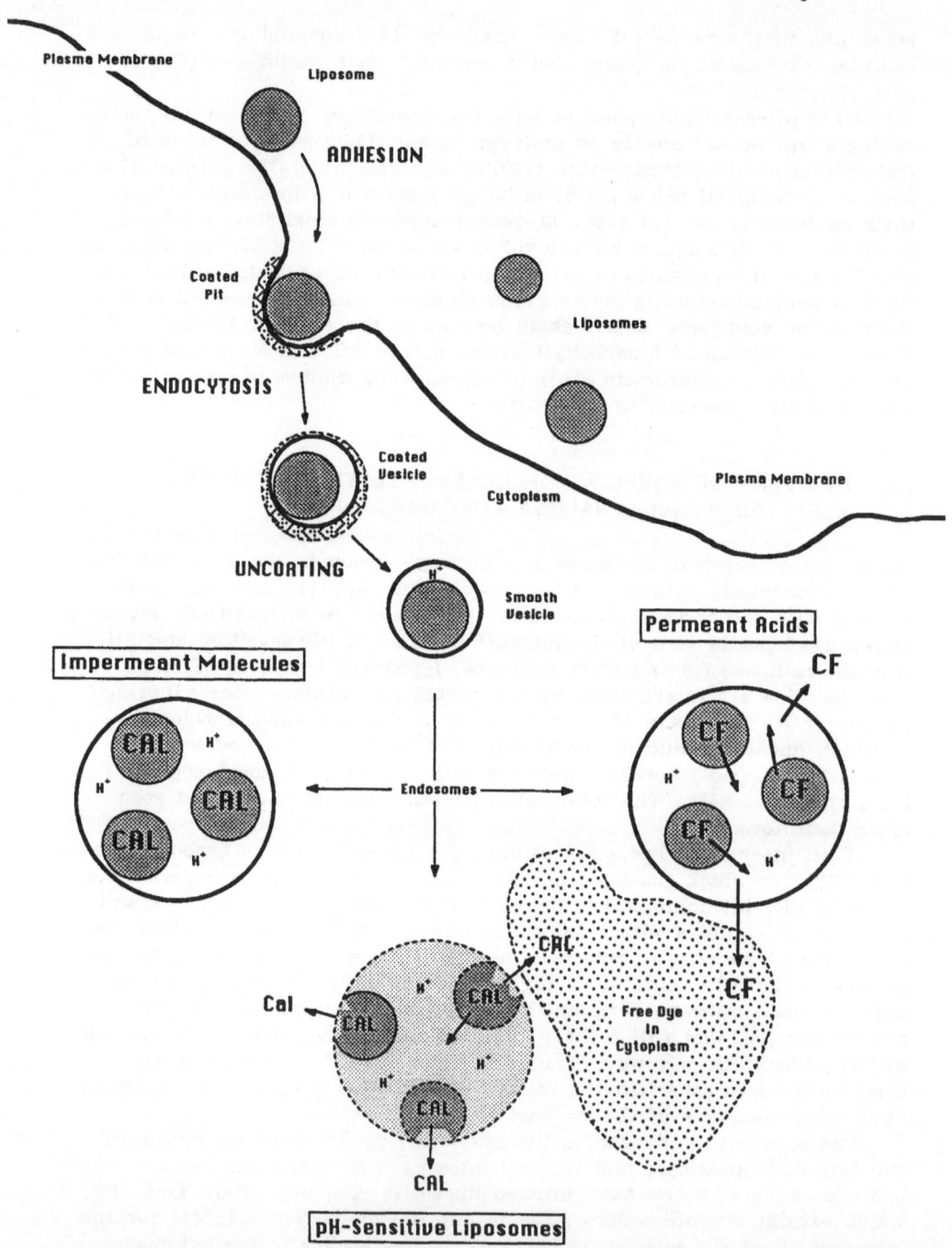

FIGURE 2 Uptake and intracellular processing of pH-sensitive and pH-insensitive liposomes.

TABLE 2 Cytoplasmic Delivery of Foreign Molecules via pH-Sensitive Liposomes[a]

Liposome composition	Cell type	Molecule delivered to the cytoplasm	Reference
DOPE/PHC anti-H2K^k(LUV)	L929	Calcein	Connor and Huang (61)
OA/TPE, OA/TPE/Chol (extruded MLV)	CV-1	Calcein F-dextran	Straubinger et al. (39)
Succinyl-PE/DOPE(LUV)	Macrophages	Calcein	Schroit et al. (66)
OA/DOPE/cholesterol (LUV)	Carrot protoplasts	Calcein	Wang et al. (67)
OA/DOPE/anti-H2K^k(LUV)	L929	ara-C Methotrexate	Connor and Huang (62)
OA/PE/anti-H2K^k(LUV)	L929	Diphtheria toxin A fragment	Collins and Huang (63)
OA/TPE (LUV)	CV-1	Pokeweed Antiviral protein	Baldwin et al. (64)
OA/DOPE/cholesterol/ anti-H2K^k (LUV)	RDM-4	DNA (chloramphenicol acetyltransferase gene)	Wang and Huang (63a)

[a]Abbreviations are defined in footnote to Table 1.

are active even at pH 6.2, which is the maximal level endosomal pH is thought to reach in the presence of chloroquine (60). This pH is within the range necessary to induce leakage and fusion of these liposomes (32,39). The inclusion of 30 mol% cholesterol in OA/TPE liposomes is not inhibitory to the delivery of calcein; in fact, the intensity of cytoplasmic fluorescence is higher. OA/TPE liposomes can also deliver fluoresceinated (F) dextran (average molecular weight 18 or 40 kD) into the cytoplasm following glycerol treatment of the cells for a brief period (Fig. 4; Ref. 39). The effect of glycerol may be attributed to the increase in internalization of the pH-sensitive liposomes rather than membrane permeabilization by osmotic shock, since cytoplasmic delivery of calcein from pH-insensitive liposomes is not increased (39,58). The mechanism of pH-sensitive liposome-mediated delivery of calcein and dextran is thought to be the destabilization of the endosome membrane by the pH-sensitive liposomes, or the fusion of the two membranes, much as in the case of influenza virus or Semliki Forest virus. At present, there are no definitive data in support of either hypothesis.

Similar observations have been made by Connor and Huang (61,62), who have used liposomes composed of DOPE and either PHC or OA to deliver calcein and the antitumor drugs ara-C and methotrexate to the cytoplasm

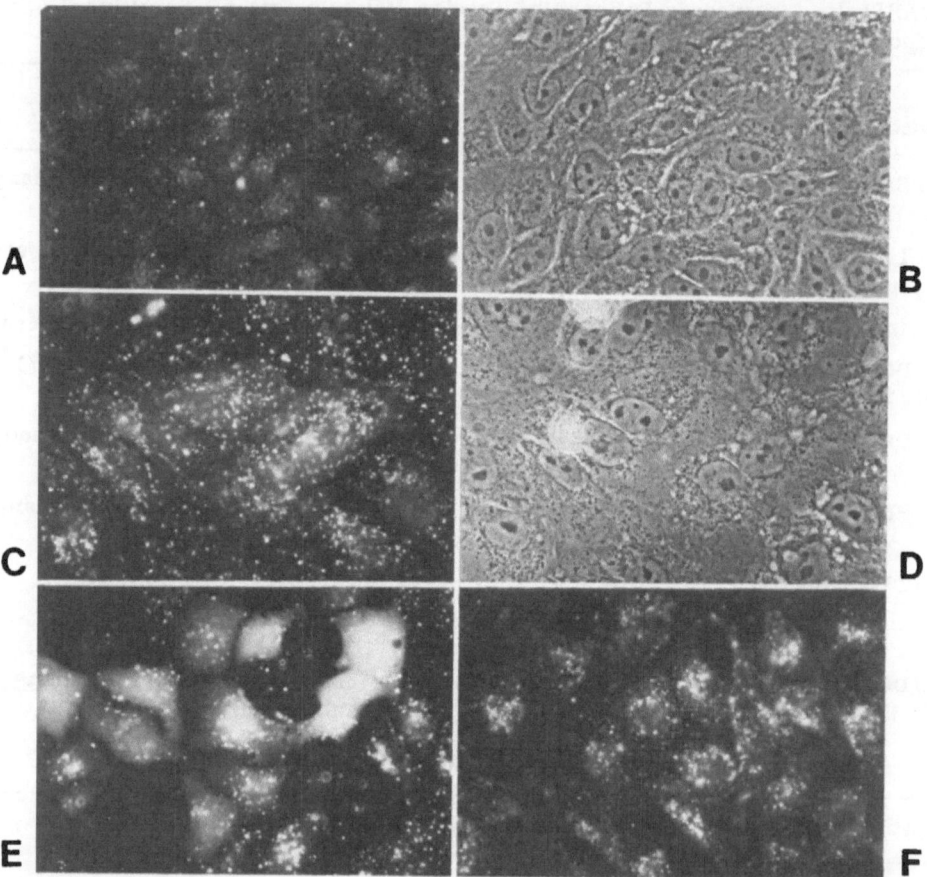

FIGURE 3   Delivery of calcein to CV-1 cells via pH-sensitive liposomes.
(A) PS/PE liposomes show punctate fluorescence. (B) Phase contrast image
of the field shown in A. (C) OA-containing liposomes reveal both cytoplasmic
and punctate fluorescence. (D) Phase contrast image of the field shown
in C. (E) Treatment of the cells with 21% glycerol results in intense diffuse
calcein fluorescence. (F) Glycerol treatment of cells incubated with PS/PE
liposomes does not result in significant cytoplasmic delivery of calcein.
Reproduced, with permission, from Straubinger et al. (39).

of L929 cells. In these studies, liposomes were targeted to cells by means
of a surface-bound antibody recognizing the H-2K^k histocompatibility antigen. Collins and Huang (63) have shown that OA/DOPE liposomes mediate
the inhibition of protein synthesis in L929 cells by the membrane-impermeable
A fragment of diphtheria toxin encapsulated in the liposomes. The effect
of the A fragment could be inhibited by pretreatment of the cells with
NH_4Cl or chloroquine, or by excess free antibody or empty liposomes.
The delivery of a plasmid containing the chloramphenicol acetyltransferase
gene to RDM-4 lymphoma cells inside mouse peritoneum, following encapsulation in pH-sensitive liposomes targeted to H-2K^k, has also been demonstrated
(63a).

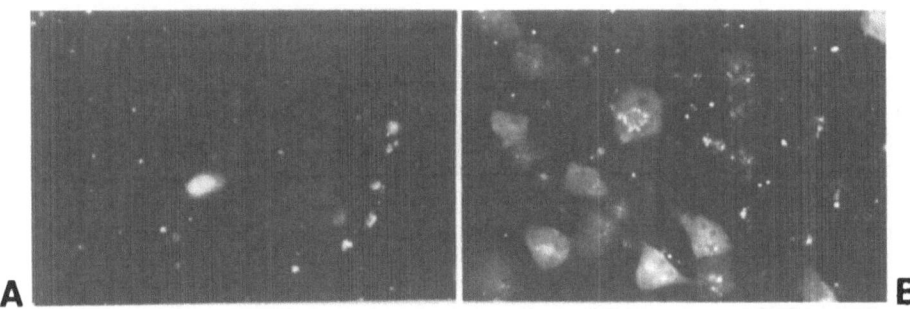

FIGURE 4 Cytoplasmic delivery of F-dextran encapsulated in pH-sensitive
liposomes incubated with CV-1 cells. Cells were incubated with the liposomes
and treated with glycerol for 5 min. (A) Punctate fluorescence arising from
intact PS/PE liposomes. (B) Diffuse fluorescence resulting from F-dextran
in the cytoplasm, delivered via OA/TPE liposomes. Reproduced, with permission, from Straubinger et al. (39).

The potent plant toxin pokeweed antiviral protein has also been encapsulated in pH-sensitive liposomes, and delivery of the intact protein to
CV-1 cells has been demonstrated by the inhibition of cell growth and
protein synthesis (64). The free toxin is relatively nontoxic, since it does
not have a receptor on the cell surface and is not readily internalized.
Delivery via pH-sensitive liposomes results in a much greater inhibition
of cell growth and protein synthesis than with pH-insensitive liposomes
or incubation of the cells with free protein. In this study, the presence
of serum in the cell growth medium was found to decrease the delivery
of the pokeweed antiviral protein by pH-sensitive liposomes, while delivery
by pH-insensitive liposomes was affected to a lesser extent. Inhibition
of cell growth by the toxin encapsulated in pH-sensitive liposomes has
also been observed with the ovarian carcinoma cell line OVCAR-3 (P. A.
Baldwin, R. M. Straubinger, and D. Papahadjopoulos, unpublished experiments). Growth inhibition was more efficient with pH-sensitive liposomes
than with pH-insensitive liposomes or free toxin (Table 3). One difficulty
with the use of OA/TPE liposomes is the extensive leakage of encapsulated
molecules in the presence of cells or serum. Strategies to improve cytoplasmic delivery should address the problem of liposome stability under
such conditions. Amphiphiles with two alkyl chains and serine or histidine
headgroups, in combination with PE, form pH-sensitive liposomes, but
appear to be less susceptible to desorption from the vesicles in the presence
of serum (65,65a).

pH-sensitive liposomes have also been used for the delivery of calcein
to peritoneal macrophages (66) and to plant protoplasts (67). Superoxide
release in neutrophils is stimulated when the inducer oleoyl-acetyl-glycerol
is incorporated in liposomes composed of PE and arachidonic acid (68).
Activation of macrophages to the tumoricidal state is enhanced when γ-
interferon and muramyl dipeptide are encapsulated in DOPE/SOPE (7:3)
liposomes, which are sensitive to low pH, in contrast to SOPE liposomes,
which are destabilized at neutral pH (69).

The above studies indicate that the cytoplasmic delivery of highly
charged molecules and macromolecules that do not cross the endosome membrane by themselves can be achieved by certain pH-sensitive liposomes.

TABLE 3 Cytotoxicity of Liposome-Encapsulated Pokeweed Antiviral Protein Against OVCAR-3 Cells[a]

Liposome composition[b]	Liposome type	IC$_{50}$
OA/PE (3:7)	pH-sensitive	1×10^{-8} μM
PS/PE (3:7)	pH-insensitive	4.2×10^{-8} μM
Free protein	—	1.4×10^{-7} μM

[a]Liposomes containing pokeweed antiviral protein were added to OVCAR-3 cells, an ovarian carcinoma cell line, in the presence of medium without serum. After 4 h, 10% fetal bovine serum was added and the cells were allowed to grow for 4 days before counting in a Coulter counter. IC$_{50}$ is defined as the concentration of pokeweed antiviral protein at which cell growth is 50% of control cells.
[b]Abbreviations are defined in footnote to Table 1.

The in vivo application of pH-sensitive liposomes may require the use of analogous pH-sensitive liposome systems that are not susceptible to destabilization by serum. Nevertheless, recent observations by Connor et al. (70) on the in vivo distribution of OA/DOPE liposomes suggest that these liposomes may be able to carry lipophilic drugs to the lungs, even though they would be ineffective as a carrier of water-soluble drugs.

ACKNOWLEDGMENTS

This work was supported by NIH Grants GM28117 (D.P. and N.D.), CA-35340 (D.P. and R.M.S.) and AI25534 (N.D.), a Grant-in-Aid from the American Heart Association (N.D.), a Grant from the State of California Universitywide AIDS Research Program (N.D.), and a Fellowship from the American Cancer Society (P.A.B.). We thank Drs. Leaf Huang and Rajiv Nayar for sending us their manuscripts before publication, and Mrs. Rose Antonucci for invaluable editorial assistance.

REFERENCES

1. Mayhew, E., and Papahadjopoulos, D. (1983). Therapeutic applications of liposomes. In: *Liposomes* (M. J. Ostro, ed.). Marcel Dekker, New York, pp. 289-341.
2. Schroit, A. J., Fidler, I. J. (1986). The design of liposomes for delivery of immunomodulators to host defense cells. In: *Medical Applications of Liposomes* (K. Yagi, ed.). Japan Scientific Societies Press, Tokyo, and Karger, Basel, pp. 141-149.
3. Alving, C. R. (1986). Liposomes as drug carriers for the treatment of leishmaniasis. In: *Medical Applications of Liposomes* (K. Yagi, ed.). Japan Scientific Societies Press, Tokyo, and Karger, Basel, pp. 105-112.
4. Knight, C. G., Bard, D. R., and Thomas, P. P. (1985). Liposomes as carriers of antiarthritic agents. *Ann. NY Acad. Sci.* 446:415-428.

5. Juliano, R. L., Lopez-Berestein, G., Hopfer, R., Mehta, R., Mehta, K., and Mills, K. (1985). Selective toxicity and enhanced therapeutic index of liposomal polyene antibiotics in systemic fungal infections. *Ann. NY Acad. Sci.* 446:390-402.

6. Gregoriadis, G., Senior, J., Wolff, B., and Kirby, C. (1985). Targeting of liposomes to accessible cells in vivo. *Ann. NY Acad. Sci.* 446: 319-340.

7. Düzgüneş, N., Perumal, V. K., Kesavalu, L., Goldstein, J. A., Debs, R. J., and Gangadharam, P. R. J. (1988). Enhanced effect of liposome-encapsulated amikacin on *Mycobacterium avium - M. intracellulare* complex infection in beige mice. *Antimicrob. Agents Chemother.* 32:1404-1411.

8. Fraley, R., Subramani, S., Berg, P., and Papahadjopoulos, D. (1980). Introduction of liposome-encapsulated SV40 DNA into cells. *J. Biol. Chem.* 255:10431-10435.

9. Fraley, R., Straubinger, R. M., Rule, G., Springer, E. L., and Papahadjopoulos, D. (1981). Liposome mediated delivery of deoxyribonucleic acid to cells: Enhanced efficiency of delivery related to lipid composition and incubation conditions. *Biochemistry* 20:6978-6987.

10. Walter, M. F., Uster, P. S., and Deamer, D. W. (1986). Liposome-mediated delivery of antibody to a *Drosophila* cell line. *Eur. J. Cell Biol.* 40:195-202.

11. Straubinger, R. M., and Papahadjopoulos, D. (1982). Liposome-mediated DNA transfer. In: *Techniques in Somatic Cell Genetics* (J. W. Shay, ed.). Plenum Press, New York, pp. 399-413.

12. Wilson, T., Papahadjopoulos, D., and Taber, R. (1979). The introduction of poliovirus RNA into cells via lipid vesicles (liposomes). *Cell* 17:77-84.

13. Ostro, M. J., and Giacomoni, D. (1983). Liposomes as a tool in molecular biology: A comparison to other methodologies. In: *Liposomes* (M. J. Ostro, ed.). Marcel Dekker, New York and Basel, pp. 145-208.

14. Papahadjopoulos, D., Straubinger, R. M., Hong, K., and Düzgüneş, N. (1987). Liposomes as a carrier system for delivery of foreign molecules into cells: Methodology for enhanced cytoplasmic delivery. In: *Beijing Symposium on Bioenergetics and Biomembranes* (S. Fleisher, T. E. King, and S. Papa, eds.). Vanderbilt University Printing Services, Nashville, TN, pp. 43-52.

15. Yatvin, M. B., Kreutz, W., Horwitz, B. A., and Shinitzky, M. (1980). pH-sensitive liposomes: Possible clinical implications. *Science* 210:1253-1255.

16. Magee, W. E., Goff, C. W., Schoknecht, J., Smith, M. D., and Cherian, K. (1974). The interaction of cationic liposomes containing entrapped horseradish peroxidase with cells in culture. *J. Cell Biol.* 63:492-504.

17. Weissmann, G., Bloomgarden, D., Kaplan, R., Cohen, L., Hoffstein, S., Collins, J., Gotlieb, A., and Nagle, D. (1975). A general method for the introduction of enzymes, by means of immunoglobulin-coated liposomes, into lysosomes of deficient cells. *Proc. Natl. Acad. Sci. USA* 72:88-92.

18. Leserman, L. D., Weinstein, J. N., Blumenthal, R., and Terry, W. D. (1980). Receptor-mediated endocytosis of antibody-opsonized liposomes by tumor cells. *Proc. Natl. Acad. Sci. USA* 77:4089-4093.

19. Poste, G., and Papahadjopoulos, D. (1978). The influence of vesicle membrane properties on the interaction of lipid vesicles with cultured cells. *Ann. NY Acad. Sci.* 308:164-184.

20. Straubinger, R. M., Hong, K., Friend, D. S., and Papahadjopoulos, D. (1983). Endocytosis of liposomes and intracellular fate of encapsulated molecules: Encounter with a low pH compartment after internalization in coated vesicles. *Cell* 32:1069-1079.

21. Straubinger, R. M., Hong, K., Friend, D. S., Düzgüneş, N., and Papahadjopoulos, D. (1985). Endocytosis of liposomes and intracellular fate of encapsulated molecules: Strategies for enhanced cytoplasmic delivery. In: *Receptor-Mediated Targeting of Drugs* (G. Gregoriadis, G. Poste, J. Senior, and A. Trouet, eds.). Plenum Press, New York, pp. 297-315.

22. Tycko, B., and Maxfield, F. R. (1982). Rapid acidification of endocytic vesicles containing alpha$_2$-macroglobulin. *Cell* 28:643-651.

23. Mellman, I., Fuchs, R., and Helenius, A. (1986). Acidification of the endocytic and exocytic pathways. *Annu. Rev. Biochem.* 55:663-700.

24. Marsh, M., Bolzau, E., and Helenius, A. (1983). Penetration of Semliki Forest virus from acidic prelysosomal vacuoles. *Cell* 32:931-940.

25. Yoshimura, A., Kuroda, K., Kawasaki, K., Yamashina, S., Maeda, T., and Ohnishi, S-I. (1982). Infectious cell entry mechanism of influenza virus. *J. Virol.* 43:284-293.

26. Marsh, M. (1984). The entry of enveloped viruses into cells by endocytosis. *Biochem. J.* 218:1-10.

27. Ohnishi, S-I. (1985). Membrane fusion: Mechanism and role in cellular response to enveloped virus. In: *Biomolecules: Electronic Aspects* (C. Nagata, M. Hatano, J. Tanaka, and H. Suzuki, eds.). Japan Scientific Societies Press, Tokyo, and Elsevier, Amsterdam, pp. 227-252.

27a. Ohnishi, S-I. (1988). Fusion of viral envelopes with cellular membranes. In: *Membrane Fusion in Fertilization, Cellular Transport, and Viral Infection* (N. Düzgüneş and F. Bronner, eds.). Academic Press, San Diego, pp. 257-296.

28. Ellens, H., Bentz, J., and Szoka, F. C. (1985). H^+ and Ca^{2+}-induced fusion and destabilization of liposomes. *Biochemistry* 24:3099-3106.

29. Connor, J., Yatvin, M. B., and Huang, L. (1984). pH-sensitive liposomes: Acid-induced liposome fusion. *Proc. Natl. Acad. Sci. USA* 81:1715-1718.

30. Ellens, H., Bentz, J., and Szoka, F. C. (1986). Destabilization of phosphatidylethanolamine liposomes at the hexagonal phase transition temperature. *Biochemistry* 25:285-294.

31. Lai, M-Z., Vail, W. J., and Szoka, F.C. (1985). Acid- and calcium-induced structural changes in phosphatidylethanolamine membranes stabilized by cholesterol hemisuccinate. *Biochemistry* 24:1654-1661.

32. Düzgüneş, N., Straubinger, R. M., Baldwin, P. A., Friend, D. S., and Papahadjopoulos, D. (1985). Proton-induced fusion of oleic acid-phosphatidylethanolamine liposomes. *Biochemistry* 24:3091-3098.

33. Wang, C-Y., and Huang, L. (1984). Polyhistidine mediates an acid-dependent fusion of negatively charged liposomes. *Biochemistry* 23:4409-4416.

34. Uster, P. S., and Deamer, D. W. (1985). pH-dependent fusion of liposomes using titratable polycations. *Biochemistry* 24:1-8.

35. Düzgüneş, N., and Bentz, J. (1987). Fluorescence assays for membrane fusion. In: *Spectroscopic Membrane Probes*, Vol. I (L. M. Loew, ed.). CRC Press, Boca Raton, FL, pp. 117-159.

36. Ellens, H., Bentz, J., and Szoka, F. C. (1984). pH-induced destabilization of phosphatidylethanolamine-containing liposomes. Role of bilayer contact. *Biochemistry* 23:1532-1538.

37. Bondeson, J., Wijkander, J., and Sundler, R. (1984). Proton-induced membrane fusion. Role of phospholipid composition and protein-mediated intermembrane contact. *Biochim. Biophys. Acta* 777:21-27.

38. Düzgüneş, N., Straubinger, R. M., and Papahadjopoulos, D. (1983). pH-dependent membrane fusion. *J. Cell Biol.* 97:178a.

39. Straubinger, R. M., Düzgüneş, N., and Papahadjopoulos, D. (1985). pH-sensitive liposomes mediate cytoplasmic delivery of encapsulated macromolecules. *FEBS Lett.* 179:148-154.

40. Kolber, M. A., and Haynes, D. H. (1979). Evidence for a role of phosphatidylethanolamine as a modulator of membrane-membrane contact. *J. Memb. Biol.* 48:95-114.

41. Baldwin, P. A., Düzgüneş, N., and Papahadjopoulos, D. (1985). Free fatty acids affect H^+ and Ca^{2+}-induced fusion of model membranes. *Biophys. J.* 47:112a.

42. Düzgüneş, N., Wilschut, J., Fraley, R., and Papahadjopoulos, D. (1981). Studies on the mechanism of membrane fusion: Role of head-group composition in calcium- and magnesium-induced fusion of mixed phospholipid vesicles. *Biochim. Biophys. Acta* 642:182-195.

43. Sundler, R., Düzgüneş, N., and Papahadjopoulos, D. (1981). Control of membrane fusion by phospholipid head groups. II. The role of phosphatidylethanolamine in mixtures with phosphatidate and phosphatidylinositol. *Biochim. Biophys. Acta* 649:751-758.

44. Ahkong, Q. F., Fisher, D., Tampion, W., and Lucy, J. A. (1973). The fusion of erythrocytes by fatty acids, esters, retinol and α-tocopherol. *Biochem. J.* 136:147-155.

45. Creutz, C. E. (1981). cis-Unsaturated fatty acids induce the fusion of chromaffin granules aggregated by synexin. *J. Cell Biol.* 91:247-256.

46. Baldwin, P. A., Düzgüneş, N., and Papahadjopoulos, D. (1990). pH-Sensitive liposomes: Role of the bilayer-hexagonal (H_{II}) phase transition and fatty acids in membrane fusion and destabilization, in preparation.

47. Düzgüneş, N., Hong, K., Baldwin, P. A., Bentz, J., Nir, S., and Papahadjopoulos, D. (1987). Fusion of phospholipid vesicles induced by divalent cations and protons. Modulation by phase transitions free fatty acids, monovalent cations, and polyamines. In: *Cell Fusion* (A. E. Sowers, ed.). Plenum Press, New York, pp. 241-267.

48. Michell, R. H. (1975). Inositol phospholipids and cell surface receptor function. *Biochim. Biophys. Acta* 415:81-147.

49. Michell, R. H., Kirk, C. J., Jones, L. M., Downes, C. P., and Creba, J. A. (1981). The stimulation of inositol lipid metabolism that accompanies calcium mobilization in stimulated cells: Defined characteristics and unanswered questions. *Phil. Trans. R. Soc. Lond. B.* 296:123-137.

50. Geuze, H. J., Slot, J. W., Strous, G. J. A. M., Lodish, H. F., and Schwartz, A. L. (1983). Intracellular site of asialoglycoprotein receptor-ligand uncoupling: Double label immunoelectron microscopy during receptor-mediated endocytosis. *Cell* 32:277-287.

51. Dautry-Varsat, A., Ciechanover, A., and Lodish, H. F. (1983). pH and the recycling of transferrin during receptor-mediated endocytosis. *Proc. Natl. Acad. Sci. USA* 80:2258-2262.

52. Brown, M. S., Anderson, R. G. W., and Goldstein, J. L. (1983).

Recycling receptors: The round-trip itinerary of migrant membrane
proteins. *Cell* 32:663-667.

53. Schneider, H., Lemasters, J. J., Höchli, M., and Hackenbrock, C. R.
    (1980). Fusion of liposomes with mitochondrial inner membranes. *Proc.
    Natl. Acad. Sci. USA* 77:442-446.

54. Chazotte, B., Wu, E-S., Höchli, M., and Hackenbrock, C. R. (1985).
    Calcium-mediated fusion to produce ultra large osmotically active mito-
    chondrial inner membrane of controlled protein density. *Biochim. Bio-
    phys. Acta* 818:87-95.

54a. Arvinte, T., and Steponkus, P. L. (1988). Characterization of the
     pH-induced fusion of liposomes with the plasma membrane of rye proto-
     plasts. *Biochemistry* 27:5671-5677.

55. Driessen, A. J. M., Hoekstra, D., Scherphof, G., Kalicharan, R. D.,
    and Wilschut, J. (1985). Low pH-induced fusion of liposomes with
    membrane vesicles derived from *Bacillus subtilis*. *J. Biol. Chem.* 260:
    10880-10887.

56. Nayar, N., and Schroit, A. J. (1985). Generation of pH-sensitive
    liposomes: Use of large unilamellar vesicles containing *N*-succinyl-
    dioleoylphosphatidylethanolamine. *Biochemistry* 24:5967-5971.

57. Yatvin, M. B., Cree, T. C., and Tegmo-Larsson, I-M. (1984). Theo-
    retical and practical considerations in preparing liposomes for the pur-
    pose of releasing drug in response to changes in temperature and pH.
    In: *Liposome Technology*, Vol. III (G. Gregoriadis, ed.). CRC Press,
    Boca Raton, FL, pp. 157-175.

57a. Tegmo-Larsson, I-M., Hofmann, K. P., Kreutz, W., and Yatvin,
     M. B. (1985). The effect of pH on vesicles composed of phosphatidyl-
     cholines and *N*-acylamino acids: A calcein release fluorescence study.
     J. Controlled Release 1:191-196.

57b. Tirrell, D. A., Takigawa, D. Y., and Seki, K. (1985). pH sensitiza-
     tion of phospholipid vesicles via complexation with synthetic
     poly(carboxylic acid)s. *Ann. NY Acad. Sci.* 446:237-247.

58. Straubinger, R. M., Düzgüneş, N., and Papahadjopoulos, D. (1983).
    pH-sensitive liposomes: Enhanced cytoplasmic delivery of encapsulated
    macromolecules. *J. Cell. Biol.* 97:109a.

59. deDuve, C., de Barsy, T., Poole, B., Trouet, A., Tulkens, P.,
    and van Hoof, F. (1974). Lysosomotropic agents. *Biochem. Pharmacol.*
    23:2495-2531.

60. Ohkuma, S., and Poole, B. (1978). Fluorescence probe measurement
    of the intralysosomal pH in living cells and the perturbation by various
    agents. *Proc. Natl. Acad. Sci. USA* 75:3327-3331.

61. Connor, J., and Huang, L. (1985). Efficient cytoplasmic delivery of
    a fluorescent dye by pH-sensitive immunoliposomes. *J. Cell Biol.* 101:
    582-589.

62. Connor, J., and Huang, L. (1986). pH-sensitive immunoliposomes as
    an efficient and target-specific carrier for antitumor drugs. *Cancer
    Res.* 46:3431-3435.

63. Collins, D., and Huang, L. (1987). Cytotoxicity of diphtheria toxin A
    fragment to toxin resistant murine cells delivered by pH-sensitive
    immunoliposomes. *Cancer Res.* 47:735-739.

63a. Wang, C-Y., and Huang, L. (1987). pH-sensitive immunoliposomes
     mediate target-cell-specific delivery and controlled expression of a
     foreign gene in mouse. *Proc. Natl. Acad. Sci. USA* 84:7851-7855.

64. Baldwin, P. A., Straubinger, R. M., and Papahadjopoulos, D. (1986).

Enhanced delivery of pokeweed antiviral protein to cultured cells by pH-sensitive liposomes. *J. Cell Biol.* 103:57a.

65. Diacovo, T., Leventis, R., and Silvius, J. R. (1986). Proton-induced fusion of lipid vesicles containing novel pH-sensitive amphiphiles. *Biophys. J.* 49:19a.

65a. Leventis, R., Diacovo, T., and Silvius, J. R. (1987). pH-dependent stability and fusion of liposomes combining protonatable double-chain amphiphiles with phosphatidylethanolamine. *Biochemistry* 26:3267-3276.

66. Schroit, A. J., Madsen, J., and Nayar, R. (1986). Liposome-cell interactions: In vitro discrimination of uptake mechanism and in vivo targeting strategies to mononuclear phagocytes. *Chem. Phys. Lipids* 40:373-393.

67. Wang, C-Y., Hughes, K. W., and Huang, L. (1986). Improved cytoplasmic delivery to plant protoplasts via pH-sensitive liposomes. *Plant Physiol.* 82:179-184.

68. Tsusaki, B. E., Kanda, S., and Huang, L. (1986). Stimulation of superoxide release in neutrophils by 1-oleoyl-2-acetylglycerol incorporated into pH-sensitive liposomes. *Biochem. Biophys. Res. Commun.* 136:242-246.

69. Nayar, R., Fidler, I. J., and Schroit, A. J. (1988). Potential applications of pH-sensitive liposomes as drug delivery systems. In: *Liposomes as Drug Carriers: Recent Trends and Progress* (G. Gregoriadis, ed.). Wiley-Interscience, New York, pp. 771-782.

70. Connor, J., Norley, N., and Huang, L. (1986). Biodistribution of pH-sensitive immunoliposomes. *Biochim. Biophys. Acta* 884:474-481.

71. Papahadjopoulos, D., Vail, W. J., Newton, C., Nir, S., Jacobson, K., Poste, G., and Lazo, R. (1977). Studies on membrane fusion. III. The role of calcium-induced phase changes. *Biochim. Biophys. Acta* 465:579-598.

72. Ellens, H., Bentz, J., and Szoka, F. C. (1986). Fusion of phosphatidylethanolamine containing liposomes and the mechanism of the L_α-H_{II} phase transition. *Biochemistry* 25:4141-4147.

73. Farias, R. N., Vinals, A. L., and Morero, R. D. (1985). Insulin-mediated fusion of negatively charged phospholipid vesicles at low pH. *Biochem. Biophys. Res. Commun.* 128:68-74.

74. Schenkman, S., Araujo, P. S., Dijkman, R., Quina, F. H., and Chaimovich, H. (1981). Effects of temperature and lipid composition on the serum albumin-induced aggregation and fusion of small unilamellar vesicles. *Biochim. Biophys. Acta* 649:633-641.

75. Garcia, L. A. M., Schenkman, S., Araujo, P. S., and Chaimovich, H. (1983). Fusion of small unilamellar vesicles induced by bovine serum albumin fragments. *Braz. J. Med. Biol. Res.* 16:89-96.

76. Hope, M. J., Walker, D. C., and Cullis, P. R. (1983). Calcium and pH-induced fusion of small unilamellar vesicles consisting of phosphatidylethanolamine and negatively charged phospholipids: A freeze-fracture study. *Biochem. Biophys. Res. Commun.* 110:15-22.

77. Bentz, J., Ellens, H., Lai, M-Z., and Szoka, F. C., Jr. (1985). On the correlation between H_{II} phase and the contact-induced destabilization of phosphatidylethanolamine-containing membranes. *Proc. Natl. Acad. Sci. USA* 82:5742-5745.

78. Blumenthal, R., Henkart, M., and Steer, C. J. (1983). Interaction of liver clathrin coat protein with lipid model membranes. *J. Biol. Chem.* 258:3409-3415.

79. Hong, K., Yoshimura, T., and Papahadjopoulos, D. (1985). Interaction of clathrin with liposomes: pH-dependent fusion of phospholipid membranes induced by clathrin. *FEBS Lett.* 191:17-23.

80. Pattus, F., Cavard, D., Crozel, V., Baty, D., Adrian, M., and Lazdunski, C. (1985). pH-dependent membrane fusion is promoted by various colicins. *EMBO J.* 4:2469-2474.

81. Maezawa, S., Yoshimura, T., Hong, K., Düzgüneş, N., and Papahadjopoulos, D. (1989). Mechanism of protein-induced membrane fusion: Fusion of phospholipid vesicles by clathrin associated with its membrane binding and conformational change. *Biochemistry* 28:1422-1428.

82. Parente, R. A., Nir, S., and Szoka, F. C., Jr. (1988). pH-dependent fusion of phosphatidylcholine small vesicles. Induction by a synthetic amphipathic peptide. *J. Biol. Chem.* 263:4724-4730.

83. Oku, N., Shibamoto, S., Ito, F., Gondo, H., and Nango, M. (1987). Low pH induced membrane fusion of lipid vesicles containing proton-sensitive polymer. *Biochemistry* 26:8145-8150.

84. Epand, R. M., Cheetham, J. J., and Raymer, K. E. (1988). Acid-induced fusion of liposomes: studies with 2,3-seco-5α-cholestan-2,3-dioic acid. *Biochim. Biophys. Acta* 940:85-92.

85. Ellens, H., Siegel, D. P., Alford, D., Yeagle, P. L., Boni, L., Lis, L. J., Quinn, P. J., and Bentz, J. (1989). Membrane fusion and inverted phases. *Biochemistry*, in press.

86. Cabiaux, V., Vandenbranden, M., Falmagne, P., and Ruysschaert, J-M. (1984). Diphtheria toxin induces fusion of small unilamellar vesicles at low pH. *Biochim. Biophys. Acta* 775:31-36.

87. Cabiaux, V., Lorge, P., Vandenbranden, M., Falmagne, P., and Ruysschaert, J. M. (1985). Tetanus toxin induces fusion and aggregation of lipid vesicles containing phosphatidylinositol at low pH. *Biochem. Biophys. Res. Commun.* 128:840-849.

32

Sendai Virus Envelopes as a Biological Carrier
Reconstitution, Targeting, and Application

ABRAHAM LOYTER, VITALY CITOVSKY, and NURIT BALLAS
The Hebrew University of Jerusalem, Jerusalem, Israel

I. INTRODUCTION

Two alternative routes have been suggested for the entry of animal viruses into recipient cells (see Chapter 13). Most enveloped virions enter cells by endocytic-like processes. Fusion of the viral envelopes with endosomal membranes is promoted by the intraorganelle low pH (1). A second mechanism of entry has been observed for members of the paramyxovirus group such as Newcastle disease virus (NDV) or Sendai virus, whose penetration occurs by fusion of the viral envelope with the cell plasma membrane (1,2). Viral membrane fusion processes have been studied mainly using electron microscopy methods (2) or by following virus-induced lysis of recipient cells (3). Recently, energy transfer or fluorescence dequenching methods have also been employed to estimate, on a quantitative bases, virus-membrane fusion (4,5).

Attachment of Sendai virions to cell surface receptors (sialic-acid-containing components) is mediated by a viral envelope glycoprotein designated as the hemagglutinin/neuraminidase (HN) polypeptide. Fusion of the viral envelope with recipient cell plasma membranes requires the presence of a second envelope glycoprotein, the F (fusion) polypeptide (6,7). Studies in our laboratory and other laboratories have shown that, in addition to its function as the viral binding protein, the HN glycoprotein also actively participates in the virus-membrane fusion process (8,9).

II. RECONSTITUTED SENDAI VIRUS ENVELOPES AS A BIOLOGICAL VEHICLE: A SHORT REVIEW

The view that the viral envelope glycoproteins, namely the HN and F polypeptides, are required for attachment of Sendai virus to cell surfaces and for promotion of virus-membrane fusion was inferred from experiments showing that reconstituted viral envelopes are as fusogenic as intact Sendai virions (7). Reconstituted Sendai virus envelopes can be obtained by

731

solubilization of intact virions with nonionic detergents such as Nonidet P-40 or Triton X-100 (7,10). After sedimentation of detergent-insoluble material, removal of the detergent from the clear supernatant results in the formation of membrane vesicles composed of the viral phospholipids and the two viral envelope glycoproteins, the HN and F glycoproteins (see also Chapter 18). Such reconstituted Sendai virus envelopes (RSVE), similar to intact virions, are able to induce cell lysis and promote cell-cell fusion (7). Virus-induced hemolysis has been shown to reflect a process of virus-membrane fusion (3), and therefore, it has been suggested that such RSVE possess the same fusogenic characteristics as those attributed to intact virions.

A. Fusion-Mediated Microinjection of Macromolecules

It has been well established that, if soluble macromolecules such as polypeptides, RNA, or DNA are added to the detergent-solubilized mixture of the viral glycoproteins and phospholipids, they are entrapped within the envelope formed after removal of the detergent (7). Such loaded envelopes have been used as biological carriers for the introduction of macromolecules into animal cells (7).

RSVE have been used as a carrier to introduce functional DNA molecules such as SV_{40} DNA into SV_{40}-susceptible or SV_{40}-resistant cell lines. Expression of the microinjected SV_{40} DNA has been studied by the appearance of the specific SV_{40} T antigen in the nucleus of the recipient cells. For example, about 20% of CV_1 cells (an African monkey kidney cell line permissive to infection by SV_{40}) expressed SV_{40} T antigen, following fusion with SV_{40}-DNA-loaded RSVE (7,11). RSVE have also been loaded with various restriction fragments of Epstein-Barr virus (EBV) DNA. Fusion-mediated microinjection of the enclosed DNA was used to study the function of the different genes located in the various fragments (12). Similarly, RSVE have been used to transfer the HSV TK gene (herpes simplex thymidine kinase gene) to TK⁻ cells with relatively high efficiency (11).

RSVE were also used as an efficient vehicle for the introduction of poly(I)·poly(C) into cultured cells to study inhibition of protein synthesis in interferon-treated cells (13). It has been observed that incubation of loaded RSVE with L and HeLa cells resulted in strong inhibition of protein synthesis, indicating fusion-mediated microinjection of the enclosed poly(I)·poly(C). Inhibition of protein synthesis in L cells but not in HeLa cells was dependent on pretreatment with interferon (IFN). Incubation of poly(I)·poly(C)-loaded viral envelopes with IFN-treated variant cells of the NIH 3T3 line, which possess small amounts of RNase L, resulted in a very low degree of inhibition of protein synthesis. The results, using this system, clearly indicate that the inhibition of protein synthesis in L cells by RSVE loaded with poly(I)·poly(C) is due to activation of the (2'-5')-oligoadenylate-synthetase-RNase L pathway (13).

Fusion-mediated microinjection was also used to study the biological function of intracellular polyamines (14). Serum amine oxidase and/or porcine kidney diamine oxidase was trapped within RSVE with full retention of their enzymatic activity. Both enzymes are involved in the catabolism of the naturally occurring polyamines, spermidine, spermine, and the diamine putrescine. During oxidation of these polyamines, aldehydes, hydrogen peroxides, and ammonia are produced (15). These products are known for their cytotoxicity. Indeed, when RSVE loaded with the above enzymes, namely serum amine oxidase and porcine kidney diamine oxidase, were incubated with cultured fibroblasts, an arrest in protein and DNA synthesis was observed (14). It appears that the microinjected enzymes catalyzed the oxidation of intracellular polyamines. The resulting oxidation products apparently caused an arrest in intracellular synthesis of macromolecules.

B. Fusion-Mediated Implantation of Membrane Proteins

Previous work has shown that, if a membrane component is added to the detergent solution of the viral envelope components, it is inserted into the viral envelope, thus resulting in the formation of hybrid RSVE containing viral glycoproteins and nonviral components within the same membrane (7). Fusion of such hybrid reconstituted vesicles, bearing nonviral exogenous polypeptides, with living cells results in fusion-mediated implantation of the nonviral polypeptides in the cell's plasma membrane (7).

Fusion-mediated implantation of the human erythrocyte band 3, the erythrocyte anion exchange system, into Friend erythroleukemia cells (FELC) results in a marked and specific stimulation of the exchange of anions in the target cells (7,16). Since band 3 can stoichiometrically be labeled with tritiated 4,4'-diisothiocyano-stilbene-2,2'-disulfonic acid ($[^3H]H_2$DIDS), its insertion into FELC could also be followed by determination of cell-associated radioactive material (7,16). Similarly, a calcium-binding protein (CBP) was implanted into variants of basophilic cells that lack this protein. Fusion-mediated implantation led to the restoration of Ca^{2+} uptake and degranulation capacity of the variant cells after IgE-mediated stimulation (17).

In addition to the use of purified membrane proteins such as the human erythrocyte band 3 and the CBP of rat basophilic leukemia cells, extracts of whole plasma membrane can also be inserted into the viral envelope, using the coreconstitution system. Membrane extracts rich in receptors for the viruses SV_{40} or EBV have been coreconstituted with Sendai virus envelope glycoproteins (7,18). Fusion of such hybrid vesicles with receptor-negative cells resulted in fusion-mediated implantation of the two receptors for these viruses. This was inferred from experiments showing that following fusion with the hybrid vesicles, the recipient cells became susceptible to infection by SV_{40} or EBV (7,18). Likewise, insertion of syngeneic H-2^b membrane components from insulin responder mice into allogeneic spleen cells from H-2^k nonresponder mice endows the latter with the capacity to present beef insulin to H-2^b responder T lymphocytes (19).

III. QUANTITATIVE DETERMINATION OF RSVE-MEMBRANE FUSION PROCESSES BY THE USE OF FLUORESCENTLY LABELED RSVE

For the efficient use of RSVE as a biological carrier, it is important to develop a method that will allow quantitative determination of its fusogenic activity. Virus- or RSVE-induced hemolysis, although possibly reflecting virus-membrane fusion, is not a direct measure of such a process, especially when various cell lines are used as recipients. Furthermore, the residual amount of detergent that, in some cases, may be left in the RSVE preparation may either cause or stimulate the lytic processes promoted by incubation of the viral envelopes with human erythrocytes or cultured cells.

Energy transfer and fluorescence dequenching methods have been used to study fusion processes occurring between enveloped viruses and biological membranes (4,5; see also Chapter 14). We have recently shown that the fluorescent probe N-(4-nitrobenzo-2-oxa-1,3-diazole)-phosphatidylethanolamine (N-NBD-PE) can be inserted into RSVE, if added during the reconstitution process (5). Insertion of N-NBD-PE into RSVE at a ratio of viral lipids to N-NBD-PE of 94:6 results in self-quenching. However, N-NBD-PE cannot be incorporated into envelopes of intact Sendai virions, thus preventing its use for study of fusion processes occurring between intact virions and biological membranes. Nevertheless, fusion of intact Sendai virions can be studied by the use of octadecylrhodamine B (R_{18}) (4,20). Since R_{18} can be inserted into intact virions as well as RSVE, its use permits comparison of the fusogenic activities of these two viral preparations. Indeed, recent experiments in our laboratory (see Table 1) have shown that incubation of fluorescently labeled intact virions or RSVE with either human erythrocytes or living cultured cells results in a relatively high degree of fluorescence dequenching (4,20,21). The extent of fluorescence dequenching obtained is dependent on the amount of cells added as well as on the pH and temperature of the incubation medium. A high degree of fluorescence dequenching was obtained between pH 6.0 and 9.0 and at 37-39°C (5,20,21).

The results in Table 1 show the degree of fluorescence dequenching obtained by incubation of intact Sendai virions with saturating amounts of human erythrocyte ghosts (HEG), FELC, and mouse L cells. Based on our previous observations (5), it should be noted that about 58% of intact Sendai virions fuse with HEG, while 51% and 48% fuse with FELC and L cells, respectively. It also appears (Table 1) that the fusogenic activity of the RSVE preparation is very similar to that of intact Sendai virions, since the degree of fluorescence dequenching obtained with RSVE is very close to that observed with intact Sendai virions.

Treatment of intact virions with trypsin or phenyl-methyl-sulfonyl-fluoride (PMSF) has been shown to effectively inhibit their infectivity and fusogenic activity (23,24). Such treated virions are still able to bind to cell-surface receptors and to cause hemagglutination, since trypsin and PMSF specifically affect the viral fusion protein F, but not the HN glycoprotein. On the other hand, reduction with dithiothreitol (DTT) (3-5 mM) affects the viral HN glycoprotein (25). DTT-treated virions are not able to bind to and, therefore, to fuse with cell plasma membranes (25). Indeed, a very low degree of fluorescence dequenching was obtained following incubation of nonhemolytic, i.e., trypsin-, PMSF-, or DTT-treated virions, with HEG, FELC, or L cells (Table 1). These results further emphasize the view that the fluorescence dequenching observed with un-treated virions indeed reflects a process of virus-membrane fusion.

The results in Table 1 (see also Ref. 5) show that similar to intact virions, also incubation of fluroescently labeled unfusogenic (trypsin-, PMSF-, or DTT-treated) RSVE with human erythrocytes or HEG neither induced hemolysis nor resulted in a high degree of fluorescence dequench-ing. A low degree of fluorescence dequenching was also obtained when RSVE were incubated with neuraminidase-treated cells, showing that sialic-acid-containing components are needed to allow RSVE-membrane fusion. All these observations unequivocally prove that RSVE-induced hemolysis, as well as the fluorescence dequenching observed, cannot be attributed to residual traces of detergent left in the RSVE preparations but are due to the authentic activity of the viral glycoproteins. Practically the same

TABLE 1 Interaction of Fluorescently Labeled, Intact Sendai Virions or RSVE with Human Erythrocytes and Living Cultured Cells[a]

Treatment	Dequenching (%)			Hemolysis (%)
	HEG	FELC	L mouse cells	
R_{18}-labeled intact Sendai virions				
None	58	51	48	88
Trypsin	3	6	4	6
PMSF	4	8	5	8
DTT	4	4	5	8
R_{18}-labeled RSVE				
None	44	48	48	95
Trypsin	8	3	6	7
PMSF	4	6	6	8
DTT	10	6	4	8
N-NBD-PE-labeled RSVE				
None	74	49	N.D.	88
RSVE + neuraminidase-treated cells	0	6	N.D.	4

N.D., not determined.

[a]Intact Sendai virions were obtained and labeled with R_{18} exactly as described before (20). RSVE were prepared as described, following removal of the detergent by direct addition of SM-2 Bio-beads (22). RSVE were labeled either with R_{18} (20) or with N-NBD-PE as described previously (5). Human erythrocyte ghosts (HEG) were prepared and Friend erythroleukemia cells (FELC) or mouse L cells were obtained as described elsewhere (20). All other experimental conditions were as described before (20).

TABLE 2 Fusion of Viral Envelopes with Biological Membranes: Effect of the Viral Glycoproteins

Membrane vesicles bearing:	Dequenching (%)		Hemolysis (%)
	HEG	FELC	
HN-polypeptide	7	7	8
F-polypeptide	3	6	8
HN- and F-polypeptides in separate vesicles (1:1)	3	6	9
HN- and F-polypeptides in the same vesicle	38	32	68
RSVE	41	39	72

RSVE or membrane vesicles bearing purified F, HN, or both F and HN (1:1, w/w) polypeptides were prepared and labeled with N-NBD-PE as described before (5). Hemolysis was determined as described previously (20,24). All other experimental conditions were as described in Table 1.

results were obtained with RSVE preparations labeled with either R_{18} or with N-NBD-PE (Table 1).

Further support for the view that the activities observed with the RSVE preparations indeed reflect a process of virus-membrane fusion is obtained from the results summarized in Table 2. Membrane vesicles, bearing only the viral HN or F glycoprotein, are not hemolytic, and their incubation with either HEG or FELC results in a very low degree of fluorescence dequenching. Membrane vesicles bearing the viral binding protein, namely the viral HN glycoprotein, readily attach to the cell plasma membrane and agglutinate human erythrocytes. A mixture of equal amounts of HN- and F-bearing vesicles is also inactive (Table 2). Only reconstituted vesicles containing both the HN and F glycoproteins within the same envelope possess fusogenic as well as hemolytic activities.

IV. CONSTRUCTION OF TARGETED RSVE

Loaded or hybrid fusogenic RSVE may serve as an excellent vehicle for the intracellular delivery of drugs, especially of macromolecules, either in vitro or in vivo. An important characteristic of an ideal carrier should be that its binding and subsequent fusion will be limited and specific to the cells of interest, and that after intravenous injection it will possess the same features that were demonstrated following interaction with cell cultures. In this regard, it should be mentioned that loaded liposomes have been used during the last few years to deliver certain drugs (26) and, recently, also DNA molecules (27) into specific organs of whole animals. However, from these and other experiments, it became clear that after intravenous injection most of the injected liposomes are taken into cells of the reticuloendothelial system (RES) by endocytosis (28). A strategy to overcome this uptake of liposomes by the RES was to construct liposomes bearing antibodies raised against specific surface antigens (29).

The use of RSVE for the delivery of drugs and macromolecules may have an advantage over the use of phospholipid vesicles. Even if loaded RSVE will, like liposomes, associate with cells of the RES, fusion with their plasma membranes may make RSVE a highly efficient vehicle. However, it has been observed that the interaction of Sendai virions with cell cultures is greatly inhibited by the presence of serum (30). It is conceivable that certain glycoproteins present in the serum prevent virus-cell interaction, either by masking the virus binding protein (the HN glycoprotein) or by blocking cell surface sialic-acid-containing components which are known to serve as virus receptors. A similar inhibition may occur after intravenous injection, thus preventing interaction and fusion of RSVE with cells of the RES. This limitation may potentially be overcome by the use of targeted RSVE, the binding of which to cell surfaces is mediated by nonviral protein.

Construction of fusogenic targeted RSVE can be achieved by covalent attachment of specific recognition proteins, such as antibodies, hormones, or plant lectins, to the envelope membrane. Essentially two methods have been developed in our laboratory for the preparation of RSVE bearing specific binding proteins, such as polypeptide hormones on antibodies raised against cell surface membrane antigens (31,32). Anti-human-erythrocyte antibodies (anti-HE-Ab) were covalently attached to the viral glycoproteins by the use of the cross-linking reagents succinimidyl-4-(p-maleimido-phenyl)-butyrate (SMPB) and N-succinimidyl-3-(2-pyridyl-dithio)propionate (SPDP) (31). Also, insulin molecules were covalently attached to the viral envelope glycoproteins by similar techniques (31). RSVE bearing anti-HE-Ab or insulin molecules were able to bind to, but not to fuse with, virus-receptor-depleted (neuraminidase-treated) human erythrocytes or cultured cells. The same results were obtained with membrane vesicles containing only the viral fusion (F) protein (31). Fusogenic targeted RSVE were obtained only after coreconstitution of Sendai viral glycoproteins, coupled to anti-HE-Ab or insulin molecules, with untreated viral glycoproteins (31). These results clearly demonstrate that even when binding of viral particles to cell surfaces is mediated by a nonviral ligand and not by the HN glycoprotein, the presence of HN is essential to allow RSVE-membrane fusion, suggesting an active role for the viral binding protein in the process of virus-membrane fusion (31). The same conclusion was drawn from experiments using targeted RSVE that were prepared by a different method that allowed attachment of a foreign polypeptide to the virus envelope, without the need to chemically modify the viral glycoproteins. The cross-linking reagent SMPB was used to covalently attach ANTI-HE-Ab to the thiol-containing paraffin dodecanethiol (32). The complex formed, dodecanethiol-MPB(maleimidophenyl-butyrate)-antibody, was then added to a detergent solution of the viral phospholipids and glycoproteins. Removal of the detergent led to the formation of vesicles containing intact, unmodified viral glycoproteins and the dodecanethiol-MPB-antibody complex within the same membrane. Such targeted RSVE are able to bind to and fuse with neuraminidase-treated human erythrocytes. Fusion was inferred from the ability of such targeted RSVE (RSVE bearing antibodies) to induce hemolysis in virus-receptor-depleted human erythrocytes (32). Addition of the complex dodecanethiol-MPB-anti-HE-Ab to a detergent solution containing only the viral F glycoprotein resulted in the formation of vesicles that were able to bind to, but not to fuse with, the neuraminidase-treated erythrocytes. These results confirm previous observations (8,9) indicating that the viral HN glycoprotein, besides mediating binding to membrane sialic-acid-containing components, also plays an active role in the fusion (hemolytic) process.

It is conceivable that the fusogenic activity of targeted RSVE is dependent on two unrelated parameters: (1) the attachment affinity of the nonviral binding protein (antibody) to its surface receptor, and (2) the expression of the fusogenic activity of the viral glycoproteins in the presence of a nonviral binding protein. In order to study the fusogenic activity of targeted RSVE following incubation with various cell lines, it is essential to use the same binding protein, thus eliminating variations in this parameter. This can be achieved by using the interaction between trinitrobenzoyl(TNB) groups and anti-dinitrophenyl(DNP) antibodies as a system (33), which allows targeting of RSVE to almost any cell in culture (Fig. 1). TNB groups can be covalently attached to cell plasma membranes by the use of TNB-sulfonic acid, without significantly affecting cell viability (33). Anti-DNP-IgG can be incorporated into RSVE following coupling to dodecanethiol, using SMPB as a crosslinking reagent (32) (Fig. 1).

The results in Table 3 clearly show that the membrane-bound TNB and RSVE bearing anti-DNP antibodies can be used to study the activity of targeted RSVE. As shown in Table 3, RSVE are able to induce hemolysis in untreated, but not in neuraminidase-treated, human erythrocytes. However, targeted RSVE bearing anti-DNP antibodies cause a relatively high degree of hemolysis when incubated with either type of erythrocytes containing exposed TNB groups. Much less hemolysis is obtained by incubating targeted RSVE with desialized erythrocytes lacking TNB groups or by

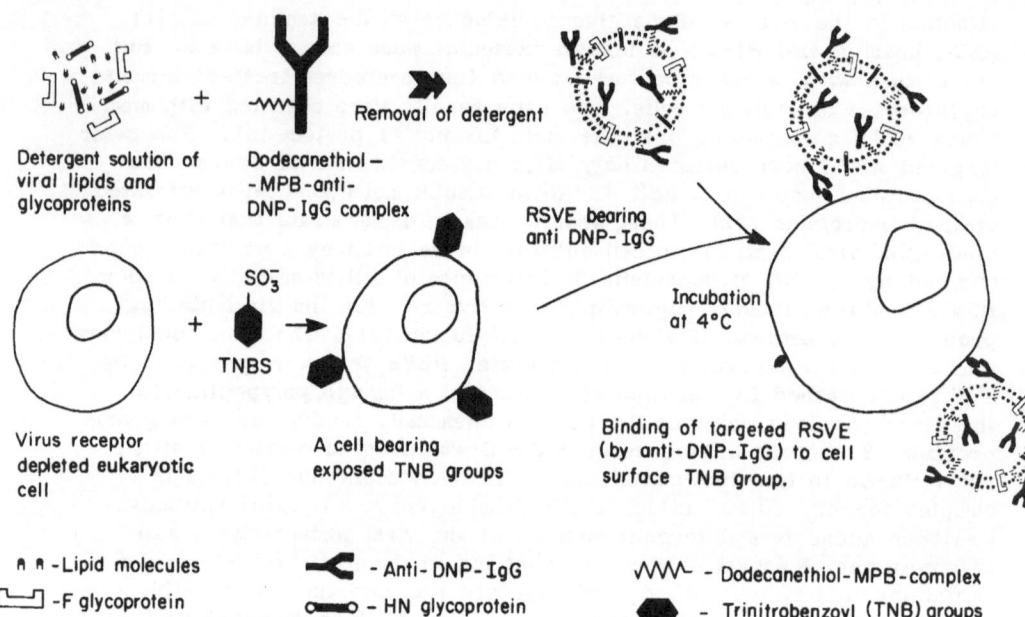

Removal of detergent

Detergent solution of
viral lipids and
glycoproteins

Dodecanethiol –
- MPB-anti-
DNP- IgG complex

RSVE bearing
anti DNP-IgG

Virus receptor
depleted eukaryotic
cell

TNBS

A cell bearing
exposed TNB groups

Incubation
at 4°C

Binding of targeted RSVE
(by anti-DNP-IgG) to cell
surface TNB group.

ᴧ ᴧ -Lipid molecules

-F glycoprotein

- Anti- DNP- IgG

- HN glycoprotein

- Dodecanethiol-MPB-complex

- Trinitrobenzoyl (TNB) groups

FIGURE 1   Preparation of targeted, fusogenic RSVE: schematic representation. (a) RSVE are reconstituted from supernatant of Triton X-100 (4%) solution of Sendai virions (7,22). (b) Preparation of dodecanethiol-MPB-IgG: The IgG fraction of either normal human serum or rabbit-anti-DNP antiserum is purified on DEAE-cellulose column (31). Dodecanethiol-MPB-IgG (1:1:0.2 initial molar ratio) complexes are prepared as described in Ref. 32, except for the presence of 4% (w/v) Triton X-100 in the antibody solution and throughout all the subsequent steps in order to prevent formation of dodecanethiol micelles. (c) Incorporation of the dodecanethiol-MPB-IgG complex into RSVE: A detergent solution of dodecanethiol-MPB-IgG complexes (100 µl in 4% (w/v) Triton X-100) formed from 6 nmol of IgG, or as desired, is mixed with 50 µl of Triton X-100 (10%, w/v) solution, obtained following solubilization of 2.5 mg of intact Sendai virions (39). Reconstitution is performed by removal of the detergent (22,32). The vesicles bearing viral components and dodecanethiol-MPB-IgG complexes are washed with 20 volumes of 150 mM NaCl, 20 mM Tricine (pH 7.4) and resuspended in the same buffer, to give a protein concentration of 2 mg/ml. (d) Preparation of trinitrobenzoyl (TNB) human erythrocytes: Intact or neuraminidase-treated human erythrocytes (type O, RH⁻) are washed three times in NaCl/Tricine, resuspended in the same buffer to a concentration of 10% (v/v), and incubated with an equal volume of a solution containing 4 mM of TNB-sulfonate (TNBS) for 30 min at 37°C. After incubation, the cells are washed three times with 5 mM of glycyl-glycine in NaCl/Tricine, followed by two washings with NaCl/Tricine. Binding of RSVE bearing anti-DNP-IgG (5 µg protein) to TNB human erythrocytes (250 µl) is performed by incubation for 10 min at 4°C.

TABLE 3 RSVE-Induced Hemolysis in Desialized Human Erythrocytes: Effect of Targeting

Viral envelopes	Hemolysis (%)		
	Untreated HE	Desialized HE	TNB-desialized HE
RSVE	80	8	10
RSVE-IgG	78	16	16
RSVE-anti-DNP-IgG	82	5	45

Control or targeted RSVE (10 µg) were incubated for 20 min at 37°C with 250 µl of 2-3% (v/v) of control, desialized, or TNB-desialized human erythrocytes (HE). At the end of the incubation period, hemolysis was determined, as described before (20,24). Preparation of RSVE and TNB-labeled erythrocytes was as described in Figure 1. RSVE-anti-DNP-IgG and RSVE-IgG are RSVE bearing anti-DNP-IgG or human IgG, respectively.

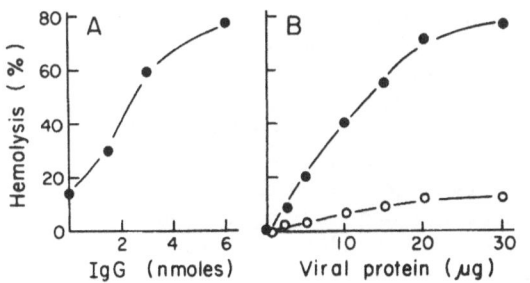

FIGURE 2 Induction of hemolysis in desialized human erythrocytes by targeted RSVE. (A) Effect of increasing amounts of anti-DNP-IgG present in the coreconstitution mixture: Targeted RSVE were prepared by core-constitution with dodecanethiol-MPB-anti-DNP-IgG complexes formed from the indicated amounts of antibodies, as described in Figure 1. RSVE (12 µg) were incubated for 40 min at 37°C with desialized TNB human erythrocytes (250 µl of 2-3%, v/v). Subsequently, hemolysis was determined from the absorbance of the supernatants at 540 nm, after dilution and sedimentation of the cells; total hemolysis was estimated by addition of Triton X-100. (B) The indicated amounts of RSVE bearing (● - ●) or lacking (○ - ○) anti-DNP-IgG were incubated with 250 µl of 2-3% (v/v) desialized TNB human erythrocytes, for 40 min at 37°C. The hemolytic activity was determined as described above.

TABLE 4   Fusion of Fluorescent, Targeted RSVE with Membranes of TNB-Desialized Human Erythrocytes Ghosts

Fluorescent viral envelopes	Pyrene DQ (%)	
	TNB-HEG	TNB-desialized-HEG
RSVE	65	10
RSVE$_{PMSF}$	6	12
RSVE$_{4°C}$	8	4
RSVE-anti-DNP-IgG	62	27
RSVE$_{PMSF}$-anti-DNP-IgG	8	10
RSVE$_{4°C}$-anti-DNP-IgG	3	7

Human erythrocytes were desialized and then incubated with TNB, as described in Figure 1. Human erythrocyte ghosts (HEG) were prepared from either TNB-treated or desialized TNB-treated human erythrocytes as previously described (5). Fluorescent RSVE were prepared by including 1-palmitoyl-2(10-pyrenyl-decanoyl)-sn-glycero-3-phosphorylcholine (5 mol% of RSVE phospholipids) into the reconstitution mixture (see legend to Fig. 1). Under these conditions, the pyrene fluorescene is self-quenched and its dequenching is proportional to the decrease of the fluorochrome surface density (not shown). To induce pyrene dequenching (DQ), fluorescent-targeted (bearing anti-DNP-IgG) or nontargeted RSVE (2 µg) were incubated with TNB-HEG or TNB-desialized-HEG (200 µg) for 40 min at 37°C. After incubation, the degree of fluorescence dequenching was estimated. Fusogenic activity of RSVE was inhibited either by pretreatment with phenyl-methyl-sulfonyl-fluoride (PMSF, 7 mM, 30 min 37°C) (RSVE$_{PMSF}$) or by incubating with HEG at 4°C (RSVE$_{4°C}$).

incubation of RSVE, bearing nonspecific human IgG, with neuraminidase-treated erythrocytes lacking or bearing TNB groups (Table 3). These results confirm previous observations (32) showing that insertion of the complex dodecanethiol-MPB-IgG into the viral envelope does not affect its hemolytic (fusogenic) activity. It is noteworthy, as already mentioned, that virus-induced hemolysis reflects a process of virus-membrane fusion (3). Hemolysis induced by targeted RSVE in desialized TNB-erythrocytes is dependent on both the amount of anti-DPN-IgG added to the reconstitution system (anti-DNP-IgG/virus) and the amount of targeted RSVE added to the erythrocyte suspension (Fig. 2, a and b).

A direct quantitative determination of the fusogenic activity of targeted RSVE can be performed by energy transfer or fluorescence dequenching methods. However, the fluorescent probe for labeling the targeted RSVE should be carefully chosen to prevent possible cross-reactions with a nonviral ligand used. The use of N-NBD-PE was therefore avoided because, in preliminary experiments, it has been observed that the addition of dodecanethiol-MPB-anti-DNP antibodies, together with N-NBD-PE molecules, to the detergent solution of the viral glycoprotein impedes the formation of targeted RSVE. A strong association probably occurs because of interaction between the anti-DNP antibody and the aromatic group of the N-

NBD-PE molecules. Hence, the fluorescent pyrene-phosphatidylcholine, which has been used to follow liposome-liposome fusion (34) or fusion between Sendai virus and liposomes (35), was used in these experiments.

As shown in Table 4, incubation of fluorescently labeled, nonmodified or targeted RSVE with untreated HEG results in a high degree of fluorescence dequenching. As expected, much less fluorescence dequenching was observed with PMSF-treated RSVE or after incubation at 4°C. A low degree of fluorescence dequenching was also observed when fluorescent RSVE were incubated with neuraminidase-treated HEG, clearly showing that binding of fusogenic, fluorescent RSVE is mediated by cell-surface virus receptors. The results in Table 4 also show that an increase in the fluorescence dequenching is observed when targeted RSVE (bearing anti-DNP antibodies) are incubated with TNB-HEG.

From a comparison of the degree of fluorescence dequenching observed after incubation of targeted RSVE with neuraminidase-treated TNB-HEG with that obtained by incubation of nonmodified envelopes and untreated TNB-HEG, it appears that the fusogenic activity of the targeted RSVE is about half of that observed with nonmodified RSVE. The results in Table 4 also show that the fusogenic activity, based on the observed degree of fluorescence dequenching, of targeted RSVE is about three times higher than that of nontargeted RSVE when incubated with desialized HEG.

V. THE USE OF TARGETED, LOADED VIRAL ENVELOPES AS A BIOLOGICAL CARRIER

Recent work in our laboratory has clearly shown that, besides specific antibodies, polypeptide hormones such as insulin may also be used for the construction of targeted, reconstituted envelopes (36). RSVE bearing insulin molecules were shown to be able to interact with neuraminidase-treated cell cultures (Table 5 and Ref. 36). Based on these observations, experiments were performed to demonstrate the ability of targeted RSVE to microinject their content into virus-receptor-depleted cells. SV_{40} DNA was used to study the fusion-mediated microinjection of specific genes, while fragment A of the plant toxin ricin was a model system to demonstrate microinjection of functional polypeptides.

The results in Table 5 show that very little, if any, binding is observed when ^{125}I-labeled RSVE, empty or loaded with SV_{40}, are incubated with neuraminidase-treated hepatoma tissue culture cells (HTC. Consequently, no T antigen is detected in the desialized cells following incubation with either empty or loaded RSVE (Table 5). On the other hand, loaded RSVE, bearing insulin molecules, are able to attach to desialized HTC as well as to induce the synthesis of T antigen (Table 5). As can be inferred from our previous work (36) and the results presented in Table 5, binding of RSVE is mediated by the virus-attached insulin molecules, whereas fusion and microinjection are promoted by the fusogenic activity of the viral envelope glycoproteins. Loaded RSVE, bearing bovine serum albumin (BSA) molecules instead of insulin, are inactive (Table 5, system 4). Addition of free insulin molecules at the end of the incubation period results in very little inhibition of the binding of RSVE-insulin and of the appearance of T antigen (Table 5, system 5). Conversely, when free insulin molecules are added prior to the incubation with RSVE-insulin, containing SV_{40} DNA, negligible binding is observed and no synthesis of the SV_{40} T antigen is

TABLE 5   Fusion-Mediated Microinjection of SV_{40} DNA into Virus-Receptor-Depleted Cells: The Use of Insulin Molecules to Mediate Binding of RSVE to Cell Membranes

Viral envelopes	^{125}I-RSVE bound (%)	T-antigen positive cells (%)
1. RSVE	7	0
2. RSVE-(SV_{40} DNA)	8	0
3. RSVE-(SV_{40} DNA)-insulin	32	41
4. RSVE-(SV_{40} DNA)-BSA	5	0
5. RSVE-(SV_{40} DNA)-insulin + insulin	25	30
6. Insulin + RSVE-(SV_{40} DNA)-insulin	5	0
7. Trypsinized RSVE-(SV_{40} DNA)-insulin	33	0.1
8. PMSF-RSVE-(SV_{40} DNA)-insulin	35	0.2
9. PMSF-RSVE-(SV_{40} DNA)-insulin + insulin	10	0.2

RSVE were prepared, loaded with SV_{40} DNA, and radiolabeled with $Na^{125}I$. Insulin and GSA were covalently attached to the viral glycoproteins to obtain RSVE-insulin and RSVE-BSA, respectively. Hepatoma tissue culture cells (HTC) were treated with neuraminidase. For inactivation of the viral fusogenic activity, intact Sendai virions (before RSVE preparation) were treated with trypsin or PMSF. For determination of virus-cell binding, 40 μg of ^{125}I-RSVE-(SV_{40} DNA) was incubated with 10^6 neuraminidase-treated HTC. After 45 min of incubation at 37°C, radioactive material associated with the cells was determined. For induction of T-antigen synthesis, 10 μg of RSVE loaded with SV_{40} DNA and bearing insulin (RSVE-(SV_{40} DNA)-Insulin) or bovine serum albumin (RSVE-(SV_{40} DNA)-BSA) was incubated for 30 min at 37°C with a monolayer containing about 5×10^4 HTC. Appearance of T-antigen was determined after 24-48 h of incubation. For experimental details, see Ref. 36.

detected (Table 5, system 6). These results clearly show that excess free insulin effectively competes with the binding of (and microinjection by) RSVE-insulin, if added before but not after the incubation of cells with the targeted RSVE. Further support to the view that binding and fusion-mediated microinjection are caused by insulin molecules and the viral glyco-proteins, respectively, was obtained from experiments using unfusogenic RSVE (Table 5, systems 8 and 9). As can be seen, a high degree of bind-ing but no induction of SV_{40} T antigen is obtained following incubation with PMSF-treated, loaded RSVE-insulin with desialized HTC (Table 5, system 8).

As already mentioned, PMSF specifically inactivates the viral fusion protein (24). Contrary to the observation with fusogenic targeted RSVE (Table 5, system 5), addition of free insulin at the end of the incubation to cells incubated with PMSF-treated RSVE-insulin results in removal of most of the bound viral envelopes (Table 5, system 9). This is expected, since PMSF-treated RSVE-insulin are bound to, but not fused with HTC plasma membranes.

Essentially the same results are obtained by the use of targeted RSVE, bearing insulin molecules, loaded with ricin A (36). Ricin is a potent toxin composed of two subunits, A and B (37). The B subunit mediates binding of the whole toxin to cell surface receptors, while the A subunit, after being transferred to the cell cytoplasm, causes inhibition of protein synthesis and eventually cell death. The fragment A (ricin A), by itself, is harmless, unless it is microinjected into the intracellular space of living cells (37). Experiments in our laboratory (36) have shown that incubation of HTC with RSVE containing enclosed ricin A causes an almost complete inhibition of protein synthesis in the recipient cultured cells (36). No inhibition of protein synthesis is observed when empty RSVE are incubated with HTC in the presence of externally added ricin A. Protein synthesis could be inhibited in desialized HTC, following incubation with targeted loaded RSVE, that is, RSVE containing ricin A and bearing covalently attached insulin molecules. These experiments (36) demonstrate the effective use of targeted RSVE as a biological carrier for microinjection of macromolecules into virus-receptor-depleted cells.

VI. DISCUSSION AND CONCLUSIONS

The development of an appropriate and efficient vehicle for the microinjection of polypeptides and polynucleotides (specific genes) into the cytoplasm of living cells is of inestimable value. Such carriers may allow the delivery of their content into specific cells of a living organism. Indeed, impermeable membrane vesicles have been used extensively as a biological carrier of both polypeptides and polynucleotides. The methodology of vesicle-mediated microinjection is based in all cases on two steps: enclosure of macromolecules within the appropriate vesicle and then interaction or fusion of the loaded vesicle with either the cell plasma membrane or membranes of intracellular organelles. Basically three kinds of vesicles have been used as a carrier of macromolecules: erythrocyte ghosts, liposomes, and reconstituted virus envelopes.

A. Erythrocyte-Mediated Microinjection

Erythrocyte-mediated microinjection is based on the fusion of loaded red blood cell ghosts with cultured cells (38,39). Human erythrocytes have been used as a carrier of macromolecules, since they are easy to obtain and readily fuse with cultured animal cells induced by either Sendai virus or by nonviral fusogenic agents, such as poly(ethylene glycol) (38,39). In addition, human erythrocytes possess a relatively large internal volume, thus allowing the trapping of substantial amounts of macromolecules. Macromolecules are trapped within human erythrocyte ghosts during hypotonic hemolysis. Pores form during this process in the erythrocyte membrane, allowing the entry of macromolecules present in the incubation medium. Following resealing, the loaded erythrocyte ghosts can be fused with cultured cells (38,39; see also Chapter 33).

Various experiments (39) have shown that about 10^6-10^7 copies of polypeptides of a molecular weight between 25,000 and 50,000 Daltons can be enclosed within a single human erythrocyte ghost. However, since the size of the pores formed in the erythrocyte membrane during the hypotonic hemolysis is limited, it appears that only polypeptides with a molecular weight lower than 200,000 Daltons can be encapsulated. Moreover, the

efficiency of trapping is highly dependent on the size of the macromolecule
to be enclosed, being high with polypeptides of 25,000-50,000 Daltons
but becoming very low with those of 100,000 Daltons and above. This is
probably the reason why the erythrocyte-mediated microinjection technique
was used extensively for the delivery of polypeptides or RNA, but rarely
for gene transfer (38-40). Recently, however, DNA molecules have been
enclosed within erythrocyte ghosts by freezing and thawing (40). The
trapping efficiency for DNA molecules appeared to be much lower than
that observed for polypeptides. About 2-3% of the added DNA was found
to be associated with erythrocyte ghosts either enclosed within or adsorbed
externally to the erythrocyte membrane (40).

Erythrocyte-ghost-mediated microinjection of polypeptides has been
used extensively to study the mechanism of intracellular protein degradation
(41). Radioactive polypeptides of various sizes are enclosed within the
erythrocyte ghosts, which are then fused with different cell lines grown
in culture (41,42). Degradation of the injected polypeptides is then monitored. The ability to introduce individually labeled proteins into large
numbers of cells permits accurate measurements of the degradation rate,
localization, and modification of the injected proteins.

Erythrocyte-mediated microinjection has also been used for injection
of specific antibodies to probe selective properties of various intracellular
systems (43). In addition, the method has been employed for injection
of nuclear proteins to study their behavior and the permeability of the
nuclear membrane (44). In summary, erythrocyte ghosts have been used
as carrier for polypeptides as well as polynucleotides such as m-RNA (45),
t-RNA (46), and, to a lesser extent, DNA (40).

B. Liposomes as Biological Carrier

Macromolecules can also be enclosed within phospholipid vesicles (liposomes).
Many methods are available for construction of liposomes with various
characteristics such as composition of phospholipids, surface charge, permeability, and size (47-49). However, it seems that the large unilamellar
vesicles are the best suited for delivery of macromolecules, as they exhibit
a high ratio of internal aqueous space per unit lipid. Loaded liposomes
have been used for the delivery of polypeptides as well as of DNA and
RNA into a variety of cultured cells (50). Liposomes loaded with DNA
have been used for the delivery of specific genes into mammalian cells
as well as plant protoplasts (50). However, for the efficient delivery of
the liposome content into living cells, the addition of facilitators, such
as dimethylsulfoxide (DMSO), glycerol, or poly(ethylene glycol), is necessary (50,51). Glycerol has been found to be the most effective, increasing
the delivery of SV_{40} DNA into cultured cells 100 to 200-fold (51). The
detailed mechanism by which the content of the liposome is released is
still unknown. It appears, however, that liposomes, like other particulate
material, are taken up by endocytosis (28). Processing of the liposomes
by intralysosomal enzymes may promote the release of their content.
Recently, in order to stimulate the release of the liposome content in the
low-pH environment of the endosomes or lysosomes, vesicles composed
of phosphatidylethanolamine (PE) and oleic acid (OA) have been constructed
(52; see also Chapter 31). These liposomes have been shown to be pH
sensitive, i.e., to release their content at low pH values, and to be more
efficient for the functional delivery of DNA molecules.

As mentioned above, following intravenous injection, liposomes are taken up mostly by the macrophages of the RES. Extent and rate of uptake by particular tissues of the RES system depend on the size, surface properties, and lipid composition of the vesicles. Coupling of specific antibodies or other molecules results in the formation of targeted liposomes, whose distribution among the cells of the RES may be different from that of untargeted liposomes (28,53).

In addition to transfer of specific drugs into the cells of the RES (26), liposomes have been recently used as carrier for gene transfer in vivo (27). Intravenous injection into rats of large liposomes carrying a recombinant plasmid, containing the rat preproinsulin I gene, resulted in a transient expression of the gene in the liver and spleen of the recipient animals. Incorporation of glycolipid with a terminal nonreduced β-galactose residue into the liposome bilayer resulted in the formation of targeted liposomes. Liposomes carrying the glycolipid interacted in vivo more readily with hepatocytes, probably owing to the interaction of the β-galactose vesicles with the galactose-specific lectin of the hepatocytes (27). These and other studies clearly show the high potential offered by the use of liposomes as a biological carrier for in vivo research.

C. Reconstituted Viral Envelopes

Macromolecules such as polypeptides or DNA can be enclosed within reconstituted viral envelopes when added during the reconstitution process. Similarly, membrane components can be inserted into the viral envelope. Fusion of loaded or hybrid RSVE, which is promoted by the viral envelope glycoproteins, results in microinjection or transfer of the RSVE content into the cell cytoplasm or cell plasma membranes, respectively (7). This method, using RSVE as a biological carrier, is based on the observation that penetration of Sendai virions (which belong to the paramyxovirus group) into living cells occurs by fusion of the viral envelope with the cell plasma membrane (6). However, most enveloped viruses penetrate living cells through a different route. Enveloped viruses belonging to the orthomyxovirus or togavirus groups, such as influenza or Semliki Forest virus (SFV), are taken up into eukaryotic cells, similar to liposomes, by an endocytotic process (1). Thus, the viral nucleocapsid is introduced into the cell cytoplasm following fusion of the viral envelope with the inner surface of the endosomal membrane. Virus-membrane fusion is activated by the intraendosomal low pH (1). Use of envelopes of these viruses as a biological carrier may be of advantage over use of reconstituted envelopes prepared from Sendai virions. The fact that virions such as influenza or SFV are taken into cells by endocytosis may allow the use of high amounts of viral envelopes without induction of cell-cell fusion or lysis. Furthermore, fusion with the endosome at low pH may avoid the release of the content of the viral envelopes or cells into the external medium, which may occur during fusion of the plasma membrane. Thus, reconstituted envelopes of influenza or SFV may potentially serve as a powerful carrier for the delivery of macromolecules into mammalian cells.

The use of viral envelopes for the delivery of drugs and, especially, of macromolecules may have an important advantage over that of phospholipid vesicles. Fusion of viral envelopes with recipient cells in vivo, similar to that observed in vitro, may make them highly efficient biological carriers, enabling the use of low doses of biological material. It has recently been observed that following intravenous injection, most of the injected RSVE accumulate in the liver within a short period of time (54). Results in our laboratory have shown that after the intravenous injection of radiolabeled RSVE into mice, about 50% of the radioactive material is found associated with the liver as soon as 10 min after injection (54). However, no information is as yet available about the fusogenic properties of the RSVE in vivo.

Endocytosis of RSVE by cells of the RES may result in irreversible inhibition of their fusogenic activity because of intraendosomal or lysosomal low pH environment. The same environment, on the other hand, may activate the fusogenic proteins of viruses such as influenza or SFV, thus leading to their fusion with the endosomal membrane, resulting in the release of their content into the cell cytoplasm. Hence. it will be of interest to study in the future the ability of reconstituted influenza or SFV envelopes to microinject their content, first into cultured cells and, eventually, into cells of whole animals.

ACKNOWLEDGMENTS

This work was supported by grants from the International Genetic Sciences Partnership, from the National Council for Research and Development, Jerusalem, Israel, and from the Gesellschaft für Strahlung and Umweltforschung, Munich, Federal Republic of Germany.

REFERENCES

1. White, J., Kielian, M., and Helenius, A. (1983). Membrane fusion proteins of enveloped animal viruses. *Q. Rev. Biophys.* 16:151-195.
2. Hosaka, Y., and Shimizu, K. (1977). Cell fusion by Sendai virus. In: *Cell Surface Reviews*, Vol. 2 (G. Poste and G. L. Nicolson, eds.). North Holland, Amsterdam, pp. 129-155.
3. Maeda, Y., Kim, J., Koseki, I., Mekada, E., Shiokawa, Y., and Okada, Y. (1977). Modification of cell membranes with viral envelopes during fusion of cells with HVJ (Sendai virus). *Exp. Cell Res.* 108: 95-106.
4. Hoekstra, D., De Boer, T., Klappe, K., and Wilschut, J. (1984). Fluorescence method for measuring the kinetics of fusion between biological membranes. *Biochemistry* 23:5675-5681.
5. Chejanovsky, N., and Loyter, A. (1985). Fusion between Sendai virus envelopes and biological membranes. *J. Biol. Chem.* 260:7911-7919.
6. Poste, G., and Pasternak, C. A. (1978). Virus-induced cell fusion. In: *Cell Surface Reviews*, Vol. 5 (G. Poste and G. L. Nicolson, eds.). Elsevier, North Holland, Amsterdam, pp. 305-367.
7. Loyter, A., and Volsky, D. J. (1982). Reconstituted Sendai virus envelopes as carriers for the introduction of biological material into animal cells. In: *Cell Surface Reviews*, Vol. 8 (G. Poste and G. L. Nicolson, eds.). North Holland, Amsterdam, pp. 215-266.
8. Miura, N., Uchida, T., and Okada, Y. (1982). HVJ (Sendai virus)-induced envelope fusion and cell fusion are blocked by monoclonal anti-HN protein antibody that does not inhibit hemagglutination activity of HVJ. *Exp. Cell Res.* 141:409-420.
9. Citovsky, V., Yanai, P., and Loyter, A. (1986). The use of circular dichroism to study conformational changes induced in Sendai virus envelope glycoproteins. *J. Biol. Chem.* 261:2235-2239.

10. Hosaka, Y., and Shimizu, K. (1972). Artificial assembly of envelope particles of HVJ (Sendai virus). I. Assembly of hemolytic and fusion factors from envelopes solubilized with Nonidet P-40. *Virology* 49:627-639.

11. Vainstein, A., Razin, A., Graessmann, A., and Loyter, A. (1983). Fusogenic reconstituted Sendai virus envelopes as a vehicle for introducing DNA into viable mammalian cells. *Methods Enzymol.* 101:492-512.

12. Volsky, D. J., Gross, T., Sinangil, F., Kuszynski, C., Bartzatt, R., Dambaugh, T., and Keiff, E. (1984). Expression of Epstein-Barr virus (EBV) DNA and cloned DNA fragments in human lymphocytes following Sendai virus envelope-mediated gene transfer. *Proc. Natl. Acad. Sci. USA* 81:5926-5930.

13. Arad, G., Hershkovitz, M., Panet, A., and Loyter, A. (1986). Use of reconstituted Sendai virus envelopes for fusion-mediated microinjection of double-stranded RNA: Inhibition of protein synthesis in interferon-treated cells. *Biochim. Biophys. Acta* 859:88-94.

14. Bachrach, U., Ash, I., Abu-Elheiga, L. A., Hershkovitz, M., and Loyter, A. (1986). Microinjection of amine- and diamine oxidases into cultured cells using Sendai virus envelopes. *J. Cell Physiol.* 131:92-98.

15. Tabor, C. W., and Tabor, H. (1984). Polyamines. *Annu. Rev. Biochem.* 53:749-790.

16. Volsky, D. J., Cabantchik, Z. I., Beigel, M., and Loyter, A. (1979). Implantation of isolated human erythrocyte anion channel into plasma membranes of Friend erythroleukemic cells. *Proc. Natl. Acad. Sci. USA* 76:5440-5444.

17. Mazurek, N., Bashkin, A., Loyter, A., and Pecht, I. (1983). Restoration of Ca^{2+} influx and degranulation capacity of variant RBL-2H3 cells upon implantation of isolated cromolyn binding protein. *Proc. Natl. Acad. Sci. USA* 80:6014-6018.

18. Loyter, A., Vainstein, A., Graessmann, M., and Graessmann, A. (1983). Fusion-mediated injection of SV_{40}-DNA. *Exp. Cell Res.* 143:415-425.

19. Prujanski-Jakobovitz, A., Volsky, D. J., Loyter, A., and Sharon, N. (1981). Alteration of lymphocyte surface properties by insertion of foreign functional plasma membrane components. *Proc. Natl. Acad. Sci. USA* 77:7247-7251.

20. Citovsky, V., Blumenthal, R., and Loyter, A. (1985). Fusion of Sendai virions with phosphatidylcholine-cholesterol liposomes reflects the viral activity required for fusion with biological membranes. *FEBS Lett.* 193:135-140.

21. Chejanovsky, A., Henis, Y. I., and Loyter, A. (1986). Fusion of fluorescently labeled Sendai virus envelopes with living cultured cells as monitored by fluorescence dequenching. *Exp. Cell Res.* 164:353-365.

22. Vainstein, A., Hershkovitz, M., Israel, S., Rabin, S., and Loyter, A. (1984). A new method for reconstitution of highly fusogenic Sendai virus envelopes. *Biochim. Biophys. Acta* 181:773-781.

23. Ozawa, M., Asano, A., and Okada, Y. (1979). Biological activities of glycoproteins of HVJ (Sendai virus) studied by reconstitution of hybrid envelope and by Concanavalin-A-mediated binding: A new function of HANA protein and structural requirements for F protein hemolysis. *Virology* 99:197-202.

24. Israel, S., Ginsberg, D., Laster, Y., Zakai, N., Milner, Y., and Loyter, A. (1983). A possible involvement of virus-associated protease in the fusion of Sendai virus envelopes with human erythrocytes. *Biochem. Biophys. Acta* 732:335-346.

25. Tomasi, M., and Loyter, A. (1981). Selective extraction of biologically active F-glycoprotein from dithiothreitol reduced Sendai virus particles. *FEBS Lett.* 131:381–385.

26. Alving, C. R., Glenn, M., and Swartz, A., Jr. (1984). Preparation of liposomes for use as drug carriers in the treatment of Leishmaniasis. In: *Liposome Technology*, Vol. II (G. Gregoriadis, ed.). CRC Press, New York, pp. 65–68.

27. Nicolau, C., Legrand, A., and Soriano, P. (1984). Liposomes for gene transfer and expression in vivo. In: *Cell Fusion* (CIBA Foundation Symp. 103). Pitman, London, pp. 254–267.

28. Gregoriadis, G. (1981). Targeting of drugs: Implications in medicine. *Lancet* 2:241–247.

29. Wolff, B., and Gregoriadis, G. (1984). The use of monoclonal anti-$Thy_1 IgG_1$ for the targeting of liposomes to AKR-A cells in vitro and in vivo. *Biochim. Biophys. Acta* 802:259–273.

30. Okada, Y. (1969). Factors in fusion of cells by HVJ. In: *Current Topics in Microbiology and Immunology*, Vol. 48. Springer Verlag, Berlin, pp. 102–128.

31. Gitman, A. G., Kahane, I., and Loyter, A. (1985). Use of virus attached antibodies or insulin molecules to mediate fusion between Sendai virus envelopes and neuraminidase-treated cells. *Biochemistry* 24:2762–2768.

32. Gitman, A. G., and Loyter, A. (1984). Construction of fusogenic vesicles bearing specific antibodies. *J. Biol. Chem.* 259:9813–9820.

33. Mangros, M., Bischoff, D., Dusch, M., and Oth, D. (1983). Differential binding of 2,4,6-trinitrobenzene sulfonate (TNBS) on RDM-4 lymphoma cells in vitro, as a function of the culture condition. *IRCS Med. Sci.* 11:989–990.

34. Owen, C. S. (1980). A membrane-bound fluorescent probe to detect phospholipid vesicle-cell fusion. *J. Membr. Biol.* 54:13–20.

35. Amselem, S., Barenholz, Y., Loyter, A., Nir, S., and Lichtenberg, D. (1986). Fusion of Sendai virus with negatively-charged liposomes as studied by pyrene-labeled phospholipid liposomes. *Biochim. Biophys. Acta* 860:301–313.

36. Gitman, A. G., Graessmann, A., and Loyter, A. (1985). Targeting of loaded Sendai virus envelopes by covalently attached insulin molecules to virus receptor-depleted cells: Fusion-mediated microinjection of ricin A and Simian virus 40-DNA. *Proc. Natl. Acad. Sci. USA* 82:7309–7313.

37. Refsnes, K., Olsnes, S., and Pihl, A. (1974). On the toxic proteins abrin and ricin. *J. Biol. Chem.* 249:3557–3562.

38. Kulka, R. G., and Loyter, A. (1979). The use of fusion methods for the microinjection of animal cells. In: *Current Topics in Membranes and Transport*, Vol. 12 (F. Bronner and A. Kleinzeller, eds.). Academic Press, New York, pp. 365–430.

39. Schlegel, R. A., and Rechsteiner, M. C. (1986). Erythrocyte-mediated transfer: Methods. In: *Microinjection and Organelle Transplantation Techniques* (J. E. Celis, A. Graessmann, and A. Loyter, eds.). Academic Press, New York, pp. 66–87.

40. Iino, T., Furusawa, M., Furasawa, I., and Obinata, M. (1983). Transformation of L-cells with virus thymidine kinase genes introduced by red cell-mediated microinjection. *Exp. Cell Res.* 148:475–450.

41. Freikopf-Cassel, A., and Kulka, R. G. (1981). Regulation of the degradation of ^{125}I-glutamine synthetase introduced into cultured hepatoma cells by erythrocyte ghost-mediated injection. *FEBS Lett.* 128:63–66.

42. Rechsteiner, M. C., and Schlegel, R. A. (1986). Erythrocyte-mediated transfer: Applications. In: *Microinjection and Organelle Transplantation Techniques* (J. E. Celis, A. Graessmann, and A. Loyter, eds.). Academic Press, New York, pp. 89-116.

43. Smith, J. H., Subbarao, M. N., and Eliceiri, G. L. (1983). Small nuclear RNAs and translation. *J. Cell Physiol.* 114:1-6.

44. Wu, L., Rechsteiner, M. C., and Kuehl, L. (1981). Comparative studies on microinjected high-mobility-group chromosomal proteins, HMG1 and HMG2. *J. Cell Biol.* 91:488-496.

45. Boogaard, C., and Dixon, G. H. (1983). Red cell ghost-mediated microinjection of RNA into HeLa cells. *Exp. Cell Res.* 143:175-190.

46. Kaltoft, K., Zeuthen, J., Engbaek, F., Piper, P. W., and Celis, J. E. (1976). Transfer of tRNAs to somatic cells mediated by Sendai virus-induced fusion. *Proc. Natl. Acad. Sci. USA* 73:2793-2796.

47. Szoka, F., and Papahadjopoulos, D. (1980). Comparative properties and methods of preparation of lipid vesicles (liposomes). *Annu. Rev. Biophys. Bioeng.* 9:467-508.

48. Poste, G., Kirsch, R., and Koestler, T. (1984). The challenge of liposome targeting in vivo. In: *Liposome Technology*, Vol. III (G. Gregoriadis, ed.). CRC Press, New York, pp. 1-28.

49. Szoka, F., and Papahadjopoulos, D. (1978). Procedure for preparation of liposomes with large internal aqueous space and high capture by reverse-phase evaporation. *Proc. Natl. Acad. Sci. USA* 75:4194-4198.

50. Straubinger, R. M., and Papahadjopoulos, D. (1986). Liposomes for microinjection of molecules into cells. In: *Microinjection and Organelle Transplantation* (J. E. Celis, A. Graessmann, and A. Loyter, eds.). Academic Press, New York, pp. 117-134.

51. Straubinger, R. M., and Papahadjopoulos, D. (1983). Liposomes as carriers for intracellular delivery of nucleic acids. *Methods Enzymol.* 101:512-527.

52. Straubinger, R. M., Düzgünes, N., and Papahadjopoulos, D. (1985). pH-sensitive liposomes mediate cytoplasmic delivery of encapsulated macromolecules. *FEBS Lett.* 179:149-154.

53. Gregoriadis, G., Senior, J., and Wolff, B. (1986). Liposomes in drug targeting. In: *Microinjection and Organelle Transplantation* (J. E. Celis, A. Graessmann, and A. Loyter, eds.). Academic Press, New York, pp. 135-155.

54. A. Loyter, A. G. Gitman, N. Chejanovsky, and O. Nussbaum. (1986). Fusion-mediated microinjection of macromolecules into living cultured cells by "targeted" Sendai virus envelopes. In: *Microinjection and Organelle Transplantation* (J. E. Celis, A. Gaesssmann, and A. Loyter, eds.). Academic Press, New York, pp. 179-197.

33

Enveloped Viruses as a Tool in the Study of Lipid Organization in the Plasma Membrane of Epithelial Cells

GERRIT van MEER

University of Utrecht, Utrecht, The Netherlands

I. INTRODUCTION

Epithelial cells are organized as single layers of cells that separate "external" from "internal" compartments in the organism. As a consequence, each cell faces two different environments, the external milieu on the apical side and the mesenchymal space and blood supply on the serosal or basolateral side. In order to maintain a tight barrier, the individual cells in the cell sheet are connected to each other by specialized junctions (1,1a). One of these, the zonula occludens or tight junction, separates the external milieu on the apical side from the basolateral tissue by forming a continuous belt of sealing contacts around the apex of each cell.

Epithelial cells perform vectorial functions that are reflected by the specialized organization of their apical and basolateral cell surfaces (2). The two surfaces are distinguishable both morpholobically and biochemically. The apical surface is covered with microvilli, while the basolateral membrane is heavily folded (3). The two plasma membrane domains also display unique protein compositions (2). The tight junction may constitute the boundary between the two domains and probably functions as a diffusion barrier to prevent mixing of diffusible membrane proteins (4).

The present chapter focuses on the lipid distribution in the plasma membrane of epithelial cells and examines the role of the tight junction as a barrier to the lateral diffusion of lipid molecules. A method is discussed which permits direct insertion of lipids into the apical surface of the cell by fusion of liposomes with plasma membrane.

II. LIPID POLARITY IN EPITHELIAL CELLS

A. Epithelial Tissues

The most careful studies concerning the lipid composition of apical and basolateral plasma membrane domains have been performed on intestinal cells (5-9). These studies involved cell fractionation and were facilitated

751

by the organization of the apical membrane into a highly differentiated
brush border, which can be easily purified. The apical membrane relative
to a basolateral plasma membrane fraction displayed a two- to fourfold
enrichment of glycosphingolipids and a two- to fourfold depletion of phosphatidylcholine (PC). The amount of the second major phospholipid, phosphatidylethanolamine (PE), as a percentage of total lipid was equivalent
in the two domains.

Apical and basolateral domains have also been purified from other
tissues, such as epithelial cells of rat kidney (10,11), dog kidney (12),
rat intestinal mucosa (13), and rat liver (14,15). Clear-cut differences
in the phospholipid compositions of the plasma membrane domains were
reported. The apical domain generally contained more sphingomyelin (SPH)
and less PC. Glycosphingolipids were not analyzed in these studies.

B. Cultured Epithelial Cells

Epithelial monolayers in cell culture provide excellent models for studies
of lipid polarity. Experimental manipulation is facilitated by growing the
cells on a permeable support (2). The system that has been best characterized is the Madin-Darby canine kidney (MDCK) cell line. MDCK cells form
a confluent cell monolayer which develops a transepithelial electrical resistance and transports ions (16,17). Two strains of MDCK cells with different
characteristics have been defined (for discussion, see Ref. 18). One of
the major differences between MDCK strain I and strain II cells is that
strain I MDCK cells form a "tight" epithelium which develops transepithelial
electrical resistances greater than 2000 ohm.cm^2, whereas a monolayer
of MDCK strain II cells is fairly leaky and typically displays a resistance
of 100-300 ohm.cm^2. Nevertheless, the distribution of plasma membrane
proteins is equally polarized in both MDCK cell strains (19).

To address the question of whether MDCK cells in culture also display
a polarized lipid distribution, we have made use of the observation by
Rodriguez Boulan and Sabatini that enveloped viruses bud from MDCK
cells in a polarized manner (20). Influenza viruses bud from the apical
surface, while vesicular stomatitis virus (VSV) obtains its envelope by
budding through the basolateral plasma membrane domain (Fig. 1). These
viruses can be used to sample the lipid composition of either domain since
there is no selection for specific phospholipids during viral budding. For
example, when fowl plague virus (FPV), an influenza virus, and VSV
were harvested from nonpolar BHK cells, they displayed identical phospholipid compositions (21-24).

In contrast, remarkable differences in phospholipid composition between
the viruses were observed when they were harvested from MDCK strain I
cells. For this, FPV was isolated from the apical medium and VSV from
the medium below the nitrocellulose filter support which, in order to allow
passage of the VSV, had a pore diameter of 3 μm. The intactness of the
tight junctions was assessed throughout the experiments by the maintenance
of a transepithelial electrical resistance of greater than 1000 ohm.cm^2
(21,24). Similar differences between the phospholipids of the viruses were
found when they were harvested from the apical and basolateral surfaces
of strain II MDCK cells grown on a solid support (25). The results are
summarized in Table 1. Both viruses display a typical plasma membrane
composition. They contain far less PC and more SPH and phosphatidylserine
(PS) than the total cells. The major difference between the apical virus
(FPV) and the basolateral virus (VSV) is a nearly fivefold depletion of PC
in the apical membrane relative to PE.

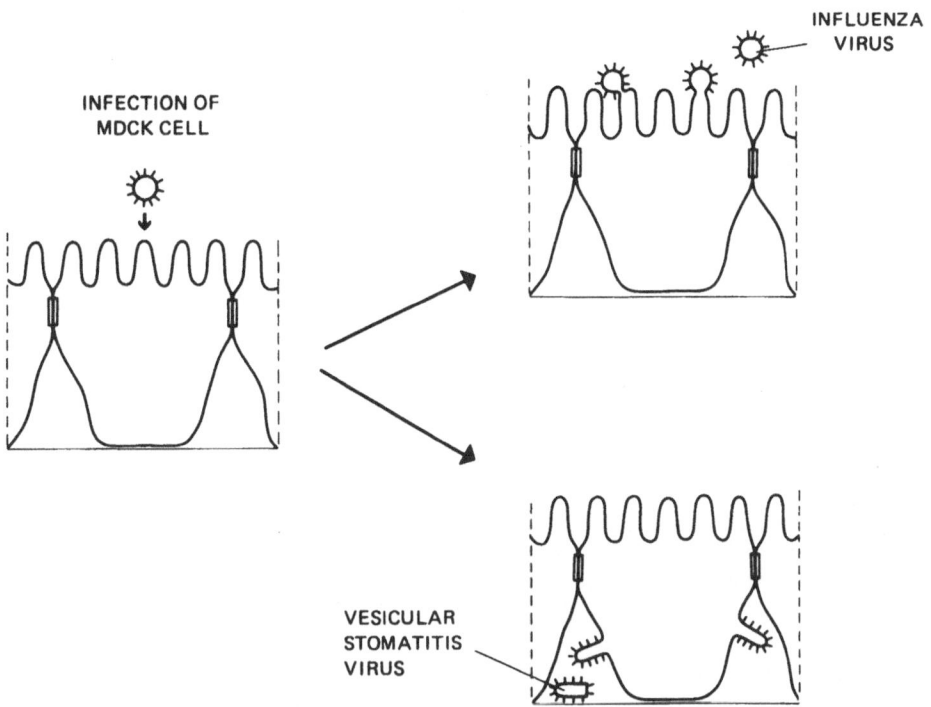

FIGURE 1 Selective budding of influenza virus from the apical surface and VSV from the basolateral surface of MDCK cells (20).

TABLE 1 Phospholipids (mol%) of Viruses Harvested from the Apical (FPV) or Basolateral Surface (VSV) of MDCK Cells (21,25)

Phospholipid[a]	FPV	VSV	Total MDCK cells
SPH	19.8	21.0	10.2
PC	6.0	20.1	48.7
PI	2.5	5.1	7.1
PS	21.6	18.1	5.7
PE	50.4	36.0	28.3

SPH, sphingomyelin; PC, phosphatidylcholine; PI, phosphatidylinositol; PS, phosphatidylserine; PE, phosphatidylethanolamine.

Since these viruses do not select for specific phospholipids, the difference in phospholipid composition between the viruses must reflect differences
in phospholipid composition between the apical and basolateral plasma membrane domain of MDCK cells. The difference in phospholipid composition
between the plasma membrane domains in MDCK cells is very similar to
that reported for intestinal cells (Section IIA). In addition, recent studies
have shown that MDCK cells (18,26,26a,27) and other kidney epithelia
(28,29) also have a very high glycosphingolipid content. As in intestinal
cells, these have been primarily localized to the apical domain. Furthermore,
virus that buds from the apical surface of Madin-Darby *bovine* kidney
cells has been shown to possess typical apical *phospholipids*. These were
localized primarily in the inner leaflet of the viral lipid bilayer, which
was attributed to a high glycolipid content (22,23,30). Studies are now
necessary to determine the total lipid compositions of the apical and the
basolateral plasma membrane domain directly. The available data for the
polarized distribution of plasma membrane lipids in epithelial cells are
summarized by the model in Figure 2. The apical cell surface appears to
be covered by glycosphingolipids (7-9,31) and/or SPH, while basolateral
plasma membranes are enriched in PC. The third major phospholipid, PE,
is distributed equally between the two domains, as is cholesterol, which
constitutes 30 mol% of the total plasma membrane lipids (8,9; reviewed
in Ref. 31a).

C. Role of Tight Junctions

The observation that the apical and basolateral membrane domains have
different lipid compositions raises the question: why don't the lipids intermix? The simplest explanation would be the presence of a diffusion barrier
in the plane of the plasma membrane. The most likely candidate for such
a barrier would be the tight junction, which is also thought to maintain
protein polarity. The tight junction forms the boundary between the apical
and the basolateral plasma membrane, and treatments that open tight junctions allow intermixing of apical and basolateral proteins (2,4).

The question of whether the tight junction acts as a diffusion barrier
for lipids has been tested using several different approaches. First, the
ability of the ganglioside G_{M1}, present on the apical surface of MDCK
cells, to equilibrate with the basolateral surface was investigated and no
net movement was observed (27). In another approach MDCK cells were
incubated with filipin, a probe that interacts with cholesterol. The surface
density of the resulting membrane lesions appeared to be different on
apical and basolateral surfaces (32). However, parameters other than
cholesterol concentration are involved in the formation of these lesions
and therefore the surface density of the lesions cannot be used as a measure of cholesterol concentration (33). Furthermore, even if an unequal
distribution of cholesterol over the two plasma membrane domains existed,
this would not necessarily be incompatible with free diffusion across the
tight junction (34).

The role of tight junctions has also been studied by the addition of
certain fluorescent amphiphiles, in ethanolic solution, to the apical surface
of an epithelial monolayer grown on a solid support (35). Some amphiphiles
were shown to remain on the apical surface, while others redistributed
to the basolateral surface. The authors suggested that basolateral redistribution has occurred following prior translocation of certain probes to
the cytoplasmic leaflet and that diffusion to the basolateral surface takes

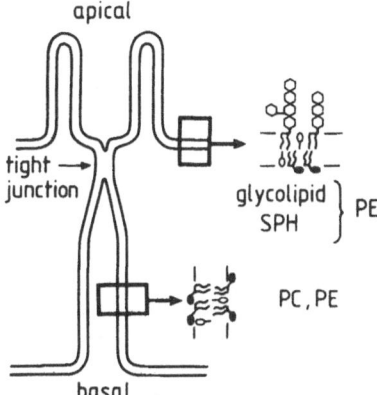

FIGURE 2 Polarized distribution of plasma membrane lipids in epithelial cells.   , Phospholipid; O , carbohydrate; ? , cholesterol.

place exclusively in this leaflet. Unfortunately, translocation of these probes in the plasma membrane could not be proven directly. A second problem in the interpretation of these experiments is that they were performed at 20°C, a temperature at which endocytosis and transcytosis are not inhibited (36). Therefore, redistribution may have occurred via an intra-cellular route. Moreover, the lipid probes used were water-soluble and it was not possible to exclude the possibility that equilibration between the plasma membrane domains had occurred through the cytosol. Finally, the intactness of the tight junctions under the conditions of the experiments was not controlled and the existence of a polarized lipid distribution was not demonstrated in the cell system used.

We have tried to overcome these problems using a more direct experimental approach. We chose as an experimental system strain I and strain II MDCK cells grown on permeable supports. Under these conditions both strains display polarized protein and lipid compositions. In order to perform these studies, it was necessary to develop a method whereby a water-insoluble, fluorescent phospholipid could be introduced into cells. A water-insoluble probe would be expected to mimic naturally occurring phospholipids more closely than the analogs used before. We also wanted to insert this phospholipid directly into either the external or the cytoplasmic leaflet of the plasma membrane. For this purpose we devised a method to fuse liposomes with the apical membrane of the MDCK cells.

III. LIPOSOME FUSION WITH THE APICAL MEMBRANE

A. Influenza Virus Hemagglutinin as a Fusogen

The interaction between liposomes and living cells has been widely studied, but thus far controlled fusion of liposomes with the plasma membrane has been achieved in only a few cases (37). In order to obtain controlled and efficient fusion of liposomes with the apical plasma membrane domain of MDCK cells, we have made use of a naturally occurring fusion protein, the hemagglutinin (HA) glycoprotein of influenza virus. The fusogenic properties of this protein have been very well characterized (see Chapter 14).

In nature it mediates the fusion of the influenza virus membrane with a
host cell membrane, whereby the viral nucleocapsid enters the host cell
cytoplasm (38; Chapter 13). The first steps in this process are binding
of the influenza virus to a receptor, containing sialic acid, on the cell
surface and internalization of the virus by endocytosis through coated
pits. Subsequently, the virus particle is delivered to an endosome. The
low pH (~5) in the endosomal compartment triggers a conformational change
in the influenza virus HA, leading to fusion of the viral membrane with
the membrane of the endosome and release of the nucleocapsid into the
cytoplasm.

We have adapted this natural event for the purpose of liposome-plasma
membrane fusion (39,40) by taking advantage of two additional observations.
First, HA expressed on the plasma membrane of cells after infection with
influenza virus can mediate cell-cell fusion when the pH is lowered (41).
Second, influenza virus in solution can be induced to fuse at low pH with
liposomes of various lipid compositions (42; see also Chapters 14 and 17).
As a first step, we infected a monolayer of MDCK cells with influenza
virus to obtain HA expression on the plasma membrane. Subsequently,
we added liposomes containing a ganglioside with a terminal sialic acid,
acting as an HA receptor, to the apical side of the cell monolayer. These
liposomes bound to the apical HA (43) at 0°C, and fusion of the liposomes
with the plasma membrane was induced by incubation for 1 min at 37°C
and the pH optimum for fusion of the particular influenza virus used
(Fig. 3). After this fusion step the monolayer was immediately cooled to
0°C and assayed for fusion and liposomal lipid distribution.

The influenza virus HA was selected as a fusion protein instead of
other viral fusion proteins for several reasons. HA is expressed exclusively
on the apical surface of MDCK cells (43) and specific binding of HA to
sialic acid has been well defined. In addition, HA-mediated fusion is very
efficient and extremely rapid (42). The $t_{1/2}$ of fusion is less than 1 min.
A distinct advantage of HA-mediated fusion over the Sendai virus system
(see Chapters 15–17) is the fact that the fusion activity can be initiated
by lowering the pH and terminated by neutralizing the medium (41,42,44–46).

B.   Quantitative Assays of Fusion

We developed or adapted four independent fusion assays to test whether
lowering the pH led to the expected HA-mediated fusion of the liposomes
with the plasma membrane (40). The fusion assays were developed to allow
quantitation in order to facilitate characterization and improvement of fusion
activity.

Our first assay for discriminating cell-associated liposomes which were
merely bound from those which had fused was based on the observation
that the initial liposome-cell binding was sensitive to neuraminidase (39).
This was due to the fact that the terminal sialic acid of the liposomal
ganglioside G_{D1a}, which mediated the liposome binding to the HA on the
cell surface, could be removed by neuraminidase. The influenza virus
neuraminidase which is also present on the apical cell surface of infected
MDCK cells (Fig. 3) was inactive during the liposome binding at 0°C but
released most of the bound liposomes within 10 s after warming to 37°C
(39). This implied that during the fusion step, 60 s at 37°C, any unfused
liposomes would automatically be removed by the viral neuraminidase. The
remaining cell-associated liposomes would therefore be considered as having
fused in this assay. Neuraminidase-resistant, cell-associated liposomes were

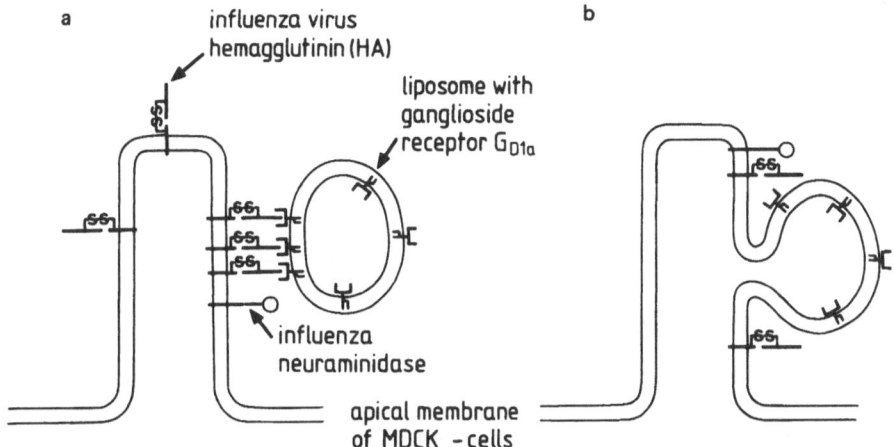

FIGURE 3 Fusion of liposomes with the apical plasma membrane of MDCK cells. (a) Liposomes are bound to the HA on the apical surface at 0°C. (b) When the pH is lowered to 5.0 for 1 min at 37°C, a conformational change in the HA molecule leads to fusion. At the same time, unfused liposomes are removed by the viral neuraminidase.

quantitated by incorporating a nonexchangeable radioactive lipid marker into the liposomes. The use of a nonexchangeable lipid marker was essential since exchange through the water phase and partitioning into the plasma membrane would have given rise to false positive results.

While the neuraminidase assay was based on a change in the nature of the liposome-cell association, our second assay measured the actual continuity of the liposomal lipid bilayer with that of the plasma membrane. Two nonexchangeable fluorescent lipids N-NBD-PE and N-Rh-PE were incorporated into the liposomes and the loss of fluorescence resonance energy transfer between the probes as a result of dilution into the plasma membrane was used as a measure of fusion (47; see Chapter 17). Resonance energy transfer describes the interaction between N-NBD-PE and N-Rh-PE that takes place when the probes are in a proximity of about 4 nm (47,48). It was measured in a monolayer of MDCK cells either by scraping the cell monolayer from the solid support and resuspending the monolayer in a fluorescence cuvette (39) or by growing the cells on a glass coverslip that could be fitted directly into a cuvette (40).

In a third fusion assay, we monitored the diffusion of the nonexchangeable fluorescent lipid N-Rh-PE over distances as large as 4000–8000 nm by fringe pattern fluorescence photobleaching (40,49). Since the mean diameter of the liposomes used was smaller than 200 nm, lateral diffusion over distances larger than 4000 nm would only be possible if N-Rh-PE were present in the plasma membrane. Due to the nonexchangeability of the probe, this could only be the case following liposome fusion. Quantitation of the mobile fraction of N-Rh-PE was used to assess the percentage of cell-associated N-Rh-PE that had fused with the plasma membrane.

The fourth and most convenient fusion assay was developed in the course of our fusion studies. We observed (40) that after fusion two non-exchangeable radioactive lipid markers in the liposomes, cholesterol-[^{14}C]oleate (51,52) and glycerol-tri[^{14}C]oleate (53), were hydrolyzed.

The amount of hydrolysis as a fraction of total cell-associated radioactivity correlated very well with results obtained by measuring the relative loss of resonance energy transfer under all conditions studied. We concluded that the hydrolysis must be due to accessibility of the radioactive lipids to one or more cellular hydrolases after insertion into the plasma membrane as a result of fusion. For analysis after fusion, the cell monolayer was incubated for 1 h at 0°C to allow enzyme-mediated hydrolysis of all accessible radioactive substrate. Quantitation of fusion was performed by lipid extraction, separation of the individual lipids by thin-layer chromatography, and quantitation of the radioactivity in the various spots.

C. Characteristics of the Fusion Process

1. Viral Infection and Liposome Binding

In order to characterize the low-pH-induced fusion of liposome with the plasma membrane mediated by HA, we infected confluent monolayers of strain II MDCK cells grown on a solid support with an avian influenza virus. The monolayers were used for liposome binding studies 4.5 h after infection. At that time influenza virus hemagglutinin is present in appreciable amounts on the apical surface of strain II MDCK cells grown on both solid and permeable supports (54,55).

First, we studied the ability of various gangliosides incorporated into the liposomes to support liposome binding to HA on the apical cell surface (39). Gangliosides possessing a terminal sialic acid, G_{D1a} and G_{T1b}, were equally efficient. The viral neuraminidase is able to remove the terminal neuraminic acid of these gangliosides while leaving the proximal ones intact (56,57), yielding G_{M1} and G_{D1b}. This resulted in liposome release in both cases. Also, a direct test of G_{D1b} demonstrated that this ganglioside does not bind to HA. Also, G_{M3}, the shortest ganglioside with a terminal neuraminic acid, but not G_{M2}, in which the neuraminic acid is turned into a proximal one, has been shown to act as an acceptor for an influenza virus (57a). These findings are contradictory to those reported by Bergelson et al. (57,58), who concluded that G_{M1} is the basic ganglioside structure that is sufficient for HA binding. The observed requirement of a terminal neuraminic acid for influenza virus HA binding has also been reported for Sendai virus binding (59,60). Based on these results, we used large unilamellar liposomes consisting of egg PC/cholesterol/G_{D1a} (1:1:0.1 mol/mol), prepared by octylglucoside dialysis. For the various fusion assays we included 1 mol% each of N-NBD-PE and N-Rh-PE and a trace amount of cholesterol-[^{14}C]oleate (40).

2. Neuraminidase-Resistant Liposome Binding

The binding of liposomes to the apical surface leveled off after 30 min at 0°C (39). If the cells were warmed to 37°C at pH 7.4 following binding, most of the liposomes were released as a consequence of the action of the viral neuraminidase (see Section IIIB). However, when this warming step was performed at a pH below 5.5, a fraction of the liposomes became resistant to release by neuraminidase. Maximal liposome fusion occurred at pH 5.0, which is identical to the pH optimum for fusion of the influenza N virus strain used in this study (45). According to the "neuraminidase assay," 50% of the bound liposomes fused with the plasma membrane of the cells.

In strain II MDCK cells the HA of influenza N virus is not proteolytically cleaved into its HA_1 and HA_2 subunits (54,55,61). This cleavage is necessary for the HA to obtain its fusogenic properties (38; Chapter 14) and can be achieved by a trypsin treatment of the apical cell surface prior to liposome binding (54,55). In this system, we were able to verify that the acquisition of neuraminidase resistance of the liposome binding at low pH occurred only when the HA had been cleaved into its fusogenic form (39). The acquisition of neuraminidase-resistant binding was very rapid. At pH 5.1 and 37°C the process was complete within 10 s. Furthermore, the rate and extent of fusion as measured by this assay were independent of the liposomal lipid composition (40).

3. Loss of Resonance Energy Transfer Between Liposomal Lipids

The second fusion assay, measuring loss of resonance energy transfer between the liposomal N-Rh-PE and N-Rh-PE (see Section III.B), exhibited roughly the same pH dependence for the fusion process as did the neuraminidase assay (40). Fusion occurred at pH 5.5 and reached a maximum at pH 5.1. However, a closer analysis of the results revealed some significant differences between the two assays. Whereas the neuraminidase assay quantitated the amount of liposomes that were cell-associated and resistant to neuraminidase, the loss of resonance energy transfer assay quantitated the fraction of neuraminidase-resistant liposomes that had actually fused with the plasma membrane. The difference between the two parameters was most evident when PC/cholesterol liposomes containing G_{D1a} were bound and fused with MDCK cells. Less than half of the liposomes that acquired neuraminidase resistance displayed loss of energy transfer under all conditions tested. At early time points during fusion, up to three-quarters of the cell-associated liposomes that were resistant to neuraminidase had not fused, as measured by loss of resonance energy transfer.

The rate of the fusion process measured by the two assays was also observed to differ. In contrast to the acquisition of neuraminidase resistance, the actual fusion as measured by the loss of resonance energy transfer was dependent on the lipid composition of the liposomes (Fig. 4). Fusion of liposomes with a matrix of PC/cholesterol had a half-time of over 60 s, whereas the replacement of 50% of the PC by egg PE increased the rate of fusion dramatically (half-time, 10-15 s). This facilitating effect of an unsaturated PE on the rate of fusion has also been reported for the fusion of liposomes to intact influenza virus (42) and for fusion in a number of other model systems. Two properties of PE might explain this effect. First, PE reduces the repulsive hydration force between two apposed bilayers (Chapter 3), and second, PE especially the more unsaturated species, exhibit a tendency to assume nonbilayer configurations which have been suggested to be intermediate structures in the fusion reaction (Chapters 2 and 8).

4. Sequential Stages in the HA-Mediated Fusion Process

As described above, liposomes could become associated with the apical plasma membrane such that they were resistant to neuraminidase, but without actually fusing with the plasma membrane (Fig. 4). The HA-mediated fusion reaction thus seems to occur in two separable stages. During the first stage a change takes place in the binding of the liposome to sialic-acid

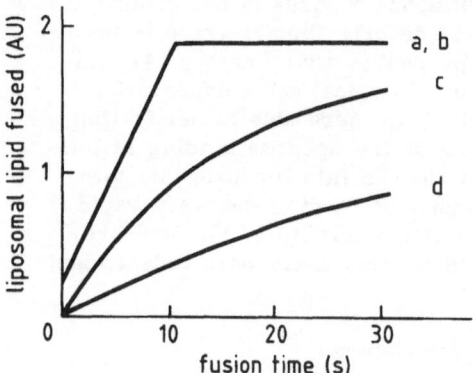

FIGURE 4 Absolute amounts of liposome-plasma membrane fusion at pH 5.0 and 37°C as measured by the neuraminidase resistance assay (a,b) or the loss of resonance energy transfer (c,d). Two types of liposomes were used: liposomes with a matrix of PC/cholesterol (1:1, mol/mol) in b and d or of PC/PE/cholesterol (1:1:2, mol/mol) in curves a and c. Units on the vertical axis are arbitrary (AU). Under optimal conditions on a plastic support 10% of the liposomes that had originally been added to the cell monolayer acquired neuraminidase resistant binding, and 85% of these actually fused with the plasma membrane (40). When the cells had been grown on filters the latter number changed to 50% (21).

residues on the plasma membrane to a condition where either sialic acid is no longer accessible to neuraminidase or in which it is no longer important for binding. This stage precedes the actual liposome-plasma membrane fusion event monitored by the loss of resonance energy transfer. An additional difference between binding of the liposomes to the plasma membrane and the actual fusion event is that only the latter is dependent on the liposomal lipid composition (Fig. 4). However, also the change in the type of binding can be considered an integral part of the fusion process since, in contrast to binding mediated by sialic acid, it is strictly dependent both on the cleavage of HA into its fusogenic HA_1 and HA_2 subunits and on the low pH treatment (39,40).

A change in the type of binding as an initial step in the HA-mediated fusion has been suggested previously (62,63) on the basis of the amino acid sequence of HA_2 (see also Chapter 14). At low pH a conformational change occurs in the HA glycoprotein whereby HA_2, the subunit of the spike that is anchored in the viral membrane, exposes a second hydrophobic segment. This segment is highly conserved in different influenza virus strains and is thought to penetrate into the target membrane (64; Chapter 14). Similarly, in our system the HA expressed on the plasma membrane might penetrate the liposomal bilayer. As a consequence, the liposome would now become linked to the plasma membrane by an HA_2 bridge anchored in both membranes by hydrophobic peptide segments. Subsequently, further bending of the HA_2 (65) would bring the two membranes in such close proximity that actual fusion would be induced (Fig. 5). This model is supported by the following observations (see Chapter 14): (1) lowering the pH induces binding of the water-soluble ectodomain of HA to liposomes, (2) the HA_2 subunit mediates low-pH binding, and (3) the binding is independent of liposomal lipid composition (63,64), while the actual fusion is not (40,42). The model shown in Figure 5 is consistent with all of our data (40). The intermediate stage of fusion would indeed be dependent on low pH and proteolytic cleavage of HA into HA_1 and HA_2, and the subsequent liposome-cell association would be insensitive to neuraminidase. The reason for the arrest of fusion between liposomes and the plasma membrane in this intermediate stage, a phenomenon that has not been previously reported, is still unclear.

5. Photobleaching and the Hydrolysis of Liposomal Lipids by Cellular Enzymes

Two additional fusion assays, the measurement of the mobility of liposomal N-Rh-PE over distances exceeding the diameter of one liposome by photobleaching and the hydrolysis of liposomal cholesterol-[^{14}C]oleate by a cellular enzyme (IIIB), displayed identical pH dependencies and time courses of fusion as the loss of resonance energy transfer (40). Therefore, these assays could also be used to monitor the actual fusion event.

In addition to providing information about the size of the fraction of N-Rh-PE that is mobile over a certain distance scale, the fringe pattern photobleaching technique gives an accurate estimate of the diffusion coefficient of this mobile fraction (49). In the apical plasma membrane domain of strain II MDCK cells, grown on a glass support, we defined a diffusion coefficient for N-Rh-PE after liposome fusion of 7×10^{-10} cm^2s^{-1} (40). This is significantly slower than the diffusion coefficient normally found for lipids in membranes (see Chapter 1) and studies are under way to correlate this behavior with the special properties of the apical membrane of MDCK cells.

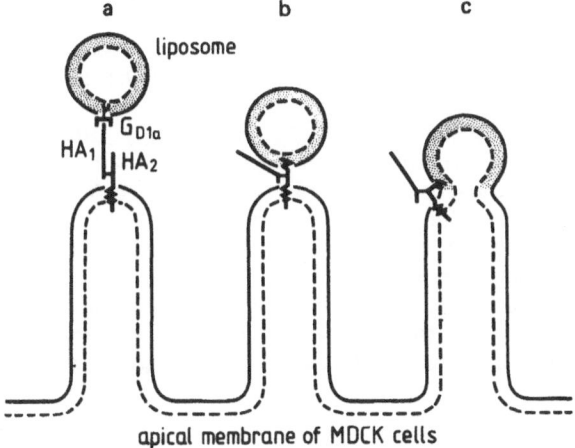

FIGURE 5 Hypothetical stages in the HA-mediated liposome-plasma membrane fusion. (a) The liposome is bound to the HA by means of the HA receptor, the ganglioside G_{D1a}. (b) Lowering the pH to 5.0 leads to a conformational change in the HA by which the second hydrophobic domain inserts into the liposomal membrane. (c) During a later step, which is also pH dependent, actual fusion of the lipid bilayers occurs (24,62-64).

The identity of the enzyme(s) responsible for the hydrolysis of cholesterol-[^{14}C]oleate and glycerol-tri[^{14}C]oleate is not known. The involvement of lysosomal hydrolases is very unlikely since this would imply endocytosis and transport of the substrate to the lysosomes within 10 s at 37°C, a process that normally takes on the order of 30 min in MDCK cells (3). Various cholesterol esterases have been localized in the plasma membranes and cytosol of several cells (66-68). These enzyme(s) are more likely to be responsible for hydrolysis of the radioactive substrates than lysosomal esterases under the assay conditions used. Furthermore, some of these cholesterol esterases have been reported to be identical with, or complexed to, a triacylglycerol lipase (69-72).

During these studies we observed that a third lipid, dipalmitoyl phosphatidic acid, was also hydrolyzed, giving rise to diglyceride at 0°C after introduction into the plasma membrane by fusion. This enzymatic activity, which was also observed by Pagano and Sleight (73) by means of a fluorescent phosphatidic acid analog, will have to be characterized in more detail. It may be relevant with respect to the regulation of the activity of protein kinase C by plasma membrane diglycerides (74). From the fact that cholesterol esters, triglycerides, and phosphatidic acid are hydrolyzed when inserted into the plasma membrane, we conclude that under normal conditions none of these lipids are present in the plasma membrane.

6. Alternative Applications of HA-Mediated Fusion

In parallel with our system of HA-mediated liposome-plasma membrane fusion, a method was developed for introducing macromolecules into living cells by fusing loaded erythrocyte ghosts with the plasma membrane (75). Erythrocyte ghosts were used instead of liposomes since they have a 1000-fold larger included volume relative to the 200-nm diameter liposomes. As in our system, influenza virus HA was expressed on the cell surface to mediate erythrocyte binding and fusion. Fusion was assayed in two ways. First, the amount of cell-associated erythrocytes resistant to neuraminidase was quantitated, and second, delivery of the contents of the erythrocyte ghosts into the cytoplasm of the HA expressing cell was assessed by light or electron microscopy.

Viral infection generally leads to cytopathic effects which might interfere with subsequent studies especially if long time courses (over 6 h) are required. Doxsey et al. (75) proposed two alternatives to the infection with influenza virus for surface expression of HA: transfection of the cells with a SV40 vector carrying the HA gene or the use of a cell line that had been transformed to permanently express HA. The latter method is especially promising (76). Alternatively, the viral fusion protein can be incorporated in a liposomal bilayer by reconstitution. For a discussion of reconstitution of viral envelopes and their application see Chapter 19 (76a,b).

D. Liposome Fusion with Filter-Grown Cells

As discussed in Section IIC, the methodology for fusion of liposomes with the plasma membrane was developed to test the properties of the tight junction as a diffusion barrier for lipids. For these experiments the MDCK cells were grown on permeable supports. On filters the cells take up nutrients from the basal side and form a monolayer of cuboidal cells, which is more stable to experimental manipulations (5,17,55,77).

We adapted the fusion method described above to filter-grown MDCK cells, monitoring fusion by the hydrolysis of cholesterol-[^{14}C]oleate. The fusion of liposomes with strain I MDCK cells did not require prior tryptic cleavage of the HA, which was necessary when MDCK strain II cells were used (IIIC). SDS gel electrophoresis demonstrated spontaneous cleavage of HA in strain I cells, and serum-free conditioned medium from strain I cells was shown to increase the level of cleaved HA (G. van Meer, unpublished observations). Apparently, MDCK strain I cells secrete a protease that is responsible for the observed HA cleavage.

In order to prevent damage of the apical membrane, we limited the amount of lipid introduced by fusion to 2% of the total lipid in the apical plasma membrane domain, as calculated from morphometrical data (5,21). The fusion efficiency on filter-grown cells was lower than the efficiency observed for cell monolayers grown on a solid support. Routinely 40-55% of the cell-associated liposomes were fused according to the cholesterol-[^{14}C]oleate hydrolysis assay.

IV. REDISTRIBUTION OF FLUORESCENT PHOSPHOLIPID OVER THE TIGHT JUNCTION

A. Transbilayer Orientation of N-Rh-PE in Liposomes

We selected the fluorescent phospholipid N-Rh-PE as a probe for the study of the properties of the tight junction. N-Rh-PE does not exchange between membranes via the aqueous phase (47) and is able to diffuse laterally in the apical membrane of MDCK cells after fusion as indicated by photobleaching (24,50). N-Rh-PE can be traced in confluent monolayers of living MDCK cells with a fluorescence microscope equipped with a water-immersion objective.

During fusion of liposomes with the plasma membrane the external leaflet of the liposome should become continuous with the exoplasmic leaflet of the plasma membrane and the internal leaflet should come into continuity with the cytoplasmic leaflet. Since the fluorescent phospholipid molecules might display a different behavior depending on whether they were inserted into the exoplasmic or into the cytoplasmic leaflet of the plasma membrane, we assayed the transbilayer orientation of N-Rh-PE in the liposomes by using phospholipase A$_2$ (78-80).

We first tested large unilamellar liposomes prepared by octylglucoside dialysis, which had been used in previous studies (24). These liposomes contained egg PC/egg PE/cholesterol/G$_{D1a}$ (25:25:50:5, mol/mol), 1mol% each of N-NBD-PE, N-Rh-PE, [^{14}C] DOPE, and cholesterol-[^{14}C]oleate. At 10°C phospholipase A$_2$ hydrolyzed 55% of the PC and PE and also of the [^{14}C] DOPE, as expected for phospholipids with a symmetrical transbilayer distribution. In striking contrast, the hydrolysis of the fluorescent phospholipids N-NBD-PE and N-Rh-PE reached levels of 95% and 85%, respectively. They are thus highly enriched in the external leaflet of liposomes prepared by octylglucoside dialysis (21). This asymmetrical arrangement of the N-NBD-PE and N-Rh-PE in the external leaflet is most likely due to insertion of the fluorescent lipids after closed bilayer structures were already formed during detergent dialysis (81).

The asymmetrical liposomes were freeze-thawed 10 times and the resulting larger structures reduced to the original liposome size by extrusion through a polycarbonate filter. Following this treatment, the asymmetrical arrangement of the fluorescent phospholipids had essentially disappeared. Alternatively, symmetrical large unilamellar liposomes (0.2 μm diameter) could be prepared directly from the lipid mixture mentioned above by reverse-phase evaporation (REV) (21,82).

B. Fusion of Asymmetrical Liposomes

The highly asymmetrical orientation of N-Rh-PE in the octylglucoside-dialysis liposomes made it possible to fuse the fluorescent phospholipid predominantly into the exoplasmic leaflet of the apical plasma membrane. When such liposomes were bound to filter-grown MDCK strain II cells, a characteristic pattern was observed under the fluorescence microscope (Fig. 6A,B). The fluorescent dots most likely represent the microvilli on the apical cell surface (cf. Ref. 5). No sharp fluorescent pattern was observed below the plane of the apical membrane. The pattern of fluorescence did not change after fusion of 40% of the cell-associated fluorescent lipids into the apical plasma membrane, as measured by the cholesterol oleate hydrolysis assay (IIIB). All fluorescence was found on the apical surface and remained there for several hours at 0°C. This fluorescence pattern changed dramatically immediately after the tight junctions were opened by an incubation in Ca^{2+} free medium (83), as shown in Figure 6C,D. When the microscope was focused below the apical plane, a honeycomb pattern of sharp fluorescent lines appeared, typical of lateral membrane staining (cf. Ref. 77). The fact that N-Rh-PE when fused into the exoplasmic leaflet of the apical plasma membrane, could not diffuse to the basolateral surface without first opening the tight junctions suggested that the tight junction functions as a barrier to lipid diffusion in the exoplasmic leaflet of the plasma membrane bilayer (21).

C. Fusion of Symmetrical Liposomes

Fusion of octylglucoside-dialysis liposomes scrambled by freeze-thawing or symmetrical REV with the apical surface should result in the insertion of N-RhPE in both leaflets of the plasma membrane. Binding of either type of liposomes produced a dotted pattern of fluorescence identical to that shown in Figure 6A,B. Upon fusion, however, the resulting fluorescence pattern was very different from that obtained after fusion of the asymmetrical liposomes. Immediately after fusion, a pattern of fluorescent lines was observed at the level of the lateral membrane, as in Figure 6C,D. The immediate diffusion of N-Rh-PE to the lateral surface following fusion of symmetrical liposomes is evidence for the absence of a diffusion barrier in the cytoplasmic leaflet of the plasma membrane (21).

D. Implications for Lipid Organization

The finding that the tight junction acts as a diffusion barrier exclusively in the exoplasmic leaflet of the plasma membrane implies that the observed difference in lipid composition between the apical and basolateral plasma membrane domain is caused by differences in the exoplasmic leaflets. If, indeed, the free diffusion of lipid molecules in the cytoplasmic leaflet leads to an identical lipid composition of the cytoplasmic leaflets of both domains, interesting predictions concerning the transbilayer lipid distribution in the plasma membrane would result. Let us assume that the exoplasmic leaflet of the apical domain is nearly entirely occupied by glycosphingolipids, as

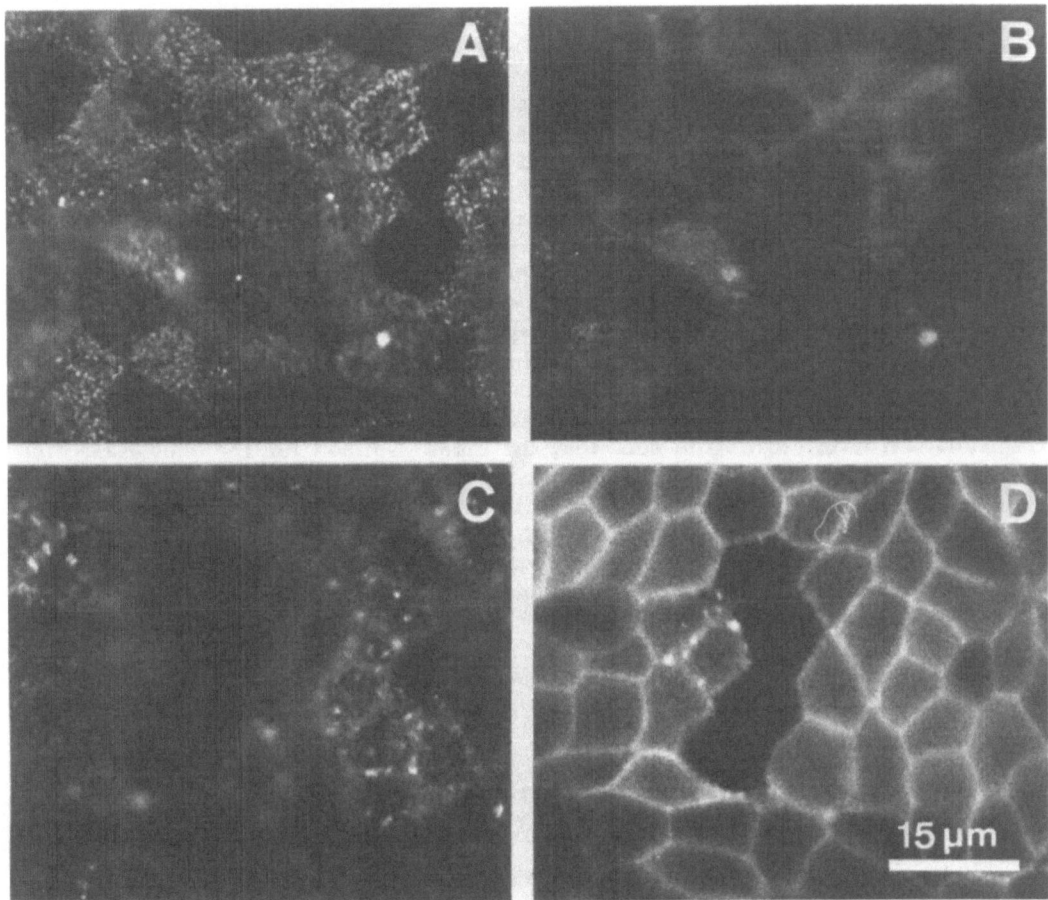

FIGURE 6 *N*-Rh-PE does not diffuse to the lateral surface when fused
into the exoplasmic leaflet of the apical plasma membrane of strain II MDCK
cells. Asymmetric liposomes were bound to the apical surface (A, B). Subsequently they were fused and the tight junctions were opened in Ca^{2+}
free medium (C, D). Panels A and C: apical surface; panels B and D:
lateral surface.

suggested in Figure 2 (7-9,31). As a result, most of the apical phospholipids would now be situated in the cytoplasmic leaflet. If the two cytoplasmic leaflets equilibrate freely, as suggested by our results, the
cytoplasmic leaflet of the basolateral membrane should have an identical
phospholipid composition. From this information, the distribution of the
individual phospholipid classes across the basolateral membrane bilayer
can be calculated from the total basolateral phospholipid composition. This
leads to the following predictions for the two major phospholipid classes:
65-90% of the PE and only 10-25% of the PC would be localized in the cytoplasmic leaflet (8,21,25). This is strikingly similar to the phospholipid
distribution in the erythrocyte plasma membrane (84-86) and may reflect
general principles underlying the phospholipid organization in mammalian
plasma membranes.

The predicted structure of the tight junction sheds some light on how the differences in lipid composition between the domains may be generated within the cell. The pathways of plasma membrane proteins destined for either domain have been fairly well elucidated (2). Both sets of proteins are transported through the same Golgi stack, after which the two pathways most likely bifurcate at the *trans* Golgi network (87). Since the membrane vesicles that transport the plasma membrane proteins to their destination necessarily transport lipids at the same time, the sorting of apical and basolateral lipids is also likely to take place in the *trans* Golgi network. This sorting process must involve the lateral segregation of unique sets of lipids to areas of the sorting compartment membrane where budding of vesicles destined for either plasma membrane domain occurs. The conclusion that the different lipid compositions of the domains reflect differences in the lipids of their exoplasmic leaflets (see above) implies that the lipids have to segregate in the sorting compartment only in its noncytoplasmic leaflet. It is interesting to note that the lipids destined for the exoplasmic leaflet of the apical domain, glycosphingolipids and sphingomyelin, have been reported to self-associate (88). Since this by itself would not be sufficient to direct them into the budding vesicles destined for the apical surface, additional factors must be postulated. However, the first "sorting protein" still (1990) awaits identification.

V.   REDISTRIBUTION OF LIPIDS FROM ONE EPITHELIAL CELL TO THE NEXT

A.   Endogenous Glycolipids

The results presented above are in agreement with the hexagonal lipid model for the tight junction (89,90; Fig. 8). The central point of this model is that the core of the tight junction is formed by a hexagonal cylinder of lipids. It predicts free diffusion between the apical and basolateral plasma membrane of one cell through the cytoplasmic leaflet, but the lipid cylinder would inhibit diffusion between the two domains through the exoplasmic bilayer leaflet. At the same time, this model predicts that lipid molecules are free to diffuse from the exoplasmic leaflet of the apical surface of one epithelial cell to that of the neighboring cell. We tested this prediction directly using endogenous glycolipids as probes.

Strain II MDCK cells express a glycosphingolipid, called the Forssman antigen, in the exoplasmic leaflet of their apical membrane (18,26,26a,b) which is absent in strain I cells. We cocultured MDCK strain II cells with strain I cells and labeled the apical surface of the mixed monolayer after 96 h with an anti-Forssman monoclonal antibody. Only 50% of the cells were labeled, suggesting that the endogenous glycosphingolipid was unable to pass from strain II cells to strain I cells. We demonstrated that tight junctions had formed between the two cell types by their impermeability to an antibody against a basolateral membrane marker (91). These experiments have also been published by Nichols and colleagues (91a).

B.   Fluorescent Lipids

In an alternative approach we fused fluorescent lipids into the apical membrane of MDCK cells as described above (see section III). However, we limited the lipid insertion to about half of the MDCK cells in a confluent monolayer and determined whether or not they could redistribute to non-

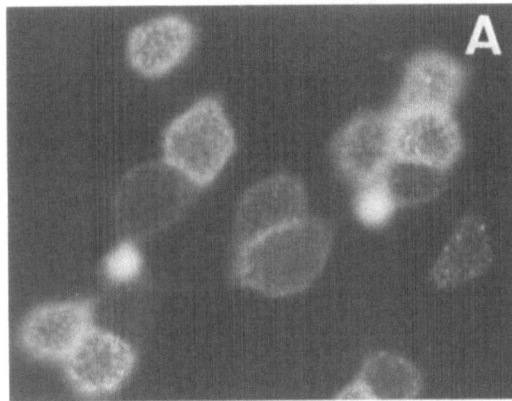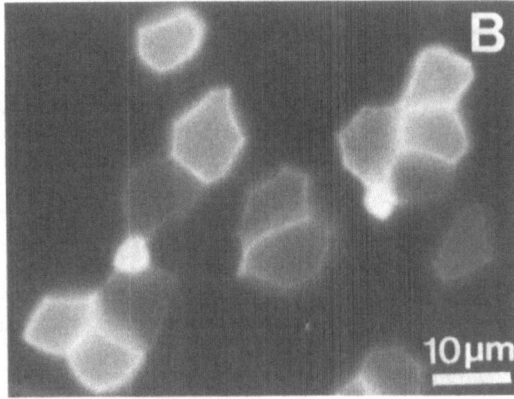

FIGURE 7   R18, fused into half of the strain II MDCK cells in a confluent monolayer, diffuses to the lateral domain but not to neighboring cells. (A) Apical surface and (B) lateral surface. The same pictures were obtained for *N*-Rh-PE when fused as REV into strain I MDCK cells (21).

fluorescent neighbor cells. This was done by using a low multiplicity of influenza virus for the infection. The liposomes subsequently bound and fused only to those cells which were infected. In this way we fused liposomes having a symmetrical transbilayer distribution of *N*-Rh-PE into MDCK strain I cells. In a parallel experiment octadecyl rhodamine, a less natural fluorescent lipid, was fused into MDCK strain II cells. The integrity of the tight junctions was confirmed by the maintenance of a high transepithelial electrical resistance for strain I MDCK cells and by the inaccessibility of a lateral membrane protein to a specific antibody when this was added from the apical surface in the case of strain II cells. In both cases, fusion could be demonstrated by the presence of fluorescent phospholipid in the lateral plasma membrane domain (Fig. 7). However, no diffusion of the fluorescent lipids to the adjacent noninfected cells was observed. Bright and dark cells remained juxtaposed in the cell monolayer for several hours at 0°C (91).

C.   Implications for Tight Junction Structure

We have presented evidence that three different types of lipids present in the apical plasma membrane domain are unable to diffuse into the apical plasma membrane of adjacent cells under conditions where the tight junctions were intact. Similar results were previously reported by Dragsten and colleagues (35). They observed that after insertion of a fluorescent amphiphile into a monolayer of epithelial cells and complete bleaching of the fluorescence in one cell by a laser beam, no fluorescence returned to the black cell from the surrounding bright cells. Assuming that the lipid molecules are mobile in the exoplasmic leaflet of the apical plasma membrane domain, these observations provide evidence that no continuity exists between the exoplasmic leaflets of the apical plasma membrane of neighboring cells. The results are thus difficult to reconcile with the hexagonal lipid model of the tight junction (89,90; Fig. 8a). The more reasonable model seems to be a protein bridge between two cells which inhibits the diffusion of proteins and lipids between the two domains in the exoplasmic

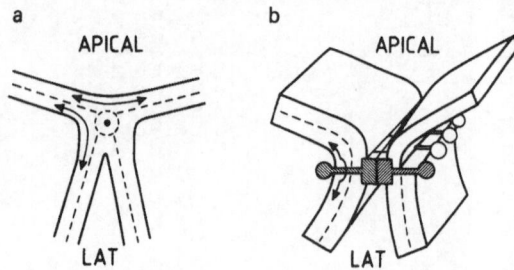

FIGURE 8 Two models for the structure of the tight junction. In model a
(89,90), the core of the tight junction consists of a hexagonal cylinder
of lipids. In model b the tight junction is formed by a protein bridge between two cells. The differences in lipid composition between the apical
and basolateral plasma membrane domain can be fully explained by both
models. However, the observation (91) that lipids do not pass from one
cell to the next favors model b. (Reprinted, with permission, from Ref. 21.)

leaflet (1,1a; Fig. 8b). Lipid molecules in the cytoplasmic leaflet would
diffuse freely between the protein subunits of the tight junction on the
cytoplasmic surface.

This model would also be more compatible with the occluding function
of the tight junction: to inhibit the passage of water-soluble molecules
through the lumen between the cells. It is well known that the tight junction displays an ion selectivity. It is virtually impermeable to anions like
Cl^- and this correlates nicely with our observation that phospholipid
molecules with negative charges on the surface of the cell are also unable
to pass this barrier. In contrast, the tight junction is permeable to cations
and it displays a defined sequence of cation conductances (17,92). How
this could be achieved in the lipid model of the tight junction is difficult
to envisage.

NOTE ADDED IN PROOF

After this chapter was written, further details of the fusion process were
provided by a study using lipid autoradiography and freeze-fracture electron
microscopy (93). The preparation of fusogenic liposomes by the functional
reconstitution of viral spike proteins has now also been accomplished (76a,b).
Further confirmation concerning the barrier properties of the tight junction
was obtained in that [^{14}C] dipalmitoyl PC, when inserted into both bilayer
leaflets of the apical membrane, was able to pass to the basolateral domain
(93), while a fluorescent SPH analog in the exoplasmic leaflet was unable
to diffuse past the tight junction in either direction (94). Finally, the
suggestion in section IV.D that lipid polarity in MDCK cells is generated
by a sphingolipid sorting event in the *trans* Golgi has now materialized
(94; for reviews see Refs. 95,96).

ACKNOWLEDGMENTS

The present work of the author is made possible by a senior fellowship
from the Royal Netherlands Academy of Arts and Sciences. The work on

which this paper is based was carried out at the European Molecular Biology Laboratory, Heidelberg, FRG, in the laboratory of Kai Simons, who co-authored many of the original papers. I am very grateful to Kai and to Angela Wandinger-Ness for critically reading the manuscript. I would like to thank my colleagues Karl Matlin, Barry Gumbiner, Stephen Fuller, and Jean Davoust for their contributions and critical comments and Thomas Gabran, Doris Hübsch, and Hilkka Virta for their excellent technical assistance. I am also grateful to Annie Steiner and Sabine Myers for typing the manuscript and to Petra Riedinger for preparing the drawings.

REFERENCES

1. Gumbiner, B. (1987). Structure, biochemistry, and assembly of epithelial tight junctions. *Am. J. Physiol.* 253:C749-C758.

1a. Stevenson, B. R., Anderson, J. M., and Bullivant, S. (1988). The epithelial tight junction: structure, function, and preliminary biochemical characterization. *Mol. Cell. Biochem.* 83:129-145.

2. Simons, K., and Fuller, S. D. (1985). Cell surface polarity in epithelia. *Annu. Rev. Cell Biol.* 1:243-288.

3. von Bonsdorff, C-H., Fuller, S. D., and Simons, K. (1985). Apical and basolateral endocytosis in Madin-Darby canine kidney (MDCK) cells grown on nitrocellulose filters. *EMBO J.* 4:2781-2792.

4. Gumbiner, B., and Louvard, D. (1985). Localized barriers in the plasma membrane: a common way to form domains. *TIBS* 10:435-438.

5. Forstner, G. G., Tanaka, K., and Isselbacher, K. J. (1968). Lipid composition of the isolated rat intestinal microvillus membrane. *Biochem. J.* 109:51-59.

6. Douglas, A. P., Kerley, R., and Isselbacher, K. J. (1972). Preparation and characterization of the lateral and basal plasma membranes of the rat intestinal epithelial cell. *Biochem. J.* 128:1329-1338.

7. Forstner, G. G., and Wherrett, J. R. (1973). Plasma membrane and mucosal glycosphingolipids in the rat intestine. *Biochim. Biophys. Acta* 306:446-459.

8. Kawai, K., Fujita, M., and Nakao, M. (1974). Lipid components of two different regions of an intestinal epithelial cell membrane of mouse. *Biochim. Biophys. Acta* 369:222-233.

9. Brasitus, T. A., and Schachter, D. (1980). Lipid dynamics and lipid-protein interactions in rat enterocyte basolateral and microvillus membranes. *Biochemistry* 19:2763-2769.

10. Hise, M. K., Mantulin, W. W., and Weinman, E. J. (1984). Fluidity and composition of brush border and basolateral membranes from rat kidney. *Am. J. Physiol.* 247:F434-F439.

11. Molitoris, B. A., and Simon, F. R. (1985). Renal cortical brush-border and basolateral membranes: Cholesterol and phospholipid composition and relative turnover. *J. Membrane Biol.* 83:207-215.

12. Carmel, G., Rodrigue, F., Carriere, S., and LeGrimellec, C. (1985). Composition and physical properties of lipids from plasma membranes of dog kidney. *Biochim. Biophys. Acta* 818:149-157.

13. Chapelle, S., and Gilles-Baillien, M. (1983). Phospholipids and cholesterol in brush border and basolateral membranes from rat intestinal mucosa. *Biochim. Biophys. Acta* 753:269-271.

14. Kremmer, T., Wisher, M. H., and Evans, W. H. (1976). The lipid composition of plasma membrane subfractions originating from the three major functional domains of the rat hepatocyte cell surface. *Biochim. Biophys. Acta* 455:655-664.

15. Meier, P. J., Sztul, E. S., Reuben, A., and Boyer, J. L. (1984). Structural and functional polarity of canalicular and basolateral plasma membrane vesicles isolated in high yield from rat liver. *J. Cell Biol.* 98:991-1000.

16. Misfeldt, D. S., Hamamoto, S. T., and Pitelka, D. R. (1976). Transepithelial transport in cell culture. *Proc. Natl. Acad. Sci. USA* 73: 1212-1216.

17. Cereijido, M., Robbins, E. S., Dolan, W. J., Rotunno, C. A., and Sabatini, D. D. (1978). Polarized monolayers formed by epithelial cells on a permeable and translucent support. *J. Cell Biol.* 77:853-880.

18. Hansson, G. C., Simons, K., and van Meer, G. (1986). Two strains of the Madin-Darby canine kidney (MDCK) cell line have distinct glycosphingolipid compositions. *EMBO J.* 5:483-489.

19. Fuller, S. D., and Simons, K. (1986). Transferrin receptor polarity and recycling accuracy in "tight" and "leaky" strains of Madin-Darby canine kidney cells. *J. Cell Biol.* 103:1767-1779.

20. Rodriguez Boulan, E., and Sabatini, D. D. (1978). Asymmetric budding of viruses in epithelial monolayers: A model system for study of epithelial polarity. *Proc. Natl. Acad. Sci. USA* 75:5071-5075.

21. van Meer, G., and Simons, K. (1986). The function of tight junctions in maintaining differences in lipid composition between the apical and the basolateral cell surface domains of MDCK cells. *EMBO J.* 5:1455-1464.

22. Klenk, H-D., and Choppin, P. W. (1970). Plasma membrane lipids and parainfluenza virus assembly. *Virology* 40:939-947.

23. Klenk, H-D., and Choppin, P. W. (1970). Glycosphingolipids of plasma membranes of cultured cells and an enveloped virus (SV 5) grown in these cells. *Proc. Natl. Acad. Sci. USA* 66:57-64.

24. van Meer, G., Fuller, S. D., and Simons, K. (1985). Sorting of (glyco-?) lipids in epithelial cells: Possible implications for tight junction structure. In: *Protein Transport and Secretion* (M-J. Gething, ed.). Cold Spring Harbor Laboratory, Cold Spring Harbor, NY, pp. 179-183.

25. van Meer, G., and Simons, K. (1982). Viruses budding from either the apical or the basolateral plasma membrane domain of MDCK cells have unique phospholipid compositions. *EMBO J.* 1:847-852.

26. Nichols, G. E., Shiraishi, T., Allietta, M., Tillack, T. W., and Young, W. W. Jr. (1987). Polarity of the Forssman glycolipid in MDCK epithelial cells. *Biochim. Biophys. Acta* 930:154-166.

26a. Nichols, G. E., Shiraishi, T., and Young, W. W. Jr. (1987). Polarity of neutral glycolipids, gangliosides, and sulfated lipids in MDCK epithelial cells. *J. Lipid Res.* 29:1205-1213.

26b. Nichols, G. E., Lovejoy, J. C., Borgman, C. A., Sanders, J. M., and Young, W. W. Jr. (1986). Isolation and characterization of two types of MDCK epithelial cell clones based on glycosphingolipid pattern. *Biochim. Biophys. Acta* 887:1-12.

27. Spiegel, S., Blumenthal, R., Fishman, P. H., and Handler, J. S. (1985). Gangliosides do not move from apical to basolateral plasma membrane in cultured epithelial cells. *Biochim. Biophys. Acta* 821:310-318.

28. Zalc, B., Helwig, J. J., Ghandour, M. S., and Sarlieve, L. (1978). Sulfatide in the kidney: How is this lipid involved in sodium chloride transport? *FEBS Lett.* 92:92-96.

29. Turner, R. J., Thompson, J., Sariban-Sohraby, S., and Handler, J. S. (1985). Monoclonal antibodies as probes of epithelial membrane polarization. *J. Cell Biol.* 101:2173-2180.

30. Rothman, J. E., Tsai, D. K., Dawidowicz, E. A., and Lenard, J. (1976). Transbilayer phospholipid asymmetry and its maintenance in the membrane of influenza virus. *Biochemistry* 15:2361-2370.

31. Lewis, B. A., Gray, G. M., Coleman, R., and Michell, R. H. (1975). Differences in the enzymic, polypeptide, glycopeptide, glycolipid and phospholipid compositions of plasma membranes from the two surfaces of intestinal epithelial cells. *Biochem. Soc. Trans.* 3:752-753.

31a. van Meer, G. (1988). How epithelial cells grease their microvilli. *TIBS* 13:242-243.

32. Miller, R. G., and Baldridge, W. H. (1985). The tight junction as a barrier to cholesterol in canine epithelial cells. *J. Ultrastruct. Res.* 90:275-285.

33. Severs, N. J., and Robenek, H. (1983). Detection of microdomains in biomembranes: An appraisal of recent developments in freeze-fracture cytochemistry. *Biochim. Biophys. Acta* 737:373-408.

34. Wattenberg, B. W., and Silbert, D. F. (1983). Sterol partitioning among intracellular membranes. Testing a model for cellular sterol distribution. *J. Biol. Chem.* 258:2284-2289.

35. Dragsten, P. R., Blumenthal, R., and Handler, J. S. (1981). Membrane asymmetry in epithelia: is the tight junction a barrier to diffusion in the plasma membrane? *Nature* 294:718-722.

36. Pesonen, M., Ansorge, W., and Simons, K. (1984). Transcytosis of the G protein of vesicular stomatitis virus after implantation into the apical plasma membrane of Madin-Darby canine kidney cells. I. Involvement of endosomes and lysosomes. *J. Cell Biol.* 99:796-802.

37. Gregoriadis, G. (1980). The liposome drug-carrier concept: Its development and future. In: *Liposomes in Biological Systems* (G. Gregoriadis and A. C. Allison, eds.). Wiley, New York, pp. 25-86.

38. White, J., Kielian, M., and Helenius, A. (1983). Membrane fusion proteins of enveloped animal viruses. *Q. Rev. Biophys.* 16:151-195.

39. van Meer, G., and Simons, K. (1983). An efficient method for introducing defined lipids into the plasma membrane of mammalian cells. *J. Cell Biol.* 97:1365-1374.

40. van Meer, G., Davoust, J., and Simons, K. (1985). Parameters affecting low-pH-mediated fusion of liposomes with the plasma membrane of cells infected with influenza virus. *Biochemistry* 24:3593-3602.

41. White, J., Matlin, K., and Helenius, A. (1981). Cell fusion by Semliki Forest, influenza, and vesicular stomatitis viruses. *J. Cell Biol.* 89:674-679.

42. White, J., Kartenbeck, J., and Helenius, A. (1982). Membrane fusion activity of influenza virus. *EMBO J.* 1:217-222.

43. Rodriguez Boulan, E., and Pendergast, M. (1980). Polarized distribution of viral envelope proteins in the plasma membrane of infected epithelial cells. *Cell* 20:45-54.

44. Maeda, T., and Ohnishi, S. (1980). Activation of influenza virus by acidic media causes hemolysis and fusion of erythrocytes. *FEBS Lett.* 122:283-287.

45. Huang, R. T. C., Rott, R., and Klenk, H-D. (1981). Influenza viruses cause hemolysis and fusion of cells. *Virology* 110:243-247.

46. Stegmann, T., Hoekstra, D., Scherphof, G., and Wilschut, J. (1985). Kinetics of pH-dependent fusion between influenza virus and liposomes. *Biochemistry* 24:3107-3113.

47. Struck, D. K., Hoekstra, D., and Pagano, R. E. (1981). Use of resonance energy transfer to monitor membrane fusion. *Biochemistry* 20:4093-4099.

48. Fung, B. K., and Stryer, L. (1978). Surface density determination in membranes by fluorescence energy transfer. *Biochemistry* 17:5241-5248.

49. Davoust, J., Devaux, P. F., and Leger, L. (1982). Fringe pattern photobleaching, a new method for the measurement of transport coefficients of biological macromolecules. *EMBO J.* 1:1233-1238.

50. Omitted in proof.

51. Kamp, H. H., Wirtz, K. W. A., and van Deenen, L. L. M. (1973). Some properties of phosphatidylcholine exchange protein purified from beef liver. *Biochim. Biophys. Acta* 318:313-325.

52. Rothman, J. E., and Dawidowicz, E. A. (1975). Asymmetric exchange of vesicle phospholipids catalyzed by the phosphatidylcholine exchange protein. Measurement of inside-outside transitions. *Biochemistry* 14:2809-2816.

53. Zilversmit, D. B. (1971). Stimulation of phospholipid exchange between mitochondria and artificially prepared phospholipid aggregates by a soluble fraction from liver. *J. Biol. Chem.* 246:2645-2649.

54. Matlin, K. S., and Simons, K. (1983). Reduced temperature prevents transfer of a membrane glycoprotein to the cell surface but does not prevent terminal glycosylation. *Cell* 34:233-243.

55. Matlin, K. S., and Simons, K. (1984). Sorting of an apical plasma membrane glycoprotein occurs before it reaches the cell surface in cultured epithelial cells. *J. Cell Biol.* 99:2131-2139.

56. Suzuki, Y., Morioka, T., and Matsumoto, M. (1980). Action of ortho- and paramyxovirus neuraminidase on gangliosides. Hydrolysis of ganglioside GM_1 by Sendai virus neuraminidase. *Biochim. Biophys. Acta* 619:632-639.

57. Slepushkin, V. A., Bukrinskaya, A. G., Prokazova, N. V., Zhigis, L. S., Reshetov, P. D., Shaposhnikova, J. I., and Bergelson, L. D. (1985). Action of influenza virus neuraminidase on gangliosides. *FEBS Lett.* 182:273-277.

57a. Suzuki, Y., Matsunaga, M., and Matsumoto, M. (1985). N-Acetylneuraminyllactosylceramide, G_{M_3}-NeuAc, a new influenza A virus receptor which mediates the adsorption-fusion process of viral infection. *J. Biol. Chem.* 260:1362-1365.

58. Bergelson, L. D., Bukrinskaya, A. G., Prokazova, N. V., Shaposhnikova, G. I., Kocharov, S. L., Shevchenko, V. P., Kornilaeva, G. V., and Fomina-Ageeva, E. V. (1982). Role of gangliosides in reception of influenza virus. *Eur. J. Biochem.* 128:467-474.

59. Holmgren, J., Svennerholm, L., Elwing, H., Fredman, P., and Strannegard, Ö. (1980). Sendai virus receptor: Proposed recognition structure based on binding to plastic-adsorbed gangliosides. *Proc. Natl. Acad. Sci. USA* 77:1947-1950.

60. Markwell, M. A. K., Svennerholm, L., and Paulson, J. C. (1981). Specific gangliosides function as host cell receptors for Sendai virus. *Proc. Natl. Acad. Sci. USA* 78:5406-5410.

61. Klenk, H-D., Rott, R., Orlich, M., and Bloedorn, J. (1975). Activation of influenza A viruses by trypsin treatment. *Virology* 68:426-439.

62. Gething, M. J., White, J. M., and Waterfield, M. D. (1978). Purification of the fusion protein of Sendai virus: Analysis of the NH_2-terminal sequence generated during precursor activation. *Proc. Natl. Acad. Sci. USA* 75:2737-2740.

63. Skehel, J. J., Bayley, P. M., Brown, E. B., Martin, S. R., Waterfield, M. D., White, J. M., Wilson, I. A., and Wiley, D. C. (1982). Changes in the conformation of influenza virus hemagglutinin at the pH-optimum of virus-mediated membrane fusion. *Proc. Natl. Acad. Sci. USA* 79: 968-972.

64. Doms, R. W., Helenius, A., and White, J. (1985). Membrane fusion activity of the influenza virus hemagglutinin. The low pH-induced conformational change. *J. Biol. Chem.* 260:2973-2981.

65. Ruigrok, R. W. H., Wrigley, N. G., Calder, L. J., Cusack, S., Wharton, S. A., Brown, E. B., and Skehel, J. J. (1986). Electron microscopy of the low pH structure of influenza virus haemagglutinin. *EMBO J.* 5:41-49.

66. Eto, Y., and Suzuki, K. (1973). Cholesterol ester metabolism in rat brain. A cholesterol ester hydrolase specifically localized in the myelin sheath. *J. Biol. Chem.* 248:1986-1991.

67. Riddle, M. C., Smuckler, E. A., and Glomset, J. A. (1975). Cholesteryl ester hydrolytic activity of rat liver plasma membrane. *Biochim. Biophys. Acta* 388:339-348.

68. Nilsson, A. (1976). Hydrolysis of chyle cholesterol esters with cell-free preparations of rat liver. *Biochim. Biophys. Acta* 450:379-389.

69. Teng, M-H., and Kaplan, A. (1974). Purification and properties of rat liver lysosomal lipase. *J. Biol. Chem.* 249:1064-1070.

70. Pittman, R. C., and Steinberg, D. (1977). Activatable cholesterol esterase and triacylglycerol lipase activities of rat adrenal and their relationship. *Biochim. Biophys. Acta* 487:431-444.

71. Cook, K. G., Yeaman, S. J., Stralfors, P., Fredrikson, G., and Belfrage, P. (1982). Direct evidence that cholesterol ester hydrolase from adrenal cortex is the same enzyme as hormone-sensitive lipase from adipose tissue. *Eur. J. Biochem.* 125:245-249.

72. Blaner, W. S., Prystowsky, J. H., Smith, J. E., and Goodman, D. S. (1984). Rat liver retinyl palmitate hydrolase activity. Relationship to cholesteryl oleate and triolein hydrolase activities. *Biochim. Biophys. Acta* 794:419-427.

73. Pagano, R. E., and Sleight, R. G. (1985). Defining lipid transport pathways in animal cells. *Science* 229:1051-1057.

74. Hokin, L. E. (1985). Receptors and phosphoinositide-generated second messengers. *Annu. Rev. Biochem.* 54:205-235.

75. Doxsey, S. J., Sambrook, J., Helenius, A., and White, J. (1985). An efficient method for introducing macromolecules into living cells. *J. Cell Biol.* 101:19-27.

76. Sambrook, J., Rodgers, L., White, J., and Gething, M-J. (1985). Lines of BPV-transformed murine cells that constitutively express influenza virus hemagglutinin. *EMBO J.* 4:91-103.

76a. Metsikkö, K., van Meer, G., and Simons, K. (1986). Reconstitution of the fusogenic activity of vesicular stomatitis virus. *EMBO J.* 5: 3429-3435.

76b. Stegmann, T., Morselt, H. W. M., Booy, F. P., van Breemen, J. F. L., Scherphof, G., and Wilschut, J. (1987). Functional reconstitution of influenza virus envelopes. *EMBO J.* 6:2651-2659.

77. Fuller, S., von Bonsdorff, C-H., and Simons, K. (1984). Vesicular stomatitis virus infects and matures only through the basolateral surface of the polarized epithelial cell line, MDCK. *Cell* 38:65-77.

78. Sundler, R., Alberts, A. W., and Vagelos, P. R. (1978). Phospholipases as probes for membrane sidedness. Selective analysis of the outer monolayer of asymmetric bilayer vesicles. *J. Biol. Chem.* 253: 5299-5304.

79. Wilschut, J. C., Regts, J., and Scherphof, G. (1979). Action of phospholipase A$_2$ on phospholipid vesicles. Preservation of the membrane permeability barrier during asymmetric bilayer degradation. *FEBS Lett.* 98:181-186.

80. Kumar, A., and Gupta, C. M. (1985). Transbilayer phosphatidylcholine distributions in small unilamellar sphingomyelin-phosphatidylcholine vesicles: Effect of altered polar head group. *Biochemistry* 24:5157-5163.

81. Helenius, A., Sarvas, M., and Simons, K. (1981). Asymmetric and symmetric membrane reconstitution by detergent elimination. *Eur. J. Biochem.* 116:27-35.

82. Szoka, F., Jr., and Papahadjopoulos, D. (1978). Procedure for preparation of liposomes with large internal aqueous space and high capture by reverse-phase evaporation. *Proc. Natl. Acad. Sci. USA* 75:4194-4198.

83. Gumbiner, B., and Simons, K. (1986). A functional assay for proteins involved in establishing an epithelial occluding barrier: Identification of a uvomorulin-like polypeptide. *J. Cell Biol.* 102:457-468.

84. Bretscher, M. S. (1972). Phosphatidyl-ethanolamine: Differential labelling in intact cells and cell ghosts of human erythrocytes by a membrane-impermeable reagent. *J. Mol. Biol.* 71:523-528.

85. Op den Kamp, J. A. F. (1979). Lipid asymmetry in membranes. *Annu. Rev. Biochem.* 48:47-71.

86. Seigneuret, M., and Devaux, P. F. (1984). ATP-dependent asymmetric distribution of spin-labeled phospholipids in the erythrocyte membrane: relation to shape changes. *Proc. Natl. Acad. Sci. USA* 81:3751-3755.

87. Griffiths, G., and Simons, K. (1986). The *trans* Golgi network: Sorting at the exit site of the Golgi complex. *Science* 234:438-443.

88. Thompson, T. E. and Tillack, T. W. (1985). Organization of glycosphingolipids in bilayers and plasma membranes of mammalian cells. *Annu. Rev. Biophys. Biophys. Chem.* 14:361-386.

89. Pinto da Silva, P., and Kachar, B. (1982). On tight-junction structure. *Cell* 28:441-450.

90. Kachar, B., and Reese, T. S. (1982). Evidence for the lipidic nature of tight junction strands. *Nature* 296:464-466.

91. van Meer, G., Gumbiner, B., and Simons, K. (1986). The tight junction does not allow lipid molecules to diffuse from one epithelial cell to the next. *Nature* 322:639-641.

91a. Nichols, G. E., Borgman, C. A., and Young, W. W. Jr. (1986). On tight junction structure: Forssman glycolipid does not flow between MDCK cells in an intact epithelial monolayer. *Biochem. Biophys. Res. Commun.* 138:1163-1169.

92. Madara, J. L., Dharmsathaphorn, K. (1985). Occluding junction structure-function relationships in a cultured epithelial monolayer. *J. Cell Biol.* 101:2124-2133.

93. Knoll, G., Burger, K. N. J., Bron, R., van Meer, G., and Verkleij, A. J. (1988). Fusion of liposomes with the plasma membrane of epithelial cells: Fate of incorporated lipids as followed by freeze-fracture and autoradiography of plastic sections. *J. Cell Biol.* 107:2511-2521.

94. van Meer, G., Stelzer, E. H. K., Wijnaendts-van-Resandt, R. W., and Simons, K. (1987). Sorting of sphingolipids in epithelial (Madin-Darby canine kidney) cells. *J. Cell Biol.* 105:1623-1635.

95. Simons, K., and van Meer, G. (1988). Lipid sorting in epithelial cells. *Biochemistry* 27:6197-6202.

96. van Meer, G. (1989). Polarity and polarized transport of membrane lipids in a cultured epithelium. In: *Modern Cell Biology: Functional Epithelial Cells in Culture* (B. H. Satir, ed.). Alan R. Liss, New York, pp. 41-68.

34

Structure and Function of Bacterial Membranes
Insertion of Exogenous Lipids and Proteins by Fusion with Lipid Vesicles

ARNOLD J. M. DRIESSEN and WILHELMUS N. KONINGS

University of Groningen, Haren, The Netherlands

I. INTRODUCTION

One of the most striking aspects of bacteria is their wide biological diversity and nutritional versatility. Microorganisms can be found in the most adverse environments. For instance, species of the sulfur-oxidizing organism *Sulfolobus* are able to grow in media with pH values as low as 1 and at temperatures as high as 85°C. On the other hand, several alkalophilic bacteria have been isolated that can grow at pH values as high as 10.5. To cope with these extreme environments the cytoplasm of a bacterial cell is protected by the cell envelope. In most bacteria this envelope is formed by a cell wall (peptidoglycan layer) and a cytoplasmic membrane located on the inner side of the cell wall. In Gram-negative bacteria a third layer, the outer membrane, is found outside the cell wall.

The cytoplasmic membrane of bacteria is functionally more diverse than the plasma membrane of eukaryotes. It constitutes a very selective barrier between the interior of the cell and the external environment, which usually has a completely different and more dilute ionic composition. Furthermore, the cytoplasmic membrane of bacteria performs a vital role in metabolic free-energy transduction. For instance, in aerobic bacteria the respiratory electron transport systems are located in the cytoplasmic membrane, while in eukaryotes the components of the respiratory electron transport chain are mainly found in the mitochondrial inner membrane. In addition to these energy-transducing systems which generate electro-chemical ion gradients across the cytoplasmic membrane (primary transport systems), a large number of solute transport systems (secondary transport systems and group translocation systems) are found in the cytoplasmic membrane (Fig. 1). These systems allow the selective translocation of solutes across the cytoplasmic membrane to which this membrane is otherwise impermeable. Most translocation processes occur by specific transport proteins at the expense of metabolic energy, the so-called "active transport processes." In addition, two different mechanisms of solute movement down a concentration gradient can occur: "passive diffusion," which takes place

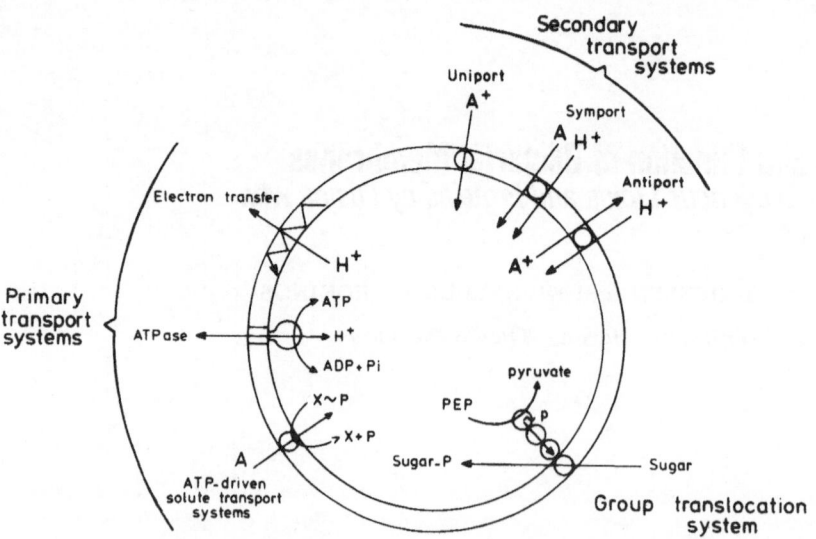

FIGURE 1   Schematic presentation of the different solute transport systems
found in bacteria. The ATP driven transport systems do not necessarily
use ATP but possibly other phosphate-bond energy.

without specific interaction with membrane proteins, and "facilitated
diffusion," which is mediated by specific membrane proteins.

Active transport processes are generally subdivided in three groups
(Fig. 1):

1. Primary transport systems. Transport is by specific enzyme systems
which convert chemical, redox, or light energy into electrochemical energy.
These transport systems include the cytochrome-linked electron transfer
chains, the ATPase complex, the light-driven proton pump bacteriorhodopsin, and the phosphate-bond-driven solute transport systems.

Proton pumps translocate protons from the cytoplasm to the external
medium, which results in the generation of an electrochemical proton
gradient ($\Delta\tilde{\mu}_{H^+}$). This $\Delta\tilde{\mu}_{H^+}$ exerts a force on the protons, the protonmotive
force ($\Delta\tilde{\mu}_{H^+}/F$, Δp), which consists of a transmembrane electrical potential
($\Delta\psi$) and a chemical gradient of protons across the membrane ($Z\Delta pH$,
expressed in mV).

In equation form:

$$\Delta\tilde{\mu}_{H^+}/F = \Delta p = \Delta\psi - Z\Delta pH \text{ (mV)} \tag{1}$$

where Z equals 2.3 RT/F and ΔpH equals ($pH_{in} - pH_{out}$).

2. Secondary transport systems. Solute transport by these systems
is driven by the components of the proton motive force and the solute
gradient (1). The driving force supplied by the protonmotive force depends
on the total charge and the number of protons that are cotranslocated in
the transport process. The following general equation holds for secondary
transport processes:

$$\text{Driving force} = \Delta\tilde{\mu}_A/F + (n+m)\,\Delta\psi - nZ\,\Delta pH \text{ (mV)} \tag{2}$$

where A is the translocated solute with charge m;

$$\Delta \bar{\mu}_A / F = Z \log \frac{(A)_{in}}{(A)_{out}} \tag{3}$$

and n is the number of translocated protons.

Major secondary transport systems found in bacteria are: electrogenic cation uniporters, sugar/H^+ and amino acid/H^+ symporters, and cation/H^+ antiporters (Fig. 1).

3. Group translocation: The solute is chemically modified during the translocation across the membrane by an enzyme system associated with the transport protein. The main group translocation systems found in bacteria are the phosphoenolpyruvate (PEP)-dependent sugar transport systems (PTS).

The difficulties in the elucidation of the properties of membrane-associated processes in intact bacteria have led to the development of procedures for the isolation of closed cytoplasmic membranes that retain the functional and structural properties of this cytoplasmic membrane. These isolated bacterial membrane vesicles have been shown to be excellent model systems for studies on energy-transducing processes (for review, see Refs. 1,2). Isolation procedures usually comprise two essential steps (Fig. 2). First, the bacterial cell is converted into an osmotically sensitive form (protoplast or sphaeroplast) in a hypertonic medium by selective removal of the cell wall (by the action of lysozyme). From Gram-negative bacteria the outer membrane must be removed first with divalent-cation-chelating agents in order to make the cell wall accessible to lysozyme. In the second step, osmotic lysis of these proto- or sphaeroplasts is induced by dilution into a hypotonic medium in the presence of deoxyribonuclease (DNase) and ribonuclease (RNase) in order to hydrolyze the liberated DNA and RNA. This procedure yields sealed membrane vesicles, predominantly with the in vivo orientation of the cytoplasmic membrane (Fig. 2). Membrane vesicles with an inside-out orientation (like submitochondrial particles) can be obtained by ultrasonic irradiation or high pressure treatment of proto- or sphaeroplasts.

The membrane-associated enzymes that remain functional in these membrane vesicles include the primary transport systems. In the presence of suitable energy donors, a $\Delta\dot{p}$ can be generated across the membrane that forms the driving force for secondary transport of solutes (for reviews, see Refs. 1-3).

Many attempts have been made to modify the chemical composition of bacterial membranes in order to examine structure-function relationships. This has been achieved either (1) by the isolation of mutants defective in a single step of the biogenesis of membrane lipids or (2) by the treatment of cells with an inhibitor of phospholipid metabolism resulting in an alteration of phospholipid composition (4). Another approach has been the fusion of bacterial membranes with membranes of defined composition. The development of rapid kinetic membrane fusion assays has made it possible to study the efficiency of fusion (for review, see Ref. 5), and this has subsequently led to the development of improved procedures to induce fusion between biological membranes and lipid vesicles. These fusion procedures are a valuable tool to analyze the role of specific protein-protein and protein-phospholipid interactions in membrane functions. In addition to the manipulation of the lipid-to-protein ratio or lipid composition, fusion

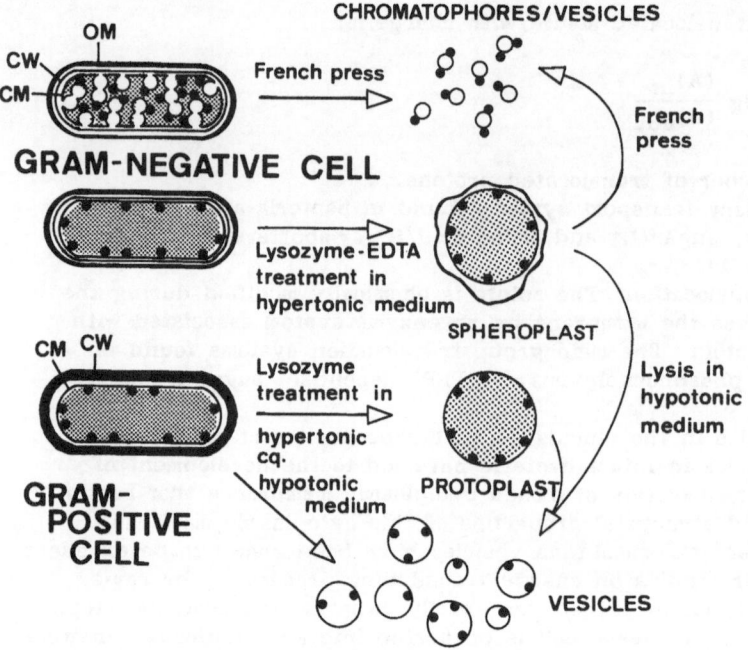

FIGURE 2 Scheme of the isolation procedure of membrane vesicles. OM,
outer membrane; CW, cell wall (peptidoglycan layer); CM, cell membrane.
Black dots indicate the location of the catalytic site of the ATPase complex.

can also be used to insert membrane proteins, reconstituted in lipid vesicles,
into biological membranes.

The recent advances in studies on structure and function of bacterial
membranes using membrane fusion techniques are discussed in this chapter,
with special emphasis on the use of fused membranes in studies on solute
transport.

II. LIPID-ENRICHED PHOTOSYNTHETIC MEMBRANES

A. The Cyclic Electron Transfer Chain and Associated Pigments

Membrane fusion techniques have been most extensively utilized in studies
on the topological arrangement of protein complexes and their interaction
in the cytoplasmic membrane of phototrophic bacteria. Many phototrophic
bacteria develop extensive intracytoplasmic membrane systems when grown
under low oxygen pressure and low light intensity (6). These intracytoplasmic membranes result from invaginations of the cytoplasmic membrane
and are typically tubular, lamellar, or vesicular structures. Closed membrane
vesicles, the so-called chromatophores, can be derived from these membrane
structures after cell disruption (Fig. 2). These chromatophores have provided a well-defined system for the investigation of the mechanism of light-
energy conversion in bacterial photosynthesis, especially in purple nonsulfur
bacteria. They contain in their membrane the light-harvesting complexes
as well as the photochemical electron transport system. Chromatophores
have retained the topology of the invaginated cytoplasmic membrane in the

intact cell. The catalytic site of the ATPase which faces the cytoplasm now faces the external medium. In chromatophores the process of ATP synthesis coupled to light-driven electron transfer, a process called photophosphorylation, can be well studied. Light energy is absorbed by bacteriochlorophyll (Bchl) and carotenoids. Most species of Rhodospirillaceae (purple nonsulfur bacteria) contain two light-harvesting systems, the B870 and the B800-850 complexes (6). Excitation energy is finally transferred to the special pair bacteriochlorophyll (P_{870}) (Bchl)$_2$ in the reaction centers (RC), or the energy is dissipated as fluorescence or heat (7). Excitation of the RC will release an electron that is subsequently transferred via a series of redox components back to the RC (Fig. 3). The redox components involved are quinones, b- and c-type cytochromes and FeS clusters. Within the RC the electron is channeled to the primary acceptor quinone (Ubiquinone$_{10}$), which in turn reduces a secondary quinone. This step probably includes the uptake and binding of a proton from the internal medium (Fig. 3). Simultaneously, the oxidized P_{870} becomes rereduced by an electron derived from cytochrome c_2, associated with the RC at the outer surface of the intact cell, e.g., the inner surface of the chromatophores.

Electron flow from the secondary quinone to oxidized cytochrome c_2 is catalyzed by the ubiquinol:cytochrome c_2 oxidoreductase (bc$_1$ complex). The overall process, which is called *cyclic electron transfer*, results in the translocation of protons from the cytoplasm to the internal medium in intact cells and from the medium to the inner-chromatophoral space in chromatophores. In chromatophores, therefore, cyclic electron transfer leads to the generation of a protonmotive force, inside positive and acid. Close interactions (collisions) of redox components must occur to facilitate

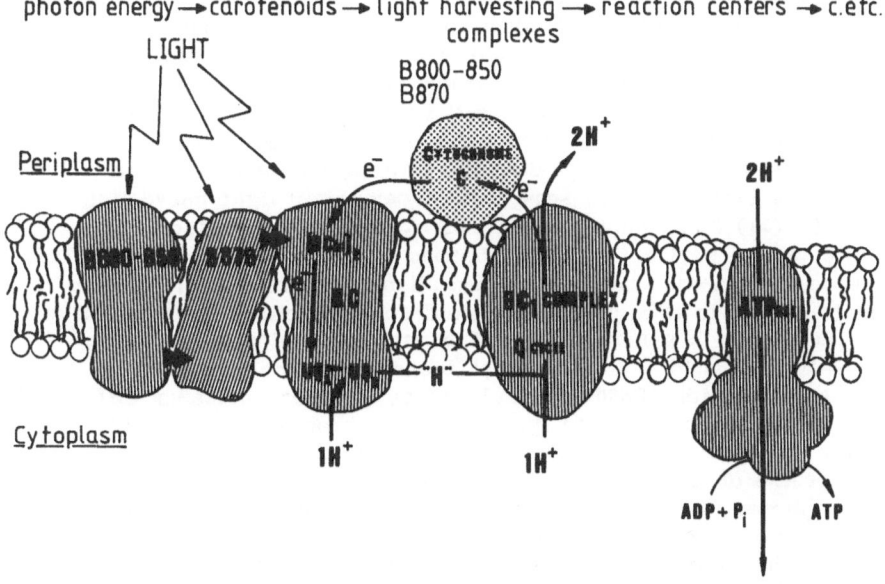

FIGURE 3 Scheme of light-harvesting complexes, cyclic electron transfer chain components, and ATPase complex in phototropic bacteria-like *Rhodopseudomonas sphaeroides*.

electron transfer between the functional groups. Interactions must also occur for energy transfer between light-harvesting complexes and RC. Therefore, the mean distance between the components in the membrane and their arrangement as complexes in the membrane will be an important factor in determining energy transfer efficiencies and electron transfer rates. Although many techniques have been used to study these interactions, a number of questions remain. One approach to investigate the lateral arrangement of components in a membrane and their interaction is to dilute the components in the membrane by lipid enrichment, followed by measurement of the resulting functional effects. The recent progress made in studies on structure-function relationships of photosynthetic membranes with lipid-enriched chromatophores will be reviewed.

B. General Properties of Lipid-Enriched Chromatophores

To increase the bilayer surface of chromatophores by fusion with lipid vesicles, several fusion-inducing techniques appear to be appropriate. Most frequently, fusion has been induced either by the low pH (8–11) or by the freeze-thaw method (10,12), followed by a brief sonication step (13). However, fusion induced by calcium phosphate precipitation (10) and calcium/poly(ethylene glycol) (14) is as efficient. Despite the large number of reports on fusion between chromatophores and lipid vesicles, little information is available about the phospholipid requirements of these fusion events. In almost all cases, lipid vesicles containing soybean phospholipids (asolectin) or phospholipids extracted from chromatophores have been used. The resultant fused membranes are relatively heterogeneous in size and density. Depending on the initial phospholipid-to-protein ratio and fusion procedure employed, three to four membrane fractions are obtained upon fractionation on discontinuous sucrose gradients. Phospholipid-to-protein ratios (w/w) can vary up to 15-fold between these fractions. Concomitant with a decrease in buoyant density of the chromatophores, a decrease in the number of intramembranous particles per membrane area is observed, as revealed by freeze-fracture electron microscopy (11,13). In fused membranes obtained by freeze-thaw sonication (13) or low pH (11), no multilamellar membrane structures were detected.

Fusion does not dramatically alter the protein composition of the chromatophores (11), although some loss of peripheral light-harvesting complex B800 (11) and soluble cytochrome c_2 has been reported (13). Apparently, also a part of B800-850 becomes denatured after fusion (14). Furthermore, no drastic changes in orientation of electron transport chain components appear to take place since lipid-enriched chromatophores have retained the ability to generate upon illumination a protonmotive force, inside positive and acid (8,10,13). Electrochromic bandshifts of the endogenous carotenoids (B800-850) induced by light or potassium diffusion potentials are still observable in lipid-enriched chromatophores (8,10,13). In line with the increased capacitance expected from an increased surface area, a decrease of the extent of this signal is observed upon fusion (13). In chromatophores fused by freeze-thaw sonication the decay of the electrochromic bandshift is still as slow as in control chromatophores (13).

Chromatophores fused by low pH show a faster decay of the electrochromic bandshift and also of pH gradients (10). This is an indication for increased ion permeability of the membranes of these preparations. In agreement with this, an increase of the apparent K_m for light, estimated

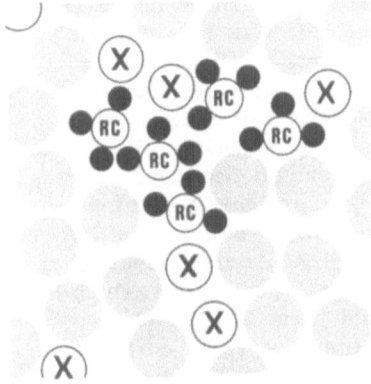

FIGURE 4 Model of the two-dimensional organization of the photosynthetic apparatus of *Rhodopseudomonas capsulata*. The lateral topography of RC, B870 (small black circles), and B800–850 light-harvesting complexes (large gray circles) is shown. The ATPase and quinole: cytochrome c_2 oxido-reductase are shown by X, but their topography has not been studied. (From Ref. 17.)

from the rates of photophosphorylation under continuous illumination, is observed in these membranes (12).

C. Interaction Between Light-Harvesting Complexes and Reaction Centers

One of the questions tackled using lipid-enriched chromatophores concerns the membrane topology of the light-harvesting complexes and their inter-action with each other and with reaction centers (RC). Kinetic studies on fluorescence emission suggest that the B870 complexes are organized in the membrane in such a way that many RC are interconnected and that the B800–850 complexes are arranged peripherally, interconnecting many photosynthetic units (15,16) (Fig. 4). The size of the photosynthetic unit, i.e., the number of light-harvesting pigment molecules per RC, and the topography of the photosynthetic units are important for the efficiency of energy transfer (17). Different variants of this lake model are proposed in which these photosynthetic units are either functionally and physically separated or energetically interconnected (17). In the latter case, lipid dilution of chromatophores is expected to disrupt downhill energy transfer to the RC, unless tight protein-protein interactions are of vital importance. Conclusions regarding these interactions can only be drawn when lipid enrichment itself does not directly affect the spectral properties of one or more of the components involved in excitation energy transfer.

Takemoto and co-workers (11) showed that lipid enrichment of chromato-phores of *Rhodopseudomonas capsulata* has a major effect on excitation energy transfer between B870 and RC. Excitation energy transfer efficien-cies, estimated from the fluorescence of the B870 complex, decreased with increasing lipid-to-protein ratio. The activity of the RC was hardly affected by the fusion procedure, while energy coupling between B800–850 and B870 or RC was not changed, in contrast to studies performed with chro-matophores of *Rps. sphaeroides* (14). A decrease in excitation energy transfer efficiency between B800–850 and B870 or RC was observed, while excitation energy transfer efficiency between B870 and RC was not altered. Apparently, *Rps. sphaeroides* and *Rps. capsulata* differ in this respect, which would mean that a fundamental difference exists in the molecular arrangement of light-harvesting complexes in these species. It should be noted that the chromatophore preparations used differ with respect to their B800-850 to B870 ratio, which was high in chromatophores of *Rps. sphaeroides* (14). Alternatively, these differences should be explained in terms of different protein-protein interactions or protein-lipid interactions. In *Rps. capsulata*, the observed effects can be explained either by assuming that B870 is physically separated from RC, or by assuming a dispersion of photosynthetic units which are normally close enough to allow interunit energy transfer. It is obvious that these studies alone do not allow discrimination of which variants of the lake model describe the interactions most precisely. The findings with lipid-enriched chromatophores of *Rps. capsulata* also argue against the possibility that RC and B870 complexes are held tightly together solely by protein-protein interactions in discrete photosynthetic units. Studies with chemical cross-linking agents could possibly reveal which of the alternative associations of complexes occurs (11) Another approach to study protein-protein interactions is to reconstitute the participating proteins. Garcia and co-workers (18) fused chromatophores of a mutant strain of *Rps. capsulata* lacking the B800-850 complex with proteoliposomes containing purified B800-850 complex. Both excitation energy transfer to B870 or RC and the electrochromic band-shift of the carotenoids of the B800-850 complex were restored.

D.   Interaction Between Redox Components

Another type of study performed with lipid-enriched chromatophores is the interaction of electron-transfer chain components, in analogy to investigations by Hackenbrock and co-workers on electron flow in lipid-enriched mitochondrial inner membranes (19; see also Chapter 35). Photosynthetic membranes have the advantage over mitochondrial membranes that a time resolution of the kinetics of light-induced electron flow can be obtained in the millisecond time scale.

Chromatophores contain ubiquinones in a large excess over the other electron transport chain components (20). These molecules behave as a large, thermodynamically homogeneous pool (20) of about 25-30 ubiquinones per reaction center (RC). Crofts and co-workers (21) proposed a model of cyclic electron transport in which the quinones of the pool act as diffusable H-carriers between RC and the ubiquinol:cytochrome c_2 oxidoreductase. In their model they challenge the actual presence of a specific ubiquinone molecule (UQ_z), structurally bound to the ubiquinol:cytochrome c_2 oxidoreductase in about a 1:1 stoichiometry with cytochrome c_1 [characterized by a midpoint potential of 150 mV at pH 7 (22)], and distinct from that of the ubiquinol pool [$E_0' = 90$ mV at pH 7 (20)]. Their kinetic model proposes that the ubiquinone pool acts uniformly and directly as a reductant at the Q_z-site of the ubiquinol:cytochrome c_2 oxidoreductase, via a second-order collisional mechanism.

Snozzi and Crofts (8) and Casadio and co-workers (13) applied membrane fusion to study the diffusion-controlled reactions of the quinone pool. Dilution of the chromatophore membrane with exogenous phospholipid led to a decrease in the rate of cytochrome b_{561} reduction upon single-flash illumination in the presence of a bc_1 complex inhibitor, antimycin A. Under those conditions the rate of cytochrome b_{561} reduction reflects the turnover

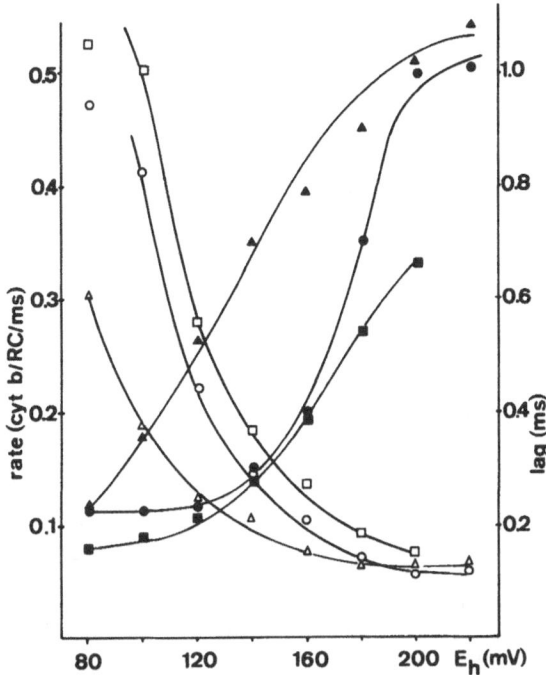

FIGURE 5 Titration curves of the rates of cytochrome b_{561} reduction versus ambient redox potential (E_h) in chromatophores of *Rhodopseudomonas sphaeroides* (■, □), fused with plain liposomes (▲, △), and fused with ubiquinone-containing liposomes (●, ○). Filled symbols were used for the lag and open symbols for the initial reduction rate. (Reprinted, with permission, from Ref. 8.)

of the ubiquinol oxidase site of the ubiquinol:cytochrome c_2 oxidoreductase complex (21). Fast electron transfer could be restored when the concentration of the ubiquinone pool in the lipid bilayer was reconstituted by the addition of exogenous quinones at redox potentials around 100 mV, where the quinone pool is approximately 50% reduced (Fig. 5). Although the rate of cytochrome b_{561} reduction following single-flash illumination could almost completely be restored when ubiquinone-containing lipid vesicles were used, the lag (delay) between the flash and onset of cytochrome b_{561} reduction was increased when ubiquinone-containing and also when ubiquinone-free lipid vesicles were used (Fig. 5). Single-flash illumination results in the reduction of a single ubiquinone by a RC, and two semiquinones have to be formed before oxidation can take place by the ubiquinol:cytochrome c_2 oxidoreductase. Snozzi and Crofts (8) suggested that this lag is most likely due to the time needed for ubiquinol molecules to diffuse from the RC to the ubiquinol:cytochrome c_2 oxidoreductase. Alternatively, this lag can be explained by the time it takes for two RC to collide in order to form two semiquinones, unless RC function as dimers and dimerization is not influenced by lipid enrichment. Since the rate of cytochrome b_{561} reduction could be restored upon addition of ubiquinones, it was concluded that cyclic electron transfer in lipid-enriched chromatophores is limited by the lateral diffusion of pool ubiquinone molecules in the plane of the membrane (8).

Casadio and co-workers (13) found that the increase of the half-time related to the cyclic electron transfer is proportional to the increase of the lipid bilayer surface area. In this case the rate of rereduction of photooxidized RC was measured. From their data, they calculated a lateral diffusion coefficient of about 2.5×10^{-10} cm^{2}s^{-1} for the diffusible species, which is in the same order as reported for membrane proteins [10^{-11} to 10^{-9} cm^{2}s^{-1} (23)] and at least 2 orders of magnitude lower than estimated for quinone molecules in lipid vesicles (about 10^{-8} cm^{2}s^{-1}). The results suggest that diffusion of cytochrome c_2 is the limiting factor for electron transfer in lipid-enriched chromatophores. Previous studies performed on electron flow in ubiquinone-extracted chromatophores indicated that almost 80% of the ubiquinone pool could be removed by solvent extraction without impairing the kinetics of cyclic electron transfer induced by single (20) or multiple turnover flashes (24). In order to reconcile these findings with lipid-enriched and ubiquinone extracted chromatophores, a role of protein-protein interaction in the efficiency of electron transfer via a collisional mechanism was suggested (13). This model involves a dynamic equilibrium between freely diffusing and associated forms of redox components in which ubiquinone molecules (possibly UQ$_z$) would still mediate the fast electron transfer within transient aggregates of protein complexes. Further studies with ubiquinone-extracted chromatophores are being performed to obtain information about the role of protein-protein interactions in controlling efficient electron transfer (B. A. Melandri, personal communication). In this respect, lipid-enriched chromatophores have the disadvantage that some soluble cytochrome c_2 is lost (13), which complicates the interpretation of the results.

III. INSERTION OF PROTONMOTIVE-FORCE-GENERATING SYSTEMS BY MEMBRANE FUSION

A. Coreconstitution of Membrane Proteins

The difficulty in drawing unambiguous conclusions about the mechanism of energy coupling in the intact bacterial cell has led to a search for an adequately defined experimental system in which the structural and functional properties of the cytoplasmic membrane are retained. For this purpose, Kaback (25) devised a procedure for the isolation of closed membrane vesicles of *Escherichia coli* with the same polarity as the cytoplasmic membrane of the intact cell. These membrane vesicles contain a large number of membrane-associated enzymes and perform several membrane-related functions, such as phospholipid synthesis, electron transfer, ATP synthesis and hydrolysis, transport by group translocation, and secondary solute transport (for reviews, see Refs. 1-3,25-27). Studies with membrane vesicles have contributed enormously to an understanding of the vital role of the protonmotive force (Δp) in bioenergetic processes, particularly with respect to secondary transport (1-3,26,27).

An even simpler system to study energy coupling are liposomes reconstituted with membrane proteins. A large number of integral membrane proteins involved in energy transduction have been extracted, purified, and reinserted into lipid vesicles, and important information has been obtained in this way about energy transduction (for reviews, see Refs. 28-31). The coreconstitution of energy-yielding and energy-consuming enzyme complexes, in some cases isolated from totally unrelated organisms, and the demonstration of coupled activities without the necessity of a structural link, has provided convincing evidence for the chemiosmotic hypothesis of Mitchell (32,33). The coupled enzyme activities in these coreconstituted systems are usually very low compared to the activities in vivo. These and other observations suggest that other factors besides the Δp could be involved in energy coupling (34-36). On the other hand, the low coupled enzyme activities observed in coreconstituted systems could well be a result of general problems connected to the reconstitution of the individual integral membrane proteins. Factors to be considered in coreconstitution studies are: (1) differences in phospholipid requirements, (2) orientation of the proteins, and, most important, (3) the heterogeneity in the membrane preparations.

An elegant strategy to deal with these problems is to reconstitute the membrane proteins separately using the most favorable reconstitution procedure and then couple the activities by fusion of the proteoliposomes. Coupling of the activities of two membrane proteins in this way was first described by Miller and Racker (37) for cytochrome c oxidase and the proton-conducting part of the ATPase, F_0. In (co)reconstitution of membrane proteins into lipid vesicles, often a fusion-inducing step (freeze-thaw) is included. Functional coreconstitution of the lactose permease and cytochrome c oxidase of *E. coli* was achieved by combining detergent dilution with freeze-thaw sonication (38).

Fusion as a method to couple membrane protein activities is not only applicable to isolated membrane proteins, but can also be used to (re)insert certain membrane proteins into biological membranes. An important application of this idea is the insertion of Δp-generating systems into biological membranes lacking such a system. In this way the functional properties of membrane proteins that depend on the Δp for their activity can be studied without the need to isolate and reconstitute these proteins. Many biological membranes, such as the plasma membrane of eukaryotic cells or the cytoplasmic membrane of fermentative bacteria, contain only the proton- (or ion)-translocating ATPase as Δp generating system. In membrane vesicles with the same polarity as the plasma membrane or cytoplasmic membrane in intact cells, the catalytic site of the ATPase is located at the inner surface and is therefore not accessible for externally added ATP. Although procedures are available to generate artificially a Δp across the membrane, e.g., by the use of ionophore-mediated ion-diffusion potentials (39) or by the use of outwardly directed diffusion gradients of weak acids (40), the transient character of the generated potentials severely limits its applicability for detailed studies on Δp-dependent membrane functions. To generate a Δp for a longer period of time, a powerful Δp-generating system that can be continuously supplied with an energy source should be present in the membrane. Functional incorporation of Δp-generating systems into membrane vesicles of the homofermentative lactic acid bacterium *Streptococcus cremoris* has been achieved by fusion of these bacterial membranes with proteoliposomes containing a Δp-generating system (41-44, 44a,b). The features of the fusion procedures involved and the properties of the fused membranes in relation to the properties of the inserted Δp-generating system will be discussed below. Experiments were performed on the incorporation of the light-induced proton pump bacteriorhodopsin (bR), bacterial reaction centers, and the redox-linked proton pump cytochrome c oxidase into membrane vesicles of *S. cremoris*.

B. Bacteriorhodopsin

The light-induced proton pump bacteriorhodopsin (bR) is a very attractive enzyme to introduce as a Δp generator by membrane fusion. Bacteriorhodopsin

TABLE 1   Properties of Two Primary Transport Systems

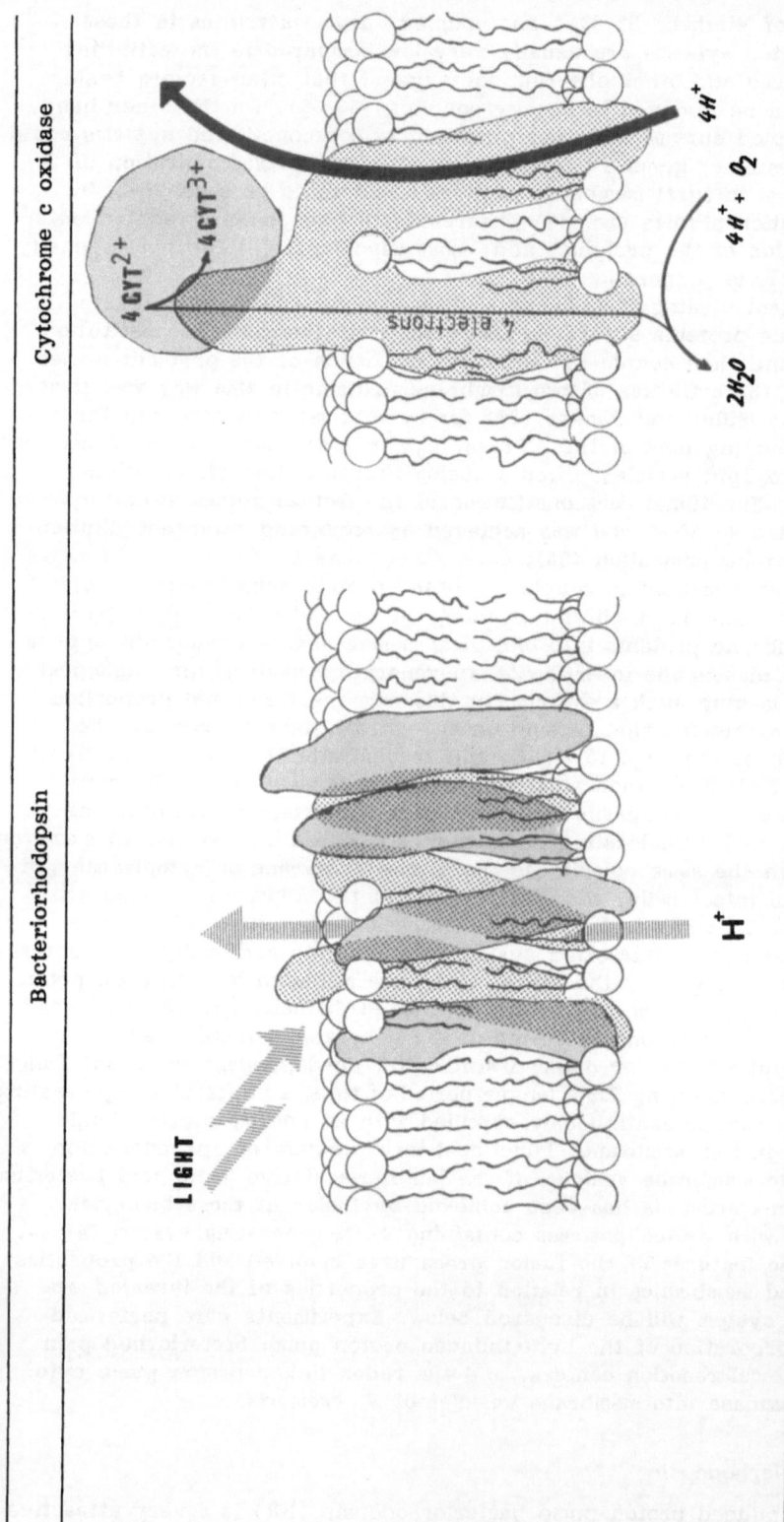

Bacteriorhodopsin

LIGHT

H⁺

Light-induced proton pump

Cytochrome c oxidase

4 CYT²⁺

4 CYT³⁺

4 electrons

4H⁺

4H⁺ + O₂

2H₂O

Redox-linked proton pump (reduced cytochrome c as electron donor)

	(Halobacterium halobium)	Beef heart mitochondria
Origin		
Mol. weight	26 kD (61)	Approx. 125 kD (60)
Structure	Single polypeptide, with retinal chromophore (62)	7 subunits, 2 haem a groups, 2 copper atoms (60,65)
Turnover (activity)	$1-2\ H^+ \cdot s^{-1}$ (47) Photocycle duration: 10 ms (63)	$300-500$ electrons $\cdot\ s^{-1}$ (60,65)
Proton motive force[a]	R.S.O.[b] $\Delta\psi - 144$ mV $-Z\Delta pH - 162$ mV (49a) I.S.O.[d] $-Z\Delta pH + 126$ mV to $+ 162$ mV (47)	$\Delta\psi - 125$ mV[c] (50)
Isolation	Osmotic lysis of cells, isolation of purple membranes by centrifugation (64)	Purification by $(NH_4)_2SO_4$ precipitation after solubilization of mitochondria (66)
Remarks	Extremely resistant to low pH, sonication, detergents, and proteolysis	Unidirectional proton translocation in proteoliposomes by the use of cytochrome c

[a] The protonmotive force is defined as the maximal magnitude of the Δp generated by the pump in the reconstituted system, under specified conditions when the ion permeability of the membrane approaches zero. The values were determined under conditions that the only component of the Δp was the Δψ or the ΔpH.

[b] Trypsin-modified bR reconstituted into cardiolipin liposomes by low-pH sonication, R.S.O., right-side-out.

[c] Cytochrome c oxidase reconstituted into asolectin by detergent dialysis. Measurement performed with cytochrome c with was reduced for 20%.

[d] bR reconstituted into egg phosphatidylcholine or asolectin by solication. I.S.O., inside-out.

can easily be isolated from the membranes of *Halobacterium halobium*, since it is found in patches, so-called purple membranes, which contain bR organized in a two-dimensional crystalline lattice (45). Moreover, it is extremely resistant to low pH, detergents, and prolonged sonication. Several procedures have been described to reconstitute bR into lipid vesicles with a unidirectional orientation (46-49,49a). Upon illumination a Δp is generated across the membrane, which can easily be varied in magnitude by varying the light intensity. A number of important properties of bR are summarized in Table 1.

One interesting point is that upon illumination, bR generates only a low electrical potential ($\Delta \psi$) but a large pH gradient (ΔpH) across the membrane. This is in contrast to cytochrome c oxidase, which maintains a large $\Delta \psi$ and only a small ΔpH under steady-state conditions (44,50). Upon addition of nigericin, an ionophore that mediates the electroneutral exchange of protons for potassium ions and therefore collapses the pH gradient across the membrane, a complete interconversion of the ΔpH into a $\Delta \psi$ is observed (49a). Valinomycin, an ionophore that mediates the electrogenic translocation of potassium ions, thus collapsing the $\Delta \psi$, induces a complete interconversion of $\Delta \psi$ into ΔpH. With bR as Δp-generating system, the magnitude of the Δp can be varied by light and also the composition of the Δp can be varied at will by ionophores.

One of the best-characterized fusion assays applicable to bacterial membranes is the membrane fusion assay based on resonance energy transfer (RET) between fluorescent lipids (51). It has been used to demonstrate that lipid vesicles fuse with membrane vesicles of *Bacillus subtilis*, *E. coli*, and *S. cremoris* upon lowering of the pH (52).

Efficient interaction of the bacterial membranes was observed only with liposomes containing negatively charged lipids such as cardiolipin or phosphatidylserine. Interaction was promoted by the inclusion of cholesterol in the lipid vesicles. Evidence for the involvement of membrane fusion in the observed interaction was presented by (1) the proportional transfer of nonexchangeable cholesteryl-[1-^{14}C] oleate, [^{14}C]egg-PC, and the RET probes from the lipid vesicles to the bacterial membranes, (2) the formation of products with an intermediate buoyant density, and (3) the appearance of colloidal gold, initially encapsulated in the lipid vesicles, in the internal volume of fused structures as revealed by thin-section electron microscopy. Transfer of aqueous contents (ferritin) from lipid vesicles into flagellated membrane vesicles from *E. coli* by incubation of the membranes at neutral pH was demonstrated by Lelkes et al. (53). At neutral pH, also, a low but significant level of interaction between *B. subtilis* membrane vesicles and lipid vesicles was observed (52).

Incorporation of bR into lipid vesicles did not decrease the fusion capacity of the negatively charged liposomes (41,42). Low-pH-induced fusion between *S. cremoris* membrane vesicles and bR proteoliposomes not only resulted in the dilution of the RET probes from the liposomal membrane into the membrane after fusion, but also a dilution of bR molecules was demonstrated (42). This is concluded from fluorescence energy transfer from the fluorescence donor, the phospholipid analog, N-(7-nitro-2,1,3-benzoxadiazol-4-yl)phosphatidylethanolamine (N-NBD-PE), to the acceptor bR, both present in the proteoliposomes.

In these studies, bR had been reconstituted into cardiolipin lipid vesicles by sonication (47). The direction of proton pumping by bR in these

TABLE 2   Light-Induced Δp Generation in bR Proteoliposomes and *S. cremoris* Membrane Vesicles Fused with bR Proteoliposomes in Relation to the Initial Ca^{2+} Uptake Rate

Ionophore	bR proteoliposomes			Fused membranes			
	$\Delta\psi^a$	$-Z\Delta pH^b$ (mV)	Δp	$\Delta\psi^a$	$-Z\Delta pH^b$ (mV)	Δp	$V_{Ca^{2+}}$ (nmol/mg/min)
No addition	+35	+90	+125	+14	+32	+46	4.8
TPT	+118	0	+118	+51	0	+51	1.9
Nonactin	0	+109	+109	0	+42	+42	6.4
S-13/dark	0	0	0	0	0	0	1.5

[a] Δψ (interior positive) was calculated from the distribution of tetraphenylboron across the membrane using a tetraphenylphosphonium selective electrode.
[b] ΔpH (interior acid) was estimated from the fluorescence quenching of 9 amino-acridine.
TPT, triphenyltin; S-13, 5-chloro-3-tert-butyl-2'-chloro-4'-nitro-salicylanilide. $V_{Ca^{2+}}$, initial Ca^{2+} uptake rate. (A. J. M. Driessen, unpublished results.)

lipid vesicles was opposite to that of bR in the intact cell and a Δp was generated, interior positive and acid (Table 2). The orientation of bR did not change upon fusion, although the magnitude of the Δp was markedly reduced (Table 2). Unequivocal evidence for a functional incorporation of bR into *S. cremoris* membrane vesicles is not provided by Δp measurements alone. Solid evidence would be provided by observation of solute transport by transport systems of *S. cremoris* energized by bR activity. Since the orientation of bR is inside-out in the fused membranes, solute uptake studies are limited to extrusion systems. Calcium (54) and sodium (55) ions are extruded by bacterial cells by cation/proton antiports (Fig. 1). In *S. cremoris* membrane vesicles fused with bR proteoliposomes, a low but significant uptake of Ca^{2+} was observed (41). Accumulated Ca^{2+} was released from the fused membranes upon addition of the ionophore A 23187, which mediates the exchange of divalent cations for protons or Mg^{2+} (56). In the presence of Mg^{2+}, release of Ca^{2+} can be explained as a result of Ca^{2+}/Mg^{2+} exchange. A collapse of the Δψ resulted in an increased rate of Ca^{2+} uptake (Table 2). From these observations it is concluded that Ca^{2+} uptake by membrane vesicles of *S. cremoris* occurs in an electroneutral exchange with protons (41,42).

These studies unequivocally demonstrate that in fused membranes coupled activities can be obtained between an inserted Δp-generating system and an endogenous secondary solute transport system. It would be experimentally more attractive if a Δp could be generated with the same polarity as the Δp in intact cells (inside negative and alkaline). bR can be inserted into lipid vesicles in the in vivo orientation by a low pH (pH 2.5-2.7)-sonication procedure (47,49), and the fraction rightside-out-oriented bR can be improved by trypsin treatment of bR (49a). However, the in vivo orientation of bR is only partly retained upon fusion. Apparently several environmental factors have a strong influence on the orientation of bR in the membrane.

C. Cytochrome c Oxidase

Cytochrome c oxidase catalyzes the oxidation of cytochrome c to the reduction of molecular oxygen to H_2O, the terminal step in the mitochondrial electron transfer chain (60). Electron transfer from cytochrome c at the outer surface of the cytoplasmic membrane to O_2 at the inside results in translocation of negative charge from the outside to the inside, while translocation of (a) proton(s) during each turnover also leads to internal alkalization. Since only oxidase molecules with their cytochrome c binding site at the outer surface are able to bind the electron donor, reduced cytochrome c, always a unidirectional generation of a Δp, inside negative and alkaline, is assured in the reconstituted or the fused membranes. The turnover of cytochrome c oxidase is an order of magnitude higher than that of bR (Table 1). A Δp can therefore be generated even when the ion permeability of the membrane is high.

Fusion between cytochrome c oxidase proteoliposomes and *S. cremoris* membrane vesicles was mainly induced by the freeze-thaw/sonication procedure, but membranes with comparative properties can be obtained by low-pH-induced fusion. Fused membranes obtained by freeze-thaw/sonication have a lower ion permeability than those obtained by the low-pH procedure (A. J. M. Driessen, unpublished results). A disadvantage of the first procedure is that a partial inversion of the orientation of integral membrane proteins takes place. In the fused membranes 55-60% of the cytochrome c oxidase molecules are oriented with their cytochrome c binding site at the outer surface, compared to 65-70% in the proteoliposomes (44). A similar observation is made for the proton-translocating ATPase, present in *S. cremoris* membrane vesicles. Prior to fusion, approximately 20% of N,N'-dicyclohexylcarbodiimide-sensitive ATP-hydrolysis activity is accessible from the outer surface, compared to 35-40% after freeze-thaw/sonication (B. Poolman and A. J. M. Driessen, unpublished results). Furthermore, freeze-thaw/sonication immediately leads to an inversion of the orientation of bR molecules that originally were incorporated into lipid vesicles in the in vivo orientation (49a).

The freeze-thaw/sonication and the low-pH procedure induce the dilution of nonexchangeable fluorescent phospholipid probes, originally present only in the bacterial membrane or only in the liposomal membrane (44). This dilution of fluorescent probes is also observed before sonication.

The buoyant density of the membranes after fusion is between those of the starting membrane preparations, while freeze-etch electron microscopy showed a less dense intramembranous particle distribution in the fused membranes than in the bacterial membranes (43). Fused membranes are tightly sealed structures with a relatively low ion permeability. These membranes show increased capacity to maintain an artificially imposed Δψ and/or ΔpH in comparison to the original bacterial membranes (44). Also, the uptake of several amino acids, driven by these artificially imposed potentials, proceeds for a longer period of time. To obtain such tightly closed membranes, the sonication step is essential.

The relatively low ion permeability makes it possible to generate a high protonmotive force (>100 mV, inside negative and alkaline) by cytochrome-c-oxidase-mediated oxidation of the electron donor system ascorbate/N,N,N',N'-tetramethyl-p-phenylenediamine/cytochrome c. Although the magnitude of the Δp generated by cytochrome c oxidase in the fused membranes is comparable to the Δp generated in the proteoliposomes, no respiratory control could be demonstrated in the fused membranes. Preliminary results with fused membranes from cytochrome c oxidase proteoliposomes, containing cardiolipin and 20 mol% cholesterol instead of soybean lipids, and *S. cremoris* membrane vesicles showed that not only was a high Δp generated, but that a low level of respiratory control (approximately 2) could be detected (G In't Veld and A. J. M. Driessen, unpublished results). Unequivocal evidence for a functional incorporation of cytochrome c oxidase into the *S. cremoris* membrane vesicles was presented by the accumulation of amino acids coupled to the cytochrome c oxidase activity (43,44).

According to the chemiosmotic theory, zwitterionic substrates with pI's near neutrality like leucine should enter the vesicles in symport with protons in response to the Δp. Under steady-state conditions the accumulation level of leucine is expected to reach thermodynamic equilibrium with the driving force. According to Eq. (2), steady-state accumulation for a neutral solute translocated in symport with n protons will be reached when:

$$\Delta \bar{\mu}_A / F = n(\Delta \psi - Z \Delta pH) = n \Delta p \qquad (4)$$

The H^+/solute stoichiometry (n) can be determined from the steady-state accumulation of the solute and the driving force (Δp). For the branched amino acid leucine a constant value for n of 0.8 was found over a broad pH range (Fig. 6) (60a-d). In the presence of nigericin, a value for n of 0.9-1.1 was found for the neutral amino acids alanine and serine at low pH (5.5-6.0). The presence of cytochrome c oxidase proteoliposomes which have not been fused with the bacterial membranes would contribute to the observed generation of a Δp and therefore an underestimation of the H^+/solute stoichiometry. The H^+/solute stoichiometry is expected to be an integer. If the real H^+/solute stoichiometry were 2, about 80-85% of the cytochrome c oxidase would be present in proteoliposomes and/or fused membranes which would not contain leucine, alanine, or serine carriers. This appears unlikely in view of the results obtained with membrane fusion assays and sucrose gradients. The most likely conclusion, therefore, is that leucine, alanine, and serine enter the vesicles in symport with one proton. In these considerations possible roles of inner leaks of the carrier (slip) or outer (passive diffusion) leaks of the substrates in attaining the steady-state level are ignored (60b).

At high external pH values (pH > 6.5) no thermodynamic steady state of alanine and serine accumulation can be reached and H^+/amino acid stoichiometries cannot be estimated (60a,b). The activity of the amino acid carriers depends strongly on the internal pH. This becomes obvious when the initial rate of alanine and serine is measured with a Δp of constant magnitude but of different composition. The measurements were done at different external pH values under conditions that $pH_{in} = pH_{out}$ (in the presence of nigericin) and when $pH_{in} = pH_{out} + \Delta pH$ (no additions) (Fig. 7). The activities of the alanine and serine carrier depend more strongly on the internal pH than the external pH, with an apparent pK of about 7.0 (60a,d).

Fused membranes are not only useful for studies on solute transport. Fused membranes containing entrapped ADP and P_i can perform oxidative phosphorylation (B. Poolman and A. J. M. Driessen, unpublished results). In this system the coupled activities of the ATPase and cytochrome c oxidase can be measured, and these studies can yield more information about the H^+/ATP stoichiometries and the thermodynamical threshold for ATP synthesis.

Insertion of cytochrome c oxidase into *S. cremoris* membrane vesicles clearly offers attractive possibilities for studies on the mechanism of solute

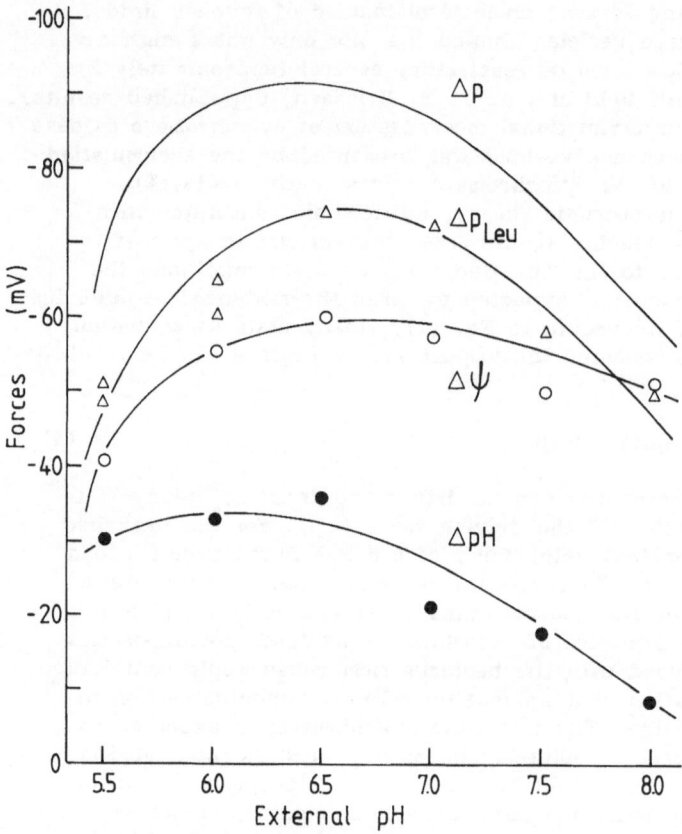

FIGURE 6 Relation between the proton motive force and leucine accumulation in *Streptococcus cremoris* membrane vesicles fused with cytochrome c oxidase proteoliposomes as a function of the external pH. (S. de Jong and A. J. M. Driessen, unpublished results.)

transport in these fermentative bacteria. The fusion procedure should in principle also be applicable to other biological membranes, which lack a suitable Δp-generating system, such as the plasma membrane of eukaryotic cells (60e).

D. Reaction Centers

Reaction centers (RC) form together with the cytochrome bc_1 complex and quinones, a cyclic electron transport chain (60f). In RC proteoliposomes, in which the cytochrome bc_1-complex is absent, cyclic electron transfer is accomplished by the use of short-chain quinones (ubiquinone-O) (60g–j). At alkaline pH, diffusion of the quinol across the membrane occurs rapidly enough to allow sufficient turnover in order to generate a considerable Δp (60k). Since RC's interact with cytochrome c_2 on one side of the membrane, and with quinones on the opposite site (60l), just like cytochrome c oxidase, an exclusive generation of a right-side-out Δp is obtained when cytochrome c is added on the outside of the proteoliposomes. With this light-dependent Δp-generating system the magnitude of Δp can be adjusted

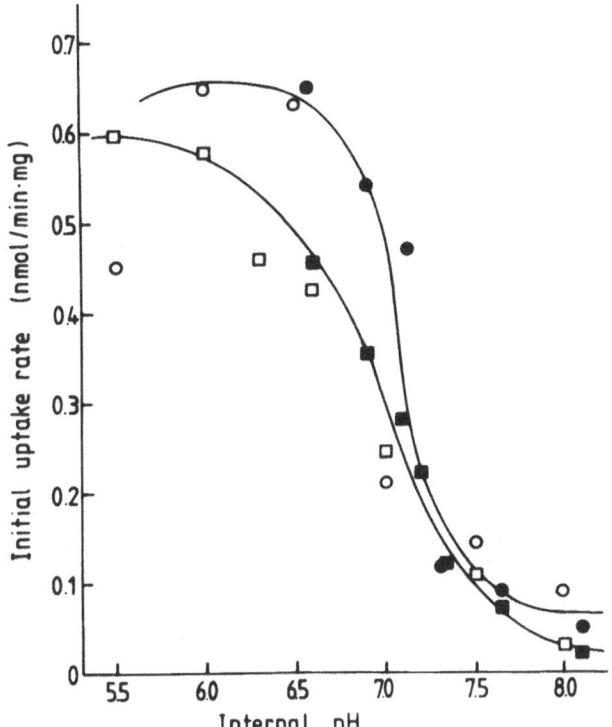

FIGURE 7 Relation between the intravesicular pH and the initial rate of serine (● , ○) and alanine (■ , □) uptake by *Streptococcus cremoris* membrane vesicles fused with cytochrome c oxidase proteoliposomes. Uptake performed in the absence (filled symbols) and presence (open symbols) of a ΔpH. (Reprinted, with permission, from Ref. 60a.)

by varying the light intensity. This system has been used to study the force-flux relationship of Δp-driven leucine transport in *S. lactis* (44b). A limitation, however, is the fact that this system can only be used at neutral or alkaline pH. Since oxygen is not involved in functioning of RC, the system can also be used in an anaerobic environment. This might be useful when studying solute transport in obligate anaerobic bacteria. RC proteoliposomes fused with membrane vesicles of the obligate anaerobe *Clostridium acetobutylicum* retain their ability to catalyze the secondary transport of several amino acids (44b).

IV. CONCLUDING REMARKS

This chapter reviews the recent progress of studies on structure and function of bacterial membranes using membrane fusion techniques. There has been no intention to give a complete description of all applications performed thus far. For instance, the effects of incorporation of exogenous phospholipids (60m,n) and ubiquinone analogs on electron transfer have not been mentioned (57). Membrane fusion procedures can also be applied to trap enzymes and small compounds in the intravesicular space (A. J. M.

Driessen and B. Poolman, unpublished results). Another interesting application is the use of lipid vesicles to promote delivery of DNA into protoplasts of bacteria.

One striking difference between lipid-enriched bacterial membranes and lipid-enriched mitochondrial inner membranes, is that only the latter have lost the energy-conserving properties (19,58). An explanation for this loss could be the introduction of contaminating outer membrane components in the fused membranes. Porins present in these membranes might influence the permeability of the fused membranes for ions and small compounds. Freeze-thaw/sonication-induced fusion between submitochondrial particles and bR proteoliposomes yields fused membranes that perform light-dependent ATP synthesis (59). In these studies, submitochondrial particles were purified by sucrose-gradient centrifugation.

The insertion of Δp-generating systems into bacterial membranes by membrane fusion offers the attractive opportunity to study mechanisms of energy transduction in membrane vesicles of strictly anaerobic fermentative bacteria, like *Clostridia* (60o). Membrane fusion techniques can also be used to manipulate the lipid composition of the bacterial membrane. This offers the possibility to study the phospholipid requirements of integral membrane proteins which are difficult to isolate and purify. Lipid-enriched bacterial membranes are excellent starting preparations for the extraction of solute carriers (A. J. M. Driessen, unpublished results) or other integral membrane proteins by the use of detergents. Long-term stabilization of the membranes might be achieved by the insertion of polymerizable lipids (67).

ACKNOWLEDGMENTS

The authors thank Dr. K. J. Hellingwerf, Dr. B. Poolman, and Dr. T. Abee for critically reading the manuscript and helpful discussions. This study was made possible by the Stichting voor Biophysica with financial support from the Netherlands Organization for Scientific Research (N. W. O.).

REFERENCES

1. Konings, W. N., and Michels, P. A. M. (1980). Electron transfer driven solute translocation across Bacterial Membranes. In: *Diversity of Bacterial Respiratory Systems* (C. J. Knowles, ed.). CRC Press, Boca Raton, FL, pp. 33-86.
2. Kaback, H. R. (1982). Membrane vesicles, electrochemical ion gradients, and active transport. *Curr. Top. Membr. Transp.* 16:393-404.
3. Konings, W. N. (1977). Active transport of solutes in bacterial membrane vesicles. *Adv. Microbiol. Physiol.* 15:175-251.
4. Ohta, T., Okuda, S., and Takahashi, H. (1977). Relationship between phospholipid compositions and transport activities of amino acids in *Escherichia coli* membrane vesicles. *Biochim. Biophys. Acta* 466:44-56.
5. Wilschut, J., and Hoekstra, D. (1984). Membrane fusion from liposomes to biological membranes. *Trends Biochem. Sci.* 11:479-483.
6. Drews, G., and Oelze, J. (1981). Organization and differentiation of membranes of phototrophic bacteria. *Adv. Microbiol. Physiol.* 22: 1-92.

7. Zankel, K. L. (1978). Energy transfer between antenna components and reaction centers. In: *The Photosynthetic Bacteria* (R. K. Clayton and W. R. Sistrom, eds.). Plenum Press, New York, London, pp. 341-347.

8. Snozzi, M., and Crofts, A. R. (1984). Electron transport in chromatophores from *Rhodopseudomonas sphaeroides* GA fused with liposomes. *Biochim. Biophys. Acta* 766:451-463.

9. Costa, B., Gulik-Krzywicki, T., Reiss-Husson, F., and Rivas, E. (1982). Fusion entre chromatophore extraits de *Rhodopseudomonas sphaeroides* et liposomes. *CR Acad. Sci. (Paris)* 295:517-522.

10. Garcia, A. F., and Drews, G. (1984). Properties of membrane fractions prepared by chromatophore-liposome fusion. *Z. Naturforsch.* 39c:1112-1119.

11. Takemoto, J. K., Schonhardt, T., Golecki, J. R., and Drews, G. (1985). Fusion of liposomes and chromatophores of *Rhodopseudomonas capsulata*: Effect on photosynthetic energy transfer between B875 and reaction center complexes. *J. Bacteriol.* 162:1126-1134.

12. Garcia, A. F., Reidl, H. H., and Drews, G. (1985). Efficiency of light conversion in photophosphorylation measured in chromatophores fused with liposomes and treated with inhibitors. *Biochim. Biophys. Acta* 808:180-185.

13. Casadio, R., Venturoli, G., DiGioia, A., Castellani, P., Leonardi, L., and Melandri, B. A. (1984). Phospholipid-enriched bacterial chromatophores. A system to investigate the ubiquinone-mediated interactions of protein complexes in photosynthetic oxidoreduction processes. *J. Biol. Chem.* 259:9149-9157.

14. Pennoyer, J. D., Kramer, H. J. M., van Grondelle, R., Westerhuis, W. H. J., Amesz, J., and Niederman, R. A. (1985). Excitation energy transfer in *Rhodopseudomonas sphaeroides* chromatophore membranes fused with liposomes. *FEBS Lett.* 182:145-149.

15. Clayton, R. K. (1966). Relations between photochemistry and fluorescence in cells and extracts of photosynthetic bacteria. *Photochem. Photobiol.* 5:807-821.

16. Monger, T. G., and Parson, W. W. (1977). Singlet-triplet fusion in *Rhodopseudomonas sphaeroides* chromatophores. A probe of the organization of the photosynthetic apparatus. *Biochim. Biophys. Acta* 460:393-407.

17. Drews, G. (1985). Structure and functional organization of light-harvesting complexes and photochemical reaction centers in membranes of phototrophic bacteria. *Microbiol. Rev.* 49:59-70.

18. Garcia, A. F., Gad'on, N., and Drews, G. (1985). Fusion of proteoliposomes containing the B800-850 light harvesting complex with membranes of the mutant strain NK3 of *Rhodopseudomonas capsulata* lacking B800-850. *Arch. Microbiol.* 141:239-243.

19. Schneider, H., Lemasters, J. J., Hochli, M., and Hackenbrock, C. R. (1980). Fusion of liposomes with mitochondrial inner membranes. *Proc. Natl. Acad. Sci. USA* 77:442-446.

20. Takamiya, K., and Dutton, P. L. (1979). Ubiquinone in *Rhodopseudomonas sphaeroides*. Some thermodynamic properties. *Biochim. Biophys. Acta* 546:1-16.

21. Crofts, A. R., Meinhardt, S. W., Jones, K. R., and Snozzi, M. (1983). The role of the quinone pool in the cyclic electron-transfer chain of *Rhodopseudomonas sphaeroides*. A modified Q-cycle mechanism. *Biochim. Biophys. Acta* 723:202-218.

22. Evans, E. H., and Crofts, A. R. (1974). In situ characterization of photosynthetic electron transport in *Rhodopseudomonas capsulata*. *Biochim. Biophys. Acta* 357:89-102.

23. Cherry, R. J. (1979). Rotational and lateral diffusion of membrane proteins. *Biochim. Biophys. Acta* 559:289-327.

24. Baccarini Melandri, A., Gabellini, N., Melandri, B. A., Jones, K. R., Rutherford, A. W., Crofts, A. R., and Hurt, E. (1982). Differential extraction and structural specificity of specialized Ubiquinone molecules in secondary electron transfer in chromatophores from *Rhodopseudomonas sphaeroides*, GA. *Arch. Biochem. Biophys.* 216:566-580.

25. Kaback, H. R. (1974). Transport in isolated bacterial membrane vesicles. *Meth. Enzymol.* 31:689-712.

26. Kaback, H. R. (1974). Transport studies in bacterial membrane vesicles. *Science* 186:882-892.

27. Konings, W. N. (1979). Energization of solute transport in membrane vesicles from anaerobically grown bacteria. *Meth. Enzymol.* 56:378-387.

28. Kaback, H. R. (1983). The Lac carrier protein in *Escherichia coli*. *J. Membr. Biol.* 76:95-112.

29. Eytan, G. D. (1982). Use of liposomes for reconstitution of biological functions. *Biochim. Biophys. Acta* 694:185-202.

30. Kagawa, Y. (1972). Reconstitution of oxidative phosphorylation. *Biochim. Biophys. Acta* 265:297-338.

31. Casey, R. P. (1984). Membrane reconstitution of the energy-conserving enzymes of oxidative phosphorylation. *Biochim. Biophys. Acta* 768:319-347.

32. Mitchell, P. (1968). *Chemiosmotic Coupling and Energy Transduction*. Glynn Research, Bodmin, England.

33. Mitchell, P. (1972). Chemiosmotic coupling in energy transduction: A logical development of biochemical knowledge. *Bioenerg.* 3:5-24.

34. Westerhoff, H. V., Melandri, B. A., Venturoli, G., Azzone, G. F., and Kell, D. B. (1984). A minimal hypothesis for membrane-linked free-energy transduction. The role of independent, small coupling units. *Biochim. Biophys. Acta* 768:257-292.

35. Ferguson, S. J. (1985). Fully delocalised chemiosmotic or localized proton flow pathways in energy coupling? A scrutiny of experimental evidence. *Biochim. Biophys. Acta* 811:47-95.

36. Slater, E. C., Berden, J. A., and Herweijer, M. A. (1985). A hypothesis for the mechanism of respiratory-chain phosphorylation not involving the electrochemical gradient of protons as obligatory intermediate. *Biochim. Biophys. Acta* 811:217-231.

37. Miller, C., and Racker, E. (1976). Fusion of phospholipid vesicles reconstituted with cytochrome oxidase and mitochondrial hydrophobic protein. *J. Membr. Biol.* 26:319-333.

38. Matsushita, K., Patel, L., Gennis, R. B., and Kaback, H. R. (1983). Reconstitution of active transport in proteoliposomes containing cytochrome o oxidase and lac carrier protein purified from *Escherichia coli*. *Proc. Natl. Acad. Sci. USA* 80:4889-4893.

39. Hirata, H., Altendorf, K., and Harold, F. M. (1973). Role of an electrical potential in the coupling of metabolic energy to active transport by membrane vesicles of *Escherichia coli*. *Proc. Natl. Acad. Sci. USA* 70:1804-1807.

40. Lancaster, J. R., and Hinkle, P. C. (1977). Studies on the galactoside transporter in everted membrane vesicles of *Escherichia coli*. I. Sym-

metrical facilitated diffusion and proton coupled transport. *J. Biol. Chem.* 252:7657-7661.

41. Driessen, A. J. M., Hellingwerf, K. J., and Konings, W. N. (1985). Light-induced generation of a proton motive force and Ca^{2+}-transport in membrane vesicles of *Streptococcus cremoris* fused with bacteriorhodopsin proteoliposomes. *Biochim. Biophys. Acta* 808:1-12.

42. Driessen, A. J. M., Hellingwerf, K. J., and Konings, W. N. (1985). Light-induced generation of a protonmotive force and Ca^{2+}-transport in membrane vesicles of *Streptococcus cremoris* fused with bacteriorhodopsin proteoliposomes. In: *Recent Advances in Biological Membrane Studies. Structure and Biogenesis, Oxidation and Energetics* (L. Packer, ed.). Plenum Press, New York, pp. 439-462.

43. Driessen, A. J. M., de Vrij, W., and Konings, W. N. (1985). Incorporation of beef heart cytochrome c oxidase as a proton-motive-force-generating mechanism in bacterial membrane vesicles. *Proc. Natl. Acad. Sci. USA* 82:7555-7559.

44. Driessen, A. J. M., de Vrij, W., and Konings, W. N. (1986). Functional incorporation of beef-heart cytochrome c oxidase into membranes of *Streptococcus cremoris*. *Eur. J. Biochem.* 154:617-624.

44a. Driessen, A. J. M., Hellingwerf, K. J., and Konings, W. N. (1987). Membrane systems in which foreign proton pumps are incorporated. *Microb. Sci.* 4:173-180.

44b. Crielaard, W., Driessen, A. J. M., Molenaar, D., Hellingwerf, K. J., and Konings, W. N. (1988). Light-induced amino acid uptake in membrane vesicles of *Streptococcus cremoris* and *Clostridium acetobutylicum* fused with reaction centre containing proteoliposomes. 170:1820-1824.

45. Stoeckenius, W. (1976). The purple membrane of salt-loving bacteria. *Sci. Am.* 243:38-46.

46. Hwang, S. B., and Stoeckenius, W. (1977). Purple membrane vesicles: Morphology and proton translocation. *J. Membr. Biol.* 33:325-350.

47. Hellingwerf, K. J. (1979). Structural and functional studies on lipid vesicles containing bacteriorhodopsin. Ph.D. Thesis, University of Amsterdam.

48. van Dijck, P. W. M., and van Dam, K. (1982). Reconstitution of bacteriorhodopsin in phospholipid vesicles. *Meth. Enzymol.* 88:17-25.

49. Happe, M., Teather, R. M., Overath, P., Knobling, A., and Oesterhelt, D. (1977). Direction of proton translocation in proteoliposomes formed from purple membrane and acidic lipids depends on the pH during reconstitution. *Biochim. Biophys. Acta* 465:415-420.

49a. Driessen, A. J. M., Hellingwerf, K. J., and Konings, W. N. (1987). The effect of trypsin treatment on the incorporation and energy transducing properties of bacteriorhodopsin in liposomes. *Biochim. Biophys. Acta.* 891:165-176.

50. de Vrij, W., Driessen, A. J. M., Hellingwerf, K. J., and Konings, W. N. (1986). Measurements of the proton motive-force generated by cytochrome c oxidase from *Bacillus subtilis* in proteoliposomes and membrane vesicles. *Eur. J. Biochem.* 156:431-440.

51. Struck, D. K., Hoekstra, D., and Pagano, R. E. (1981). Use of resonance energy transfer to monitor membrane fusion. *Biochemistry* 20:4039-4099.

52. Driessen, A. J. M., Hoekstra, D., Scherphof, G., Kalicharan, R. D., and Wilschut, J. (1985). Low pH-induced fusion of liposomes with membrane vesicles derived from *Bacillus subtilis*. *J. Biol. Chem.* 250: 10880-10887.

53. Lelkes, P. I., Klein, L., Marikousky, Y., and Eisenbach, M. (1984). Liposome-mediated transfer of macromolecules into flaggelated cell envelopes from bacteria. *Biochemistry* 23:563-568.

54. Rosen, B. P. (1982). Calcium transport in micro organisms. In: *Membrane Transport of Calcium* (C. Carafoli, ed.). Academic Press, London, pp. 187-216.

55. Lanyi, J. K. (1979). The role of Na^+ in transport processes of bacterial membranes. *Biochim. Biophys. Acta* 559:377-398.

56. Ambudkar, S. V., Zlotnick, G. W., and Rosen, B. P. (1984). Calcium efflux from *Escherichia coli*. *J. Biol. Chem.* 259:6142-6146.

57. Bergsma, J., Meihuizen, K. E., van Oeveren, W., and KOnings, W. N. (1982). Restoration of NADH oxidation with menaquinones and menaquinone analogue in membrane vesicles from the menaquinone-deficient *Bacillus subtilis* aro D. *Eur. J. Biochem.* 125:651-657.

58. van de Bend, R. L. (1985). Studies on the functional coupling of bacteriorhodopsin and mitochondrial ATP synthase. Ph.D. Thesis, University of Amsterdam.

59. Seren, S., Casadio, R., and Sorgato, M. C. (1985). Fusion of bacteriorhodopsin liposomes with submitochondrial particles yields a new system with retention of energy coupling and acquisition of photophosphorylation activity. *Biochim. Biophys. Acta* 810:370-376.

60. Azzi, A. (1980). Cytochrome c oxidase. Towards a clarification of its structure, interactions and mechanisms. *Biochim. Biophys. Acta* 594:231-252.

60a. Driessen, A. J. M., Kodde, J., de Jong, S., and Konings, W. N. (1987). Neutral amino acid transport by membrane vesicles of *Streptococcus cremoris* is subjected to regulation by internal pH. *J. Bacteriol.* 169:2748-2754.

60b. Driessen, A. J. M., Hellingwerf, K. J., and Konings, W. N. (1987). Mechanism of energy coupling to entry and exit of neutral and branched chain amino acids in membrane vesicles of *Streptococcus cremoris*. *J. Biol. Chem.* 262:12438-12443.

60c. Driessen, A. J. M., de Jong, S., and Konings, W. N. (1987). Transport of branched chain amino acids in membrane vesicles of *Streptococcus cremoris*. *J. Bacteriol.* 169:5193-5200.

60d. Poolman, B., Driessen, A. J. M., and Konings, W. N. (1987). Regulation of solute transport in Streptococci by the external and internal pH values. *Microbiol. Rev.* 51:498-508.

60e. Opekarovà, M., Driessen, A. J. M., and Konings, W. N. (1987). Proton-motive-force-driven leucine uptake in yeast plasma membrane vesicles. *FEBS Lett.* 213:45-48.

60f. Feher, G., and Okamura, M. Y. (1978). Chemical composition and properties of reaction centres. In: *The Photosynthetic Bacteria* (R. K. Clayton and W. R. Siström, eds.). Plenum Press, New York, pp. 349-386.

60g. Darszon, A., Vandenberg, C. A., Schönfeld, M., Ellisman, M. H., Spitzer, N. C., and Montal, M. (1980). Reassembly of protein-lipid complexes into large bilayer vesicles. Perspectives for membrane reconstitution. *Proc. Natl. Acad. Sci. USA* 77:239-243.

60h. Schönfeld, M., Montal, M., and Feher, G. (1979). Functional reconstitution of photosynthetic reaction centers in planar lipid bilayers. *Proc. Natl. Acad. Sci. USA* 76:6351-6355.

60i. Hellingwerf, K. J. (1987). Reaction centers from *Rhodopseudomonas sphaeroides* in reconstituted phospholipid vesicles. I. Structural studies. *J. Biomembr. Bioenerg.* 19:203-224.

60j. Hellingwerf, K. J. (1987). Reaction centers from *Rhodopseudomonas sphaeroides* in reconstituted phospholipid vesicles. II. Light-induced proton translocation. *J. Biomembr. Bioenerg.* 19:225-238.

60k. Molenaar, D., Crielaard, W., and Hellingwerf, K. J. (1988). Characterization of protonmotive force generation in liposomes reconstituted from phosphatidylethanolamine, reaction centers with light-harvesting complexes isolated from *Rhodopseudomonas palustris*. *Biochemistry* 27:2014-2023.

60l. Prince, R. C., Baccarini-Melandri, A., Hauska, G. A., Melandri, B. A., and Crofts, A. R. (1975). Assymetry of an energy transducing membrane: the location of cytochrome c_2 in *Rhodopseudomonas sphaeroides* and *Rhodopseudomonas capsulata*. *Biochim. Biophys. Acta* 387:212-227.

60m. Driessen, A. J. M., Tan Zheng, In't Veld, G., Op den Kamp, J. A. F., and Konings, W. N. (1988). The lipid requirements of the branched chain amino acid carrier of *Streptococcus cremoris*. *Biochemistry* 27:865-872.

60n. Tan Zheng, Driessen, A. J. M., and Konings, W. N. (1988). Effect of cholesterol on the branched chain amino acid transport system in *Streptococcus cremoris*. *J. Bacteriol.* 170:3194-3198.

60o. Driessen, A. J. M., Ubbink-Kok, T., and Konings, W. N. (1988). Amino acid transport by membrane vesicles of an obligate anaerobic bacterium, *Clostridium acetobutylicum*. *J. Bacteriol.* 170:817-820.

61. Oesterhelt, D., and Stoeckenius, W. (1974). Rhodopsin-like protein from the purple membrane of *Halobacterium halobium*. *Nature* 233:149-152.

62. Stoeckenius, W., and Bogolmolni, R. (1982). Bacteriorhodopsin and related pigments of Halobacteria. *Annu. Rev. Biochem.* 51:587-616.

63. Stoeckenius, W., Lozier, R., and Bogolmolni, R. (1979). Bacteriorhodopsin and the purple membrane of halobacteria. *Biochim. Biophys. Acta* 505:215-278.

64. Danon, A., and Stoeckenius, W. (1974). Photophosphorylation in *Halobacterium*. *Proc. Natl. Acad. Sci. USA* 71:1234-1238.

65. Wikstrom, M. K. F., Krab, K., and Saraste, M. (1981). Proton-translocating cytochrome complexes. *Annu. Rev. Biochem.* 50:623-655.

66. Yu, C. A., Yu, L., and King, T. E. (1975). Studies on cytochrome c oxidase. *J. Biol. Chem.* 250:1383-1392.

67. Bader, H., Dorn, K., Hupfer, B., and Ringsdorf, H. (1985). Polymeric Monolayers and Liposomes as Models for Biomembranes in *Polymer Membranes* (M. Gordon, ed.). Springer-Verlag, Berlin, pp. 2-62.

35

Modulation and Analysis of Structure and Function of the Inner Mitochondrial Membrane Through the Application of Phospholipid Enrichment and Membrane Fusion Techniques

BRAD CHAZOTTE and CHARLES R. HACKENBROCK

School of Medicine, University of North Carolina—Chapel Hill, Chapel Hill, North Carolina

I. INTRODUCTION

Experimental and theoretical studies of membrane fusion have been undertaken mainly in an attempt to understand the mechanism of the fusion process itself and how it may relate to physiological fusion events, e.g., endocytosis, exocytosis, fertilization, myogenesis, virus infection, etc. Studies toward this end have utilized native membranes and liposome model systems, and recently there has been considerable interest in using liposome fusion as a drug delivery method (1). In contrast, our laboratory has pursued membrane fusion techniques to investigate structure-function relationships in biological membranes. In our studies of structure-function relationships, we have developed methods of membrane engineering, reconstitution, and modification as specific approaches to the analysis of the activities and interactions of catalytic membrane components and we have found the mitochondrial inner membrane to be ideal for such studies.

It is our intention here to examine the use of membrane lipid enrichment and fusion techniques in the assessment of structure-function relationships in the mitochondrial inner membrane. The acquisition and analyses of data derived from modulated inner membranes are presented. We include the following aspects: (a) the contribution of membrane modulation techniques to the evolution of a random-collision model of mitochondrial electron transport; (b) low-pH lipid enrichment (2) and calcium-mediated fusion techniques (3,4), including methods, compositional and ultrastructural analyses, possible molecular bases for fusion, and current and future applications; (c) kinetic and physical measurements of modulated inner membranes interpreted in terms of mitochondrial structure-function relationships; (d) applications of lipid enrichment and fusion techniques to other membranes.

II. OVERVIEW

The idea of using fusion techniques to study structure-function relationships in the mitochondrial inner membrane evolved gradually from early

experimental results from this laboratory. These results engendered a
rethinking of the fundamental basis of mitochondrial electron transport
from which developed the Random-Collision Model (5-7). The essence of
this model is that electron transfer occurs by diffusion-based random collisions between the electron-transferring (redox) components of the inner
membrane (Fig. 1). This model is in contrast to the previous concept that
mitochondrial electron transport occurs via a permanent, physically associated macromolecular assembly, or respiratory chain (see, e.g., Ref. 8).

The existence of *lateral mobility* in the inner membrane was suggested
by a number of early experimental observations. Immunospecific probes
in conjunction with electron microscopy revealed a random distribution
of cytochrome oxidase in the plane of the inner membrane (9). Calorimetric
studies showed that the inner membrane has a broad, subzero, lipid-phase
transition temperature consistent with a highly fluid membrane (10). The
coupling of freeze-fracture and calorimetric analyses revealed that the
integral proteins of the inner membrane are capable of reversible, long-
range movement inducible by thermotropic lateral-phase separations. Significantly, it was found that inner membranes subjected to a subzero thermotropic
lateral phase separation could, upon warming back to room temperature,
conduct oxidative phosphorylation with virtually the same efficiency as
membranes not subjected to cooling (11). The latter finding suggested
that the inner membrane components did not require a highly ordered
structure and that the mobility of its integral proteins may be a natural
occurrence (11). Polycationic ferritin latticing of inner-membrane proteins
dramatically inhibits protein mobility (12), suggesting that the mitochondrion
has no cytoskeleton-like proteins affecting the free lateral motion of membrane proteins. However, these integral proteins may not normally be
physically associated, since the mobility of smaller integral proteins was
preserved in the presence of ferritin latticing. In terms of specific membrane
proteins, it was determined that cytochrome oxidase diffuses laterally in the
membrane plane from studies combining immunospecific labeling, freeze-
fracture, and thermotropic lateral-phase separations (13). At this juncture
it remained uncertain whether cytochrome oxidase diffuses independently
or in association with other integral proteins.

It was clear that if *all* inner membrane proteins were naturally mobile,
then (a) a new concept of electron transport would be required and (b)
new techniques to determine mobility of the inner membrane components
would have to be applied and/or developed. These techniques would have
to permit an assessment of the two-dimensional environment of the inner-
membrane bilayer as a solvent in analogy to more traditional studies of
chemical reactions in solution. It was thought that if independent diffusion
of the redox components is intrinsic to electron transport, increasing the
distance between such components should decrease the observed electron
transport rates. In solution chemistry, more solvent is added to dilute
reactant concentrations; this could be done for membranes by adding more
lipid. Therefore, we developed a method of incorporating exogenous lipid
into the bilayer of the inner membrane in a controlled fashion. This was
done by developing a low-pH enrichment (fusion) technique, which is detailed
in Section III. Also, it was essential to measure the actual rate of diffusion
of redox components in the inner membrane. The lateral diffusion of proteins
and lipids in a membrane can be measured using fluorescence recovery
after photobleaching (FRAP), which can quantitate the rate of lateral diffusion as well as the proportion of the component population that is diffusible.
FRAP requires membranes with a minimum diameter of 5 μm. Since the

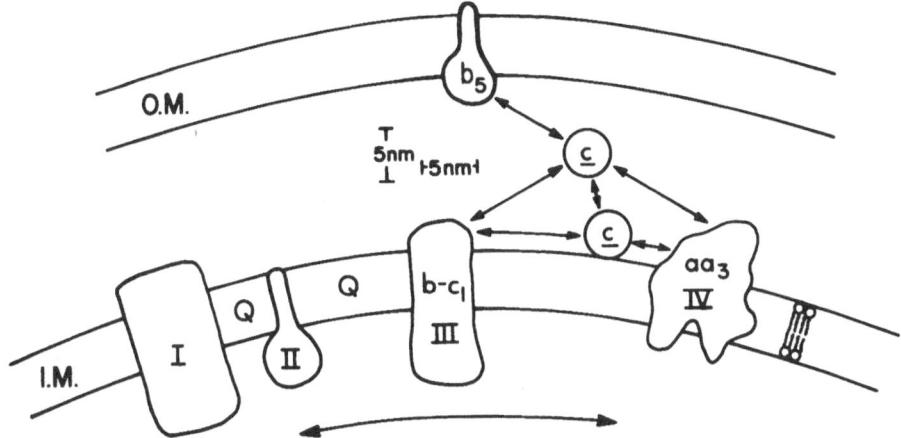

FIGURE 1 The random collision model of mitochondrial electron transport (to scale, except for component separations): I.M., inner membrane; O.M. outer membrane; I, II, III, IV; Complexes I, II, III, IV, respectively; c, cytochrome c; Q, ubiquinone; b_5, cytochrome b_5.

average, swollen, spherical, rat liver mitochondrial inner membrane is 1.5 μm in diameter, we developed techniques to fuse spherical inner membranes to a size large enough for FRAP. These fusion techniques are detailed in Section IV. Finally, we combined the lipid enrichment and membrane fusion techniques. This permitted us to use inner membranes of controlled protein density to compare the rates of diffusion of electron-redox components measured by FRAP, with the kinetic rates of electron transport as a function of protein concentration. This combined technique is presented in Section IV.

III. LOW-pH-MEDIATED FUSION OF PHOSPHOLIPID VESICLES WITH MITOCHONDRIAL INNER MEMBRANES: LIPID ENRICHMENT

A. Description of the Technique

1. Preparation of Inner Membranes and Liposomes

Rat liver mitochondrial inner membranes (mitoplasts) prepared in a simple spherical configuration are routinely used (2). Small unilamellar vesicles (SUV) are used for the low-pH lipid enrichment procedure (2). Preparation of SUV is typically done using asolectin, which is rich in negatively charged phospholipids and has a composition similar to that of the mitochondrial inner membrane bilayer (14). We have used asolectin enriched with specific membrane constituents, such as cholesterol (15), ubiquinone (16), cardiolipin (3), or mixtures of specific phospholipids, in the absence or presence of lipid-soluble probes or other agents. It is possible to use SUV containing selected integral membrane proteins, but the maintenance of protein orientation ("sidedness") must be assessed. Sonication of SUV should always be carried out at a temperature high enough to ensure the liquid-crystalline state of the lipids and prepared just before use in membrane enrichment.

2. Induction of Membrane Enrichment (Fusion)

The low-pH method of Schneider et al. (2) is used to enrich spherical
inner membranes with SUV-lipid mixtures. Briefly, this process entails
incubation of the inner membranes with SUV at pH 6.35 and 30°C. Divalent
cations are *not* used. The rate of enrichment tends to increase at lower
pH values, but this also tends to increase irreversible aggregation of inner
membranes. At higher pH values the rate of enrichment progressively
slows, with no significant enrichment occurring at pH 7.0.

3. Separation of Enriched Membrane Populations

Following lipid enrichment, the inner membranes can be purified by discontinuous sucrose density gradient centrifugation. According to their
degree of lipid enrichment, four sharp, density-distinct, inner-membrane
populations sediment into the sucrose gradient (2). From the least to the
most dense, these membrane populations have been designated Band 1,
Band 2, Band 3, and Pellet, with residual SUV remaining atop the gradient.
The yield of membrane protein per enriched membrane population varies
with a number of factors, including SUV and inner-membrane concentrations,
incubation time, pH, temperature, and lipid composition. Longer incubations
and negatively charged phospholipids tend to shift the distribution in all
bands progressively toward Band 1.

B. Analysis of Phospholipid-Enriched Inner Membranes

In order to assess the occurrence and the extent of the fusion-based enrichment, a number of methodologies may be applied. In recent years various
fluorescence assays for membrane fusion have been developed (see Chapter 4), such as the examination of aqueous contents (17-19) or lipid mixing
(20,21,23). Since we were interested in the enrichment of inner membranes
with phospholipid, we tested for the incorporation of lipid into the bilayer.
We used biochemical and ultrastructural analyses to determine the extent
of this incorporation (2,22).

1. Biochemical Analysis

The relative amounts of mitochondrial inner-membrane components, e.g.,
electron transport proteins, integral proteins, phospholipids, etc., can
be quantitated by chemical analysis of bulk membrane suspensions. Protein,
phospholipid, and heme contents are analyzed as described in Refs. 2
and 22. Since each cytochrome oxidase (an integral protein) contains two
heme *a* cytochrome groups, an increase in the lipid phosphorus to heme *a*
ratio from native inner membranes to Band 1 membranes is indicative of
exogenous phospholipid associating with the inner membrane (Table 1,
column 1). The constant heme *a* to total protein ratio (Table 1, column 2)
in all four enriched membrane bands shows that cytochrome oxidase is
not lost as a result of the enrichment process. While cytochrome *c*, a peripheral protein, is lost to a significant extent (Table 2, column 1), the content
of the integral membrane heme proteins and ubiquinone remains constant
(Table 2, columns 2-4). Mitochondrial matrix protein is released as a result
of the enrichment process, as can be deduced from the heme *a* to protein
ratio of native membranes compared to that of the phospholipid-enriched
membranes (Table 1, column 2). This loss of internal protein during the
fusion process has been noted more recently in other studies using low-pH
fusion and will be dealt with in more detail in Section III.C.

TABLE 1 Composition of Phospholipid-Enriched Mitochondrial Inner Membranes

Fraction	Lipid P. (μmole/mg protein)	Heme a (nmole/mg protein)	Lipid P/heme a (mole/mole)	% increase bilayer surface area	Average surface area[a] (cm^2*10^8)
Native	0.24	0.27	900	0	7.07
Pellet	0.61	0.51	1200	30	9.10
Band 3	0.90	0.57	1600	80	12.72
Band 2	1.65	0.54	3100	240	16.97
Band 1	3.76	0.52	7200	700	49.49

[a]Average surface area refers to inner membrane bilayer surface area based on an average rat liver mitochondrion.

TABLE 2 Mitochondrial Heme Proteins (Cytochromes) and Ubiquinone to Heme a Ratios in Native and Phospholipid-Enriched Inner Membranes

Fraction	c/a	c_1/a	b/a	Q/a
Native	0.69	0.25	0.53	8.5
Pellet	0.13	0.31	0.63	7.5
Band 3	0.13	0.32	0.61	7.6
Band 2	0.18	0.31	0.63	8.1
Band 1	0.21	0.31	0.58	8.6

Assuming that all exogenous lipid associated with the membrane enters the bilayer, the extent of lipid enrichment can be expressed in terms of the percent increase in the bilayer surface (Table 1, column 4) and bilayer surface area in cm^2, based on an average rat liver mitochondrion (Table 1, column 5). Although our results have not shown selective enrichment or loss of integral membrane components during the enrichment process (2,15, 16,22), it is nonetheless advisable to quantitatively analyze enriched membranes when dealing with other membrane systems and also to ensure that the discontinuous sucrose gradient maximizes separation.

2. Ultrastructural (Electron Microscopic) Analysis

Biochemically based compositional analysis can only reveal that exogenous lipid associates with the inner membrane, whereas freeze-fracture electron microscopy can demonstrate that the associated lipid is in fact incorporated into the bilayer. Freeze-fracture comparison of native and phospholipid-enriched inner membranes shows a clear, progressive increase in the average surface area (2). In addition, a progressive decrease in the integral protein density distribution occurs from native to Band 1 inner membranes (Fig. 2). It is clear, therefore, that exogenous lipid incorporates into the inner membrane bilayer to increase the separation between randomly distributed

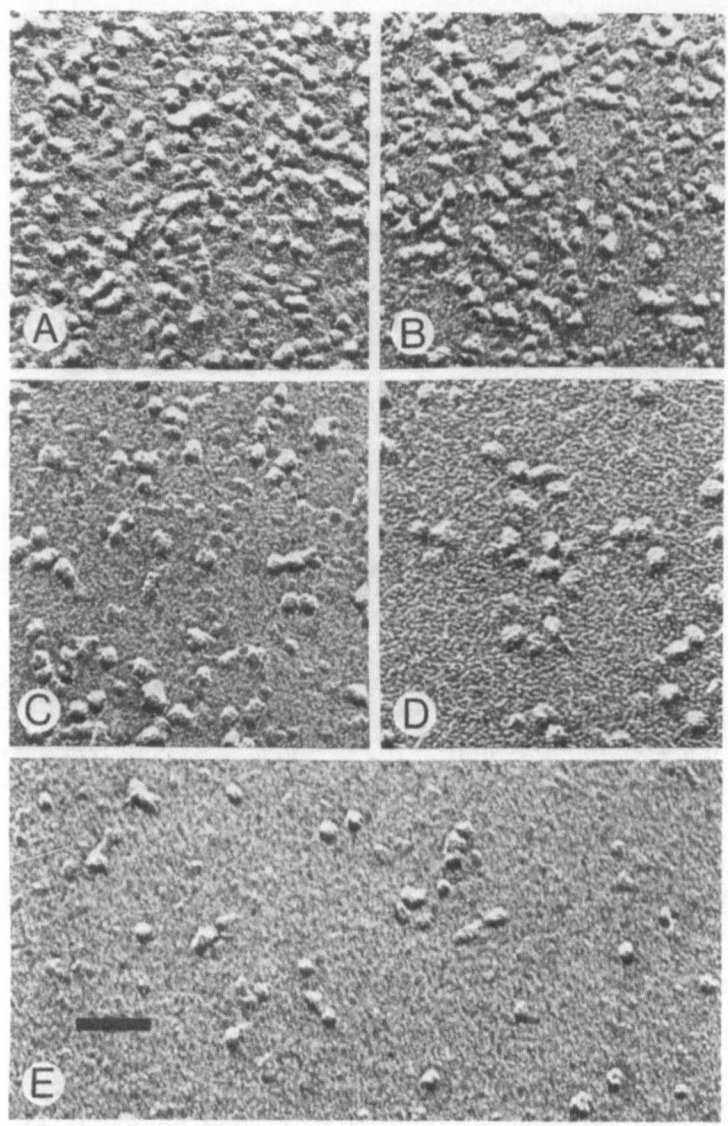

FIGURE 2   Freeze-fracture electron micrographs of native (control) and
phospholipid (asolectin)-enriched mitochondrial inner membranes. Bar = 0.04
μm. (A) Native as nonenriched control; (B) pellet, 30% phospholipid-enriched;
(C) Band 3, 80% phospholipid-enriched; (D) Band 2, 240% phospholipid-
enriched; (E) Band 1, 700% phospholipid-enriched. The two-dimensional
integral protein density (IMP) decreases in proportion to the extent of
phospholipid enrichment.

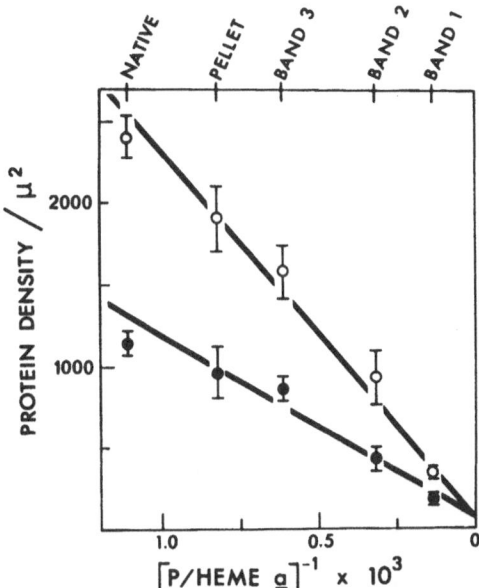

FIGURE 3 Integral membrane protein (IMP) density on the convex (○) and concave (●) freeze-fracture faces of native (control) and phospholipid-enriched inner membranes as a function of phospholipid to heme a mole ratio. Protein density on both faces decreases proportional to the extent of enrichment.

integral proteins proportional to the degree of enrichment. Further evidence for fusion is found in the fact that the relative density distribution in *both* halves of the fractured bilayer decreases proportionately (Fig. 3). In some instances the density distribution of proteins resulting from enrichment can be nonrandom. This may occur when a foreign exogenous agent is incorporated in sufficiently high amounts. In the case of inner membranes, incorporation of 20 mol% or more of cholesterol [not normally present in the inner membrane (24,25)] results in clustering and aggregation of integral proteins (Fig. 4). When enrichment is performed for functional studies of a native membrane, it is advisable to ensure, by freeze-fracture analysis, that the resulting density distribution of integral proteins is unaltered. Conversely, alteration of the density distribution can also be used as a tool in functional studies (15). Functional analyses are presented in Section III.D.

C. Low-pH-Mediated Fusion: Mechanisms

The original impetus in developing the low-pH fusion technique was to dilute the protein in the mitochondrial inner membrane while working in bulk solution without irreversible aggregation and precipitation of the membranes. Such controlled incorporation would allow the use of standard electron transport kinetic assays of the resultant membranes. It has not been our goal to exhaustively examine the molecular mechanisms of the incorporation process. However, we have made a number of relevant observations on the nature of the low-pH fusion process, which we present here.

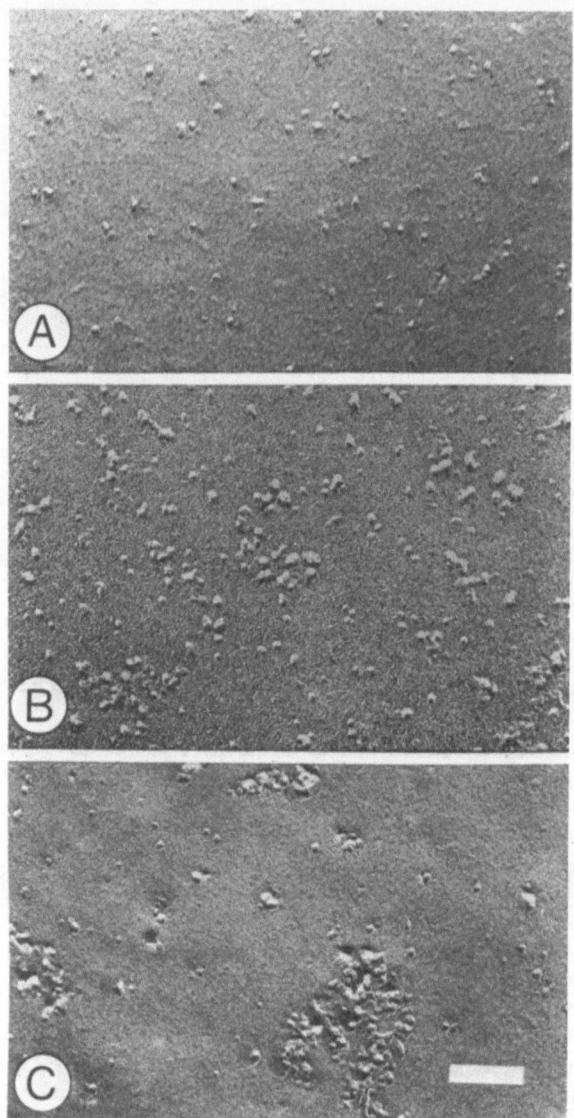

FIGURE 4 Freeze-fracture electron micrographs of the convex faces of
a Band 1 inner membrane after enrichment of inner membranes with liposomes
containing no cholesterol (A), 103 mg cholesterol/g phospholipid (B), and
172 mg cholesterol/g phospholipid (C). The cholesterol concentration expressed in mol% for each of the membrane fractions was determined to
be 0 (A), 27 (B), and 34 (C). Bar = 0.1 μm. Clustering and aggregation
of proteins (IMP) increases with the mol% of cholesterol in the enriched
membrane.

In order for fusion to occur between two membranes (1) interbilayer contact must be established, (2) contact must be followed by a transient destabilization of the two bilayers, and (3) destabilization must lead to a merging of the bilayers (fusion) (27-30). In order to obtain an understanding of membrane fusion, it is appropriate to examine the various factors and/or forces that could contribute to this process. Organizationally, these individual factors will be dealt with separately, keeping in mind that membranes exist as a result of an equilibrium of forces, and fusion follows some perturbation of this equilibrium.

1. Role of Lipid Composition

The lipid compositions of the target (inner membrane) and the source (SUV) membranes affect the susceptibility to and the extent of fusion. The phospholipid composition of the inner membrane has been documented by Colbeau et al. (24) and reviewed by Daum (25). The membrane phospholipid is highly unsaturated (65%) and contains 38-45% phosphatidylcholine (PC), 32-39% phosphatidylethanolamine (PE), and 14-23% cardiolipin (CL) as its major phospholipid classes; cholesterol is virtually absent. Asolectin consists of 36.5% PC, 36.9% PE and phosphatidylserine (PS), and 8.8% CL as the major classes (14); it is highly unsaturated (62% polyunsaturated, 14% monounsaturated).

Our earlier results (2) indicate that fusion of asolectin SUV to mitochondrial inner membranes is supported by the high proportion of PE and CL in both the SUV and inner membranes. Comparison of the constituent phospholipids of the inner membrane with recent model studies of low-pH fusion (31-36) provides an indication of which lipid classes may be more involved in the fusion process and also highlights the "leaky" nature of this fusion, i.e., loss of internal aqueous contents (2). Bondeson et al. (33), using glycolipid-phospholipid vesicles containing phosphatidic acid (PA) and a relatively high proportion of PE, reported that lowering the pH to ~6.5 induces fusion followed rapidly by leakage. PC was found to be inhibitory toward fusion. Düzgünes et al. (35) found that vesicles composed of oleic acid (OA) and PE become destabilized and fuse below pH 6.5, whereas OA/PC or PS/PE vesicles do not fuse or even aggregate at pH values as low as 4. Connor et al. (31), using vesicles containing PE and palmitoylhomocysteine, reported rapid fusion upon lowering of the pH to 5; the presence of PC, but not PE, inhibited fusion. Considering these model system studies, it may well be that the success of the low-pH technique in fusing SUV to inner membranes without concomitant, indiscriminant, and irreversible aggregation of the inner membranes, is due in part to the balance of phospholipid classes as well as the omission of divalent cations. In this regard, we have previously reported (3) that when the CL content of asolectin SUV is increased to be roughly equivalent to that determined for the native inner membrane, the low-pH fusion is enhanced, resulting in a pronounced and progressive shift in the distribution of the enriched membranes toward Band 1 membranes. This indicates an enhancement of the fusion process, with more SUV fusing per native inner membrane. We have also successfully incorporated a PS/CL mixture into inner membranes producing results similar to asolectin. Based on these results, more extensive model system studies involving CL might prove fruitful. In this regard, we note that Stegmann et al. have studied low-pH fusion of vesicles with influenza virus (37).

In our mitochondrial inner membrane low-pH fusion, cholesterol-supplemented asolectin SUV do not noticeably affect the fusion process itself, although cholesterol does markedly affect protein distribution in the inner membrane (15). A low-pH fusion that has an absolute requirement for cholesterol has been reported in the literature for Semliki Forest virus; however, this fusion is mediated by a viral glycoprotein at low pH (26). In this case PC and sphingomyelin were found to inhibit the fusion process.

2. Role of Nonbilayer Phases

Nonbilayer lipids, such as lipids preferring the hexagonal (H_{II}) organization in isolation, have been proposed to be involved in membrane fusion processes (27,30,36,38-41; see also Chapters 2, 4, and 6]; particularly, inverted micellar intermediates might function as dynamic fusion intermediates. It is well established that PE can adopt the H_{II} phase, and so can CL in the presence of divalent cations. Both of these phospholipid classes are contained within the mitochondrial inner membrane and asolectin. In addition, low pH may induce the H_{II} phase in PE systems (32), as Ca^{2+} is known to do in CL systems. Presently, a role for nonbilayer structures in the low-pH enrichment method (2) is not confirmed.

3. Role of Membrane Dehydration

Based on model system studies (30,33,42,43), it has been suggested that membrane dehydration plays a role in membrane fusion. With regard to the low-pH fusion technique, a case for a role for membrane dehydration can only be made by analogy to these model studies. The hydration force provides a high degree of repulsion at close range (44; see also Chapter 3) and has to be overcome to allow fusion to occur. In this context there are, perhaps, two possibilities: first, a general dehydration of the membrane, and second, a localized decrease in the water of hydration of the phospholipid headgroups. Studies by MacDonald (42), using poly(ethylene glycol), dextran, or sucrose, indicate that dehydration can produce membrane fusion over a period of time, i.e., minutes, at room temperature. Gibson and Strauss (43) argue that removal of water associated with the bilayer is an effect common to all membrane fusion processes, irrespective of whether they are induced by calcium, temperature changes, or freezing and thawing, and that different extents of fusion can be correlated with the amount of water removed. How this effect might be promoted in the case of low-pH fusion requires further study.

4. Role of Proteins

Studies regarding the role of proteins in membrane fusion have considered (a) movement of integral proteins away from the bilayer contact (fusion) site, (b) movement of integral proteins to the contact site, (c) integral proteins intimately involved in the fusion process (viral fusion), (d) use of exogenous proteins to mediate fusion. In light of these possibilities, the influence of proteins, when present, on fusion may not be singular.

The fusion of native inner membrane with exogenous SUV occurs in the presence of inner-membrane integral proteins, which comprise one-third to one-half of the inner membrane's surface area. A specific requirement for these proteins in fusion is not yet known. It is known in the case of low-pH fusion of influenza virus with liposomes that the hemagglutinin protein of the virus is essential for fusion (37,59). Similarly, the membrane glycoprotein of Semliki Forest virus is required for fusion with liposomes at low pH (26). It has also been found that bovine serum albumin can induce fusion of PC vesicles at pH 3.5 (45). Conversely, vesicles containing a high mole fraction of PE can be made to undergo rapid fusion at low pH in the absence of proteins (31).

The role of membrane proteins in the fusion of inner membranes is not known. It is well documented that these integral membrane proteins are mobile, undergoing lateral (6,46-48) and rotational (49) diffusion. We have found from our work on low-pH fusion of erythrocyte membranes with asolectin SUV that the cytoskeleton (spectrin) must be removed in order for fusion to occur. It is thought that this inhibitory effect is due to suppression of lateral mobility. Volsky and Loyter (50) used cationic ferritin to lattice chicken erythrocyte membrane proteins and reported an inhibition of fusion by calcium or by Sendai virus. They argue that a steric hindrance effect of cationic ferritin is not likely and mobility is the important factor. However, other effects cannot be ruled out due to the presence of cationic ferritin on the membrane surface, which may inhibit fusion. Proteins may facilitate fusion directly or indirectly as a result of their mobility. It is not clear whether proteins aggregate during fusion at a contact site, as proposed earlier by Poste and Allison (51), or migrate away from the contact site, as observed by Ahkong et al. (52). Indeed, it is not clear whether there is one specific mechanism for all membranes.

5. Role of Vesicle Size

The degree of bilayer curvature can play a role in membrane fusion. The low-pH method (2) is found to proceed best with extensively sonicated SUV that have a minimal radius of curvature. This sharp curvature may strain the packing order of the bilayer lipids (53). This strain on the lipid packing may reduce the degree of perturbation (energy) necessary to transiently destabilize the vesicle bilayer. A size dependence for vesicle fusion has been found by many other researchers in studies of liposome fusion induced by divalent cations, smaller vesicles tending to be more susceptible to fusion than larger ones (27,54-56). Nir et al. (27) state that the observed resistance to fusion of large unilamellar vesicles (LUV) may be due to an enhanced stability upon reaching a certain size and/or an increase in the potential barrier for close approach with increasing vesicle size. Theoretical analyses (27,44,57) also show that the repulsion between approaching small vesicles is less than that of vesicles of larger diameter. Hence, small vesicles have the greatest probability for contact and subsequent fusion.

6. Role of Low pH and Membrane Charge

Fusion-based lipid enrichment increases as the pH is lowered below 7.0. Mechanistically, a critical question is how membrane electrostatics, Van der Waals attraction, and hydration repulsion are affected by hydrogen ions. Rand (44), Gingell and Ginsburg (57), and Traüble (58) have addressed this question in a quantitative manner (see also Chapters 3, 4, and 5). We shall present only a qualitative description related to our observations.

Typically, the fusion is performed at pH 6.35-6.5; lower pH values (<6.0) produce an irreversible aggregation of inner membranes. Calcium also produces irreversible aggregation and precipitation of inner membranes and SUV. It appears that, given the composition of the membranes involved, hydrogen ions permit a more "gentle" fusion than calcium. Interpreted via the kinetic conceptualization of Nir et al. (27, see also Chapter 5), the aggregation of vesicles is readily reversible at a mildly acidic pH, yet

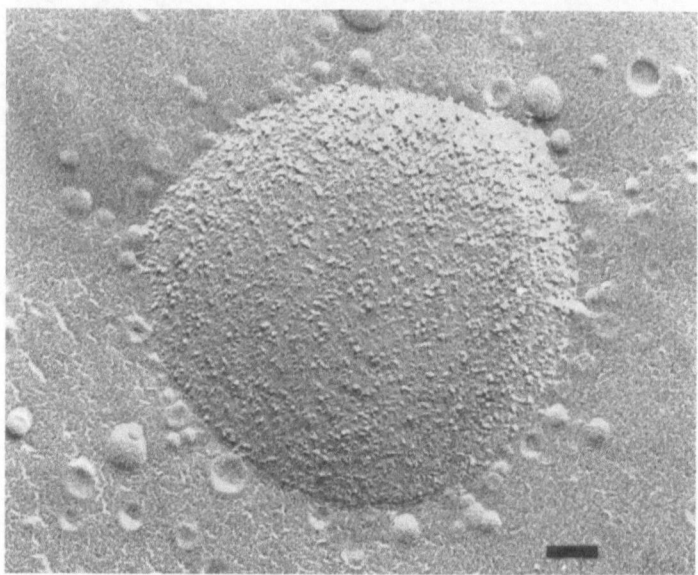

FIGURE 5   Asolectin SUV-mitochondrial inner membrane mixture during
incubation at pH 6.5, propane jet frozen without cryoprotection. SUV are
shown in close apposition with a protein-rich inner membrane during the
fusion process. Bar = 0.1 μm.

the fusion rate constant is sufficiently favorable to support fusion during
the transient binding of the SUV to the inner membrane. Freeze-fracture
analysis of inner membranes in the presence of SUV at pH 6.5 shows the
SUV sequestered near and on the inner membrane surface (Fig. 5), a
result not seen at pH 7.4 (2). The interface between the surfaces of the
inner membranes and the sequestered SUV may have a localized pH lower
than that of the bulk solution, and the pK_a's of ionizable groups on the
membranes' surfaces could be different in these localized environments.
Either of these two phenomena could facilitate fusion.

We have noted that the fusion of SUV to inner membranes causes "leakiness." Enriched inner membranes are not osmotically active, are permeable
to NADH, lose much of the mitochondrial matrix protein, and lose their
ability to carry out oxidative phosphorylation (2). We conjecture that the
major portion of the fusion occurs during, and/or immediately (1-2 min)
following, the lowering of the bulk solution pH. This can be based on
two related observations. First, we have found that the addition of SUV
in three aliquots, each followed by a low pH adjustment, is more effective
than one addition of SUV followed by one pH adjustment. Second, following
the adjustment of the pH to the desired low value, we observed a rise
in the pH, which required addition of acid to maintain the low pH. This
rise is greatest and most rapid in the first 2 min after the lowering of
the pH. It may be due to leakage of matrix proteins from the inner membranes and/or of pH 7.4 solution from the SUV. Regardless of specific
origin, we suppose that this pH rise indicates the most active phase of
the fusion process. Rapid fusion following the pH jump and the "leaky"
nature of low-pH fusions are consistent with studies on model systems
(32,33). However, the duration of the low-pH incubation does affect the

distribution (yield) of the phospholipid-enriched membrane populations, with longer incubation times increasing the amount, although more slowly, of highly enriched membranes (2).

D. Analysis of Structure-Function Relationships

In Section III.B.2, we showed that incorporation of exogenous bulk phospholipid into the inner membrane results in an enrichment of the bilayer and an increase in the separation of integral proteins. In this section we present kinetic (functional) data on the electron transport activities of some of these proteins as a function of enrichment. These data led to the development of the Random-Collision Model for mitochondrial electron transport (Fig. 1).

It is possible to study individual segments of electron transport, as well as the complete sequence, by judicious use of natural and artificial electron donors and acceptors combined with appropriate specific inhibitors. This permits assessment of interactions between specific redox components. Kinetic studies of electron transport activities in native and phospholipid-enriched inner membranes reveal a distinct relationship between the catalytic rate and the two-dimensional density of redox components (22). Electron-transferring activities by redox component interactions decrease in proportion

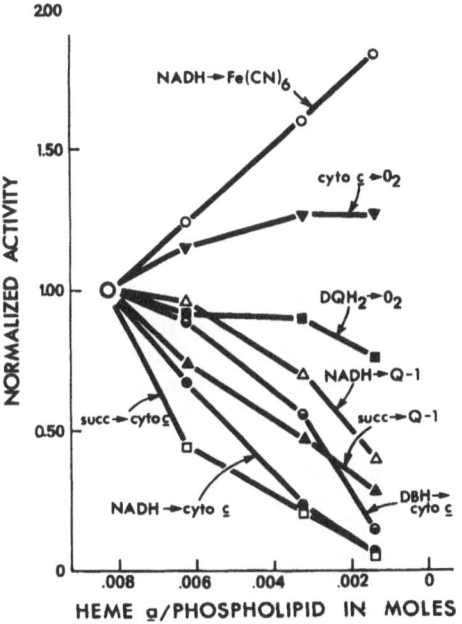

FIGURE 6 Relative electron transfer activities of phospholipid-enriched inner membranes plotted as a function of heme a to phospholipid mole ratios. Activities requiring collisions of the redox partners decrease in proportion to phospholipid enrichment. Assay conditions are given in Ref. 22. NADH dehydrogenase (NADH $\rightarrow$ Fe(CN)$_6$); cytochrome c oxidase (cyto $c \rightarrow O_2$); duroquinol oxidase (DQH$_2 \rightarrow O_2$); NADH-ubiquinone reductase (NADH $\rightarrow$ Q-1); succinate-ubiquinone reductase (succ $\rightarrow$ Q-1); ubiquinol-cytochrome c reductase (DBH $\rightarrow$ cyto c); succinate-cytochrome c reductase (succ $\rightarrow$ cyto c).

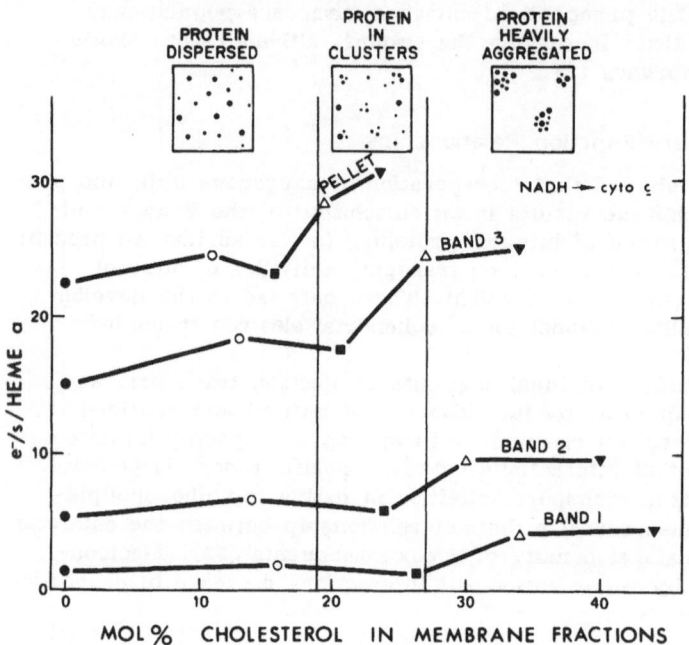

FIGURE 7 Specific activity of NADH-cytochrome c reductase transport as a function of cholesterol concentration (mol%) in lipid-enriched membrane fractions. Cholesterol content in mg/g phospholipid of liposomes used for enrichment: 0 (●), 26 (○), 103 (■), 172 (△), and 240 (▼). Clustering and aggregation of IMP due to cholesterol incorporation moderates the decrease in activity due to enrichment.

to the degree of membrane enrichment (Fig. 6), which coincides with the two-dimensional dilution of redox components and all integral proteins (Fig. 2). In addition, when ubiquinone, a small lipoidal redox component, is incorporated with phospholipid into inner membranes, a significant increase in activity occurs (16). Furthermore, incorporation of cholesterol induces a proportionate clustering of integral proteins (Fig. 4) along with an increase in electron transfer (Fig. 7). The cholesterol study shows that a decrease in protein-protein distance due to clustering results in an increase in electron transport rates, while the surface area of the membrane increases markedly.

These functional studies, based on structural modulation of the inner membrane by phospholipid enrichment, reveal that redox components of the inner membrane are independent lateral diffusants. Clearly, the kinetic data reveal that the rate of electron transfer between interacting redox partners is affected by the distance between such partners.

IV. CALCIUM-MEDIATED FUSION OF NATIVE AND PHOSPHOLIPID-ENRICHED MITOCHONDRIAL INNER MEMBRANES: PRODUCTION OF ULTRALARGE MEMBRANES

A. Description of the Techniques

Two techniques have been developed in our laboratory to prepare ultralarge inner membranes for use in FRAP measurements. These membranes are also amenable to other single-membrane analysis techniques, as will be discussed later. The fusion of spherical, native inner membranes to obtain nonosmotically active, ultralarge, spherical inner membranes is performed on a glass slide, as described by Höchli et al. (4). The protocol produces glass-attached, spherical inner membranes with diameters greater than 10 μm. In general, larger inner membranes are produced by increasing the initial concentration of the native inner membranes. The method described by Chazotte et al. (3) can be used not only with native inner membranes, but also with inner membranes that have been previously enriched with lipid. Unlike the method of Höchli et al. (4), ultralarge membranes produced by this methodology are osmotically active.

B. Analysis of Calcium-Fused Membranes

1. Microscopic Studies

Light Microscopy. The occurrence and the extent of inner-membrane fusion are routinely determined by phase-contrast microscopy; a dry objective lens is preferred. The nonosmotically active fused inner membranes (4) range from 10 to 50 μm in diameter and are attached directly to the glass surface. The osmotically active fused inner membranes (3) range from 20 to 50 μm in diameter, with some up to 200 μm in diameter. Figure 8 shows fused, phospholipid-enriched inner membranes. Most of the ultralarge fused membranes are attached to aggregated nonfused membranes which are attached to the glass surface. Some of the large fused membranes break away from aggregates as independent, intact, spherical membranes. Observation during the fusion process shows the ultralarge fused membranes expanding outward from the (unfused) aggregated membranes.

Freeze-Fracture Electron Microscopy. Two essential properties of the calcium-mediated fusion procedure of Chazotte et al. (3) have been verified by freeze-fracture electron microscopy. First, in all fused membranes examined it was determined that the integral membrane proteins (IMP) are present and randomly distributed (Fig. 9a and b). Second, it was determined that when inner membranes were first lipid-enriched and then fused, the protein density in the ultralarge fused membranes was in very close agreement with the average density in the unfused enriched membranes (see Figs. 2 and 9).

2. Osmotic Activity

One of the salient feature of the native as well as phospholipid-enriched inner membranes fused by the method of Chazotte et al. (3) is that they are osmotically active. The existence of osmotic activity in these fused membranes was unexpected since both the phospholipid-enriched inner membranes, produced by the low-pH method (2), and the fused native inner membranes, produced by the method of Höchli et al. (4), are not osmotically active. The decrease in membrane permeability following fusion by the method of Chazotte et al. (3) was noted in a number of ways. Osmotic activity can be seen in a sequence of photographs of (a) a spherical membrane in 40 mosM buffer, (b) shrinkage of the membrane induced by addition of a high-osmolarity sucrose solution (700 mosM), and (c) restoration of the spherical shape upon returning to 40 mosM buffer, as shown by phase-contrast images (Fig. 10a, b, and c, respectively). It was found that various fluorescent dyes do not penetrate the membrane aqueous interior. Specific studies with Rhodamine 123 indicated that the osmotically active

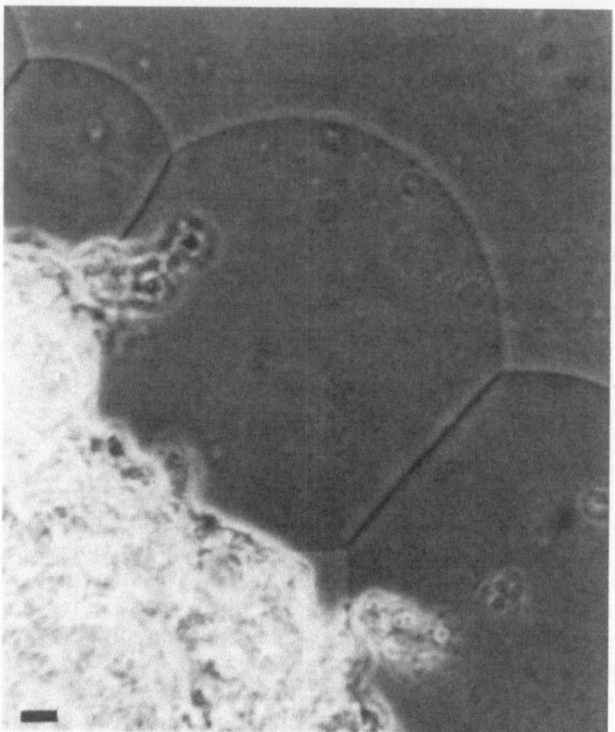

FIGURE 8 Calcium-fused, 30% asolectin-enriched mitochondrial inner membranes in a phase contrast image. The fusion method (3) results in the
aggregation of the small, enriched inner membranes seen in the lower left
and the fusion of such small membranes into osmotically active, ultralarge,
enriched, intact inner membranes typified in the center. Bar = 5 μm.

fused membranes can be energized with succinate and deenergized with
CCCP, which would argue for the existence of a transmembrane electrochemical potential comparable to that in whole intact mitochondria, and,
hence, for a high degree of impermeability to protons. Apparently, some
change in the phospholipid-enriched and the native membranes is effected
by calcium, pH, and elevated temperature, as used in the fusion technique
of Chazotte et al., to render the fused membranes impermeable. It is well
documented that erythrocyte ghosts are commonly "sealed" by incubation
in NaCl at 37° for 1 h (75). Therefore, it is not an anomalous phenomenon
for the permeability of a membrane to decrease as a result of increased
temperature in the presence of ions.

C.   Calcium-Mediated Fusion: Mechanisms

Although our purpose in developing calcium-mediated fusion procedures
was utilitarian to obtain individual, ultralarge, inner membranes for diffusion
studies using the FRAP technique, we shall briefly discuss and compare
our observations with other relevant studies on membrane fusion. In making
these comparisons, although we primarily refer to the method of Chazotte
et al., many of the comparisons pertain to the nonosmotically active, ultralarge

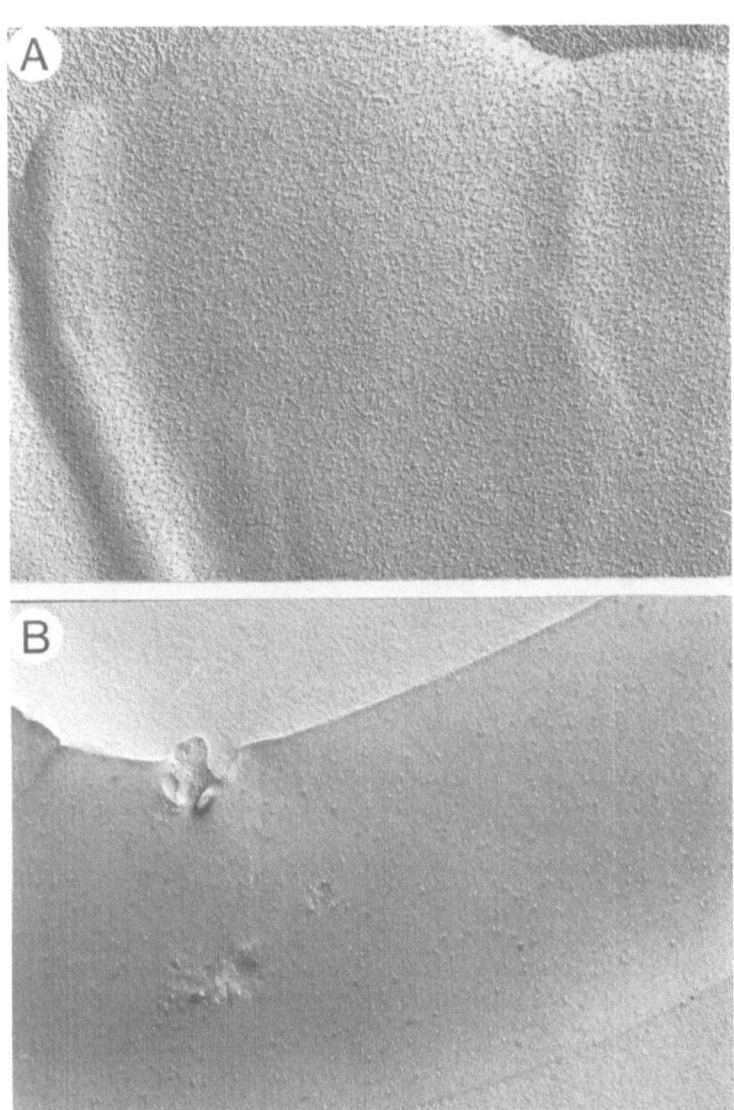

FIGURE 9   Freeze-fracture electron micrographs of membranes fused by
the method in Ref. 3. (A) Part of a freeze-fracture face of a calcium-fused,
30% asolectin-enriched mitochondrial inner membrane that as a 20-μm minimum
diameter. (B) Part of a freeze-fracture face of a calcium-fused, 240%
asolectin-enriched inner membrane that has a 40-μm minimum diameter.
Bar = 0.4 μm.

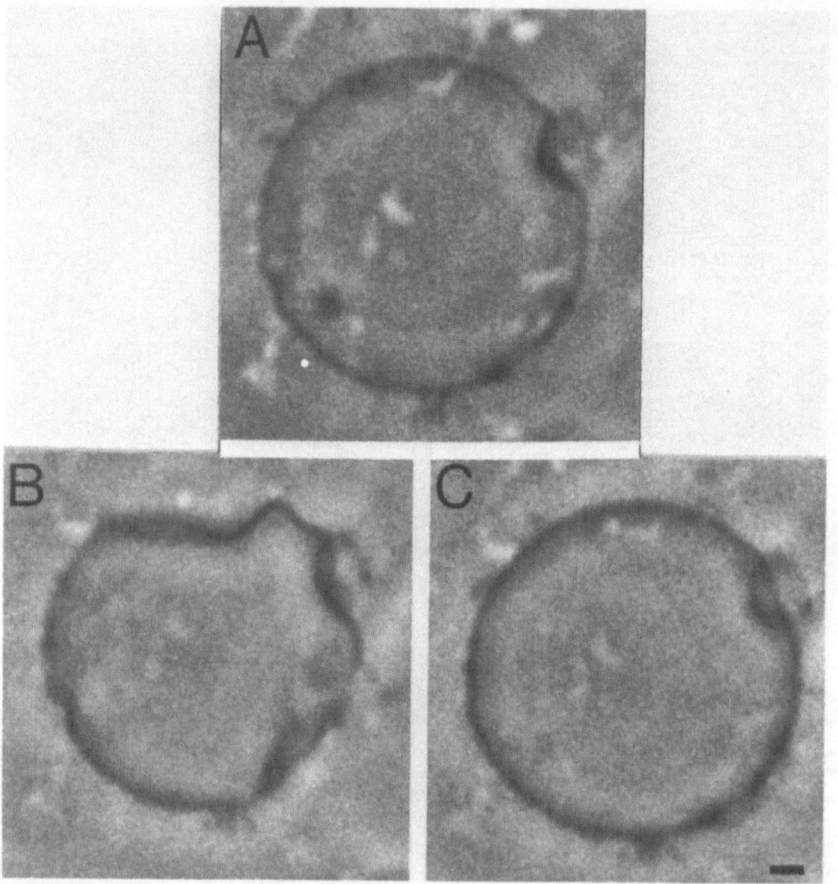

FIGURE 10   A sequence of phase contrast images showing osmotic sensitivity
of a calcium-fused, 30% asolectin-enriched inner membrane. (A) Initial conditions of 40 mosM buffer. (B) After addition of a high-osmolarity (700
mosM) sucrose solution to the slide chamber resulting in rapid crenation.
(C) After returning to 40 mosM buffer the membrane resumes its original
spherical shape. Bar = 2 µm.

membranes of Höchli et al. (4) as well. Fusion mediated by divalent cations,
in particular calcium, has been studied extensively and is reviewed in
this volume (Chapter 4) and elsewhere (27-29).

   For osmotically active membranes, we utilize calcium (10 mM), low pH
(6.5), and elevated temperature (37°C) as an optimal condition to induce
fusion of native and phospholipid-enriched inner membranes (3). Each
of these three parameters has been used independently to induce fusion
in different membranes. While these different parameters act in concert
to achieve fusion, we shall discuss each parameter separately and describe
interrelationships or synergistic effects where applicable.

1.   Role of Lipid Composition

CL is a major negatively charged phospholipid in the inner membrane and
may play an important role in fusion. This speculation is partly based on

our observations of low-pH enrichment of inner membranes (3), as discussed above. There is an extensive literature on fusion of negatively charged vesicles in model systems (27-29; see also Chapter 4 and 5]. Most of these studies have been carried out with PS vesicles or vesicles containing PS in combination with other phospholipids (see Chapter 4 and 5). However, there have also been reports on fusion of CL-containing vesicles (60,61) or vesicles containing PA (62,63). Part of the reason for the emphasis on PS has been the fact that it is a major component of most mammalian plasma membranes (Ref. 64, and references therein).

With regard to calcium fusogenicity, it is established that the lipid composition of the bilayer has a significant effect. The rat liver mitochondrial inner membrane is composed primarily of PE (32-39%), PC (38-45%), and CL (14-23%) (24,25). Both CL- and PE-containing vesicles undergo calcium-mediated fusion in model systems. It is known from model studies that membrane aggregation and fusion due to CL, PE, PS, or PA are inhibited by PC, related to its relative concentration (21,60). We have found that inner membranes fuse more readily when they contain a higher CL-to-phospholipid ratio. Calcium is known to bind with high affinity to CL (65) and is known to induce fusion of CL-containing vesicles (39,60,66). Considering all these data, we conclude that CL plays a major role in the calcium-mediated fusion of inner membranes.

2. Role of Calcium Concentration and pH

The role of calcium in the induction of fusion is fundamental and is related to the lipid composition of membranes. It has long been conjectured that divalent cations help reduce the electrostatic repulsion of approaching membranes. The concentration dependence of calcium on fusion in model systems has been considered (28,29,60,67). Calcium binds stoichiometrically to the PS headgroup (68), and this interaction can have a number of effects, e.g., charge neutralization, dehydration, nonbilayer phase formation, alteration of phase transition behavior, etc. (see also Chapter 4). Other divalent cations have been studied as well, such as Mg^{2+}, Ba^{2+}, Sr^{2+}, and Mn^{2+} (69). Calcium has been found to be the most effective fusogenic divalent cation.

With respect to the calcium-mediated fusion procedure of Chazotte et al. (3), inner membrane aggregation and fusion occur to varying degrees over a pH range of 5.0-7.0 and a calcium concentration range of 0.5-100 mM. Aggregation of membranes is greatly decreased and fused membranes are barely detectable at the lowest calcium concentrations (0.5-1.0 mM), whereas, the use of lower pH values (5.0-5.75) at lower calcium concentrations results in a small degree of improvement in the yield of aggregated and subsequently fused membranes. As the pH is progressively increased above 6.5, or the final concentration of calcium above 10 mM, the amount of fused membranes decreases. It was found that high calcium concentrations (>10 mM) produce marked membrane aggregation but little fusion.

The combination of 10 mM calcium and pH 6.5 induces fusion of native or of phospholipid-enriched mitochondrial inner membranes. Both of these membrane types normally have a minimum diameter of 1-2 μm, far greater than that of sonicated vesicles, and are thus less likely to fuse. The success of these two "fusogens," i.e., calcium and pH, in combination with elevated temperature (37°C) may be due to a number of factors, which are discussed in the following sections.

3. Role of Nonbilayer Phases and Lateral Phase Separations

A role for nonbilayer phases, such as the H_{II} phase, in calcium-mediated fusion has been suggested by some researchers (36,70, see also Chapter 2, 4, and 6). This view is not universally accepted. Hoekstra and Martin (40) argue against a role for nonbilayer phase in the case of PS/PE SUV fusion, since extensive or virtually complete bilayer fusion can be shown to occur before significant lipid phase separation occurs. It has been reported that calcium can trigger the transition to the H_{II} phase in multilamellar mixtures composed of different acidic phospholipids with unsaturated PE (Ref. 36 and references therein). Also, it is well established that CL adopts the H_{II} phase in the presence of calcium. It follows that nonbilayer phase formation could play a significant role in both of our calcium-mediated fusion techniques (3,4). During fusion, a lateral-phase separation may be at least temporarily induced in the inner membrane, since both the protein and lipoidal components of this membrane diffuse laterally (7,46,47) and this membrane is capable of exhibiting lateral phase separations (10). However, following the completion of calcium-mediated fusion no lateral-phase separations are detected (3).

4. Role of Membrane Dehydration

A number of research groups believe dehydration plays an important role in the process of calcium-mediated fusion (33,43,62) and in fusion in general (42-44). This is based on the observation that hydration pressure is the major repulsive force between bilayers (44). Sundler and co-workers (33,67), using steric probes, have found that the phospholipid headgroup hydration is significantly reduced by calcium in cases where fusion is induced. Ohki and Ohshima (62), comparing ion binding to PA headgroups, found that there is no difference between Ca^{2+} and Mg^{2+}, as opposed to the more bulky PS headgroup, where Ca^{2+} binds more effectively and is a stronger fusogen. Ohki and Ohshima attribute these results to the small size and hydration of calcium, permitting it to penetrate into the polar group region. Wilschut et al. (Ref. 54, and references therein; see also Chapter 4) point out that the ability of calcium to induce fusion is likely related to its capacity to form dehydrated *trans* complexes between PS molecules in opposing bilayers, which brings the membranes into close proximity. Gibson and Strauss (43) studied fusion induced by calcium, free thawing, and passage through the lipid-phase transition. These authors contend that water effectively locks small unilamellar vesicles in a metastable state and that the extent of fusion can be correlated to the amount of water removed by each of these three methods.

These model studies show that dehydration may be an underlying mechanism for membrane fusion. A definitive case for dehydration in the technique of Chazotte et al. (3) cannot be made based on current experimental knowledge, although we have noted that there is a tendency for fusion to be more extensive for phospholipid-enriched inner membranes near the open edges of the microscope slide assembly. The possibility that more extensive fusion of inner membranes is due to some evaporation of solution from the open edge of the slide chamber during heating and a resultant increase in dehydration is purely subjective at this time. It is not unreasonable, however, that various agents (e.g., calcium) or conditions (e.g., evaporation) may induce or promote a dehydration instability necessary for fusion of inner membranes.

5. Role of Proteins

Two related points should be addressed regarding the role of proteins in calcium-mediated fusion. First, it appears that when proteins are present in membranes undergoing fusion, they must be laterally mobile. Second, with regard to site(s) of intermembrane contact for fusion, it is not clear whether proteins are present at, or rather excluded from, these sites.

The mitochondrial inner membrane has mobile integral proteins. By measurement we know that nonfused, inner membranes (13,48,71), lipid-enriched inner membranes (22), and calcium-fused inner membranes (6,7,47) all have mobile integral proteins. We also know that the low-pH fusion is inhibited by integral protein immobility (Section III.C). Volsky and Loyter (50) have also reported that calcium-mediated fusion of chicken erythrocytes is inhibited when integral protein mobility is restricted. Following calcium-mediated fusion, freeze fracture reveals the presence of randomly distributed proteins, which suggests that any lateral-phase separations between proteins and lipids that may be induced during fusion are transient. It has been suggested (Ref. 72 and references therein) that fusion is preceded by a lateral-phase separation where integral proteins are excluded from the interbilayer contact site, leaving free lipid domains that are the sites of subsequent fusion. The argument for pure lipid domain involvement in the calcium-mediated fusion of inner membranes is strengthened by the fact that calcium-mediated fusion also occurs with pure lipid membranes.

6. Role of Temperature

Temperature plays a significant role in calcium-mediated membrane fusion. The procedure of Chazotte et al. (3) utilizes an elevated temperature (37°C). Inner membranes left at room temperature (21°C) in the presence of 10 mM Ca^{2+} at pH 6.5 for 24-48 h produce only few fused membranes, whereas 15 min at 37°C under the same conditions results in a significant amount of fused membranes. Incubation for 1-2 h at 50-60°C at pH 7.4 in the absence of calcium results in aggregation of native inner membranes and clustering of IMP, whereas in phospholipid-enriched inner membranes, clustering of proteins occurs, but not membrane aggregation. The membrane aggregation effect most likely reflects a difference in the surface adhesion properties of these two membrane types attributable to the altered lipid composition, such as the lower CL content of the lipid-enriched inner membranes. Clearly, inner-membrane aggregation is enhanced by increased temperature alone, while fusion, i.e., membrane destabilization and membrane intermixing, is enhanced in the presence of calcium with elevated temperature.

The effect of elevated temperature (37°C) has been investigated in model system studies. In agreement with our findings, a number of researchers have also concluded that bilayer phospholipids must be in a fluid state for fusion to occur. The mitochondrial inner-membrane lipids contain 65% unsaturated fatty acid, and the lipid bilayer is highly fluid with its main phase transition temperature at -4°C (10). It has been pointed out that each of the discrete steps in fusion exhibits a specific temperature dependence (27,54). Bilayer destabilization, for example, may exhibit distinct characteristics in specific membrane systems, which may be maximal at a particular temperature (27), leading to a maximum in the fusion rate constant. In model studies all rate constants involved in fusion increase with temperature (73), with the fusion rate constant increasing 3- to 10-fold and the aggregation and vesicle dissociation rate constants 1.5- to 2-fold per 10°C. Bentz et al. (73) point out that were it not for fusion occurring between most membrane pairs before dissociation occurs, overall aggregation of vesicles would decrease with increasing temperature. A decrease in overall aggregation with increasing temperature was noted in a vesicle system that could aggregate, but not fuse (73).

7. Summary

The conditions of pH 6.5, 10 mM final calcium concentration, and 37°C for 15 min are optimal for fusion of rat liver mitochondrial native and phospholipid-enriched inner membranes. Each of these parameters has been used independently to induce fusion. It is apparent, however, that the fusion of these membranes does not occur at one critical set of conditions; rather, fusion occurs over a range of conditions that has an optimum. This finding is consistent with results of studies on various other systems where fusion was found to occur over a range of conditions (27-29,57,74).

The molecular mechanisms of membrane fusion are still not known, but membranes must come into close apposition and then be transiently destabilized. In this regard, analyses suggest that calcium and hydrogen ions permit the close apposition of two membranes by decreasing or neutralizing repulsions due to electrostatic charge and short-range hydration pressure while elevated temperature can provide the energy through Brownian (kinetic) motion to randomly drive two membranes together. Once in close apposition, contact-mediated modification of the bilayer structure can occur (44). Analyses indicate that calcium, pH, and/or bilayer contact, supplied with energy due to the elevated temperature, can induce some transient destabilization (e.g., lateral-phase separation, nonbilayer-phase formation, dehydrated complex formation, etc.) to precipitate the fusion event(s). The kinetic treatment of Nir and co-workers (27,73) may clarify how a number of parameters interact over a range of conditions in the fusion process.

D. Analysis of Structure-Function Relationships

In order to elucidate a diffusion-based collisional mechanism for electron transport, it was necessary to directly establish lateral diffusion of specific redox components, subsequently measure the rate of the diffusion, and compare the number of successful collisions (electron transfers) between redox partners with the total number of collisions (theoretical) that occur between such partners. The technique of FRAP is a powerful, direct approach for the measurement of lateral diffusion. As discussed in Section II, it requires membranes with a minimum diameter of 5 μm, thus necessitating the development of fusion techniques to faithfully and competently produce enlarged inner membranes. In this section we present the results of diffusion measurements made possible by the use of ultralarge inner membranes and discuss these results in terms of structure and function related to the Random-Collision Model of mitochondrial electron transport (5-7,76).

1. Lateral Diffusion in Calcium-Fused Ultralarge Inner Membranes

All Redox Components Diffuse. Direct lateral diffusion measurements were made via FRAP on specific redox components in ultralarge inner membranes produced by calcium-induced fusion (3,4) and in ultralarge inner membranes of megamitochondria (47) isolated from livers of cuprizone-fed mice (i.e., no fusion). The diffusion of transmembranous redox complexes gave measured diffusion coefficients (D) in the mid 10^{-10} cm^2/s range (Table 3), independent of ionic strength (47). Measurement of Complex III in fused membranes and megamitochondrial inner membranes yields similar D values (Table 3), revealing that the diffusion of redox components in fused membranes is not affected by the fusion process compared to unfused membranes. Diffusion of a fluorescent ubiquinone analogue has a D in the mid 10^{-9} cm^2/s range (Table 3). The diffusion of the peripheral redox component, cytochrome *c*, is dependent on ionic strength (76a,b). Cytochrome *c* at 56 mM ionic strength has a D of 1.9×10^{-9} cm^2/s (Table 3) and at 100 mM ionic strength diffuses primarily in three dimensions. Phospholipids, like the fluorescent lipid *N*-NBD-PE or the fluorescent lipid analog DiI, have a D in the mid 10^{-9} cm^2/s range (Table 3). Direct measurements have also been made for the D values of Complex I (77) and Complex V (ATP synthetase) (78) in calcium-fused, native membranes.

The measurements of D, in concert with earlier studies mentioned in Section II, verified that all redox components are lateral diffusants. In addition to determining these D values, FRAP measurements reveal that more than 90% of the population of each redox component is mobile and that all members of each population diffuse in a common pool. These findings are in complete agreement with the ultrastructural observations that (a) cytochrome oxidase (Complex IV) is randomly distributed in the plane of the inner membrane (9), (b) all integral proteins in the membrane diffuse laterally as a result of thermotropic lateral phase separations (10-12), (c) cytochrome oxidase aggregates into planar clusters as a function of time in the presence of its monospecific antibody (13) and (d) all integral proteins migrate into a single patch in the inner membrane plane under an electric field (48,71).

Diffusion as a Function of Membrane Protein Concentration. In an effort to study the effect of two-dimensional protein concentration on diffusion via the kinetic measurements of Schneider et al. (22), we utilized the low-pH enrichment method (2), followed by calcium-induced fusion (3) to produce ultralarge, inner membranes of controlled, protein density for FRAP measurements. It was determined that lateral diffusion is affected by the integral protein concentration, such that D increases as the protein concentration decreases (46,78a). The D of Complex III increases from 6.9×10^{-10} to 1.25×10^{-8} cm^2/s as the protein concentration decreases from that of native membranes, while the lipid D increases from 3.9×10^{-9} to 1.3×10^{-8} cm^2/s (Fig. 11). Also, we found that the diffusion of ubiquinone is approximately 5 times faster in pure DMPC bilayers than in calcium-fused native (nonenriched) inner membranes (6,7,76,78a). These results show that collisional interactions decrease the rate of diffusion and that the extent of this decrease depends on the number and size of the obstructions as well as the size of the diffusant. This follows from the observed increase in D with decreasing protein concentration and the observation that protein (large) and lipid (small) D values change to different degrees as a function of protein concentration. These results are in agreement with the theory of Eisinger et al. (79) for obstructed, long-range diffusion. Further, the similar D values of phospholipids and the larger-diameter Complex III in the highly enriched inner membranes agree with the prediction of Saffman and Delbrück (80) for a weak dependence of D on the diffusant radius in dilute, two-dimensional solutions. Furthermore,

TABLE 3 Lateral Diffusion Coefficients of Mitochondrial Inner-Membrane Components[a]

Component	Membrane/source	Probe	$D*10^{-10}$ (cm^2/s)	% recovery	Temp. (°C)	μ	Ref.
Complex III	FIM (RAT LIVER)	TRITC-IgG	4.4	>90	23	73	47
	FIM	TRITC-Fab	4.4	>90	23	73	47
	MEGA (M. LIVER)	TRITC-IgG	3.6	>90	23	73	47
	FIM	TRITC-IgG	6.9	>90	25	0.3	6,78a
	F-PELLET	TRITC-IgG	13	>90	25	0.3	6,78a
	F-BAND 3	TRITC-IgG	67	>90	25	0.3	6,78a
	F-BAND 2	TRITC-IgG	75	>90	25	0.3	6,78a
	F-BAND 1	TRITC-IgG	125	>90	25	0.3	6,78a
Complex IV	FIM	TRITC-IgG	3.6	>90	23	73	47
	FIM	TRITC-Fab	3.7	>90	23	73	47
Cytochrome c	FIM	FITC-CYT. c	0.59	>85	23	0.3	47
	FIM	FITC-CYT. c	2.7	>93	23	23	47
	FIM	FITC-CYT. c	19	>83	23	56	47
Ubiquinone	FIM	$Q_0C_{10}NBD$	26	>95	21	0.3	6,76
	MEGA	$Q_0C_{10}NBD$	37	>90	21	0.3	6,76
	DMPC	$Q_0C_{10}NBD$	550	>95	30	0.3	6,76
	DMPC	$Q_0C_{10}NBD$	4	>90	15	0.3	6,76

Phospholipid						
FIM	DII	39	>95	21	0.3	46
F-PELLET	DII	60	>95	21	0.3	46
F-BAND 3	DII	90	>95	21	0.3	46
F-BAND 2	DII	110	>95	21	0.3	46
F-BAND 1	DII	130	>95	21	0.3	46
ASOLECTIN	DII	300	>95	21	0.3	46
FIM	N-NBD-PE	68	>95	21	0.3	98
DMPC	DII	600	>95	30	-	6
DMPC	DII	0.65	>95	15	-	6

[a] μ, ionic strength in mM units. FIM, fused ultralarge mitochondrial inner membrane from rat liver. TRITC-IgG or -Fab, tetramethylrhodamine isothiocyanate conjugated to rabbit IgG or Fab. MEGA, inner membranes from megamitochondria isolated from livers of cuprizone-fed mice. F-PELLET, fused, ultralarge 30% asolectin-enriched rat liver inner membranes. F-BAND 3, fused, ultralarge 80% asolectin-enriched rat liver inner membranes. F-BAND 2, fused, ultralarge 240% asolectin-enriched rat liver inner membranes. F-BAND 1, fused, ultralarge 700% asolectin-enriched rat liver inner membranes. FITC-Cyt. c, fluorescein isothiocyanate conjugated to cytochrome c from horse heart. $Q_0C_{10}NBD$, fluorescent ubiquinone analogue; hexanoic acid conjugated to 2,3-dimethyl-5-methyl-6-(10-hdyroxydecyl)quinone. DMPC, dimyristoyl-phosphatidylcholine. DII, fluorescent lipid analogue; dioctyl- or dihexyldecylcarbocyanine. ASOLECTIN, soy bean phospholipids. N-NBD-PE, N-4-Nitrobenz-2-oxa-1,3-diazole conjugated to phosphatidylethanolamine.

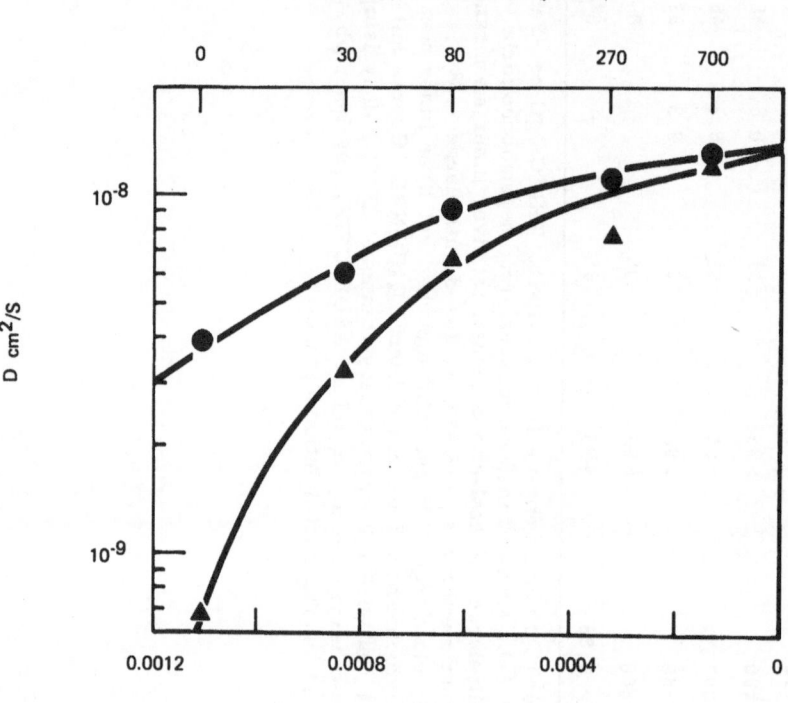

FIGURE 11 Effect of mitochondrial inner membrane integral protein density
on lateral diffusion as a function of heme *a* to phospholipid ratio. D at
23°C, measured by FRAP, in native (nonenriched) and phospholipid-enriched
inner membranes. Complex III diffusion (▲) was reported by TRITC-
conjugated anti-Complex-III IgG bound to inner membrane Complex III
and lipid diffusion (●) was reported by DiI, a fluorescent lipid analog,
incorporated into the inner membrane. D values increase as the protein
density decreases.

these results are in agreement with the ultrastructural observation of Sowers
and Hackenbrock (71), showing that the D for all integral membrane proteins varies with the naturally occurring variation in protein concentration
among individual mitochondria. These results offer a clear explanation
for the more rapid protein diffusion in (pure) lipid model systems and
the slower diffusion in native biomembranes. The measured low microviscosity, 0.9 P, and the absence of cholesterol from the inner membrane
reflect the need for a low resistance to motion in this lipid bilayer to offset
the effect of all the collisional interactions that generate catalytic functions,
as well as the nonproductive collisional interactions.

   Diffusion as a Function of Temperature. Lateral diffusion of protein
redox complexes and phospholipid as a function of temperature was studied
using FRAP in membranes of controlled protein density (79a). It was found
that D increased with temperature (Fig. 12), as is typical for diffusion
generally. The coefficient for the temperature increase of D can be expressed in terms of an apparent activation energy E_a by plotting log D vs.

1/T. E_a reflects the energy required for diffusion. Thus, the energetic requirements for diffusion in membranes of different protein density can be ascertained. It was found that the degree of increase in D as the temperature is raised depends also on the protein density of the membrane. Membranes with the greatest protein density exhibit the greatest temperature dependency for D, as illustrated by the comparison of the apparent E_a for diffusion of Complex III in inner membranes of different phospholipid to protein ratios (Table 4). The D for phospholipid was found to have a greater temperature dependence than for Complex III at a given membrane protein density (Table 4). The latter result can be expected since the smaller-diameter phospholipid molecules will diffuse in more restricted small "channels" between proteins, thus experiencing more collisions and a longer path length for the same degree of displacement per unit time.

2. Comparison of Electron Transport with Lateral Diffusion

In order to test the random-collision model for electron transport, it is essential to determine the frequency of collision of randomly diffusing redox components and compare this with observed electron transport rates. This comparison can be accomplished using the mathematical and conceptual framework developed by Hardt (81), which describes the diffusion-controlled collision rates of diffusants for a given dimensionality (e.g., two-dimensional). The equations utilize D values, reactant concentrations, and reactant collision radii parameters, which we have adapted for use in mitochondrial electron transport (47). Additionally, the Hardt equations can be used to run

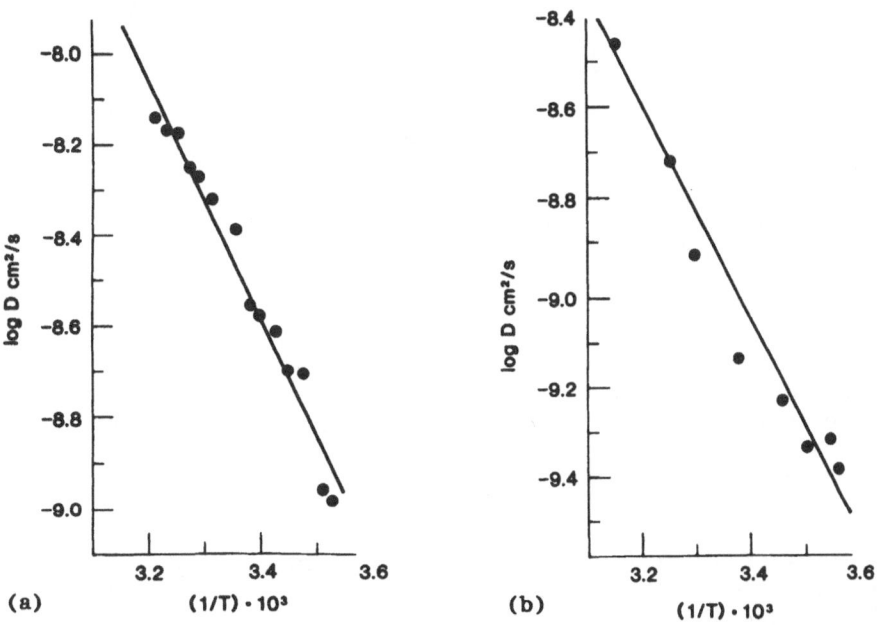

FIGURE 12 Temperature dependence of lateral diffusion in inner membrane. Diffusion coefficients measured by FRAP using probes in Figure 11. Native inner membranes: (a) DiI for ubiquinone or phospholipid, (b) Complex III. Note that D increases markedly as a function of temperature.

TABLE 4   Diffusion Coefficient Apparent Activation Energies

Membrane	E_a of diffusant (kcal/mole)	
	Complex III	Phospholipid
Native inner membrane	10.8	12.04
30% phospholipid-enriched	9.6	10.1
80% phospholipid-enriched	8.6	8.9

simulations to test the limits of validity for the random-collision model of mitochondrial electron transport (82,83).

Diffusion-Coupled Electron Transport. Following the approach outlined above, it was determined by Gupte et al. (47) that mitochondrial electron transport is *diffusion-coupled*; i.e., all bimolecular redox reactions (between appropriate redox partners at effective redox concentrations) are preceded by lateral diffusion and random collision. This followed from the finding that, in all cases examined, there are more collisions occurring between redox partners, as determined via the Hardt equation, than electron transfers at *maximum* rates of respiration. Based on this analysis, it was concluded that no physically associated, macromolecular assembly, respiratory chain, or permanent or transient aggregate is necessary to account for maximal rates of electron transport.

Electron Transport and Multicollisional, Obstructed, Long-Range Diffusion. Experimental, structural, and functional observations have established that mitochondrial electron transport is a multicollisional, long-range (>10 nm) diffusion process. From a structural perspective, redox components are randomly distributed (9), redox complexes are on average widely dispersed (7), integral proteins occupy less than 50% of the bilayer surface area of native inner membranes (48), and all inner membrane components are mobile (47,77,78). From a functional perspective, every potentially reactive redox component undergoes one or more diffusion-based collisions with its potentially reactive redox partner to effect one turnover (7,47), D values increase as the protein density decreases (6,7,46,78a), and multi-molecular electron transport rates decrease as the protein density decreases (6,7). Collectively, these data demonstrate that (a) electron-transferring reactions between all redox components and their respective redox partners at their effective redox concentrations occur via a long-range diffusional process, (b) the rates of diffusion as well as the rates of reaction are affected by nonproductive, obstructive multicollisions among redox partners themselves and among redox components and other nonredox proteins, (c) the degree of obstructive multicollisions is proportional to the protein concentration of the inner membrane.

Electron Transport, Collision Efficiency, and Protein Density. The kinetic analysis of Schneider et al. (22) and the diffusion analysis in Gupte et al. (47) have been extended to encompass a comparison of the rates of electron transport and diffusion in inner membranes of controlled protein concentration (78a). Electron transport activities involving multimolecular interactions decrease as a function of phospholipid enrichment (22). Using the Hardt equation for two-dimensional systems and experimentally determined electron transfer rates, it was calculated that the collisions/turnover, i.e., per electron transfer, increase for those activities that include ubiquinone diffusion; thus, efficiency decreases as the protein density decreases (Table 5). Whether this is a real increase in the collisions/turnover (decrease in collision efficiency) or an anomalous result due to the nature of the Hardt equation is not immediately clear. It may well be that more nearest-neighbor collisions occur per unit time in the more crowded native inner membrane, compared to the phospholipid-enriched inner membrane, due to a "cage effect," analogous to the Franck-Rabinowitz effect (84,85) in condensed phases. This effect in classical collision theory gives a three- to fivefold difference between theory and experimental results for condensed phases. Thus, the Hardt equation would tend to underestimate the true (greater) number of collisions in the protein-dense native inner membranes and give a more true value in lower-protein-density, phospholipid-enriched inner membranes. The observed increase in D as the protein density decreases may be a natural way of minimizing the difference in net electron transport rates in a population of mitochondria, which are known to have a naturally occurring variation (71) in inner-membrane protein density. Other data, which we will present, suggest that collision efficiency *is retained* as protein concentration is varied. Clearly, utilization of fusion techniques to modulate the inner membrane in order to compare rates of diffusion, collision, and electron transfer has helped to elucidate the relationships between structure, function, the theory for electron transport.

   The Direct Influence of Diffusion Rates on Electron Transport; Diffusion as a Rate-Limiting Step. Having established that mitochondria electron transport is diffusion-coupled (47), we turned our attention to ascertain whether diffusion is controlling or rate-limiting, or whether electron transport is controlled by the chemical reaction of the electron transfer itself. In this regard we have set forth criteria to define these possibilities (7).

   Since mitochondrial electron transport is a sequence of consecutive reactions comprising many individual rate constants, it is difficult to determine a rate-limiting step. Therefore, we have applied concepts put forth

TABLE 5 Collisions/Turnover for Q-III Redox Partners as a Function of Membrane Protein Density During Maximal Respiration

Activity[a]	Mitochondrial inner membrane (% enriched)				
	Native	30%	80%	240%	700%
Succinate oxidase	35	35	76.5	130	103
Succinate cyt. c reductase	18.2	31.1	71.7	152	150
DBH cytochrome c reductase	3.4	4.8	7.2	8.8	9.1
Duroquinol oxidase	11.1	15.0	19.2	16.1	6.1

[a]Assay conditions as in Ref. 22.

in rate process theory (86) and the theory of absolute reaction rates (87)
as a conceptual framework for our experimental approach (6,7,76,79a).
This conceptualization requires the determination of temperature dependences,
in terms of apparent activation energies (E_a), for physical and/or chemical
rate processes (e.g., diffusion and electron transfer) that constitute the
overall kinetic rate process. It can be predicted, via this rate process
theory, that the E_a for the rate-limiting step (diffusion or electron transfer)
will constitute the most significant contribution (largest portion) to the
E_a of the overall electron transport process (i.e., diffusion plus electron
transfer).

We compared the temperature dependence of the overall diffusive steps
to the temperature dependence of the overall steps (diffusion and electron
transfer) in the Complex II to ubiquinone to Complex III (II-Q-III) electron
transport process in the inner membrane. This was accomplished by experimental measurement of the temperature dependence of the D values of
the appropriate redox components by FRAP as well as the temperature
dependence of the overall succinate-linked, maximum electron transport
rates at various temperatures over the 5-40°C range and plotting the results
as log rate vs. 1/T, after Arrhenius. The Hardt equation for two-dimensional
diffusion was utilized to calculate the E_a values of the collision frequencies
for II-Q and Q-III redox partners (Table 6, column 2) by substituting
the appropriate D values of Q and Complex II or Q and Complex III, at
various temperatures. In this rate process approach the specific two-
dimensional, reaction-diffusion equation used in this calculation is not
consequential since no explicit temperature-dependent term is found in
any of these equations other than that implicit in the experimentally determined Ds. Furthermore, the relationship among the individual points as
a function of temperature in an Arrhenius plot will not be equation dependent; the slope and consequently the E_a will be independent of the equation
used. In accordance with the treatment of Gutman (88) of assigning separate
rate constants to the diffusion steps in electron transport and applying
the E_a values of the collision frequencies obtained for the II-Q and Q-III
redox partners, the overall E_a for the diffusive steps for the II-Q-III
sequence was determined (Table 6, column 3). E_a values for succinate-
cytochrome *c* reductase and succinate oxidase activities were also determined
(Table 6, column 4).

Comparison of the temperature dependencies in the native (nonenriched)
membrane, for the overall succinate oxidase electron transport process
(E_a = 12.8 kcal/mole, Fig. 13a) and the overall succinate-cytochrome *c*
reductase electron transport process (E_a = 14.3 kcal/mole, Fig. 13b) with
the temperature dependence of the overall diffusion steps (E_a = 12.2
kcal/mole), shows that the diffusion steps of Q and its redox partners
comprise the largest portion of the E_a of the overall kinetic process (cf.
Table 6, column 3 vs. column 4). Thus, the corresponding chemical reaction
steps, i.e., the actual electron transfer during or after collision, require
a smaller portion of the total E_a. This finding is compatible with diffusion
as a rate-limiting step (7).

Additionally, it can be postulated that if diffusion is, in fact, rate-
limiting, any affect on the E_a of diffusion will be approximately the same
on the E_a of electron transport. Consistent with diffusion as a rate-limiting
step, the E_a values of both lateral diffusion and electron transport decrease
approximately proportional to the decrease in the protein concentration
(by lipid enrichment) of the inner membrane (Table 6, columns 3 and 4).
This result also suggests that the collisions/turnover ratio does not change

TABLE 6 Comparison of Apparent Activation Energies

Membrane	Redox sequence	E_a (kcal/mole)		
		Collision frequency	Diffusion	Uncoupled electron transport activity
Native	II–Q	11.8		
	Q–III	11.9		
	II–Q–III		12.2	12.87 succinate cytochrome *c* reductase
				14.3 succinate oxidase
30% enriched	II–Q	9.07		
	Q–III	9.2		
	II–Q–III		9.55	10.8 succinate cytochrome *c* reductase
				9.3 succinate oxidase
80% enriched	II–Q	8.6		
	Q–III	8.9		
	II–Q–III		9.22	10.5 succinate cytochrome *c* reductase
				9.0 succinate oxidase

See text for details.

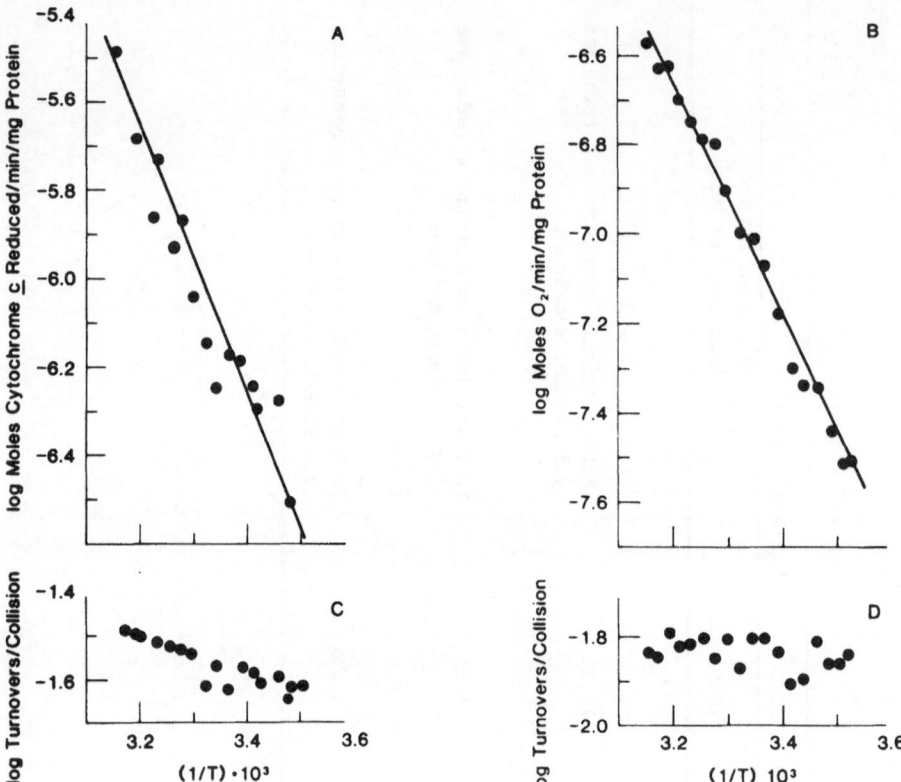

FIGURE 13   Temperature dependence of electron transport and collision
efficiencies in native inner membranes. Assay conditions are given in Refs.
7 and 22.   (A) Maximum succinate-cytochrome c reductase activity.   (B)
Maximum succinate oxidase activity.   (C) Collision efficiency as a function
of temperature of Q-III redox partners based on turnovers during maximum
succinate-cytochrome c reductase activity. Turnovers derived from panel A
and theoretical collisions calculated using the Hardt equation [viz. Gupte
et al. (47)].   (D) Same as in panel C, but using maximum succinate oxidase
activity.

with decreasing protein density; i.e., the true collision efficiency does
not change as a function of protein density (see Section IV.D.2).

     Finally, one would expect that for diffusion as a rate-limiting step
the collision efficiencies of the redox partners will be virtually independent
of temperature since any effect on the diffusion steps will be almost the
same as for the overall electron transport process. Accordingly, the collision
efficiencies of II-Q and Q-III were determined at various temperatures by
plotting, in Arrhenius fashion, the ratio of the actual turnovers to theoretical collisions (at various temperatures) of succinate-cytochrome c reductase
or succinate oxidase. This analysis shows a very weak or no temperature
dependence for collision efficiencies of Q-III (Fig, 13C and D) or of II-Q
(not shown). This indicates that any increase in the electron transport
rate can be accounted for by the increase in the rate of diffusion-based
collisions. These results on the temperature dependencies in native and

phospholipid-enriched inner membranes provide a third argument in favor of diffusion as a rate-limiting step (7).

3. Summary

These studies of structure-function relationships in mitochondrial electron transport were made possible by the application of membrane enrichment and fusion techniques. The random-collision model of mitochondrial electron transport, reviewed recently (7), conceptualizes electron transfer resulting from diffusion-based collisions between correct redox partners in the membrane plane. In this model collisional interactions occur concurrently and nonproductively (i.e., no electrons transferred), among all membrane components. The study of such a dynamic system was in large part dependent on the development of the low-pH fusion (enrichment) technique, which permitted the two-dimensional concentration of redox components to be varied in a controlled manner. It was determined that electron transport is dependent on the two-dimensional concentration of redox components, i.e., proportional to the average lateral separation of these components. The development of a calcium-mediated membrane fusion technique to produce ultralarge inner membranes made it possible to use FRAP to measure the diffusion rates of redox components in the native inner membrane and relate the diffusion rates to electron transport rates. The enrichment and fusion techniques also permitted the correlation of electron transport and diffusion as a function of concentration of membrane components and temperature, as well as an assessment of fundamental properties of diffusion in a biological membrane.

V. PERSPECTIVES AND APPLICATIONS

Membrane fusion can be a powerful tool in studies of membrane function and structure-function relationships. For such studies, membrane fusion techniques permit the modulation of a membrane's size and intrinsic properties. This makes possible the study of membrane properties and functions through the creation of membranes that are not normally amenable to instrumental analysis.

A. Criteria for the Application of Membrane Enrichment and Fusion Techniques

A study of function or structure-function relationships in a biological membrane can only be pursued by the application of enrichment and/or fusion techniques, if a number of criteria are met, relating to the following questions. (a) Structural order: Is a function predicated on a particular order or repeat of components that could be altered to vary performance? (b) Two-dimensional concentration dependence: Is a function dependent on the membrane planar concentrations of components such that performance could be altered? (c) Mobility of membrane components: Is a function dependent on the mobility of membrane components such that chemical reactions result after diffusion and collision, which may or may not involve transient, i.e., other than instantaneous, aggregation of membrane components? (d) Lipid-protein interactions: Is a function markedly dependent on the bilayer environment of a membrane-bound protein, such as phospholipid acyl chain dynamics, length, degree of saturation, headgroup composition, phase behavior, etc.? If one or more of these criteria are met by the system to be studied, then membrane modulation via enrichment and/or fusion techniques may be a viable approach.

B.   Applications of the Low-pH Enrichment Technique

A number of examples meeting one or more of the above criteria can be enumerated. The mitochondrial inner membrane, the major focus of our work and the membrane used to develop these techniques, is, of course, a prime example. Millner et al. (89) used low-pH fusion (2) to enrich thylakoid membranes from peak chloroplasts to study electron transport. Similarly, Snozzi and Crofts (90) enriched chromatophores from *Rhodopseudomonas sphaeroides* with liposomes to study quinone electron transport function. Driessen et al. (91) also used low pH to enrich vesicles derived from *Bacillus subtilis* and *Streptococcus cremoris*, and inner membranes of *Escherichia coli*, to examine the utility of low-pH fusion for bacterial membranes (see also Chapter 34). Siegel et al. (92) adapted the basic principles of the low-pH technique for use with the thylakoid membranes of spinach chloroplasts to produce membranes having different densities of integral membrane particles in order to study electron transport function. In a similar fashion, Casadio et al. (93) produced phospholipid-enriched bacterial chromatophores from *R. sphaeroides* to study quinone electron transport.

Receptor-mediated endocytosis, which is believed in some cases to involve simple diffusion and clustering of the receptors, may be tractable by enrichment and/or fusion methods (94). Clustering of ligands and receptors on the plasma membrane is believed to initiate endocytosis. Examples are IgG on B lymphocytes (94), C3b receptor on neutrophils (94), and the asialoglycoprotein receptor on hepatocytes (95). It has been suggested that membrane localization of receptors to coated pits may occur either as a result of a membrane receptor's diffusion in the bilayer or through a ligand-binding-induced slowing of the rate of diffusion through coated pits. Apparently, the slowing of diffusion results in a net accumulation at pits without triggering coat formation (94). Various drug and hormone interactions might involve receptor diffusion and clustering. Calculations by Hanski et al. (96) have suggested that adenylate cyclase is activated by β-receptors via a diffusion-controlled process. Whether diffusion is involved in the coupling of the β-adrenergic receptors and adenylate cyclase is not presently clear, but on the basis of reconstitution studies, the membrane environment is known to be important (97).

It may be possible to study a number of cellular functions via modulation by enrichment techniques. If studies are attempted in vivo, the normal turnover of the plasma membrane could be a problem in maintaining a stable membrane composition. Fusion-induced leakage might also present a problem. Additional lipids as well as receptors might be added to a membrane to determine their affects on function. It might be possible to isolate cellular membranes, modify them, then microinject them back into living cells to determine if function is altered. Various intracellular transport and processing functions might be studied in this manner. It would be feasible to modulate the plasma membrane of platelets and monitor changes in various aspects of their clotting function.

Thus, the low-pH method of Schneider et al. (2) provides a number of valuable possibilities. It should be emphasized that the development of this technique in our laboratory was engendered by a desire to modulate membrane properties rather than to study fusion. The low-pH method permits the incorporation of exogenous lipids in a controlled fashion into liquid-crystalline biological membranes having some negatively charged phospholipids and mobile proteins. Lipid compositional changes may be chosen to possibly modulate the activity of membrane-bound enzymes or transport proteins as expressed in their Arrhenius plot behavior, to modulate membrane fluidity, or to alter the phase behavior of the membrane. Accordingly, one may choose to specifically enrich or diminish specific lipid components of the membrane relative to others.

The incorporation of exogenous phospholipids may be used to dilute the integral membrane protein concentration (density) to modulate structure and/or function. This has been so far the singularly unique contribution of the enrichment technique. With regard to structure, the incorporation of exogenous phospholipids of the same or similar composition compared to the intact, native membrane can be used to lower the membrane concentration of the proteins, while still preserving their random distribution. Conversely, the introduction of certain lipoidal molecules may permit the graduated induction of nonrandom protein distributions. Thus, membrane proteins may be induced to aggregate to different extents until an almost complete lateral-phase separation of protein and lipid components has been achieved, as with the use of cholesterol in the mitochondrial inner membrane (15). With regard to structure-function relationships, if the mobility of one or more membrane components is essential or of major importance for function, function should be affected proportional to the decrease in the concentration of the mobile component. Concomitantly, there may be an increase in long-range mobility in a native membrane due to a decrease in obstructive collisions, e.g., in membranes where significant obstructive collisions occur, such as the mitochondrial inner membrane. This increase in mobility may mitigate some of the decrease in function due to concentration. Alteration of the distribution of membrane components from random to nonrandom could also be interpreted as support for structural mobility, although definitive proof should be a change in function.

The low-pH enrichment technique may be applied to the selective incorporation of specific integral proteins into a native membrane. This can be accomplished by utilizing vesicles carrying these proteins in the membrane. However, the correct and/or uniform orientation of the proteins once introduced into the biological membrane should be verified by independent means, such as antibody probing.

C. Application of Calcium-Mediated Fusion Techniques

The calcium-mediated fusion method(s) developed in our laboratory (3,4) should be considered when a significantly larger-diameter vesicle is required by experimental design or instrumental constraints. It is considered essential for fusion to have a membrane in a liquid-crystalline state, some negatively charged lipids present, and probably mobile proteins. It should be clear from Section IV that the specific conditions of calcium concentration, low pH, and temperature necessary to achieve fusion will be dependent on the lipid composition of the membranes. Therefore, it is necessary to determine the appropriate condition(s) for a particular membrane system. It can be speculated that a combination of calcium and low pH may facilitate fusion of various vesicles and cellular membranes. However, successful application of these fusion techniques to a given membrane system may entail considerable experimentation with pH, calcium concentration, temperature, and duration of incubation.

There are a number of possible applications for the calcium-mediated fusion techniques. First is the expansion of a membrane to a diameter sufficient for the FRAP technique (5 μm minimum). Membranes increased to these larger sizes also become amenable to microelectrode studies and possibly microinjection-based studies. Some membrane functions may be possible to study in enlarged membranes using optical microscopy. Since it may also be possible to control the permeability of these 'fused membranes by the choice of fusion conditions (see, e.g., Ref. 3 vs. Ref. 4), functions related to permeability may be studied. From a more strictly functional perspective, it may be possible to vary the surface-area-to-volume ratio of a membrane vesicle to assess effects on transport, or other membrane functions. The coupling of enrichment and fusion techniques, as we have used to produce ultralarge mitochondrial inner membranes of controlled protein density, may also be applied to other cell membranes.

D. Conclusion

In conclusion, we can think of enrichment and fusion techniques from a number of important perspectives. While a physical mechanism(s) is still being sought for fusion, it is possible to effectively use fusion-based techniques to solve other important biological problems. It is hoped that in the latter respect this article will stimulate other applications of enrichment and fusion to study structure-function relationships in a variety of membranes.

NOTE ADDED IN PROOF

We have updated only references to our work that have appeared after the 1986 date this chapter was submitted.

Electron transfer refers to the actual transmission of reducing equivalents between redox components. Electron transport is inclusive of the overall process of electron transfer and lateral diffusion of redox components.

ACKNOWLEDGMENT

The research presented in this review was supported in part by funding from NIH GM28704, and NSF PCM79-10968 and PCM84-05269. The authors acknowledge the contributions of Drs. En-Shinn Wu, Heinz Schneider, Matthias Höchli, John J. Lemasters, Sharmila S. Gupte, Arthur E. Sowers, Luzia Höchli, and Kenneth A. Jacobson to various aspects of the work presented here.

REFERENCES

1. Juliano, R., and Layton, D. (1980). Liposomes as a drug delivery system. In: *Drug Delivery Systems* (R. Juliano, ed.). Oxford University Press, Oxford, pp. 189-236.
2. Schneider, H., Lemasters, J. J., Höchli, M., and Hackenbrock, C. R. (1980). Fusion of liposomes with mitochondrial inner membranes. *Proc. Natl. Acad. Sci. USA* 77:442-446.

3. Chazotte, B., Wu, E-S., Höchli, M., and Hackenbrock, C. R. (1985). Calcium-mediated fusion to produce ultra-large osmotically active mitochondrial inner membranes of controlled protein density. *Biochim. Biophys. Acta* 818:87-95.

4. Höchli, M., Höchli, L., and Hackenbrock, C. R. (1985). Independent lateral diffusion of cytochrome bc1 complex and cytochrome oxidase in mitochondrial inner membrane. *Eur. J. Cell Biol.* 38:1-5.

5. Hackenbrock, C. R. (1981). Lateral diffusion and electron transfer in mitochondrial inner membrane. *Trends Biochem. Sci.* 6:151-154.

6. Hackenbrock, C. R., Gupte, S. S., and Chazotte, B. (1985). The random collision model of mitochondrial electron transport and diffusion control. In: *Achievements and Perspectives of Mitochondrial Research*, Vol. I: *Bioenergetics* (E. Quagliariello, E. C. Slater, F. Palmieri, C. Saccone, and A. M. Kroon, eds.). Elsevier, Amsterdam, pp. 83-101.

7. Hackenbrock, C. R., Chazotte, B., and Gupte, S. S. (1986). The random collision model and a critical assessment of diffusion and collision in mitochondrial electron transport. *J. Bioenergetics Biomemb.* 18:331-368.

8. Green, D. E. (1974). The electromechanochemical model for energy coupling in mitochondria. *Biochim. Biophys. Acta* 346:27-78.

9. Hackenbrock, C. R., and Hammon, K. M. (1975). Cytochrome c Oxidase in liver mitochondria. Distribution and orientation determined with affinity purified immunoglobulin ferritin conjugates. *J. Biol. Chem.* 250:9185-9197.

10. Hackenbrock, C. R., Höchli, M., and Chau, R. M. (1976). Calorimetric and freeze fracture analysis of lipid phase transitions and lateral translational motion of intramembrane particles in mitochondrial membranes. *Biochim. Biophys. Acta* 455:466-484.

11. Höchli, M., and Hackenbrock, C. R. (1976). Fluidity in mitochondrial membranes: Thermotropic lateral translational motion of intramembrane particles. *Proc. Natl. Acad. Sci. USA* 73:1636-1640.

12. Höchli, M., and Hackenbrock, C. R. (1977). Thermotropic lateral translational motion of intramembrane particles in the inner mitochondrial membrane and its inhibition by artificial peripheral proteins. *J. Cell Biol.* 72:278-291.

13. Höchli, M., and Hackenbrock, C. R. (1979). Lateral translational diffusion of cytochome c oxidase in the mitochondrial energy-transducing membrane. *Proc. Natl. Acad. Sci. USA* 76:1236-1240.

14. Kagawa, Y., and Racker, E. (1966). Partial resolution of enzymes catalyzing oxidative phosphorylation. *J. Biol. Chem.* 241:2467-2474.

15. Schneider, H., Höchli, M., and Hackenbrock, C. R. (1982). Relationship between the density distribution of intramembrane particles and electron transfer in the mitochondrial inner membrane as revealed by cholesterol incorporation. *J. Cell Biol.* 94:387-393.

16. Schneider, H., Lemasters, J. J., and Hackenbrock, C. R. (1982). Lateral diffusion of ubiquinone during electron transfer in phospholipid- and ubiquinone-enriched mitochondrial membranes. *J. Biol. Chem.* 257:10787-10793.

17. Wilschut, J., and Papahadjopoulos, D. (1979). Ca^{2+}-induced fusion of phospholipid vesicles monitored by mixing of aqueous contents. *Nature* 281:690-692.

18. Ellens, H., Bentz, J., and Szoka, F. C. (1985). H^+- and Ca^{2+}-induced fusion and destabilization of liposomes. *Biochemistry* 24:3099-3106.

19. Sundler, R., Düzgüneş, N., and Papahadopoulos, D. (1981). Control of membrane fusion by phospholipid headgroups II. *Biochim. Biophys. Acta* 649:751–758.

20. Owen, C. S. (1980). A membrane-bound fluorescent probe to detect phospholipid-cell fusion. *J. Memb. Biol.* 54:13–20.

21. Düzgüneş, N., Wilschut, J., Fraley, R., and Papahadjopoulos, D. (1981). Studies on the mechanism of membrane fusion. Role of head-group composition in calcium- and magnesium-induced fusion of mixed phospholipid vesicles. *Biochim. Biophys. Acta* 642:182–195.

22. Schneider, H., Lemasters, J. J., Hochli, M., and Hackenbrock, C. R. (1980). Liposome-mitochondrial inner membrane fusion. Lateral diffusion of integral electron transfer components. *J. Biol. Chem.* 255:3748–3756.

23. Struck, D. K., Hoekstra, D., and Pagano, R. E. (1981). Use of resonance energy transfer to monitor membrane fusion. *Biochemistry* 20:4093–4099.

24. Colbeau, A., Nachbaur, J., and Vignais, P. M. (1971). Enzymic characterization and lipid composition of rat liver subcellular membranes. *Biochim. Biophys. Acta* 249:462–492.

25. Daum, G. (1985). Lipids of mitochondria. *Biochim. Biophys. Acta* 822:1–42.

26. White, J., and Helenius, A. (1980). pH dependent fusion between the Semliki Forest virus membrane and liposomes. *Proc. Natl. Acad. Sci. USA* 77:3273–3277.

27. Nir, S., Bentz, J., Wilschut, J., and Düzgüneş, N. (1983). Aggregation and fusion of phospholipid vesicles. *Prog. Surface Sci.* 13:1–124.

28. Papahadjopoulos, D. (1977). Calcium induced phase changes and fusion in natural and model membranes. In: *Membrane Fusion*, Cell Surface Reviews, Vol. 5. Elsevier, Amsterdam, pp. 765–790.

29. Papahadjopoulos, D., Poste, G., and Vail, J. W. (1979). Studies on membrane fusion with natural and model membranes. *Meth. Memb. Biol.* 10:1–121.

30. Hoekstra, D. (1982). Role of lipid phase separations and membrane hydration in phospholipid vesicle fusion. *Biochemistry* 21:2833–2840.

31. Conner, J., Yatvin, M. B., and Huang, L. (1984). pH-sensitive liposomes: Acid-induced fusion. *Proc. Natl. Acad. Sci. USA* 81:1715–1718.

32. Ellens, H., Bentz, J., and Szoka, F. C. (1984). pH-induced destabilization of phosphatidylethanolamine-containing liposomes: Role of contact. *Biochemistry* 23:1532–1538.

33. Bondeson, J., Wikander, J., and Sundler, R. (1984). Proton-induced membrane fusion. Role of phospholipid composition and protein-mediated intermembrane contact. *Biochim. Biophys. Acta* 777:21–27.

34. Kolber, M. A., and Haynes, D. H. (1979). Evidence for a role of phosphatidylethanolamine as a modulator of membrane-membrane contact. *J. Memb. Biol.* 48:95–114.

35. Düzgüneş, N., Straubinger, R. M., Baldwin, P. A., Friend, D. S., and Papahadjopoulos, D. (1985). Proton induced fusion of oleic acid/phosphatidylethanolamine liposomes. *Biochemistry* 24:3091–3098.

36. Hope, M. J., Walker, D. C., and Cullis, P. R. (1983). Ca^{2+} and pH induced fusion of small unilamellar vesicles consisting of phosphatidylethanolamine and negatively charged phospholipids: A freeze fracture study. *Biochim. Biophys. Res. Commun.* 110:15–22.

37. Stegmann, T., Hoekstra, D., Scherphof, G., and Wilschut, J. (1985). Kinetics of pH-dependent fusion between influenza virus and liposomes. *Biochemistry* 24:3107–3113.

38. Cullis, P. R., Verkleij, A. J., and Ververgaert, P. H. J. Th. (1978). Polymorphic phase behavior of cardiolipin as detected by 31P NMR and freeze fracture techniques. Effects of calcium, dibucaine and chlorpromazine. *Biochim. Biophys. Acta* 513:11-20.

39. Verkleij, A. J., van Echteld, C. J. A., Gerritsen, W. J., Cullis, P. R., and De Kruijff, B. (1980). The lipidic particle as an intermediate structure in membrane fusion processes and bilayer to hexagonal H_{II} transitions. *Biochim. Biophys. Acta* 600:620-624.

40. Hoekstra, D., and Martin, O. C. (1985). Transbilayer redistribution of phosphatidylethanolamine during fusion of phospholipid vesicles. Dependence on fusion rate, lipid phase separation and formation of nonbilayer structures. *Biochemistry* 24:6097-6103.

41. Cullis, P. R. and De Kruijff, B. (1979). Lipid polymorphism and the functional roles of lipids in biological membranes. *Biochim. Biophys. Acta* 559:399-420.

42. MacDonald, R. I. (1985). Membrane fusion due to dehydration by polyethylene glycol, dextran, or sucrose. *Biochemistry* 24:4058-4066.

43. Gibson, S. M., and Strauss, G. (1984). Reaction characteristics and mechanisms of lipid bilayer vesicle fusion. *Biochim. Biophys. Acta* 769:531-542.

44. Rand, R. P. (1981). Interacting phospholipid bilayers: Measured forces and induced structural changes. *Annu. Rev. Biophys. Bioeng.* 10:277-314.

45. Schenkman, S., De Araujo, P. S., Sesso, A., Quina, F. H., and Chaimovich, H. (1981). A kinetic and structural study of two-step aggregation and fusion of neutral phospholipid vesicles promoted by serum albumin at low pH. *Chem. Phys. Lipids* 28:165-180.

46. Chazotte, B., Wu, E-S., and Hackenbrock, C. R. (1984). Effect of varied membrane protein density on the lateral diffusion of lipids in the mitochondrial inner membrane. *Biochem. Soc. Trans.* 12:463-464.

47. Gupte, S. S., Wu, E-S., Höchli, L., Höchli, M., Jacobson, K. A., Sowers, A. E., and Hackenbrock, C. F. (1984). Relationship between lateral diffusion, collision frequency, and electron transfer of mitochondrial inner membrane oxidation-reduction components. *Proc. Natl. Acad. Sci. USA* 81:2606-2610.

48. Sowers, A. E., and Hackenbrock, C. R. (1981). Rate of lateral diffusion of intramembrane particles: Measurement by electrophoretic displacement and rerandomization. *Proc. Natl. Acad. Sci USA* 78:6246-6250.

49. Kawato, S., Lehner, C., Müller, M., and Cherry, R. J. (1982). Protein-protein interactions of cytochrome oxidase in inner mitochondrial membranes. The effect of liposome fusion on protein rotational mobility. *J. Biol. Chem.* 257:6470-6476.

50. Volsky, D. J., and Loyter, A. (1978). Inhibition of membrane fusion by suppression of lateral movement of membrane proteins. *Biochim. Biophys. Acta* 514:213-224.

51. Poste, G., and Allison, A. C. (1973). Membrane fusion. *Biochim. Biophys. Acta* 300:421-465.

52. Ahkong, Q. F., Tampion, W., and Lucy, J. A. (1975). Promotion of cell fusion by divalent cation ionophores. *Nature* 256:208-209.

53. Sheetz, M. P., and Chan, S. I. (1973). Effect of sonication on the structure of lecithin bilayers. *Biochemistry* 11:4573-4581.

54. Wilschut, J. N., Duzgunes, N., Hoekstra, D., and Papahadjopoulos, D. (1985). Modulation of membrane fusion by membrane fluidity: Temperature

dependence of divalent cation induced fusion of phosphatidylserine vesicles. *Biochemistry* 24:8-14.

55. Morris, S. J., Gibson, C. C., Smith, P. D., Greif, P. C., Stirk, C. W., Bradley, D., Haynes, D. H., and Blumenthal, R. (1985). Rapid kinetics of Ca2+-induced fusion of phosphatidylserine/phosphatidyl-ethanolamine vesicles. The effect of bilayer curvature on leakage. *J. Biol. Chem.* 260:4122-4127.

56. Bentz, J., and Düzgünes, N. (1985). Fusogenic capacities of divalent cations and effect of liposome size. *Biochemistry* 24:5436-5443.

57. Gingell, D., and Ginsberg, L. (1978). Problems in the physical inter-pretation of membrane interaction and fusion. In: *Membrane Fusion*, Cell Surface Reviews, Vol. 5 (G. Poste and G. Nicholson, eds.). Elsevier, New York, pp. 792-830.

58. Traüble, H. (1977). Membrane electrostatics. In *Structure of Biological Membranes*. Nobel Foundation Symposium 34. Plenum Press, New York, pp. 509-549.

59. Van Meer, G., Davoust, J., and Simons, K. (1985). Parameters affecting low-pH-mediated fusion of liposomes with the plasma membrane of cell infected with influenza virus. *Biochemistry* 24:3593-3602.

60. Wilschut, J., Holsappel, M., and Jansen, R. (1982). Ca^{2+}-induced fusion of cardiolipin/phosphatidylcholine vesicles monitored by mixing of aqueous contents. *Biochim. Biophys. Acta* 690:297-301.

61. Verkleij, A. J., Mombers, C., Gerritsen, W. J., Leunissen-Bijvelt, L., and Cullis, P. R. (1979). Fusion of phospholipid vesicles in association with the appearance of lipidic particles as visualized by freeze fracturing. *Biochim. Biophys. Acta* 555:358-361.

62. Ohki, S., and Ohshima, H. (1985). Divalent cation-induced phosphatidic acid membrane fusion. Effect of ion binding and membrane surface tension. *Biochim. Biophys. Acta* 812:147-154.

63. Sundler, R., and Papahadjopoulos, D. (1981). Control of membrane fusion by phospholipid headgroups. I. Phosphatidate/phosphatidylinositol specificity. *Biochim. Biophys. Acta* 649:743-750.

64. Rothman, J. E., and Lenard, J. (1977). Membrane asymmetry. *Science* 195:743-753.

65. Sokolove, P., Brenza, J. M. and Shamoo, A. E. (1983). Ca^{2+}-cardiolipin interaction in model system. Selectivity and apparent high affinity. *Biochim. Biophys. Acta* 732:41-47.

66. De Kruijff, B., and Cullis, P. R. (1980). Cytochrome *c* specifically induces nonbilayer structures in cardiolipin containing membranes. *Biochim. Biophys. Acta* 602:477-490.

67. Sundler, R. (1984). Studies on the effective size of phospholipid head-groups in bilayer vesicles using lectin-glycolipid interaction as a steric probe. *Biochim. Biophys. Acta* 771:56-67.

68. Portis, A., Newton, C., Pangborn, W., and Papahadjopoulos, D. (1979). Studies on membrane fusion: Evidence for an intermembrane Ca^{2+}-phospholipid complex. Synergism with Mg^{2+} and inhibition by spectrin. *Biochemistry* 18:780-790.

69. Bentz, J., Düzgüneş, N., and Nir, S. (1983). Kinetics of divalent cation induced fusion of phosphatidylserine vesicles: Correlation between fusogenic capacities and binding affinities. *Biochemistry* 22:3320-3330.

70. Cullis, P. R., De Kruijff, B., Hope, M. J., Nayar, R., Reitveld, A., and Verkleij, A. J. (1980). Structural properties of phospholipids in the rat liver mitochondrial inner membrane. *Biochim. Biophys. Acta* 600:625-635.

71. Sowers, A. E., and Hackenbrock, C. R. (1985). Variation in protein lateral diffusion coefficients is related to variation in protein concentration found in mitochondrial inner membranes. *Biochim. Biophys. Acta* 821:85–90.

72. Zimmermann, U. (1982). Electric field-mediated fusion and related electrical phenomena. *Biochim. Biophys. Acta* 694:227–277.

73. Bentz, J., Düzgüneş, N., and Nir, S. (1985). Temperature dependence of divalent cation induced fusion of phosphatidylserine liposomes: Evaluation of kinetic rate constants. *Biochemistry* 24:1064–1072.

74. Lucy, J. A. (1978). Mechanisms of chemically induced cell fusion. In *Membrane Fusion*, Cell Surface Reviews, Vol. 5 (G. Poste and G. L. Nicholson, eds.). Elsevier, New York, pp. 268–304.

75. Minelli, M., and Ceccabini, M. (1982). Protein-dependent lipid lateral phase separation as a mechanism of human erythrocyte ghost resealing. *J. Cell Biochem.* 19:59–75.

76. Hackenbrock, C. R., Chazotte, B., and Gupte, S. S. (1986). Lateral diffusion of coenzyme Q and its redox partners in the control of mitochondrial electron transport. In: *Biomedical and Clinical Aspects of Coenzyme Q*, Vol. 4 (K. Folkers, and Y. Yamamura, eds.). Elsevier, Amsterdam, pp. 25–38.

76a. Gupte, S. S., and Hackenbrock, C. R. (1988). Multidimensional diffusion modes and collision frequencies of cytochrome *c* with its redox partners. *J. Biol. Chem.* 263:5241–5247.

76b. Gupte, S. S., and Hackenbrock, C. R. (1988). The role of cytochrome *c* diffusion in mitochondrial electron transport. *J. Biol. Chem.* 263:5248–5253.

77. Chazotte, B., and Hackenbrock, C. R. (1988). Physical constraints on lateral diffusion and ubiquinone mitochondrial electron transport. In: *Integration of Mitochondrial Function* (J. J. Lemasters, C. R. Hackenbrock, R. G. Thurman, and H. V. Westerhoff, eds.). Plenum Press, New York, pp. 53–61.

78. Gupte, S. S., and Hackenbrock, C. R. (1988). Lateral diffusion of mitochondrial F_1F_0ATP synthase: A random collision model. In: *Integration of Mitochondrial Function* (J. J. Lemasters, C. R. Hackbrock, R. G. Thurman, and H. V. Westerhoff, eds.). Plenum Press, New York, pp. 95–101.

78a. Chazotte, B., and Hackenbrock, C. R. (1988). The multicollisional, obstructed, long-range, diffusional nature of mitochondrial electron transport. *J. Biol. Chem.* 263:14359–14367.

79. Eisenger, J., Flores, J., and Peterson, W. P. (1986). A milling crowd for local and long-range obstructed lateral diffusion-mobility of excimer probes in the membrane of intact erythrocytes. *Biophys. J.* 49:987–1001.

79a. Chazotte, B., and Hackenbrock, C. R. (1989). Lateral diffusion as a rate-limiting step in ubiquinone-mediated mitochondrial electron transport. *J. Biol. Chem.* 264:4978–4985.

80. Saffman, P. G., and Delbrück, M. (1975). Brownian motion in biological membranes. *Proc. Natl. Acad. Sci. USA* 72:311–314.

81. Hardt, S. L. (1979). Rates of diffusion controlled reactions in one, two and three dimensions. *Biophys. Chem.* 10:239–243.

82. Chazotte, B., and Hackenbrock, C. R. (1985). Experimental and theoretical relationships of lateral diffusion and electron transport in mitochondrial inner membranes of controlled, varied protein density. *Proc. XIII International Congress of Biochemistry*, Amsterdam, The Netherlands. Elsevier, Amsterdam, p. 352.

83. Chazotte, B., and Hackenbrock, C. R. (1985). Experimental and theoretical analysis of the random collision model of mitochondrial electron transport. *Proc. International Symposium on Achievements and Perspectives in Mitochondrial Research*, Bari, Italy. Adriatica Editrice, Bari, p. 56.

84. Franck, J., and Rabinowitz, E. (1934). Some remarks about free radicals and the photochemistry of solutions. *Trans. Faraday Soc.* 30:120-125.

85. Rabinowitz, E. (1937). Collision, co-ordination, diffusion and reaction velocity in condensed systems. *Trans. Faraday Soc.* 33:1225-1230.

86. Johnson, F. H., Eyring, H., and Stover, B. J. (1975). *Theory of Rate Processes in Biology and Medicine*. Wiley, New York.

87. Eyring, H. (1935). The theory of absolute reaction rates. *J. Chem. Phys.* 3:107-115.

88. Gutman, M. (1980). Electron flux through the mitochondrial ubiquinone. *Biochim. Biophys. Acta* 594:53-84.

89. Millner, P. A., Grouzis, J. P., Chapman, D. J., and Barber, J. (1983). Lipid enrichment of thylakoid membranes. I. Using soybean phospholipids. *Biochim. Biophys. Acta* 722:331-340.

90. Snozzi, M., and Crofts, A. R. (1984). Electron transport in chromatophores from *Rhodopseudomonas sphaeorides* GA fused with liposomes. *Biochim. Biophys. Acta* 766:451-463.

91. Driessen, A. J. M., Hoekstra, D., Scherphof, G., Kalicharan, R. D. and Wilschut, J. (1985). Low pH-induced fusion of liposomes with membrane vesicles derived from *Bacillis subtilis*. *J. Biol. Chem.* 260: 10880-10887.

92. Siegel, C. O., Jordan, A. E., and Miller, K. R. (1981). Addition of lipid to the photosynthetic membrane: Effects on membrane structure and energy transfer. *J. Cell Biol.* 91:113-125.

93. Casadio, R., Venturoli, G., Di Gioia, A., Castellani, P., Leonardi, L., and Melandri, B. A. (1984). Phospholipid-enriched bacterial chromatophores. A system suited to investigate the ubiquinone mediated-interactions of protein complexes in photosynthetic oxidoreduction processes. *J. Biol. Chem.* 259:9149-9157.

94. Silverstein, S. C., Steinman, R. M., and Cohen, Z. A. (1977). Endocytosis. *Annu. Rev. Biochem.* 46:669-722.

95. Ashwell, G., and Harford, J. (1982). Carbohydrate-specific receptors of the liver. *Annu. Rev. Biochem.* 51:531-554.

96. Hanski, E., Rimon, G., and Levitzki, A. (1979). Adenylate cyclase activation by the beta-adrenergic receptors as a diffusion-controlled process. *Biochemistry* 18:846-853.

97. Lefkowitz, R. J., Stadel, J. M., and Caron, M. G. (1983). Adenylate cyclase-coupled beta-adrenergic receptors: Structure and mechanisms of activation and desensitization. *Annu. Rev. Biochem.* 52:159-186.

98. Chazotte, B., Wu, E-S., and Hackenbrock, C. R. (1983). Rates of lipid diffusion in mitochondrial inner membranes of varying protein density. *Fed. Proc.* 42:2170.

36

Liposomal Drug Delivery
Current Status and Future Prospects

FRANCIS C. SZOKA, JR.

University of California—San Francisco, San Francisco, California

I. INTRODUCTION

It has been 19 years since Gregoriadis and Ryman (1) proposed the liposome as a carrier to treat lysosomal storage disorders. This early suggestion to use liposome-encapsulated enzymes for enzyme replacement therapy was soon followed by a plethora of potential medical applications (2,3). Unfortunately, progress in gene therapy with liposome-encapsulated enzymes or DNA has been meager. This is primarily due to the fact that most liposome compositions used for drug delivery do not fuse with cellular membranes and hence do not efficiently introduce their contents into the cytoplasm. Therefore, understanding of viral membrane fusion at the biochemical and biophysical level has an obvious relevance for future improvements in cytoplasmic delivery using liposomes.

In the intervening years, the perception has arisen in some circles that liposomes would not be widely useful as a vehicle for drug delivery (4). Indeed, the progression of the liposome from the laboratory to the clinic has been slow. Until recently, clinical trials of liposome-encapsulated agents were limited to one or, at most, a few patients (5). This is partially due to the reluctance of large pharmaceutical companies to become involved with an unproven system and the attendant unpredictable return on investment (6). It is also partially due to the fact that many of the proposed applications for liposomes require injection (2,3). This route of administration demands that the system be sterile, pyrogen free, and of a controlled particle size. Given these criteria, not many academic research teams had the resources to bring a new liposomal dosage form to the clinic. The persistence of a few groups has led to the use of liposome-encapsulated drugs in preliminary clinical trials (7-13) (Table 1).

Prior to 1980 about 30 patients were treated with liposome-encapsulated agents; since then over 140 patients have participated in published studies (14,15). This acceleration in the use of liposomes in humans continues (Table 2) and is catalyzed by liposome drug delivery companies formed

TABLE 1   Preliminary Clinical Studies of Liposome-Encapsulated Drugs, 1983-1989

Drug	Disease state	Number of patients	Ref.
Doxorubicin	Cancer	10	7
Doxorubicin	Cancer	20	8
Doxorubicin	Cancer	6	9
Doxorubicin	Cancer	15	13
Amphotericin B	Fungal infections	12	10
Amphotericin B	Fungal infections	15	11
NSC 251635	Cancer	14	12

TABLE 2   Actual and Projected Human Trials of Liposome-Encapsulated Agents in the United States in 1990

Drug	Application	Current status
Doxorubicin	Cancer	Phase I
Doxorubicin	Cancer	Phase II
Liposomal platinum	Cancer	Phase I
Daunorubicin	Cancer	Phase I
Muramyl tripeptide	Cancer	Phase II
Indium	Tumor imaging	Phase II/III
Amphotericin B	Antifungal	Phase I 1990
Gentamicin	Antibacterial	Phase I 1990
Metaproterenol sulfate	Asthma	Phase II

Source: Modified from Refs. 15 and 16.

in the 1980s. Such companies raised between 120 and 150 million dollars (16) to commercialize liposomes as drug carriers. The fruits of their labor can be seen in Table 2. Ten different liposomal drug formulations and one in vivo diagnostic formulation were projected to be in human clinical trials in the United States by 1990.

From Tables 1 and 2 it is clear that two drugs, amphotericin B and doxorubicin, have received the most attention as liposome-encapsulated agents for use in humans. One reason for this interest is that both agents exhibit an improved therapeutic index in animal studies when administered as the liposomal form compared to the nonencapsulated or "free" form (vide infra). Thus amphotericin B and doxorubicin serve as excellent paradigms for one mechanism by which liposomes can improve the therapeutic index of a drug.

In this chapter, I briefly discuss how a drug carrier can improve drug therapy, how liposomes are processed after intraveneous administration, and how liposomes are handled on a cellular level. These topics have been the focus of numerous reviews (2,5,15,17-19) and only the most salient features are mentioned here. Then, with doxorubicin and amphotericin B as examples, the current status of liposomal drug delivery in humans is reviewed. The extensive information from animal studies with liposomal doxorubicin (L-dox) can help to delineate the mechanism of the improvement in the therapeutic index and to explicate the role lipid composition and size play in the therapeutic effect.

Finally, I emphasize the importance of selecting agents that can escape from the lysosomal compartment and the need to improve methods to fuse the liposome with cellular membranes so that future macromolecular therapeutic agents can be delivered in liposomes.

II. MECHANISM OF DRUG CARRIER IMPROVEMENTS IN DRUG THERAPY

The encapsulation and delivery of a drug in liposome has the potential to improve drug therapy by (1) rate control of drug input, (2) site-specific drug delivery, (3) site-avoidance drug delivery (Fig. 1). These three distinct mechanisms can be inferred from basic principles of therapeutics.

The first principle, or what might be designated the zeroth law of therapeutics, is that the effect is caused by the drug. A further assumption, which can be considered to be the first law of therapeutics, is that the drug concentration at the target site is related to the magnitude of the effect. Based on these postulates, the simplest model to describe carrier-mediated drug delivery is illustrated in Figure 2. It is apparent from this figure, that the drug must be released from the carrier to have an effect. Thus even the simplest drug carrier systems must initially be viewed as two-compartment pharmacokinetic models. In evaluating a drug as a candidate for delivery in a carrier system, consideration must be given to how, when, and where the drug will be released from the carrier.

When liposomes are used to deliver drugs, a complete kinetic description of the system must include at least: the elimination rate of the liposome, and the elimination rate of the nonencapsulated drug. The possibility that drug can return to the blood compartment after intracellular release from the liposomes must also be considered (Fig. 3) (vide infra).

When drug concentration is measured in the blood or plasma compartment after intravenous administration in the liposomal form, the quantity measured includes both liposome-encapsulated and nonencapsulated drug. This is particularly true at early time points in the elimination curve. Therefore, when the plasma level of a liposome-encapsulated drug is found to be greater than the corresponding plasma level of the nonencapsulated

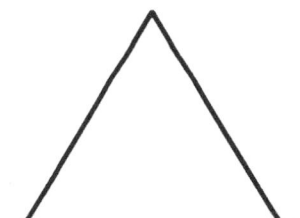

FIGURE 1 Mechanisms by which drug carriers improve drug therapy.

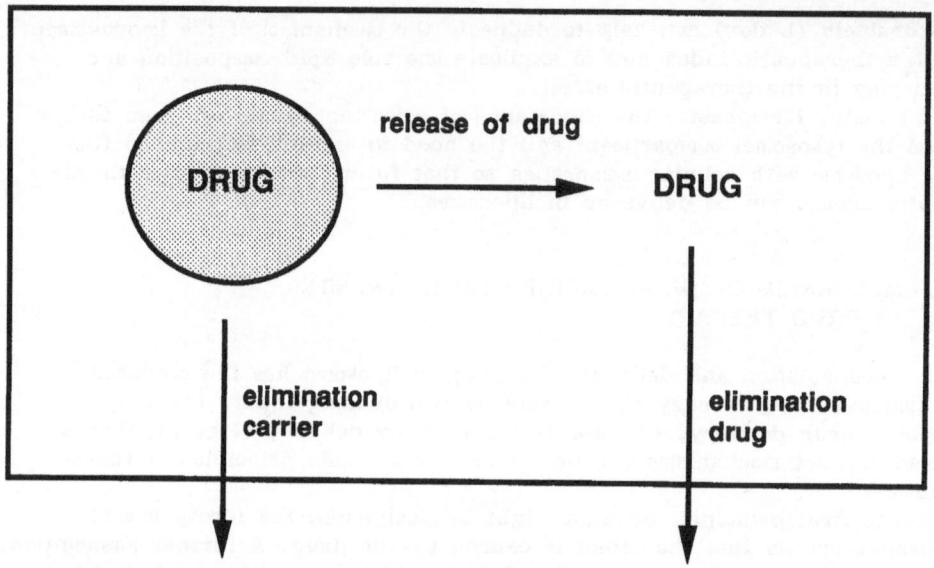

FIGURE 2 Simple pharmacokinetic scheme to describe the initial behavior
of a liposome drug carrier following administration.

compound, it cannot be assumed that drug activity is greater with the
liposomal form than the nonencapsulated form. If the drug remains in the
liposome, it may not be available and it may ultimately be metabolized so
that it never gains access to the target site.

All therapeutic agents have both desirable and undesirable effects
in an organism; this can be considered the second law of therapeutics.
The therapeutic and toxic effects can be related to drug concentrations
in the body, and usually the therapeutic effect occurs at a lower drug
concentration than the toxic effect. A therapeutic index for a compound
can be computed as the ratio of the drug concentration that causes a toxic
effect in 50% of a group of experimental animals to the concentration that
elicits a therapeutic effect in 50% of the animals (20). Controlling drug
input rates can be used to maintain the drug concentration above a therapeutic level but below a toxic level (21). This is the rationale for controlled
drug delivery and is the guiding principle for a number of oral and trans-
dermal drug delivery systems (22)

When the therapeutic effect occurs at a different locus in the body
than the toxic effect, carriers can improve the therapeutic index of a
drug by redirecting the drug to the therapeutic site or away from the
toxic site.

A. Controlled Drug Delivery Using Liposomes

When drug input is controlled, the drug concentration in the plasma compartment is regulated, but the ratio of the drug concentration at the target

and toxic sites is not altered. This improves drug therapy by minimizing fluctuations of drug concentrations above the maximum tolerated concentration or below the minimum effective concentration but does not change the therapeutic index of the compound. In the simplest case, where elimination of the liposome-drug complex is ignored, liposomes fulfill the criteria of a first-order rate controller and an adjustable drug release rate can be provided by selection of the liposome composition (23). For instance, liposome-encapsulated cytosine arabinoside is significantly more effective than the nonencapsulated compound against L1210 murine leukemia (24,25). The therapeutic effect of this compound can be modulated by varying the mole ratio of cholesterol in the liposome; liposomes with high ratios of cholesterol/phospholipid are more stable in plasma and release encapsulated cytosine arabinoside more slowly (23,24). Thus one can obtain an optimal therapeutic effect per unit dose, which is a more economical and convenient form of drug administration.

The mechanism of the liposome improvement of the therapeutic effect for cytosine arabinoside was studied by Mayhew and colleagues (26). They compared the therapeutic index of cytosine arabinoside in mice with the

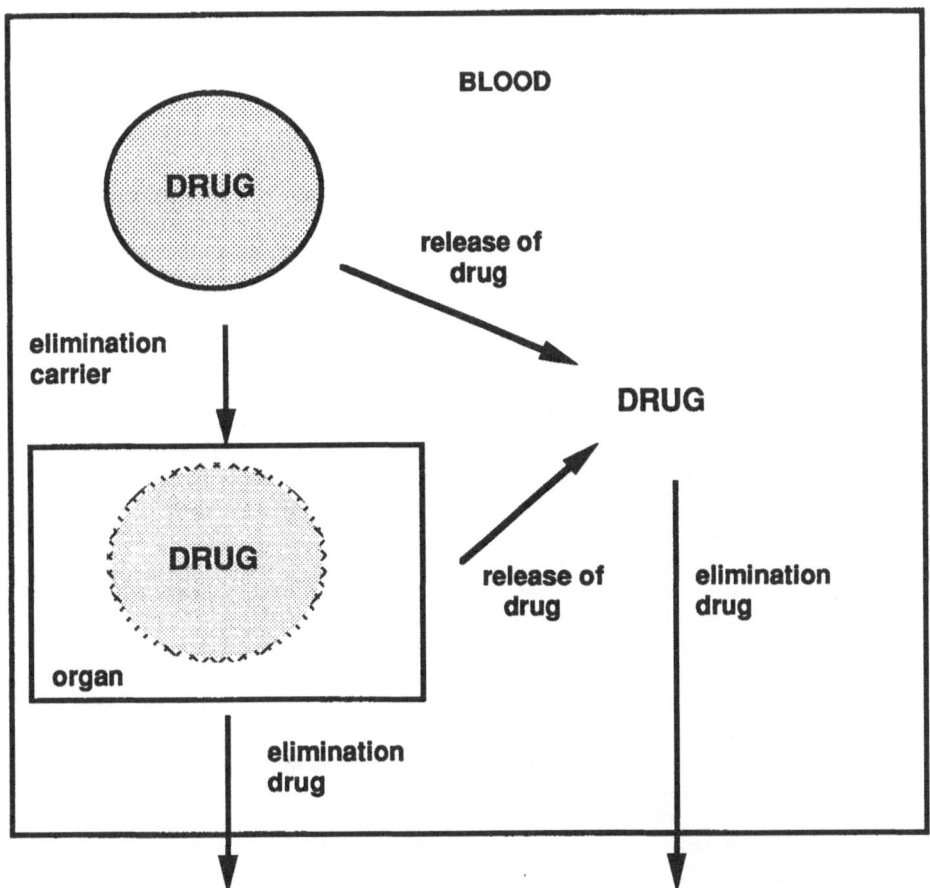

FIGURE 3 Redistribution of entrapped drug following intracellular release from a liposome.

L1210 murine tumor when the drug was administered as an intermittent multiple injection, a continuous infusion, or a single dose of liposome-encapsulated compound (26). In all cases a single dose of liposome-encapsulated drug has a therapeutic index that was at least equivalent to the continuous mode of administration. Cytosine arabinoside is rapidly deaminated in the blood and the liposome protected the compound until it was released. Mayhew et al. concluded that the liposome was behaving as a circulating rate controller for the input of cytosine arabinoside.

When the tumor was inoculated intraperitoneally (ip) and the drug given intravenously (iv), the liposomal drug had a better therapeutic index than a continuous iv infusion of the free drug. This improvement of the therapeutic index for the liposomal cytosine arabinoside, when tested against the ip tumor, was because a portion of the liposome-encapsulated drug was released in the vicinity of the tumor (26). Thus in the case of the ip tumor, the iv-administered liposome exhibited both controlled drug input and a site-specific delivery effect.

Other examples of controlled drug input applications of liposome-encapsulated compounds include intramuscular injections (27,28), ocular depots (29), as a delivery system for the lung (30), and as topical formulations (31). Although the therapeutic index of the compound is not usually improved by such systems, the decrease of adverse effects, improved patient convenience, and decreased frequency of drug administration significantly benefit the patient.

B.   Site-Specific Drug Delivery Using Liposomes

To improve the therapeutic index of a compound it must be delivered in either a site-specific or site-avoidance manner. The former approach is the classical concept of the "magic bullet" proposed by Paul Ehrlich at the turn of the century (32). The carrier promotes delivery of the compound to the target site; thus at the same total drug dose as the nonencapsulated compound, a greater fraction of the dose reaches the target site. Since the liposome is removed from circulation by the mononuclear phagocytic system (MPS) (17), liposomes are a particularly attractive vehicle for conveying drugs to cells of the MPS (33). However, anatomical barriers impede targeting to sites outside the vascular system (4). When infections of the MPS system are the target, orders-of-magnitude improvements in the therapeutic efficiency occur (34) as well as substantial improvements in the therapeutic index of the compound.

Activation of macrophages to a tumoricidal or virucidal state is greatly augmented by encapsulating muramyl dipeptides or other immunostimulatory compounds in liposomes and delivering them to macrophages (35). This type of site-specific delivery depends on phagocytosis of the liposome by the macrophage, and the internalized drug must be able to reach and interact with its molecular target from inside the cell. Compounds that interact with cell surface receptors, compounds that are degraded in the lysosome, or compounds that cannot escape the lysosome may not exhibit an increase in their therapeutic index (36).

In addition, such efforts are most successful for improving the therapeutic index if the compound does not readily redistribute once it reaches the target site (37,38). The upshot of this selection criteria is that many compounds that are already "good" drugs might not exhibit significant improvements in their therapeutic index when administered in a liposome. Thus compound selection is critical to achieve the maximal benefit from site-specific delivery.

C. Site-Avoidance Delivery Using Liposomes

An improvement to the therapeutic index of the encapsulated compound can also be achieved by site-avoidance delivery (39). In this mechanism, less drug reaches the sites of toxicity but drug levels at the therapeutic site are unchanged. At first this may seem unusual; however, often only a small fraction of the administered dose goes to either the target or toxic sites, and thus redistributing drug to the therapy site or away from the toxic site may have only a diminishingly small effect on drug concentration at other sites in the body. For instance, reducing the fraction of the injected dose of drug in the toxic site from 0.5 to 0.1% would have no appreciable effect on drug levels in other locations in the body, yet the toxicity would be dramatically reduced. Examples of a decrease in toxicity with no change in efficacy following drug administration in liposomes have been reported for actinomycin D (40), amphotericin B (41,42), primaquine (43), and doxorubicin (reviewed below). When this occurs with liposomal delivery, the drugs can interact with phospholipid membranes, and for at least two of them the sites of chronic toxicity are the heart (doxorubicin) or kidney (amp B), organs to which iv-administered liposomes do not readily distribute. Thus the mechanism of site-avoidance delivery may relate to a slowing of the rate of transfer of the liposomal drug to sites of toxicity. If this is generally true, then rate control via dissociation from the liposome is important. This can be modulated by changing lipid composition (44).

If the three mechanisms of carrier-mediated improvement in drug therapy are illustrated as the apexes of a triangle (Fig. 1), any point within the triangle then represents a mixture of the three mechanisms. For instance, intramuscular (im) injection of a liposomal drug is a good example of controlled drug input whereas iv-administered cytosine arabinoside has elements of both controlled drug input and site-specific delivery. Thus a point off the apex representing controlled drug delivery and toward the apex representing site-specific delivery more accurately describes the mechanism of action of liposomal cytosine arabinoside.

The conceptual scheme in Figure 1 can account for drug delivery in general. Although this discussion has focused on improvements in the therapeutic index or therapeutic efficacy of a compound, it should also be apparent that the liposome may decrease the therapeutic index of a drug and hence make it more toxic. In these instances, sites of toxicity may be in organs that remove liposomes from the circulation. In fact liposome-encapsulated dichloromethylenediphosphonate has been used as a research tool to oblate macrophages in studies of the immune response to liposome-encapsulated antigens (45). Thus poisoning of the MPS is also a consideration when evaluating liposomes as drug carriers (4,45,46).

For a similar reason, liposomal delivery may decrease the efficacy of a drug, hence requiring a greater or more frequent administration of the liposomal dose than for the free drug (36). An example of this is the decrease in the antitumor efficacy of liposome-encapsulated cyclophosphamide compared to the free compound (47). When this occurs, it usually means that the drug is being released from the liposome in a location, such as the lysosome, where it cannot reach the target.

III. LIPOSOME DISTRIBUTION AT THE ORGAN, CELLULAR, AND SUBCELLULAR LEVEL

Liposome composition and size can be manipulated so that organ, cellular, and subcellular delivery can be controlled. Liposomes administered iv rapidly redistribute in the vascular compartment and are eliminated from circulation by uptake into cells in the liver and spleen [extensively reviewed by Senior (19) and Hwang (48)]. The rate and extent of uptake are affected by vesicle size, charge, liposome dose, and lipid composition. Vesicles larger than 0.15 μ are primarily removed by cells of the mononuclear phagocytic system (19), while vesicles with smaller diameters can also be removed by hepatocytes (19,48). Up to 80% of vesicles in the 0.2-3.0-μ-diameter range may be eliminated by the liver. It must be strongly emphasized that with vesicles in the 0.03-0.1-μ-diameter range, lipid compositions can be selected so that only 10-20% of the injected dose is cleared by the liver (19,48). The remainder of the vesicles distribute throughout the vascular compartment. Liposomes composed of high-transition-temperature lipids (rigid liposomes) (49) or that also contain ganglioside GM1 have a prolonged circulation time in mice (50,51). Liposomes with the latter composition have been suggested to avoid the reticuloendothelial system (RES) and proposed as a prolonged circulatory reservoir (50).

At the cellular level surface adsorption, lipid exchange, and endocytosis/phagocytosis are the predominant modes of liposome cell interactions. In the early literature liposome fusion with cells was often discussed (17,18). The self-quenched fluorescence dyes, such as carboxyfluorescein, gave an indication of significant cytoplasmic delivery into cells which was interpreted as liposome-cell fusion (52). However, carboxyfluorescein undergoes a pH-induced rapid leakage from liposomes at pH 6.0 (53). Internalization of the liposome via endocytosis places the liposome into a low-pH compartment (pH 4.5-6.0) and would mediate transfer of the dye from the endosome into the cytoplasm. Hence interpretation of liposome-cell fusion based on this technique was called into question (53). Fusion of liposomes with most cell types is a low-frequency event, as demonstrated with a variety of biophysical (54), pharmacological (55), and electron microscopic techniques (56). The fact that most liposome compositions tested to date do not fuse with cells in culture is now well accepted (57). Adsorption, endocytosis, and/or phagocytosis of liposomes following intravenous injection in animals is also the principal mode of liposome interaction with cells in vivo (19,36,58).

Endocytosis places the liposome in an intracellular compartment that undergoes a progressive decline in pH from 7.4 to between 4.5 and 5.5. The ultimate fate of the liposome and its contents in this pathway is the lysosome. Here the liposome and its contents are exposed to a variety of hydrolytic and proteolytic enzymes involved in the degradation of lipids, proteins, and nucleic acids (58). Degradation of both lipid and proteins commences shortly after liposome internalization and can reach 90% of the internalized dose within 6 h of internalization both in vitro (2,3,59) and in vivo (36,60). For drug delivery, this implies that the contents of liposomes must be resistant to degradation in the lysosome and capable of escaping this compartment to be active at sites other than the lysosome (36). The low-pH environment presents an opportunity for creating liposomes or virosomes that are capable of fusing with the early endosomal compartment and delivering their contents into the cytoplasm, thus avoiding delivery of contents into the lysosome (61-63). Clearly, a better understanding of membrane fusion in general and viral-cell fusion in particular would assist in this pursuit.

IV. HUMAN CLINICAL TRIALS OF LIPOSOME-ENCAPSULATED DOXORUBICIN

Doxorubicin (Adriamycin) is an anthracycline antibiotic with a wide spectrum of activity against human neoplasms (64). It is one of the most commonly used antineoplastic agents. Cardiotoxicity is a major dose-limiting factor in doxorubicin therapy. Cardiotoxicity is severe, cumulative, and in most cases irreversible. This necessitates that doxorubicin administration be discontinued after a cumulative dosage of 550 mg/M^2 (65). Bone marrow depression, stomatitis, alopecia, nausea, and vomiting all are found as acute toxicities following administration of doxorubicin (64,65). In addition, extravasation of the drug causes a fulminating necrosis, which can result in a pitting ulceration. Although the drug has been extensively studied, the biochemical basis of its toxicity and antitumor efficacy has not been fully resolved (66).

Since 1979, when the first report on the in vivo use of doxorubicin encapsulated in liposomes appeared (67), more than 40 original articles have been published on this topic. Six different groups of investigators have reached a unanimous conclusion that the cardiotoxicity of doxorubicin is significantly diminished when the drug is administered in liposomes (68-73). In addition, liposomal doxorubicin (L-dox) has been found in most cases to have equal or slightly superior antitumor efficacy compared to the nonencapsulated drug in animal tumor models (vide infra).

The extensive evidence of an improvement in the therapeutic index of L-dox in animal studies formed an excellent rationale for human clinical trials of L-dox. The use of L-dox in humans has been reported in four clinical trials (Table 3).

Kumai and co-workers (7), at the Keio University Medical Center in Japan, treated 10 cancer patients with doxorubicin encapsulated in MLV composed of PC/Chol/dipalmitoylphosphatidic acid (DPPA):10/5/1. Non-encapsulated drug was removed from the liposomal drug by centrifugation and the L-dox was administered via hepatic arterial infusion at cumulative doses of 8-30 mg/patient. The status of the patients and the previous or concurrent therapeutic history were not presented in this brief report. The authors claim that no serious adverse reaction was observed except slight fever in 3 of the 10 patients and slight leukocytopenia (nadir) 2700 in one patient. The latter patient also received liposomes that had an anti-alpha-fetoprotein monoclonal antibody attached to its surface. Thus the significance of the slight leukocytopenia is not evaluable.

The second clinical trial was undertaken at the Hadassah University Hospital in Jerusalem, Israel, under the direction of Dr. Alberto Gabizon (8). This phase I trial was designed to test the toxicity of L-dox after iv administration in cancer patients. L-dox was formulated in a liposome composition of PC/Chol/PG/tocopherol succinate:7/3/3/0.1 molar ratio. The liposomes were prepared in 0.9% sterile, pyrogen-free saline to which deferoxamine mesylate was added to a final concentration of 50 µM. The mean diameter of the liposome preparations was 0.16 µ. Final concentrations of phospholipids in all batches were in the range of 10-20 µmoles/ml, which corresponded to 0.5-1.0 mg/ml doxorubicin and 2.5-5.0 mg/ml lactose. Extensive quality control, pyrogenicity, and sterility tests were run on the L-dox preparations prior to administration to patients. Nonencapsulated doxorubicin ranged from 10 to 30% in these preparation, with batches of liposomes prepared in the latter phase of the trial having the lowest percentage of free doxorubicin.

TABLE 3 Summary of Clinical Trials of Liposomal Doxorubicin

	Reference			
	7	8	9	13
Type of patient	Cancer, various	Primary and hepatic metastases	Hepatic metastases	Ovarian
Number of patients	10	20	6	15
Liposome type	MLV	EMLV	EMLV	SUV
Liposome composition	EPC/DPPA/Chol: 10/5/1	EPC/EPG/Chol/TS: 7/3/3/0.1	EPC/EPG/Chol/TS: 7/3/3/0.1	CDL/PC/Chol/SA: 1/5/3.5/2
Percent non-encapsulated drug	N.S.	10-30%	10%	N.S.
Maximum single dose	30 mg	70 mg/M^2	20 mg/M^2	100 mg/2 liters
Dose schedule	N.S.	3-week interval	weekly × 3	3-week interval
Maximum lipid single dose	N.S.	About 2 g/M^2	About 650 mg/M^2	100 mg/2 liters
Adverse effects	Fever, 3 patients	Fever, 3 patients; myelosuppression, 12/31 courses; hair loss, 7/16 evaluable patients	None	Increased bowel movement, mild to moderate abdominal distress, 1 peritonitis
Responses	3	3/9 evaluable with 2 courses of therapy	1 reduction of hepatomegoly	3/10 evaluable
Pharmacokinetics	No	No	No	Plasma levels

CDL, Cardiolipin; Chol, cholesterol; DPPA, dipalmitoylphosphatic acid; EPC, egg phosphatidylcholine; EPG, egg phosphatidylglycerol; MLV, multilamellar vesicles; N.S., not specified; SUV, sonicated unilamellar vesicles; TS, tocopherol succinate.

Twenty patients with primary and metastic liver cancer refractory to conventional therapy were admitted to the study. A total of 35 courses were administered by fast-drip intravenous infusion of a suspension of L-dox in physiological saline. The L-dox and phospholipid doses were escalated from 20 mg/m^2 and 0.3 g/m^2 to 70 mg/m^2 and 1.2 g/m^2, respectively. A cumulative L-dox dose of 210 mg/m^2 with a 3-week intermittent schedule was reached. Treatment was reported to be generally well tolerated, and acute toxic effects such as nausea and vomiting were mild and infrequent. Myelosuppression was observed in 12 of 31 evaluable courses. Agranulocytosis and fever requiring intravenous antibiotics were observed in three patients. No changes in liver function tests indicative of hepatotoxicity were found, nor was any evidence of acute cardiotoxicity detected. Significant hair loss was noticed in 7 of 16 evaluable patients who completed at least two courses of treatment. The authors suggested that the available data support the feasibility of systemic administration of L-dox at therapeutically effective dose levels and point out the significant improvement in tolerance with regard to the acute toxicity of the soluble doxorubicin. A maximum tolerated dose had not yet been established in this trial.

Sells and co-workers (9) examined the effects of weekly intraveneous injections of doxorubicin in extruded MLV composed of EPC/EPG/Chol/TS: 7/3/3/0.1 in six patients with hepatic metastases from primary gastrointestinal adenocarcinoma. The concentration of the entrapped drug ranged from 2.6 to 2.8 mg/ml and the amount of free drug in the preparations was less than 10% of the total. The maximum dose reported in this trial was 20 mg/M^2 given weekly for 3 weeks. A total of 40 patient weeks of treatment were given. No alopecia, ulceration, nausea, vomiting, or phlebitis was seen in any patient. This is significantly better than that reported for a 20 mg/M^2 course with nonencapsulated drug where 34% of the patients experienced nausea and 8% suffered severe alopecia (24). Three patients had five transient episodes of mild pyrexia which occurred about 5 h after injection. Although this study had not yet reached a maximum tolerated dose, a significant reduction of malignant hepatomegaly in one patient and the apparent arrest of tumor growth in a second patient were reported. The trial is continuing until a maximum tolerated dose for the L-dox is established.

Liposome-encapsulated doxorubicin administered intraperitoneally was evaluated in a combined phase I/II clinical trial in patients with advanced ovarian cancer (13). Liposomes composed of cardiolipin, phosphatidylcholine, cholesterol, and stearylamine, containing about 5 mol% of doxorubicin with a diameter of approximately 1 μ, were infused into the peritoneal cavity through an infusaport. The liposome suspension, in 2 liters of saline, was infused over a period of 1 h and allowed to remain in the peritoneal cavity for 4 h. This protocol was repeated every 21 days, and the dose of liposomal doxorubicin was escalated to 100 mg/2 liters. Forty-two cycles of therapy were administered in 15 patients. The highest dose was well tolerated, with no evidence of myelosuppression, abnormalities of liver function tests, or alopecia. Nausea and vomiting were minimal and the liposomal doxorubicin was extravasated in two patients without harmful sequelae. Increased bowel movements with mild to moderate abdominal distress were encountered during the initial 24 h after infusion in 2/3 of the patients and a single patient had an elevated temperature (39.5°C) with a presumed chemically induced peritonitis. The maximum tolerated dose had not yet been attained in this study, yet 3 of 10 evaluable patients showed evidence of a treatment response (13).

The peak intraperitoneal concentration was 28.6 µg/ml, which was reduced to 23.6 µg/ml by 2 h postinfusion. Concurrent plasma levels were in the range of 0.2-0.5 µg/ml. The preliminary pharmacokinetic data demonstrate the advantage of the intraperitoneal route of administration. Although it is too early to predict if the doxorubicin in liposomes will be efficiently transferred to the ovarian tumors, in the instances when neoplastic cells were withdrawn from the peritoneal cavity, they contained high levels of cell-associated doxorubicin (13). This promising finding highlights the potential for liposomes as a local sustained-release system for antineoplastic drug delivery into various body cavities.

In summary, preliminary clinical results with L-dox show that a number of formulations are well tolerated and the acute signs of toxicity, with the exception of myelosuppression, appear to be significantly reduced. Considerably more work is required to document the effect of L-dox on the cardiomyopathy induced by doxorubicin. Although only a small number of patients have been treated at suboptimal doses in these studies, preliminary responses already reported (7-9,13) suggest that the therapeutic effect of doxorubicin is not abrogated by incorporation into liposomes. We will return to this important point when results from the animal studies are described.

V.   HUMAN CLINICAL TRIALS OF LIPOSOME-
ENCAPSULATED AMPHOTERICIN B

A second important milestone in the movement of liposomal dosage forms out of the laboratory and into the clinic has been work accomplished with liposomal amphotericin B. Amphotericin B (amp B), a polyene antibiotic, is the major drug used for the treatment of systemic fungal infections (74,75). Its mechanism of action involves the binding to membrane sterols and subsequent permeabilization of the membrane through which ions and other molecules can leak from the cell (76). This interaction is reversible, and amphotericin B can transfer between membranes. The rate of transfer is controlled by the lipid and sterol content of the membrane (76); amp B binds more avidly to ergosterol than to cholesterol (76). This is the basis for the selectivity of the drug for fungal membranes, which contain ergosterol, as opposed to mammalian membranes, which contain cholesterol. However, the avidity of amp B for cholesterol is sufficiently high that intraveneous administration of amp B solubilized in deoxycholate (Fungizone) leads to partitioning of the drug into mammalian membranes, which results in immediate and chronic toxicities in humans. Chills, fever, vomiting, and nephrotoxicity result from its administration. These toxicities are sufficiently unpleasant for the patient that in the past physicians administered amp B only for confirmed fungal infections. This situation is changing with the introduction in the clinic of more aggressive antifungal protocols (77).

The observation by New and colleagues (41) that incorporation of amp B in liposomes decreased the toxicity in animals without diminishing the efficacy opened up the possibility of improved antifungal therapy with liposomal amp B. Through the efforts of Lopez-Berestein and his collaborators at M. D. Anderson Hospital, Texas, liposomal amp B was administered under a Compassionate Investigational New Drug Permit (IND) to cancer patients with systemic fungal infections (10; reviewed in 41; Table 4). All the patients had failed antifungal treatment with the conventional formu-

TABLE 4 Summary of Clinical Trials of Liposomal Amphotericin B

	Reference	
	10	11
Type of patient	Documented fungal infections	Documented fungal infections
Number of patients	12	15
Liposome type	MLV	SUV
Liposome composition	DMPC/DMPG:7/3	EPC/Chol/SA:4/3/1
Maximum single dose drug	1.5 mg/kg	100 mg
Maximum single lipid dose	32.5 mg/kg	About 6.6 g
Maximum cumulative drug	3.1 g	1 g
Maximum cumulative lipid	66.5 g	66 g
Adverse effects	No common side effects of amphotericin B, very well tolerated compared to current formulation of amp B in both studies	
Responses	3 cures/5 partial responses	All fungemias eliminated
Pharmacokinetics	No	Yes

Chol, Cholesterol; DMPC, dimyristoylphosphatidylcholine; DMPG, dimyristoyl-phosphatidylglycerol; EPC, egg phosphatidylcholine; SA, stearylamine; SUV, small unilamellar vesicle.

lation of amp B, Fungizone. In addition, nine of the patients were granulo-cytopenic. An MLV preparation of amp B composed of DMPC/DMPG:7/3 with a size range between 0.2 and 5 μ, containing about 5 mol% amp B, was infused intraveneously over a 15-45-min period. The infusion rate is considerably more rapid than is used with Fungizone. The amp B dose ranged from 0.4 to 1.5 mg/kg. The maximum cumulative dose of drug was 3.1 g. The maximum cumulative dose of lipid administered was 66.5 g. Eight of the twelve patients entered into the study responded to the lipo-somal amp B, and three of these were cured of the fungal infection (10).

Follow-up studies on two of the cured patients revealed no recurrence of the infection and no evidence of central nervous system or renal toxicity (41). Moreover, there was no evidence of toxicity attributable to the satu-rated lipid used to prepare the liposomes. In most of the patients the liposomal amp B was well tolerated, unlike the clinical experience with Fungizone. Based on the response rate and the minimal toxicity of the formulation, it appears that the therapeutic index for the liposomal amp B is superior to that for the Fungizone preparation.

Pharmacokinetic studies of the liposomal amp B were not performed owing to the rapid clearance of the liposome drug complex. However, the investigators examined the clearance and disposition of ^{99m}Tc-labeled MLV in cancer patients (78). The liposomes were prepared from the same lipids but lacked the amp B (78). The liposomes were cleared from circulation

with a half-life of less than 15 min with >90% of the dose appearing in
the liver, spleen, and lung. These researchers attributed the rather large
fraction of the dose (about 14%) lodging in the lung to the large-diameter
liposomes in the preparation (78).

The promising increase in the therapeutic index of amp B by encapsulation in liposome encouraged a second group to initiate a clinical trial
of a different liposomal form (11). Coune and his colleagues at the Jules
Bordet Institute in Brussels were interested in the ability of liposomes
to solubilize hydrophobic drugs as a vehicle to administer them to humans
(79). They had previously used SUV composed of EPC/Chol/SA:4/3/1 to
deliver a quinazolone anticancer compound NSC 251635 to 14 cancer patients
(12; 80; Table 1) and had studied the distribution of ^{111}In-labeled SUV
containing this drug in humans (81).

Investigations with the NSC 251635 SUV demonstrated that large volumes
(up to 1 liter) and high lipid doses (up to 20 g) were well tolerated when
infused into humans. Side effects observed in some patients during infusion
included mild sedation, fever, chills, lumbar pain, and bronchospasm. In
all patients a significant activation of the complement system was observed
(12), which may be related to the presence of stearylamine in the liposomes.

Elimination of the NSC 251635 from the blood was quantitated by a
high-performance liquid chromatography (HPLC) assay. The blood levels
of drug peaked at the end of the infusion period and gradually declined
to a plateau, which was followed by a prolonged terminal phase. Plasma
phospholipid concentration was used to estimate the amount of the liposomes
remaining in the blood. Although there was a large background level from
the endogenous phospholipids, the authors were able to demonstrate that
serum phospholipid levels peaked at the end of the infusion period and
appeared to return to preinfusion values within 10 h. In patients who
received the highest doses (20 g), the concentration of phospholipid remained elevated for 24 h. The peak concentration of lipid was proportional
to the amount and rate of infusion. The peak concentration of drug was
proportional to the amount infused. Based on the noncorrespondence of
the lipid and drug profiles, it was concluded that NSC 251635 rapidly
dissociates from the liposomes following infusion. No objective regressions
of the tumors were observed, which, as pointed out by the authors, may
be more a function of the compound that of the delivery system (12).
A maximum tolerated dose was not reached, and the authors commented
that the limiting factor in the phase I study was the ability to prepare
sufficient quantities of liposomes.

Using the ^{111}In-labeled drug containing SUV, the authors demonstrated
that there was no significant accumulation in cancerous masses in the lung
and only a slight accumulation (<3%) in the lung itself; that there was a
relatively important uptake of labeled liposomes by the liver, spleen, and
bone marrow; and finally, that there was a relatively high blood level
of the drug for a prolonged period (81). These results using SUV must
be contrasted with the experience of the M. D. Anderson group, who
observed both a rapid clearance and a high lung level for the MLV liposomes. This points out how liposome distribution in humans, as has been
found in other animals (19), can be controlled by physical parameters.

Based on the clinical experience with the EPC/Chol/SA:4/3/1 SUV,
the Jules Bordet group incorporated amp B into liposomes of the same
composition and type and administered them to 15 cancer patients with
fungal infections (11, Table 4). The total dose of amp B ranged from 20
to 1004 mg per patient, and infusion doses of up to 1.8 mg/kg were well

tolerated. The number of infusions varied from 1 to 20, and the duration of treatment ranged from 1 to 29 days.

None of the common side effects of Fungizone occurred, and it was particularly noteworthy that no renal impairment was observed (11). As noted in the M. D. Anderson trial (10), the liposomal amp B could be administered much more rapidly (0.5-1 h compared to 4-6 h) than the Fungizone suspension and no limiting toxicity was observed in this trial. No late neurological toxicity was seen nor was there delayed nephrotoxicity. One patient, who received a very rapid infusion, had symptoms and transient serum enzyme activity consistent with acute pancreatitis. A second patient experienced hyperkalemia, a complication also reported with the rapid infusion of Fungizone (11).

Serum amp B concentrations were much higher than could be achieved with Fungizone. This fact permitted measurement of the pharmacokinetics of the liposomal amp B, a study the M. D. Anderson group was unable to do because of the rapid clearance of the MLV liposomes used in the Texas clinical trial (10). With a daily schedule, peak and trough serum amp B concentrations were 10-20 µg/ml and 5-10 µg/ml, respectively. Serum concentrations of amp B given as Fungizone did not exceed 2 µg/ml. Amp B administered in SUV had a prolonged serum half-life (25.3 ± 16 h), and serum antifungal activity was considerably higher when liposomal amp B was administered than when Fungizone was administered (11). A concomitant increase of serum phospholipid levels over the 24-h period suggests that the SUV provide a circulating reservoir of the drug.

The authors conclude that liposomal amp B has a better therapeutic index than Fungizone because it is better tolerated and possibly because of a higher antifungal activity due to the prolonged elevation of the amp B serum levels (11).

Thus, in both clinical trials, liposomal amp B has a better therapeutic index than the current Fungizone preparation (10,11). This occurs in spite of the fact that drastically different liposome formulations were used, with differing pharmacokinetic and distribution properties. The pharmacokinetic analysis of the liposomal amp B by Sculier and co-workers (11) is the most comprehensive examination of the kinetics of a liposome-encapsulated drug in humans reported to date. Although the authors were unable to differentiate between liposomal and nonliposomal amp B in the plasma, their study clearly documents the ability of the liposome to alter the pharmacokinetic profile of a lipid-soluble drug in a manner that retains efficacy and diminishes toxicity.

Liposomal delivery of amphotericin B and doxorubicin is similar in a number of respects: both show reduced toxicity, both maintain efficacy, both can interact with membranes and transfer into cells readily. Thus, in the ensuing sections we will focus on studies of doxorubicin delivery in liposomes as an example of drugs of this nature.

VI. PHARMACOKINETICS OF LIPOSOME-ENCAPSULATED DOXORUBICIN IN LABORATORY ANIMALS

A major consequence of drug entrapment in liposomes is a change in the tissue distribution and pharmacokinetics (19). Tissue distribution following intravenous administration of liposome encapsulated agents demonstrates (1) increased drug levels in liver and spleen, and (2) decreased drug levels in tissues separated from the intravascular compartment by a continuous capillary endothelium and intact basement membrane (brain, heart, kidney). This general pattern has been demonstrated for liposome-encapsulated doxorubicin (Table 5).

A.   Intravenous Administration of Liposome Doxorubicin

Forssen and Tökes (67), using negatively charged, small, unilamellar liposomes composed of egg phosphatidylcholine (EPC), bovine brain phosphatidylserine (PS), and cholesterol (Chol) of a respective molar ratio 7/3/1, altered the distribution of radiolabeled doxorubicin following iv injection in Swiss-Webster mice. Doxorubicin levels were increased in liver and significantly decreased in brain, heart, kidney, and lung at 1 and 4 h after administration in liposomes.

Rahman and colleagues (68) administered iv doxorubicin encapsulated in positively charged SUV composed of EPC/Chol/stearylamine(SA):10/2/3 to DBA/2 mice. Doxorubicin tissue levels were quantitated by a fluorescence method. Liposomal doxorubicin resulted in lower concentration × time levels from 5 min to 24 h in the heart and kidney but greater levels in the liver, lung, and spleen, than the nonencapsulated compound. Doxorubicin encapsulated in negatively charged SUV composed of PC/Chol/PS:10/2/1 and given iv resulted in greater levels in heart, liver, kidney, and spleen and comparable levels in lung as the free drug. This finding of increased heart levels of the liposomal form is not in agreement with results from Forssen and Tökes (67) and seven other groups that have examined the distribution of doxorubicin encapsulated in small, negatively charged liposomes (Table 5). There is no obvious reason for this discrepancy, and based on the evidence from other studies (Table 5), it is clear that delivery of doxorubicin in negatively charged SUV results in lower drug levels in the heart, kidney, and lungs.

The tissue distribution of L-dox in tumor-bearing mice following iv administration was measured using an HPLC method (82). Higher levels were observed in serum and liver at 20 and 60 min after administration of doxorubicin in neutral liposomes composed of PC/Chol:7/2, negatively charged liposomes PC/Chol/dicetylphosphate:7/2/1, and positively charged liposomes composed of PC/Chol/SA:7/2/1, compared to nonencapsulated doxorubicin. The negatively and positively charged liposomes gave higher liver levels of doxorubicin than the nonencapsulated compound at 20 and 60 min. This group found comparable or lower levels of doxorubicin in the heart when administered in liposomes, regardless of charge.

Forssen and Tökes (69) administered doxorubicin encapsulated in PC/Chol/PS:6/3/2 SUV to mice and measured the heart levels by a fluorescence technique. They observed a significant decrease in heart levels of the L-dox compared to the nonencapsulated drug. Olson and co-workers (70) administered doxorubicin encapsulated in extruded multilamellar vesicles (EMLV) composed of PC/Chol/phosphatidylglycerol (PG):4/5/1 to mice. They measured significantly greater levels in plasma yet significantly lower doxorubicin levels in heart for 24 h following iv injection when compared to the nonencapsulated doxorubicin. Higher liver levels from this same formulation were found in mice out to 48 h after injection when compared to the nonencapsulated doxorubicin (83).

When L-dox MLV composed of PC/Chol/PS:7/10/3 were injected into mice, higher liver and spleen levels and lower levels in heart were found (84). Liposomes composed of cardiolipin (DCP)/PC/Chol:1/4/5 resulted in increased liver and spleen levels but no significant decrease in heart levels compared to the nonencapsulated compound (84).

TABLE 5 Pharmacokinetics and Distribution of Liposome-Encapsulated Doxorubicin in Laboratory Animals

Ref.	Liposome type/composition[a]	Route	Species	Detection[b]	Outcome[c]
Forssen and Tökes (67)	SUV:EPC/Chol/PS:7/3/1	iv	Mouse	Radioactivity	Decrease heart, brain, kidney, lung, increase liver
Rahman et al. (68)	SUV:EPC/Chol/SA:10/2/3	iv	Mouse	Fluorescence	Decrease heart, kidney; increase liver, lung, spleen
	SUV:EPC/Chol/PS:10/2/1	iv	Mouse	Fluorescence	Comparable lung; increase heart, liver, spleen, kidney
Shinozawa et al. (82)	SUV:EPC/Chol:7/2 SUV:EPC/Chol/DCP:7/2/1 SUV:EPC/Chol/SA:7/2/1	iv	Mouse	HPLC	Decrease heart; increase serum, liver
Forssen and Tökes (69)	SUV:EPC/Chol/PS:6/3/2	iv	Mouse	Fluorescence	Decrease heart
Olson et al. (70)	EMLV:EPC/Chol/PG:4/5/1	iv	Mouse	Fluorescence	Decrease heart; increase serum
Mayhew et al. (83)	EMLV:EPC/Chol/PG:4/5/1	iv	Mouse	Fluorescence	Increase liver
Gabizon et al. (84)	MLV:EPC/Chol/PS:7/10/3	iv	Mouse	Fluorescence	Decrease heart; increase liver, spleen
	MLV:CDL/EPC/Chol:1/4/5	iv	Mouse	Fluorescence	Comparable heart; increase liver, spleen
Rosa and Clementi (85)	EPC/Chol/DCP:7/4/2	iv	Mouse	Fluorescence	Decrease heart; increase liver, spleen, plasma
Forssen and Tökes (86)	SUV:EPC/Chol/PS:6/3/2	iv	Mouse	Fluorescence	Decrease heart; increase plasma
Gabizon et al. (87)	MLV:EPC/Chol/PS:7/10/3	iv	Mouse	Fluorescence	Decrease heart; increase liver and spleen

(continued)

(Table 5, Continued)

Ref.	Liposome type/composition[a]	Route	Species	Detection[b]	Outcome[c]
Van Hoesel et al. (71)	SUV:EPC/Chol/PS:10/4/1	iv	Rat	HPLC	Decrease heart, kidney, muscle, lung; increase liver and spleen
Rahman et al. (88)	SUV:CDL/EPC/Chol/SA: 1/5/3.5/2	iv	Mouse	Fluorescence	Decrease heart, lung, kidney; increase liver and spleen
Rahman et al. (110)	SUV:CDL/EPC/Chol/SA: 1/5/3.5/2	iv	Rat	Fluorescence	$T_{1/2}$; dox = 17 h; L-dox = 69 h AUC 42-fold higher with L-dox; less in heart; more in liver
Kojima et al. (89)	SUV:EPC/Chol/SULFATIDE: 7/2/1	iv	Mouse	Fluorescence	Decrease in heart, kidney; increase blood, liver, and spleen
Konno et al. (109)	SUV:EPC/Chol/DPPA 9/5/1:monoclonal attached	iv	Mouse	Fluorescence	Decrease in heart, kidney, lung; increase liver, spleen, and serum
Storm et al. (95)	EMLV:EPC/Chol/PS:10/4/1 EMLV:DSPC/DPPG/Chol: 10/1/10	iv	Rat	Radioactivity	Compared 2 L-dox compositions, more rapid clearance and greater dox release from EPC; greater liver levels from EPC
Mayer et al. (90)	E LUVET various compositions, remote loaded	iv	Mouse	Fluorescence	Decrease heart, kidney, lung; increase liver, spleen, and blood

	Composition	Route	Animal	Method	Effect
Gabizon et al. (91)	EMLV:HPI/HPC/CHOL remote loaded	iv	Mouse	Fluorescence	Decrease in heart; increase in liver; blood clearance greatly decreased
Parker et al. (98)	LUV:EPC/Chol/SA:3/3/1	ip	Rat	Radioactivity	Decrease heart, kidney, lung; increase liver, spleen, lymph nodes, and peritoneal cavity
Rosa and Clementi (85)	MLV:EPC/Chol/DCP:7/4/2	ip	Mouse	Fluorescence	Decrease heart, liver, spleen; comparable kidney intestine and testicle
Forssen and Tökes (99)	SUV:EPC/Chol/PS:6/3/2	id	Mouse	Fluorescence	Slower clearance of L-dox than free dox from site

aNumbers following composition represent molar ratio of components.

bMethod used to detect doxorubicin levels in tissues.

cIncrease means a greater concentration of doxorubicin in the respective tissue compared to the nonencapsulated drug given at the same dose and via the same route. Decrease means a lesser concentration in the respective tissue compared to the nonencapsulated doxorubicin.

CDL, Cardiolipin; Chol, cholesterol; DCP, dicetylphosphate; DPPG, dipalmitoylphosphatidylglycerol; DSPC, distearoyl-phosphatidylcholine; EPC, egg phosphatidylcholine; id, intradermal; ip, intraperitoneal; iv, intravenous; PG, egg phosphatidylglycerol; PS, bovine brain phosphatidylserine; SA, stearylamine; EMLV, multilamellar vesicles sized via membrane extrusion; MLV, multilamellar vesicle; SUV, small unilamellar vesicle, size less than 100 nm.

Doxorubicin administered to mice in liposomes composed of PC/Chol/DP: 7/4/2 gave increased doxorubicin levels in liver and spleen and decreased heart levels of the drug compared to the nonencapsulated compound (85). L-dox also had a longer plasma half-life than the nonencapsulated compound. Forssen and Tökes (86) found that doxorubicin encapsulated in liposomes composed of PC/Chol/PS:6/3/2 and injected into Swiss-Webster mice had an increased area under the plasma concentration × time curve and a decreased area under the heart concentration × time curve compared to the nonencapsulated doxorubicin.

Gabizon and co-workers (87) measured the tissue distribution of doxorubicin encapsulated in PC/Chol/PS:7/10/3-sonicated MLV in tumor-bearing mice. They observed increased levels in the liver and spleen and decreased doxorubicin levels in the heart when compared to the free drug. Van Hoesel and colleagues (71) examined the tissue distribution of doxorubicin encapsulated in SUV composed of PC/Chol/PS:10/4/1 and administered to LOU/M Ws1 rats. Using a HPLC assay there were lower drug levels at 4 h after injection in plasma, heart, kidney, muscle, and lung and greater doxorubicin levels in liver and spleen when the compound was administered in liposomes than when given as the nonencapsulated form. Rahman and colleagues (88) compared the tissue disposition of nonencapsulated doxorubicin to that of doxorubicin encapsulated in DCP/PC/Chol/SA:1/5/3.5/2 SUV injection in mice. Using a fluorescent assay, they found lower tissue levels in the heart, lung, and kidney and elevated levels in the liver and spleen compared to the nonencapsulated compound. Sulfatide-containing L-dox also had a higher blood level and a lower level in the heart than the free compound (89).

Two particularly interesting approaches to L-dox delivery involve entrapping the drug via a pH trapping method to achieve a high encapsulation efficiency in liposomes (90) and using a liposome composition that yields a long-circulating liposome (91). Mayer and co-workers used a pH trapping method (92) to load doxorubicin into preformed liposomes of various compositions. They were able to attain doxorubicin/lipid weight-to-weight ratios of 0.3 with a 99% trapping efficiency. This is a higher ratio that previous methods have been able to attain and could be reached in liposome compositions lacking charged phospholipids. Doxorubicin levels in the blood were greatly increased for all liposome compositions tested, as were the drug levels in the liver and spleen. Conversely, drug levels in the heart, lung, and kidney were decreased (90). This comprehensive study confirmed results from previous L-dox animal distribution and pharmacokinetic studies, but with a formulation where the doxorubicin was predominantly in the aqueous compartment.

Liposomes with a prolonged circulatory half-life have been used to deliver doxorubicin in mice (91). The liposomes were composed of hydrogenated phosphatidylinositol (HPI), hydrogenated phosphatidylcholine, and cholesterol and contained ammonium sulfate. Doxorubicin was loaded into the preformed liposomes by an exchange reaction of the encapsulated ammonia for the doxorubicin, so that, as in the report of Mayer and co-workers (90), the drug was predominantly in the aqueous compartment. Administration of the HPI formulation into Balb/c mice resulted in a terminal half-life more than 10 times longer than for doxorubicin delivered in liposomes composed of EPG/EPC/Chol and 100-fold longer than when given as the free drug. Significantly more drug was observed in the liver and spleen and significantly less drug was found in the heart than when the nonencapsulated drug was given. One of the interesting features of this study was that liposome-encapsulated doxorubicin in plasma was separated from the free drug (93) and doxorubicin remained encapsulated in the HPI liposomes until the liposome was cleared from circulation more than 72 h, which was significantly longer than for liposomes composed of EPG/EPC/Chol = 5 h (91).

The animal studies reviewed in this section provide convincing evidence that administration of L-dox significantly influences the pharmacokinetics and tissue distribution of the drug. Liposome encapsulation prolongs the plasma concentration, increases drug levels in the liver and spleen, and reduces drug levels in the heart, lung, and kidney. Liposome composition, size, and drug/lipid ratio influence the extent of the pharmacokinetic modulation.

An unresolved question concerns the fate of liposome-encapsulated doxorubicin delivered into macrophages and other cells of the MPS. It has been suggested that doxorubicin is released from cells of the MPS back into the circulation (Fig. 3) and that macrophages may serve as a reservoir of doxorubicin in the vicinity tumors (71,83,94,95). Macrophages store and subsequently release free doxorubicin or its active metabolites into the surrounding medium (96). Storm and co-workers (97) addressed this question by harvesting macrophages from the peritoneal cavity of rats after ip administration of L-dox. The cytostatic potential of the macrophages in culture depended on the type of L-dox injected and was directly related to the amount of doxorubicin in the macrophage. Doxorubicin was chemically stable during this process, as determined by HPLC. Moreover doxorubicin was released from the macrophages into the culture medium at cytotoxic concentrations when the supernatant was placed on cultured tumor cells. Thus macrophages can serve as a reservoir of doxorubicin after internalizing L-dox.

B. Intraperitoneal Administration of Liposomal Doxorubicin in Laboratory Animals

Parker and colleagues (98) examined the effect of liposome encapsulation on the clearance and tissue disposition of doxorubicin in Sprague-Dawley rats when given ip. Radiolabeled doxorubicin was encapsulated in large liposomes composed of PC/Chol/SA:3/3/1, and tissue levels were determined by scintillation counting of oxidized samples. Doxorubicin and L-dox appeared in and were cleared from the plasma at a similar rate when given ip. L-dox led to decreased tissue levels in the heart, kidney, and lung and increased tissue levels in the liver, spleen, and lymph nodes draining the peritoneal cavity. These results indicate that doxorubicin left the peritoneal cavity associated with the liposome.

Rosa and Clementi (85) measured tissue levels of doxorubicin following ip injection into C57BL6 mice either as the nonencapsulated drug or encapsulated in PC/Chol/DP:7/4/2 MLV. The L-dox led to decreased levels in the heart, liver, and spleen out to 12 h and comparable levels in the kidney, intestine, and testicles compared to the nonencapsulated doxorubisin.

C. Intradermal Injection of Liposomal Doxorubicin

Forssen and Tökes (99) injected doxorubicin and liposome doxorubicin (PC/Chol/PS:6/3/2) intradermally into mice and observed a slower clearance of the L-dox from the site of injection than the nonencapsulated compound.

In spite of the longer residence time, the encapsulated compound caused significantly less tissue damage than the nonencapsulated compound.

D. Summary of Liposomal Pharmacokinetics in Animals

Entrapment of doxorubicin in liposomes results in a prolonged plasma concentration, an increase in the levels associated with the liver and spleen, and a decrease in the doxorubicin levels in the heart, kidney, and lung (Table 5). L-dox internalized in macrophages is able to escape from this cell and remains bioactive. The extent of liposome modulation of doxorubicin pharmacokinetics depends on the liposome size, composition, amount of lipid injected, and route of administration.

VII. TOXICITY OF LIPOSOMAL DOXORUBICIN IN ANIMALS

A summary of the various studies to evaluate the toxicity of L-dox preparations is given in Tables 6 and 7. In general, the liposome compositions tested exhibit a higher LD_{50} or a decrease in adverse effects when compared to the nonencapsulated drug.

Forssen and Tökes (67) were the first to point out the reduction in toxicity of doxorubicin when administered in liposomes. Intravenous administration of doxorubicin encapsulated in SUV composed of PC/Chol/PS:7/3/1 at a dose of 10 mg/kg body weight into Swiss mice, followed a week later by a second dose of 10 mg/kg, produced a less severe loss of body weight immediately following each injection than did the nonencapsulated drug. Mice receiving the L-dox also showed a more rapid recovery of body weight and a greater final weight gain at the end of the 10-week experiment.

Rahman and co-workers (68) examined the effect of intravenous administration of L-dox on myocardial structure in DBA/2 mice. Mice received nonencapsulated doxorubicin or doxorubicin encapsulated in positively charged liposomes (PC/Chol/SA:10/2/3) or negatively charged liposomes (PC/Chol/PS:10/2/1) at a dose of 4 mg/kg for 3 consecutive days. Mice were sacrificed on the first or fifth day after the last injection and samples of the left ventricle were taken for electron microscopic examination. Nonencapsulated drug and drug encapsulated in the negatively charged liposomes produced qualitative damage of cardiac myocytes. Such alterations were infrequent in myocytes treated with the positive charged L-dox. This report of no difference in myocyte damage when doxorubicin is administered encapsulated in negatively charged liposomes compared to nonencapsulated doxorubicin is at variance with results obtained by five other groups (69-72,89) that also used microscopic techniques to assess myocardial damage. There is no identifiable reason for this discrepancy, and based on the weight of the evidence, L-dox containing negatively charged lipids must be considered less cardiotoxic than free doxorubicin (Table 6).

Forssen and Tökes (69) administered iv doxorubicin encapsulated in PC/Chol/PS:6/3/2 SUV at a dose of 5 mg/kg for either 4 or 8 weeks to Swiss mice. Mice were sacrificed at either 12 weeks (low dose) or 13 weeks (high dose), and artrial and ventricle portions of the heart were stained and examined by light microscopy. In the low-dose groups, mice receiving nonencapsulated drug had a mean toxicity index of 2.65 ± 0.81 (mean ± SEM). Those receiving liposome-encapsulated doxorubicin had a mean toxicity index of 0.99 ± 0.25, which was indistinguishable from that of saline-injected controls. In the high-dose group, mice receiving nonencapsulated

TABLE 6 Toxicity of Liposomal Doxorubicin Administered Intravenously in Mice

Ref.	Liposome[a]	Outcome[b]
Forssen and Tökes (67)	SUV -	Less severe loss of body weight, more rapid recovery, and greater body weight at 10 weeks after dosing
Rahman et al. (68)	SUV-, SUV+	Decrease in myocardial damage assessed by electron microscopy with positive-charged liposomes, no difference with negative charge
Forssen and Tökes (69)	SUV-	Significantly less myocardial damage assessed by histology and no loss in weight as found with nonencapsulated drug
Olson et al. (70)	E-MLV-	Acute and chronic LD_{50} for L-dox was 2-3-fold greater than that of the nonencapsulated dox; significant decrease in myocardial damage assessed by electron microscopy
Mayhew et al. (83)	E-MLV-	2-3-fold decrease in chronic toxicity
Forssen and Tökes (86)	SUV-	L-dox depressed response to antigen and suppressed leukocyte numbers less than nonencapsulated drug
Gabizon et al. (87)	SUV-	No difference in liver histology between nonencapsulated and L-dox although L-dox concentration was 3-4-fold higher than free dox
Ganapathi and Krishan (101)	SUV	L-dox or continuous infusion of free dox superior to bolus free dox in animal survival or weight gain measures; cholesterol content in L-dox between 0.1 and 0.5 mole ratio had no effect on outcome
Rahman et al. (88)	SUV+/-	L-dox significantly reduced histopathological lesions in cardiac tissue, L-dox less toxic to hematopoetic system
Mayhew and Rustum (102)	E-MLV	iv bolus L-dox less toxic than iv infusion or iv bolus of free dox; maximum tolerated dose L-dox = 30 mg/kg, infusion = 20 mg/kg, bolus dox = 10 mg/kg
Gabizon et al. (72)	SUV-	L-dox superior (less toxic) in all toxicological criteria; significantly less myocardial histopathology with L-dox

(continued)

(Table 6, Continued)

Ref.	Liposome[a]	Outcome[b]
Rahman et al. (123)	SUV+/–	Suppression of antigen-specific cellular cytotoxicity and response to mitogens after a single dose are altered in timing but not in magnitude and duration of immunosuppressive effects is shorter with L-dox
Rahman et al. (124)	SUV+/–	Immunosuppressive effect of multiple doses of L-dox is the same as with free drug but damage is repaired more rapidly with L-dox

[a]E-MLV, Extruded MLV; MLV multilamellar vesicle; SUV, small unilamellar vesicle; (–), negatively charged, (+) positively charged; (+/–) contains positively and negatively charged lipids; no symbol following liposome type, neutral. Composition is given in Table 5.

[b]dox, Doxorubicin; L-dox, liposome-associated doxorubicin; free dox, nonencapsulated doxorubicin; comparisons of L-dox are related to equivalent doses of nonencapsulated doxorubicin given via the same route.

doxorubicin had a high toxicity score of 3.98 ± 0.87. Animals receiving L-dox exhibited a markedly diminished toxicity score of 1.05 ± 0.5, which was slightly greater than that of the saline-injected control of 0.69 ± 0.23. Nonencapsulated doxorubicin plus non-drug-containing liposomes had a toxicity score of 3.73 ± 0.94, confirming that the drug had to be encapsulated in order for the liposome to confer a protective effect. Forssen and Tökes (69) found that animals receiving the L-dox in both the low- and high-dose regimes had a similar weight gain as saline-injected control mice over the course of the experiment, whereas the nonencapsulated-doxorubicin-treated groups had significantly lower ($p < 0.05$) weight gain.

Olson and co-workers (70) compared the inherent toxicities of non-drug-containing liposomes administered as an iv bolus in C57/BL 6 mice. They observed LD_{50} values of 1.1 g/kg body weight for positively charged liposomes composed of PC/Chol/SA:5/4/1, an LD_{50} of 7.2 g/kg body weight for liposomes composed of PC/Chol:1/1, and an LD_{50} of 7.5 g/kg body weight for liposomes composed of PC/Chol/PG:5/4/1. Based on these results, doxorubicin was encapsulated in negatively charged liposomes and the LD_{50} of the liposome-encapsulated doxorubicin was compared to the nonencapsulated compound administered as a single iv bolus into DBA/2 mice. Two phases of toxicity were observed: an acute phase, with animals dying in the first 14 days after injection, and a chronic phase, with deaths occurring between 45 and 70 days after injection. Nonencapsulated doxorubicin had an LD_{50} in the acute phase of 20 mg/kg, which decreased to 10-15 mg/kg in the chronic phase. Liposome-encapsulated doxorubicin had an LD_{50} in the acute phase of about 50 mg/kg, which decreased to 25-30 mg/kg in the chronic phase. This group also studied the effects of L-dox on the morphology of the murine heart. As contrasted to the findings of Rahman and co-workers (68) using electron microscopy, they observed the absence of electron-dense bodies, a significant decrease in the incidence of enlarged perivascular space, and a diminished enlargement

TABLE 7 Toxicity of Liposomal Doxorubicin in Species Other than Mice or via Routes Other than Intravenous Administration

Ref.	Liposome[a]	Route[b]	Species	Outcome[c]
Herman et al. (100)	SUV+/-	iv	Dog	Significantly less cardiomyopathy, decrease in alopecia, comparable bone marrow suppression
Van Hoesel et al. (71)	MLV-, SUV+	iv	Rat	L-dox less histopathology in the heart, less kidney damage than free dox for both liposome types
Litterst et al. (103)	LUV+	ip	Rat	Peritonitis marginally less with L-dox but L-dox systemic toxicities were significantly reduced
Forssen and Tökes (99)	SUV-	id	Mouse	Highly significant attentuation of local tissue damage with L-dox
Storm et al. (107)	E-MLV-	iv	Rat	L-dox less toxic, cardio- and nephrotoxicity, weight loss, survival; infusion of free dox could reduce its cardiotoxicity but not the nephrotoxicity, whereas liposome encapsulation could reduce both

[a]See Table 6 for abbreviations.
[b]id, intradermal; ip, intraperitoneal; iv, intravenous.
[c]Comparison is liposomal doxorubicin to nonencapsulated doxorubicin at the same dose and via the same route.

of the sarcoplasmic reticulum when doxorubicin was administered in liposomes. Thus the L-dox was markedly less cardiotoxic than the nonencapsulated drug.

Mayhew and colleagues (83) examined the lethality of MLV liposomes composed of PC/Chol/PG/tocopherol:4/4/1/0.1 containing doxorubicin in C57BL/6 mice. L-dox given as a single injection at 25 mg/kg resulted in no animal deaths (10/10 survivors) out to 140 days postinjection whereas nonencapsulated doxorubicin killed 100% of the animals (0/10 survivors). In addition nonencapsulated doxorubicin at 15 mg/kg resulted in only 30% animal survival at 140 days (3/10 survivors). When administered every day, five doses of the L-dox at 5 mg/kg caused no fatalities at 140 days (8/8 survivors) whereas five doses of the nonencapsulated drug resulted in 70% fatalities (3/10 survivors). When administered every 2 weeks, five doses of 10 mg/kg of L-dox resulted in no fatalities at 140 days (10/10 survivors) whereas five doses of the nonencapsulated caused 100% fatalities (0/10 survivors).

Herman and co-workers (100) examined the cardiotoxicity of L-dox in beagle dogs. L-dox SUV prepared with a lipid composition of PC/Chol/cardiolipin/stearylamine:5/3.5/1/2 and administered iv at a dosage of 1.75 mg/kg at 3-week intervals for seven doses was significantly less cardiotoxic than the same amount of nonencapsulated drug. They could demonstrate no cardiomyopathy with the liposomal form. In addition, they observed that weight loss was more severe and that alopecia was extensive in animals given the nonencapsulated drug. No alopecia was noted in animals given the liposomal doxorubicin. At the doses given, the liposomal and free compounds exhibited comparable levels of bone marrow suppression.

Forssen and Tökes (86), using SUV composed of PC/Chol/PS:6/3/2, demonstrated no immunosuppression due to L-dox as measured by the ability of Swiss Webster mice to respond to human red blood cells. They administered human red blood cells sc followed by treatment with either the free or L-dox at 5 mg/kg at 24 and 48 h after immunization. The mice were reimmunized 7 days later and the drug treatments were repeated. Antibody titers against human red blood cells were determined 4 weeks after the second immunization. Treatment with the nonencapsulated doxorubicin depressed antibody titers 64-fold whereas treatment with the L-dox had no significant effect on antibody titers. Free doxorubicin significantly reduced leukocyte numbers whereas L-dox-treated animals had similar numbers of leukocytes as the nontreated controls two weeks after dosing. On this dosing schedule, L-dox was considerably less immunosuppressive than nonencapsulated doxorubicin.

Gabizon and colleagues (87) used SUV composed of PC/Chol/PS:7/10/3 to examine the toxicity and efficacy of L-dox in BALB/c mice bearing the J-6456 T-cell lymphoma. At 100-200 μg doxorubicin per mouse they found increased drug levels in the liver and spleen and decreased levels in the heart when the drug was administered as the liposomal form. With light microscopy no differences were detected in the histopathology of the liver between the free and L-dox-treated groups, in spite of the fact that 3-4 times more drug distributed into the liver when given as the liposomal than as the free form.

Van Hoesel and colleagues (71) compared the toxicity of L-dox to that of nonencapsulated doxorubicin administered to rats bearing an immunocytoma. Small L-dox MLV composed of either PC/Chol/PS:10/4/1, PC/Chol/SA:10/4/3, or nonencapsulated doxorubicin were administered iv at doses up to 2 mg/kg for 5 consecutive days to LOU M/Ws1 rats. Grade 3 cardiomyopathy was observed 47 days after treatment with free doxorubicin, whereas at the same dose, L-dox resulted in only grade 1 cardiomyopathy. Free doxorubicin caused a significantly greater nephrotoxicity and a greater decline in serum albumin than L-dox when compared at the same dose levels. These researchers conclude that treatment with doxorubicin in either the positive or negative charged liposomes resulted in a prolonged survival of tumored animals, less albuminuria, and higher serum albumin levels than animals treated with the nonencapsulated drug. Also, fewer lesions in the heart and kidney were observed by microscopy. This decrease in adverse effects correlated with the lower doxorubicin levels in these organs when L-dox was administered.

The survival times of DBA/2 mice receiving free doxorubicin or L-dox by various routes of administration were examined by Ganapathi and Krishan (101). Free doxorubicin given ip as either a single (20 mg/kg) or multiple (5 mg/kg, QD × 4) doses resulted in 100% lethality (0/8 survivors) at day 40. Mice in these groups showed a 10–15% loss in body weight after treatment and never recovered to pretherapy weights. In contrast, no mortality was observed in 60 days in mice treated with a single dose (20 mg/kg) of L-dox composed of EPC/Chol:10/1 or when the free drug was administered as a continuous 96-h infusion via the tail vein. The latter two groups experienced a 10% loss in body weight, but they recovered to pretherapy weights by the second week and weight gain at later times was similar to that of controls. No difference was observed in the survival of mice given L-dox when the lipid composition of the liposome was changed to PC/Chol:1/0.5 or 1/1. Thus, decreasing the PC/cholesterol ratio did not have a significant effect on the ability of liposomes to further reduce the lethality of encapsulated doxorubicin.

Rahman and colleagues (88) examined the cardiotoxic properties of L-dox in DBA/2 mice. Doxorubicin was encapsulated in SUV composed of PC/Chol/cardiolipin/SA:5.1/3.5/1/2 and administered to mice iv at a dose of 15 mg/kg. Mice were sacrificed at 3, 5, and 7 days after injection and tissue sections were prepared for electron microscopic evaluation. These researchers concluded that, compared to free drug, the L-dox significantly reduced the histopathological lesions in cardiac tissue of mice. They also observed that at a dose of 6 mg/kg the nadir of white blood cell counts occurred on day 3 postinjection and was reduced by 50% compared to controls, whereas in animals given L-dox the nadir occurred on day 7 and was reduced only 23%. Thus they conclude that L-dox was less toxic than the free compound to both the heart and the hematopoietic system.

Mayhew and Rustum (102) were concerned with the mechanism of the reduction of toxicity brought about by encapsulation of the drug in negatively charged EMLV liposomes composed of PC/Chol/PG/tocopherol:4/4/1/0.1. Specifically, they were interested in learning if a constant slow infusion of the free drug could reduce the toxicity as much as encapsulation of the drug in liposomes. Mice (C57BL/6) were injected iv with free or L-dox or were administered the drug by constant infusion via the tail vein for 24 or 72 h. Mice were weighed at regular intervals and survival of the mice was determined. An iv bolus of L-dox was superior in its lack of toxicity compared to an iv infusion of free drug or an iv bolus of free drug given as a single injection. The maximum tolerated doses for the different forms or routes of administration for a single injection were L-dox, 30 mg/kg; 24-h infusion free drug, 20 mg/kg; and bolus free drug, 10 mg/kg. The L-dox also had a higher maximum tolerated dose than the free drug when both were administered for five consecutive daily injections or five injections at 2-week intervals. In addition, the L-dox had a superior efficacy compared to the bolus and infused free drug, at equitoxic doses against L1210 leukemia model, and an equal efficacy as the free drug given by iv bolus against the sc form of the M5076 tumor. The L-dox had a significantly enhanced efficacy against the liver metastasis of the M5076 compared to the other two forms of administration.

Gabizon and colleagues (72) published an extensive study of the toxicology of L-dox SUV composed of PC/Chol/PG:7/4/3 administered iv to BALB/c or outbred Sabra mice. Encapsulation of doxorubicin in liposomes reduced the body and organ weight losses, reduced the severity of pathological changes, and reduced the incidence and severity of blood biochemical changes compared to the free doxorubicin. Nephrotoxicity was extremely frequent and severe in mice treated with free doxorubicin but was insignificant among L-dox-treated mice. Mice treated with the free doxorubicin at 7.5 mg/kg had a 100% mortality (0/15 survivors) while at the same dose L-dox caused no mortality (15/15 survivors) at 180 days postinjection. At 180 days at a dose of 5 mg/kg, mice given the free drug had a cardiotoxicity index of 4.5 ± 0.5 while animals administered the L-dox had a cardiotoxicity index of 0.6 ± 0.2. This extensive toxicity study confirmed that liposome encapsulation significantly reduces the toxic manifestations of doxorubicin when administered to animals.

The LD_{50} of various L-dox formulations loaded by the pH trapping method were examined in DBA/2 (73) and CD1 mice (90). All of the lipid compositions tested had a greater LD_{50} than the free drug. For the EPC composition with a 0.25/1 doxorubicin/lipid ratio, the L-dox had a 53% higher LD_{50} than the free drug (73). Liposome size influenced the LD_{50}, particularly with the distearoyl PC(DSPC)-containing compositions, and liposome composition had a pronounced effect. L-dox composed of distearoyl PC had a lower toxicity than liposomes composed of EPC. An LD_{50} value could not be determined for large-diameter DSPC/cholesterol compositions since only 20% of the animals died at the highest dose that could be administered. A high drug/lipid ratio was also correlated with a lower toxicity (90). The authors conclude that changes in liposome properties that result in a more stable drug retention by the liposome or lead to reduced residence time of the doxorubicin in circulation can greatly decrease the toxicity (90).

A. Toxicity of Liposomal Doxorubicin Administered Intraperitoneally

Litterst and co-workers (103) studied the toxicity of doxorubicin administered in large unilamellar liposomes composed of PC/Chol/SA:3/3/1 via the ip route to Sprague-Dawley rats. They observed that the peritonitis was marginally less with L-dox than with the free drug (Table 7). However, they found that the systemic effects of doxorubicin were significantly reduced when the compound was given as the liposomal form. As previously indicated, L-dox is well tolerated when administered to humans via the intraperitoneal route (13).

B. Toxicity of Liposomal Doxorubicin Administered Intradermally

Forssen and Tokes (99) compared the toxic effects of doxorubicin administered as either the free or liposome-encapsulated form intradermally to mice. They found that doxorubicin encapsulated in SUV composed of PC/Chol/PS:6/3/2 significantly attenuated local tissue damage when injected intradermally as compared to the free drug. They raise the possibility that liposome encapsulation of doxorubicin could significantly reduce local tissue trauma caused by extravasation of doxorubicin during administration to humans.

The vesicant properties of L-dox were also assessed by Balazsovits and co-workers (73) in DBA/J mice using the EPC/Chol pH loaded composition. Mice receiving the free drug as a single sc injection (0.4 mg) immediately developed erythema and edema at the injection site, which became ulcerated. Mice receiving L-dox developed a slight edema and erythema, but no ulceration occurred. Moreover, all signs of irritation had subsided by 3 weeks postinjection in the L-dox group (73) (Table 7).

C. Summary of Toxicity of Liposomal Doxorubicin

Animal studies in a number of academic laboratories document the decrease in both acute and chronic lethality in mice and rats of L-dox using a variety of liposomal types and lipid compositions compared to the nonencapsulated drug. Significant decreases in cardio- and nephrotoxicity in mice, rats, and dogs treated with L-dox are also evident (Tables 6, 7). These data formed the basis for proceeding with the phase I clinic trials (vide supra). These clinical trials, although still in the early phase, indicate that acute toxicity of doxorubicin is significantly reduced when the drug is administered as a liposome-encapsulated formulation.

Lowering of toxicity by L-dox appears to be due to a reduction in the rate of entry of drug into the circulation. The lower free doxorubicin level in the blood results in a lower drug level in the heart and hence reduced toxicity. This effect is similar to that achieved by a slow infusion of doxorubicin (104) with the additional advantage that extravasation of the liposome at the site of infusion will cause much less local tissue trauma than free doxorubicin. Liposome formulations that retain drug in the liposome and/or lead to rapid clearance of the drug from the circulation into the liver and spleen have a lower toxicity. Does the reduced toxicity translate into a reduced efficacy against tumors? Studies presented in the following section suggest that L-dox retains its antitumor efficacy.

VIII. EFFICACY OF LIPOSOMAL DOXORUBICIN AGAINST ANIMAL TUMOR MODELS

The efficacy of L-dox has been shown in at least 14 studies to be equal or superior to the free doxorubicin against a variety of animal tumor models (Table 8).

Shinozawa and co-workers (82) compared the efficacy of L-dox to that of nonencapsulated doxorubicin in ICR male mice with the Ehrlich ascites tumor inoculated on the back. Mice were treated via the ip route of administration on days 7, 8, and 9, sacrificed on day 11, and the solid tumor weighed. Neutral, positively charged, and negatively charged L-dox-treated animals all had a significantly greater decrease in tumor weight than mice treated with free drug at a dose of 1.25 mg/kg body weight.

The efficacy of doxorubicin encapsulated in EMLV liposomes composed of PC/Chol/PG:4/5/1 was compared to that of the nonencapsulated drug against the L1210 leukemia (70). Female mice DBA 2/J were innoculated with tumor cells via the ip route. L-dox or free doxorubicin was administered on day 1 via the iv route at 10 or 20 mg/kg body weight. At 10

TABLE 8 Antitumor Effects of Liposomal Doxorubicin in Animal Models

Ref.	Liposome[a]	Species	Tumor[b]	Drug route	Outcome[c]
Rahman et al. (68)	SUV+	Mouse	P388 ip	iv X 4	L-dox similar antitumor effect to dox
Shinozawa et al. (82)	SUV+/-neutral	Mouse	Ehrlich sc	ip	All L-dox compositions superior antitumor effect
Forssen and Tökes (69)	SUV-	Mouse	L1210 iv P388 ip	iv	Enhanced antitumor effect with L-dox
Gabizon et al. (84)	MLV-neutral SUV-	Mouse	J6456 iv	iv	Similar antitumor effect Superior to free drug
Olson et al. (70)	EMLV-	Mouse	L1210 ip	iv X 4	Similar antitumor effect
Mayhew et al. (83)	EMLV-	Mouse	M5076 iv	iv	Equitoxic dose, L-dox more effective
Forssen and Tökes (86)	SUV-	Mouse	Sarcoma 180 sc Lewis lung sc iv	iv	Similar antitumor effect Greater antitumor effect with L-dox
Gabizon et al. (87)	SUV-	Mouse	J6456 iv	iv	Enhanced effect versus metastases in liver
Abra et al. (105)	LMLV-	Mouse	EMT-6 iv	iv	No significant difference in lung tumor nodules, not better than free drug
Van Hoesel et al. (71)	EMLV- SUV+	Rat	IgM immunocytoma sc	iv	L-dox similar reduction in tumor volume as free drug but enhanced survival; decreased antitumor effect for SUV+
Rahman et al. (88)	SUV neutral SUV(+/-)	Mouse	P388 ip	iv	Enhanced antitumor effect
Mayhew and Rustum (102)	EMLV	Mouse	L1210 iv	iv bolus/infusion	Similar at equivalent doses; L-dox superior at equitoxic doses
		Mouse	M5076 sc M5076 iv	iv iv	Equivalent survival L-dox and dox L-dox is superior

Reference	Liposome type	Species	Tumor/route	Route	Comments
Gabizon et al. (94)	SUV-	Mouse	J-6456 iv	iv	L-dox superior, type of negative lipid or presence of cholesterol did not effect outcome
			J-6456 im	iv	L-dox decreased antitumor effect
Storm et al. (95)	EMLV-	Rat	IgM immunocytoma sc	iv	Similar antitumor effect but solid liposome composition was less effective for first 3 days, but final antitumor effect similar
Kojima et al. (89)	SUV-	Nude mouse	Ovarian tumor sc	iv	L-dox equipotent free dox
Mayhew et al. (106)	EMLV-	Mouse	Colon carcinoma 26 spleen	iv	L-dox superior versus intrasplenic implanted tumor
		Mouse	Colon carcinoma 38 colon	iv	L-dox superior versus intracecum implanted tumor; both are models for liver metastases
Konno et al. (109)	SUV	Mouse	Human hepatoma sc	iv	Antibody L-dox superior to L-dox or free dox
Storm et al. (107)	EMLV-	Rat	IgM immunocytoma sc	iv	Antitumor effect of bolus free, infusion free, and L-dox is similar but toxicity for L-dox is less than for other two modalities
Mayer et al. (90)	EMLV-/remote loaded	Mouse	L1210 iv	iv	EPC and DSPC/Chol L-dox equipotent; EPC/Chol 1 µM less potent and 0.1 µm more potent than free dox
Balazsovits et al. (73)	EMLV	Mouse	P815 iv	iv	L-dox equipotent to free dox

aSee Tables 5 and 6 for abbreviations.

bRoute of tumor inoculation given after or under tumor. ip, intraperitoneal; im, intramuscular; iv, intravenous, sc, subcutaneous.

cComparison is made between liposomal doxorubicin and the nonencapsulated doxorubicin administered by the same route and at the same dose unless otherwise indicated.

mg/kg L-dox increased the mean survival time of the tumored animals to the same extent as the free drug. At 20 mg/kg the L-doxorubicin was more efficacious than the free drug, because the toxicity of the free drug but not the L-dox was manifested.

Gabizon and co-workers (84) studied the antitumor activity of L-dox and free doxorubicin against the J-6456 lymphoma model in BALB/c mice. Large MLV composed of either PC/Chol/PS:7/10/3 or PC/Chol:4/1 given as a single iv injection (18 mg/kg) were as effective at prolonging survival as equivalent doses of free doxorubicin. Repeated iv injections (four doses) at 6 mg/kg of SUV composed of PC/Chol/PS:7/10/3 resulted in a significantly improved survival of the animals compared to free doxorubicin.

The effect of L-dox against tumors that can metastasize to the liver was studied by the Roswell Park group (83) in C57B1/6 mice. Small MLV liposomes composed of PC/Chol/PG/tocopherol:4/4/1/0.1 were administered to the animals with the M5076 colon carcinoma inoculated via the iv route on day 8 after tumoring. L-dox at an equitoxic dose to the free doxorubicin gave a fourfold increase in survival time. Long-term survivors were observed only in the L-dox-treated groups. The free and L-dox forms of the drug had equal efficacy against the tumor when the tumor was inoculated sc, but the L-dox was more effective against the liver metastases, indicating the differential antitumor effect of the L-dox is correlated with the increased uptake of L-dox in the liver.

Forssen and Tökes (86) compared L-dox to free doxorubicin in both the sarcoma 180 tumor and the Lewis lung tumor in mice. Swiss mice had the sarcoma 180 tumor implanted sc in the flank and treatment was started 7 days after implantation. Doxorubicin encapsulated in SUV composed of PC/Chol/PS:6/3/2 or free drug was administered iv at a dose of 5 mg/kg at weekly intervals. At 23 and 33 days postimplantation of the tumor L-dox was as active as the free drug in decreasing tumor volume. Female C57BL/6 mice received sc implants of Lewis lung tumor in the flank. Two days after implantation mice received an iv treatment with either the free or liposome-encapsulated doxorubicin (same composition as used in the sarcoma 180 tumor study) at a dose of 5 mg/kg for three injections at weekly intervals. Both forms of the drug produced a significant decrease in tumor volume; however, the L-dox demonstrated greater antitumor activity than the free drug.

Gabizon and colleagues (87) examined the antitumor effect of L-dox against the J-6456 lymphoma inoculated iv in BALB/c mice. Doxorubicin encapsulated in SUV composed of PC/Chol/PS:7/10/3 or free drug at a dose of 200 µg/mouse was administered and tumor cells were isolated from the liver 24 h after treatment. Tumor cells recovered from L-dox-treated mice had a significantly greater decrease in growth than tumor cells recovered from mice treated with the free doxorubicin. The in vivo growth of tumor cells from mice treated with the L-dox was also significantly more depressed than from mice treated with the free drug. Thus, as in the Mayhew study (83), L-dox was more effective against tumor residing in the liver than the free drug.

Abra et al. (105) examined the antitumor effect of L-dox against the EMT-6 tumor in BALB/c mice. Doxorubicin in large MLV liposomes composed of PC/Chol/PS/tocopherol:4/5/1/0.2 or free drug was administered iv at a dose of 2.2 mg/kg to irradiated animals 24 h after tumoring and 13 days later animals were sacrificed and lung tumors quantitated. No significant difference was found between the number of tumors in the lungs of animals treated with the free drug or the L-dox.

Van Hoesel and co-workers (71) tested the antitumor effect of L-dox and free doxorubicin against an IgM immunocytoma inoculated sc in rats. Doxorubicin was encapsulated in small MLV with a 1000-nm diameter composed of PC/Chol/PS:10/4/1. Seventeen days after tumor implantation treatment was started. Tumored animals received either free or L-dox iv for 5 consecutive days at a dose of 2 mg/kg for 8 consecutive weeks. Animals treated with L-dox had a significant increase in survival time when compared to the free doxorubicin. L-dox was as efficacious as the free drug in reducing the tumor volume at the site of implantation. The authors conclude that L-dox resulted in prolonged survival and less toxicity than the free drug.

The antitumor effect of L-dox and free doxorubicin was compared in the P388 tumor inoculated ip in male BALB/c X DBA/2 F_1 mice (88). Doxorubicin encapsulated in SUV composed of PC/Chol/cardiolipin/stearylamine: 5/3.5/1/2 or free drug was administered iv on days 1, 3, and 7 after tumor implantation at a dose of 7.5 mg/kg. Mice receiving the L-dox had a significantly increased survival time compared to mice receiving free doxorubicin.

Mayhew and Rustum (102) compared the antitumor effect of doxorubicin and L-dox in DBA/2 mice tumored with L1210 or C57B1/6 mice tumored with M5076. They compared free drug given as a bolus or slow infusion to a bolus injection of L-dox. Doxorubicin was encapsulated in small MLV composed of PC/Chol/PG/tocopherol:4/4/1/0.1. L1210 cells were innoculated iv into mice and treatment was started at 24 h postinoculation. Doxorubicin was administered at doses from 5 to 30 mg/kg. The free drug was administered either as an iv bolus or as a tail vein infusion over a 24-h interval and the L-dox was administered as an iv bolus. At equivalent doses all three dosage forms gave the same increase in survival. At equitoxic doses the L-dox gave a significant increase in survival times over the two forms of the free drug. This is because the L-dox was less toxic and more could be infused.

The M5076 tumor was inoculated into the C57B1/6 mice by either the sc or iv route (102). The iv inoculation served as a model for metastatic disease. Animals tumored via the sc route were treated by iv bolus on days 21 and 28 at a dose of 10 mg/kg. The L-dox and free doxorubicin gave comparable increases in animal life span. Animals tumored via the iv route received treatments on days 8, 14, and 21 by iv-administered doxorubicin or L-dox. Animals receiving the L-dox had a significant increase in life span compared to animals that received the free doxorubicin. Thus the authors conclude that L-dox is equal in efficacy to free doxorubicin against primary tumors implanted sc and superior in efficacy to free doxorubicin against metastatic tumors (102). The Mayhew group (106) extended the above studies concerning the efficacy of L-dox (EMLV, EPG/EPC/Chol/tocopherol:1/4/4/0.1, 1500 nm diameter) against metastases to the liver by using the mouse colon carcinomas 26 and 28. These tumors are implanted in either the iliocolic vein (CT38LD) or the spleen (CT26) and metastasize to the liver. The CT38LD line was about 30 times more sensitive to doxorubicin in cell culture than the CT26 line. Free doxorubicin had no therapeutic effect against the CT26 tumor line in vivo. When the tumors were inoculated sc, the free doxorubicin had a greater effect on the growth of the CT38LD tumor than did the L-dox and neither form was effective against the CT26 tumor. L-dox was more effective against the liver metastases for both tumor lines and less toxic than free doxorubicin. These results reinforce the potential of L-dox for passive targeting to metastatic tumors in the liver and spleen.

Gabizon and co-workers (94) compared the efficacy of L-dox to that of free doxorubicin in the J-6456 lymphoma tumor in BALB/c mice. Doxorubicin was encapsulated in SUV composed of PC/Chol/PS:7/10/3 and 7/1.5/3 or PC/Chol/PS:7/10/3 and 7/2.5/3. In mice inoculated iv with the tumor, L-dox administered as a single injection at a dose of 8 mg/kg significantly improved animal survival compared to the free doxorubicin. There was no difference in the antitumor effect of the L-dox when phosphatidylserine was replaced by phosphatidylglycerol or when the cholesterol/phospholipid ratio was reduced from 1/1 to 1/4. Only in the L-dox-treated group were there long-term survivors. The authors estimated that the cytoreductive effect of the L-dox was about 100 times greater than that of the free doxorubicin.

The effect of L-dox was also compared to that of the free doxorubicin in prolonging animal survival in animals where the tumor was inoculated im into the thigh. Mice were treated iv on days 3, 10, and 17 after tumor inoculation. Both L-dox and free doxorubicin increased animal survival; however, free doxorubicin was superior to the L-dox in this tumor model. Thus Gabizon and colleagues (94) conclude that free doxorubicin is better than L-dox against tumor cells growing in the im site while the superior antitumor effect of L-dox was expressed on tumors located in the liver and spleen. Efficacy of L-dox to human tumors implanted sc in nude mice has also been studied (89). In this case the negative charge on the SUV was provided by a sulfatide and the L-dox inhibited the growth of the AMOC-I human tumor in nude mice to a similar degree as the free drug (89).

Storm and co-workers examined the influence of lipid composition on the antitumor effects of L-dox in the rat IgM immunocytoma tumor implanted sc (97). A delay in the antitumor effect was evident in the DPPC/DPPG/Chol L-dox formulation (diameter 800 nm) compared to the antitumor effect elicited by the EPC/PS/Chol L-dox formulation (diameter 300 nm) although the final activity of the two formulations was similar. It was not clear if the delay in antitumor effect was caused by the size of the liposomes or the composition. These researchers further showed that the antitumor effect of the bolus free, infusion free, and bolus L-dox were equipotent in the rat IgM immunocytoma, but the L-dox was the least toxic of the three dosage forms (107).

The antitumor efficacy of negatively charged L-dox preparations composed of EPC/EPG/Chol or EPC/PS/Chol with a diameter of 200 nm was tested in the J-6456 tumor or the BCL1 tumor in Balb/c mice (108). The various L-dox formulations were more active on tumors infiltrating the liver and spleen and equipotent on tumors in the bone marrow compared to the free drug. Free doxorubicin given iv was more potent than L-dox administered via the same route when the tumors were implanted in sc sites. When administered ip to treat a peritoneal tumor, L-dox was twofold less toxic and considerably more effective (108).

Passive targeting of doxorubicin to tumors in the liver and spleen by encapsulating the drug in liposomes, as Mayhew (83,102,106) and Gabizon (87,94,108) have shown, results in increased potency for the compound. Attempts to increase the antitumor efficacy of L-dox by attaching a tumor-specific antibody to the liposome have also been undertaken (109). The human alpha-fetoprotein-producing hepatoma strain Li-7 growing in nude mice was treated with antibody-targeted L-dox. Antibody-targeted L-dox (7.5 mg/kg) caused a greater decrease in the weight of the sc-implanted tumor than either L-dox (7.5 mg/kg) or free doxorubicin (4 mg/kg). The antitumor effect occurred in the absence of a clear increase in drug levels in the tumor compared to the nontargeted L-dox with an equivalent amount of nonattached antibody. This suggests that specifically targeting L-dox using a monoclonal antibody may confer additional benefits above tumor localization; the liposome-associated drug may have enhanced access to the tumor cell.

Recent studies using L-dox prepared by the pH trapping technique (90) also demonstrate that L-dox can be as effective as the free drug but considerably less toxic. For instance the EPC/Chol composition had a 53% higher LD_{50} compared to the free drug but was equipotent to the free drug when tested in the p815 mastocytoma murine tumor model. Tumor and drug were inoculated via the iv route (73). However, the potency of the L-dox was a function of the liposome size (90). In general, L-dox with 100-nm diameters were as potent as or more potent than the free drug in the L1210 model, whereas L-dox with diameters of approximately 1000 nm were less potent that the free drug. L-dox composed of DSPC/Chol with a 230-nm diameter were more potent than the same composition with an 820-nm diameter. The authors suggest that differences in the antitumor effect observed in their studies, as well as in the studies of Storm and colleagues (95) where large vesicles composed of more rigid lipids had a delayed antitumor effect compared to smaller vesicles composed of EPC, are most likely due to the size of the vesicles and not to the difference in composition (90). A comprehensive examination of the role of drug/lipid ratio in both negative and neutral vesicles was undertaken in this study. This was possible because the pH strapping method is relatively unaffected by vesicle charge, unlike prior methods used to prepare doxorubicin-containing liposomes, which need a negatively charged lipid to obtain high lipid to obtain high trapping efficiencies (Table 8). The drug/lipid ratio did not influence the potency of the L-dox although it did influence the toxicity (90).

In summary, in a number of rodent tumors liposome-encapsulated doxorubicin has equal or superior efficacy compared to the free drug. This equivalent antitumor efficacy to the free drug occurs with a variety of liposome types, lipid compositions, and drug entrapment methods. In general, small liposomes of about 100 nm were found to have a superior potency to larger liposomes (circa 1000 nm). In the J-6456 tumor implanted im or sc and treated via the iv route, free doxorubicin gave a greater increase in animal survival time than the L-dox. In three other sc implanted tumor models, the sarcoma 180, Lewis lung, and M5076, L-dox had either equal or superior efficacy to the free drug. L-dox was also more effective than free drug against tumors that had metastasized into the liver. The increased liver level of doxorubicin delivered by the liposomes was capable of suppressing tumor growth even in tumors that were not particularly sensitive to the drug.

IX. LIPOSOMAL DOXORUBICIN AN EXAMPLE OF AN IMPROVED THERAPEUTIC INDEX BY SITE-AVOIDANCE DELIVERY

The equal potency and decreased toxicity of the L-dox compared to the free drug establish that the therapeutic index for doxorubicin in animals is significantly enhanced (twofold) when delivered in liposomes. A model of the potential pathways of drug disposition following intravescular administration of L-dox is given in Figure 4. The decreased toxicity is due to the fact that L-dox significantly decreases the initial high plasma level of free doxorubicin. This results in lower levels of doxorubicin in the heart and other tissues that can be adversely affected.

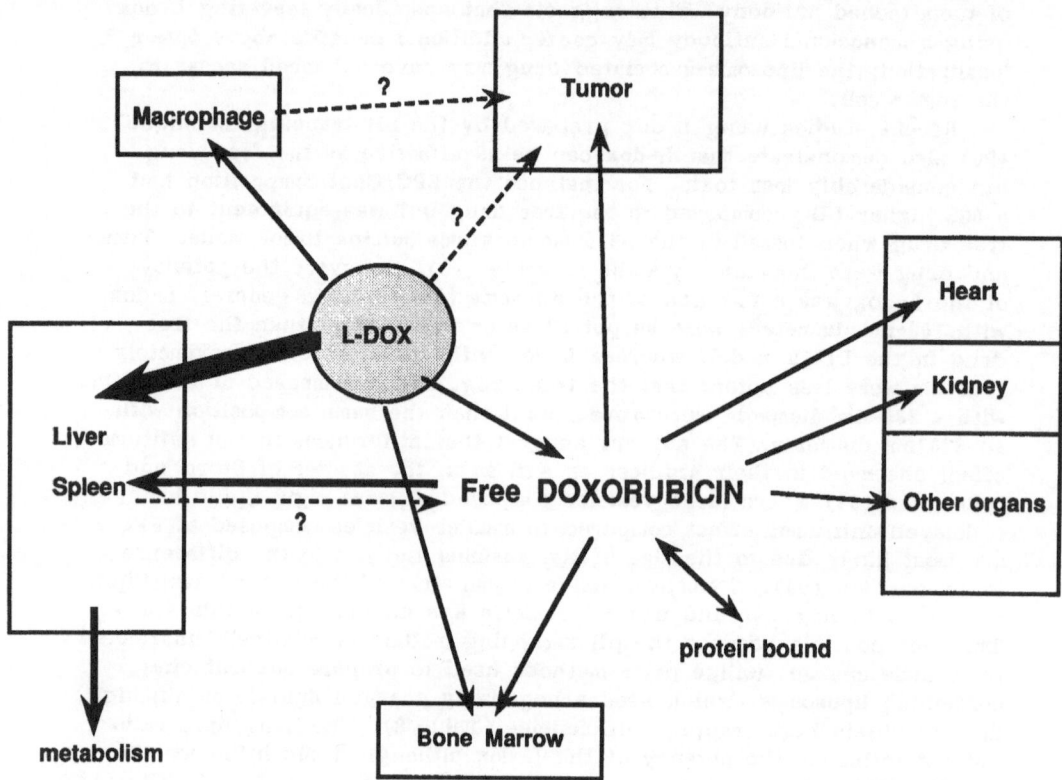

FIGURE 4 Pathways for doxorubicin disposition following administration
in a liposome. Dotted lines represent drug that redistributes after it is
released from the liposome intracellularly. In this scheme toxicities in the
heart and kidney occur via free doxorubicin. Toxicities in the bone marrow
arise from either free or L-dox in the bone marrow. The antitumor effects
are either due to free doxorubicin in the plasma or from doxorubicin released
from the liposome intracellularly in the tumor or extracellularly near the
tumor.

The maintenance of potency of the L-dox implies that a comparable
doxorubicin concentration is attained in the tumor site by the free drug
and L-dox. A significant fraction of the L-dox is taken up by the liver
and spleen, which would remove the drug from the circulation and hence,
in principle, reduce its availability to the tumor. Therefore, the drug
must either be able to exit its initial distribution site and access the tumor
or distribute to the tumor in the liposome so that a high local doxorubicin
concentration is attained.

Three posibilities exist: (1) Doxorubicin is able to return to the central
vascular compartment in a prolonged slow-release fashion. This could account
for the decrease in toxicity and possibly provide a rationale for the maintenance of efficacy in the cases where L-dox is equipotent. (2) Macrophages
that ingested L-dox could provide a source of drug in the vicinity of the
tumor. Macrophages exposed to either free (96) or L-dox (97) can provide
a source of bioactive drug that is cytocidal. Macrophages that have internalized L-dox either because they are already in the vicinity of the tumor or

have migrated into the tumor after ingesting L-dox could provide a local reservoir of drug. This mechanism could also explain the examples of equal and superior potency, yet decreased toxicity, of L-dox. (3) The liposomes provide a slow-release reservoir in the vicinity of the tumor. This mechanism could account for the decrease in toxicity since cardiac tissue is shielded from close approach of liposomes by the endothelial barrier. This would protect the heart from the local release of doxorubicin from liposomes, whereas vascular-accessible tumors could be exposed to a high local concentration of doxorubicin from surface-bound or tumor-internalized liposomes. Thus toxicity would be decreased and efficacy enhanced. The fact that small liposomes (100 nm) have a better antitumor effect than large liposomes (1000 nm) is compatible with this mechanism (84,90), as are the experiments that demonstrate an equal potency and less toxicity for L-dox compared to continuously infused doxorubicin (102,107). This mechanism would also be consistent with the suggestion that liposomes containing cytosine arabinoside act as a depot local release (26). If this is the case, then small liposomes with enhanced circulatory properties, such as the so-called stealth liposome (50,51), could be a significantly improved antitumor L-dox formulation.

The important element for the liposome improvement of doxorubicin therapy is that the encapsulated doxorubicin is able to exit the liposome and its initial distribution site to reach the tumor.

X. IMPROVEMENTS OF LIPOSOMAL DRUG DELIVERY BASED ON FUSION MECHANISMS

Unlike the successful examples of liposomal drug delivery described above, a variety of potential macromolecular therapeutic agents, whose site of action is the cytoplasm, such as oligonucleotides, ribozymes, and DNA, cannot efficiently be delivered into the cytoplasm with current liposome technology. When delivered in the majority of liposome compositions used to date, they would be conveyed to the lysosome and degraded. The therapeutic prospects for these agents would be substantially enhanced if they could be designed to transfer through membranes in a low pH environment, as is the case with *Pseudomonas* exotoxin (111), or if the liposomes could be engineered to fuse with cellular membranes and transfer their contents into the cytoplasm.

The advantage of the latter approach is that a single liposome composition could be employed to deliver a variety of encapsulated molecules. This would be more efficient than redesigning each molecule for enhanced delivery. Engineering of fusogenic liposomes has been attempted by alterations of liposome composition (61-63) and by incorporation of viral fusion proteins into the liposome to mimic the fusion action of viruses with cells (114-116).

Attempts to create pH-sensitive fusogenic liposomes based on modifications of lipid composition have combined phosphatidylethanolamine (PE) with lipids containing acid titratable groups. At neutral pH the ionized lipids stabilize the PE in a bilayer, whereas when the pH is lowered and the ionizable groups are protonated, the PE prefers a hexagonal II phase. Such liposomes are designed to destabilize membranes or become fusogenic when they are exposed to an acidic environment (61-63). In the process of endocytosis the pH is reduced in the endosome, a compartment that precedes the lysosome. The appropriately designed pH-sensitive liposome might then transfer its contents into the cytoplasm before the liposome can be conveyed to the lysosomes.

Straubinger and co-workers (62) demonstrated that a liposome composed of oleic acid/phosphatidylethanolamine can deliver membrane impermeant calcein and fluoresceinated dextran (FITC-D) to the cytoplasm. Connor and Huang (61) incorporated monoclonal antibodies with the fatty-acid-containing, pH-sensitive liposomes to construct pH-sensitive immunoliposomes and were able to deliver various chemotherapeutic agents and DNA to the cytoplasm of target cells.

We have demonstrated that cholesterylhemisuccinate (CHEMS) behaves like cholesterol and stabilizes PE vesicles at neutral pH (112,113) and that protonated CHEMS accelerates the destabilization of PE vesicles at low pH (< 6.0) by catalyzing the formation of the hexagonal H_{II} phase (113). Thus, the CHEMS/PE composition is sensitive to the pH change that occurs along the endocytic pathway and can deliver macromolecules into the cytoplasm of cells in culture (63).

The advantage of liposomes with a pH-sensitive lipid composition is that they are relatively simple to prepare and can be made from inexpensive components. However, the liposome-cell fusion using pH-sensitive compositions is not yet a high-efficiency process.

An alternative approach is to incorporate viral fusogenic proteins or their fragments into the bilayer to create virosomes (114-116). This approach, underway in a number of laboratories, is based on the hypothesis that viral fusogenic proteins will be highly efficient at mediating the fusion between the liposome and the target cell. The disadvantage of this tactic is that high densities of envelope proteins are needed to obtain efficient fusion (117-119). It is unlikely that this will be feasible for drug delivery in the foreseeable future; fusion proteins are difficult to obtain in high quantities and their reconstitution into liposomes in a functional state depends on a large number of variables.

Nonetheless, an understanding of the key elements in the mechanism of protein-induced fusion may permit the creation of small peptide analogs of viral fusion proteins that can induce liposome-cell fusion in a defined environment. Attempts are underway to create pH-sensitive fusogenic peptide sequences (120-122) that can be attached to liposomes and induce liposome-cell fusion in a controlled fashion. If such attempts are to succeed, a much more thorough understanding of the role of protein amino acid sequences in membrane fusion is required.

Liposome drug delivery has achieved significant progress with compounds, such as amphotericin B and doxorubicin, that can transfer from the liposome into the target site. In the next few years commercial products with an improved therapeutic index over the currently used drug will enter the clinic. This will be a welcomed and important milestone for liposomal drug delivery. However, significant improvements in their ability to deliver macromolecules to the cytoplasm are needed before liposomes can fulfill their initially projected therapeutic potential (1).

ACKNOWLEDGMENTS

Portions of this review were written during my sabbatical year. I am grateful for the kindness of Professor Helmut Hauser for providing me with space, lively discussions, and resources in his laboratory at the Department of Biochemistry, ETH, Zurich. I also thank Professor Daan Crommelian for providing me with space and support during a 3-month stay in his group at the school of Pharmacy at the State University of Utrecht. Daan and the members of his laboratory, particularly Drs. Herre Talsma, Pierre Peters, and Gert Storm, provided me with many hours of interesting scientific discussions on liposome drug delivery in general and doxorubicin drug delivery in particular.

REFERENCES

1. Gregoriadis, G., and Ryman, B. E. (1971). Liposomes as carriers of enzymes or drugs: A new approach to the treatment of storage diseases. *Biochem. J.* 124:58P.

2. Gregoriadis, G. (1976). The carrier potential of liposomes in biology and medicine. *N. Engl. J. Med.* 295:704-710.

3. Papahadjopoulos, D., ed. (1978). Liposomes and their uses in biology and medicine. *Ann. NY Acad. Sci.* 308:1-462.

4. Poste, G., and Kirsh, R. (1983). Site-specific (targeted) drug delivery in cancer therapy. *Biotechnology* 1:869-878.

5. Mayhew, E. M., and Papahadjopoulos, D. (1983). Therapeutic applications of liposomes. In: *Liposomes* (M. J. Ostro, ed.). Marcel Dekker, New York, pp. 289-341.

6. Fildes, F. J. T. (1981). Liposomes: The industrial viewpoint. In: *Liposomes from Physical Structure to Therapeutic Applications* (C. G. Knight, ed.). Elsevier Biomedical Press. Amsterdam, pp. 465-485.

7. Kumai, K., Takahashi, R., Tsubouchi, K., Yoshino, K., Ishibiki, K., and Abe, O. (1985). Selective hepatic arterial infusion of liposomes containing anti-tumor agents. *Jpn. J. Cancer Chemother.* 12:1946-1948.

8. Gabizon, A., Dagan, A., Goren, D., Barenholz, Y., and Fuks, Z. (1986). Phase I study with liposome associated adriamycin: Preliminary report. *Proc. Ann. Meeting Am. Soc. Clin. Oncol.* 5:43.

9. Sells, R. A., Owen, R. R., New, R. R. C., and Gilmore, I. T. (1987). Reduction in toxicity of doxorubicin by liposomal entrapment. *Lancet* 2:624-625.

10. Lopez-Berestein, G., Fainstein, V., Hopfer, R., Mehta, F., Sullivan, M. P., Keating, M., Luna, M., Hersh, E. M., Reuben, J., Juliano, R. L., and Bodey, G. P. (1985). Liposomal amphotericin B for the treatment of systemic fungal infections in patients with cancer: A preliminary study. *J. Infect. Dis.* 151:704-710.

11. Sculier, J. P., Coune, A., Meunier, F., Brassinne, C., Laduron, C. Hollaert, C., Collette, N., Heymans, C., and Klastersky, J. (1988). Pilot study of amphotericin B entrapped in sonicated liposomes in cancer patients with fungal infections. *Eur. J. Cancer Clin. Oncol.* 24:527-538.

12. Sculier, J. P., Coune, A., Brassinne, C., Laduron, C., Atassi, G., Ruysschaert, J. M., and Fruhling, J. (1986). Intravenous infusion of high doses of liposomes containing NSC251635, a water-insoluble cytostatic agent. A pilot study with pharmacokinetic data. *J. Clin. Oncol.* 4:789-797.

13. Delgado, G., Potkul, R. K., Treat, J. A., Lewandowski, G. S., Barter, J. F., Forst, D., and Rahman, A. (1989). A phase I/II study of intraperitoneally administered doxorubicin entrapped in cardiolipin liposomes in patients with ovarian cancer. *Am. J. Obstet. Gynecol.* 160:812-819.

14. Zonneveld, G. M., and Crommelin, D. J. A. (1988). Liposomes: parenteral administration to man. In: *Liposomes as Drug Carriers: Trends and Progress* (G. Gregoriadis, ed.). Wiley, Chichester, pp. 795-817.

15. Ostro, M. J., and Cullis, P. R. (1989). Use of liposomes as injectable-drug delivery systems. *Am. J. Hosp. Pharm.* 46:1576-1587.

16. Klausner, A. (1988). Is 1988 the year of the liposome? *Biotechnology* 6:20.

17. Kimelberg, H. K., and Mayhew, E. G. (1978). Properties and biological effects of liposomes and their uses in pharmacology and toxicology. *Crit. Rev. Toxicol.* 6:25-79.

18. Pagano, R. E., and Weinstein, J. N. (1978). Interactions of liposomes with mammalian cells. *Annu. Rev. Biophys. Bioeng.* 7:435-468.

19. Senior, J. H. (1987). Fate and behavior of liposomes in vivo: A review of controlling factors. *Crit. Rev. Therap. Drug Carrier Syst.* 3:123-193.

20. Goldstein, A., Aronow, L., and Kalman, S. M. (1968). *Principles of Drug Action.* Harper & Row, New York.

21. Urquhart, J., Fara, J. W., and Willis, K. L. (1984). Rate-controlled delivery systems in drug and hormone research. *Annu. Rev. Pharmacol. Toxicol.* 24:199-236.

22. Prescott, L. F., and Nimmo, W. S. (eds.). (1985). *Rate Control in Drug Therapy.* Churchill Livingstone, London.

23. Hunt, C. A. (1982). Liposome disposition in vivo. V. Liposome stability in plasma and implications for drug carrier function. *Biochim. Biophys. Acta* 719:450-463.

24. Mayhew, E. G., Rustum, Y. M., Szoka, F. C., and Papahadjopoulos, D. (1979). Role of cholesterol in enhancing the antitumor activity of cytosine arabinoside entrapped in liposomes. *Cancer Treat. Rep.* 63:1923-1928.

25. Kataoka, T., and Kobayaski, T. (1978). Enhancement of chemotherapeutic effect by entrapping 1-β-D-arabinofuranosylcytosine in lipid vesicles and its mode of action. *Ann. NY Acad. Sci.* 308:387-393.

26. Mayhew, E. G., Rustum, Y. M., and Szoka, F. C., Jr. (1983). Therapeutic efficacy of cytosine arabinoside entrapped in liposomes. In: *NATO Symposium ASI: Targeting of Drugs* (G. Gregoriadis, ed.). New York, Plenum Press, pp. 249-260

27. Arakawa, E., Imai, Y., Kobayashi, H., Okumura, K., and Sezaki, H. (1975). Application of drug-containing liposomes to the duration of the intramuscular absorption of water-soluble drugs in rats. *Chem. Pharm. Bull.* 23:2218-2222.

28. Schreier, H., Levy, M., and Mihalko, P. (1987). Sustained release of liposome-encapsulated gentamicin and fate of phospholipids following intramuscular injection. *J. Controlled Release* 5:187-192.

29. Barza, M., Baum, J., and Szoka, F. C., Jr. (1984). Pharmacokinetics of subconjunctival liposome-encapsulated gentamicin in normal rabbit eyes. *Invest. Ophthalmol. Vis. Sci.* 25:486-490.

30. Mihalko, P. J., Schreier, H., and Abra, R. M. (1988). Liposomes: A pulmonary perspective. In: *Liposomes as Drug Carriers* (G. Gregoriadis, ed.). Wiley, Chichester, pp. 679-694.

31. Mezei, M. (1988). Liposomes in the topical application of drugs: A review. In: *Liposomes as Drug Carriers* (G. Gregoriadis, ed.). Wiley, Chichester, pp. 663-677.

32. Ehrlich, P. (1906). *Collected Studies on Immunity.* Vol. 2. Reprinted. Wiley, New York, pp. 442-447.

33. Alving, C. R. (1983). Delivery of liposome-encapsulated drugs to macrophages. *Pharmacol. Ther.* 22:407-422.

34. Popescu, M. C., Swenson, C. E., and Ginsberg, R. S. (1987). Liposome-mediated treatment of viral, bacterial and protozoal infections. In: *Liposomes: From Biophysics to Therapeutics* (M. J. Ostro, ed.). Marcel Dekker, New York, pp. 219-251.

35. Fidler, I. J., and Poste, G. (1982). Macrophage-mediated destruction of malignant tumor cells and new strategies for the therapy of metastic disease. *Springer Semin. Immunopathol.* 5:161-187.

36. Szoka, F. C., Jr. (1986). The cellular availability of liposome encapsulated agents: Consequences for drug therapy. In: *Medical Applications of Liposomes* (K. Yagi, ed.). Karger, New York, pp. 21-30.

37. Stella, V. J., and Himmelstein, K. J. (1985). Site-specific drug delivery via prodrugs. In: *Design of Prodrugs* (H. Bundgaard, ed.). Elsevier, Amsterdam, pp. 177-198.

38. Hunt, C. A., MacGregor, R. D., and Siegel, R. A. (1986). Engineering targeted in vivo drug delivery. I. The physiological and physicochemical principles governing opportunities and limitations. *Pharm. Res.* 3:333-344.

39. Mufson, D., and Szoka, F. C., Jr. (1986). The application of liposome technology to targeted delivery systems. In: *The Latest Developments in Drug Delivery Systems*, Proceedings of the Robert S. First, Inc., Conference on Drug Delivery Systems. Springfield, IL, Aster, pp. 16-21. Erratum *Pharm. Tech.* 10:18.

40. Rahman, Y. E., Hanson, W. R., Bharucha, J., Ainsworth, E. J., and Jaroslow, B. N. (1978). Mechanisms of reduction of antitumor drug toxicity by liposome encapsulation. *Ann. NY Acad. Sci.* 308:325-341.

41. New, R. R., Chance, M. L., and Heath, S. (1981). Antileishmanial activity of amphotericin B and other antifungal agents entrapped in liposomes. *J. Antimicrob. Chemother.* 8:371-381.

42. Lopez-Berestein, G., and Juliano, R. L. (1987). Application of liposomes to the delivery of antifungal agents. In: *Liposomes: From Biophysics to Therapeutics* (M. J. Ostro, ed.). Marcel Dekker, New York, pp. 253-276.

43. Pirson, P., Steiger, R. F., Trouet, A., Gillet, J., and Herman, F. (1980). Primaquine liposomes in the chemotherapy of experimental murine malaria. *Ann. Trop. Med. Parasitol.* 74:383-392.

44. Szoka, F. C., Jr., Milholland, D., and Barza, M. (1987). Effect of Liposome composition and size on toxicity and in vitro fungicidal activity of liposome-intercalated amphotericin B. *Antimicrob. Agents Chemother.* 31:422-429.

45. Van Rooijen, N., and Classen, E. (1988). In vivo elimination of macrophages in spleen and liver, using liposome encapsulated drugs. In: *Liposomes as Drug Carriers* (G. Gregoriadis, ed.). Wiley, Chichester, pp. 131-143.

46. Allen, T. M., Murray, L., MacKeigan, S., and Shah, M. (1984). Chronic liposome administration in mice: Effects on reticuloendothelial function and tissue distribution. *J. Pharmacol. Exp. Therap.* 229:267-275.

47. Fichtner, I., Arndt, D., and Reszka, R. (1986). Antineoplastic activity and toxicity of some alkylating cytostatics (cyclophosphamide, CCNU, Cytostasan) encapsulated in liposomes in different murine tumour models. *J. Microencapsulation* 3:77-87.

48. Hwang, K. J. (1987). Liposome pharmacokinetics. In: *Liposomes: From Biophysics to Therapeutics* (M. J. Ostro, ed.). Marcel Dekker, New York, pp. 109-156.

49. Senior, J., Crawley, J. C., and Gregoriadis, G. (1985). Tissue distribution of liposomes exhibiting long half-lifes in the circulation after in vivo injection. *Biochim. Biophys. Acta* 839:1-17.

50. Allen, T. M., and Chonn, A. (1987). Large unilamellar liposomes with low uptake into the reticuloendothelial system. *FEBS Lett.* 223:42-46.

51. Gabizon, A., and Papahadjopoulos, D. (1988). Liposome formulations with prolonged circulation time in blood and enhanced uptake by tumors. *Proc. Natl. Acad. Sci. USA* 85:6949-6953.

52. Weinstein, J. N., Yoshikami, S., Henkart, P. A., Blumenthal, R., and Hagins, W. A. (1977). Liposome-cell interaction: Transfer and intracellular release of a trapped fluorescent marker. *Science* 195:489-492.

53. Szoka, F. C., Jr., Jacobson, K., and Papahadjopoulos, D. (1979). The use of aqueous space markers to determine the mechanism between phospholipid vesicles and cells. *Biochim. Biophys. Acta* 557:9-23.

54. Szoka, F. C., Jacobson, K., Derzko, Z., and Papahadjopoulos, D. (1980). Fluorescence studies on the mechanism of liposome-cell interactions in vitro. *Biochim. Biophys. Acta* 601:559-571.

55. Rustum, Y. M., Mayhew, E., Szoka, F. C., and Campbell, J. (1981). Inability of liposome encapsulated 1-β-D-arabinofuranosylcytosine nucleotides to overcome drug resistance in L1210 cells. *Eur. J. Cancer* 17:809-817.

56. Straubinger, R. M., Hong, K., Friend, D., and Papahadjopoulos, D. (1983). Endocytosis of liposomes and itnracellular fate of encapsulated molecules: Encounter with a low pH compartment after internalization in coated vesicles. *Cell* 32:1069-1079.

57. Weinstein, J. N. (1987). Liposomes in the diagnosis and treatment of cancer. In: *Liposomes: From Biophysics to Therapeutics* (M. Ostro, ed.). Marcel Dekker, New York, pp. 277-338.

58. Scherphof, G. L. (1986). Liposomes in biology and medicine (a biased review). In: *Lipids and Membranes: Past, Present and Future* (J. A. F. Op den Kamp, B. Roelofsen, and K. W. A. Wirtz, eds.). Elsevier Biomedical Press, Amsterdam, pp. 113-136.

59. Dijkstra, J., Van Galen, W. J. M., and Scherphof, G. (1984). Effects of ammonium chloride and chloroquine on endocytic uptake of liposomes by Kupffer cells in vitro. *Biochim. Biophys. Acta* 804:58-67.

60. Gregoriadis, G., and Neerunjum, E. D. (1974). Control of the rate of hepatic uptake and catabolism of liposome-entrapped proteins injected into rats. Possible therapeutic implications. *Eur. J. Biochem.* 47:179-186.

61. Connor, J., and Huang, L. (1985). Efficient cytoplasmic delivery of a fluorescent dye by pH-sensitive immunoliposomes. *J. Cell Biol.* 101:582-589.

62. Straubinger, R. M., Düzgünes, N., and Papahadjopoulos, D. (1985). pH-sensitive liposomes mediate cytoplasmic delivery of encapsulated macromolecules. *FEBS Lett.* 179:148-154.

63. Chu, C-J., Dijkstra, J., Lai, M-Z., Hong, K., and Szoka, F. C., Jr. (1990). Efficiency of cytoplasmic delivery by pH-sensitive liposomes to cells in culture. *Pharm. Res.* 7, in press.

64. Blum, R. H., and Carter, S. K. (1974). Adriamycin. A new anticancer drug with significant clinical activity. *Ann. Intern. Med.* 80:249-259.

65. Young, R. C., Ozols, R. F., and Myers, C. E. (1981). The anthracycline antineoplastic drugs. *N. Engl. J. Med.* 305:139-153.

66. Gianni, L., Corden, B. J., and Myers, C. E. (1983). The biochemical basis of anthracycline toxicity and antitumor activity. *Rev. Biochem. Toxicol.* 5:1-82.

67. Forssen, E. A., and Tökes, Z. A. (1979). In vitro and in vivo studies with adriamycin liposomes. *Biochem. Biophys. Res. Commun.* 91:1295-1301.

68. Rahman, A., Kessler, A., More, N., Sikic, B., Rowden, G., Woolley, P., and Schein, P. S. (1980). Liposomal protection of adriamycin-induced cardiotoxicity in mice. *Cancer Res.* 40:1532-1537.

69. Forssen, E. A., and Tökes, Z. A. (1981). Use of anionic liposomes for the reduction of chronic doxorubicin-induced cardiototoxicity. *Proc. Natl. Acad. Sci. USA* 78:1873-1877.

70. Olson, F., Mayhew, E., Maslow, D., Rustum, Y., and Szoka, F. (1982). Characterization, toxicity and therapeutic efficacy of adriamycin encapsulated in liposomes. *Eur. J. Cancer Clin. Oncol.* 18:167-176.

71. Van Hoesel, Q. G., Steerenberg, P. A., Crommelin, D. J., van Dijk, A., vab Oort, W., Klein, S., Douze, J. M., de Wildt, D. J., and Hillen, F. C. (1984). Reduced cardiotoxicity and nephrotoxicity with preservation of antitumor activity of doxorubicin entrapped in stable liposomes in the LOU/MWsl rat. *Cancer Res.* 44:3698-3705.

72. Gabizon, A., Meshorer, A., and Barenholz, Y. (1986). Comparative long-term study of the toxicities of free and liposome-associated doxorubicin in mice after intravenous administration. *J. Natl. Cancer Inst.* 77:459-469.

73. Balazsovits, J. A. Z., Mayer, L. D., Bally, M. B., Cullis, P. R., McDonell, M., Ginsberg, R. S., and Falk, R. E. (1989). Analysis of the effect of liposome encapsulation on the vesicant properties, acute and cardiac toxicities, and antitumor efficacy of doxorubicin. *Cancer Chemother. Pharmacol.* 23:81-86.

74. Graybill, J. R., and Craven, P. C. (1983). Antifungal agents used in systemic mycoses. Activity and therapeutic use. *Drugs* 25:41-62.

75. Medoff, G., Brajtbury, J., and Kobayashi, G. S. (1983). Antifungal agents useful in therapy of systemic fungal infections. *Annu. Rev. Pharmacol. Toxicol.* 23:303-330.

76. Bolard, J. (1986). How do the polyene macrolide antibiotics affect the cellular membrane properties? *Biochim. Biophys. Acta* 864:257-304.

77. Meunier, F. (1989). New methods for delivery of antifungal agents. *Rev. Infect. Dis.* 11 (Suppl. 7):S1605-S1612.

78. Lopez-Berestein, G., Ksia, L., Rosenblum, M. G., Haynie, T., Jahns, M., Glenn, H., Mehta, R., Magvligit, G. M., and Hersh, E. M. (1984). Clinical Pharmacology of ^{99m}Tc-labeled liposomes in patients with cancer. *Cancer Res.* 44:375-378.

79. Coune, A. (1984). Lipids and liposomes for improving efficacy of cancer chemotherapy; commentary. *Eur. J. Cancer Clin. Oncol.* 20:443-445.

80. Coune, A., Sculier, J. P., Fruhling, J., et al. (1983). IV administration of a water-insoluble antimitotic compound entrapped in liposomes. Preliminary report on infusion of large volumes of liposomes to man. *Cancer Treat. Rep.* 67:1031-1033.

81. Fruhling, J., Coune, A., Ghanem, G., Sculier, J. P., Verbist, A., Brassinne, C., Laduron, C., and Hildebrand, J. (1984). Distribution in man of ^{111}In-labelled liposomes containing a water-insoluble antimitotic agent. *Nuclear Med. Commun.* 5:205-208.

82. Shinozawa, S., Araki, Y., and Oda, T. (1981). Tissue distribution and antitumor effect of liposome-entrapped doxorubicin (Adriamycin) in Ehrlich solid tumor-bearing mouse. *Acta Med. Okayama* 34:395-405.

83. Mayhew, E., Rustum, Y., and Vail, W. J. (1983). Inhibition of liver metastases of M5076 tumor by liposome entrapped Adriamycin. *Cancer Drug Deliv.* 1:43–58.

84. Gabizon, A., Dagan, A., Goren, D., Barenholz, Y., and Fuks, Z. (1982). Liposomes as in vivo carriers of Adriamycin: Reduced cardiac uptake and preserved antitumor activity in mice. *Cancer Res.* 42:4734–4739.

85. Rosa, P., and Clementi, F. (1983). Absorption and tissue distribution of doxorubicin entrapped in liposomes following intravenous or intraperitoneal administration. *Pharmacology* 26:221–229.

86. Forssen, E. A., and Tökes, Z. A. (1983). Improved therapeutic benefits of doxorubicin by entrapment in anionic liposomes. *Cancer Res.* 43:546–550.

87. Gabizon, A., Goren, D., Fuks, Z., Barenholz, Y., Dagan, A., and Meshorer, A. (1983). Enhancement of Adriamycin delivery to liver metastatic cells with increased tumoricidal effect using liposomes as drug carriers. *Cancer Res.* 43:4730–4735.

88. Rahman, A., White, G., More, N., and Schein, P. S. (1985). Pharmacological, toxicological, and therapeutic evaluation in mice of doxorubicin entrapped in cardiolipin liposomes. *Cancer Res.* 45:796–803.

89. Kojima, N., Ueno, N., Takano, M., Yabushita, H., Noguchi, M., Ishihara, M., and Yagi, K. (1986). Effect of adriamycin entrapped by sulfatide-containing liposomes on ovarian tumor-bearing nude mice. *Biotechnol. Appl. Biochem.* 8:471–478.

90. Mayer, L. D., Tai, L. C. L., Ko, D. S. C., Masin, D., Ginsberg, R. S., Cullis, P. R., and Bally, M. B. (1989). Influence of vesicle size, lipid composition, and drug-to-lipid ratio on the biological activity of liposomal doxorubicin in mice. *Cancer Res.* 49:5922–5930.

91. Gabizon, A., Shiota, R., and Papahadjopoulos, D. (1989). Pharmacokinetics and tissue distribution of doxorubicin encapsulated in stable liposomes with long circulation times. *J. Natl. Cancer Inst.* 81:1484–1488.

92. Mayer, L. D., Bally, M. B., and Cullis, P. R. (1986). Uptake of adriamycin into large unilamellar vesicles in response to a pH gradient. *Biochim. Biophys. Acta* 857:123–126.

93. Druckmann, S., Gabizon, A., and Barenholz, Y. (1989). Separation of liposome-associated doxorubicin in human plasma: Implications for pharmacokinetic studies. *Biochim. Biophys. Acta* 980:381–384.

94. Gabizon, A., Goren, D., Fuks, Z., Meshorer, A., and Barenholz, Y. (1985). Superior therapeutic activity of liposomes-associated adriamycin in a murine metastatic tumour model. *Br. J. Cancer* 51:681–689.

95. Storm, G., Roerdink, F. H., Steerenberg, P. A., de Jong, W. H., and Crommelin, D. J. A. (1987). Influence of lipid composition on the antitumor activity exerted by doxorubicin-containing liposomes in a rat solid tumor model. *Cancer Res.* 47:3366–3372.

96. Martin, F., Caignard, A., Olson, O., Jeanin, J. F., and Leclerc, A. (1982). Tumoricidal effect of macrophages exposed to Adriamycin in vivo and in vitro. *Cancer Res.* 42:3851–3855.

97. Storm, G., Steerenberg, P. A., Emmem, F., van Borssum Waalkes, M., and Crommelin, D. J. A. (1988). Release of doxorubicin from peritoneal macrophages exposed in vivo to doxorubicin-containing liposomes. *Biochim. Biophys. Acta* 965:136–145.

98. Parker, R. J., Priester, E. R., and Sieber, S. M. (1982). Effect of route of administration and liposome entrapment on the metabolism and disposition of adriamycin in the rat. *Drug Metabol. Dispos.* 5:499–504.

99. Forssen, E. A., and Tökes, Z. A. (1983). Attenuation of dermal toxicity of doxorubicin by liposome encapsulation. *Cancer Treat. Rep.* 67:481-484.

100. Herman, E. H., Rahman, A., Ferrans, V. J., Vick, J. A., and Schein, P. S. (1983). Prevention of chronic doxorubicin cardiototoxicity in beagles by liposomal encapsulation. *Cancer Res.* 43:5427-5432.

101. Ganapathi, R., and Krishan, A. (1984). Effect of cholesterol content of liposomes on the encapsulation, efflux and toxicity of adriamycin. *Biochem. Pharmacol.* 33:698-700.

102. Mayhew, E., and Rustum, Y. M. (1985). The use of liposomes as carriers of therapeutic agents. In: *Molecular Basis of Cancer*, Part b: *Macromolecular Recognition, Chemotherapy, and Immunology.* Alan R. Liss, New York, pp. 301-310.

103. Litterst, C. L., Sieber, S. M., Copley, M., and Parker, R. J. (1982). Toxicity of free and liposome-encapsulated Adriamycin following large volume, short-term intraperitoneal exposure in the rat. *Toxicol. Appl. Pharmacol.* 64:517-528.

104. Robert, J. (1987). Continuous infusion or intravenous bolus: What is the rationale for doxorubicin administration? *Cancer Drug Deliv.* 4:191-199.

105. Abra, R. M., Hunt, C. A., Fu, K. K., and Peters, J. H. (1983). Delivery of therapeutic doses of doxorubicin to the mouse lung using lung-accumulating liposomes proves unsuccessful. *Cancer Chemother. Pharmacol.* 11:98-101.

106. Mayhew, E. G., Goldrosen, M. H., Vaage, J., and Rustum, Y. M. (1987). Effects of liposome-entrapped doxorubicin on liver metastases of mouse colon carcinomas 26 and 38. *J. Natl. Cancer Inst.* 78:707-713.

107. Storm, G., Van Hoesel, Q. G. C. M., Degroot, G., Kop, W., and Crommelin, D. J. A. (1989). A comparative study of the antitumor effect, cardiotoxicity and nephrotoxicity of doxorubicin given as a bolus, continuous infusion or entrapped in liposomes in the Lou/MWSL rat. *Cancer Chemother. Pharmacol.* 24:341-348.

108. Gabizon, A., Goren, D., and Barenholz, Y. (1988). Investigations on the antitumor efficacy of liposome-associated doxorubicin in murine tumor models. *Isr. J. Med. Sci.* 24:512-517.

109. Konno, H., Suzuki, H., Tadakuma, T., Kumai, K., Yasuda, T., Kubota, T., Ohta, S., Nagaike, K., Hosokawa, S., Ishibiki, K., Abe, O., and Saito, K. (1987). Antitumor effect of adriamycin entrapped in liposomes conjugated with anti-human alpha-fetoprotein monoclonal antibody. *Cancer Res.* 47:4471-4477.

110. Rahman, A., Carmichael, D., Harris, M., and Roh, J. K. (1986). Comparative pharmacokinetics of free doxorubicin and doxorubicin entrapped in cardiolipin liposomes. *Cancer Res.* 46:2295-2299.

111. Pastan, I., and FitzGerald, D. (1989). *Pseudomonas* exotoxin: Chimeric toxins. *J. Biol. Chem.* 264:15157-15160.

112. Ellens, H., Bentz, J., and Szoka, F. C., Jr. (1984). pH-induced destabilization of phosphatidylethanolamine containing liposomes: Role of bilayer contact. *Biochemistry* 23:1532-1538.

113. Lai, M-Z., Vail, W. J., and Szoka, F. C., Jr. (1985). Acid- and calcium-induced structural changes in phosphatidylethanolamine membranes stabilized by cholesterylhemisuccinate. *Biochemistry* 24:1654-1661.

114. Uchida, T., Kim, J., Yamaizumi, M., Miyake, Y., and Okada, Y. (1979). Reconstitution of lipid vesicles associated with HVJ (Sendai

virus) spikes: Purification and some properties of vesicles containing
non-toxic fragment A of diphtheria toxin. *J. Cell Biol.* 80:10-20.

115. Loyter, A., and Volsky, D. J. (1982). Reconstituted Sendai virus
     envelopes as carriers for the introduction of biological material into
     animal cells. In: *Cell Surface Reviews, Membrane Reconstitution*
     (G. Poste and G. L. Nicolson, eds.). Elsevier Biomedical Press,
     Amsterdam, pp. 215-266.

116. Mannino, R. J., and Gould-Fogerite, S. (1988). Liposome mediated
     gene transfer. *Biotechniques* 6:682-690.

117. Ohnishi, S. (1988). Fusion of viral envelopes with cellular membranes.
     Curr. Topics Membranes Transport 32:257-295.

118. Hoekstra, D., and Kok, J. W. (1989). Entry mechanism of enveloped
     viruses. Implications for fusion of intracellular membranes. *Biosc.
     Rep.* 9:273-305.

119. Stegmann, T., Doms, R. W., and Helenius, A. (1989). Protein-mediated
     membrane fusion. *Annu. Rev. Biophys. Biophys. Chem.* 18:187-211.

120. Lear, J. D., and Degrado, W. F. (1987). Membrane binding and
     conformational properties of peptides representing the NH2 terminus
     of influenza HA-2. *J. Biol. Chem.* 262:6500-6505.

121. Subbarao, N. K., Parente, R. A., Szoka, F. C., Jr., Nadasdi, L.,
     and Pongracz, K. (1987). pH-dependent bilayer destabilization by
     an amphipathic peptide. *Biochemistry* 26:2964-2972.

122. Parente, R. A., Nir, S., and Szoka, F. C., Jr. (1988). pH-dependent
     fusion of phosphatidylcholine small vesicles; induction by a synthetic
     amphipathic peptide. *J. Biol. Chem.* 263:4724-4730.

123. Rahman, A., Ganjei, A., and Neefe, J. R. (1986). Comparative immuno-
     toxicity of free doxorubicin encapsulated in cardiolipin liposomes.
     Cancer Chemother. Pharmacol. 16:28-34.

124. Rahman, A., Fumagalli, A., Barbieri, B., Schein, P. S., and Casazza,
     A. M. (1986). Antitumor and toxicity evaluation of free doxorubicin
     and doxorubicin entrapped in cardiolipin liposomes. *Cancer Chemother.
     Pharmacol.* 16:22-27.

Index